Biochemistry and Molecular Biology

Biochemistry and Molecular Biology

Fifth edition

Despo Papachristodoulou

Alison Snape

William H. Elliott

Daphne C. Elliott

OXFORD
UNIVERSITY PRESS

Great Clarendon Street, Oxford, OX2 6DP,
United Kingdom

Oxford University Press is a department of the University of Oxford. It furthers the University's objective of excellence in research, scholarship, and education by publishing worldwide. Oxford is a registered trade mark of Oxford University Press in the UK and in certain other countries

Impression: 1

British Library Cataloguing in Publication Data
Data available

Library of Congress Control Number: 2013956989

ISBN 978–0–19–960949–9
Printed in Italy by L.E.G.O. S.p.A. —Lavis TN

Preface

Though perhaps not as intimidating as writing a book from scratch, taking on a new edition of an established text is a daunting task. William and Daphne Elliott, in their preface to the fourth edition, described this as a book 'designed for undergraduates taking science and health-related courses in biochemistry and molecular biology' and said that their aim was 'to provide a text suitable for students meeting those subjects for the first time, but of sufficient depth to be intellectually satisfying ... written in an approachable style, but with lots of explanation to promote understanding.' Our experience of using this book for undergraduate courses is that they succeeded in their aim, achieving what one reviewer has described as a style that is 'refreshingly different more personal and more direct' that 'seems to take the reader on a one-to-one tutorial'. It is this admirable quality of the earlier editions that we have sought to retain. The book is aimed at students in the early years of their undergraduate programmes. It assumes little prior knowledge, and seeks to provide accessible explanations of terms and concepts that may be unfamiliar, while building up to a level that should enable students to tackle short review articles of the kind found in many journals. It also avoids giving the impression that biochemistry and molecular biology are closed subjects. Undergraduates at the beginning of their course are sometimes uncomfortable with the notion of uncertainty and ask to be given 'just the facts'. However, by emphasizing the relevance of this material to current medical and biological research and pointing out areas where our knowledge is still imperfect, we hope to give students an appreciation of the interest and importance of continuing developments and to inspire some of them to pursue their studies to a higher level.

It has been noted many times that adding new material to a textbook creates the dilemma of which older material should be discarded in order to keep the size of the book manageable. While we have tried to be up to date we have been relatively cautious about leaving anything out that was covered in previous editions, recent developments in metabolomics and cancer research having shown that those who suggested that certain metabolic topics or even the whole of metabolism were no longer worth learning were speaking prematurely. In genomics and molecular biology the vast quantity of new data and novel methodology presents particular challenges, and we have included some new material in that area. However, we have considered it worthwhile to retain coverage of techniques such as Southern blotting and library screening for now, even though they may be rarely used in modern laboratories, because we feel that knowledge of these methods provides an historic framework that allows students to build up a thorough understanding of our present position. In order to save space and allow for these additions and retentions we have taken the step of moving online the suggestions for further reading and the web based bioinformatics exercise. This has the added advantage that these resources can be more easily updated.

About the book

The book is organized into six main parts:

1 Basic concepts of life
2 Structure and function of proteins and membranes
3 Metabolism and nutrition
4 Information storage and utilization
5 Cells and tissues
6 Protective mechanisms against disease.

The chapters are arranged to give a seamless progression through the subject but we recognize that the order in which topics are taught varies with the teacher. There is, therefore, extensive cross-referencing between chapters in order to help students with their learning.

What is new in this edition?

- All the chapters have been reviewed to update them. The extent of the update varies in the different chapters from extensive rewriting and major additions to minor changes. A number of complex areas, which some students find difficult, have been rewritten for increased clarity. These include X-ray crystallography and nuclear magnetic resonance (Chapter 5); mechanisms of transport, storage and mobilization of dietary components (Chapter 11); mechanisms of metabolic control and their applications to metabolic integration (Chapter 20); homologous recombination (Chapter 23); cell signalling, particularly the insulin receptor and insulin signalling (Chapter 29); and vesicle transport of proteins (Chapter 27).
- A new chapter has been added, General principles of nutrition (Chapter 9) to introduce the subject and the terminology before dealing with the handling of dietary components.

- Material on regulation of gene expression, both transcriptional and post-transcriptional, has been reorganized and moved into a single new chapter (Chapter 26). This chapter includes an updated section on the role of chromatin in eukaryotic gene control with a new box on Genomic imprinting disorders (Box 26.1). It also incorporates material on microRNAs and RNA interference, which was given a separate chapter in the previous edition.
- The last section of the book has been reorganized. Part 5, *Cells and tissues*, is designed to illustrate how cell communication and coordination of cell division and cell death take place in eukaryotes, and then to illustrate how perturbation of these processes may lead to cancer. Part 6 then deals briefly with some special topics relating to protective mechanisms against disease.
- Cell division (mitosis and meiosis) is dealt with in the same chapter as the cell cycle (Chapter 30), which also includes cell death (apoptosis).

In metabolism, attention has been paid to the clinical relevance of the biochemical events. New material includes:

- Lipid transport, including cholesterol homeostasis and the hyperlipoproteinaemias.
- An extensive review of the regulation of metabolism in the fed state, fasting state, starvation and diabetes mellitus.
- New boxes: The Warburg effect (Box 13.1); Glucose-6-phosphate dehydrogenase deficiency (Box 15.1); Alcohol and the oriental flushing syndrome (Box 16.1); The alpha and the omega of fatty acids and diet (Box 17.1).

Important developments covered in Chapter 28 on DNA manipulation include the following:

- An introduction to Next Generation Sequencing methods for DNA and the potential medical application of rapid sequencing. No doubt these topics will gain still greater prominence in future editions.
- The principle of Genome Wide Association Studies used for elucidation of genetic causes for common complex diseases such as type 2 diabetes and coronary heart disease.
- Recent promising clinical trials using human embryonic stem cells, and the production of induced pluripotent stem cells from patients and their use to model diseases.

Using the book

This book includes a number of features to help make it easy to use, and to make learning from it as effective as possible.

- Index of diseases. A separate index of diseases and medically relevant topics helps students on health-related courses to identify relevant topics.
- Medical boxes. These illustrate the direct relevance of biochemistry and molecular biology to medicine and health-related issues. A separate list of these boxes is shown on the Contents pages.
- Questions and answers. Questions at the end of each chapter (with answers at the back of the book) are designed to support student learning.
- Chapter summaries. Summaries at the end of each chapter highlight the key concepts presented and aid revision.
- Further reading references. References online direct the reader mainly to review articles of the shorter type found in Trends journals. References can be accessed by scanning the QR code image at the end of each chapter. QR Code images are used throughout this book. QR Code is a registered trademark of DENSO WAVE INCORPORATED. If your mobile device does not have a QR Code reader try this website for advice www.mobile-barcodes.com/qr-code-software.

Online Resource Centre

The Online Resource Centre, which has been built to accompany this text, contains a number of useful teaching and learning resources for lecturers and students. Visit the site at **www.oxfordtextbooks.co.uk/orc/snape_biochemistry5e/**

For registered adopters of the book:

- Figures from the book available to download, for use in lectures and presentations.

For students:

- Comprehensive further reading lists, organized by chapter, and linked to relevant sections in the book via QR codes.
- An extensive bank of multiple-choice questions for self-directed learning. Each question has feedback keyed to the book so that students can review the relevant concepts easily.
- Links to three-dimensional structures of key biological molecules featured in the book.

Acknowledgements

Our first and enormous debt is, of course, to Bill and Daphne Elliott, who developed and wrote the four previous editions of this book. Their breadth and depth of knowledge, their skill in organizing and explaining complex material in an accessible way, and their enthusiasm and capacity for keeping up to date with new developments across a range of subject areas are truly inspirational. They wrote to us most generously on hearing that we would take on the fifth edition, and we can only hope that we have learned from them sufficiently to maintain the standard they set. It was especially sad to hear that Bill Elliott died in July 2012, before we had a chance to meet, particularly since it turns out that he and AS also have their native county, Durham (UK) in common. The Elliotts acknowledged several colleagues for help with previous editions, and as we have retained much of the earlier structure and material, we have also without knowing them personally benefitted from their expertise and advice.

Our second major thanks go to the staff at Oxford University Press, Jonathan Crowe, who decided to let us loose on the book, and our editors, Holly Edmundson and Alice Mumford, who have guided us with patience, firmness, and good humour. We would also like to thank the members of the advisory and reviewing panels appointed by OUP who made many constructive suggestions for the development of this edition and helpful comments on draft versions of some of our chapters.

Finally we would like to thank staff and students at King's College London. Our colleagues have contributed greatly to our knowledge and understanding of numerous topics, in particular Dr Liz Andrew, who reviewed Chapter 33 on the immune system. Professor Guy Tear, Dr Paul Brown, Dr Renee Tata and others have provided invaluable support and encouragement. Our students have persistently, and often perspicaciously, made comments and asked questions that have helped us develop our ability to teach.

> It may metaphorically be said that natural selection is daily and hourly scrutinising, throughout the world, the slightest variations; rejecting those that are bad and adding up all that are good; silently and insensibly working, whenever and wherever opportunity offers, at the improvement of each organic being in relations to its organic and inorganic conditions of life.
>
> *Charles Darwin*

> Evolution is a tinkerer.
>
> *Francois Jacob*

Brief contents

Contents

Part 2 Structure and function of proteins and membranes

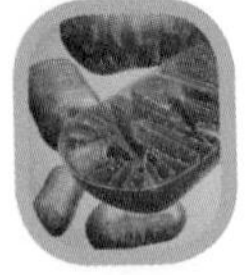

Part 3 Metabolism and nutrition

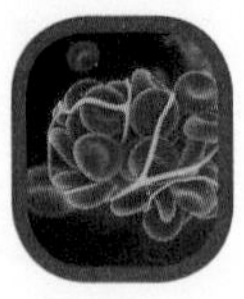

Part 6 Protective mechanisms against disease

Diseases and medically relevant topics

Abbreviations

A	adenine
AA	aminoacyl group
ABC	ATP-binding cassette
ACAT	acyl-CoA:cholesterol acyltransferase
ACP	acyl carrier protein
ACTH	adrenocorticotrophic hormone (also called corticotrophin)
ADH	antidiuretic hormone (also called vasopressin)
ADP	adenosine diphosphate
AgAp	agouti-related appetite stimulant
Akt	protein kinase (*see also* PKB)
AIDS	acquired immunodeficiency syndrome
ALA	5-aminolevulinic acid
ALA-S	aminolevulinate synthase
ALT	alternative mechanism of lengthening telomeres
AMP	adenosine monophosphate
AMPK	AMP-activated protein kinase
AP	apurine or apyrimidine
APC	antigen-presenting cell
APR	anaphase promoting complex
Apaf-1	apoptopic protease mediating factor
A site	acceptor site (or amino acyl site) of ribosome
ATCase	aspartyl transcarbamylase
ATM	ataxia telangiectasia mutated
ATP	adenosine triphosphate
ARE	AU-rich element (in mRNA)
AZT	azidothymidine
BAC	bacterial artificial chromosome
bp	base pair
BPG	2,3-bisphosphoglycerate
BSE	bovine spongiform encephalopathy (mad cow disease)
C	cytosine
c-	'cellular', denotes protooncogene (c-*ras*, c-*myc*, etc.)
cal	calorie
CAM	calmodulin
cAMP	adenosine-3′,5′-cyclic monophosphate
CAK	Cdk activating kinase
CAP	catabolite gene-activator protein
CBP	CREB-binding protein
CD	cluster of differentiation (proteins)
Cdk	cyclin-dependent kinase
cDNA	complementary DNA
CETP	cholesterol ester transfer protein
CGRP	calcitonin gene related peptide
CJD	Creutzfeldt-Jakob disease
CDP	cytidine diphosphate
Cki	Cdk inhibitor protein
COP	coat protein (of transport vesicles)
CoA	coenzyme A (A = acyl)
CoQ	ubiquinone (*see also* Q, UQ)
COX	cyclooxygenase
CRE	cAMP-response element
CREB	CRE-binding protein
CSF	colony-stimulating factor
CTD	carboxy-terminal domain (of eukaryotic RNA polymerase)
CTP	cytidine triphosphate
CVD	cardiovascular disease
d-	deoxy (as in deoxyribonucleotides: dATP, dCTP etc.)
DAG	diacylglycerol
Da	Dalton (unit of atomic or molecular mass: one twelfth of the mass of a carbon 12 atom)
dd-	dideoxy (as in dideoxyribonucleotides: ddATP, ddCTP etc.)
DHAP	dihydroxyacetone phosphate
DNA	deoxyribonucleic acid
DNase	deoxyribonuclease
DPE	downstream promoter element
ds	double stranded (DNA, RNA)
DNP	dinitrophenol
DSB	double stranded break
E_0'	redox potential value at pH 7.0
ECM	extracellular matrix
EF	elongation factor (in translation)
EF_2	elongation factor 2 (eukaryotic ribosomal translocase)
EF-G	elongation factor-G (*E. coli* ribosomal translocase)
EF-Tu	elongation factor temperature unstable
EGF	epidermal growth factor
eIF	eukaryotic initiation factor (in translation)
ELISA	enzyme-linked immunoabsorbent assay

ENCODE — Encyclopoedia of DNA elements (project)
ER — endoplasmic reticulum
ES cell — embryonic stem cell
ESI — electrospray ionization
E site — exit site of ribosome

F_o — membrane rotary subunit of ATP synthase
F_1 — catalytic subunit of ATP synthase
F — Faraday constant (96.5 kJ V^{-1} mol^{-1})
FAD — flavin adenine dinucleotide
FADD — Fas-associated protein with a death domain
$FADH_2$ — reduced form of FAD
Fd — ferredoxin
FFA — free fatty acid
FH_2 — dihydrofolate
FH_4 — tetrahydrofolate
fMet — formylmethionine
FMN — flavin mononucleotide
FSH — follicle-stimulating hormone

G_1 *or* G_2 — 'gap' phases of cell cycle
G — guanine
G — free energy (Gibbs)
$G^{0\prime}$ — standard free energy (Gibbs) at pH 7.0
G-1-P — glucose-1-phosphate
G-6-P — glucose-6-phosphate
G6PD or G6PDH — glucose-6-phosphate dehydrogenase
GAG — glycosaminoglycan
GAP — GTPase-activating protein
GDP — guanosine diphosphate
GEF — guanine nucleotide exchange factor
GLUT — glucose transporter
GMO — genetically modified organism
GroEL — multisubunit molecular chaperone (chaperonin) of Hsp 60 class
GroES — 'lid' structure of groEL chaperonin complex
GRB — growth receptor-binding protein
GRK — G-protein receptor kinase
GSH — reduced glutathione
GSK3 — glycogen synthase kinase 3
GSSG — oxidized glutathione
GTP — guanosine triphosphate

H — enthalpy
HAT — histone acetyltransferase
Hb — haemoglobin
HbO_2 — oxyhaemoglobin
HDAC — histone deacetylase
HDL — high-density lipoprotein
hESC — human embryonic stem cell
HGPRT — hypoxanthine–guanine phosphoribosyltransferase
HIF — hypoxia-inducible factor
HIV — human immunodeficiency virus
HLH — helix-loop-helix
HMG-CoA — 3-hydroxy-3-methylglutaryl-CoA
HNPCC — hereditary nonpolyposis colorectal cancer
hnRNP — hetero-ribonucleoprotein complex
HPLC — high-pressure (or high performance) liquid chromatography
Hsp — heat-shock protein
HTH — helix–turn–helix (DNA-recognition motif)

I — inosine
IDL — intermediate-density lipoprotein
IF — initiation factor (e.g. IF1, IF2, IF3) in translation
IF — intermediate filament
Ig — immunoglobulin (IgG, IgG1, IgA, etc.)
IGF — insulin-like growth factor (IGFI, IGFII)
Il — interleukin
Inr — initiator (eukaryotic transcription)
IP_3 — inositol trisphosphate
iPSC — induced pluripotent stem cell
IRE — iron-responsive element
IRP — IRE-binding protein
IRS — insulin receptor substrate

JAK — type of tyrosine kinase (*J*anus *k*inase)
J — Joule

K — Kelvin
K_a — acid dissociation constant
K_{cat} — turnover number of an enzyme (number of molecules of substrate converted to product by a molecule of enzyme at saturating levels of substrate per second)
K_{eq} — equilibrium constant of a reaction
K'_{eq} — equilibrium constant at pH 7.0
K_m — Michaelis constant:the substrate concentration at which a Michaelis–Menten enzyme works at half-maximal velocity
kb — kilobase

LCAT — lecithin:cholesterol acyltransferase
LDL — low-density lipoprotein
LINES — long interspersed elements
LTR — long terminal repeat (sequences in retroviruses and certain retrotransposons)

M — Molar (moles dm^{-3} or moles $litre^{-1}$)
MALDI — matrix-assisted laser-desorption ionization

MAP mitogen-activated protein (kinase)
MHC major histocompatibility complex
miRNA microRNA
mRNA messenger RNA
MS mass spectrometry
MTOC microtubule-organizing centre
m/z mass-to-charge ratio

N unspecified base in a nucleotide (e.g. NTP)
NAD^+ nicotinamide adenine dinucleotide (oxidized form)
NADH reduced form of NAD
$NADP^+$ nicotinamide adenine dinucleotide phosphate (oxidized form)
NADPH reduced form of NADP
N-CAMS nerve cell adhesion proteins
NEFA non-esterified fatty acid
NES nuclear export signal
NF-κB nuclear factor family of eukaryotic transcription factors
NGS next generation sequencing
NK natural killer cells
NLS nuclear localization signal
nm nanometer (10^{-9} metres)
NMR nuclear magnetic resonance spectroscopy
NPY neuropeptide Y; appetite stimulant
NSAID nonsteroidal anti-inflammatory drugs
NTF nuclear transport factor

—Ⓟ high-energy phosphoryl group
P450 cytochrome P450
PAGE polyacrylamide gel electrophoresis
PBG porphobilinogen
PC phosphatidylcholine (lecithin)
Pc plastocyanin
PCNA proliferating cell nuclear antigen (eukaryotic sliding clamp protein)
PCR polymerase chain reaction
PDB protein database
PDGF platelet-derived growth factor
PDH pyruvate dehydrogenase
PDI protein disulfide isomerase
PE phosphatidylethanolamine (cephalin)
PEP phosphoenolpyruvate
PEP-CK PEP carboxykinase
PET positron emitting tomography
PFK phosphofructokinase (PFK1, PFK2)
PG prostaglandin
3-PGA 3-phosphoglycerate
PH pleckstrin homology (domain)
P_i inorganic phosphate
pI isoelectric point
PIAS protein inhibitor of activated STATS
PIC preinitiation complex
PI 3-kinase phosphatidylinositide 3-kinase
piRNA piwi-interacting RNA
PK protein kinase (PKA, PKB, PKC, etc.)
PK pyruvate kinase
pK_a the pH at which there is 50% dissociation of an acid
PKB mammalian homologue of Akt
PKU phenylketonuria
PLC phospholipase C
PLP pyridoxal-5′-phosphate
Pol DNA or RNA polymerase
POMC pro-opiomelanocortin; appetite repressor
PP_i inorganic pyrophosphate
PPI peptidylproline isomerase
Pq plasto-quinone
PRE polypyrimidine (C-rich) element (in mRNA)
pri-miRNA primary microRNA
PrP^c prion protein (constitutive)
PrP^{sc} prion protein (scrapie)
PRPP 5-phosphoribosyl-1-pyrophosphate
PS phosphatidylserine
PS photosystem (PSI, PSII)
P site peptidyl site (of ribosome)
PTGS posttranscriptional silencing (plants)
PTS peroxisome-targeting signal
PYY-3–36 neuropeptide appetite inhibitor

Q ubiquinone (*see also* CoQ, UQ)
Q quadrupole (in mass spectrometry)
qPCR quantitative PCR (polymerase chain reaction)

R gas constant (8.315 J mol^{-1} K^{-1})
Rb retinoblastoma
RFLP restriction fragment length polymorphism
RF release factor (in translation)
RISC RNA-induced silencing complex
RNA ribonucleic acid
RNase ribonuclease
RNAi RNA interference
ROS reactive oxygen species
R-5-P ribose-5-phosphate
RPA replication protein A (detects single-stranded DNA)
rRNA ribosomal RNA
Rubisco ribulose-1,5-bisphosphate carboxylase

S Svedberg unit
S ‘synthesis’ (DNA replication) phase of cell cycle
S entropy
SAM *S*-adenosylmethionine

SCID	severe combined immunodeficiency disease
SCNT	somatic cell-nuclear transfer
SECIS	selenocysteine insertion sequence
SDS	sodium dodecylsulfate
SDS-PAGE	SDS polyacrylamide gel electrophoresis
SH2	Src homology region 2
SINES	short interspersed elements
siRNA	small interfering RNA
Sn	stereospecific numbering
SNP	single nucleotide polymorphism
snRNAs	small nuclear RNAs
snoRNA	small nucleolar RNA
SOCS	suppressors of cytokine signalling
SOS	'son of sevenless'
SR	sarcoplasmic reticulum
SRP	signal-recognition particle
Ss	single-stranded (DNA or RNA)
SSB	single-strand binding protein
STAT	signal transducer and activator of transcription
STR	short tandem repeats (microsatellites)
T	thymine
T_3	triiodothyronine
T_4	thyroxine
TAF	TBP-associated factor
TAG	triacylglycerol
TBP	TATA-binding protein
TCA	tricarboxylic acid
TCR	T cell receptor
TF	transcription factor
TFIID	transcription factor D for RNA polymerase II
TIM	translocator of the inner mitochondrial membrane
TNF-α	tumour necrosis factor-α
TOF	time of flight
TOM	translocator of the outer mitochondrial membrane
TPA (t-pa)	tissue plasminogen activator
TPP	thiamin pyrophosphate
tRNA	transfer RNA
$tRNA^{phe}$	tRNA specific for phenylalanine (by analogy, $tRNA^{Leu}$, $tRNA^{Met}$, etc.)
$tRNA_f$	tRNA formyl (bacterial translation initiation)
$tRNA_i$	tRNA for eukaryote translation initiation
TSH	thyroid-stimulating hormone
U	uracil
UDP	uridine diphosphate
UDPG	uridine diphosphoglucose
UDP-Gal	uridine diphosphogalactose
UTP	uridine triphosphate
UQ	ubiquinone (*see also* CoQ, Q)
UTR	untranslated region (of mRNA)
UV	ultraviolet (light)
v-	'viral', denotes oncogene (v-*ras*, v-*myc*, etc.)
V	velocity of reaction
$V0$	initial velocity of reaction
V_{max}	maximum velocity of reaction
VLDL	very-low-density lipoprotein
VNTR	variable number of tandem repeats
X-5-P	xylulose-5-phosphate
YAC	yeast artificial chromosome

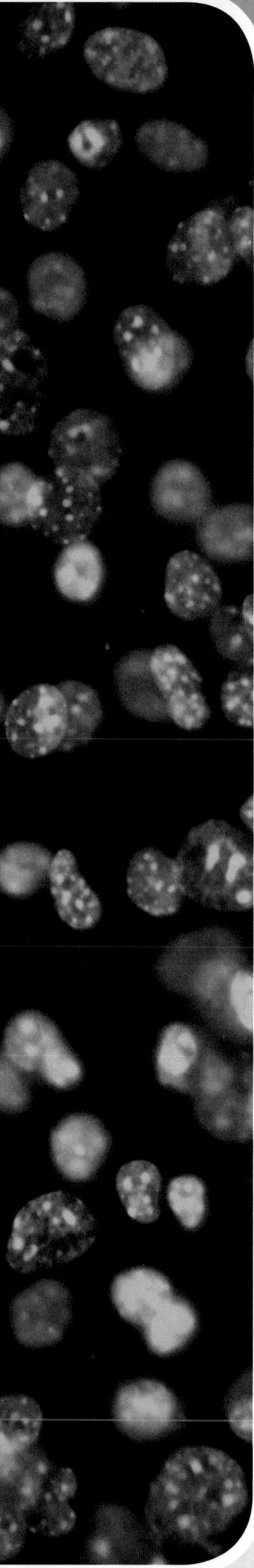

Part 1

Basic concepts of life

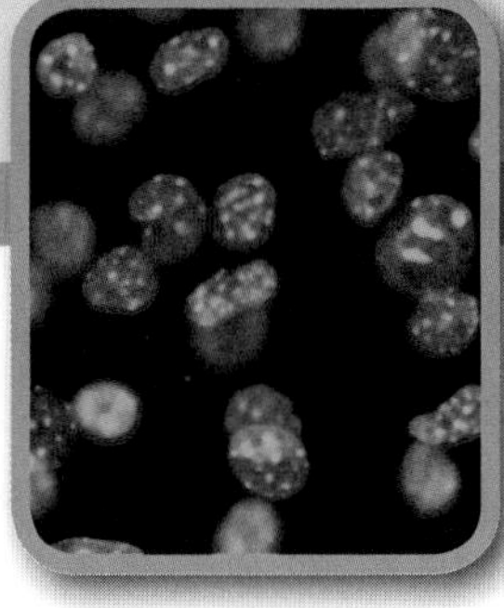

The basic molecular themes of life

Living systems have the property of being able to reproduce themselves by a process of self assembly from nonliving materials in the environment. Life is a molecular process in which ordinary chemical compounds are able to achieve this. It is a complex process in detail but is based on a few basic molecular themes common to all life forms on this planet.

A reviewer commented that all writers of biochemical texts face the problem that everything should come first because most aspects are interdependent. This chapter is as near as we could get to achieving what should come first. It gives a preliminary survey of the concepts, and of the laws of the universe that determined them. All of the topics discussed here are dealt with at greater length in later chapters to which references are given. It is hoped that this preliminary survey will help students to understand where each topic fits into the overall picture as they come to them in more detail in later chapters.

Biochemistry and **molecular biology** are the disciplines concerned with understanding the mechanism of life in molecular terms. Biochemistry is the name for the earliest studied aspects in which the metabolism of food and small molecules is a principal focus. Molecular biology developed later, from the 1950s, and was the name given to the study of biological macromolecules, particularly proteins and DNA, and how these function in the genetic mechanism. The distinction between biochemistry and molecular biology has become blurred because the two are interdependent, using much the same techniques, but the terms are still used as convenient broad labels. Many biochemical departments and biochemical societies have added 'molecular biology' to their titles. The joint subject is of ever increasing importance in medicine, agriculture, and all aspects of biology. It is an exciting subject with seemingly never-ending discoveries of molecular mechanisms by which life operates. Research progresses at an exhilarating pace in almost all areas of biological science; the medical potential of discoveries at the molecular level has given rise to the biotechnology boom.

All life forms are similar at the molecular level

All living organisms at the molecular level have a basic unity; whether this will extend to life on other planets, or wherever it might be discovered, is a fascinating question since the same physical and chemical laws that have governed the development of life on earth apply throughout the universe.

To return to earth, a famous dictum of the French Nobel prize winner, Jacques Monod, is that 'what holds for the Coli bacterium is true for an elephant', meaning that the similarities between a bacterium living in the human gut and an elephant far exceed the differences between the two organisms when viewed at the molecular level. A mass of evidence shows that all current life had a single origin. Initially life must have been extremely simple, but it would have had to involve a molecule with the capability of directing its own replication and of determining the characteristics of 'offspring' of the replicating unit. This is where **nucleic acids** come into the picture. **DNA** and its close relative, **RNA**, are nucleic acids. DNA is the basis of all cellular life, in that it carries the information necessary for reproduction. It carries the 'genetic code' that determines the characteristics of offspring. RNA has the same role in certain viruses. The structure of DNA and RNA lends itself to self-directed replication, as explained later. It is believed that life originated with RNA as the 'genetic' material, but RNA has been replaced in all cellular life with DNA. DNA is chemically more stable and therefore more suitable for storing genetic information.

At the origin of life, once the primordial form of a self-replicating cell-like 'unit' had evolved, many of the fundamental biochemical processes must have already been established and life was locked into these. This has given all life forms that evolved from this initial replicating unit their basic unity. Diversity in the detail has arisen through natural selection. Mistakes in replication of the DNA are inevitable; small random variations continually occur in the form of mutations, so that offspring have slight variations from the parents. Any variation that increases the chances of the organism reproducing itself is preserved and any that reduces it is likely to be eliminated by natural selection. This leads to the evolution of new life forms better adapted to the environment, though their underlying molecular processes remain much the same. The essential unity of life explains why, in biochemical research, a variety of organisms is often used to elucidate a given biochemical process. To understand how a process works, for example in humans, the best strategy may be first to study the bacterium, *Escherichia coli*, or a virus, as the basic information might be more easily obtained from these simpler systems. The yeast cell is a eukaryote and has become an important model system for investigating human diseases such as cancer. There are differences in molecular processes between bacterial and human cells but these are more matters of detail rather than of principle. Most biochemical knowledge is applicable to all life forms.

The energy cycle in life

Living cells obey the laws of physics and chemistry. To grow and reproduce, cells take in simple molecules such as sugars and nitrogenous compounds from the external medium and build them up into the large organized complex molecules of cells. The synthesis of complex molecules from simpler ones involves an increase in energy content of the cell, so chemical work must be done (see Chapter 3). A living cell is at a higher energy level than the random collection of molecules in its external environment from which it was assembled. It is far from being in thermodynamic (energetic) equilibrium with its surroundings; this is achieved only by decomposition after the death of the cell. The state of a cell is somewhat analogous to that of a flying aeroplane in which the high gravitational potential energy state is maintained by constant fuel oxidation, which releases energy. If this stops, the aeroplane crashes to a minimum energy state on the ground.

Most cells get their fuel from food molecules, in which the energy initially came from the sun and was converted into chemical energy in the form of sugars by photosynthetic plants. For some organisms, such as bacteria living around hydrothermal vents in the ocean, the 'foods' that supply their energy are chemicals such as hydrogen sulphide from the earth's crust. These organisms are known as **chemotrophs**.

The laws of thermodynamics deal with energy

Thermodynamics is the study of energy transformations. It is a daunting subject to many students, partly because it usually has to deal with systems such as steam engines in which temperature and pressure changes are mechanistically involved, thus complicating the subject. Living organisms are, however, mostly systems functioning at constant temperature and pressure. They do not depend on temperature and pressure gradients within the organism. This fact greatly simplifies thermodynamics as applied to life processes. Indeed, many biochemists seldom use thermodynamic equations. However, they need to understand the simple concepts presented here and in Chapter 3.

The **first law of thermodynamics** states that energy can be neither created nor destroyed – the total energy content of the universe remains constant. We generally believe that we know what energy is, but it is difficult to define. A useful concept is that it is what makes things happen. Although energy cannot be destroyed, not all energy is useful in the sense of being capable of performing work. When you burn fuel, the liberated heat may be made to perform work as in a steam engine, but part of it becomes heat that cannot be used to drive the engine. For example, the heat in a hot car engine block is still energy, but it cannot be used to propel the car. An important concept is that if you consider the total amount of energy released by any process, such as the oxidation of food, only part of it can be used to do work. The rest increases the total **entropy** of the universe. Entropy is the degree of randomness or disorder in any system. High entropy equals relatively low energy and low entropy relatively high energy. Heat increases the random motion of molecules and hence increases entropy. When a molecule breaks down into smaller ones, or anything increases the number of particles in a system, this also increases entropy. To anyone unfamiliar with the concept, it will probably seem odd that the vitally important **second law of thermodynamics** specifies that all processes must increase the total entropy of the universe. It seems, however, that increasing entropy is a major driving force in the universe and all processes must contribute to it. That is why no process can ever be energetically 100% efficient. It would seem that the ultimate fate is a dark, silent universe of infinite entropy and maximum stability where nothing whatsoever can happen. The universe appears to have a relentless 'drive' to achieve this state.

At first sight, living cells appear to be magically defying the second law of thermodynamics. Cells reproduce by dividing into two after doubling in size. To do so they have to convert small randomly arranged compounds of high entropy from the environment into the large highly organized structures of life (low entropy). They thus might appear to be unique entities in which the drive to increase randomness is reversed and the second law defied, since a living cell is at a lower entropy and higher energy level than the starting materials from which it was built. The answer to the paradox is that in oxidizing foodstuff molecules to release energy to drive the process, there is a large increase in entropy due to the liberation of heat and the breakdown of molecules to smaller ones such as CO_2, which escape into the cell's surroundings. This entropy increase exceeds the decrease in entropy that occurs in the production of cells, so that if we consider the cell, plus its surroundings, there is a net increase in total entropy and the second law is obeyed.

Energy can be transformed from one state to another

A familiar example of energy transformation is the conversion of **kinetic energy** (movement) into heat. The kinetic energy of a rock crashing to the ground is converted into heat, by friction on the way down, and when it hits the bottom the conversion is complete. There is also **potential energy**; a rock perched up on the cliff has **gravitational potential energy**, which converts to kinetic energy when it falls. Each molecule has a certain amount of **chemical potential energy** built into it depending on its structure. The molecules of the food that you eat are rich in potential chemical energy, while molecules such as H_2O and CO_2 have, in this context, none. When glucose is oxidized to carbon dioxide and water, its potential energy is released. Food is taken in by organisms where it is oxidized to carbon dioxide and water, and the energy so released is used to drive all the reactions of a living cell. This is summarized in the energy cycle of life shown in Fig. 1.1.

ATP (adenosine triphosphate) is the universal energy currency in life

As already emphasized, so far as life is concerned, heat from food oxidation is 'waste' energy; it keeps you warm but that is different from it doing work. It has, however, the important role of increasing the entropy of the universe as required by the second law. Life is a chemical process and cells must harness the useful energy from food in a form of chemical energy. There are several different classes of food molecules to be oxidized – carbohydrates, fats, proteins, and a variety of other compounds including alcohol. There are also different uses to which the energy is coupled – chemical work, osmotic work, mechanical work, electrical work. It would be impossibly complicated to link all the different type of foodstuff oxidations to all the individual uses of the energy without a degree of flexibility.

This problem has an ingeniously simple solution. Virtually all processes releasing energy from all food molecules trap it in a single compound, **adenosine triphosphate (ATP)**, shown in diagrammatic form in Fig. 3.3. With trivial exceptions, all processes needing energy use ATP to supply it. ATP is the universal energy currency of life. To give a simple illustration, when you contract a muscle, ATP is converted into adenosine diphosphate (ADP) and releases inorganic phosphate. To state an elementary point that may be obvious, for ATP to supply energy to a process, its breakdown must, in some manner, be tightly coupled to the process. If ATP were simply hydrolysed to ADP and phosphate, the energy would simply be released as heat. You will, in this book, come across a great many examples of the ways in which enzymes couple ATP breakdown to the performance of work. The breakdown supplies the requisite energy for muscular activity, for example, by the mechanism described in Chapter 8. Food breakdown processes then immediately leap into action and replenish the ATP reserve by resynthesizing it from ADP and phosphate. This one molecule is the immediate source of energy that drives processes in all living creatures. Dinosaurs roamed on it, electric fishes generate electricity on it, fireflies flash on it (to emphasize its universal role). There is nothing magical about ATP; it is an ordinary chemical with a structure that allows its remarkable role (see Chapter 3). It is obtainable as a white powder that can be stored in the freezer. ATP cannot penetrate into cells, so they must generate it themselves; it cannot be supplied to them from the outside.

Trapping the energy released by food oxidation rather than dissipating it as waste heat is a complex task in detail and is a subject that occupies a large proportion of Part 3 of this book.

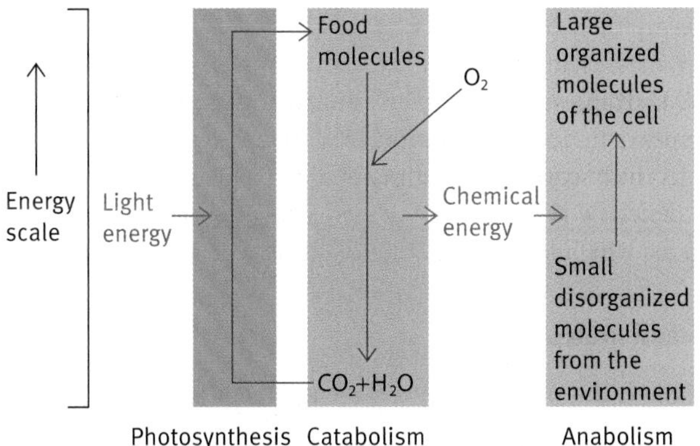

Fig. 1.1 The energy cycle in life. Catabolism is the breakdown of complex molecules releasing energy in the cell. Anabolism is the energy-requiring transformation of simple molecules into more complex ones.

Types of molecules found in living cells

Molecules are formed by the chemical bonding of atoms. There are several types of chemical bonds (see Chapter 3); the most familiar and the strongest is the **covalent bond**. It is formed by two atoms sharing a pair of electrons, a simple example being the formation of a hydrogen molecule from two hydrogen atoms (see Box 1.1). Each electron is attracted to the positive

BOX 1.1 Covalent bond in formation of the hydrogen molecule

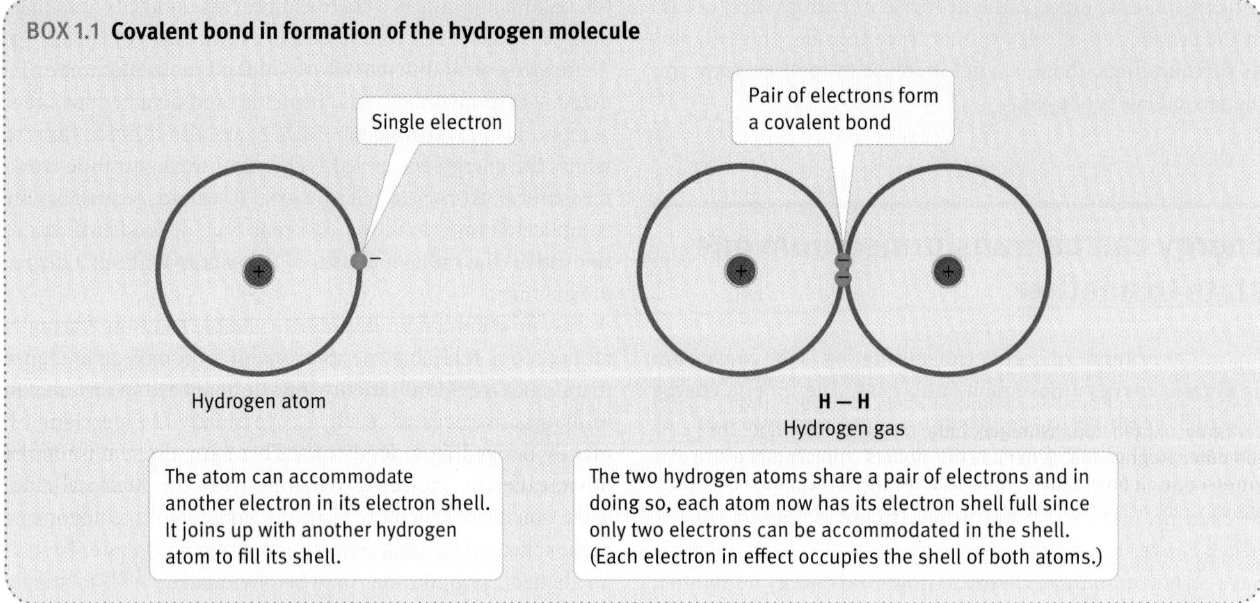

nucleus of both atoms and this holds them together. Hydrogen atoms react in this way to form molecules so that the proportion of free hydrogen atoms in hydrogen gas is negligible. The reaction is energetically strongly 'downhill'; energy from the reaction is liberated as heat. Chemical systems tend to achieve the lowest possible energy state – to achieve the lowest chemical potential energy. The lower this is, the less reactive chemicals are – they are more stable.

Single electron
Pair of electrons form a covalent bond
Hydrogen atom
H—H
Hydrogen gas
The atom can accommodate another electron in its electron shell. It joins up with another hydrogen atom to fill its shell.
The two hydrogen atoms share a pair of electrons and in doing so, each atom now has its electron shell full since only two electrons can be accommodated in the shell. (Each electron in effect occupies the shell of both atoms.)

Atoms are maximally stable when their outer electron shells are filled, as is the case with the inert noble gases such as helium, neon, argon, and krypton. Chemical reactions appear to achieve atomic structures as close as possible to those of the noble gases closest to them in the periodic table. They might be regarded as part of the tendency of the universe to achieve the most stable state and maximum entropy. Most biochemical reactions do not involve free atoms but rather rearrangements or exchanges of chemical groupings between or within molecules. However, the same energetic principle applies in that a biochemical reaction can occur only if it liberates energy and increases entropy as required by the second law of thermodynamics.

Biological molecules are based on the carbon atom bonded mainly to carbon, hydrogen, oxygen, and nitrogen atoms. These four atoms constitute something like 99% of the total number in the body. They can all form strong covalent bonds, the strength of these being inversely proportional to the weights of the atoms involved. The four bonds of carbon are tetrahedrally arranged allowing branching structures to form (Fig. 1.2). This together with its ability to form long chains by C—C bonds enables it to form a wide variety of structures of different shapes and properties. In this respect carbon is unique among the 92 natural elements. Silicon can form chains but has nowhere near the versatility of carbon. Other elements are important in life, including phosphorus and sulphur, which are also strong covalent bond formers. Sodium, potassium, and calcium are also of great importance.

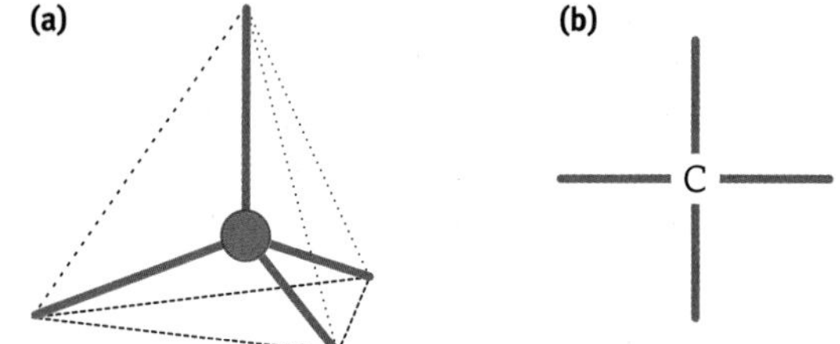

Fig. 1.2 (a) The tetrahedral arrangement of the four single bonds of the carbon atom. (b) As represented in structural formulae.

We arbitrarily divide cellular molecules into two categories, **small molecules** and **macromolecules**. Small molecules are exemplified by glucose, fats, and amino acids, with molecular masses in the range of a few hundred Daltons or less. (A **Dalton** or Da is a unit of **atomic or molecular mass** formally defined as one twelfth of the mass of a carbon 12 atom, effectively equal to the mass of a hydrogen atom. The size of a molecule is commonly described by quoting its molecular mass, which is also sometimes called its 'molecular weight'.)

Small molecules

Water is the most prevalent of the small molecules

Water constitutes about 70% of a typical cell. Although the water molecule has no charged groups and is overall electrically neutral, it is a polar molecule (one that has unequally distributed electrical charge). This is because the two hydrogen atoms are bonded to the oxygen at an angle as shown in Fig. 1.3, giving the molecule an asymmetric shape with the

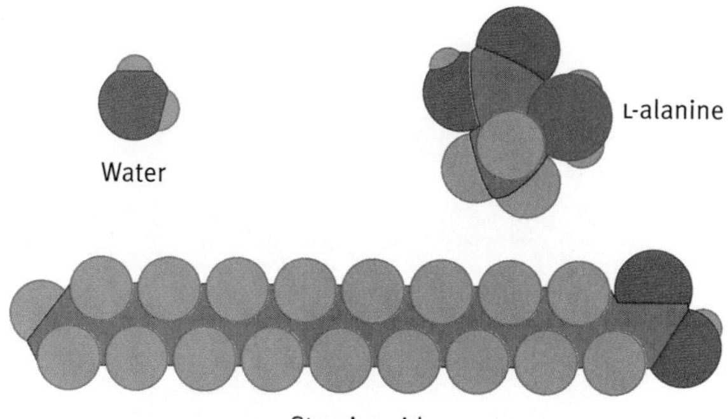

Fig. 1.3 Space-filling models of water, the amino acid L-alanine, and a lipid, stearic acid. The colours of the atoms are: carbon, dark grey; oxygen, red; hydrogen, blue; nitrogen, dark blue. The computer program that generates the models represents the size of the electron cloud of atoms, which is affected by the nature of their attached atoms. In the case of hydrogen with a single electron, the represented size is greatly reduced when attached to an electrophilic atom such as oxygen or nitrogen.

oxygen at one end and the hydrogens at the other. The oxygen atom is electronegative – it attracts electrons and so the electrons in the O—H covalent bonds are closer to the oxygen than they are to the hydrogen. The hydrogens therefore have **partial** positive charges and the oxygen a partial negative one giving an overall polarity to the molecule (see structure in Chapter 3).

Hydrogen bonds have a central role in life

The polarity of water has important repercussions because it allows the molecules to be linked by **hydrogen bonds** causing bulk water to have a cohesive structure, without which the world would be a different place. Hydrogen bonds are also important in cellular structures. They belong to the class of chemical bonds known as **noncovalent or weak secondary bonds**. Despite their description as 'weak' and 'secondary', these bonds lie at the heart of life processes. While covalent bonds give molecules stability and are broken and formed in biochemical reactions, many molecules interact in life processes by noncovalent bonds, as will become evident. Hydrogen bonds form between water molecules due to weak electrostatic attractions between the partial positive charges on the hydrogen atoms and the partial negative charge on the oxygen of adjacent molecules. They can also form between appropriate hydrogen atoms and oxygen or nitrogen atoms of other molecules (see Chapter 3) and play a vital role in biological structures. The genetic apparatus of all organisms is dependent on hydrogen bonding.

Water is a good solvent for polar compounds

The polar nature of water makes it an excellent solvent for polar substances such as sugars. Molecules of sugars in solution become separated by weakly attached water molecules. Such soluble molecules are known as **hydrophilic** (water loving). Nonpolar **hydrophobic** (water hating) molecules such as benzene cannot form bonds with water and so do not dissolve in it. The tendency of water to reject hydrophobic molecules has a profound effect on the structure of macromolecules such as proteins (see Chapter 4) and of biological membranes (see Chapter 7).

Water molecules in the pure state are only minutely dissociated into hydrogen ions and hydroxyl ions both being at 10^{-7} M.

Small molecules other than water

Most of the rest of the small molecules found in cells are monomers (single molecules that can be linked together to form larger structures), such as sugars, fats, and amino acids, mainly derived from foodstuffs. There are also thousands of other, different, molecules resulting from the chemical reactions going on in the cells. The sugars such as glucose and sucrose are called carbohydrates because they have the elements of carbon and water, many conforming to the general formula $C_m(H_2O)_n$. They are important energy stores. Amino acids are short carbon chains with basic amino groups and acidic carboxyl groups (see Chapter 4). Lipids or fats have various roles, the two most prominent being the formation of cell membranes (see Chapter 7) and the storage of energy in animals (see Chapter 10).

Fig. 1.3 shows molecular models of water, l-alanine (a typical amino acid) and stearic acid (a typical lipid), which has a long hydrocarbon chain.

Macromolecules are made by polymerization of smaller units

Glycogen, **starch**, and **cellulose** are large **polymers** known as **polysaccharides** formed by joining together glucose units in a slightly different manner in each of the three cases (see Chapter 9). Glycogen and starch function as energy stores in animals and plants respectively and cellulose provides structural strength in plants. Cellulose is broken down by bacteria in ruminants, for whom it is their major food source, but cannot be used as food by other animals. Only glucose monomers are involved in the synthesis of these molecules (though more complex polysaccharides exist); all that is needed for their synthesis is a mechanism to link them together in the appropriate way. There is, therefore, no information content in them. They resemble long strings of a single unit.

Protein and nucleic acid molecules have information content

Proteins, DNA, and RNA are macromolecules built up from a variety of monomers that are linked together in a specific order in each molecule. Therefore they have information content based on the sequence of monomers. The cell contains template instructions on the correct sequences for each protein, RNA, and DNA molecule. These can be compared with meaningful messages composed from alphabets.

The flow of information is DNA→RNA→Protein.

Proteins

It has been said that proteins may be the most remarkable molecules in the universe. The word protein is derived from the Greek meaning 'primary'; proteins are of primary importance in life; genes and the genetic system exist to make their production possible. As a generalized statement, proteins do everything in the mechanism of life. They are built up from a menu of 20 different amino acids, which are polymerized into long chains, known as **polypeptides** (Fig. 1.4). The DNA of the genes encodes the amino acid sequence for each protein, one gene for each polypeptide chain. Life depends on these sequences being correct. A single incorrect amino acid among hundreds or thousands in a protein molecule may result in a genetic disease. After synthesis, the chains fold up into three-dimensional compact shapes determined by the particular sequence of amino acids. Fig. 1.5 shows a space-filling molecular model of human **deoxyhaemoglobin**, an average-sized protein of 574 amino acids and molecular mass of 64,500 Da. Proteins range in size from the small insulin molecule (molecular mass 5,733 Da), which is comprised of 51 amino acids linked together, to large ones of several thousand amino acids.

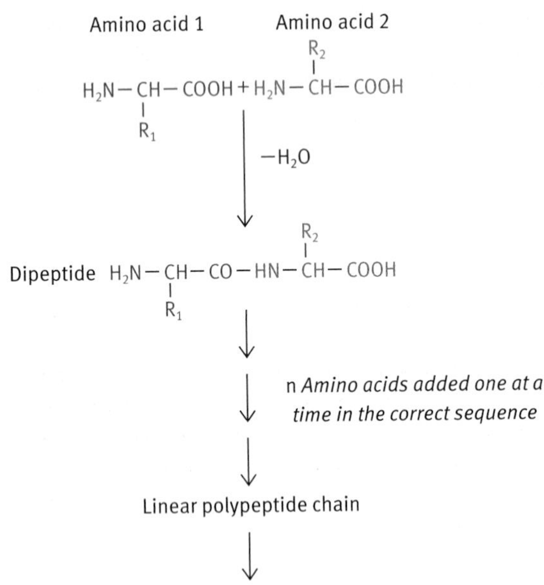

Fig. 1.4 Outline of protein synthesis. Note that although peptide synthesis involves the removal of water molecules overall, the process in the cell is not a direct condensation. Protein synthesis is carried out by cellular structures called ribosomes. The sequence of amino acids linked to form the polypeptide chain is specified by a molecule of messenger RNA, which is a copy of the base sequence of the gene coding for the protein.

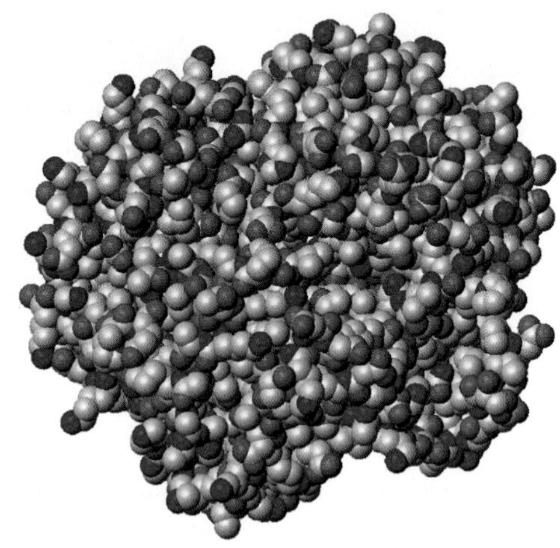

Fig. 1.5 Space-filling model of haemoglobin. The CPK (CoreyPauling-Koltun) colour scheme is used: carbon, light grey; oxygen, red; nitrogen, blue; sulphur, yellow. The Protein Data Bank accession code for haemoglobin is 1A3N.

Catalysis of reactions by enzyme proteins is central to the existence of life

Enzymes are catalytic proteins. Thousands of different chemical reactions occur in a living cell even though the mild conditions there are not such that would facilitate chemical reactions – almost neutral pH, low temperature, no especially reactive substances and chemicals present in dilute aqueous solution. In the chemistry laboratory, reactions are commonly brought about by high temperatures, extreme pH and high concentrations of reactants. A sugar such as glucose is stable at body temperature and left in air in a bowl will undergo no change for many years. If you eat the sugar, it is involved in rapid chemical reactions in the cell due to enzymes combining with the molecules and catalysing the reactions. Enzymes are specific protein catalysts – usually one enzyme, one reaction. Without the ability of proteins to bind precisely to their target molecules (in this case, enzyme substrates) and catalyse specific reactions, life would be impossible. Since there are thousands of different reactions in a cell it follows that there are thousands of different enzymes catalysing them. They are efficient catalysts. One molecule of the enzyme carbonic anhydrase, important in red blood cells, catalyses the conversion of 600,000 molecules of substrate per second. The mechanisms by which proteins are so catalytically efficient are described in Chapter 6.

What is the function of enzymes?

Earlier, we described how the second law of thermodynamics determines whether a reaction may be able to proceed, not whether a given reaction *will* actually take place. This is a separate question involving the nature of chemical reactions. There is an energetic barrier to chemical reactions occurring, without which everything that could react would have done so long ago. Everything combustible would burst into flames in the presence of oxygen if there was no barrier. Enzymes cannot change the thermodynamics of a reaction – they cannot tinker with the laws of thermodynamics – but they can lower the barrier to chemical reactions. Enzyme catalysis, explained in Chapter 6, is one of the wonders of life.

Proteins work by molecular recognition

As stated, in order to act catalytically on specific substrates, enzymes must 'recognize' their target molecule and bind to it. They do this by virtue of noncovalent bonds between the protein and the target molecule. This can be exquisitely specific. The ability of proteins to recognize other molecules is central to almost everything in life. Cellular structures, muscle contraction, nerve impulses, hormone action, chemical signalling, and regulation of metabolism all depend on it. Proteins are very versatile, ranging from delicate enzymes and exquisite molecular machines to the tough proteins of cartilage, hair, and horses' hooves.

Life is self-assembling due to molecular recognition by proteins

Life is based on a one-dimensional linear coding system (DNA) and yet the information is translated into three-dimensional organisms. The linear code is initially translated into linear, one-dimensional polypeptide chains, but these fold up to form three-dimensional proteins that have uniquely specific recognition sites on them. The folding is determined by the amino acid sequences of the polypeptides, which themselves are determined by nucleotide sequences in the genes. The evolution of genes has therefore determined the specific abilities of proteins to recognize other molecules, which in turn determine the development of all living organisms. Rabbit genes direct production of rabbit proteins, and these will cause a rabbit to develop from a fertilized egg because of a multitude of molecular recognitions.

Many proteins are molecular machines

We come now to one of the most remarkable properties of many or most proteins. They can undergo microscopic conformational changes in their three-dimensional shape on binding to their cognate molecules. (The term 'cognate', in this context, means 'having affinity to'. A cognate molecule is one that it is appropriate for the protein to bind.) When most or all enzymes bind to their substrate they change shape very slightly; it is part of the catalytic mechanism. Also, the activity of enzymes must be regulated. Fuels need to be metabolized more rapidly during exercise than when asleep. For example, key control enzymes change their shape on detecting that ATP supplies are adequate, and switch off the oxidation of food (see Chapter 20). Muscle contraction depends on vast numbers of protein molecules undergoing conformational change and interaction with protein fibres. Other proteins literally walk along special protein tracks in the cell pulling loads; they are molecular motors (see Chapter 8). Haemoglobin does not just carry oxygen; it responds to signals and changes its shape so as to maximize oxygen pick up in the lungs and oxygen surrender in the tissues (see Chapter 4). A vast network of cell–cell communication regulates cell activities (see Chapter 29). This signalling is dependent on protein conformational changes, as are gene regulation (see Chapter 26) and control of cell division (see Chapter 30).

How can one class of molecule carry out so many tasks?

If we regard amino acids as an alphabet of 20 letters, proteins can be regarded as words, typically several hundred or more letters long; the number of theoretically possible different amino acid sequences is virtually infinite. It is the sequence that determines the three-dimensional shape of a protein and the shape that gives it a specific function. The thousands of protein sequences that exist today have evolved over billions of years of random mutation and natural selection and are stored encoded in the genes.

Evolution of proteins

As stated, evolution is a process in which natural selection can preserve those mutations that increase the chance of progeny reaching reproductive age. Deleterious mutations are likely to be eliminated. Since genes code for proteins, evolution depends on the synthesis of new proteins that give a selective advantage (or, at least, give no disadvantage). The chance of a random change in the amino acid sequence of a protein being advantageous is small, so that evolution is a slow, chancy business, but because vast timescales are involved it has resulted in a myriad of different proteins and hence in the complexity and variety of life.

Development of new genes

The evolution of proteins requires the development of new genes. The problem of how an essential gene can change into a different one without eliminating the function of the original can often be explained by chance gene duplication. One of the duplicates can then mutate into a new gene that results in the production of a new protein, while the other continues to code for the original protein. There is much evidence in the sequences of genes indicating that this has often happened; families of related genes and proteins exist that have evolved from common ancestors (see Chapter 4).

DNA (deoxyribonucleic acid)

It was established in the 1940s that DNA (**deoxyribonucleic acid**) is the substance of genes. Complete DNA molecules are present in cells as **chromosomes**, with protein components present as structural support. The *E. coli* chromosome has a molecular weight of 12 million Daltons, and the largest human chromosome is several billion Daltons. As already explained, DNA constitutes the chemical message that instructs the cell how to assemble amino acids in the correct order to produce the protein encoded by each gene. The information is contained in the sequence of the monomers called **nucleotides**, which make up DNA. A nucleotide has the structure:

base – sugar – phosphate

There are four different nucleotides in DNA, differing in the base components, and linked together by a 'backbone' of alternating sugar-phosphate residues, with the bases projecting from the sugars. The sequence of different bases carries the information of the gene.

DNA exists in the form of a double strand held together by secondary bonds of which hydrogen bonds are critical, as illustrated in Fig. 1.6. Two of the four species of bases in DNA, **adenine** and **thymine** (A and T) hydrogen bond together because their shapes are complementary. The same is true for the other pair, **guanine** and **cytosine** (G and C). This pairing is known as **Watson–Crick base pairing**, after its discoverers. It is specific; base pairing in this way in the double helix occurs only between G and C, and A and T respectively. The two strands in a DNA molecule are not parallel, as indicated for simplicity, in Fig 1.6, but rather wind around each other to form the well-known double helix, shown more realistically in the space-filling model shown in Fig. 1.7. Chapter 22 deals with DNA structure.

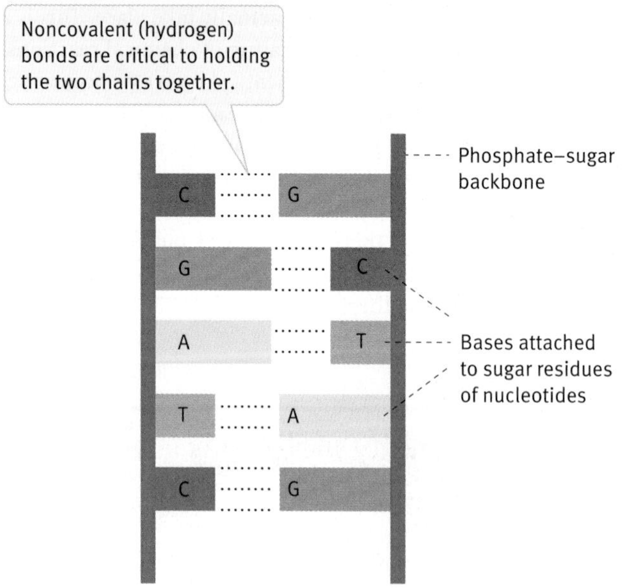

DNA directs its own replication

The central requirement of any genetic system is that the hereditary information can be passed on to daughter cells. Nucleic acids fulfil this requirement, as they have the capacity to direct their own replication as well as performing the function of directing protein synthesis. The two strands of DNA are separated and each strand acts as a template for the assembly of new partner strands. An A on the template strand matches a T on the new strand, G is matched to a C and *vice versa*. Fig. 1.8 illustrates the principle. This results in two new double helices identical to the original. DNA synthesis is discussed in detail in Chapter 23.

Fig. 1.6 Diagram of the structure of double-stranded DNA. The backbone consists of alternating sugar–phosphate residues to which the four types of base are attached. The base pairs are always between G and C or between A and T. Note that each base pair always includes one larger and one smaller base so that all base pairs are of the same size. The two chains are held together by noncovalent bonds; there are three between G and C, and two between A and T. The two strands are shown as being parallel for clarity but in fact they form a double helix as shown in the model in Fig. 1.7.

Genetic code

Triplets of bases in the DNA known as codons specify amino acids of polypeptide chains. The 'dictionary' that describes which triplet represents which amino acid is known as the **genetic code**. With 4 different bases, 64 different triplet combinations (**codons**) are possible (4×4×4).

The DNA of the human genome (the complete collection of **genes**) contains 3.2 billion nucleotide pairs, The complete sequencing of these has been achieved by the **Human Genome Project** completed in 2003.

Genes are part of the continuous chromosomal DNA molecule. In general, each gene is distinct from the next, separated by spacer sequences between them.

Proteins are synthesized on cellular structures known as **ribosomes**. These take instructions (indirectly) from the gene.

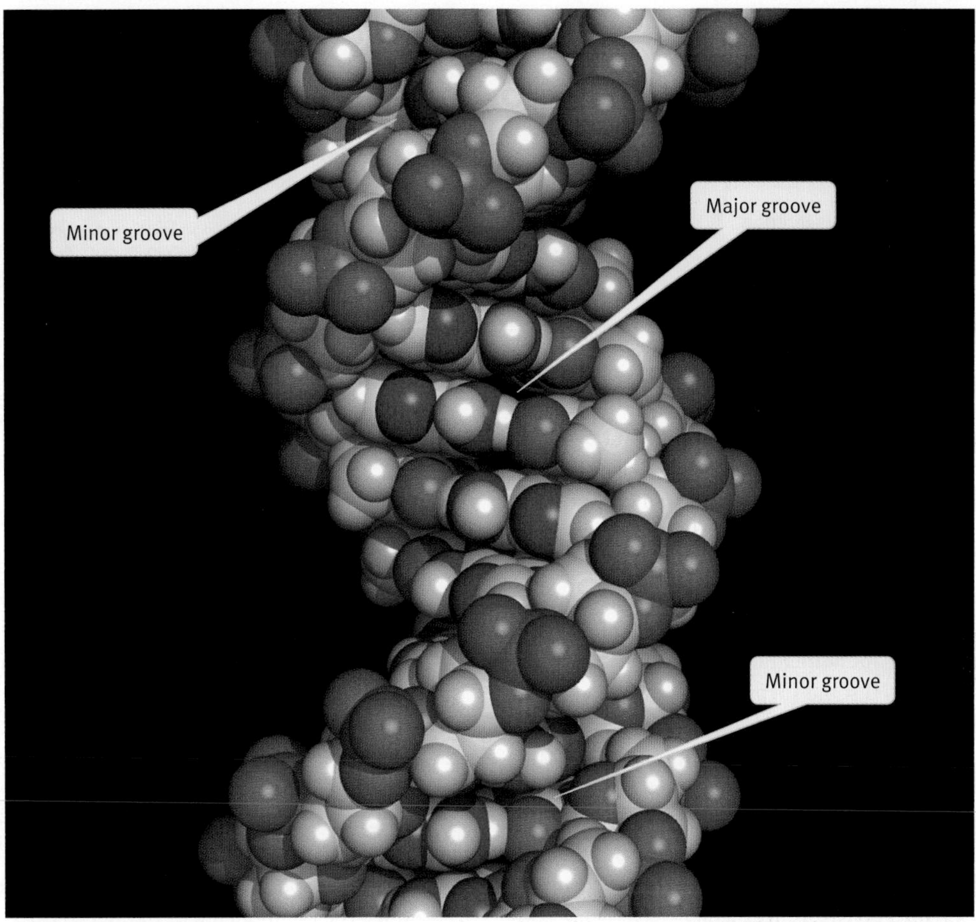

Fig. 1.7 A model of B DNA, Protein Data Bank code 1BNA. Space-filling atomic model of a DNA segment with one major groove and part of two minor grooves.

Each gene is independently copied into a different relatively short nucleic acid called **messenger RNA (mRNA)**, which delivers the message of coded instructions from the gene to the ribosome. RNA has almost the same structure as a single strand of DNA with small chemical differences. The bases in RNA specifically base pair much as in DNA. The sequence of information flow is as shown below.

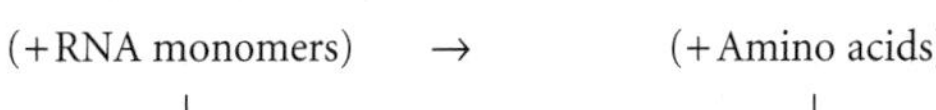

DNA of gene 1 → messenger RNA 1 → polypeptide 1 → folded protein 1
DNA of gene 2 → messenger RNA 2 → polypeptide 2 → folded protein 2

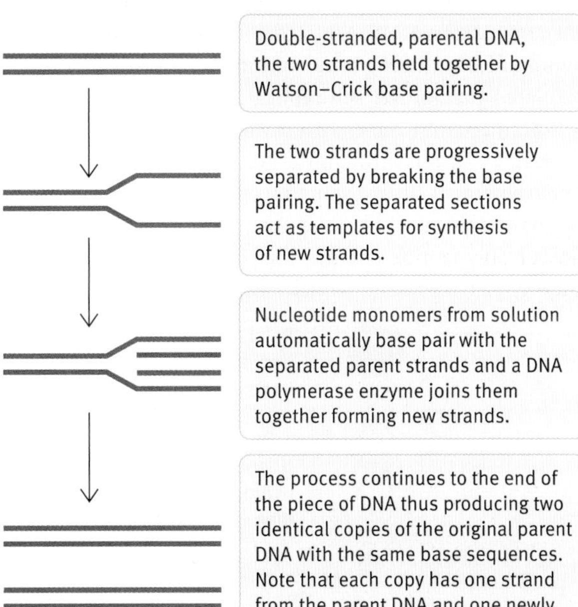

Double-stranded, parental DNA, the two strands held together by Watson–Crick base pairing.
The two strands are progressively separated by breaking the base pairing. The separated sections act as templates for synthesis of new strands.
Nucleotide monomers from solution automatically base pair with the separated parent strands and a DNA polymerase enzyme joins them together forming new strands.
The process continues to the end of the piece of DNA thus producing two identical copies of the original parent DNA with the same base sequences. Note that each copy has one strand from the parent DNA and one newly synthesized strand. This is known as semi-conservative replication.

Fig. 1.8 Principle of DNA replication. The two strands of the double helix are held together by hydrogen bonds between bases A and T, and G and C respectively. When the strands are separated the single strands are now available for base pairing by incoming monomer nucleotides. The nucleotides thus lined up are linked together to give two identical daughter double helices.

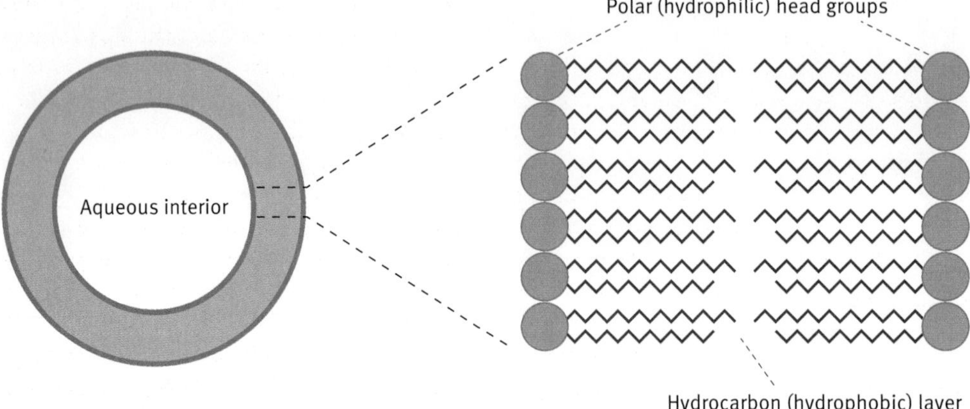

Fig. 1.9 A synthetic liposome made of a lipid bilayer structure.

Organization of the genome

The completion of the monumental task of sequencing the human genome delivered some surprises. The genes are scattered about within the linear DNA sequence and over half the DNA has no obvious function, and is sometimes referred to as 'junk DNA'. Bacteria have very little of this 'extra' DNA and look more orderly in their organization, with genes adjacent to one another in the DNA molecule. However, the concept of junk DNA has been reassessed following the discovery of large numbers of **microRNA genes**, which have been conserved over long periods of evolution. They code for short **microRNAs** that do not code for proteins, and are potentially highly significant in gene regulation. This area will be dealt with in Chapter 26.

How did life start?

Living organisms consist of one or more cells. Each cell is surrounded by a cell membrane, a thin sheet composed mainly of lipid molecules that holds the contents of the cells together, as well as carrying out other functions.

As already stated, at some time in the establishment of life there must have been a primordial self-replicating molecular system from which living cells developed. Hypotheses have been formulated of how such systems might have been established on a mineral surface or in a drop of liquid or sea pool, but, at an early stage, the system had to be contained by a membrane or it presumably would have dispersed. A striking fact is that when molecules of a suitable substance are simply agitated in water they form small spherical vesicles (see Chapter 7). The boundary of these vesicles is a structure known as a **lipid bilayer**, which is virtually identical to the basic structure found in the membranes of modern cells (Fig. 1.9). Such vesicles may have enclosed a drop of the first self-replicating system. From such a primordial cell-like structure all life is postulated to have originated. The requirements for a molecule to be capable of forming a lipid bilayer are not demanding; it needs to have amphoteric properties, by which we mean one part of a molecule is water-insoluble (hydrophobic) and the other water-soluble (hydrophilic), and be of a roughly suitable shape, as illustrated in Fig. 1.10.

What was the source of the molecular building blocks needed to produce the components of living cells? Experiments have been done in which electrical discharges were passed through a mixture of gases (hydrogen, methane, and ammonia, in the presence of water) intended to resemble the atmosphere believed to exist on the primitive earth, as suggested by geologists and astronomers. A mixture of potential precursors of biomolecules including some amino acids was produced. The postulated primordial self-replicating cell must have taken in molecules from the environment to produce new cellular material. Diffusion through the containing membrane before the development of transport mechanisms would have been slow, and replication likewise slow, but vast time scales were involved.

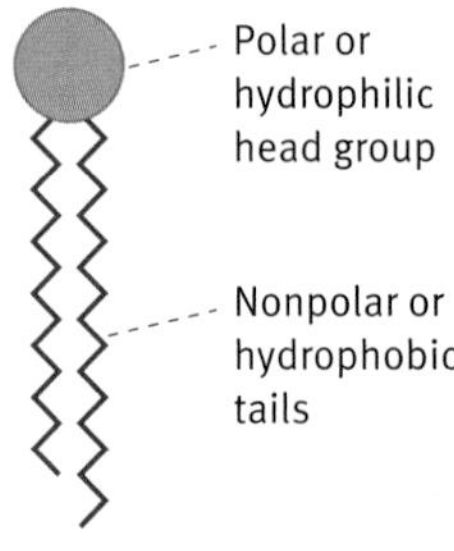

Fig. 1.10 An amphipathic molecule of the type found in cell membranes.

The RNA world

A more difficult problem in the establishment of a self-replicating system is to identify the initial catalysts and the primitive 'genetic system' that would ensure faithful replication. In short a chicken and egg problem; which came first, proteins to catalyse reactions or nucleic acids to direct the synthesis of primitive proteins? This dilemma received a possible answer with the discovery that RNA can catalyse some chemical reactions including conversion of short polynucleotides into longer sequences. Such catalytic RNA molecules were given the name of '**ribozymes**' (not to be confused with ribosomes). It was the first time that biological molecules other than proteins had been found to catalyse specific reactions. RNA has the same potentiality for acting as a template in its own replication as explained for DNA. RNA was probably both the catalyst and the primitive 'genetic system' for self-replication in the origin of life, thus avoiding the chicken and egg dilemma. It may be speculated that the first short polynucleotides were formed from nucleotide monomers by heat chemically condensing the nucleotides together by driving off water. From this stage, evolution of more efficient catalysts, namely proteins, to replace RNA catalysts is postulated to have occurred, though the first 'proteins' must have been primitive and presumably were short peptides of low catalytic efficiency.

The concept of an RNA-based biological world that preceded the DNA world is generally accepted, as there is much supporting evidence. In modern cells, although protein enzymes bring about almost all catalysed reactions, the displacement of RNA from this role is not complete. What might be regarded as a few 'fossil' catalysts – hangovers from the RNA world – exist in cells as ribozymes. One of these in ribosomes is involved in the synthesis of all proteins (see Chapter 25), providing an interesting link between one type of catalytic system (RNA) and a more efficient one (proteins). Ribosomes are giving us a glimpse into the ancient RNA world, somewhat akin to astronomers viewing the past universe through long-distance telescopes.

Proteomics and genomics

From what has been said in this chapter, it will be clear that sequences of amino acids in proteins and those of the nucleotides in DNA underlie just about everything in life. As these sequences were determined, it was realized that the flood of molecular information would be of little avail without an efficient retrieval system. To this end, protein and DNA computer databases were established in various centres around the world in which information on proteins and genes is recorded. Details of the sequences of thousands of genes and proteins together with the three-dimensional structures of many proteins are now available. Software in the public domain is available to search the databases and analyse the information contained in them. This area of science, known as **bioinformatics**, has become of immense importance in biochemistry and molecular biology. These developments have made computers with their remarkable software programs essential tools in molecular research.

Parallel to this there has been development of methods for the automatic sequencing of DNA that have resulted in the completion of the human genome project and the sequencing of the genomes of other species, such as those of the mouse, the rice plant, and *Drosophila*, the fruit fly, to cite only a few. Other technical developments, using DNA microarrays (see Chapter 28), allow the simultaneous study of which genes are active. In the protein field, the relatively recent application of mass spectrometry to proteins, a development of immense importance (see Chapter 5), makes it feasible to investigate many proteins at once.

These developments are sometimes referred to informally as the '**omics revolution**'. The entire collection of proteins in a cell (in any one state, for it varies from time to time) is called the **proteome**, and that of genes, the **genome**. The collective studies of these are called **proteomics** and **genomics** respectively. The essential aspect is that large numbers of proteins and genes are analysed simultaneously. An apt analogy has been put forward to illustrate the meaning of these terms to the effect that many of the instruments in an orchestra (genes and proteins) have been identified. The next stage is to listen to the music the orchestra plays with them. In other words, the proteins and genes in a cell function as a collective whole, and will need to be considered as such to achieve a full understanding of life processes and the misfunctions that occur in disease. As a simple illustrative example, comparing the collection of genes, gene activities and proteins in cancer cells with their normal neighbours, could give leads on what has caused the cancer and suggest therapeutic strategies.

Summary

Unity of life

Despite the diversity of life forms, at the molecular level all life is basically the same, with variations being modifications. This suggests a single origin of life.

Living cells obey the laws of physics and chemistry

Energy is derived in cells from breaking down food molecules (ultimately produced by plants using sunlight energy). The energy must be released in a form that can drive chemical and other work. Heat cannot do work in the cell.

ATP is the universal energy currency in life

Energy from food breakdown is used to synthesize ATP (adenosine triphosphate) from ADP (adenosine diphosphate) and phosphate. ATP breakdown can then be coupled to do biochemical work.

Molecules found in living cells

These include small molecules such as water, food molecules and their breakdown products. Macromolecules, among which proteins and DNA are pre-eminent, are large molecules formed by polymerization of smaller units.

Proteins

These are the cell's workhorses and the basis of most living structures. They are long chains of amino acids, typically hundreds long, but folded up into a precise three-dimensional structure. Twenty different amino acids are found in proteins; each protein has a unique amino acid sequence.

Enzyme catalysis

Enzymes are proteins that catalyse virtually all the thousands of chemical reactions of life with great specificity: one enzyme, one reaction. Relatively recently, however, it has been discovered that RNA (ribonucleic acid) can have catalytic activity and that a small number of cellular enzymes are RNA molecules.

DNA

The cell must have an instruction manual for the sequence of each of the thousands of proteins it synthesizes. This is the function of DNA in the form of genes, each gene specifying the amino sequence of one polypeptide. DNA consists of two strands of polynucleotides in a double helix. A nucleotide has the structure: base–sugar–phosphate. The bases are paired by hydrogen bonds, the base A linked to T, and G to C. This automatic pairing is the basis of self-directed replication. The base sequences also act as a code, specifying individual amino acids. Groups of three bases, known as codons, each represent an amino acid. The genetic code correlates codons to the amino acids they specify.

Ribosomes translate the base sequences of genes into proteins. Each gene is copied into messenger RNA (a polymer resembling DNA and with the same information content), which attaches to and instructs a ribosome.

Evolution of genes and proteins

DNA is the record of sequences needed to synthesize proteins, the information having developed through billions of years of evolution. Mistakes in replicating DNA inevitably occur. These are mutations that result in incorrect amino acid sequences in proteins, which may in turn result in genetic diseases. Random mutations are also the material on which evolution, *via* natural selection, develops new genes and proteins.

Molecular recognition by proteins

Apart from recognition of substrates by enzymes, proteins recognize (bind to) other molecules such as hormones and growth factors, thus directing development, growth and metabolic processes. The binding is by multiple weak bonds whose formation depends on atoms being close enough for the bonds to form. This means that only molecules closely complementary to one another bind. Weak bonding in molecular recognition confers flexibility and reversibility. It is the basis of biological specificity.

How did it all start?

Life presumably must have originated by the spontaneous formation of a molecule capable of self-replication. It is generally believed that life originated with RNA, which has the information to direct its own replication and can also form catalytic molecules, thus fulfilling the functions of both DNA and proteins. The 'RNA world' is still seen in all cells in the form of ribosomes, which have a high content of structural RNA and an RNA enzyme – a 'ribozyme'. DNA replaced RNA because it is a chemically more stable repository of genetic information.

The new 'omics' phase of biochemistry and molecular biology

Since the 1990s an explosion of new technologies has revolutionized biochemistry and molecular biology. Prominent among these are automated DNA sequencing, mass spectrometry for the study of proteins and DNA microarrays for gene and gene expression studies. They are having enormous effects on biological science, medicine, and agriculture. The branches of science using these are termed proteomics and genomics, which are collective terms to specify that large numbers of proteins and genes can be examined together.

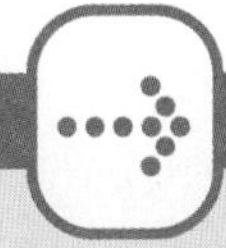

Further reading

To access the further reading, please scan the QR code image or go to http://global.oup.com/uk/orc/biosciences/molbiol/snape_biochemistry/student/reading/ch01/

Problems

1 How is the entropy level related to the energy level of a system?

2 What is energy?

3 List the various forms of energy.

4 What do we mean by saying that proteins and DNA have information content?

5 What is messenger RNA in broad terms?

6 Which properties enabled nucleic acids to become the basis of life?

Chapter 2

Cells and viruses

This chapter offers a broad survey of the structures and properties of cells and viruses, the aim being to provide a biological background for the more detailed mechanisms of cell biology, biochemistry, and molecular biology that follow in subsequent chapters. Viruses, it should be noted, are not living organisms in the sense that cells are. They are incapable of replication themselves and have virtually no biochemical activities of their own. However, they possess genomes and have the ability to enter living cells, where they cause the host's molecular biology machinery to replicate them, as a result of which they cause diseases. We will deal first with cellular life and then with viruses.

Cells are the units of all living systems

All living organisms consist of one or more cells and, with rare exceptions, cells are microscopic in size. Bacteria are the smallest; *Escherichia coli (E. coli),* is rod-shaped with dimensions of about 2 μm in length and 1 μm in diameter. Animal and plant cells typically have dimensions about 10 times larger. A delimiting lipid membrane and a DNA genome are common features of all cells.

Bacteria are single cell organisms that live independently of each other, or occasionally in small aggregations. Their 'strategy' of life is, essentially, to grow as rapidly as possible and outstrip the growth of competitors in the struggle for survival. At the other end of the biological scale are multicellular organisms which, in the case of more complex ones such as mammals, are composed of vast numbers of cells. In a human it is estimated that there are around 10^{11}–10^{13} cells. In such organisms there are many different types of cells specializing in different functions, and this requires mechanisms by which their activities are controlled so that they are appropriate to the needs of the organism as a whole.

What determines the size of cells?

Physical considerations can account for the small size of almost all cells, the surface area/volume ratio being probably the most important. Cells take in molecules from the environment, such as food, and release unwanted molecules such as carbon dioxide, so that there is constant molecular traffic across the membrane. Adequate surface area is needed to allow a rate of exchange sufficient to support the needs of the cell. In addition, incoming molecules and ions have to reach all parts of the cell. Diffusion in solution is a relatively slow process and would be inadequate for the cell's needs over more than tiny distances. The advantages of small cells are that the ratio of surface to volume is maximized, and the length of diffusion paths within the cells minimized. As well as incoming molecules, macromolecules synthesized in one part of a cell often have to reach their functional sites in other parts. Bacteria are small enough to rely on diffusion for this transport but in animal and plant cells, 1,000 times larger in volume, it would be inadequate, and energy-requiring transport systems are also used.

The necessity of accommodating the essential quota of macromolecules, particularly the DNA and proteins, places a minimum size on cells.

Classification of organisms

There are three main evolutionary branches of organisms, **bacteria** (or **eubacteria**), **archaea** (or **archaebacteria**), and **eukaryotes**. The bacteria and archaea were previously classed together as **prokaryotes**, and although it is now considered that archaea constitute a separate evolutionary group, their cell structure resembles that of bacteria. Prokaryotic and eukaryotic cells are therefore the two main types.

Bacteria include 'typical' bacteria such as *E. coli* and the photosynthetic cyanobacteria. Archaea include **extremophiles**, which are found in hostile environments such as acid hot springs and around hydrothermal vents in the ocean floor, possibly representing conditions on primitive earth. Extremophiles live on chemical 'food' such as hydrogen sulphide emanating from the earth's crust. Possibly they represent survivors from life on primitive earth. Some extremophiles have enzymes of unusual thermal stability, which have found important applications in recombinant DNA manipulation (see Chapter 28).

Prokaryotic cells have no nuclear membrane

Eukaryotes are the so-called higher plant and animal cells, which range from unicellular yeasts and protozoa to humans. 'Karyon' in Greek refers to a nucleus. Prokaryotes ('before nucleus') do not have a distinct nucleus bounded by a membrane enclosing their DNA genome. Eukaryotes ('true nucleus') have a nucleus bounded by a membrane. Although the presence or absence of a membrane around the DNA of a cell might seem like a trivial difference, in fact it represents a major evolutionary divide. In eukaryotes, the separation of genes issuing instructions from the cytoplasm where those instructions are carried out has profound repercussions on the molecular biology of the cell, as will be explained later in the book. We will now describe the main characteristics of both classes of cells.

Prokaryotic cells

E. coli, a bacterium that lives in the human gut, is a typical prokaryotic cell. It is structurally simple, being a single compartment with no **internal** membranes. Stated differently, it has no **organelles**, these being the membrane-bounded 'small organs', such as mitochondria, that are so important in eukaryotic cells (Fig. 2.1).

The *E. coli* cell is surrounded by the plasma membrane, which is constructed of lipid molecules (collectively called membrane or polar lipids) into which proteins that carry out various functions are inserted (see Chapter 7). Membrane proteins include systems that transport molecules into the cell, and the membrane is also the site of the vitally important system for generating ATP. Outside the membrane is a protective strong cell wall made of a meshwork of chains of modified sugars (**glycans**) cross-linked by short peptides. The structure resembles a seamless string bag, the whole cell wall being a single molecule. (The antibiotic **penicillin** inactivates the enzyme that catalyses the final step in the formation of the peptide cross-links. A weakened cell wall results, causing the cell to burst due to the high osmotic pressure inside.) A second

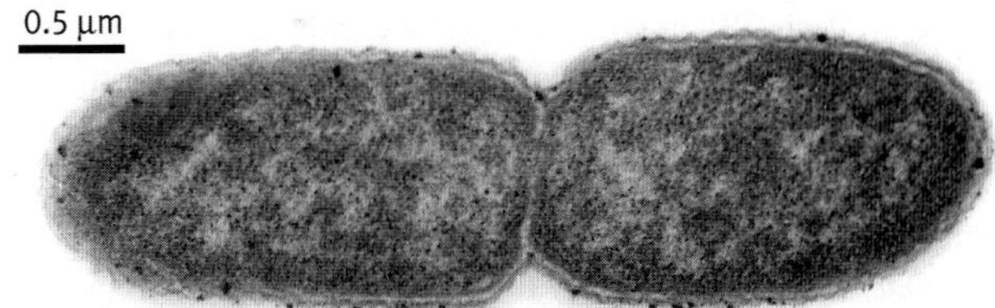

Fig. 2.2 Electron micrograph of an *Escherichia coli* cell undergoing septum formation. Septum formation occurs at the middle of the cell and involves ingrowth of the cell membrane and cell wall layers. Reproduced from Wang, Smith & Davis Thrive in Cell Biology (2013) Oxford University Press, Oxford.

membrane exists outside the cell wall in *E. coli* but is not present in all bacteria.

The prokaryotic chromosome resides in the cell as a nucleoid, a diffuse, tangled, thread-like structure visible in the electron microscope (Fig. 2.2). The main chromosome of *E. coli* is a circle of double-stranded DNA, though in the tangled form in which it is packed into the cell it cannot be discerned to be circular. The *E. coli* chromosome is estimated to comprise just over 4,000 genes. The cell contains only a single copy of the main chromosome but, in addition, it often contains small circular DNA molecules, variable in number, known as **plasmids**. Plasmids contain a variety of genes that often include those conferring resistance to antibiotics. They replicate independently of the main chromosome. Plasmids play a major role in recombinant DNA technology (genetic engineering), dealt with in Chapter 28.

Cell division in prokaryotes

Some prokaryotes such as *E. coli* can grow in simple media containing inorganic salts and a source of energy such as glucose. Some can replicate in about 20 minutes under ideal conditions, which means that a single cell can, overnight, multiply to hundreds of millions.

The prokaryotic cell cycle is uncomplicated in that the replication of the chromosome goes on continuously while the cell enlarges, and once a critical size is reached cell division occurs by binary fission (Fig. 2.2). In this process, a cell wall and new membranes begin to form around the circumference of the cell,

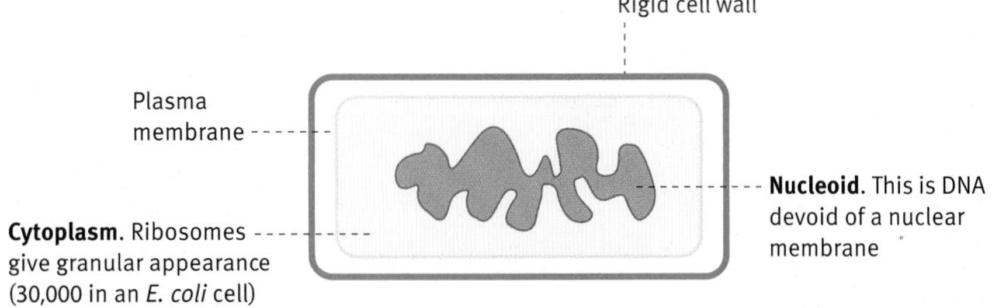

Fig. 2.1 Diagram of a prokaryotic cell. Most are either rod-shaped or spherical. Note the complete absence of any membranous structures inside the plasma membrane. (Plasma membrane is the term usually used for the cell membrane.) Some bacteria like *Escherichia coli* have an additional membrane on the outside of the cell wall.

at the midline. When complete, these divide the cell into two. Each daughter cell must receive a copy of the chromosome, so the two DNA molecules arising from the chromosome replication have to be segregated into what will become the two daughter cells. In the most intensively studied bacteria (*E. coli* and *Bacillus subtilis*) the favoured model links segregation to chromosome replication. The replication apparatus (**replisome** or **segrosome**) remains at the mid-cell position for most of the cell cycle. As the circular chromosome is replicated, the two copies of DNA are extruded to either side of the mid-cell line in the direction of the two ends. The process is physically complex and not completely understood.

E. coli cells also undergo conjugation, in which genes are exchanged between two genomes; this **recombination** increases genetic variability, which is important for evolution. A cell that contains a 100 kb (kilobase) circular plasmid known as the **F plasmid** (an F^+ cell) can form a temporary bridge to a recipient F^- cell lacking the plasmid, and a copy of the plasmid is transferred across. In some F^+ cells the F plasmid becomes integrated into the main chromosome, so that a section of chromosomal DNA is also transferred into the F^- cell where it replaces the corresponding section in the recipient's chromosome. The recipient thus receives new genes, creating genetic diversity. The donor cell's chromosome remains unchanged by the process because only a single-strand of DNA is transferred, and in both donor and recipient cells DNA replication restores the second strand.

Eukaryotic cells

A **plasma membrane** surrounds the eukaryotic cell. It is a lipid structure, as in bacteria, with proteins inserted into it used for molecular transport, receipt of cell signals and other functions (see Chapter 7). In contrast to prokaryotes, however, eukaryotic cells contain membrane-enclosed structures, known as **organelles** (Fig. 2.3). The cell **nucleus** is the most prominent organelle inside the cell and contains the chromosomes. The typical eukaryotic **genome** is **diploid**, there being two copies of each chromosome, one from each of the two parents (see Chapter 30). The nucleus is surrounded by a double membrane supported on the inside by a layer of filamentous proteins (see Chapter 8) and studded with **nuclear pores** (see Fig. 27.15), through which there is both diffusion and active transport of specific molecules in both directions. An elaborate mechanism controls the movement of molecules through the pores (see Chapter 27).

There is some variation in the way the term **cytoplasm** is used when referring to eukaryotic cells. Strictly, it is all the contents of a cell excluding the nucleus but including the other membrane bound organelles described below, while **cytosol** is

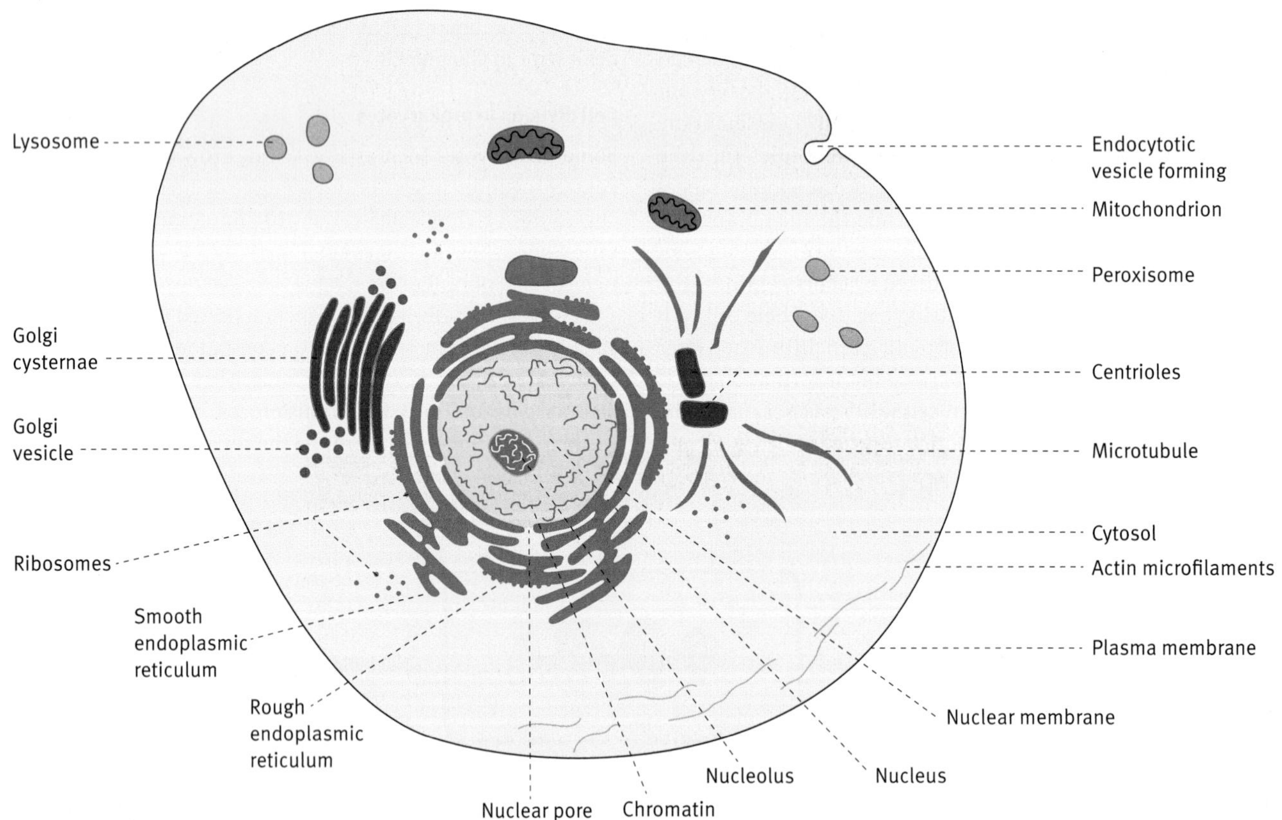

Fig. 2.3 Diagrammatic representation of a generalized animal cell, showing different types of organelles.

the term for the contents of the cell excluding the nucleus and other membrane bound organelles.

The **endoplasmic reticulum (ER)** is a membranous structure that forms a completely enclosed convoluted tubular sac. It surrounds the nucleus, the membrane of which is continuous with that of the ER. The inner **lumen** of the ER tubules form a separate cellular compartment, with the cytosol outside. In some cells the ER is present in large amounts and in cross section almost fills the cell (Fig. 2.4). In other cells with different functions there is much less. The ER is involved in the synthesis and delivery of proteins to their various cellular destinations or, in the case of plasma proteins and digestive enzymes, to the plasma membrane for secretion. The membrane of part of the ER, known as the **rough ER**, is studded with **ribosomes**, the protein–RNA complexes that synthesize proteins. Other sections (continuous with the rough ER) have no ribosomes and constitute the **smooth ER**, which is the site of synthesis of new membrane lipids, and also processes and packages proteins into vesicles for transport to the Golgi apparatus. The smooth ER also has other metabolic roles. It is important to note that only those ribosomes that happen to be synthesizing proteins destined for secretion are attached to the rough ER membrane. They detach and join the general pool of ribosomes in the cytoplasm when synthesis of each protein molecule is complete, and re-attach if they again initiate synthesis of a secretory protein. Additionally, large numbers of ribosomes that are free in the cytosol are synthesizing cytosolic proteins and proteins destined to be transferred into organelles.

The **Golgi apparatus**, named after its discoverer, plays the role of a mail sorting room, the 'mail' in this case being newly synthesized proteins. The Golgi consists of closed membranous flattened structures enclosing the **Golgi cisternae** (Fig. 2.5). They resemble in cross section a stack of large platelike vesicles placed near the nucleus. Membrane-bounded transport vesicles carrying newly synthesized proteins from the lumen of the smooth ER are budded off, and travel to fuse with the Golgi membranes and deliver their contents into the cisternae where proteins are given 'destination labels', packaged into appropriately labelled vesicles and released into the cytosol to be carried to their cellular destinations. The complex process is described fully in Chapter 27 on protein delivery, or targeting as it is called. It is an astonishing sorting and delivery mechanism.

Mitochondria are the powerhouses of the cell. They are present in multiple copies in all eukaryotic cells except the mature red blood cells; they are about the size of an *E. coli* cell. They have a double membrane enclosing a dense collection of proteins. Mitochondria are responsible for most of the ATP production of eukaryotic cells, the energy required being derived from the terminal oxidation of food molecules such as carbohydrates and fat. The inner mitochondrial membrane is

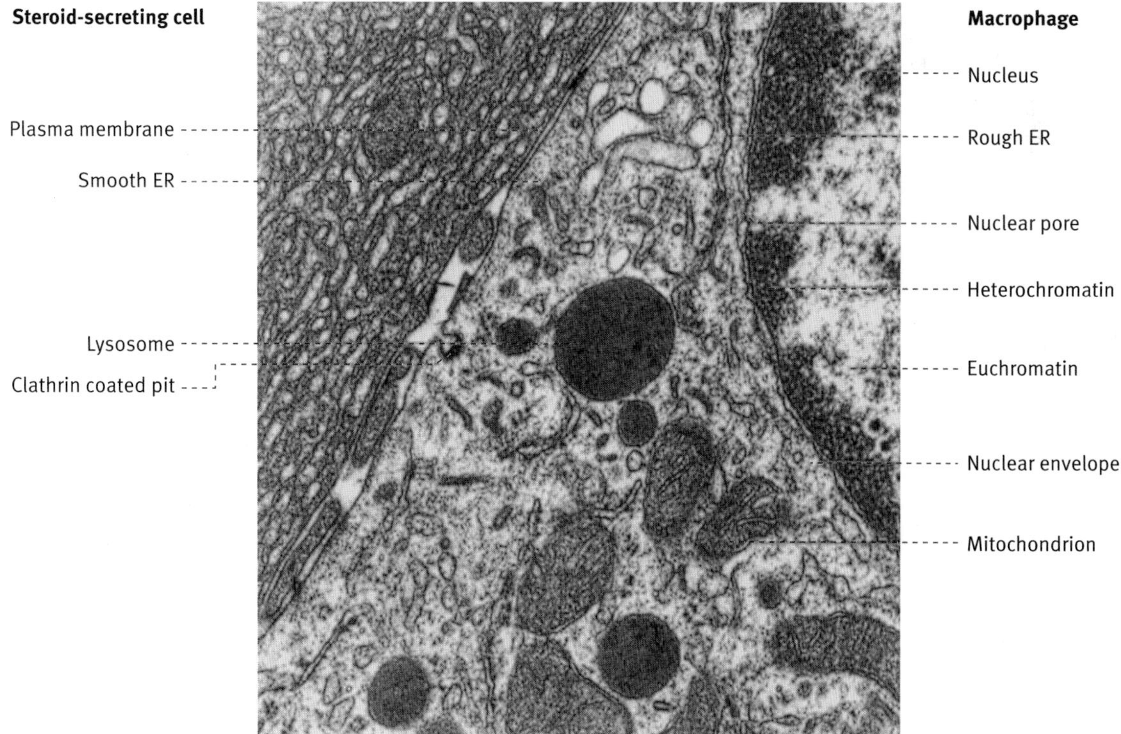

Fig. 2.4 Electron micrograph of part of sectioned testis showing a steroid-secreting cell (upper left) packed with smooth endoplasmic reticulum (ER), alongside a macrophage (right). Photograph kindly provided by Professor W. G. Breed, Department of Anatomy, University of Adelaide, Australia.

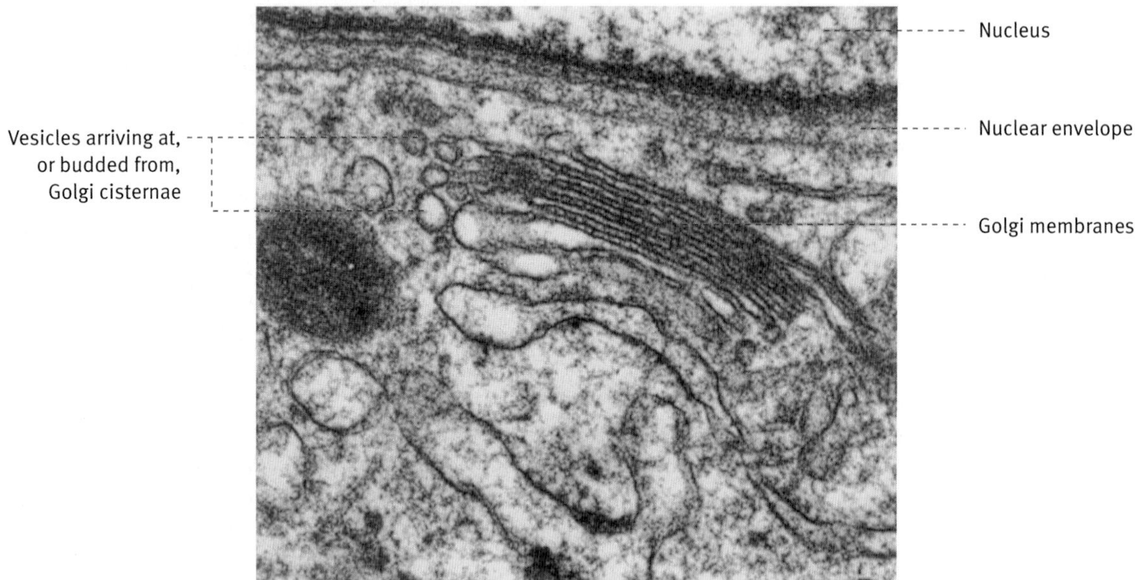

Fig. 2.5 Higher magnification electron micrograph of Fig. 2.4 in which a prominent Golgi apparatus is evident. Photograph kindly provided by Professor W. G. Breed, Department of Anatomy, University of Adelaide, Australia.

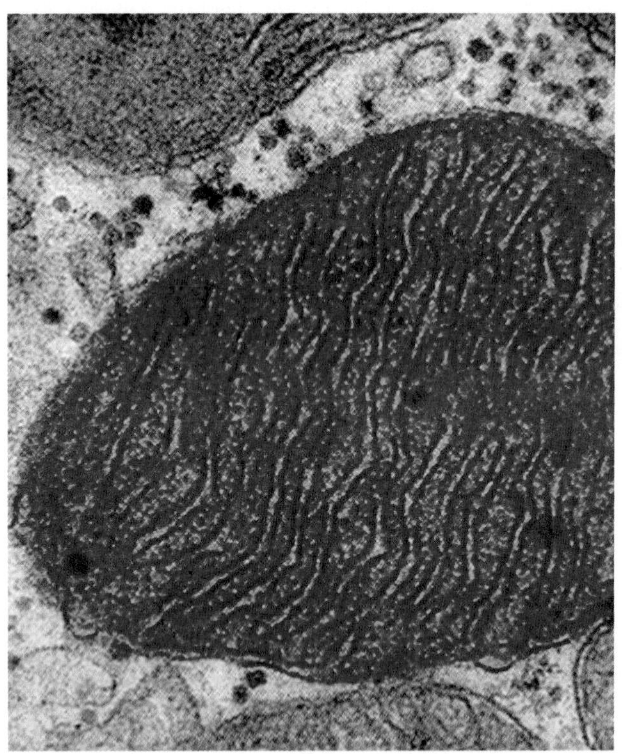

Fig. 2.6 Coloured transmission electron micrograph of a heart muscle mitochondrion, showing cristae. © Robert Harding.

the site of ATP generation. Its area is increased by invagination to form folds or **cristae** whose extent reflects the level of ATP demand. Heart muscle mitochondria, for example, have closely packed cristae in keeping with the high ATP demand (Fig. 2.6).

It is generally accepted that mitochondria originated in the evolutionary sense from prokaryotic cells being engulfed by the precursor of eukaryotic cells and becoming established inside the cells. In modern eukaryotes, all aerobic (oxygen-dependent) energy release from food is performed by mitochondria. The acquisition of aerobic metabolism by eukaryotic cells was of very great evolutionary importance since the yield of ATP from complete glucose oxidation is 18–19 times greater than from its anaerobic breakdown.

The mitochondria have their own circular DNA molecule and undergo replication so that daughter cells have the full complement as they grow, but they have surrendered much of their genetic independence. Most of the proteins in a mitochondrion are coded for by genes in the nucleus of the cell, synthesized in the cytosol and transported into the organelle. The mitochondria do, however, contain their own apparatus for synthesis of mitochondrially encoded proteins. The ribosomes within mitochondria are prokaryotic in type; they are smaller than those of eukaryotes and have certain different characteristics, reflecting their evolutionary origin.

Chloroplasts are present in plant cells. They contain the photosynthetic apparatus (see Fig. 21.2). It is believed that these originated from photosynthetic prokaryotes (cyanobacteria) engulfed by precursors of eukaryotic cells, and became established inside the cells much the same as occurred with mitochondria. They are analogous to mitochondria in having their own circular DNA and limited protein synthesis, and replicate by simple division.

Lysosomes are small spherical membrane vesicles, present in the cytoplasm of eukaryotic cells in large numbers. They contain a collection of about 50 hydrolytic enzymes capable of destroying biological molecules. They do not have their own genome. Their function is to break down some material taken in from outside the cell by endocytosis (see Chapter 7) or cellular material due for destruction. It is essential that this occurs within closed membrane vesicles to protect the cell itself from the destructive enzymes. Several lethal diseases are associated with lysosomal abnormalities (see Box 27.1).

Peroxisomes are small membrane-enclosed vesicles in which a number of molecules not metabolized elsewhere are oxidized by enzymes that use molecular oxygen directly and produce hydrogen peroxide. Hydrogen peroxide is destroyed by further reactions catalysed by catalase and peroxidases. Oxidation in peroxisomes does not generate ATP as it does in mitochondria. Peroxisomes have no genetic system and do not synthesize their enzymes, which are transported into them from the cytosol.

Glyoxysomes are small membrane-bounded vesicles found in plants. They contain enzymes not present in animals. They make it possible for fat reserves to be converted into carbohydrate, which is important, for example, in the germination of seeds where fat provides the main source of energy. Importantly, animals cannot convert fat to carbohydrate.

The **cytoskeleton** (see Chapter 8) is an internal structure of protein fibres that, in most eukaryotic cells, pervades all of the cytosol. Many of the fibres are attached to the cell membrane and influence the shape of the cell. However the term 'skeleton' gives a misleading impression of its nature. The fibres are mainly tracks on which molecular motors operate. They are involved in cell movement, separation of chromosomes at cell division, and transport of molecules within the cell. They are also mainly ephemeral, being set up and demolished rapidly as needed. Prokaryotes, being about 1,000 times smaller than eukaryotes in volume, do not have a cytoskeleton as most transport is by simple diffusion.

Eukaryotic cell growth and division

Single celled eukaryotes such as yeast can be grown in simple media, as anyone who has made bread or home-brewed beer will know. Mammalian cells can be cultured in the laboratory but they are more demanding. They are usually grown in culture dishes (flat circular plastic dishes with lids) that require a specially treated surface to allow cell attachment. In the whole animal, cells mutually control proliferation; that is to say one cell's divisional activity is strongly influenced, even controlled, by other cells, often its neighbours. (Independent growth of mammalian cells is characteristic of cancer.) Coordination is achieved through cell communication via protein or large peptide signalling molecules known as **growth factors** and **cytokines**, which are produced by many types of somatic cells. To culture mammalian cells in the laboratory, these signalling molecules are supplied in the tissue culture medium, often as a mixture in fetal blood serum.

When cells of tissues such as liver are cultured, after a certain number of cell divisions (perhaps 40 in the case of skin cells known as fibroblasts) proliferation ceases. It seems that each cell has a limitation on the number of divisions that it can undergo, which may be a factor in ageing. The limitation is due to a structure, the **telomere**, at the end of each chromosome, which acts like a cell division counter, shortening every time DNA is replicated (see Chapter 23). At less than critical telomere length, normal cell division ceases. However, cells may mutate and escape this control; for example in cancerous cells, which can divide and be cultured indefinitely to establish immortal cell lines, telomeres are continually replenished. Telomere shortening does not limit division of prokaryotic cells because the DNA is typically circular and so there are no ends.

Cell division in eukaryotes is more complex than in prokaryotes, as might be expected. The process in somatic cells is called **mitosis**, in which the resultant daughter cells are exact diploid copies of the parent cell. **Germ cells** undergo a special type of division called **meiosis**, which produces the **haploid** sperm and eggs containing only one copy of each chromosome. The mechanism of these cell divisions is described in Chapter 30.

Basic types of eukaryotic cells

If you consider an animal such as a mammal, its development starts with a single fertilized egg – a single diploid cell. From this, embryogenesis produces all the different types of cells found in the adult body. Most of these are known as **somatic cells**, which are all the cells of the body except the **germ cells** that give rise to eggs and sperm. Somatic cells of an adult are mainly **differentiated cells** – they have developed the specific characteristics required for their specialized functions. For example a liver cell has properties quite different from those of a nerve or muscle cell, even though they all have the same DNA sequences. The process of cells changing in this way is known as **differentiation** and it is generally an irreversible process. If a liver cell divides it will only produce more liver cells. In fact somatic cells are in a nondividing state most of the time. Their division replaces dying cells or repairs wounds and hence maintains a constant number of cells appropriate to the organ they belong to. This keeps the organ proportionate to the size of the organism. Most somatic cells can, however, divide quite rapidly if given the signal to do so, muscle cells, erythrocytes (red blood cells) and some nerve cells being exceptions. If you experimentally remove two thirds of a mouse liver it will regain its full size in days and then abruptly stop growing. Control of cell division is elaborate and vital (see Chapter 30). In cancer, the control has gone wrong (see Chapter 31).

Certain groups of cells are, however, constantly renewed throughout the life of an organism. Skin cells are an example, as are the cells lining the intestine, that experience great wear and tear, and are the most rapidly replaced of all. Blood cells are renewed at a great rate. The life span of an erythrocyte is 120 days in a human, where 2–3 million red blood cells are made per second to replace old cells destroyed in the liver and spleen by macrophages.

You can see that since differentiated cells can only replace themselves, but all cells in an organism arise from a single one, the zygote, there must be a class of cells that can divide but give rise to other types of cells. These are known as **stem cells**. They are cells that divide, and the two daughter cells can do one of two things: they can remain as stem cells or they can differentiate into somatic cells. There are different types of stem cells. **Embryonic stem cells** are those initially produced from the fertilized egg during embryogenesis. They are **pluripotent** meaning that they can give rise to any cell type. **Adult stem cells** are a halfway house between embryonic stem cells and somatic cells in their capacity to form different cell types. If you consider, for example, the renewal of skin, there are various types of skin cells that have to be replaced, so underlying skin there is a pool of stem cells that can give rise to all of these, but cannot give rise to non-skin cells. There are other adult stem cells that give rise to the various blood cells but only to blood cells, and so on. The process by which all blood cells have their origin in bone marrow stem cells that continually divide is known as haemopoiesis (Fig. 2.7).

Stem cells, as is widely known, are of major medical interest because of their potential to replace diseased or dead cells. Moreover, although differentiation of a stem cell into a somatic cell is naturally irreversible in mammals, technologies have been developed to reverse differentiation. It is possible to reprogramme the nucleus back to its pluripotent state as was shown by the celebrated work on Dolly the sheep.

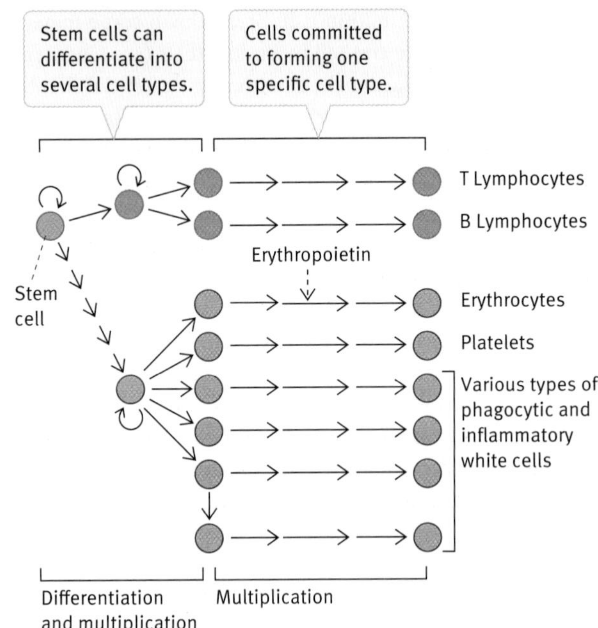

Fig. 2.7 Simplified diagram of haemopoiesis. At each arrow, cell multiplication can occur, and this may be controlled specifically by various protein cytokines – the colony-stimulating factors, interleukins, and erythropoietin. For example, erythropoietin stimulates erythrocyte (red blood cell) production at the step indicated. Differentiation of stem cells into committed cells is likewise controlled. A. J. Paul, J. C. Vickerman, M. Grasserbauer et al; Organics at Surfaces, their Detection and Analysis by Static Secondary Ion Mass Spectrometry [and Discussion]; Philosophical Transactions A; 1990, 333, 1648; by permission of the Royal Society.

BOX 2.1 Some of the organisms used in experimental biochemical research

In Chapter 1 we briefly alluded to the fact that it is best when trying to elucidate a molecular process to choose an organism that gives you the best chance of success. The unity of basic life processes in all life forms often means that discoveries in one organism are to a considerable degree relevant to others. A brief survey of some of the most popular 'model' organisns used in biochemical research may help to make scientific literature more comprehensible.

Viruses can often be easily grown in bacteria and animal cells. Their genomes are very small and reproductive times are measured in minutes. Phage lambda (λ) has been used extensively in gene cloning (see Chapter 28).

Bacteria: Although a variety of bacteria is used for specific purposes in research, ***E. coli*** (nonpathogenic strain) was for a long time the main workhorse of biochemists. It is small, easily cultivated, and replicates in 20 minutes, and has only 4,000 genes as compared with the 20–25,000 of human cells. It is still a key organism in genetic engineering work.

Yeast is a single cell eukaryote and is much used for genetic studies. It has most of the molecular characteristics of more complex eukaryotes but with a very much smaller genome.

C. elegans (*Caenorhabditis elegans*) is a small easily grown nematode roundworm that has become the simplest model in studies on animal cells. The adult worm has only 959 somatic cells plus germ cells. The lineage of every somatic cell back to the fertilized egg has been traced in a remarkable piece of work that earned the Nobel Prize in 2002 for Sydney Brenner, Robert Horvitz, and John Sulston. *C. elegans* is important for animal developmental studies and for the detection and isolation of genes. Genes identified in *C. elegans* have been found to occur widely in animals.

Drosophila melanogaster, the fruit fly, is one of the most used models for animal developmental studies and has been important in molecular biology in general. It has a short reproductive time and the existence of a larval stage makes it a convenient model for identifying genes that function in embryonic development. Work on *Drosophila* has revealed developmental and other control genes that also exist throughout vertebrates.

The **mouse** (and also the **rat**) are the most used mammalian models for studies on vertebrates.

Cultured mammalian cells are also important for molecular biology studies since they give direct access to cells under controlled experimental conditions.

Arabidopsis thaliana is a small cress plant that is the workhorse for genetic and developmental studies on plants.

Viruses

As viruses lack a cell membrane and the ability to reproduce independently they are not classed as cells. However, they have a genome, which can be DNA or RNA (all cells have a DNA genome), and they have the capacity to make use of a host cell's processes for their own reproduction. Hence, quite apart from their biological and medical importance, viruses are important in biochemical research, as they provide relatively simple models for studying gene function and replication.

The chief (but not the only) protection mammals have against viral infections is the immune system (see Chapter 33), which produces antibodies against viruses leading to their destruction, or causes the destruction of infected cells, thus aborting viral replication. Viruses have often developed mechanisms that outwit the immune defence systems, and knowledge of these mechanisms can be an important guide in research for unraveling highly complex immune strategies.

Viruses are much smaller than cells, an electron microscope being needed to see them rather than the light microscope that is adequate for most bacteria. They have no metabolism by themselves – a virus particle or **virion**, as it is called, on its own does nothing. It is an inert, organized, complex of molecules that may sometimes be crystallized. Nonetheless, viruses are reproduced when they infect living cells. Different viruses infect specific animal and plant cells, and also bacteria (viruses that infect the latter are known as **bacteriophages**). They transmit their genetic material into cells and use the host cell machinery for their own replication.

A virus particle comprises a small amount of nucleic acid surrounded by a protective shell of protein molecules. The total number of genes may be about a hundred for a large virus, such as **vaccinia** (used for smallpox vaccination), or three or four for the smallest. The protein shell surrounds the genome and is called the **nucleocapsid**. In some viruses there is an additional membrane coat.

Viruses enter the cells they infect and release their genetic material (though bacteriophages inject their genome without the whole assembly entering). The viral genes then direct the synthesis of the components required for the assembly of new virus particles, which escape from the cell. The membrane of a cell is a barrier to virus entry but there are ways past it. For example, animal cells have mechanisms for engulfing molecules (endocytosis, see Chapter 7) and some viruses exploit this; they hitch a ride on a normal cellular import mechanism. A second route available to viruses with a membrane is direct fusion with the plasma membrane of the host cell, causing release of the nucleocapsid into the cell. **Human immunodeficiency virus**, (**HIV**) uses this route.

In **bacteriophages**, a different route is followed. The bacterial cell has a rigid cell wall around it. **Phage lambda** (λ) (Fig. 2.8) has a **head** capsule, formed by protein molecules, in which resides the viral DNA. The phage attaches to the cell wall by its tail fibre and injects its DNA molecule into the bacterial cell, almost like the action of a hypodermic syringe.

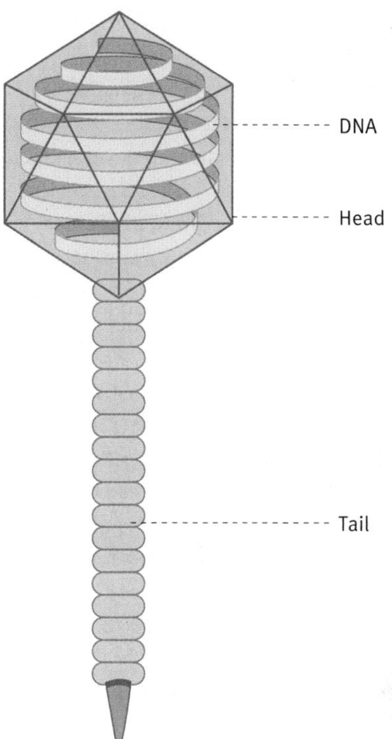

Fig. 2.8 A phage λ particle. The λ chromosome – some 50,000 base pairs of DNA – is wrapped around a protein core in the head. Ptashne, M (1992); A Genetic Switch: Phage 1 and Higher Organisms, 2nd edition; Reproduced by permission of John Wiley & Sons.

Genetic material of viruses

The genetic material of viruses may be DNA, or RNA. RNA occurs in different viruses in single-stranded and double-stranded form. Among viruses with a DNA genome double-stranded DNA is most common; single-stranded DNA also occurs, but it is converted to the double-stranded form for replication.

Why has DNA superseded RNA as the medium for storing genetic information in all cells? The answer almost certainly is that DNA is chemically more stable than RNA (see Chapter 22). However, the continued existence of viruses with RNA genomes is a cause of their rapid evolution, and hence their capacity to evade host immunity. Cells contain enzymes that repair DNA if a mistake is made in their genome replication or if the genome is damaged in some way (see Chapter 22). RNA damage, however, is not repaired and RNA viruses therefore mutate rapidly. Thus, by constantly changing the proteins that the immune system recognizes, new viral strains escape immune attack. So primitive molecular instability, coupled with lack of repair, is an advantage for viruses even in the modern world: human immunodeficiency virus (HIV), influenza, poliomyelitis, mumps, foot and mouth, measles, and rubella viruses are a few familiar RNA viruses that infect animals and humans. Plant viruses also exist.

Some examples of viruses of special interest

Influenza is an RNA virus consisting of a membrane with proteins inserted in it enclosing its genetic material, which is split into eight separate packages (Fig. 2.9). The influenza virus mutates rapidly, so that new strains develop each year requiring annual vaccinations for protection against the current strain. Immune protection is directed against the surface proteins but these change because of the mutations. This means that protection is usually gradually lost as new strains appear (known as **antigenic drift**), but residual protection means that attacks may be comparatively mild. However, on fortunately rare occasions, the coat protein may be changed completely due to recombination between two strains of virus infecting the same cell. This is known as **antigenic shift**. In this situation, immunological protection of the populace due to previous exposure to the influenza virus is almost completely lost and a **pandemic** (a widespread, possibly worldwide, infection) may result. Pandemics of the past have killed many millions of people. The 1918 outbreak is believed to have originated from a reassortment of blocks of genes (recombination) between human and bird strains of the virus when an animal (possibly a pig) became infected by both types.

Antibiotics are not effective against viruses

Antibiotics are molecules produced by microorganisms and fungi, which they release to kill competing organisms in their neighbourhood, contributing to their reproductive success. Although all cells are to a large extent basically the same, there are differences, often subtle, that make processes susceptible to molecular 'spanners' thrown into the works. From a human standpoint, the ideal antibiotic is one that attacks these biochemical systems of infectious organisms that have no counterpart in human biochemistry. Penicillin, for example, prevents susceptible bacteria making cell walls, which animal cells lack. As resistance to antibiotics develops, it has been necessary to create new antibiotics by modifying natural ones and to seek additional pathogen specific processes as targets.

Viruses infecting humans present a different problem because they have essentially no biochemical machinery of their own, and reproduce using host cell machinery. It is difficult, therefore, to inhibit the virus reproduction without damaging the host. No antiviral 'antibiotics' have evolved. The main human protection against viral infection is the immune system. This relies on distinguishing the amino acid sequence of viral proteins from human ones. This is a highly sophisticated task and makes the immune system almost incredibly complex. However, chemically made antiviral drugs have been successful in controlling a number of viruses, such as influenza and HIV, by exploiting subtle differences or finding targets in the viral processes.

The influenza virus has, in its outer shell, molecules of an enzyme called **neuraminidase**, which play some part in the release of the virus from infected cells, but may have other roles. A therapeutic strategy has been designed in the form of commercially available drugs – '**Relenza**' was the first–that bind to the active site of this enzyme with high affinity and block its action. The active site of an enzyme (see Chapter 6) is less likely to mutate and to become drug resistant because any change may interfere with its catalytic activity. The unchanging active site of neuraminidase might seem to expose the virus to antibody attack but it is not accessible to the relatively large antibody molecules. However, the drugs are small enough to enter the site and are reported to lessen the severity of the disease.

It is important for medical professionals to understand that although antibiotics may be useful in controlling secondary bacterial infections following a viral disease, they are of no use in combating viral infection themselves, and that their overuse may lead to the development of resistant bacteria.

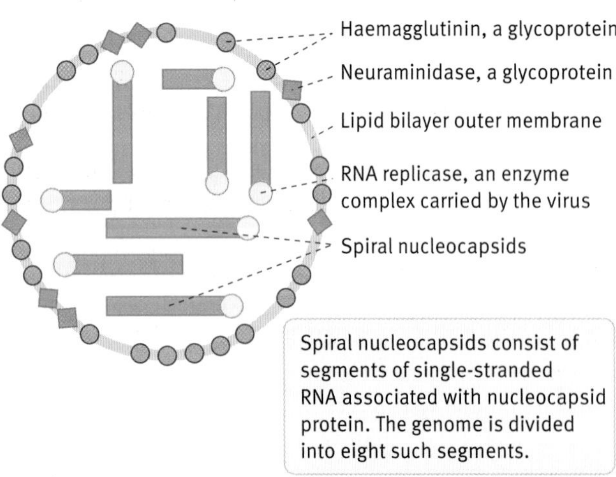

Fig. 2.9 Structure of the influenza virion. A layer of matrix protein underlying the outer membrane is not shown.

Retroviruses

Retroviruses are RNA viruses of immense current interest, partly, but not only, because HIV is of this type. The retrovirus particle carries within itself a few molecules of an enzyme, the discovery of which caused initial disbelief, to be followed by the award of a Nobel Prize in 1975 to its discoverers David Baltimore, Renato Dulbecco, and Howard Temin; it is called **reverse transcriptase**. Before its discovery, it was a dogma that DNA could direct RNA synthesis, but not the reverse. However, viral reverse transcriptase copies the viral RNA into DNA. Another enzyme, carried in the virus, causes integration of the DNA into the host chromosome where it becomes, in essence, an extra set of genes carried by the cell. Once in, the viral genome is replicated along with host DNA for cell division. For the production of new retrovirus particles, the proviral genes (that is, viral genes in the host chromosomes) are transcribed (copied) into RNA transcripts. The retroviral life cycle is described later (see Chapter 23). The viral genome integration is of relevance to cancer (see Chapter 31), because retroviruses may pick up a host gene fragment or gene control sequence that when transmitted into a new host may be oncogenic (cancer-producing). Several examples of cancers in animals, and a single leukaemia in humans are known to have this origin.

HIV, like influenza, mutates very rapidly due to its error-prone replication mechanism, so that any immunity that is developed by the host is quickly bypassed. Also new mutant strains develop even during a single infection, giving little chance of any significant immunity developing. In addition the virus attacks helper cells of the immune system. These are essential in developing antibodies against the virus so that the patient's antibody production against the virus and any other pathogen is severely impaired leading, potentially, to multiple infections.

The drug **AZT (azidothymidine)**, a nucleoside analogue, is used as one component of therapy against the AIDS virus (HIV); it works by inhibiting the viral reverse transcriptase by becoming incorporated into, and so terminating the synthesis of, viral DNA chains. The drug is specific to the virus because human cells do not use the reverse transcriptase enzyme. Other components of an antiviral drug 'cocktail' may be a non-nucleoside compound that binds directly to reverse transcriptase and prevents RNA being copied to DNA, and a protease inhibitor that binds to, and blocks, an essential HIV protease.

BOX 2.2 **Structure of the drug azidothymidine**

O
H_3C
NH
N
O
$HOCH_2$
O
N_3

Azidothymidine (AZT)

Viroids

Viroids are the smallest infectious particles known – even smaller and simpler than viruses. They are very small, naked RNA molecules without any protein or other type of coat,that can cause some plant diseases. Infection of plant cells is dependent on mechanical damage to the host plant. The effect of viroid infection varies from some impairment of the health of the plant to certain death of, for example, palm trees. The really remarkable feature is that no genes coding for proteins are present and it is not certain how viroids produce disease.

Summary

Cells are the units of living systems. Each cell is surrounded by a lipid membrane and has a DNA genome. They are with rare exceptions microscopic in size to maximize the surface area to volume ratio. The two main classes of cellular organism are prokaryotes and eukaryotes.

Prokaryotes include modern bacteria. They lack a nuclear membrane or any internal membrane-bounded organelles. Their small size makes it possible for all molecules, including macromolecules, to diffuse to parts of the cell where they are needed. *E. coli*, the typical prokaryote, has a single circular chromosome. It can replicate in about 20 minutes.

Eukaryotic cells are about 1,000 times larger in volume and replicate in about 24 hours. They have a nucleus containing the DNA genome, and other membrane-bounded organelles. The **cytoplasm** can be defined as the content of the cell excluding the nucleus, while the **cytosol** refers to the soluble constituents of the cytoplasm from which membrane-bounded organelles have been removed.

The **endoplasmic reticulum (ER)** is an extensive continuous membranous structure enclosing a lumen, separated from the rest of the cell. The ER modifies newly synthesized proteins and buds off transport vesicles to deliver them to the Golgi. The ER also synthesizes new lipid membrane. **Golgi vesicles** are a series of enclosed sacs, involved in the delivery of newly synthesized proteins to their destinations in the cell. **Mitochondria** are the site of most of the ATP generation from oxidation of food molecules and are believed to have originated in evolutionary terms from engulfed prokaryotic cells. **Chloroplasts** are the site of photosynthesis in plants. **Lysosomes** are bags of enzymes needed for the destruction of selected cellular material. **Peroxisomes** oxidize certain molecules forming peroxide but do not generate ATP.

The **DNA** of prokaryotes resides in the cytoplasm as an extended tangled thread-like molecule. For cell division this is duplicated and the duplicates are pulled apart, followed by separation of the daughter cells. In eukaryotic cells the division process is more elaborate, involving mitosis in which highly condensed chromosomes are segregated into daughter cells. The eukaryotic cell cycle is more highly controlled than that of prokaryotes.

Cells contain large numbers of **ribosomes**. These are large complex structures of RNA and proteins found in the cytosol. They synthesize proteins by the process of translation of messenger RNA.

In animals, **cell growth** is controlled by growth factors and cytokines. Normal cells can only divide for a limited number of times because the telomeres of their chromosomes shorten at each replication and eventually become so short that replication stops. In cancerous cells the telomeres are replaced continually, allowing unlimited cell division.

Most of the **differentiated cells** in the body are known as somatic cells. These can only give rise to their own type of cell. The germ cells are those that give rise to sperm and eggs. **Embryonic stem cells** are pluripotent and can give rise to all cell types, while **adult stem cells** can give rise to a limited number of cell types.

Viruses have genetic material that can be either DNA or RNA. When they infect a cell their genetic material is released into the cell and they utilize the metabolism of the host to reproduce. Influenza virus is an RNA virus. It mutates rapidly and its antigenic properties change, reducing the resistance of its host animals. Occasionally recombination generates a totally new influenza strain, giving rise to a pandemic.

Retroviruses, which include HIV, are RNA viruses that after infection copy their RNA to form DNA. This then integrates into the host cell genome and the viral genes can be copied into new RNA viruses.

Viroids are small naked RNA molecules that can infect plants, entering cells through sites of mechanical damage.

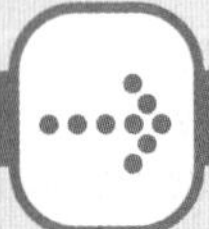

Further reading

To access the further reading, please scan the QR code image or go to http://global.oup.com/uk/orc/biosciences/molbiol/snape_biochemistry/student/reading/ch02/

Problems

1 Why are cells, with a few exceptions, microscopic in size?

2 What sets the lower size limit of cells?

3 Briefly summarize the main characteristics of prokaryotic and eukaryotic cells.

4 Comment on the nature of somatic cells and stem cells.

5 Why do biochemists often use many different types of organisms to elucidate a problem?

6 Influenza epidemics sweep the world at long intervals. Most are mild but occasionally, as in 1918, a highly lethal one occurs. Why is this so?

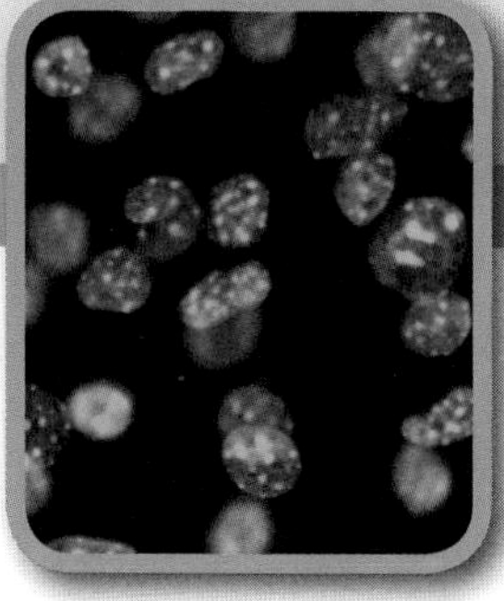

Chapter 3

Energy considerations in biochemistry

> If your theory is found to be against the second law of thermodynamics, I can give you no hope; there is nothing for it but to collapse in deepest humiliation.
>
> Sir Arthur Eddington

In Chapter 1 there is a more general survey of the basic energy concepts that apply to life, which it will be helpful to read. As explained there, the relevant thermodynamic considerations are greatly simplified by the fact that life operates at constant temperature and pressure.

The consideration underlying all life processes is energy. A printed chemical equation on its own lacks a vital piece of information and that is the energy change involved. It may not be obvious what is meant by this, as energy change in chemical reactions occurring in solution is not something that is easy to see or detect unless very violent, and such spectacular reactions do not occur in biology. But energy considerations in biochemical reactions determine whether a reaction is possible on a significant scale, and whether the reverse reaction can occur to a significant degree. They are of prime importance in determining the chemical activities in cells.

An example of how this applies directly to living processes is the conversion in muscle, during vigorous exercise, of glycogen (a polymer of glucose) into lactic acid by the metabolic pathway called **glycolysis** (see Chapter 13). This pathway consists of a series of a dozen chemical reactions. After exercise, during the ensuing rest period, accumulated lactic acid is converted back into glycogen, but not through the reverse of exactly the same set of chemical reactions by which it was formed. There are important differences between the forward and reverse pathways. This can seem to be pointlessly complicated chemistry, but a knowledge of simple energy considerations makes it at once apparent why this mechanism has to be followed by the cell. An understanding of simple energy considerations throws a flood of light onto the biochemistry of cellular processes.

Energy considerations determine whether a chemical reaction is possible in the cell

As already implied, energy considerations mean that certain chemical reactions can occur while others cannot. Note that in biochemistry we are concerned with reactions occurring to a *significant* extent. In principle, energy considerations do not completely prevent a reaction occurring (unless reactants and products are precisely at equilibrium concentrations), but if the reaction ceases after a minute amount of conversion has taken place, then, for the practical purposes of life, it has not occurred.

First, what do we understand by energy change in a chemical 'system', by which we mean an assemblage of molecules in which chemical reactions can take place, such as that that exists inside a cell? The concept is not as self-evident as say, gravitational potential energy change when a weight falls. A chemical system involves a huge number of individual molecules each of which contains a certain amount of energy, dependent on its structure. This energy can be described as the heat content or **enthalpy** of the molecule and depends on the chemical bonds within it. Enthalpy values for many compounds are available in physical chemical tables. When a molecule is converted into a different structure in a chemical reaction, its energy content usually changes; the change in the enthalpy is written as ΔH (delta H), which stands for the *change* in heat content. The ΔH may be negative (heat is lost from molecules and released, so raising the temperature of the surroundings) or positive (heat is taken up from the surroundings, which, correspondingly, cool down).

At first sight it may seem surprising that reactions with a positive ΔH can occur at all since it might seem to represent an energy uptake analogous to a weight simply raising itself from the floor and cooling the surrounding air as it does so. This is the point at which physical analogies such as weights falling become inadequate as models for chemical reactions. In chemical reactions, a negative ΔH favours the reaction and a positive ΔH has the opposite effect, but ΔH is not the final arbiter as is gravitational energy with a weight system. **Entropy change** (the drive to randomness), known as ΔS, also has a say in the matter.

Entropy is the degree of randomness of a system (see Chapter 1 for a simple explanation of entropy if you are not familiar with the term). In a chemical system, this can take three forms: first, a molecule is not usually rigid or fixed – it can vibrate, twist around bonds, and rotate. The more this occurs the higher the entropy. Secondly, the vast numbers of individual molecules may be scattered in a random manner (higher entropy) or they may have a more ordered arrangement (lower entropy). A living cell has a more ordered arrangement of molecules than that of molecules in the outside-cell surroundings and therefore a lower entropy level, analogous to a finished house having a lower entropy level than that of the materials from which it was constructed. Thirdly, the number of individual molecules or ions may change as a result of a chemical change. The greater the number of individual particles (molecules and ions), the greater the randomness; the greater the randomness, the greater the entropy. These three factors can all contribute to entropy change. Thus, a chemical reaction may change the entropy of the system. Increasing entropy lowers the energy level of the system; decreased entropy (increased order) increases the energy level.

As already stated above, both ΔH and ΔS play a part in determining whether a chemical reaction may occur. A negative ΔH and a positive ΔS both reinforce the 'yes' decision; a positive ΔH and negative ΔS both reinforce the 'no' decision. A negative ΔH and negative ΔS have opposite impacts on the decision, as do a positive ΔH and a positive ΔS, and whether the outcome is yes or no depends on which is quantitatively the larger.

The driving force of increasing entropy is illustrated by the example of the melting of a block of ice in warm water. Heat is taken up as the ice melts (ΔH is positive), but the scattering of the organized molecules of the ice crystal, as it dissolves, increases the entropy so the process proceeds. If energies are similar, the disordered state occurs more readily, and is more probable, than the ordered state; this is an everyday experience. In a collection of molecules, or any objects, there are vastly more possible random arrangements than organized ones. It is not easy, however, to intuitively accept that entropy change can affect processes in the way one can visualize gravitational energy causing a weight to fall or a car to roll downhill. If you find it difficult, rather than have a mental block, it would be best to accept the concept and it will become familiar as you progress through the book. As pointed out in Chapter 1, despite entropy perhaps seeming a nebulous concept, achieving maximum entropy appears to be an irresistible tendency of the universe. Another wording of the second law of therrmodynamics given in Chapter 1 specifies that all happenings must increase the total entropy of the universe. For example the released heat of a reaction increases entropy.

The situation we have described between ΔS and ΔH is not convenient; we have two terms of variable sizes that may reinforce or oppose each other in determining whether a reaction is possible. Moreover, in biological systems it is difficult or impossible to measure the ΔS term directly. The situation was greatly ameliorated by the concept of **Gibbs free energy**, which combined the two terms into a single one. The change in free energy (**ΔG** after J. Willard Gibbs) is given by the famous equation

$$\Delta G = \Delta H - T\Delta S$$

where T is the **absolute temperature**. This equation applies to systems where the temperature and pressure remain constant during a process, which is the case for biochemical systems. Note that we are concerned with the **change of free energy** in a reaction. The ΔG of a reaction is an all-important thermodynamic term; in its application to chemical reactions the rule specified by the **second law of thermodynamics** is as follows: the products have lower free energy than the reactants.

The term 'free' in free energy means free in the sense of being *available* to do useful work, not free as in something for nothing. ΔG represents the *maximum* amount of energy available from a reaction to do useful work. It is somewhat like available cash being 'free' for you to use to make purchases, or a shop assistant being 'free' or available to serve you. Useful work includes muscle contraction, chemical synthesis in the cell, and osmotic and electrical work. ΔG values are expressed in terms of calories (cal) or joules (J) per mole (1 cal=4.184 J); the joule is now the official unit (and the one used in this book) but calories are still frequently used in biochemical texts, especially those originating in the United States. Since the values are large, the terms kilocalories (kcal) or kilojoules (kJ) per mole are used. Kilojoules per mole is usually abbreviated to kJ mol^{-1}.

The fact that a chemical reaction, if it were to occur, would comply with the second law of thermodynamics does not mean that the reaction actually does occur, though noncompliance guarantees that it will not. A negative ΔG is a necessary but not sufficient condition for a reaction to occur. We have explained in Chapter 1 that enzyme catalysis is also needed for biochemical reactions to take place. The reason, in summary, is due to the fact that there is an energy barrier to chemical reactions occurring; molecules must be raised to a transition state before they can react (see Chapter 6). This is why sugar in the bowl on the table does not burn despite the fact that the energy considerations of the sugar oxidizing to CO_2 and water are highly favourable.

Reversible and irreversible reactions and ΔG values

Strictly speaking, *all* chemical reactions are reversible. This might imply that the ΔG value must be negative in both directions – remembering that a reaction cannot occur unless the ΔG value is negative. The answer to this apparent paradox is that the ΔG of a reaction is not a fixed constant but varies with the reactant and product concentrations. The relationship is given later in this chapter. Thus in the reaction $A \rightleftharpoons B$, if A is at a high concentration and B at a low one, the ΔG may be negative in the direction $A \rightarrow B$ and, of course, positive from $B \rightarrow A$. Reverse the concentrations and the ΔG can be negative for the reverse direction. A reaction will proceed to the point at which A and B are in concentrations at which the ΔG is zero in both directions and then no further net reaction can occur. This is the **chemical equilibrium point**.

If the ΔG of a biochemical reaction $A \rightleftharpoons B$ is small, significant reversibility may be possible in the cell because changes in reactant concentrations may be sufficient to reverse the sign of the ΔG. If the ΔG is large, for all practical purposes the reaction is irreversible. In cellular reactions there is relatively little scope for concentration change in the metabolites (as reactants and products are called). The concentrations are always relatively low: 10^{-3}–10^{-4} M would be typical. The net result is that most reactions with large negative ΔG values are irreversible because concentration changes are insufficient to reverse the sign of those values. (We later explain, in Chapter 13, why certain reactions may appear to contradict this.)

As a general guide, hydrolytic reactions in the cell – reactions in which a bond is split by water – are irreversible, in the sense that the synthesis of substances in the cell does not occur by the reverse of the hydrolytic reactions. Conversely reactions in which molecules become linked together by the elimination of a water molecule from between them will require input of energy from adenosine triphosphate (ATP). As you proceed through the chapters on metabolism you will become familiar with which reactions are reversible and which are irreversible.

To summarize, a reaction with a small ΔG is likely to be reversible in the cell, the direction being determined by small changes in reactant and product concentrations. A reaction with a large ΔG value will, in cellular terms, proceed in one direction only and, moreover, will proceed to virtual completion because the equilibrium point is so far to that side of the reaction. Put in another way, in the latter case, the ΔG of the reaction does not become zero until virtually all of the reactant(s) have been converted into product(s).

The importance of irreversible reactions in the strategy of metabolism

The major chemical processes of the cell usually involve not single reactions, but series of reactions organized into metabolic pathways in which the products of the first reaction are the reactants of the next and so on. In the example of the glycogen → lactic acid conversion in muscle mentioned earlier, a dozen successive reactions are involved.

An important general characteristic of metabolic pathways is that they are, as a whole, irreversible. Many of the individual reactions in a pathway may be freely reversible, but it virtually always contains one or more reactions that cannot be directly reversed in the cell. Such irreversible reactions act as one-way valves and ensure that from a thermodynamic viewpoint the pathway can proceed to completion. This is not the same as saying that overall physiological chemical processes are irreversible. Lactic acid formed from glycogen *can be* converted back into glycogen in the body but by a different reaction pathway to that which produced it and, while many steps in the process are the simple reversal of those in the forward process, there are steps that cannot be reversed directly and alternative reactions are necessary. These involve the input of energy, which makes the alternative reactions also irreversible, but in the opposite direction. So the forward pathway (glycogen → lactic acid) is directly irreversible as in the reverse pathway (lactic acid → glycogen). A typical metabolic situation is:

$$A \rightleftharpoons B \circlearrowright C \rightleftharpoons D \rightleftharpoons E \circlearrowright F \rightleftharpoons \text{products.}$$

The red arrows represent irreversible reactions with large negative ΔG values.

A general biochemical principle emerges. *Whenever the overall chemical process of a metabolic pathway has to be reversed, the reverse pathway is not exactly the same as the forward pathway – some of the reactions are different in the two directions.*

What is the significance of irreversible reactions in a metabolic pathway?

The alternative to the situation we have outlined is that *all* the reactions of a metabolic pathway are reversible, that is,

$$A \rightleftharpoons B \rightleftharpoons C \rightleftharpoons D \rightleftharpoons E \rightleftharpoons F \rightleftharpoons \text{products.}$$

The major drawback of this arrangement is that the whole process is subject to mass action. If the concentration of A increases (perhaps due to ingested food), the reaction would shift to the right and more products would be formed. If substance A decreased in amount, some of the products would revert to A to maintain the equilibrium. Imagine that the pathway is for the synthesis of molecules such as the DNA of genes or of vital proteins, and you can see how impossible this scenario is. It would be rather like building the walls of a house with the laying of bricks being a freely reversible process. The walls would rise and fall to maintain a constant equilibrium with the number of bricks lying on the ground.

There is a second important, and related, significance in having dissimilar reactions in the forward and reverse direction of metabolic pathways. Metabolism must be controlled. As already stated, during vigorous exercise, muscles convert glycogen into lactic acid. During rest, the lactic acid is converted back into glycogen and the forward conversion switched off. To independently control the two directions (that is, to switch one on and the other off) there must be separate reactions to control; otherwise it would be possible only to switch both directions on and off together. Thus, the irreversible reactions are usually the control points. Metabolic control is a major subject that we will deal with in Chapter 20.

How are ΔG values obtained?

The ΔG value of a reaction, as explained, is *not* a fixed constant, but the ΔG value of a reaction *under specified standard conditions* is fixed. The conditions defined as standard for biological reactions are with reactants and products at 1.0 M,

25 °C, and pH 7.0; that is, the free energy difference between separated one-molar solutions of reactants and products. The ΔG value, *under these conditions*, is called the **standard free energy change of a reaction**. It is denoted as $\Delta G^{0\prime}$, the prime being used in biochemical systems to indicate that the pH is 7.0, rather than pH 0 which is used in physical sciences.

The $\Delta G^{0\prime}$ value may often be calculated from readily available standard free energies of formation. For many reactions, the $\Delta G^{0\prime}$ value can be calculated by adding up the free energies of formation of the reactants and (separately) those of the products. The difference is the $\Delta G^{0\prime}$ value. An alternative is to experimentally determine the equilibrium constant for the reaction and from this the $\Delta G^{0\prime}$ value is easily calculated.

There is a simple direct relationship between $\Delta G^{0\prime}$ values and ΔG values at given reactant and product concentrations. If, therefore, we know the relevant metabolite concentrations in the cell, the ΔG value for the reaction in the cell is readily determined (see Box 3.2, for an illustrative calculation). This has been done for many biochemical reactions so that such values are often quoted.

There is a snag – determining the actual concentrations of the thousands of metabolites in a cell is not a trivial matter. They are present at low concentrations and are changing anyway due to metabolic activities in the cell. So, for many biochemical reactions, we do not have this data and therefore do not have the ΔG values. However, it is found that the more easily obtained $\Delta G^{0\prime}$ values usually correlate well with known cellular happenings so that they are a useful guide in understanding metabolic reactions. Thus, although such values are not directly applicable to cells because metabolite concentrations are never 1.0 M, they are frequently quoted to explain why certain reactions behave as they do. It is a compromise but a very useful one.

Standard free energy values and equilibrium constants

A particularly useful aspect of knowing the $\Delta G^{0\prime}$ value of a reaction is that, from it, the **equilibrium constant** of that reaction is readily determined. The equilibrium constant K'_{eq} of a reaction represents the ratio at equilibrium of the products to reactants. (The prime indicates that it is the K_{eq} at pH 7.0.) Thus, if one considers the reaction $A+B \rightleftharpoons C+D$, the K'_{eq} is calculated from the concentrations of A, B, C, and D present after the reaction has come to equilibrium – that is, when there is no net change in their concentrations,

$$K'_{eq} = \frac{[C][D]}{[A][B]}$$

The relationship between the value of the $\Delta G^{0\prime}$ and the K'_{eq} for a reaction is

$$\Delta G^{0\prime} = -RT \ln K'_{eq} = -RT\, 2.303 \log_{10} K'_{eq}.$$

In the expression 2.303 $\log_{10}$, 2.303 is a conversion factor to convert from $\log_{10}$ to the natural log, ln, also known as $\log_e$. R is the gas constant (8.315 J mol^{-1} K^{-1}) and T is the absolute temperature in Kelvin (298 K = 25 °C). At 25 °C, RT = 2.478 kJ mol^{-1}.

Thus, if the K'_{eq} is determined, the $\Delta G^{0\prime}$ for the reaction can be calculated and *vice versa*. Table 3.1 shows the relationship between $\Delta G^{0\prime}$ value and the K'_{eq} value for chemical reactions.

Approximate $\Delta G^{0\prime}$ (kJ mol^{-1})	K'_{eq}
+17.1	0.001
+11.4	0.01
+5.7	0.1
0	1.0
−5.7	10.0
−11.4	100
−17.1	1000

Table 3.1 Relationship of the equilibrium constant (K_{eq}) of a reaction* to the $\Delta G^{0\prime}$ value of that reaction.

* For a reaction $A+B \rightleftharpoons C+D$, the equilibrium constant is the molar concentration of C×D divided by that of A×B.

$\left(K'_{eq} = \frac{[C][D]}{[A][B]}\right)$; K'_{eq} is the K_{eq} at pH 7.0.)

The release and utilization of free energy from food

The conversion of simple precursor molecules to larger cellular molecules (such as DNA and proteins) involves increases in energy, and therefore cannot occur without energy input. The required energetic assistance comes ultimately from food breakdown. Chemical conversions involving positive free energy changes – the synthetic or 'building-up' processes – are collectively called **anabolism** or anabolic reactions. (The anabolic steroids of sporting ill-repute promote increase in body mass, hence their name.) The other half of metabolism consists of the 'breaking-down' reactions with negative free energy changes – the catabolic reactions, or collectively, **catabolism. Metabolism** is composed of catabolism and anabolism. Catabolism of food liberates free energy, which is used to drive the synthesis of ATP from adenosine diphosphate (ADP) and inorganic phosphate. ATP is used to drive the energy-requiring processes of anabolism by the mechanism explained below.

We can summarize the overall situation as in Fig. 3.1. Food oxidation releases free energy, captured in ATP, which is then used to drive energetically unfavourable processes. To keep the system going on a global scale, CO_2 and H_2O are reconverted during photosynthesis to food molecules (such as glucose or its derivatives) using light energy. These food molecules are

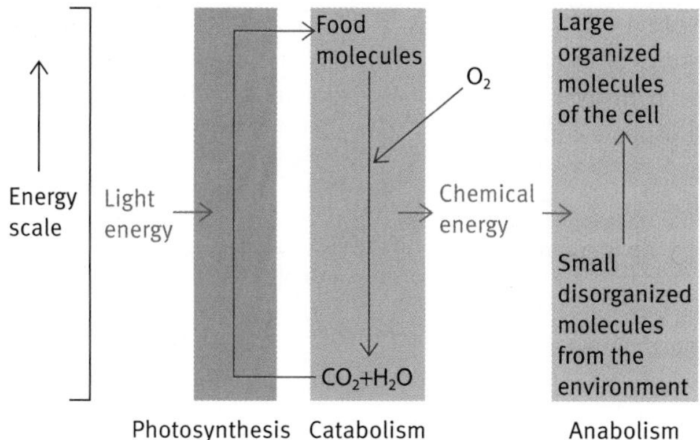

Fig. 3.1 The energy cycle in life.

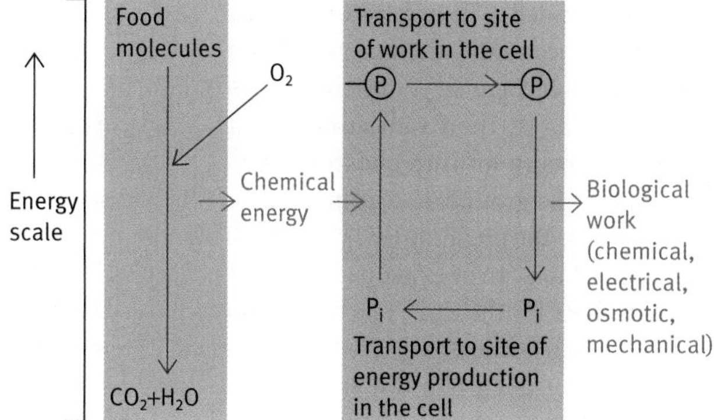

Fig. 3.2 The role of the phosphoryl group in the energy economy of a cell. —Ⓟ represents the phosphoryl group in a molecule whose hydrolysis to liberate P_i is associated with a $\Delta G^{0\prime} > 30$ kJ mol^{-1}.

converted by organisms into other food molecules such as fat. Although the assembly of large cellular structures involves a decrease in entropy (unfavourable), the oxidation of food molecules involves a greater entropy increase (favourable). The entropy change of the total system (cell and surroundings) is positive, and so the second law of thermodynamics is obeyed.

ATP is the universal energy intermediate in all life

As already explained in Chapter 1, oxidation of energy-supplying food molecules in the cell without appropriate coupling to the energy-requiring reactions would simply liberate heat and this cannot be used to do chemical or other work in the cell. Instead, the free energy change involved in food breakdown must be coupled to the energy-requiring processes. This occurs by converting ADP plus inorganic phosphate to ATP, which is the universal energy intermediate of life.

ATP is a '**high-energy phosphate compound**' (defined in 'High and low-energy phosphates'). It is transported to wherever work is to be performed in the cell, where the attached phosphates, known as **high-energy phosphoryl groups**, are converted back into inorganic phosphate ions with the liberation of the free energy that went into the formation of the groups in ATP. This does not mean that *direct* hydrolysis of ATP occurs, as this could only liberate the energy as useless heat. The mechanisms by which the energy is harnessed for work will be described shortly. We now have to deal with how ATP has this central role. Before we get to its structure in detail we will first explain something about the chemistry of phosphates.

High and low-energy phosphates

If we consider cellular compounds containing a phosphoryl group, they can be divided into two categories: **low-energy phosphate** compounds, the hydrolysis of which to liberate inorganic phosphate (**P**$_i$) is associated with negative $\Delta G^{0\prime}$ values in the range of about 9–20 kJ mol^{-1}, and **high-energy phosphate** compounds with corresponding negative $\Delta G^{0\prime}$ values larger than about 30 kJ mol^{-1}. The concept of a high-energy phosphate compound used to be described in terms of the 'high-energy phosphate bond', but this use has largely been abandoned because a high-energy bond in chemical terms refers to a bond where the breakage requires a large input of energy, the reverse of the intended biochemical concept. The high-energy phosphoryl group can be regarded as the universal energy currency of the living cell. This concept is illustrated in Fig. 3.2, which is a refinement of Fig. 3.1.

What are the structural features of high-energy phosphate compounds?

Phosphoric acid, H_3PO_4, is an oxyacid of phosphorus with three dissociable protons as shown below.

$$\mathrm{HO{-}P({=}O)(OH){-}OH} \underset{pK_{a_1}=2.2}{\rightleftharpoons} \mathrm{HO{-}P({=}O)(OH){-}O^-} + \mathrm{H^+}$$

$$\mathrm{HO{-}P({=}O)(OH){-}O^-} \underset{pK_{a_2}=7.2}{\rightleftharpoons} \mathrm{HO{-}P({=}O)(O^-){-}O^-} + \mathrm{H^+}$$

$$\mathrm{HO{-}P({=}O)(O^-){-}O^-} \underset{pK_{a_3}=12.3}{\rightleftharpoons} \mathrm{{}^-O{-}P({=}O)(O^-){-}O^-} + \mathrm{H^+}$$

At normal cell pH, a mixture of both single negatively and double negatively charged phosphate ions exists in the solution. In biochemistry, this mixture of phosphate ions is symbolized as P_i, the i indicating 'inorganic'; it represents the lowest energy form of phosphate in the cell and can be regarded as the ground state of phosphate when considering its energetics. The predominance in the cell of these two forms of phosphate ion can be explained by considering the dissociation constants for the three reactions shown above. These are expressed as pK_a values; a pK_a is the pH at which there is 50% dissociation of the group. An explanation of pK_a values and buffers is given in the appendix to this chapter; this should be studied if you are not already familiar with the subject. At physiological pH (say 7.4), the group with a pK_a of 2.2 will be fully ionized and that with a pK_a of 12.3 will be undissociated. The group with a pK_a value of 7.2 will be partially dissociated. The actual degree of dissociation of the latter can be calculated from the **Henderson–Hasselbalch equation** (Box 3.1).

Inorganic phosphate, esterified with an alcohol, is a **phosphate ester**.

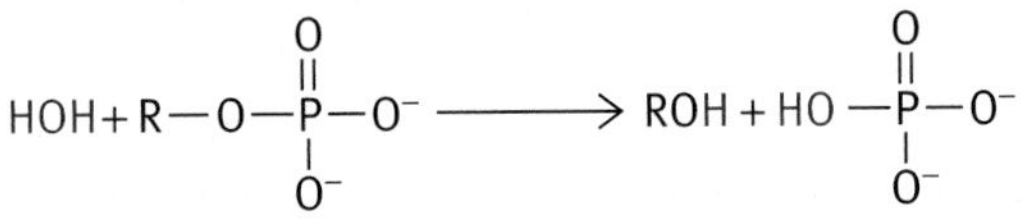

$$HOH + R{-}O{-}\overset{\overset{O}{\|}}{\underset{\underset{O^-}{|}}{P}}{-}O^- \longrightarrow ROH + HO{-}\overset{\overset{O}{\|}}{\underset{\underset{O^-}{|}}{P}}{-}O^-$$

Phosphate ester Alcohol Inorganic phosphate (P_i)

The $\Delta G^{0\prime}$ for the hydrolysis of this class of ester is roughly of the order of $-12.5\,\text{kJ mol}^{-1}$, resulting in the equilibrium for the hydrolytic reaction being strongly to the side of hydrolysis. In the cell, the reverse reaction does not occur. However, relative to ATP, this phosphate ester is a low-energy phosphate compound.

As well as forming phosphate esters with alcohols, P_i can form a **phosphoric anhydride** called **inorganic pyrophosphate** (*pyro* means fire; pyrophosphate can be made by driving off water from P_i at high temperatures) and in biochemistry it is often written as $\mathbf{PP_i}$. The $\Delta G^{0\prime}$ for the hydrolysis of this compound is $-33.5\,\text{kJ mol}^{-1}$, which is much higher than the value for hydrolysis of the ester phosphate. PP_i is a high-energy phosphate compound.

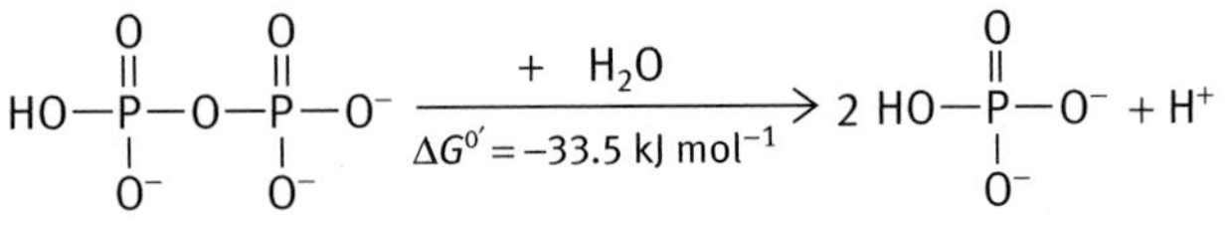

$$HO{-}\overset{\overset{O}{\|}}{\underset{\underset{O^-}{|}}{P}}{-}O{-}\overset{\overset{O}{\|}}{\underset{\underset{O^-}{|}}{P}}{-}O^- \xrightarrow[\Delta G^{0\prime} = -33.5\ \text{kJ mol}^{-1}]{+\ H_2O} 2\ HO{-}\overset{\overset{O}{\|}}{\underset{\underset{O^-}{|}}{P}}{-}O^- + H^+$$

Inorganic pyrophosphate (PP_i) Inorganic phosphate (P_i)

We will explain the high-energy nature of the anhydride by referring to inorganic pyrophosphate because it makes the explanation simpler. (PP_i is not commonly an energy donor in the cell in the sense that ATP is). The very high free energy release associated with the hydrolysis of the phosphoric anhydride group is due to several factors. When the pyrophosphoryl group is hydrolysed, the electrostatic repulsion between the two negatively charged phosphoryl groups is relieved. This factor is made clear by the reflection that, in the reverse reaction to synthesize a pyrophosphoryl bond, the electrostatic repulsion of the two ions has to be overcome in bringing them together. The phosphoric anhydride bond formation might be likened to compressing a spring.

Secondly, the products of the reaction (in this case, two P_i ions) are **resonance stabilized** – they have a greater number of possible resonance structures (see below) than has the pyrophosphate structure. This increases the entropy and therefore decreases the energy level of the products so that on breaking the bond there is a bigger yield of energy.

In an inorganic phosphate ion, all of the P–O bonds are partially double-bonded in character rather than the proton being associated with any one oxygen, resulting in an increase in entropy and a lowering of their energy level. The main resonating forms of the phosphate ion are shown in the following figure:

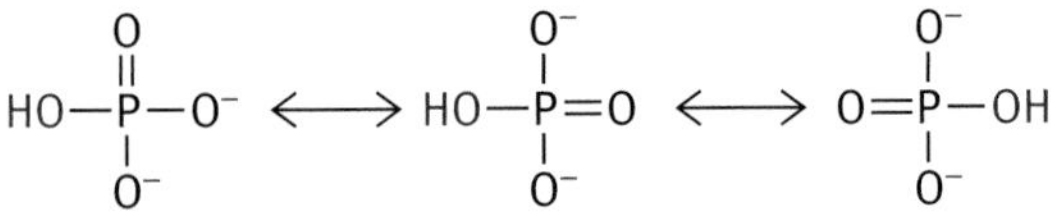

$$HO{-}\overset{\overset{O}{\|}}{\underset{\underset{O^-}{|}}{P}}{-}O^- \longleftrightarrow HO{-}\overset{\overset{O^-}{|}}{\underset{\underset{O^-}{|}}{P}}{=}O \longleftrightarrow O{=}\overset{\overset{O^-}{|}}{\underset{\underset{O^-}{|}}{P}}{-}OH$$

Note that the $\leftrightarrow$ symbol has a special meaning in chemistry; it is not the same as $\rightleftharpoons$. It does not imply that the different ionized forms are interconverting but that the structure that exists is not any of the forms shown but is an intermediate in which all oxygens have a partial negative charge and the proton is not associated with any one form. The same comments apply to the structures of the resonating forms of carboxylic acid and guanidino compounds described in the following text.

These factors mean that phosphoric anhydrides have equilibrium constants decisively in favour of P_i formation (equals a large negative $\Delta G^{0\prime}$ value). These considerations apply to any phosphoric anhydride group and in particular to those

BOX 3.1 Henderson–Hasselbalch equation calculation

$$pH = pK_{a_2} + \log_{10}\frac{[\text{salt}]}{[\text{acid}]}$$

$$7.4 = 7.2 + \log_{10}\frac{[HPO_4^{2-}]}{[HPO_4^-]}$$

Therefore,

$$\log_{10}\frac{[HPO_4^{2-}]}{[HPO_4^-]} = 0.2$$

so that

$$\frac{[HPO_4^{2-}]}{[HPO_4^-]} = 1.58$$

in ATP. However, the factors we have discussed above do not apply to hydrolysis of ester phosphates, which explains why the energy release from hydrolysing them is much less.

The phosphoric anhydride structure discussed above is not the only biological high-energy phosphate compound (though it is the predominant one). Three other types are found.

The first is the mixed anhydride between phosphoric acid and a carboxyl group (often called **acylphosphates**), hydrolysis of which by the reaction shown has a very large negative $\Delta G^{0\prime}$ value (−49.3 kJ mol^{-1} in a typical case). The large free energy change is associated with the resonance stabilization of both products, namely, P_i and the carboxy acid (the latter illustrated below). This type of molecule occurs in metabolic reactions.

$$R{-}C(=O){-}O{-}P(=O)(O^-){-}O^- + H_2O \longrightarrow R{-}C(=O){-}O^- + HO{-}P(=O)(O^-){-}O^- + H^+$$

Acylphosphate anhydride

$$R{-}C(=O){-}O^- \longleftrightarrow R{-}C(-O^-){=}O$$

Resonance stabilization

The second structure is that of **guanidino-phosphate**, the hydrolysis of which to produce P_i also has a very large negative $\Delta G^{0\prime}$ value (−43.0 kJ mol^{-1} in a typical case). Again this is due to resonance stabilization of both products. An example is creatine phosphate in muscle (see Chapter 8).

$$\overset{+}{N}H_2{=}C(NH{-}R){-}NH{-}P(=O)(O^-){-}O^- + H_2O$$

Guanadino-phosphate

$$\downarrow$$

$$HO{-}P(=O)(O^-){-}O^- + \overset{+}{N}H_2{=}C(NH_2)(NH{-}R) \longleftrightarrow NH_2{-}C(NH_2){=}\overset{+}{N}H{-}R \longleftrightarrow NH_2{-}C(=\overset{+}{N}H_2)(NH{-}R)$$

Resonance stabilization

A third type is in a different category – an enol-phosphate structure. This is found in the metabolite, **phosphoenolpyruvate**. This looks an unlikely candidate for high-energy status, but, on removal of the phosphate group, the **enol pyruvate** structure so formed, spontaneously rearranges into the keto form of pyruvate, the equilibrium of this being far to the right.

$$CH_2{=}C(COO^-){-}O{-}P(=O)(O^-){-}O^- + H_2O \longrightarrow CH_2{=}C(COO^-){-}OH + HO{-}P(=O)(O^-){-}O^-$$

Phosphoenolpyruvate (Enol phosphate) → Unstable

$$CH_2{=}C(COO^-){-}OH \longrightarrow CH_3{-}C(=O){-}COO^-$$

Stable

As a result of this the conversion of phosphoenolpyruvate to P_i and the keto form of pyruvate has a $\Delta G^{0\prime}$ value of −61.9 kJ mol^{-1}. Phosphoenol pyruvate is a component of the glycolytic pathway, which harvests energy during the breakdown of glucose.

The structure of ATP

We have so far spoken of the phosphoryl groups present in high-energy phosphate compounds in general terms. Thus, in Fig. 3.2, P_i is shown as being elevated to a 'high-energy phosphoryl group' and then transported around the cell to where it is needed to supply the energy for work. Clearly, the phosphoryl group —Ⓟ must be covalently attached to another molecule that acts as its carrier within the cell. Which brings us to the next question.

What transports the —Ⓟ around the cell?

The general carrier is **adenosine monophosphate (AMP)** (the structure is shown in Fig. 3.3). AMP is, by itself, a relatively low-energy phosphate ester. This figure gives a diagrammatic representation of adenosine and its derivatives without any

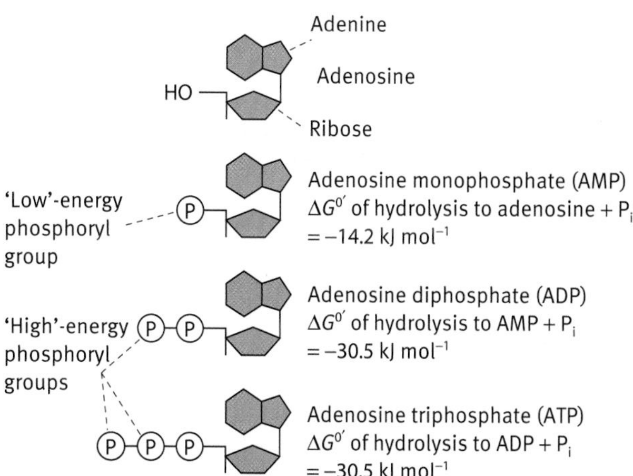

Fig. 3.3 Diagrammatic representation of adenosine and its phosphorylated derivatives. Adenosine is a nucleoside with ribose and adenine as its components.

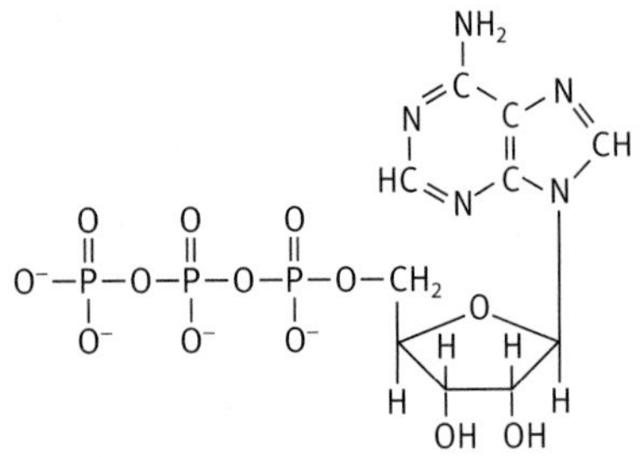

Fig. 3.4 The structure of adenosine triphosphate (ATP).

detailed structures since they are not needed in this context. (However, for reference purposes, Fig. 3.4 gives the full structure of ATP.) AMP carrying a single —Ⓟ group is called ADP or **adenosine diphosphate**. AMP carrying two —Ⓟ groups is ATP or **adenosine triphosphate**. ATP therefore can be written as AMP —Ⓟ—Ⓟ. The two terminal phosphates are attached to AMP by phosphoric anhydride linkages of the same type as already met in PP_i, and the reasons given earlier for the large $\Delta G^{0'}$ for hydrolytic removal of these two groups apply equally here. Note that, as stated, AMP itself is a low-energy ester phosphate and does not itself directly participate in the energy cycle except as carrier for the two —Ⓟ groups. You will see from Fig. 3.3 that the $\Delta G^{0'}$ of hydrolysis of each of the two terminal groups is $-30.5\,\text{kJ mol}^{-1}$; that is, of the high-energy type. When food is oxidized or otherwise broken down, the free energy released, as explained, is coupled to the conversion of ADP (or AMP—Ⓟ) to ATP:

$$ADP + P_i \xrightarrow[+\ \text{energy}]{} ATP + H_2O$$

Each cell contains only a small quantity of ATP at any one time. The amount would last only a very short time and the cell cannot get any ATP from the outside, since ATP, ADP, or AMP cannot diffuse through the cell membrane because they are highly charged. Each cell has to synthesize the molecule itself. ATP thus 'turns over' or cycles very rapidly in the cell, by which we mean it breaks down to ADP and P_i and is resynthesized to ATP. We can now modify the diagram given in Fig. 3.2 to include ATP and ADP (Fig. 3.5).

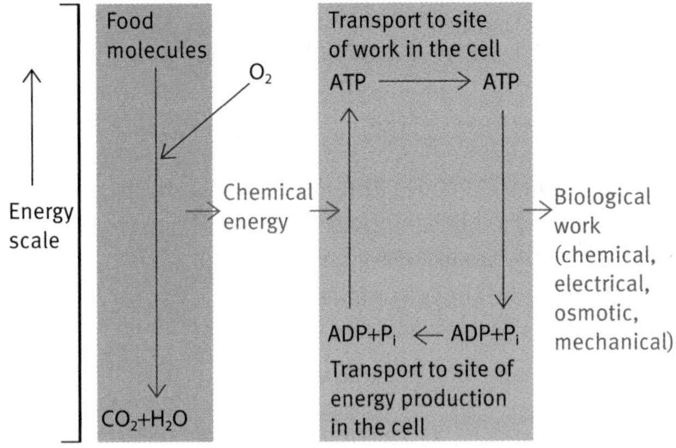

Fig. 3.5 The role of ATP in the energy economy of a cell. Note that some types of work involve breakdown of ATP to AMP, but, as described in the text, this does not change the concept given here.

How does ATP drive chemical work?

Suppose the cell needs to synthesize X—Y from the two reactants, X—OH+Y—H, and the $\Delta G^{0'}$ change involved in the conversion is $12.5\,\text{kJ mol}^{-1}$. The simple reaction $XOH + YH \rightarrow XY + H_2O$ cannot occur to any significant extent because the $\Delta G^{0'}$ is positive and the equilibrium is far to the left. The usual solution is to use **coupled reactions** involving ATP breakdown, as shown below. Coupled reactions are two or more reactions in which the product of one becomes the reactant for the next. No physical coupling is necessarily involved – they simply have to be present in the same chemical system, for example the same organelle in the same cell. Note that all reactions in the cell involving ATP must be enzymically catalysed – the fact that hydrolysis of each of the two phosphoric anhydride groups is strongly exergonic (giving out energy) does not mean that ATP is an unstable or highly reactive molecule. For clarity, ATP will be represented as AMP—Ⓟ—Ⓟ below. The coupled reactions to synthesize XdY are as follows:

Reaction 1: XOH+AMP—Ⓟ—Ⓟ→X–P+AMP—Ⓟ

Reaction 2: X–P+YH→X–Y+P_i

Sum of 1+2: XOH+YH+AMP—Ⓟ—Ⓟ→X–Y+AMP—Ⓟ+P_i

In reaction 1, a phosphoryl group is transferred from ATP to X—OH forming X—P. In the second reaction the phosphoryl group is replaced by Y, liberating inorganic phosphate and forming X—Y. Both reactions may occur on a single enzyme without the X—P ever being free, but in other cases X—P may leave the first enzyme and diffuse to a second enzyme, which carries out the second reaction. From an energetic viewpoint the two possibilities are the same. The overall $\Delta G^{0'}$ for coupled reactions is the arithmetic sum of the $\Delta G^{0'}$ values of the component reactions. The $\Delta G^{0'}$ for the $XOH + YH \rightarrow X{-}Y + H_2O$ we take to be $12.5\,\text{kJ mol}^{-1}$; the $\Delta G^{0'}$ for the reaction $ATP + H_2O \rightarrow ADP + P_i$ is $-30.5\,\text{kJ mol}^{-1}$. Therefore, the $\Delta G^{0'}$ of the overall process is $-18.0\,\text{kJ mol}^{-1}$, a strongly exergonic process that will proceed essentially to completion – the equilibrium constant will be about 10^3 – which means that the reactions can proceed to approximately 99.9% to the side of X—Y formation (see Table 3.1). It is to be noted that the use of ATP here involves phosphoryl group transfer from ATP to one of the reactants and only then is P_i liberated. *Direct* hydrolysis of ATP to ADP and P_i is not occurring.

In the cell, the actual ΔG value of the ATP breakdown to ADP and P_i is considerably larger than the $\Delta G^{0'}$ value, because ΔG values are affected by reactant concentrations, and cellular levels of ATP, ADP, and P_i are, of course, nowhere near 1.0 M (the standard concentrations specified in $\Delta G^{0'}$

BOX 3.2 Calculation of ΔG value

$$\Delta G = \Delta G^{0\prime} + RT2.303\log_{10}\frac{[\text{products}]}{[\text{reactants}]}$$

where R is the gas constant (8.315×10^{-3} kJ mol^{-1} K^{-1}); and T is the temperature in Kelvin (298 K = 25 °C).

The concentrations of ATP, ADP, and P_i will vary from cell to cell. ATP 'levels' generally fall in the range 2–8 mM, ADP levels are about one-tenth of this, and P_i values are similar to those of ATP. However, for simplicity let us assume a situation in which all three are at 10^{-3} M. (For reactions in dilute aqueous solution, the concentration of water is considered to have a value of 1.) Substituting these values into the equation:

$$\Delta G = \Delta G^{0\prime} + RT2.303\log_{10}\frac{[\text{ADP}][\text{P}_i]}{[\text{ATP}]}$$

we obtain:

$$\Delta G = -30.5\,\text{kJmol}^{-1} + (8.315\times10^{-3})(298)$$
$$\times 2.303\log_{10}\frac{[10^{-3}][10^{-3}]}{[10^{-3}]}$$
$$= -30.5 - 17.1 = -47.6\,\text{kJmol}^{-1}$$

determinations). If the actual cellular concentrations of these components are known, the ΔG value for the hydrolysis of ATP to ADP + P_i can be calculated (see Box 3.2) and is even more energetically favourable than is $\Delta G^{0\prime}$.

The reaction sequence given above is used for many biochemical reactions coupled to ATP breakdown. But, more commonly, for the synthesis of molecules such as nucleic acids and proteins, the cell uses an even more energetically effective trick to guarantee direct irreversibility of the reaction. Instead of breaking off one —Ⓟ of ATP to P_i it releases two. The negative $\Delta G^{0\prime}$ value that results from this is so large that the equilibrium is such that the reaction is completely irreversible in the cell. The way this is done is as follows, again taking XOH + YH → X—Y + H_2O as the example:

Reaction 1: XOH + AMP—Ⓟ—Ⓟ → X—AMP + PP_i

Reaction 2: X—AMP + YH → X—Y + AMP

Reaction 3: $PP_i + H_2O \rightarrow 2P_i$

Sum: XOH + YH + AMP—Ⓟ—Ⓟ + H_2O → X—Y + AMP + $2P_i$

In the first reaction, instead of a phosphoryl group being transferred to X—OH, the AMP group is attached, displacing the two terminal phosphoryl groups as inorganic pyrophosphate. The X—AMP reacts in a second reaction with X—OH, displacing the AMP and forming the desired X—Y. The two reactions occur on the surface of a single enzyme. So far this is little different energetically from the first scheme in which X—OH is phosphorylated, for only one phosphoric anhydride group has been broken. However a different enzyme hydrolyses the inorganic pyrophosphate. Assume that the $\Delta G^{0\prime}$ for the reaction from XOH + YH → X—Y + H_2O is 12.5 kJ mol^{-1}: the $\Delta G^{0\prime}$ for the reaction ATP + H_2O → AMP + PP_i is −32.2 kJ mol^{-1}; that for PP_i hydrolysis is −33.5 kJ mol^{-1}.

The $\Delta G^{0\prime}$ for the reaction ATP + H_2O → AMP + $2P_i$ is −65.7 kJ mol^{-1}. The overall process of synthesizing X—Y at the expense of ATP breakdown thus has a $\Delta G^{0\prime}$ of −65.7 + 12.5 = −53.2 kJ mol^{-1}, a very large negative value indeed.

The mechanism depends on the fact that inorganic pyrophosphate, PP_i, is degraded rapidly in the cell. We have stated earlier that ATP hydrolysis must be coupled to a reaction in order to drive it, and that simple direct hydrolysis would be useless. Although it may now seem contradictory to state that PP_i hydrolysis helps to drive a reaction, the way in which PP_i hydrolysis helps to drive the formation of X—Y is that it removes the product (PP_i) of the reaction. This has the effect of shifting the reaction to the right. It is an equilibrium mass action effect. This explains why cells contain an enzyme, called **inorganic pyrophosphatase**, that catalyses PP_i hydrolysis.

This enzyme, once regarded as unimportant, is a driving force in biochemical syntheses and is widely distributed. As already implied, the enzyme simply has to be present in the same chemical system (in the same cell) without physical association with the synthesis of X—Y. It is the free energy change for the overall conversion scheme that is important. The total free energy change is the arithmetic sum of that of all the reactions.

How does ATP drive other types of work?

As well as performing chemical work, ATP breakdown powers muscle contraction, the generation of electrical signals, and pumping of ions and other molecules against concentration gradients, and much else. The mechanisms are, in principle, the same. Whatever the process, provided ATP breakdown is coupled into the mechanism the resultant free energy liberation will drive it. Many of these processes will be dealt with in subsequent chapters.

High-energy phosphoryl groups are transferred by enzymes known as kinases

Many ATP-requiring chemical syntheses in the cell produce AMP + $2P_i$, not ADP + P_i. AMP cannot, itself, be converted into ATP by the foodstuff-oxidation system; only ADP is accepted as shown in Fig. 3.5. However, AMP is rescued by an enzyme that transfers —Ⓟ from ATP to AMP. The reaction is:

AMP + AMP—Ⓟ—Ⓟ ⇌ 2AMP—Ⓟ

or, as written more conventionally:

AMP + ATP ⇌ 2ADP.

The enzyme is called **AMP-kinase**. Kinase is the term for an enzyme that carries a phosphoryl group from ATP to elsewhere; AMP-kinase means it transfers the group to AMP (AMP is also called adenylic acid; hence AMP-kinase may also be called adenylate kinase). Note that, in this case, the —Ⓟ group is transferred directly from one molecule to another without hydrolysis or significant release of energy. You will find as you go through the book that such freely reversible 'shuffling' at the 'high-energy level' occurs frequently. The ADP (AMP—Ⓟ) so produced is now accepted by the food-oxidation system and converted to ATP.

In addition to these transfers at a high-energy level a multitude of specific kinases transfer phosphoryl groups to other molecules resulting in the formation of relatively low-energy phosphate esters. Such transfers are not reversible because of the large negative ΔG value involved. You will come across these kinases in almost every aspect of biochemistry.

To give perspective to where we are, we have not in this account dealt with the mechanisms by which ATP is synthesized from ADP and P_i at the expense of food catabolism. This is a very large topic that forms a substantial part of the chapters on metabolism later in this book. Also, we have dealt with the utilization of ATP energy to perform work only in general terms so far – as you progress through the book you will encounter example after example, as ATP utilization is involved in virtually all biochemical systems.

We come now to a change of topic, though it is still concerned with free energy changes in chemical processes.

Energy considerations in covalent and noncovalent bonds

The chemistry we have been discussing so far in this chapter involves covalent bonds, which are strong. They are formed by two atoms sharing a pair of electrons, a simple example being the formation of a hydrogen molecule from two hydrogen atoms:

$$H^{\cdot} + H^{\cdot} \rightarrow H{:}H$$

Each electron is attracted to the positive nucleus of both atoms, which holds the two together. When any chemical bond is formed, energy is liberated as required by the second law of thermodynamics. To break a chemical bond the same amount of energy has to be provided as was liberated in its formation. Covalent bonds are needed to form stable molecules such as glucose. They are very stable because large amounts of energy are liberated in their formation. To give an example, in forming a molecule of O_2 from two oxygen atoms the standard free energy change is about $-460\,\text{kJ mol}^{-1}$, so that the equilibrium is such that oxygen gas has a negligible number of free atoms. Collisions between molecules in solution can provide the energy required to break some chemical bonds. However, the average kinetic energy of molecules in solution at 25 °C is only in the range of about $4\text{–}30\,\text{kJ mol}^{-1}$, far below the range needed to *destroy* a covalent bond. In biochemical reactions, the breakage of a covalent bond is accompanied by the simultaneous formation of another bond via the transition state so that the net energy change involved is far less than the very large value required to *destroy* a covalent bond.

Now we come to a different type of molecular interaction involving **noncovalent** bonds, also referred to as secondary or weak bonds. The latter two terms belie their importance in life. There are few, if any, processes in the mechanism of life that do not depend on them.

Noncovalent, weak, or secondary bonds do not involve sharing a pair of electrons between atoms; they are electrostatic attractions of several types. Their free energies of formation are about $0.5\text{–}40\,\text{kJ mol}^{-1}$; this means that the kinetic energy of thermal motion of water molecules is sufficient to disrupt these bonds, so that they are continually being spontaneously formed and disrupted. Their energies are so small that there is not the same energy barrier that there is to molecules interacting *via* noncovalent bonds. Hence there is (usually) no need for enzyme catalysis to make or break them.

Noncovalent bonds would not be any good at forming stable discrete single molecules such as glucose. However, if a sufficient number of such bonds are present they can hold molecules together to form larger molecular structures.

Noncovalent bonds are the basis of molecular recognition and self-assembly of life forms

In Chapter 1, we explained that one of the fundamental themes essential to life was that protein molecules can recognize other chemical structures and bind to them. There is virtually no biochemical process in which this does not play a vital part. It is, for example, the reason why life is a self-assembling process. Produce the correct proteins inside a cell in the correct order and quantities, and they will interact to allow the organism to develop. As we phrased it earlier, produce rabbit proteins and a rabbit will develop. This is the remarkable way in which a linear one-dimensional code in DNA can produce such a variety of three-dimensional living organisms.

How is this achieved? Noncovalent bonds form between appropriate atoms of molecules provided they are close enough to each other to do so. Because of their weakness many such bonds must be formed to hold the molecules together. Two different protein molecules can get close enough only if there are patches of structural complementarity; in other words only if they have complementary shapes so that they can fit together and form noncovalent bonds at the attachment points. There must also be large enough numbers of bonds formed to hold the molecules together. Thirdly, the attachment points must have the appropriate chemical groups to favour weak bond formation. Put all these requirements together and the chance of random protein molecules being joined is remote. Only proteins that have evolved the appropriate structural complementarities will associate. Thus weak bond dependent associations can be highly specific. Only appropriate associations will occur and therefore only molecular assemblies appropriate to the particular organism can form. The same applies to enzymes recognizing their substrates and to gene control proteins recognizing their genes, to give two further examples (see Chapters 6 and 26).

Noncovalent bonds are also important in the structures of individual protein molecules and other macromolecules

So far we have been discussing the role of weak bonds in causing specific associations of protein molecules, but they are also important in the structures of individual proteins. Primary protein structure is the covalent linking of a long sequence of amino acids (see Chapter 4) to form the long polypeptide chain. However, an extended polypeptide chain is rarely a functional protein. The chains fold up into compact three-dimensional shapes. The precise folding and therefore the external functional groups and their spatial arrangement in a protein are determined entirely in most cases, or mainly in some, by noncovalent interactions between amino acid residues of the polypeptide. (A few proteins that must face the rigours of the extracellular world, such as insulin released into the blood or digestive enzymes released into the gut, are stabilized in their folded state by a small number of covalent S—S bonds, analagous to a few steel rivets in the structure.)

There is a good reason for using weak bonds to produce the final compact protein structures. Many or most proteins are molecular machines that need to change their shape somewhat as they function. They have to undergo conformational change in response to ligand binding. Weak bonds allow this. To give another illustration, DNA depends on hydrogen bonding (see below) of the two strands; these must be separated for replication and gene activity. The weak hydrogen bonds are sufficiently strong to hold the strands together and sufficiently weak to be broken as needed.

Types of noncovalent bonds

The three relevant types of bond or force are shown in Table 3.2. All are electrostatic attractions between appropriate atoms of covalent molecules, each with a different strength.

	Bond strengths (kJ mol^{-1})
van der Waals interactions	0.4–4
Hydrogen bonds	12–30
Ionic bonds	20

Table 3.2 Noncovalent bonds and their characteristics

Ionic bonds

An ionic bond is the attraction between negative and positive groups of ions. A typical case is the attraction between a negatively charged carboxylate ion and a positively charged amino group. The bond has no intrinsic directionality but since the groups are present in defined locations in the molecule(s) they can be involved in specific molecular recognition.

$$-COO^{-}---H_3N^{+}-$$

For comparison, covalent bonds typically have bond strengths in the region of a few hundred kJ mol^{-1}. The value for a C—C bond is about 350 kJ mol^{-1}.

Additionally hydrophobic forces or interactions are not bonds, but are important in determining the molecular structures of many biological molecules (see text).

Hydrogen bonds

These are electrostatic attractions between atoms of polar molecules, but the attractions are not due to fully separated ionized groups but to a much weaker form of positive and negative charge separation, the electric dipole moment. Although a covalent bond is overall electrically neutral, the bond can have a weakly polar nature. An example is the water molecule.

$$^{\delta+}H-O^{\delta-}-H^{\delta+}$$

The hydrogen atoms are attached to the oxygen with an angle of 104.5° between them, giving the molecule an asymmetric shape with the oxygen at one end and the two hydrogens at the other. Although the covalent O—H bonds are the result of a shared pair of electrons, the electrons are not shared exactly equally because oxygen is an electronegative atom. It attracts the shared electrons in the bond to be closer to it than they are to the hydrogen. This gives the hydrogens a partial positive charge (δ^+) and the oxygen a partial negative charge (δ^-) so that a weak attraction occurs between the O and H atoms of adjacent water molecules. This attraction is known as a hydrogen bond. Each water molecule is bonded to four other molecules because the oxygen atom can form two hydrogen bonds and the hydrogen atoms one each. This means that bulk water has a cohesive structure. The partial charges on water molecules also have a profound effect on electrical attractions between other polar molecules because they partially neutralize the charges.

A hydrogen atom of any molecule can participate in hydrogen bond formation provided it is attached to an electronegative atom. The main relevant electronegative atoms in biochemistry are nitrogen and oxygen. The bond must involve also an electronegative acceptor atom – also usually nitrogen or oxygen.

The types of hydrogen bond common in biochemistry are illustrated here with hydrogen bonds shown as dashed lines.

$$-O-H---O-$$
$$-O-H---N-$$
$$-N-H---O-$$
$$-N-H---N-$$
$$-N^{+}-H---O-$$

In contrast to ionic bonds, *the hydrogen bond is highly directional* and of maximal strength when all of the atoms are in a straight line.

van der Waals forces

This is the collective term used to describe a group of weak interactions between closely positioned atoms. Atoms are electrically neutral overall but very weak transitory polarities exist. The electrons of any atom are in constant motion; at any one time they may not be evenly distributed so that transitory fluctuations in electron density around the nucleus of the atoms occur. This means that the negative charge distribution in the atom can fluctuate, which causes, at any instant, one part to have a slight positive charge and the other a slight negative one. This in turn can affect electron distribution in neighbouring atoms. A negative charge on one atom will tend to repel electrons in its neighbour, thus inducing a local positive charge and resulting in an electrostatic attraction between the two atoms.

Van der Waals forces can operate between *any* two atoms, which must be positioned close together so that their electron shells are almost touching. The attraction between them is inversely proportional to the sixth power of the distance, so precisely close positioning is essential. If atoms tend to become too close together, their electron shells overlap and a repulsive force is generated. In short, precise positioning is a prerequisite for van der Waals attractions to arise. This means that for them to form between, say, two different proteins these must have precise complementarity of structure at contact points for van der Waals forces to be generated. Although the attractions are extremely weak they become significant if they exist in large numbers.

Hydrophobic force

This is also termed **hydrophobic interactions**. The force does not involve bonds between hydrophobic molecules (see Chapter 1), but it causes hydrophobic molecules to associate together. A hydrophilic (polar) molecule such as sugar that can form hydrogen bonds is soluble in water because, although it disrupts hydrogen bonding between water molecules (energy-requiring), it can itself form bonds with the latter (energy-releasing), making the process energetically feasible. Salts such as NaCl are soluble because the Na^+ and Cl^- ions become surrounded by hydration shells in which the ion-water attractions exceed those between the ions themselves and, when separated, there is a large increase in entropy from the crystalline state. If by contrast an attempt is made to dissolve a nonpolar substance such as olive oil in water, the oil molecules get in the way of hydrogen bonding between water molecules. Since hydrogen bonding is highly directional, the water molecules around the oil molecules rearrange themselves so that none of their bonding sites are aimed at the oil molecules. This more highly ordered arrangement is at a lower entropy level than that of randomly arranged water molecules (a higher energy level) and so the solubilization of the oil is opposed by the second law of thermodynamics. The nonpolar molecules are forced to associate together so as to present the minimum oil/water interface area. The olive oil forms droplets and then a separate layer, which is the minimum free energy state. Hydrophobic groups occur in proteins, DNA, and other cellular molecules, and hydrophobic forces play a crucial role in the structure of these molecules, as will be seen later in this book; it is remarkable how the necessity of 'hiding' these hydrophobic groups from water determines so much in living cells.

With that introduction to energy considerations in life we come to the next five chapters, in which protein structure and function are the main themes – protein structure, methods in protein investigation, enzymes, membranes, molecular motors, and the cytoskeleton. These topics are necessary for understanding metabolism, which is the major area covered in the middle section of the book.

Appendix: Buffers and pK_a values

It is very important in biochemistry to understand what buffers and pK_a values are. We have placed this is an appendix to avoid disrupting the text and because many will have dealt with it in their chemistry studies.

The pH of a living cell is maintained in the range 7.2–7.4. (pH is the negative logarithm to base 10 of the H^+ concentration expressed in moles per litre). Special situations occur, such as in the stomach where HCl is secreted, and in lysosomes into which protons are pumped to maintain an acid pH, but, otherwise, the pH of cells and of circulating fluids is maintained within narrow limits. This is despite the fact that metabolic processes producing acid,such as lactic acid and ethanoic (acetic) acid, and CO_2 conversion to carbonic acid (H_2CO_3) in the blood, occur on a large scale.

This pH stability is largely due to the buffering effect of weak acids. An acid in this context is defined as a molecule that can release a hydrogen ion (a proton), and a base is defined as a proton acceptor.

A carboxylic acid dissociates, liberating a proton:

$$\text{R COOH} \rightleftharpoons \text{R COO}^- + \text{H}^+$$

Written in a more general form, the equation for acids is:

$$\text{HA} \rightleftharpoons \text{H}^+ + \text{A}^-$$

Acids vary in their tendency to dissociate. Stronger acids do so more readily than weaker ones – this is why, say, a 0.1 M solution of methanoic (formic) acid (HCOOH) has a lower pH than a 0.1 M solution of ethanoic acid (CH_3COOH). The tendency to dissociate is quantitated as a dissociation constant,

K_a for each acid: the larger the value of K_a, the greater the tendency to dissociate and the stronger the acid:

$$K_a = \frac{[H^+][A^-]}{[HA]}.$$

For ethanoic acid $K_a = 1.74 \times 10^{-5}$. This value is not much used, as such, by biochemists, as there is another way of expressing the strength of the acid by a much more convenient term – the pK_a value. The two are related by the equation:

$$pK_a = -\log_{10} K_a$$

Thus, ethanoic acid has a pK_a of 4.76 and methanoic acid a pK_a of 3.75. These values represent the pH at which an acid is 50% dissociated. As the pH increases (i.e as H^+ concentration decreases), the acid becomes more dissociated; as the pH decreases (H^+ concentration increases), the reverse occurs. This is because the $HA \rightleftharpoons H^+ + A^-$ equilibrium is affected by the H^+ concentration as would be expected (Fig. 3.6).

An amine base also has a pK_a value because the ionized form can dissociate to liberate a proton and, in this sense, is an acid:

$$R\,NH_3^+ \rightleftharpoons R\,NH_2 + H^+$$

(a) for the molecule RCOOH with a pK_a of 4.2:

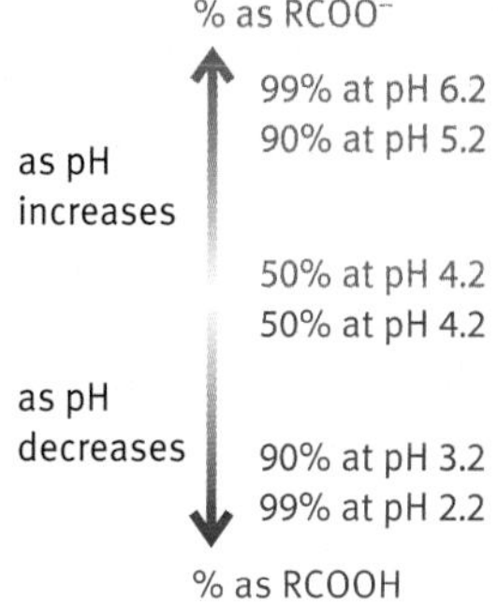

(b) for the molecule RNH_2 with a pK_a of 9.2:

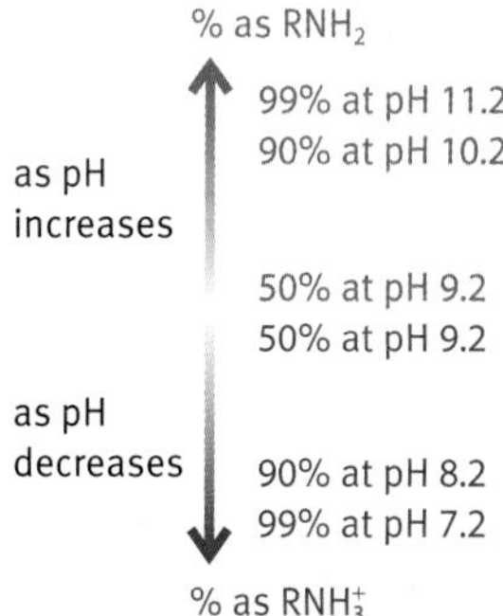

Fig. 3.6 Effect of pH on the ionization of **(a)** –COOH and **(b)** –NH_2 groups. Kindly provided by R. Rogers.

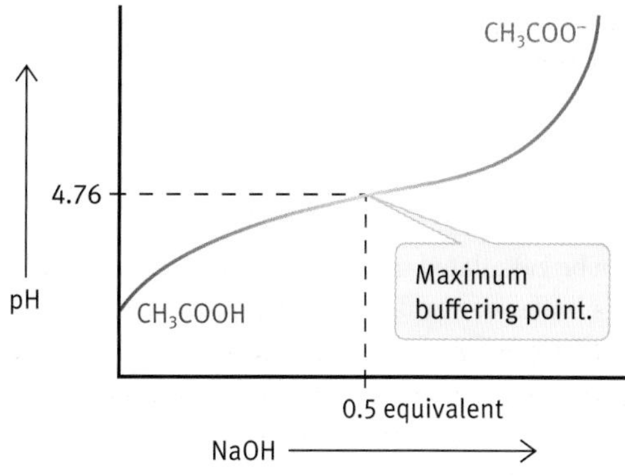

Fig. 3.7 Titration curve for ethanoic acid (pK_a = 4.76)

In this case, as the pH increases (H^+ concentration decreases), the dissociation of the amine base increases and its ionization decreases. With both carboxylic acids and amine bases, an increase in H^+ concentration (decreased pH) causes increased protonation. The difference is that protonation of a carboxylic acid reduces the amount of ionized form while, with amine bases, the proportion in the ionized form increases (Fig. 3.6).

pK_a values and their relationship to buffers

If you take a 0.1 M solution of ethanoic acid and gradually add 0.1 M NaOH, measuring the pH at each step, the curve shown in Fig. 3.7 is obtained.

At the beginning of the titration, the added OH^- neutralizes existing H^+ (the ethanoic acid is slightly dissociated) and the pH rises rapidly. However, as the pH begins to approach the pK_a value of ethanoic acid, as more NaOH is added, the ethanoic acid dissociates into the ethanoate ion liberating hydrogen ions, which neutralize the hydroxide ions so that the pH change is relatively small until the ethanoic acid is all ionized. The reaction is:

$$CH_3COOH + OH^- \rightarrow CH_3COO^- + H_2O.$$

(Note that sodium ethanoate is fully dissociated.)

Similarly, from the other side of the pH scale, additions of H^+ cause little change to the pH because of the reaction:

$$CH_3COO^+ + H^+ \rightarrow CH_3COOH.$$

This is the pH-buffering effect of ethanoic acid. Buffering is maximal at the pK_a so that an equimolar mixture of ethanoate and ethanoic acid gives its maximum buffering effect at that pH. As a rule of thumb, in biochemistry, a useful buffer covers a range of about 1 pH unit on either side of the pK_a. The pH of a solution containing an acid-base conjugate pair can be calculated from the Henderson–Hasselbalch equation:

$$pH = pK_a + \log_{10} \frac{[\text{proton acceptor}]}{[\text{proton donor}]}$$

or, as often expressed:

$$pH = pK_a + \log_{10} \frac{[salt]}{[acid]}.$$

The pH of any mixture of ethanoic acid and sodium ethanoate can be calculated from this equation so that the composition of a buffer of any desired pH can be determined.

As an example for a solution containing 0.1 M ethanoic acid and 0.1 M sodium ethanoate:

$$pH = 4.76 + \log_{10} \frac{[0.1]}{[0.1]}$$
$$= 4.76 + 0 = 4.76.$$

If the mixture contains 0.1 M ethanoic acid and 0.2 M sodium ethanoate:

$$pH = 4.76 + \log_{10} \frac{[0.2]}{[0.1]}$$
$$= 4.76 + \log_{10} 2 = 4.76 + 0.30 = 5.06.$$

Suppose that to the buffer containing 0.1 M ethanoic acid and 0.1 M sodium ethanoate you added NaOH to a concentration of 0.05 M; the pH of the resultant solution would be:

$$4.76 + \log_{10} \frac{[0.15]}{[0.05]}$$

(since half of the ethanoic acid would be converted to sodium ethanoate)

$$= 4.76 + \log_{10} 3 = 4.76 + 0.48 = 5.24.$$

Ethanoic acid/ethanoate mixture has no significant buffering power at physiological pH values, but compounds with pK_a values close to pH 7 exist in the body, and these are effective buffers. Among the most important are the phosphate ion and its derivatives. Phosphoric acid has three dissociable groups as shown earlier, the second of which ($H_2PO_4 \rightleftharpoons HPO_4 + H^+$) has a p$K_a$ value of 6.86, so that phosphate is an excellent buffer at cellular pH.

Another buffering structure is the imidazole nitrogen of the histidine residue in proteins, which has a pK_a around 6. The dissociation involved is shown below.

$H_3^+N—CH(CH_2\text{-imidazole, NH / N}^+H)—COO^- \rightleftharpoons H_3^+N—CH(CH_2\text{-imidazole, NH / N})—COO^- + H^+$

The phenomenon of buffering can be illustrated by a very simple observation. If you take a test-tube containing distilled water and (as dissolved CO_2 normally makes it slightly acidic) adjust the pH to 7.0 with NaOH, and then add a few drops of 0.1 M HCl, there will be a precipitous drop in pH. If you take a test-tube containing 0.1 M sodium phosphate, also adjusted to pH 7.0, and add the same amount of HCl, the pH will hardly change, due to the buffering action of the phosphate ion.

The buffering action of compounds with pK_a values near 7 thus protects the cell and body fluids against large pH changes.

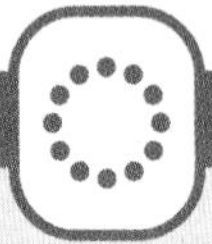

Summary

Energy changes in chemical reactions are reliable guides to the biochemistry of cells. The most useful value is the free energy change (ΔG), which is an expression of the amount of energy change in a reaction available to perform useful work. The ΔG value can be used to determine the equilibrium constant of a reaction and whether it is likely to be reversible in cells.

Free energy (a technical term meaning useable energy) made available from food breakdown is used to synthesize ATP, the universal energy carrier of life.

Catabolism (breakdown) of food molecules drives anabolism (synthesis of molecules) with ATP being the energy-carrying go-between. ATP is termed a high-energy phosphate molecule; the release of the two terminal phosphate groups liberates large quantities of free energy, which is used to perform work of all kinds in coupled cellular reactions. There are other high-energy compounds in the cell that can transfer their phosphate groups to ADP to form ATP.

Weak noncovalent chemical bonds play a crucial part in life. These are hydrogen bonds, ionic bonds and van der Waals forces. Unlike covalent bonds, their formation and breakdown involves only small free energy changes, and occur without the need for enzyme catalysis. They are important because binding between molecules is dependent on a number of weak bonds forming. This will occur only if the atoms involved are sufficiently physically close, which means that only molecules with complementary shaped binding sites will bind by these forces. This complementarity of shape is the basis of molecular recognition (biological specificity) by proteins on which all life depends.

Hydrophobic forces or hydrophobic interactions are not bonds, but they contribute to molecular structures and molecular interactions by causing hydrophobic groups and molecules to associate together in the aqueous environment of the cell.

Noncovalent bonds are essential for molecules to specifically recognize and bind to each other, and are essential in most biochemical processes. They are also crucially involved in molecules such as proteins and DNA, and in the formation of protein complexes. They have the advantage that because they are easily broken they confer flexibility on structures; for example the two DNA strands of a double helix are held together by hydrogen bonds. To replicate DNA the strands have to be separated and reformed.

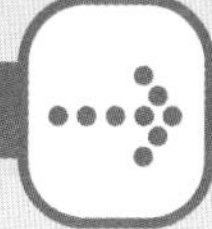

Further reading

To access the further reading, please scan the QR code image or go to http://global.oup.com/uk/orc/biosciences/molbiol/snape_biochemistry/student/reading/ch03/

Problems

1 Suppose that a 70 kg man has a food intake for his energy needs of 10,000 kJ per day. Assume that the free energy available in his diet is used to form ATP from ADP and P_i with an efficiency of 50%. In the cell, the ΔG for the conversion of ADP + P_i is approximately 55 kJ mol^{-1}. Calculate the total weight of ATP the man synthesizes per day, in terms of the disodium salt (molecular mass = 551 Da).

2 The $\Delta G^{0\prime}$ for the hydrolysis of ATP to ADP and P_i is 30.5 kJ mol^{-1}. Explain why, in Problem 1, a ΔG value of 55 kJ mol^{-1} was suggested as the amount of free energy required to synthesize a mole of ATP from ADP and P_i in the cell.

3 The hydrolysis of ATP to ADP + P_i and that of ADP to AMP + P_i have $\Delta G^{0\prime}$ values of −30.5 kJ mol^{-1}, while the hydrolysis of AMP to adenosine and P_i has a value of −14.2 kJ mol^{-1}. What are the reasons for the large difference?

4 Explain

(a) why benzene will not dissolve in water to any significant extent

(b) why a polar molecule such as glucose is soluble in water

(c) why NaCl is soluble in water.

5 In the cell, ADP is converted to ATP using the energy derived from food catabolism, but AMP is not utilized by the ATP-synthesizing system; many synthetic processes convert ATP to AMP. How is AMP brought back into the system?

6 An enzyme catalyses a reaction that synthesizes the compound XY from XOH and YH, coupled with the breakdown of ATP to AMP and PP_i. The $\Delta G^{0\prime}$ of the reaction XOH + YH → XY + H_2O is 10 kJ mol^{-1}. Determine the $\Delta G^{0\prime}$ of the reaction: (a) in the cell; and (b) using a completely pure preparation of the enzyme. Explain your answer. (You are told that the $\Delta G^{0\prime}$ values for ATP hydrolysis to AMP and PP_i, and for PP_i hydrolysis are −32.2 and −33.4 kJ mol^{-1}, respectively.)

7 What are the different types of weak bonds of importance in biological systems and their approximate energies. Why is it that their formation does not require enzymic catalysis? If they are so weak, why are they of importance?

8 The pK_a values for phosphoric acid are 2.2, 7.2, and 12.3.

(a) Write down the predominant ionic forms at pH 0, pH 4, pH 9, and pH 14.

(b) What would be the pH of an equimolar mixture of NaH_2PO_4 and Na_2HPO_4?

(c) Suppose that you wanted a buffer fairly close to physiological pH and all you had available were the 20 amino acids found in proteins, which one would be the most suitable?

9 Virtually all biological processes involve specific interactions between proteins and other molecules

(which may also be proteins). Explain how the specific interactions are achieved.

10 The use of the term 'high-energy phosphate bond' (indicated by a squiggly bond) by Lipmann in 1940 has been of great importance in the development of the concept of biological energy. Although sometimes still used, because it is a convenient shorthand notation, it has fallen out of favour because it is not chemically correct. Discuss this.

11 What is the fundamental driving force that causes chemical reactions to occur under appropriate conditions? In other words why should reactions ever occur?

12 The second law of thermodynamics specifies that all processes must increase the total entropy of the universe and yet living cells are at a lower entropy level than the randomly arranged molecules in the environment from which the living cells are assembled. Does this mean that living cells are islands in the universe exempt from the second law? Explain your answer.

13 What is meant by 'free energy?

14 What is entropy? Discuss its significance briefly.

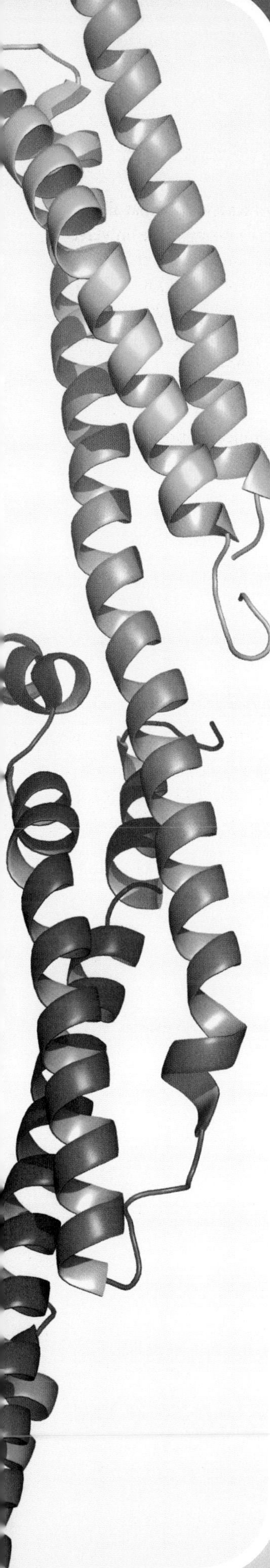

Part 2

Structure and function of proteins and membranes

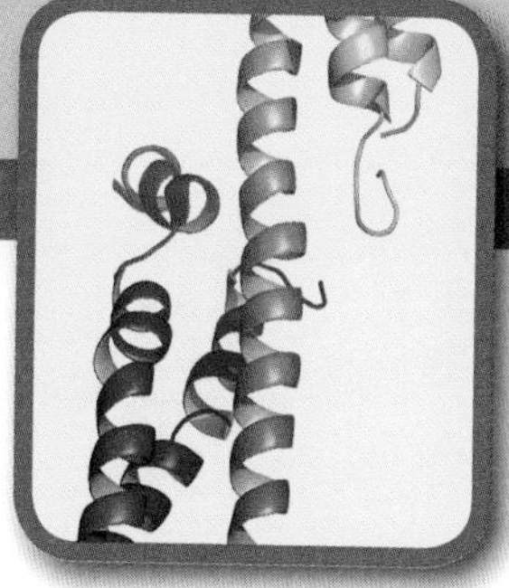

Chapter 4

The structure of proteins

It has been said, with considerable justification, that it is improbable that there exists in the universe any type of molecule with properties more remarkable than those of proteins. Life depends on thousands of different proteins whose structures are fashioned so that individual protein molecules combine, with exquisite precision, with other molecules. Chemical reactions in the cell and just about all cellular activities depend on them. Proteins essentially 'do' everything in life. They are the workhorses and the basis of most biological structures. The elaborate genetic machinery based on DNA coding is there so that the correct proteins are there at the right time and in the correct quantities.

Proteins are, in basic chemical structure, more complex molecules than DNA, though they do not reach its immense size. They are made of long strings of 20 different species of amino acids, while DNA has only four variable 'building blocks'.

This chapter deals with the structures of these giant molecules, which carry out such a range of activities. Their versatility is based on the fact that a virtually unlimited number of different protein molecules can theoretically exist. A remarkable property of proteins is that while they have definite structures and shapes, many are flexible molecules that can change their conformations – their arrangements in three-dimensional space – in the process of performing their functions. Such proteins may be regarded as molecular machines rather than inert chemical molecules. Haemoglobin is a classic example. Muscle contraction (see Chapter 8) is an extreme example of a biological function resulting from protein molecules changing their conformation. It is remarkable that the miniscule atomic movements of groups of atoms can result in the massive contractions of animal muscles. In this chapter we will deal with the basic structures of proteins in general and how the structures of various classes of protein have evolved to fulfill their particular functions. Enzymes are, of course, an important class of proteins and will be discussed in more detail in Chapter 6. Membrane proteins have special characteristics and we will discuss these in Chapter 7.

Proteins are composed of **polypeptide** chains. A protein molecule may have more than one such chain. A polypeptide chain consists of a large number of amino acids linked together – anything from 50 to several thousand. Twenty different amino acid structures are used for the assembly of the polymers.

Structures of the 20 amino acids used in protein synthesis

These amino acids are the building blocks with which evolution works to produce the vast variety of different proteins needed for the life of organisms. More than 20 different amino acids are found in nature, but only what Francis Crick has called the 'magic 20' are used for protein synthesis. (Although a few proteins incorporate selenocysteine by an exceptional mechanism – see Chapter 25). The same 20 are used for all life forms on earth; their different shapes, sizes, chemical properties, and polarity characteristics gives evolution flexibility in 'trying out' different protein structures, with natural selection deciding whether variants should be adopted *via* the genetic system.

An α-amino acid, written in the nonionized form has the structure (a), but in aqueous neutral solution it exists as the zwitterionic form (b).

(a) $H_2N-\overset{\alpha}{C}H(R)-COOH$ **(b)** $^{+}H_3N-\overset{\alpha}{C}H(R)-COO^{-}$

Every amino acid, with the exception of proline which is actually an **imino acid**, has the same $H_2N-CH-COOH$ part – only the R group attached to the α-carbon atom varies. The R group or **side chain** is shown below in red.

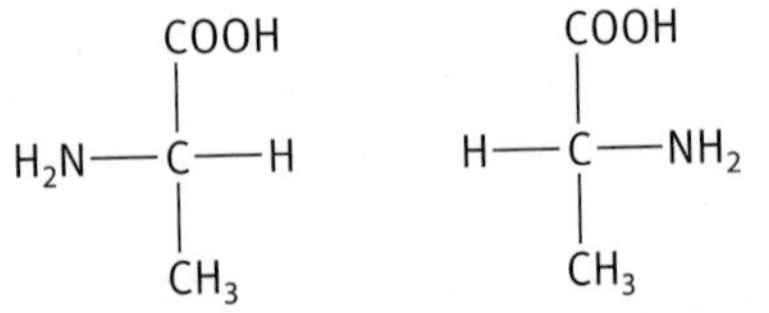

$$\begin{array}{cc} \begin{array}{c} COOH \\ | \\ H_2N{-}C{-}H \\ | \\ CH_3 \end{array} & \begin{array}{c} COOH \\ | \\ H{-}C{-}NH_2 \\ | \\ CH_3 \end{array} \\ \text{L-Alanine} & \text{D-Alanine} \end{array}$$

Fig. 4.1 Stereoisomers of L- and D-alanine. In this projection, vertical lines represent bonds that project below the plane of the paper, and horizontal lines represent bonds that project outward from the paper. Note that the two stereoisomers are mirror images of one another.

With the exception of glycine, which has no asymmetric carbon atom, amino acids in proteins are of the L-configuration. Fig. 4.1 gives the structures of L and D amino acids, the two being mirror images of one another. It is not necessary to specify that an amino acid in any biological context is of the L-configuration since D-amino acids are only rarely encountered (mainly in certain microbial structures) and where they occur these are always specified.

Symbols for amino acids

There are two types of abbreviations for amino acids as shown in Table 4.1. The old three letter system is often still used when short sequences are represented because they have the advantage of being self-evident. More usually now and for longer sequences, the single letter abbreviations are less cumbersome and save a lot of storage space in databases, though less easy to remember.

Amino acids are classified according to the structures and properties of their side chains. Although the criteria used and hence the groupings may vary somewhat depending on where you look them up, the **hydrophobic** or **hydrophilic** ('water hating' or 'water loving') nature of the side chain is usually emphasized, as it is a key determinant of how the amino acid in question interacts with other chemical groups. Hydrophilic amino acids have side chains that are charged or polar (with charge distributed unevenly) at physiological pH. You may wish to revisit the subject of pK_a (see Chapter 3) to remind you of the relationship between pH and the charge of a chemical group. The pK_a values for amino acids with ionizable side chains are given in Table 4.1.

Aliphatic amino acids

The aliphatic amino acids (aliphatic means open noncyclic structures) are **glycine, alanine, valine, leucine,** and **isoleucine.**

Glycine ,with H as its side chain, is the smallest amino acid and also the only one with no L- and D-isomer. Its small size and lack of strong hydrophobic or hydrophilic properties means that glycine fits flexibly into small spaces in protein structures.

Amino acids	One letter	Three letter	Side chain pK_a
Alanine	A	Ala	
Arginine	R	Arg	12.5
Asparagine	N	Asn	
Aspartic acid	D	Asp	3.9
Cysteine	C	Cys	8.4
Glutamine	Q	Gln	
Glutamic acid	E	Glu	4.1
Glycine	G	Gly	
Histidine	H	His	6.0
Isoleucine	I	Ile	
Leucine	L	Leu	
Lysine	K	Lys	10.5
Methionine	M	Met	
Phenylalanine	F	Phe	
Proline	P	Pro	
Serine	S	Ser	
Threonine	T	Thr	
Tryptophan	W	Trp	
Tyrosine	Y	Tyr	10.5
Valine	V	Val	
Unspecified or unknown	X	Xaa	

Table 4.1 Single letter and three letter symbols for amino acids. Approximate pK_a values are given for ionizable side chains. Note that pK_a values are affected by the microenvironment of the side chains (temperature, ionic strength, surrounding chemical groups) and therefore may vary somewhat when the amino acids are incorporated into proteins.

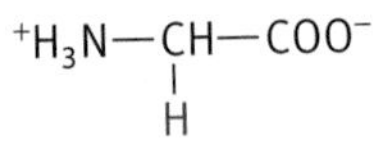

$$\begin{array}{c} {}^{+}H_3N{-}\underset{\displaystyle H}{\underset{|}{CH}}{-}COO^{-} \\ \text{Glycine} \end{array}$$

Alanine, valine, leucine, and isoleucine have side chains of increasing hydrophobicity. The latter three are known as the **branched-chain aliphatics. Methionine** is a rather special hydrophobic aliphatic amino acid (special more because of its role in methyl group metabolism than its role in protein structure).

Hydrophobic amino acids

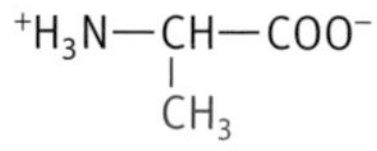

$^{+}H_3N-CH-COO^-$ with side chain CH_3 — Alanine

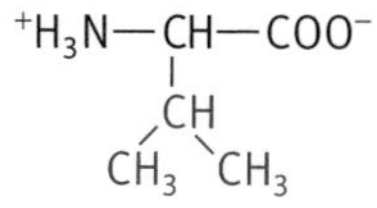

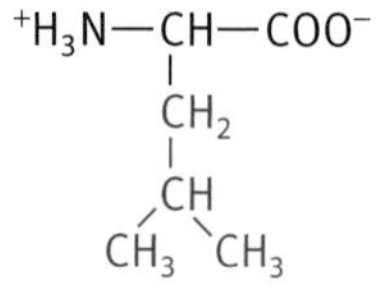

$^{+}H_3N-CH-COO^-$ with side chain $CH(CH_3)CH_3$ — Valine

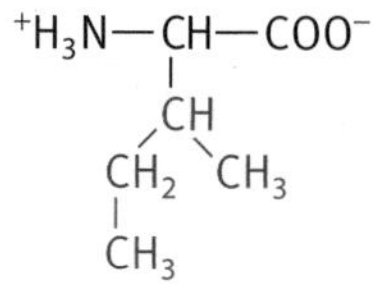

$^{+}H_3N-CH-COO^-$ with side chain $CH_2-CH(CH_3)CH_3$ — Leucine

$^{+}H_3N-CH-COO^-$ with side chain $CH(CH_3)-CH_2-CH_3$ — Isoleucine

$^{+}H_3N-CH-COO^-$ with side chain $CH_2-CH_2-S-CH_3$ — Methionine

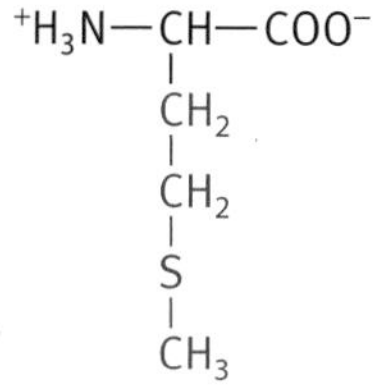

Aromatic amino acids

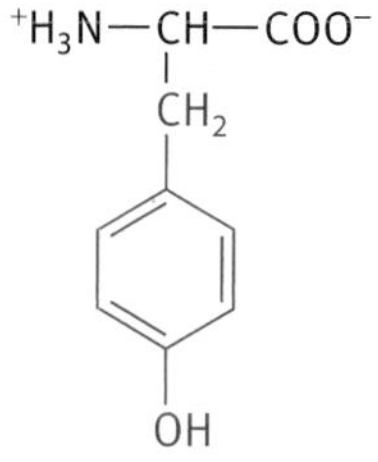

We next come to the large side chains – aromatic ones (those with cyclic planar rings):

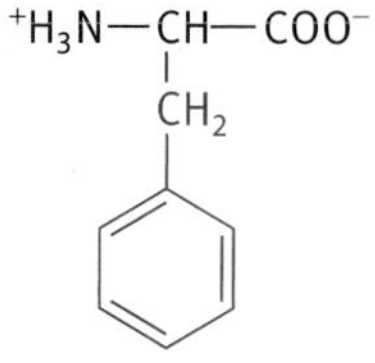

$^{+}H_3N-CH-COO^-$ with side chain CH_2–(benzene ring) — Phenylalanine

$^{+}H_3N-CH-COO^-$ with side chain CH_2–(benzene ring)–OH — Tyrosine

$^{+}H_3N-CH-COO^-$ with side chain CH_2–(indole ring, NH) — Tryptophan

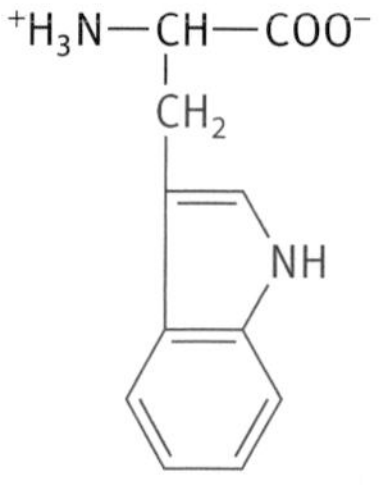

While phenylalanine is clearly hydrophobic, tyrosine with its polar —OH group can form a hydrogen bond. However, the aromatic ring of tyrosine is large and hydrophobic so it is of somewhat mixed classification. Tryptophan can also form a hydrogen bond with its —NH group but is large and mainly hydrophobic.

Ionized hydrophilic amino acids

An ionized group is hydrophilic. Acidic amino acids (with an extra $-COO^-$ on the side chain) are negatively charged at the pH of the cell, and thus hydrophilic. Basic amino acids have an extra positive charge and are also hydrophilic. (The term basic here means a side chain that acts as a proton acceptor as opposed to an acidic proton donor at cellular pH). **Aspartic** and **glutamic acids** both have acidic side chains, and are called **aspartate** and **glutamate** when negatively charged at physiological pH. **Lysine** and **arginine** have strongly basic side chains, and histidine is a weakly basic amino acid. The pK_a of the **imidazole ring** structure of histidine is around 6, so it easily gains or loses a proton at physiological pH. The side chains of the charged amino acids can form hydrogen bonds and salt bridges, the latter being any ionic bond between positive and negative groups.

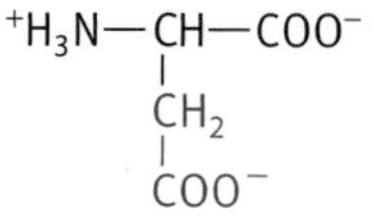

$^{+}H_3N-CH-COO^-$ with side chain CH_2-COO^- — Aspartate

$^{+}H_3N-CH-COO^-$ with side chain $CH_2-CH_2-COO^-$ — Glutamate

$^{+}H_3N-CH-COO^-$ with side chain $CH_2-CH_2-CH_2-CH_2-{}^{+}NH_3$ — Lysine

$^{+}H_3N-CH-COO^-$ with side chain $CH_2-CH_2-CH_2-NH-C(NH_2)=NH_2^+$ — Arginine

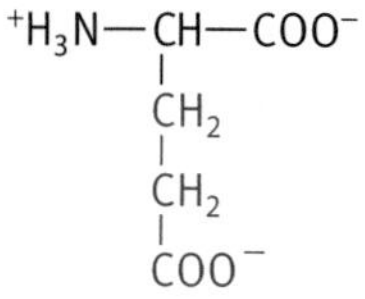

$^{+}H_3N-CH-COO^-$ with side chain CH_2–(imidazole ring: C–NH⁺=CH–NH–HC=C) — Histidine

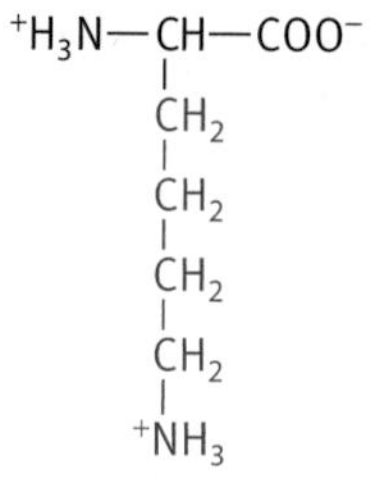

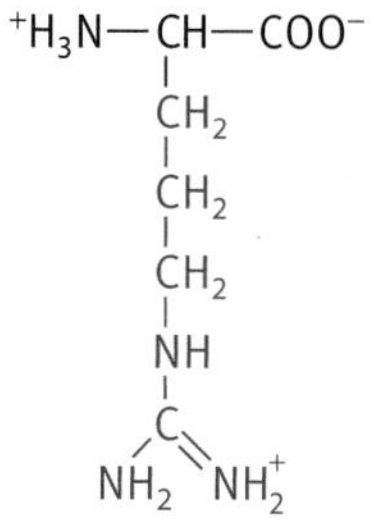

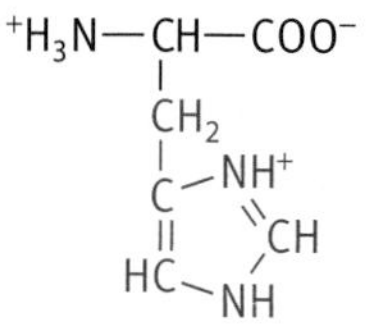

Uncharged polar hydrophilic amino acids

Amide derivatives of aspartic and glutamic acid also exist: **asparagine** and **glutamine**, respectively. The amide group is polar but does not ionize, so these two amides are less hydrophilic and form much weaker hydrogen bonds than the parent compounds.

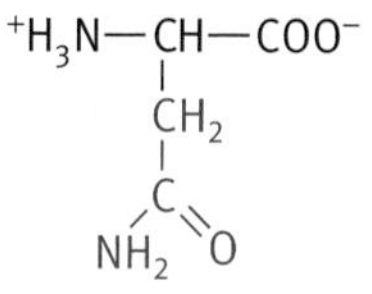

$^{+}H_3N-CH-COO^-$ with side chain $CH_2-C(=O)NH_2$ — Asparagine

$^{+}H_3N-CH-COO^-$ with side chain $CH_2-CH_2-C(=O)NH_2$ — Glutamine

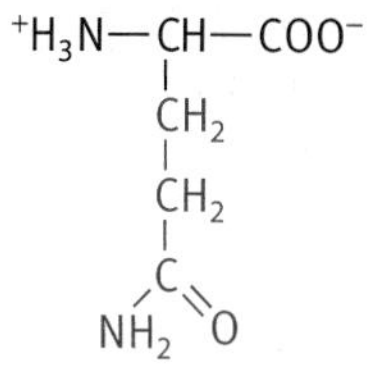

Serine and threonine side chains are nonionized and weakly hydrophilic. They can both form hydrogen bonds with their polar —OH groups.

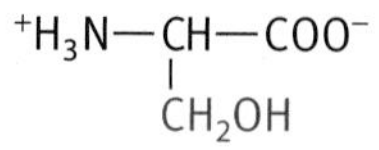

$^{+}H_3N-CH-COO^-$ with side chain CH_2OH — Serine

$^{+}H_3N-CH-COO^-$ with side chain $CHOH-CH_3$ — Threonine

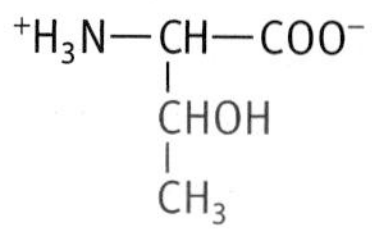

Two amino acids with unusual properties

Cysteine is similar to serine but with $-SH$ (a thiol or sulphydryl group) instead of $-OH$. As sulphur is less electronegative than oxygen the thiol group is not strongly polar and cysteine is usually classed as hydrophobic. It plays two special roles in protein structure – supplying external $-SH$ groups such as in the active centres of enzymes, and forming covalent $-S-S-$ bonds or disulphide bonds internally. Cysteine that has formed a disulphide bond (through an oxidation reaction) is called **cystine**.

$$^{+}H_3N-CH(CH_2SH)-COO^-$$

Cysteine

Proline is the oddity – literally a kinky amino acid in that it puts a kink into the conformation of polypeptide chains. If you have trouble in memorizing proline remember that it is an ordinary amino acid except that the side chain forms a loop by bonding at one end to the α-carbon and at the other end to the nitrogen, giving it an $-NH_2^+$ imino rather than an $-NH_3^+$ amino group. The looped side chain forbids rotation about the bond between the nitrogen and the α-carbon, causing proline to have a large effect on protein structure.

Imino nitrogen ----> $^{+}H_2N-CH-COO^-$ (ring: CH_2, CH_2, CH_2)

This is the unusual bond (between N and CH_2)

(a) Primary

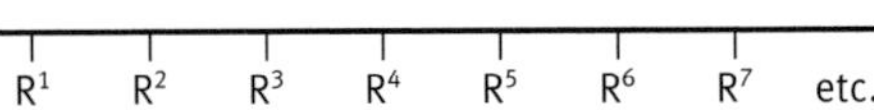

This structure will be represented below as a simple line.

(b) Secondary

The polypeptide **backbone** exists in different sections of a protein either as an α helix, a β-pleated sheet, or random coil.

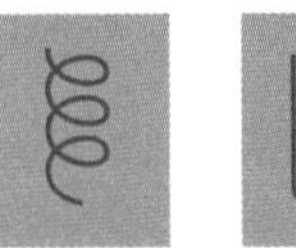

α helix

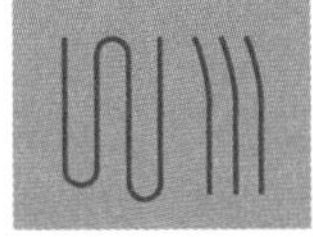

β-pleated sheet

Random coil or loop region

(c) Tertiary

The secondary structures above are folded into the compact globular protein.

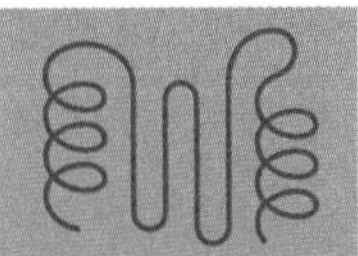

This protein will be represented below as:

(d) Quaternary

Protein molecules known as subunits assemble into a multimeric protein held together by weak forces.

Fig. 4.2 Diagrammatic illustration of what is meant by primary, secondary, tertiary, and quaternary structures of proteins.

The different levels of protein structure – primary, secondary, tertiary, and quaternary

We first give a brief overview of this topic, to be followed by a more detailed treatment. The sequence of amino acids that are linked together covalently in a polypeptide chain is the **primary** structure (Fig. 4.2(a)). It says nothing about how that polypeptide is arranged in a three-dimensional space, just the order of the amino acids. In the first level of protein folding the polypeptide backbone itself is arranged in a particular conformation known as the **secondary** structure, which includes sections of regular repeating structures known as α helices (helices being the plural of helix) and β sheets (Fig. 4.2(b)). The secondary structure is folded on itself to give the **tertiary** structure (Fig. 4.2(c)). The complete molecule so formed by the primary, secondary, and tertiary structures may be the final functional protein or it may be a **protein monomer** or **subunit**, which associates with other protein monomers (which may be the same or different) to form a functional protein. This is the **quaternary** structure (Fig. 4.2(d)).

With this overview, we will now deal in more detail with the different levels of protein structure.

Primary structure of proteins

Two amino acids can be linked together by a condensation reaction with the removal of H_2O (though note that protein synthesis does not occur so simply in the cell, as this would be thermodynamically impossible). The result is a **dipeptide**, as shown below (structures are written in the nonionized form for clarity). The $-CO-NH-$ bond is the **peptide bond** or link, which you will note results in formation of an amide group.

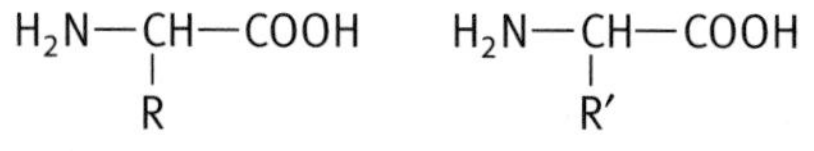

```
H2N—CH—COOH      H2N—CH—COOH
    |                |
    R                R'
```

Two amino acids

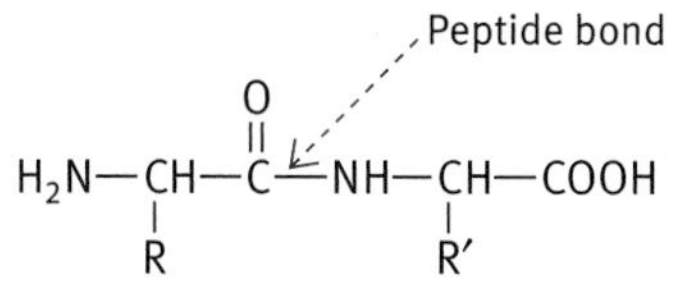

```
                 Peptide bond
        O       ↙
        ||
H2N—CH—C—NH—CH—COOH
    |        |
    R        R'
```

A dipeptide

The dipeptide is the simplest 'peptide unit'. Multiple identical peptide bonds are formed in the **polypeptide** structure shown:

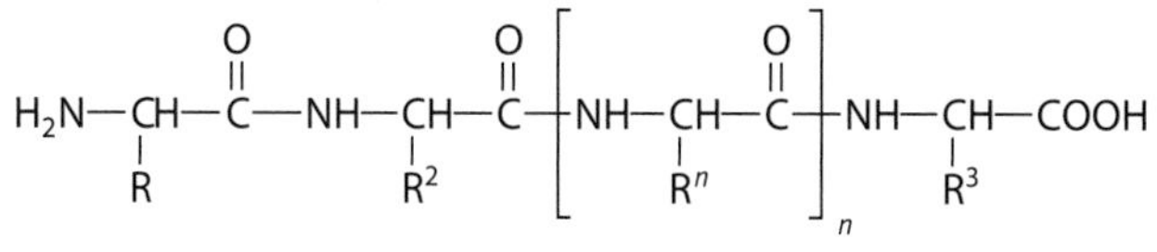

```
        O          O         O
        ||         ||        ||
H2N—CH—C—NH—CH—C—[NH—CH—C]—NH—CH—COOH
    |        |        |     n     |
    R        R2       Rn          R3
```

A polypeptide

The order in which the 20 different amino acids are arranged in the polypeptide is the **amino acid sequence**. The sequence, as mentioned, is the **primary structure** of a protein. Determining the sequence is referred to as **protein sequencing**, or amino acid sequencing, dealt with in the next chapter. The terminology used for peptides of different lengths is rather arbitrary, but as a rough guide a short chain of a few amino acids, perhaps a dozen or so, is referred to as an **oligopeptide** (*oligo* – few), a **polypeptide** has 'many' amino acids, and the terms polypeptide and protein are often used interchangeably: a protein may have several polypeptide chains, or consist of just one very long polypeptide. A few more terms are useful at this point: the $-NH_3^+$ end of a protein is referred to as the **amino terminal** or **N-terminal** end, and the other $-COO^-$ end as the **carboxy terminal** or **C-terminal** end. The central chain, without the R groups, (−CH−CO−NH −CH−CO−NH−CH−, etc.) is called the **polypeptide backbone.** An amino acid in a peptide is referred to as an **amino acid residue** or **amino acyl residue**; the R groups are variously referred to as amino acid side chains, protein side chains, or simply as **side chains.**

By convention the amino acid sequence is written from the N-terminal end to the C-terminal end. Polypeptide chains have direction. Consider a given amino acid sequence such as Ala−Gly−Leu−Phe. If the N-terminal amino acid is Ala, the molecule Ala−Gly−Leu−Phe is biologically different from its inverse, Phe-Leu-Gly-Ala.

Ionization of amino acids in polypeptide chains

As already mentioned, free amino acids in aqueous solution have the zwitterionic structures in which both the α-amino and α-carboxyl groups are ionized (the pK_a values of these groups being around 9–10 and 2–3 respectively). However, when incorporated into a polypeptide chain, these groups are no longer ionizable, except for the terminal amino and carboxyl groups. The ionized state of a polypeptide, and therefore of a protein, is therefore almost entirely dependent on the side chains of aspartic and glutamic acid, lysine, arginine, and histidine residues, since their ionizable side chain groups are not blocked by peptide bond formation, as shown for glutamic acid and lysine in the illustration below.

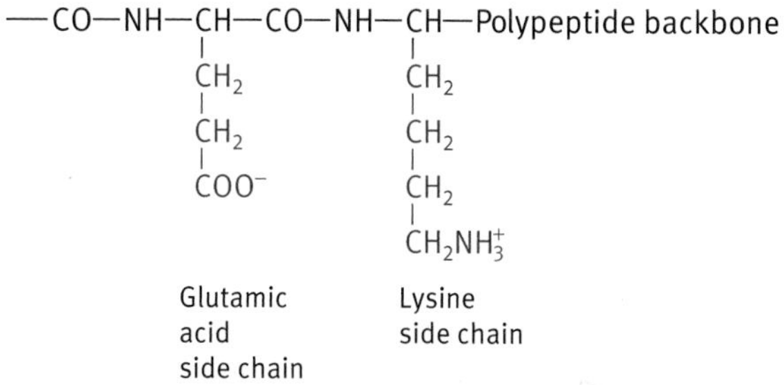

```
—CO—NH—CH—CO—NH—CH—Polypeptide backbone
        |         |
        CH2       CH2
        |         |
        CH2       CH2
        |         |
        COO-      CH2
                  |
                  CH2NH3+

   Glutamic     Lysine
   acid         side chain
   side chain
```

The side chain $-COO^-$ groups of aspartic and glutamic acids have pK_a values around 4, so they are virtually fully dissociated at pH 7.4. These provide the negatively charged side chains of proteins. The basic amino acids, lysine and arginine, with pK_a values for their basic side chain groups of 10.5 and 12.5 respectively, are fully ionized by an additional proton at physiological pH. The third basic amino acid, histidine, has an imidazole ring as the side chain group whose pK_a value is near neutrality (around 6 in proteins), so that histidine is often found in active sites of enzymes where movement of a proton is involved in the reaction catalysed; the imidazole group can accept or donate a proton at a pH near that existing in the cell (see the chymotrypsin enzyme mechanism described in Chapter 6 for an example of this).

The distribution of charged amino acid residues in a protein has an important effect on the conformation that a polypeptide chain can adopt. Charges of the same sign close to each other repel each other. Closely positioned positive and negative charges will attract each other.

The peptide bond is planar

The simple polypeptide structure as depicted previously does not convey an important feature of a polypeptide chain Although the CO−NH peptide bond is written as an ordinary single bond (about which rotation might be expected), in fact the peptide bond is a hybrid between two structures: (1) in which the bond between the carbon and nitrogen atom is a single bond; and (2) in which it is a double bond.

```
(1) —C       H          (2) —C       H
      \     /                 \     /
       C—N                     C=N+
      //    \                 /     \
     O       C—              O-      C—
```

The electron density is between the two, giving the C—N bond about 40% double-bonded character. This is sufficient to prevent rotation about it, making the polypeptide chain more rigid. In theory, there are two possible configurations for the peptide bond, *cis* and *trans*. In practice, peptide bonds in proteins are almost always in the *trans* configuration as the side chains prevent *cis* peptide bond formation due to steric hindrance (i.e. the side chains would get in each other's way in the *cis* configuration).

trans *cis*

The architecture of a polypeptide chain is shown in Fig. 4.3. The successive α-carbon atoms lie above and below the plane of the paper. As shown, rotation of the N—C and C—C bonds adjacent to the α-carbon atoms can occur, and their angles of rotation (known as *dihedral* or *torsion angles*) determine the configuration of the chain. These angles are known by the Greek letters phi (φ) and psi (ψ) as shown. The omega (ω) angle is the angle around the peptide bond and is 180° for a *trans* peptide bond.

The structural biologist G. N. Ramachandran showed that in practice only certain combinations of phi and psi angle values are found in proteins, presumably because of steric hindrances that prevent other combinations. In this famous piece of work he plotted the pairs of phi and psi angles adopted by

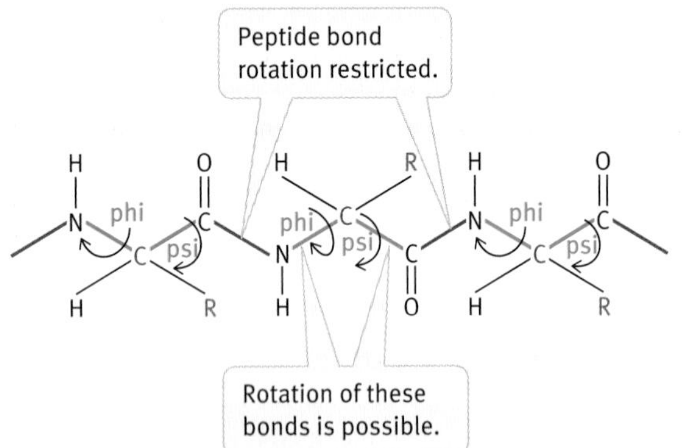
Peptide bond rotation restricted.
phi
psi
Rotation of these bonds is possible.

Fig. 4.3 Section of a polypeptide. Successive α-carbon atoms (shown in red) of amino acid residues lie above and below the plane of the paper as also do the R groups (shown in blue). The peptide dCOdNHd bond (shown in red) has a partial double-bonded character that prevents rotation and gives the group a rigid planar structure. However, the adjacent bonds (shown in green) are capable of rotation with a single-bonded structure. The angle adopted by these bonds are known respectively by the Greek letters phi (φ) and psi (ψ). The conformation of a polypeptide chain is determined by the value of these angles for each amino acid in the chain.

each amino acid in a large number of proteins. They clearly formed two tight clusters, which correspond to the angles found in the two common secondary structures, the α helix and the β-pleated sheet.

Secondary structure of proteins

Most proteins have a compact globular shape rather than the extended configuration of a polypeptide, though fibrous proteins exist. The interior of globular proteins is a strongly hydrophobic environment. This imposes limitations on the possible types of secondary structures since in order to fold the polypeptide backbone into a compact shape it has to criss-cross the hydrophobic interior. The problem is that a polypeptide has polar groups capable of hydrogen bonding – two bonds per amino acid unit, since the C—O and N—H of the peptide bond are capable of participating in hydrogen bond formation. This bonding potentiality must be satisfied by bond formation as far as possible to produce a stabilized structure. The backbone groups in the interior of the molecule cannot hydrogen bond to the hydrophobic side chain groups. So what can these hydrogen atoms bond with? The answer is with groups on the same, or an adjacent, polypeptide backbone.

There are two main classes of secondary structures – the **α helix** in which the backbone is arranged in a spiral and the **β-pleated sheet** in which extended polypeptide backbones are side by side. These structures are stable; and as in both cases the side chains extend outwards from the structure made by the peptide backbone, in a protein that is found in an aqueous environment such as the cytosol they can occur at the exterior of the protein with appropriate hydrophilic side chains or they can occur in the hydrophobic interior with appropriate hydrophobic side chains.

The α helix

In the α helix, the polypeptide backbone is twisted into a right-handed helix which, for L-amino acids, is more stable than a left-handed one (Fig. 4.4(a)). You can visualize the direction of twist of the right-handed helix; if you look down the axis either way the helix turns clockwise. You can also imagine tightening a conventional screw with your right hand to give you the direction. The shape is shown in Fig. 4.4(a).

The α helix structure has 3.6 amino acid units per turn, which results in the C—O of each peptide bond being aligned to form a hydrogen bond with the peptide bond N—H of the fourth distant amino acid residue. The C—O groups point in the direction of the axis of the helix and are nicely aimed at the N—H groups with which they hydrogen bond, giving maximum bond strength. All the C—O and N—H groups of the polypeptide backbone are hydrogen bonded in pairs, forming a cylindrical, rod-like structure (Fig. 4.4(a)). In cross-section an α helix is a virtually solid cylinder, with all the side chains projecting outwards (Fig. 4.4(b)).

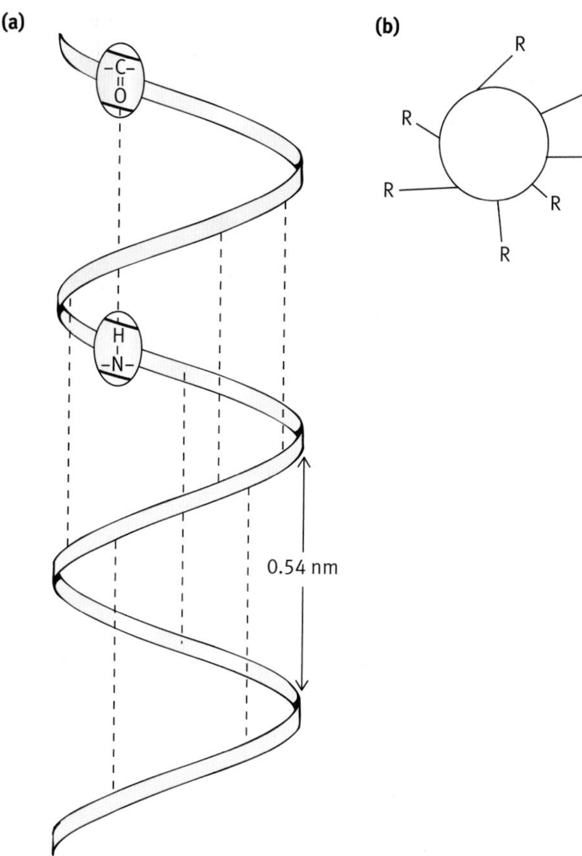

Fig. 4.4 The α helix form of a polypeptide chain. **(a)** Hydrogen bonding between C=O and N–H groups of the polypeptide backbone (side chains not shown). The hydrogen bonds (broken lines) are shown in approximate positions only. The pitch (vertical length of one complete turn) of the helix is 0.54 nm. **(b)** Looking down the axis of an α helix, with the amino acid side chains projecting from the cylindrical structure (each at a different distance below the plane of the paper). The R groups are not drawn in their exact orientation from the axis. Since there are 3.6 residues per turn, each residue occurs every 100° around a circle (360°/3.6=100°).

Not all of the polypeptide chain of a globular protein is in the α helix form. The α helical sections average 10 amino acid residues in length, but the range of α helix lengths varies a great deal from this average in different proteins, as does the number of helical sections. Some proteins, like haemoglobin, are composed almost entirely of α helices, while other proteins have very few, but all variations can be found.

Amino acids vary in their tendency to form α helices

Some amino acids, such as leucine and methionine, are excellent helix formers, some indifferent, and a few are α helix breakers or terminators – a proline residue in particular forces a bend in the structure, so where a proline is found in the polypeptide chain that will be the end of any α helix. With proline in peptide linkage, there is no hydrogen atom on the nitrogen available for hydrogen bonding, and the structure of the residue restricts rotation, so that it cannot assume the conformation needed to fit into an α helix.

—NH—CH(R)—C(=O)—N—CH—C(=O)—NH—
(CH_2–CH_2–CH_2 ring closing on N)

Imino nitrogen in peptide linkage not able to hydrogen bond

We will return shortly to how runs of amino acids in α helices fit into protein structures, but before that we must deal with the alternative to the α helix, namely the β-pleated sheet. Proteins often contain mixtures of the two, with each constituting different sections of the polypeptide chain, but some have only one of these two structures.

The β-pleated sheet

The β-pleated sheet also forms a stable structure in which the polar groups of the polypeptide backbone are hydrogen bonded to one another, thus forming a stable structure that is often in the hydrophobic interior of a globular protein.

The principle is simple. The polypeptide chain lies in an extended or β form with the C–O and N–H groups hydrogen bonded to those of a neighbouring chain (which may be formed by the same chain folding back on itself, or by a separate chain lying alongside, Fig. 4.5). Several chains can thus form a sheet of polypeptide. It is 'pleated' because successive α-carbon atoms of the amino acid residues lie alternately slightly above and below the plane of the β sheet (Fig. 4.3). The side chains also alternate on either side of the plane of the sheet.

The adjacent polypeptide chains bonded together can run in the same direction (parallel) or opposite directions (antiparallel). In the latter case, the polypeptide makes tight 'β turns' or hairpin bends to fold the chain back on itself (Fig. 4.6(c)). Four amino acid residues constitute the turn; a hydrogen bond between the C–O and N–H of residues 1 and 4 stabilize the hairpin bend. Parallel β sheets are connected by a longer motif, such as an α helix or connecting loop (Fig. 4.6(b)).

Connecting loops

In a protein, the α helices and β sheet sections are connected together by unstructured polypeptide known as **connecting loops**. A connecting loop may be in any conformation (other than a recognizable α helix or β sheet) as determined by the various group interactions in the protein structure. Since the structure of a loop may not satisfy the hydrogen bonding potentials of the C–O and N–H groups of the backbone, or those of the side chains, such loops are often found at the exterior of proteins, in contact with water. In a given protein, connecting loops will have a conformation determined by their interactions with other groups.

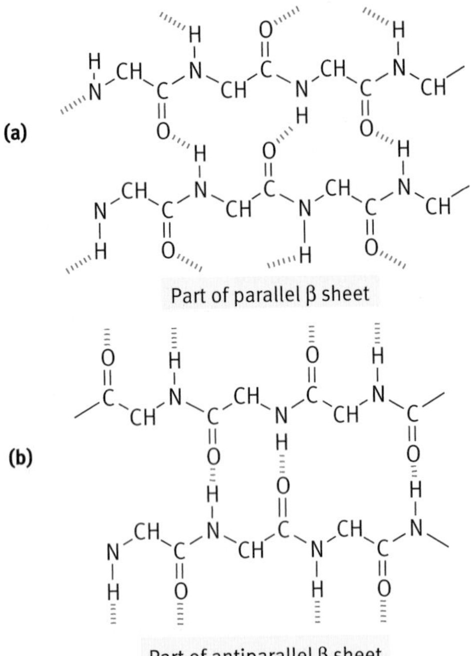

Fig. 4.5 **(a)** Hydrogen bonding between two polypeptide chains running in the same direction forming a parallel β sheet. The R groups attached to the —CH— groups are alternately above and below the plane of the paper. **(b)** Adjacent polypeptide chains running in opposite directions can also mutually hydrogen bond forming an antiparallel β sheet. This enables a single chain to form a β sheet by folding back on itself. Hydrogen bonds are denoted by dotted lines.

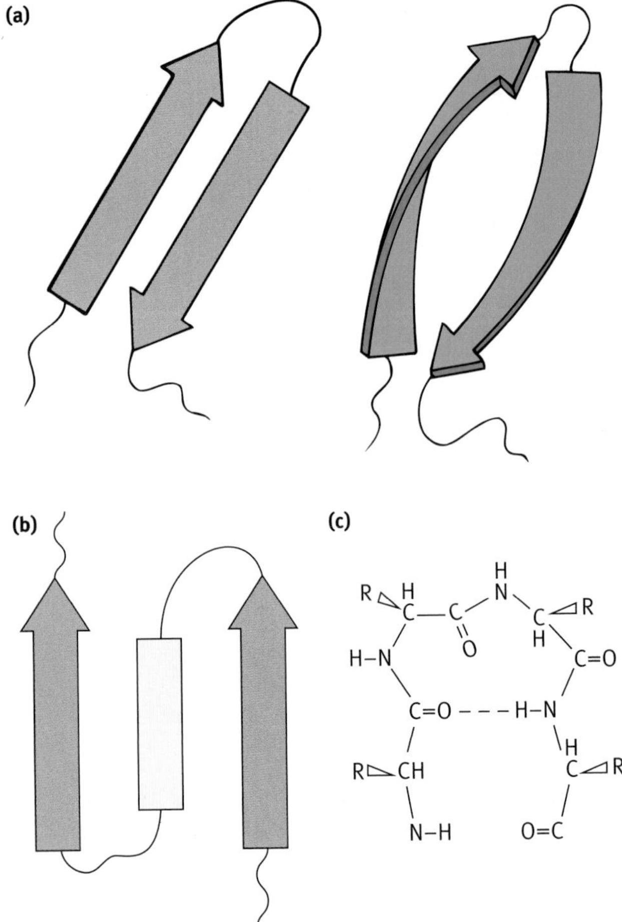

Fig. 4.6 (a) Symbols used in depictions of protein structure to indicate a pair of polypeptide sequences forming an antiparallel β-pleated sheet. Because of the slight right-handed twist the arrows are represented as on the right. The single lines connecting structures represent random coil or loop sections. **(b)** Symbols used to indicate a parallel β-pleated sheet, with the two sections connected by loops and an α helix. **(c)** Molecular structure of a typical β turn, stabilized by hydrogen bonding between the first and fourth amino acid residues.

Tertiary structure of proteins

A single polypeptide chain in a protein can be arranged as a mixture of the various secondary structures constituting different parts of the chain, which themselves are folded up and packed together to form the protein molecule. This arrangement of the various secondary structures into the compact structure of a globular protein is referred to as the tertiary structure.

To simplify structural diagrams of the tertiary structure of proteins, conventions have been adopted to depict the arrangements of the secondary structures of which they are composed. An α *helix* is represented either as a solid cylinder, sometimes with a helix inside it, or alternatively as a helical ribbon as seen in the structures in Figure 4.7. The individual sections of polypeptide chains or β strands which participate in β-pleated sheet formation are represented as broad arrows (Fig. 4.6(a)). An extended polypeptide chain in the β configuration has been shown to twist slightly to the right so that the arrows are usually drawn in protein structures with this twist. Connecting loops are shown as any sort of line. It should always be remembered, however, that proteins are solid structures with tightly packed atoms, not the open structures of springs and wires these convenient diagrammatic motifs might suggest. Space-filling models (such as that of the haemoglobin molecule shown in the first chapter (Fig. 1.5)) are more realistic, but their internal structures cannot be seen. They are particularly useful where we are interested in interactions of protein surfaces with other molecules.

Large numbers of proteins have had their three-dimensional structure determined by X-ray diffraction studies (see Chapter 5). In Figure 4.7, representations of a few illustrative protein structures are given. Myoglobin, (Fig. 4.7(a)) described later in this chapter, has only α helices connected by short loop sections with its haem group inserted into a cleft. The staphylococcal nuclease (an enzyme hydrolysing nucleic acid) has a mixture of an antiparallel β sheet and three α helices connected by unstructured polypeptide (Fig. 4.7(b)). A common arrangement is the so-called **α/β barrel**, in which there is a central core of β strands arranged like the staves of a wooden

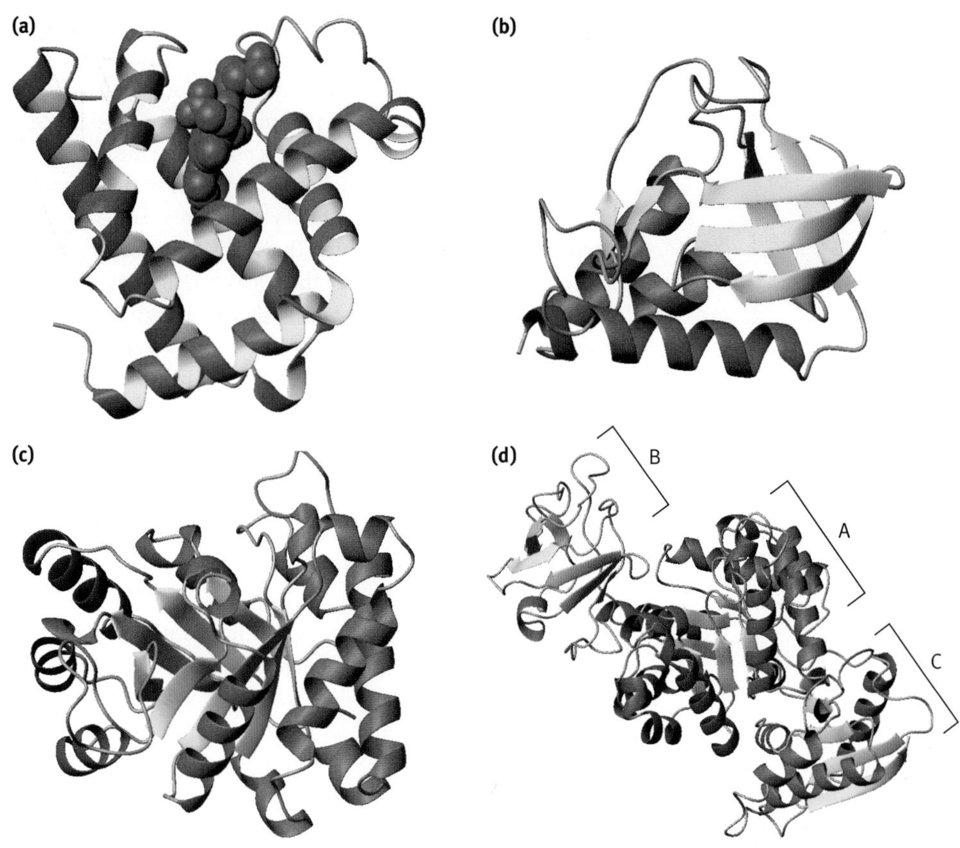

Fig. 4.7 Ribbon diagrams of the structures of different proteins. Numbers in brackets are identification codes for the online Protein Database (PDB) resource where details of the structure are stored. **(a)** myoglobin (1MBO), showing the haem molecule in blue; **(b)** staphylococcal nuclease (1A2T); **(c)** triosephosphate isomerase (1AG1); **(d)** pyruvate kinase of yeast (1A3W) has three structural and functional domains. A is the catalytic domain and has an α/β barrel structure. B is a small β barrel domain that forms a 'cap' over the active site of the catalytic domain. C is the regulatory domain and has an α/β open sheet motif. (b)–(d) Colours indicate α helices in red, β strands in yellow, and all else in grey.

barrel except that they are twisted. The barrel encloses tightly packed hydrophobic side chains. Surrounding this barrel are α helices. The diagram of the triosephosphate isomerase in Fig. 4.7(c) illustrates this. Another good example of a β barrel is shown in Figure 7.9. The enzyme pyruvate kinase, shown in Fig. 4.7(d), shows a structure with three sections, illustrating the domain organization of proteins, discussed below.

When the structures of numbers of proteins were determined, it emerged that certain patterns of tertiary folding are frequent. Families of proteins exist, different in detail but with the same basic arrangement of secondary structures. Myoglobin illustrates the 'globin fold', a structure of eight α helices which in this case forms a pocket into which the haem group fits. This fold is found in a large family of proteins. It seems that old structural patterns are often preserved as new proteins are evolved.

Noncovalent bonds are mainly responsible for stabilizing the tertiary structures of proteins

The bonds involved in protein tertiary structure are predominantly noncovalent – hydrogen bonds and van der Waals forces with some salt bridges between charged amino acid side chains. Hydrophobic interactions that bury nonpolar side chains in the interior are are also important.

The folding of a protein can be regarded as a chemical reaction, with an associated free energy change, ΔG, which must have a negative value if folding is energetically favourable and hence spontaneous. The equation $\Delta G = \Delta H - T\Delta S$ (see Chapter 3) tells us that ΔH, the enthalpy change associated with bond formation, and ΔS, changes in entropy that occur during folding, contribute to the value of the associated free energy change. When a cytosolic protein folds, bonds that form between different sections of the protein replace those between the unfolded (denatured) protein and its surroundings. Since an extended, unfolded protein can form multiple hydrogen bonds with water, it has been calculated that the net enthalpy change associated with folding is not terribly favourable. However, we must also consider entropy. While folding a protein into a single 'correct' structure represents a reduction of entropy when compared to the multiple possible conformations of a denatured protein, there is a compensatory increase in the entropy of water due to the hydrophobic effect: that is, an unfolded protein has a large surface area and creates an ordered 'shell' of water molecules around it, while a folded, more compact protein has a reduced surface area and thus orders fewer water molecules. It can be calculated that the overall outcome of combined enthalpy and entropy changes is that the ΔG associated with protein folding is small and negative, i.e. most folded proteins are only marginally stable. You can show this for yourself by frying an egg: the heat is sufficient to denature the egg white albumin protein (i.e. break the noncovalent bonds), creating a tangled

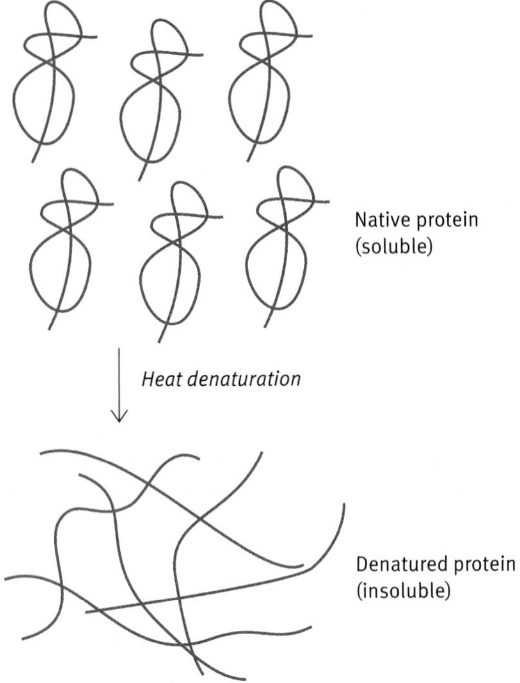

Fig. 4.8 Hypothetical representation of egg albumin denaturation. (The folded configuration is drawn arbitrarily.)

insoluble mass as opposed to the clear, ordered and soluble protein (Fig. 4.8).

The folded structure of a protein is determined by the amino acid sequence of the polypeptide chain

This was proved in the classical experiment of Anfinsen. He inactivated the enzyme ribonuclease by exposing it to high concentrations of urea, a hydrogen bond breaker. He also reduced the disulphide bonds. This treatment denatures a protein – its polypeptide chain is unfolded. On removal of the urea by dialysis, the ribonuclease protein refolded itself and enzyme activity was restored. Anfinsen's experiment showed that proteins can spontaneously fold into the correct way, and the amino acid sequence is sufficient in itself to determine the final form. This is of great importance. It establishes that the simple one-dimensional linear code of genes is sufficient to specify the folded functional form of proteins and hence the three-dimensional form of thousands of different proteins, and hence of all life forms. What makes this even more remarkable is that it is still not fully understood how it happens, as we will now discuss.

How does protein folding take place?

In the cell a newly synthesized polypeptide chain folds up in a minute or two. The seemingly obvious mechanism is that the polypeptide 'tries out' different folding conformations until the lowest energy state is found. However, this is not tenable, for it has been calculated that there are so many possible conformations even for a moderately sized protein of one or two hundred residues, it would take billions of years to fold correctly. The earth has not existed long enough for a single protein molecule to have folded up by this mechanism, so the process must proceed differently in the cell. A postulated mechanism is that as the polypeptide is synthesized, sections of the polypeptide rapidly assume their secondary structures. This would be a stepwise mechanism in which the series of secondary structures occur and finally arrange themselves in the correct form.

The Anfinsen experiment was with a small protein present in low concentration so that incorrect interactions between chains would be minimized. Conditions in a cell are very different, with tightly packed large polypeptides giving every chance of incorrect associations. Special proteins have evolved to cope with this problem. They prevent improper associations and like their Victorian human counterparts are called **chaperones** or **molecular chaperones**. There are various ways that chaperone proteins work. One way is to enclose the unfolded molecule in an isolated box that provides an environment favourable to proper folding, and then let it out once it has achieved this. The chaperone in a sense does what Anfinsen did for ribonuclease, allowing it to refold under favourable conditions, and is therefore sometimes described as an 'Anfinsen cage'. It is important to note, however, that chaperones cannot give *directions* on how proteins fold; they can simply make it more likely that they will fold according to their amino acid sequences by stopping them interacting with other unfolded proteins. We will return to a fuller discussion of chaperones in Chapter 25.

Covalent –S–S– bonds stabilize some proteins

Although their tertiary structures are largely the result of noncovalent bonds, some protein structures are 'locked' or strongly stabilized by **disulphide bridges** (S–S), which, being covalent, are very strong. Examples where these occur are proteins liberated into the blood (insulin is one example, see Fig. 4.9), or the intestine (digestive enzymes). This stabilization is achieved by pairs of thiol (S–H) groups of the cysteine side chains, brought together by polypeptide folding. An oxidase enzyme forms the S–S link between them by the following reaction:

$$2RSH + O_2 \rightarrow RS{-}SR + H_2O_2$$

The trivial name for two cysteine molecules which have become linked by an S–S bond is **cystine** and the symbol for it is Cys–Cys. It provides a very strong cross-linking bond – more or less like a steel rivet in the structure. A few of these between different sections of a chain make the folded shape more stable. Insulin has three disulphide or S–S bridges (Fig. 4.9). Proteins with disulphide bonds are often less easily denatured by heat. Few intracellular proteins have disulphide bonds in their structure, possibly because the interiors of cells are strongly reducing environments and might be sufficient to disrupt them; most S–S cross-linked proteins are extracellular.

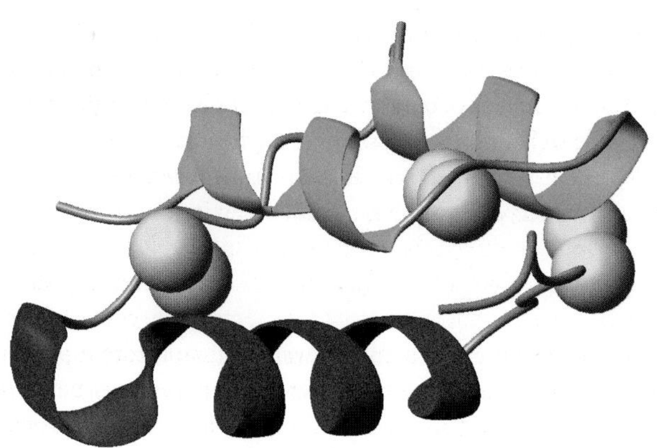

Fig. 4.9 Ribbon model of insulin (3INS) showing the two polypeptide chains joined by two disulphide bridges, with a third disulphide bridge internally in the A chain. The S—S bonds are visible in yellow.

An extreme example of stabilization of a protein by disulphide bridges is the **α-keratin** protein of hair. The long polypeptides are interlinked by many disulphide bonds, which are important in locking in the configuration of the hair. In permanent waving, these are broken by thiol reduction, heat and moisture being used to disrupt the original hydrogen bonding, followed by setting the hair into a new configuration. On cooling, hydrogen bonds reform the α-helical structure, and then the new configuration is made permanent by the 'neutralizer', which reoxidizes cysteine—SH groups to reform disulphide bonds between the multiple polypeptide chains of the hair structure. These bonds are new ones between —SH groups that have been brought together in the new stable (permanent) curled configuration of the hair.

Quaternary structure of proteins

With the tertiary structure, we now have a protein molecule, and for many proteins that is the final stage. Many functional proteins, however, have more than one such protein molecule (or protein monomer) held together in a single complex by noncovalent bonds. These monomers may be the same or different, but the molecules have to be structured so as to fit *via* complementary surface patches, so that only the correct subunits complex together. The resultant multi-component molecules are variously called **oligomeric**, **polymeric**, or **multi-subunit proteins**, while proteins with two, three and four subunits may be more specifically termed dimeric, trimeric and tetrameric. The term **homodimer** is used to describe a protein consisting of two identical monomers, and **heterodimer** to describe one where the two monomers are different. Allosterically regulated enzymes, discussed in Chapter 6, are mostly multi-subunit proteins, as is haemoglobin. The arrangement of subunits into a single functional complex is referred to as the quaternary structure of a protein (Fig. 4.2(d)).

Quaternary structure greatly increases the number of functional proteins. For example, if the active form of a protein is a dimer, by forming a range of heterodimers in which different but related monomers combine, many different functional proteins are possible. This feature is commonly found in DNA-binding proteins (see Chapter 26), but is not limited to those.

Protein homologies and evolution

Evolution involves the development of new protein structures and therefore modification of existing genes by mutation. It may seem puzzling that an existing gene coding for an essential protein can be modified into a different one without losing the original. Gene duplication (see Chapter 22) followed by evolution of one copy avoids elimination of existing genes.

The amino acid sequence of a protein is a consequence of its evolution and can be used by researchers to gain insights into the past. The analysis of proteins in this way is part of the discipline of **bioinformatics**. Amino acids essential for function tend to be **conserved** in evolution – they are not substituted, or only by very similar amino acids. Mutational changes in a particular amino acid residue resulting in another one with similar properties (for instance substitution of asparagine with glutamine) are termed **conservative** changes. Less functionally important amino acids tend to change, and this is known as **nonconservative** substitution. It is observed repeatedly that different proteins with different functions have such close amino acid sequence similarities that they must have had a common ancestral protein. Proteins that have evolved from a common ancestral protein are said to be **homologous**.

When the amino acid sequence of a protein has been determined, protein databases containing data for all known proteins can be searched for homologous sequences. Methods are available for aligning the sequences while allowing for deletions and insertions of residues having occurred through evolution. Sequence similarities can then be quantitatively assessed and the statistical probability of resemblances being due to chance can be determined. In this way protein and gene families that evolved from common ancestors can be identified and information on evolutionary relationships obtained.

Tertiary structural resemblances between proteins can also be used to detect homologies, for example the arrangement of helices and β sheets can be compared. These protein structures tend to be conserved because they are most intimately related to function, and can be more diagnostic than sequences of evolutionary relationships between proteins. For example, the existence and relative spatial arrangement of an α helix and a β sheet may be conserved between two proteins even if the amino acid sequences contributing to the two secondary structures are different.

Protein domains

If we consider a protein molecule consisting of a single polypeptide chain, its folded structure may be a single compact entity, no part of which could exist on its own. However, especially in proteins larger than about 200 amino acids in length, it is often seen that there are two or more regions that form compact **domains** of folded structure usually linked together by an unstructured polypeptide chain. It is possible to use experimental techniques to obtain some of these domains as separate entities, without the rest of the protein, and their integrity of structure and, in some cases, catalytic activity confirmed. A suggested definition of a protein domain is that it is a sub-region of the polypeptide that possesses the characteristics of a folded globular protein. The structure of the enzyme pyruvate kinase illustrated in Fig. 4.7(d) illustrates three domains joined together to form a single protein. Further examples are found among DNA-binding proteins (see Chapter 26), which usually have a domain that even when isolated from the rest of the protein can bind specific DNA sequences, as well as other domains that may have protein-protein interactions or catalytic functions.

Thus, protein domains are often associated with different partial activities of a protein that together enable it to carry out its function. For instance, many enzymes with a single catalytic function combine with at least two substrates. The NAD^+ dehydrogenases (see Chapter 12) are a typical case. Different NAD^+ dehydrogenases all bind NAD^+, but each binds a different oxidizable substrate and all catalyse the reaction:

$$AH_2 + NAD^+ \rightleftharpoons A + NADH + H^+$$

(where A is any substrate molecule).

It is found that several enzymes catalysing such reactions have separate domains for binding NAD^+ and the substrate (AH_2). Together the two binding sites form the active site of the enzyme. However, the NAD^+ binding domains of the different enzymes examined have a similar structure, suggesting homology and common ancestry, while the AH_2 binding domains are different. This suggested that once evolution had developed an NAD^+ binding domain, it was duplicated and combined repeatedly with different substrate-binding domains that evolved separately. Many similar examples of protein evolution through repeated use of a single functional domain are now known. One particularly striking example is the SH3 domain of a protein involved in signal transduction (see Chapter 29); in the human genome sequence over 300 DNA sequences exist that would be translated into amino acid sequences homologous with that of the SH3 domain.

In some cases, however, structural similarities may be the result of **convergent evolution** in which a particular sequence of amino acids is independently evolved in proteins of different ancestry. An example is the catalytic triad of proteases (see Chapter 6), which occurs in a family of eukaryotic proteases with clearly homologous structures, but also in the bacterial protein subtilisin, which is ancestrally unrelated to the eukaryotic enzymes but has a similar function.

Domain shuffling

Domain shuffling (or domain swapping) is the name given to the evolutionary process in which new genes are assembled from sections of DNA coding for pre-existing protein domains. It leads to the synthesis of novel proteins made up of new mixtures of domains. Modular construction of enzymes and other proteins permits more rapid evolution of new functional proteins than would occur from single amino acid substitutions.

Protein domains are often (though not always) coded for in eukaryotes by specific separate gene subsections, called exons, described in Chapter 22. The divided structure of eukaryotic genes and localization of DNA sections coding for functional domains increases the chances of recombination leading to new functional proteins.

Membrane proteins

We have so far concentrated on globular proteins that are cytosolic and hence water soluble, but membrane proteins necessarily have slightly different structural properties. Many have a water soluble section at each end with one end extending into the exterior and the other into the interior of the cell. These are linked by a hydrophobic section in the middle that is in contact with the hydrocarbon layer in the centre of the lipid bilayer. The structure of membrane proteins is dealt with in more detail in Chapter 7.

Conjugated proteins and post-translational modifications of proteins

Many proteins need nothing but the folded polypeptide chain(s) for their function. However, many enzymes require a metal ion for activity; some have attached a complex molecule called a **prosthetic group**, which forms part of the active site. The protein part in such cases is called an **apoenzyme** (*apo*= detached or separate) and the complete enzyme a **holoenzyme**.

Other proteins have carbohydrate attachments and are called **glycoproteins**. Most membrane proteins have oligosaccharides attached to the sections of the polypeptide on the outside surface of the membrane. Attachment is via the —OH of serine or threonine side chains (*O*-linked) or to an asparagine side chain (*N*-linked). The latter attachment is illustrated below.

Polypeptide backbone
Asparagine side chain
Sugar unit

Many secreted proteins such as blood and saliva proteins are glycoproteins, with the carbohydrate groups serving a variety of functions. The degradation of carbohydrate attachments to serum proteins and to erythrocyte membrane proteins marks them for uptake and destruction by liver cells – these carbohydrates thus act as indicators of age of the components. Glycosylation of some proteins makes them effective lubricants, as in saliva, or protects them from proteolytic attack. In some cases the different sugars are involved in recognition. They play this role in the sorting of proteins by the Golgi apparatus (see Chapter 27).

Post-translational modifications of proline residues are important in extracellular proteins (see Extracellular matrix proteins).

Extracellular matrix proteins

The free soluble proteins we have discussed so far have been globular. In structural terms we now come to a different class: proteins of the extracellular matrix are mainly elongated fibrous proteins that are usually partly immobilized by being bound into larger structures. Extracellular matrix proteins play important organizational roles, and they are a subject of medical interest.

A general description of the extracellular matrix (ECM) may be helpful here to give perspective on its role. All cells are embedded or bathed in ECM; even in tissues such as liver when the cells are in close contact and the layer is thin. Additionally, between specialized tissues the bodies of animals contain spaces filled with connective tissue, which is particularly rich in ECM proteins and the cells producing them. There are different types of connective tissue in different parts of the body. The **dense connective tissues** include bone and tendons, the latter linking muscles to bone in order to transmit the tension of contraction. Both bone and tendons are predominantly **collagen**. In the case of bone, the tissue is calcified. At the other end of the spectrum are the **loose connective tissues** found under all epithelial layers; wherever there is a bodily cavity, such as the intestine and the blood vessels, there is a layer of epithelial cells lining the cavity. This has no mechanical strength but is supported by a layer of protein fibres known as the **basal lamina**. Underneath the basal lamina is a layer of loose connective tissue as illustrated in Fig. 4.10 where skin is used as an example. The connective tissue joins the epithelial layer to the underlying tissue. It is flexible and resists compression. The background substance that resists compression is a soft, highly hydrated gel formed by **proteoglycans** (see Structure of proteoglycans), which on its own has no mechanical strength against tension but is reinforced with collagen and **elastin** fibres. Some fibres link the basal lamina to the epithelial cells above it and to the connective tissue below. Further links join the underlying tissue cells to components of the connective tissue so that the whole structure is stable. There are also adhesive proteins that help to interconnect everything, the best known of which is **fibronectin**.

The components of connective tissue are secreted by cells known as **fibroblasts** dotted around in the background matrix and occupying little of the volume of connective tissue. Bone and cartilage have special fibroblasts known as osteoblasts and chondroblasts respectively.

With that general introduction we will now describe the structures of the reinforcing proteins, collagen and elastin, the structures of proteoglycans which form the jelly-like ground substance of connective tissue. Finally we will briefly describe some of the proteins (fibronectin and integrins) that connect the ECM to intracellular components.

Structure of collagens

Collagen is the most plentiful protein in the mammalian body. It is a secreted protein and therefore occurs outside cells. The protein from which it is assembled is secreted by cells in the form of **procollagen**, which is subjected to a variety of chemical changes catalysed by enzymes, resulting in the mature collagen.

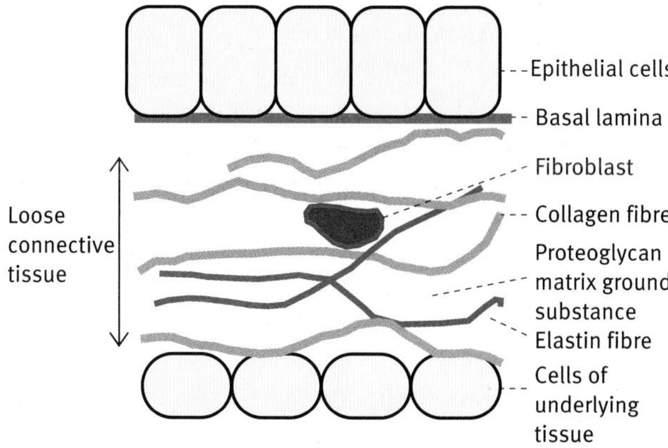

Fig. 4.10 Components of loose connective tissue that underlies epithelial cell layers (e.g. of skin or intestinal lining). The structures of the collagen and elastin fibres and of the proteoglycans are described later in the text.

Procollagen contains a triple superhelix – three helical polypeptides twisted around each other (Fig. 4.11(a)). Each of the polypeptides in the triple superhelix of procollagen is an *unusual left-handed helix* not the right-handed α helix of globular proteins. However, *the three polypeptides are twisted around each other in a right-handed manner* to form the triple helix. About one in three amino acid residues is proline and every third residue is glycine. At the ends of the procollagen helices are extra peptides with a different structure, which, after secretion of the procollagen, are cleaved off leaving triple helical **tropocollagen** molecules from which collagen fibrils are assembled.

Many of the proline and also lysine residues of tropocollagen are hydroxylated to form **hydroxyproline** and **hydroxylysine**. The amino acid residue are hydroxylated in the endoplasmic reticulum after synthesis of the tropocollagen molecule. Hydroxylation of proline in the polypeptide requires ascorbic acid (vitamin C), which keeps an essential Fe^{2+} atom in the enzyme prolyl hydroxylase in the reduced form. In deficiency of this vitamin, connective tissue is not properly formed, resulting in the painful consequences of **scurvy** – bleeding gums and failure of wound healing.

—N—CH—C(=O)— with CH_2, CH_2, CH_2 ring → (O_2, Ascorbic acid; Prolyl hydroxylase) → —N—CH—C(=O)— with CH_2, CH_2, CH–OH ring

Proline residue in polypeptide → Hydroxyproline residue in polypeptide

(Note that the hydroxylation reaction is more complex than shown here and also involves 2-oxoglutarate.)

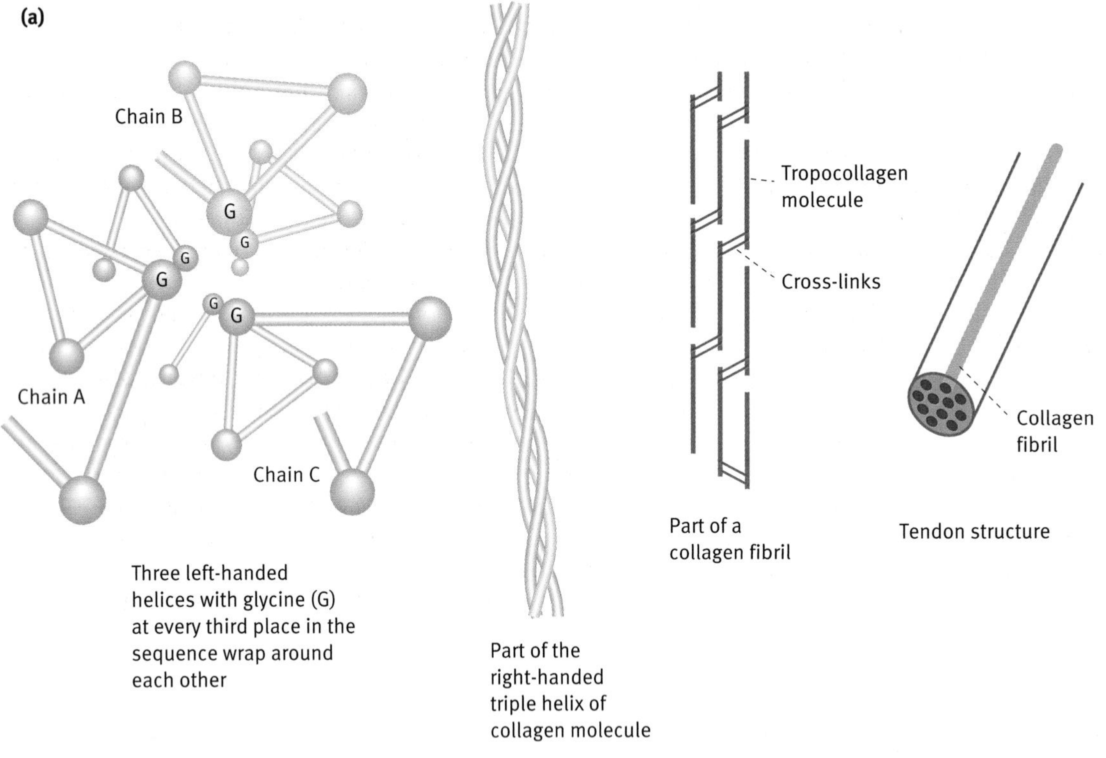

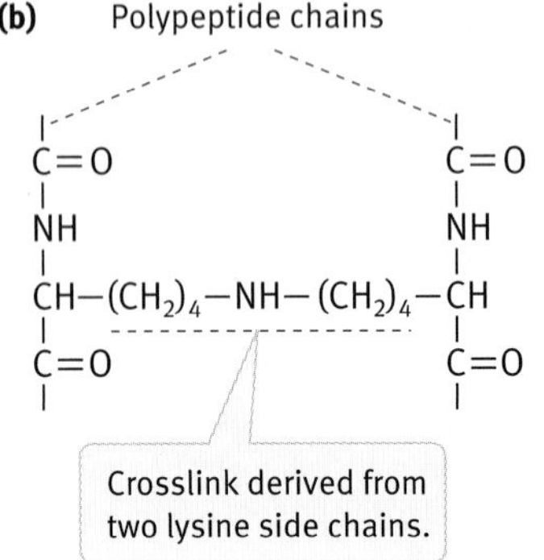

Fig. 4.11 **(a)** Assembly of collagen fibres. Three left-handed helices are wound around each other to form a right-handed triple helix, which is processed to form the tropocollagen molecule. Tropocollagen triple helices are cross-linked to form collagen fibrils. The bonds in red are covalent links formed between lysine residues. **(b)** One type of cross-link formed between two adjacent lysine residues. Note that several specific types of collagen exist for individual functions. Adapted from van der Rest M, Garrone R.; Collagens as multidomain proteins; Biochimie; 1990 Jun–Jul; 72 (6–7): 473–84; Elsevier.

The three strands of the triple superhelix are in close association, forming a very strong structure. The side chains of polypeptide chains would normally prevent such close association but the structure of collagen allows this. The proline residues mean that the left-handed helix is more extended than an α helix, with three amino acid residues per turn. Every third residue is glycine and, in the triple superhelix, the contacts between the chains occur always at glycine residues, the side 'chain' of which is only a hydrogen atom that does not get in the way of close contact (Fig. 4.11(a)). Hydroxylated lysines and prolines form hydrogen bonds between the three chains, thus stabilizing the super helix.

The tropocollagen molecule so far described has about 1000 amino acid residues. These assemble into collagen fibrils by staggered head to tail arrangement as shown in Fig. 4.11(a). The 'holes' in this structure are believed to be sites where, in bone, crystals of hydroxyapatite ($Ca_{10}(PO_3)_6(OH)_2$) are laid down to initiate mineralization. In tendons additional strength is achieved by the formation of unusual covalent links between the tropocollagen units; adjacent lysine side chains are modified in various ways to form the links, an example of which is shown in Fig. 4.11(b). The collagen fibrils so formed aggregate to form tendons by a parallel arrangement. In skin, it is more of a two-dimensional network. Different subclasses of collagen exist, dependent on the precise structure of the three polypeptides in the triple superhelix. Variations between the different types of chains include the content of hydroxylysine and hydroxyproline. In some types of collagen, these hydroxylated residues are glycosylated to varying degrees by covalent bonding of glucose and galactose to their hydroxyl groups.

Structure of elastin

The elastin molecule has a unique structure, different from that of collagen. It is a major component of connective tissue such as the lung and arteries. It is a hydrophobic insoluble protein which forms a three-dimensional elastic network, which can reversibly stretch in any direction with a structure more elastic than rubber (Fig. 4.12(a)). The network is assembled from a soluble protein unit called proelastin, which is secreted from cells and then cross-linked to form the elastic network. Proelastin is rich in glycine, alanine, and lysine. Lysine occurs every few residues and in between are short stretches of helical or random coil sections. It is these sections that can reversibly stretch in an elastic manner. The proelastin is assembled into the three-dimensional network of elastin by the formation of covalent links between lysine residues of four polypeptide chains forming desmosine, as shown in Fig. 4.12(b). Desmosine formation requires the enzymic oxidation of lysyl residues.

Structure of proteoglycans

The proteoglycans provide the jelly-like matrix substance of the connective tissues. Hydrated jellies in nature are based on negatively charged carbohydrate polymers. The chains of polar sugars are highly hydrophilic and the mutual repulsion of their charged groups ensures that the chains are fully extended and occupy a large volume, thus entrapping a lot of water. A proteoglycan consists of chains of charged sugars attached to the serine side chains of **core protein** molecules (Fig. 4.13). The protein chain is fully extended, as are the carbohydrate chains. The negative charges localize a cloud of cations, which contribute to the osmotic pressure of the matrix, which is important in resisting compression.

The carbohydrate chains are all made of repeating disaccharide units each of which has either ***N*-acetylglucosamine** or ***N*-acetylgalactosamine** (Fig. 4.14) as one component so that the polysaccharides are known as **glycosaminoglycans** (GAGs). The *N*-acetylglucosamine or *N*-acetylgalactosamine is often modified by one or more sulphate groups at varying positions. The second sugar usually has a carboxyl group as shown. Sulphate and carboxylic acid groups increase the negative charge of the molecule. The general pattern of a repeating GAG disaccharide is shown in Fig. 4.15. The main GAGs are known as **chondroitin sulphate, dermatan sulphate, heparan sulphate,** and **keratan sulphate**. (Keratan sulphate should not

BOX 4.1 Genetic diseases of collagen

The functions of collagens are always to provide tough reinforcing filaments in connective tissues, but there are different needs in different situations. This is reflected by the existence of almost 20 types of collagen (identified by roman numerals); there are about 30 different genes coding for the constituent polypeptides of these collagen types. All contain the triple helix but this can vary from occupying the entire length of the collagen molecule to chains in which it is quite short. Not surprisingly with so many genes for the collagen chains and for the enzymes carrying out post-translational modification, there are many associated genetic diseases, with a wide variety of results: for example, weakened blood vessels leading to aortic rupture, hyperelastic skin, and hypermobility of joints.

In another disease the small filaments that attach the basal lamina to the underlying collagen fibres in the connective tissue of skin are deficient. In people with this condition the skin forms blisters as a result of the slightest provocation as it becomes detached from the connective tissue.

The enzyme lysyl oxidase, which forms the covalent collagen cross-links (see main text), is deficient in children with the genetic disorder Menkes disease. The result is connective tissue abnormalities causing aortic aneurisms, fragile bones, and loose skin. The genetic lesion causes faulty copper homoeostasis in the body; lysyl oxidase is a copper-requiring enzyme. Copper-deficient animals show some characteristics resembling those seen in Menkes disease.

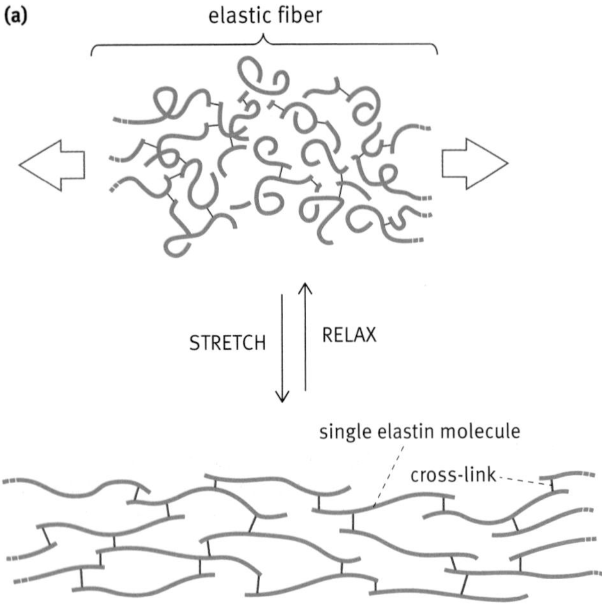

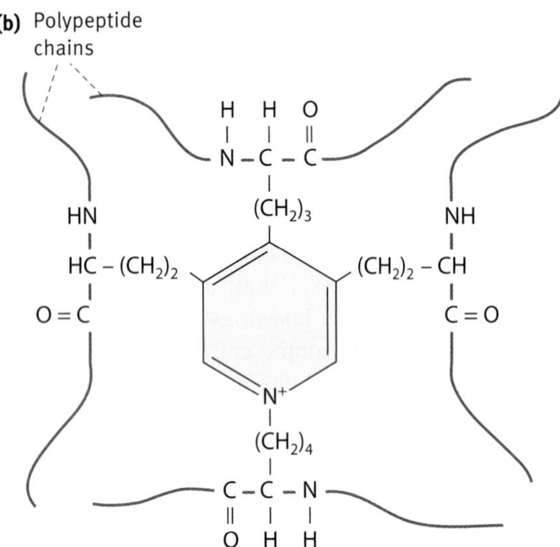

Fig. 4.12 (a) Representation of the general structure of elastin showing how 'stretching' of the protein occurs. (Bruce Alberts, Dennis Bray, Karen Hopkin and Alexander D Johnson (2009); Essential Cell Biology, 3rd Edition; Reproduced by permission of Taylor and Francis Group.) **(b)** Desmosine cross-link between four polypeptide chains of elastin. The structure is formed by enzyme modification of four lysine residues.

be confused with the fibrous protein, keratin, found in skin and hair.) **Heparin** is structurally similar to heparan sulphate, but the latter has more sulphate groups. Heparin has a different role, however, as its exists as free GAG in blood vessels, and is important in controlling blood clotting (see Chapter 32).

There are many different proteoglycans; the basic design is extremely flexible. The length of the core protein varies from about 1000 to 5000 amino acid residues, and the number of polysaccharide chains attached to the core protein varies up to about 100; the length of the polysaccharide chain varies but typically is about 80–100 sugar residues long. Finally, the number and type of charged groups can vary. **Aggrecan,** the chief proteoglycan of cartilage, provides an example to illustrate this. Cartilage has to withstand very large compressive forces and be very tough. In knee joints, for example, it is needed to prevent direct contact of leg bones and has to withstand enormous pressures. Aggrecan (Fig. 4.16) consists of a protein core with two different GAGs, keratan sulphate and chondroitin sulphate, attached to serine side chains. Large numbers of these proteoglycan molecules are complexed noncovalently to yet another long GAG called hyaluronan, forming a gigantic molecule (Fig. 4.16) resembling a bottle brush as seen in the electron microscope. Hyaluronan is also called hyaluronate or hyaluronic acid. This matrix is heavily reinforced with collagen fibres that resist tension so that the cartilage resists both compression and tearing. The GAG hyaluronan is also widely found in soft extracellular matrices and synovial fluid, where it exists free rather than associated with protein.

Fibronectin and integrins connect the extracellular matrix to the interior of the cell

Fibronectin is an important and widespread member of a large and diverse class of proteins known as cell adhesion molecules or CAMs. While some CAMs mediate direct cell-cell interactions, fibronectin is found in the extracellular matrix and mediates attachment of cells to the ECM. It consists of two flexible elongated protein chains joined at one end by two disulphide bridges. It has several different sites on it that bind to proteoglycans, collagen, and cell surfaces, respectively (Fig. 4.17). Fibronectin binds to cell surfaces via dimeric transmembrane proteins known as **integrins,** so called because they integrate the attachment of cells to extracellular matrix proteins (Fig. 4.18). There is a very large family of different integrin heterodimers that can be found on different cell types. Integrins selectively attach to different ECM molecules including collagen and proteoglycans, as well as to fibronectin. Their intracellular domains are attached to actin fibres of the cytoskeleton and thus form a link between the extracellular matrix and the inside of the cell (Fig. 4.18), giving a strong interaction as required in the attachment of muscle cells to tendon, for instance. However, integrins are not simply structural proteins: binding to their ligands can trigger intracellular signalling responses (see Chapter 29), thus enabling cells to respond to changes in their environment. Integrins are therefore a class of cellular receptors. Integrin signalling is of profound importance in many aspects of cell life including embryonic development, blood platelet aggregation in response to injury, and cellular responses to infection.

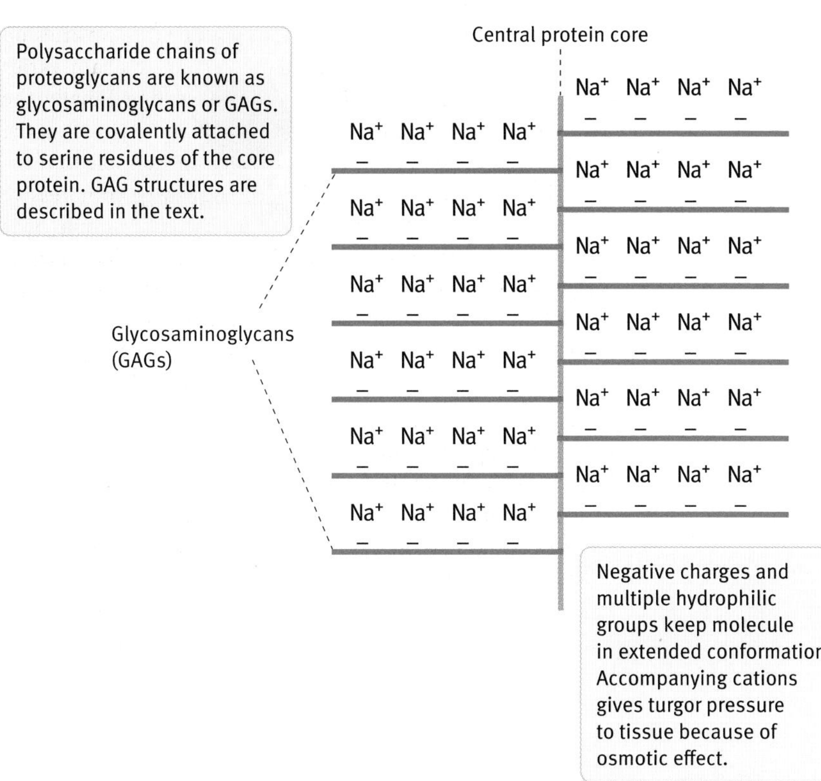
Polysaccharide chains of proteoglycans are known as glycosaminoglycans or GAGs. They are covalently attached to serine residues of the core protein. GAG structures are described in the text.
Central protein core
Glycosaminoglycans (GAGs)
Na^+ Na^+ Na^+ Na^+
Negative charges and multiple hydrophilic groups keep molecule in extended conformation. Accompanying cations gives turgor pressure to tissue because of osmotic effect.

Fig. 4.13 The design of proteoglycans. Note that there can be many variations on the common theme: the size of the central protein core can vary; the number of glycosaminoglycan (GAGs) attached to it can vary; the GAGs can vary in length, in the number, position, and nature of the charged groups, and in the details of their chemistry. We show the accompanying cloud of cations as Na^+ but other cations could be involved. The structures of the GAGs are described in the text.

BOX 4.2 Smoking, elastin, emphysema, and antiproteinases

We will see in Chapter 10 how the digestive system escapes the actions of its own proteolytic enzymes. However, proteases exist elsewhere in the body (see A major question in digestion – why doesn't the body digest itself?). A particularly important one, in the present context, is the **elastase** of **neutrophils**. Neutrophils are phagocytic white cells attracted to sites of infection or irritation. When activated at such sites, they secrete elastase, which clears away connective tissue from the site. Elastin is the elasticity-conferring connective tissue protein from which elastase gets its name.

In the lung, air passages lead to minute pockets, the alveoli, which result in the lung having the very large surface area needed for the diffusion of gases between blood and air. Neutrophils attracted to the lung liberate elastase. In normal circumstances this is prevented from destroying the lung structure by **α_1-antitrypsin** (**α_1-antiproteinase**), a protein that is produced by the liver and secreted into the blood. The α_1-antitrypsin inhibits several proteases, including trypsin as the name implies, but is especially effective on elastase which, *in vivo*, is probably the most important of the proteases in this context. It inhibits by combining tightly with the enzyme and blocking its catalytic site so tightly that it is known as a 'suicide inhibitor' because once it binds, both the enzyme and the inhibitor molecule are not recoverable.

α_1-Antitrypsin in adequate levels in the blood is essential for the protection of the lung. The molecule diffuses from the blood into the alveoli. If, due to a genetic defect, the level of α_1-antitrypsin is subnormal, neutrophil elastase may destroy alveoli, resulting in much larger pockets in the lung structure and consequent reduction of surface area available for gaseous exchange. The result is emphysema, a symptom of which is extreme shortness of breath.

Smokers are prone to emphysema for two reasons. The smoke irritation attracts neutrophils to the lungs, with a consequent increased release of elastase. Secondly, oxidizing agents in the smoke destroy α_1-antitrypsin; they chemically oxidize the sulphur atom of a crucial methionine side chain to a sulphoxide group (S→SdO). This prevents the α_1-antitrypsin from inactivating elastase, and may result in proteolysis of lung tissue, thus resulting in emphysema.

This is not the only physiological role of antiproteinases in the body. Trypsin, chymotrypsin, and elastase are secreted by the pancreas into the intestine in an inactive zymogen form. They are activated by proteolysis in which an initial small activation becomes an autocatalytic cascade. This makes even a small amount of premature activation in the pancreas cells potentially dangerous, because any active proteinase may activate all of the zymogen in a proteolytic cascade and cause pancreatitis. The three digestion proteases all depend on an active serine residue (see Chapter 6) and the antiproteinases are collectively called **serpins** (**serine protease inhibitors**). As with α_1-antitrypsin they work by very tightly associating with the active sites of the enzymes and blocking their activity.

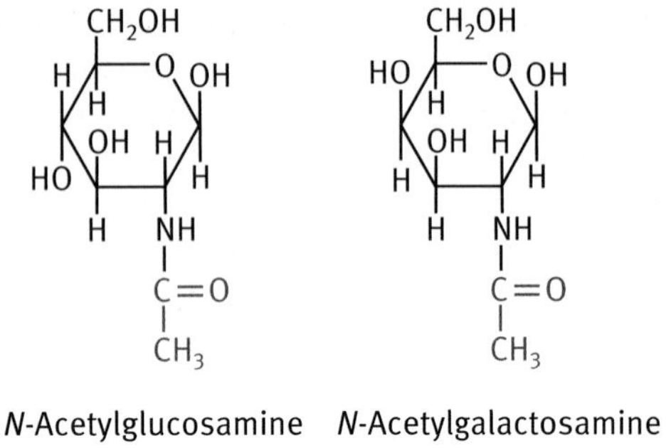

N-Acetylglucosamine *N*-Acetylgalactosamine

Fig. 4.14 *N*-Acetylglucosamine and *N*-acetylgalactosamine.

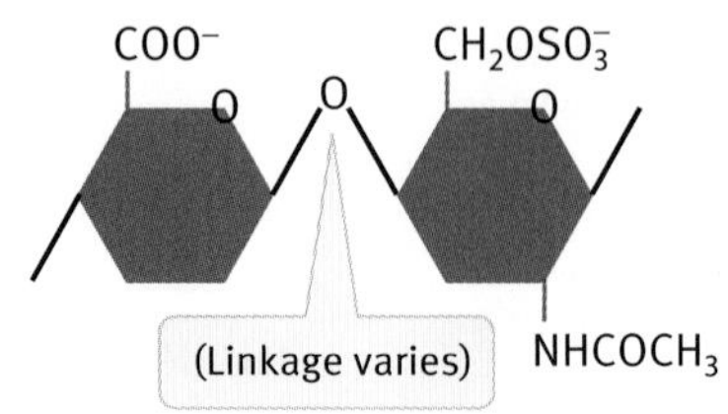

Fig. 4.15 A disaccharide unit of glycosaminoglycan (GAG). For simplicity all bonds and substituents except those characteristic of GAGs have been omitted. The polysaccharide portion of proteoglycans are made of long unbranched chains of these disaccharides. Different GAGs vary in the number and positions of sulphate and carboxyl groups and in other details such as the nature of the glycosidic link between the sugars.

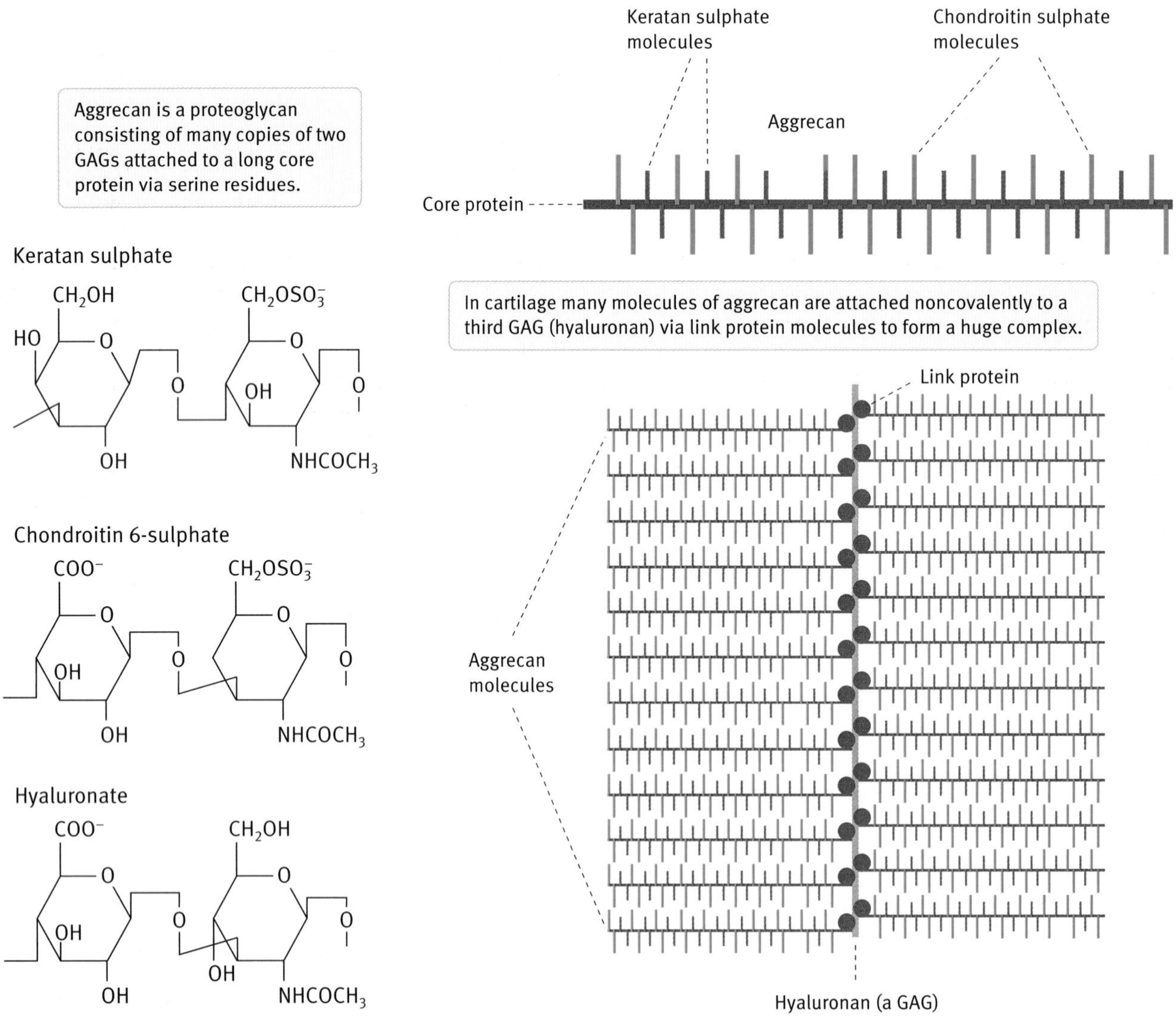

Aggrecan is a proteoglycan consisting of many copies of two GAGs attached to a long core protein via serine residues.

Keratan sulphate

Chondroitin 6-sulphate

Hyaluronate

Keratan sulphate molecules

Chondroitin sulphate molecules

Aggrecan

Core protein

In cartilage many molecules of aggrecan are attached noncovalently to a third GAG (hyaluronan) via link protein molecules to form a huge complex.

Link protein

Aggrecan molecules

Hyaluronan (a GAG)

Fig. 4.16 The main proteoglycan found in cartilage. Note that hyaluronan is simply a glucosaminoglycan (GAG) molecule. It forms a huge noncovalent complex with multiple copies of the proteoglycan aggrecan, the attachment being via link protein molecules. Hyaluronate is also widely found in soft extracellular matrices where it exists free, not linked to proteins — often called hyaluronic acid in this context.

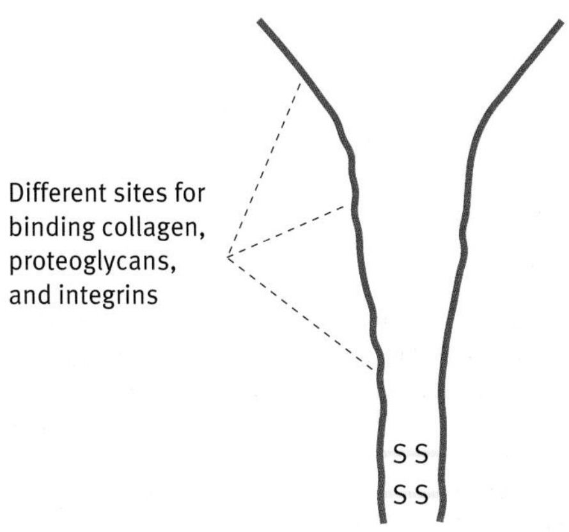

Fig. 4.17 A fibronectin molecule consisting of two polypeptides linked by two disulphide bridges (yellow). The integrins are transmembrane receptors on cells. The binding of fibronectin to collagen fibres, proteoglycans, and integrins links together the extracellular matrix structure.

Myoglobin and haemoglobin illustrate how protein structure is related to function

The oxygen-binding proteins myoglobin and haemoglobin have been studied intensively and form the classic illustration of the way in which knowledge of protein structure leads to an understanding of biological function. Tissues of the body have to be supplied with oxygen. The most primitive animals and some cold-water fish transport oxygen in solution in their blood, but the solubility of the gas is too low for this to be adequate for more active animals. Much the same is true for the removal of CO_2 from the tissues. A specialized transport protein such as haemoglobin is required.

Myoglobin is the red pigment found in striated muscle (see Chapter 8), where it has the role of an oxygen store to be used during intense muscular activity that results in the consumption of more oxygen than the blood can deliver. Haemoglobin is the oxygen carrier in blood. The two proteins are closely related, both in structure and evolutionary ancestry, but myoglobin is a relatively simple molecule while haemoglobin is a sophisticated molecular machine superbly evolved for its complex roles. A comparison of the two proteins is helpful in understanding haemoglobin.

Myoglobin

The myoglobin protein is monomeric — the single polypeptide chain of 153 amino acid residues is arranged entirely in the form of α helices connected by loops (Fig. 4.23 (a)). Inserted into a pocket formed by the helices is a molecule of haem. Haem is a ferrous iron — tetrapyrrole (Fig. 4.19) with hydrophobic side chains on three sides and the fourth with hydrophilic carboxyl radicals. The molecule is buried in the hydrophobic interior of the molecule with the hydrophilic side oriented to the exterior. Haem has an intense red colour resulting from its **conjugated system** of alternating single and double bonds around the molecule. The iron in haem can bond to six ligands, four being taken up by the pyrrole nitrogen atoms, the fifth by attachment to a histidine residue of the protein, and the sixth is available for the reversible attachment of oxygen (Fig. 4.20).

Fig. 4.21 shows the percentage saturation of myoglobin at increasing oxygen tensions. The curve is hyperbolic, much the same as that of a 'classic' enzyme-substrate-binding response (see Chapter 6). The molecule is almost fully saturated at the low oxygen tension of 20 Torr present in capillaries. (A torr is a unit of pressure named after Torricelli. 20 Torr is equivalent to 2.7 kPa). This indicates that myoglobin has a high oxygen affinity and does not give it up at normal oxygen tensions. It

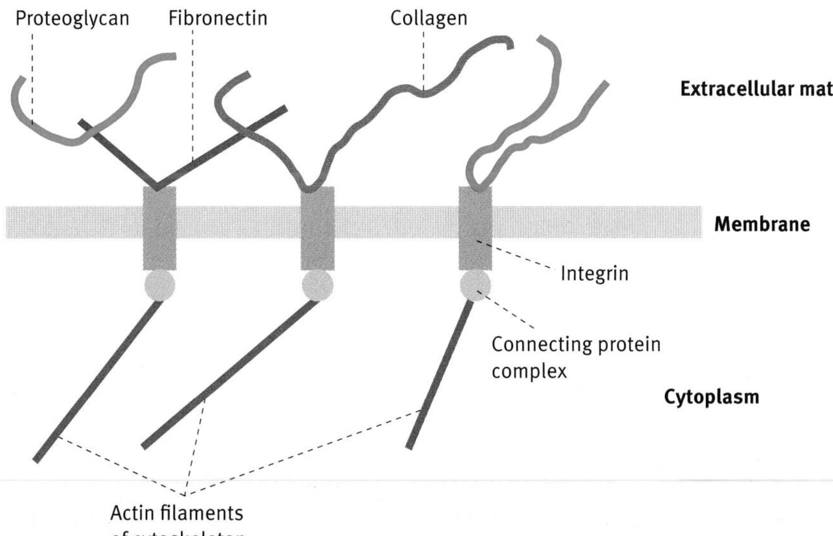

Fig. 4.18 The role of integrins in connecting components of the extracellular matrix to the cytoskeleton. A family of fibronectins exist. The fibronectin is represented as a single line rather than its dimer shape (Fig. 4.17) for simplicity. The cytoskeleton is described in Chapter 8.

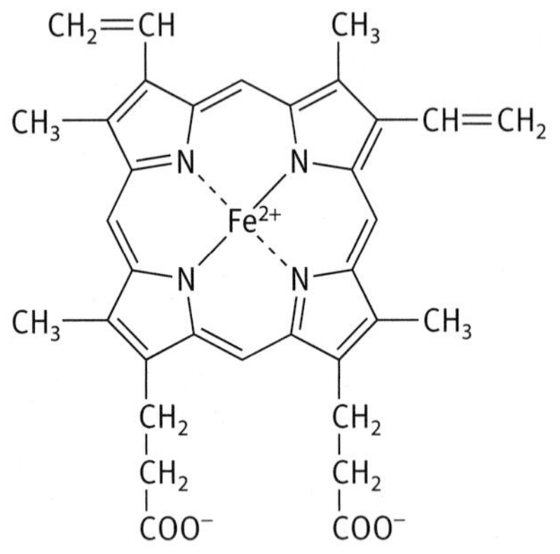

Fig. 4.19 The structure of haem. This form is found in haemoglobin; other haems differing in their side chains are found in cytochromes. Note that, at one side of the molecule, there are two hydrophilic propionate groups ($-CH_2CH_2COO^-$) while the remaining side chains are all hydrophobic. In myoglobin, haem sits in a cleft of the molecule with the hydrophilic side pointing out towards water and the hydrophobic groups buried into the nonpolar interior of the protein.

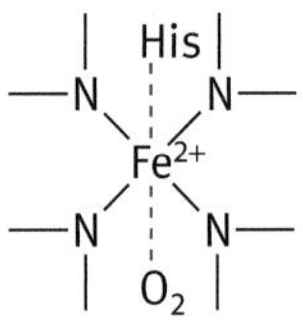

Fig. 4.20 Binding ability of Fe^{2+} in haem. The iron atom in haem can bond to six ligands in total: four bonds to the flat pyrrole nitrogen atoms as shown and the other two above and below the plane of the page. One of the perpendicular bonds is bound to the nitrogen atom of a histidine residue, the other is the binding site for an oxygen molecule.

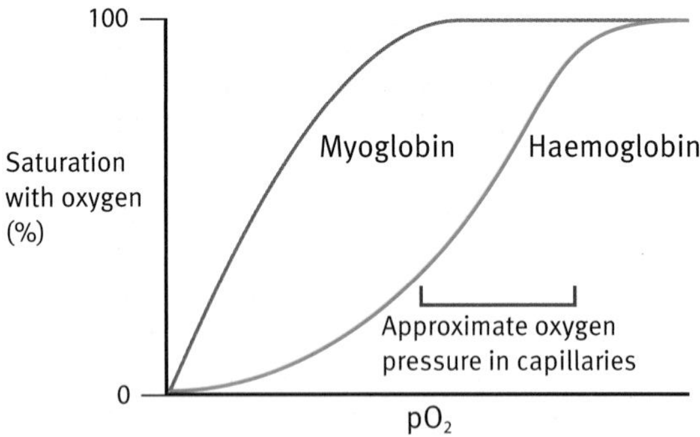

Fig. 4.21 Oxygen–haemoglobin saturation curve. The higher affinity for oxygen of myoglobin as compared to haemoglobin means that myoglobin in the muscles readily accepts oxygen from the blood.

does so only when intense muscular activity further lowers the oxygen tension in the tissue so that it is acting as a reserve store of oxygen in times of need. It has a higher affinity for oxygen than has haemoglobin and so can extract oxygen from the blood to top up its store as required. Since myoglobin does not surrender oxygen at the levels normally found in tissues it would be unsuitable as the oxygen carrier in the blood: this is the role of haemoglobin.

Structure of haemoglobin

Haemoglobin is a tetramer of protein subunits (Fig. 4.22) each of which has a haem molecule capable of binding oxygen. The two α subunits are identical as are the two β ones; they are known as α_1 and α_2, β_1 and β_2. The subunits closely resemble myoglobin in structure, being composed of α helices, and each subunit has a haem unit similarly located between the helices. Myoglobin and the β-haemoglobin subunit are virtually identical in structure, as shown in Fig. 4.23(a) and (b).

Binding of oxygen to haemoglobin

In the body, the binding of oxygen to haemoglobin and its release in the tissues is not a passive reversible process. The haemoglobin molecule works in a way that demonstrates an exquisite match of structure and function.

A molecule of haemoglobin binds four molecules of oxygen, one per subunit. When an oxygen-saturation curve is plotted, instead of the hyperbolic curve seen with myoglobin, there is a sigmoid curve that is well to the right of that of myoglobin (Fig. 4.21). Higher oxygen concentrations are required to 50% saturate haemoglobin than is the case for myoglobin, indicating, as already stated, that myoglobin has a higher affinity for oxygen.

Haemoglobin needs to readily pick up as much oxygen as it can in the lungs but then readily surrender it in the tissue capillaries. The sigmoidal oxygenation curve is most steep (that is, haemoglobin surrenders the most oxygen) at oxygen pressures encountered in the capillaries (Fig. 4.21) but, nonetheless, it is still capable of becoming virtually saturated at the oxygen pressures encountered in the lungs.

How is the sigmoidal oxygen saturation curve achieved?

Haemoglobin is an **allosteric protein** – a term that we have not introduced previously. Allosteric proteins, of which there are many, are proteins that have more than one binding site for ligands and in which binding of a ligand at one site influences the interaction of the protein and ligand at another site. Most allosteric proteins are, like haemoglobin, multi-subunit proteins. Each of haemoglobin's four subunits is capable of binding an oxygen molecule. The oxygen-binding curve of haemoglobin is sigmoid because binding of the oxygen at low pressures when few of the sites are occupied is more difficult than when more are occupied. There is a progressive increase in affinity of haemoglobin for oxygen as site occupancy increases, so that at the higher oxygen pressures the affinity is increased 20-fold. Although the haem groups in haemoglobin are distant from one another, the initial

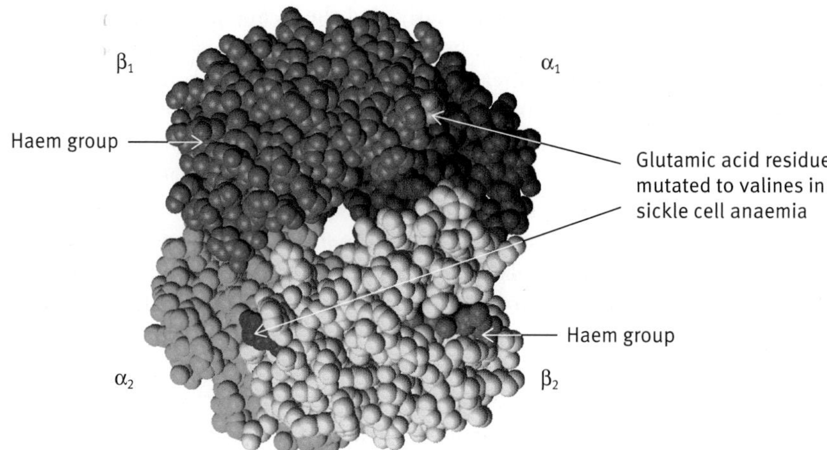

Fig. 4.22 Space-filling model of haemoglobin (1A3N). Model of the subunits and their arrangement in haemoglobin showing the central cavity into which 2,3-bisphosphoglycerate fits in the deoxygenated state. Haem groups are shown in red. In sickle cell disease the glutamic acid residues at position 6 in the β chains (shown in contrasting colours) are mutated to valines, creating hydrophobic patches on the molecule.

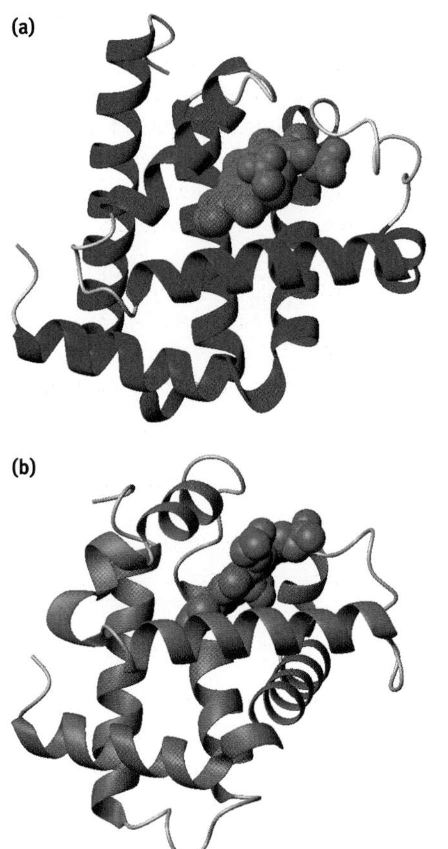

Fig. 4.23 Models of myoglobin (1MBO) **(a)** and a single chain (β chain) of haemoglobin (1HHO). **(b)** Computer-generated diagrams showing the folding of the polypeptide in myoglobin and haemoglobin and the positioning of haem in the molecule.

binding of oxygen to one subunit facilitates the binding of further molecules of oxygen to the other subunits. This is known as a **homotropic positive cooperative effect** (homotropic, because the ligand bound at each site is the same i.e. oxygen).

It is possible to estimate cooperative interactions by a graphic method which gives a value called the **Hill coefficient**. For a protein with a single binding site such as myoglobin, or where there is no cooperativity between sites if there is more than one, the value is 1. Values greater than 1 indicate the degree of cooperativity. For haemoglobin the value is 2.8.

Theoretical models to explain protein allostery

There are two theoretical models used to explain the mechanism of cooperative oxygen binding. In both of these, increase in oxygen binding results in progressively more subunits in a given collection of haemoglobin molecules being in a high affinity state, so that the initial binding of oxygen facilitates the further binding of more oxygen.

In the **concerted model** (also known as the MWC model after its discoverers, Monod, Wyman, and Changeaux), either *all* the subunits of haemoglobin bind oxygen with low affinity (known as the tense or T state) or *all* bind with high affinity (known as the relaxed or R state), the two forms being in spontaneous equilibrium (Fig. 4.24), but with the equilibrium to the low affinity side. Binding of oxygen to a single subunit swings this equilibrium towards the high affinity state. As oxygen pressure increases more of the binding sites become occupied and this swings the equilibrium further to the high affinity state. There is no precise number of oxygen molecules that need to be bound to cause the change but increased binding increases the statistical probability of the change and with it a progressive increase in the affinity.

The **sequential model** (Fig. 4.25) differs in that it assumes that, in the absence of oxygen, *all* haemoglobin molecules are in the low affinity state; there is no equilibrium with high affinity state molecules. When one molecule of oxygen binds to one of the subunits, this single unit changes its conformation from tense (T) to relaxed (R) so that unlike the postulated situation in the concerted model a single molecule can

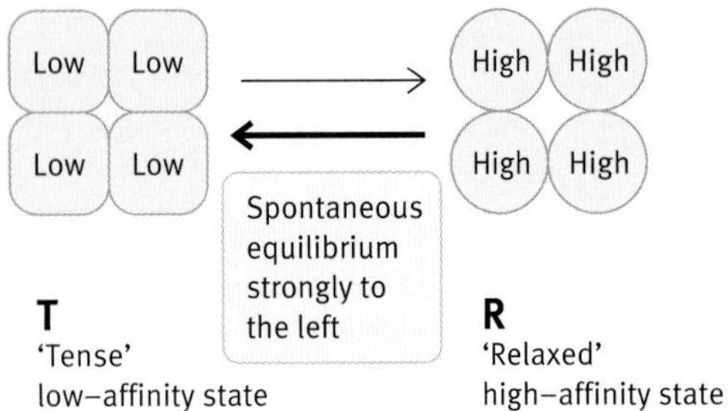

Fig. 4.24 Concerted model of cooperative binding of oxygen to haemoglobin. In this model the haemoglobin exists in two forms, T and R, the two being in spontaneous equilibrium. Binding of an oxygen molecule to the R state swings the equilibrium to the right, thus increasing the affinity of all of the subunits in that molecule. 'High' and 'low' refer to affinities of the haemoglobin for its ligand, oxygen.

have mixtures of subunits in the two states. The change from T to R has the effect of facilitating a similar change in an adjacent subunit so that a second molecule of oxygen binds more easily. Binding of the second molecule of oxygen increases still further the ease with which a third adjacent subunit can make the conformational change with a resultant increase of affinity for oxygen and so on. Since, in a study of the oxygenation of haemoglobin, large numbers of individual molecules are involved, as more and more oxygen binds, more molecules of haemoglobin are in the relaxed state and, therefore, the observed oxygen affinity in the solution increases.

The concerted model explains many but not all of the observed properties of haemoglobin, as also does the sequential model. It may be that the actual mechanism is somewhere in between the two models and individual allosteric proteins may differ in the exact mechanism that operates. There are large numbers of allosteric proteins, many of which are enzymes, as discussed further in Chapter 6. As will be described in Chapter 20, allosteric proteins play a tremendously important role in virtually all aspects of biochemical regulation.

Mechanism of the allosteric change in haemoglobin

By means of X-ray diffraction studies the conformation of the haemoglobin tetramer has been determined in the oxygenated and deoxygenated states. A useful way to view haemoglobin is that the α_1 and β_1 subunits are firmly associated as a dimer and, similarly, that the α_2 and β_2 form a dimer. It is the interaction between the two dimers in the tetramer that undergoes rearrangement in the T $\rightleftharpoons$ R conversion. Figure 4.26 illustrates the relative rotation between the dimers in the conversion from the T to R state caused by binding of oxygen.

What causes this allosteric change to the haemoglobin molecule when oxygen molecules bind? On binding of oxygen to haem, the iron atom moves slightly and, in doing so, brings about the T→R conversion. Although haem, as usually written, appears as a planar molecule, in the deoxygenated state the iron atom lies above the plane of the molecule because it is too big to fit into the tetrapyrrole (Fig. 4.27(a)) and so the tetrapyrrole is not quite flat. The iron atom itself is bonded to a histidine residue in one of the α helices of which the protein subunit is composed.

On binding of oxygen, the iron atom of haem becomes effectively smaller in diameter and moves into the plane of the tetrapyrrole, thus flattening the molecule. The protein rearranges itself as a result of the movement of the iron atom (Fig. 4.27(b)). This causes a relative movement at the point where the α unit of one dimer interacts with the β unit of the other dimer (α_1–β_2/α_2–β_1 interfaces), and thus the relative rotation shown in Figure 4.26. This movement results in the T→R change.

The essential role of 2,3-bisphosphoglycerate (BPG) in haemoglobin function

$$\begin{array}{l} COO^- \\ | \\ CH-O-P(=O)(O^-)-O^- \\ | \\ CH_2-O-P(=O)(O^-)-O^- \end{array}$$

2,3-Bisphosphoglycerate

2,3-Bisphosphoglycerate (BPG) plays an important physiological role in oxygen transport by lowering the affinity of haemoglobin in red blood cells for oxygen, and thus increasing the unloading of oxygen to the tissues; it moves the dissociation curve of oxyhaemoglobin to the right.

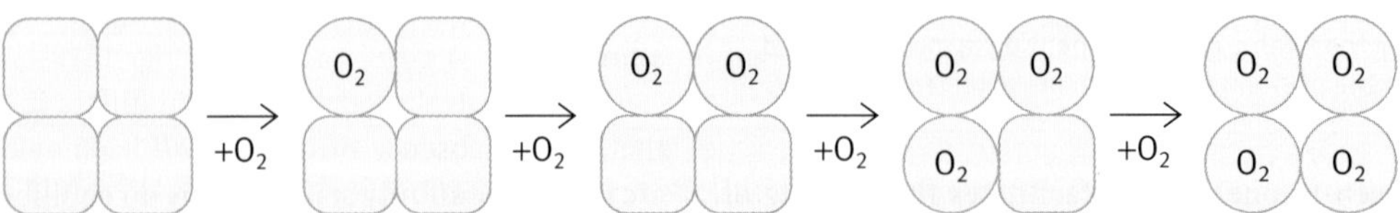

Fig. 4.25 The sequential model for the combination of haemoglobin with its substrate. The binding of a single oxygen molecule to a subunit causes a conformational change in the subunit. This facilitates the conformational change of a second subunit when it combines with an oxygen molecule and so on for the next subunit. The net effect is to make it 'easier' for successive subunits to undergo the conformational change, which is seen as increasing affinity for oxygen by successive subunits.

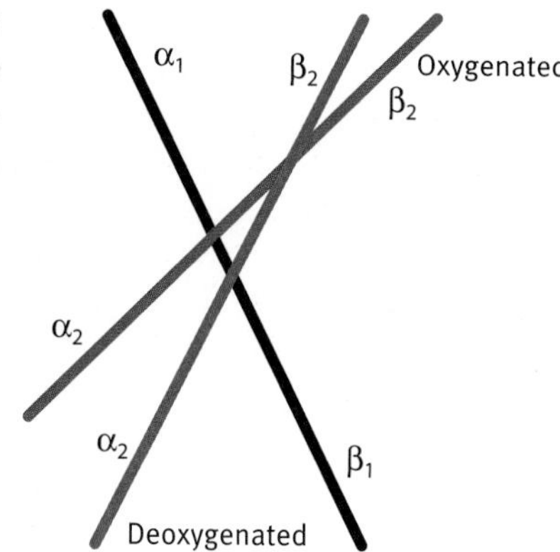

Fig. 4.26 The relative subunit positioning in deoxygenated and oxygenated haemoglobin molecules, the axes of which are represented by straight lines. The α_1, β_1 pair and the α_2, β_2 pair should be regarded as single dimer units. On oxygenation, these dimers rotate and slide, relative to each other, by about 15°. The black and blue lines represent the relative positions of the dimers in the T ('tense') state. In the oxygenated (R, 'relaxed') state the red line shows the rotation of the α_2/β_2 dimer relative to the α_1/β_1 dimer, which is represented as fixed. The change alters the contacts between the dimers.

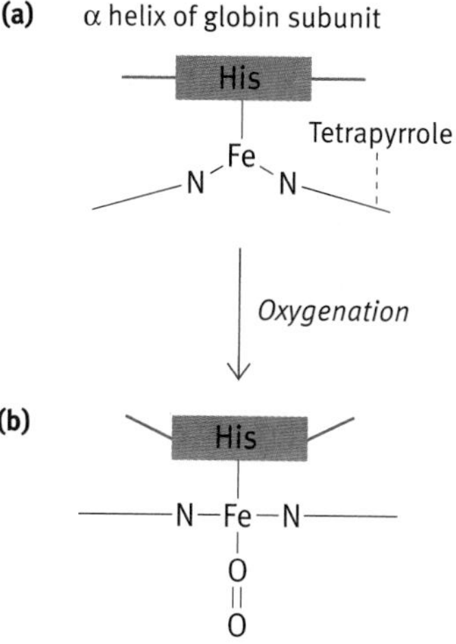

Fig. 4.27 The changes in the haem of haemoglobin upon oxygenation. **(a)** The haem molecule in deoxyhaemoglobin with the tetrapyrrole structure strained into a slightly domed shape. **(b)** Attachment of haem in oxyhaemoglobin. The movement makes the iron atom a microswitch which, by its attachment to an α helix of the haemoglobin, alters the conformation of the protein.

The haemoglobin tetramer, looked at from the appropriate viewpoint, has a cavity running through the molecule (see Fig. 4.22). Projecting into this cavity are amino acid side chains with positive charges. The BPG molecule has five negative charges at the pH of the blood, and has just the correct size and configuration to fit into the cavity of deoxygenated haemoglobin, and to make ionic bonds with these positive charges. This helps to hold haemoglobin in its deoxygenated position; in effect it cross-links the β units in that position. In the deoxygenated (T) state, the haemoglobin can accommodate a molecule of BPG. However, on oxygenation, because of the conformational change in the protein, the cavity of the R state becomes smaller and is unable to accommodate BPG. If we consider oxyhaemoglobin in the capillaries, the ability of BPG to strongly bind to, and stabilize the deoxygenated state favours unloading of the oxygen. In effect, the process is:

(a) Hb—oxygen $\rightleftharpoons$ Hb + oxygen
(b) Hb + BPG $\rightleftharpoons$ Hb – BPG

Reaction (b) with BPG will tend to pull the equilibrium of reaction (a) to the right and favour oxygen release.

If blood cells are stripped of all of their BPG, the haemoglobin remains virtually saturated with oxygen even at oxygen concentrations below that encountered in the tissue capillaries. It would be incapable, therefore, in that state, of delivering oxygen to the tissues efficiently. The effect of BPG on oxygen binding by haemoglobin is illustrated in Fig. 4.28. As it binds to haemoglobin at a site other than the oxygen binding site, but affects its affinity for oxygen, BPG acts as a **heterotropic allosteric modulator** (or **allosteric effector**) of haemoglobin.

BPG is synthesized in red blood cells from 1,3-bisphosphoglycerate, an intermediate in the glycolysis pathway (see Chapter 13). The normal molar concentration of BPG is roughly equivalent to that of tetrameric haemoglobin. The higher the concentration of BPG, the more the deoxygenated form is favoured. This constitutes a regulatory system – if oxygen tension in the tissues is low, synthesis of more BPG in the red blood cells favours increased unloading of oxygen. Acclimatization at high altitudes involves, in part, the establishment of higher BPG levels in red blood cells. It is to be noted that while BPG causes greater delivery of oxygen to the tissues, the decreased oxygen affinity has little effect on the degree of oxygenation in the lungs.

There is yet another refinement of the BPG-based regulatory system that illustrates how small changes to proteins

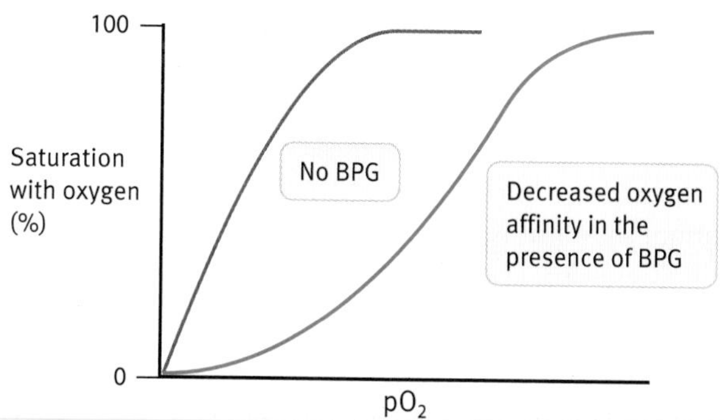

Fig. 4.28 Oxygen saturation curves for haemoglobin, illustrating the effect of 2,3-bisphosphoglycerate (BPG).

can have major physiological effects. For a mother to deliver oxygen to a foetus, it is necessary for the foetal haemoglobin to extract oxygen from the maternal oxyhaemoglobin across the placenta. This requires the foetal haemoglobin to have a higher oxygen affinity than that of the maternal carrier. This is achieved by a foetal haemoglobin subunit (called γ) replacing the adult β chains. Each γ chain lacks one of the positive charges of the β subunit. The missing charges are those that, in adult haemoglobin, line the cavity into which BPG fits; therefore foetal haemoglobin has two fewer ionic groups to bind BPG, the latter therefore being held less tightly. BPG is therefore less efficient in lowering the oxygen affinity, giving foetal haemoglobin a higher O_2 affinity than that of maternal haemoglobin. Thus, the maternal haemoglobin readily transfers its oxygen load to the foetus.

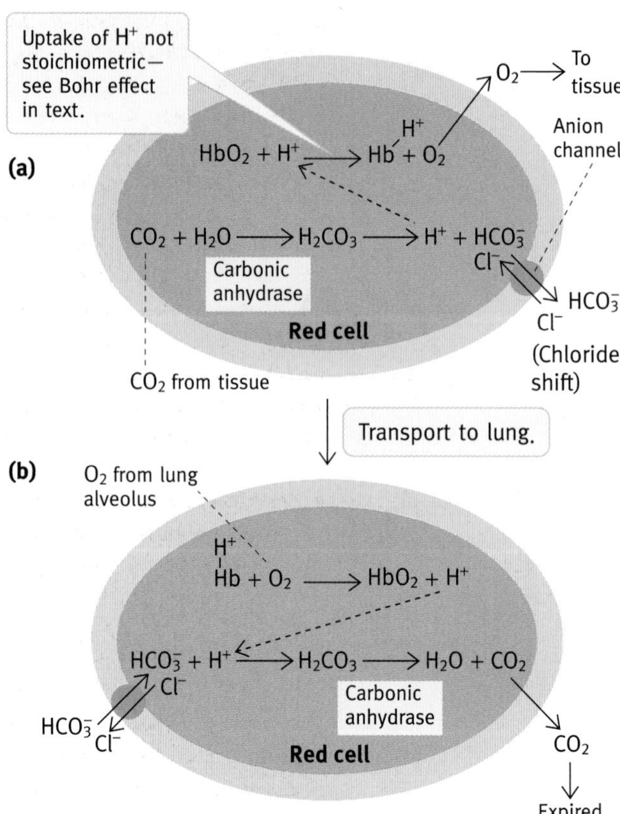

Fig. 4.29 Transport of CO_2 in the blood. (a) Reactions in the tissue capillaries. (b) Reactions in the lungs. The diagrams omit transport of CO_2 as carbamino groups of haemoglobin.

Effect of pH on oxygen binding to haemoglobin

Haemoglobin in the deoxygenated state has a higher binding affinity for protons than has oxyhaemoglobin. Recall that the pK_a of a dissociable group can be altered somewhat by its immediate chemical environment. The T→R conformational change on oxygenation reduces the pK_as of certain histidine side chains and of the terminal amino groups of the α chains, causing them to become deprotonated. Put in another way, the oxygenated form of haemoglobin is a stronger acid than is the deoxygenated form, resulting in dissociation of protons from the molecule when oxygen binds, a phenomenon known as the Bohr effect:

$$(1)\ Hb + 4O_2 \rightleftharpoons Hb(O_2)_4 + (H^+)_n$$

(where n is somewhere around 2; the number depends on a complex set of parameters).

Role of pH changes in oxygen and CO_2 transport

The Bohr effect, described in equation (1), has important physiological repercussions. In the tissues (Fig. 4.29(a)), CO_2 is produced and must be transported to the lungs. It enters the red blood cell where the enzyme **carbonic anhydrase** converts it to H_2CO_3, which dissociates into the bicarbonate ion and a proton:

$$(2)\ CO_2 + H_2O \rightleftharpoons H_2CO_3 \rightleftharpoons H^+ + HCO_3^-$$

The increase in proton concentration will drive the equilibrium shown in equation (1), to the left, causing the HbO_2 to unload its oxygen in the red cells, the effect thus being in harmony with physiological needs.

The HCO_3^- in the red cells passively moves out via an anion channel down the concentration gradient into the serum. The HCO_3^- movement is not accompanied by H^+ movement because there is no channel allowing passage of protons across the membrane of the red cell. To electrically balance the exit of HCO_3^-, Cl^- moves into the cell via the same anion channel. The dual movement is known as the chloride shift.

The HCO_3^- travels in solution in the serum of the venous blood, back to the lungs. Here (Fig. 4.29(b)), the proton concentration changes again help to achieve physiologically useful results. The release of protons from haemoglobin on oxygenation produces H_2CO_3 from HCO_3^- by a simple equilibrium effect:

$$(3)\ HCO_3^- + H^+ \rightleftharpoons H_2CO_3$$

H_2CO_3 (not HCO_3^-) is the substrate of the enzyme carbonic anhydrase, so this permits carbonic anhydrase to form CO_2 (reaction (4)) which is expired:

$$(4)\ H_2CO_3 \rightleftharpoons H_2O + CO_2$$

The decrease of HCO_3^- in the red blood cell causes HCO_3^- in the serum to enter the cell, down the concentration gradient, and Cl^- exits, that is, a reverse chloride shift occurs in the lungs, resulting in CO_2 expiration.

A small amount of CO_2 is transported in simple solution in the blood, but by far the greatest amount (about 75%) is transported as HCO_3^-. Additionally, about 10–15% of the CO_2 is bound to the haemoglobin itself, as CO_2 chemically and spontaneously reacts with uncharged $-NH_2$ groups of the globin to form carbamino groups:

$$(5)\ RNH_2 + CO_2 \rightleftharpoons RNHCOOH \rightleftharpoons RNHCOO^- + H^+$$

The RNH_2 groups available are mainly the terminal amino groups – lysyl and arginine side chains have too high a pK_a to be significantly uncharged.

pH buffering in the blood

(The principles of buffering are described in Chapter 3).

From the above, it is clear that there are major changes in the hydrogen ion concentration of blood associated with CO_2 and oxygen transport. When acid is produced in the tissues, the buffering power of HCO_3^-, phosphates, and haemoglobin itself is important in maintaining a physiological pH. The Bohr effect, described in equation (1), also buffers. When oxygen is unloaded, haemoglobin takes up protons. This carries roughly half of the H^+ ions generated by CO_2 in the tissues, and this again helps to prevent a drop in the pH of the red blood cell to unphysiological levels.

BOX 4.3 **Sickle cell disease and thalassaemias**

Sickle cell disease illustrates how a single amino acid change in a protein can have a profound effect. Working out the cause of the condition gave rise to the concept of a molecular disease. In the β chains of the normal human haemoglobin tetramer (haemoglobin A), amino acid number 6 is glutamic acid, whose side chain is negatively charged and highly hydrophilic. In the haemoglobin of sickle cell patients (haemoglobin S), this glutamic acid is replaced by a hydrophobic valine residue. The change requires only a single base mutation from T to A in the DNA coding for this particular glutamic acid residue of haemoglobin. The hydrophobic valine on the haemoglobin S binds to a hydrophobic pocket on another haemoglobin tetramer and so on, resulting in the formation of long rigid rods. In oxygenated haemoglobin, because of its different conformation, the hydrophobic pocket is not exposed and the haemoglobin tetramers do not bind to each other. In deoxygenated blood of people with sickle cell disease long deoxyhaemoglobin rods build up and distort the normal biconcave red blood cells into sickle-shapes that tend to block capillaries, causing tissue damage. The abnormal red cells also break up, causing anaemia.

Despite being fatal if untreated, sickle cell disease is prevalent in geographical areas where malaria is or was common and also in the descendants of people who migrated from those areas, such as African Americans. This high incidence can be explained by positive selection of the mutated gene. The haemoglobin abnormality is unfavourable for the development of the malarial parasite, even in individuals who carry one normal and one mutated copy of the gene and therefore do not suffer from sickling of their red blood cells. This so-called **heterozygote advantage** protects unaffected genetic carriers against death from malaria and the mutation is therefore preserved in the population by natural selection.

The thalassaemias are a family of genetic diseases caused by different mutations that affect haemoglobin production. In α and β thalassaemias there is a deficiency of the corresponding subunits of haemoglobins. The disease is prevalent around the Mediterranean Sea (*thalassa* is the Greek for sea) and as with sickle cell disease the mutations appear to give some protection against malaria, thus accounting for its prevalence in malarial areas. The β thalassaemias occur in varying degrees of severity.

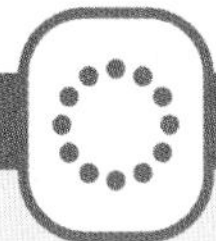

Summary

Proteins are made up of one or more polypeptide chains (amino acids linked by peptide bonds) constructed from 20 species of amino acids. The length and sequence of each polypeptide is specified by its gene. The peptide bond is the CO—NH linkage between two amino acids. The 20 different amino acids are of differing sizes and degrees of hydrophobicity, hydrophilicity, and electrical charge. They present the possibility of a vast variety of different proteins.

The primary structure is the linear amino acid sequence. Secondary structure involves folding of the polypeptide backbone. The main secondary structure motifs are the α helix and the β-pleated sheet, which are stabilized by hydrogen bonding. Proteins are built up of various combinations of these structures linked by connecting loops.

Tertiary structure involves the further folding of the secondary structure motifs into the three-dimensional form of the protein. The folded structure is determined by the amino acid sequence of the polypeptide chains so that all proteins have unique structures, but common folding patterns are recognizable. The association of protein molecules to form multi-subunit proteins produces the quaternary structure.

Larger proteins contain domains, which are sections of polypeptide chains folded into three-dimensional structures that can function independently if experimentally separated from the rest of the chain. Proteins have evolved via mutations and also by domain shuffling, in which the coding regions of genes have been reassembled to code for new combinations of domains to produce new proteins.

In globular proteins, hydrophobic residues are mainly inside the molecule and hydrophilic ones outside in contact with water. Membrane proteins have hydrophobic amino acids on the outside in contact with the hydrophobic membrane interior.

Some proteins have groups such as ions or nonprotein molecules firmly attached to them. In enzymes, such groups are often required for activity and are termed prosthetic groups: the protein alone is termed an apoenzyme and the active complex a holoenzyme. Other proteins have carbohydrate attachments to one of the amino acids serine or threonine (via the —OH group of the side chain) or to asparagine (via the —N atom of the side chain). These glycoproteins are often secreted. Depending on its nature the carbohydrate group may act as a marker, either protecting the protein or labelling it for degradation.

Extracellular matrix proteins are mainly fibrous rather than globular, and include collagens, which confer toughness on tendons, cartilage, and bone. Collagen has a strong triple superhelix structure, which is stabilized by hydrogen bonding of post-translationally modified proline residues. Elastin is the elastic protein of lungs. Proteoglycans, which contain protein and carbohydrate, form the ground substance of loose connective tissues such as the flexible layer underlying skin. The extracellular matrix also contains the cell adhesion protein, fibronectin, which binds proteoglycans and collagen and connects them to the interior cell cytoskeleton via integrin transmembrane proteins. Integrins also transmit signals from the exterior to the interior of the cell.

Myoglobin and haemoglobin are intensively studied examples that relate protein structure to function. Myoglobin is a monomeric protein that acts as an oxygen reserve in muscles. It has a haem prosthetic group to which the oxygen attaches. It surrenders its oxygen when the muscle has a low oxygen tension such as occurs in vigorous exercise.

Haemoglobin is the oxygen carrier of red blood cells. It is a tetramer of two α chains and two β chains, each of which resembles myoglobin in structure; each subunit has a haem prosthetic group.

Haemoglobin undergoes conformational changes during the attachment and release of oxygen, which are essential for its function. It is amazingly adapted to optimally perform its complex physiological functions, which require it to pick up the maximum amount of oxygen in the lungs and surrender as much as possible in the capillaries.

Haemoglobin is an allosteric protein, which changes its shape on ligand binding. Large numbers of allosteric proteins exist, many of them enzymes. The conformation change caused by binding of oxygen molecules to haemoglobin makes it easier for successive oxygens to attach. This gives a sigmoid oxygen saturation curve as compared with the hyperbolic curve of myoglobin. The concerted and sequential models can each account for aspects of haemoglobin's cooperative binding to oxygen. The true mechanism may lie somewhere between the two.

2,3-Bisphosphoglycerate (BPG) plays an important physiological role in oxygen transport: it binds allosterically and reduces the affinity of haemoglobin for oxygen, thus increasing release of oxygen in the tissues. Decreased pH (i.e. increased proton concentration) caused by production of CO_2 in the tissues also increases the release of oxygen (the Bohr effect). Most CO_2 is carried to the lungs as HCO_3^- in serum, but some CO_2 binds to amino groups of haemoglobin for transport.

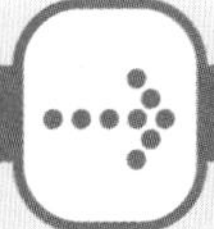

Further reading

To access the further reading, please scan the QR code image or go to http://global.oup.com/uk/orc/biosciences/molbiol/snape_biochemistry/student/reading/ch04/

Problems

1 What is the primary structure of a protein?

2 What is meant by denaturation of a protein?

3 Write down the structure and name of an amino acid with each of the following side chains:
(a) H
(b) aliphatic hydrophobic
(c) aromatic hydrophobic
(d) acidic
(e) basic.

4 Give the approximate pK_a values of:
(a) acidic amino acid side chains
(b) basic amino acid side chains
(c) the histidine side chain.

5 Which amino acids are the major determinants of the charge of a polypeptide chain containing all 20 amino acids?

6 What are the four levels of protein structure?

7 What is the peculiar structural feature of elastin that gives it its elastic properties?

8 In collagen, sections of the polypeptide chains have glycine as every third amino acid residue. What is the significance of this?

9 The activities of proteins in general are readily destroyed by mild heat. The peptide bond is quite heat stable.
(a) Why are proteins so inactivated by mild heat?
(b) A few proteins, particularly extracellular ones, are more stable than usual. What structural feature is probably responsible for this?

10 What is a protein domain and why are they of interest?

11 In a globular protein where would you statistically expect to find most of the residues of (a) phenylalanine, (b) aspartic acid, (c) arginine, and (d) isoleucine?

12 Explain why the molecular structure of proteoglycans is very suitable for creating mucins and gels with a very high water content.

13 The α helix and β-sheet structures are prevalent in proteins. What is the common feature that they have that makes them suitable for this role?

14 Compare the oxygen dissociation curves of myoglobin and haemoglobin. Discuss the rationale for the differences.

15 Binding of an oxygen molecule to haemoglobin causes a conformational change in the protein. Describe the mechanism of this.

16 Explain how the foetus is able to oxygenate its haemoglobin from the maternal carrier.

17 Explain the significance of the chloride shift in red blood cells.

18 In the context of protein tertiary structures it has been commented that when nature is onto a good thing it sticks with it. Discuss this briefly.

19 Sickle cell disease is so prevalent in certain areas of the world that there must be an explanation for its prevalence. Explain the nature of the disease and discuss why it is so prevalent in these areas.

20 Why does smoking cause emphysema?

21 The peptide bond is said to be planar. Explain briefly what is meant by this, give the structural basis of it and state the consequences of it.

22 Which of the following is out of place? Isoleucine, alanine, phenylalanine, proline, leucine.

Chapter 5

Methods in protein investigation

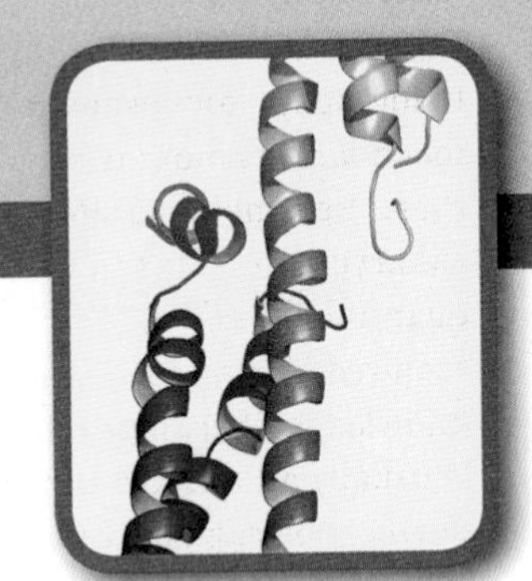

Methods for investigating proteins have undergone a revolution in the past few years, though the more traditional methods, especially in protein separation, are still important and will be described in this chapter. The main changes have resulted from the coming together of several technologies. One of the major developments has been the application of **mass spectrometry** to proteins, which allows investigations to be carried out with speed and sensitivity. Mass spectrometry has been used for a long time in organic chemistry but was initially applicable only to volatile molecules. New methods developed from the 1980s onwards have allowed its application to proteins, with spectacularly successful results. A second major change has resulted from DNA research. The amino acid sequence of proteins is coded by triplets of nucleotides in the genome. Since we know which triplets specify each amino acid, it is possible to deduce the amino acid sequence for a protein if its gene sequence is known. The sequencing of the entire human genome and the genomes of many other organisns mean that the amino acid sequences for any human protein and for thousands of others can be deduced. This means that limited information obtained about a protein by experimentation can be rapidly matched up with full sequence information from genomic databases.

The accumulated information on protein and DNA sequences has thus increased explosively and would be overwhelming had it not been for the development of protein and DNA databases into which the information is deposited as it is obtained. The databases established by international cooperation are freely available, as is computer software that permits the retrieval and analysis of the information in many different ways. This area of databases and their computer-assisted use is referred to as **bioinformatics** and has become a major tool in protein research. The direct linking up of technologies such as mass spectrometry and DNA sequencing with bioinformatics, with synergistic effects on research, has resulted in what is sometimes called the '**omics**' revolution in which large numbers of proteins and genes can be studied in parallel, the two areas being known as **proteomics** and **genomics** respectively.

We will now describe both some 'traditional' methods of studying proteins and some more recent advances. The complementary studies on DNA are dealt with in Chapter 28.

Purification of proteins

Most biochemical and molecular biology investigations require proteins to be isolated and, except for extracellular proteins, this means initially breaking open the cells – using mechanical homogenization, or enzymic methods to break down the robust walls of bacterial cells. A eukaryotic cell extract contains organelles and large protein complexes and if your protein is soluble the first step is to remove these unwanted items by centrifugation. Conversely, if your protein is present in an organelle or membrane, it may be purified by differential centrifugation followed by methods to solubilize the protein, such as the use of detergents for membrane proteins.

Cells contain thousands of different proteins and the individual protein of interest will usually constitute a very small fraction of the total. Typical protein purification protocols involve multiple steps to gradually increase the proportion of a protein extract that consists of the desired protein. An assay is needed to follow your protein through the process and measure its increasing **specific activity** (the proportion of total protein it constitutes). If the protein is an enzyme, it can often be assayed by its catalytic activity. For other proteins some other specific property, such as colour for haem proteins or an immunological assay involving recognition by an antibody, may be used. Total protein can be measured through absorbance of UV light, at a wavelength of 280 nm, by aromatic acids.

The method of purification adopted depends on the amount of the pure protein you need. Mass spectrometry can handle minute, nanogram amounts, which might be obtained in a day or so by electrophoretic methods. There are situations, however, in which relatively large amounts of the pure protein are necessary. Structural determination by X-ray crystallography, for example, requires protein crystals, and much larger amounts of pure protein are needed to produce these. Recombinant DNA technology is often used to express eukaryotic proteins in bacteria, which can be grown in large quantities (see Chapter 28), but subsequent purification is still required. A purification protocol may be developed based on knowledge of the desired protein's characteristics, but there is usually an element of trial and error involved. Large-scale purification can take a long time, so precautions such as working at low temperature are needed to prevent denaturation of the protein.

The methods that are used to begin purification generally divide the crude protein extract into **fractions**, each of which contains a subset of the original protein mix. The fraction(s) containing the protein of interest are identified using a suitable assay and then subjected to further purification steps. A common preliminary fractionation method is **precipitation** of proteins by adding increasing amounts of a highly soluble salt like ammonium sulphate to the crude extract. Different proteins precipitate at different salt concentrations. Precipitated fractions are collected by centrifugation, redissolved and assayed for the wanted protein. After such a procedure, **dialysis** removes salt and puts the proteins into suitable buffers for further purification. Dialysis involves putting the solution of proteins into a semi-permeable dialysis bag immersed in a large volume of the desired buffer. Salts and other small molecules freely diffuse out of the bag, which retains the protein molecules.

More sophisticated methods then become practical. Proteins may be separated on the basis of their differing sizes, electrical charges or chemical binding properties. Column chromatography is commonly used for these, even in quite large-scale purification.

Column chromatography

The term chromatography is widely used for analytical and preparative techniques in which components of a mixture are separated on the basis of partition between two 'phases'. A solid matrix acts as the stationary phase and the mobile phase, usually liquid rather than gas when separating proteins, flows through it. In column chromatography, the stationary matrix is packed into a column, the sample to be treated is applied to the top and washed through by the mobile phase. Fractions of the eluate are collected as they emerge from the column and are analysed for the protein of interest.

A common method is to separate on the basis of molecular size using **size exclusion** (or **molecular exclusion**) chromatography, also known as **gel filtration.** A gel filtration column is packed with microscopic beads made of an inert polymer such as agarose or Sephadex (both polysaccharides). The beads, also known as resin or **gel** (hence the commonly used name for this technique), are available commercially and contain pores of known size running through them. The protein solution is applied to the top of the column and then washed through with an appropriate buffer solution. Protein molecules too large to enter the pores of the beads flow unimpeded around the beads but those small enough to enter the beads are retarded (Fig. 5.1). Beads with small pore sizes can be used simply for separating proteins from smaller molecules and salts, as all proteins are excluded from the beads. Beads with larger pore sizes can be used to achieve fractionation of proteins: large proteins are excluded from the beads and take the most rapid route through the column, while smaller proteins enter the pores and are somewhat retarded, and small molecules are caught up in the network of pores within the beads and greatly retarded. Individual components of the starting extract are thus eluted from the column in order of decreasing molecular weight. This method will generally not achieve purification of a single protein from a complex mixture, but after assaying for the desired protein the fractions with the highest specific activity can be selected for further purification. Gel filtration may also be used as a final purification step to remove unwanted salts from the buffer in which the protein has been purified.

Additional column chromatography methods that depend on properties of individual proteins more specific than just size may be used to achieve higher levels of purification. **Ion exchange chromatography** is based on charge. Here the column is packed with beads to which positively or negatively charged groups are covalently attached. The mix of proteins is then loaded onto the column in a buffer that maintains a stable pH. If the column matrix is positively charged then proteins that have an overall negative charge at the buffer pH will be retained on the column while uncharged or positively charged proteins are washed through. The protein of interest can then be recovered from the column by disrupting its interaction with the matrix, for example by changing the pH in order to alter the charge on the protein, or by increasing the salt concentration of the buffer so that excess ions compete for binding to the column.

Another powerful variation is to use **affinity chromatography.** Suppose the protein of interest is known to bind specifically and strongly to compound X; if the inert column matrix has X attached to it by chemical means, then the protein you want will be retarded by binding to the column while other proteins in the mix are washed through. The desired protein can then be eluted, perhaps by disrupting its interaction with substance X using a buffer of different ionic strength, or by washing through a solution containing a high concentration of free substance X. In the latter method, the free substance X competes with X attached to the matrix for binding to the specific protein, which is thus washed through the column and recovered.

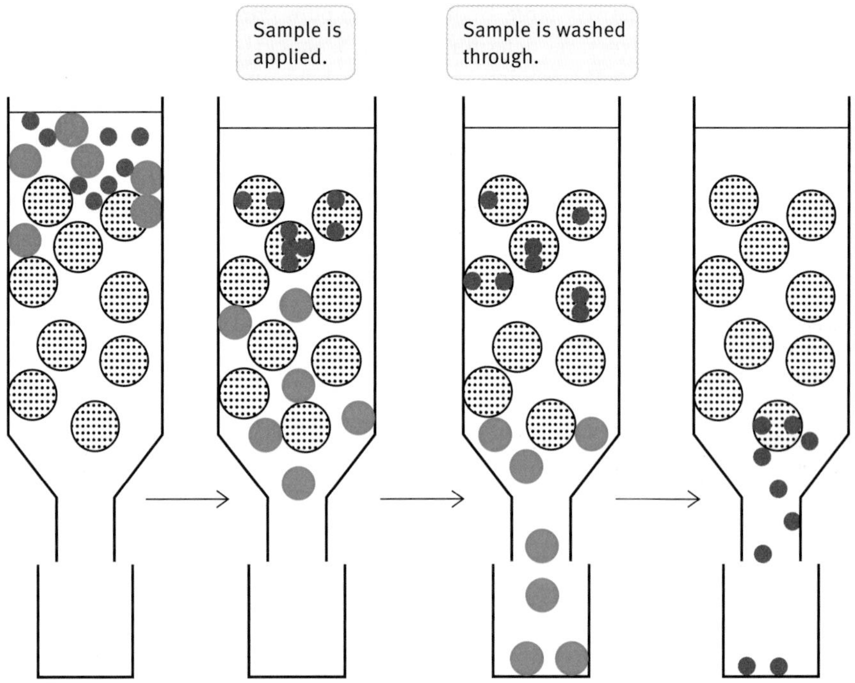

Fig. 5.1 Protein separation by gel filtration. The column is packed with beads of a gel that has pores of defined size. A mixture of large and small protein molecules is allowed to enter the column. The green molecules are too large to enter the beads but the small ones can do so. The column is washed with a suitable buffer to move the molecules down the column. The blue molecules are retarded behind the green ones because they enter the beads and so emerge from the column later than the green ones. Pure samples of each can be collected as separate fractions.

Affinity chromatography can be an extremely powerful purification technique. It makes use of noncovalent interactions of proteins with their natural molecular ligands, for example enzymes with substrates or inhibitors, antibodies with antigens and DNA binding-proteins with specific nucleotide sequences. The specificity of these interactions means that a protein of interest may be purified away from others in a single step. However, creation of the column matrix with a specific ligand attached may be technically challenging and expensive. A variation on this technique that avoids these problems can be used where a wanted protein is produced in *Escherichia coli (E. coli)* by the recombinant DNA procedures described in Chapter 28. Here an easily selectable 'tag' can be engineered into the protein. For example, a group of six histidine residues added to the end of a protein has a very high affinity for nickel. This makes it possible to isolate the pure protein from the *E. coli* cell homogenate using a column on which Ni^{2+} ions are immobilized on the matrix. Following removal of unwanted proteins, the 'His-tagged' protein can be recovered from the column by washing with a buffer containing excess histidine analogue, which out-competes the protein for binding to the column, or by altering the pH, which alters the charge on the histidine residues and hence disrupts their interaction with nickel. In many cases, the His-tag does not interfere with subsequent use of the purified protein, but if this is a concern the histidine residues may be removed by controlled use of a peptidase enzyme.

The speed and efficiency of column chromatography separations may be increased by using **high performance liquid chromatography** (HPLC), also sometimes known as **high-pressure liquid chromatography**. Here the column material is packed in a steel tube and the liquids forced through at high pressure. The method allows a more finely divided immobile phase material to be used, which increases the surface area and efficiency of separations.

SDS polyacrylamide gel electrophoresis (SDS-PAGE)

Column chromatography methods are generally used for relatively large-scale protein purification, for instance to prepare samples for determining the three-dimensional structure of the protein. For analytical separation when only small amounts of protein, in the microgram range, are involved, **gel electrophoresis** is often used. Electrophoretic analysis of column fractions is often utilized to follow the progress of a purification scheme, as illustrated in Figure 5.3.

The principle of gel electrophoresis is that molecules with a net charge migrating along an electrical field to the opposite pole are sorted according to size, shape and charge. As in column chromatography a 'gel' is used that consists of an inert polymer, usually **polyacrylamide**. In this case the gel is constituted not as beads but as a slab set between two glass plates (Fig. 5.2). The gel contains pores of controlled size and, as the proteins migrate in the electrophoresis buffer through the pores of the gel, larger proteins are impeded by the gel while smaller proteins or those with a more compact shape move faster. This mode of separation by varying degrees of 'entanglement' in the gel is known as **molecular sieving**.

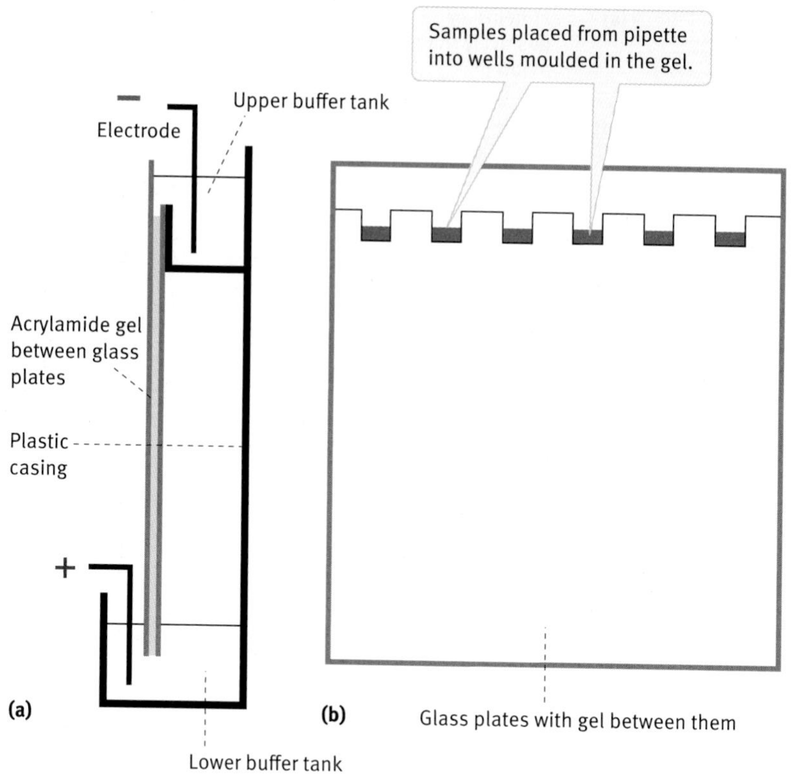

Fig. 5.2 View of polyacrylamide gel electrophoresis apparatus **(a)** and a front view of the gel between the plates **(b)**. The samples are injected into the wells through the buffer solution with a syringe or pipette. To prevent mixing of the sample in the wells with the buffer, the samples contain glycerol to make them dense. A blue dye makes it easy to see what is happening in the loading.

It is often desirable to negate the effects of three-dimensional shape and native charge of proteins and achieve separation purely on the basis of size. Thus, the most commonly used variant of polyacrylamide gel electrophoresis (PAGE) is to use a **denaturing gel** which contains a detergent, **sodium dodecylsulphate (SDS)**. SDS has a hydrophobic tail and a negatively charged sulphate group ($CH_3(CH_2)_{10}CH_2OSO_3^-Na^+$) and the protein sample is dissolved in a solution of SDS, which denatures it. The SDS inserts its hydrophobic tail into the proteins, which are thus covered with negative charges. Large amounts of the SDS attach, roughly one molecule per two amino acid residues, swamping whatever charge the native protein had so that all proteins have a strong negative charge proportionate to their size. Disulphide bonds are disrupted by including a reducing agent. The detergent also solubilizes water-insoluble hydrophobic membrane proteins so that these can be studied on gels, a major advantage of the technique. In the denaturing gel the SDS-coated proteins move towards the anode, but as the size-charge ratio is the same for all of them they are mainly separated by the **molecular sieving** effect mentioned above; small proteins moving fastest. The SDS coating also maintains the denatured proteins in a linear conformation, so that shape differences are not a factor in the separation.

In practical terms the apparatus used for PAGE is simple (Fig. 5.2). The plates with the gel between them are held vertically with the top and bottom edges of the gel exposed to tanks of the SDS buffer solution and a voltage is applied across the two tanks. When the gel is cast between the glass plates, before polymerization, a plastic 'comb', is inserted into the top edge so that, when this is removed following solidification, the gel has separate wells into which different samples are introduced. The gel is set up in the apparatus with buffer in the tanks and the samples are applied with a pipette into the wells under the buffer. The samples contain a dense substance such as glycerol to make them settle into the wells without mixing with the buffer. A blue dye in the sample allows loading to be observed.

For further analysis the separated proteins are commonly visualized as 'bands' in the gel by staining, for example with a dye, Coomassie blue. The denaturing gel can be used as a means to estimate the molecular weight of a protein by comparison of its migration distance with that of a set of pure proteins of known molecular weight, known as 'markers'. The width and intensity of a band can give an indication of the amount of protein present in a sample, although quantitation by this method is not precise. A typical result is shown in Fig. 5.3, which also illustrates the purification of a protein achieved by selective elution from an ion exchange column.

Nondenaturing polyacrylamide gel electrophoresis

Nondenaturing gels do not use SDS, so that proteins are not denatured. In this case the separation is based partly on

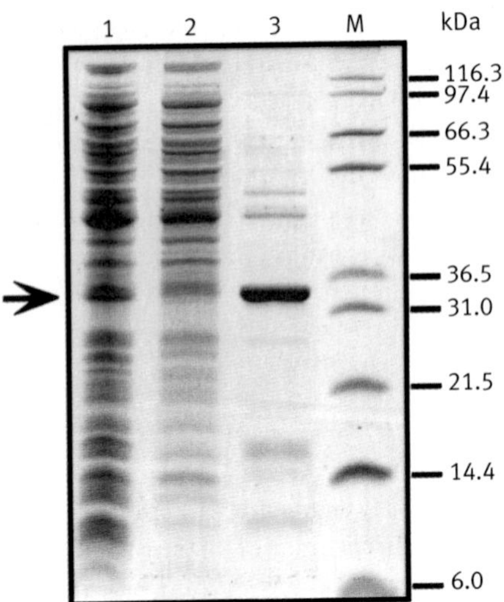

Fig. 5.3 SDS polyacrylamide gel electrophoresis. The figure also shows the purification of a protein from *Escherichia coli*. Samples from each stage of the purification were reduced and denatured, then electrophoresed on a 10% polyacrylamide gel in the presence of sodium dodecylsulphate (SDS). Separated proteins were visualized by staining with Coomassie brilliant blue. Lane 1 shows proteins extracted from *E. coli*; the arrow indicates the protein of interest in this experiment. The mixture was chromatographed on a column of cation-exchange resin. Lane 2 shows the proteins that did not bind to the resin and were recovered in the flowthrough. Lane 3 shows the protein of interest in a fraction eluted from the column. Lane M (for markers) shows proteins of known molecular weights. The protein of interest had a molecular weight of 35.3 kDa. (kDa is the molecular weight unit, kiloDaltons). Photograph courtesy Dr Anne Chapman-Smith, Department of Molecular Biosciences, University of Adelaide, Australia.

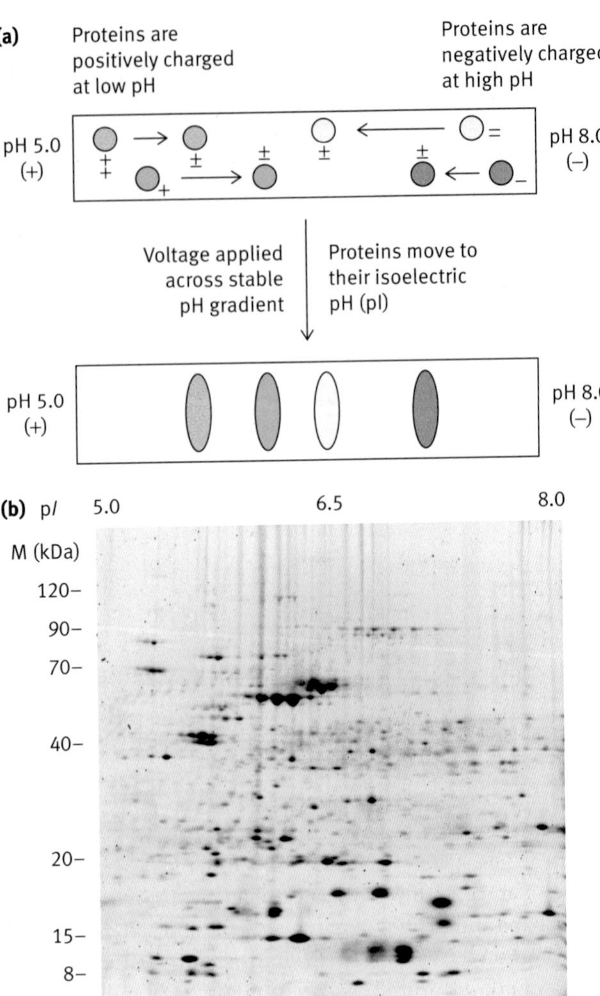

Fig. 5.4 **(a)** Principle of isoelectric focusing. A stable pH gradient is established in a narrow tube of polyacrylamide gel. Proteins subjected to electrophoresis in the gel move according to their charge. The charge on amino acid side chains alters with varying pH and when each protein reaches the pH at which its overall charge is zero (its pI) it stops moving. The figure shows four proteins, but often a complex mix is analysed of which several proteins will have the same pI. For two-dimensional electrophoresis the narrow tube of gel is then subjected to SDS-PAGE with the electric field applied at right angles to that used for the first separation, giving further separation by size. **(b)** Representative two-dimensional electrophoresis gel of whole-cell proteins from the gut pathogen *Helicobacter pylori*. Proteins were separated using immobilized pH gradient isoelectric focusing in the first dimension and a second-dimension slab gel containing SDS buffer. Gels were stained with fluorescent Sypro Ruby. pI is the isoelectric point (see text). M is the molecular weight. Image courtesy Dr Stuart Cordwell, Australian Proteome Analysis Facility, Sydney, Australia.

the net charge and varying electrophoretic mobilities of the proteins and partly on size. Positively charged particles will go in the opposite direction to negatively charged particles. Although less widely used than SDS-PAGE, nondenaturing gels have the advantage that separated proteins can be tested for a biological activity such as enzyme activity.

Isoelectric focusing

Another electrophoresis variant is **isoelectric focusing**. The **isoelectric point (pI)** of a molecule with different ionizing groups (–COOH, $-NH_2$ and certain amino acid side chains in the case of proteins) is the pH at which the positive and negative charges exactly balance so the net charge on the molecule is zero. The principle is illustrated in Figure 5.4(a). The gel first has a stable pH gradient established across it using a commercially available mixture of many small polymeric **ampholytes** (amphoteric electrolytes i.e. molecules that can act as either acid or base). Native proteins migrate electrophoretically in such a gel towards the point where the pH is such that the net charge on the protein is zero (its isoelectric point). The molecule then remains at that point: if it moves within the gel the difference in pH alters its net charge and the electric field will cause it to migrate back again to its pI.

Two-dimensional gel electrophoresis

For maximum resolving power that can be used to separate a large number of proteins from a complex mixture a combination of isoelectric focusing and SDS–PAGE is used. This technique is called **two-dimensional (2-D) gel electrophoresis.** A sample is first separated by isoelectric focusing on a gel strip with an established pH gradient (available commercially) and the 'strip' is transferred to an SDS gel and electrophoresed at right angles to the first direction. Thus, the first separation is by pI and the second by size. A crude cell extract so treated can give rise to many hundreds of separate protein spots (Fig. 5.4(b)). The result looks complex, but the method is powerful because it shows up differences in the whole collection of proteins found in a cell or tissue type, for instance by comparing the patterns produced by a normal versus a cancer cell population. Modern techniques such as mass spectrometry can be used to identify an individual protein even from the small amount available in a spot on a 2-D gel.

Immunological detection of proteins

Antibodies against specific proteins are produced in the blood of animals, or in the laboratory. This process is described in Chapter 33. The antibody has exquisite specificity for the protein it was raised against and to which it tightly binds. This property can be used to detect proteins in minute amounts in gels and in solution. For detection of a particular protein among a complex mixture that has been separated using gel electrophoresis, a method known as **Western blotting** is used (named by analogy to the Southern blotting method for detecting DNA, described in Chapter 28). The proteins on a gel are transferred to a plastic sheet, which is then soaked in a solution containing a specific antibody. Binding of the antibody and hence the location of its antigen, the protein of interest, can be detected directly if the antibody has been covalently coupled to a radioactive or fluorescent group. More frequently detection is indirect, using a secondary antibody that recognizes the first. The secondary antibody is labelled, or is coupled to an enzyme that allows it to be detected by generating a coloured or fluorescent product. A major advantage of indirect detection is that suitably coupled secondary antibodies with wide applications are commercially available; for example if the antibody specific for the protein of interest was raised in a rabbit, a second antibody that recognizes all rabbit antibodies could be used for detection.

Another use of antibodies in protein chemistry is **ELISA** (enzyme-linked immunosorbent assay), which is widely used in clinical biochemistry (see Chapter 33 for a description).

The principles of mass spectrometry

Mass spectrometry (MS) has become crucial in so many aspects of protein investigation that it will be best if we describe the various related methods in some detail before we get to its applications. Its usefulness is because of its speed and extreme sensitivity. The amount of protein in a single spot on a 2-D gel is often sufficient for an analysis, the results of which permit rapid searching of protein databases for entries that match the 'unknown'. Computer software enables this to be done using data directly fed in from mass spectrometry. The main uses of the technique are as follows (you may be unfamiliar with some of the terms used but they are explained in due course):

- A protein can be identified by **peptide mass analysis** (fingerprinting) or by **limited sequence analysis** followed by comparison to a database.
- The molecular weight of a protein can be determined rapidly with an accuracy of 1 Da in 10,000 Da.
- A protein can be **sequenced**, partly or fully.
- Post-translational modifications of proteins can be investigated.

Mass spectrometers consist of three principal components

These are:

- an ion source that converts the protein or peptide into charged particles
- one or more mass analysers that separate the ions thus produced on the basis of their mass-to-charge ratios (m/z)
- an ion detector.

Ions are separated in the analyser(s), and are collected in the detector. A computer then displays a spectrum of ion intensity *versus* mass-to-charge ratio. In the simplest case, ions are singly charged ($z=1$) so that m/z values give molecular ion masses. Spectra from multiply charged ions are often produced from proteins and the data for their m/z values can be converted into mass values from inspection, or automatically by computer. To give a very simple example, a peptide with a molecular weight of 6,000 Da might give a spectrum showing m/z values of 6,000 (representing the peptide carrying a single positive charge), 3,000 (peptide carrying a charge of +2) and 2,000 (peptide carrying a charge of +3). These multiple values obtained for a single peptide ultimately increase the accuracy of the mass measurement.

Ionization methods for protein and peptide mass spectrometry

The mass spectrometer works with ions in the gas phase. Proteins and peptides are large nonvolatile molecules and for many years this presented a barrier to analysing them using MS. The breakthrough in the 1980s that opened up the field was development of two 'soft' methods for generating gas phase ions from proteins ('soft' because the proteins are ionized without being degraded). One is **matrix-assisted laser-desorption ionization (MALDI)**. Here the protein material to be analysed (the analyte) is mixed with a UV-light-absorbing chemical matrix and deposited on a solid target surface. The target is placed in the mass spectrometer and pulsed with UV laser light, which is absorbed by the matrix causing an explosive ejection of matrix molecules. These carry with them vaporized protein or peptide molecules, usually singly charged, by accepting ions such as H^+ from the matrix.

The second method is the **electrospray ionization (ESI)** technique. In this, the protein analyte is in solution, which is raised to a high electrical potential (4 kV) and sprayed from a capillary. Fine droplets containing the peptide ions travel to the inlet of the mass analyser along a potential gradient as the solvent evaporates, leaving single peptide molecules suspended in a vacuum. ESI usually produces highly multiply charged peptide ions. An advantage of ESI over MALDI is that the analyte is treated in solution so it can be used for the direct ionization of protein fractions that have been pre-separated by HPLC. MALDI is better suited to analysis of individual protein 'spots' that are eluted from 2-D electrophoresis gels and mixed with the matrix material. Linking liquid chromatography to ESI is more amenable to automation than linking 2-D gel analysis to MALDI.

Types of mass analysers

This is a brief description of the different ways by which ions are separated to allow a spectrum of ions with different m/z values to be recorded.

Time of flight (TOF)

The TOF analyser is conceptually the simplest. Here the protein or peptide ions are propelled from the ion chamber by a high voltage applied to a grid through which the ions move into the flight tube. The flight tube has no electrical or magnetic field – the ions simply move passively to the detector. During the flight, the ions are separated solely on the basis of the m/z values. All the ions with the same z charge have the same kinetic energy, so that heavier ions move more slowly than light ones and take more time to reach the detector. A mass accuracy of 1 in 100,000 is commonly achieved.

Ion-traps and quadrupoles (Q)

Other types of analyser include ion-traps and quadrupoles. These mass analysers work by using voltage and radiofrequency fields. Ions are first retained within an ion-trap and are then sequentially ejected towards the detector, in order of increasing mass-to-charge ratios, by applying increments to the field strength. The quadrupole has four pencil-like steel rods arranged in parallel, creating a 'channel' that can be tuned to permit only ions of specific m/z values to pass through it. By tuning the quadrupole to different ranges of m/z values, and recording the numbers of ions that pass through at each step, a spectrum is generated. The mass accuracy of the spectrum is typically 1 in 10,000. A quadrupole may also be used to select a single peptide ion for further analysis, such as in peptide sequencing.

Types of mass spectrometers

The types of analysers that have been described are assembled in various combinations by manufacturers to produce different types of mass spectrometers designed for different applications.

Single-analyser mass spectrometers

Single-analyser mass spectrometers are the simplest types, which produce the spectra of peptides and proteins using either ESI or MALDI ionization methods. Examples are ESI-single-quadrupole and MALDI-TOF spectrometers. The MALDI-TOF (Fig. 5.5(a)), is now extremely important in proteomic studies because of its accuracy and relative simplicity of operation.

Tandem mass spectrometers

A tandem mass spectrometer (MS/MS) can be used to generate amino acid sequence data from a peptide. The description 'tandem' refers to the arrangement of two mass analysers in series separated by a central collision cell as shown in Fig. 5.5(b). The first analyser can be set by the computer to allow only a single peptide ion of a given m/z ratio to progress into the collision cell. Other ions are scattered to the sides of the first analyser. In the collision cell the selected ions collide with argon gas molecules and are fragmented. The fragment (daughter) ions move into the second mass analyser, which gives their m/z spectrum. More detail on the use of the data generated to determine the sequence of the peptide and localize post-translational modifications is given in the Applications of mass spectrometry section.

Applications of mass spectrometry

Molecular weight determination of proteins

The MS method determines the molecular weight of a protein in a minute or so with an accuracy of 1 Da in 10,000 Da or 100 ppm. Molecular weight determinations are important in industrial biotechnology for quality control of proteins produced by recombinant DNA technology (see Chapter 28) and also help to identify proteins. A single spot on a 2-D gel can give sufficient protein for this application.

Identification of proteins using mass spectrometry without sequencing

When studying proteins it is not always necessary to determine the amino acid sequence. A common application of MS that makes use of peptide mass measurements but not sequence is for analysing protein expression: that is determining which proteins are present in a particular cell or tissue type under certain conditions. We have already mentioned the comparison of proteins found in normal versus cancer cells using 2-D gel electrophoresis: if a spot on a 2-D gel is noted to appear specifically in cancer cells, it is obviously desirable to find out which protein that spot contains, and MS technology combined with computer analysis make this task relatively straighforward.

The simplest method to identify an 'unknown' protein is to see if the protein is already recorded on a protein database, in which case a lot more information about it may be readily available. This can be done for 2-D gel spots by using MS to carry out **peptide mass analysis** (also known as **peptide mass fingerprinting**). The protein is enzymically digested into peptides by trypsin. MALDI-TOF MS analyses the mixture and records the m/z values of individual peptide ions as peaks, as shown in Fig. 5.5(a). As discussed further in Chapter 6, trypsin is specific in cutting a polypeptide at certain amino acid residues (lysine and arginine), which means that it is possible to predict the pattern of tryptic peptides produced from a given protein for which the amino acid sequence is already known. Computer software compares the actual pattern obtained from the 'unknown' protein by MS, with the patterns predicted by theoretical (notional) trypsin 'digestion' of every protein in the database. Any protein that would produce a set of peptides on trypsin digestion corresponding to ones found in the analysis is reported. Modern protein databases contain thousands of amino acid sequences generated from genomic sequence data, so the chances of finding a match are very good, and the process is extremely rapid and sensitive; a few hundred femtomoles of protein are adequate. (One fmol equals 10^{-15} moles; a fmol of a protein of 100,000 molecular weight therefore equals 10^{-10} g.)

Identification of proteins by limited sequencing and database searching

This differs from peptide mass analysis in that it produces some sequence data with which to match the 'unknown' to a protein in databases; MS/MS is used. The method of protein sequencing is described in the next major section. Even small stretches of amino acid sequences from an unidentified protein are sufficient to match it with one in the database with a high degree of confidence. Partial but not exact sequence matches between the protein under investigation and database entries can be of interest because they may be obtained from homologues of the protein. This can lead to identification of new protein families or a new member of a known family.

Analysis of post-translational modification of proteins

Many proteins are covalently modified after synthesis, such as by the addition of glycosyl (carbohydrate) or phosphoryl groups. The mass spectrometer can measure modifications at the level of changes in the mass of the whole protein or a peptide fragment. Within a peptide the specific modified amino acid residue may be identified using MS/MS. Techniques are available that detect characteristic 'reporter' fragmentation products from specific modifications. Alternatively phosphoryl and glycosidic groups can be enzymically removed (enzyme kits for this are commercially available) and MS analysis before and after removal can both identify the modified peptides and give an indication of the type of modification.

Methods of sequencing protein

Classical methods

The original method pioneered by Fred Sanger of Cambridge, UK, was to cut up the protein into peptides and label the *N*-terminal amino acid of each with a detectable group, allowing its identification. This is very laborious and requires a lot of protein. It took Sanger years to be the first to sequence a small protein, insulin, which earned him his first Nobel Prize. The method is not used now. An advance was made by Edman, who used the principle of deducting and identifying amino acids one by one from the ends of peptides (the Edman degradation process). This allowed proteins to be automatically sequenced in a 'sequenator'. Runs were limited to about 30 amino acids so larger proteins had to be broken down into smaller peptides for sequencing. It took months to sequence a whole protein. Although the efficiency and speed of Edman degradation have been greatly increased it is no longer used routinely.

Sequence prediction of proteins from gene DNA sequences

DNA codes for the amino acid sequences of proteins. Each amino acid is coded for by a triplet of nucleotides in the DNA. The genetic code relates the triplets to amino acids so that gene sequences can be interpreted as amino acid sequences of the proteins they encode. Advances in gene isolation and DNA sequencing technology mean that the DNA sequences of many proteins are known, and it is often easier to isolate the

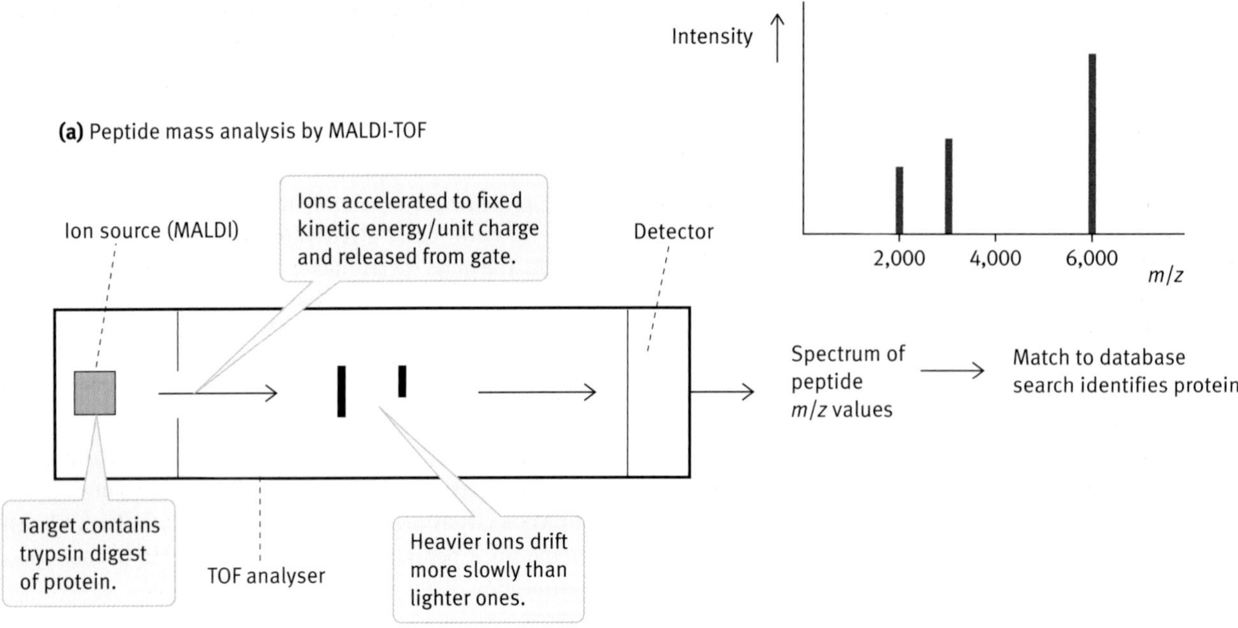
(a) Peptide mass analysis by MALDI-TOF
Intensity
2,000
4,000
6,000
m/z
Ion source (MALDI)
Ions accelerated to fixed kinetic energy/unit charge and released from gate.
Detector
Spectrum of peptide m/z values
Match to database search identifies protein
Target contains trypsin digest of protein.
TOF analyser
Heavier ions drift more slowly than lighter ones.

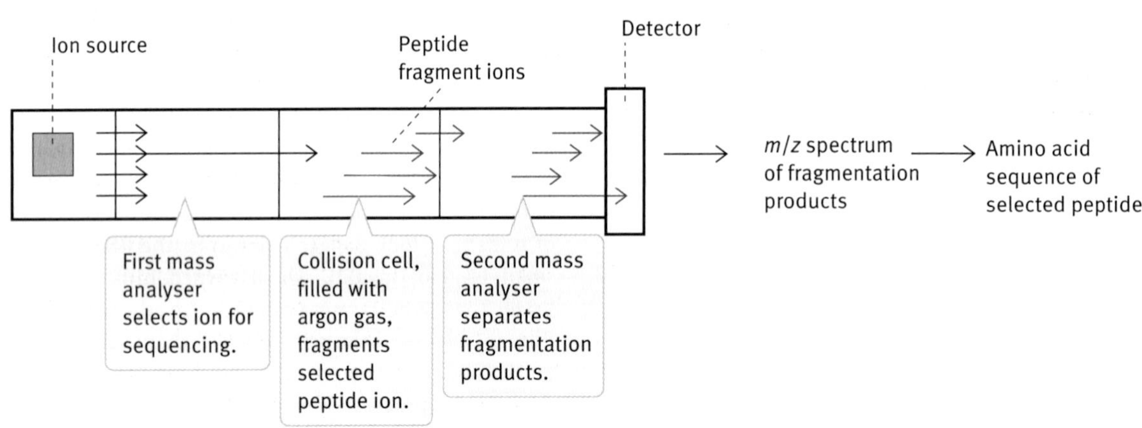
(b) Amino acid sequencing by tandem mass spectrometry (MS/MS)
Ion source
Peptide fragment ions
Detector
m/z spectrum of fragmentation products
Amino acid sequence of selected peptide
First mass analyser selects ion for sequencing.
Collision cell, filled with argon gas, fragments selected peptide ion.
Second mass analyser separates fragmentation products.

Fig. 5.5 Simplified diagram of the methods of peptide mass analysis and peptide sequencing. **(a)** Peptide mass analysis of a protein. The protein is trypsin-digested and the peptides subjected to matrix-assisted laser-desorption ionization time of flight (MALDI-TOF) 'fingerprinting'. Matching the mass analysis pattern to databases can identify a protein. **(b)** Amino acid sequencing. The protein is digested by, say, trypsin and then ionized, often by electrospray ionization. The first analyser selects a peptide ion of defined *m*/*z* value and releases it into a central collision cell where it is fragmented in a collision with argon gas. The fragmentation products pass into the final mass analyser and the *m*/*z* spectrum of the fragments is recorded. This can be used to directly deduce the sequence of the selected peptide. The key to sequencing is that peptide fragmentation occurs in a predictable manner so that the 'fragmentation ladder', as the spectrum is called, can be interpreted in terms of amino acid sequences. The method is often coupled to high-pressure liquid chromatography separation of the tryptic digest and the peptides sequentially fed into the mass spectrometer.

gene for a protein and determine its nucleotide sequence than to directly sequence the protein. The human genome project has determined the 3.2 billion base pairs of human DNA so that if the gene for any human protein can be identified in the base sequence, the amino acid sequence of its protein is easily deducible. The genomes of several other species have also been determined.

Sequencing by mass spectrometry

For sequencing by MS, tandem MS (MS/MS) is used. This is illustrated in Fig 5.5(b). In MS/MS, the individual peptides from a tryptic (or other protease) digest of the protein are analysed by the first mass analyser of a tandem mass spectrometer, which is set so that only one selected peptide ion species of given m/z value is allowed to proceed into the central collision cell. There the selected peptide molecules are fragmented into a pattern of ions, which are separated in the second mass analyser, giving a spectrum known as a fragmentation ladder. The key to the method is that *the fragmentation occurs in a predictable manner*; one amino acid at a time is removed from either end of the peptide. Thus a set of fragments is produced that differ from each other by the mass of one or more amino acids and the fragmentation spectrum can be translated into sequence. A limitation is that the isoleucine and leucine residues are not readily distinguished.

A refinement is to couple the MS/MS to reverse phase HPLC of peptides to be analysed. The separated or partially separated components are fed into the mass spectrometer as they emerge from the column. ESI is the most suitable for this. This permits sequencing of many or all the peptides obtained from a small sample of protein. About 100 fmoles of a protein from a gel are required for the analysis.

To obtain a complete protein sequence it is necessary to work out the order in which the sequence of individual peptides should be joined together. As mentioned earlier, trypsin cleaves proteins at specific amino acids. If a fresh sample of the protein is now digested with a protease of different specificity, such as chymotrypsin, a set of peptides is created that overlap those produced by tryptic digest. Sequencing this second set of peptides and comparing them with the first set permits assembly of the complete sequence.

Determination of the three-dimensional structure of proteins

The linear sequence of amino acids tells us little about the protein as a functional unit because an unfolded protein has no biological activity. Biological activity depends on the folding of the polypeptide into a three-dimensional (3-D) structure, as specified in the amino acid sequence of the protein. Determining the 3-D structure of proteins is really the ultimate goal, for these permit elucidation of the molecular mechanisms by which they function. There is a practical importance too; therapeutic drugs bind to specific sites on proteins, only recognizable in the 3-D structure, which are therefore of prime importance to developing new therapies. 3-D structures of many proteins have been determined and are available in the protein databases, but for novel proteins the amino acid sequence is obtained and then two other methods are used. These are X-ray diffraction and nuclear magnetic resonance.

X-ray diffraction

In simplistic terms, X-ray crystallography maps the electron densities around the atoms of a protein. The key to the method is that X-rays directed at the protein mostly pass through, but they are scattered (diffracted) when they encounter electrons.

The determination of protein structures by **X-ray diffraction** requires that the protein first be crystallized to produce a three-dimensional array of identical molecules all oriented in the same way. The protein used for crystallization experiments is typically produced by recombinant DNA technology and then purified by some of the chromatographic methods described earlier. Crystallization is achieved by gradually increasing the salt concentration of a highly concentrated solution of the protein. Crystallization can be one of the most difficult phases of the work, though robotic methods for screening large numbers of crystallization recipes have a high success rate.

Onced formed, the crystal is mounted in an apparatus that bombards it with X-rays with a wavelength of the same order as the distance between atoms (~1.5 Å or 0.15 nm). The diffracted X-rays are measured directly because there is no lens suitable for focusing them into an image, as one does in a camera. Where diffracted waves meet and interact they mostly cancel each other out, but where they meet 'in phase' they reinforce each other, producing a spot ('reflection') at the detector. The crystal is rotated, producing a characteristic array of spots – the diffraction pattern. The position and intensity of the spots gives information about the arrangement of electrons and hence atoms in the protein. As no direct image is possible, the measurements made must be manipulated mathematically in order to produce a three-dimensional electron density map onto which the deduced molecular structure is then superimposed.

Experiments to measure the X-ray diffraction pattern from crystals are now commonly performed with **synchrotron radiation**. The extreme brightness of this X-ray source allows measurements to be made from very small crystals (~10 μm in diameter), an enormous advantage if these are the only crystals available. Further, the tuneability of synchrotron radiation to different wavelengths simplifies certain experimental manipulations (not described here) that are required in order to generate structures from X-ray diffraction patterns.

Nuclear magnetic resonance spectroscopy

A second method of protein 3-D structure determination, based on the magnetic properties of certain atomic nuclei, and of increasing importance, is **nuclear magnetic resonance spectroscopy** (NMR). Some atomic nuclei, of which protons are the most important in the present context, have a property known as spin and behave as minute magnets. These nuclei can be affected by a constant powerful magnetic field that causes the magnets to be aligned in one of two possible spin states, either with or against the field direction. The spin state that is aligned against the field is of slighly higher energy than is the other. Pulses of selected radiofrequency can cause the nuclei to change from the lower to the higher energy state, as they absorb radiation if it has a frequency that exactly coincides with the energy difference between the two. Thus, a pulse of appropriate radiofrequency sets up a resonance between the two states and produces a measurable emission signal as the nucleus relaxes to its lower energy level. Either an absorption or an emission spectrum can be obtained, but emission is normally measured by modern NMR machines.

The crucial factor that allows NMR to be used to determine molecular structure is that the atomic environment surrounding individual nuclei affects the strength of the magnetic field to which they are effectively subjected. This in turn affects the energy difference between the two spin orientations and hence the frequency at which an emission signal is produced. The externally applied magnetic field is held constant while the frequency of the radiation pulses applied to the protein is varied, and this gives a spectrum of emission peaks, which can be correlated with amino acid residues in a protein primary sequence, and, most importantly, gives information on the distance between pairs of atoms. An additional useful property is that spin on one nucleus can affect the spin state of another that is 'coupled' to it either by covalent bonding between the two atoms or by physical proximity within the molecular structure. Thus, analysis of NMR spectra can identify atoms that are close together in a folded protein but distant from each other in the primary structure, and enables secondary and tertiary structures to be inferred.

A major advantage of NMR over X-ray diffraction is that it is performed in solution, thus removing the necessity to crystallize the protein, which is frequently difficult or unachievable. Another advantage of determining structures of proteins in solution is that this correlates better with their physiological state. A disadvantage of NMR is that high concentrations of protein are needed. The method is most easily used on proteins below 100 amino acid residues in size but, with extra refinements in methodology, has been used for proteins three times larger than this. As of 2012, approximately 12% of the roughly 80,000 protein structures that have been experimentally determined and recorded in the Protein Data Bank were determined by NMR, with the overwhelming majority of the rest determined by X-ray crystallography.

Homology modelling

Despite our accumulated knowledge of protein structures and protein folding, it is still not possible, from theoretical principles, to predict the 3-D structure of a protein from its amino acid sequence. However, if two different proteins have similar sequences and the 3-D structure of one is known from, say, X-ray crystallography, then it may be possible to make a fairly reliable estimate of the structure of the other. The method, known as homology modelling, requires at least a 50% sequence homology (see 'Protein homologies and evolution' in Chapter 4) between the two proteins for useful predictions.

An exercise in obtaining a 3-D structure from a protein database

The use of protein databases is important in searching for proteins for which the 3-D structure has been determined. On the Online Resource Centre, we give step-by-step instructions on how to obtain the 3-D structure of human haemoglobin from the Protein Data Bank, if you feel interested in trying it.

Proteomics

A living cell or organism contains a large collection of proteins. Differential splicing of mRNAs (see Chapter 24) and post-translational modifications (see Chapter 4) can result in it producing even more proteins than it has protein-coding genes. To refer to these collectively, the term '**proteome**' was coined. This is analogous to the term '**genome**', which refers to the entire collection of DNA in an organism. The two terms are not exactly comparable because, with a tiny number of exceptions, the genome of all cells in an organism is the same whereas the proteome varies from cell to cell and within a given cell from time to time. At any one time, the proteome represents the functional portion of the genome. For example, a liver cell makes liver-specific proteins but not brain-specific or muscle-specific proteins and *vice versa*. Thus the proteomes of liver, brain, and muscle cells overlap as both will contain essential structural proteins and metabolic enzymes, for example, but they also differ to a considerable extent. The proteins of a given cell may change during differentiation (the development of specialized cell types) or in response to physiological needs. In a liver cell, some enzymes are needed in greater amounts after eating than during fasting so that the proteome may vary from hour to hour. A comparison of the proteomes of normal and diseased cells may reveal differences in individual proteins, correlated with the disease state. As already mentioned in the context of 2-D gel electophoresis, it may be desirable, therefore, to look at whole collections of proteins to see how the proteome varies in development, in response to physiological needs, and in disease such as cancer. The study of whole proteomes or large collections of proteins such as those in complexes or organelles is known as '**proteomics**'. Proteomics differs from conventional 'protein chemistry' essentially in being concerned with large numbers of proteins at once. By analogy the large-scale study of genes is called '**genomics**'. Other terms such as the '**transcriptome**' (the full set of RNA transcripts) and the '**metabolome**' (the full set of small molecule metabolic intermediates) have also been coined.

Proteomic studies are technically extremely challenging (much more so than genomics, which is discussed in Chapter 28, due to the chemical complexity of proteins compared to nucleic acids). However, modern methods are opening up the field. As described, 2-D gel electrophoresis made it possible to separate hundreds or thousands of proteins in a crude cell extract. We have explained that MS methods can be applied to single protein spots on the 2-D gels so that each of the spots can be studied and in many cases identified in databases. The speed of MS makes rapid throughput and automation possible. Modern mass spectrometers can characterize over 1,000 proteins per day so that they can be identified from databases by the procedures described above. An important use of MS is to study the proteins present in organelles and complexes in the cell: mitochondria and nuclear pore complexes are two examples. Many or most proteins do not function in isolation but participate in larger assemblies to perform a specific task. If the organelle complexes can be isolated by methods that do not dissociate their proteins, then their components can be identified by MS analysis. The components of the nuclear pore complex protein were determined in this way. MS is also opening up the study of post-translational protein modifications on a larger scale than was hitherto possible.

The combination of genomics and proteomics represents a major new approach to the study of biology by examining how entire collections of genes and proteins function together to constitute the living process, rather than by examining the constituent parts. Its development in a few years has been spectacular.

Bioinformatics and databases

Bioinformatics is the (relatively new) branch of science that deals with the storage, retrieval and utilization of the vast amount of data generated by genomic, proteomic and other '-omic' studies. This brief section can only give a general idea of the rapidly growing importance of the discipline and let you have some idea of what it is about. Some database website addresses are given in Box 5.1. Most biochemists and molecular biologists nowadays will need to learn and make use of at least some bioinformatic skills. However, specific training is required to practice bioinformatics as a principal activity, as it requires multiple skills and intersects the fields of biochemistry and molecular biology (including molecular genetics and molecular evolution), computer science, mathematics, and statistics, depending on the project.

The most obvious function of bioinformatics is to allow the storage and retrieval of nucleic acid and protein sequence data. The development of automated DNA sequencing, which

BOX 5.1 Database of website addresses

A useful introduction for students is at the National Centre for Biotechnology Information website (**www.ncbi.nlm.nih.gov/About/primer/index.html**), which contains educational material on a range of topics.

A few selected database website addresses are given below although it is not expected that many users of this book will find them immediately useful without further formal instruction.

Primary sequence databases containing publicly available DNA sequences

GenBank
http://www.ncbi.nlm.nih.gov/genbank/

EMBL – the European Molecular Biology Laboratory
http://www.ebi.ac.uk/embl/

DDBJ – the DNA DataBank of Japan
http:// www.ddbj.nig.ac.jp/

Protein databases

UniProt – combines data from Swiss-Prot and TrEMBL to give high quality protein sequence and functional information
http://www.uniprot.org/

PIR – Protein Information Resource
http://pir.georgetown.edu/

PDB – Database of 3-D protein structures
Structure data determined by X-ray crystallogaphy and NMR.
http://www.rcsb.org/pdb/

BLAST – basic local alignment search tool
Tool to probe sequence data banks with sequence.
http://blast.ncbi.nlm.nih.gov/Blast.cgi

Clustal
Tool for multiple sequence alignment.
http://www.clustal.org/

ExPASy – Expert Protein Analysis System
A resource portal operated by the Swiss Institute of Bioinformatics that provides access to hundreds of biological databases and software tools – not only for protein analysis!
http://www.expasy.org/

has allowed the sequencing of the entire human genome and more and more plant, animal and microbial genomes each year, has contributed to a data 'explosion'. The primary open access nucleotide sequence databases are GenBank in the USA, the European Molecular Biology Laboratory (EMBL), and the DNA Data Bank of Japan (DDBJ), which together collect and share all publicy available sequences. Sequence records in these databases are annotated with key information such as the source of the sequence, publication references, and biologically significant features, for example, which sections actually code for protein. Nucleotide databases are intimately involved in protein bioinformatics since, increasingly, amino acid sequences are generated from DNA sequences, and these three databases also publish the amino acid translations of all coding sequences deposited in them.

In addition to these, there are numerous databases that specialize in specific molecules, specific organisms, or specific methodologies – for example data from 2-D gel electrophoresis studies. Of these, databases of protein sequences, many derived from translations of the DNA, are probably the most widely used. These include the European Swiss-Prot and TrEMBL (Translated EMBL Nucleotide Sequence Data Library) resources and the US-based PIR (Protein Information Resource). Redundancy (the same sequence being submitted and appearing more than once with slight variations and differing annotations) can create difficulties with data retrieval and utilization, and the three resources have joined forces to create UniProt, a merged database that aims to maintain a high standard of curation and annotation to deal with such issues. To cope with the large volume of new data the initial annotation of a sequence is done automatically by computer, but this is followed up by laborious manual reviews. Another key protein resource is PDB (the Protein Data Bank), which is the central repository for protein structures, and also contains some three-dimensional structures determined for other biological macromolecules.

Retrieval of sequence data may involve searching by gene, protein name, function, or by organism. However, some of the most useful searches are those that look for sequence homology. Finding that the gene or protein you are studying is already in the database, perhaps identified in a different cell type, or that genes and proteins of similar sequence that are related through evolution (gene and protein 'families') have already been studied, is a rapid way of gaining information about the structure and possible function of 'your' molecule. The most widely used tool for homology searches is the Basic Local Alignment Search Tool (BLAST). With the BLAST family of programs you can use a sequence of interest of either DNA or amino acids to search a DNA (BLASTN) or protein (BLASTP) sequence database for similar sequences. Having found them, is also potentially useful to make more extensive comparisons to infer the evolutionary history, function, or structure of the sequence of interest. To accomplish such analyses, the sequences, which usually vary in length, must first be aligned to ensure that comparisons are made among homologous sites along each sequence. The most commonly used publicly available tool for multiple sequence alignment is Clustal.

It is impossible to detail all the possible uses of bioinformatic searches and analyses, but some examples are given below.

- As proteins have evolved from a few ancestral species, their amino acid sequences diverged. Some changes may have little effect on the function of a protein but those amino acids which are essential to activity are usually strongly conserved since any alteration could inactivate it. Comparisons of homologous protein sequences can identify conserved sequences, suggesting which ones are essential for activity. This information can give clues to help elucidate the mechanism of action of the protein.
- It is often found that a particular activity or role of a protein is associated with a particular structural domain or motif. To give an example, one large class of membrane proteins has a helix of hydrophobic residues that just spans the hydrophobic section of the lipid bilayer. It is possible, therefore, to scan possible databases for all proteins with this characteristic and thereby identify putative membrane proteins. Another example from the important field of cell signalling (see Chapter 29) is that a large family of enzymes, the protein kinases, all have a domain that effects transfer of a phosphoryl group from ATP to proteins. Consequently it is possible to ask the question of how many such protein kinases exist by searching databases for proteins with domains with close homology to this.
- Proteins which lack obvious sequence homology may share homologies of tertiary structure, giving clues to both functional and evolutionary relationships.
- Another application is to locate protein-coding genes in DNA sequences. It is relatively easy to identify a prokaryotic gene by searching for an open reading frame: that is a nucleotide sequence long enough to code for a protein that is not interrupted by a 'nonsense codon', which does not represent any amino acid. By contrast, in eukaryotes, genes are interrupted by noncoding introns (see Chapter 22), so this approach is not applicable. However, the ends of introns have specific base sequences and these can be searched for to locate genes in new sequences.

The above summary can only scratch the surface and does not touch on the many novel and sophisticated uses of bioinformatics that are in developement, for example in modelling and predicting molecular interactions and complex cellular processes.

Summary

The classical methods of protein purification are often relatively simple procedures that can be applied on a large scale and are therefore used as preliminary purification steps. The simplest is differential centrifugation, which sediments organelles on the basis of mass; these can be discarded or collected depending on the location of the protein of interest. Purification methods for soluble proteins include selective precipitation by high concentrations of ammonium sulphate. The salts are then removed by dialysis.

More specific separations can be achieved by column chromatography. Molecules are separated as they run through column packings. Separations are variously based on molecular (size) exclusion, ion exchange or specific molecular affinities. Column chromatography is used on a preparative scale but also is important analytically on a small scale.

Electrophoresis separates proteins by movement in an electric field on the basis of size, charge and shape. However, the most commonly used method, sodium dodecylsulphate-polyacrylamide gel electrophoresis (SDS-PAGE) denatures proteins and eliminates the effects of charge and shape, giving separation on the basis of size alone and hence allowing determination of molecular weight. In isoelectric focusing, a stable pH gradient is established in which proteins migrate to the pH at which they have zero charge (isoelectric point). Two-dimensional gel electrophoresis separates proteins by isoelectric focusing in the first dimension and SDS-PAGE in the second. It can separate many proteins in a single run.

Since the 1980s a revolution in protein technology has superseded many of the earlier methodologies. The newer methods work on minute amounts of protein and are rapid. The protein databases, which record vast amounts of information on all aspects of proteins, now complement these technologies. Amino acid sequencing of a protein was a laborious task. It is often now easier to determine the base sequence of the gene for a given protein and translate this into amino acid sequence using the genetic code. The completion of the human genome project means that the sequence of every human protein for which the gene can be identified is obtainable in this way.

Mass spectrometry (MS) has revolutionized much of protein methodology. It permits rapid identification of proteins in databases. Using the minute amount of protein from a spot in a two-dimensional gel, a protein can be identified in a database by its molecular mass and its peptide 'fingerprint'. These can be obtained rapidly by MS. It can also perform amino acid sequencing more quickly than by any other method.

Methods for determining the three-dimensional structures of proteins are X-ray diffraction, now commonly performed with synchrotron radiation, and nuclear magnetic resonance spectroscopy.

The ability to rapidly identify minute amounts of proteins has lead to proteomics in which large numbers of proteins are studied at once. Bioinformatics, which involves the computer-assisted use and analysis of database information is itself a major science and 'mining the databases' is a fruitful research activity.

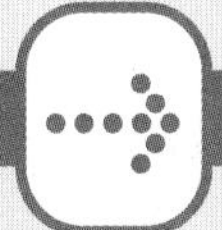

Further reading

To access the further reading, please scan the QR code image or go to http://global.oup.com/uk/orc/biosciences/molbiol/snape_biochemistry/student/reading/ch05/

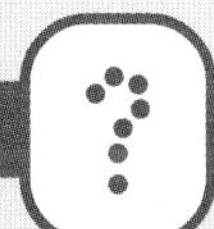

Problems

1. Describe, without chemical detail, four methods for determining the primary structure of a protein.
2. What is meant by the term proteome?
3. What is meant by the term proteomics? It has come into research prominence only relatively recently. What has been a major factor in this?
4. When proteins are separated by polyacrylamide gel electrophoresis, it is common to include sodium dodecylsulphate (SDS) in the gel and reagents. What is the reason for this?
5. List the various types of column-chromatographic separation of proteins.

6 Suppose you have a minute amount of an unidentified protein as a spot on a gel. How could it be sufficiently characterized rapidly to identify a corresponding protein in a protein database?

7 Mass spectrometry has been known for a long time but only relatively recently has it been applied to proteins. What caused this change?

8 Protein databases have assumed great importance. Briefly explain their relevance and use.

9 Briefly explain the basis of protein sequencing by mass spectrometry.

10 List the methods by which the three-dimensional structures of proteins are determined.

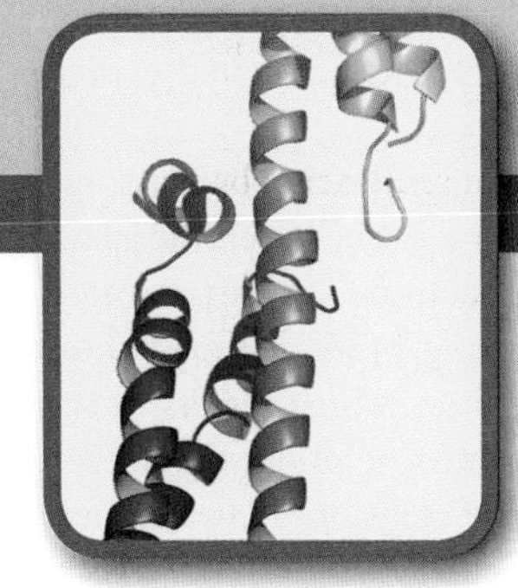

Chapter 6

Enzymes

Given that a reaction has a negative ΔG value, what determines whether it actually takes place at a perceptible rate in the cell? This question follows on from Chapter 3, where we explained that on the question of whether a reaction is theoretically possible (*may* occur), energy considerations have absolute authority. If the free energy change involved in the reaction under the prevailing conditions is negative, it may occur; if it is zero or positive it cannot occur and nothing in the universe can alter that. Note, however, that chemical *conversions* (such as the synthesis of compound X–Y from reactants XOH and YH) involving increases in energy levels do occur in the cell, by incorporating ATP breakdown into the process resulting in a negative ΔG for the overall process. It is also important to appreciate that the rule that a reaction must have a negative ΔG refers to *net* chemical change in the reaction. There may be some confusion generated by the observation that at equilibrium the free energy change of a reaction is zero, but this is consistent since there is also no *net* chemical change at equilibrium; the reaction in one direction is exactly balanced by that in the reverse direction.

To go back to what determines whether a reaction actually occurs, we have touched briefly on this in Chapter 3, but we will elaborate on it here, as it is of basic importance. Energy considerations determine whether a reaction *may* occur, not whether it *does* occur. This is just as well, because energy considerations say that everything combustible around you may burst into flame. But petrol does not ignite by itself, and sugar in the bowl on the table does not burn. Nonetheless, petrol in a car cylinder burns and sugar inside the body is oxidized to CO_2 and H_2O.

From these considerations you can see that there is a barrier to the occurrence of chemical reactions and, in the case of biochemical reactions, it is sufficiently large to prevent them occurring at a finite rate. Even though a reaction has a strong negative ΔG something restrains the reaction from happening. In the case of petrol in a car cylinder, a spark from the spark plug overcomes that barrier. This brings us to one of the most fundamental problems that had to be solved before life could exist. How can chemical reactions be caused to occur in the cell? The answer lies in **enzyme catalysis**. It is worth reflecting on what a formidable obstacle this problem posed for the development of life, for it leads to an appreciation of what an astonishing phenomenon enzyme catalysis is. It was necessary to develop a means whereby at low temperatures, at almost neutral pH, in aqueous solutions, and at low reactant concentrations, otherwise stable molecules undergo the rapid chemical conversions needed for life.

Enzyme catalysis

An enzyme is a catalyst that brings about (usually) one particular chemical reaction but itself remains unchanged at the end of the reaction, as required in the definition of a catalyst. There are thousands of biochemical reactions, each catalysed by a separate enzyme. Multifunctional enzymes with several catalytic activities on the same molecule and multienzyme complexes exist for special situations, but the general principle remains. It automatically follows from what we have said about energy considerations that an enzyme cannot affect the equilibrium or direction of a reaction; it can only promote a reaction subject to the limitations of energy considerations. This is little different from saying that you cannot get energy for nothing.

An enzyme is a protein molecule. Proteins are built up of 20 different species of amino acids linked together to form one or more long chains. Enzymes are therefore large molecules – a molecular weight of 10,000 Da corresponds to a small enzyme, and they range in size up to hundreds of thousands of daltons in molecular weight. One of the reasons why they are so large is that the long chain(s) of which they are constituted must be folded such that there is an **active site** on the surface so shaped that it is a three-dimensional pocket or cleft into which the compounds attacked by the enzyme (known as **substrates**) fit with exquisite precision. The active site is also called an **active centre** or **catalytic site**. This site occupies a very small part of the protein in most cases. As with all such specific protein–ligand binding (a ligand is any binding molecule including an enzyme substrate), the substrate attaches reversibly by noncovalent, or weak, bonds. As explained earlier (see Chapter 3), the specificity of all such attachments arises from the fact that several weak bonds are needed. It follows that unless there is a precise fit between the interacting groups on the substrate and enzyme, the attachment will not occur. Life processes depend on this simple principle by which specific interactions of proteins with other molecules (which may also be proteins) are achieved.

The nature of enzyme catalysis

To explain enzyme catalysis, we must look at the nature of chemical reactions. A reaction occurs in two stages. Consider a reaction converting substrate to product (S→P). In this reaction, S must first be converted to the **transition state**, $S^‡$, which might be thought of as a 'halfway house' in which the molecule is distorted to an electronic configuration that readily converts to P. The transition state has an exceedingly brief existence of 10^{-14}–10^{-13} of a second. The overall reaction S→P must have a negative free energy change; otherwise it could not occur. An important thermodynamic principle is that the free energy change of a reaction is determined solely by the free energy difference between the starting and final products. The 'energy pathway' or energy profile, the route by which the reaction takes place, cannot affect the overall free energy change of the reaction. Hence it is in no way a contradiction that the transition state ($S^‡$) is at a higher free energy level than S. The free energy change for S→$S^‡$ is positive. It is called the **energy of activation** for that reaction (Fig. 6.1).

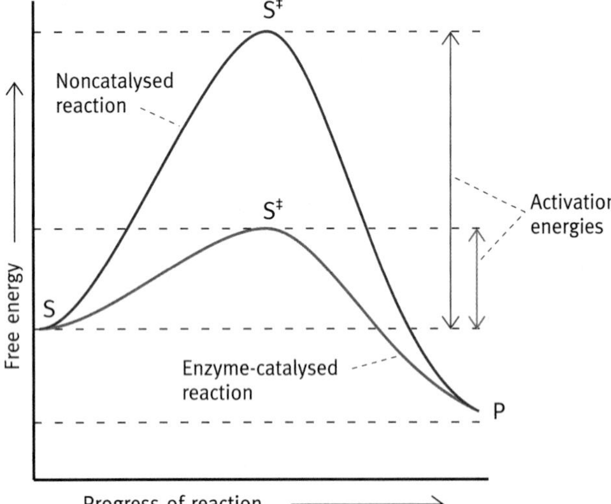

Fig. 6.1 Energy profiles of noncatalysed and enzyme-catalysed reactions. The inverse relationship between the rate constant and the activation energy of the reaction is exponential so that the rate of the reaction is extremely sensitive to changes in the activation energy. S, substrate; $S^‡$, transition state; P, products.

This energy hump constitutes a barrier to chemical reactions occurring. If it were not present and the energy profile S→P were a straight downward slope then, as stated earlier, everything that could react would do so.

It follows that energy of activation must be supplied in some way to permit a reaction to occur. In a car cylinder, the spark causes a few molecules of petrol to be activated to the transition state, which then react, and in oxidizing, produce enough heat to activate further molecules and so on, which results in the explosion. In a noncatalysed reaction in solution, the energy to surmount the hump is supplied by collisions between molecules. Provided that colliding molecules are appropriately oriented, and of sufficient kinetic energy, reactant molecule(s) can be distorted into the appropriate higher energy transition state, which enables the reaction to occur. The rate of formation of the transition state therefore determines the overall reaction rate. High temperatures, which increase molecular motion and increase collision frequency between molecules, facilitate this. Hence the organic chemist usually employs high temperatures to promote reactions. At physiological temperatures, however, around 37 °C, most biochemical reactions, uncatalysed, proceed at imperceptible rates.

As stated, each enzyme has an active site to which the substrate(s) bind. The binding has several effects.

- It positions substrate molecules in the most favourable relative orientations for the reaction to occur.
- The active site is perfectly complementary, not to the substrate in its ground state (as the unactivated substrate molecule is referred to), but to the **transition state**, which is intermediate between the reactant molecules and products.

A substrate in its transition state, for this reason, binds to the enzyme more tightly than does the substrate in its ground state. This is due to the structure of the active site, which has evolved to do this. A transition state tightly bound to the enzyme is at a lower energy level than the same transition state in free solution as occurs in a noncatalysed reaction. (The binding liberates energy). The transition state has too ephemeral an existence to measure just how tightly it binds to the enzyme, but **transition state analogues** have been synthesized. These are stable molecules similar in structure to the transition state. It has been found that these bind to enzymes with remarkably high affinity – in one case, binding to the enzyme thousands of times more tightly than the substrate.

Another way of looking at this is that the formation of the transition state involves a partial redistribution of electrons. The amino acid groups in the structure of the active centre of the enzyme are of such a nature, and so positioned, that they stabilize the electron distribution of the transition state. You will find this explained in greater detail when you come to the mechanism of the enzyme chymotrypsin, later in this chapter. The fact that the active site is a less perfect fit to the substrate than it is to the transition state results in the substrate being strained on binding to the active site, and this favours transition state formation. The net effect is to lower the activation energy of the reaction; the energy hump barrier to the reaction is reduced (Fig. 6.1), and the reaction rate is increased. This is the central principle of enzyme catalysis on which life depends. Once formed, the transition state rapidly converts to products. The products bind less tightly to the enzyme and diffuse away. The catalysis rate is very sensitive to changes in the activation energy, there being an inverse exponential relationship. A very small reduction in the activation energy, equivalent to the small amount of energy released on formation of a single average hydrogen bond, can increase the rate of the reaction by a factor of 10^6. Enzymes increase the rate of chemical reactions sometimes by a factor of 10^7–10^{14}. The enzyme urease, which catalyses the hydrolysis of urea to ammonia and carbon dioxide, reduces the energy of activation by 84 kJ mol^{-1} and increases the reaction rate 10^{14} times.

Enzyme catalysis has an additional beneficial effect. Most molecules are capable of participating in many different chemical reactions, each with its own transition state. In an uncatalysed reaction, promoted by high temperatures, molecules collide unpredictably and different transition states are formed, resulting in a variety of side reactions. By contrast, an enzyme catalyses a specific reaction for which there are only a limited number of well-defined products.

From what has been said, it might be deduced that if one could produce a protein with a high affinity for the transition state of a given chemical reaction, it would catalyse that reaction; such has been found to be the case. Antibodies (described in Chapter 33) are proteins which bind tightly to specific structures in molecules, and moreover they can be produced to bind to selected molecules. It was found that an antibody raised against a stable analogue of the transition state involved in the hydrolysis of a synthetic ester behaved as a hydrolytic enzyme towards that ester. The term **abzyme** has been coined for such proteins, *ab* denoting 'antibody'.

The induced-fit mechanism of enzyme catalysis

The binding of an enzyme to its substrate implies that the relationship between an enzyme and its substrate(s) is simply a 'lock-and-key' model in which the active site is envisaged to be a rigid structure in an unchanging form and the substrate (the key) fits to it (Fig. 6.2(a)). A more recent concept, however, is the '**induced-fit**' mechanism, which is based on the view that the enzyme is not a rigid structure analogous to a lock, but rather is a flexible structure capable of changing its conformation slightly in an interactive way when its substrate binds. ('Conformation' refers to the particular arrangement of the protein chain(s) in three-dimensional space.) In other words, it changes its shape slightly, which has the important effect of altering the spatial arrangements of groups on the molecule. Conformational changes in proteins, as will become clear

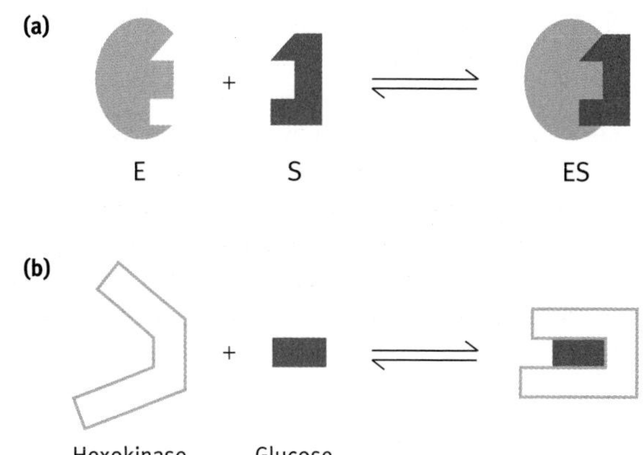

Fig. 6.2 (a) Lock-and-key model of enzyme mechanism. E, enzyme; S, substrate. **(b)** Induced-fit model of the hexokinase mechanism.

later in the book, are of central importance to biological processes. This induced-fit mechanism was first established for the enzyme **hexokinase**. The enzyme, discussed further in Chapter 11, catalyses the transfer of a phosphoryl group from ATP to glucose (a hexose sugar). The enzyme has two 'wings' to its structure. In the absence of glucose, these have an 'open' conformation, but on binding of glucose the wings close in a jaw-like movement that results in the creation of the catalytic site (Fig. 6.2(b)). The postulated conformational change has been shown to occur by X-ray crystallographic studies (Fig. 6.3).

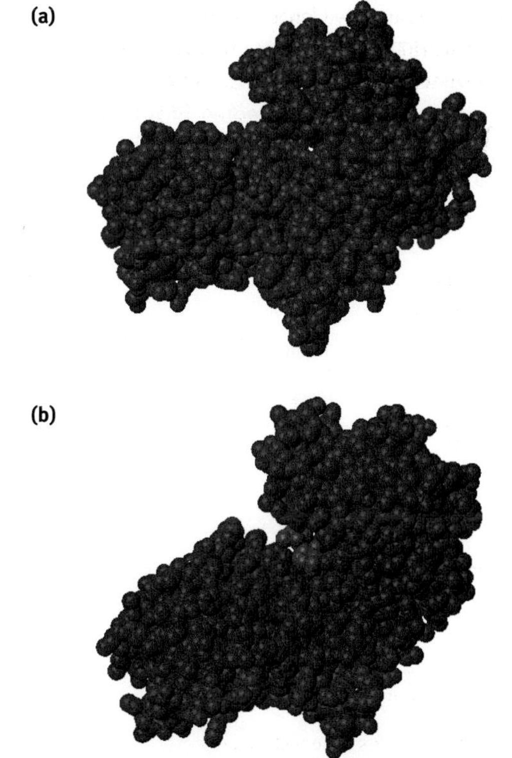

Fig. 6.3 Space-filling models of yeast hexokinase **(a)** unbound (Protein Data Bank Code 1HKG) and **(b)** with a glucose analogue (red) bound (2YHX).

Enzyme kinetics

Enzyme kinetics refers to the study of enzymes by determining their reaction rates. Measurement of enzyme activities is routine in much of biochemistry and other biological sciences, and also in medicine, where the presence and activity of specific enzymes in clinical samples is often used as a diagnostic tool. In biochemistry, kinetic studies can aid the understanding of the mechanisms by which enzymes work, and this is also of importance in pharmacology and drug design. Many drugs work by inhibiting enzymes and it is essential in developing these to understand the different types of inhibition by which compounds may work and how they affect reaction kinetics. The subject is also important in considering aspects of metabolic control discussed in Chapter 20.

In a typical case, an enzyme rate is measured by incubating it with substrate(s) at a defined temperature and pH and following the production of a reaction product with time, or alternatively following the disappearance of substrate. As seen in Fig. 6.4, the reaction, with a fixed amount of enzyme, gradually diminishes in rate with time; this may be due to several factors including depletion of substrate, conversion of product back to substrate by the reverse reaction (which occurs at a significant rate when the product accumulates), and, in the case of a delicate enzyme, denaturation of the enzyme. In order to avoid these complications and obtain meaningful quantitative assays of enzyme activity, it is necessary to ensure that **initial reaction velocity**, designated V_0, is measured, which typically means using short time periods before there is a significant amount of product formed. In this situation the reaction velocity is linearly proportional to the amount of enzyme added (assuming that substrate is in excess over enzyme).

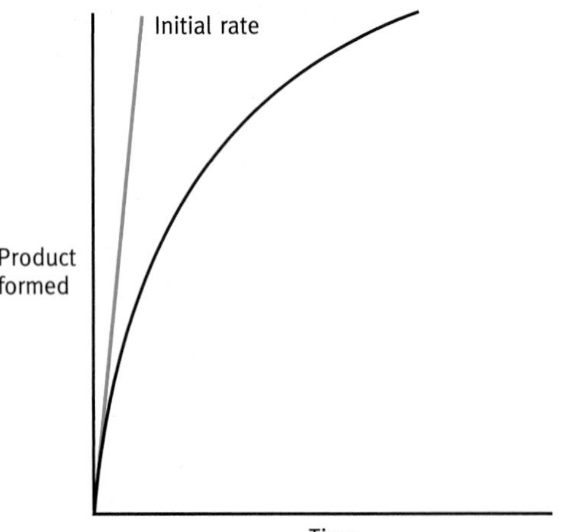

Fig. 6.4 Time course of a typical enzyme reaction in which the amount of enzyme is constant.

Hyperbolic kinetics of a 'classical' enzyme

In 1913 Leonor Michaelis and Maud Menten proposed a model of the way in which enzymes act, to fit the observed kinetics of enzyme catalysis for a single substrate enzyme. As the starting point for their model they used the reaction scheme:

$$E + S \underset{k_{-1}}{\overset{k_1}{\rightleftarrows}} ES \underset{k_{-2}}{\overset{k_2}{\rightleftarrows}} E + P$$

where a single substrate, S, binds reversibly to the active site of the enzyme, E. The complex, ES can either dissociate to E and S, or the substrate can be converted to product, P, which dissociates and diffuses away from the enzyme. The reactions are reversible, and k_1, k_{-1}, etc., denote rate constants.

The scheme is then simplified still further to consider a situation where conversion of product back to substrate does not occur at an appreciable rate, as is the case if the initial velocity of the reaction, V_0, is what is measured.

$$E + S \underset{k_{-1}}{\overset{k_1}{\rightleftarrows}} ES \xrightarrow{k_2} E + P$$

The outcome of an experiment in which V_0 is measured at increasing substrate concentrations is shown in Fig. 6.5. At low [S], the rate of formation of ES is relatively low and will vary with changes in [S], so that the rate of reaction varies accordingly. As [S] increases, so will the proportion of enzyme in the form of ES increase and therefore the rate of reaction increases proportionately to [S], displaying **first order** kinetics. However, as [S] further increases, the point is reached at which the the enzyme is **saturated** with substrate. The rate of the reaction is then called V_{max} (for maximum velocity), and further increases in the substrate have no effect on the rate of catalysis (the reaction displays **zero order** kinetics). V_{max} is a function of the amount of enzyme present in a given experiment and the rate at which each molecule of enzyme catalyses the reaction. In the case of many or most enzymes, when the velocity of enzyme activity is plotted against substrate concentration a **hyperbolic curve** as shown in Fig. 6.5 is obtained. An enzyme displaying these kinetics is referred to as a **Michaelis–Menten** enzyme and the kinetics as Michaelis–Menten kinetics or **hyperbolic kinetics**.

An equation can be derived which describes this hyperbolic relationship between the velocity of an enzyme reaction and substrate concentration. The derivation depends on certain simplifying assumptions:

- Only a single substrate is involved. This may seem rather limiting since many enzymes catalyse reactions with two or more substrates. However, in practice, if all substrates bar one are present in excess the reaction will display Michaelis–Menten kinetics on varying the concentration

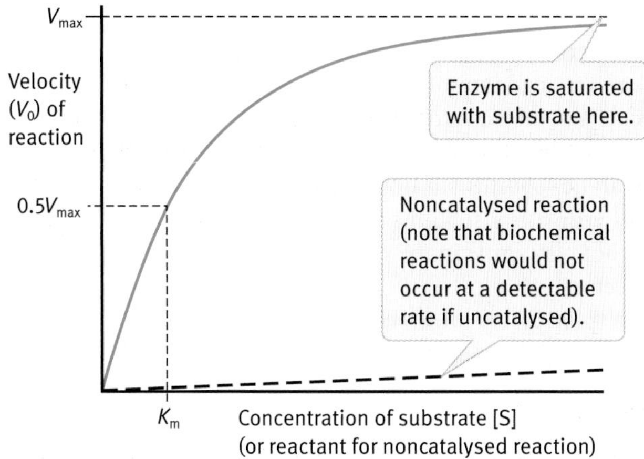

Fig. 6.5 Effect of substrate concentration on the reaction velocity catalysed by a classic Michaelis–Menten type of enzyme. K_m, Michaelis constant. The dashed line shows, for comparison, the effect of reactant concentration on a noncatalysed chemical reaction. Note that the two lines are drawn to illustrate their shapes, not their relative rates.

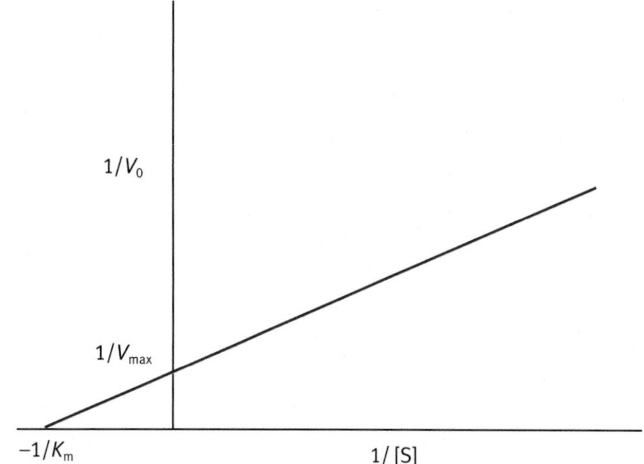

Fig. 6.6 Double reciprocal plot of an enzyme reaction.

of the single rate-limiting substrate, and useful information can be gained from studying the kinetics of the reaction.

- As already mentioned, initial velocities (V_0) are measured so that the concentration of product is negligible compared with that of the substrate.
- It is also assumed that the system is in a **steady state** in which the rate of formation of ES is exactly balanced by the rate of its removal.
- The substrate is in vast molar excess over the enzyme.

The latter two conditions are usually met because steady state kinetics are established almost instantly and the molar amount of enzyme is usually negligible. The equation that describes the relationship between the velocity of an enzyme reaction and substrate concentration and gives the hyperbolic curve shown in Fig. 6.5 is known as the **Michaelis–Menten equation**:

$$V_0 = \frac{[S]V_{max}}{[S]+K_m}.$$

The **Michaelis constant (K_m)** is a very useful constant. It is derived by simplifying a term in the equation involving the rate constants:

$$K_m = \frac{k_{-1}+k_2}{k_1}.$$

By rearranging the Michaelis–Menten equation it can be shown that the K_m value, expressed in units of molar concentration, is equal to the concentration of S at which the enzyme is working at half maximal velocity. At the K_m value of substrate concentration, half the total number of enzyme active sites are occupied and half vacant. The K_m value of an enzyme is an inherent property of the enzyme and does not depend on enzyme concentration. As can be seen in Fig. 6.5, it can be estimated from an experimentally obtained Michaelis–Menten plot by first obtaining a value for V_{max} and then reading off the value of [S] at which V_0 is $0.5V_{max}$.

In the Michaelis–Menten plot the hyperbolic curve approaches, but never quite reaches the true value of V_{max}. Other methods of plotting graphs are used to determine K_m values more precisely; if instead of plotting reaction velocity (V_0) against [S], $1/V_0$ is plotted against 1/[S] (Fig. 6.6) this gives a straight line, which, if extrapolated back, intercepts the horizontal axis at the negative reciprocal of the K_m. This is known as a **double reciprocal plot**, or a **Lineweaver–Burk** plot after its authors. The intercept of the line with the vertical axis gives a value for $1/V_{max}$.

What does the value of K_m tell us about an enzyme? It is [S] at which the enzyme is 50% saturated with substrate. It depends on how tightly the substrate binds to the enzyme, or as it is usually expressed, what the **affinity** of the enzyme is for its substrate. This depends on the precise relationship of the substrate to its enzyme – the nature and number of the weak bonds established between them. *The higher the K_m, the lower is the affinity of the enzyme for its substrate.* K_m values can be used to compare the affinities of different enzymes for their substrates. They are of interest from the viewpoint of metabolism for they tell us how an enzyme will respond to changes in the concentration of substrate in the cell.

In the case of many enzymes, the K_m represents a true affinity constant, but only for those in which the rate of dissociation of ES back to E+S is much faster than the catalytic step of ES to E+P. The reason is that since an affinity constant represents the position of the equilibrium $E+S \rightleftharpoons ES$, if ES is rapidly removed by conversion to E+P, a true equilibrium is not established. The K_m is thus a true affinity constant only in the case where k_2 is much smaller than k_{-1}, in which case k_2 can be ignored in our expression for K_m, which becomes:

$$K_m = \frac{k_{-1}}{k}.$$

This is true for many enzymes. However, even where it is not, the *apparent* affinity of the enzyme for its substrate, which is in all cases reflected in the K_m value, is a useful way of comparing how different enzymes will respond to substrate concentration changes. As a general statement, most enzymes have K_m values such that they are working in the cell at sub-saturating substrate levels. For many enzymes the K_m is in the range of 10^{-4}–10^{-6} M.

The **turnover number** or K_{cat} for an enzyme is another useful value, and can be calculated if the molar concentration of the enzyme in an experiment is known. This is the number of molecules of substrate converted to product by a molecule of enzyme at saturating levels of substrate per second. K_{cat} values have been measured ranging up to 4×10^7/s for the enzyme **catalase**, which converts toxic hydrogen peroxide to oxygen and water.

The **specificity constant** of an enzyme, K_{cat}/K_m is a value that takes account of both the rate at which the enzyme converts a substrate to product (K_{cat}) and the affinity of the enzyme for that substrate (K_m). It is therefore a good overall measure of the catalytic efficiency of an enzyme with respect to a particular substrate.

Allosteric enzymes

Enzymes that display Michaelis–Menten kinetics are sometimes known as 'classical' enzymes. **Allosteric enzymes** are another very important class of enzymes. They do not have hyperbolic kinetics. These are regulatory enzymes whose activities are controlled by chemical signals in the cell and which in turn control metabolism, a major subject in Chapter 20. The prefix *allo* means 'other'; it refers to the existence on an enzyme of one or more binding sites other than for substrate. The ligands which bind to the allosteric sites are called **allosteric activators or inhibitors** and collectively allosteric modulators. They do not have to have any structural relationship to the substrates of the enzymes. *At a given substrate concentration*, an allosteric modulator increases or inhibits the activity of the enzyme when it combines at the allosteric site.

Allosteric modulators typically alter the K_m or apparent affinity of the enzyme for its substrates (S). (We have already explained that the K_m of an enzyme may not be a true affinity constant.) Most enzymes work in the cell at subsaturating levels of [S], so that an *increase in their affinity will increase their activity and a decrease in their affinity has the reverse effect.* At saturating levels of [S], the activity is unchanged, even if the affinity is changed, but this situation does not (usually) occur in the cell.

The mechanism of allosteric control of enzymes

Allosteric enzymes mainly have a multisubunit structure; they are made up of more than one catalytic protein molecule assembled into a single enzyme complex by noncovalent bonds. In some cases the enzyme is a collection of only catalytic subunits but in others there is a complex of catalytic and regulatory subunits. The latter have no catalytic activity but play a role in the response of the enzyme to allosteric effectors.

A plot of reaction velocity versus substrate concentration of allosteric enzyme is shown in Fig. 6.7. It differs from Michaelis–Menten enzyme kinetics in that the response of the enzyme velocity to changes in substrate concentration is sigmoidal, rather than hyperbolic. In fact, the plot looks similar in shape to the oxygen saturation curve of haemoglobin, and in what follows you will see many similarities between allosteric enzymes and haemoglobin, an allosteric protein.

When an allosteric activator binds to the enzyme, the sigmoid curve is moved to the left, and with an allosteric inhibitor, it moves to the right, as shown in Fig. 6.8. The allosteric activator increases the binding affinity of the enzyme for its substrate; the allosteric inhibitor reduces it. The sigmoidal shape of the response to substrate concentration means that over a range of substrate concentrations (centred at that giving half maximum velocity), the rate of enzyme catalysis is more sensitive to substrate concentration change than with a Michaelis–Menten type of enzyme. By the same token, it means that increasing or decreasing the substrate affinity of an enzyme displaying sigmoidal kinetics has a greater effect on reaction velocity at a given substrate concentration than with hyperbolic kinetics. Since, as stated, enzymes in the cell usually are working in the sensitive range of substrate

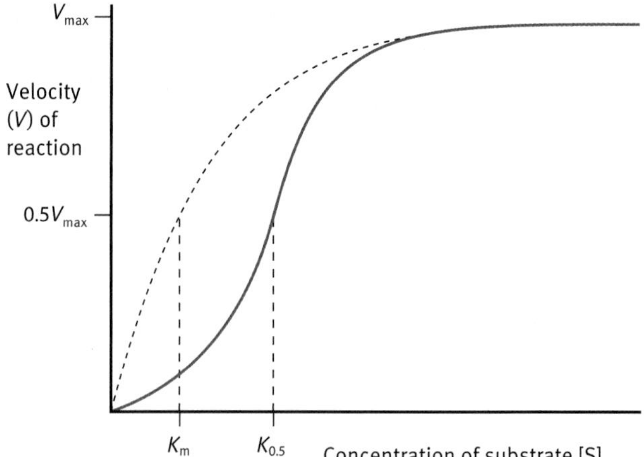

Fig. 6.7 Effect of substrate concentration on the rate of reaction catalysed by a typical allosteric enzyme. The dashed line shows, for comparison, the corresponding curve for a classical Michaelis–Menten enzyme. The term K_m is not strictly applicable to a non-Michaelis–Menten enzyme, and the term $K_{0.5}$ is used instead.

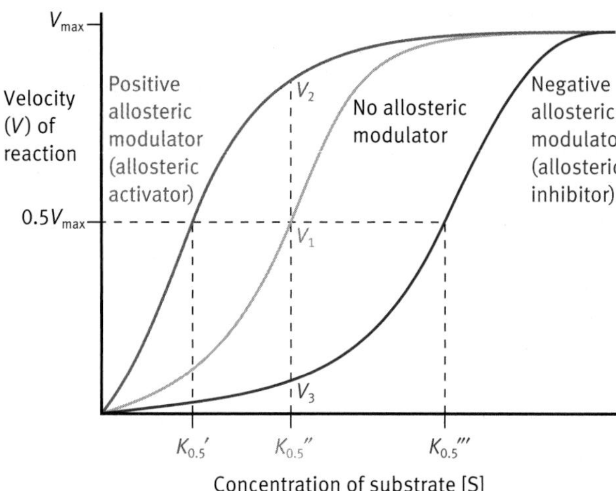

Fig. 6.8 Effect of substrate concentration on a typical allosterically regulated enzyme. V_1, V_2, and V_3 are, respectively, the reaction velocities observed with no allosteric modulator, with an allosteric activator, and with an allosteric inhibitor, at a fixed concentration of S ($K_{0.5''}$). The curves are drawn arbitrarily; actual shapes and positions will be a function of the particular system.

concentration, the sigmoidal response to substrate concentration maximizes the effect of an allosteric modulator.

What causes the sigmoidal response of reaction velocity to substrate concentration?

Binding of a substrate molecule to *one* of the catalytic subunits in an allosteric enzyme has the result that binding of subsequent substrate molecules to the other subunits of the enzyme occurs more readily. This is known as **positive homotropic cooperative binding** of substrate: 'positive' because affinity increases; 'homo' because only one type of ligand, the substrate, is involved; and cooperative because the different catalytic subunits are interacting. There are two models to explain the sigmoid kinetics. These were described when we dealt with haemoglobin, the classical allosteric protein. We will use here the concerted model for illustrative purposes.

In this model, either *all* of the subunits in an enzyme bind substrate with low affinity or *all* with high affinity, the two forms being in spontaneous equilibrium (Fig. 6.9), but with the equilibrium strongly to the low-affinity side. In incorporating the effects of allosteric modulators, the model further assumes that allosteric modulators alter the position of the equilibrium between the low-affinity (tense or T) state and the high-affinity (relaxed or R) state. If the allosteric modulator displaces the $T \rightleftharpoons R$ equilibrium to the right it results in a higher proportion of molecules being in the R state; it would, at a given substrate concentration, activate the enzyme. As the concentration of the positive allosteric modulator is increased and more of the enzyme converts to the high-affinity state, the substrate-reaction velocity curve tends towards the hyperbolic type. If, by contrast, the allosteric effector is of the negative type, it is assumed that it stabilizes the low-affinity T state.

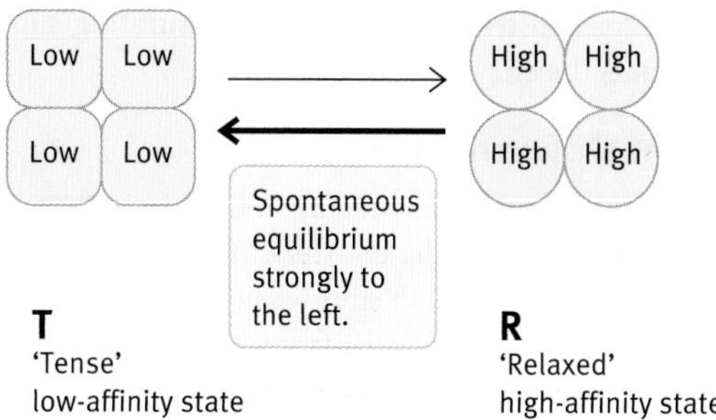

Fig. 6.9 Concerted model of cooperative binding of substrate (S). In this model the enzyme exists in two forms, T and R, the two being in spontaneous equilibrium. Binding of a substrate molecule to the R state swings the equilibrium to the right, thus increasing the affinity of all of the subunits in that molecule. 'High' and 'low' refer to affinities of the enzyme for its substrate. The model assumes that allosteric modulators that move the equilibrium towards the R form will activate. If a modulator moves the equilibrium to the T form, it will inhibit.

This results in more enzyme molecules being in this state and a decreased reaction rate in the collection of enzyme molecules as a whole – it inhibits. Allosteric activators or inhibitors of enzymes are, **heterotropic allosteric modulators** of enzymes, as 2,3-bisphosphoglycerate is for haemoglobin.

Aspartate transcarbamylase is the classical model of an allosteric enzyme

Aspartate transcarbamylase (ATCase) of bacteria was one of the earliest allosterically controlled enzymes to be studied in very great structural detail. It is useful to describe here because it illustrates the principle of feedback control of metabolism and the concerted mechanism of allosteric control. It also illustrates the principle that allosteric effectors do not have to have any structural resemblance to the substrate(s) of the enzyme. ATCase catalyses the first metabolic step committed to pyrimidine nucleotide synthesis (described in Chapter 19). One of the products of the pathway is cytidine triphosphate (CTP), which is the allosteric feedback inhibitor of bacterial ATCase. CTP shuts down the pathway when sufficient amounts of pyrimidine nucleotides are present. The enzyme is a large one consisting of six catalytic subunits and six regulatory subunits. These are arranged in the quaternary structure as two catalytic trimers linked by three regulatory dimers. The regulatory dimers do not interact with the catalytic active sites but are the sites for the CTP binding.

When the enzyme is combined with substrate, a massive quaternary structural change occurs. In the absence of substrate, the enzyme is in a compact T (tense) state with a low affinity for substrate but when combined with substrate it expands into a relaxed state with a higher affinity for substrate. The two states are in equilibrium amounts, dependent on the amount of substrate bound – the higher this is, the greater the proportion of enzyme in the high-affinity relaxed state. Binding of CTP, the allosteric inhibitor, to the regulatory subunits shifts the equilibrium to the tense state, lowers the affinity for substrate and thus at subsaturating substrate levels, inhibits the enzyme.

Reversibility of allosteric control

Allosteric control is virtually instantaneous in both its application and reversibility. The allosteric modulator attaches to its site by noncovalent bonds and when the concentration of ligand is reduced, it dissociates from combination with the enzyme and everything goes into reverse.

General properties of enzymes

Nomenclature of enzymes

Usually enzyme names end in '-ase', preceded by a term which most often indicates the nature of the reaction it catalyses and/or indicates the substrate. Thus *amylase* catalyses the hydrolysis of *amylose*. A *dehydrogenase* removes hydrogen atoms from a substrate: lactate dehydrogenase is an example. This is not always the case: proteolytic enzymes discussed later in the chapter often end in '-in'; *pepsin, chymotrypsin, plasmin,* and *thrombin* are examples. We will explain names more fully when the enzymes are considered in biochemical systems. An international committee has systematized all enzyme names, but as the systematic names are necessarily rather long, there are also shorter recommended names for everyday use. The systematic names and reference numbers are usually given once in published research papers and reference works to avoid any ambiguity.

Isozymes

It is quite common to find that the same reaction is catalysed by a number of distinguishable different enzymes, called isozymes or isoenzymes. There is considerable similarity in the amino acid composition of the different isozymes catalysing a given reaction, suggesting that they all have the same evolutionary ancestor but the genes coding for them have diverged somewhat to suit particular roles of the isozymes in the body. Isozymes are usually found in different tissues or in different locations in cells. The reason for multiple versions of the same enzyme is to tailor them to the specific needs of the cell. Thus in some cases the isozymes differ in substrate affinities, or have different regulatory mechanisms or other properties. This is not too surprising because different tissues have quite different roles. A classic example is that of hexokinase and glucokinase (see Chapter 11) – catalysing the same reaction (though with a different range of specificities), with different tissue distributions, different physiological roles and different K_m values.

Isozymes often have a number of different subunits, or protein molecules, which join together to form the complete enzyme. Thus another enzyme that you will meet later, **lactate dehydrogenase**, (see Chapter 12), has four subunits of two different types, one known as H because it is the main one in the heart enzyme, and the other known as M because of its association with muscle. Various combinations of H and M subunits produce the different isozymes in different tissues. Isozymes, including those of lactate dehydrogenase, can often be separated by electrophoresis because of their different electrical charge. Placed in a gel with a voltage gradient across it, they migrate at different rates.

Enzyme cofactors and activators

In the simplest case, the enzyme protein combines with a single substrate, the reaction occurs, and the products leave the active site to make way for another molecule. Frequently the enzyme requires a **cofactor** for activity. This may be a metal ion such as Mg^{2+} or Zn^{2+}, which participates in the reaction mechanism (e.g. carboxypeptidase, see A brief description of other types of protease Section in this chapter), or it may be an organic molecule attached to the enzyme in which case it is known as a **prosthetic group**. The protein part is then called an **apoenzyme** (*apo*=detached or separate) and the complete enzyme with the prosthetic group attached, a **holoenzyme**. The prosthetic group is sometimes a vitamin derivative, and some vitamins activate several enzyme species. Since each vitamin molecule in this case activates an enzyme molecule, which can bring about reactions in vast numbers of substrate molecules, it explains why small amounts of vitamins can have a huge effect on the body. **Coenzymes** also have vitamins as components and behave somewhat similarly but are not firmly attached to the enzyme (details will be given at appropriate places later in the book). They are very much like substrates in that in most cases they react and leave the enzyme. Their role is to couple enzymic reactions together. For example, the coenzyme nicotinamide adenine dinucleotide (NAD^+; see Chapter 12) is reduced to $NADH+H^+$ (equivalent to NAD^+ with two hydrogen atoms attached) in dehydrogenation reactions.

$$AH_2 + NAD^+ \rightarrow NADH + H^+$$

The reduced coenzyme leaves the enzyme and becomes the substrate for a second one.

$$B + NADH + H^+ \rightarrow BH_2 + NAD^+$$

The overall result is that the coenzyme acts as the intermediary in transferring hydrogen atoms from one substrate to another. This type of activity plays a prominent role in the release of energy from foodstuffs as described in later chapters.

Covalent modification of enzymes

Covalent modification of enzymes, for instance by transfer of phosphate groups, is a common mechanism by which their activity is regulated in cells. This is discussed in detail in the chapters on metabolic regulation (see Chapter 20) and cell signalling (see Chapter 29).

Effect of pH on enzymes

The activity of an enzyme is influenced by pH in several ways. The protein structure is influenced by the state of its ionizable groups and the function of the active site may be likewise dependent on this. The ionization of the substrate itself may also be affected. The rate of catalysis is therefore dependent on the pH. Enzyme pH activity profiles vary from one to another but the optimum is often around neutral pH; a typical plot is shown in Fig. 6.10(a). Exceptions occur such as the case of the digestive enzyme, pepsin, which functions in the acidic stomach contents; its pH optimum is near 2.0.

Effect of temperature on enzymes

Temperature also affects enzyme activity rates. As the temperature increases, the rate of most chemical reactions increases

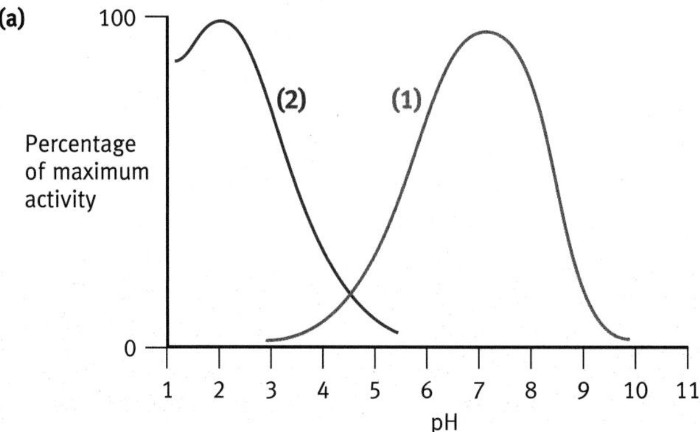

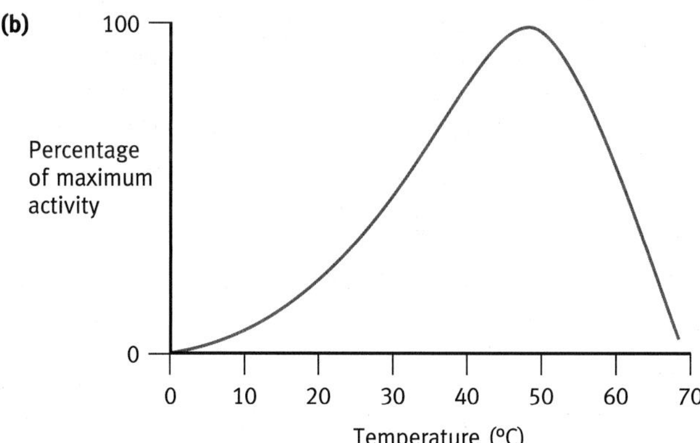

Fig. 6.10 Effect of pH and temperature on enzyme activity. **(a)** Effect of pH on enzyme activity. Curve 1 is typical of the majority of enzymes with maximal activity near physiological pH. Curve 2 represents pepsin, an exceptional case, since the enzyme functions in the acidic stomach contents. **(b)** Effect of temperature on a typical enzyme. The precipitous drop at high temperatures is due to enzyme destruction (though a number of heat-stable enzymes are known). Note that, as described in the text, the temperature 'optimum' of an enzyme has little significance, since the shape of the curve will depend on the length of time the enzyme is maintained at a given temperature before measuring. Adapted from Figure 1 in Dodson, G. and Wlodawer, A.; Catalytic triads and their relatives; Trends Biochem Sci.; 1998; 23, 347–52; Elsevier.

(approximately two-fold for each 10 °C) but, because of the inherent instability of most protein molecules, the enzyme is inactivated at higher temperatures. Thus a typical enzyme optimum temperature plot would appear as shown in Fig. 6.10(b) but, although perhaps useful in a practical sense, this has little absolute significance since the optimum will depend on the experimental time period used in measuring the rates; the shorter the time the less will be the destructive effect of higher temperatures. A few enzymes are stable to high temperatures but, in general, temperatures over 50 °C are destructive and some enzymes are more labile still.

Effect of inhibitors on enzymes

The activity of an enzyme can be affected by inhibitory compounds. In this section we are not so much considering physiological inhibitors as drugs and toxins that can be used experimentally or therapeutically, or may have poisonous effects if they are ingested or encountered in the environment. Such inhibitors may be reversible or irreversible. In the former case, the enzyme and inhibitor exist in a reversible equilibrium ($E+I \rightleftharpoons EI$). Irreversible inhibitors bind to the enzyme and do not dissociate from it to an appreciable extent; the extreme case of this is where the inhibitor becomes covalently attached. The effect of aspirin on an enzyme, described in the following section, is one such case. Penicillin is another, as penicillin drugs covalently bind and inhibit an enzyme that cross-links and strengthens bacterial cell walls.

Competitive and noncompetitive inhibitors

Reversible inhibition of an enzyme may be of different types. One class, called **competitive inhibitors**, simply mimic the substrate and compete with the latter for binding to the active site. The degree to which the latter is occupied by the inhibitor will determine the degree of inhibition. The inhibition will be a function of the relative affinities for the enzyme of substrate and inhibitor and their relative concentrations. For enzymes showing Michaelis–Menten kinetics, it is possible to distinguish between competitive and noncompetitive inhibition using the double reciprocal plot already described and which is shown in Fig. 6.11. A competitive inhibitor will have no effect at infinite substrate concentration since the substrate will completely win in the competition to bind to the active site. The intersection of the reciprocal plot with the vertical axis (which represents infinite [S]) will be the same whether inhibitor is present or not (i.e. *V_{max} is unchanged*). Since the inhibitor interferes with binding of the substrate to the active site, it will, however, appear to reduce the affinity and change the K_m. The intersection with the horizontal axis, which gives the reciprocal of the K_m value, is changed by the inhibitor (*K_m is increased*).

Inhibitors of enzymes play an important part in the treatment of diseases. For example, physostigmine is a competitive inhibitor of acetylcholinesterase and is used for patients with myasthenia gravis (see The problem of autoimmune reactions

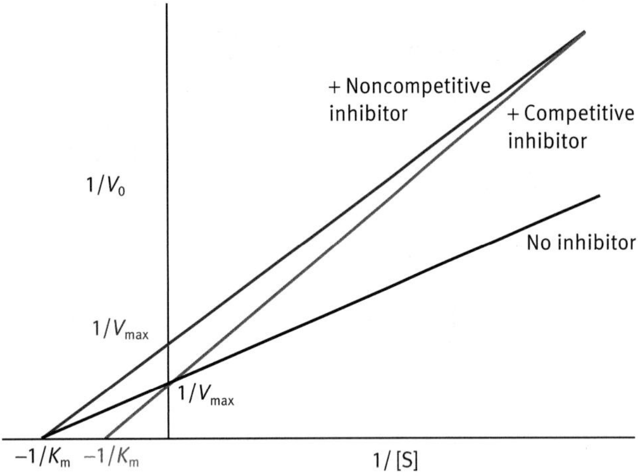

Fig. 6.11 Double reciprocal plot of enzyme reactions in the presence of competitive and noncompetitive inhibitors respectively. See text for explanation of inhibitors.

in Chapter 33). The so-called '**statin**' drugs, which are used to treat people with high cholesterol levels, are competitive inhibitors of the HMG-CoA reductase enzyme that controls the synthesis of cholesterol (see Box 11.2).

A **noncompetitive inhibitor** binds to the enzyme at a position separate from the active site so that there is no competition with the substrate; this is readily seen in a double reciprocal plot, as shown in Fig. 6.11, in which it is seen that *the V_{max} at infinite substrate concentration is reduced while the K_m is unchanged.* Noncompetitive inhibitors work in various ways. For example, a heavy metal such as mercury may react with a thiol group essential for catalytic activity. Removing the metal with a thiol compound or a metal-chelating agent may reverse the inhibition.

In other cases, a noncovalent inhibitor may covalently acylate the active site of the enzyme in an irreversible manner. Thus aspirin inactivates an enzyme (cyclooxygenase) involved in prostaglandin synthesis (see Box 17.2) as follows:

ENZ—OH + Aspirin (COOH, O—C(=O)—CH_3)

Cyclooxygenase Aspirin

↓

ENZ—O—C(=O)—CH_3 + 2-Hydroxybenzoic acid (COOH, OH)

Acylated enzyme 2-Hydroxybenzoic acid

Prostaglandins are involved in pain and inflammation, so inhibition of their synthesis by aspirin ameliorates these symptoms.

Although competitive inhibitors are often encountered, in practice purely noncompetitive enzyme inhibitors that have no effect on enzyme-substrate interactions are rare. Instead **mixed inhibition** may be observed, in which both V_{max} and K_m are affected to some extent.

A third type of inhibition is known as ***uncompetitive*** (subtly different from noncompetitive). In this, the inhibitor binds to the enzyme and does not affect substrate binding, but it binds only to the ES form of the enzyme (whereas a noncompetitive inhibitor bind both E and ES). Like noncompetitive inhibitors, uncompetitive inhibitors reduce V_{max} since their effect cannot be ameliorated by increasing [S], but unlike noncompetitive inhibitors they reduce K_m. The reduction in K_m may seem surprising, but what happens is that the inhibitor effectively removes ES complex, so more is formed to restore the equilibrium of the $E+S \rightleftharpoons ES$ reaction. The enzyme's affinity for its substrate appears to be increased.

Mechanism of enzyme catalysis

We can illustrate how one class of enzymes works in structural terms, using the enzyme **chymotrypsin** as an example. Enzymes may increase reaction rates by 10^{14} times. A 10^{14}-fold increase means that an enzyme catalyses in 1 second an amount of chemical reaction that without the enzyme would require hundreds or thousands of years.

The chemistry of the chymotrypsin reaction that follows involves little more than a proton jumping from one group to another and back again. It is one of the most satisfying illustrations of the remarkable abilities of proteins to perform specific chemical tasks at incredible speeds and by mechanisms, which are, in concept, very simple.

Mechanism of the chymotrypsin reaction

Chymotrypsin is a digestive enzyme produced by the pancreas; it hydrolyses specific peptide bonds of proteins in the diet. Its digestive role is described in Chapter 10.

The **active site** of an enzyme is usually in a cleft of the protein, into which fits the substrate to be attacked, as already stated. In the case of chymotrypsin, the natural substrate is a polypeptide and it hydrolyses those peptide bonds whose carbonyl group is donated by a large hydrophobic amino acid residue (mainly the aromatic ones phenylalanine, tyrosine, and tryptophan but also methionine). The active centre has a large hydrophobic pocket to accommodate the bulky hydrophobic group. The enzyme is an **endopeptidase** – it hydrolyses internal peptide bonds of proteins, in contrast to an **exopeptidase**, which hydrolyses terminal peptide bonds (*endo*=within;

exo=without). The reaction given here in nonionized structures for clarity is as follows:

$$\text{R-CONH-R}' + H_2O \longrightarrow \text{RCOOH} + \text{R}'\text{N}\begin{smallmatrix}\text{H}\\ \text{H}\end{smallmatrix}.$$

R represents a large hydrophobic residue of a polypeptide substrate that provides the carbonyl group of the peptide bond. R′ represents the rest of the peptide.

Chymotrypsin is one of a group of proteases known as **serine proteases** because the active site contains a serine residue. Hydrolysis of the peptide bond takes place in two stages. In the first, the serine-OH becomes acylated and the first product ($R'NH_2$) is released. In the second stage the acyl enzyme is hydrolysed and the second product (RCOOH) released. (Nonionized structures are used for clarity; R and R′ are as defined in the equation already given above.)

Stage 1:

$$\underset{\text{Peptide substrate}}{\text{RCONHR}'} + \underset{\text{Enzyme with serine–OH group}}{\text{Enz-OH}} \longrightarrow \underset{\text{Intermediate acyl enzyme}}{\text{RCOO-Enz}} + \underset{\text{First product}}{\text{R}'\text{N}\begin{smallmatrix}\text{H}\\ \text{H}\end{smallmatrix}}.$$

Stage 2:

$$\text{RCOO-Enz} + \text{HOH} \longrightarrow \underset{\text{Second product}}{\text{RCOOH}} + \underset{\text{Enzyme in original state}}{\text{Enz-OH}}.$$

The ester bond in the acyl-enzyme intermediate is hydrolysed by water, releasing the second product (RCOOH) and restoring the enzyme to its original state.

There are two main questions to be considered.

1 Why is the –OH of the serine residue so reactive in stage 1 of the enzymic reaction? Serine itself is a very stable molecule and its hydroxyl group is unreactive at neutral pH; other serine residues in the chymotrypsin molecule are inert, so why is that one reactive? The peptide bond of the substrate to be attacked is likewise quite stable at neutral pH and on its own does not react at a perceptible rate.

2 Why does water so readily hydrolyse the ester bond in stage 2 of the reaction? A carboxylic ester is relatively stable at neutral pH in aqueous solution. To answer these questions, we must look at the structure of the catalytic centre of the enzyme.

The catalytic triad of the active site

Projecting into the active site of the enzyme are the side chains of three amino acid residues that are part of the polypeptide chain comprising the enzyme – aspartate, histidine, and serine, known as the **catalytic triad**. Although quite widely separated in the polypeptide chain of the enzyme, folding of the chain brings them together in the active centre as diagrammatically shown in Fig. 6.12. This is an important role of enzyme proteins – to bring together and fix in optimal relationships the reactive groups involved in catalysis. This is a reason why enzymes are large.

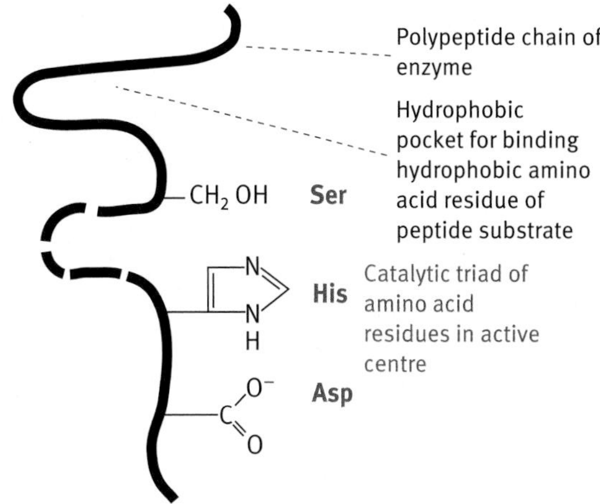

Fig. 6.12 The active site of chymotrypsin. Folding of the polypeptide brings together the three amino acid residues of serine, histidine, and aspartate. In the unfolded chain they are quite widely separated. The substrate specificity of chymotrypsin for peptides whose carbonyl group is donated by a hydrophobic amino acid derives from the specific hydrophobic binding site of the active site.

A few points about these groups first.

- The side-chain carboxyl group of aspartate (the ionized form of aspartic acid) has a pK_a of about 4 and is therefore dissociated at physiological pH.
- The serine –OH group with a pK_a of about 14 is not significantly dissociated.
- The imidazole side chain of histidine is the interesting one with a pK_a in its protonated form of about 6. This means that at physiological pH, there is a rapid equilibrium between the protonated and unprotonated states. On the Brønsted–Lowry definition of an acid being a proton donor and a base being a proton acceptor, in its protonated form histidine is an acid and in its unprotonated form it is a base.

Thus the histidine side chain in its protonated form can readily function as a proton donor, in which case it is acting as a **general acid**. In its unprotonated form, it can accept a proton and function as a general base. It can thus promote **general acid–base catalysis**, as will shortly be explained.

It might be helpful to remember some chemical principles. A covalent bond involves two atoms sharing a pair of electrons between them. Each electron is attracted to both atoms and this mutually holds the two atoms together. Formation of the bond releases energy, which is why it forms. A simple example is the formation of a covalent bond between two hydrogen atoms to form a hydrogen molecule (H_2):

$$H^{\cdot} + H^{\cdot} \rightarrow H{:}H$$

Each hydrogen atom has a single electron so that the two atoms contribute equally to formation of the bond. In some molecules or ions, an atom has an unshared pair of electrons that can be donated to another atom to form a covalent bond. The chemical entities containing such donor atoms are called **nucleophiles**. The acceptor of the pair of electrons is called an **electrophile**. The process of forming such bonds is known as a **nucleophilic attack**. For nomenclature reasons in chemistry, in the one specific case where the electron acceptor is a proton (H^+) the chemical entity supplying the electrons is called a base and not a nucleophile, but the principle is unchanged.

$$\underset{\text{Nucleophile}}{X{:}} + \underset{\text{Electrophile}}{Y} \rightarrow X^{+}{:}\,Y^{-}$$

One of the nitrogen atoms of the imidazole side chain of histidine has such an unshared pair of electrons that can interact with a proton. The dissociation of the imidazole group of histidine occurs as follows:

HN⌒N⁺H ⇌ HN⌒N: + H^+.

In donating two electrons to H^+ to form the covalent bond in the reverse reaction above, the nitrogen atom acquires a positive charge. Note that the imidazole ring of histidine can exist in two tautomeric forms. In the unprotonated form, the single hydrogen atom can be on either of the two nitrogen atoms. With that background we can proceed to the mechanism of chymotrypsin catalysis.

The reactions at the catalytic site of chymotrypsin

All three amino acid residues of the catalytic triad – aspartate, histidine, and serine – are essential for the enzyme to function properly but actual chemical changes (reactions) occur only between the histidine and serine, so we will deal with these first to keep it simple. The role of aspartate will be explained later.

A peptide substrate molecule attaches to the catalytic site of chymotrypsin by the binding of its hydrophobic group to a specific nonpolar pocket such that the carbonyl carbon (C=O) atom of the peptide bond to be attacked is close to the serine –OH. Simultaneously the hydrogen atom of the serine –OH transfers to the histidine nitrogen atom and the oxygen atom forms a bond with the carbonyl carbon atom of the substrate as shown in step 1 of Fig. 6.13. This produces the first tetrahedral intermediate shown in the yellow box in the figure. (The term tetrahedral refers to the organization of bonds of the carbon atom of the bond to be broken; a tetrahedral carbon has its four bonds pointing to the vertices of a tetrahedron.) In step 1 the formation of the tetrahedral intermediate requires the C=O of the carboxyl group to become single-bonded thus forming an oxyanion as shown. This is stabilized by interacting with an oxyanion 'hole' in the enzyme.

That is what happens, but how can serine react in this way? The serine –OH group is normally unreactive at neutral pH. For the oxygen atom to form the bond with the carbon atom of the substrate, it has to lose its hydrogen atom as a proton, but, with a pK_a of 14, the –OH group does not significantly dissociate except in strongly alkaline solutions whose pH is near the pK_a of the group. The answer lies in the ability of the histidine N atom to abstract the hydrogen – it is acting as a **general base** (defined, we remind you, as a group that accepts a proton). It does so because the serine and histidine of the catalytic triad are oriented so that the hydrogen of the OH group is perfectly positioned to interact with the nitrogen atom of the histidine.

The histidine with its acquired proton is now a **general acid** – it can donate a proton. (You can now see why the process is referred to as **general acid–base catalysis**.) The proton is transferred to the tetrahedral intermediate causing it to break down as shown in step 2 of Fig. 6.13, to liberate the first product ($R'NH_2$) and form the acylated enzyme.

We are halfway there; the next step is to hydrolyse the ester bond of the acyl-enzyme intermediate and thus liberate the second product of peptide hydrolysis (RCOOH) and restore the enzyme to its original state ready for reaction with the next substrate molecule. It is basically a repeat of the strategy used in the first stage. What is required is for the oxygen atom of water to make a nucleophilic attack on the carbonyl carbon atom of the ester bond of the acyl-enzyme intermediate. When a molecule of water enters the active site (Fig. 6.13, step 3) the histidine group, acting as a general base abstracts a proton (just as it did from serine in the previous stage of the reaction). The water oxygen atom makes a nucleophilic attack forming the second tetrahedral intermediate (Fig. 6.13, step 4). Exactly as before, the protonated histidine is now a general acid; it donates its acquired proton back to the tetrahedral intermediate (Fig. 6.13, step 5) resulting in completion of the reaction and liberation of the second product (RCOOH). The serine –OH is restored to its original state ready for reaction with the next substrate molecule.

The actual mechanisms by which the chemical changes are brought about are thus remarkably simple, involving little more than histidine acquiring a proton and giving it back at each of the two stages.

What is the function of the aspartate residue of the catalytic triad?

If, by genetic engineering, the aspartate residue is deliberately converted to an asparagine residue in which the carboxyl

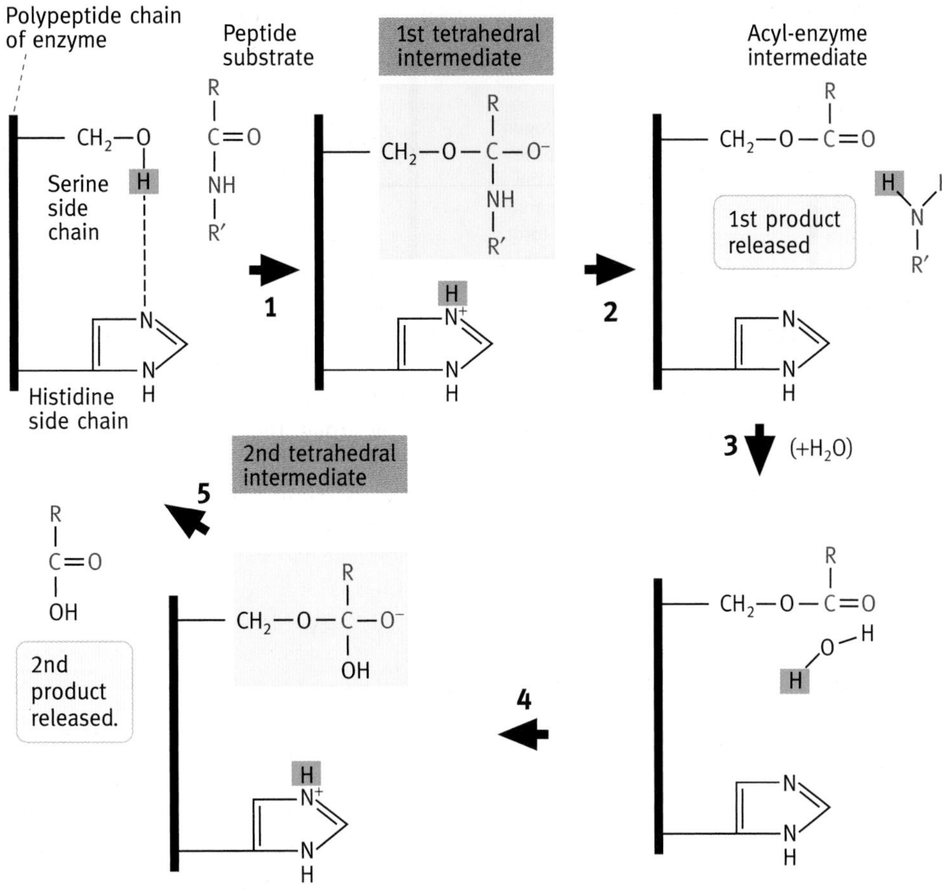

Fig. 6.13 Reactions at the catalytic centre of chymotrypsin involved in the hydrolysis of a peptide substrate. For ease of presentation the enzyme polypeptide chain is given as a straight line with only two of the catalytic triad residues shown. The role of the third residue (aspartate) is described later. The broken line represents a hydrogen bond. The steps are described in the text.

group now becomes an amide group that does not significantly dissociate, the catalytic activity of chymotrypsin falls by a factor of 10,000. The aspartate carboxylate anion is thus essential but nevertheless it does not undergo any chemical reaction during the catalytic process. Why then is it needed? The aspartate carboxylate anion forms a strong hydrogen bond with the histidine side chain as shown in Fig. 6.14. Its main function is to hold the histidine residue in the orientation and tautomeric form shown in the figure, so that the nitrogen atom which accepts the proton from the serine residue is always facing the latter, optimally positioned to abstract the proton from the -OH group. If the aspartate is converted into asparagine, the hydrogen-bonding potentiality is very much weaker and this immobilizing effect on the histidine residue is missing. It is historically interesting that initially it was believed that this residue in the catalytic triad was in fact asparagine. When it was appreciated that aspartate would make more sense mechanistically, the structure determination was re-checked revealing that this residue is indeed aspartate.

Other serine proteases

The catalytic triad mechanism has been adopted by a variety of hydrolytic enzymes. The serine proteases chymotrypsin, trypsin, and elastase have the same mechanism, all with the aspartate, histidine, and serine residues. They all hydrolyse peptides but have different specificities for the component amino acid residues forming the peptide bond. The active sites differ in the pockets in the active sites required for the binding of the substrates; they will accept only the particular amino acid side chains of their specific substrates. The pocket in **chymotrypsin** is hydrophobic (Fig. 6.15(a)), whereas that of **trypsin** has a negatively charged aspartate residue (different from the one in the catalytic triad) to which binds the partially charged basic side chains of trypsin-specific substrates (Fig. 6.15(b)). That of **elastase** is smaller and access to it is restricted by threonine and valine residues so that the enzyme is specific for peptide bonds whose carbonyl group is contributed by amino acid residues with small side chains (Fig. 6.15(c)). The three enzymes described above are structurally closely related to one another

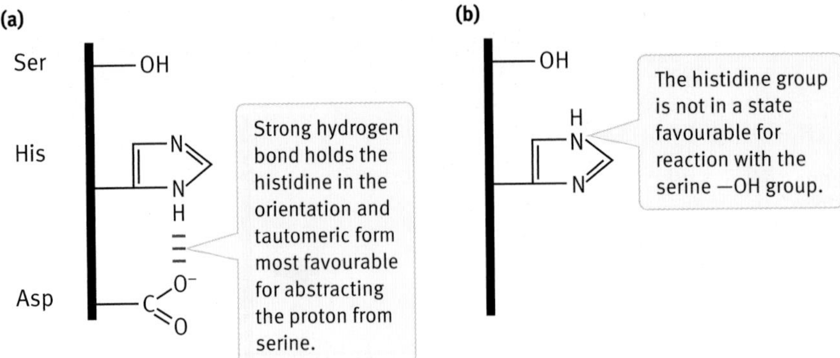

Fig. 6.14 The function of aspartate in the catalytic triad. In situation **(a)** the histidine is held in the form shown, by a strong hydrogen bond with aspartate. This results in the nitrogen with the unshared pair of electrons facing the serine proton, which it can abstract. If the histidine is not held by being hydrogen bonded to aspartate, the situation shown in **(b)** could result with the protonated nitrogen facing the serine and therefore the serine proton would be less efficiently abstracted.

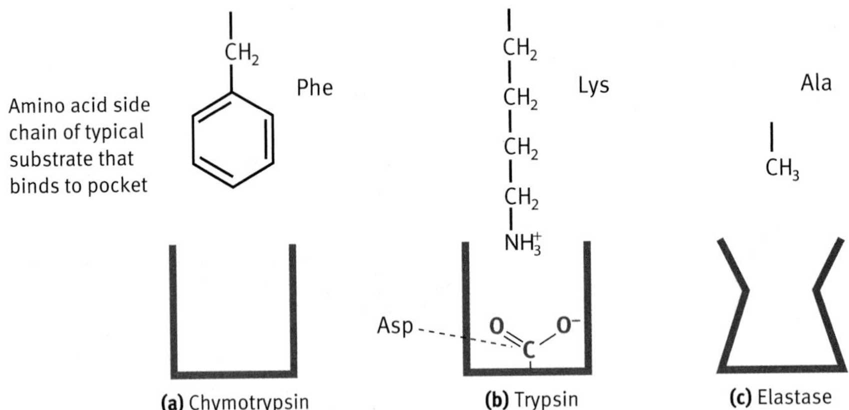

Fig. 6.15 Simplified diagram of the pockets in the active sites of chymotrypsin, trypsin, and elastase into which fit the amino acid side chains of their respective substrates. That of chymotrypsin accommodates a bulky hydrophobic group such as the side chains of phenylalanine or tryptophan, that of trypsin accommodates the positively charged side chain of lysine or arginine that bind to the negatively charged aspartic acid residue present in the binding site. The elastase pocket accepts only smaller amino acid side chains of substrate molecules, the entrance to the binding site being restricted by the side chains of valine and threonine residues.

and clearly have an evolutionary relationship. The bacterial proteolytic enzyme, subtilisin (from *B. subtilis*) is a totally different protein but still has the identical catalytic triad; the independent convergent evolution of this emphasizes its basic importance. Several enzymes unrelated in function to the proteinases, also have the catalytic triad. Acetylcholinesterase (see the Nerve-impulse transmission Section in Chapter 7) is an example.

A brief description of other types of protease

As well as the serine proteases, there are three other classes of protease in terms of the structures of their catalytic sites. These are the thiol, aspartic, and zinc proteases. The **thiol proteases** are very similar to the serine types but instead of an activated serine hydroxyl group as in chymotrypsin, there is an activated thiol group of cysteine. An intermediate thioester (RCO—S—Enz) is formed instead of the carboxylic ester RCO—O—Enz) as in chymotrypsin. The plant proteolytic enzyme papain (found in the latex juice of the papaya) is an example.

Pepsin, whose digestive role is described in Chapter 10, belongs to the **aspartic protease** class. It has a catalytic diad of two aspartate residues: one is unprotonated and can accept a proton, the other is protonated and can donate one. The two act in turn as a general acid and a general base respectively and reverse their roles with each round of reaction. Several enzymes of this class are known, such as the kidney enzyme **renin**, which is involved in blood-pressure control. The finding that HIV (the AIDS virus) has an aspartic protease required for its replication has heightened interest in the group. Inhibition of this enzyme is a site for therapeutic attack on the disease.

A third class of proteases is the metalloproteases, which depend on a metal ion, usually zinc, in the active site. An example of the class of **zinc proteases** is **carboxypeptidase A**, a digestive enzyme with a preference for hydrophobic amino acid residues, which hydrolyses off the terminal carboxyl residue from peptides. It exhibits a remarkably large conformational change on substrate binding, a good example of induced fit. Mechanistically the catalytic process has strong resemblances to that of chymotrypsin in that general acid–base catalysis is involved – a proton is removed from water by transfer, in this case to a glutamate residue. The activated water molecule then attacks the carbonyl carbon of the peptide, forming a tetrahedral transition state, as shown in Fig. 6.13, step 4). Activation of the water molecule in this case is promoted by its binding to a zinc atom, which is bound to the enzyme active site.

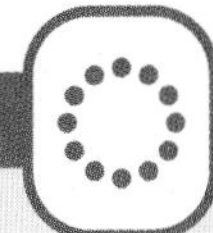

Summary

An enzyme is a catalyst that brings about a specific reaction but is unchanged in the process. It cannot affect the equilibrium of a reaction. The free energy change of a reaction determines whether it may proceed but says nothing about whether it will occur.

For a chemical reaction to happen, an energy barrier has to be overcome. The reactants must be activated to a higher energy transition state, which rapidly converts to products. The energy for this is the activation energy, supplied in uncatalysed reactions by molecular collisions.

An enzyme is a protein that binds substrates and lowers the activation energy of the reaction it catalyses. It does so because the active site binding the substrate is perfectly complementary not to the substrate itself but to the transition state. It also aligns substrates and provides a catalytic site. This is a small region of the protein that contains chemical groupings essential for the catalysis. On binding of substrates to enzymes the protein often changes shape, a process known as the induced-fit mechanism.

Enzyme kinetics explores enzymes by measuring reaction rates. The Michaelis–Menten equation explains the kinetics of many enzyme-catalysed reactions in terms of the substrate binding to the enzyme and the enzyme-substrate complex then breaking down to liberate products. An enzyme displaying Michaelis–Menten kinetics gives a hyperbolic curve when initial reaction velocity is plotted against increasing substrate concentration. An important constant is the K_m or Michaelis constant, which is the substrate concentration at which the enzyme has half its maximal activity. It is useful in comparing enzyme affinities for substrates. Other kinetic constants that give useful insight into enzyme activities are turnover number or K_{cat} and the specificity constant of an enzyme, K_{cat}/K_m.

Allosteric enzymes are multisubunit proteins with allosteric sites, to which molecules (modulators) attach and affect the activity. The effect of substrate concentration on their reaction rates is sigmoidal rather than hyperbolic. Attachment of modulators usually alters the affinity of the enzyme for its substrate(s) and this affects its rate of activity at a given sub-optimal substrate concentration. Two theoretical models that account for the properties of allosteric enzymes are the concerted model and the sequential model. Both involve conformational changes. Allosteric control is virtually instantaneous and reversible. It is a crucial mechanism for regulation of diverse metabolic pathways.

Enzymes are affected by temperature, pH, and inhibitors. Inhibitors may compete with the substrate for the active site or be noncompetitive and inhibit by binding elsewhere. A graphical procedure, the double reciprocal plot, which is an adaptation of the Michaelis–Menten plot, distinguishes between the two.

Enzymes often need activators, prosthetic groups, or coenzymes attached to them. Activators may be metal ions; prosthetic groups are permanently attached small organic molecules; coenzymes differ from these in that they attach transitorily.

Different enzymes catalysing the same reaction are known as isoenzymes or isozymes. They have characteristics tailored to their particular roles, usually in different tissues.

Chymotrypsin is a digestive enzyme that hydrolyses proteins and the mechanism of which is understood in detail. Its active site consists of a hydrophobic substrate-binding site and a catalytic triad of amino acid residues (serine, histidine, and aspartate). The serine –OH group is reactive because the histidine group is perfectly positioned to abstract the serine –OH proton. During the hydrolysis of the peptide bond an intermediate is formed that is decomposed into an acylated serine group by the donation of the abstracted proton from the histidine group. The function of the aspartate residue is to hydrogen bond the histidine residue and fix it in the favourable orientation. The second stage of the reaction is to hydrolyse the acylated serine complex liberating the bound peptide-acyl group. A water molecule is activated by the abstraction of a proton by the histidine group, forming an intermediate that is also decomposed by the donation of the abstracted proton from the histidine residue. The mechanism of this catalysis is astonishingly simple, involving essentially a histidine residue activating the serine -OH group by abstracting a proton in the first half of the reaction and then activating a water molecule in a similar manner. Other proteases use different but related mechanisms.

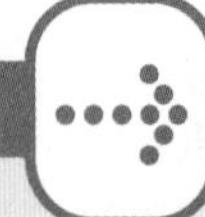

Further reading

To access the further reading, please scan the QR code image or go to http://global.oup.com/uk/orc/biosciences/molbiol/snape_biochemistry/student/reading/cho6/

Problems

1 What information can be drawn from a plot of the rate of activity of an enzyme against temperature?

2 The oxidation of glucose to CO_2 and water has a very large negative $\Delta G^{0\prime}$ value, and yet glucose is quite stable in the presence of oxygen. Why is this so?

3 Explain how an enzyme catalyses reactions.

4 In the case of some enzymes a K_m value is a true measure of the affinity of the enzyme for the substrate. In other cases it is not. Explain this.

5 (a) Compare the relationship between substrate concentration and rate of enzyme catalysis in a nonallosteric enzyme and a typical allosteric enzyme.
(b) What is the advantage of the allosteric enzyme substrate/velocity relationship?

6 If a typical allosterically controlled enzyme is exposed to saturating levels of substrate, what would be the effects of allosteric activators on the reaction velocity?

7 Describe how you would determine whether an inhibitor of an enzyme reaction is a competitive or noncompetitive one. Draw a graph to illustrate your answer.

8 A competitive inhibitor for a specific enzyme works by combining with its active site and blocking access to it by its substrate. Transition state analogues have been found to be very effective in some cases. Why would you expect such a molecule to be more effective than a competitive analogue of the substrate of the same enzyme?

9 If you want to compare the amount of an enzyme in different preparations by measuring reaction rates what precaution must you take if the measurements are to be meaningful?

10 In the active site of chymotrypsin, there is a serine residue that makes a nucleophilic attack on the carbonyl carbon atom of the peptide substrate forming a covalent bond to it. For this, the proton of the serine –OH has to be removed at the same time. Other serine residues in the protein are totally inert in this respect, as is free serine. Explain how this reaction is triggered in the catalytic centre of the enzyme.

11 The active site of chymotrypsin contains an aspartate residue; although this does not participate in the chemical reaction involved in catalysis, its conversion to asparagine lowers the activity of the enzyme 10,000 fold. Why is this so?

12 Chymotrypsin, trypsin, and elastase all have the same catalytic mechanism but have different specificities. Explain the reason for this.

13 Explain the meaning of the terms thiol protease and aspartyl protease.

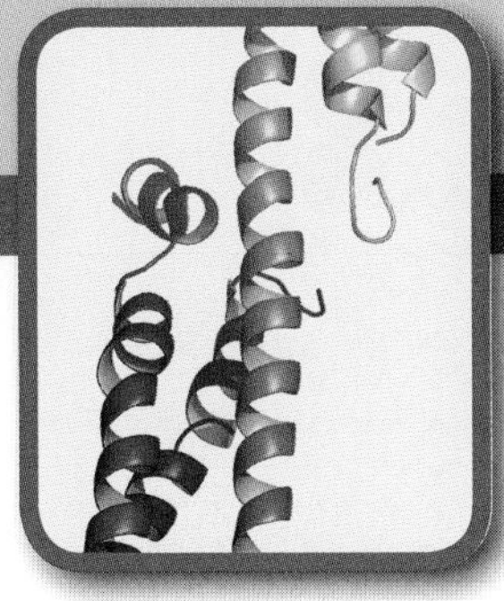

Chapter 7

The cell membrane and membrane proteins

In Chapter 3 we discussed the importance of three types of weak, noncovalent bonds as well as the hydrophobic force resulting from water molecules rejecting nonpolar molecules. Formation of weak bonds (as with all bonds) involves a decrease in free energy so that only when formation of such bonds is maximized do you have the preferred most stable structure. This forms a link to the topic now to be dealt with – the cell membrane, which is an assembly of molecules held together mainly by noncovalent bonds.

Other links are to Chapter 1 where the need for the first self-replicating system to be contained by a membrane was discussed and to Chapter 2 where the variety of membranous structures inside most eukaryotic cells was described.

Basic lipid architecture of membranes

The basic structure of all biological membranes is the lipid bilayer, the constituents of which are **polar lipids**. A polar lipid (Fig. 7.1) is an **amphipathic** or **amphiphilic** molecule (*amphi*=both kinds of) because it has a polar part, the so-called head group, and a pair of nonpolar hydrophobic tails. In the presence of water, such molecules are forced into structures in which the hydrophilic head groups have maximum contact with water, which maximizes noncovalent bond formation. Conversely the hydrophobic tails are forced into minimum contact with water because such contact forces water into a higher-energy arrangement (see Chapter 3) and therefore is resisted. Also, the contact between hydrophobic tails maximizes van der Waals interactions (see Chapter 3), again minimizing the energy level of the structure.

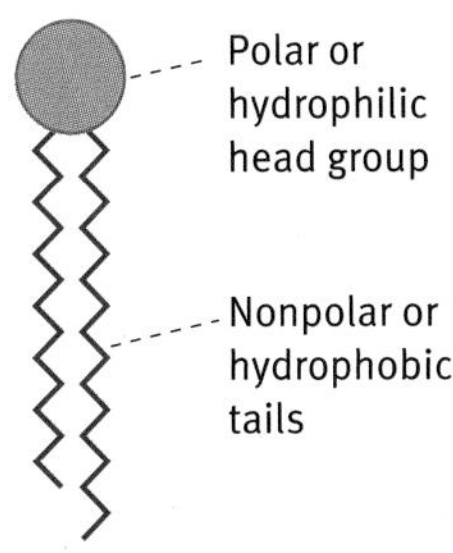

Fig. 7.1 An amphipathic molecule of the type found in cell membranes.

When polar lipids of the type found in membranes are agitated in an aqueous medium, one of the structures formed is called a **liposome** – a hollow spherical lipid bilayer. As mentioned in Chapter 1 the formation of liposomes may be relevant to the origin of cells. A lipid bilayer has two layers of polar lipids with their hydrophobic tails pointing inwards and their hydrophilic heads outwards in contact with water and each other as shown in Fig. 7.2. The presence of two hydrophobic tails favours this arrangement rather than that of a micelle, which is a solid sphere with all the tails pointing to its centre. Single tails favour micelle formation.

If a synthetic liposome is sectioned and stained with a heavy metal that attaches to the polar heads, the lipid bilayer appears in the electron microscope as a pair of dark parallel 'railway lines' owing to absorption of electrons by the metal stain. A living cell membrane treated in the same way has an identical appearance.

The polar lipid constituents of cell membranes

There are a number of membrane polar lipids, structures that look quite different from one another on paper, but in space-filling models they all have the same basic shape as the molecule shown in Fig. 7.1, with a polar head and two hydrophobic tails. Before dealing with the structures of the various polar lipids, a brief discussion of the nature of lipids in general may be useful.

A lipid is a fat. **Neutral fat** is derived from fatty acids; *neutral fats never occur in membranes.* A fatty acid has the structure RCOOH where R is a long hydrocarbon chain. The commonest fatty acids have 16 or 18 carbon atoms (C_{16} and C_{18}). These

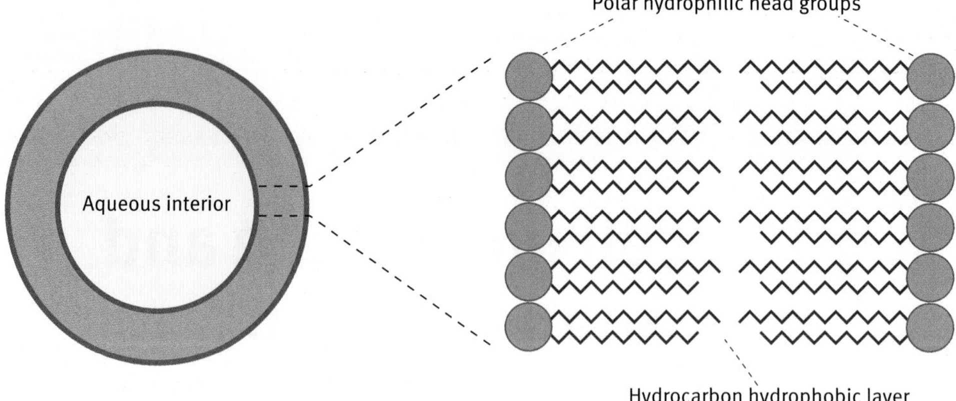

Fig. 7.2 A synthetic liposome made of a lipid bilayer structure.

may be called hexadecanoic and octadecanoic acids, respectively, if fully saturated or, by their common names, palmitic and stearic acids. They can also be represented as 16:0 and 18:0, indicating the number of carbon atoms and the number of double (unsaturated) bonds.

Stearic acid (C_{18}) has the structure:

$$H_3C-(CH_2)_{16}-C(=O)-OH.$$

For convenience, saturated fatty acids are often represented simply as:

$$\text{(zig-zag chain)}-C(=O)-OH$$

or $CH_3(CH_2)_{16}COOH$.

At neutral pH the sodium salt of a fatty acid is called soap:

$$\text{(zig-zag chain)}-C(=O)-O^-\ Na^+.$$

Humans eat large quantities of fats, which provide a substantial proportion of their food, but soaps are not edible. They are unpalatable and their detergent action could disrupt cell membranes. Instead we eat mainly neutral fats, known as **triacylglycerols (TAGs)**, in which three molecules of fatty acids are esterified to the three hydroxyl groups of glycerol as shown in Fig. 7.3. A carboxylic acid attached by an ester linkage is an acyl group, and there are three acyl groups attached to one glycerol molecule, hence the name triacylglycerol. The term triglyceride is sometimes used but is not chemically correct.

If neutral fat is boiled with NaOH or KOH, the ester bonds are hydrolysed, forming soaps and glycerol (this is how soap is made). Neutral fat is suitable as a food; it is not a detergent.

Glycerol	Three molecules of fatty acid	One molecule of neutral fat (or triacylglycerol (TAG))
CH_2-OH	$^-O-C(=O)-R$	$CH_2-O-C(=O)-R$ (Ester bond)
$CH-OH$	$^-O-C(=O)-R$	$CH-O-C(=O)-R$
CH_2-OH	$^-O-C(=O)-R$	$CH_2-O-C(=O)-R$

Fig. 7.3 A triacylglycerol and its component parts.

As stated, *neutral fat does not occur in membranes.* We have discussed it to help in understanding what a polar lipid is by contrast. A TAG molecule has no polar groups and therefore forms oily droplets or particles in water – it could never produce a lipid bilayer structure. It is, however, the most efficient storage form of metabolic fuel, as we will see in the chapters on metabolism. The membrane polar lipids can be classified according to their structures.

Glycerophospholipids

Membrane lipids containing glycerol are the most abundant form and are known as **glycerophospholipids**. They are based on glycerol-3-phosphate.

$$^{1}CH_2-OH$$
$$HO-{}^{2}C-H$$
$$^{3}CH_2-O-P(=O)(OH)-OH$$

sn-Glycerol-3-phosphate

The prefix sn, for stereospecific numbering, refers to the nomenclature system used here. The central carbon atom of this compound is asymmetric; in glycerol-3-phosphate as found in the cell, the secondary hydroxyl group is represented in this projection to the left and the carbon atoms numbered from the top. If two fatty acids are esterified to the primary (C-1) and secondary (C-2) hydroxyl groups of glycerol-3-phosphate, the product is **phosphatidic acid**. It does not occur in this highly reactive form in biological systems, but it is the parent compound of the glycerophospholipids found in membranes.

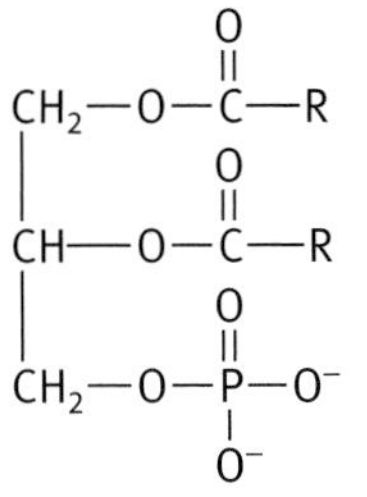

$$\begin{array}{l} CH_2-O-\overset{O}{\overset{\|}{C}}-R \\ | \\ CH-O-\overset{O}{\overset{\|}{C}}-R \\ | \\ CH_2-O-\overset{O}{\overset{\|}{P}}-O^- \\ \qquad\qquad\ \ | \\ \qquad\qquad\ O^- \end{array}$$

Phosphatidic acid

We can attach other polar molecules to the phosphoryl group. If phosphatidic acid is attached to something else (e.g. 'X') it becomes a **phosphatidyl group** (e.g. phosphatidyl-X) with the structure:

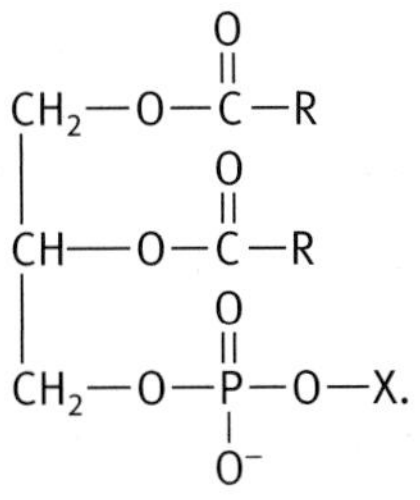

$$\begin{array}{l} CH_2-O-\overset{O}{\overset{\|}{C}}-R \\ | \\ CH-O-\overset{O}{\overset{\|}{C}}-R \\ | \\ CH_2-O-\overset{O}{\overset{\|}{P}}-O-X. \\ \qquad\qquad\ \ | \\ \qquad\qquad\ O^- \end{array}$$

If X is also a highly polar molecule we now have an extremely polar head group.

The structure above may give a misleading impression of the shape of the molecule. A space-filling model would look somewhat like this:

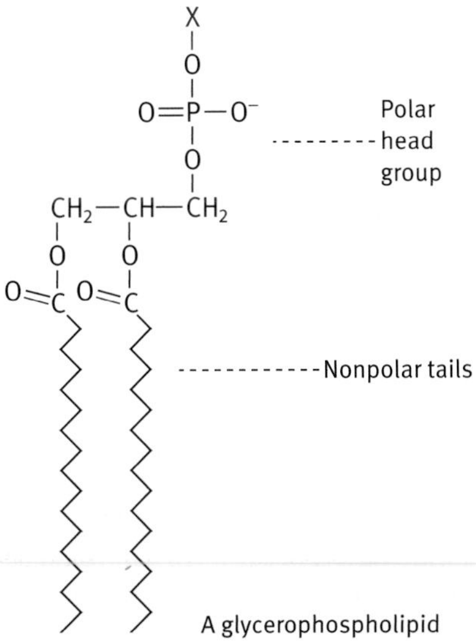

X–O–P(=O)(–O⁻)–O–CH_2 of glycerol: $CH_2-CH-CH_2$, with O–C(=O) on two carbons attached to two zig-zag chains

Polar head group

Nonpolar tails

A glycerophospholipid

What are the polar groups attached to the phosphatidic acid?

Living cells have a variety of structures in their membranes, some quite complex, but all structures used have *the same overall shape and amphipathic properties*. Several different polar groups are attached to different phosphatidic acid molecules.

One polar substituent is **ethanolamine** ($HOCH_2CH_2NH_3^+$) giving the phospholipid **phosphatidylethanolamine** (PE for short, or the trivial name is **cephalin**). Its structure is:

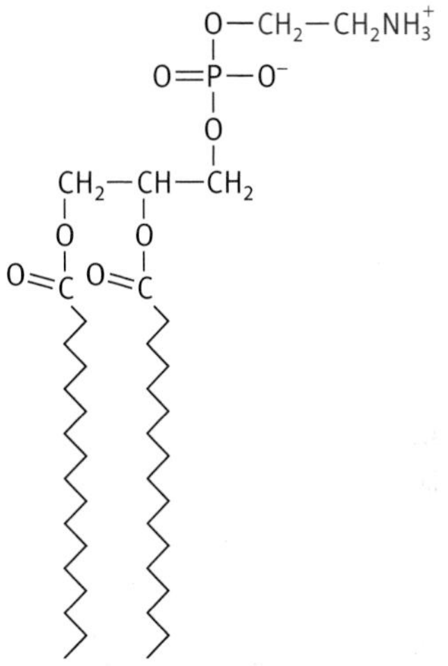

$$\begin{array}{c} O-CH_2-CH_2NH_3^+ \\ | \\ O=P-O^- \\ | \\ O \\ | \\ CH_2-CH-CH_2 \\ | \qquad | \\ O \qquad O \\ | \qquad | \\ O=C \quad O=C \end{array}$$

(with two zig-zag hydrocarbon tails)

If the nitrogen of ethanolamine is trimethylated, it becomes **choline**:

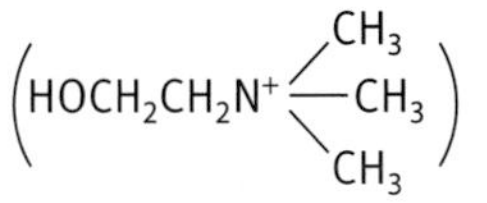

$$\left(HOCH_2CH_2N^+\begin{array}{l} CH_3 \\ CH_3 \\ CH_3 \end{array}\right)$$

and the phospholipid derived from it is **phosphatidylcholine** (PC) or **lecithin**. Its structure is the same as that given for PE above, apart from the three methyl groups on the nitrogen atom.

If the ethanolamine is carboxylated, we have a **serine** substituent giving **phosphatidylserine** or PS (serine is $HOCH_2CHNH_2COOH$). The attachment is like this:

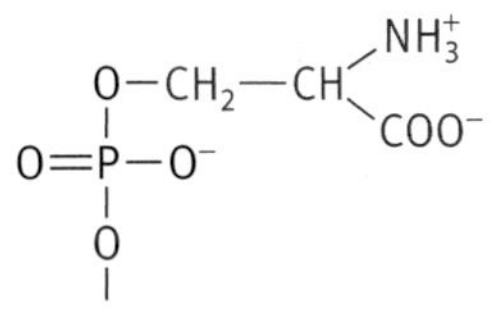

$$\begin{array}{l} O-CH_2-CH\begin{array}{l} NH_3^+ \\ COO^- \end{array} \\ | \\ O=P-O^- \\ | \\ O \\ | \end{array}$$

Another quite different polar substituent is the hexahydric alcohol, **inositol**:

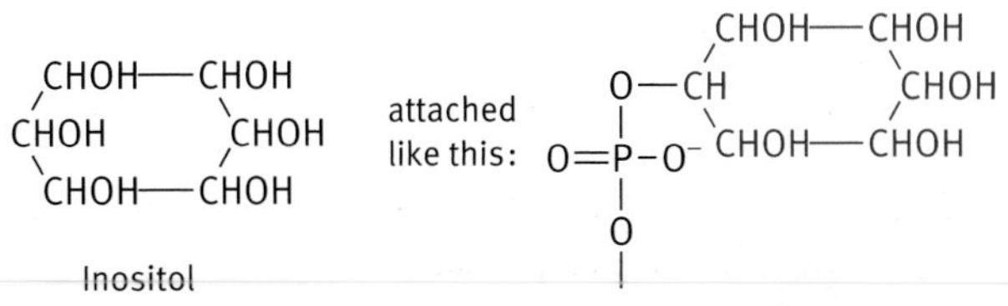

Inositol: six-membered ring of CHOH groups (CHOH—CHOH, CHOH, CHOH, CHOH—CHOH)

attached like this: $O=P(-O^-)(-O-)$ linked via O–CH to the ring of five CHOH groups

Inositol

giving **phosphatidylinositol**, usually abbreviated to PI.

Another glycerol based phospholipid is **Cardiolipin**, or **diphosphatidylglycerol**, which has two phosphatidic acids linked by a third glycerol unit.

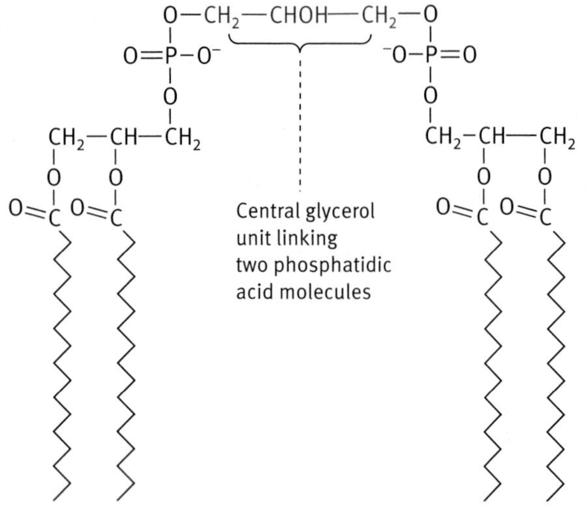

Central glycerol unit linking two phosphatidic acid molecules

The resultant structure still has the amphipathic shape to fit into lipid bilayers. It occurs mainly in the inner mitochondrial membrane and in bacterial cell membranes.

All of the above polar lipids are based on glycerol. However, the process of evolution has produced molecules of almost identical overall shape derived from a different structure called **sphingosine**.

Sphingolipids

The basic structure of **sphingosine** is not all that different from glycerol; the middle (C-2) hydroxyl group of glycerol is replaced by an $-NH_2$ group and a hydrogen on carbon atom 1 is replaced by a C_{15} hydrocarbon group – more or less a permanently 'built-in' hydrocarbon tail. One tail is not enough for a bilayer constituent. A second tail, a fatty acid, is attached to the central $-NH_2$ group by means of a $-CO-NH-$ linkage forming a **ceramide**, similar in shape to a diacylglycerol.

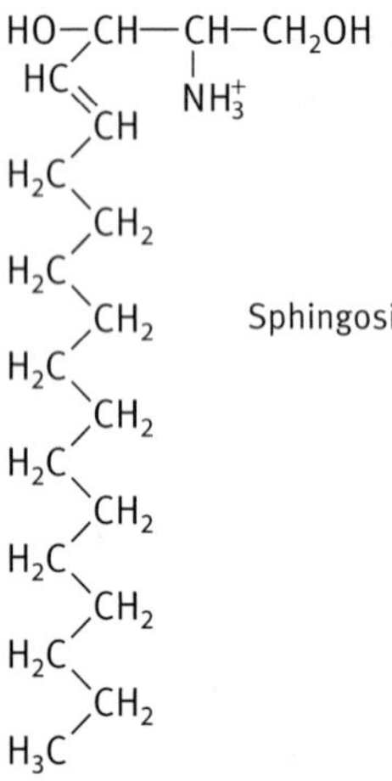

Sphingosine

and

A ceramide

If now we add a phosphorylcholine group, as in lecithin, the product is **sphingomyelin**:

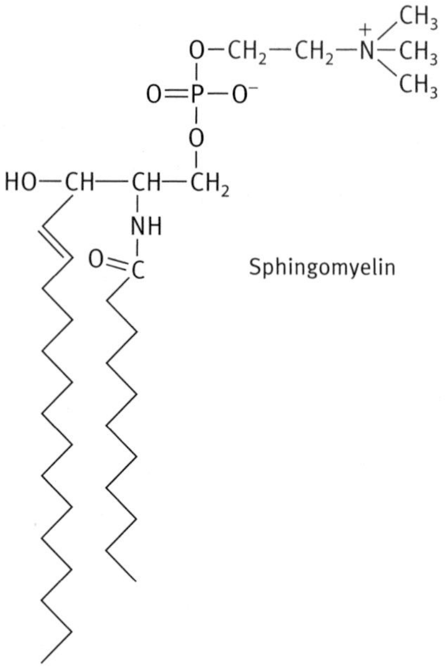

Sphingomyelin

which is similar in shape to lecithin. It is prevalent in the myelin sheath of nerve axons.

A number of different polar groups are found attached to the free $-OH$ of the ceramide molecule. Sugars, which are highly polar molecules, are good candidates for this purpose and the resulting molecules are called **glycosphingolipids**.

If a single sugar is the polar group, we get a **cerebroside**, important in brain cell membranes. Cerebrosides contain either glucose or galactose. The latter is a stereoisomer of glucose in which the C-4 hydroxyl is inverted.

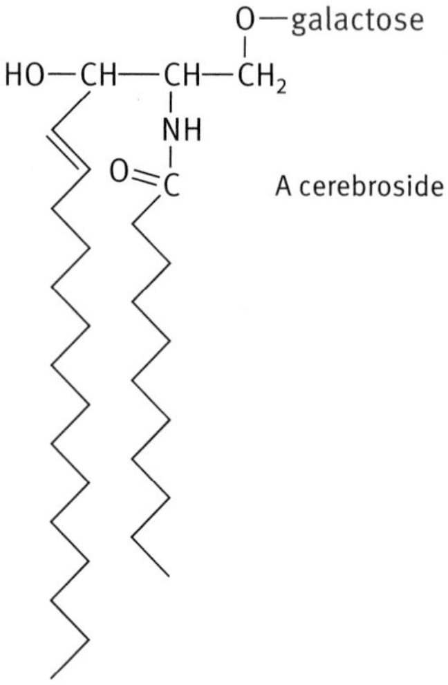

A cerebroside

There is a wide variety of sugars in nature. Some are **amino sugars** such as glucosamine (whose structure is given below) or derivatives of them.

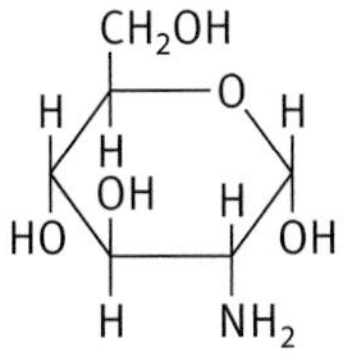

Combinations of the different sugars (**oligosaccharides**) can produce a variety of molecules in which a small number of different sugars are linked together in a branched oligosaccharide structure. (*Oligo* = few, so an oligosaccharide is a small polymer of sugars – perhaps 3–20 sugars rather than the hundreds or thousands found in polysaccharides.) Such an oligosaccharide is highly polar. When one of these is attached to a ceramide the resulting molecule is a **ganglioside**.

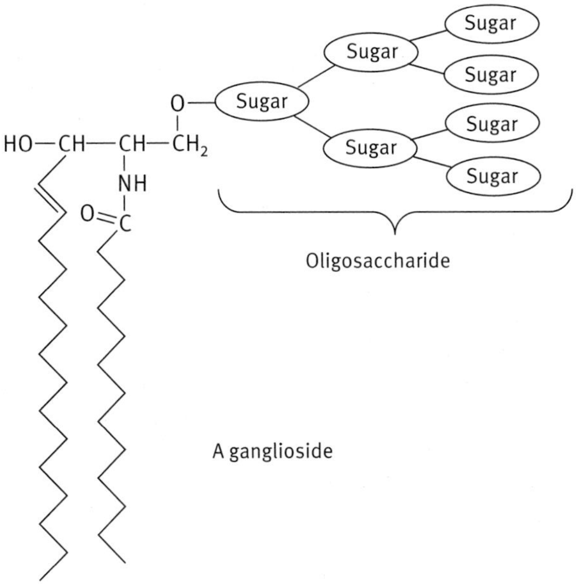

Oligosaccharide

A ganglioside

***N*-Acetylglucosamine** and **sialic acid** are shown here because they are of special interest as components of gangliosides.

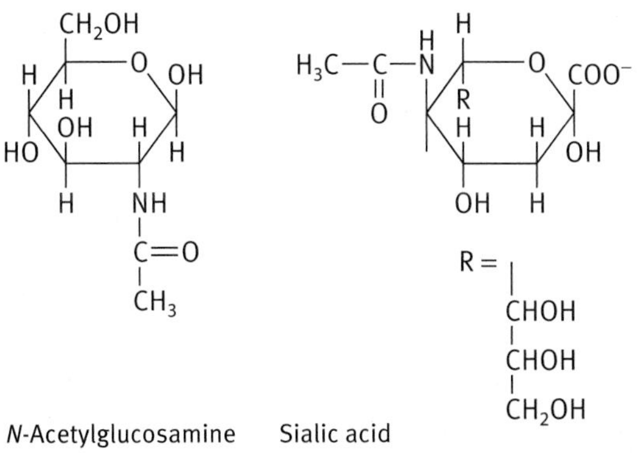

N-Acetylglucosamine

Sialic acid (also called *N*-acetylneuraminic acid)

Sialic acid is involved in infections by the influenza virus. Gangliosides, based on sphingosine, are what distinguishes the human blood groups O, A, and B.

Membrane lipid nomenclature

The names of individual polar lipids have been given above, but alternative collective terms are sometimes used. The terms **membrane lipids** and **polar lipids** include any lipid found in cell membranes. A **phospholipid** is any lipid containing phosphorus. Those based on glycerol are **glycerophospholipids**, as opposed to sphingomyelin which is a single specific sphingosine-based phospholipid. The ceramide-based membrane lipids with carbohydrate polar groups and lacking any phosphoryl group are called **glycolipids** or **glycosphingolipids** to indicate that they are based on sphingosine. A plasmalogen is a glycerophospholipid in which one of the hydrophobic tails is linked to glycerol by an ether bond (not shown). (Cholesterol, another membrane component described below, is also classified as a lipid.)

What is the advantage of having so many different types of membrane lipid?

The different membrane lipids confer different properties on the membrane surface. The choline substituent of lecithin (PC) has a positive charge, the serine of PS is a zwitterion (see Chapter 4), and the carbohydrate of a cerebroside has no charge. Different cells have quite different membrane lipid compositions. Cerebrosides and gangliosides are common in brain cell membranes, and the cell membranes of the myelin sheath of nerve axons are rich in glycosphingolipids. As well as different cells having different lipid compositions, the outer and inner halves of the bilayer of the one membrane are different from one another and different membranes within the cell have different compositions. For example, glycolipids are always on the outer side of the cell membrane so that their sugars point outwards from the cell into the external aqueous environment. This asymmetry is preserved by the fact that transverse movement of lipids, known colloquially as 'flip-flop' (from one side of the membrane to another), is severely restricted; such movement would involve transferring the polar heads through the central hydrocarbon layer to get to the other side. This is energetically unfavourable. Proteins catalysing energy-dependent membrane flip-flop are involved in the creation and maintenance of this asymmetry.

The limited flip-flop movement of membrane lipids contrasts with their potential for rapid lateral movement within the plane of the bilayer. The assembly of lipids into the bilayer structure does not involve covalent bonds and, in general, they move around freely.

Some particular membrane lipids (e.g. PI) have a special role as the source of chemical intracellular signals. Also cell signalling in some cases is dependent on specific proteins associating with patches of the cytosolic membrane rich in particular lipids (see Chapter 29). Specific protein–lipid associations, in general, appear to be responsible for the diversity of membrane lipids. Indeed the association and dissociation of proteins from membranes is increasingly recognized to play a vital part in the regulation of the activities of cells. Given the wide variety of different membrane proteins and their functions, it is not surprising that different types of membrane lipids are needed.

Some membrane lipids are of special medical interest. For example, there is a clinical interest in gangliosides, because of genetic diseases called **glycosphingolipidoses**. One example is Tay–Sachs disease (see Box 27.1), in which there is mental retardation and early death; another is Gaucher's disease. In these, the glycosphingolipids are not broken down properly in lysosomes (see Chapter 27) and their residues accumulate causing severe brain disorders.

The fatty acid components of membrane lipids

The fatty acyl 'tail' components of a membrane lipid, such as lecithin, vary. In length they may range from C_{14} to C_{24} (almost always an even number), but C_{14}–C_{18} are most common. A variety of fatty acids, often unsaturated with one or more double bonds, are present in membrane lipids. The usual ones are C_{18} and C_{16} with one double bond in the middle (oleic and palmitoleic acids, respectively), linoleic (C_{18} with two double bonds), and arachidonic acid (C_{20} with four double bonds). The **nomenclature** of such acids is dealt with in a later chapter (see Chapter 17).

The two fatty acyl residues in a single phospholipid molecule may be the same or different, but usually in a glycerophospholipid, the fatty acid attached to the C-1 —OH group is saturated and that attached to the C-2 —OH group is unsaturated. The degree of saturation of fatty acid tails is of great importance because the central hydrocarbon core of the bilayer must be a two-dimensional fluid rather than solid. Unsaturated hydrocarbon tails reduce the temperature at which a bilayer loses its fluidity.

COOH — Saturated fatty acid

COOH — Unsaturated fatty acid with *cis* double bond

COOH — Unsaturated fatty acid with *trans* double bond

Membrane fluidity is essential to allow lateral movement of transmembrane proteins so that they can interact with one another. In addition, such proteins undergo conformational changes needed for ion channels, transporters, and receptors to function. Without a fluid bilayer such changes might be difficult to accommodate. (The proteins referred to are described later.) A saturated fatty acid tail is straight, but a double bond in the *cis* configuration introduces a kink into it as illustrated, and kinks increase fluidity. Natural unsaturated fatty acids are almost always in the *cis* configuration (Box 7.1).

The straight saturated chains pack together comfortably and interact, but the kinked ones cannot do so and stay fluid at lower temperatures. This physical effect of unsaturation can be seen by comparing hard mutton fat with olive oil. Both are TAGs but the olive oil has unsaturated fatty acyl tails.

The importance of maintaining membrane fluidity is illustrated by the fact that bacteria adjust the degree of unsaturation of the fatty acid components of their membrane bilayers according to the ambient temperature. In special situations, such as during hibernation, animals modulate the degree of saturation of cell membrane components to cope with lower body temperature.

What is cholesterol doing in membranes?

From its structure, as conventionally drawn (Fig. 7.4(a)), cholesterol seems an improbable membrane constituent. However, the conformation of the ring system, shown in Fig. 7.4(b), shows that the molecule is elongated, the steroid nucleus being rigid and the hydrocarbon chain flexible. It is an amphipathic molecule, the —OH group being weakly polar.

Cholesterol in membranes acts as a 'fluidity buffer'. At temperatures above the melting point of a lipid bilayer, it reduces fluidity because of its rigid structure, but, at lower temperatures it increases fluidity because it prevents close-packing of the flexible fatty acid tails. It thus 'buffers' fluidity. The observed effect is that cholesterol 'blurs' the melting point of a

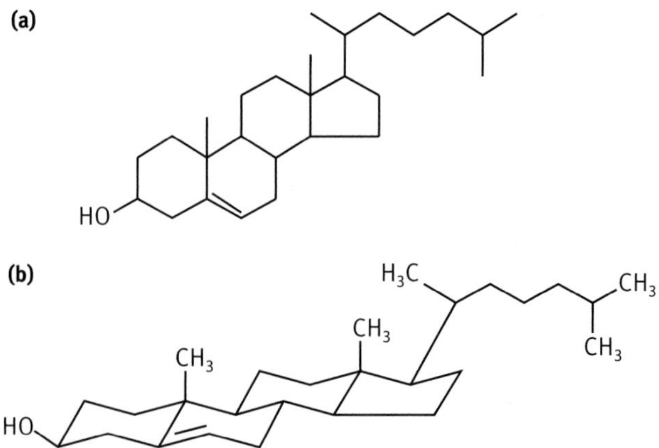

Fig. 7.4 **(a)** Structure of cholesterol as conventionally drawn. **(b)** Structure drawn to give a better indication of the actual conformation of cholesterol.

BOX 7.1 ***Trans* fatty acids**

As is widely known, there is considerable concern that the intake of *trans* unsaturated fatty acids is harmful to health. Natural foods have very little of these acids present. They originate mainly from ruminant bacteria in cattle and sheep and are associated with the meat. The main source in commercial foods comes from partial hydrogenation of unsaturated oils, used to solidify the oils to varying extent for use in the food industry such as manufacture of pastries. Dietary regulations or guidelines are being introduced in various countries to minimize or eliminate partially hydrogenated foods.

The structures of *cis* and *trans* unsaturated fatty acids have already been shown. It can be seen that the *cis* variety is kinked while the *trans* isomer is a straight chain and, in this respect, resembles saturated fatty acids. There is good evidence that *cis* unsaturated acids are essential to maintain the liquidity of cell membranes presumably by preventing the tight side-by-side packing of the fatty acid tails of membrane polar lipids.

Since *trans* fatty acids have the straight chain structure it might be presumed that they are similar to saturated fatty acids in this respect, but there is no direct evidence that this is a significant effect.

There is evidence, however, that *trans* fatty acids have undesirable physiological effects and may contribute to cardiovascular disease, though the biochemical mechanism(s) by which these occur are not understood. Dietary studies indicated that *trans* fats, compared with the same amount of saturated or *cis* unsaturated fats, increase the concentration of **low-density lipoproteins (LDLs)** in the blood and lowers the concentration of **high-density lipoproteins (HDLs)**. LDLs are sometimes referred to as 'bad' cholesterol, and HDLs as 'good' cholesterol. The reduction in the ratio of HDL cholesterol to total cholesterol is well known to be associated with an increased risk of cardiovascular disease. The *trans* fats were also found to increase the concentration of serum triacylglycerols as compared with the effects of comparable amounts of saturated or *cis* fats.

A number of studies have concluded that near elimination of manufactured *trans* fats from the diet could prevent many thousands of cardiovascular events each year in the USA.

lipid bilayer. Without cholesterol, the transition from solid to liquid is sharper than when cholesterol is present. A red blood cell membrane may be about 25% cholesterol. On the other hand, bacterial membranes have no cholesterol and animal cell mitochondria have very little. Plants contain other sterols, known as **phytosterols**, but no cholesterol. Fungal membranes contain ergosterol, not cholesterol. (See Box 7.5 where amphotericin B is discussed. It has affinity for ergosterol over cholesterol.)

The self-sealing character of the lipid bilayer

The lipid bilayer is effectively a two-dimensional fluid. The bilayer gives cells flexibility and a self-sealing potential, the latter being essential when a cell divides (Fig. 7.5(a)).

This versatility is also essential in **endocytosis** – the process by which cells can engulf material (Fig. 7.5(b)). The reverse also happens. If the cell needs to secrete a substance such as a digestive enzyme, it is synthesized inside the cell and enclosed within a membrane-bounded vesicle that migrates to the plasma membrane, fuses with it, and releases its contents to the outside. The process is called **exocytosis**. A good example of this process is the secretion of digestive enzymes into the intestine by pancreatic cells (Fig. 7.5(c)). Another, is the release of peptide hormones such as insulin, from the β cells of the endocrine part of the pancreas into the circulation.

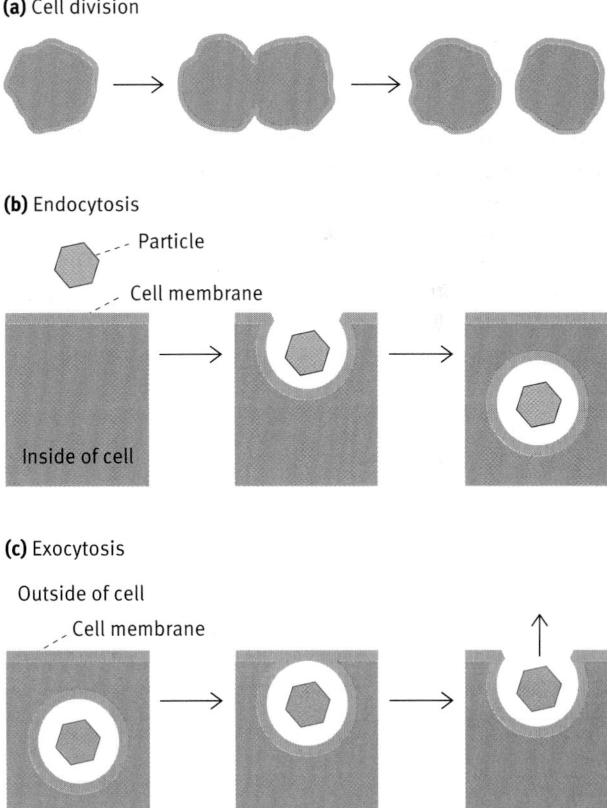

Fig. 7.5 (a) Cell division; **(b)** endocytosis; **(c)** exocytosis for the release of, for example, a digestive enzyme from cells.

Permeability characteristics of the lipid bilayer

The ability of small molecules to diffuse through a lipid bilayer is related to their fat solubility. Strongly polar molecules such as ions traverse the bilayer extremely slowly if at all. Large, more weakly polar molecules such as glucose penetrate very slowly, although smaller ones such as ethanol or glycerol diffuse across more readily. Ionized groups of polar molecules and inorganic ions are surrounded by a shell of water molecules, which must be stripped off for the solute to pass through the lipid bilayer hydrocarbon centre, and this is energetically unfavourable. The lipid bilayer is, therefore, almost impermeable to such molecules and to ions, but slow leaks inevitably occur.

Water molecules, despite being polar, apparently pass through the lipid bilayer with sufficient ease for the needs of some cells. Presumably this is due to the small size of the molecule and its lack of charge, but there is some uncertainty as to how it traverses the bilayer. Cells involved in water transport, such as those of renal tubules and secretory epithelia, have a transmembrane protein, **aquaporin**, which allows free movement of water. Many cells have **porins** (see Fig. 7.9). The lipid bilayer also readily allows gases in solution such as oxygen to diffuse through it.

A high proportion of molecules of biochemical interest are polar in nature and cannot pass through the lipid bilayer at rates commensurate with cellular needs. This means that although the bilayer structure is ideal for holding in the contents of a cell, special arrangements must be made to permit rapid movement of molecules across the membrane as needed. Membrane proteins are responsible for this transport.

Membrane proteins and membrane structure

Different membranes have different functions and it is not surprising that they contain different protein molecules. In the cell, membrane proteins are inserted into the membrane as the proteins are synthesized. Proteins experimentally isolated from membranes have been incorporated into synthetic phospholipid liposomes where they function exactly as in their parent cell membrane.

The membrane structure containing proteins is called a **lipid fluid mosaic** (Fig. 7.6). In this, laterally mobile protein molecules are present in a two-dimensional lipid layer. Such proteins are called **integral proteins** (labelled I in Fig. 7.6). Other proteins are called **peripheral proteins** (labelled P in Fig. 7.6) because they associate with the periphery of the membrane.

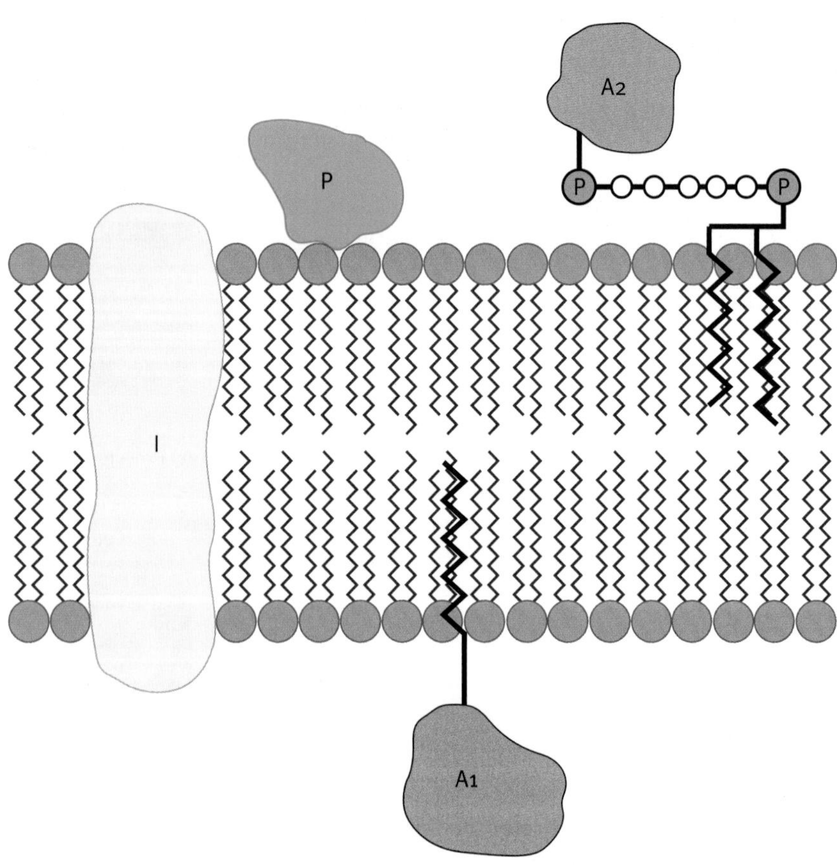

Fig. 7.6 Lipid fluid mosaic model for membrane structure. I, integral protein; P, peripheral protein; A, anchored proteins: A1 linked to FA chains and A2 (shown on the top here) linked to a glycolipid.

Another class of membrane proteins are the so-called anchored proteins. They are not deeply embedded in the membrane but are covalently linked to fatty acyl chains or to glycolipids on the cell surface.

Structures of integral membrane proteins

Most integral membrane proteins are so structured that they are held in the lipid bilayer by noncovalent forces. Integral membrane proteins with α helical structures are amphipathic with ends made of hydrophilic amino acids in contact with water and the middle section of hydrophobic ones as shown diagrammatically in Fig. 7.7(a). The essential feature of such integral membrane proteins is that the hydrophobic section corresponds in length with the hydrocarbon middle zone of the membrane lipid bilayer. The parts of the protein projecting from the membrane that are in contact with water, or the polar head groups of membrane lipids, have hydrophilic residues as illustrated in Fig. 7.7(b). This is a diagram of **glycophorin**, a major component of the erythrocyte membrane. On the external surface of the membrane is a large *N*-terminal section of polypeptide, rich in hydrophilic amino acids; attached to serine, threonine, and asparagine residues are carbohydrates (see Chapter 4, Protein domains), making this part of the protein extremely hydrophilic. On the inside face of the bilayer is a shorter C-terminal section devoid of carbohydrate attachments but also containing hydrophilic amino acids. Connecting the two external sections and spanning the lipid bilayer is a stretch of 19 amino acid residues which, in the form of an α helix, are sufficient to just span the hydrocarbon central layer of the membrane, which is about 30 nm wide. In this section of the chain, isoleucine, leucine, valine, methionine, and phenylalanine, all strongly hydrophobic and good α-helix-formers, predominate. There are no strongly hydrophilic residues, except in those cases where the protein is a transmembrane aqueous pore, lined with hydrophilic residues.

In some proteins, the polypeptide chain loops back and crosses the bilayer several times – for this, alternating hydrophilic and hydrophobic sections are required, each of the hydrophobic sections being about 19 amino acids long and forming α helices, which nicely span the bilayer. One of the best-studied examples is **bacteriorhodopsin**, a protein present in the membrane of the purple bacterium *Halobacterium halobium* found in brine ponds. The protein crisscrosses the membrane seven times, forming a cluster of seven α helices spanning the membrane and connected by hydrophilic loops (Fig. 7.8). The cluster has a light-absorbing pigment at its

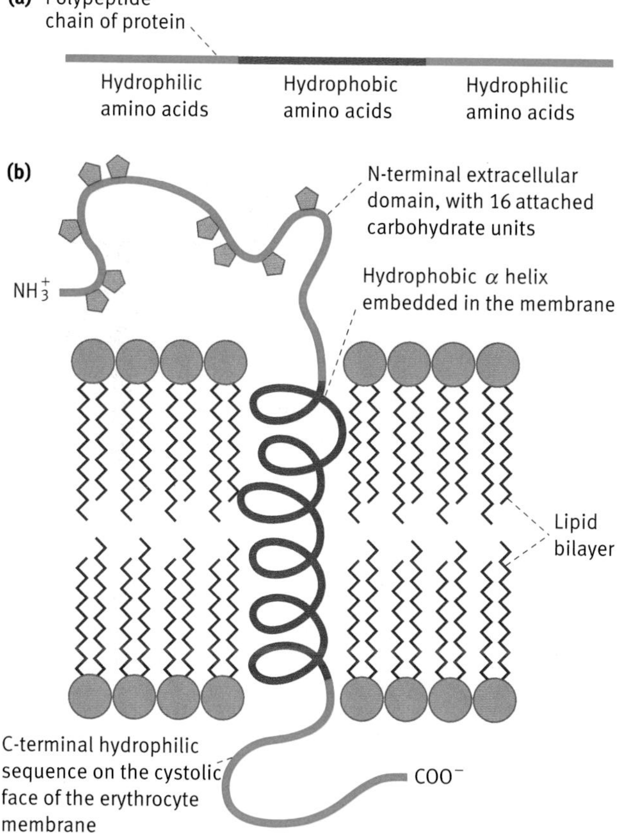

Fig. 7.7 (a) The structural plan of an integral membrane protein. **(b)** Glycophorin, a protein of the erythrocyte membrane. The α helix, containing about 19 hydrophobic amino acid residues, is approximately 30 Å in length, which is sufficient to span the nonpolar interior of the lipid bilayer.

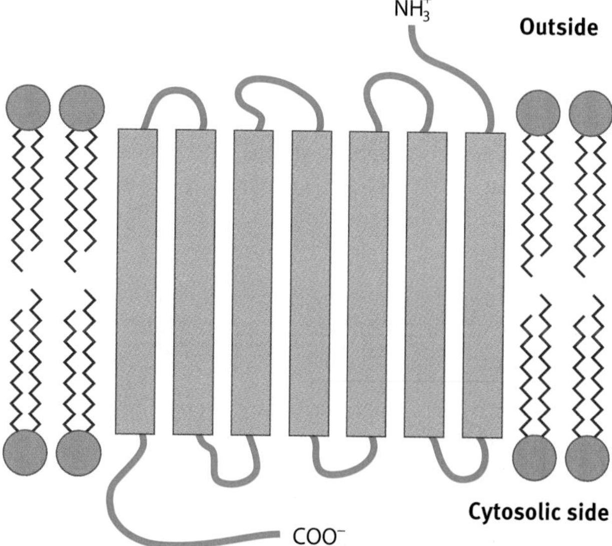

Fig. 7.8 Topological representation of bacteriorhodopsin, with seven α helices spanning the lipid bilayer. The α helices are not actually arranged linearly, as shown here for convenience, but are clustered compactly together.

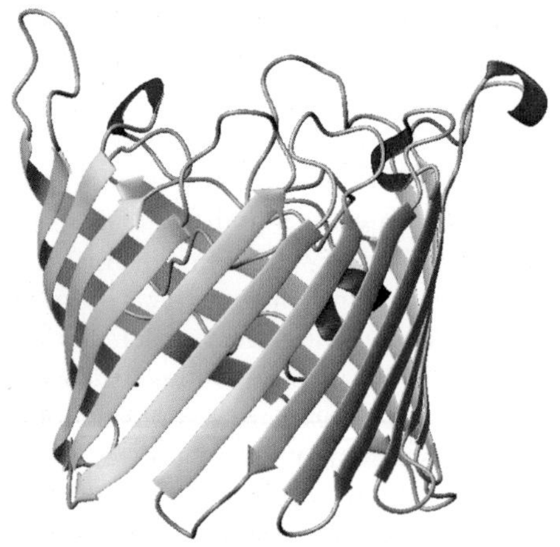

Fig. 7.9 The β-barrel structure of a bacterial porin (Protein Data Bank code 1BH3). β strands are yellow, α helices are red, and everything else is grey.

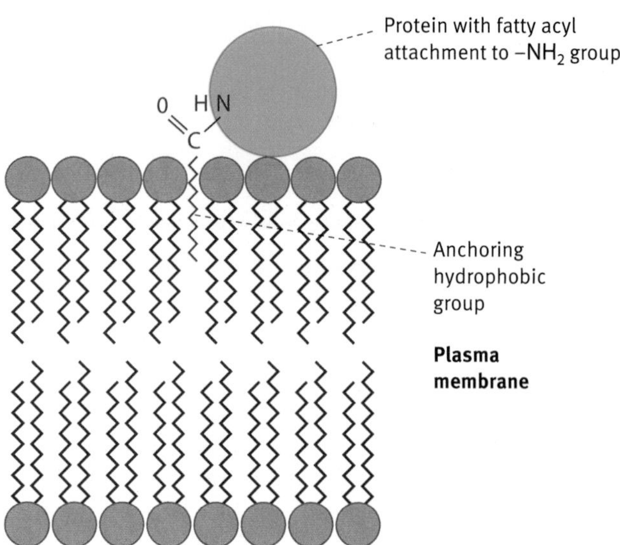

Fig. 7.10 A fatty acid anchoring molecule joined by CO—NH linkage to the amino group of a protein. Alternative linkages such as ester and thiol ester can occur with appropriate side groups of the protein.

centre, which captures light energy and drives the pumping of protons from the cell to the outside. The energy of the proton gradient so formed is used to drive the synthesis of ATP by a mechanism described in Chapter 13.

The transmembrane **porins** are different. These are proteins found in most cell plasma membranes, including the outer membrane of certain bacteria and in the outer mitochondrial and chloroplast membranes. These outer membranes are relatively porous to small molecules due to porins. These are water-filled channels with a β barrel structure (Fig. 7.9) varying in different cases from 8 to 20 antiparallel strands with alternating hydrophilic and hydrophobic residues, arranged so that the hydrophobic residues are outside, in contact with the hydrophobic lipid bilayer components, while the hydrophilic residues point inwards. There is some selectivity in what the porin channels conduct; in bacteria some allow anions to cross and others conduct sugars.

If the protein tended to move out of the membrane, the hydrophobic groups of the protein would come into contact with water and hydrophilic groups into contact with the hydrocarbon layer. Both are energetically resisted and the protein is fixed in the transmembrane sense. Unless otherwise restrained it can move laterally in the bilayer.

Anchoring of peripheral membrane proteins to membranes

A peripheral water-soluble membrane protein may be associated with a membrane by hydrogen bonding and ionic attractions. However, an alternative arrangement exists in the case of certain proteins. In these, the proteins have attached to them a fatty acid whose hydrocarbon chain is inserted into the lipid bilayer, thus anchoring the protein to the membrane (Fig. 7.10).

C_{14}, C_{16}, and C_{18} fatty acids may anchor proteins by linking to amino acid residues such as glycine, to serine or threonine groups by ester linkage, or to cysteine by thioester linkage. More complex acids are also found in ether linkages.

An example of a fatty acyl linked protein is a signalling G protein called RAS, which relays information to the interior of the cell from the cell membrane. RAS in fact refers to a family of related proteins and the name is an abbreviation of 'rat sarcoma' indicating the origins of discovery of this family of proteins. When RAS is activated by incoming signals it activates various other components of the signalling system resulting in activation of genes involved in cell growth and differentiation. An overactive RAS may lead to the development of cancer (see Chapters 29 and 33).

As mentioned above, some anchored proteins are attached to glycolipid on the cell surface. An example of a glycolipid anchored protein is the enzyme alkaline phosphatase, an enzyme responsible for removing phosphate groups from molecules such as proteins and nucleotides.

Glycoproteins

Many membrane proteins have attached to them branched oligosaccharides, on the external surface. The carbohydrates are covalently attached to the side chains of asparagine or serine (see Chapter 4). The sugars which make up the oligosaccharides include various isomers of glucose and amino sugars.

In glycophorin of the red blood cell (Fig. 7.7(b)), half the weight of the protein molecule is carbohydrate. The exact function of these carbohydrate attachments is an area of uncertainty but may, in some cases, have recognition roles on cell surfaces. Glycoproteins are not only found in membranes, but are also found elsewhere, such as in the blood, as one example.

Functions of membranes

Membranes have a wide variety of functions; for the moment we will deal with the plasma membrane surrounding the cell. We will come to the internal membranes in later chapters.

The main functions of such membranes, apart from retaining cell contents, are:

- transport of substances in and out of the cell
- ion transport and nerve-impulse conductance
- cell signalling (a major topic dealt with in Chapter 29)
- maintaining the shape of the cell
- cell–cell interactions.

Membrane transport

Many, or most, of the substances that must be taken up by the cell or have to be removed from the cell, cannot diffuse across the lipid bilayer. The lipid bilayer has to be impermeable to most molecules other than small hydrophobic ones because it must retain the necessary cell constituents against leakage. For transport of polar structures such as sugars, amino acids and inorganic ions, specific membrane proteins are needed. Although inorganic ions are very small, they cannot diffuse through a cell membrane because they are surrounded by a shell of water molecules and, as explained earlier, removal of these is energetically unfavourable. Simple holes in the membrane will not do – they would allow nonspecific leakage in and out and would be lethal. Usually transport systems handle only specific molecules so there are many different transport systems and therefore many different transport proteins.

The problem is a major one for all life forms. Bacteria must take in essential nutrients, so must all cells but the problem for, say, a mammal is greatly compounded because the chemical environment in the blood is very different from that inside cells. (The sodium and potassium ion concentrations of blood and intracellular space are different and if equilibrated would be lethal. A remarkable situation is that animals use the Ca^{2+} ion as one of the most potent regulators of much of cell chemistry, and yet Ca^{2+} is present extracellularly at high levels. For this to work, calcium levels inside the cell must be kept extremely low. Because there is such a steep gradient of the ion across the membranes, signals can cause instant fluxes of the ion to be transported into the cell. Pumps then immediately withdraw the calcium to terminate the signal as required. The result is that an animal as complex as a mammal is a mass of never-ending membrane pumping. Bearing in mind that nerve activity is also a matter of ion movement across membranes (described shortly) the brain is energetically an expensive organ to maintain. The energy ultimately derives from ATP and perhaps 30% of one's total energy expenditure is on membrane pumping in general.

Active membrane transport

We can separate transport systems into active and passive types. An active system is transport that requires the performance of work, which usually involves the hydrolysis of ATP directly or indirectly. It requires work to be done because the movement of substances occurs against concentration gradients. However, rapid influxes of ions into a cell can occur by simply opening a channel to allow the ion to rush through driven by a concentration gradient. This happens in nerve-impulse transport. The maintenance of the gradient in these cases requires work.

The amount of energy required to transport a solute into a cell, against a gradient, can be calculated from the equation for chemical reactions given in Chapter 3 (see Box 7.2).

We will now deal with various types of active transport.

The sodium/potassium pump

A good example of active transport is the Na^+/K^+ pump present in animal cells. This pumps Na^+ out of cells and K^+ into cells using ATP energy. Animal cells have a high internal K^+ concentration (140 mM) and a low Na^+ concentration (12 mM) as compared with those of the blood (4 mM and 145 mM respectively). The ion gradients are necessary for electrical conduction in excitable membranes and, in some cases, for driving solute transport across membranes.

BOX 7.2 Calculation of energy required for transport

$$\Delta G = \Delta G^{0\prime} + RT\,2.303\log_{10}\frac{[\text{products}]}{[\text{reactants}]}$$

For transport of a solute whose structure does not change, $\Delta G^{0\prime}$ is zero and the equation becomes:

$$\Delta G = RT\,2.303\log_{10}\frac{[C_2]}{[C_1]}$$

where C_1 is the concentration outside and C_2 is the concentration inside, taken in this example to be in the ratio of 10/1 so that

$$G = (8.315\times10^{-3}\ \text{kJ mol}^{-1}\,\text{K}^{-1})(298\,\text{K})2.303\ \log_{10}10/1$$
$$= 5.706\ \text{kJ mol}^{-1}.$$

Thus the transport of 1 mol, under these conditions (at 25 °C), requires 5.706 kJ. Note that the above calculation applies to the transport of an uncharged solute. With a charged solute, generation of an electrical potential requires an additional term to correct for this.

Mechanism of the Na^+/K^+ pump

The Na^+/K^+ pump is also called the Na^+/K^+ ATPase because ATP is hydrolysed to ADP + P_i as Na^+ is pumped out of the cell and K^+ in. The protein is a complex of four polypeptides or subunits. There are two identical α subunits and two β subunits. Some proteins have the property of slightly changing their shape when specific ligands bind to them. This may be due to a change in the conformation of a protein, or to a relative change in position of subunits of a protein complex. The covalent attachment of a phosphoryl group to the Na^+/K^+ ATPase by transference from ATP also causes a **conformational change**. Such a change can alter the ability of a protein to bind a given ligand. Thus, in one form, the protein of the pump binds Na^+ but not K^+ and another form K^+ but not Na^+.

In the model shown in Fig. 7.11, the Na^+/K^+ ATPase protein exists in two conformations. Form (a) is open to the interior of the cell and binds Na^+ but not K^+. ATP is used to phosphorylate this, (on an aspartyl residue) yielding ADP + phosphorylated protein. In this form it is effectively open to the outside, no longer binds Na^+ (which diffuses away), but does bind K^+ which therefore attaches. This gives form (b) (Fig. 7.11) (see also Box 7.3). The –P group is now hydrolysed from the protein giving P_i and the protein reverts to form (a) to which the K^+ no longer binds. The latter therefore enters the cell and more Na^+ attaches to the pump (Fig. 7.11(c)). Evidence from antibody studies shows that the whole protein does not revolve in the process as might seem an obvious mechanism.

The net result is that hydrolysis of ATP pumps Na^+ out and K^+ in. The ratio is 3 Na^+ out and 2 K^+ in.

The overall equation is:

$$3Na^+(in) + 2K^+(out) + ATP + H_2O \rightarrow 3Na^+(out) + 2K^+(in) + ADP + P_i$$

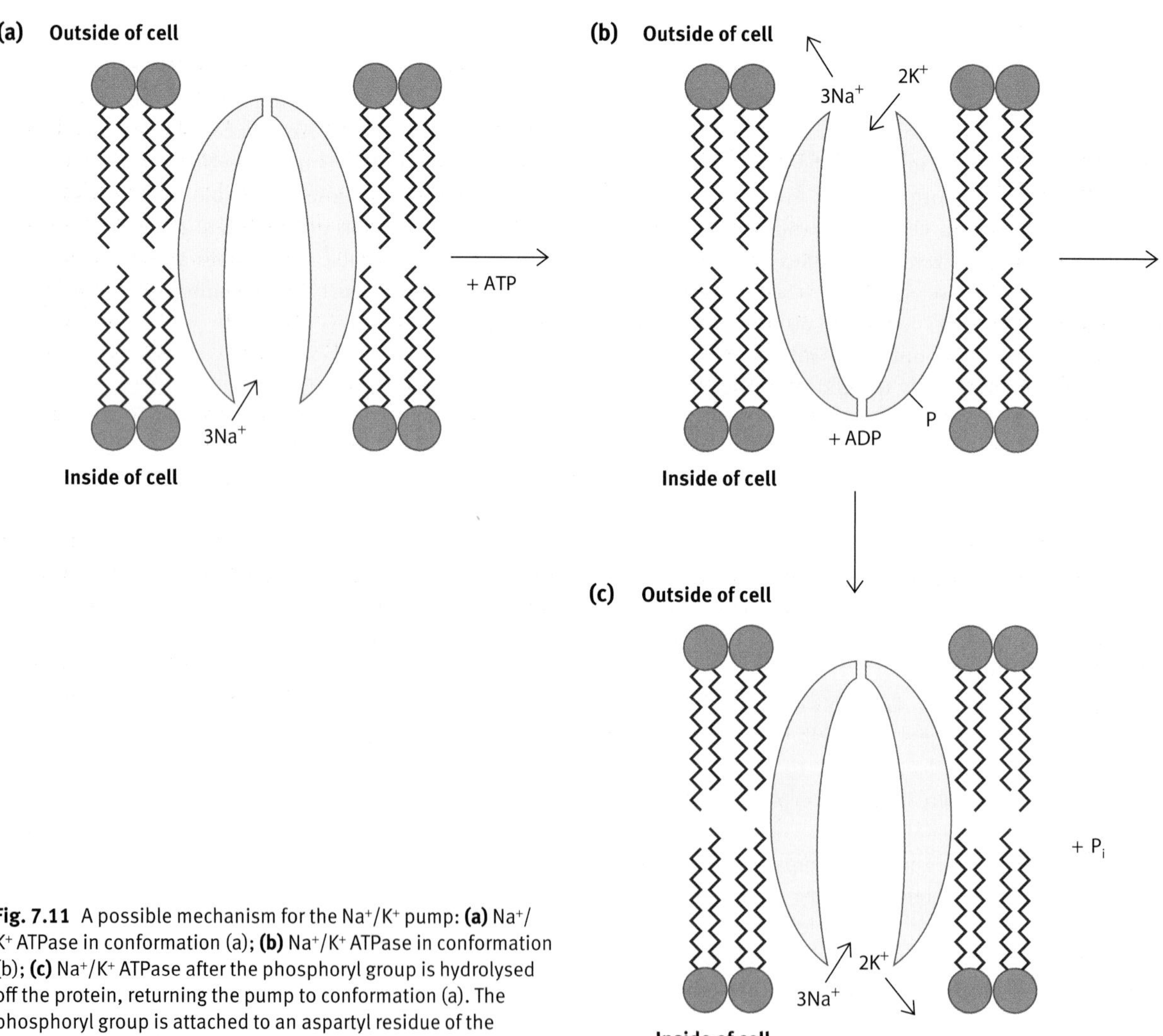

Fig. 7.11 A possible mechanism for the Na^+/K^+ pump: **(a)** Na^+/K^+ ATPase in conformation (a); **(b)** Na^+/K^+ ATPase in conformation (b); **(c)** Na^+/K^+ ATPase after the phosphoryl group is hydrolysed off the protein, returning the pump to conformation (a). The phosphoryl group is attached to an aspartyl residue of the enzyme.

BOX 7.3 Cardiac glycosides

Cardiac glycosides are a group of compounds found in the foxglove plant (*Digitalis purpurea*). They are steroids or steroid glycosides; steroids resemble cholesterol in structure; glycosides are molecules with sugars attached. Such compounds inhibit the Na^+/K^+ ATPase by preventing the removal of the phosphoryl group from the transport protein. This 'freezes' the pump in one form (Fig. 7.11(b)) and stops the ion transport. The cardiac glycosides have long been used clinically as treatment for congestive heart failure. The compounds are lethal in excess, but in appropriate doses, the partial inhibition of the Na^+/K^+ ATPase increases the Na^+ concentration inside heart muscle cells and thus lowers the Na^+ gradient from outside to inside. This has the effect of raising the cytosolic Ca^{2+} level because there is another system which transports Na^+ into the cell and Ca^{2+} out of the cell. The ejection of Ca^{2+} is driven by the Na^+ gradient; if the latter is lowered by cardiac glycosides then ejection of Ca^{2+} is reduced and the internal level of Ca^{2+} rises which stimulates heart muscle contraction; the role of Ca^{2+} in contraction is dealt with in Chapter 8. Ouabain, the African arrow-tip poison, has similar effects to cardiac glycosides.

A large family of related ATP-dependent pumps transport specific solutes across membranes

The mechanism is much the same as the sodium/potassium pump. One very important example is the Ca^{2+}-ATPase of muscle. As described in the next chapter, vertebrate striated muscle is triggered to contract by a neuronal signal that causes release of Ca^{2+} into the muscle cytoplasm (known as the sarcoplasm). The ATPase almost instantly withdraws the ion into a reservoir thus terminating the contraction. Another example is provided by the ATP-driven proton pump, which causes acidification of lysosomal vesicles.

ATP-binding cassette (ABC) proteins

One of the largest groups of transport systems is known as **ATP-binding cassette (ABC) proteins**. These are multi-domain transport proteins all of which have two cytosolic ATP-binding domains (the unit being called a cassette) and two transmembrane ones. At the expense of ATP breakdown, small molecules like drugs are transported across the membrane. A wide variety of such systems exist and are found in both bacteria and eukaryotes. In eukaryotes a protein known as the multidrug resistance protein (MDR protein) is an ABC transporter. It ejects pharmacological agents used in cancer treatment. The cancer cell may increase its amount of MDR protein and become resistant to the therapy. An increase of such resistance to one drug may result in multiple drug resistance – hence its name. The protein is also referred to as **P (permeability) glycoprotein**.

A number of genetic diseases are associated with deficiencies in ABC transporters. The most important is **cystic fibrosis**, which affects 1 in 2500 Caucasians. The responsible gene called the **CFTR** gene has been isolated. It codes for an ABC transporter protein known as the **C**ystic **F**ibrosis **T**ransmembrane conductance **R**egulator protein. A number of different mutations of the CFTR gene are known to cause the disease but most often the regulatory domain of the CFTR is affected. This protein functions as a channel across the membrane of cells that produce mucus, sweat, saliva, tears, and digestive enzymes. The channel transports negatively charged chloride ions into and out of cells. The transport of chloride ions regulates the movement of water across membranes, which allows the production of thin, freely flowing mucus. Mucus is a slippery substance that lubricates and protects the lining of the airways, digestive system, reproductive system, and other organs and tissues.

In patients with the disease, chloride secretion from the cells is decreased in the lungs and the ducts of the pancreas and other glands. In the lungs, this leads to the formation of viscous mucus which blocks the bronchioles of the lungs. Death may result from lung infection or heart failure.

Cotransport systems

The examples of ATP-dependent transport proteins given above all involve a solute simply being transported across a membrane at the expense of ATP hydrolysis. The ATP is *directly* involved with the transport. Such systems are called **uniports**. There is, however, a thermodynamically feasible alternative way. When the uniport establishes an ion or other gradient, this gradient has potential energy and ions can flow down the gradient. If appropriately harnessed, the flow of ions can be made to perform work. The energy still comes from ATP but not directly. There are two basic types of cotransport systems, symport and antiport.

The Na^+/K^+ ATPase, described above, produces a steep Na^+ gradient across the cell membrane, which has potential energy and, given a chance, the accumulated exterior Na^+ will flow back into the cell.

Let us consider glucose absorption from the intestine, an example of symport, as both glucose and Na^+ are transported together by the same transporter, in the same direction (Fig. 7.12; see also Chapter 10). The glucose symport protein permits movement of Na^+ and glucose across the luminal side of the intestinal epithelial cell membrane *when both are present*

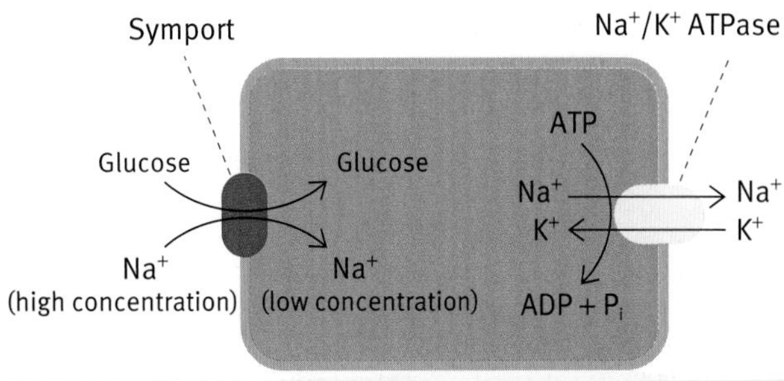

Fig. 7.12 The Na^+/glucose cotransport system – a symport.

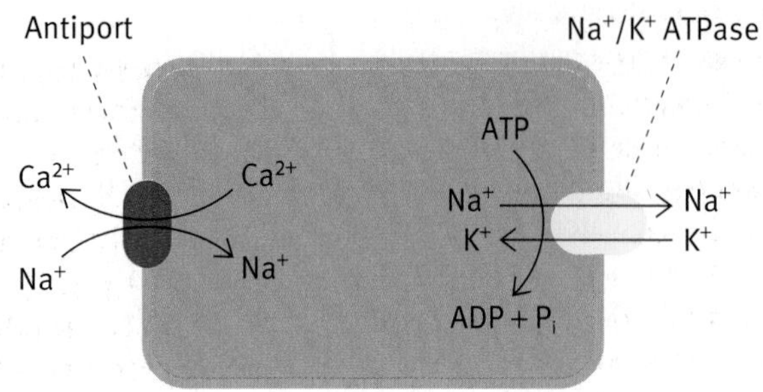

Fig. 7.13 The Na^+/Ca^{2+} cotransport system – an antiport.

in the lumen of the intestine, but not when only one is present. It can thus transport glucose against a concentration gradient utilizing the potential energy of the Na^+ gradient. The Na^+ gradient is maintained by the Na^+/K^+ ATPase pumping Na^+ out of the opposite side of the cell so that it is ATP hydrolysis that indirectly supplies the energy for glucose uptake.

Absorption of amino acids from the gut occurs by a similar mechanism, separate symport proteins being required for the different substances transported.

Antiport systems also exist. As mentioned previously (Box 7.3), in connection with cardiac glycoside action on the heart, the cotransport of Na^+ can be used in another way – to pump out Ca^{2+}. This is an **antiport** system (Fig. 7.13). Na^+ is transported into the cell only if at the same time Ca^{2+} is transported out. Again the energy in the Na^+ gradient established by the Na^+/K^+ ATPase system is the driving force.

Passive transport or facilitated diffusion

In passive transport, a protein permits the movement of a substance across the membrane so that the movement will be down whatever concentration gradient exists across the membrane; no energy is involved in the actual transport process. This is **facilitated diffusion**. We have already described **porins**, the hydrophilic β barrel channels which allow this transport process. Another good example is the anion-transport protein in red blood cells which lets$HCO_3^-$and Cl^- ions pass through the membrane in either direction (Fig. 7.14). The function of this was discussed earlier (see Chapter 4). Another important example of facilitated diffusion is the glucose transporters possessed by many animal cells. They allow glucose to diffuse passively across the membrane. Their importance will become obvious when we deal with the uptake and utilization of glucose in various tissues (see Chapters 10 and 20).

Gated ion channels

An important class of passive-transport systems is that of the gated ion channels. These are aqueous pores, highly selective for specific ions, that open and close on receipt of a signal. They are found in large numbers and of varying specificities in most membranes, and the different channels respond to different signals. Figure 7.15 illustrates ligand-gated and voltage-gated channels. The most important **gated pores** are those for **Na^+**, **K^+**, and **Ca^{2+}** (each selective for one ion) and the **Na^+/K^+ channel**. The passage of ions through these channels, when open, is the result of concentration gradients across the membrane so that the flow proceeds in either direction as determined by these gradients. Since they are steep in respect of ions such as Na^+, K^+, and Ca^{2+}, the rate of movement of these ions through open channels is very much faster, perhaps 1000-fold, than the rate of transport of Na^+ and K^+ ions brought about by the Na^+/K^+ ATPase pump. This becomes an

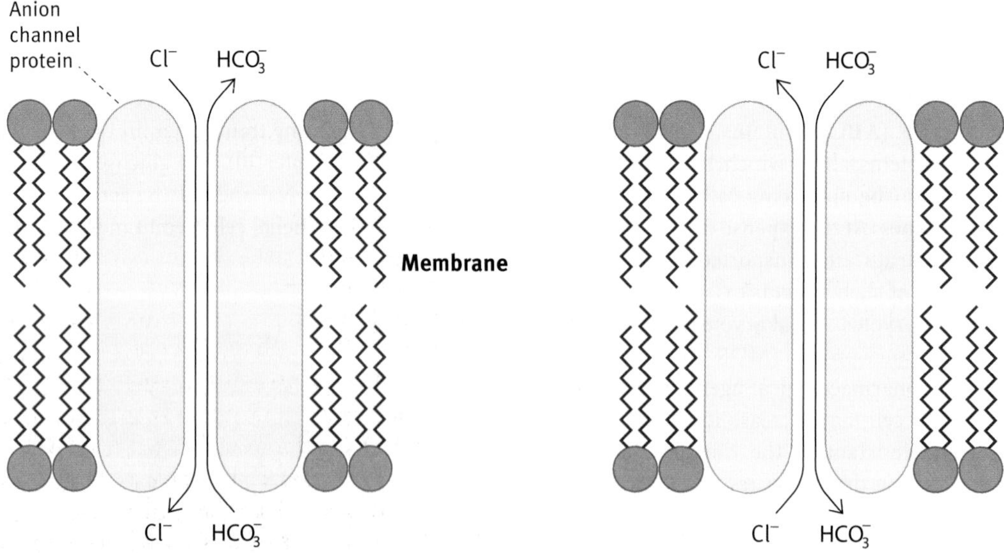

Fig. 7.14 An anion channel of red blood cells. Cl^- and HCO_3^- may move in either direction, according to concentration gradients across the membrane. The counterflow of the ions prevents any electrical potential developing across the membrane.

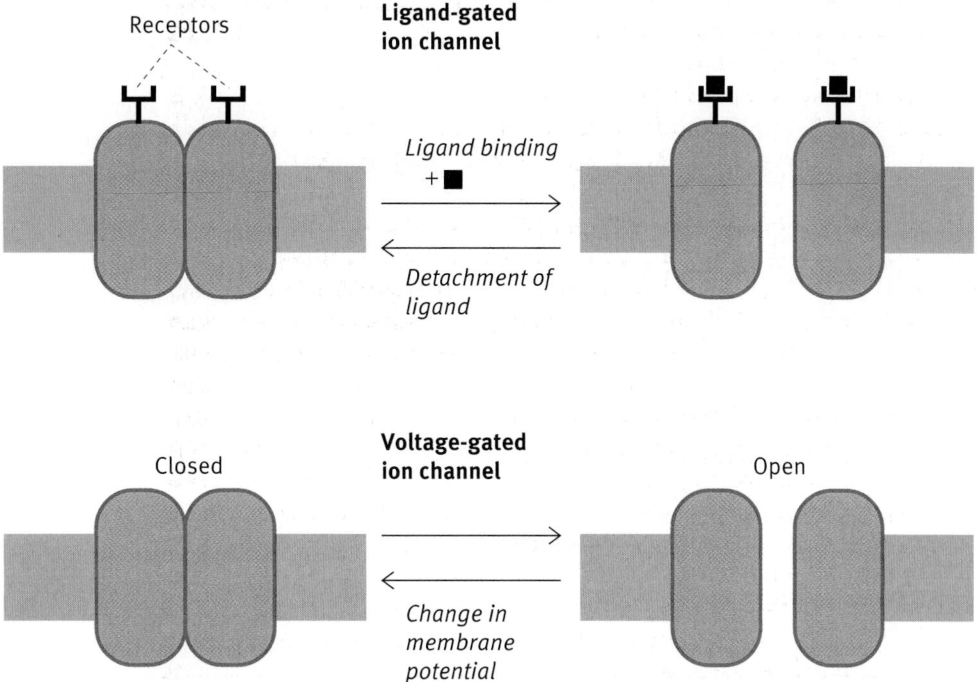

Fig. 7.15 Ligand-gated and voltage-gated ion channels. Examples of ligand-gated channels are the acetylcholine-gated Na^+/K^+ channels found on postsynaptic membranes and on muscle membranes at neuromuscular junctions (also known as the acetylcholine receptors), but the principle applies to other neurotransmitters and to a wide variety of chemical signals in other types of cells. Examples of voltage-gated channels are the Na^+ and K^+ channels found in nerve axons. Movement of ions is passive, driven solely by concentration gradients.

important factor in nerve-impulse generation (see the Nerve-impulse transmission Section).

Mechanism of the selectivity of the potassium channel

The most intensively studied ion channels are the gated channels of the nerve cells, described below. However, a major advance in understanding how different channels exert their strict selectivity for certain ions has come from the crystallization of a bacterial potassium selectivity channel protein which permitted the determination of its three-dimensional structure by X-ray diffraction. The membrane-spanning units of this protein are homologous with subunits of eukaryotic sodium/potassium channels.

The selectivity filter of the channel consists of four identical transmembrane protein subunits, creating a cone-shaped channel. Figure 7.16 shows the tetrameric structure in end-on view. The extracellular end of the pore is at the pointed end of the cone (lower end, Fig. 7.17). Here hydrated ions can enter the cavity where a potassium ion (green) is shown surrounded by eight water molecules (red). (Remember that the cavity is completed by two additional monomers, above and below those two shown side-by-side in the figure.) The channel then narrows into the selectivity part of the pore where there are several sites. The narrow section is formed by four connecting polypeptide loops belonging to each of the four protein subunits (two are shown in Fig. 7.17). The polypeptide chains of these loops are oriented so that the peptide carbonyl oxygens (C=O groups shown red in the figure) point into the channel. There are groups of eight of these for each site in the complete channel. The selectivity loops consist of a sequence, Thr—Val—Gly—Tyr—Gly, which is highly conserved and found in all K^+ channels.

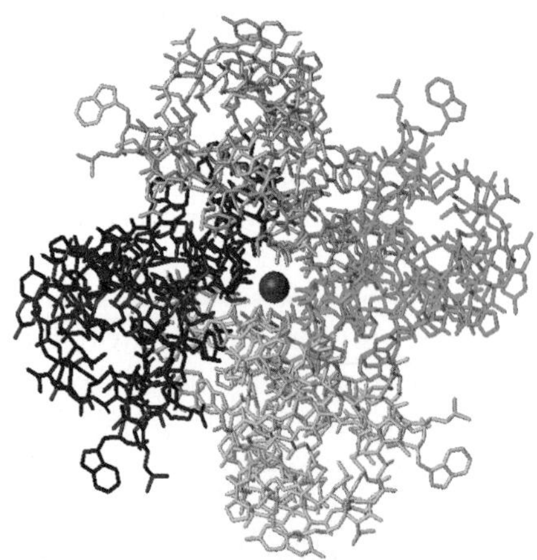

Fig. 7.16 Potassium channel (Protein Data Bank code 1BL8) stick diagram, showing an end view of the four proteins making up the channel, looking from the inside of the cell, with K^+ ions making their way in the pore.

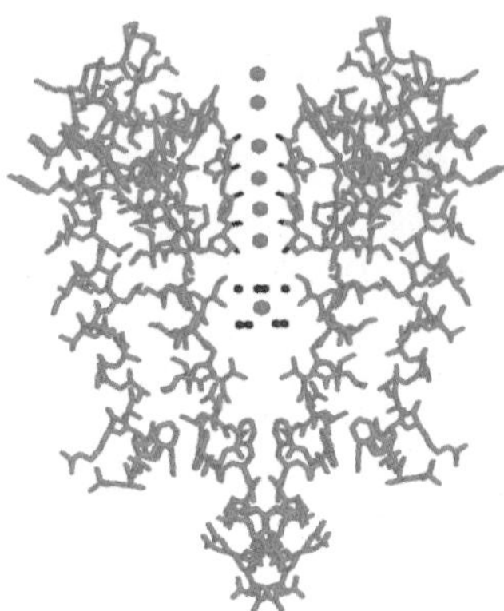

Fig. 7.17 Potassium channel (Protein Data Bank code 1K4C) diagram. Two of the four protein molecules that make up the pore are shown, with K^+ ions (green spheres) passing through. The K^+ at the bottom, in aqueous solution, is surrounded by eight water molecules (red), and in this form is too large to traverse the pore. The K^+ ion has to shed the water molecules in order to pass through the channel, but this is energetically unfavourable. Carbonyl oxygen atoms of the polypeptide chain that line the pore (in red) facilitate this by substituting for the eight water molecules. Once through the pore each K^+ ion becomes hydrated again (not shown). The hydrated sodium ion is too large to pass through the selectivity filter, while the filter cannot facilitate the shedding of its water molecules, since unhydrated sodium ions, being smaller than potassium, cannot interact with the oxygen atoms lining the pore. Illustration taken from PDB Molecule of the Month series.

The channel allows the free movement of potassium ions through the pore but almost completely blocks the passage of sodium ions. For each 10,000 K^+ ions only one Na^+ ion is allowed to pass despite the fact that the sodium ion is smaller than the potassium ion. The selectivity is based on thermodynamic principles. The potassium ion in solution is comfortably surrounded by eight water molecules, which must be removed for the ion to traverse the pore. Removal of these water molecules is energetically unfavourable since it means breaking their noncovalent attractions to the ion. The peptide carbonyl groups lining the selectivity filter channel are spaced so that they exactly mimic the arrangement of eight water molecules around the ion. The potassium ion can therefore easily slip from the water molecules into the thermodynamically similar situation within the filter channel so that there is no energy barrier. In Fig. 7.17 before the selectivity filter, in the large cavity, as mentioned, you see one potassium ion (green) surrounded by the eight water molecules (red). The ions passing through the narrow selectivity filter can be seen to be surrounded by four oxygen atoms of the peptide-bond carbonyl oxygens (eight for the four subunit chains) thus mimicking the structure of a hydrated ion. The ions move through the filter from one site to the next. Why can sodium ions not do the same? Being smaller, a dehydrated ion cannot attach to the oxygens in the channel for the latter are too far apart to allow this. The hydrated sodium ion is too large to pass through the selectivity filter, while the filter cannot facilitate the shedding of its water molecules. The barrier, the energy requirement for shedding the hydrating molecules, is not removed in the case of sodium ions as it is for potassium. The high selectivity of the channel for K^+ is coupled with a very rapid flow. It is believed that this results from the K^+ bound to the selectivity sites being sequentially displaced from one site to the next by incoming K^+ ions. The simplicity of the mechanism for achieving such selectivity is astonishing.

It is likely that the selectivity filters of other K^+ channels are similar to the bacterial channel described above, but have different gating mechanisms attached to them. K^+ channels in nerve cells are voltage-gated – they open and close as described below.

Nerve-impulse transmission

Transmission of nerve impulses constitutes one of the most elegant applications of molecular biology. Mechanistically it is an astonishingly simple and intellectually satisfying system.

The ionic gradients established across cell membranes are used by cells that have gated ion channels in their membranes. When channels open, rapid flows (fluxes) of ions occur through the membrane, as appropriate to their function. The most spectacular use of this principle is in nerve conduction, which we will use as an important and best-studied example. The gated ion channels involved in nerve conduction are those for Na^+/K^+, Na^+ alone, K^+ alone, and Ca^{2+}.

A nerve impulse is transmitted by a series of neurons or nerve cells, which have a central cell body and a thin axon; axons may be very long, even metres in length, and terminate in branches. The gap between one neuron and another is called a **nerve synapse** (Fig. 7.18); the signal is transmitted chemically across synapses by **acetylcholine** (or other neurotransmitter), which is released from the neuron ending to stimulate the next neuron. The signal molecules combine with receptors on the postsynaptic membrane – the starting membrane of the next neuron. We confine our discussion here to acetylcholine as the transmitter substance. The acetylcholine released into the synapse is rapidly hydrolysed by **acetylcholinesterase** bound to the synaptic membrane and the neuron becomes ready to accept a new signal. The anticholinesterase

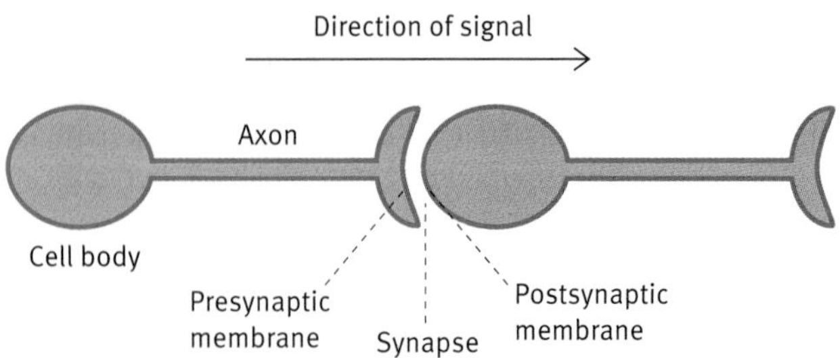

Fig. 7.18 A synapse between two neurons. A nerve impulse in the first one causes the liberation of a neurotransmitter such as acetylcholine into the cleft. This stimulates the succeeding neuron.

organophosphate nerve gases and insecticides inhibit the hydrolysis. Snake venoms such as cobra toxin attach to the channels in the postsynaptic membrane and inactivate them.

The same type of receptor occurs at neuromuscular junctions; acetylcholine liberated by motor nerve endings binds to the receptors, and triggers muscle contraction. **Curare**, the arrow poison, blocks the action of acetylcholine here and paralyses muscles. To relax voluntary muscles during surgery, anaesthetists use modern derivatives of curare-like nondepolarizing muscle relaxants such as **vecuronium**, which block the signal from motor nerves to skeletal muscles. All act essentially like curare, but are much shorter acting and have less propensity for causing histamine release and hypotension than curare. Fig. 7.19 gives a simplified overview of how a neuron conducts a signal (see also Box 7.4).

The acetylcholine-gated Na^+/K^+ channel or acetylcholine receptor

The channel consists of a pore created by five protein subunits arranged in a circle. Two of the subunits have a binding site for acetylcholine. Each subunit contains an α helix kinked in the middle thus restricting the channel size (Fig. 7.20). In the closed form, hydrophobic groups project into the channel but when two molecules of acetylcholine bind to the receptor the helices tilt slightly and cause the groups to swing out of the way leaving the channel open to Na^+ and K^+ ions. The gate remains open only for about a millisecond, even if acetylcholine is still bound, because the gate rapidly becomes desensitized and closes. The channel is highly selective for Na^+ and K^+ ions; anions are repelled by negatively charged carboxyl groups of amino acid side chains.

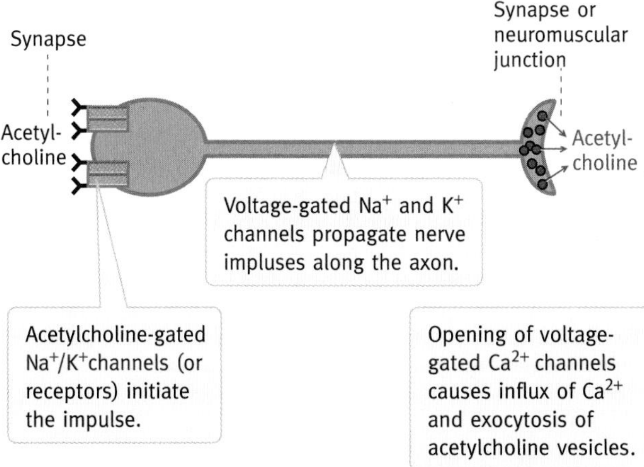

Fig. 7.19 Simplified diagram of the events in the transmission of a nerve impulse along a neuron, starting with the acetylcholine stimulation of the postsynaptic membrane and ending with the release of acetylcholine into a nerve synapse or into a neuromuscular junction. The acetylcholine-gated channel conducts both Na^+ and K^+ ions; there are two separate voltage-gated channels for Na^+ and K^+ respectively for the propagation of the impulse along the axon.

BOX 7.4 Cholinesterase inhibitors and Alzheimer's disease

Alzheimer's disease (AD) is the most common form of dementia involving the parts of the brain that control thought, memory, and language. Amyloid plaques (abnormal extracellular protein clumps) and neurofibrillary tangles (tangled bundles of intracellular protein fibres) in the brain are considered hallmarks of AD.

Ageing is associated with some neuronal cell loss and subtle neurotransmitter changes whereas, in AD, cholinergic neurotransmitter function is markedly depleted. Neurons secreting acetylcholine degenerate, the brain levels of this neurotransmitter fall sharply, and the function of some acetylcholine receptors is changed. The effect of AD is much more severe than the effect of ageing. Acetylcholine is a critical player in the process of forming, storing, and retrieving memories. Regions of the brain critical to memory and thinking include the hippocampus and cerebral cortex – two regions devastated by AD. These findings led naturally to the idea that increasing levels of acetylcholine, replacing it, or slowing its breakdown may ameliorate the disease.

Four prescription drugs are currently approved by the US Food and Drug Administration to treat the symptoms of mild-to-moderate AD and three are licensed and used by the NHS in the UK. These medications act by inhibiting the action of acetylcholinesterase, an enzyme that normally breaks down acetylcholine. The drugs are aimed at increasing the concentration of acetylcholine in the synapse, thereby facilitating neurotransmission. They can help improve some of the symptoms (including behavioural changes) and/or delay deterioration. None of these medications stops the disease itself, as they do not prevent ongoing brain cell degeneration. As AD progresses, the medications may eventually lose their effect, probably due to decreased acetylcholine production.

AD is a heterogeneous disease, and a number of other lines of research into its causes and treatment are being undertaken.

How does acetylcholine binding to a membrane receptor result in a nerve impulse?

Consider a nerve synapse at which acetylcholine is released by the presynaptic neuron (the membrane at the end of the neuron carrying the impulse to the synapse). It diffuses the short distance to the postsynaptic membrane of the next neuron where it triggers a nerve impulse in that neuron. As with other cells, neurons have a high concentration of K^+ and a low concentration of Na^+ inside relative to the levels outside as a result of the Na^+/K^+ pump.

In the resting stage, K^+, high in concentration inside, leaks out because of special K^+ 'leak' channels thus creating a negative charge inside and a positive one outside, since the membrane is impermeable to anions. The K^+ leakage is self-limiting, because the internal negative charge so created holds K^+ ions back and an equilibrium is established in which the resting cell potential across the membrane is about −60 mV (more negative inside). This results in a separation of electric charges at the membrane (which is an electrical insulator)

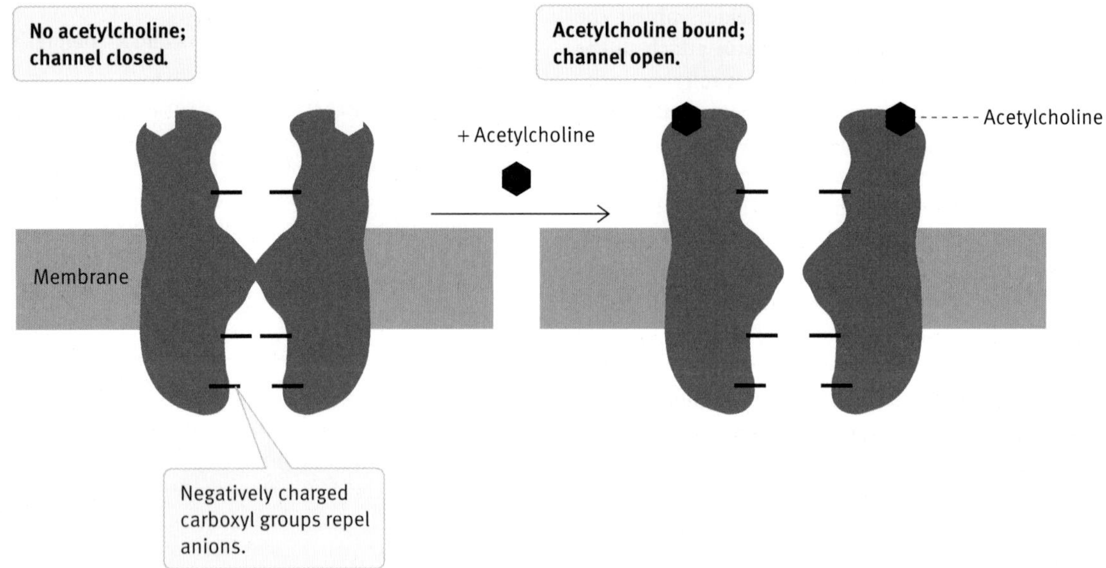

Fig. 7.20 Diagram of the acetylcholine-gated Na^+/K^+ channel of postsynaptic membrane and the receptor at neuromuscular junctions. The channel is composed of five subunits arranged to form a pore but only two are shown for simplicity; two are identical and each has an acetylcholine-binding site, which induces the allosteric change to open the channel on ligand binding. It opens for only a fleeting period.

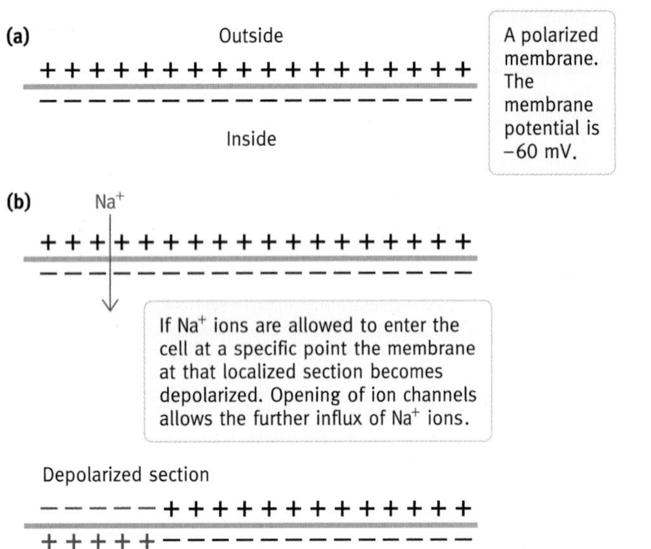

Fig. 7.21 Diagram to illustrate the meaning of the term 'polarized membrane' and its depolarization. In talking about depolarization in the context of neuronal function, it is important to remember that only localized sections of membrane are referred to, not the whole cell membrane, and also that the depolarization is transitory as explained in the text. **(a)** A polarized membrane; **(b)** depolarization by Na^+ entry.

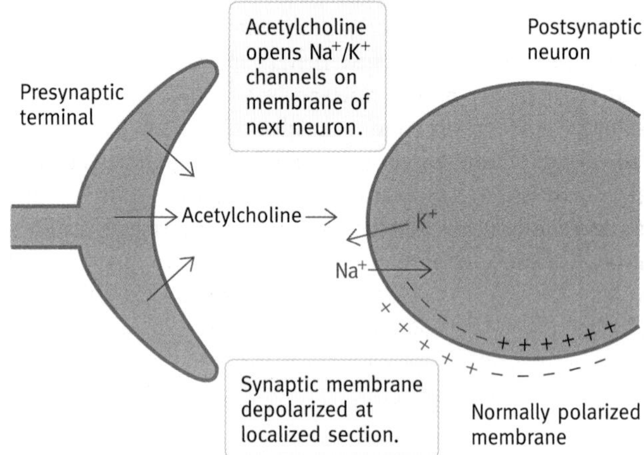

Fig. 7.22 Transmission of a nerve impulse across a nerve synapse. The presynaptic membrane liberates acetylcholine (or other neurotransmitter) on arrival of a nerve impulse. The acetylcholine binds to the receptors (the Na^+/K^+-gated channels) on the postsynaptic membrane of the next neuron. The resultant channel opening causes an influx of Na^+ ions and a smaller efflux of K^+ ions. This causes a local depolarization of the postsynaptic membrane. This depolarization triggers the propagation of the nerve impulse along the axon of the neuron as described in the text.

with a surplus of negative charges inside and positive outside. They attract one another across the lipid bilayer as shown in Fig. 7.21(a), creating an electrical or **membrane potential**. The membrane is then said to be **polarized**. If you placed an electrode inside the cell and another outside, a voltage would be recorded by a meter placed in the circuit connecting the two.

When Na^+/K^+ channels in the postsynaptic membrane open in response to acetylcholine binding, the resulting Na^+ influx (Fig. 7.21(b)) is greater than the efflux of K^+ because the negative charge inside the membrane opposes the latter. While the ion fluxes are large, they only occur for about a millisecond and there is negligible effect on the overall Na^+ and K^+ gradients between the inside and the outside of the cell. The membrane polarization in the postsynaptic membrane in the vicinity of the channels is reversed – it changes from about −60 mV to −65 mV and the section is then said to be **depolarized** (Fig. 7.21(b)). The channels now close. The synaptic transmission events are summarized in Fig. 7.22.

Nerve-impulse propagation is driven by ion gradients

Nerve-impulse propagation uses ion gradients that exist across the cell membranes to supply the required energy, the propagation mechanism involving nothing more than the opening and closing of voltage-gated channels in the correct sequence. Nothing physically moves along the length of the axon in the sense of a flow of molecules or ions. All that is needed is that, at the end of the neuron, the presynaptic membrane becomes locally depolarized.

The mechanism of nerve-impulse propagation along the axon was elucidated by the classic work of Hodgkin and Huxley in Cambridge in 1952. The acetylcholine-gated channel, as described, creates a small patch of depolarized membrane at the postsynaptic membrane (Fig. 7.23(a)). This, in turn, partially depolarizes the adjacent section of membrane due to the diffusion of ions along the axon for a short distance (Fig. 7.23(b)). In the axon membrane, there are separate **voltage-gated Na^+ channels** and **K^+ channels**. When the membrane in which they are located is partially depolarized by this, more Na^+ channels open. The resultant influx of Na^+ fully depolarizes the local section of membrane. The Na^+ channels are inactivated after about a millisecond. (This will be returned to shortly.) At this point, the depolarization peaks and the membrane potential reverts to that of the resting stage – it is repolarized. The repolarization results from opening of voltage-gated K^+ channels, which open slightly later than the Na^+ channels and allow efflux of K^+ ions. The K^+ channels also are rapidly inactivated so that the resting potential is restored. The whole thing takes about 3 ms.

The depolarization of a small section of membrane followed by its repolarization can be measured experimentally by having electrodes placed across the membrane linked to a cathode ray oscillograph. The result is shown in Fig. 7.24 (black curve). The electrical 'spike' of voltage change is known as an **action potential**. At the risk of overexplaining what may be obvious, this spike is the experimental measurement of the changes in membrane polarization. It is the changes in membrane polarization that conduct the impulse; the action potential spike is not something in addition to this but simply a demonstration of it. The conductivity of the patch of membrane to Na^+ ions and to K^+ ions during an action potential indicates the opening of the relevant channels (red and blue curves in Fig. 7.24). What we have achieved so far is the depolarization of a short section of membrane adjacent to the synapse. This depolarization propagates itself along the axon, one small section at a time, until finally the distal synaptic membrane is depolarized causing the release of the transmitter substance to carry the impulse across the synaptic cleft. How does this travelling

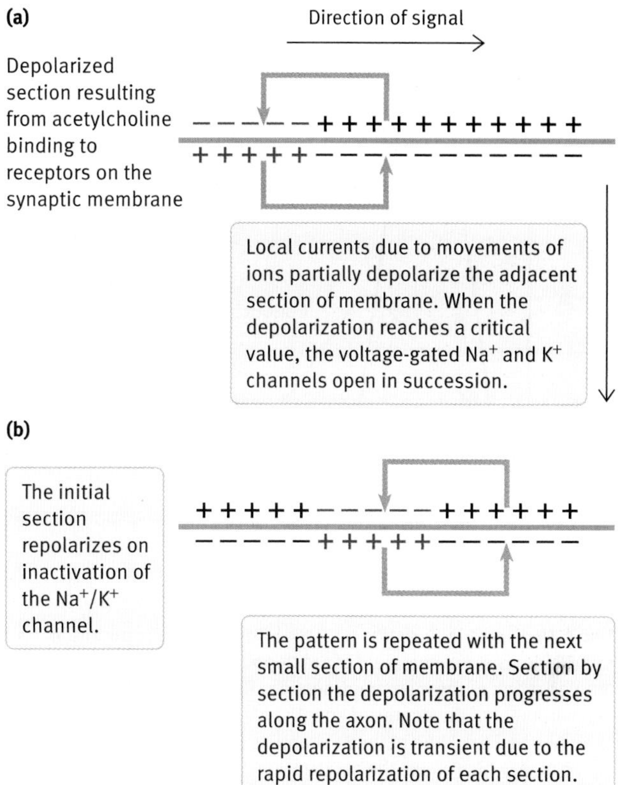

(a)
Direction of signal
Depolarized section resulting from acetylcholine binding to receptors on the synaptic membrane
– – – – – + + + + + + + + + + + +
+ + + + + – – – – – – – – – – – –
Local currents due to movements of ions partially depolarize the adjacent section of membrane. When the depolarization reaches a critical value, the voltage-gated Na+ and K+ channels open in succession.
(b)
The initial section repolarizes on inactivation of the Na+/K+ channel.
+ + + + + – – – – – + + + + + + +
– – – – – + + + + + – – – – – – –
The pattern is repeated with the next small section of membrane. Section by section the depolarization progresses along the axon. Note that the depolarization is transient due to the rapid repolarization of each section.

Fig. 7.23 How a nerve impulse is propagated along the axon. **(a)** Starting with the small depolarized section caused by acetylcholine at the synaptic membrane as shown in the previous figure, **(b)** when a small section of membrane is depolarized, local currents due to ion movements (brown arrows) cause partial depolarization of the adjacent section of membrane. When the membrane potential in the latter has fallen to the threshold value of −40 mV, opening of voltage-gated Na^+ channels causes rapid further depolarization of that section. Closing of the Na^+ channels and opening of the K^+ channels restores the polarization but meanwhile the next section has become sufficiently depolarized to trigger channel opening in that section. The same cycle is repeated until the end of the neuron is reached. The impulse is prevented from going backwards by a mechanism described in the text.

of the depolarization along the axon occur? At each section exactly the same is repeated so that the next small patch of membrane is depolarized by spread of the Na^+ ions along the inside face of the membrane so there is another action potential generated; this spreads to the next and so on thus producing an action potential at one small section of membrane after another, very rapidly, until at the end of the neuron the synaptic membrane is likewise depolarized.

Mechanism for ensuring the nerve impulse only goes forward

The Na^+ ions from the depolarized section can spread in both directions. What is to stop the nerve impulse travelling in the backwards direction as well? The Na^+ and K^+ channels have a special property which copes with this. There is a **refractory period** during which the channels which have opened and

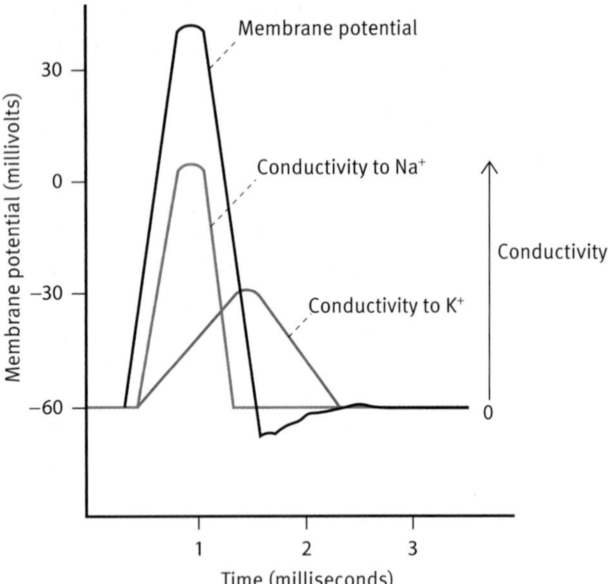

Fig. 7.24 Events in an action potential. The membrane potential (black line) is a measurement of the depolarization and repolarization of the membrane section. The conductivity is a measurement of the degree of opening of the voltage-gated Na^+ channels (red line) and of the K^+ channels (blue line) in the axon membrane, which cause the polarization changes. These curves represent experimental measurements of the depolarization and repolarization events at the membrane. It is the latter that propagates the nerve impulse along the axon.

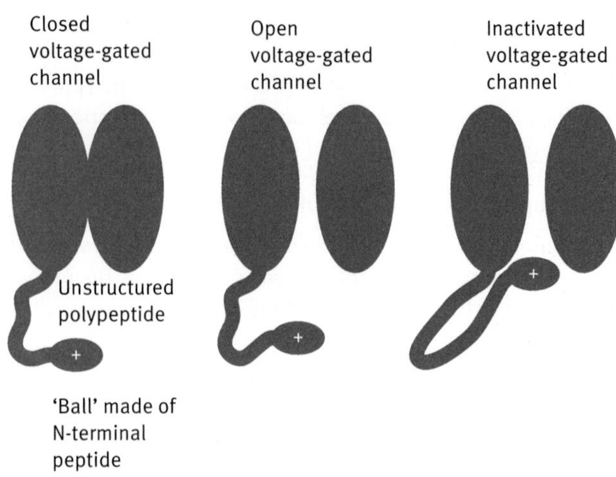

Fig. 7.25 The general principle of the 'ball-and-chain' inactivation of the voltage-gated K^+ channels. Only two of the subunits forming the pore are shown. Restoration of the original closed state occurs after the inactivated state. The Na^+ channels are believed to be inactivated by a mechanism similar in principle but the blocking peptide is a cytosolic loop of the protein connecting two transmembrane domains rather than an N-terminal 'ball'.

closed cannot be re-opened until after a slight delay. By the time they are capable of re-opening, the depolarization has moved too far along the axon for it to affect the channels; they will open only when the next impulse arrives. Therefore the impulse can only go forwards.

Mechanism of control of the voltage-gated Na^+ and K^+ channels

The Na^+ and K^+ channels show considerable structural homology, suggesting that they have a common ancestor. The Na^+ channel is a protein with four transmembrane domains linked on the cytoplasmic side by extensive hydrophilic peptide loops and smaller loops on the outside. The four domains are arranged in the membrane to form the channel. Each of the four is comprised of six transmembrane α helices making it a large protein. In each of the transmembrane domains, one of the transmembrane helices is a voltage sensor rich in basic residues, which slightly changes its position in response to membrane-potential changes and this is what causes channel opening and closing. However, we have pointed out above that the channels, opened by a depolarization of the membrane, are inactivated after about 1 ms. Although in this state they do not allow passage of ions, this does not involve restoration of the original closed state until after the refractory period mentioned above. How does this inactivation occur?

In the case of the K^+ channel, a 'ball-and-chain' mechanism illustrated in Fig. 7.25 involves an N-terminal peptide (the ball) tethered to the channel by a section of polypeptide such that the ball is free to move. When the channel opens, the positively charged ball, attracted by a negative charge, fits into and blocks the channel. The Na^+ channel may have a similar inactivation mechanism, but in this case it is one of the cytosolic loops connecting two of the domains that is believed to be involved rather than a terminal peptide 'ball'.

At the next synapse (or at neuromuscular junctions) the membrane depolarization activates voltage-gated Ca^{2+} channels. The resultant inflow of Ca^{2+} causes acetylcholine-filled vesicles to discharge the neurotransmitter from the synaptic membrane (see Fig. 7.19).

The central feature of nerve-impulse propagation mechanism is that the signal strength is maintained right along the lengths of axons. There is no modulation of the strength of a nerve impulse: it is all or nothing. If the initial signal is sufficient to trigger the initial depolarization, the impulse will travel. It is the frequency of the impulses that is controlled. The signal strength is maintained by the electrochemical gradient across the neuronal membrane, generated by the Na^+/K^+ ATPase; it is boosted every time an action potential is generated.

Myelinated neurons permit more rapid nerve-impulse transmission

In motor neurons, which trigger muscle contraction, there is another refinement. The axon, instead of being bare, is insulated by a myelin sheath (Fig. 7.26), which prevents ion movement (signal leakage) across it. The myelin sheath is interrupted every couple of millimetres at places called the **nodes of Ranvier** and it is here that the voltage-gated Na^+ and

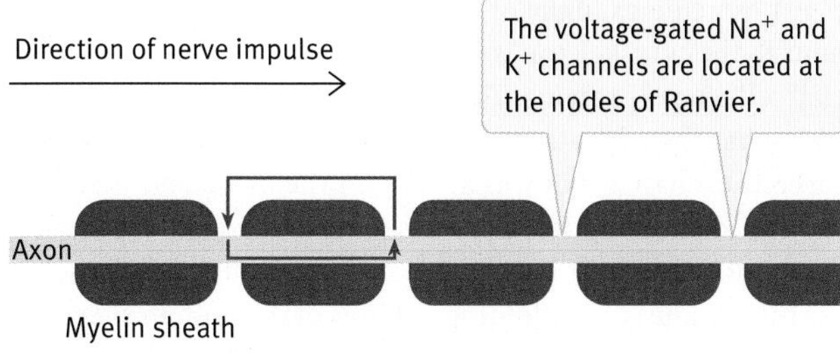
Direction of nerve impulse
The voltage-gated Na+ and K+ channels are located at the nodes of Ranvier.
Axon
Myelin sheath

Fig. 7.26 Myelinated nerve found in the motor and many other neurons. Breakdown of the insulating myelin sheath occurs in multiple sclerosis, and the resulting defect in transmission of action potentials causes the symptoms of this disease. The red arrows represent the local currents caused by depolarization at the nodes from one node to the next so that depolarization and triggering of action potentials leaps along the axon in a saltatory (jumping) fashion: this greatly speeds up the rate of transmission.

K^+ channels are located so that active depolarization of the membrane occurs at the nodes. The axon between the nodes is heavily insulated by the myelin sheath. This enables the depolarization to spread passively and rapidly to the next node. The result is that the action potentials leap from node to node in what is called saltatory conduction. In **multiple sclerosis**, breakdown of the myelin sheath insulation impairs the rapid conduction, causing nerve impulses to travel along the nerve much more slowly. (See the section on Sphingolipids for the structure of a main myelin component.)

Why doesn't the Na^+/K^+ pump conflict with the propagation of action potentials?

Since the action potential propagation depends on migration of Na^+ ions across the neuron membrane, it might be thought that the Na^+/K^+ pump in continually restoring the balance would be confusing the mechanism. This does not occur because the local ion movements in nerve conduction are much faster than those caused by the Na^+/K^+ pump, which acts as a slow 'trickle charger', keeping the ion gradient topped up. The membrane potential derives from the relatively small number of ions near the membrane surfaces, and the movements of ions in generating action potentials involve only a minute proportion of the ions in the bulk of the cells and extracellular medium. There is little concentration change involved.

Role of the cell membrane in maintaining the shape of the cell

Eukaryotic cells have an internal scaffolding which maintains the shape of the cell and is involved in amoeboid motility. It

BOX 7.5 Membrane-targeted antibiotics

In the unceasing battle for survival, microorganisms throw chemical missiles at one another. One of the targets is the ionic gradients across cell membranes. Membrane antibiotics destroy these gradients by allowing their equilibration across the membrane. Two classes of antibiotic are used to achieve this, mobile ion carriers (**ionophores**) and channel formers.

Valinomycin is the best-known ionophore, produced by a *Streptomyces* species, and is active against *Mycobacterium tuberculosis*. It is an unusual 12-membered ring containing D-valine and L-valine, and two hydroxy acids linked by peptide and ester linkages. The outside is hydrophobic while the carbonyl oxygens point inwards and in this way they can chelate monovalent ions, especially K^+. K^+ in solution is attached to water molecules by weak bonds, breaking of which requires energy, so there is a thermodynamic barrier to the loss of the attached water molecules. Valinomycin offers the K^+ an equally stable thermodynamic environment chelated to its carbonyl groups so that the ion can slip from its water cage into the valinomycin cage easily, the free energies of the two states being similar. (The same mechanism is used in the selectivity filter of potassium channels discussed in this chapter). The folded antibiotic enclosing the ion with its lipid-soluble exterior carries the ion across the membrane. Since there is an equilibrium between a hydrated ion and a valinomycin-enclosed ion, the effect is to equilibrate the ions across the membrane. It picks up K^+ inside the cell where its level is high and releases it outside where it is low. Several such ionophores are known. Nonactin, produced by the *Streptomyces* species, is selective for K^+. The antibiotic A23187 exchanges Ca^{2+} for H^+; calcium ions are extremely important in cell regulation so disruption of normal transport is an effective weapon. It is a useful biochemical research tool.

The second type of membrane antibiotic is the channel formers, of which the gramicidins, produced by *Bacillus brevis*, are the best known. Gramicidins A, B, C, and D are dimers of linear peptides, 15 residues long, consisting of alternating L- and D-amino acids. They insert themselves into lipid bilayers, each dimer spanning only one half of the layer. Two of these temporarily associate end-to-end to form a contiuous channel across the membrane, and allow the free passage of H^+, K^+, and Na^+ ions. This destroys the ion gradient essential to the life of the cell. Gramicidin S is quite different. It is a cyclic 10-residue peptide. Polymixins are of this type and active against Gram-negative bacteria. Clinical use of these agents is usually only in the form of ointments for superficial applications since they may affect animal cell membranes.

A different type of action is shown by amphotericin B. This is a polyene antibiotic acting against many fungal infections. It allows loss of low-molecular-weight substances from cells by creating pores formed by several molecules of antibiotic complexed with membrane cholesterol. Amphotericin B affects any cholesterol containing membrane, so that the patient's own tissues are under threat. As the side effects are serious, it will, for example lyse blood cells, it is only used for potentially life-threatening fungal infections. It is ineffective against bacteria as their membranes do not contain cholesterol and it is equally ineffective against viruses. Mitochondria are not affected since the antibiotic is active only against sterol-containing membranes. Candidin, nystatin, and fungichromin have similar activities.

is called the **cytoskeleton** (see Chapter 8). It is made of protein microfilaments, which pervade the cytosol and, at various places, are attached to integral proteins in the cell membrane. Such membrane proteins are not then free to move laterally in the lipid bilayer, as are other proteins.

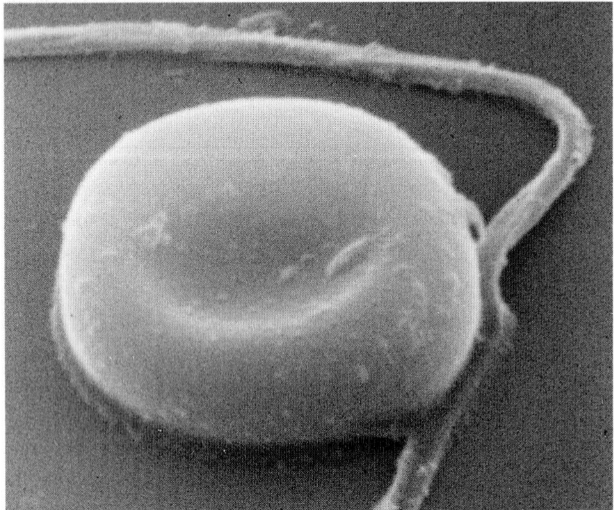

Fig. 7.27 Scanning electron micrograph of an erythrocyte. Courtesy Professor W. G. Breed, Department of Anatomy, University of Adelaide.

An extreme example of the attachment of membrane proteins to the cytoskeleton is in the red blood cell, which has special 'cell-shape' proteins. The cell is a biconcave disc (Fig. 7.27), which has the advantage of presenting a large surface area for gaseous exchange, but it is always on the move and therefore subject to shearing forces as it squeezes through capillaries, demanding a robust but flexible cell membrane. Underneath the membrane is a scaffolding of fibres of the protein **spectrin**, anchored to the anion-channel protein (Fig. 7.28) by a protein appropriately called **ankyrin**. The name spectrin comes from the fact that it is possible to release the red cell contents but retain the empty membrane with its cytoskeleton, producing a red cell ghost (spectre). The anion-channel protein is a large protein which projects into the cytosol providing a cytoskeleton attachment point (Fig. 7.28). Spectrin is also linked to glycophorin by other specific linking proteins.

In a small number of people the cytoskeleton of red blood cells is deficient because of faulty spectrin or ankyrin due to a genetic defect. The most common is a defect in forming spectrin tetramers, which are necessary to form the long filaments of spectrin. Fig. 7.29 shows the structure of the human erythrocyte spectrin tetramerization domain. The cells are abnormally shaped and tend to be destroyed by the spleen. The diseases are called **hereditary spherocytosis** and **hereditary elliptocytosis**.

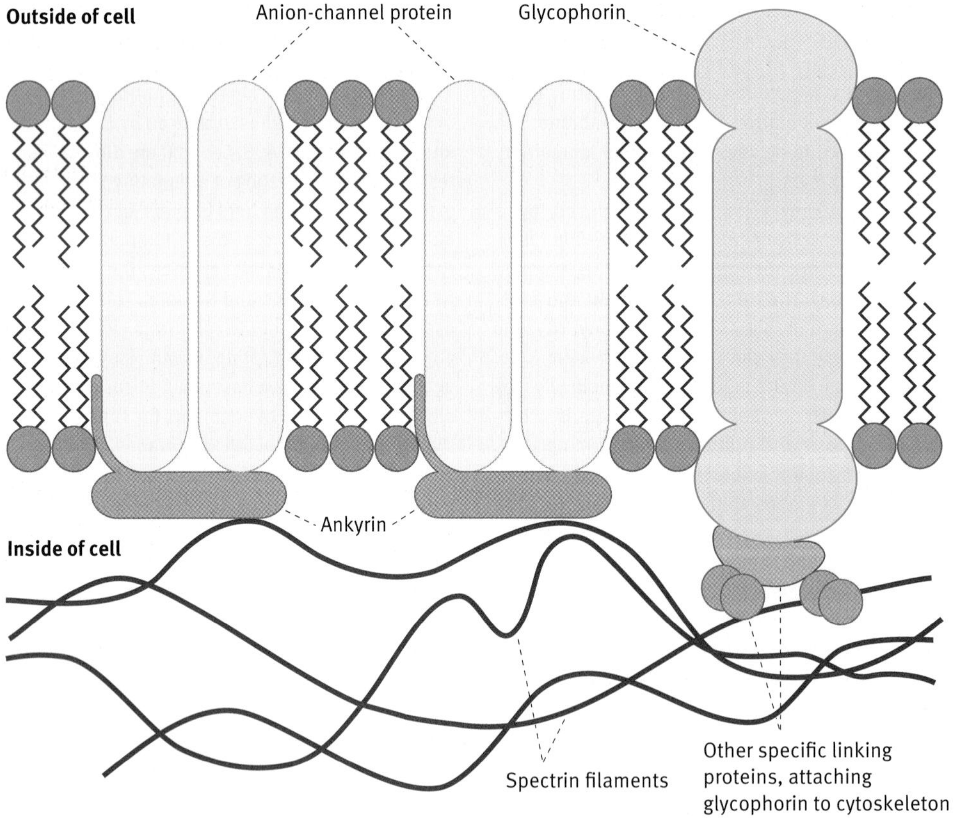

Fig. 7.28 Diagram to illustrate attachment of anion-channel protein and glycophorin to the cytoskeleton.

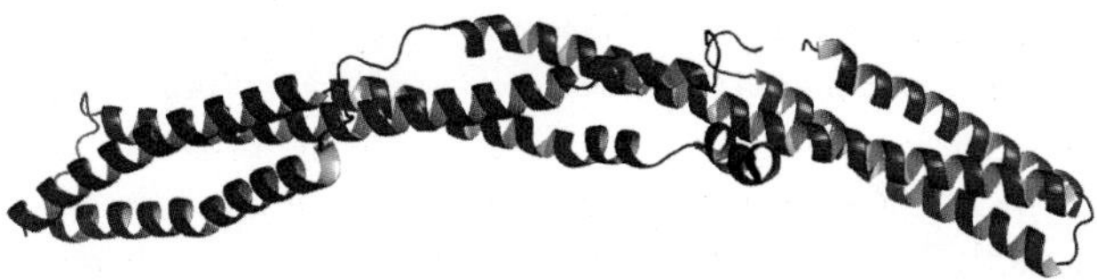

Fig. 7.29 The structure of the human erythrocyte spectrin tetramerization domain complex (Protein Data Bank code 3LBX).

Cell–cell interactions – tight junctions, gap junctions, and cellular adhesive proteins

In an epithelial tissue such as the lining of the intestine, the products of food digestion are selectively taken up by the cells and then transported from the cell by appropriate systems in the membrane of the opposite side of the cell facing the blood vessels. Special bands of membrane proteins encircle the cells and these bind to corresponding proteins of the neighbouring cells, creating **'tight' junctions** or nonleaky cell–cell contacts.

Adjacent cells in some cases exchange molecules, which allow coordination of chemical activities throughout a tissue. This is achieved by proteins forming a tunnel between cells, known as **gap junctions**, which will allow smaller molecules such as ATP to pass, but not proteins. The coordination of heart cell contractions, for example, depends on gap junctions.

Another function of membrane proteins is to promote adhesion between cells of a tissue to form that tissue. If different types of embryo cells are mixed together they will re-associate with cells of their own type – kidney cells to kidney cells and so on. This is a function of tissue-specific cell–cell adhesion proteins called **cadherins** (see also Chapter 4). Another family of this class of proteins, the glycoproteins known as ***N*-CAMS** (for nerve cell adhesion molecules), is important in nervous tissue formation.

Summary

Biological membranes have a lipid bilayer structure made up of a variety of different amphipathic lipids held together by noncovalent bonds. They are arranged with their hydrophobic tails pointing to the middle of the bilayer and their hydrophilic sections to the outside. Lipid bilayers are two-dimensional fluids that can self-seal. This permits endocytosis, a process by which cells take in material, and enables cells to eject molecules in a reverse process called exocytosis.

The fatty acid components may be saturated or unsaturated; the latter, in a *cis configuration*, are essential in maintaining the bilayer in a fluid condition. *Trans* unsaturated fatty acids resemble saturated fatty acids in that they are straight chain, not kinked. Cholesterol also plays a moderating influence on membrane fluidity.

Proteins are embedded in the bilayer as required for the functions of various membranes and often retain their lateral mobility. The membrane proteins have structures in keeping with their location in the hydrophobic membrane interior with hydrophobic amino acid residues in the central region and hydrophilic residues in domains that will be on either side of the membrane.

Porin proteins form hydrophilic channels across the membrane and these have hydrophobic amino acids on the outside face of the protein that is in contact with the hydrophobic layer of the membrane and hydrophilic ones pointing into the channel. Most membrane proteins are glycosylated on their exterior domains.

Lipid bilayers are largely impermeable to hydrophilic molecules and transport systems are needed for molecular traffic. These may be active, driven by ATP breakdown, or passive, driven by concentration gradients only. Nearly all cells have the Na^+/K^+ ATPase, which pumps sodium out and potassium in. The ion gradient so formed is used to drive transport of other molecules in symport and antiport mechanisms.

Many channels are selective gated pores, which are controlled by ligand attachments or membrane potential. The structure and mode of action of the potassium channel have been elucidated. It is based on an ingenious device basically thermodynamic in concept that allows the K^+ ions to shed their attached water molecules and pass through the pore. Na^+ ions, although smaller, cannot do so.

Nerve-impulse conduction is the most spectacular membrane activity involving gated ion channels opening and closing at specific times. The energy for the nerve impulse derives from the ion gradient established by the Na^+/K^+ pump.

Membranes have other functions; one of the most important of these in eukaryotes is cell signalling (see Chapter 29). The cell membrane has a role in maintaining the shape of the cell by interactions between the cytoskeleton (see Chapter 8) and integral membrane proteins. Cell-cell interactions such as tight junctions, gap junctions, and cell-cell adhesions (cadherins, Chapter 4), all involve special membrane proteins.

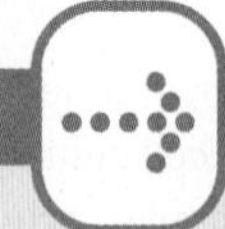

Further reading

To access the further reading, please scan the QR code image or go to http://global.oup.com/uk/orc/biosciences/molbiol/snape_biochemistry/student/reading/ch07/

Problems

1 Describe the lipid bilayer structure of membranes.

2 What is the role of cholesterol in eukaryotic membranes?

3 Outline the structure of phosphatidic acid.

4 Name three glycerophospholipids based on phosphatidic acid. Name the substituent in each case attached to the latter.

5 What is the significance of *cis* unsaturated fatty acyl tails in membrane phospholipids?

6 Why do inorganic ions not readily pass through membranes?

7 What is meant by facilitated diffusion through a membrane? Give an example.

8 Compare the structure of a triacylglycerol and a polar lipid. Why can the former never be found in lipid bilayers?

9 Explain how an ion gradient across a cell membrane can be harnessed to provide the energy for the active transport of an unrelated molecule or ion.

10 What is a gated ion channel? Name two types.

11 Describe the mechanism by which digitalis is used to strengthen the heartbeat in patients with congestive heart failure.

12 Discuss the principle of the mechanism of the selectivity filter of a potassium ion channel.

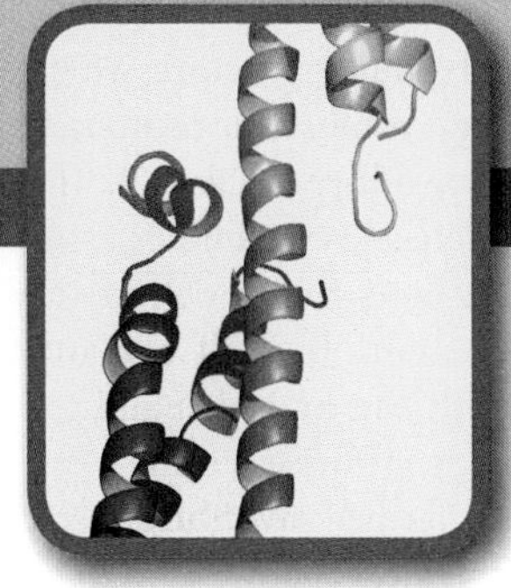

Chapter 8

Muscle contraction, the cytoskeleton, and molecular motors

A surprising amount of mechanical work goes on in almost all eukaryotic cells. This is driven by adenosine triphosphate (ATP)-fuelled molecular motors (**motor proteins**), which act on proteins of the **cytoskeleton**: microfilaments (polymers of actin) and microtubules (polymers of tubulin). Muscle contraction is a specialized case in which motor proteins are collected together in vast numbers and organized so that together they can produce the remarkable forces that muscles develop. In this case the motors are fixed in position and act on actin filaments that are forced to slide. Motor proteins present in nonmuscle cells, such as liver and most other cells, are in motion, running along fixed cytoskeletal 'tracks' and pulling loads of macromolecules, vesicles, and organelles between various parts of the cell. The intracellular motor activity resembles that of a busy city. As described later, the cytoskeleton has several important roles in cellular structure and motility, in addition to providing transport tracks.

Prokaryotes, because of their small size, do not need an internal cytoskeleton. Some do, however, have external flagellae equipped with molecular motors to propel cells through liquid.

In this chapter we will first deal with the mechanism of muscle contraction, which provides a good basis for understanding the action of motor proteins. Following this we will describe the nonmuscle cytoskeleton and the molecular motors that operate on its tracks.

Muscle contraction

A reminder of conformational changes in proteins

Muscle contraction is a remarkable phenomenon. Biological processes depend wholly on molecules and, for contraction or movement, individual molecules must move in a directional fashion. What type of molecular activity is available for this? The only type we know of is conformational change in proteins – minute changes in shape of individual protein molecules that occur as a result of different ligands binding to them. Each conformational change is minute, but collectively they can add up to the gross movement of muscles.

For movement to occur, whatever exerts force must have something to exert it against. A 'molecular motor' exerting force by conformational change must always have a partner structure to react against – if the motor is fixed, the partner molecule will move; if the partner molecule is fixed, the motor molecule will move. To anticipate the mechanism of muscle contraction, the motor protein is **myosin**, which is fixed, and the structure it acts against, and slides, is the **actin filament**.

Types of muscle cell and their energy supply

The two main classes of muscles are **striated** and **smooth**. Striated muscle is found in skeletal muscle, which is under voluntary nerve control and contracts rapidly. It derives its name from its appearance under the microscope (Fig. 8.1). Heart muscle is striated but not identical in structure to skeletal muscle, and is under involuntary control. Smooth muscle is found in the intestine and blood vessels, which are typically under involuntary nervous control and also frequently under control by hormones. It contracts slowly and can maintain the contraction for extended periods. It is not striated.

In all muscles the reserve of ATP, on which contraction depends, is only enough for a brief period of intensive contraction. This explains a role for a fast-acting reserve in skeletal muscles of creatine phosphate, a high-energy phosphate compound. When ATP is converted to adenosine diphosphate (ADP) and inorganic phosphate (P_i) by contraction, ATP is

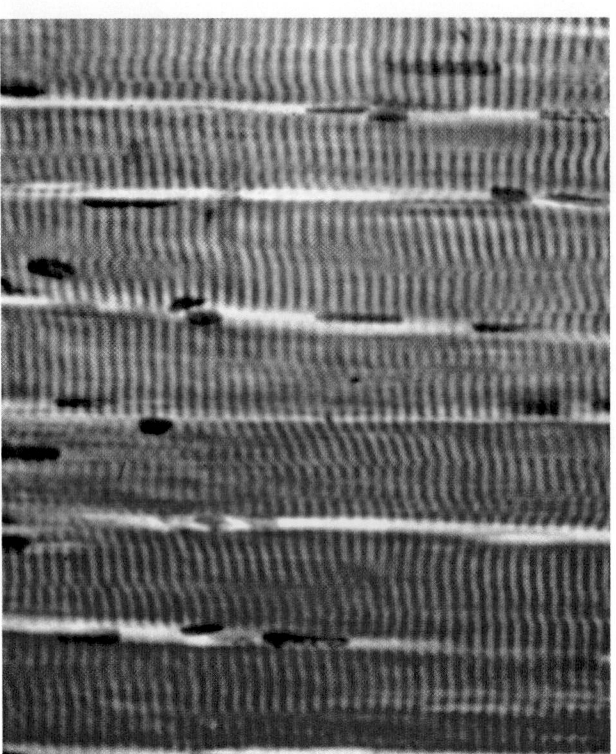

Fig. 8.1 Photomicrograph of myofibrils of striated skeletal muscle. © Robert Harding.

rapidly regenerated by a reaction catalysed by the enzyme, creatine kinase, using the phosphoryl group of creatine phosphate.

$$\text{Contraction event: ATP} + H_2O \rightarrow \text{ADP} + P_i$$

$$\text{ATP-regeneration event: ADP} + \text{creatine}(-Ⓟ) \rightarrow \text{ATP} + \text{creatine}$$

During muscle recovery, ATP formed by oxidative metabolism replenishes the pool of creatine phosphate by the creatine kinase reaction, which is reversible in the presence of high levels of ATP and low levels of ADP.

Creatine phosphate has the structure shown below. Because of the reactivity of the phosphoryl group, there is a spontaneous slow (noncatalysed) formation of **creatinine**; this has no function and is excreted in the urine at a daily rate proportional to muscle mass. Since creatinine is formed at a relatively constant rate in the body and its only fate is to be excreted in the urine, plasma (or serum) creatinine is used as an indicator of kidney function; a rise in plasma creatinine indicates problems with the kidneys.

```
                O⁻
                |
          NH—P=O
         /      |
H₂⁺N=C        O⁻
         \
          N—CH₂—COO⁻
          |
          CH₃

     Creatine phosphate
```

```
            O
            ||
      NH—C
     /     |
HN=C       |
     \     |
      N—CH₂
      |
      CH₃

   Creatinine
```

Structure of skeletal striated muscle

Skeletal muscle has a very specialized cellular structure, which gives rise to some specific terminology. A muscle is composed of long multinucleated cells called **myofibres** (Fig. 8.2). The plasma membrane (the **sarcolemma**) has nerve endings associated with it at neuromuscular junctions, which deliver the nervous signal that triggers contraction. The cell contains many mitochondria, in keeping with the high demand for ATP to drive contraction. Running lengthwise within it are multiple elongated protein structures termed **myofibrils**, each surrounded by a membranous sac, the **sarcoplasmic reticulum**. The sarcoplasmic reticulum acts as a repository for calcium ions.

Structure of the myofibril

The myofibril is the structure that does the contracting. Each myofibril is divided into segments termed **sarcomeres** (Fig. 8.3(a)) bounded by **Z discs** of proteins (Z for *zwischen* or *between*). On contraction, the Z discs are pulled closer together, thus shortening the individual sarcomeres and hence also the myofibrils (Fig. 8.3(b)). This in turn shortens the myofibre, causing muscle contraction.

How does the sarcomere shorten?

The Z discs of protein at each end of the sarcomeres have attached to them thin **actin filament 'rods'** pointing to the centre of the sarcomere. The filaments are attached to the Z discs only at one of their ends. In a vertebrate muscle cross-section, the **thin filaments** are in multiple hexagonal arrays on the two discs (Fig. 8.3(c)). Inside each array is a **thick filament**, which has finger-like projections that do the actual work of contraction. By a ratchet-like mechanism the finger-like projections 'claw' the thin filaments towards the centre

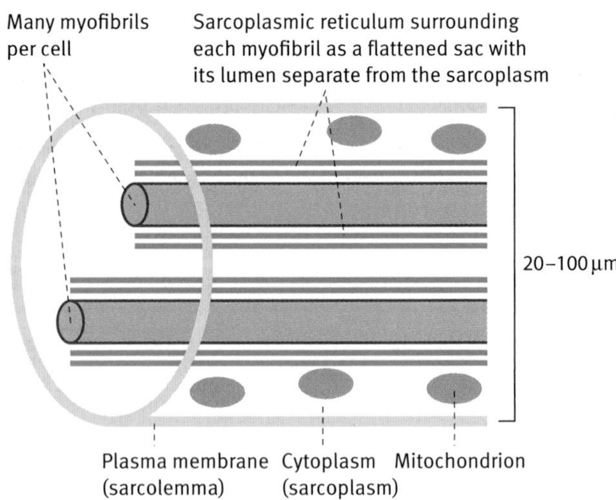

Fig. 8.2 A myofibre (muscle fibre) or muscle cell from striated muscle. Bundles of myofibres make up a muscle.

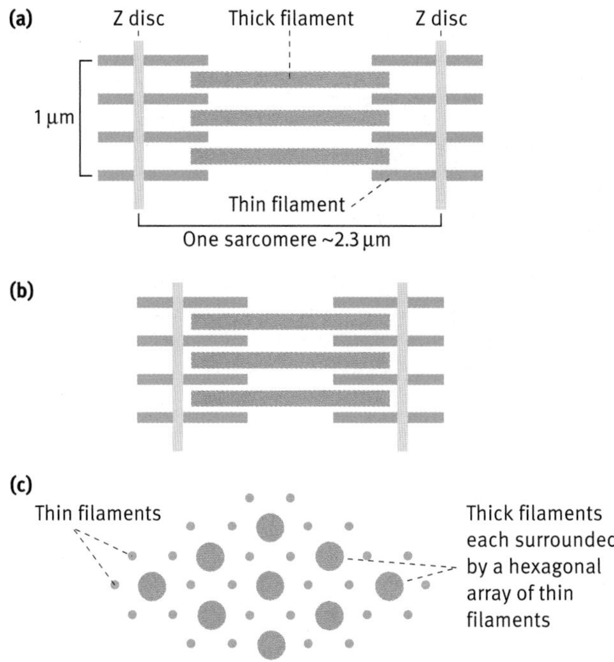

Fig. 8.3 Arrangement of thick and thin filaments in a sarcomere. **(a)** Relaxed. The striated appearance of myofibril sections in electron microscopy is caused by the amount of protein the beam traverses; **(b)** contracted; **(c)** arrangement in cross-section. The diagram shows only a few filaments but each sarcomere has a large number, all in lateral register.

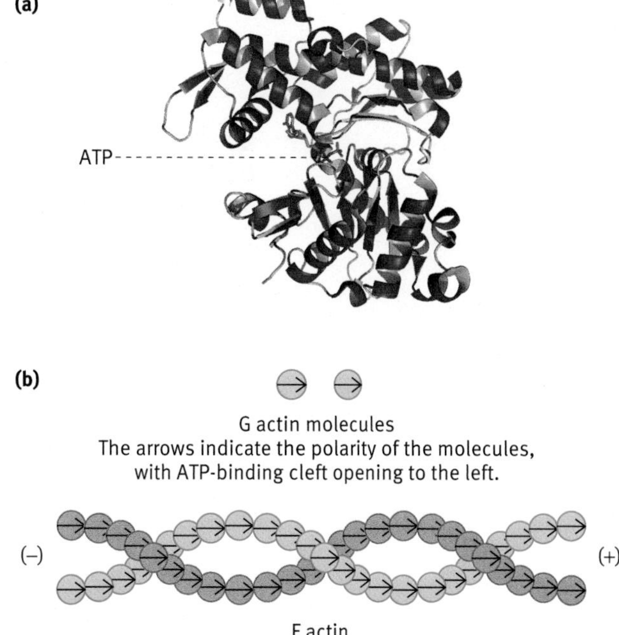

Fig. 8.4 **(a)** Protein structure of globular (G) actin. **(b)** Fibrous (F) actin made from the polymerization of G actin. The two strands are actually closely apposed forming a rod-like structure, but are presented here and in Fig. 8.10 in more open form for clarity.

and in doing so pull the Z discs closer together (Fig. 8.3(b)). The process is known as the **sliding-filament model.**

That, then, is the overall picture of muscle contraction. To understand how contraction happens we must look at the molecular structures involved.

Structure and action of thick and thin filaments

The thin filaments are made of polymerized actin. The monomeric **G actin** protein molecule (G for globular) (Fig. 8.4(a)) has two lobes separated by a deep cleft on one side of the molecule, giving it a polarity. The cleft is actually an ATP-binding site, a point that we will return to later, but which is not important for explaining muscle contraction. G actin polymerizes with a head-to-tail polarity into long filamentous structures,termed **F actin**, (Fig. 8.4(b)). A **thin filament** consists of two F actin strands coiled around each other with a long pitch, the ends being referred to as (+) and (–).

Myosin, the protein of the **thick filament**, contains two identical 'heavy' polypeptide chains. An α helical section of each chain wraps around the other in a coiled-coil configuration to form a straight rod structure (Fig. 8.5(a)). Each α helix has regularly spaced residues that form hydrophobic attachments to its partner, giving strength and rigidity. Each heavy chain terminates in a globular head. Two other small polypeptide chains, called myosin light chains, are wrapped around the neck of each myosin head. Myosin light chains are concerned with regulation, as discussed in Control of smooth muscle contractions, later in this chapter. The myosin of muscle cells is more precisely designated myosin II, because it is part of a family of related molecules. The function of some other myosins is mentioned later in the chapter.

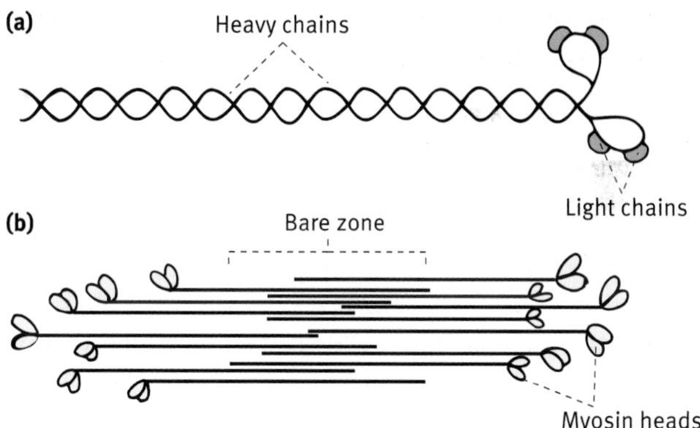

Fig. 8.5 **(a)** A myosin molecule consisting of a dimer structure of two heavy chains, each terminating in a myosin head. The latter has two dissimilar light chains attached to it. **(b)** A thick filament made of about 300 myosin molecules arranged in a bipolar fashion.

Thick filaments are formed from several hundred myosin molecules arranged in a bipolar fashion as shown in Fig. 8.5(b). They are held in a central position relative to the thin

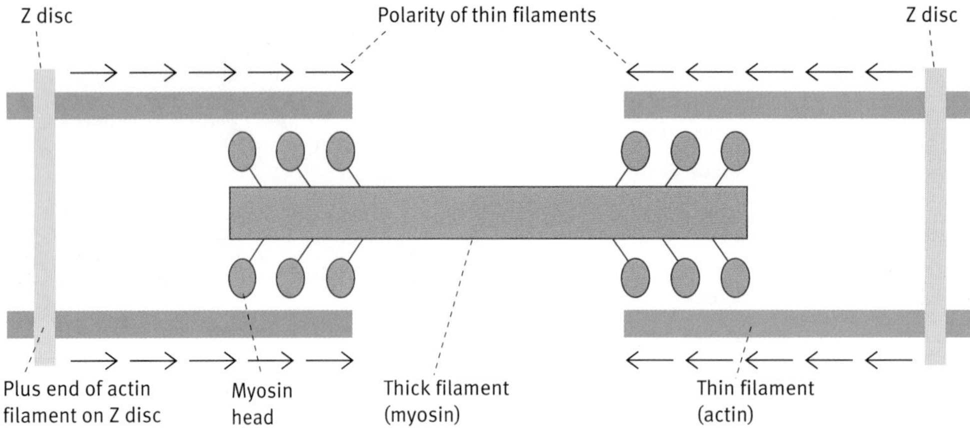

Fig. 8.6 Diagram showing the arrangement of thick and thin filaments in a sarcomere. The arrows in the latter are to show the polarity of the actin filaments. In a contraction event, in effect, the myosin heads track along the actin filaments towards their (+) ends and, in doing so, claw the two discs together. The heads are shown as simple shapes for clarity but their actual structures are described below.

filaments by other proteins involved in sarcomere structure. One is **titin**, a huge protein. The thin filaments are anchored to the Z discs at their (+) ends so that the myosin heads, at both ends of the thick filament, are oriented the same way, relative to the thin-filament polarity (Fig. 8.6).

How does the myosin head convert the energy of ATP hydrolysis into mechanical force on the actin filament?

The myosin head is an ATPase enzyme. It hydrolyses ATP to ADP and P_i. (Note that this is not the ATP attached to actin). Myosin undergoes conformational changes at the expense of ATP hydrolysis, which results in it exerting a force on the actin filament. The actual 'power stroke' in contraction does not occur on ATP hydrolysis, as you might expect. Instead the myosin head–ADP/P_i complex takes up a different conformational state; you might think of it as adopting a 'high-energy' conformation. It is when the P_i+ADP leave the protein that the liberation of free energy occurs, so that the 'power stroke' in muscle contraction correlates, not with ATP hydrolysis *per se*, but with release of the products of the hydrolysis reaction from the myosin head. The energy transfer is mediated by conformational changes in the myosin head.

Mechanism of the conformational changes in the myosin head

A myosin head has a compact region that attaches to the actin filament (the right-hand side in Fig. 8.7). Its orientation to the actin filament is always perpendicular. There also is an extended α helix forming the 'lever arm' (shown to the left-hand side of the diagram) which, at its distal end, joins to the rod part of the myosin molecule. During contraction, when the ATP is hydrolysed on the myosin head, with the ADP+P_i still bound, a small conformational change occurs in the head. This causes the α helix to adopt the primed, 'high-energy' conformation ready for the power stroke which occurs on the release of ADP+P_i. When this happens, the lever arm swings relative to the actin-binding part and this forces the head to move. Since the head is attached to the actin filament the latter is forced to slide as illustrated in Fig. 8.7. Figure 8.8 shows the three-dimensional structures of the myosin head before and after the power stroke. On the right of each structure is the site that attaches to the actin. You will recall that the myosin rod section is encircled by the two myosin light chains (not represented in Figs 8.7 and 8.9). These stabilize the lever arm.

Let us now look more closely at the steps involved in this cycle, starting with Fig. 8.9(a) in which the head has just finished its power stroke in a contraction. (Only one of the two heads of myosin is represented in the diagram in the interests of clarity.) Vast numbers of individual heads are involved in any contraction.

The following sequence of events then occurs as depicted in Fig. 8.9.

- In (a) the myosin head is attached to the actin filament having just completed its previous power stroke in which a force was applied to the actin filament. The head cannot detach from the actin until a molecule of ATP binds. This resembles the state of muscles in rigor mortis in which muscles are contracted.
- In (b) a molecule of ATP has bound, causing detachment of the myosin head from the actin filament. This is the state existing in relaxed muscle.
- In (c) ATP hydrolysis occurs but the products, P_i and ADP, are still attached to the myosin head. A conformational change occurs in the head causing the lever arm to adopt the 'primed' conformation.
- In (d) the head attaches to the actin filament resulting in state (e) (the ADP and P_i are still attached).

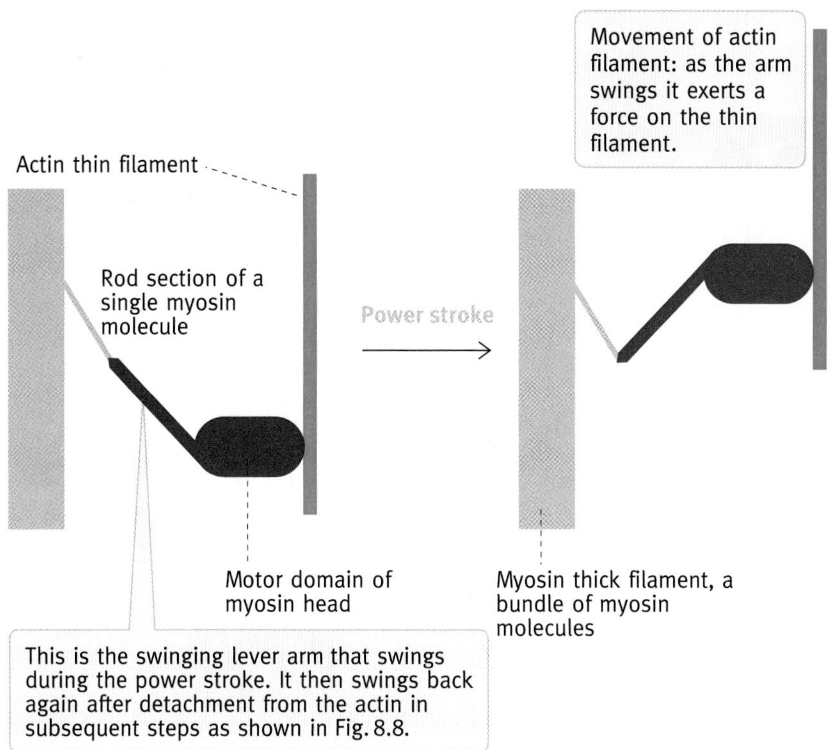
Movement of actin filament: as the arm swings it exerts a force on the thin filament.
Actin thin filament
Rod section of a single myosin molecule
Power stroke
Motor domain of myosin head
Myosin thick filament, a bundle of myosin molecules
This is the swinging lever arm that swings during the power stroke. It then swings back again after detachment from the actin in subsequent steps as shown in Fig. 8.8.

Fig. 8.7 Diagrammatic representation of the swinging-lever-arm mechanism of muscle contraction. The myosin head has a motor domain that binds to the actin filament without changing its angle of attachment. The motor domain is the site of nucleotide (ATP/ADP) attachment. It is connected to the myosin rod by an α helix which forms the lever arm; this is surrounded by the myosin light chains (not shown). When the power stroke is induced by P_i+ADP dissociation from the motor domain (see text) the small resultant conformational change is amplified by the lever arm which swings through an arc of 70°, sufficient to displace the actin filament by about 10 Å. Note that for clarity only a single myosin head is shown; each myosin molecule has two heads.

- In (e) the power stroke occurs in which P_i and ADP are released and the actin filament is forced to slide, thus causing a contractile force. This returns the head to our starting point shown in (a).

Control of voluntary striated muscle

In skeletal muscles, contraction is initiated by a nerve impulse causing Ca^{2+} ions to be liberated into the myofibril from the sarcoplasmic reticulum which surrounds each myofibril within the muscle cell. The reticulum membrane has **voltage-gated Ca^{2+} channels**, normally closed. On receipt of a nerve impulse to the muscle cell, they open and release Ca^{2+} from the lumen of the reticulum into the myofibril, causing contraction. The reticulum membrane is rich in a Ca^{2+} ATPase (an ATP-dependent pump) that pumps Ca^{2+} from the cytosol (known as the sarcoplasm) surrounding the myofibril into the lumen of the reticulum. This depletes the myofibril of Ca^{2+} and terminates muscle contraction.

How does Ca^{2+} trigger contraction?

Thin filaments have additional protein molecules called **tropomyosin** associated with them. This is an elongated molecule that lies along the two helical grooves between the actin fibres of a thin filament (Fig 8.10). A tropomyosin molecule has on it seven actin attachment sites, each binding to an actin monomer within the groove, and successive molecules overlap to form continuous threads along the thin filament.

Each tropomyosin molecule has attached to it, at one end, an additional complex of three more globular proteins called **troponin**, in which Ca^{2+} causes a conformational change. Tropomyosin impedes the interaction of the myosin head with actin, but the conformational change of troponin in response to Ca^{2+} slightly shifts the position of tropomyosin. This allows the myosin head access to the actin filament and starts the **myosin-actin power cycle**. Withdrawal of Ca^{2+} terminates the contraction event.

For a long muscle to contract, all of the sarcomeres in a myofibril and all the myofibres in the muscle need to respond to a motor nerve impulse essentially simultaneously, or otherwise the contraction will be uncoordinated. The nerve impulse causes liberation of acetylcholine from the nerve ending at the neuromuscular junction; this causes a local depolarization of the plasma membrane that rapidly propagates throughout the latter. Depolarization causes the voltage-gated Ca^{2+} channels of the sarcoplasmic reticulum to open and release the ion on to the myofibril. In order to ensure that the signal reaches all of the sarcoplasmic reticulum within a cell very rapidly, the plasma membrane is invaginated into transverse (T) tubules that enter the cell at Z discs and make direct contact with the sarcoplasmic reticulum membrane. This permits the electrical signal to reach, virtually simultaneously, all of the contractile units controlled by that nerve impulse (Fig. 8.11).

Fig. 8.8 The three-dimensional structure of the myosin head: (upper) before the power stroke with the swinging lever arm in the upright position; (lower) the head immediately after the power stroke. Adapted from Figs 7 & 8 in Geeves, M. A. and Holmes, K. C. Annu. Rev. Biochem. (1999) 68, 687–728; Reproduced by permission of Annual Reviews Inc.

Smooth muscle differs in structure and control from striated muscle

Smooth muscle is found in the walls of blood vessels, in the intestine, and in the urinary and reproductive tracts. The long spindle-shaped cells have a single nucleus and associate together to form a muscle in patterns appropriate to their function, such as an annular arrangement in blood vessels and a criss-cross network in the bladder.

The basic principles of contraction are the same as in striated muscle but the contractile components are not so highly organized. There are no myofibrils or sarcomeres, the latter being the reason for the absence of a striated appearance under the microscope. Instead, actin filaments run the length

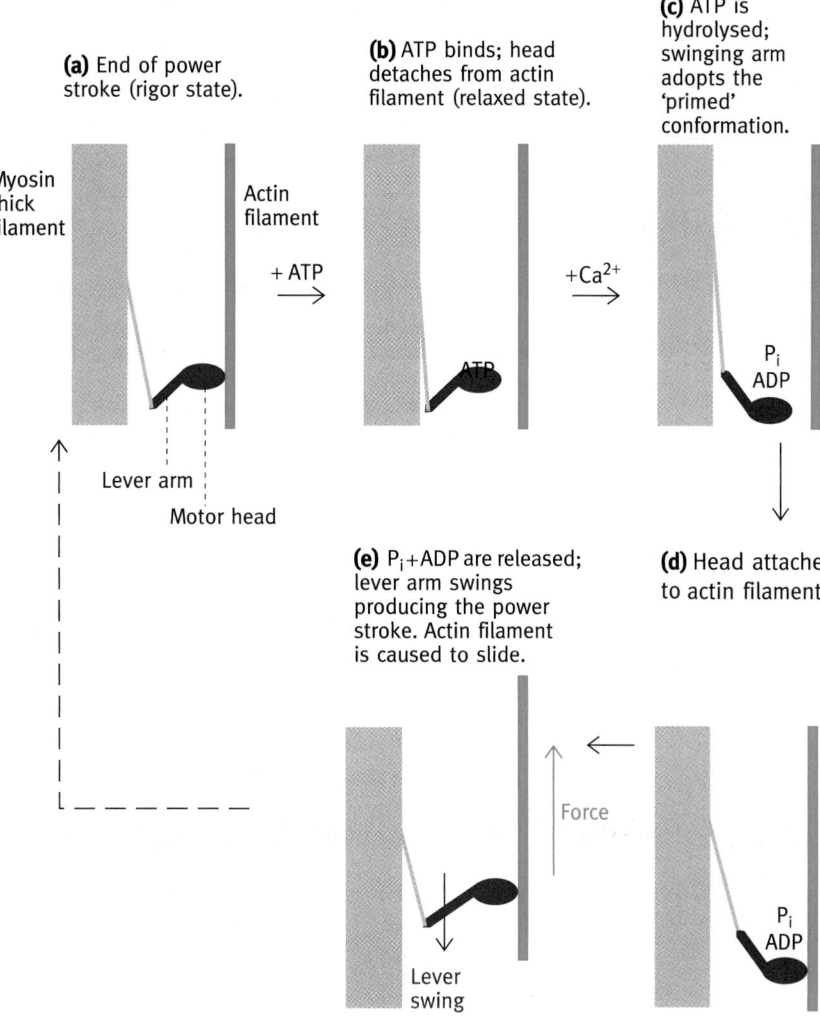
(a) End of power stroke (rigor state).
(b) ATP binds; head detaches from actin filament (relaxed state).
(c) ATP is hydrolysed; swinging arm adopts the 'primed' conformation.
Myosin thick filament
Actin filament
+ ATP
ATP
$+Ca^{2+}$
P_i
ADP
Lever arm
Motor head
(e) P_i+ADP are released; lever arm swings producing the power stroke. Actin filament is caused to slide.
(d) Head attaches to actin filament.
Force
Lever swing
P_i
ADP

Fig. 8.9 Swinging-lever-arm model of muscle contraction. The pink bar is the thick filament made of multiple myosin molecules; the thin pink line represents an individual myosin molecule; only one of two heads is shown (red) for clarity. The green line is the thin actin filament. ATP, P_i, and ADP represent molecules bound to the myosin head. The essence of the model is that the power stroke is associated with P_i and ADP release, and the force is exerted by the proximal subunit of the myosin head (the lever arm) swinging in a lever-like fashion. See text for individual steps.

BOX 8.1 **Muscular dystrophy**

An incompletely understood aspect of muscle contraction relates to muscular dystrophy. This term covers a group of related inherited diseases in which there is a progressive muscular weakness and muscle wasting. There are many different forms of muscular dystrophy, with age of onset varying from birth (congenital muscular dystrophy) through to adulthood (some types of limb girdle muscular dystrophy). The clinical course may be static or progressive. In some patients there is involvement of the respiratory and cardiac muscles, which can result in respiratory failure (the major cause of death) or cardiomyopathy, respectively.

The best known and most common form is Duchenne muscular dystrophy. This is an X-linked disorder that predominantly affects boys. Weakness becomes clinically obvious in the pre-school age group and is progressive; affected boys usually lose the ability to walk by their teenage years and there is progressive weakness of respiratory muscles, resulting in death in the second or third decade of life.

The affected gene in this disease has been cloned; it encodes a very large protein called dystrophin. This protein is attached at one end to structural elements of the muscle cell cytoskeleton and at the other to a transmembrane complex of proteins, which are, in turn, linked to proteins of the extracellular matrix that surrounds muscle cells. The linkage of intracellular structural components to the membrane and thence to proteins of the extracellular matrix (through the dystrophin-associated-protein complex) is thought to play a part in stabilizing, or strengthening the muscle cell membrane to withstand the contractile forces involved in muscle contraction. In Duchenne muscular dystrophy, dystrophin is absent so that the connection of the intracellular cytoskeleton is broken, resulting in progressive muscle cell damage and necrosis. Some other forms of muscular dystrophy (inherited in an autosomal fashion) result from mutations in the genes encoding other members of the dystrophin-associated-protein complex.

BOX 8.2 **Malignant hyperthermia**

An inherited condition known as **malignant hyperthermia** exists in some humans, pigs, dogs, and poultry. Following the administration of so-called triggering agents, such as a depolarizing muscle relaxant (suxamethonium) or one of the volatile anaesthetics such as halothane, a hypermetabolic process is initiated within the skeletal muscle. The pathological process is thought to be an increase in the opening time and conductance of the calcium channel of the sarcoplasmic reticulum, the **ryanodine receptor protein**. In approximately 50–80% of sufferers the faulty calcium channel is associated with a mutation in the gene for this protein. The increased myoplasmic calcium levels catalyse increased muscle contraction, increased mobilization of energy through glycogen breakdown and glycolysis, resulting in increased aerobic and anaerobic metabolism. If unrecognized or left untreated, respiratory and metabolic acidosis, breakdown of muscle fibres (rhabdomyolysis), renal failure, and death may occur. Treatment includes withdrawal of the triggering agents, hyperventilation with 100% oxygen, administration of a specific antidote, dantrolene, and active cooling. The clinical incidence is 1:15,000 children and 1:50,000 adults where triggering agents are used. The faulty gene prevalence is unknown, but estimated to be about 1:5000.

of the cell (which is small, compared with a striated muscle cell) and are anchored into the cell membrane.

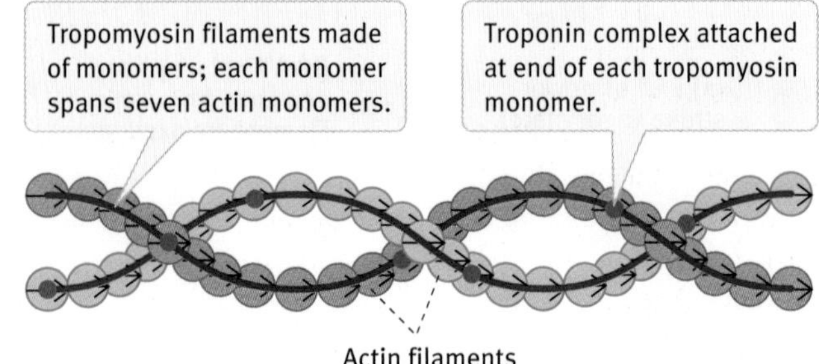

Fig. 8.10 The relationship of the actin and tropomyosin molecules to each other. Each tropomyosin molecule binds to seven actin monomers forming a continuous filament in each of the actin filament grooves. Each tropomyosin molecule also has a troponin complex bound at one end.

Control of smooth muscle contractions

Although Ca^{2+} controls smooth muscle also, the mechanism is different from that in striated muscle. A smooth muscle contracts typically about 50 times more slowly than a striated muscle. There is no requirement for contraction to be almost instantaneous throughout the structure. The contraction signal to cells can spread at a more leisurely pace. A nerve impulse from the autonomic system causes Ca^{2+} gates in the **plasma membrane** to open and allow an inrush of the ion into the cells from the outside. There is no sarcoplasmic

(a)

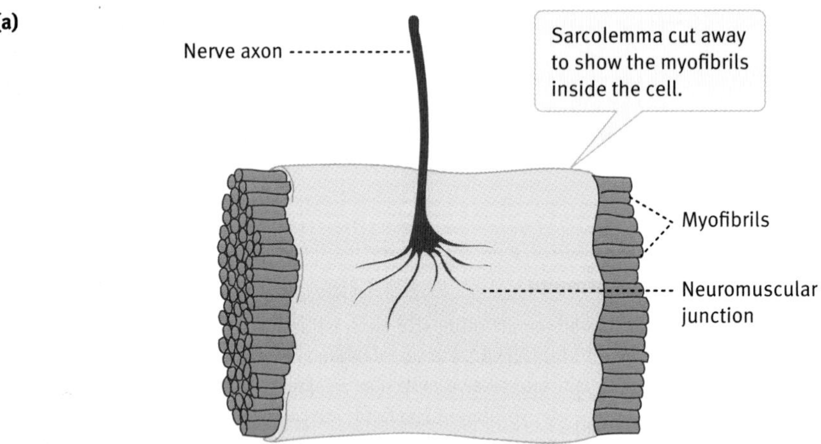

(b)

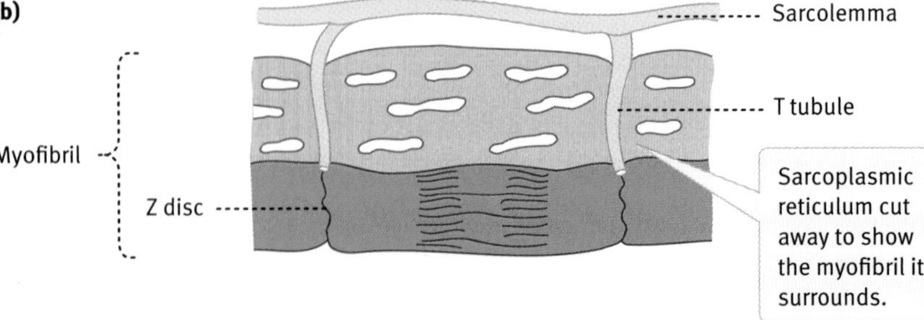

Fig. 8.11 **(a)** Plasma membrane (sarcolemma) of the myofibre, showing the neuromuscular junction. **(b)** The transverse (T) tubules that carry the plasma membrane depolarization signal to the sarcoplasmic reticulum (SR). It is postulated that the plasma membrane depolarization is directly transmitted to the SR Ca^{2+} channels causing rapid Ca^{2+} release from the SR.

reticulum. The relatively slow diffusion of Ca^{2+} throughout the cell can be tolerated because of the small distances involved and the slow response requirements. Special junctions between cells allow a neurological signal to spread throughout the muscle.

At the neck of each myosin molecule, as already described, are two small polypeptides known as **myosin light chains.** In resting smooth muscle, one of these (designated the **regulatory light chain**) inhibits the binding of the myosin head to the actin fibre, and thus prevents contraction. Ca^{2+} activates a myosin kinase that catalyses phosphorylation of the regulatory light chain by ATP, and abolishes its inhibitory effect, thus triggering contraction.

The Ca^{2+} does not directly activate the kinase; but does so via the protein, **calmodulin.** When the Ca^{2+} levels fall due to pumping out from the cell by a Ca^{2+} ATPase in the cell membrane, the process reverses; a phosphatase enzyme catalyses dephosphorylation of the myosin light chain abolishing the binding between myosin and actin, and causing muscle relaxation. The scheme is summarized in Fig. 8.12.

As well as neurological control of smooth muscle contractions, several hormones also exert control; some prostaglandins cause muscle contraction (see Fig. 17.10 and Box 17.2). Noradrenaline (norepinephrine) causes contraction of certain blood vessel muscles.

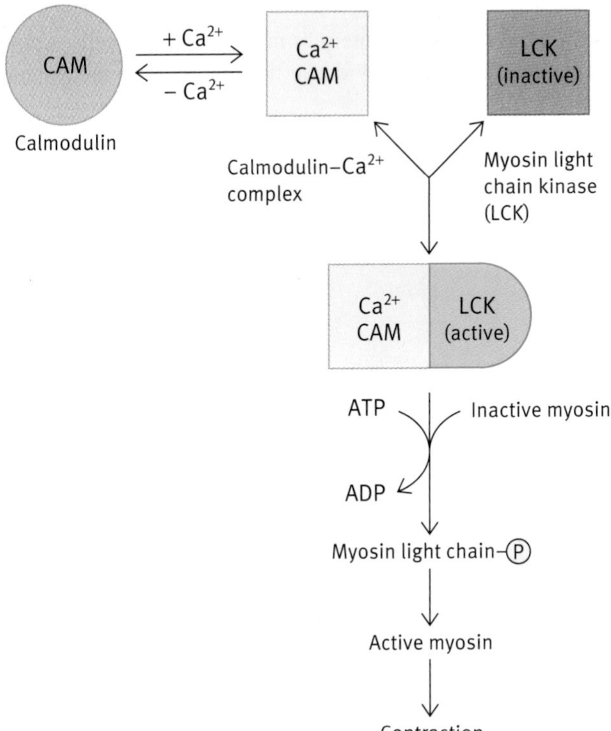

Fig. 8.12 Mechanism of activation of smooth muscle contraction by Ca^{2+}.

The cytoskeleton

An overview

The eukaryotic cell **cytoskeleton** is a complex scaffolding of protein filaments that pervades all parts of the cell. It has several roles: it connects the interior of cells to the extracellular matrix; it confers shape on animal cells; it is involved in transport within cells; it is involved in cytokinesis and in chromosome separation at cell division; it also enables cells to move. Motility is not something we associate with most cells in the body, but, quite apart from cells such as macrophages and leukocytes, which migrate through tissues by amoeboid-like action, motility is a widespread property of animal cells. In embryonic development for example, cells must move at various stages in order to establish the final body pattern. Embryonic cells may migrate as single cells over surfaces by a crawling action, or as coordinated sheets. Cell migration also occurs in normal wound healing. Sperm are another example of motile cells as they use flagellae for swimming. Yet another form of cell movement (though not migration) is due to cilia, whose beating causes movement of surface mucous layers in the respiratory passages.

Perhaps the most remarkable role of the cytoskeleton is that it provides transport tracks within cells for molecular motors to run along pulling loads to various parts of the cell. There is constant internal transport in eukaryotic cells, for which the cytoskeleton fibres provide the tracks. There has to be active transport of materials within eukaryotic cells. The distances are too big for diffusion to be adequate, as it is in prokaryotes. A nerve axon may be half a metre or more in length (several metres, in giraffes and whales); synthesis of proteins and vesicles occurs in the nerve cell body and may have to be transported to the tip of the axon. Membrane-enclosed transport vesicles carry proteins to their cellular destinations pulled along cytoskeletal tracks by ATP-powered motor proteins. These motors also have other roles in causing movement. For example they are needed to separate chromosomes at the time of cell division so that at mitosis, chromosomes on the spindle move apart into daughter cells.

There are three major classes of cytoskeletal components. The first two are the **actin microfilaments** and **tubulin microtubules.** Fig. 8.13(a) and (b) show typical microfilament and microtubule cellular networks stained so that they are visible in the light microscope. The third group, **intermediate filaments,** are quite different structurally and in function. They are named through being intermediate in size between the other two, the approximate fibre diameters being 7–9 nm (microfilaments), 24 nm (microtubules) and 10 nm (intermediate filaments). Microfilaments and microtubules are involved in the cell movement and intracellular transport functions of the cytoskeleton. The intermediate filaments play

(a)

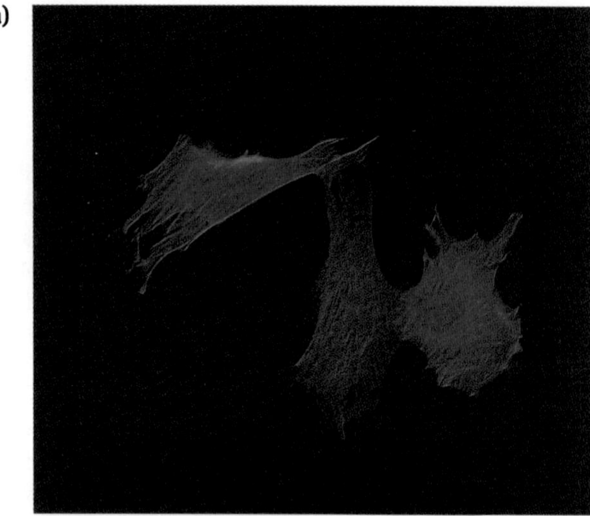

(b)

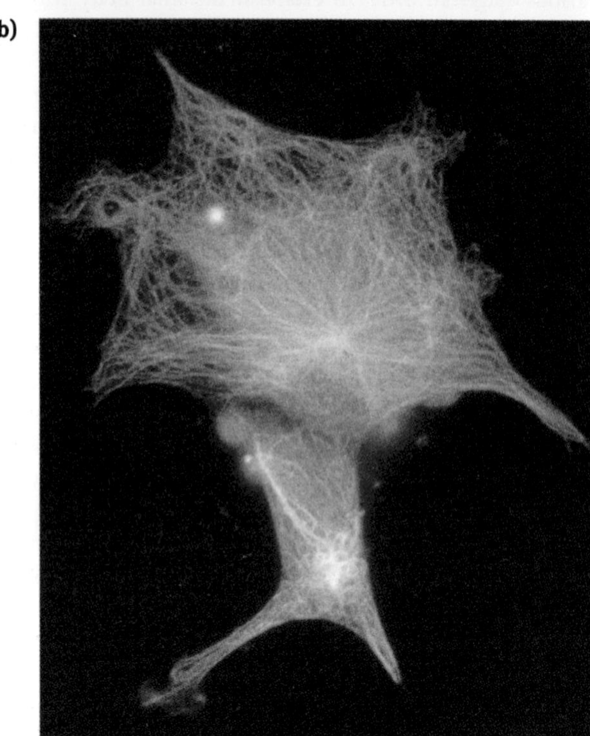

Fig. 8.13 **(a)** Actin filaments in mouse myoblasts visualized with an actin antibody. The filaments pervade the cell and bundles of them (called stress fibres) are attached at focal points to the cell membrane, often at points at which the membrane makes contact with a solid surface. The fibres are rapidly disassembled and assembled. Photograph kindly provided by Dr P. Gunning, Children's Medical Research Institute, Sydney, Australia. **(b)** Microtubules in cytosol of a cell radiating from the microtubule-organizing centre (MTOC) or centrosome (see text). Adapted from Fig. 2.13 in Lewin B. (1994) Genes V, Oxford University Press, Oxford. Photograph was provided by Frank Solomon.

a rather more limited structural role, conferring strength and toughness.

The cytoskeleton is in a constant dynamic state

In one sense the cytoskeleton is a bewildering structure because actin fibres and microtubules are subject to continual assembly and collapse. We are not used to the concept that a highly organized transport system can be based on such a flickering, apparently random, ephemeral set-up. Rather than a nicely laid out road system, it is more like a complex of microscopic tracks that may disappear at any moment and reappear pointing in a different direction. It is hardly possible to envisage how such a system can handle such complex traffic with such an organized outcome.

There are a few exceptions to the dynamic instability of actin fibres and microtubules. Muscle thin filaments are permanent; another example is the brush border microvilli (Fig. 8.14) of the intestinal cells, which are stabilized by an actin framework and increase the absorptive area of the cells. Microtubules running the length of cilia and flagellae are also stable, permanent structures.

We will now deal in turn with the actin filaments, then the microtubules, and finally with the intermediate filaments.

The role of actin and myosin in nonmuscle cells

Actin is an abundant protein in most eukaryotic cells. Actin filaments are particularly dense near the plasma membrane

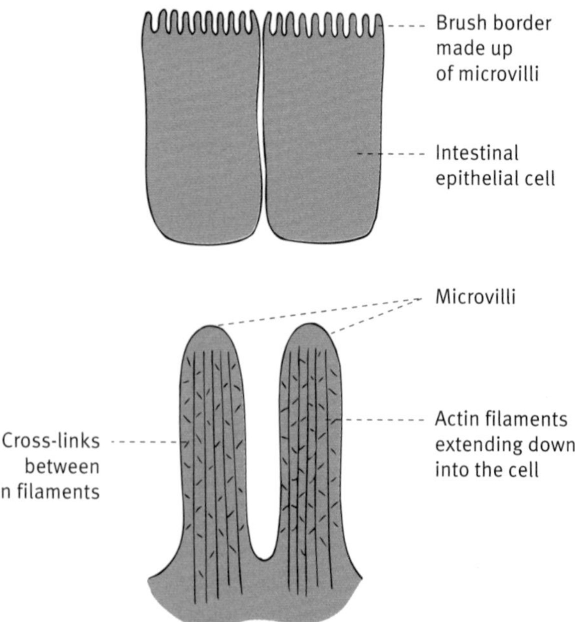

Fig. 8.14 Diagram of actin filaments in microvilli. Note that the filaments have a complex cross-linking and anchoring arrangement with other proteins to form a robust cytoskeletal structure.

where bundles of them are anchored into the membrane to form **stress fibres**, involved in maintaining cell shape (Fig. 8.13(a)) and in causing amoeboid movement. Myosin homologues also are an almost universal constituent of such cells but are present in smaller amounts than myosin II in muscles.

Assembly and collapse of actin filaments

Actin filaments of the cytoskeleton reversibly assemble and disassemble rapidly from G actin monomers, which polymerize by noncovalent bonding. As described for muscle actin, cytoskeletal actin monomers have head-to-tail asymmetry, and they polymerize in a head-to-tail orientation by a self-assembly process so that the F actin filaments have a polarity with ends referred to as plus (+) and minus (–).

As mentioned earlier, free actin monomers have a molecule of ATP bound within a nucleotide-binding cleft; this is hydrolysed to ADP shortly after incorporation into the growing end of an actin polymer. ATP hydrolysis is not required for polymerization: the actin monomer acts as an ATPase but only after it has been incorporated into a growing actin filament. Actin-ATP polymerizes much more readily than does actin-ADP so that a terminal monomer in the ADP form is likely to dissociate from an actin filament. Continued elongation of the filament requires new actin-ATP monomers to be added before the ATP on the previously added unit is hydrolysed. Unless this happens the filament will tend to depolymerize.

Addition of actin-ATP monomers can occur at either end of an actin filament, but it occurs more readily at the (+) end than at the (–) end. The concentration of free G actin monomers required to drive the reaction towards polymerization, the so-called **critical concentration**, is thus lower for the (+) end than for the (–) end. The consequence in the cell can be that (+) end grows while the (–) end shrinks, a phenomenon known as treadmilling. The constantly changing arrangement of actin filaments is associated with cell shape changes, cell movement and internal transport. It is a very dynamic situation.

Actin-binding proteins

A bewildering variety of actin-binding proteins play different, sometimes multiple, roles in regulating the cytoskeleton. Some influence the dynamics of polymerization/depolymerization, while others affect the spatial organization of the microfilament network. A list of some of their activities with a few examples is given below.

G actin binding

By binding monomeric G actin, **thymosin β_4** sequesters it and hence inhibits microfilament polymerization. **Profilin**, on the other hand, binds G actin and stimulates ADP-ATP exchange, thus increasing polymerization. Since it also prevents G actin from binding to thymosin and promotes G actin monomer addition to the (+) end, profilin assists microfilament growth via several mechanisms.

Capping

CapZ binds the (+) end of microfilaments, hence stabilizing them but preventing further growth. Similarly **tropomodulin** binds and stabilizes the (–) end. The thin filaments of skeletal muscle cells are permanently stabilized by CapZ and tropomodulin binding, but these proteins also function in a more dynamic fashion in other cell types. **Formins** have a slightly different function. They cap the (+) end, but they are processive: that is they move along as the filament grows and actually promote addition of new G actin monomers.

Severing

Gelsolin severs microfilaments, cutting them in the middle. It then remains bound to the newly formed (+) ends, but leaves the new (–) ends to rapidly depolymerize. The name gelsolin thus derives from the observation that the gel-like consistency of preparations of polymerized actin becomes much less viscous if gelsolin is added (a 'gel to sol' transformation).

Branching

The **Arp2/3** protein complex binds to the side of a microfilament and promotes growth of a new filament from that point, creating a branched network.

Cross-linking

There are multiple cross-linking proteins, which give rise to a variety of three-dimensional arrangements. **Fascin** binds microfilaments close together in parallel bundles, in microvilli for example (Fig. 8.14). **Filamin** allows a greater distance between the cross-linked filaments so they can be arranged at angles, creating a loose network with a gel-like consistency.

Mechanism of contraction in nonmuscle cells

Myosin II, which is the 'conventional' muscle-like myosin molecule (with two heads) found in skeletal muscle, is also present in nonmuscle cells. When contracting mechanisms are needed in these cells, myosin II molecules aggregate into bipolar filaments of about 16 myosin molecules, analogous to but much smaller than striated muscle thick filaments. In an action very similar to that described for muscle, these then exert a contracting force on adjacent actin filaments (Fig. 8.15) and exert a pull on the cell membrane to which the filaments are anchored. Control of contraction is by phosphorylation of myosin as in smooth muscle.

Cytokinesis, or the constriction of a dividing cell into two daughter cells, provides a good example of this type of contractile system. An annulus (ring) of actin fibres assembles at the equatorial plane where constriction occurs. Actin filaments are anchored at the plasma membrane. Overlapping of these filaments (with opposite polarities) allows cytosolic myosins to exert a contraction force, rather like that in muscle. The set-up is disassembled as the contraction progresses. The action constricts the cell and ultimately separation into two daughter cells occurs.

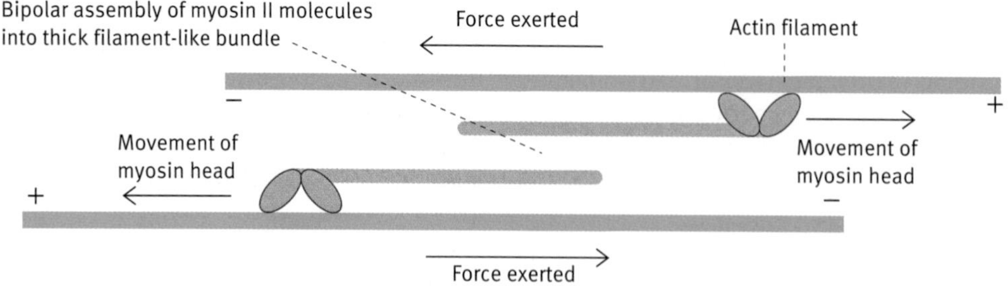

Fig. 8.15 Diagram of the pulling action of myosin II on actin filaments. Note that ATP hydrolysis supplies the energy. The achievement of useful work by this system will depend on appropriate anchoring of the actin filaments to the cell membrane so as to change cell shape. Each myosin molecule has a region capable of interacting with other molecules and thus allowing self-assembly of bipolar filaments. The diagram is a cross-sectional representation – the bipolar assemblies are cylindrical. In the arrangement of actin fibres and myosin bundle shown, a contraction results from the relative sliding of the actin filaments.

The role of actin and myosin in cell movement

Cell movement depends on both the dynamic polymerization and depolymerization and the contraction of microfilaments. When cells 'crawl' over a surface, such as the extracellular matrix, they push out broad flat extensions termed **lamellipodia** (singular lamellipodium), or long thin extensions called **filopodia** (singular filopodium) from their front leading edge. These extensions are formed by directed (by actin-binding proteins) polymerization of actin filaments, which form networks in lamellipodia or parallel bundles in filopodia. The forward extensions attach to the underlying substrate and the body of the cell is then pulled up behind through myosin-mediated contraction of the elongated stress fibres.

The role of actin and myosin in intracellular transport of vesicles

The rod-like 'tail' of the muscle-type 'conventional' myosin molecule is designed so that, as described, myosin can assemble itself into bipolar filaments both in muscle and nonmuscle cells and cause contractions to occur. A family of other types of myosin exist, numbered in order of their discovery. Instead of the long rod-like tail of myosin II, the tail attached to the motor head is small. These molecules cannot form bipolar bundles needed for contraction but single motor molecules move along actin filaments, at the expense of ATP hydrolysis, by the same basic mechanism as described for muscle. The tail of these myosins is designed to attach to another structure (the cargo) such as the membrane of a vesicle, which is moved along the actin filament. The members of the family have different tails for binding to specific vesicle membranes or other molecular cargoes. They have been found in almost all tissues including brain. With one exception, they move towards the (+) end of the actin fibre.

We now turn to a different class of transport systems in which both the tracks and the motors are different from the actin-myosin ones.

Microtubules, cell movement, and intracellular transport

Cytosolic microtubules are made by polymerization of **tubulin** protein subunits, which are dimers of α and β tubulin molecules that give the dimer a polarity. Tubulin dimers reversibly polymerize to form a hollow tube bounded by 13 longitudinal rows of subunits (Fig. 8.16). The polymerization of the dimers occurs in a head-to-tail fashion so that a microtubule has a polarity, with the ends referred to as plus (+) and minus (–). Microtubules are basically unstable in the cell. They can assemble and disassemble very rapidly, a phenomenon known as **dynamic instability**. The instability is associated with unprotected ends – where these occur, the microtubules undergo collapse by depolymerizing into free tubulin subunits. In the cell, the (–) ends are associated with a **microtubule-organizing centre** (**MTOC**) or **centrosome**, a structure near the nucleus. In animal cells the centrosome contains a pair of small bodies, the centrioles, made of fused microtubules (Fig. 8.17). The function of the centrioles is not well understood as

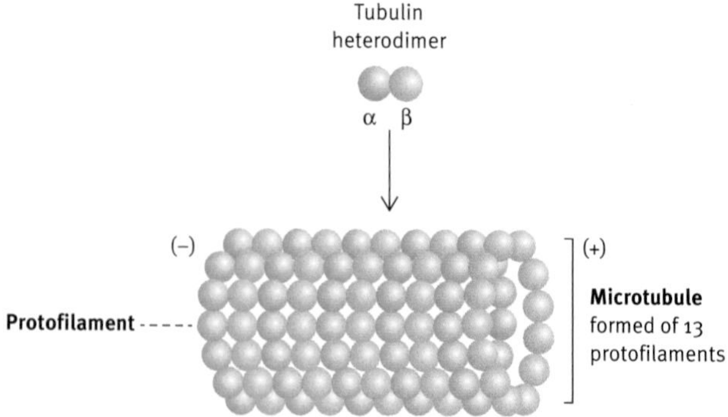

Fig. 8.16 Structure of a section of microtubule. Heterodimers of α and β tubulin polymerize to form cylindrical tubules of 13 protofilaments. Polymerization occurs more readily at the (+) end, which is stabilized by GTP-bound β tubulin.

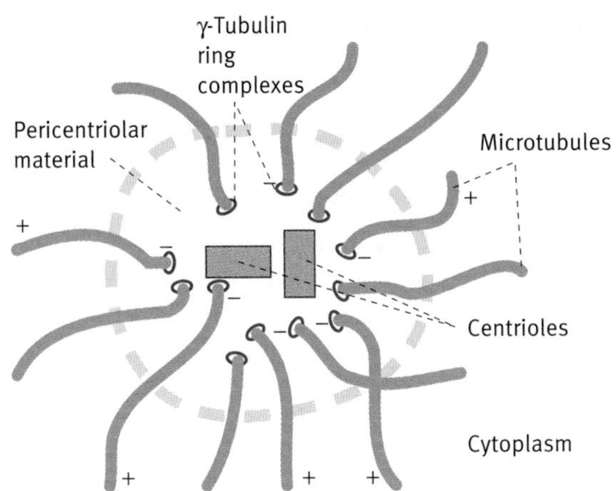

Fig. 8.17 Diagram of microtubules radiating out from the microtubule-organizing centre (MTOC). The (+) and (–) indicate the polarity of the microtubules. The centrioles are a pair of tube-like structures made of fused microtubules; the radiating microtubules originate in the centrosome material surrounding the centrioles. (There is no precise boundary to this.) Each microtubule is initiated at a γ-tubulin complex, which includes several other proteins.

it is the pericentriolar material of centrosomes that acts as the nucleation point for microtubules (the point from which they begin to grow outwards). The cell is pervaded by microtubules radiating out from the MTOC (Fig. 8.13(b)).

The MTOC protects the (–) end. Microtubules grow (and collapse) in the cell from the (+) ends. It is thought that when the microtubule reaches an appropriate destination, target proteins 'cap' and protect the microtubule. Microtubules grow out of the MTOC in random directions. Those that make contact with an appropriate component of the cell will, on this model, be capped and stabilized while the remainder collapse.

But what protects a growing microtubule *before* it reaches a target protein? Free tubulin subunits have GTP attached to them; a GTP-tubulin subunit added to a growing microtubule protects the (+) end. β-Tubulin is a latent GTPase; its GTP is hydrolysed to GDP and P_i after addition of a subunit to a growing tubule, but this does not happen until after a slight delay. Thus newly added subunits will be in the GTP form, so temporarily capping the microtubule, which protects the end from collapse. A tubulin-GDP subunit is not protective so that unless a new tubulin-GTP monomer is added before the GTP on the last added monomer is hydrolysed, the microtubule depolymerizes. The GTP hydrolysis in this model is acting as a clock. Note that this situation parallels that in actin-filament polymerization, described earlier, except that there it is ATP rather than GTP. In cilia and flagella, where permanent microtubules occur (see below), covalent modification of tubulin with cross-links between the microtubules occurs after assembly and stabilizes the structure.

Molecular motors: kinesins and dyneins

Two types of microtubule-associated motor proteins have been identified, **kinesin** and **dynein**, both of which are dependent on ATP for their movement. The motor domains of kinesins and myosin show structural similarities, which suggest that they evolved from a common ancestor. Kinesin and dynein molecules (supplied with ATP) will 'walk' along microtubule fibres immobilized on a solid, just as myosin heads will move along immobilized actin fibres. Most kinesins travel along a microtubule in the (–)→(+) direction, and dyneins travel in the opposite direction. There are families of kinesin and dynein molecules specialized for different functions. The heads, which perform the actual movement along the microtubule, are probably a constant motif within each family, while different tail structures attach to specific cargoes such as vesicles (Fig. 8.18(b)).

The best known kinesin, known as **conventional kinesin**, is illustrated in Fig. 8.18(a). There are two heads that power the movement along microtubules. The motor heads are about half the size of those of the myosins. The movement differs from that of the myosins in that the two heads indulge in a walking-like action along the microtubule, one attached at a time with the heads swivelling past one another at each step. The neck of the molecule is a flexible section that allows this. The movement is driven by ATP hydrolysis. The other end of the molecule binds to the cargo, such as a vesicle.

As stated, the kinesin travels towards the (+) end of a microtubule, which means in general towards the periphery of the cell. In a nerve axon they pull vesicles from the cell body outwards along the axon. A genetic condition exists in which neuronal kinesin is defective, resulting in a **peripheral neuropathy**, which causes progressive stiffness and weakness of the lower limbs.

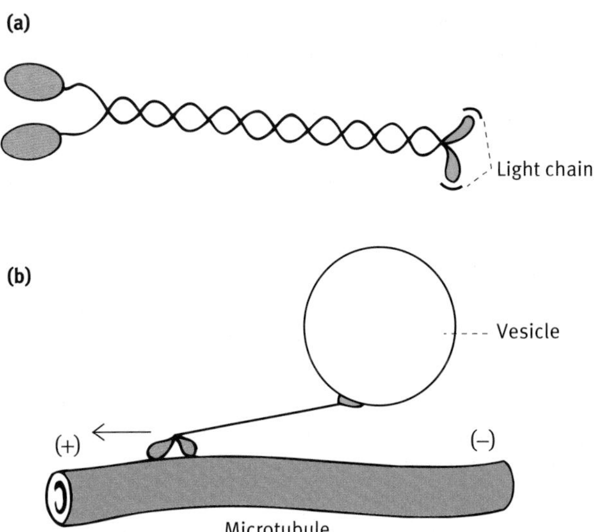

Fig. 8.18 **(a)** Diagram of the structure of a kinesin molecule. **(b)** Diagram illustrating how a kinesin molecule can transport vesicles along a microtubule. Also see Fig. 8.19 for an electron micrograph of vesicles being transported within a cell. © 1988 Rockefeller University Press. J. Cell. Biol. 106: 111–125.

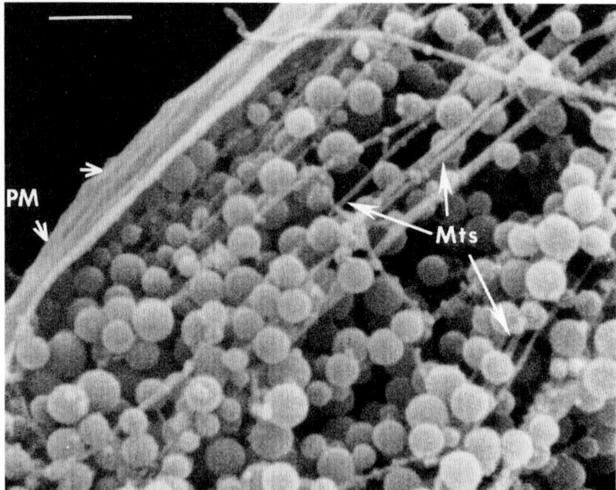

Fig. 8.19 Vesicle transport by microtubules. Scanning electron micrograph of pigment-containing vesicles being transported along microtubules in a chromatophore of a squirrel fish. The change of colour of the cell is effected by the movement of the vesicles to and from the cell centre. Mts, microtubules; PM, plasma membrane. Scale bar=0.5 μm.

Dynein is a much larger molecule, again with two motor heads. It travels along microtubules in the direction opposite to that of kinesin. Dynein motors are found in almost all eukaryotic cells and also have the task of transporting vesicles.

A striking example of vesicle transport along microtubules is shown in the beautiful scanning electron micrograph in Fig. 8.19. Some fish and amphibia have a camouflage mechanism, which rapidly changes the colour of the skin. This is achieved by vesicles containing pigment either moving along microtubules to the periphery of the cells or becoming evenly distributed giving the background appearance.

Role of microtubules in cell movement

The role of microtubules in cell movement is relatively limited in comparison to cytoskeletal actin, being confined to cilia and flagella in eukaryotes. Cells lining the respiratory passages of lungs have large numbers of cilia whose beating motion sweep along mucus and its entrapped foreign particles. Sperm propel themselves with flagellae. Cilia are smaller than flagella, but other than that the organelles are basically similar in structure. Microtubules, in this case permanent structures arranged in pairs that are fused in parallel along their length, run down the length of the organelle, originating in a basal body closely resembling a centriole. Associated with the microtubules are dynein molecules. Using ATP energy, they 'walk' along microtubules (towards the (–) end) with their tails attached to an adjacent microtubule pair. However, the microtubule pairs are cross-linked and cannot slide relative to one another, so in this case dynein movement causes a bending wave motion in the cilium or flagellum, instead of a sliding one. It should be noted that although many bacterial species possess flagella, bacterial flagella are different (nonmicrotubule) structures, which propel the bacterium by rotating, rather than a wave motion.

Role of microtubules and molecular motors in mitosis

Chromatids, as the pairs of new chromosomes are called in mitosis (see Chapter 30), are initially held together and have to be separated (Fig. 8.20). The nuclear membrane disappears and the centrosome divides, one copy of each migrating to opposite ends of the cell. From these, microtubules grow out to form the mitotic spindle. The duplicated chromosomes become arranged in the central, equatorial plane of the spindle. There are three types of microtubule in the spindle. The first type, called **kinetochore** microtubules, attach to each chromatid (Fig. 8.20) at the kinetochore, a protein complex at the centromere of the chromatid. Tension from the kinetochore microtubules aligns the chromosomes at the equator. The second type are **polar** microtubules, which overlap at their positive ends, and the third type are **aster** microtubules, which attach the spindle to the cell membrane at each pole of the dividing cell.

Three actions are involved in chromosome segregation. First the chromosomes are pulled towards the poles. This is caused by shortening of the kinetochore fibres attached to the centromeres. Microtubules cannot contract; the shortening is due to depolymerization of the microtubule at the attachment point to the chromosome. Despite the progressive loss of tubulin dimers the (+) end of the microtubule remains attached to the kinetochore and the two chromatids are pulled apart.

In the second action, kinesins, which operate between the overlapping polar fibres, pull them past each other and drive the centrosomes to opposite ends of the cell. Kinesin tails are bound to the one fibre and their motor heads to the overlapping one of opposite directionality (Fig. 8.20). In moving towards the (+) end of the latter they reduce the overlap of the two fibres and in doing so drive the centrosomes to the opposite sides of the cell. In the third action, dyneins attached to the inner cortex of the cell membrane move along the aster microtubules towards the (–) end at the centrosomes, pulling them apart.

Intermediate filaments

The third type of filaments in eukaryotic cells average 10 nm in diameter, between the dimensions of microfilaments and microtubules – hence their name of **intermediate filaments (IFs)**. These have a different role from actin filaments and microtubules. In short, the role of IFs is to confer mechanical strength to cells. They are found in vertebrates, but not in all cells, and only in a few nonvertebrates.

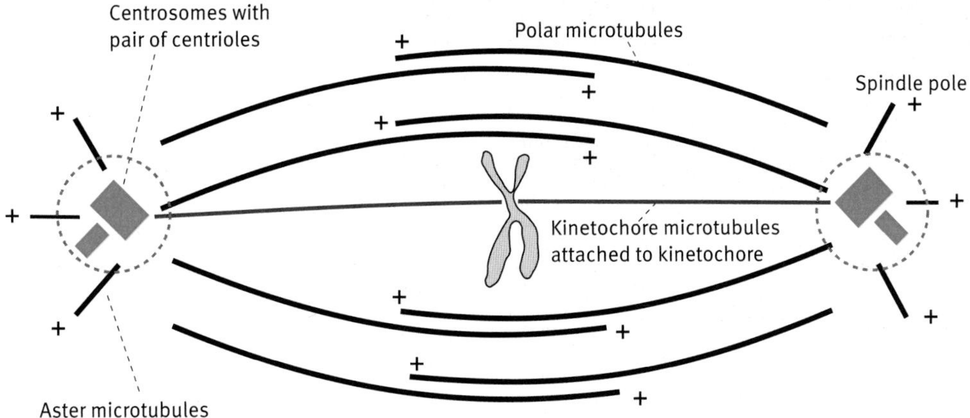

Fig. 8.20 Mitotic spindle at the metaphase state. The microtubules originate at the centrosomes. The duplicated chromosomes (only one duplicate is shown) are arranged at the centre of the spindle attached to the kinetochore fibres at the complex of proteins known as the kinetochore. The overlapping polar fibres propel the centrosomes apart; it is believed that kinesins in the overlapping regions of the polar fibres cause the separation. The aster microtubules assist in the separation of the spindle poles, (+)-directed kinesin motors being involved. The kinetochore fibres are progressively shortened at the kinetochore attachment site by disassembly. This pulls the duplicated chromosomes apart to opposite ends of the cell in preparation for cell division.

BOX 8.3 Effects of drugs on the cytoskeleton

Actin and microtubule filaments are assembled and disassembled with great speed. The correct organization of these processes is essential to the development, multiplication, and life of the cell. This is illustrated dramatically by the effects of a number of plant and sponge-derived drugs that these organisms produce as a defensive measure against predators. The drugs work by binding either the monomer forms of the filaments or the polymerized form, which perturbs the equilibrium to one side or the other. Some work on actin and some on microtubules.

The poisonous mushroom *Amanita phalloides*, or death cap, produces the deadly toxin **phalloidin**; this binds to F actin and stabilizes actin filaments so preventing their turnover. **Cytocholasin** is a toxin of fungal origin that binds to the growing (+) ends of actin filaments; it prevents the assembly and disassembly of the filaments and so inhibits, for example, the formation of the contractile ring needed to separate daughter cells after mitosis. It acts on mammalian cells.

Turning to microtubules, **colchicine** is an alkaloid produced by the autumn crocus; it binds tightly to tubulin monomers, preventing microtubule assembly and therefore spindle formation. It promotes depolymerization of microtubules and freezes mitosis at metaphase. **Vinblastine** and **vincristine** are alkaloids from *Vinca rose*, the Madagascar periwinkle, which also freeze mitosis at metaphase by binding to spindle fibres. These drugs are used to treat some cancers. **Paclitaxel**, first isolated from the bark of the Pacific yew tree, but since synthesized chemically, interferes with normal microtubule growth during cell division. Sold as **Taxol®**, it is also used in cancer chemotherapy.

IFs comprise a diverse group of homologous proteins. Typically there is a core filament about 350 amino acid residues in length, with the ends varying in the different types of IF. They are elongated molecules whose central α helical sections form coiled-coil dimers. These in turn laterally pack together forming robust structures.

Various types of IF occur in different eukaryotic cells and are expressed at specific stages of development and differentiation, suggesting that they play important roles. These include **keratin** IFs in epidermal cells, which confer skin toughness. Keratins from dead epithelial cells form hair, fingernails, and horses' hooves. **Neurofilaments** exist in nerve cells to give mechanical support to the long axons. **Desmin** filaments are located in the Z discs of sarcomeres. A number of **lamin** proteins form a network associated with the inner surface of the inner nuclear membrane. Intermediate filaments in general are not ephemeral, as most actin and microtubules are, but lamins are the exception in that at mitosis the nuclear membrane is disassembled after phosphorylation of the lamins, and is reformed at the end of the process.

IFs are not essential for cell growth and division; cells in which IF formation does not occur due to mutations still grow and divide in laboratory tissue culture. This survival is possibly because these cells are not subject to the mechanical stresses experienced by cells in functional tissues. The role of IFs in protecting cells from mechanical stress is demonstrated by mutation of skin keratin in humans, which causes a pathological condition, **epidermolysis bullosa**, in which the skin blisters due to weakness in the basal layer of the epidermis.

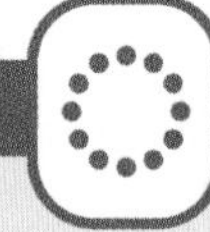

Summary

Muscle contraction and all molecular motors depend on conformational changes in proteins. Skeletal muscle cells have multiple myofibrils running through their length, each divided into sarcomeres, which are the contractile units. The sarcomeres have at each end a strong Z disc and projecting from these are thin filaments made of the fibrous protein actin pointing to the centre of the sarcomere. The thin filaments form a hexagonal 'cage' inside of which is a thick filament: a bundle of hundreds of myosin molecules arranged in a bipolar fashion. Myosins are rod-like molecules, each a dimer arranged as a coiled-coil and terminating in a pair of globular heads.

Contraction is explained by the sliding-filament model in which sarcomeres are made to shorten by thick filaments. The myosin heads contact the actin filaments. A cycle of events takes place driven by hydrolysis of myosin-bound ATP to ADP and P_i, in which the heads exert a force on the thin filaments pulling the Z discs towards the centre of the sarcomere, thus shortening it. ATP hydrolysis, however, does not coincide with the power stroke. The power stroke occurs when P_i and ADP are released and the myosin heads exert their force through a swinging-lever-arm mechanism in which the myosin rod swings from one position to another, causing the attached actin filament to slide.

Contraction is triggered by nerve impulses, which cause release of calcium ions from the sarcoplasmic reticulum, a sac surrounding the myofibrils. The ions initiate the contractile cycle of the myosin heads by causing a conformational change in a tropomyosin complex attached to the actin filaments, which shifts its position to allow myosin attachment. The contraction is terminated by removal of the Ca^{2+} ions by an ATP-driven pump, which returns the ions to the sarcoplasmic reticulum. Smooth muscle also utilizes the actin and myosin contraction mechanism, but lacks the sarcomere structure and has a different control system.

Muscle is a specialized case, but almost all eukaryotic cells contain actin and myosin as components of the cytoskeleton. The other main components are tubulin microtubules and intermediate filaments. The cytoskeleton has important functions in cell structure, cell movement, and transport of membrane vesicles containing newly synthesized macromolecules to parts of the cell that they would not reach by diffusion.

Microfilaments are polymers of globular G actin, as are thin filaments of muscle. Unlike thin filaments, which are permanent structures, cytoskeletal microfilaments are assembled and disassembled as needed. This dynamic process involves hydrolysis of actin-bound ATP and is regulated by a large number of actin-binding proteins with different properties and functions. Rapid growth of microfilaments drives cells forward in cell migration. Members of the myosin family are involved in the cell transport function of microfilaments. They have a pair of globular heads as in muscle myosin but the long rod-like tail is replaced by a short one that attaches to the vesicles to be transported.

Microtubules are hollow tubes made of polymerized tubulin protein dimers. They also develop and collapse as needed, the process involving hydrolysis of tubulin-bound GTP. ATP-driven molecular motors known as kinesins and dyneins move along the microtubules in opposite directions pulling cargoes. The role of microtubules in cell movement is limited to cilia and flagella, but they play an important role in chromosome movements during cell division as well as in intracellular transport.

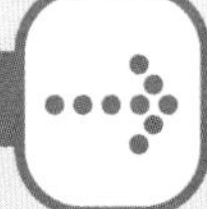

Further reading

To access the further reading, please scan the QR code image or go to http://global.oup.com/uk/orc/biosciences/molbiol/snape_biochemistry/student/reading/ch08/

Problems

1 How is contraction of a voluntary striated muscle sarcomere controlled?

2 How is smooth muscle contraction controlled?

3 Can you see any similarities in principle between the mechanisms of ATP synthesis by ATP synthase and its utilization by myosin for contraction?

4 Explain with simple diagrams the mechanism by which myosin causes movement of the actin filament.

5 Actin is found in nonmuscle cells. What are its roles there?

6 What are microtubules? What controls their assembly and collapse?

7 Microtubules with unprotected ends undergo collapse. What protects the ends as they form?

8 What are kinesin and dynein? How does their movement differ from that of myosins?

9 At cell division, chromosomes on the metaphase equatorial plate move apart. Microtubules are attached to the kinetochores and shorten as the chromosomes move apart. Does this mean that microtubules contract? Explain your answer.

10 What are intermediate filaments? What are their functions?

11 What do the proteins G actin and tubulin have in common?

Part 3

Metabolism and nutrition

Chapter 9

General principles of nutrition

This chapter will introduce the subject of nutrition and introduce various concepts which will help us understand the chapters on metabolism which will follow.

We need energy and specific nutrients in our diet and we oxidize or modify and store what we eat, but essentially we and the food we eat are made up of the same components. We consume carbohydrates, proteins and fats, which act as metabolic fuels, and we also take in the diet essential nutrients, such as vitamins and minerals.

Proteins carbohydrates and fats are known as macronutrients as they constitute the bulk of the diet. Vitamins and minerals are consumed in much smaller amounts and are known as micronutrients. All of them are needed for the correct functioning of the body. Let us define and explain some terms used with regard to nutrition:

Nutrition is the science of food and the substances contained in it, and the importance of nutrition was already recognized in antiquity. Hippocrates, for example, advised the consumption of liver for people who suffered from night blindness (vitamin A deficiency cured by consumption of vitamin A), without actually knowing the relationship between the defect and the cure. Doctors throughout history have tried to heal or cure patients by making changes to their diets. The first recognition of nutrition as a science came from Lavoisier in the 18th century, who concluded, after a number of experiments, that 'life is a combustion'.

Health is a state of optimal function of an organism and **disease** a state of impaired function.

Diet is the course of regular consumption of food and drink adopted by an individual or a group of people.

Nutritional status is the state of the organism with respect to the consumption, utilization and storage of energy and nutrients.

A **nutrient** is an essential component of the diet, which cannot be synthesized by the body. Nutrients include essential amino acids, essential fatty acids, vitamins, and minerals. Metabolic fuels such as carbohydrates, fats, and proteins, in general, are not classified as nutrients, but as sources of energy. It is also desirable to include inert substances in the diet, such as non-starch carbohydrates (fibre), which cannot be digested or metabolized but are valuable for the correct functioning of the gastrointestinal system.

Malnutrition is the term used most often to denote inadequate nutrition, either through general lack of food or referring to a specific nutrient, for example, vitamin C deficiency manifesting itself as scurvy. Malnutrition could also be used to describe excessive consumption of food leading to obesity and disorders such as diabetes mellitus and cardiovascular disease.

In this chapter we will look at dietary components and their function, the effects of deficiency or excess, and the control of food intake and body weight; Chapter 10 will deal with the digestion, absorption, and distribution of dietary components in the body; Chapters 11–19 with the synthesis and degradation of metabolic fuels; and Chapter 20 with the integration of metabolic processes.

The requirement for energy and nutrients

Energy is constantly needed for muscular contraction, maintenance of ionic equilibria, transport processes, and synthesis of macromolecules. In theory, it does not matter whether the energy is provided by oxidation of carbohydrate, fat, or protein but there are practical considerations. In Chapter 18 we will see how a carbohydrate, such as glucose, can be produced from protein, but most diets do not contain enough protein to supply the glucose requirements of the organism. Most diets provide 10–15% of the energy as protein and the rest is made up of carbohydrate and fat. In the developed world, carbohydrate accounts for about 30% of the energy of the diet but this can be as high as 90% in less economically developed countries. There is no dietary requirement for fat except for small amounts of essential fatty acids (see Chapter 17), but a fat free diet is extremely unpalatable and it would be difficult to eat enough of it to meet one's energy requirements, as protein or carbohydrate provide only half as much energy as fat per unit

Metabolic fuel	kJ/g	kcal/g (1 kcal=4.186 kJ)
Carbohydrate	17	4
Protein	17	4
Alcohol	29	7
Fat	37	9

Table 9.1 The approximate energy yields of metabolic fuels in kilojoules and kilocalories per gram. Note that 1 kilocalorie is equal to 4.186 kilojoules.

weight (Table 9.1). A fat free diet would also lead to deficiencies in fat soluble vitamins such as A and D, as these nutrients are present in oily foods and also need a certain amount of fat for absorption from the gut.

Moderately active females (19–30 years old) need 6.3–10.5 MJ per day (1500–2500 kcal) while males of the same age need 10.5–13.8 MJ (2500–3300 kcal).

Protein

Unlike fat, carbohydrate, and alcohol, there is a definite dietary requirement for protein. Only about half of the amino acids found in proteins can be synthesized by the body, the so-called **nonessential amino acids**, and the rest have to be provided in the diet, so they are classified as **essential amino acids**. The need for protein is the need for the essential amino acids, that is, those whose carbon skeleton cannot be synthesized in the body.

The essential amino acids are: histidine, isoleucine, leucine, lysine, methionine and/or cysteine, phenylalanine and/or tyrosine, threonine, tryptophan, and valine.

Apart from the obvious need for dietary protein to synthesize new body protein as in a pregnant woman or a growing child, there is also the need in adults to replace protein, which is constantly degraded. There is also the need for nitrogen for the synthesis of nucleic acids and certain neurotransmitters. Very little nitrogen comes into the body in a form other than protein. Strict vegetarians should ensure that their diets provide enough protein to cover the need for lysine and tryptophan, which are low in plant proteins.

Protein/energy malnutrition (PEM) is seen in many less-economically developed countries, particularly in sub-Saharan Africa, the Indian subcontinent, and parts of Central America. The two extreme forms are marasmus (lack of food in general) and kwashiorkor (thought to be more specifically a protein deficiency). The situation is more complex than that, as these conditions are exacerbated by poor sanitation, infection, and deficiency in antioxidant nutrients as well. In developed countries PEM is seen in people with a medical condition such as cancer, AIDS, and eating disorders, for example, anorexia nervosa.

Fats

Humans can synthesize fat from carbohydrate but most of our fat is of dietary origin and is modified to forms specific to humans. The storage form of fat in mammals is triacylglycerol composed of glycerol esterified with three fatty acids.

Saturated fatty acids, that is, those with no double bonds in the fatty acid chain, are derived mainly from animal sources. **Unsaturated** or polyunsaturated **fatty acids**, with a number of double bonds in the chain, are mainly of plant origin, such as sunflower, soya bean, or corn oil. **Monounsaturated fatty acids** have only one double bond and are found mainly in olive oil and rapeseed oil.

***Trans* fatty acids** do not occur naturally but are hydrogenated commercially and in the process the double bonds are converted from the natural *cis*- to the *trans*-configuration (also see Box 7.1). *Trans* fats are useful in the food industry as they improve the shelf life of a lot of manufactured foods, but caution is necessary as they may be as harmful as, or worse than, saturated fats with respect to cardiovascular disease. Epidemiology shows that mono- and polyunsaturated fats are not harmful and may be beneficial with respect to cardiovascular disease.

The **essential fatty acids**, which cannot be synthesized by the body and need to be taken in the diet, are **linoleic** (18 carbon atoms, 2 double bonds), **linolenic** (18 carbon atoms, 3 double bonds) and **arachidonic** (20 carbon atoms, 4 double bonds) acids. The first two are needed mainly as structural components of cell membranes and the third is the precursor of prostaglandins.

Cholesterol is taken in the diet if meat or animal products are consumed, so it is not present in strict vegetarian diets (vegan diets), but it can be synthesized in the body.

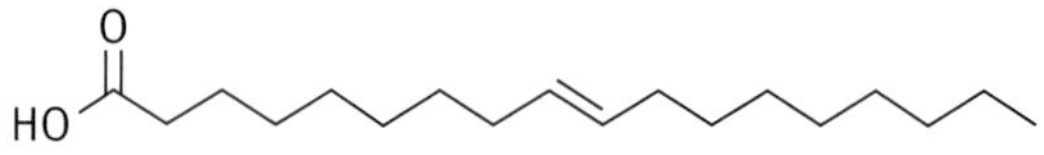

O
HO
trans-Oleic acid

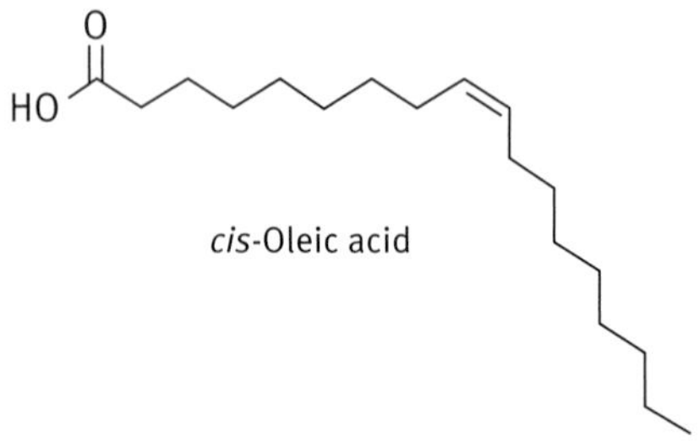

O
HO
cis-Oleic acid

Fig. 9.1 The structure of *trans* and *cis*-oleic acid.

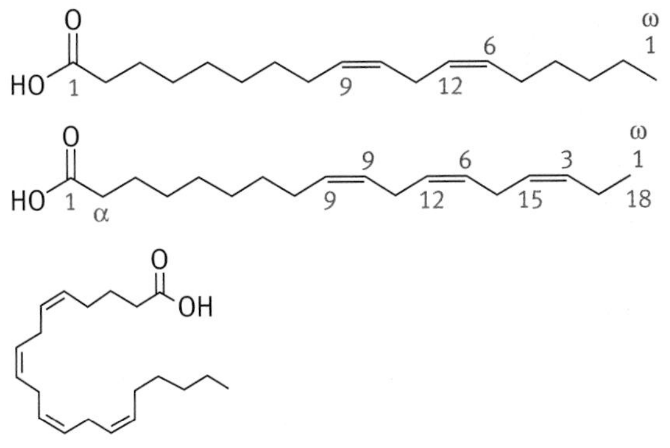

Fig. 9.2 The structure of linoleic, linolenic, and arachidonic acids.

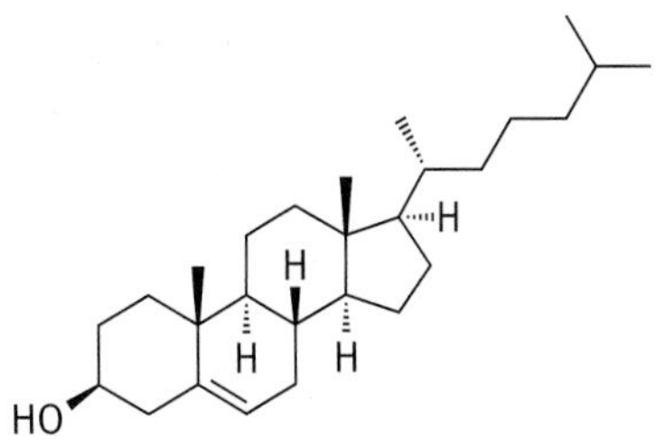

Fig. 9.3 The structure of cholesterol.

Carbohydrates

The bulk of carbohydrate in the diet is made up of **starch** and sugar (**sucrose**). The presence of carbohydrate in the diet has a protein sparing effect. The brain, nerves, and erythrocytes need glucose as a metabolic fuel and cannot use fatty acids. If no carbohydrate is present in the diet, all the glucose needs of the body have to be met by protein and as it is unlikely that any diet is so high in protein to provide enough glucose, body protein would be compromised as well.

Non-starch polysaccharides (NSP, dietary fibre) are not digested by the human gut as there are no human enzymes that can hydrolyse the β glucosidic bond (β acetal bond), but they are considered beneficial. Their main source is whole grain cereals and fruit and vegetables. Diets high in refined carbohydrates and low in NSP are associated with disorders of the colon such as cancer and diverticular disease and also with increased blood cholesterol. Fig. 9.4 shows the difference in the structure of cellulose and starch. Cellulose is made up of β glucose units, whereas starch is made up of α glucose units joined by α glycosidic (α acetal) bonds, which can be hydrolysed by human enzymes.

Vitamins

The concept of a deficiency disease is fairly new in medicine. For hundreds of years diseases were thought to be caused by the presence of a toxic factor, not the lack of an essential factor. Lind recognized in 1757 that scurvy, which was decimating the British navy, could be cured by supplying the sailors with lemons and limes. Eijkman, a Dutch physician working in Java in the 1890s treated his beriberi patients, whose diet was largely based on white rice, causing deficiency of thiamin (vitamin B1) by adding the rice polishings to their food, thus supplying the missing element. The studies of Gowland Hopkins in the early 1900s, looking for essential dietary factors, helped to lay the foundations of biochemistry as a science. These complex essential factors were called vitamins as they were originally thought to be 'vital amines'.

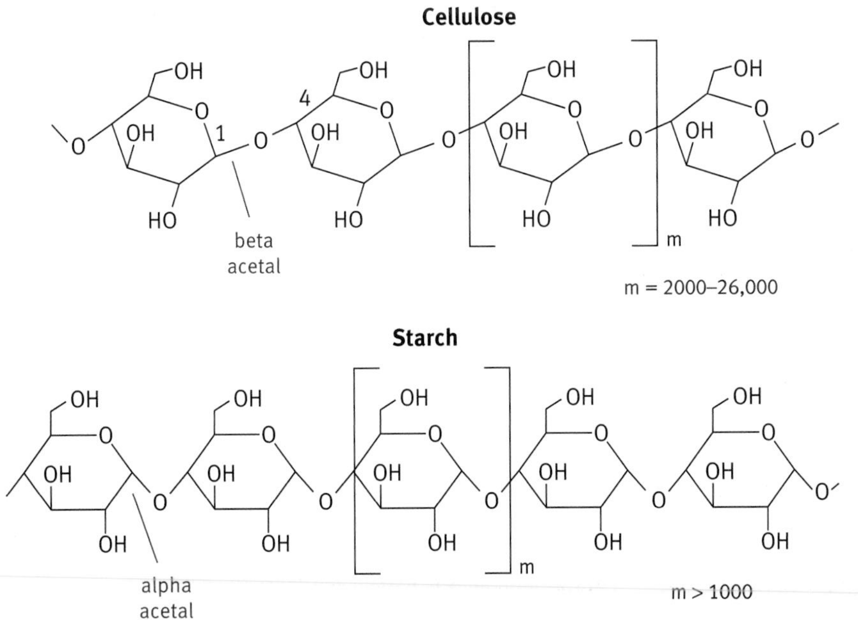

Cellulose

beta acetal

m = 2000–26,000

Starch

alpha acetal

m > 1000

Fig. 9.4 Structure of cellulose and starch.

The definition of a vitamin now is 'a complex organic substance required in the diet in small amounts, compared to other dietary components, such as protein, carbohydrate, and fat, and whose absence leads to a deficiency disease'.

Traditionally, vitamins are divided into **water soluble** and **fat soluble**. Although there is no good theoretical reason for this classification, as there is no similarity in structure or function within either group, there are some practical implications. Water-soluble vitamins are excreted if taken in excess, so toxicity is low, but on the other hand they are not stored extensively, so they need to be taken in the diet more frequently than the fat soluble vitamins. Fat soluble vitamins, include vitamin A, D, K and E, and the water soluble ones are the B group vitamins and vitamin C.

Some vitamins need to be converted into an active form before they can have biological activity.

A summary of the vitamins, their main dietary sources, active form, function, deficiency disease, and an indication of whether they can be toxic if taken in excess is shown below. The more detailed involvement of the vitamins in metabolism will be considered in the relevant chapters.

Water-soluble vitamins

The B group vitamins

All vitamins belonging to the B group act as cofactors or coenzymes in metabolic pathways.

Thiamin (vitamin B1)

Thiamin is found in meat, yeast and unpolished cereals. The active form is thiamin pyrophosphate and is a coenzyme in carbohydrate metabolism (see Chapter 13). Deficiency of thiamin results in a condition known as beriberi, mainly seen in South Asia, and associated with diets high in polished white rice. It is characterized by a peripheral neuropathy, with or without congestive heart failure. In the developed world it is mainly seen as an alcohol induced dementia (Wernicke–Korsakoff syndrome) characterized by loss of recent memory and disorientation in time and space. Thiamin is water soluble and easily excreted so it is nontoxic if taken in excess.

Riboflavin (vitamin B2)

Riboflavin is found in eggs, dairy products, and in general in high protein diets. The active forms are FAD and FMN and they are involved in cellular respiration in oxidation/reduction reactions (see Chapter 13). Deficiency is rare and the symptoms are not severe, consisting of cracked lips (angular stomatitis) and inflammation of the tongue (glossitis). Riboflavin is water soluble and any excess is easily excreted, so it is not toxic.

Niacin (nicotinic acid or nicotinamide)

Niacin is found in meat, yeast and dairy products. The active forms are NAD and NADP. They are involved in cellular respiration in oxidation/reduction reactions (see Chapter 13). Deficiency is associated with diets based on untreated maize and causes pellagra. Pellagra is characterized by dermatitis on sun exposed areas of the skin, diarrhoea and dementia. Niacin is easily excreted so it is not toxic in excess.

Biotin

Biotin is found in eggs and milk and is produced by intestinal bacteria. The active form is enzyme bound biotin; it is involved in carboxylation reactions (see Chapter 14). Deficiency is rare, as biotin is widely available and a large percentage of the requirement is supplied by intestinal flora. Eating raw eggs may cause a deficiency as they contain avidin, which binds biotin and makes it unavailable. The characteristics of deficiency are dermatitis, glossitis and nausea. It is nontoxic in excess.

Pyridoxine (pyridoxamine, pyridoxal, vitamin B6)

Pyridoxine is widespread in plant and animal products. The active form is pyridoxal phosphate. It is a cofactor mainly involved in amino acid metabolism (see Chapter 18). Deficiency is rare except when using the drug isoniazid to treat tuberculosis. Isoniazid binds pyridoxine and makes it unavailable so supplementation is necessary. Deficiency is characterized by **anaemia** and convulsions. Unlike other water-soluble vitamins excessive intake causes sensory neuropathy. It has been observed in women overdosing on pyridoxine to treat premenstrual tension syndrome.

Folic acid (folate)

It is found, as the name implies, in green leafy vegetables and also in liver. The active form is tetrahydrofolic acid. It is involved in 1-carbon transfer reactions, particularly in amino acid, and purine and pyrimidine synthesis (see Chapters 18 and 19). Deficiency causes megaloblastic anaemia where blood contains large immature erythrocytes. Deficiency in pregnancy can cause neural tube defects in the fetus such as spina bifida. Excess is not toxic.

Cobalamin (Vitamin B12)

It is found in meat and animal products only. A glycoprotein, known as **intrinsic factor**, which is secreted by the stomach, is necessary for the absorption of vitamin B_{12}. The active forms are methyl cobalamin or adenosyl cobalamin. It is involved in methyl group transfer reactions particularly in methionine synthesis and in purine and pyrimidine metabolism (see Chapters 18 and 19). The deficiency disease is **pernicious anaemia** caused primarily by defective intrinsic factor secretion. The characteristics of deficiency are megaloblastic anaemia, as in folate deficiency, but with additional neurological complications such as central and peripheral neuropathy. It is not toxic in excess.

Pantothenic acid

Pantothenic acid is widely available (from the Greek 'pantothen' meaning from everywhere). It forms part of coenzyme

A and it functions as an acyl group carrier. Deficiency is rare and toxicity is not known.

Vitamin C (ascorbic acid)

Vitamin C is present in citrus fruits, tomatoes and some berries. The active form is ascorbic acid. It is a reducing agent necessary for hydroxylation reactions in the synthesis of collagen (see Chapter 4). Deficiency leads to scurvy, characterized by impaired wound healing, gastrointestinal bleeding, loose teeth and sore bleeding gums. A number of people take large doses of vitamin C to prevent diseases such as colds and cancer. There is no harm except for the appearance of oxalate kidney stones in some susceptible people nor is there evidence that these megadoses are effective against these diseases.

Fat soluble vitamins

Vitamin A (retinol, retinal, retinoic acid, β-carotene)

Good sources of vitamin A are fish liver oils and butter. Plant sources include β-carotenes taken in the diet in carrots and other yellow and orange vegetables. β-carotenes can be converted into retinol in the liver. The active forms of vitamin A are retinol, retinal and retinoic acid. Retinol is involved in reproduction, retinal in vision (see Chapter 29) and retinoic acid in growth, gene expression and differentiation of epithelial tissues (see Chapter 29). Deficiency of vitamin A leads to growth failure in children, infertility later in life, and night blindness, xerophthalmia (keratinization of the cornea), keratomalacia (degeneration of the cornea) and finally blindness. Excess is stored in the liver and can reach toxic levels with supplementation; it can predispose the individual to bone fractures later in life. Pregnant women should not take supplements as it is teratogenic, that is, it can cause birth defects, nor should they use the chemically related isotretinoin for treatment of acne.

Vitamin D (cholecalciferol)

Vitamin D (Fig. 9.5) is strictly speaking a hormone rather than a vitamin as it can be synthesized in the skin from dehydrocholesterol under the action of UV light.

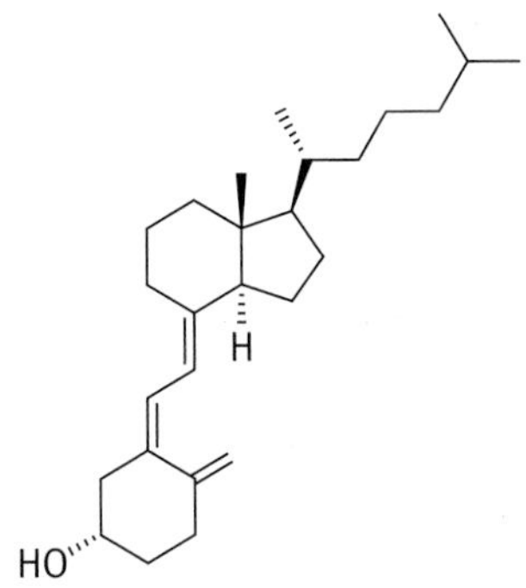

HO H

Fig. 9.5 The structure of vitamin D, or cholecalciferol, synthesized in the skin or taken in the diet.

Although a hormone, it is still classified as a vitamin because a number of people do not synthesize enough and it therefore becomes a dietary essential. Fish oils are good dietary sources. The active form is 1,25-dihydroxycholecalciferol. It is needed for the absorption of calcium from the intestine and proper mineralization of bones. Deficiency causes **rickets** in children and **osteomalacia** in adults. In both cases, the bone mineral to matrix ratio is reduced, resulting in soft and fragile bones. Groups of the population at risk include the elderly, particularly if they are housebound, and people from any culture that dictates covering most of the body by clothing and whose diet is not enriched in vitamin D. Rickets is reappearing in some northern countries in some breast fed babies of mothers of low vitamin D status and because of overprotection from exposure to the sun. It is toxic in excess. Excessive intake from natural sources is unlikely but it can occur from excessive supplementation. Ectopic calcification and mental retardation result, and large doses are lethal.

Vitamin E (α-tocopherol)

Good sources of vitamin E are wheat germ, vegetable oils and nuts. The active form consists of a number of tocopherol derivatives. It is an antioxidant and acts as a free radical scavenger in cells preventing the peroxidation of unsaturated fatty acids in membranes (see Chapter 32). Deficiency causes haemolytic anaemia because of fragile erythrocyte membranes. It is very rare but has been seen in low birth weight premature babies. There is no known toxicity.

Vitamin K (menadione, menaquinone, phylloquinone)

Vitamin K is found in green leafy vegetables but is also synthesized in the intestine by bacteria. The active forms are menadione, menaquinone and phylloquinone. It is involved in γ-carboxylation of glutamate residues in clotting factors and other proteins (see Chapter 32). In deficiency, there is a prolonged clotting time and haemorrhages. Deficiency is rare in adults, except on long term antibiotic treatment, but common in neonates. Vitamin K administration is advised shortly after birth to prevent haemorrhagic disease of the newborn. High doses may be toxic in babies.

Minerals

A great number of minerals are needed in the diet and the requirements are known for many of them, but for some we have no information as they are needed in such small amounts that no natural diet is deficient in them.

In this chapter we will deal with some major minerals whose deficiency is common.

Calcium

Good sources of calcium are milk and dairy products. Calcium is less well absorbed from plant sources. Calcium is a structural component, which forms 18% of bone, is involved in blood clotting (see Chapter 32), and is an intracellular signalling molecule (see Chapter 29). It is particularly important in pregnancy and growth. Calcium homeostasis is achieved by the concerted action of parathyroid hormone, vitamin D and calcitonin. Deficiency leads to bone fragility and fractures.

Iron

Meat is a good source of haem iron, and nuts and pulses contain non-haem iron. Iron is a component of haemoglobin (see Chapter 4), myoglobin, and cytochromes (see Chapter 13). Deficiency leads to anaemia, which is particularly common in women of reproductive age as iron is lost during menstruation. The requirement for iron also increases in pregnancy and lactation as there is transfer of iron to the fetus and infant. Excretion of iron is very poor, except in cases of blood loss, so caution should be taken with iron supplements, as excessive intake can be toxic, particularly in children, leading to liver failure and death.

Iodine

Iodine is found in seafood, plants grown on iodine-rich soil and meat of animals eating iodine-rich plants. It can also be taken in the diet as iodized salt. It is a component of thyroxine and triiodothyronine involved in regulation of metabolism. Deficiency gives rise to goitre (swelling of the thyroid gland), reduced metabolic rate and weight gain. Deficiency in pregnancy leads to mental retardation of the child. Excessive intake is unlikely from food sources but excessive supplementation can cause gastrointestinal disturbances.

Zinc

Zinc is found in protein rich foods such as nuts, meat, and pulses. Bioavailability of zinc is lower in plant than in animal sources. It is a component of many enzymes, for example, carbonic anhydrase (see Chapters 4 and 10) and alcohol dehydrogenase (see Chapter 10). Several proteins that bind to DNA and affect transcription have special zinc binding domains known as zinc fingers (see Chapter 24). About a third of the world is at risk of zinc deficiency, primarily because of vegetarian diets low in bioavailable zinc.

Acrodermatitis enteropatica is a genetic condition where zinc cannot be absorbed from the diet. Skin is dry and scaly and susceptible to bacterial infections. Deficiency of zinc results in gastrointestinal disorders, loss of smell and hypogonadism in men. Excessive intake is rare, usually arising from exposure to paints and dyes. It leads to intestinal irritation and convulsions.

Guidelines for a healthy diet

The recommendations for a healthy diet are very similar in developed countries. Both the Food Standards Agency of the UK and the USA National Research Council give similar guidelines

- Reduction of dietary fat to 30% of the energy of the diet with saturated fat less than 10% and alcohol less than 5%
- Five portions of fruit and vegetables per day
- Moderate intake of protein
- Intake of energy to maintain an appropriate body weight
- Reduction of salt to less than 6 g per day
- Adequate calcium and iron intake
- Optimal intake of fluoride and restriction of dietary sucrose to less than 60 g/day to prevent dental caries.

Regulation of food intake

The epidemic of **obesity** in many developed countries with its association with type 2 diabetes and other health problems has greatly stimulated research interest into the normal controls on food intake, weight homeostasis, and energy balance. If the intake of energy exceeds the expenditure of energy, the excess will be stored in the form of fat. Even a small imbalance between food intake and energy output would, over long periods, cause excessive weight changes. A number of controls exist to achieve this balance; they are very complex and involve signals being integrated in the brain as well as more direct effects on metabolism in peripheral tissues. It seems that the known control systems are not sufficient to prevent the current epidemic of obesity. This may be due to the temptations of readily available food and leisure in today's society, where the energy intake has increased and the energy expenditure has diminished. Over the millions of years of human existence, death by starvation was a more likely danger than over-abundance of food. Survival of the species may have depended on the ability to store excess food efficiently as fat.

A useful criterion for assessing normality, underweight or obesity in an individual is the Body Mass Index (BMI). It is calculated by dividing the subject's body weight by the square of his/her height.

$$\mathrm{BMI} = \frac{\mathrm{mass(kg)}}{(\mathrm{height(m)})^2}$$

Table 9.2 shows the classification of adults as underweight, normal healthy, overweight and obese. These are guidelines and are not absolutely accurate, as, for example a very muscular person may fall under the category of overweight or obese while having a low percentage of body fat. They are, however, useful general guidelines and easily calculated.

Despite considerable advances, in our understanding in the last decade, there is still no widely applicable therapeutic remedy to the problem of obesity, other than reduction in food intake and increase in activity.

Hunger, appetite and satiety

Hunger describes the feeling of the need to eat. **Appetite** describes the desire to eat a particular type of food. Sometimes these terms are used interchangeably but in this chapter we use

Category	BMI range (kg/m^2)
Underweight	less than 18.5
Normal (healthy weight)	from 18.5 to 24.9
Overweight	from 25 to 29.9
Obese Class I (Moderately obese)	from 30 to 34.9
Obese Class II (Severely obese)	from 35 to 39.9
Obese Class III (Very severely obese)	over 40

Table 9.2 Classification of normal body weight, underweight, overweight and obesity based on calculations of Body Mass Index.

the definitions above. **Satiety** is the feeling of having consumed an adequate amount of food and is experienced somewhere between the feeling of relief from hunger and the uncomfortable feeling of having eaten in excess. The perception of the energy content of food is poor in humans and not immediate. Eating habits and customs override physiological control mechanisms.

There are several hormones produced by the digestive tract in response to the absence or presence of food, which act as hunger or satiety signals.

Ghrelin is a peptide hormone produced by the stomach when it is empty of food; it acts as a hunger signal and stimulates eating. Its concentration in the blood increases rapidly in fasting and falls just as quickly after a meal.

Cholecystokinin (CCK) is secreted by the intestine in the presence of food and acts as a satiety signal. It is not known whether its physiological effect is exerted through gastrointestinal CCK acting on the brain or CCK released within the central nervous system.

PYY-3-36, a pro-opiomelanocortin related peptide, is produced, in the presence of food, by endocrine epithelial cells of the small intestine and the colon, and is released into the bloodstream, which carries it to the brain. This neuropeptide is also a satiety signal. Injection of PYY-3-36 into experimental animals or humans, in amounts sufficient to give normally achieved blood levels, suppresses hunger for about 12 hours.

The digestive tract is not the only source of hormones that control hunger. The cells of adipose tissue are also involved.

Leptin is produced by adipocytes and its concentration in the blood is proportional to the size of the adipose stores. Leptin is involved in long-term control of eating and thus of maintaining constancy of body weight. It is transported to the brain where it binds to its receptors and acts as a satiety signal. It has also been found to have more direct effects on fatty acid metabolism in muscles.

Leptin was a discovered in 1994 when its gene was pinpointed in a mutant strain of extremely obese mice known as ob-ob mice, which did not produce the hormone. The mice became obese through uncontrollable eating. Injection of leptin caused weight loss in the mice. Attempts to treat human obesity with leptin injections failed as the obese have high concentrations of leptin but seem to be leptin resistant. In very rare cases, where children had a mutation affecting the production of leptin, injection of the hormone corrected the obesity. There are very few cases of leptin deficiency described worldwide.

Adiponectin is also produced by adipocytes, with low concentrations reported in obesity. Adiponectin and leptin are two of several **adipokines** or **adipocytokines** (see Chapter 29 for cytokines) secreted by adipose tissue.

Insulin is produced by the pancreas when blood glucose is high as occurs after meals. Its best known role is metabolic fuel homeostasis but it also has effects on hunger and satiety, resembling those of leptin.

Amylin is also produced by the pancreatic β-cells, and is co-secreted with insulin, but in much lower concentrations. It is a peptide consisting of 37 amino acids. It exerts a short-term satiety effect *via* receptors in the posterior part of the brain. Administration of amylin to rats caused a decrease in food intake, and some weight loss.

It has been shown that injection of both leptin and amylin into diet-induced obese rats caused weight loss greater than either hormone alone (see Roth et al. in Further reading). The loss was mainly in the fat content, and the effect on leptin responsiveness occurred only with amylin. Further clinical studies suggested that this applied also to human obese or overweight subjects, in which 12.7% greater weight loss occurred with leptin and an amylin analogue, than with either alone. It was suggested that amylin may restore sensitivity to leptin. The authors suggest that a multifactorial approach with different hormones may be advantageous in obesity control studies.

Integration of hunger and satiety signals by the hypothalamus

An important hunger control centre is located in the **hypothalamus**, a small area at the base of the brain. In a region known as the **arcuate nucleus**, there are two subsets, or groups, of neurons with opposing effects on appetite. They are controlled by some of the circulating hormones described previously. One of these groups (the **NPY/AgRP** producing set) produces two neuropeptides (**Neuropeptide Y** and **Agouti-related peptide**, the latter discovered in agouti mice), which stimulate hunger. The other group, known as the **pro-opiomelanocortin (POMC)** set, produces neuropeptides, which suppress hunger. The NPY/AgRP neuropeptides block the action of the POMC neuropeptides. It is another example of a push–pull mechanism so common in homeostatic control. Fig. 9.6 shows the way in which the various circulating hormones interact with the arcuate nucleus to control hunger.

- At times of the day when the stomach is empty, ghrelin is secreted and stimulates the NPY/AgRP subset to produce their neuropeptides and stimulate eating.

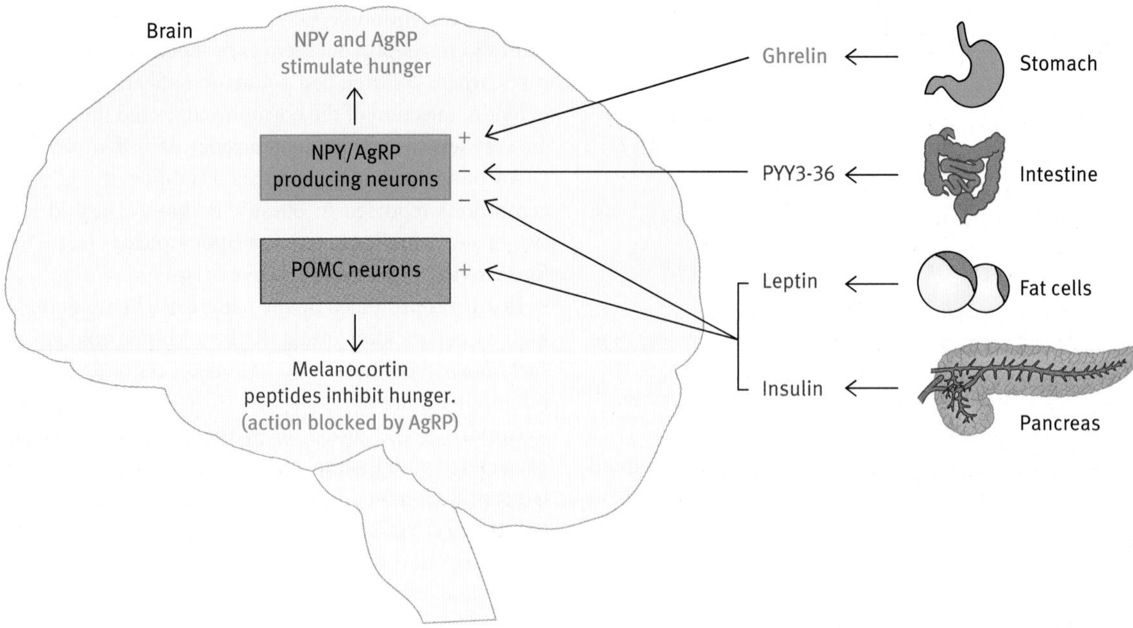

Fig. 9.6 Simplified diagram of appetite control via hypothalamic neurons. See text for explanation. The green plus sign represents stimulation of the target neurons and the red minus sign, inhibition. NPY, neuropeptide Y; AgRP, agouti related peptide. POMC, pro-opiomelanocortin.

- After intake of food, ghrelin secretion stops and PYY-3-36 is secreted in the presence of food in the intestine and inhibits the NPY/AgRP group of neurons from producing its hunger stimulating peptides.
- Leptin, produced when fat reserves are high, inhibits production of the hunger stimulating NPY/AgRP neuropeptides and stimulates production of the hunger inhibiting POMC neuropeptides.
- Insulin has similar effects to leptin.
- In fasting or starvation, the reverse happens. Ghrelin is produced by the stomach, stimulating hunger. In the absence of food in the intestine, PYY-3-36 is not produced so its inhibitory influence is not present. Leptin production falls as fat reserves decline so its inhibitory influence is diminished. Insulin concentrations will be low. The stimulation of hunger by ghrelin is, therefore, unopposed.

Other controls on eating exist, though less is known of the way in which they work. Maintenance of body weight is achieved by a balance between food intake and energy expenditure. The subject is complex, as energy intake affects energy expenditure and vice versa. The metabolic rate and hence energy expenditure is also affected by the concentration of thyroid hormones, as well as the amount of physical activity. It has also been shown that leptin increases energy expenditure and there is evidence that it increases the metabolism of fat rather than allowing it to be deposited in adipocytes. We will deal with the major metabolic controls on energy metabolism in Chapter 20.

Summary

Components of the diet

Nutrition is the science of food and the substances contained in it. The main components of the diet are macronutrients, that is, protein, carbohydrate, and fat, which provide the bulk of the diet, and micronutrients, which include essential vitamins and minerals. Protein is needed to supply essential amino acids and fat is needed to supply essential fatty acids and also to provide sufficient energy in the diet. Most diets provide 10–15% of the energy in the form of protein and the remainder is made up of carbohydrate and fat. Consumption of saturated fat is one of the risk factors for cardiovascular disease and some cancers whereas consumption of polyunsaturated fat is considered healthy. A high consumption of sucrose can lead to dental caries. Non-starch polysaccharides are not digested but are desirable components of the diet as they may prevent a number of diseases of the colon and may lower blood cholesterol.

Vitamins are complex essential dietary factors consumed in small quantities compared to protein carbohydrate and fat. Their inadequacy in the diet results in deficiency diseases. There are a number of minerals of practical importance such as iron, calcium, iodine, and zinc, and there are minerals whose precise requirements in the diet are not known as they are ubiquitous and only needed in minute amounts. Vitamins are classified as water- or fat- soluble. Water-soluble vitamins are, on the whole, required frequently in the diet as they are not stored and they are generally nontoxic as any excess is excreted. Fat-soluble vitamins require the presence of fat in the diet for their absorption , they are stored so they do not need to be eaten frequently but can be toxic in excess.

Guidelines for a healthy diet

Guidelines for a healthy diet in the developed world include reduction of saturated fat intake and increase of polyunsaturated fat (as long as the energy intake does not lead to obesity), reduction in the consumption of sucrose and salt, and a moderate protein intake.

Regulation of food intake

This is an area of great interest owing to the epidemic of obesity particularly in the developed world. Several hormones are involved in regulating food intake. Ghrelin stimulates hunger and is produced when the stomach is empty. A number of other hormones suppress hunger. These are the peptide hormone PYY-3-36 (produced by the intestinal epithelial cells in the presence of food), leptin (produced by fat cells with the rate of release increasing as the fat stores increase). Insulin (produced by the pancreas when blood glucose concentrations are high) has a role similar to leptin. Ghrelin and the opposing hormones work on hunger and satiety centres in the brain.

There is great interest in the possibility of hormones being used to control obesity though experiments with leptin have not been successful except in a few obese children, who lacked the hormone completely.

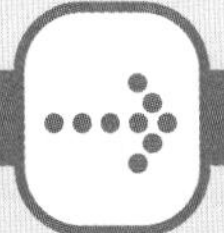

Further reading

To access the further reading, please scan the QR code image or go to http://global.oup.com/uk/orc/biosciences/molbiol/snape_biochemistry/student/reading/ch09/

Problems

1 What makes protein an essential component of the diet?

2 It is said that carbohydrate is not an essential part of the diet as it can be synthesized from protein. What would be the consequences of consuming a diet devoid of carbohydrate?

3 Why is an excessive intake of salt and sucrose undesirable?

4 What is the function of nondigestible components of the diet such as non-starch polysaccharides?

5 Describe the role of leptin. Is it likely to be of therapeutic use in human obesity? Explain your answer.

6 Outline the systems by which hunger and satiety are regulated.

Chapter 10

Food digestion, absorption, and distribution to the tissues

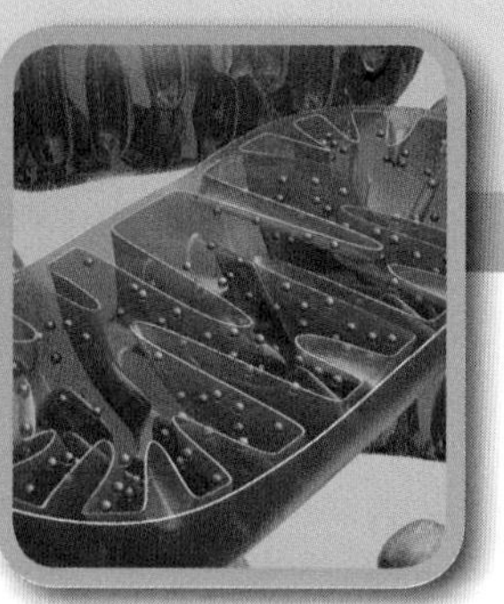

In this chapter we are going to deal with the metabolism of the macronutrients provided by food. We will start with **digestion**, which is the breakdown of macromolecules in the gastrointestinal tract into smaller units that can be **absorbed** into the bloodstream, and the **distribution** and uptake of these units, by cells, to be used for cellular metabolism.

Metabolism refers to the set of chemical reactions that take place in living organisms to sustain life. Metabolism can broadly be divided into **catabolism**, which is the breakdown of organic compounds by cells, usually to provide energy, and **anabolism**, which is the synthesis of complex organic molecules from simpler units.

The **roles of metabolism** can be summed up as follows:

- A major role of metabolism is to oxidize food to provide energy in the form of adenosine triphosphate (ATP).
- Food molecules are converted into new cellular material and essential components.
- Waste products are processed to facilitate their excretion in the urine.
- Some specialized cells in human babies and hibernating animals oxidize food to generate heat. This is exceptional since heat is, in general, a byproduct of metabolism.
- Excess metabolic fuel, that is, that which does not need to be oxidized immediately, is converted into a storage form and stored providing reserves for times of need.

In this chapter we will not give any details of the metabolic pathways or the metabolites involved, but will concentrate on the broad picture.

In this chapter we will discuss:

- how food is prepared for absorption into the bloodstream – what the chemical bonds are that need to be broken in these compounds
- how food molecules reach the bloodstream
- how food molecules are transported between the blood and tissues, and between different tissues
- how this traffic is regulated to satisfy different physiological needs – in essence, the broad logistics of fuel movements in the body
- how the body responds to a plentiful supply of food, fasting, starvation, and emergency situations.

Chemistry of foodstuffs

As discussed in Chapter 9, there are three main classes of food – proteins, carbohydrates, and fats.

- **Proteins** are large polymers of amino acids linked together to form polypeptide chains, which, in turn, may assemble into dimers or larger aggregates.
- **Carbohydrates** are the sugars and their derivatives; the name comes from their empirical formulae with carbon atoms and the elements of water in the ratio of 1:1 (CH_2O). Although simple monosaccharide sugars, such as glucose, occur in food, most of the carbohydrate in food is in the form of disaccharides, such as sucrose and lactose, or polysaccharides such as starch.
- **Fats** in the diet are mainly in the form of triacylglycerols (TAGs), sometimes called neutral fats. Polar lipids are also present, derived from the digested cellular membranes of animal and plant material. Small amounts of cholesterol are also taken in from diets containing animal products.
- The diet also contains small amounts of vitamins and minerals, essential nutrients needed in small quantities and whose absence in the diet leads to specific deficiency diseases. Vitamins and minerals provide negligible, if any, energy in the diet.

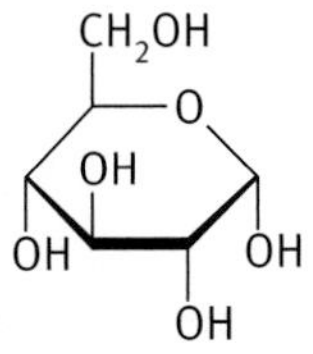

glucose

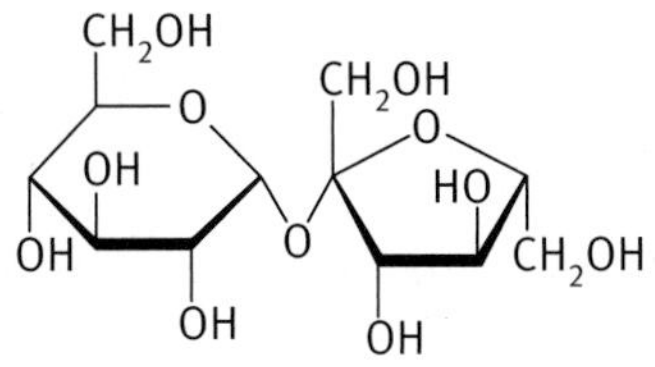

sucrose

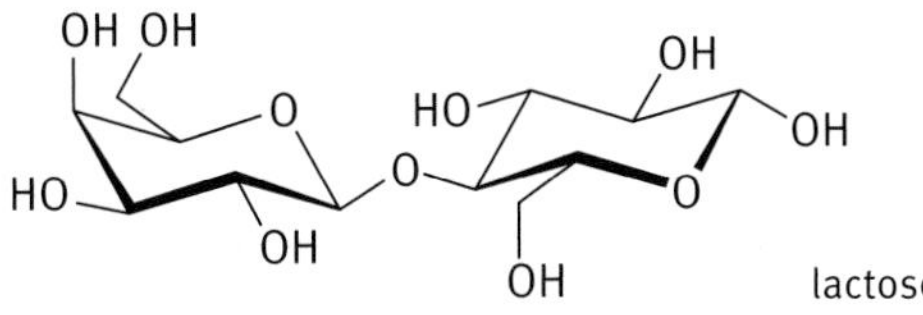

lactose

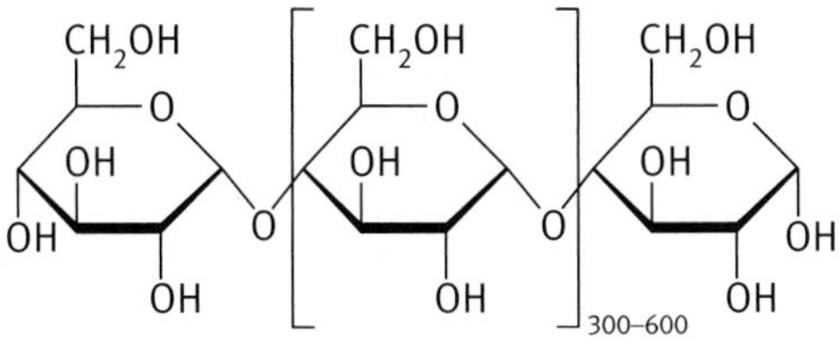

$[\ldots]_{300-600}$ amylose component of starch

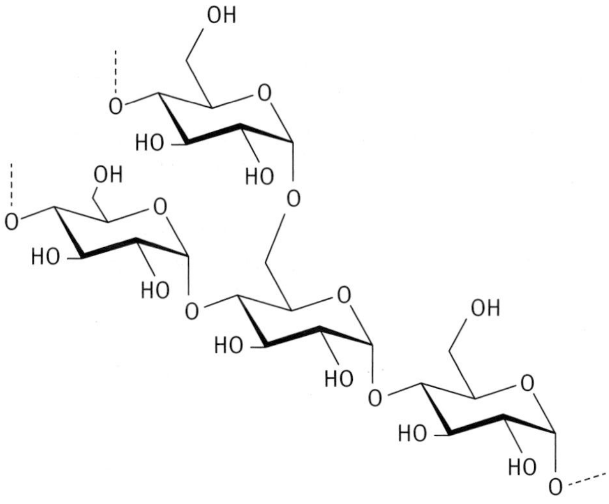

amylopectin component of starch

Digestion and absorption

With the exception of monosaccharides such as glucose, all of the foodstuffs mentioned above are digested by hydrolysis into their constituent parts in the small intestine (a minor exception is the absorption of dipeptides and tripeptides). To be absorbed, substances must cross membranes to enter the mucosal cells lining the intestine. TAGs cannot cross cellular membranes as they are large and neutral, nor can proteins, and among the carbohydrates only monosaccharides are absorbed. Thus, digestion consists mainly of the conversions:

- proteins, held together by peptide bonds→amino acids
- carbohydrates, held together by various glycosidic bonds→sugar monomers (monosaccharides)
- TAGs, held together by ester bonds→fatty acids and monoacylglycerol, and eventually glycerol.

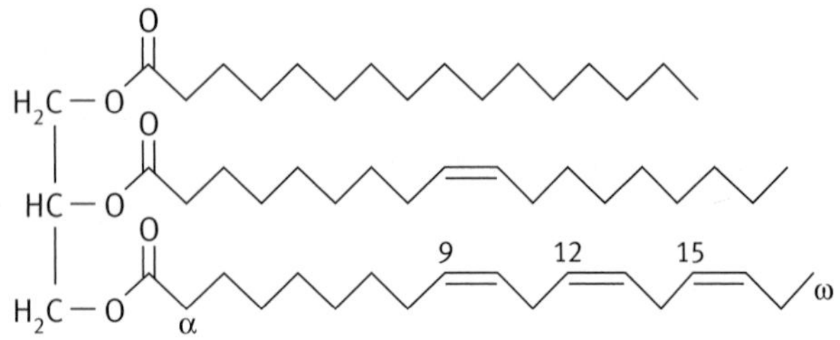

a triacylglycerol (neutral fat)

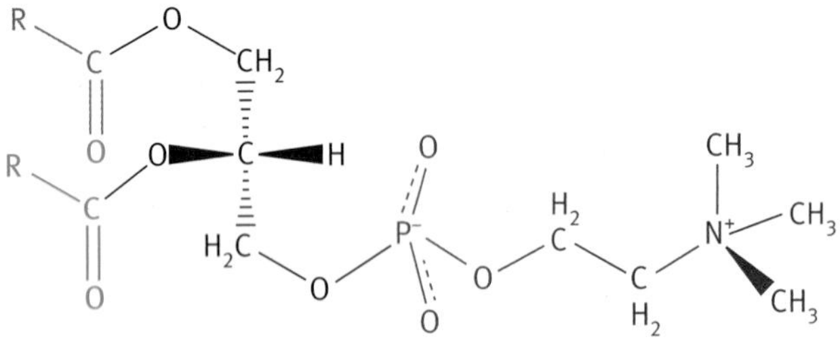

phosphatidylcholine (a polar lipid)

Anatomy of the digestive tract

The following regions (of nonruminants) are involved.

- In the **mouth**, food is masticated and lubricated for swallowing; limited starch digestion occurs.
- The **stomach** contains hydrogen chloride (HCl), which 'sterilizes' food and denatures proteins; partial digestion of proteins occurs.
- The **small intestine** is the major site of digestion and absorption of all classes of food. It is lined by fine finger-like processes, the villi, which are covered by epithelial cells – these are known as brush border cells because the microvilli of the epithelial cells resemble bristles on a brush (see Fig. 8.14). The microvilli are on the external membrane of the epithelial cells of the villi, giving the large surface area needed for absorption.
- The **large intestine** is involved in the removal of water. It is also the site of bacterial fermentation of some fibre and other components resistant to normal digestion.

What are the energy considerations in digestion and absorption?

So far as digestion goes, there are no thermodynamic problems. Hydrolytic reactions such as degradation of proteins to amino acids, of disaccharides and polysaccharides to monosaccharides, and of fats to fatty acids and monoacylglycerols are all exergonic processes – they have negative ΔG values sufficient to push the equilibrium entirely to the side of hydrolysis. Hydrolytic reactions (lysis by water) in biochemistry are invariably of this type. Absorption is a different matter since it is often an active process in which molecules are absorbed against a concentration gradient, and energy is needed.

A major question in digestion – why doesn't the body digest itself?

Food is, chemically, little different from the tissues of the animal that eats it. A fearsome array of enzymes is produced by the digestive system to completely digest the food into its components, and those enzymes have to be produced inside living cells, which, if exposed to their action, would be destroyed. There are two major types of defence against this.

Zymogen or proenzyme production

The enzymes are produced as inactive **proenzymes** or **zymogens** that are activated only when they reach the stomach and small intestine. Glands producing digestive enzymes secrete most of them as inactive proteins and are never themselves exposed to the destructive processes. In the case of enzymes that pose no threat (amylase, the enzyme that hydrolyses starch, is one), proenzymes are not involved. The question also arises as to how cells selectively secrete digestive enzymes or other proteins, but this is best left until we deal with protein targeting (see Chapter 27) for the mechanism is complex and not directly related to digestion. We will deal with the mechanisms of zymogen activation when we come to particular enzymes.

Protection of intestinal epithelial cells by mucus

The cells lining the intestinal tract are protected from the action of activated digestive enzymes, the main line of defence being the layer of mucus that covers the epithelial lining of the gut. The essential components of mucus are the **mucins**. These are large glycoprotein molecules – proteins with large amounts of carbohydrate attached to their polypeptide chains in the form of oligosaccharides that contain a mixture of sugars you have already met in membrane glycoproteins (see Protein domains in Chapter 4): glucosamine, fucose, and sialic acid to name a few. The mucins form a network of fibres, interacting by non-covalent bonds and resulting in a gel containing more than 90% water due to the hydrophilic carbohydrates that protect intestinal cells. The carbohydrates may protect the mucin proteins themselves from digestion. The mucin gel is quite permeable to low molecular weight digestion products, but much less permeable to digestive enzymes. The mucins are synthesized and secreted by special goblet cells in the epithelial lining of the gut. The amount of mucin secreted is controlled.

Digestion of proteins

In a normal or 'native' protein, the polypeptide chain is folded up, the shape being largely determined by weak bond interactions. In this compact folded form, many of the peptide bonds are hidden away inside the molecule where they are not accessible to hydrolytic enzymes. An important early step in digestion is to denature the native proteins. This is done by the acid in the stomach where, owing to HCl secretion, the pH is about 2.0. This partially disrupts the polypeptide folding, making the polypeptide chain susceptible to proteolysis.

HCl production in the stomach

The stomach epithelial lining contains **parietal** or **oxyntic** cells that secrete acid. In essence, the process consists of secreting H^+ – this is similar to the ejection of Na^+ from cells against a concentration gradient (see Cotransport systems in Chapter 7), which is achieved by a Na^+/K^+ ATPase in the membrane. In a similar manner, the acid-secreting cells have a H^+/K^+ ATPase. Using energy from ATP hydrolysis, they eject H^+ and import K^+, the latter recycling back to the exterior of the cell. The process is shown in Fig. 10.1. Where do the protons come from? An enzyme, **carbonic anhydrase**, converts CO_2 inside parietal cells to carbonic acid, which dissociates as shown.

$$CO_2 + H_2O \rightleftharpoons H_2CO_3 \rightleftharpoons H^+ + HCO_3^-$$

carbonic anhydrase

The resultant protons are pumped into the stomach lumen and the bicarbonate ions then exchange with Cl^- in the blood via an anion-transport protein (Fig. 10.1). We have already met this type of exchange in the anion channel of the red blood cell. The Cl^- ions then exit to the stomach lumen (cavity) to form HCl with the secreted protons, as shown in the figure.

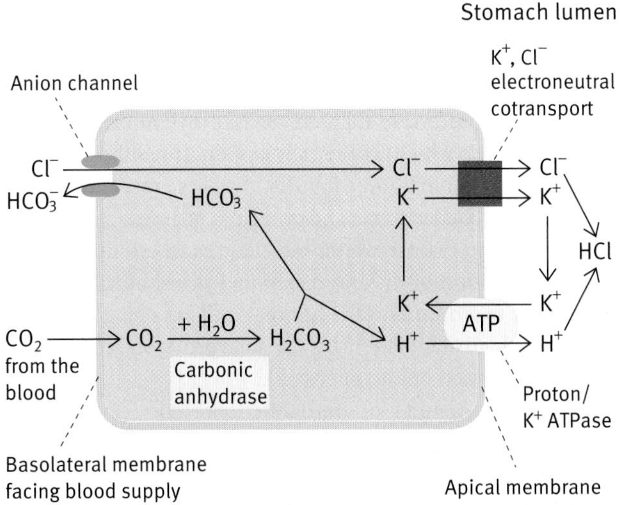

Fig. 10.1 Mechanism of gastric hydrochloric acid (HCl) secretion.

Pepsin, the proteolytic enzyme of the stomach

To indicate an inactive precursor of an enzyme, the suffix '-ogen' is used; for example, pepsinogen is the inactive form of pepsin. Alternatively, the term proenzyme is used in some cases.

Cells of the stomach epithelium secrete pepsinogen. The secretion is stimulated by the hormone **gastrin** released by stomach cells into the blood in response to food. Pepsinogen is pepsin with an extra stretch of 44 amino acids on the polypeptide chain. This additional segment covers, and blocks, the active site of the enzyme. When pepsinogen meets HCl in the stomach, a change in conformation exposes the catalytic site, which self-cleaves – it cuts off the extra peptide from itself to form active pepsin. As soon as a small amount of active pepsin is produced in this way, it converts the rest of the secreted pepsinogen to pepsin.

Pepsin is unusual for an enzyme in that it works optimally at acid pH (Fig. 6.10(a)), most enzymes requiring a pH near neutrality. It hydrolyses peptide bonds of proteins within the molecule, producing a mixture of peptides. It is therefore an ***endo*enzyme** or ***endo*peptidase**; that is, it does not attack terminal peptide bonds at the end of molecules, but only those within the molecule. The derivation is from the Greek *endo*, meaning within.

Proteolytic enzymes (proteases) are usually specific in their action – they hydrolyse only peptide bonds adjacent to certain amino acid residues. Pepsin has this characteristic so that only partial digestion of proteins can occur in the stomach. Pepsin produces peptides with C-terminal amino acid residues derived from aromatic amino acids (tyrosine, phenylalanine, or tryptophan) or long-chain neutral amino acids.

In ruminants, another stomach enzyme, **rennin**, acts on the casein of milk, causing clotting. As a result, casein does not pass too rapidly through the stomach and it undergoes gastric digestion.

Completion of protein digestion in the small intestine

The chyme, as the partially digested stomach contents are called, enters the duodenum at the start of the small intestine. The acid stimulates the duodenum to release hormones (**secretin** and **cholecystokinin**) into the blood, which stimulate the pancreas to release pancreatic juice. This is alkaline and (together with bile juice) neutralizes the HCl giving a slightly alkaline pH suitable for pancreatic enzyme action and terminating pepsin activity.

A battery of pancreatic proteases is produced from clusters of cells in the pancreas, and secreted into the intestine as pancreatic juice via the main pancreatic duct. There are three endopeptidases – **trypsin**, **chymotrypsin**, and **elastase** – all entering the small intestine in the form of the inactive proenzymes, trypsinogen, chymotrypsinogen, and proelastase, respectively. Trypsin hydrolyses peptides in such a way as to produce peptides with basic residues (arginine or lysine) at the C-terminal end. Chymotrypsin, has the same specificity as pepsin and produces peptides with C-terminal residues derived from aromatic or long-chain neutral amino acids. Elastase produces peptides with the C-terminal end derived from small neutral amino acid residues.

Two exopeptidases, **carboxypeptidases** A and B, secreted as inactive proenzymes, remove amino acids from the C-terminal end of peptides. Carboxypeptidase A attacks peptides with aromatic C-terminal residues. Carboxypeptidase B attacks peptides with basic C-terminal residues.

$$H_2N{-}\underset{R}{CH}{-}CO{-}NH{-}\underset{R'}{CH}{-}CO{-}NH{\cdots\cdots}NH{-}\underset{R''}{CH}{-}COOH$$

Amino terminal end or N-terminal end

Carboxy terminal end or C-terminal end

The mechanism by which these enzymes work has been given in the section on Mechanism of enzyme catalysis in Chapter 6.

Activation of the pancreatic proenzymes

As with pepsinogen activation, pancreatic proenzymes are activated by proteolytic cleavage of the proenzymes. The activation process is triggered by a specialized enzyme secreted by the cells of the small intestine, an active enzyme (not a proenzyme in this case) called **enteropeptidase**, which hydrolyses a single specific peptide bond of trypsinogen, activating it to trypsin. The initial amount of active trypsin produced in this way, now activates all proenzymes (including trypsinogen itself) so that all are rapidly activated in a proteolytic cascade (Fig. 10.2). The activation details differ for different enzymes. This is an elegant mechanism, which ensures that the array of *active* enzymes is present only in the intestinal lumen. Their premature activation in the pancreas is deleterious – if it occurs, the disease pancreatitis ensues. Blockage of the pancreatic duct, or damage to the gland, can trigger this disease. After synthesis, the proenzymes are stored in the cells producing them in membrane-bound secretory vesicles. On hormonal or neurological stimulation these fuse with the cell membrane and release their contents by exocytosis. The cells contain a **trypsin-inhibitor protein** capable of inactivating any trypsin that might accidentally leak from the vesicles before secretion into the cytosol. The inhibitor protein fits the trypsin active site so perfectly that a sufficient number of weak bonds are formed to make the combination of the two, almost completely irreversible. The mechanism by which enzymes to be secreted are produced and enveloped in the secretory vesicles is more suitably dealt with later (see Chapter 27).

Proteolysis resulting from pancreatic secretions is not the only force at work. Additional enzymes, known as **aminopeptidases (exopeptidases)**, located on the microvilli on the luminal side of intestinal cells, progressively hydrolyse

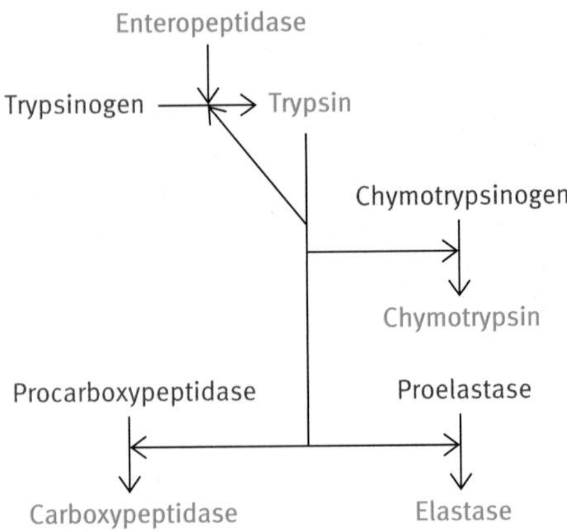

Fig. 10.2 Activation of the pancreatic proteolytic enzymes. Proenzymes (zymogens) are shown in red; activated proteolytic enzymes are shown in green.

N- terminal amino acids from peptides. In this way, with the three endopeptidases (trypsin, chymotrypsin, and elastase) acting in the middle of polypeptides, each with a different specificity for the peptide bonds they attack, and the carboxypeptidases and aminopeptidases acting at each end, proteins are finally converted into free amino acids in the lumen of the intestine or on microvilli on the lumen surface.

Absorption of amino acids into the bloodstream

Amino acids are transported from the intestine across the cell membrane of epithelial cells (the brush border cells; see Fig. 10.6) and into the cell. The brush border cells actively concentrate amino acids inside them from where they diffuse into blood capillaries inside the microvilli. There are different transport pumps in the membrane dealing with different amino acids. The cotransport mechanism is one in which the Na^+ gradient is used to drive the uptake of some of the amino acids. Figure 7.12 shows glucose being transported by a cotransport mechanism but it applies equally well to amino acids using different transport proteins. **Hartnup disease** is an autosomal recessive disorder caused by impaired amino acid transport, which affects absorption from the intestine and reabsorption in kidney tubules. Patients present with skin eruptions, cerebellar ataxia and gross loss of amino acids in the urine. Hartnup disease causes deficiencies in essential amino acids. Absorption of di- and tripeptides is not affected as the former are taken up into the luminal cells by a specific transporter and hydrolysed there by cytosolic proteases, and the latter are hydrolysed on the microvilli.

Moderate amounts of undigested proteins can also be absorbed in some cases. Maternal antibodies (IgA) secreted in milk can be absorbed from the intestine in infants.

We will take up the subject of what happens to the amino acids in the blood later in this chapter.

Digestion of carbohydrates

Structure of carbohydrates

The main carbohydrates in the diet are starch and other polysaccharides, and the disaccharides **sucrose** and **lactose** (the latter from milk). Free **glucose** and **fructose** are relatively minor components of the diet. Glycogen, the storage form of glucose in animal liver and muscles is found in human diets in insignificant amounts, even in meat containing diets as usually animals are not fed before slaughter. The function of digestion is to hydrolyse the dietary carbohydrates into monosaccharides.

In polysaccharides and disaccharides, monosaccharides are linked together by **glycosidic bonds** to form glycosides. This is an important bond from several viewpoints and needs to be described. Glucose has the following structure:

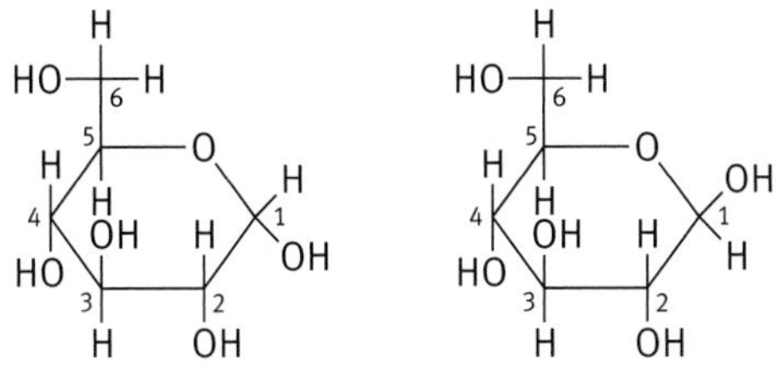

α-D-Glucose β-D-Glucose

In α-D-glucose (in the pyranose six-membered ring configuration), the -OH on carbon atom 1 points below the plane of the ring (imagine it is below the plane of the paper), and in β-D-glucose it points above it. In free monosaccharides, the two are in free equilibrium in solution by **mutarotation** (via an open-chain structure). The mutarotation occurs as shown in Fig. 10.3.

The glycosidic bond

Suppose we have two glucose molecules.

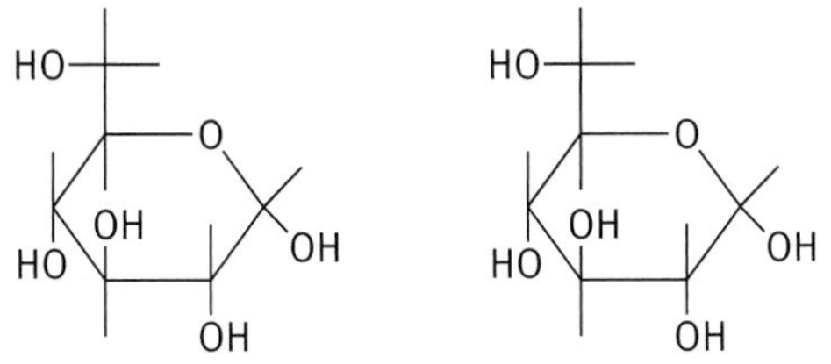

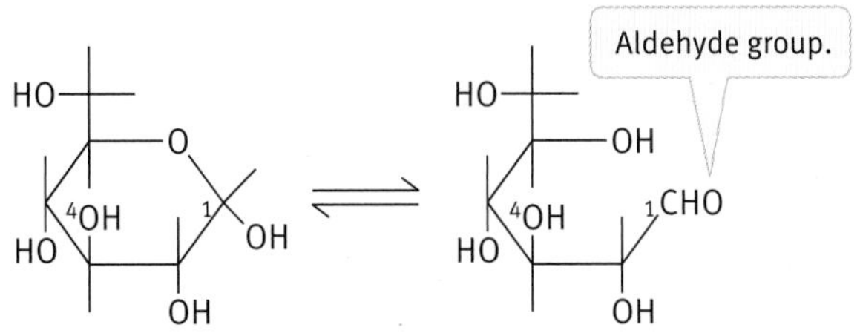

Aldehyde group.

Glucose (ring form) Glucose (open-chain form)

Fig. 10.3 Mutarotation of glucose.

They can be joined together by a glycosidic bond (by removal of the elements of water).

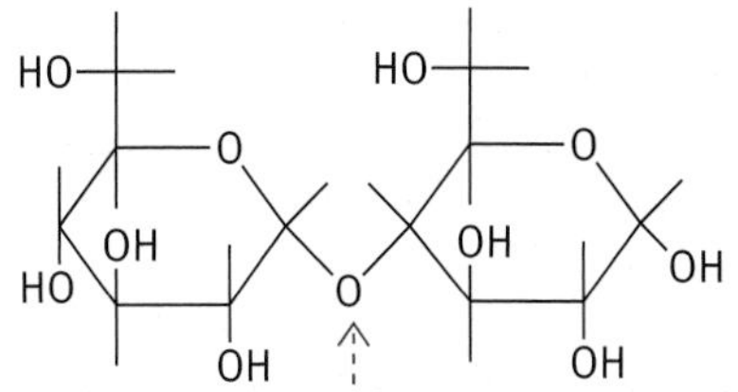

A glycosidic bond (α-configuration)

The glycosidic bond fixes the configuration on carbon atom 1 of the first monosaccharide into one form – it no longer mutarotates. In the example shown, the glycosidic bond is between carbon atoms 1 and 4 of the two units and is in the **α-configuration.** The compound is therefore glucose-(1→4)-α-glucose. It is a disaccharide that is plentiful in malted barley and has the trivial name **maltose.** As will be described shortly, disaccharides with α-glycosidic bonds also exist.

Digestion of starch

From the structure of maltose, it can be seen that glucose units can be joined together indefinitely to form a huge polysaccharide molecule. The amylose component of starch, a molecule hundreds of glucose units (glucosyl units) long, is precisely this. If we represent an α-glucosyl unit as ⌂₀ then amylose is:

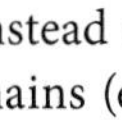

$(\ldots)_n$, etc.

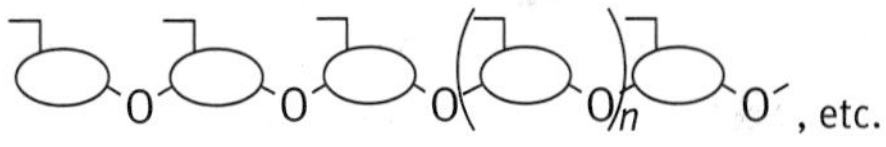

where *n* is a large number.

The second component of starch is **amylopectin.** This molecule is also a huge polymer of glucose, but instead of it being one long chain, it consists of many short chains (each about 30 glucosyl units in length) cross-linked together. The glucosyl units of the chains are (1→4)-α-glycosidic bonds as in amylose but the cross-linking between chains is by (1→6)-α-glycosidic bonds, as shown below. (We omit bonds and groups not relevant to the topic in hand.)

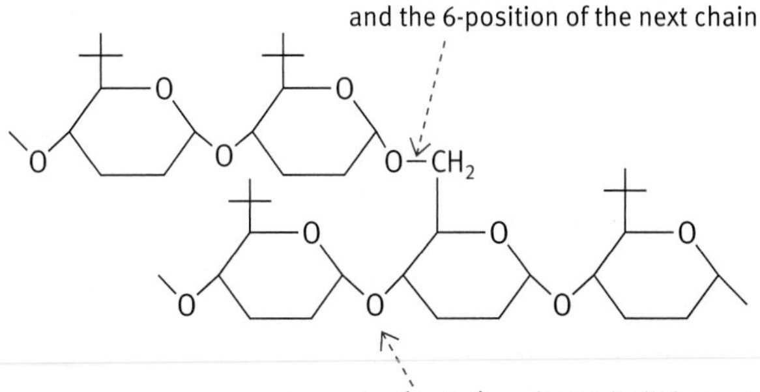

(1→6)-α-glycosidic linkage between end sugar of one chain and the 6-position of the next chain

(1→4)-α-glycosidic linkage

Many chains are linked together in this way. Amylopectin looks like this:

Digestion of starch is initiated by α-**amylase**, which is present in the saliva and pancreatic juice. It is an **endoenzyme**, hydrolysing glycosidic bonds anywhere inside the molecule except in amylopectin, where it cannot attack a bond near the (1→6) linkage. The amylase thus trims off the molecule but leaves cores containing the (1→6) bonds plus the nearby glucosyl units. This limit dextrin, as it is called (the limit of hydrolysis), is hydrolysed by an intestinal enzyme (an amylo(1→6)-α-glucosidase) that hydrolyses the (1→6) linkages. Salivary amylase has only a brief time to attack starch while food is in the mouth, and stomach acid destroys the enzyme. Most of the starch digestion, therefore, occurs in the small intestine as does the digestion of the other two main dietary carbohydrates, sucrose and lactose. The end products of α-amylase digestion are the oligosaccharides maltose, maltotriose, and α-dextrins, which are polymers of about eight glucose units containing α(1→6) linkages. The enzyme responsible for further digestion is located on the microvilli of the brush border of the small intestine and is known as α-dextrinase (isomaltase) and is primarily responsible for hydrolysing α(1→6) linkages. Together with sucrose and maltase (also present on the microvilli), they break down maltotriose and maltose.

Digestion of sucrose

Sucrose is a dimer of glucose and fructose. It is hydrolysed in the small intestine by the enzyme, **sucrase**, as shown in Fig. 10.4. Large amounts of sucrose can be taken in the diet. The absorption of large amounts of fructose presents some special metabolic problems discussed on page 308.

Sucrase and isomaltase are synthesized as a single glycoprotein which is hydrolysed by pancreatic proteases into sucrase and isomaltase. Deficiency of sucrase-isomaltase may cause diarrhoea, bloating, and flatulence after ingestion of sugar. Diarrhoea is due to the presence of osmotically active oligosaccharides in the intestinal lumen, which increase the volume of the intestinal contents. Bloating and flatulence result from the carbon dioxide and hydrogen gas produced by bacterial fermentation in the colon.

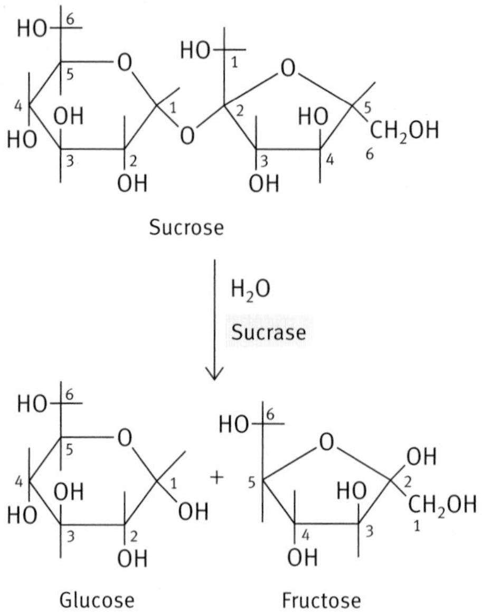

Sucrose

H_2O

Sucrase

Glucose + Fructose

Fig. 10.4 Sucrose conversion into glucose and fructose by sucrase.

Digestion of lactose

The other major disaccharide of food is lactose. Lactose is galactose-(1→4)-β-glucose, the principal sugar in milk. Galactose has the carbon 4-OH of glucose inverted.

The curved bond shown in Fig. 10.5 is used to simplify the presentation of the structure; it avoids having to invert the second glucosyl unit. An intestinal enzyme attached to the external membrane of the epithelial cells, **lactase**, hydrolyses lactose to the free monosaccharides. Because lactose is a β-galactoside, lactase is also known as **β-galactosidase**. Most people, to varying degrees, lose the ability to produce lactase as they leave childhood, and non-Europeans particularly may develop **lactose intolerance**. They are unable to hydrolyse the sugar and, since the disaccharide is not absorbed, it passes to the large intestine where it is fermented by bacteria giving rise to severe discomfort (due to gas production), and diarrhoea. In fact, most mammals lose the ability to produce lactase already when they are young, but some human populations show **lactase persistence**. 75% of the world population are lactose intolerant with figures of 5% in Northern Europe and 90% in Asia and Africa.

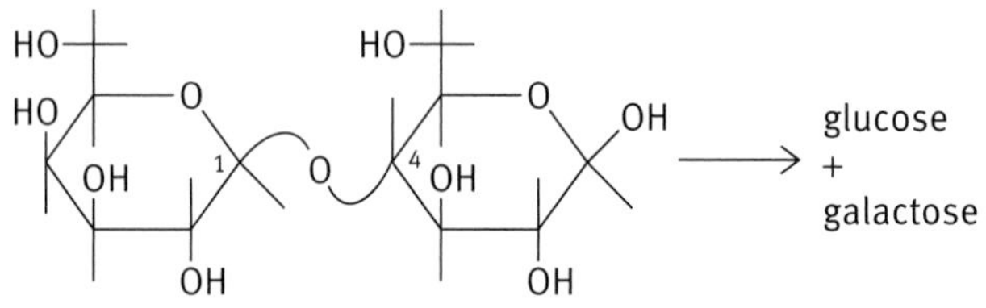

→ glucose + galactose

Fig. 10.5 Lactose conversion into glucose and galactose by lactase.

If products containing lactose are avoided, the disease symptoms disappear. Yoghurt, which has less lactose than milk, may be tolerated and cheese, having little lactose, usually presents no problem. Modern food science, however has introduced lactose into all sorts of foodstuffs, such as soups and yoghurt, and lactose intolerant people have to watch out for lactose added to manufactured foods.

Absorption of monosaccharides

Absorption of glucose into epithelial cells occurs by the Na^+ cotransport mechanism, as shown in Fig. 10.6, which uses the S-GLUT (sodium-glucose transporter). The Na^+ is continually pumped to the outside by the Na^+/K^+ ATPase (see Chapter 7) so this maintains the Na^+ concentration difference and drives the cotransport. The glucose has to exit the cells on the side opposite that of the lumen so as to enter the blood capillaries for transport to the liver by the hepatic portal vein. The glucose transporter for this movement is the facilitated-diffusion type. A full list of glucose transporters is shown in Chapter 11 (control of glucose transport into cells); the monosaccharide moves down the cell/blood concentration gradient created by active uptake from the intestine (Fig. 10.6). Fructose is absorbed from the gut by a Na^+-independent passive-transport system.

Amino acids and sugars, and other nonlipid molecules, absorbed from the intestine are collected by the portal blood system, which delivers digestion products directly to the liver. This arrangement means that these digestion products are transported to the liver before being released into the general circulation – an advantageous arrangement since the liver is responsible for removing most toxic foreign compounds entering the body via the intestine, and for processing many of the absorbed nutrients.

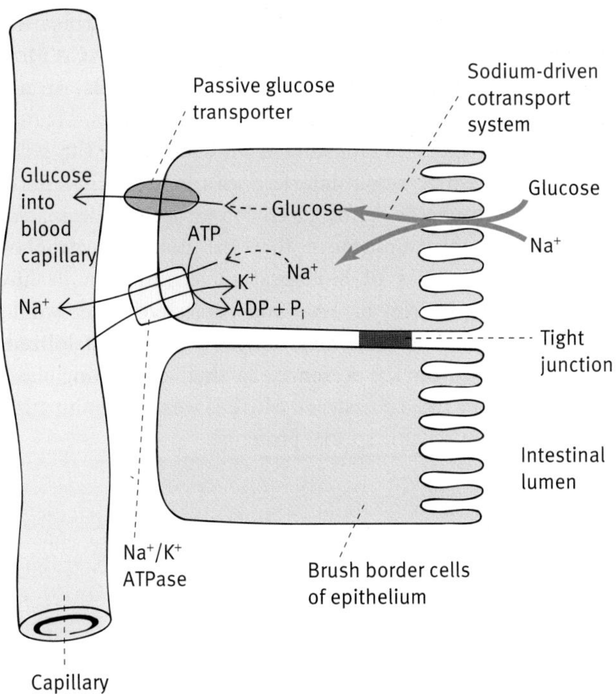

Fig. 10.6 Absorption of glucose from the intestinal lumen by cotransport with Na^+.

Digestion and absorption of fat

In Chapter 7, we described what fat is – TAG or neutral fat. Since there are no polar groups, liquid fat in water forms insoluble droplets with minimum contact between the lipid and water, and cannot be absorbed as such.

The main digestion of fat occurs in the small intestine by the action of the pancreatic enzyme, **lipase**, which mainly attacks *primary* ester bonds (red arrows in Fig. 10.7); the middle ester bond is a secondary ester bond and is not significantly attacked. Lipase is secreted as a proenzyme, and in the small intestine the first step in activation is by trypsin which hydrolyses off a specific small peptide. But then, another pancreatic protein, **colipase**, is also needed to activate lipase; it binds to the enzyme in a 1:1 ratio.

Simple as this digestion reaction is, there are problems. Fat is physically unwieldy in an aqueous medium, and as lipase can attack only at the oil/water interface, the area of this in a simple fat/water mixture is insufficient for the required rate of digestion. The available surface area is increased by emulsifying the fat. The monoacylglycerol and free fatty acids produced by lipase action, together with bile salts acting as biological detergents, help to emulsify the oily liquid into droplets.

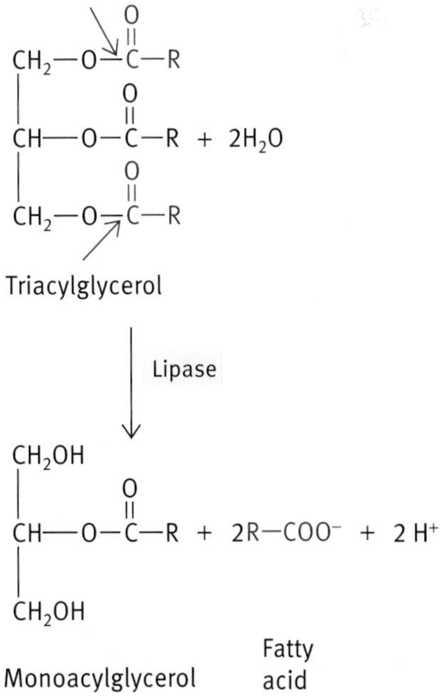

$$\begin{array}{l} CH_2{-}O{-}\overset{O}{\overset{\|}{C}}{-}R \\ | \\ CH{-}O{-}\overset{O}{\overset{\|}{C}}{-}R \;+\; 2H_2O \\ | \\ CH_2{-}O{-}\overset{O}{\overset{\|}{C}}{-}R \end{array}$$

Triacylglycerol

$\downarrow$ Lipase

$$\begin{array}{l} CH_2OH \\ | \\ CH{-}O{-}\overset{O}{\overset{\|}{C}}{-}R \;+\; 2R{-}COO^- \;+\; 2H^+ \\ | \\ CH_2OH \end{array}$$

Monoacylglycerol   Fatty acid

Fig. 10.7 Digestion of triacylglycerol to monoacylglycerol and fatty acid by lipase.

Bile acids (strictly speaking, bile salts) are synthesized in the liver and stored in the gall bladder until discharged into the duodenum. They are produced from cholesterol, whose structure they resemble (Fig. 10.8). The main change is that the hydrophobic side chain of the cholesterol molecule is converted into a carboxyl group, and -OH groups may be added.

The main bile acid is cholic acid but others varying in the number and position of hydroxyl groups also exist. The cholic acid mostly has either glycine or the sulphonic acid, taurine attached to it. Glycine is $NH_3^+CH_2COO^-$, while taurine is $NH_3^+CH_2SO_3^-$ (it does *not* occur in proteins). If cholic acid is represented as $RCOO^-$ then glycocholic acid is $RCONHCH_2COO^-$ and taurocholic acid $RCONHCH_2SO_3^-$. These conjugated acids have lower pK_a values (approximately 3.7 and 1.5, respectively) than the parent cholic acid (approximately 5.0).Conjugation ensures full ionization in the intestinal contents and makes them better detergents – the ionized forms are called bile salts.

The products of lipase action are still relatively insoluble in water but they must be moved from the emulsion to the cells lining the intestine to be absorbed. This movement is facilitated by bile salts.

The monoacylglycerols and fatty acids, in the presence of bile salts, form **mixed micelles**. These are disc-like particles in which bile salts are arranged around the edge of the disc surrounding a more hydrophobic core, which contains the digestion products of fat, together with cholesterol and phospholipids. The particles are smaller than emulsion droplets, giving a clear suspension. They carry higher concentrations of lipid digestion products than is possible in true solution. In this form, the lipid digestion products diffuse to enter the epithelial cells; probably the micelle breaks down at the cell surface and the free lipids diffuse in. Bile salts are also partially reabsorbed and transported back to the liver.

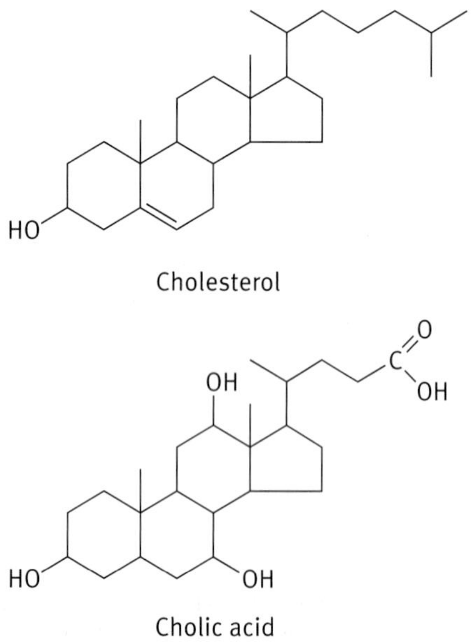

HO
Cholesterol
OH
C
O
OH
HO
OH
Cholic acid

Fig. 10.8 Structures of cholesterol and a bile acid, cholic acid.

Resynthesis of TAG in intestinal cells

The products of digestion of fat absorbed into the intestinal cells are resynthesized into fat, the fatty acids being re-esterified producing TAGs (Fig. 10.9).

The mechanism of this resynthesis is not given here because it would divert too much from the topic in hand – it is dealt with later. The TAG, together with absorbed cholesterol in esterified form (see Fig. 10.11), must now be transported from the intestinal cell to the tissues of the body. TAG cannot diffuse out of the cell through membranes and, in any case, it would be insoluble in the blood – it cannot simply be ejected into the circulation.

The solution is that TAG and cholesterol are arranged inside the epithelial cells into fine particles, called **chylomicrons**. They are released from the cell by exocytosis (see Fig. 7.5(c)), since they are too large to traverse the membrane in any other way.

Chylomicrons

Chylomicrons (Fig. 10.10) are spherical particles, in the middle of which are hydrophobic molecules – TAG and cholesterol esters. Cholesterol has a hydrophilic —OH group and a hydrophobic part. To form chylomicrons, cholesterol is converted into **cholesterol ester** by attachment of a fatty acid to its —OH group, as shown in the simplified diagram of Fig. 10.11. The enzyme which catalyses the esterification is called **acyl-CoA:cholesterol acyltransferase (ACAT)**. It requires coenzyme A (see Chapter 12, but details are not needed here).

In making a cholesterol ester from cholesterol, the polar —OH group, which would interfere with packing cholesterol into the hydrophobic centre of chylomicrons, is eliminated. It illustrates the importance that the polarity/hydrophobicity characteristics of biological molecules play in life. Hydrophobic particles of TAG and cholesterol ester would coalesce into an insoluble mass unless they were stabilized. Stability comes from the presence of a 'shell' containing phospholipids, some free cholesterol, which is weakly amphipathic and, importantly, some special proteins.

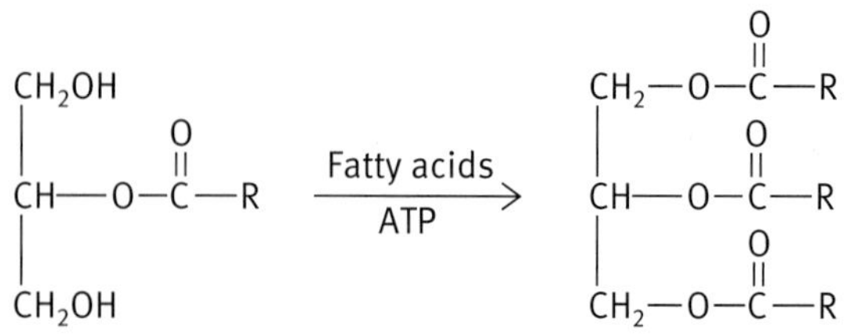

$$\begin{array}{l} CH_2OH \\ | \\ CH-O-\overset{O}{\overset{\|}{C}}-R \\ | \\ CH_2OH \end{array} \xrightarrow[\text{ATP}]{\text{Fatty acids}} \begin{array}{l} CH_2-O-\overset{O}{\overset{\|}{C}}-R \\ | \\ CH-O-\overset{O}{\overset{\|}{C}}-R \\ | \\ CH_2-O-\overset{O}{\overset{\|}{C}}-R \end{array}$$

Fig. 10.9 Summary of resynthesis of fat.

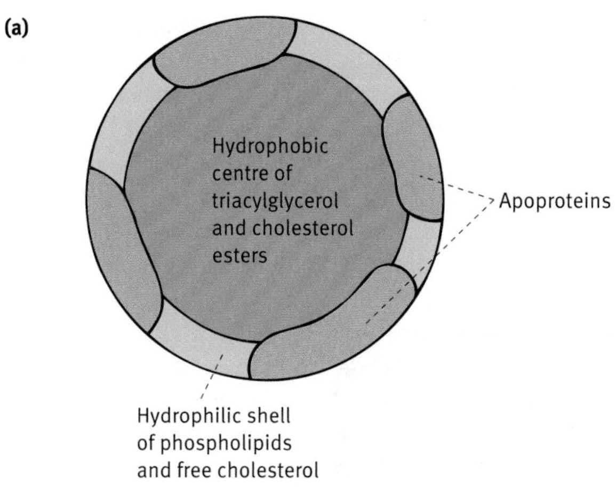

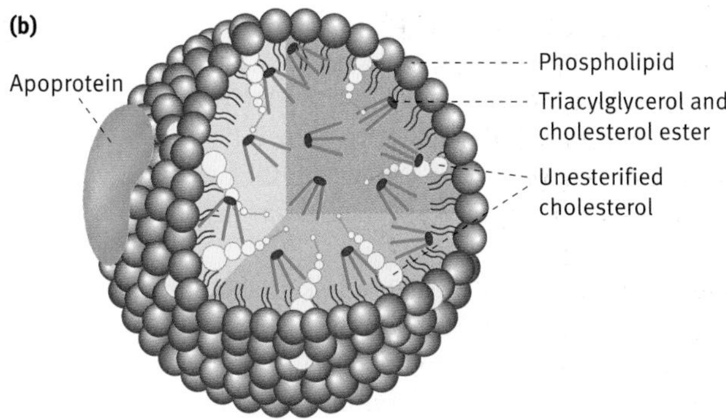

Fig. 10.10 **(a)** Cross-section of a chylomicron. **(b)** 3D structure of a chylomicron.

There are several proteins involved, a major one being **apolipoprotein B (apoB-48)**, necessary for chylomicron synthesis. ApoB is a glycoprotein, the carbohydrate attachment providing a highly polar group. The prefix *apo* means 'detached' or 'separate'. Thus an apolipoprotein is a protein normally found in lipoproteins but now is detached or separate. Apoproteins do not function in isolation, but as part of a protein-lipid complex. The whole chylomicron structure is called a **lipoprotein**. The stabilization by the hydrophilic shell allows it to remain as a suspended particle in the lymph chyle and blood. The overall composition of chylomicrons is about 90% or more TAG, giving them a low density.

The chylomicrons are not released directly into the blood (unlike absorbed amino acids and monosaccharides), but into the lymph vessels. The suspension of chylomicrons in **lymph** fluid is called chyle. A few words on lymph might be useful here. As blood circulates through the capillaries, a clear lymph fluid containing protein, electrolytes, and other solutes, filters out and bathes cells in an interstitial fluid. All tissues have a fine network of lymph capillaries closed at the fine ends into which lymph drains. The lymph capillaries join into lymphatic ducts and discharge the lymph back into the major veins of the neck via the thoracic duct. It is not actively pumped but movements of the body propel it along. After a fat containing meal, chylomicrons, entering the bloodstream

HO
Cholesterol
Fatty acid
ATP energy
R—C(=O)—O
Cholesterol ester

Fig. 10.11 Simplified diagram of the esterification of cholesterol.

via lymph, give the blood a milky appearance. They circulate in the blood and their contents are used by tissues. The traffic of fat in the body in the fed and fasting state is dealt with in Chapter 11.

Digestion of other components of food

There are components in food other than the ones dealt with so far and digestive enzymes exist, which hydrolyse them into their constituent parts. Thus phospholipids in food, which come from cell membranes of ingested plant or animal material are hydrolysed by phospholipases; nucleic acids are hydrolysed by ribonuclease (RNase) and deoxyribonuclease (DNase). These enzymes are in the pancreatic juice. Plant material contains fibre – carbohydrate molecules such as cellulose, which in humans constitute the material that is not hydrolysed but is important in the diet. Its components are cellulose, lignin, hemicelluloses, pectin, and gum. Fibre has several important beneficial effects. It provides bulk and speeds up movement through the intestine, and may provide protection against carcinogens by speeding up the rate of their elimination, and also by binding some of them. Fibre can lower blood cholesterol levels in some cases by binding bile acids, which are cholesterol derivatives, and reducing their reabsorption from the intestine, so that more cholesterol is taken up from the blood by the liver for synthesis of bile acids. Bacterial fermentation metabolizes part of the fibre in the large intestine.

Herbivores, for which cellulose is the main food component, have microorganisms in the rumen digestive tract, which produce cellulases that hydrolyse the fibre. The microorganisms convert the liberated carbohydrate into short fatty acids such as acetate and propionate, which are the main source of energy for the animals.

We now have the various products of digestion just having reached the blood – fat and cholesterol go into the general circulation via the lymph system as chylomicrons, and everything else is headed for the liver in the portal vein and thence into the general circulation.

Storage of food components in the body

Animals take in food at periodic intervals; they do not eat continuously. The body is, in effect, continually exposed to cyclical fed and fasting conditions. The periods between meals can for humans vary from short intervals during the day to longer periods during sleep and very long periods during fasting and starvation. The biochemical machinery in the body has to cope with these situations.

After a meal, the blood is loaded with digestion products absorbed from the intestine. This material is not held in the blood until it is all used up by metabolism, but rather it is rapidly cleared from the blood by uptake into the tissues so that blood levels quickly return to normal. After a fatty meal, the lipaemia (presence of milky blood plasma) is cleared within a few hours. Similarly, blood glucose may increase after a meal from a fasting concentration of about 5 mM (90 mg dL^{-1}) up to 10 mM (180 mg dL^{-1}), but it reverts back to the fasting concentration in a nondiabetic person, within 2 hours. In fact, the return of blood glucose concentration back to normal fasting concentrations within two hours after a meal, is one of the criteria of normality. The tissues are not using all of this food at once; most of it is stored. There is an advantage in employing the mechanisms of transport and storage. If glucose, fat, and cholesterol were allowed to circulate until metabolized, there would be dire results for the organism as high blood concentrations of these metabolites are associated with diabetes, and its complications, and cardiovascular disease.

How are the different food components stored in cells?

Glucose storage as glycogen

It would not be practicable for cells to store glucose as the free monosaccharide – the osmotic pressure of the glucose at high concentrations would be excessive. The osmotic pressure of a solution is proportional to the number of particles of solute in the solution. If, therefore, large numbers of glucose molecules are joined together to form a single macromolecule, the osmotic pressure, exerted by the store of glucose residues, is accordingly reduced. The resultant polymerized molecule may fall out of solution as a granule. In animals, glucose is polymerized into a highly branched so-called 'animal starch', or glycogen, with the same chemical bondings as in amylopectin (as shown previously in this chapter), but more highly branched. When needed, glycogen is degraded again. A very important fact is that *glycogen storage in animals is limited.* In humans, the glycogen reserves of liver, which are used to supply glucose to the blood for utilization by other tissues, especially the brain, are exhausted after about 24 hours without food. This has a profound effect on the biochemistry of an animal, as will become apparent later in the chapter. Muscle also has large glycogen reserves: these are used to provide energy for ATP production, needed in muscle contraction, but it only serves the muscle – free glucose cannot be produced from glycogen in muscle and so, is not released from it to the rest of the body, as happens in the liver. Other tissues do not store glycogen to any significant degree (kidney is a minor exception).

In all of the above, we have talked about glucose. What about other monosaccharides, such as galactose and fructose, present in lactose and sucrose respectively? Glucose is the monosaccharide of central metabolic importance; other monosaccharides are converted into glucose (or glycogen), or else to compounds on the main glucose-metabolizing pathways.

Storage of fat in the body

Fat is stored as triacylglycerols (TAG) by cells. The bulk of it is stored in the fat cells of adipose tissue (**adipocytes**), which is distributed in many areas of the body. A loaded fat cell under the microscope looks like droplets of oil surrounded by a thin layer of cytosol and membrane as depicted in Fig. 10.12.

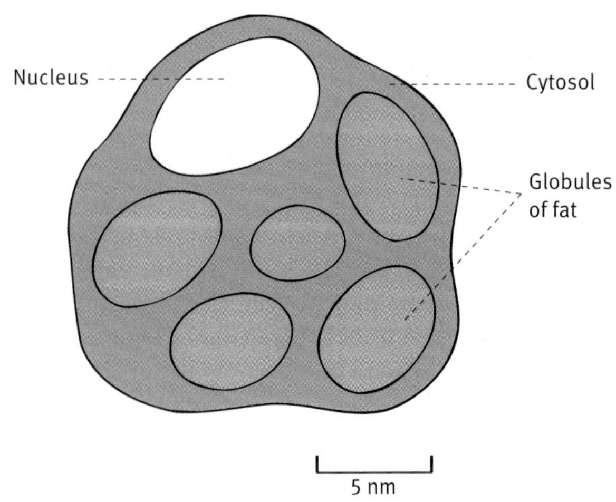

Fig. 10.12 A fat cell (adipocyte).

Unlike glycogen storage, that of fat is essentially unlimited and large reserves of energy are stored in the body as triacylglycerol in the adipose cells. In an adult of normal weight, the reserves of fat, in terms of stored energy, are much higher than those of the glycogen reserves, and can amount to 15 kg or more in total.

Since this limited glycogen storage causes the metabolic problems described later, it may be asked why the body stores so much fat, and so little glucose, when the latter is so essential to life in animals? Fat is more highly reduced (less oxidized) than carbohydrate is and hence contains more energy per unit mass, and, additionally, glycogen in the cell is hydrated while fat is not. This means that the latter occupies much less weight and volume per unit of stored potential energy than glycogen does. If the energy equivalent of fat stored in our bodies was in the form of glycogen we would need to be much larger, perhaps twice the size, than we are (migrating birds might then be too heavy to take off). Since glucose storage is so limited, and starch and sucrose intake in the diet potentially almost unlimited in humans of modern societies, it follows that glucose in excess of that used for glycogen synthesis must be stored in another way – in fact as fat. The position then is as shown in Fig. 10.13.

An important fact is that, while glucose is readily converted into fat, in the human body there is no net conversion of fatty acids into glucose. We will see why that is so in Chapters 11 and 20 (The glycerol moiety of TAG *can* be converted into glucose but this represents a small fraction of the energy available in the three long-chain fatty acids.)

Are amino acids stored by the body?

Digestion of the third main class of food, protein, results in amino acids being absorbed and taken into the blood. There is no dedicated storage form of amino acids in animals whereas plants have storage proteins in their seeds, whose only function appears to be the provision of amino acids in a convenient form to supply the developing embryo.

In animals, dietary amino acids in the blood are taken up by tissues as needed for the synthesis of cellular proteins, neurotransmitters, and other nitrogenous components, and then those in excess of immediate needs are degraded, the amino group ($-NH_2$) being removed and, in mammals, converted into urea and excreted in the urine.

In one limited sense, there is a storage of amino acids in the body – in the proteins of all cells. Muscle proteins are quantitatively of greatest importance. However, they are all functional proteins mainly involved in contraction and are not dedicated to storage. When amino acids must be made available to the body in general, these proteins are degraded, and the result is muscle wasting.

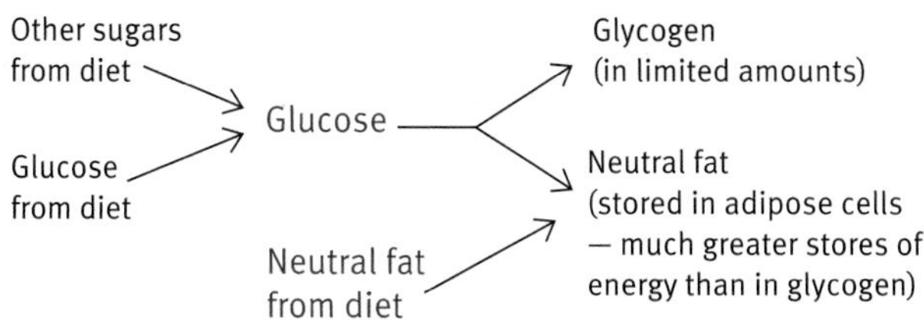

Fig. 10.13 Stroage of sugars and fat from the diet in the postabsorptive period.

Although there is no specific storage protein in animals, there is what is known as the 'free amino acid pool'. This is a collective term referring to the sum total of free amino acids present in the circulation and inside cells. The free amino acid pool supplies the appropriate amino acids for protein synthesis after a meal and can maintain tissue protein stores, or provides amino acids for oxidation when protein is not ingested. The size of the pool is limited, about 100 g in normal adults. During the postprandial period, net whole body protein synthesis takes place. Both the free amino acid pool size and amino acid oxidation rates also increase. Amino acids are consequently used as substrates to provide energy.

Urea is highly water soluble, neutral, and nontoxic. This leaves the carbon-hydrogen 'skeletons' of the amino acids, which contain chemical energy. These are converted into glycogen or fat, depending on the particular amino acid, or they can be oxidized to release energy or used to provide other metabolites (Fig. 10.14), according to the physiological needs at the time.

Characteristics of different tissues in terms of energy metabolism

The different tissues in the body have special biochemical characteristics. However, in our present context of looking at overall food traffic in the body, several organs are of overriding importance. These are the liver, the skeletal muscles, the brain, the adipose tissue (fat) cells, and the red blood cells. (We are excluding regulatory organs such as the pancreas and adrenal glands from the list though their roles are crucial.)

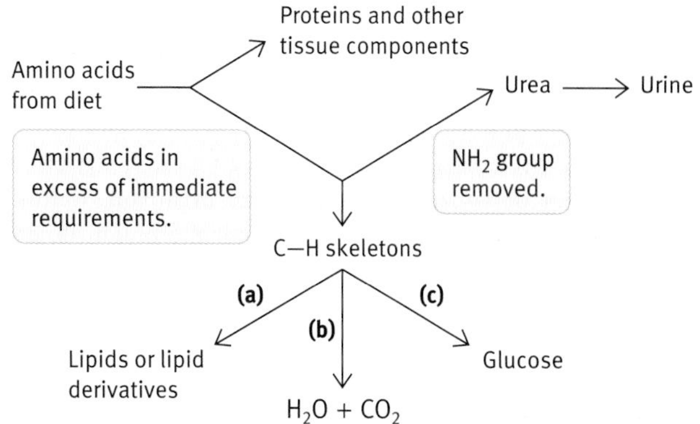

Fig. 10.14 Summary of the fate of amino acids in the diet. Which of the metabolic routes **(a)**, **(b)**, or **(c)** is followed depends on the particular amino acid, the physiological state, and biochemical control mechanisms.

- The liver has a central role in maintaining the blood glucose concentration; it is the glucostat of the body. When blood glucose is high, such as after a meal, the liver takes it up and stores it as glycogen. When blood glucose is low, it degrades the glycogen and releases glucose into the blood.

After about 24 hours of fasting , the reserves of glycogen in the liver are exhausted; the concentration of glucose in the blood would decline to lethal levels if there were no mechanism to release glucose into the bloodstream, for the brain cannot function without an adequate supply. During fasting or prolonged starvation, reserves of fat are released into the blood from the fat cells of adipose tissue. This will keep the muscles and other tissues supplied with fuel, but not the brain and red blood cells, which cannot use fatty acids. The liver can, however, convert amino acids into glucose, a process called **gluconeogenesis** (see Chapter 15), release the glucose into the bloodstream, and thus provide brain and red blood cells with the fuel they can metabolize. The main sources of amino acids in this situation are muscle proteins, which are broken down to provide them. This means destruction of functional muscle proteins, causing muscle wasting, but this is, of course, preferable to death resulting from low blood glucose.

The liver also plays an important role in fat metabolism. In starvation, the fat cells release fatty acids into the blood that can be directly used by most tissues – brain and red blood cells excepted, as stated. Brain and red cells can metabolize glucose but not fatty acids. The blood –brain barrier prevents fatty acids entering the brain cells, at least in any significant quantities to allow them to function as fuel, and the red cells have no mitochondria, which are needed to metabolize fatty acids. However, in such conditions, where massive fat utilization is occurring, the liver converts some of the fatty acids in the blood to small compounds called **ketone bodies** and releases them into the blood for use by other tissues (red blood cells again excepted, as they have no mitochondria, which are necessary to metabolize ketone bodies). Of special importance, the brain can utilize ketone bodies which, during starvation, can supply up to two thirds of that organ's energy needs with only one third coming from glucose. This is of profound importance as it spares muscle protein, which would otherwise be degraded at much higher rates to provide glucose, with detrimental effects to the organism. Ketone bodies are small and can diffuse through the blood –brain barrier whereas fatty acids cannot. The rest of the energy has to be derived from glucose. The liver itself cannot metabolize ketone bodies as it lacks the necessary enzymes. This ensures that they are released into the blood stream and used by other tissues. The liver is also the major site of fatty acid synthesis from glucose and other foods that are in excess of those required to replenish glycogen stores. The fat synthesized by the liver is not stored in the liver, but is exported to adipose cells and other tissues as TAG in the form of very-low-density lipoproteins (VLDLs described in Chapter 11). Accumulation of fat in liver is pathological.

So, energy-wise, the liver stores excess glucose as glycogen until the stores have been filled and releases glucose when needed, as long as there are glycogen reserves left. In starvation, it produces glucose mainly from amino acids released by muscle protein breakdown and ketone bodies from fat, which are turned out into the blood for use by the brain and, with respect to glucose, by red blood cells (Fig. 10.15). When food intake is plentiful, it synthesizes fatty acids from glucose and other dietary components and exports them. It preferentially oxidizes fatty acids for energy supplies. The liver has many other functions but for now we will concentrate on energy supplies in the body.

It might be useful, at this point, to summarize the fairly complex metabolic characteristics of tissues.

- The brain must, as stated, have a continuous supply of glucose. It has no significant fuel reserves. If the person becomes hypoglycaemic (low blood glucose concentration), the rate of glucose entry into the brain is reduced since it involves facilitated but not active transport, and brain and nerve cell function is impaired. Convulsions and coma result. However, as already emphasized, in starvation the brain can adapt to use ketone bodies produced by the liver from fat for up to two thirds of its energy needs. This helps to economize on scarce glucose.

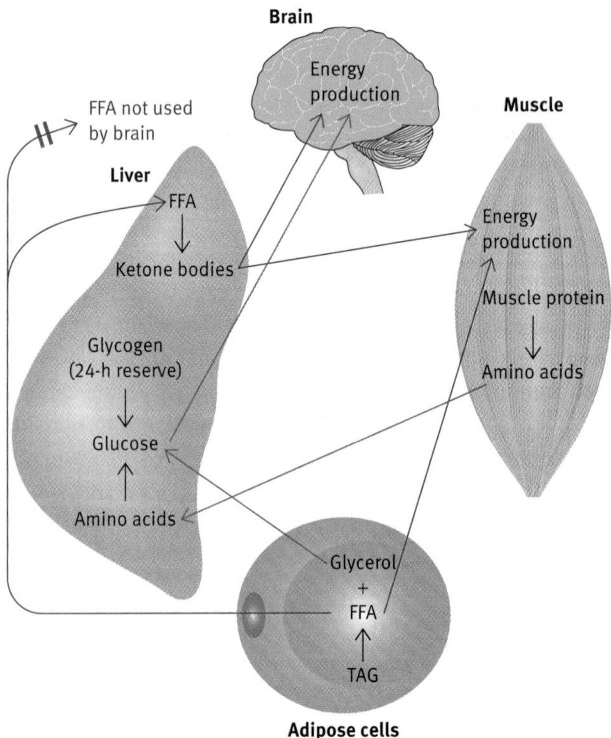

Fig. 10.15 Main fuel movements in the body during starvation. TAG, triacylglycerols; FFA, free fatty acids.

Both glucose and ketone bodies are converted by the brain into the same compound, acetyl CoA which enters the TCA cycle for further oxidation.

- Skeletal muscles require large-scale ATP production for muscle contraction. They can utilize most types of energy source. They take up glucose from the blood and store it as glycogen but, *in contrast to the liver, they do not release any glucose back into the blood.* They can oxidize fatty acids, ketone bodies, and amino acids, as well as glucose. In the presence of high concentrations of free fatty acids and ketone bodies in the blood, these are preferentially oxidized. (In starvation, muscles degrade their protein and supply the liver with amino acids for glucose synthesis.) In vigorous contraction when the energy needs may outstrip the oxygen supply, muscle can metabolize glucose (or glycogen) anaerobically resulting in lactate accumulation. This is very inefficient in ATP generation, but in an emergency it can occur on a large scale and may make the difference between survival or not.
- The role of the fat cells of adipose tissue can be stated simply. After a meal, they take up fatty acids from lipoproteins supplied in the diet or synthesized by the liver. These are stored as TAG. Glucose is needed to supply the glycerol (phosphate) moiety. As soon as the blood glucose level falls, as in fasting (or in starvation), they release reserves of fat into the blood as fatty acids.
- Red blood cells can utilize only glucose, producing lactate, which is released from the cell. They are terminally differentiated cells (that is, fully formed and do not divide) without mitochondria, as stated before, and cannot oxidize foodstuffs as can other cells. But their Na^+/K^+ ATPase pumps must run and they have other energy-requiring processes.

Overall control of fuel distribution in the body by hormones

With that general description of the major organs involved in food logistics in the body, let us now see in overview how food distribution is directed – how the various organs are coordinated in their activities to cope with different physiological situations in humans. The main controls are the hormones **insulin**, **glucagon**, and **adrenaline (epinephrine)** as will be described below. These situations in humans are as follows:

- excess food available – the fed state, or postprandial condition, or absorptive state after a meal when anabolism exceeds catabolism
- the postabsorptive state – this refers to the condition a few hours after a meal. There are no precise timings for these descriptive terms
- fasting – no food for over 12 hours or so, for example, before breakfast after the night's fasting
- starvation – fasting (or nonvoluntary deprivation of food) that goes on for longer than about 1 or 2 days
- emergency responses – this relates to conditions, for example, where vigorous muscular action is needed to avoid a threat.

The overall picture of food traffic in the body in these nutritional situations is summarized below.

Postprandial condition

In the absorptive stage, dietary components are at a high concentration in blood. Glucose is taken up by the liver and muscle and used to replenish glycogen stores. Beyond this, excess glucose is taken up mainly by the liver and converted into TAG. The healthy liver does not retain fats in large amounts, as the adipose cells do, but transfers it to other tissues *via* very-low-density lipoprotein (VLDL) (as discussed earlier in this chapter).

Amino acids are taken up by all tissues and used for the synthesis of protein or other components. Any in excess of immediate needs are converted into fat or glycogen and the nitrogen excreted in the form of urea.

Cells take up fatty acids from the TAG in chylomicrons as needed (see Chapter 11 for the mechanism). Cells of lactating mammary glands are active in this to supply milk fats, and adipose cells take them and convert them into stored TAG.

It is clear that with all of this uptake and release of foodstuffs going on there has to be a system to control it all, otherwise there would be metabolic chaos. This is achieved by hormones produced by endocrine glands, which release their chemical signals into the blood where they reach all cells and instruct target cells (those designed to receive the signals) on what they should be doing in terms of food logistics.

The main 'signal' for storage of food components to take place is the pancreatic hormone, **insulin**, whose release from the pancreas into the blood occurs in response to high blood glucose concentrations. The high insulin level is the signal to the tissues that food has entered the system and it can be stored both as glycogen and fat. Correlated with this is the low level of glucagon, another pancreatic hormone, whose release is *inhibited* by high blood glucose levels. Glucagon has the opposite effect to insulin. It signals that the tissues are short of fuel and that storage organs should release some into the blood. Note that the brain and red blood cells are not affected by insulin; they go on using glucose from the blood all the time.

Fasting condition

The insulin-stimulated storage activity in the postabsorptive period lowers the blood glucose and amino acids to normal fasting levels and clears the chylomicrons. With time, the blood glucose level begins to fall and, with it, insulin secretion. The hormone is degraded after release, so its blood level falls in a few minutes when secretion stops. In concert with these events, the pancreas releases glucagon, whose level rises in response to low blood glucose. This stimulates the liver to break down glycogen and release glucose into the blood. Glucagon also stimulates adipose tissue to hydrolyse triacylglycerol and release free fatty acids and glycerol into the blood. Muscles and liver, and other tissues, use the fatty acids to provide energy and this conserves glucose for the brain. Glycogen synthesis and fat synthesis from glucose cease due to the low insulin/high glucagon ratio. This avoids synthesizing and degrading them at the same time (futile cycling).

Prolonged fasting and starvation

After about 24 hours, the liver has no glycogen left to provide blood glucose and no other tissue, other than the quantitatively unimportant kidney, is capable of releasing glucose. The insulin level is low and that of glucagon high. In this situation the adipose cells release free fatty acids into the blood, for use by muscle and other tissues; the fat reserves in adipose cells are sufficient perhaps for weeks. Muscles break down their proteins to amino acids. The liver uses them to synthesize glucose, a process activated by glucagon. (Gluconeogenesis, as the process is called, also occurs in other circumstances; see Chapters 16 and 20.) In times of stress, the steroid **glucocorticord hormones** are released from the adrenal cortex: the main one in humans is **cortisol**. It stimulates gluconeogenesis and promotes muscle protein breakdown.

As starvation proceeds, the high glucagon causes the adipose tissue to release more and more free fatty acids into the blood, and the liver metabolizes some of them into ketone bodies, which it releases into the blood. There are two components, acetoacetate and β-hydroxybutyrate.

$CH_3COCH_2COO^-$ Acetoacetate
$CH_3CHOHCH_2COO^-$ β-Hydroxybutyrate

The name 'ketone bodies' is a misnomer; they are not 'bodies', nor is β-hydroxybutyrate a ketone (it has no C=O group), but the term is an old one and still used. As mentioned, the brain in starvation adapts to obtain part of its energy from them, which economizes on glucose.

The emergency situation – fight or flight

When an animal (such as a human) is presented with a dangerous situation, the adrenal glands release **adrenaline (epinephrine)** from the adrenal medulla into the blood in response to a neurological signal from the brain. Adipose tissue is innervated and can also be stimulated by adrenaline and noradrenaline (norepinephrine) released from nerve endings. Adrenaline, as it were, presses the biochemical panic button. It overrides normal control and stimulates the liver to release glucose into the blood and the adipose cells to release free fatty acids so that muscles have no shortage of fuel. It also stimulates glycogen breakdown in skeletal muscle enabling the cells to produce ATP, at the maximal rate. This pattern of metabolism ensures that muscles can react maximally and instantly to escape the threat.

Summary

Digestion of food

In digestion, the polysaccharides and proteins of food are hydrolysed into their monomer subunits (simple sugars and amino acids). This is required for them to be absorbed by the intestinal epithelial cells into the bloodstream. TAGs (triacylglycerols or fats) are hydrolysed into fatty acids and monoacylglycerol, the process being aided by bile salts, which emulsify the fats to give a large surface area for the enzyme lipase, secreted by the pancreas, to attack.

The body has to guard against self-digestion by the proteases released by the stomach lining and pancreas. Glycoproteins secreted from epithelial cells of the intestine form mucins, which coat and protect cells. The proteolytic enzymes are secreted in an inactive zymogen form and are activated only when they reach the lumen of the intestine.

In the stomach, pepsin partly digests protein but the main digestion and absorption occur in the small intestine. Pepsin is unusual in working optimally at about pH 2. This is maintained in the stomach by secretion of HCl from parietal cells by an ATP-dependent system.

Starch is hydrolysed by pancreatic amylase, which is secreted into the intestine as an active enzyme. It is also present in the saliva.

Sugars and amino acids are absorbed into the bloodstream and taken by the hepatic portal vein to the liver and thence into the general circulation.

Digestion products of fat (monoacylglycerol and fatty acids) are resynthesized into TAGs in the epithelial cells and sent into the lymphatics assembled as chylomicrons and thence into the blood, to be distributed around the body. Chylomicrons are lipoproteins, each a complex of phospholipids, cholesterol, and TAG together with a specific collection of protein molecules. The phospholipids keep the particle in suspension for transport.

In the large intestine, water is absorbed and some digestion of dietary fibre occurs due to the presence of bacteria.

Distribution of absorbed digestion products to the tissues

Storage of glucose: the blood after a meal becomes loaded with absorbed food components. These are rapidly cleared from the blood. Glucose is stored in muscle and liver as glycogen, a polymer of glucose. Muscles use it to supply energy but liver has the important function of using it to release free glucose when needed such as in starvation, into the bloodstream. Unless the blood glucose levels are adequate, the brain cannot function normally. Storage of glycogen in the liver is limited to about a 24-h supply of glucose during starvation. When this is exhausted it synthesizes glucose. Muscle does not release glucose into the circulation.

Storage of fat is primarily in the fat cells of adipose tissue where it occurs in large amounts as TAG. Fat storage is virtually unlimited. It has a higher energy content than glycogen and is not hydrated. If the energy stores of fat were replaced by the equivalent calorific value as glycogen, we would be much larger in size. Excess glucose is converted into fat by the liver and other sugars are converted into glucose. Glucose can be converted into fat but the reverse cannot happen in animals.

There is no dedicated storage form of protein but there is a free amino acid pool in blood and cells, which maintains protein homeostasis.

Characteristics of tissues in terms of energy metabolism

Liver has a central role in maintaining blood glucose levels. It synthesizes fats and distributes it to other tissues. Muscles store glycogen but only for their own use.

Hormonal regulation of food distribution

Insulin, released in response to high glucose levels, stimulates fat and glycogen storage. Glucagon released from the pancreas when blood glucose is low stimulates release of glucose from the liver and of fatty acids from fat cells. Insulin and glucagon have opposite effects.

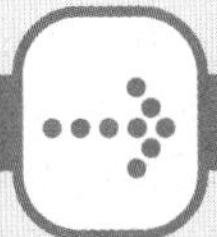

Further reading

To access the further reading, please scan the QR code image or go to http://global.oup.com/uk/orc/biosciences/molbiol/snape_biochemistry/student/reading/ch10/

Problems

1. Which digestive enzymes are produced in an inactive zymogen form? What is the advantage of having zymogens? Why is it not a disadvantage that amylase is produced as an active enzyme and not as an inactive zymogen?
2. Explain how the inactive proteases are activated in the intestine.
3. If pancreatic proenzymes are prematurely activated due to physical and chemical damage or pancreatic duct blockage, what would the result be?
4. Why is digestion of foods necessary?
5. A large number of people, particularly of non-European origin, suffer severe intestinal distress on consuming milk or milk products. Why is this so? How can it be remedied?
6. Write down the structure of a neutral fat. What is the alternative name for this? Indicate which groups in the molecule are primary esters.
7. Fat in water forms large globules with little fat/water interface where lipases can attack. Explain how the

digestive system copes with the physical intractability of lipid.

8 Amino acids and sugars absorbed into the intestinal cells move into the portal bloodstream and are carried via the liver to the rest of the body. What happens to the absorbed digestion products of fat?

9 What is the advantage of storing glucose as glycogen rather than as individual glucose molecules?

10 Triacylglycerol (TAG) is stored in much greater quantities in the body than is glycogen. What makes this possible?

11 Can glucose be converted into fat? Can fatty acids be converted in a net sense to glucose? Do amino acids have a special dedicated long-term storage form in the body? Give brief answers with explanations.

12 In terms of food logistics, describe the chief metabolic characteristics of the liver.

13 Can the brain use fatty acids for energy generation?

14 What are the chief metabolic characteristics of fat cells?

15 What is the main energy source used by red blood cells?

16 What are the main hormonal controls on the logistics of food movement in the body in different nutritive states?

17 At one time ketone bodies were regarded as pathological. Is that so? Explain your answer.

Mechanisms of transport, storage, and mobilization of dietary components

In this chapter we will deal with the handling of fuel derived from food, its distribution and storage in tissues, and subsequent mobilization from the stores in times of need, such as fasting and starvation. We will deal with carbohydrate, fat, and protein storage and release, and will also include cholesterol. Although cholesterol is not an energy yielding fuel, it is worth including it in the chapter as it is absorbed from the intestine and transported using the same system of lipoproteins as fat, and its metabolism is clinically important.

Glucose traffic in the body

The structures of carbohydrates are given in the Digestion of carbohydrates section of Chapter 10.

Mechanism of glycogen synthesis

As stated earlier, after a meal when blood glucose concentrations and the insulin/glucagon ratio are high, glucose is taken up by liver and muscle (and, to a minor extent, by kidney tubules), and used to replenish their glycogen stores. Glycogen is not stored appreciably in any other cells of the body.

The synthesis of glycogen from glucose is an **endergonic process** – it requires energy. Hydrolysis of the glycosidic bond joining glucose units in glycogen has a $\Delta G^{0'}$ value of about $-16\,kJ\ mol^{-1}$, which would give an equilibrium totally to the side of glucose for the simple reaction glycogen $+H_2O \rightarrow$ glucose. Therefore energy must be supplied from high-energy phosphoryl groups to form the glycosidic bond.

Glycogen synthesis occurs by enlarging pre-existing glycogen molecules (called 'primers') by the sequential addition of glucose units. The synthesis of a glycogen granule is initially primed by the protein, **glycogenin**, which transfers eight glucosyl residues to a tyrosine—OH on itself. The protein remains inside the glycogen granule. The new units are always added to the nonreducing end of the polysaccharide. Glucose is an aldose sugar. This is seen clearly in the open-chain form, which is in spontaneous equilibrium in solution with the ring form (Fig. 11.1), the latter predominating.

An aldehyde is a reducing agent, so that the carbon 1 end of the sugar ring is the reducing end and carbon 4 the nonreducing end. The glycogen chain thus always has a nonreducing end.

Nonreducing end of chain → OH (4, 1) O O O O

(The glucose unit is represented as O.)

The process of glycogen synthesis in essence, therefore, is as follows:

Glucose —OH + This is known as the glycogen 'primer' molecule (O O O etc.) ← ATP energy → O O O O etc.

The process is repeated over and over again, elongating the polysaccharide chain.

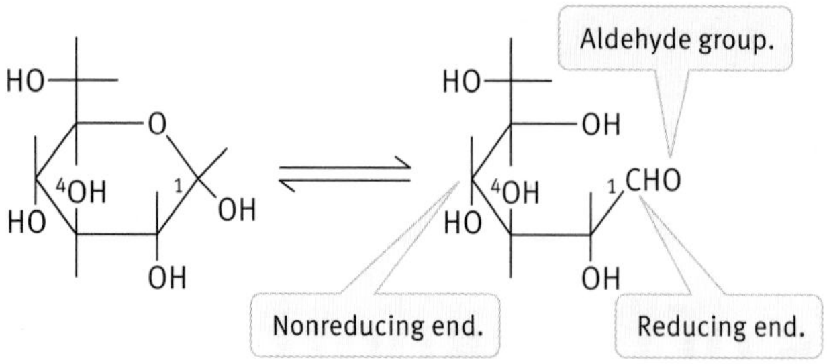

Fig. 11.1 The spontaneous equilibrium reaction of the two forms of glucose in solution.

How is glycogen synthesis driven energetically?

When glucose enters the cell it is phosphorylated by adenosine triphosphate (ATP) (Fig. 11.2). The reaction is catalysed in brain and muscle by the enzyme hexokinase. A **kinase** transfers a phosphoryl group from ATP to something else – in this case, glucose, which is a hexose sugar – hence the name, **hexokinase**. In liver, there is a different enzyme catalysing the same reaction called glucokinase, essentially active only at high glucose concentration, but hexokinase also is present and works when blood glucose levels are low (this is explained in more detail later in this chapter).

The $\Delta G^{0'}$ of this reaction is strongly negative, making it irreversible. The charged glucose-6-phosphate molecule is unable to traverse the cell membrane so that glucose phosphorylation has the effect of trapping the molecule inside the cell and causing entry of more glucose from the blood, since it enters by facilitated diffusion (see Functions of membranes in Chapter 7). The phosphoryl group is now switched around to the 1 position by a second enzyme, phosphoglucomutase, a freely reversible reaction that mutates (changes) phosphoglucose (Fig. 11.3).

The $\Delta G^{0'}$ of hydrolysis of the phosphoryl ester group of glucose-1-phosphate (G-1-P) is –21.0 kJ mol^{-1}, which is about the same as that of hydrolysis of the glycosidic bond in glycogen. You might expect from these values that the glucose unit would be transferred from G-1-P to a glycogen primer and the synthesis would be done. It was thought years ago that this was what happened. But it is not so. If glycogen synthesis occurred directly from G-1-P, the process would be freely reversible and hence uncontrollable.

G-1-P is converted into the activated form, UDPG

There is an additional step that renders glycogen synthesis thermodynamically irreversible. A compound similar to ATP (see Fig. 3.4), uridine triphosphate (UTP), exists in which the adenine of ATP is replaced by uracil (U). (The structure of U is not important here, but is given in Chapter 19 where it is relevant. Uridine is uracil-ribose, a structure analogous to adenosine.) The cell synthesizes UTP (the energy coming from ATP). In glycogen synthesis, **uridine diphosphoglucose (UDP-glucose or UDPG)** is made by a reaction between UTP and G-1-P as shown in Fig. 11.4. The enzyme for this reaction, for systematic nomenclature reasons, takes its name from the reverse reaction (which does not occur in the cell). This reverse reaction (were it to occur) would cleave UDPG

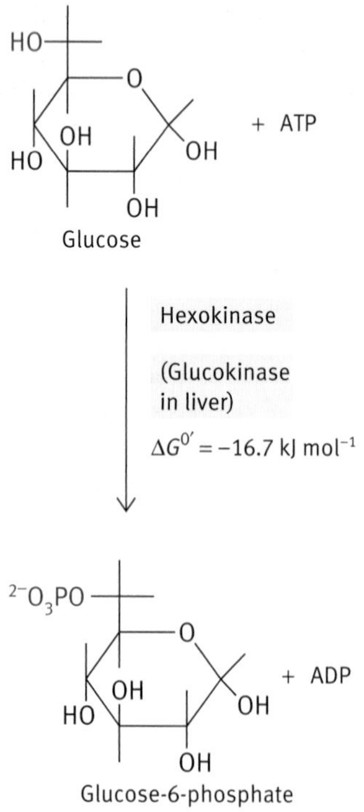

Fig. 11.2 Phosphorylation of glucose by adenosine triphosphate (ATP).

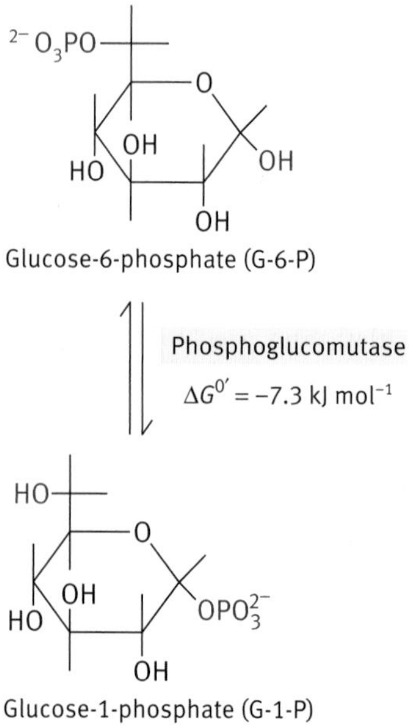

Fig. 11.3 Interconversion of glucose-1-phosphate and glucose-6-phosphate.

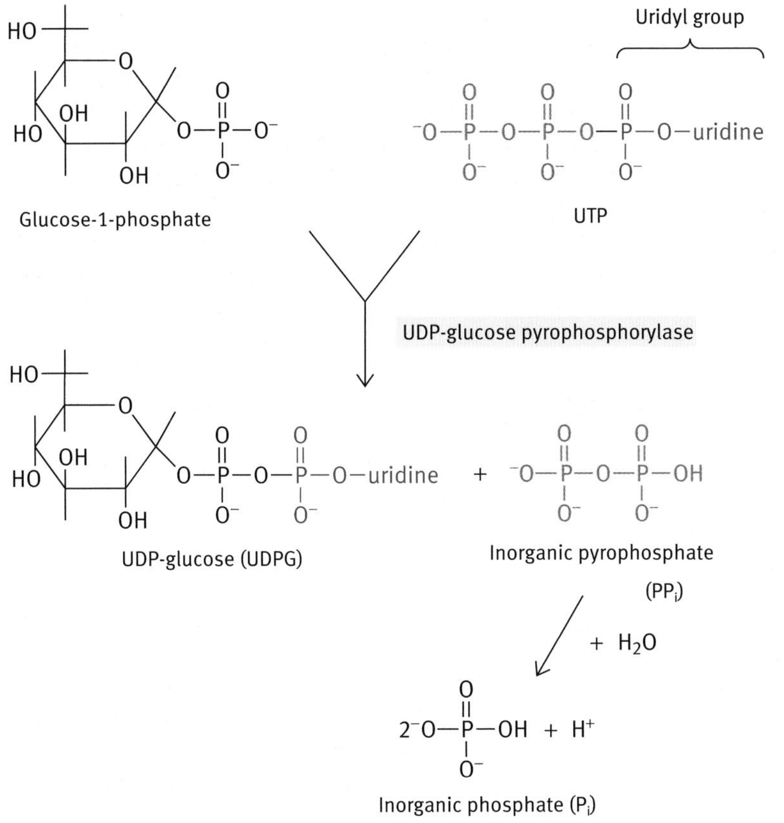

Fig. 11.4 Formation of uridine diphosphoglucose (UDPG) by a reaction between uridine triphosphate (UTP) and glucose-1-phosphate (G-1-P) catalysed by UDPG pyrophosphorylase. See text for an explanation of the name of this enzyme.

with pyrophosphate. The enzyme is therefore a pyrophosphorylase and is called **UDP-glucose pyrophosphorylase**; inorganic pyrophosphate is not available in the cell because it is rapidly destroyed and therefore the reverse reaction is inoperative.

UDPG is the 'activated' or reactive glucose compound that donates its glucosyl group to glycogen. Wherever in animals an 'activated' sugar has to be produced for a chemical synthesis, as a general rule it is a UDP derivative. (Plants use the corresponding adenine compound for starch synthesis.) Synthesis of glycogen using UDPG as a glucosyl donor occurs as shown in Fig. 11.5, catalysed by the enzyme **glycogen synthase** ('synthase' was used wherever a synthesizing enzyme does *not* directly use ATP and 'synthetase' where ATP is involved, but now either can be used). The UDPG pyrophosphorylase produces inorganic pyrophosphate. This is hydrolysed by inorganic pyrophosphatase, which, as explained in Chapter 3, has a $\Delta G^{0\prime}$ value of $-33.5\,\text{kJ mol}^{-1}$, and thus pulls the process of UDPG synthesis from G-1-P and UTP completely to the right. Glycogen synthesis in the cell is not therefore directly reversible by the route of its synthesis. The synthesis is summarized in the scheme shown in Fig. 11.6.

Adding branches to glycogen

If more and more glucosyl groups were added to primer molecules, the result would be very long polysaccharide chains – which is precisely what plants synthesize as the amylose starch component. Glycogen is different – instead of consisting of long chains of glucosyl units, it is a highly branched molecule. This structure is achieved by another enzyme, called the **branching enzyme**. When the glycogen synthase has made a 'straight-chain' extension more than 11 units in length, the branching enzyme transfers a short block of terminal glucosyl units from the end of an $(1\rightarrow4)$-α-linked chain to the C(6)–OH of a glucose unit on the same or another chain (Fig. 11.7). The energy in the $(1\rightarrow6)$-α link is about the same as in the $(1\rightarrow4)$ link so the reaction is a simple transfer one. It creates more and more ends both for glycogen synthesis and breakdown.

Thus, to return to the starting point, blood glucose, after a meal, is taken up by tissues and converted into glycogen by the reactions outlined.

Breakdown of glycogen to release glucose into the blood

The liver, as mentioned, stores glycogen, which is not used by the liver itself (as is the case in muscle), but by other tissues, and especially the brain and red blood cells. In fasting, such as between meals, it breaks down glycogen (high glucagon/low insulin are the signals for this) and releases glucose into the bloodstream. This allows the brain to take up

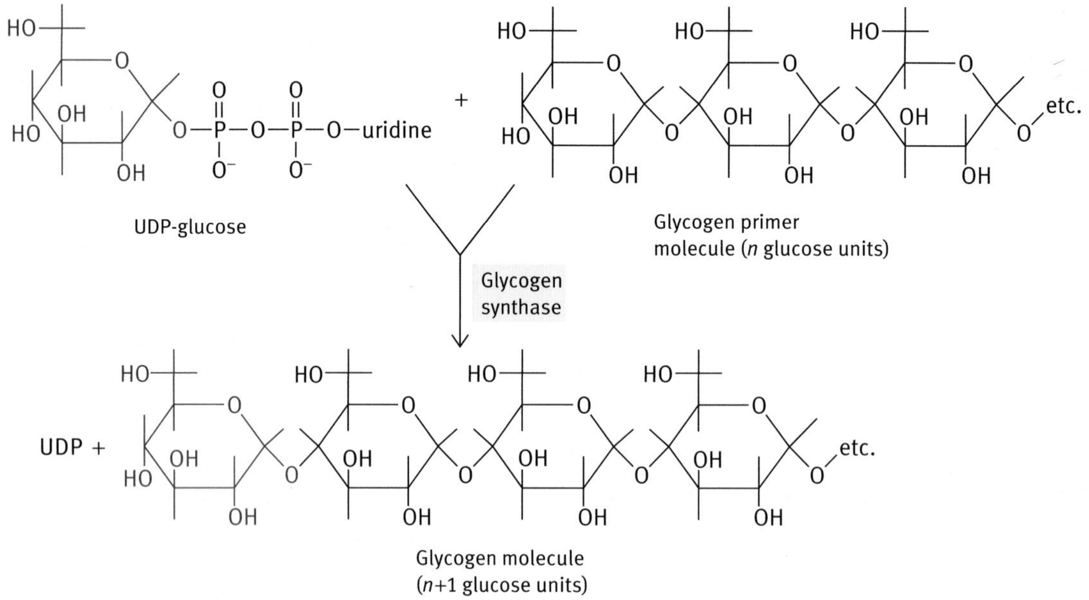

UDP-glucose + Glycogen primer molecule (n glucose units)

Glycogen synthase

UDP + Glycogen molecule (n+1 glucose units)

Fig. 11.5 Synthesis of glycogen by the elongation of a glycogen primer molecule using uridine diphosphate (UDP)-glucose (UDPG) as a glucosyl donor.

glucose and continue to function normally. Hypoglycaemia is said to exist when blood glucose concentration falls below 4 mM, but symptoms do not usually occur until blood glucose is below 2.5 mM when mental efficiency begins to decline. The liver is the glucose provider for the brain and red blood cell.

As with synthesis, **glycogen breakdown** occurs at the non-reducing end. Instead of cleaving (lysing) glucose with water (*hydrolysis*), it cleaves it with phosphate (*phosphorolysis*). The enzyme is called **glycogen phosphorylase** (Fig. 11.8). The reaction involves the coenzyme, pyridoxal phosphate, which we will meet later in transamination reactions. It acts as a general acid-base catalyst.

The G-1-P so formed is converted by **phosphoglucomutase**, working in the reverse direction, to glucose-6-phosphate, and this is hydrolysed (in liver and kidney only) by **glucose-6-phosphatase** to give free glucose, which is released into the blood. Muscle cells break down glycogen by the phosphorylase reaction but do not release glucose into the bloodstream as they do not possess the enzyme glucose-6-phosphatase. G-6-P from muscle glycogen is metabolized by muscle itself by entering the glycolytic pathway.

$$\text{Glucose-1-phosphate} \rightleftharpoons \text{glucose-6-phosphate}$$
$$\text{Glucose-6-phosphate} + H_2O \rightarrow \text{glucose} + P_i$$

A summary of the production of glucose from glycogen by liver and kidney is shown in Fig. 11.9.

Glucose-6-phosphatase is located in the membrane of the endoplasmic reticulum (ER; see Chapter 2). It is a remarkable enzyme in that its active site is exposed to the lumen of the ER, not the cytosol . Glucose-6-phosphate is transported through the ER membrane (by a transport protein) where it is hydrolysed. The products, glucose and P_i, are transported back into the cytosol by separate transport systems.

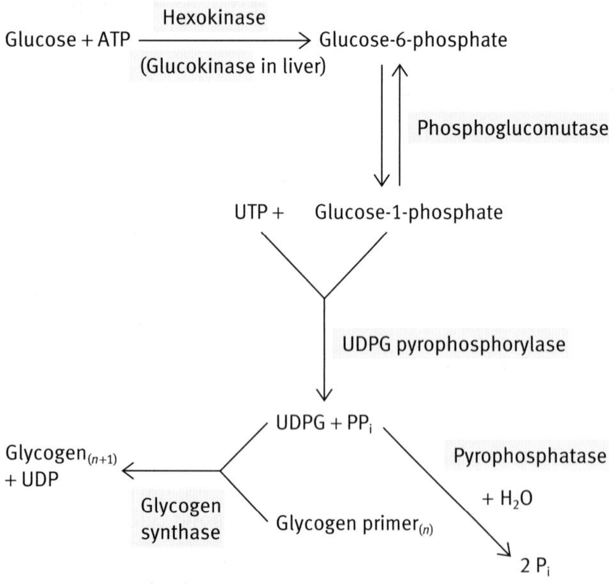

Glucose + ATP → (Hexokinase; Glucokinase in liver) → Glucose-6-phosphate

Glucose-6-phosphate ⇅ (Phosphoglucomutase) Glucose-1-phosphate

UTP + Glucose-1-phosphate → (UDPG pyrophosphorylase) → UDPG + PP_i

PP_i + H_2O → (Pyrophosphatase) → 2 P_i

UDPG + Glycogen primer$_{(n)}$ → (Glycogen synthase) → Glycogen$_{(n+1)}$ + UDP

Fig. 11.6 Summary of glycogen synthesis from glucose.

Removing branches from glycogen

The branch points of glycogen create a problem, for phosphorylase cannot function within four glucose units of such points – the enzyme is big and presumably the branch gets in the way of it attaching to the site of the glucosyl bond to be attacked. A pair of enzymes takes care of this problem, so that the breakdown can go on. The first of these, the **debranching**

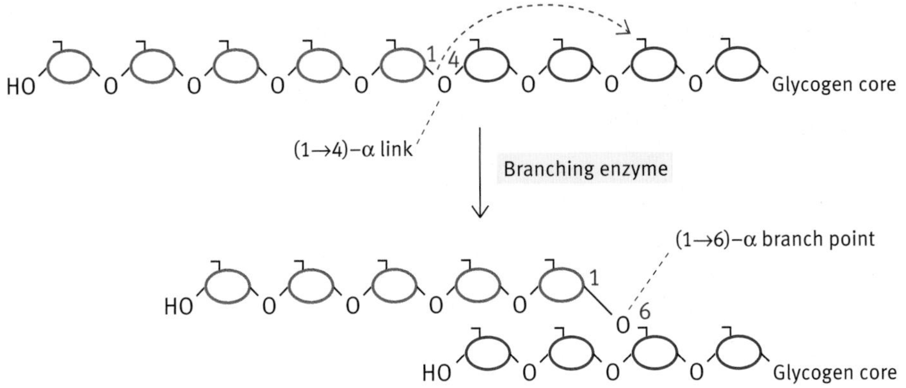

Fig. 11.7 Action of the branching enzyme in glycogen synthesis.

enzyme, transfers three of the glycosidic units of the branch to the 4-OH of another chain. This makes them part of a chain long enough for phosphorylase to work on. The last unit, in (1→6) linkage, is hydrolysed off by the (1→6)-α-glucosidase activity of *the same enzyme* (which has two activities) and this opens up the chain for further attack by phosphorylase (Fig. 11.10). Glucose uptake, storage, and release in the liver are summarized in Fig. 11.11.

Key issues in the interconversion of glucose and glycogen

There are some important points to note, especially the following.

- The glycogen synthesis and degradation pathways are different and therefore independently controllable. (Control is a major topic: see Chapter 20.)
- Glucose-6-phosphatase is found in liver and kidney only, allowing them to release glucose into the blood.
- Most importantly, insulin activates glycogen synthesis, appropriate when glucose is plentiful in the blood. The effect of glucagon is the reverse.

Thus, insulin promotes glycogen synthesis in muscle and liver; glucagon promotes release of glucose from liver. How these controls are achieved is the subject of Chapter 20.

The liver has glucokinase and the other tissues, hexokinase

Once glucose enters the cell it is phosphorylated to glucose-6-phosphate. Glucokinase in liver and hexokinase in brain and other tissues catalyse the same reaction. What are the differences between the two enzymes and how do they affect glucose homeostasis?

Fig. 11.12 shows the activities of glucokinase and hexokinase plotted against glucose concentrations. The K_m for glucokinase is 10 mM and for hexokinase 0.05 mM. Both enzymes will be saturated with substrate at concentrations of 5–10 times their respective K_ms. This means that hexokinase will

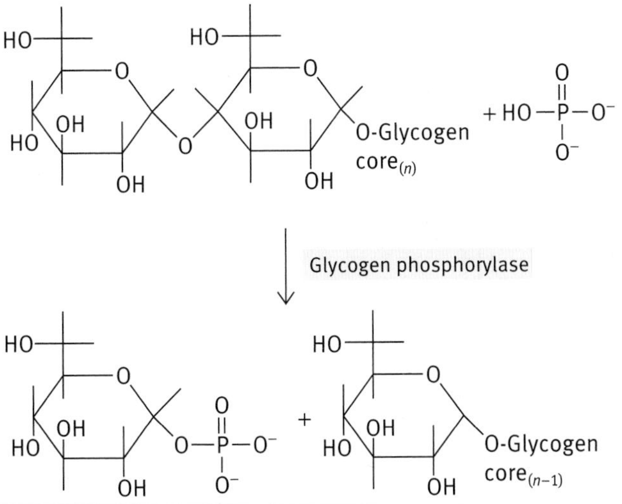

Fig. 11.8 Action of glycogen phosphorylase.

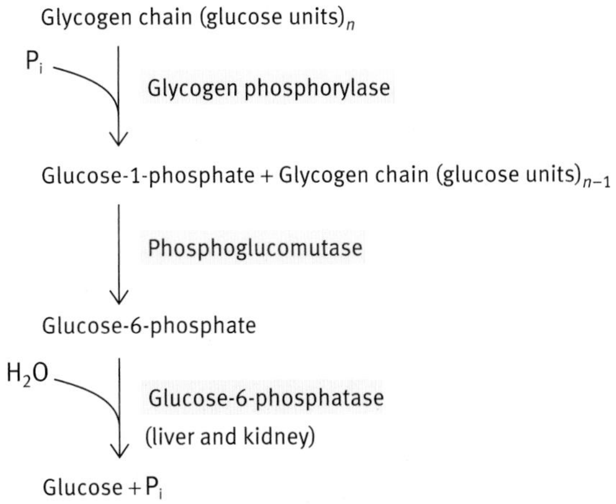

Fig. 11.9 Degradation of glycogen by phosphorolysis and the ultimate release of free glucose into the blood by the liver and kidney.

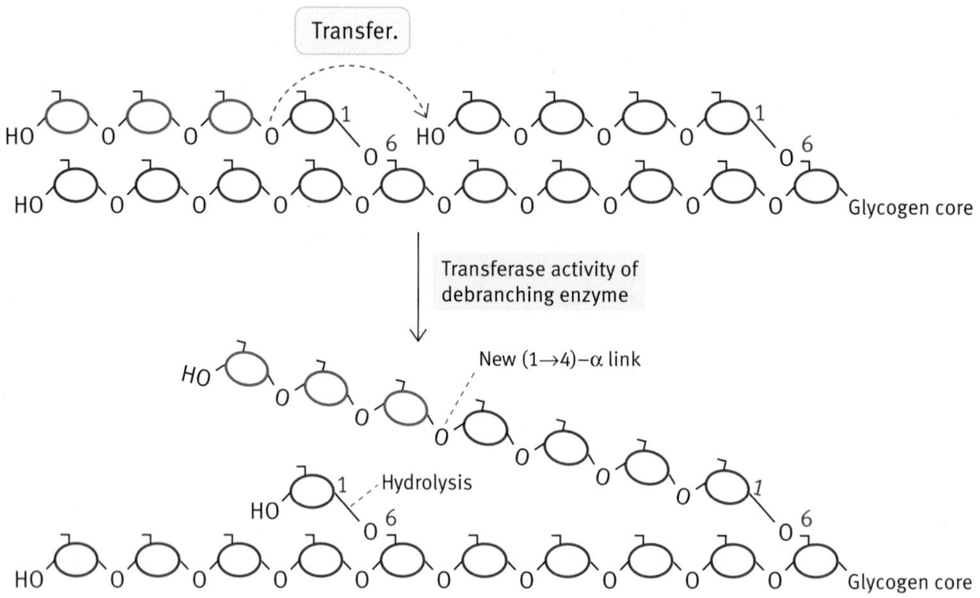

Fig. 11.10 The debranching process. Before debranching, the structure above cannot be attacked by glycogen phosphorylase. After the transferase and hydrolysis actions of the debranching enzyme, both chains are now open to phosphorylase attack.

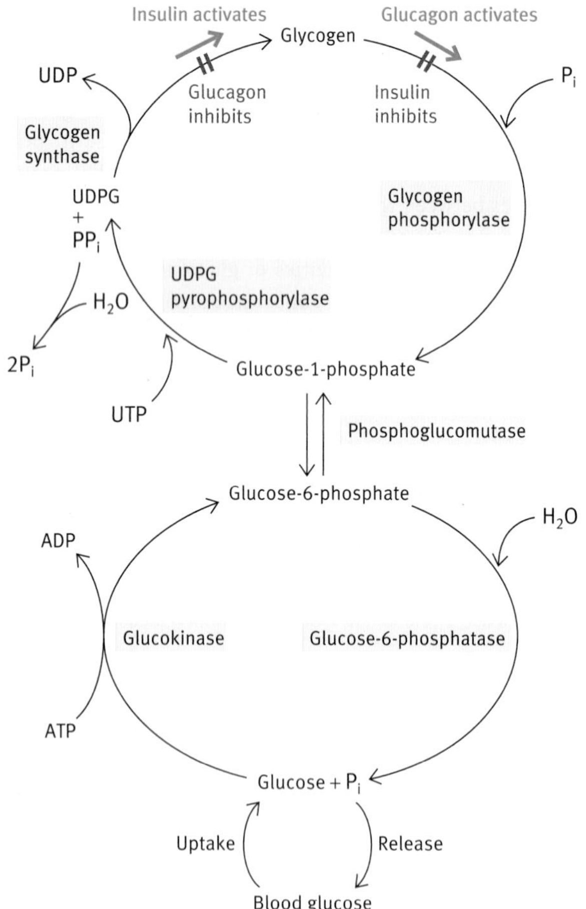

Fig. 11.11 Summary of glucose uptake, storage, and release in the liver. Note that glucagon and insulin do not act directly on glycogen metabolism enzymes (see Chapter 20).

be saturated with glucose at all physiological concentrations of glucose, whether high or low, but glucokinase only reaches saturation at concentrations of glucose of 50–100 mM. The concentration of glucose in the hepatic portal vein can reach 20–50 mM during digestion of a carbohydrate meal, so it is clear that glucokinase can function in this range and is physiologically adapted to deal with high concentrations of glucose.

There is another difference in the properties of the two enzymes. Hexokinase is inhibited by its product, G-6-P, whereas glucokinase is not, which means that metabolism of glucose in the liver can take place even at high concentrations of G-6-P inside the cell.

What happens in fasting and starvation? Hexokinase is still saturated with glucose at concentrations as low as 2–3 mM, whereas glucokinase is much less active. This means that the brain and the erythrocyte, which have hexokinase can still

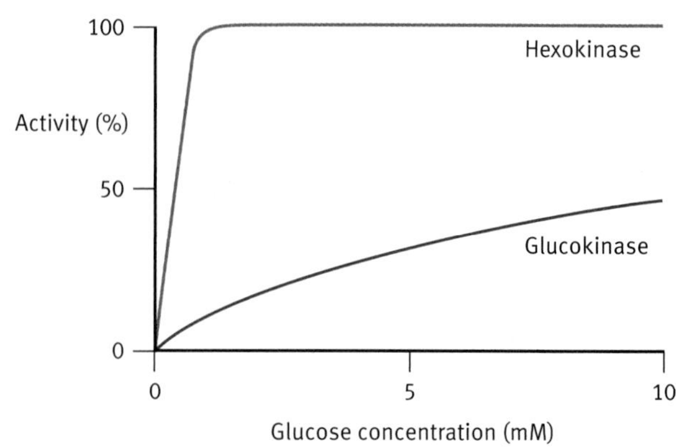

Fig. 11.12 The response of hexokinase and glucokinase to glucose concentration.

metabolize glucose whereas the liver will not, and in this way the liver does not compete with the brain and erythrocytes for scarce glucose supplies.

What about muscle which also contains hexokinase? Does it compete with the brain? The answer is no, because the entry of glucose into the cell is inhibited when glucose and insulin are low.

Brain, erythrocyte, and liver cells have glucose transporters independent of insulin. So they can transport glucose whether, its concentration is high or low. Liver will not take up glucose when glucose is low as its uptake depends on the activity of glucokinase. Muscle has an insulin dependent transporter, so although it has hexokinase, which can phosphorylate glucose at low glucose concentrations, this does not happen, as, in fasting, the glucose tranporter (GLUT4) is not present on the cell membrane. The same is true of adipose cells and in this way muscle and adipocyte do not compete with the brain and erythrocytes when glucose is low.

In summary, the brain and erythrocytes have priority when glucose is low and the liver, muscle, and adipose tissue can metabolize glucose when it is plentiful.

The glucose transporter of the liver, GLUT2, has a lower affinity for glucose than glucose transporters of most other cells. The glucose transporters of the brain (GLUT3 and GLUT1) are fully saturated, that is to say, they are working at maximal rate, at physiological blood glucose concentrations as they have a much higher affinity for glucose than the transporter in the liver (Table 11.1).

What happens to other sugars absorbed from the intestine?

The diet of animals usually results in large amounts of other sugars being absorbed into the portal bloodstream. The sugar in milk is lactose, which on hydrolysis in the small intestine yields galactose and glucose. Sucrose yields fructose and glucose. Fructose has a separate metabolic route. So far as energy metabolism goes, these other sugars are converted either into glucose or into compounds on the main glucose metabolic pathways in the liver. What happens to galactose is of special medical interest.

Galactose metabolism

To convert galactose into glucose, the conformation of the H and OH on carbon atom 4 are inverted or epimerized.

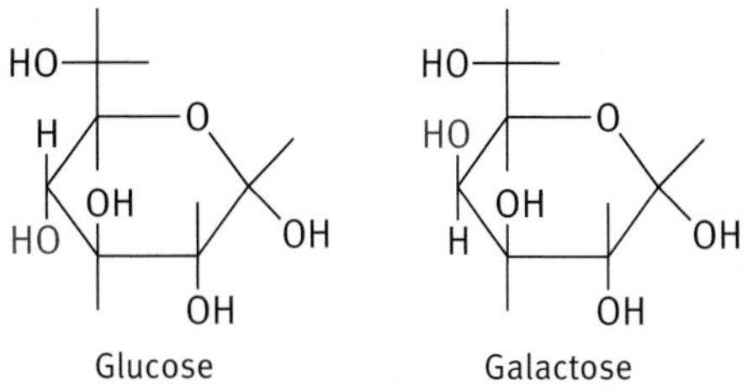

Glucose Galactose

This is done by an epimerase that attacks not galactose itself, but UDP-galactose to form UDPG. (UDPG is also involved in glycogen synthesis as described above.)

First, galactose is phosphorylated by galactokinase to galactose-1-phosphate. (This contrasts with hexokinase and glucokinase, which phosphorylate glucose on the 6-OH.)

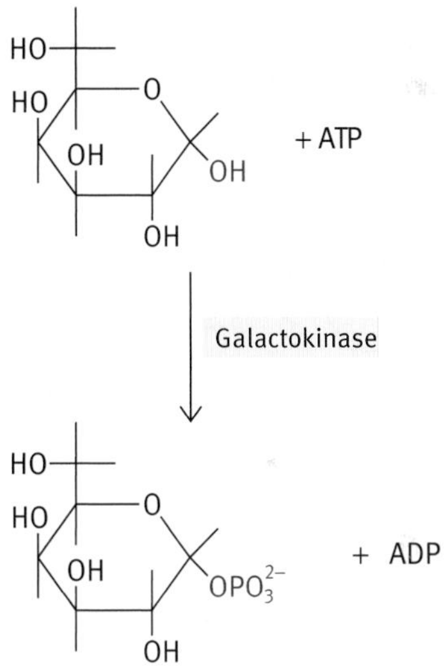

+ ATP

Galactokinase

+ ADP

You might have expected this to react with UTP by analogy with UDPG formation but that is not so. Instead, galactose-1-phosphate displaces G-1-P from UDPG forming UDP-galactose and G-1-P as follows.

	Tissue distribution	Properties/function
GLUT1	Brain, erythrocytes	High affinity, numbers increase at low blood glucose concentrations
GLUT2	Liver, pancreatic β-cell	Low affinity, high capacity
GLUT3	Brain, nerve	High affinity
GLUT4	Adipose tissue, muscle	Insulin dependent
SGLUT	Intestinal mucosa	Cotransport of sodium and glucose

Table 11.1 A summary of the well-characterized glucose transporters 1–4 and SGLUT, and their properties.

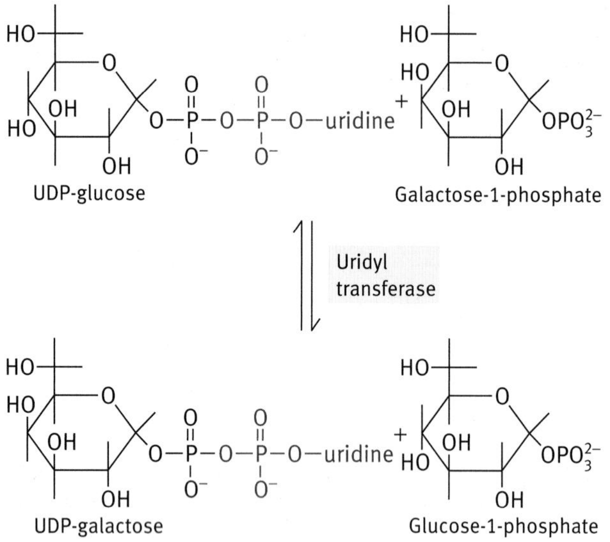

UDP-glucose + Galactose-1-phosphate

⇌ Uridyl transferase

UDP-galactose + Glucose-1-phosphate

You will see that what happens in this reaction is that the –P–uridine group (uridyl) is transferred from UDPG to galactose-1-phosphate and the enzyme is therefore a **uridyl transferase**. Now epimerization of UDP-galactose occurs.

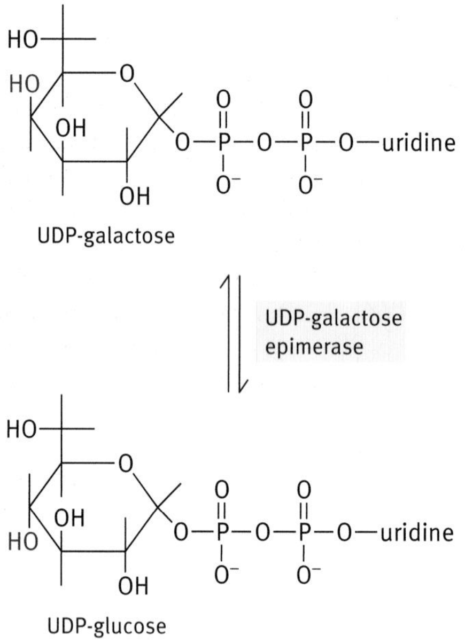

UDP-galactose

⇌ UDP-galactose epimerase

UDP-glucose

Putting the three reactions together, the following sequence occurs.

$$\text{Galactose} + \text{ATP} \xrightarrow[\text{Galactokinase}]{} \text{Galactose-1-P} + \text{ADP}$$

$$\text{Galactose-1-P} + \text{UDP-glucose} \underset{\text{Uridyl transferase}}{\rightleftharpoons} \text{Glucose-1-P} + \text{UDP-galactose}$$

$$\text{UDP-galactose} \underset{\text{UDP-galactose epimerase}}{\rightleftharpoons} \text{UDP-glucose}$$

The net effect is to convert galactose into G-1-P (as summarized in Fig. 11.13).

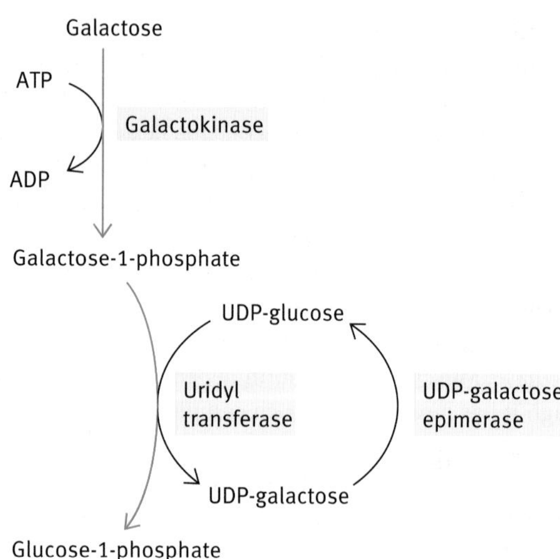

Fig. 11.13 How galactose enters the main metabolic pathways by conversion of galactose to glucose-1-phosphate. The green arrows represent the net effect of the reactions.

Amino acid traffic in the body (in terms of fuel logistics)

As there is no dedicated storage form of amino acids, apart from the free amino acid pool and tissue proteins, there is no whole-body story comparable with that of glycogen and fat traffic. The free amino acid pool is a mixture of amino

BOX 11.1 Uridyl transferase deficiency and galactosaemia

Deficiencies in enzymes involved in galactose metabolism give rise to galactosaemias. The best known and most severe one is the infant genetic disease in which the uridyl transferase is missing and galactose cannot be converted into glucose. Accumulation of galactose-1-P leads to impairment of brain development and blindness. Elimination of galactose from the diet during the 2 months after birth avoids this. UDP-galactose is needed for glycolipid and glycoprotein synthesis (see Chapter 17), but there is no problem since, in this type of galactosaemia, the epimerase is normal. UDPG, synthesized from UTP and G-1-P, can be converted into UDP-galactose as needed. The patient can therefore survive without any intake of galactose in the diet. The galactosaemic children treated in this manner develop normally.

Another hereditary defect is caused by a deficiency of galactokinase. This type of galactosaemia is a milder disorder; it causes early cataracts. The reason for this difference is that galactose can be metabolized by other pathways, whereas galactose-1-P cannot and therefore it accumulates. Galactose can be metabolized by the polyol pathway into galacticol which accumulated in the lens of the eye causing cataracts but does not affect brain development.

β-Galactosidase (lactase) deficiency has been dealt with in Chapter 10.

acids available in the cell derived from dietary sources, or the degradation of protein but it is less than 200 g in total in the average person. Amino acid traffic does occur and the movement of greatest importance is that occurring during muscle wasting in starvation. As explained, muscle proteins break down to produce amino acids, some of which are transported to the liver to provide substrates for glucose synthesis. Amino acids are not found in the circulation in the proportions that they occur in proteins, but the main amino acid traffic from muscle to liver in fasting is in the form of alanine, glutamate, glutamine, and aspartate. This topic will be dealt with in Chapter 16 when we are going to deal with amino acid metabolism.

We will therefore move on to the traffic of fat in the body – a very important subject as disorders are associated with one of the biggest killers of Western Society, cardiovascular disease.

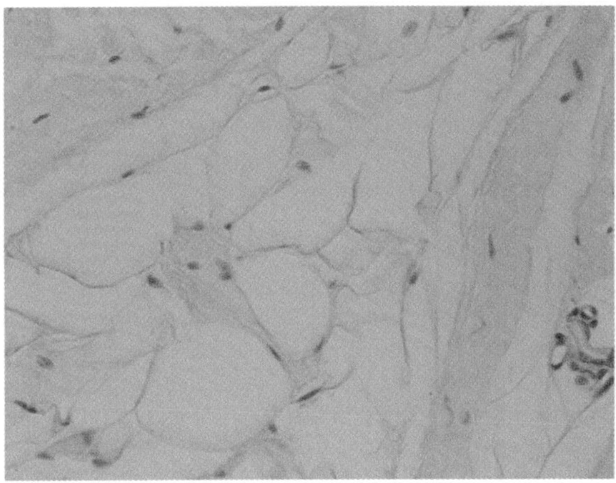

Fig. 11.14 Light micrograph of subcutaneous adipose tissue (kindly provided by Barbara Webb, Dept of Anatomy, King's College London School of Medicine).

Fat and cholesterol movement in the body: an overview

The movement of lipid and cholesterol in the body is a subject of great importance, particularly from a medical point of view. Cholesterol is essential in the body in membranes, and for bile acid and steroid hormone synthesis, but it is dangerous in excess and increased concentrations in the blood are associated with cardiovascular disease.

The liver and the intestine are the major sources of lipid and cholesterol circulating in the blood in the form of **lipoproteins**. The liver synthesizes TAG from glucose and other metabolites, and receives about 10% of the absorbed fat. It is not a storage organ for lipid – as mentioned, a 'fatty liver' in which there are extensive depositions of fat is pathological. The liver exports its TAG to peripheral tissues. As a major provider of fat, the liver supplies this to major users or storers of fat, such as muscle cells, which oxidize it as a source of energy, and adipose cells, which store it (Fig. 11.14) .

The liver is the major site of cholesterol synthesis in the body, and also receives dietary cholesterol. This also is exported to peripheral tissues resulting in an outward flow of lipid and cholesterol, from liver to peripheral tissues. There is also a reverse flow of cholesterol from peripheral tissues to the liver. This flow is summarized in Fig. 11.15. Cholesterol is picked up from peripheral cells, and returns to the liver, either as cholesterol or cholesterol esters. This 'equilibration' of cholesterol between the liver and the rest of the body ensures that all cells have adequate cholesterol supplies with any excess returned to the liver, which excretes part of it into the intestine in the form of bile salts.

With that introduction we shall proceed to an account of the mechanisms involved, but it should be kept in mind that, in detail, it is an extraordinarily complex business, is incompletely understood, and is the subject of a massive continuing research activity.

Utilization of cholesterol in the body

Cholesterol is an important constituent of animal cell membranes. It is also used in the adrenal glands and gonads for the synthesis of steroid hormones. (Examples of the structures of the latter are shown in Fig. 29.3(b).) Because of the effects of its oversupply in causing cardiovascular disease, mechanisms for its removal from the body are of great interest.

The main route for disposing of cholesterol is bile acid formation by the liver. (The structures of cholesterol and of a bile acid are shown in Fig. 10.8.) About 0.5 g per day is disposed of in humans by this route. However bile acids are partly reabsorbed from the gut and re-used. One therapeutic approach to lowering blood cholesterol levels in patients is to administer a compound that complexes bile acids in the intestine and

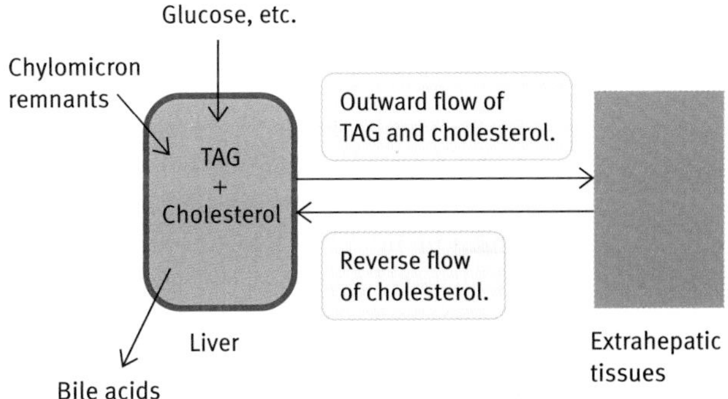

Fig. 11.15 Overview of the major movements of fat (triacyglycerol, TAG) and cholesterol to and from the liver. The liver synthesizes both components from metabolites and receives them from chylomicron remnants. It also converts cholesterol to bile salts.

BOX 11.2 **Inhibitors of cholesterol synthesis – the statins**

Therapeutical drugs, known as statins, have been developed to reduce cholesterol concentrations in the blood. The rate limiting step in cholesterol synthesis is the conversion of HMG-CoA into mevalonic acid, a reaction catalysed by the enzyme HMG-CoA reductase.

Statins resemble mevalonic acid (see Fig. 17.12) in structure, and competitively inhibit this enzyme very effectively by combining with its active site. Simvastatin, lovastatin, and atorvastatin are examples, sold under proprietary names.

Statins reduce total and LDL cholesterol concentration in the blood by decreasing its synthesis in the cells. This allows the liver and peripheral cells to take up more cholesterol from the circulation.

Other approaches for reducing total and LDL cholesterol levels are inhibitors of cholesterol absorption and bile acid-sequestering agents, which prevent their reabsorption from the intestine.

prevents their reabsorption; another approach, is to inhibit its production, as described in Box 11.2.

Cholesterol esters, whose structure and synthesis are given in Figure 11.24, is the storage form of cholesterol in cells and is the form in which much of the cholesterol is carried.

Fat and cholesterol traffic in the body: lipoproteins

Fat (TAG) and cholesterol are both synthesized from the same precursor, acetyl CoA, but they are very different from each other, both in structure, biochemistry and medical significance. As mentioned in the introduction, cholesterol is absorbed from the diet and transported by the same lipoproteins as is fat, probably because they both present the same problem of insolubility in water. It is therefore convenient to discuss the two together. Fat and cholesterol do not circulate as such in the blood, as their solubility in aqueous media is low. They are transported in the form of complexes known as lipoproteins consisting of varying amounts of TAG, cholesterol, cholesterol esters, phospholipids, and proteins, known as apoproteins or apolipoproteins. There are four main types of lipoprotein found in the circulation, **chylomicrons (CM)**, **very-low-density lipoprotein (VLDL)**, **low-density lipoprotein (LDL)** and **high-density lipoprotein (HDL)**.

In summary, chylomicrons transport dietary fat to the periphery, VLDL transport endogenously synthesized fat to the periphery, LDL is the main carrier of cholesterol from the circulation to all tissues including the liver, and HDL transports cholesterol from the periphery to the liver (known as reverse transport).

The approximate composition of plasma lipoproteins is shown in Fig 11.16. It is worth noting that there is an inverse relationship between lipoprotein size and density.

Apolipoproteins

Each type of lipoprotein has its own particular associated set of **apolipoproteins** (the name given to proteins of lipoproteins when detached). A dozen or more are already known but the functions of all of them have not yet been elucidated. We do, however, know of several roles:

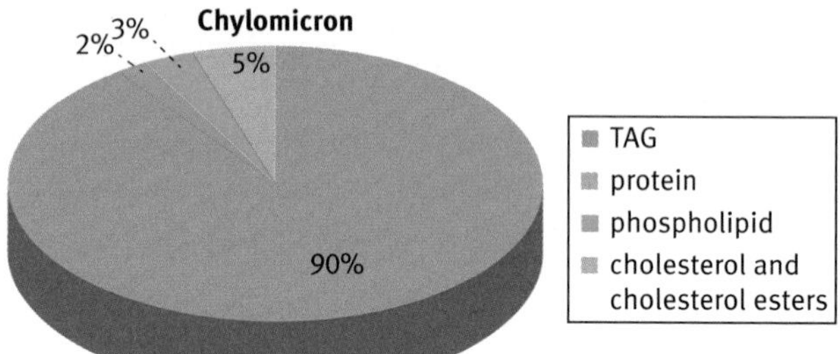

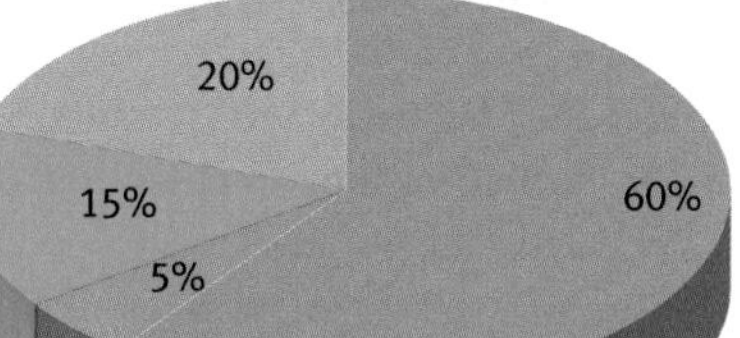

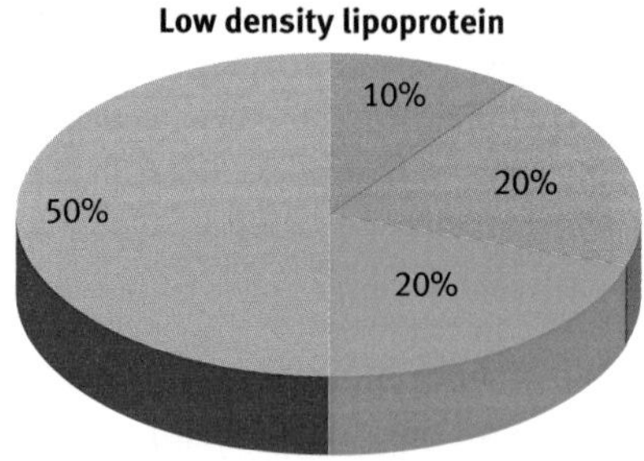

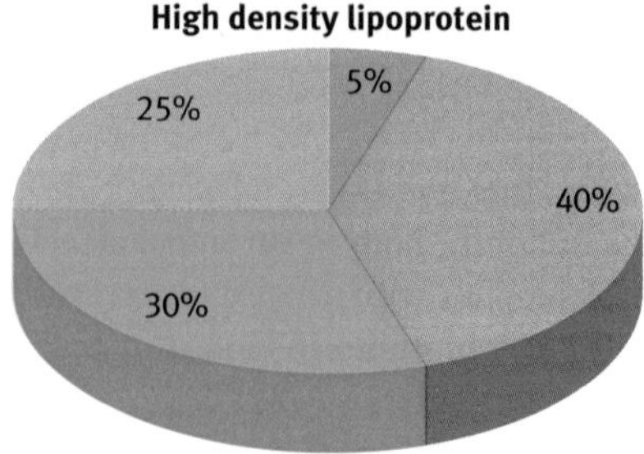

Fig. 11.16 Composition of the major classes of lipoprotein.

- Some are required as structural components for the production of lipoproteins. In the case of chylomicrons the main one is apo B-48 and, in that of VLDL, it is apo B-100.
- Some are required as destination-targeting signals – apoproteins designed to bind to specific receptors on the surface of cells. Binding leads to uptake of the bound lipoprotein by receptor-mediated endocytosis (see Fig. 11.17). In this way individual lipoproteins are taken up only by designated cells. Examples of this targeting function are apo B-100 on LDL, which binds to LDL receptors, and apo-E on chylomicron remnants, which binds to liver remnant receptors.
- Some apolipoproteins activate enzymes. Apo C-II on chylomicrons activates lipoprotein lipase, which removes fatty acids from TAG. Apo A-1 on HDL activates the enzyme LCAT (or PCAT), which forms cholesterol esters from cholesterol in peripheral cells and phospholipid on HDL itself, and carries the ester to the liver.

Lipoproteins involved in fat and cholesterol movement in the body

Chylomicrons are formed in the intestine. There are two lipoproteins produced by the liver, **very-low-density lipoprotein (VLDL)** and **high-density lipoprotein (HDL)**. The production of lipoproteins involves the ER and the Golgi apparatus, which packages them into membrane-bound vesicles for release by exocytosis. The VLDL is converted into **intermediate-density lipoprotein (IDL)** and then into **low-density lipoprotein (LDL)** by the removal of TAG by cells, so that, including chylomicrons, five different lipoproteins are found in circulation. Fig. 11.18 shows electron micrographs of different lipoproteins.

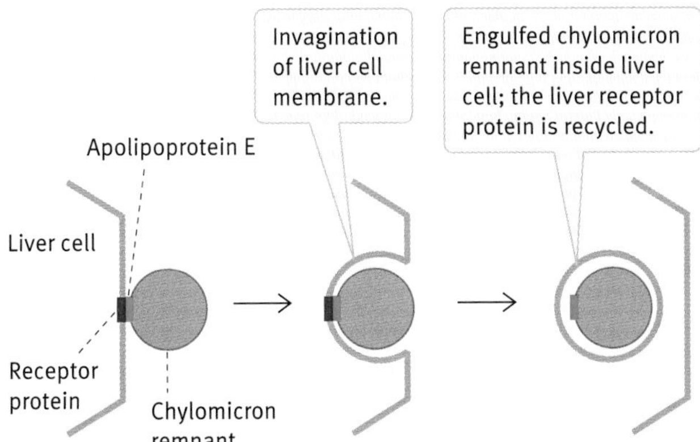

Fig. 11.17 Uptake of chylomicron remnants into a liver cell by receptor-mediated endocytosis. The receptor on the liver cell is specific for apolipoprotein E present on the chylomicron remnant.

Metabolism of chylomicrons

A summary of chylomicron metabolism is shown in Fig. 11.19.

After a meal containing fat, the blood is loaded with **chylomicrons**. The chylomicron (see Fig. 10.10) has a shell of phospholipids, cholesterol, and proteins surrounding a core of triacylglycerol (TAG) and cholesterol esters. The nascent, or newly formed chylomicron, synthesized in the intestine, contains mainly TAG and small amounts of cholesterol and cholesterol esters; it also contains apoprotein B-48. It finds its way into the circulation via the lymphatics. In the blood, it picks up further apoproteins from HDL; these are apo C-II and apo E.

In the capillaries, TAG is removed by the adipose cells for storage by muscle and other tissues, as an energy source (and also by the lactating mammary gland for secretion in milk). The TAG of chylomicrons is hydrolysed in the capillaries by **lipoprotein lipase** attached to the *outside* of endothelial cells lining the blood capillaries, to produce glycerol, which is transported to the liver, and **free fatty acids** (FFA), which enter the adjacent cells (Figs. 11.19 and 11.20). FFA, not esterified with glycerol, readily pass through cell membranes; TAG cannot do so as it is a large neutral fat. The fate of glycerol in liver is described in Chapter 16.

The amount and level of activity of lipoprotein lipase present in the capillaries of a particular tissue determines the amount of FFA released in that tissue and hence the amount of FFA uptake by it. Adipose tissue is rich in the enzyme, as is lactating mammary gland, while tissues that utilize little fat have less of it in the capillaries. The amount of lipoprotein lipase varies according to physiological need; for example, a high-insulin, low-glucagon ratio after a meal causes an increase in the synthesis of the enzyme as well as its activity. Apo C-II also activates lipoprotein lipase.

The progressive removal of TAG from chylomicrons reduces them in size to chylomicron remnants, containing all the cholesterol and its esters, and about 10% of the TAG of the original chylomicron. Apo C-II is returned to HDL in the circulation but the chylomicron remnants retain apo E, which allows them to be taken up by the liver by receptor-mediated endocytosis, through binding of apo E to the apo E receptor (**remnant receptor**) on liver cells Fig. 11.17. In this way, fat is delivered to the adipose tissue, and cholesterol and cholesterol esters, to the liver.

Metabolism of VLDL: TAG and cholesterol transport from the liver

A summary of the metabolism of VLDL is shown in Fig 11.21.

Let us start with VLDL and the outward flow of TAG and cholesterol from the liver. VLDL is synthesized in the liver and contains mainly endogenously synthesized TAG. It also contains apoprotein B-100. Apo B-100 is essential for the formation of VLDL and the subsequent export of TAG from the

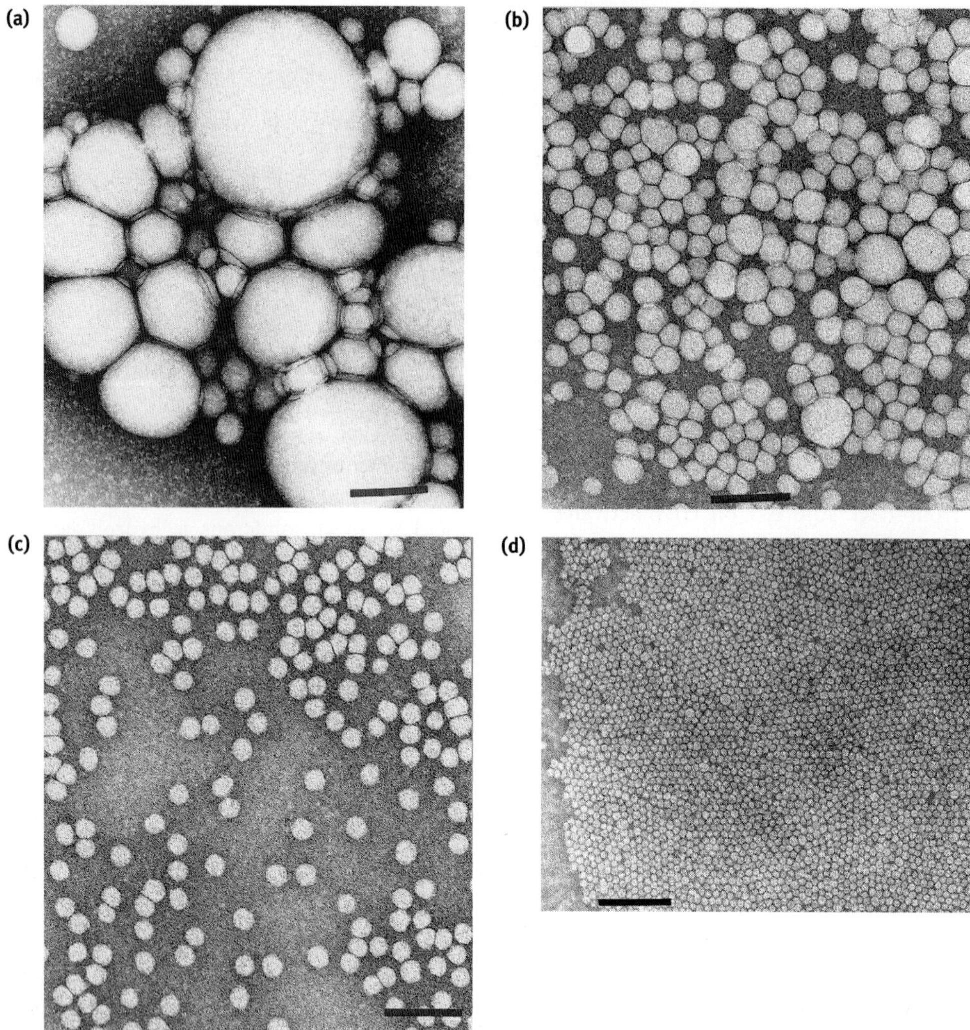

Fig. 11.18 Electron micrographs of **(a)** chylomicrons; **(b)** very-low-density lipoproteins (VLDL); **(c)** low-density lipoproteins (LDL); and **(d)** high-density lipoproteins (HDL). Scale bars, 1000 Å. Kindly provided by Dr Trudy Forte of the Lawrence Berkeley Laboratory, University of California, USA.

liver to the periphery. Defects in the synthesis of this apoprotein result in accumulation of TAG in the liver due to the failure of its export via VLDL.

Once VLDL is released, it picks up apoproteins apo E and apo C-II from HDL in the same way as chylomicrons do. In the capillaries, TAG is progressively removed by lipoprotein lipase action, again in the way already described for chylomicrons. As the amount of fat diminishes, the percentage of cholesterol and its esters rises, causing an increase in density, and the lipoproteins decrease in size. The end product of the process is LDL, formed via the intermediate state, IDL (intermediate-density lipoprotein). Apo E and apo C-II are returned to HDL. Some TAG is also transferred to HDL and some cholesterol esters are transferred from HDL to IDL by an enzyme known as cholesterol ester transfer protein (CETP). We will see the significance of this transfer later in this chapter when we deal with HDL metabolism.

About half of IDL is taken up by peripheral tissue cells via receptor-mediated endocytosis and half is converted into LDL. The latter contains mainly cholesterol and cholesterol esters and only one apoprotein, apo B-100. About half of LDL is taken up by peripheral tissues and half by the liver, by receptor-mediated endocytosis involving the apo B-100 receptor (also known as LDL receptor). In this way, by the metabolism of VLDL and LDL, TAG is delivered to extrahepatic tissues, and cholesterol is delivered to all tissues including the liver.

The number of LDL receptors on cells is regulated and controls LDL uptake. When LDL concentrations are high the receptor is downregulated, that is, fewer receptors appear on the cell surface. When the receptor-LDL complex is endocytosed, the receptors are recycled to the cell surface and the LDL is hydrolysed in lysosomes to release cholesterol, cholesterol esters, phospholipids, and amino acids (Fig. 11.22). Free

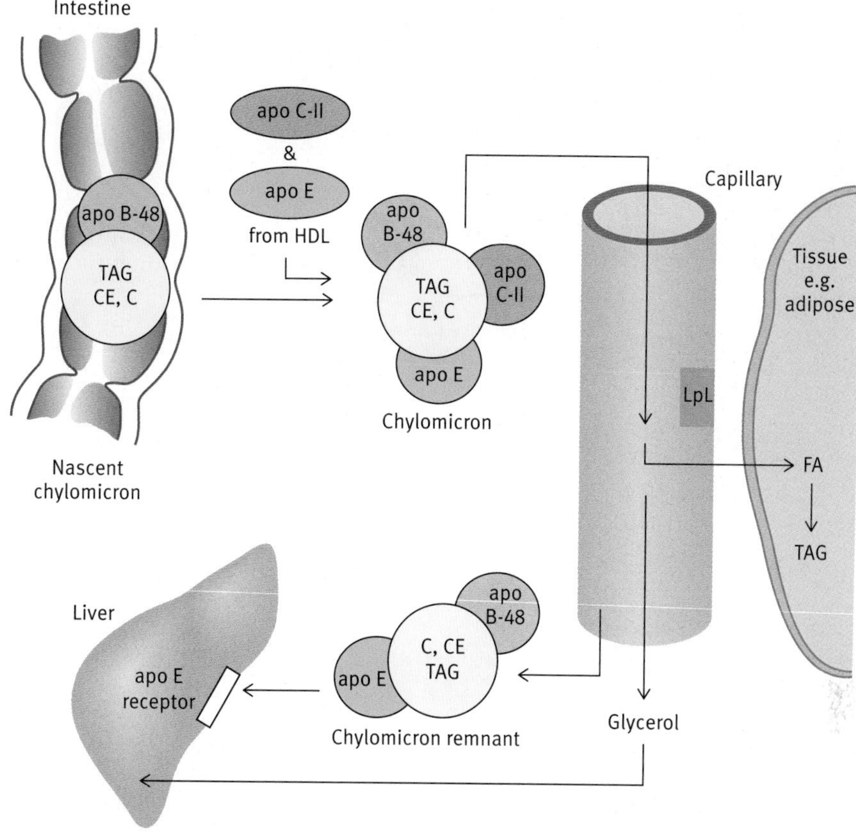

Fig. 11.19 Metabolism of chylomicrons. TAG, triacylglycerol; CE, cholesterol ester; C, cholesterol; HDL, high-density lipoprotein; apo B-48, apo C-II, apo E, apoproteins; LPL, lipoprotein lipase; FFA, free fatty acid; (not to scale).

cholesterol has a number of effects on its own synthesis and metabolism in the cell. If the concentration of cholesterol in the cell is low, synthesis and uptake are increased; if the level is high, synthesis and uptake are inhibited so that cholesterol homeostasis is achieved. How do these control mechanisms work? HMG-CoA (3-hydroxy-3-methylglutaryl-coenzyme A) is a metabolite you have not met yet in this book; it is described in Chapter 17 where we briefly deal with cholesterol synthesis, but at this point it is sufficient that it is needed for cholesterol synthesis. The first metabolite committed to cholesterol synthesis is mevalonic acid formed from HMG-CoA by the action of **HMG-CoA reductase** and this is the main control point

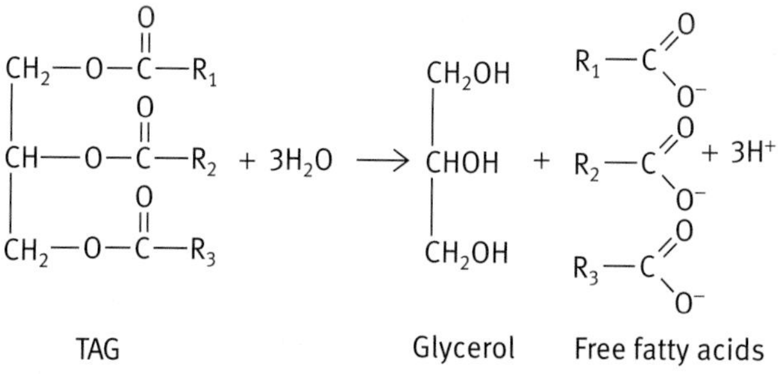

$$\begin{array}{l} CH_2{-}O{-}C({=}O){-}R_1 \\ CH{-}O{-}C({=}O){-}R_2 \\ CH_2{-}O{-}C({=}O){-}R_3 \\ \text{TAG} \end{array} + 3H_2O \longrightarrow \begin{array}{l} CH_2OH \\ CHOH \\ CH_2OH \\ \text{Glycerol} \end{array} + \begin{array}{l} R_1{-}C({=}O){-}O^- \\ R_2{-}C({=}O){-}O^- \\ R_3{-}C({=}O){-}O^- \\ \text{Free fatty acids} \end{array} + 3H^+$$

Fig. 11.20 Breakdown of triacylglycerol (TAG) by lipoprotein lipase.

(see Box 11.2). At low cholesterol levels in the cell, the genes coding for this enzyme and for LDL receptor protein are activated so that cholesterol synthesis and uptake of cholesterol by cells are stimulated. At high cellular cholesterol concentration, the genes are not activated and the level of reductase and the number of receptors diminish, and so does cholesterol synthesis and uptake via the LDL receptors. There are further controls. The HMG-CoA reductase is inhibited at high cholesterol levels; the enzyme becomes phosphorylated, in which form it is inactive. Also at high cholesterol levels the enzyme is more rapidly destroyed, so we can see that a number of separate controls operate. We will see how this control is used pharmacologically in order to reduce the amount of cholesterol in the circulation (see Box 11.2). LDL cholesterol is colloquially known as 'bad cholesterol' because high concentrations are associated with increased risk of **atherosclerosis**. A high **LDL/HDL ratio** correlates with the incidence of **coronary heart disease**. This ratio is considered a predictor of atherosclerosis but risk factor analysis and LDL concentrations are now considered more accurate predictors.

The high LDL levels lead to the development of plaques in blood vessels, a complex process in which cholesterol is involved. If LDL concentrations are high in the blood, various oxidants in the circulation are likely to modify LDL by oxidation, in a way

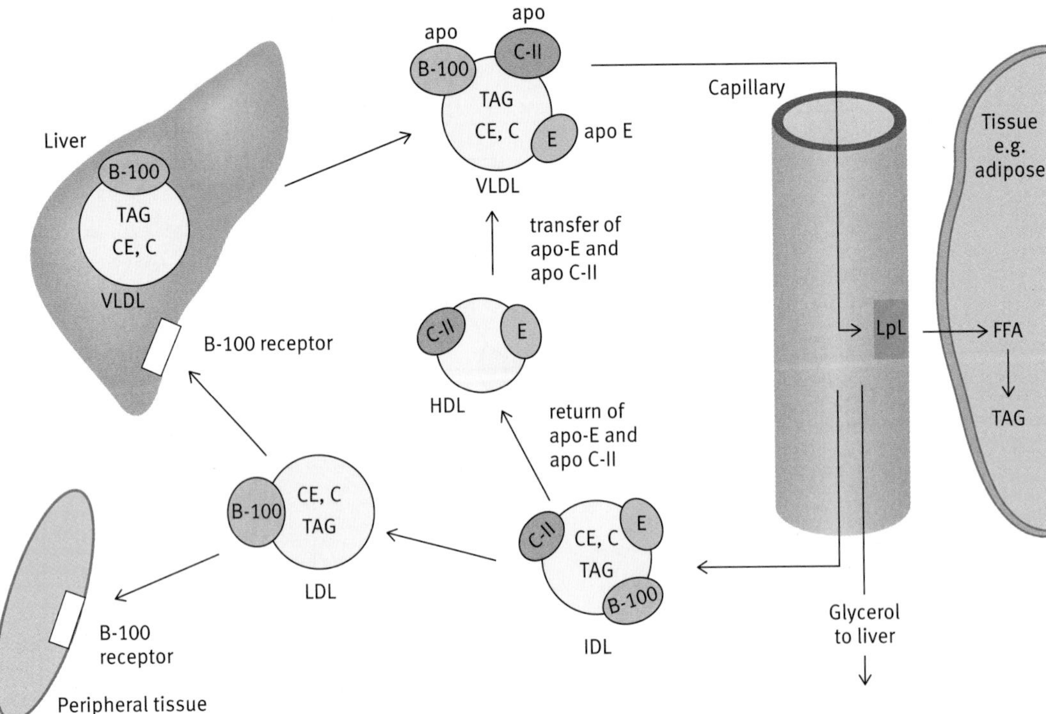

Fig. 11.21 Metabolism of VLDL and LDL. TAG, triacylglycerol; CE, cholesteryl ester; C, cholesterol; HDL, high-density lipoprotein; apo B-100, apo C-II, apo E, apoproteins; LPL, lipoprotein lipase; FFA, free fatty acid; IDL, intermediate-density lipoprotein; LDL, low-density lipoprotein (not to scale). Kindly provided by Dr Trudy Forte of the Lawrence Berkeley Laboratory, University of California, USA.

that it is not recognized by the B-100 receptors which normally exist in cells. Damaged (oxidized) LDL can, however, be taken up by scavenger receptors in macrophages in the arterial wall, eventually becoming foam cells laden with cholesterol, collagen, and lipid, which contribute to the formation of plaques. These, if present in coronary arteries, may burst causing blockages and result in heart attacks. An extreme case of this is found in the genetic disease **familial hypercholesterolaemia**, in which people have defective LDL receptors leading to very high levels of LDL cholesterol and early vascular disease even without other risk factors, such as smoking or obesity.

A lipoprotein that resembles LDL is also associated with increased risk of cardiovascular disease. Lipoprotein (a) (known as lipoprotein little a) is identical to LDL but with an additional apoprotein (apo A) covalently bonded to apo B-100. Its atherogenic properties are probably related to its structural homology with plasminogen, which is the precursor of an enzyme that hydrolyses fibrin in blood clots. It is thought that lipoprotein (a) competes with plasminogen and inhibits the breakdown of fibrin clots so contributing to blockage of arteries.

Metabolism of HDL (reverse cholesterol transport)

A summary of HDL metabolism is shown in Fig. 11.23.

HDL is produced mainly by the liver in an 'immature' disc-like form, known as **nascent HDL**, made of phospholipid

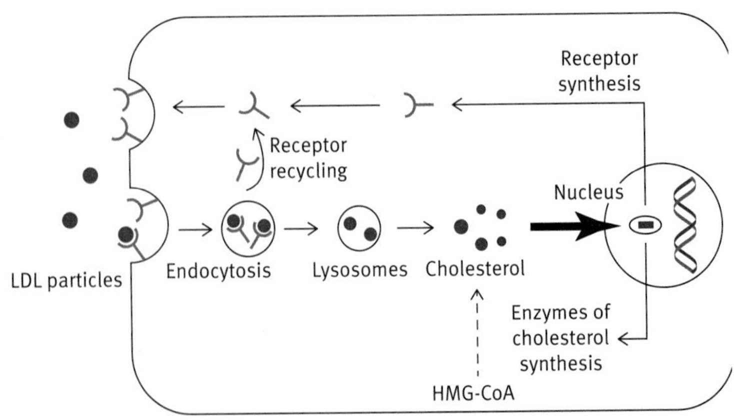

Fig. 11.22 Receptor-mediated endocytosis of LDL.

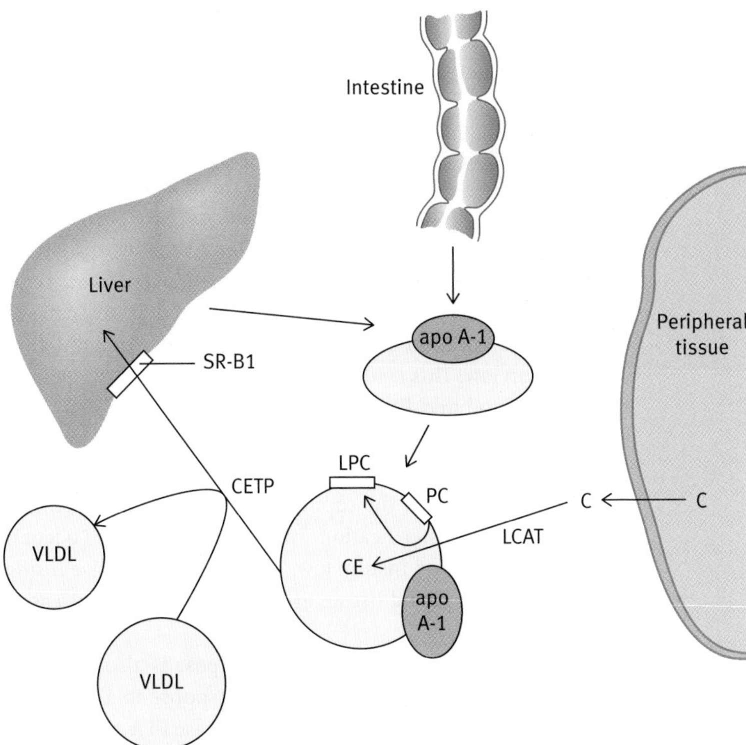

Fig. 11.23 Metabolism of HDL. C, cholesterol; CE, cholesteryl ester; LCAT, lecithin-cholesterol acyl transferase (also known as PCAT, phosphatidyl choline-cholesterol acyl transferase; CETP, cholesterol ester transfer protein; PC, phosphatidyl choline (lecithin); LPC, lysophosphatidyl choline (lysolecithin); SR-B1, scavenger receptor class B1

and apoproteins but containing little TAG or cholesterol. Nascent HDL also contains apo A-1, its principal apoprotein, as well as apo E and apo C-II, which, as we have seen earlier, can shuttle between HDL and other lipoproteins. The nascent HDL picks up cholesterol from extrahepatic cells. Free cholesterol picked up by HDL as such, would remain in the peripheral layer of the lipoprotein with its –OH group in contact with water, but, once in the HDL, it is converted into cholesterol ester. Hydrophobic force (Chapter 7) causes the latter to migrate to the centre of the HDL, away from contact with water. The accumulation of the ester causes the HDL to grow into a spherical particle. Cholesterol esterification *in cells* requires ATP but in HDL where none of this is available, a fatty acyl group of lecithin (phosphatidyl choline), which is present in the hydrophilic shell, is transferred to cholesterol in an energy-neutral reaction by an enzyme associated with HDL, called **lecithin: cholesterol acyltransferase (LCAT**, also known as **PCAT**, phosphatidyl choline-cholesterol acyl transferase) (Fig. 11.24). (The ATP-requiring synthesis, which occurs within cells is catalysed by an enzyme called acyl-CoA: cholesterol acyltransferase (ACAT).

The HDL transfers its cholesterol ester to VLDL, IDL, and LDL and also to chylomicrons by means of a **cholesterol ester transfer protein (CETP)**. For clarity, only VLDL is shown in Fig. 11.23. Although IDL and LDL deliver their contents to extrahepatic tissues as described above, a proportion return to the liver (they both use the LDL receptor). This is known as the reverse flow of cholesterol. The transfer of cholesterol ester between lipoproteins occurs reversibly. Transfer of TAG also takes place between the various components, catalysed by CETP, which exchanges cholesterol ester for TAG. It may seem bizarre that TAG are taken from VLDL and IDL back into HDL, but the exchange of TAG and cholesterol esters (CE) relieves product inhibition of LCAT by cholesteryl esters in HDL. Apart from unloading cholesteryl esters to other lipoproteins, HDL also delivers cholesteryl esters directly to the liver by means of a scavenger receptor known as SR-B1. It is not known whether the whole HDL is internalized or only the cholesteryl esters are taken up.

How does cholesterol exit cells to be picked up by HDL?

How does cholesterol get out of the peripheral donor cell to be picked up? The mechanism is now known. In an isolated island in Chesapeake Bay (Tangier Island) some of the inhabitants exhibited a condition known as **Tangier disease** in which high concentrations of cholesterol accumulate in lymphatic organs, such as tonsils and spleen, with the spleen becoming enlarged. Genetic studies pinpointed the deficient gene responsible for the condition. It codes for the enzyme ABCA1 transporter (ATP-binding cassette A1 transporter). This protein belongs to a large super family of membrane glycoproteins that transport molecules through membranes (described in Chapter 7), but its main function was in doubt. It now appears that an important role is to transport cholesterol out of cells to be picked up by HDL. The ABCA1 transporter transports cholesterol and phospholipid to HDL particles. As well as accepting cholesterol from intestine and other peripheral tissues, it also enables macrophages to unload excess cholesterol, so that ABCA1 is anti-artherogenic.

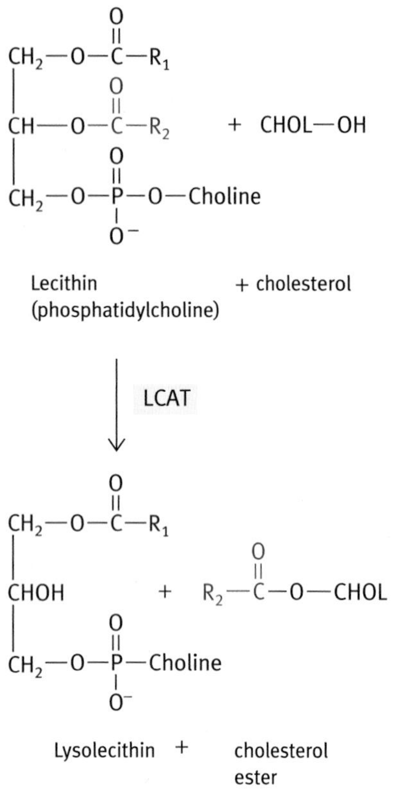

```
            O
            ||
CH2—O—C—R1
|           O
|           ||
CH—O—C—R2          + CHOL—OH
|           O
|           ||
CH2—O—P—O—Choline
            |
            O−

Lecithin                    + cholesterol
(phosphatidylcholine)

        |
        |  LCAT
        ↓

            O
            ||
CH2—O—C—R1
|                       O
|                       ||
CHOH          +   R2—C—O—CHOL
|           O
|           ||
CH2—O—P—Choline
            |
            O−

Lysolecithin  +    cholesterol
                   ester
```

Fig. 11.24 Reaction catalysed by lecithin: cholesterol acyltransferase (LCAT), where R_1 and R_2 are fatty acyl groups; CHOL–OH is cholesterol.

This results in cholesterol being sent back to the liver where, as mentioned, some of it can be disposed of as bile salts.

It has long been known that high levels of HDL cholesterol (colloquially known as 'good cholesterol' as opposed to LDL, bad cholesterol) correlate with low risk of coronary disease and clinical syndromes are known in which low levels of HDL in the blood lead to increased risk of coronary occlusions. The terms 'good' and 'bad' cholesterol as applied to HDL and LDL refer to the direction of cholesterol traffic. HDL is 'good' because it delivers cholesterol from the periphery to the liver for excretion. LDL is 'bad' because it transports cholesterol to the periphery.

The control of cholesterol levels is very important, and several factors are involved apart from the ones discussed earlier, including other dietary components. Cholesterol levels in liver increase with high levels of saturated fat in the diet. However, **unsaturated fats**, especially the **omega-3 (ω-3)** and **omega-6 (ω-6) fatty acids** (see Box 17.1) lead to decreased levels of cholesterol and also of TAG. The biochemical mechanism is unknown. Another dietary component, the vitamin niacin given in pharmacological doses increases HDL and reduces lipoprotein (a) levels.

Release of FFA from adipose cells

After a meal, adipose cells store fat. In fasting, they release free fatty acids into the blood, which supply the body with a source of energy. A lipase releases FFA from stored TAG, but it is present inside the adipose cell and not in the capillary walls as the lipoprotein lipase that we have encountered before. It is known as *hormone-sensitive lipase, is activated by a glucagon signal to the cells and inhibited by an insulin signal to the cells.* The mechanism of this is described in Chapter 20. You can see how it all fits in. After a meal, the insulin level is high and that of glucagon low. This results in adipose cells taking up glucose from the blood and FFA from chylomicrons and VLDL, and storing them as TAG; release of FFA is inhibited. The location of lipoprotein lipase in the capillaries allows the hydrolysis of TAG to take place in the capillary and the FFA to enter the adipocyte for re-esterification. In fasting, the reverse occurs; glucagon is high and insulin low. The lipoprotein lipase *inside the cell* is activated and tissues are supplied with FFA released in the blood.

The adipose cell hormone-sensitive lipase is also activated by adrenaline (epinephrine). Thus, in response to an adrenaline-releasing alarm, the adipose cells release FFA, which supply the muscles and support any required muscular activity. How adrenaline does this is an important topic, but it is more suitable to deal with it in Chapter 20. Noradrenaline (norepinephrine) released from the sympathetic nervous system that innervates adipose tissue has the same effect.

How are FFA carried in the blood?

FFA are not actually free, the word here means nonesterified. It would be unsatisfactory for relatively high levels of FFA to be carried in the blood since, at neutral pH, they are detergents. Instead they are carried adsorbed to the surface of the blood protein, **serum albumin**. This protein has hydrophobic patches on its surface and is the carrier for many molecules with hydrophobic sections. By FFA we generally mean long chain fatty acids, which are the major fatty acids in the body. (Short chain fatty acids are not bound to albumin.) The adsorbed FFA are in equilibrium with FFA in free solution so that, as cells take them up, in order to maintain the equilibrium, more dissociate from the protein carrier into the serum where they are available to cells. The serum albumin-bound FFA cannot traverse the blood–brain barrier so the brain cannot use FFA. However, ketone bodies, derived from the partial metabolism of fats, are carried in simple solution. These can enter the brain to provide, in starvation, part of the energy, as already described. The rest must be supplied by glucose. The topic will be dealt with in greater detail in Chapter 20.

Summary

After a meal, glucose is stored as glycogen. The glucose is 'activated' as uridine diphosphoglucose (UDPG) and then polymerized into glycogen by glycogen synthase. Glycogen is broken down in times of need by glycogen phosphorylase and supplies the cells with an energy source.

In liver, glycogen phosphorylase has another important function; it is used to liberate free glucose into the blood stream, which maintains sufficient levels for the use, primarily, of brain and red blood cells. Other sugars absorbed from the intestine are converted into glucose or its derivatives and are also stored as glycogen.

Glycogen metabolism is controlled by two hormones; insulin, which stimulates synthesis, and glucagon, which stimulates breakdown (described in Chapter 20). Adrenaline promotes breakdown in muscle.

Galactose conversion to glucose involves a pathway involving UDP-galactose. A genetic disease, galactosaemia, in infants affects brain development. The disease is due to the absence of an enzyme, uridyl transferase, which is needed to convert galactose-1-P to UDPgalactose, a process involving UDPG. Early detection of the disease and dietary restriction to avoid galactose intake avoids the tragic repercussions, and normal development occurs.

Amino acids from the diet are not stored but are utilized at once for protein synthesis and other tissue components, the excess being disposed of. The carbon skeletons are converted into fat or glycogen or are oxidized depending on the particular amino acid and the metabolic controls operating at the time. In terms of energy, the main 'traffic' in amino acids is in the synthesis of glucose during starvation. The topic of amino acid metabolism is dealt with in Chapter 18.

Absorbed fat present in the blood as chylomicrons, is hydrolysed by a surface enzyme on cells lining the capillaries, to release fatty acids, which are taken up by cells of adjacent tissues. Substantial storage of fat as triacylglycerol occurs in the cells of adipose tissue after eating. This is released into the blood during fasting as free fatty acids for use by other tissues, a hormone-controlled process (see Chapter 20). The chylomicron remnants remaining, after tissues have taken up much of the fat, are taken up by the liver by receptor-mediated endocytosis. In the liver, the fat and cholesterol are liberated from these, the latter being stored as cholesterol ester.

The liver also synthesizes cholesterol and fat from glucose, and other food molecules, and exports these to other tissues. This occurs in the form of chylomicron-like structures called very-low-density lipoproteins (VLDLs). The fat is removed by tissues from VLDL by the same process as described for chylomicrons. As the fat removal continues, the VLDLs become low-density lipoproteins (LDLs), which now have a high proportion of cholesterol. These are taken up by the tissues by receptor-mediated endocytosis. This outward delivery of cholesterol from liver to extrahepatic tissues is counterbalanced by a reverse flow back to the liver. High-density lipoproteins (HDLs) pick up cholesterol from cells and transfer it to LDL, part of which is taken up by the liver. The cholesterol removal as HDL is brought about by the ABCA1 transporter.

The pathway of cholesterol movement is complex and the mechanisms of the two-way traffic not completely understood. It is however of great medical importance because the liver converts some of the cholesterol to bile acids and is a route for cholesterol removal. Excess LDL in the blood is associated with atherosclerosis, causing heart attacks. The 'statin' drugs inhibit cholesterol synthesis and are widely used therapeutically to reduce cholesterol levels in blood.

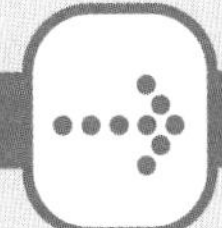

Further reading

To access the further reading, please scan the QR code image or go to http://global.oup.com/uk/orc/biosciences/molbiol/snape_biochemistry/student/reading/ch11/

Problems

1. Starting with glucose-1-phosphate, explain how the synthesis of glycogen is made thermodynamically irreversible.
2. Which tissues release glucose into the blood as a result of glycogen breakdown? How is this made possible?
3. What reaction do the two enzymes glucokinase and hexokinase catalyse? What is the physiological significance of the liver having glucokinase while brain and other tissues have hexokinase?
4. In the genetic disease galactosaemia, infants are unable to metabolize galactose properly. Removal of galactose from the diet can prevent the deleterious effects of the disease. UDP-galactose epimerase catalyses a reversible reaction; why is this important in connection with the above statement?
5. How is triacylglycerol (TAG) removed from chylomicrons and utilized by tissues?
6. Explain what very-low-density lipoprotein (VLDL) is and what is its function.
7. What is the main route of cholesterol removal from the body? Why is this of medical interest?
8. Cholesterol is esterified in high-density lipoprotein (HDL). Conversion of cholesterol to its ester is an energy-requiring process but HDL has no access to ATP. How is the esterification achieved?
9. How is fat released from adipose cells and under what conditions? How is it transported in the blood for transportation to other tissues? Compare this to the transport of fat from the liver to other tissues.
10. A genetic disease exists in which cholesterol levels in the blood are very high. What might be the cause of this?
11. How is fat from the adipose tissue transported and utilized by cells? How is fat from the liver transported and utilized by cells?
12. What is the reverse flow of cholesterol and what is its significance?

Chapter 12

Principles of energy release from food

The release of energy in the form of adenosine triphosphate (ATP) from glucose, fats, and amino acids involves fairly long and somewhat involved metabolic pathways. It is possible that, if we go into these pathways straight away, the overall strategy of energy generation by cells may be lost among the detail. We will, therefore, at this stage, deal with major stages of the processes, picking out landmark metabolites only, and view the strategies on a broad basis. When this has been done, subsequent chapters will deal in more detail with the pathways; in short, we would like you to see the wood before the trees.

Overview of glucose metabolism

Before we go into the stages of glucose oxidation we need to say a little about biological oxidation in general.

Biological oxidation and hydrogen-transfer systems

Oxidation does not necessarily involve oxygen; in fact, very few biological oxidations are direct additions of oxygen. Oxidation involves the removal of electrons, and chemically this is the definition of oxidation. It may involve *only* electron removal such as in the ferrous/ferric system:

$$Fe^{2+} \rightarrow Fe^{3+} + e^-$$

or it may be the removal of electrons accompanied by protons from a hydrogenated molecule. In biological systems it commonly involves enzymic removal of two hydrogen atoms from a metabolite molecule such as:

$$-CH_2-CH_2- \rightarrow -CH{=}CH- + 2H^+ + 2e^- \text{ or}$$
$$-CHOH-CH_2- \rightarrow -CO-CH_2- + 2H^+ + 2e^-$$

In such chemical oxidation systems, the electrons must be transferred from the electron donor to an electron acceptor. Depending on the particular electron acceptor, the electron transferred may be accompanied by the proton, in which case it is a hydrogen atom being transferred, or alternatively the proton may be liberated into solution, and only the electron transferred.

The ultimate electron acceptor in the aerobic cell is oxygen. Oxygen is **electrophilic** – it has an avidity for electrons. When it accepts four electrons, it also accepts four protons from the solution and forms water molecules:

$$O_2 + 4e^- + 4H^+ \rightarrow 2H_2O$$

However, oxygen is only the *ultimate* electron acceptor in the cell – there are other electron acceptors, which form a relay system, carrying electrons from metabolites to oxygen by a chain of intermediate carriers handing electrons from one to the next, until finally they are delivered to oxygen. They accept electrons and hand them on to the next acceptor, and thus are electron carriers. This is the **electron transport chain**, which plays a predominant part in ATP generation. The essential concept is given in Fig. 12.1. Two of the electron carriers involved in energy production are of such central importance in metabolism that they merit description here. It is essential for you to be completely familiar with these electron carriers.

NAD^+ - an important electron/hydrogen carrier?

The first carrier involved in the oxidation of many metabolites is NAD^+, which stands for **nicotinamide adenine dinucleotide**. You have already met a nucleotide in the form of adenine monophosphate (AMP) (see Fig. 3.3); a **nucleotide** has the general structure, base–sugar–phosphate, the base in AMP being adenine. (The structures of adenine and other bases are dealt with later in Chapter 22 and need not concern us now.) In the case of NAD^+, a dinucleotide is formed by linking the two phosphate groups of two nucleotides (which is unlike the way nucleotides are linked together in nucleic acids, as will be evident in later chapters). The general structure of NAD^+ is:

Base—sugar—phosphate—phosphate—sugar—base

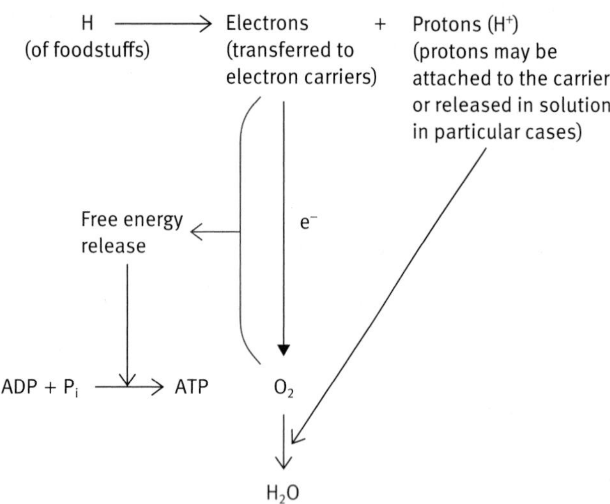

Fig. 12.1 The concept of electron transfer to oxygen and ATP generation.

The two bases are adenine (as in ATP) and nicotinamide, respectively – giving the structure of NAD below.

Adenine — Nicotinamide
| |
Ribose — Ribose
| |
Phosphate–Phosphate

NAD$^+$ is a **coenzyme** – a small organic molecule that participates in enzymic reactions. It differs from an ordinary enzyme substrate only in that its reduced form leaves the enzyme and attaches to a second enzyme where it donates its reducing equivalents to a second substrate. NAD$^+$ thus acts catalytically by being continually reduced and reoxidized, and in doing so transfers electrons from one molecule to another. It also transfers a proton, and so transfers one hydrogen atom and one electron (see following structures).

The 'business end' of the molecule is the nicotinamide group. It is derived from the vitamin, nicotinic acid or niacin. Niacin is an essential vitamin for some animals, but humans can also make it from tryptophan, which is an essential amino acid. As long as they consume a sufficiently high protein diet, they can convert some of the tryptophan into niacin. The rest of the molecule fits the coenzyme to appropriate enzymes but does not undergo any chemical change.

Nicotinamide has the structure:

H, $CONH_2$, N (pyridine ring).

When linked in NAD$^+$ the structure is:

H, $CONH_2$, N^+–R

where R is the rest of the NAD$^+$ molecule.

NAD$^+$ can be reduced by accepting two electrons from two hydrogens on a metabolite. One proton is transferred as a hydride ion (H:$^-$), to give the following structure (the second proton being liberated into solution):

H H, $CONH_2$, N–R $+ H^+$

Reduced NADH + H$^+$

NAD$^+$ is the coenzyme for several dehydrogenases that catalyse this type of reaction:

$$AH_2 + NAD^+ \rightleftharpoons A + NADH + H^+.$$

The reduced NAD$^+$ can then diffuse to a second enzyme and participate in a reaction such as:

$$B + NADH + H^+ \rightleftharpoons BH_2 + NAD^+.$$

In this way NAD$^+$ acts, effectively, as the carrier for the transfer of a pair of hydrogen atoms from A to B even though it carries only one proton:

$$AH_2 + B \rightarrow A + BH_2.$$

In biochemistry such reactions are often presented using curved arrows.

AH_2 → A; NAD$^+$ ⇄ NADH + H$^+$; B → BH_2

In case it requires extra emphasis, *NADH+H$^+$ can add two H atoms to a substrate* since the second electron it transfers to an acceptor molecule is joined by a proton from solution as shown above. In equations, reduced NAD$^+$ is written as NADH+H$^+$. In the text, the term NADH is used, but it always implies NADH+H$^+$.

FAD and FMN are also electron carriers

Another (hydrogen) carrier is **flavin adenine dinucleotide (FAD)**. In this case the electrons are transferred as hydrogen atoms. FAD is synthesized from vitamin B_2 or riboflavin. The important feature of this molecule is that it can (in combination with appropriate proteins) accept two H atoms to become

$FADH_2$. FAD is a prosthetic group – it is a permanent attachment to its apoenzyme, unlike NAD^+, which moves from one dehydrogenase to another. Its role will become clear; for the present it is sufficient that FAD can be reduced. The chemistry of this reduction is given below, where a related carrier, flavin mononucleotide (FMN), is also described.

FAD has the structure: isoalloxazine ring structure–ribitol–phosphate–phosphate–ribose–adenine. It is very unusual in having the linear ribitol molecule instead of ribose.

Oxidation–reduction reactions occur in the isoalloxazine structure:

FAD (oxidized form)

$FADH_2$ (reduced form)

FMN has the structure: isoalloxazine–ribitol–phosphate. It is reduced in the same way as FAD.

Energy release from glucose

Glucose is stored in liver and muscle as glycogen in animals and then broken down to glucose-1-phosphate, which is oxidized. This does not affect the overall account given below except that, as described in later chapters, the initial steps of the metabolism of glycogen and glucose are slightly different and for these the ATP yield from glycogen is three per glucosyl unit, not two as from glucose. (In the case of glucose an ATP molecule is used to form glucose-6-phosphate. Of course, a UTP molecule per molecule of glucose was used to synthesize glycogen in the first place.) For convenience, in this section, we will refer to glucose oxidation.

The main stages of glucose oxidation

Overall, glucose is oxidized as follows:

$$C_6H_{12}O_6 + 6O_2 \rightarrow 6CO_2 + 6H_2O$$

The $\Delta G^{0\prime}$ for this reaction is $-2820\,kJ\,mol^{-1}$, liberating large amounts of energy.

In the cell, this oxidation process is accompanied by the synthesis of more than 30 molecules of ATP from ADP and P_i. The entire process of glucose oxidation to CO_2 and H_2O can be divided up into three stages. If you cannot fully understand the summaries below do not be concerned – they will become clear shortly.

- Stage 1: **glycolysis**. This results in the lysis or splitting of glucose into two C_3 fragments ultimately yielding pyruvate, accompanied by reduction of NAD^+. *This occurs in the cytosol of cells.*
- Stage 2: the **tricarboxylic acid (TCA) cycle** or, strictly speaking tricarboxylate cycle, also known as the **Krebs cycle** (after its discoverer). For consistency, we will refer to it as the TCA cycle throughout the book. In mitochondria the carbon atoms of pyruvate are converted into an acetyl group and CO_2, and in the cycle, electrons from the acetyl groups are transferred to electron carriers. No oxygen is involved at this stage. Carbon atoms are released as CO_2, the oxygen largely from water. The cycle is located inside mitochondria in eukaryotes and in the cytosol of bacteria.
- Stage 3: the **electron transport system (or chain)**. Electrons are transported from the electron carriers to oxygen where, with protons from solution, water is formed. It is in Stage 3 that most of the ATP is generated. This occurs in the inner mitochondrial membrane in eukaryotes and in the cell membrane in prokaryotes.

Stage 1 in the release of energy from glucose: glycolysis

Glycolysis does not involve oxygen, and only two ATP molecules per molecule of glucose lysed are produced from ADP. The end products are pyruvate and NADH as shown in Fig. 12.2. In **aerobic glycolysis**, when the oxygen supply is plentiful, the NADH is reoxidized to NAD^+ *via* mitochondria in eukaryotic cells; the pyruvate is taken up by the mitochondria where it is metabolized to CO_2 and H_2O *via* the TCA cycle.

Anaerobic glycolysis

In animals, oxygen is not always plentiful since the delivery in blood may not keep up with its usage, especially in muscle during vigorous activity. Since NAD^+ acts catalytically and is present in cells in small amounts, for glycolysis to proceed the NADH must be recycled back to NAD^+; no NAD^+, no glycolysis. If the capacity of mitochondria to do this is inadequate at high glycolytic rates and/or there is insufficient oxygen availability, glycolysis needed to generate ATP for contraction would be impaired unless another way were found to regenerate NAD^+ from NADH. An 'emergency' system comes into play in this situation. The NADH is reoxidized by reducing pyruvate to produce lactate. The reaction is:

$$\underset{\text{Pyruvate}}{CH_3COCOO^-} + NADH + H^+ \underset{\text{Lactate dehydrogenase}}{\rightleftharpoons} \underset{\text{Lactate}}{CH_3CHOHCOO^-} + NAD^+.$$

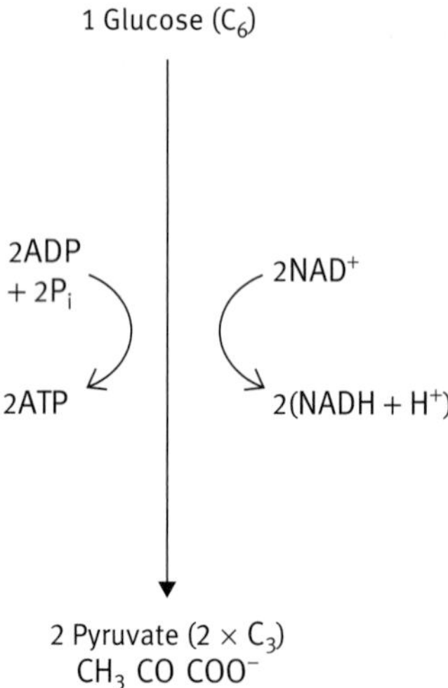

Fig. 12.2 The net result of aerobic glycolysis of glucose. NADH is oxidized by mitochondria.

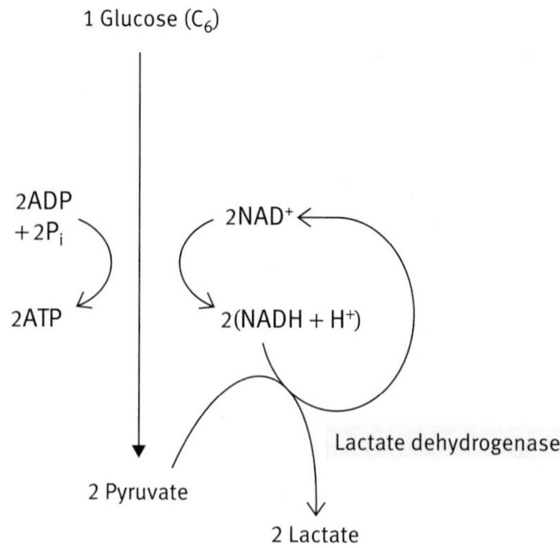

Fig. 12.3 The net result of anaerobic glycolysis of glucose.

The enzyme catalysing this reaction is **lactate dehydrogenase**. The production of lactate from glucose is known as **anaerobic glycolysis**, as opposed to aerobic glycolysis which forms pyruvate, which is then converted into acetyl-CoA and channeled into the TCA cycle. What is achieved in anaerobic glycolysis is not so much the production of lactate but the reoxidation of NADH, which permits continued ATP production from glycolysis (Fig. 12.3). Since there is a lot of lactate dehydrogenase in muscle, the NADH can be rapidly reoxidized in this way and this in turn permits glycolysis to proceed at a very fast rate. The advantage of this is that, although only two ATP molecules are generated per molecule of glucose, relatively large amounts of glucose can be degraded, which could be important when muscles are very active. If you are being chased by a tiger, the extra ATP could be very important for survival. The lactate formed enters the blood stream but is not wasted; it is used mainly by the liver to resynthesize glucose and glycogen, as described later.

Erythrocytes also depend on anaerobic oxidation as they contain no mitochondria. In this case, there is plenty of oxygen in the cell but the machinery of the TCA cycle and oxidative phosphorylation is absent.

As a parenthetic note, yeast can live entirely on anaerobic glycolysis in the absence of oxygen by using an analogous mechanism to reoxidize NADH. The pyruvate is converted into acetaldehyde+CO_2; a second enzyme, alcohol dehydrogenase, reduces the acetaldehyde to ethanol.

$$CH_3COCOO^- + H^+ \longrightarrow CH_3CHO + CO_2$$

Pyruvate decarboxylase

$$CH_3CHO + NADH + H^+ \rightleftharpoons CH_3CH_2OH + NAD^+$$

Alcohol dehydrogenase

Because of the low yield of ATP, large quantities of glucose are degraded, producing alcohol and CO_2. To illustrate the scale of this by a homely example, the latter causes the explosion of bottles if, during beer-making, it is bottled too early while glycolysis is still vigorous. Pyruvate decarboxylase is not present in animals, which is just as well perhaps, since vigorous exercise would be literally intoxicating.

As already mentioned in the introduction, in the above account, for simplicity, we have talked about glucose being the carbohydrate that is oxidized by glycolysis. It can indeed be free glucose, but in muscle it is mainly glycogen, the storage form of glucose, that is degraded. We also again remind you that in the next chapter the detailed mechanisms of glycolysis, the TCA cycle, and the electron transport system will be given. At this stage the main thing is to get an overview.

Stage 2 of glucose oxidation: the TCA cycle

The **mitochondria**, as described in Chapter 2, are small organelles located in the cytosol of the cell. They are where most of the ATP is produced. The inner membrane is the site of ATP generation. Its area is increased by invagination to form **cristae**, whose extent reflect the level of ATP demand in different tissues (Fig. 12.4). The mitochondrion is filled with a concentrated solution of enzymes – it is called the **matrix**. It is here that Stage 2 of glucose metabolism mainly occurs, only one reaction being located in the inner membrane.

Aerobic glycolysis, as stated, produces pyruvate and NADH in the cytosol. For further oxidation to take place, the pyruvate must enter the mitochondrion. For NADH oxidation, the electrons are transported in and oxidized, leaving the NAD^+ in the cytosol to participate in glycolysis. A transport system in the inner mitochondrial membrane takes the pyruvate from the cytosol into the mitochondrion.

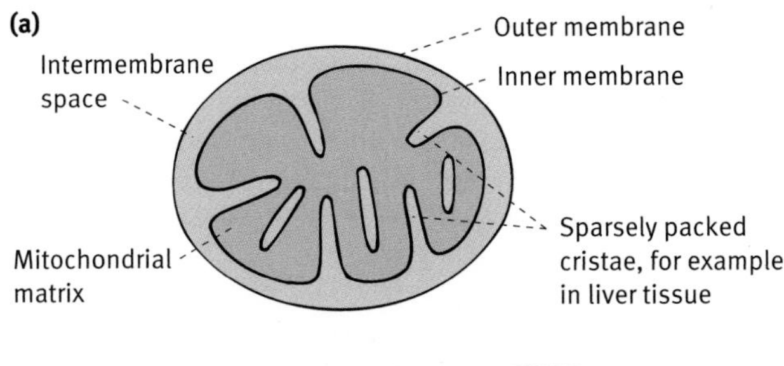

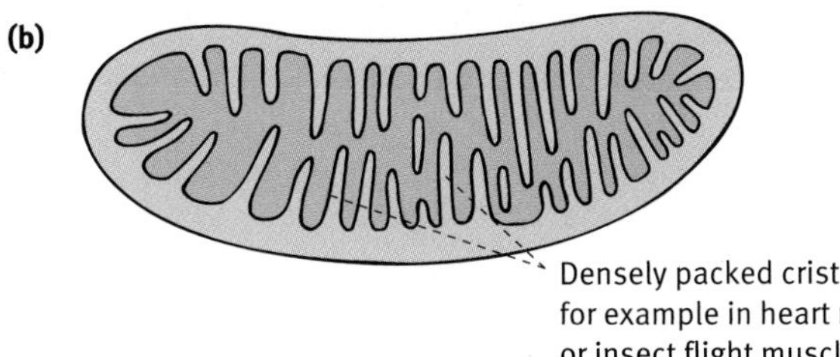

Fig. 12.4 Mitochondria from **(a)** liver and **(b)** heart tissue. The number of cristae reflects the ATP requirements of the cell. See Fig. 2.6 for an electron micrograph of a mitochondrion.

How is pyruvate fed into the TCA cycle?

We now come to an enzyme reaction of major importance in which pyruvate, transported into mitochondria, is converted into a compound that is at the crossroads of energy metabolism. This is acetyl-coenzyme A (acetyl-CoA), a compound not previously mentioned in this book.

What is coenzyme A?

Coenzyme A is usually referred to as CoA for short, but is written in equations as CoA—SH because its thiol group is the reactive part of the molecule.

Unlike NAD^+ and FAD, CoA is *not an* electron carrier, but an acyl group carrier (A for acyl). Like NAD^+ and FAD, it is a dinucleotide and, as so often happens with cofactors, incorporates a water-soluble vitamin in its structure, in this case **pantothenic acid**. Pantothenic acid has an odd sort of structure.

$$HO{-}CH_2{-}\overset{CH_3}{\underset{CH_3}{C}}{-}\overset{H}{\underset{OH}{C}}{-}\overset{O}{\overset{\|}{C}}{-}NH{-}CH_2{-}CH_2{-}C\begin{smallmatrix}{\nearrow}O\\ \searrow OH\end{smallmatrix}$$

Pantothenic acid

You would normally expect the vitamin moiety of a coenzyme to play a part in the reaction for which the coenzyme is required (as with, for example, riboflavin in FAD and nicotinamide in NAD^+), but the pantothenic acid moiety is just 'there' and apparently quite inert. It presumably provides a recognition group to help bind the CoA to appropriate enzymes but why this particular structure should be used is not obvious. It may be a quirk that happened early in evolution and has remained ever since.

The structure of CoA is:

```
         Adenine        β-Mercaptoethylamine
            |                  |
Phosphate—Ribose        Pantothenic acid
            |                  |
         Phosphate——————Phosphate
```

The **β-mercaptoethylamine** part is the business end of the molecule.

$$\underbrace{RCO{-}NHCH_2CH_2{-}SH}_{\beta\text{-Mercaptoethylamine moiety}}$$

The CoA molecule carries acyl groups as thiol esters. For example, acetyl-CoA can be written as $CH_3CO{-}S{-}CoA$. *The thiol ester is a high-energy compound* (unlike a carboxylic ester). It has a $\Delta G^{0\prime}$ of hydrolysis of approximately $-31\,kJ\,mol^{-1}$ as compared with about $-20\,kJ\,mol^{-1}$ for a carboxylic ester. The difference is due to the fact that the latter is resonance stabilized and so has a lower free-energy content than a thiol ester, which is not resonance stabilized. With that description of CoA, we can now return to the question of what happens to pyruvate transported into mitochondria.

Oxidative decarboxylation of pyruvate

The pyruvate is subjected to an *irreversible* oxidative decarboxylation in which CO_2 is released (decarboxylation), a pair of electrons is transferred to NAD^+ (oxidation), and an acetyl group is transferred to CoA. (Note that this is different from the **nonoxidative decarboxylation** carried out by yeast pyruvate decarboxylase, described earlier. In the latter, there is, as implied, no oxidation, NAD^+ is not involved, and the product is acetaldehyde, not an acetyl group.) The large negative free-energy change means that the oxidative decarboxylation reaction is irreversible. This has important metabolic repercussions namely that fatty acid cannot be converted into glucose in fasting, and starvation and body protein has to be degraded instead. The reaction, catalysed by **pyruvate dehydrogenase**, is as follows:

$$\text{Pyruvate} + \text{CoA}{-}\text{SH} + NAD^+ \rightarrow \text{Acetyl}{-}\text{S}{-}\text{CoA} + NADH + H^+ + CO_2$$
$$\Delta G^{0\prime} = -33.5\,kJmol^{-1}$$

The acetyl group of acetyl-CoA is now fed into the TCA cycle. At this stage we will not be concerned with the reactions in the cycle. The essential point is that the carbon atoms of the acetyl group of acetyl-CoA produced from pyruvate are converted into CO_2 while three molecules of NAD^+ are reduced to NADH. In addition, a molecule of FAD is reduced to $FADH_2$, the electrons coming indirectly, in part, from water (see Stage 2 – the TCA cycle in Chapter 13). The cycle generates one 'high-energy' phosphoryl group from P_i for each acetyl group fed in. Stage 2 of glucose oxidation is illustrated in Fig. 12.5.

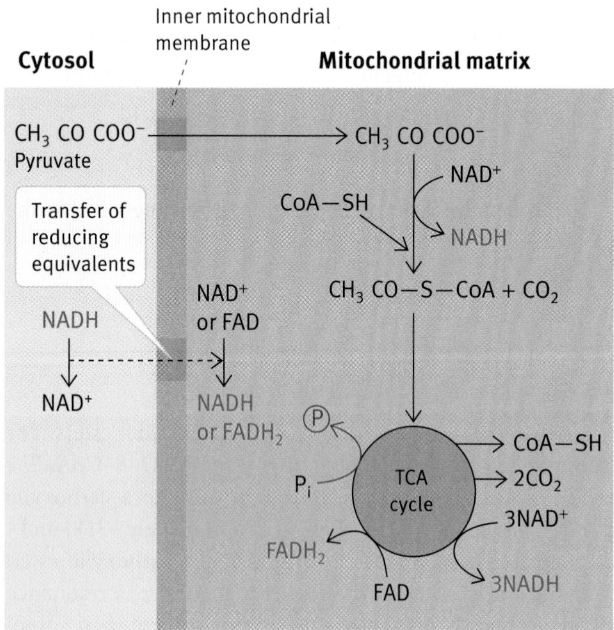

Fig. 12.5 This shows what happens to pyruvate and NADH generated in glycolysis and the generation of further NADH and $FADH_2$ inside mitochondria. The FAD is always attached to an enzyme. The source of the reducing equivalents is dealt with in the next chapter. The NADH and $FADH_2$ are oxidized by the electron transport pathway of the inner membrane, described shortly.

In summary, a molecule of pyruvate from the cytosol is converted in the mitochondria by pyruvate dehydrogenase and the TCA cycle into three molecules of CO_2 and, in the process, three molecules of NAD^+ and one molecule of FAD are reduced. We have hardly started to make ATP – only two molecules in glycolysis and two in the cycle (per starting glucose molecule) but there are almost 30 still to be made.

So far, it has been mainly preparation of fuel – the big return in the form of ATP generation comes in Stage 3.

Stage 3 of glucose oxidation: electron transport to oxygen

The oxidation of NADH and $FADH_2$ takes place in the inner mitochondrial membrane, which contains the chain of electron carriers.

The electron transport chain – a hierarchy of electron carriers

Redox potentials

The problem we are now concerned with is the transfer of electrons from NADH and $FADH_2$ to oxygen with the formation of water:

$$NADH + H^+ + \tfrac{1}{2}O_2 \rightarrow NAD^+ + H_2O$$

The $\Delta G^{0'}$ of this reaction is $-220\,kJ\,mol^{-1}$. To understand the process, we need to discuss the **redox potential** (also known as the **oxido-reduction potential**) of compounds.

Compounds capable of being oxidized are, by definition, electron donors; in any oxido-reduction reaction there must be an electron acceptor (oxidant) and a donor (reductant). The reaction:

$$X^- + Y \rightleftharpoons X + Y^-$$

can be considered to occur by two theoretical half-reactions, as follows:

$$\text{(a) } X^- \rightleftharpoons X + e^-;$$
$$\text{(b) } Y + e^- \rightleftharpoons Y^-.$$

Each of these involves the reduced and oxidized forms of each reactant, which are called **conjugate pairs** or reduction-oxidation couples or, the most convenient term, **redox couples**. X and X^- are such a couple and Y and Y^-, another couple. Clearly, any real-life oxidation-reduction reaction must involve a pair of redox couples, because one must donate electrons and the other must accept them.

Different redox couples have different affinities for electrons. One with a lesser affinity will tend to donate electrons to another of higher affinity. The redox potential value (E_0') is a measurement of the electron affinity or electron-donating potential of a redox couple. This is of importance in biochemistry for it is an indicator of the direction in which electrons will tend to flow between reactants. Equally important E_0' values are directly related to free-energy changes.

E_0' values are expressed in volts – the more negative, the lower the electron affinity, the greater the tendency to hand on electrons, the greater the reducing potential, and the higher the energy of the electrons.

Determination of redox potentials

The reason why this chemical property is expressed in volts is due to the method of redox potential determination. As stated above, two redox couples must be involved in an oxido-reduction reaction but, since electron transfer is involved, they can be in two separate vessels (half-cells) if they are connected by a copper wire to conduct electron flow. The method is to use the $2H^+ + 2e^- \rightleftharpoons H_2$ equilibrium (catalysed by the platinum black electrode) in one half-cell as the reference redox couple, and compare the unknown sample in the second half-cell with it. Electrons will flow through the wire according to the relative electron affinities of the two systems. The positive ions in each half-cell (H^+ in the hydrogen electrode half-cell and, for example, ferrous and ferric ions in the sample half-cell) are accompanied by anions. Change in the positive ions due to loss or gain of electrons in the half-cells necessitates a compensating anion migration from one half-cell to the other. The agar-salt bridge shown in Fig. 12.6 provides the route for this.

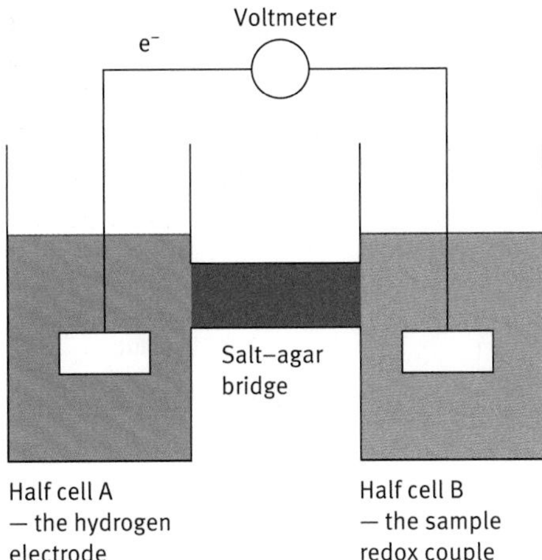

Fig. 12.6 Apparatus for measurement of redox potentials. The reference hydrogen electrode in A contains the redox couple $H_2/2H^+$ catalysed by platinum black on the electrode. The sample half-cell B contains equimolar amounts of the two components of the redox couple whose redox potential is being determined. If B is more reducing than A, electrons will flow from B to A and reduce $2H^+$ to H_2 if the half-cells are connected by a wire. The E_0' value is therefore more negative than that of the hydrogen electrode whose E_0' is assigned the value of −0.42 V. As protons in A are reduced to hydrogen atoms, anions will flow from A to B via the agar-salt bridge to preserve charge neutrality. If the sample in B is less reducing than the reference hydrogen electrode, a reverse series of events will occur and the sample redox couple will have an E_0' value more positive than −0.42 V.

The electrical potential between the half-cells is measured by a voltmeter inserted into the copper wire connecting electrodes in each half-cell. The reference (hydrogen electrode) half-cell is arbitrarily assigned the value of zero and the relative value of the sample cell is the redox potential of the redox pair in it. In physics, the convention is that electrical current flows in the opposite direction to electron flow and, therefore, the half-cell that is donating electrons has the more negative voltage. Thus, if the sample half-cell is more reducing than the reference half-cell, its redox potential value is more negative.

The E_0 values (written without a prime) are standard values measured at 1.0 M concentrations of components, or H_2 gas at 1 atmosphere pressure and 1.0 M HCl (pH 0). In biochemistry, values are adjusted to pH 7.0 instead of pH 0 and this brings the redox potential of the reference half-cell to a value of −0.42 V. Redox potentials corrected in this manner are written as E_0' values. The half-reaction $NAD^+ + 2H^+ + 2e^- \rightarrow NADH + H^+$ has an E_0' value of −0.32 V, and that of the half-reaction $\frac{1}{2}O_2 + 2H^+ + 2e^- \rightarrow H_2O$, an E_0' value of +0.82 V. The very large difference indicates that NADH has the potential to reduce oxygen to water but the reverse will not occur. The relationship between the $\Delta G^{0\prime}$ value and the E_0' value is shown in Box 12.1.

Electrons are transported in a stepwise fashion

In the cell, the transport of electrons to oxygen does not happen in a single step. In the electron transport system of the mitochondria there is a chain of electron carriers of ever-increasing redox values (decreasing reducing potentials) terminating in the ultimate electron acceptor, oxygen. In effect, electrons from NADH and $FADH_2$ bump down a staircase, each step being a carrier of the appropriate redox potential and each fall releasing free energy (Fig. 12.7). The free energy is thus liberated in manageable parcels – manageable in the sense that it can be harnessed into mechanisms that (indirectly) result in ATP generation from ADP and P_i rather than being wasted as heat, as occurs in simple burning of glucose. The oxidation of NADH and $FADH_2$ thus drives the conversion of ADP and P_i to ATP. Hence the complete process is called **oxidative phosphorylation**. Thirty or more ATP molecules are synthesized from the oxidation of one molecule of glucose (the exact number is discussed in the next chapter).

In Fig. 12.8, all three stages are put together in one scheme.

BOX 12.1 Calculation of the relationship between the $\Delta G^{0\prime}$ value and the E_0' value

There is a direct relationship between the $\Delta G^{0\prime}$ value and the E_0' value of an oxido-reduction reaction, quantified by the Nernst equation:

$$\Delta G^{0\prime} = -nF\Delta E_0',$$

where n equals the number of electrons transferred in the reaction, F is the Faraday constant (96.5 kJ V⁻¹ mol⁻¹), and E_0' is the difference in redox potential between the electron donor and electron acceptor.

If we consider the oxidation of NADH, the half-reactions are:

$$(1)\ NAD^+ + 2H^+ + 2e^- \rightarrow NADH + H^+ \quad E_0' = -0.32\ V,$$

$$(2)\ \tfrac{1}{2}O_2 + 2H^+ + 2e^- \rightarrow H_2O \quad E_0' = -0.816\ V.$$

For the overall reaction, subtracting (1) from (2), we get:

$$NADH + H^+ + \tfrac{1}{2}O_2 \rightarrow NAD^+ + H_2O,$$

$$\Delta E_0' = +0.816 - (-0.32) = +1.136\ V.$$

Therefore:

$$\Delta G^{0\prime} = -2(96.5\ kJ V^{-1} mol^{-1})\ (+1.136\ V) = -219.25\ kJ mol^{-1}.$$

Energy release from oxidation of fat

In addition to glucose and glycogen breakdown to supply the energy for ATP generation, the body oxidizes fat for the same purpose. In terms of energy production, it is the fatty acid components of neutral fat (triacylglycerol, TAG) that are quantitatively important, the glycerol portion being less significant. Chemically, fatty acids are quite different from glucose and you

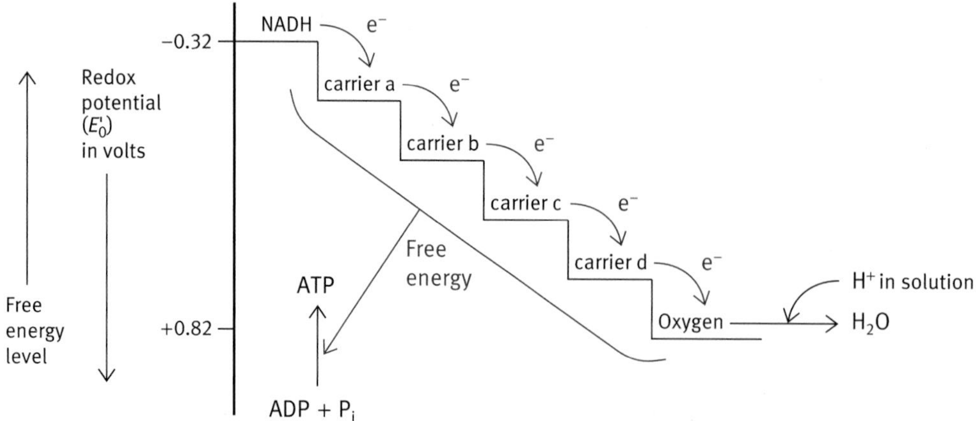

Fig. 12.7 Principle of the electron transport chain. (The number of carriers shown is arbitrary; each is a different carrier.) With some carriers only electrons are accepted, protons being liberated into solution, while in the case of other carriers, protons accompany the electrons. The final reaction with oxygen involves protons from solution.

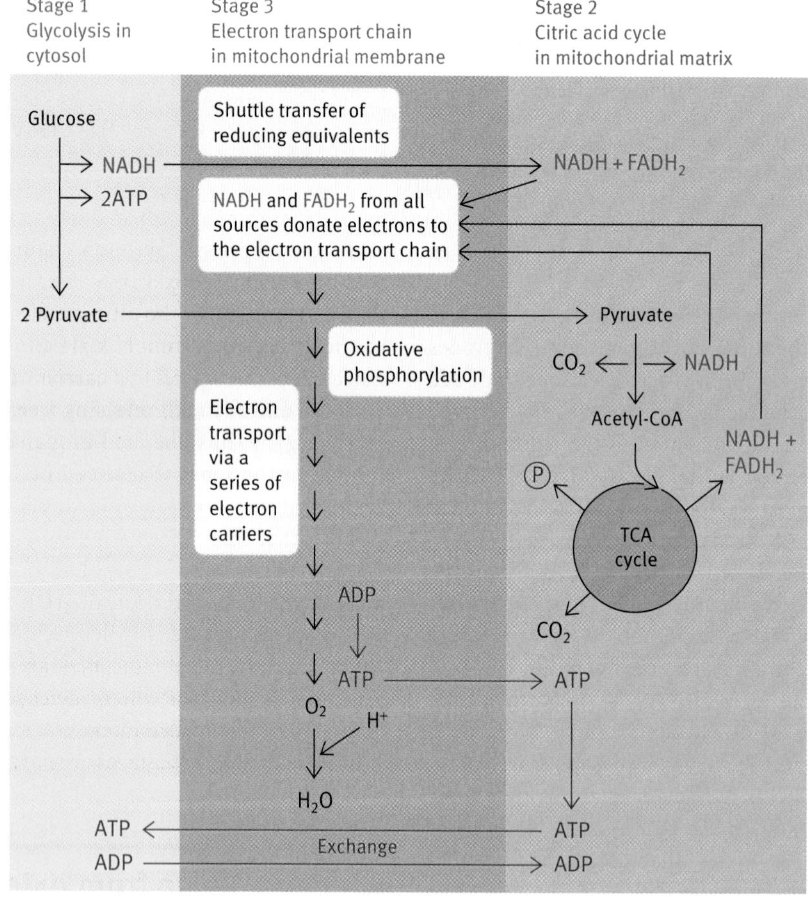

Fig. 12.8 Summary of the oxidation of glucose. For ease of presentation, only products are shown; obviously, reduced NAD is formed from NAD^+, etc. The same scheme applies to glycogen except that three ATP molecules are produced per glucose unit. We indicate the production of one from (Ⓟ) the TCA cycle rather than as one ATP because, in this case, GDP rather than ADP is the acceptor. Energetically, it amounts to the same thing and is explained in the next chapter. The ATP generated is transported to the cytosol by a mechanism involving ADP intake. An important point to note is that NAD^+ and FAD collect electrons from different metabolic systems including fatty acid oxidation and not just from glucose (see Fig. 12.10). NADH and $FADH_2$ donate electrons to different points in the electron transport chain. $FADH_2$ donates electrons to the chain at a point lower down than does NADH. See Figs 13.28 and 13.29 for details of shuttles.

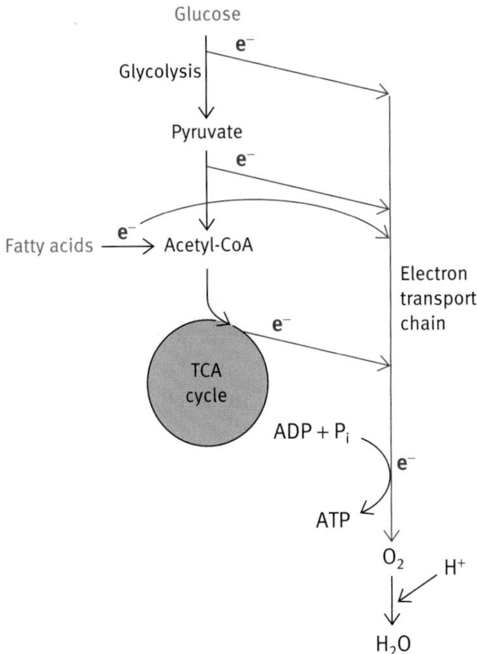

Fig. 12.9 Relationship of fat and glucose oxidation. This figure emphasizes the collection of electrons for the electron transport pathway.

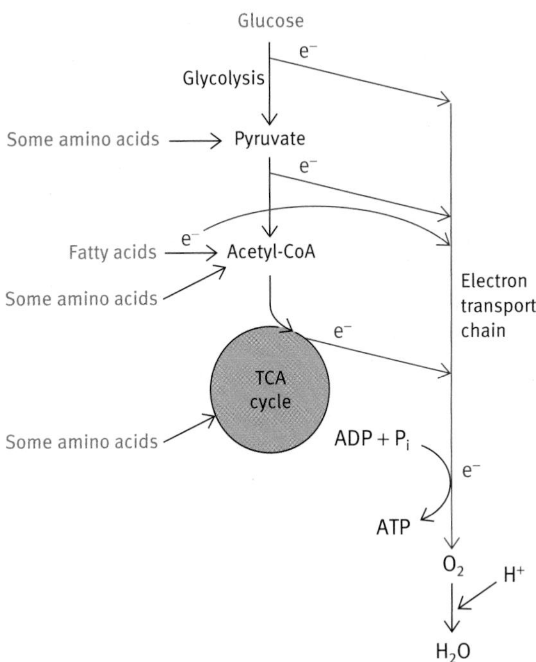

Fig. 12.10 Relationship of glucose, fat, and amino acid oxidation for energy generation.

might well expect their oxidation to be correspondingly different. It is at this point that you might get your first glimpse of how, despite complexity in detail, the metabolism of different foodstuffs dovetails together with majestic simplicity. Glucose, as described, is oxidized so that acetyl-CoA is formed and fed into the TCA cycle. Fatty acids are also metabolized by detaching two carbon atoms at a time as acetyl-CoA, which is also fed into the TCA cycle. In the preliminary metabolic steps, NAD^+ and FAD are reduced in both systems, and the electrons they carry are fed into the electron transport system as in glucose oxidation. Figure 12.9 gives the relationship between glucose and fat oxidation.

Energy release from oxidation of amino acids

The situation with regard to amino acid oxidation is more complex in detail but similar in concept. There are 20 different amino acids and, as explained earlier, if these are present in excess of immediate requirements their amino groups are removed and carbon-hydrogen skeletons used as fuel, (immediately or they may be stored as liver glycogen). The same happens in starvation when the supply is certainly not in excess, but muscle protein needs to be broken down and the amino acids used to provide glucose necessary for the brain's metabolic survival. Once again, the metabolism of the latter shows a simplicity of concept. The carbon-hydrogen skeletons are converted into pyruvate, or into acetyl-CoA, or into intermediates in the TCA cycle, so that they also join the same metabolic path (as shown in Fig. 12.10). The TCA cycle thus plays a central role in metabolism.

The interconvertibility of fuels

Although this chapter is concerned primarily with energy generation by food oxidation, a brief note here can throw light on the fuel logistics already described in earlier chapters. We described how glucose, in excess, can be converted into fat. This is because fatty acids can be synthesized from acetyl-CoA (see Chapter 17).

$$\text{Glucose} \rightarrow \text{pyruvate} \rightarrow \text{acetyl-CoA} \rightarrow \text{fatty acids}$$

In animals, pyruvate can be converted into glucose by a process known as gluconeogenesis, crucially important in starvation. Fatty acids, however, *cannot be* converted, in a net sense, into glucose because acetyl-CoA cannot be converted into pyruvate Pyruvate dehydrogenase catalyses the irreversible conversion of pyruvate into acetyl-CoA and, although fatty

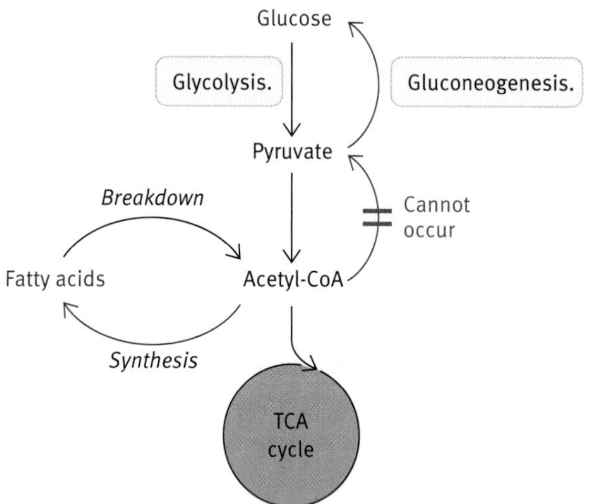

Fig. 12.11 Why glucose can be converted into fats but fats cannot be converted into glucose in animals. The reverse pathways in red are not completely the same as the forward pathways and are described in later chapters. In plants and bacteria, fat can be converted into glucose but not by the reversal of the pyruvate dehydrogenase reaction.

acids can be used as fuel in fasting and starvation, they cannot produce glucose (Fig 12.11).

Those amino acids that give rise to pyruvate or a TCA cycle intermediate, can be converted into glucose (glucogenic amino acids). In starvation, muscle proteins are degraded and the amino acids so produced transported to the liver, which converts them into glucose. There are organisms such as bacteria and plants that do have the capacity to convert fat and C_2 compounds, such as acetate to glucose, but this involves a special cycle known as the **glyoxylate cycle** (a modified TCA cycle) that is *not* present in animals. This is also described later (see Fig. 16.6).

The directions of metabolism, whether fat and glucose are oxidized or synthesized, are the result of control mechanisms to be described in Chapter 20.

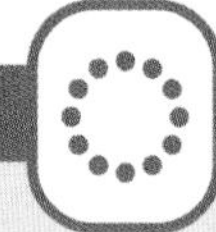

Summary

Biological oxidation involves removal of electrons and transfer to another acceptor molecule, which need not be oxygen. In biochemical systems it commonly involves enzymic removal of two hydrogen atoms from a metabolite molecule. In the cell a variety of electron/hydrogen carriers exist in the transfer of electrons to oxygen. Nicotinamide adenine dinucleotide (NAD^+) is an important one. It is a dinucleotide containing the vitamin nicotinamide, which can accept two electrons and a hydrogen atom, forming NADH. It functions as a coenzyme for dehydrogenases. FAD is of similar structure but the accepting group is the vitamin riboflavin. Flavin mononucleotide (FMN) is a single-nucleotide form. They are hydrogen carriers being reduced respectively to $FADH_2$ and $FMNH_2$.

Glucose oxidation occurs in three stages. Stage 1, glycolysis, occurs in the cytosol; Stage 2, the TCA cycle, occurs in the mitochondrial matrix; and Stage 3, the electron transport system, is in the inner mitochondrial membrane, the latter being the main site of ATP generation.

Stage 1 Glycolysis produces pyruvate from glucose or glycogen or, in the absence of sufficient capacity to reoxidize NADH by mitochondria, lactic acid.

Stage 2 Pyruvate enters the mitochondria and is converted into acetyl-CoA. Coenzyme A (CoA) is of central importance; it is a dinucleotide containing the vitamin pantothenic acid. The TCA cycle extracts high-energy electrons from the acetyl groups in the form of reduced electron carriers, NAD^+ and FAD, while the carbon atoms of the acetyl group are converted to CO_2.

Stage 3 The electron transport system is a hierarchy of carriers of different energy potentials and the electrons move from one to the other down the energy gradient to oxygen. The energy potential of the transfer of electrons from one carrier to the next is quantified by the redox value, which can be directly measured in a simple apparatus. The energy released during the electron transport is ultimately trapped as ATP.

The release of energy from the oxidation of fatty acids differs only in their preliminary conversion to acetyl-CoA, which is oxidized in mitochondria, as is the case for pyruvate. The conversion of pyruvate to acetyl-CoA is irreversible so that, in animals, fat cannot be converted into glucose as acetyl-CoA cannot be converted into pyruvate.

Energy release from oxidation of amino acids is more complicated since there are 20 different amino acids, each with its own pathway of metabolism. However they are all converted into pyruvate, acetyl-CoA or into TCA cycle intermediates.

The detailed mechanisms by which these processes occur are given in the next chapter.

Problems

1 What are the three major phases involved in the oxidation of glucose and where do they occur?

2 Write down the overall structure of NAD^+ in words; show the structure of the electron-accepting group in the oxidized and reduced form. Explain how NAD^+ acts as a hydrogen carrier between substrates.

3 What is FAD and what is its role?

4 Outline the difference between aerobic and anaerobic glycolysis in muscle and the circumstances in which they occur. What is achieved by anaerobic glycolysis?

5 Write down the structure of coenzyme A in words; also give the structure of its acyl-accepting group. What is the $\Delta G^{0\prime}$ of hydrolysis of a thiol ester? How does this compare with that of a carboxylic ster?

6 'The pyruvate dehydrogenase reaction is of central importance'. Explain this statement.

7 What normally happens to the acetyl-CoA generated in the pyruvate dehydrogenase reaction?

8 Glycolysis and the TCA cycle produce NADH and $FADH_2$. What happens to them?

9 The redox couple $FAD+2H^++2e^- \rightarrow FADH_2$ has an E_0' value of −0.219 V. That of $\frac{1}{2}O_2+2H^++2e^- \rightarrow H_2O$. Calculate the $\Delta G^{0\prime}$ value for the oxidation of $FADH_2$ by oxygen to water. The Nernst equation is $\Delta G^{0\prime}=nF\Delta E_0'$ where $F=96.5\,kJ\,V^{-1}\,mol^{-1}$.

10 What is the major source of acetyl-CoA other than the pyruvate dehydrogenase reaction?

11 Can glucose be converted into fat? Explain your answer.

12 Can fatty acids be converted into glucose in animals? Explain your answer.

Chapter 13

Glycolysis, the TCA cycle, and the electron transport system

In Chapters 10–12, you saw the overall pattern of the way in which various foodstuffs are metabolized and we outlined the three stages of the oxidation of food components absorbed from the diet.

We now want to described more fully the mechanisms of the metabolic pathways, starting with carbohydrate oxidation. In subsequent chapters we will deal with how fatty acids are metabolized to join up to the glucose oxidation pathway at acetyl-CoA and then do the same with amino acid metabolism. A potential problem in studying these pathways is to forget their overall physiological significance and to get lost in the detail. If necessary, keep on going back to the previous chapter to refresh your memory on where pathways are heading.

Regulation of these pathways is dealt with in Chapter 20. We have chosen to have a separate chapter on regulation, because it enables us to deal with the general strategies of control that apply to all aspects of metabolism. It also allows a more integrated approach.

Stage 1 – glycolysis

This, we remind you, results in the division (lysis) of glucose or a glucosyl unit of glycogen (glyco, six carbon compounds) into two molecules of pyruvate (three carbon compound) with the concomitant reduction of two molecules of NAD^+.

Glucose or glycogen?

So far, for simplicity, we have talked mainly of glucose catabolism. However, in liver, skeletal muscle, and parts of the kidney, glucose is stored as glycogen, and glycolysis may be proceeding from this rather than from free glucose. There is a difference between the two.

When glycogen is degraded, glucose-6-phosphate is produced via glucose-1-phosphate. In the liver, this can be hydrolysed to release free glucose into the blood. However, glucose-6-phosphate is on the glycolytic pathway also and in all tissues can be broken down to pyruvate. In muscle in particular, there is no possibility for the glucose-6-phosphate to release free glucose as there is no glucose-6-phosphatase. The muscle, therefore, cannot supply glucose into the bloodstream for use of other tissues but will metabolize the glycogen in situ.

Free glucose, obtained from the blood, is also converted into glucose-6-phosphate by phosphorylation using adenosine triphosphate (ATP). You have met this reaction already (see Chapter 11), as it is the same as that involved in glycogen synthesis.

Whether glucose-6-phosphate goes to glycogen or to pyruvate or to blood glucose, in the case of liver, depends on how the metabolic control switches are set, according to physiological needs, and this is a major later topic (see Chapter 20). The relationship between glycolysis, glucose, and glycogen is shown in Fig. 13.1.

ATP is needed at the beginning of glycolysis

It may seem odd that a pathway designed to produce ATP should start by using up ATP. Glycolysis involves phosphorylated compounds and ATP must be used to phosphorylate glucose – it has the necessary energy potential. Glucose-6-phosphate is a low-energy phosphoryl compound so certainly we have lost a high-energy phosphoryl group in using ATP. Think of it as an investment for, as you will see, profit is made on the ATP used in glycolysis of glucose.

Conversion of glucose-6-phosphate into fructose-6-phosphate

The next step is to convert glucose-6-phosphate, an aldose sugar, into fructose-6-phosphate, its ketose isomer.

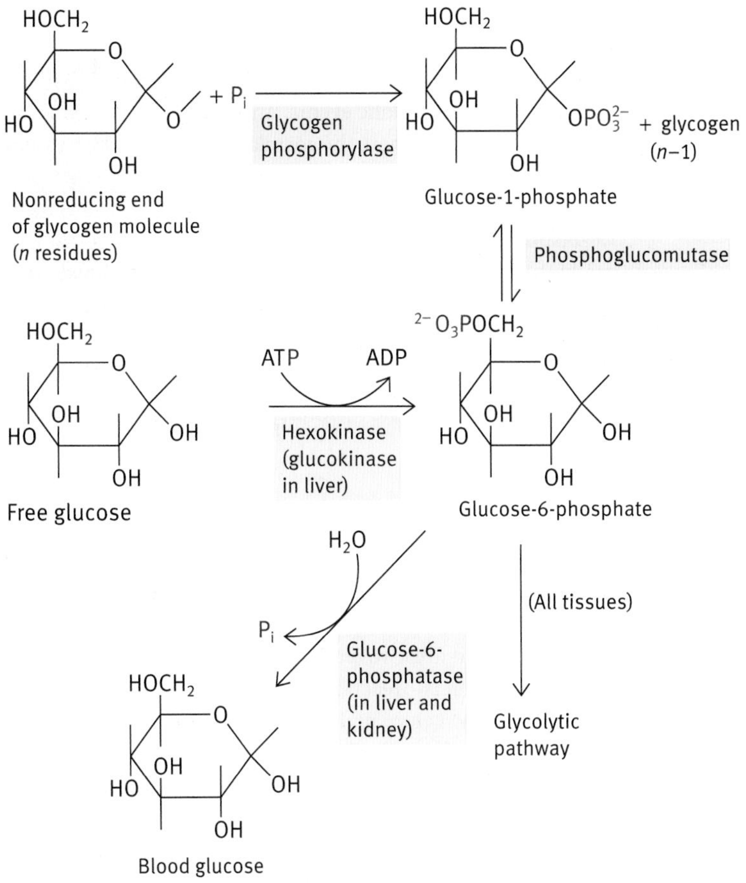

Fig. 13.1 Production of glucose-6-phosphate from glycogen or free glucose and its fate. Which routes are operative depends on control mechanisms described in Chapter 20.

To digress for a moment, organic chemistry textbooks describe a test-tube reaction called the aldol condensation. In this, an aldehyde and a ketone (or another aldehyde) condense together as shown in Fig. 13.2. The reverse of the reaction can be used to divide an aldol into two parts – an aldol being the β-hydroxycarbonyl compound shown. Glucose-6-phosphate is not an aldol, but fructose-6-phosphate is, as seen in the straight-chain formulae in Fig. 13.3. By forming the fructose isomer, the sugar phosphate can be divided into two by the aldol reaction. The glucose-6-phosphate is isomerized into fructose-6-phosphate by the enzyme **phosphohexose isomerase**. Before dividing , another phosphoryl group from ATP is transferred to the fructose-6-phosphate by the enzyme phosphofructokinase (PFK), yielding fructose-1,6-bisphosphate.

Dividing fructose bisphosphate into two C_3 compounds

The fructose-1,6-bisphosphate is now split by the enzyme **aldolase** catalysing the aldol reaction; the second phosphate means that each of the two C_3 products has a phosphoryl group giving glyceraldehyde-3-phosphate and dihydroxyacetone phosphate (Fig. 13.4). Later on, both of these C3 fragments of glucose can produce the same end compound, pyruvate. In Fig. 13.5, the same reactions are presented with the more commonly used ring structures for the sugars; the fructose-6-phosphate is in the five-membered ring (furanose) configuration. The $\Delta G^{0'}$ for the aldolase reaction is $-24.3\,\text{kJ mol}^{-1}$, which would seem to preclude its ready occurrence. There are, however, special considerations applying to this reaction (see below). In cellular conditions, the ΔG is small and the reaction freely reversible.

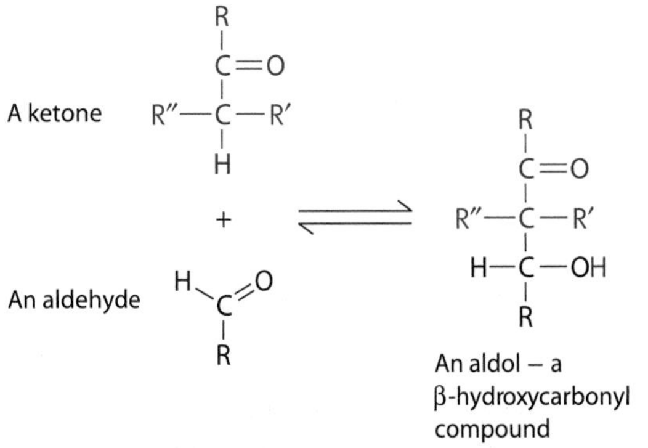

Fig. 13.2 The aldol condensation. A chemical reaction between an aldehyde and a ketone (or aldehyde).

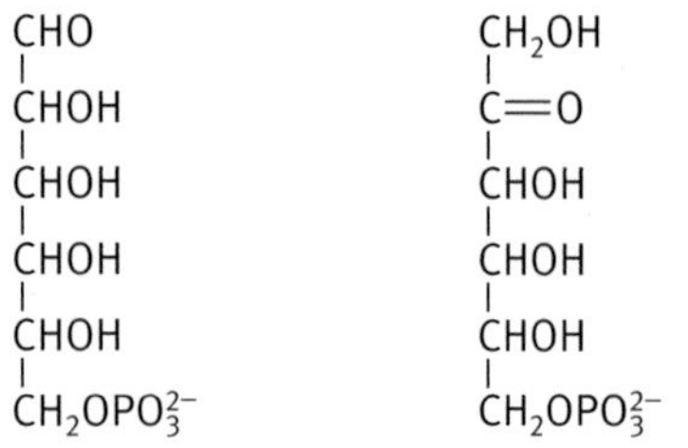

CHO	CH_2OH
CHOH	C=O
CHOH	CHOH
CHOH	CHOH
CHOH	CHOH
$CH_2OPO_3^{2-}$	$CH_2OPO_3^{2-}$
Glucose-6-phosphate, an aldose sugar. This is not an aldol.	Fructose-6-phosphate, the ketose isomer. This *is* an aldol.

Fig. 13.3 The straight-chain formulae of an aldose sugar and its ketose isomer. The glucose-6-phosphate is in equilibrium with the six-membered ring (pyranose) form and the fructose-6-phosphate with the five-membered ring (furanose) form. The ring structures are shown in Fig. 13.5.

A note on the $\Delta G^{0'}$ and ΔG values for the aldolase reaction

The reaction catalysed by aldolase has a $\Delta G^{0'}$ value of +24.3 kJ mol^{-1} and is freely reversible while reactions with smaller $\Delta G^{0'}$ values paradoxically are irreversible. The explanation is that $\Delta G^{0'}$ values are determined at 1 M concentrations of reactants and products; since concentrations in the cell are more likely to be at 10^{-3}–10^{-4} M actual ΔG values are always different from $\Delta G^{0'}$ values. Nonetheless, the latter are usually useful guides to metabolic events. This is not true, however, in the case of aldolase, where the correlation between $\Delta G^{0'}$ values and ΔG values in the cell is very poor. The reason is that, in the aldolase reaction, one molecule of the reactant, fructose-1,6-bisphosphate (F-1,6-BP), gives rise to two molecules of product, glyceraldehyde-3-phosphate (GAP) and dihydroxyacetone phosphate (DHAP). The relationship between $\Delta G^{0'}$ and ΔG values is given in the equation:

$$\Delta G = \Delta G^{0'} + RT \ln \frac{[\text{products}]}{[\text{reactants}]},$$

that is,

$$\Delta G = \Delta G^{0'} + RT \ln \frac{[\text{GAP}] \times [\text{DHAP}]}{[\text{F-1,6-BP}]}.$$

Because there are two products with low concentrations, the $RT\ln([\text{products}]/[\text{reactants}])$ moiety of the equation has a large negative value, giving a ΔG compatible with ready reversibility in the cell. To illustrate this, from the actual

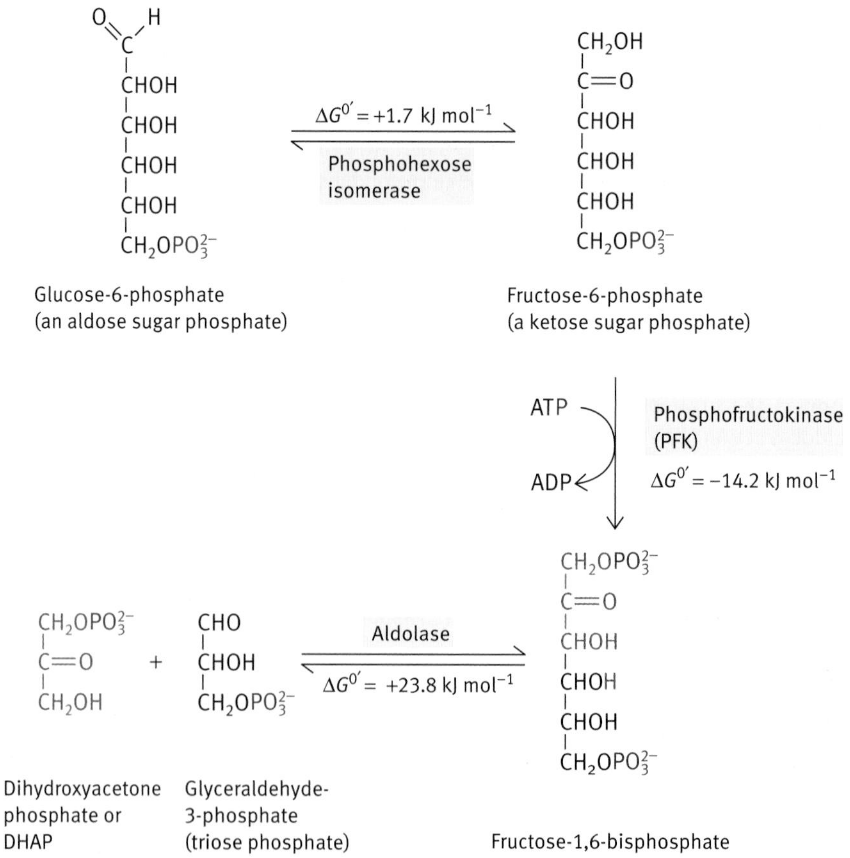

Glucose-6-phosphate (an aldose sugar phosphate): O=C(H)–CHOH–CHOH–CHOH–CHOH–$CH_2OPO_3^{2-}$

$\Delta G^{0'}$ = +1.7 kJ mol^{-1}

Phosphohexose isomerase

Fructose-6-phosphate (a ketose sugar phosphate): CH_2OH–C=O–CHOH–CHOH–CHOH–$CH_2OPO_3^{2-}$

ATP → ADP

Phosphofructokinase (PFK)

$\Delta G^{0'}$ = −14.2 kJ mol^{-1}

Fructose-1,6-bisphosphate: $CH_2OPO_3^{2-}$–C=O–CHOH–CHOH–CHOH–$CH_2OPO_3^{2-}$

Aldolase

$\Delta G^{0'}$ = +23.8 kJ mol^{-1}

Dihydroxyacetone phosphate or DHAP: $CH_2OPO_3^{2-}$–C=O–CH_2OH + Glyceraldehyde-3-phosphate (triose phosphate): CHO–CHOH–$CH_2OPO_3^{2-}$

Fig. 13.4 Conversion of glucose-6-phosphate into two C_3 compounds. The $\Delta G^{0'}$ value for the aldolase reaction in the forward direction would appear to preclude its occurrence, but see text for explanation of this. Straight-chain formulae for the sugars are used here for clarity; the reactions are commonly presented as in Fig. 13.5.

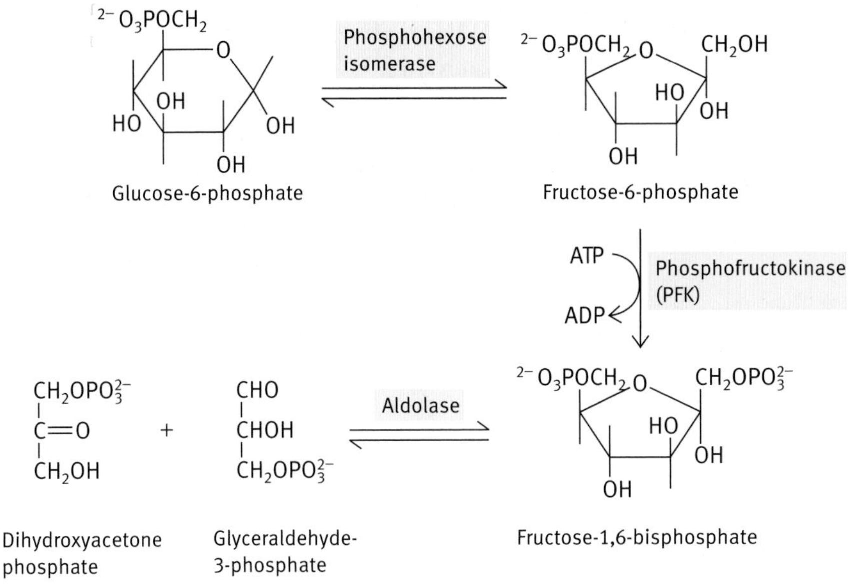

Glucose-6-phosphate ⇌ (Phosphohexose isomerase) Fructose-6-phosphate

Fructose-6-phosphate → (Phosphofructokinase (PFK); ATP → ADP) Fructose-1,6-bisphosphate

Fructose-1,6-bisphosphate ⇌ (Aldolase) Dihydroxyacetone phosphate + Glyceraldehyde-3-phosphate

Fig. 13.5 This is the same scheme as in Fig. 13.4 but presented in the more usual form with sugars as ring structures.

intracellular concentrations of fructose-1,6-bisphosphate, glyceraldehyde-3-phosphate, and dihydroxyacetone phosphate, as determined in rabbit skeletal muscle, a small ΔG value of -1.3 kJ mol^{-1} can be calculated.

Interconversion of dihydroxyacetone phosphate and glyceraldehyde-3-phosphate

Glyceraldehyde-3-phosphate and dihydroxyacetone phosphate are isomeric molecules. An enzyme, **triose phosphate isomerase**, interconverts these two compounds.

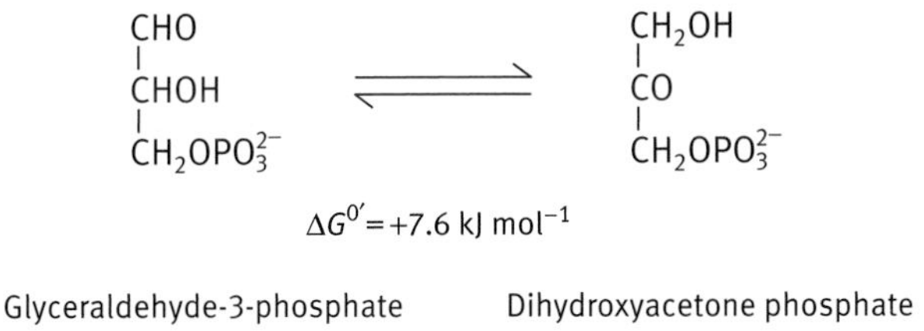

$$\text{CHO—CHOH—CH}_2\text{OPO}_3^{2-} \rightleftharpoons \text{CH}_2\text{OH—CO—CH}_2\text{OPO}_3^{2-}$$

$$\Delta G^{0'} = +7.6 \text{ kJ mol}^{-1}$$

Glyceraldehyde-3-phosphate   Dihydroxyacetone phosphate

The two compounds are in equilibrium but, since glyceraldehyde-3-phosphate is continually removed by the next step in glycolysis, all of the dihydroxyacetone phosphate is progressively converted to glyceraldehyde-3-phosphate.

Glyceraldehyde-3-phosphate dehydrogenase – an oxidation linked to ATP synthesis

The aldehyde group of glyceraldehyde-3-phosphate is oxidized by **glyceraldehyde-3-phosphate dehydrogenase**, using NAD^- (see Chapter 12) as an electron acceptor. You would expect this to produce a carboxyl group (and so it does, ultimately) but oxidation of a —CHO group to $—COO^-$ has a large negative ΔG value, sufficient in fact to generate a high-energy phosphate compound from inorganic phosphate (P_i) on the way (Fig. 13.6).

The mechanism by which this is achieved is as follows. At the active site of the enzyme there is the amino acid cysteine which has a thiol or sulphydryl group (—SH) on its side chain. The aldehyde glyceraldehyde-3-phosphate condenses with the thiol to form a thiohemiacetal.

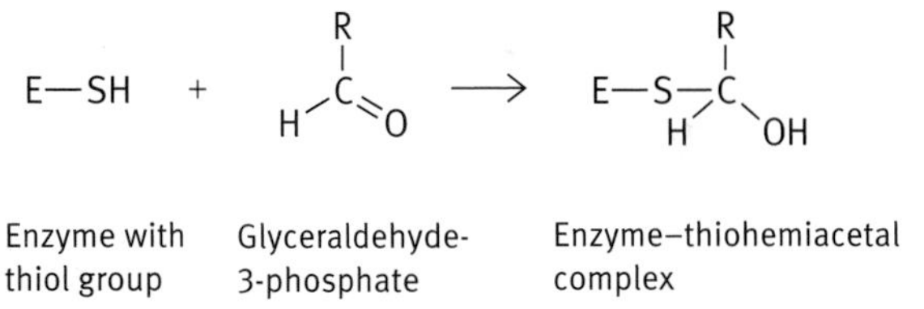

$$\text{E—SH} + \text{R—C(H)=O} \longrightarrow \text{E—S—C(R)(H)—OH}$$

Enzyme with thiol group   Glyceraldehyde-3-phosphate   Enzyme–thiohemiacetal complex

The complex is now oxidized on the enzyme active site, the electrons being accepted by NAD^+, and a thiol ester is formed with the enzyme thiol group.

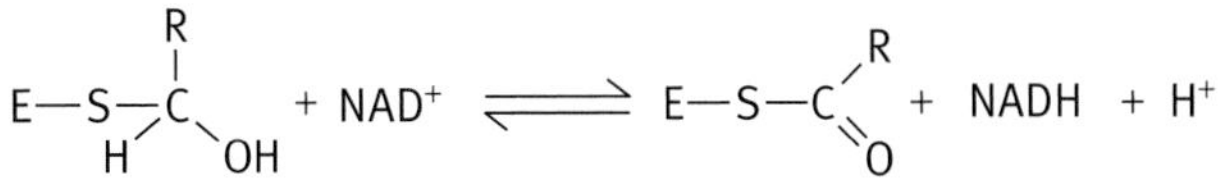

$$\text{E—S—C(R)(H)—OH} + \text{NAD}^+ \rightleftharpoons \text{E—S—C(R)=O} + \text{NADH} + \text{H}^+$$

A thiol ester (R—CO—S—), as explained earlier, is a high-energy compound – of the same order as that of a high-energy phosphate compound. It is thermodynamically feasible, therefore, for P_i to react as follows.

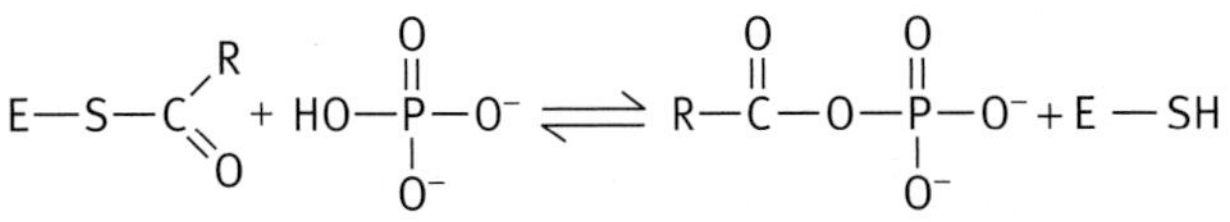

$$\text{E—S—C(R)=O} + \text{HO—P(=O)(O}^-\text{)—O}^- \rightleftharpoons \text{R—C(=O)—O—P(=O)(O}^-\text{)—O}^- + \text{E—SH}$$

The $RCO—O—PO_3^{2-}$ group is also high energy and so its phosphoryl group can be transferred to ADP forming ATP.

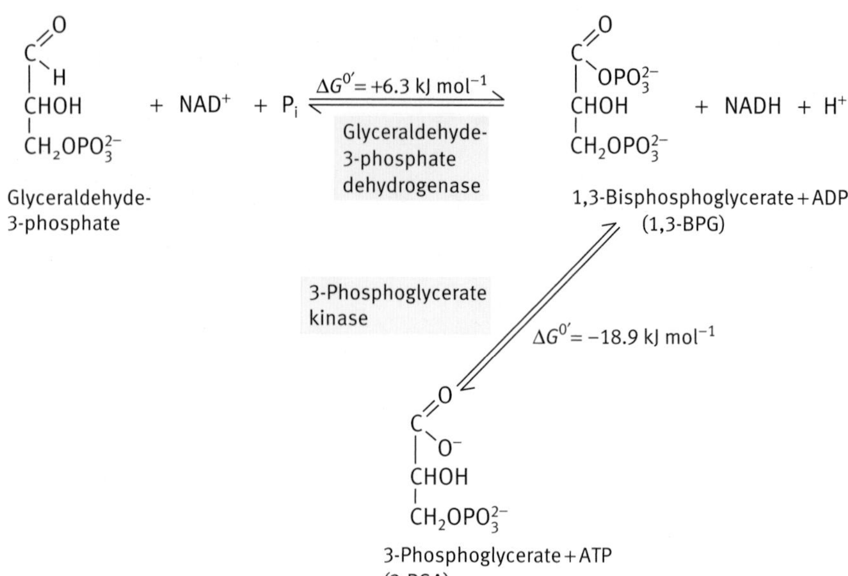

Fig. 13.6 Conversion of glyceraldehyde-3-phosphate to 3-phosphoglycerate.

The responsible enzyme is **phosphoglycerate kinase** because, in the reverse direction, it transfers a phosphoryl group from ATP to **3-phosphoglycerate (3-PGA)**. By convention, kinases are always named from the side of the reaction that involves ATP.

The phosphoryl group generated in this process is attached to the actual substrate (1,3-bisphosphoglycerate) of an enzyme. For this reason it is called **substrate-level phosphorylation**, a point to which we will refer later. 3-Phosphoglycerate is a low-energy phosphate compound and cannot phosphorylate ADP.

We will see how later on in glycolysis, the molecule is manipulated in such a way, that this low-energy phosphate ester becomes a high-energy phosphoryl group, transferable to ATP. This happens in a thermodynamically legitimate way.

The final steps in glycolysis

The phosphoryl group of 3-phosphoglycerate is transferred from the 3 to the 2 position as shown.

$$\underset{\text{3-Phosphoglycerate}}{COO^- - CHOH - CH_2OPO_3^{2-}} \rightleftharpoons \underset{\text{2-Phosphoglycerate}}{COO^- - CHOPO_3^{2-} - CH_2OH} \qquad \Delta G^{0\prime} = +4.4\ \text{kJ mol}^{-1}$$

This is called the **phosphoglycerate mutase** reaction. The reaction is not really an *intramolecular* transfer of the phosphoryl group (though it is in the enzyme from plants). The enzyme from rabbit muscle contains a phosphoryl group that it donates to the 2-OH group of 3-phosphoglycerate forming 2,3-bisphosphoglycerate. The 3-phosphoryl group is now transferred to the enzyme and replaces the donated phosphate, so that the net effect is the reaction shown above.

The next step in glycolysis is that a water molecule is removed from the 2-phosphoglycerate. Enzymes catalysing such reactions are usually called **dehydratases** but in this particular case in glycolysis, the old established name is **enolase** (because it forms a substituted enol).

$$\underset{\text{2-Phosphoglycerate}}{COO^- - CHOPO_3^{2-} - CH_2OH} \xrightleftharpoons[-H_2O]{\text{Enolase}} \underset{\text{Phosphoenolpyruvate (PEP)}}{COO^- - C(-O-PO_3^{2-}) = CH_2}$$

$$\underset{\text{Phosphoenolpyruvate (PEP)}}{COO^- - C(-O-PO_3^{2-}) = CH_2} \xrightarrow[\text{ADP}]{\text{Pyruvate kinase}} \underset{\text{Enolpyruvate}}{\left[COO^- - C(-OH) = CH_2\right]} + ATP \longrightarrow \underset{\text{Pyruvate}}{COO^- - C{=}O - CH_3}$$

The enolase reaction has a $\Delta G^{0\prime}$ of only −1.8 kJ mol^{-1}, but the enolphosphate compound is of the 'high-energy' type, with a $\Delta G^{0\prime}$ of hydrolysis of −62.2 kJ mol^{-1}; a reason for this is that the immediate product of the reaction, the enol form of pyruvate,

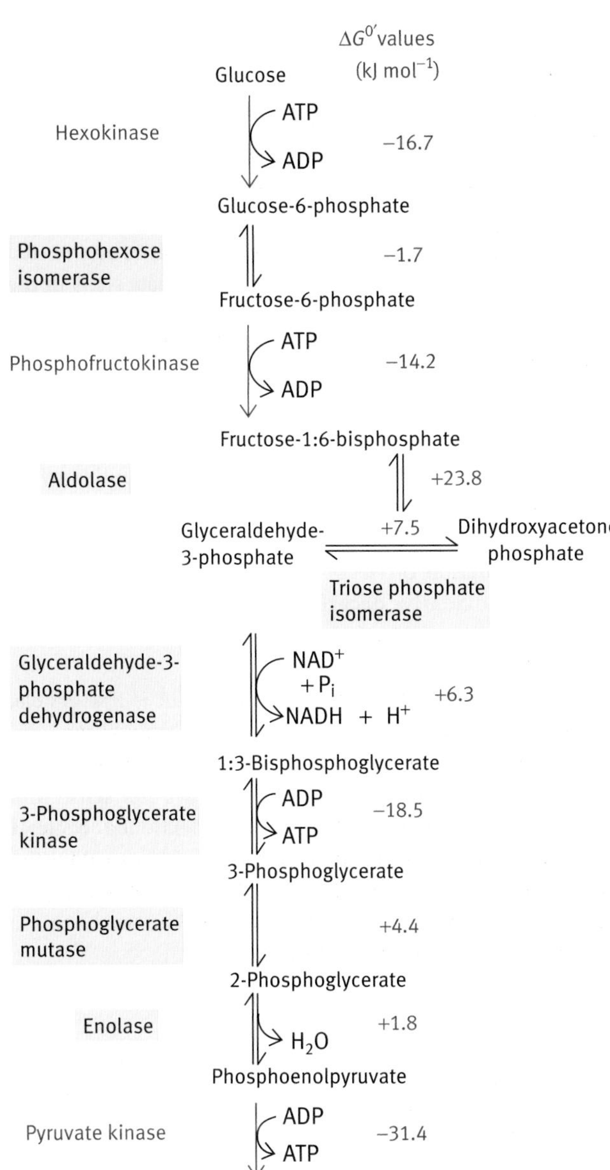

Fig. 13.7 The glycolytic pathway. Irreversible reactions are indicated in red. The free reversibility of the aldolase reaction would appear to be inconsistent with such a large $\Delta G^{0\prime}$ value (see text for explanation).

spontaneously converts to the keto form, a reaction with a large negative $\Delta G^{0\prime}$ value.

The phosphoryl group is transferred to ADP by the enzyme **pyruvate kinase**; this name might misleadingly imply that pyruvate can be phosphorylated by ATP by reversal of the reaction; the name of the enzyme derives from the convention, mentioned earlier, that a kinase is named from the reaction involving ATP on the substrate side even though, in this case, that reaction never occurs because the substrate, enol-pyruvate, does not appear, except fleetingly. It spontaneously changes to pyruvate. The irreversibility of the conversion of phosphoenolpyruvate to pyruvate has important metabolic repercussions as you will see when we come later to gluconeogenesis. (There is a potential source of confusion arising from the fact that, in certain plants and microorganisms, pyruvate *is* directly converted to phosphoenolpyruvate by a quite different enzyme that utilizes two phosphoryl groups from ATP. However, this reaction does *not occur* in animals.)

The complete glycolytic pathway is shown in Fig. 13.7.

Anaerobic glycolysis

In vigorously exercising muscle, and in red blood cells (which do not have mitochondria) , glycolysis continues to produce lactic acid rather than pyruvate. This has already been explained in Chapter 12.

The ATP balance sheet from glycolysis

Starting with glucose, two molecules of ATP were used to form fructose-1,6-bisphosphate. The phosphoglycerate kinase produced two ATP molecules per original glucose and the pyruvate kinase two – a total of four and a net gain of two.

In muscle, glycolysis may start with glycogen. In this case, the energy in the glycosidic bonds of glycogen is preserved, because the initial reaction is to release glucosyl units by using inorganic phosphate (phosphorolysis) producing glucose-1-phosphate. This saves one ATP so that the yield of ATP per glycogen unit is three. The same is true of liver, but the glucose-6-phosphate formed from glucose-1-phosphate is largely converted into free glucose which is released into the bloodstream rather than channeled into the glycolytic pathway. Most tissues are glucose users but the liver is a glucose provider (see Chapter 16).

Transport of pyruvate into the mitochondria

The other product of glycolysis besides NADH is pyruvate. Unless it is reduced to lactate in the cytosol (see Chapter 12) the pyruvate is transported into the mitochondrial matrix by an antiport type of membrane transport protein (see Chapter 7), which exchanges it for OH^- inside the matrix.

Conversion of pyruvate into acetyl-CoA – a preliminary step before the TCA cycle

Before we get to the cycle itself we must deal with the preparation of pyruvate to enter the cycle, by which we mean its conversion into acetyl-CoA.

As outlined earlier, pyruvate in the mitochondrial matrix is converted into acetyl-CoA, which feeds the

BOX 13.1 The Warburg effect

The Nobel laureate Otto Warburg observed that tumour cells mainly generated energy by anaerobic respiration, that is, conversion of glucose into pyruvate and subsequently into lactate as opposed to 'healthy' cells which would produce energy by the conversion of glucose into pyruvate and subsequent oxidation of pyruvate aerobically. This observation is referred to as the 'Warburg effect'. He proposed that cancer was a result of mitochondrial dysfunction which would not allow the tumour cells to function normally.

In recent years, the Warburg effect has re-attracted attention as an approach to cancer detection and monitoring of treatment, particularly in solid tumours. Now we know that the difference in metabolism, anaerobic as opposed to aerobic in tumour cells is not the cause of cancer but the metabolic effects of cancer causing mutations. Malignant cells can carry out glycolysis at rates up to 200 times those of normal cells. If glycolysis can be inhibited in these cells, or the oxidative capacity of mitochondria activated, then cancer growth may be halted. Studies using glycolytic inhibitors such as dichloroacetic acid (DCA) and 2-deoxy-d-glucose (2DG) have shown promising results, destroying cancer cells in vitro and also in some animal studies. Diagnosis and monitoring of treatment success is carried out by positron emitting tomography (PET) by detecting the uptake of a radioactive modified hexokinase substrate, 2-18F-2-deoxyglucose (FDG).

acetyl group into the TCA cycle (the structure of CoA is described on page X).

The overall reaction catalysed by pyruvate dehydrogenase is:

$$\text{Pyruvate} + \text{NAD}^+ + \text{CoA–SH} \rightarrow \text{Acetyl–S–CoA} + \text{NADH} + \text{H}^+ + \text{CO}_2$$

Pyruvate dehydrogenase, the responsible enzyme, is a very large complex composed of many polypeptides. It essentially consists of three different enzyme activities aggregated together, each catalysing one of the intermediate steps in the process (Fig. 13.8). The aggregation of these units increases the efficiency of catalysis. The first step is decarboxylation of pyruvate to produce CO_2 and a hydroxyethyl group CH_3CHOH– attached to the cofactor **thiamin pyrophosphate (TPP)**. TPP is derived from thiamin, or vitamin B_1, deficiency of which impairs the ability to metabolize pyruvate, among other effects. The hydroxyethyl group, in a series of steps, is converted into the acetyl group of acetyl-CoA with the reduction of NAD^+ (see Fig. 13.8). The process is known as an **oxidative decarboxylation** for obvious reasons. (Detailed structures of cofactors are given below for reference purposes.)

This conversion of pyruvate into acetyl-CoA is irreversible; the $\Delta G^{0\prime}$ of the reaction is $-33.5\,\text{kJ mol}^{-1}$. As you will see later, this is of profound significance for irreversibility and means that fatty acids can never be converted in a net sense into glucose in the animal body, though, as mentioned, bacteria and plants have a special mechanism for achieving this. The acetyl-CoA now enters the TCA cycle.

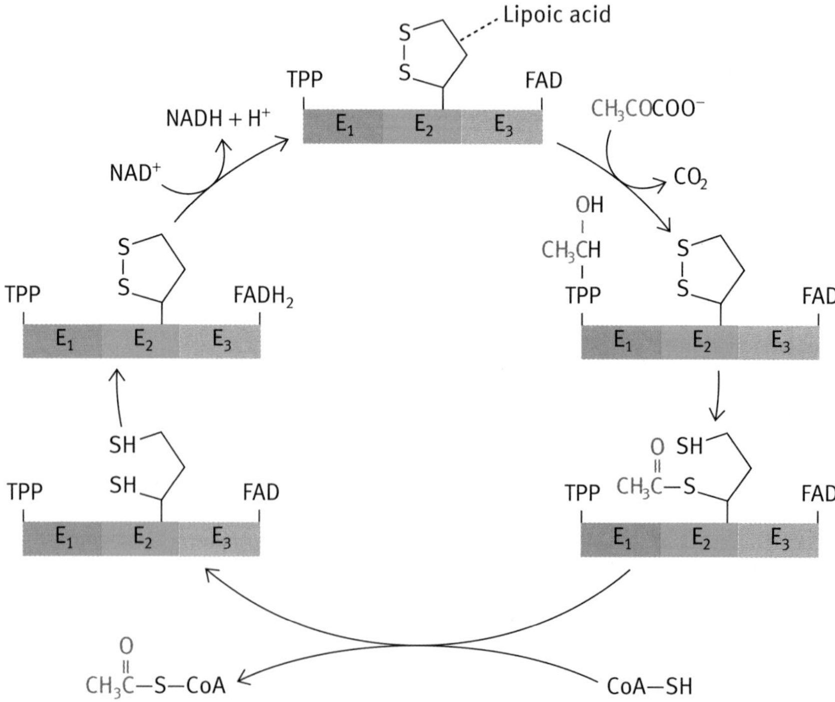

Fig. 13.8 Mechanism of the pyruvate dehydrogenase reaction. TPP, thiamin pyrophosphate; E1–E3, enzyme regions of complex. (The $FADH_2$ is at an unusually low redox potential in this enzyme and can reduce NAD^+.)

Components involved in the pyruvate dehydrogenase reaction

TPP has the structure:

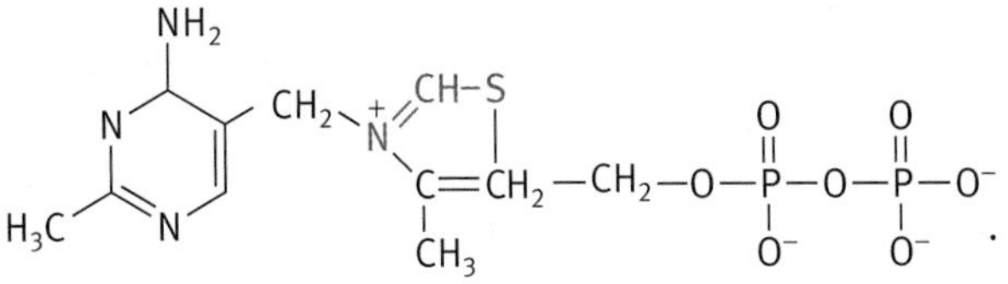

NH_2; N; N; H_3C; CH_2; $\overset{+}{N}$; CH–S; $C{=}CH_2{-}CH_2{-}O{-}\overset{O}{\overset{\|}{P}}{-}O{-}\overset{O}{\overset{\|}{P}}{-}O^-$; O^- O^-; CH_3 .

TPP-hydroxyethyl has the structure:

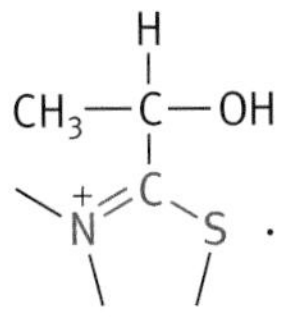

$CH_3{-}\underset{|}{\overset{H}{\overset{|}{C}}}{-}OH$; $\overset{+}{N}{=}C{-}S$.

Lipoic acid in its reduced form has the structure:

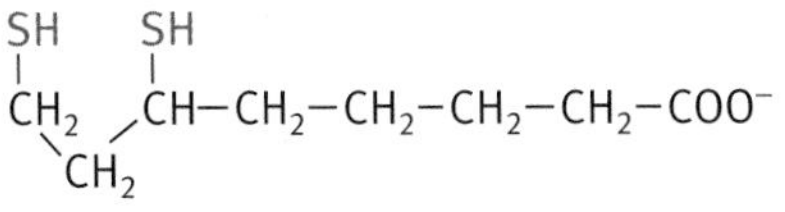

$$\overset{SH}{\overset{|}{CH_2}}\ \ \overset{SH}{\overset{|}{CH}}{-}CH_2{-}CH_2{-}CH_2{-}CH_2{-}COO^-$$
CH_2

and in its oxidized form the structure is:

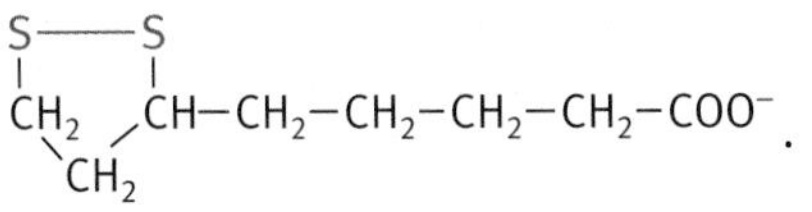

$$\overset{S{-}{-}{-}S}{CH_2\ \ CH}{-}CH_2{-}CH_2{-}CH_2{-}CH_2{-}COO^-.$$
CH_2

Lipoic acid is attached to a lysine side chain of the enzyme by –CO–NH– linkage. In Fig. 13.8 it is represented by the disulphide structure. The three enzymes represented by E_1, E_2, and E_3 are part of a very large protein complex. Its complex regulation is given in Chapter 20.

Stage 2 – the TCA cycle

A preliminary overview of the cycle is given in Chapter 12. The TCA cycle produces a combustible fuel in the form of reducing equivalents of NADH and $FADH_2$, part of which, in effect, is derived from the components of water. This fuel is burnt in the next stage, the electron transport system, to produce ATP from ADP and P_i. The cycle is thermodynamically sound because it uses the free energy made available from the destruction of the acetyl group of acetyl-CoA to drive the process.

It must be noted that the contribution of water is not a direct 'head on' process as occurs in photosynthesis, and oxygen is not liberated as such, but as CO_2. It is an indirect process that can easily be overlooked and is seldom referred to in texts. We need to explain this aspect more clearly for it is central to appreciating what the cycle is all about. It may also remove potential confusion about how eight high-energy electrons (plus a ninth hydrogen atom on CoA–SH) can arise from the destruction of an acetyl group with only three hydrogen atoms.

Acetyl-CoA enters the cycle by a reaction with oxaloacetate to produce citrate. This is described later. The reaction involves the input of a molecule of water. With one complete 'turn' of the cycle (again explained later), oxaloacetate is re-formed and the acetyl group of acetyl-CoA has disappeared. As a result of the cycle reactions, the products from one acetyl group are as follows: two molecules of CO_2; the reduction of three molecules of NAD^+ to NADH; and that of one molecule of FAD to $FADH_2$. In addition, the CoA–S– of acetyl-CoA becomes CoA–SH (Fig. 13.9). (A single high-energy phosphoryl group, as GTP, is produced from P_i.) If you add up all the reducing equivalents of the three NADH and one $FADH_2$, there are eight. (Remember that NAD^+ accepts two electrons.) The formation of the thiol group of CoA–SH (from CoA–S–) requires one more – a total of nine reducing equivalents (effectively equivalent to nine H atoms).

The CH_3CO–S–CoA supplies three of these so that there is a shortfall of six. There is no involvement of oxygen in the cycle so there is also a shortfall of three oxygen atoms to produce the two molecules of CO_2. The source of these 'missing' atoms includes two molecules of H_2O. You will see below that, as well as the input of H_2O into citrate synthesis, a second molecule of water enters the cycle. But, this still leaves a shortfall of two hydrogen atoms and one oxygen atom. It will be more convenient to explain the source of these later.

Thus, we see the remarkable feat – electrons of H_2O are raised up the energy scale to reduce NADH and $FADH_2$ and, of course, electrons from the acetyl group also are utilized for this purpose. Again, it is emphasized that this does not mean that the components of H_2O go *directly* to these

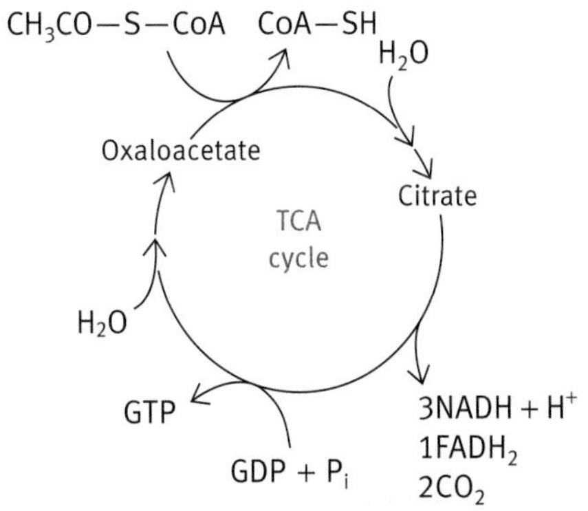

Fig. 13.9 The inputs and outputs of the TCA cycle. Individual cycle reactions are not indicated.

products. The reactions involved in converting the acetyl group into its products provide the free energy to convert H_2O into reducing equivalents. In fact, the whole cycle has a negative ΔG value and so proceeds thermodynamically 'downhill'.

With that preamble we can now turn to the reactions of the cycle.

A simplified version of the TCA cycle

Possibly one obstacle to learning the cycle is that the progression of reactions does not make much sense until you have completed it, so we will first look at a simplified version devoid of detail.

Be sure you know the structure of oxaloacetic acid, for that is where it all starts and finishes. Perhaps the easiest way to think of oxaloacetate is as a pyruvate with an added carboxyl group. Acetate is CH_3COO^-, the oxalo group is $^-OOC—C(=O)—$, so oxaloacetate is:

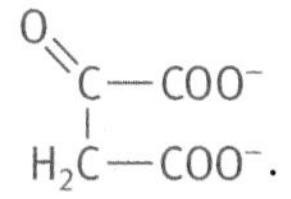

O=C—COO⁻
|
$H_2C—COO^-$.

The acetyl group of acetyl-CoA is joined to oxaloacetate to form citrate; it is easy to see how citrate can be derived from oxaloacetate.

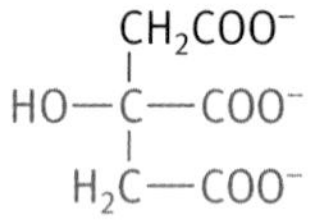

CH_2COO^-
|
HO—C—COO⁻
|
$H_2C—COO^-$

It is helpful to remember that citrate is a C_6 *symmetric* tricarboxylic compound. This is converted to its asymmetric isomer isocitrate, which in the cycle is then progressively converted to 2-oxoglutarate (C_5), succinate (C_4), fumarate (C_4), malate (C_4), and oxaloacetate (Fig. 13.10). This means that one turn of the cycle eliminates the acetyl group fed into it.

We suggest that you make yourself completely familiar with these acids of the cycle. With that preparation, a more detailed consideration of this major route of carbon-compound metabolism can be given.

Mechanisms of the TCA cycle reactions

We might, for convenience, divide the reactions into three groups.

1 The synthesis of citrate is the reaction feeding acetyl groups into the cycle.

2 The 'top part' of the cycle involves conversion of C_6 citrate into C_5 2-oxoglutarate

3 The 'lower part' of the cycle involves the conversion of succinate into oxaloacetate.

The synthesis of citrate

The name of the enzyme involved here is **citrate synthase**. The enzyme catalyses the condensation of acetyl-CoA with oxaloacetate by an aldol reaction to give citryl-CoA. This is hydrolysed to citrate. Citrate formation has a large negative $\Delta G^{0\prime}$ value (–32.3 kJ mol⁻¹), due to the hydrolysis of a thiol ester, and hence the reaction is irreversible.

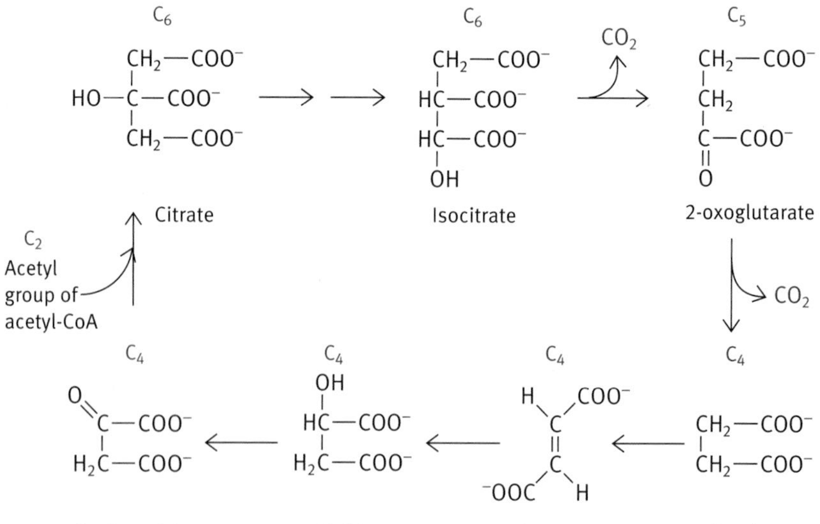

C_6 Citrate: $CH_2—COO^-$ / HO—C—COO⁻ / $CH_2—COO^-$ → → C_6 Isocitrate: $CH_2—COO^-$ / HC—COO⁻ / HC—COO⁻ / OH → (CO_2) → C_5 2-oxoglutarate: $CH_2—COO^-$ / CH_2 / C(=O)—COO⁻ → (CO_2) → C_4 Succinate: $CH_2—COO^-$ / $CH_2—COO^-$ → C_4 Fumarate: H(COO⁻)C=C(H)⁻OOC → C_4 Malate: OH / HC—COO⁻ / $H_2C—COO^-$ → C_4 Oxaloacetate: O=C—COO⁻ / $H_2C—COO^-$ → Citrate (C_2 Acetyl group of acetyl-CoA enters)

Fig. 13.10 Simplified TCA cycle showing the component acids and their sequence. The purpose of this diagram is for you to learn the structures of the main cycle acids and their interrelationships without complicating details. When you are familiar with these it will be easier to appreciate the full cycle.

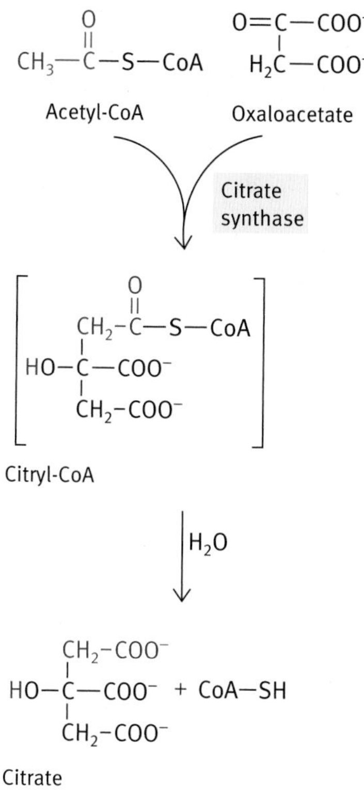

$$CH_3-\overset{O}{\overset{\|}{C}}-S-CoA \quad + \quad O{=}C(COO^-)-H_2C-COO^-$$

Acetyl-CoA   Oxaloacetate

Citrate synthase

$$\left[\begin{array}{l} CH_2-\overset{O}{\overset{\|}{C}}-S-CoA \\ HO-C-COO^- \\ CH_2-COO^- \end{array}\right]$$

Citryl-CoA

$\downarrow H_2O$

$$\begin{array}{l} CH_2-COO^- \\ HO-C-COO^- \\ CH_2-COO^- \end{array} + CoA-SH$$

Citrate

Conversion of citrate into 2-oxoglutarate

Citrate→isocitrate

In this reaction, the hydroxyl group of citrate (a symmetric molecule) is switched from the 3-position to the 2-position, giving isocitrate (an asymmetric molecule). The significance of this will become apparent. This isomerization of citrate is catalysed by a single enzyme that reversibly removes water and adds it back across the double bond in either direction.

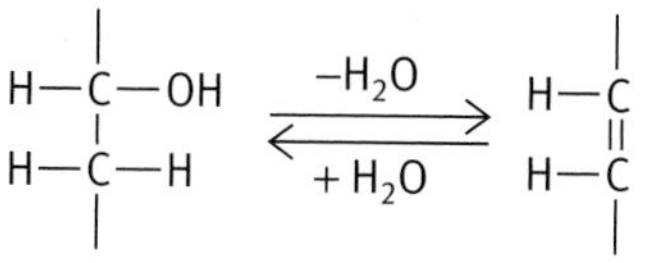

$$\begin{array}{l} H-C-OH \\ H-C-H \end{array} \underset{+H_2O}{\overset{-H_2O}{\rightleftharpoons}} \begin{array}{l} H-C \\ \quad \| \\ H-C \end{array}$$

Enzymes catalysing such reactions are, as mentioned, usually called dehydratases but in this case, due to long tradition, its name is aconitase because the unsaturated intermediate product is *cis*-aconitate (found first in the plant genus *Aconitum*).

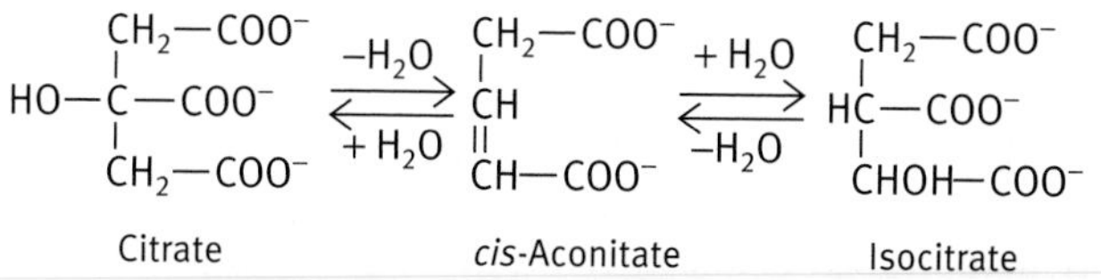

$$\begin{array}{l} CH_2-COO^- \\ HO-C-COO^- \\ CH_2-COO^- \end{array} \underset{+H_2O}{\overset{-H_2O}{\rightleftharpoons}} \begin{array}{l} CH_2-COO^- \\ CH \\ \| \\ CH-COO^- \end{array} \underset{-H_2O}{\overset{+H_2O}{\rightleftharpoons}} \begin{array}{l} CH_2-COO^- \\ HC-COO^- \\ CHOH-COO^- \end{array}$$

Citrate   *cis*-Aconitate   Isocitrate

Since isocitrate is now metabolized further in the cycle the net effect is that aconitase catalyses the reaction sequence:

Citrate → *cis* – aconitate → isocitrate

Isocitrate dehydrogenase

You have already met one example of NAD^+-requiring dehydrogenases in lactate dehydrogenase (see Chapter 12). This again is a common reaction of the type:

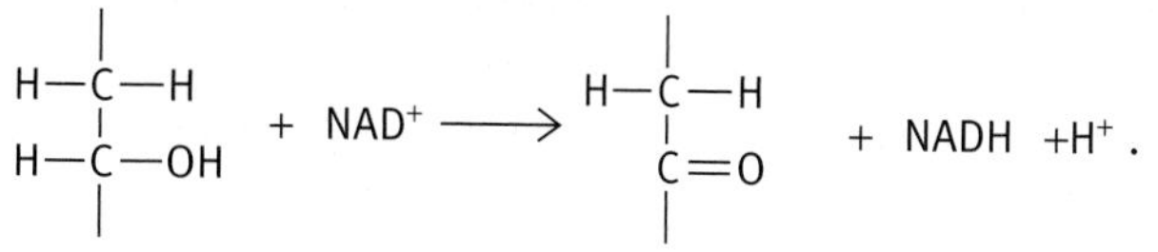

$$\begin{array}{l} H-C-H \\ H-C-OH \end{array} + NAD^+ \longrightarrow \begin{array}{l} H-C-H \\ \quad C{=}O \end{array} + NADH + H^+ .$$

In the cycle isocitrate dehydrogenase catalyses the reaction:

$$\begin{array}{l} CH_2-COO^- \\ CH-COO^- \\ CHOH-COO^- \end{array}$$

Isocitrate

$\downarrow + NAD^+$

$$\left[\begin{array}{l} CH_2-COO^- \\ \alpha\, CH-COO^- \\ \beta\, C-COO^- \\ \quad \| \\ \quad O \end{array}\right] \longrightarrow \begin{array}{l} CH_2-COO^- \\ CH_2 \\ C-COO^- \\ \| \\ O \end{array} + CO_2 .$$

$+ NADH + H^+$

Oxalosuccinate   2-oxoglutarate

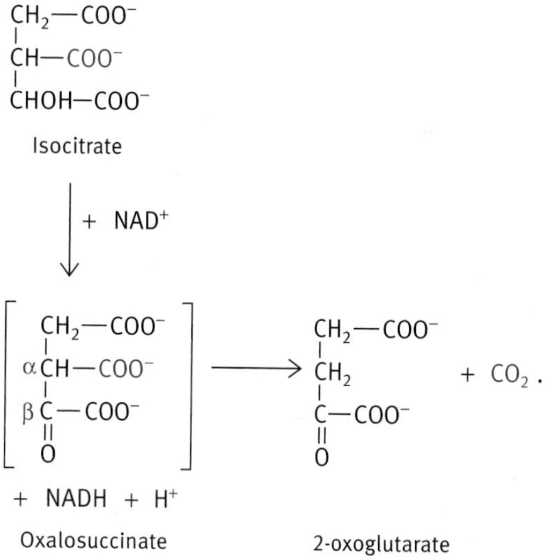

The immediate product is oxalosuccinate which is a β-keto acid (the keto group is β to the centre COOH group). Such acids are unstable and readily lose the carboxyl group as CO_2. This happens on the surface of the isocitrate dehydrogenase so the product is the C_5 acid 2-oxoglutarate as shown.

The C_4 part of the cycle

2-oxoglutarate resembles pyruvate in that both are α-keto acids. We can write both as:

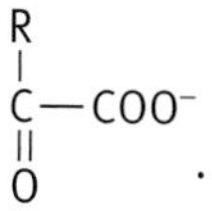

$$\begin{array}{l} R \\ C-COO^- \\ \| \\ O \end{array} .$$

For pyruvate, $R{=}-CH_3$; for 2-oxoglutarate, $R{=}-CH_2CH_2COO^-$. We have already seen that pyruvate dehydrogenase converts pyruvate into acetyl-CoA and CO_2. This reaction format is repeated here. The enzyme requires **thiamin pyrophosphate** (the active form of vitamin B1).

$$R-C(=O)-COO^- + NAD^+ + CoA-SH \rightarrow R-C(=O)-S-CoA + CO_2 + NADH + H^+$$

An equivalent enzyme complex exists, using the same set of cofactors as in pyruvate dehydrogenase, for 2-oxoglutarate and precisely the same equation applies as above except that R=$-CH_2CH_2COO^-$ and the enzyme complex attacks 2-oxoglutarate rather than pyruvate. The product of **2-oxoglutarate dehydrogenase** is therefore succinyl-CoA, analogous to acetyl-CoA. Thiamin deficiency in the diet greatly reduces energy release from foodstuffs. This is particularly noticeable in people who consume high carbohydrate diets and leads to a condition known as beriberi (see Chapter 9).

However, in the cycle, whereas acetyl-CoA is used to form citrate, succinyl-CoA is broken down to free succinate plus CoA—SH. In principle, the simplest way to do this would be to hydrolyse succinyl-CoA. However, the $\Delta G^{0\prime}$ of this hydrolysis of the thiol ester is $-35.5\,kJ\,mol^{-1}$, enough energy to raise P_i to a high-energy phosphate compound and the energy is trapped rather than wasted.

Generation of GTP coupled to splitting of succinyl-CoA

The reaction is:

$$\text{Succinyl-CoA} + GDP + P_i \rightleftharpoons \text{succinate} + GTP + CoASH;\quad \Delta G^{0\prime} = -2.9\,KJmol^{-1}.$$

The enzyme is named for the reverse reaction: hence **succinyl-CoA synthetase**. It synthesizes succinyl-CoA from succinate and GTP but, in the TCA cycle, it works in the other direction, of course. GDP is used in liver and kidney and the GTP is used in gluconeogenesis. Other tissues that do not synthesize glucose for the rest of the body use ADP, as do plants. They use a different isoenzyme.

The mechanism of the reaction is as follows: P_i displaces CoA producing succinyl phosphate:

$$CH_2(COO^-)-CH_2-C(=O)-S-CoA + HO-P(=O)(O^-)-O^- \longrightarrow CH_2(COO^-)-CH_2-C(=O)-O-P(=O)(O^-)-O^- + CoA-SH$$

Succinyl-CoA; P_i; Succinyl phosphate

The phosphoryl group is now transferred to GDP, giving succinate and GTP:

$$CH_2(COO^-)-CH_2-C(=O)-O-P(=O)(O^-)-O^- + {}^-O-P(=O)(O^-)-O-GMP \rightarrow CH_2(COO^-)-CH_2-COO^- + {}^-O-P(=O)(O^-)-O-P(=O)(O^-)-O-GMP$$

Succinyl phosphate; GDP; Succinate; GTP

Earlier we pointed out that, to balance the output of CO_2 and reducing equivalents from the cycle with the input components, in addition to the two H_2O molecules entering the cycle we need two more hydrogen atoms and one oxygen atom. These arise from the involvement of inorganic phosphate in the breakdown of succinyl-CoA as described above. In case this is not clear, the *actual* reactions given above can be notionally regarded for balance-sheet purposes as being *equivalent* to the two reactions:

$$GDP + P_i \rightarrow GTP + H_2O;$$
$$\text{Succinyl-CoA} + H_2O \rightarrow \text{succinate} + CoASH.$$

We emphasize that the reaction *does* not proceed in that way but it illustrates how the 'missing' elements of H_2O are supplied to the cycle to put the balance sheet in order.

Conversion of succinate to oxaloacetate

First, succinate is dehydrogenated by **succinate dehydrogenase** whose electron acceptor is FAD (see Chapter 12), firmly bound to the enzyme and which can reversibly accept a pair of hydrogen atoms. Why is NAD^+ used for the other dehydrogenation reactions of the cycle and FAD here? It is a question of redox potentials. In Chapter 12, we described how electrons will flow from a lower redox potential (higher reducing potential or energy level) to electron acceptors of higher redox potential (lower reducing potential or energy level). In the case of succinate dehydrogenase the reaction is of the *desaturation* type:

$$-\text{H-C-H} \; / \; \text{H-C-H}- \xrightarrow{-2H} -CH=CH- .$$

The redox potential or reducing potential of this system is such that it cannot reduce NAD^+ but can reduce FAD (which is more strongly oxidizing than NAD^+). The reaction is therefore a *dehydrogenation:*

$$\underset{\text{Succinate}}{\begin{array}{l}CH_2{-}COO^-\\ |\\ CH_2{-}COO^-\end{array}} + \text{Enzyme—FAD} \xrightarrow[\text{dehydrogenase}]{\text{Succinate}} \underset{\text{Fumarate}}{^-OOC{-}CH{=}CH{-}COO^-} + \text{Enzyme—FADH}_2.$$

The rest of the cycle, conversion of fumarate into oxaloacetate, is plain sailing because you have already met the reaction types involved. A molecule of water is added to fumarate (*cf.* aconitase, above). The enzyme should logically be called fumarate hydratase but, through long-term usage, it is called **fumarase**. The *hydration* reaction is:

$$\underset{\text{Fumarate}}{^-OOC{-}CH{=}CH{-}COO^-} + H_2O \overset{\text{Fumarase}}{\rightleftharpoons} \underset{\text{Malate}}{\begin{array}{l}HO{-}CH{-}COO^-\\ \quad |\\ H_2C{-}COO^-\end{array}}$$

The malate so produced is *dehydrogenated* by malate dehydrogenase, an NAD^+ enzyme, and so the cycle is back to the starting point, oxaloacetate.

$$\underset{\text{L-Malate}}{\begin{array}{l}HO{-}CH{-}COO^-\\ \quad |\\ H_2C{-}COO^-\end{array}} + NAD^+ \overset{\text{Malate dehydrogenase}}{\rightleftharpoons} \underset{\text{Oxaloacetate}}{\begin{array}{l}O{=}C{-}COO^-\\ \quad |\\ H_2C{-}COO^-\end{array}} + NADH + H^+$$

The $\Delta G^{0'}$ of this reaction is $-29.7\,kJ\,mol^{-1}$, which is very unfavourable; the reaction proceeds because the next reaction, the conversion of oxaloacetate to citrate, is strongly exergonic and pulls the reaction over.

The complete cycle is shown in Fig. 13.11.

What determines the direction of the TCA cycle?

The cycle operates in a unidirectional way, as shown in Fig. 13.11. This is due to three of the reactions having a

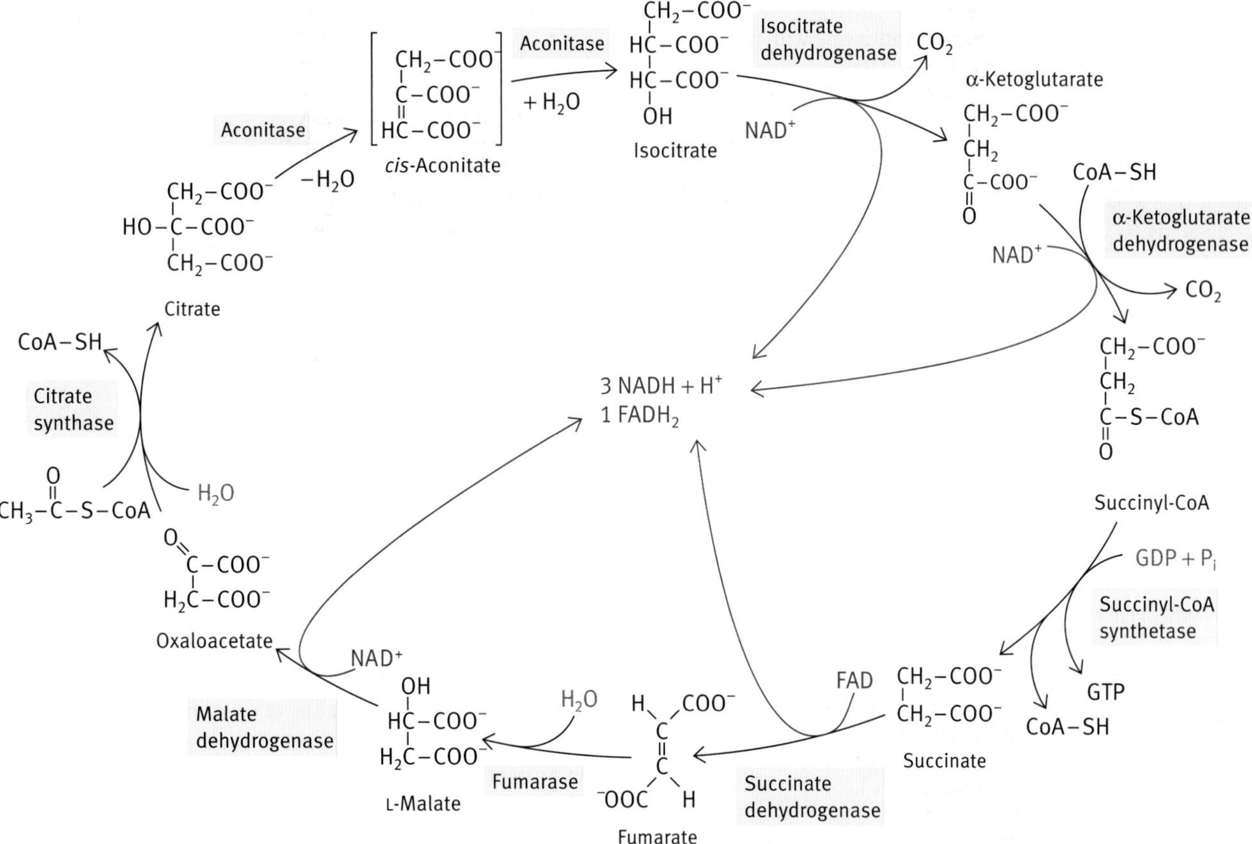

Fig. 13.11 Complete TCA cycle. Red highlights the production of reducing equivalents from the cycle. Blue highlights the supply of the elements of H_2O to the cycle. (The conversion of citrate to isocitrate involves removal and addition of H_2O but there is no net gain.) Note that the FAD is not free but is attached to the succinate dehydrogenase protein. The involvement of water in the synthesis of citrate is explained earlier in this chapter.

sufficiently large negative ΔG value to be irreversible. These are the synthesis of citrate from acetyl-CoA and oxaloacetate ($\Delta G^{0\prime}=-32.3\,kJ\,mol^{-1}$), the decarboxylation of isocitrate to 2-oxoglutarate ($\Delta G^{0\prime}=-20.9\,kJ\,mol^{-1}$), and the 2-oxoglutarate dehydrogenase reaction ($\Delta G^{0\prime}=-33.5\,kJ\,mol^{-1}$). This results in the cycle operating in one direction even though the equilibrium of the malate dehydrogenase reaction is in favour of the reverse direction ($\Delta G^{0\prime}=+29.7\,kJ\,mol^{-1}$). The overall operation of the cycle reactions has a negative ΔG value. Control of the cycle is given in Chapter 20.

Stoichiometry of the cycle

The daunting overall equation for the process is (for reference purposes only):

$$CH_3CO{-}S{-}CoA + 2H_2O + 3NAD^+ + FAD + GDP + P_i \rightarrow$$
$$2CO_2 + 3NADH + 3H^+ + FADH_2 + CoA{-}SH + GTP;$$
$$\Delta G^{0\prime} = -40\,kJ\,mol^{-1}$$

How is the concentration of TCA cycle intermediates maintained?

The cycle starts with oxaloacetate condensing with acetyl-CoA and ends with oxaloacetate so that the latter component is not used up. The cycle acids occupy a special place in metabolism in that they are not necessarily available from the diet in large amounts. Carbohydrates in the diet give rise to large quantities of C_3 acid in the form of pyruvate production but cycle acids (C_4, C_5, and C_6) are not available in such quantities. Fats provide large amounts of the C_2 (acetyl groups), but since these are completely eliminated during the cycle, they do not make any net contribution to cycle intermediates. Certain amino acids can provide cycle acids but, by the same token, cycle acids are withdrawn to synthesize some amino acids (described in Chapter 18) and other metabolites. A method of topping up cycle acids to keep the energy-generating mitochondrial reactions running properly is essential and there is such a provision in cells. An important reaction for this, called an **anaplerotic** or 'filling-up' reaction, is that of pyruvate plus CO_2 being converted into oxaloacetate, using energy from ATP hydrolysis.

$$ATP + CH_3{-}CO{-}COO^- + HCO_3^- \longrightarrow {}^-OOC{-}CH_2{-}CO{-}COO^- + ADP + P_i + H^+$$

The enzyme is called **pyruvate carboxylase** (and quite different from pyruvate decarboxylase of yeast, please note). This is the crucial point at which C_3 acids can be converted into C_4 acids, so pyruvate carboxylase is an enzyme of central importance.

Pyruvate carboxylase: biotin, the cofactor for CO_2 activation

Pyruvate carboxylase requires the vitamin biotin, one of the B group of vitamins, for its function. Wherever 'activated' CO_2 is needed for synthetic reactions catalysed by a group of carboxylase enzymes, biotin is the cofactor (see Chapters 9 and 12). It becomes covalently bound to its enzyme where it accepts a carboxy group from bicarbonate to form carboxybiotin, the reaction being thermodynamically driven by the conversion of ATP to ADP and P_i (see reaction below). Biotin is anchored to the enzyme on the ε-amino group of a lysine residue. Carboxybiotin is a reactive, but stable, form of CO_2 that can be transferred to another molecule that is to be carboxylated. The $\Delta G^{0\prime}$ for the cleavage of CO_2 from carboxybiotin is $-19.7\,kJ\,mol^{-1}$. In this case pyruvate is the substrate, but other carboxylation enzyme systems are also biotin-dependent.

$$\text{Biotin}{-}(CH_2)_4{-}CO{-}\text{Enzyme} + HO{-}CO{-}O^- \xrightarrow{ATP \;\rightarrow\; ADP + P_i} {}^-O{-}CO{-}N\text{(biotin)}{-}(CH_2)_4{-}CO{-}\text{Enzyme}$$

Pyruvate carboxylase has two catalytic sites – one to carboxylate the biotin and the other to transfer the carboxy group from biotin to pyruvate. (In some bacteria, the two activities reside on separate enzymes.) The attachment of the biotin to the long lysyl side chain of the protein provides a flexible arm to permit the biotin to oscillate between the two sites (Fig. 13.12).

Despite the stated importance of pyruvate carboxylase, some bacteria e.g. *Escherichia coli* do not possess it. Since the TCA cycle operates in these cells, how do they maintain the concentrations of the TCA cycle intermediates? The answer is that *E. coli*, in addition to the normal TCA cycle, has a modified form (the glyoxylate cycle) described in Chapter 16, which obviates the need for pyruvate carboxylase.

We will return to pyruvate carboxylase later, for it has metabolic importance other than its anaplerotic role for the TCA cycle.

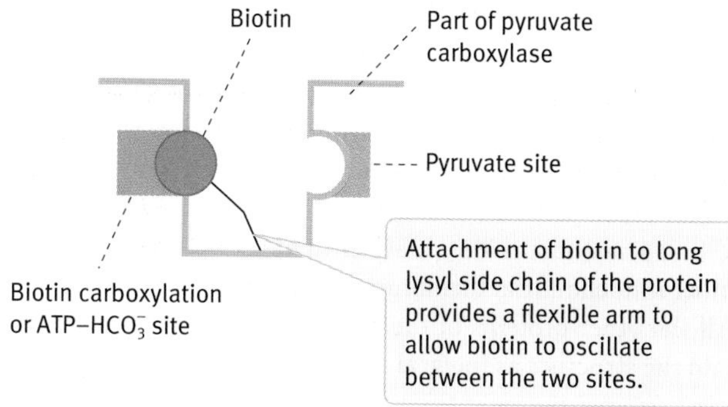

Fig. 13.12 The role of biotin in the active site of enzymes catalysing carboxylation reactions. The example shown is pyruvate carboxylase.

Stage 3 – the electron transport chain that conveys electrons from NADH and $FADH_2$ to oxygen

A preliminary overview of this stage is given in Chapter 12. Remember that what we are doing is looking at the three major stages involved in glucose (or glycogen) oxidation. Stage 1 was glycolysis, Stage 2 was the TCA cycle, and now we come to the final stage. Energy-wise we have not achieved much yet: per starting molecule of glucose only a trivial yield of four ATP molecules – two from glycolysis and two from the cycle (via GTP), with one extra if a glucosyl unit of glycogen is the starting compound. The other products per mole of starting glucose are 10 NADH (two from glycolysis, two from pyruvate dehydrogenase, six from the cycle), and 2 $FADH_2$ from the cycle. (Do not forget that one molecule of glucose produces two pyruvate molecules and hence supports two turns of the cycle.)

The oxidation of the NADH and $FADH_2$ will produce most of the ATP (from ADP and P_i) generated by the oxidation of glucose.

The electron transport chain

The basic principles of the electron transport chain are given in Chapter 12. It would probably be useful to read this again. For reasons that will become apparent, we will now discuss electron transport, pure and simple. Its purpose *is*, definitely, ATP production from ADP and P_i, but just for the time being forget all about this and concentrate on electron movement to oxygen.

Where does it take place?

Electron transport carriers exist in or on the inner mitochondrial membrane. As we have already seen, the inner membrane is folded into cristae. This increases the amount of inner membrane present, the density of cristae in a mitochondrion being related to the energy requirements of the cell.

Nature of the electron carriers in the chain

Haem is the prosthetic group of several electron carriers – called **cytochromes** because of their colour (red). The different cytochromes are called c_1, c, a, and a_3 (in order of their participation in the chain; the role of two b cytochromes is given later). The essentials of the haem structure are shown in Fig. 13.13, and its full structure in Fig. 4.19.

The important thing about the haem molecule is that, as the prosthetic group of the cytochrome electron carriers, the Fe atom oscillates between the Fe^{2+} and Fe^{3+} states as it accepts an electron from the preceding carrier and donates it to the next carrier in the chain. Note the difference from haem in haemoglobin which remains in the Fe^{2+} form. The characteristics of the haem molecule are modified by the specific protein to which it is attached, and variations in the haem side groups occur in different cytochromes and in their precise attachment to their apoproteins. Thus, it is not a contradiction that different cytochromes have different redox potentials (electron affinities) and yet have haem as their prosthetic group.

Another type of electron carrier, based on iron, is the so-called non-haem iron proteins. In these, the iron is bound to the thiol side group of the amino acid cysteine of the protein and also to inorganic sulphide ions forming **iron-sulphur complexes**, or iron-sulphur centres. The simplest type is shown in Fig. 13.14. As with the cytochromes, the iron atom in these can accept and donate electrons in a cyclical fashion, oscillating between the ferrous and ferric state. Such iron-sulphur centres are associated with flavin enzymes. They accept electrons from FAD – enzymes such as succinate dehydrogenase and the dehydrogenase involved in fat oxidation (described in Chapter 14). Another type of carrier is an FMN-protein. FMN (flavin adenine mononucleotide) consists of the flavin half of FAD (see Chapter 12). It carries electrons from NADH to an iron-sulphur centre. All of the iron-sulphur centres transfer electrons to ubiquinone.

As well as these protein-bound electron carriers, there is one carrier not bound to a protein. This is a molecule, illustrated in Fig. 13.15, called **ubiquinone**, because it can exist as a quinone and is found ubiquitously. It is often referred to as coenzyme Q (CoQ), UQ, or Q. Q is an electron carrier because

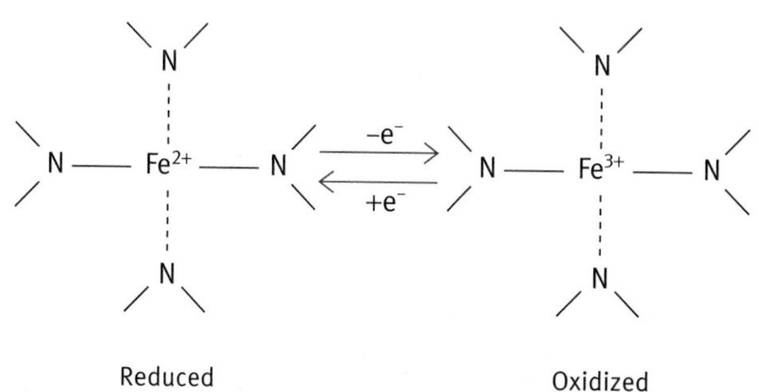

Fig. 13.13 Diagrammatic representation of haem (but take a look at the actual structure in Fig. 4.19).

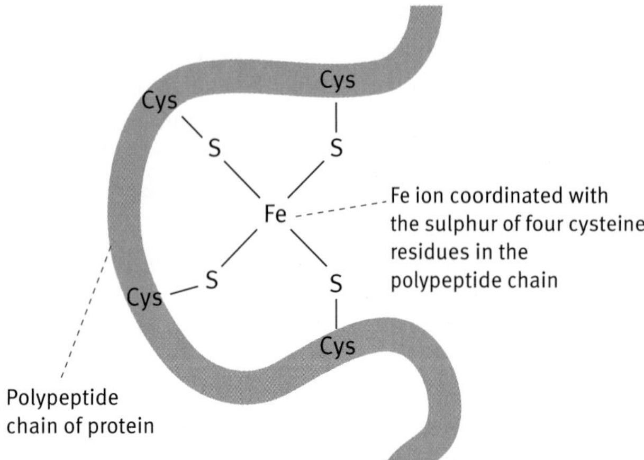

Fig. 13.14 An iron-sulphur centre. There are several types of these, increasing in complexity and numbers of Fe and S atoms. The simplest form is shown here.

it can accept protons as shown in Fig. 13.15; it can exist as the free radical, semiquinone intermediate, thus permitting the molecule to hand over a single electron to the next carrier rather than a pair of electrons. The very long hydrophobic tail on the molecule (as many as 40 carbon atoms long, in 10 isoprenoid groups) makes it freely soluble *and mobile* in the non-polar interior of the inner mitochondrial membrane.

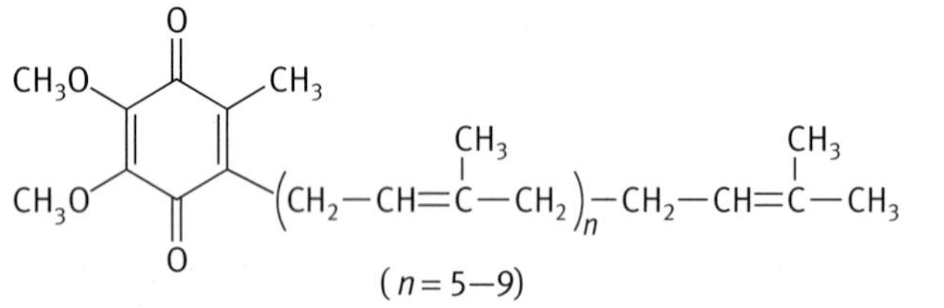

CH_3O, CH_3O, CH_3, $(CH_2-CH{=}\overset{CH_3}{C}-CH_2)_n-CH_2-CH{=}\overset{CH_3}{C}-CH_3$

($n = 5$–9)

Structure of ubiquinone (coenzyme Q)

To summarize, we have in the electron transport chain an FMN-protein, non-haem iron-sulphur proteins, Q not bound to a protein and freely mobile in the membrane, and haem proteins known as cytochromes. An important point is that one of the latter, **cytochrome *c***, is a small water-soluble protein molecule (molecular weight ~12.5 kDa, just over 100 amino acids) that is loosely attached to the outside face of the inner mitochondrial membrane so that it also is free to move. All the other proteins of the respiratory complexes are built into the structure as integral proteins.

Arrangement of the electron carriers

In Chapter 12 we discussed the redox potentials of electron acceptors and explained that electrons flow from a carrier of higher reducing potential (low redox potential or lower electron affinities) to one of lower reducing potential (more oxidizing, or higher redox potential or higher electron affinities). The electron carriers in the chain are of different redox potentials. Their electron affinities increase progressively down the chain.

The redox potentials are directly related to $\Delta G^{0'}$ values, as already discussed in Chapter 12. The electron carriers are arranged in the electron transport chain such that there is a continuous progression down the free-energy gradient (increasing redox potentials) with the corresponding release of free energy as the electrons move from one carrier to the next (Fig. 13.16). They form, a relay carrying electrons down the hill. In considering glucose oxidation, the task in this stage of metabolism is to transfer electrons from NADH and $FADH_2$ to oxygen. The whole scheme involves a somewhat formidable list of steps but (fortunately) the carriers are grouped into the four complexes shown in Fig. 13.17 so that will not give the detailed carrier list. These complexes are built into the structure of the inner mitochondrial membrane, interconnected by the mobile electron carriers, ubiquinone and

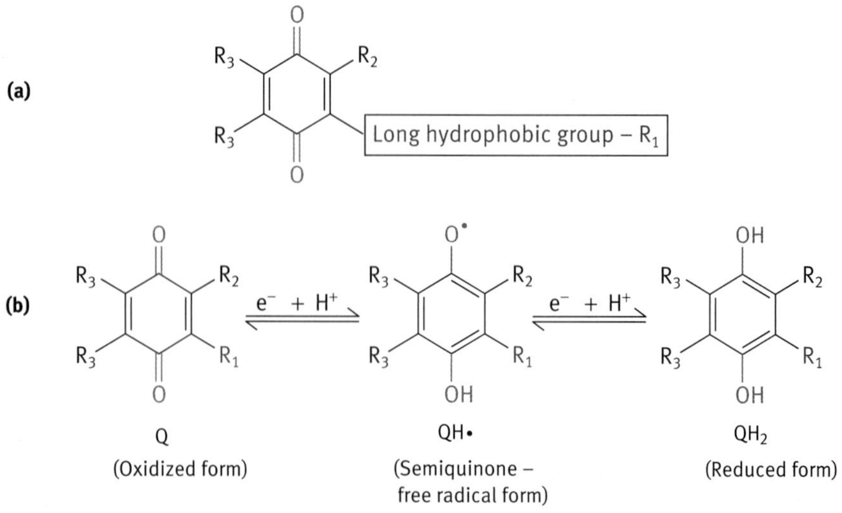

(a) R_3, R_2, R_3, Long hydrophobic group – R_1

(b) Q (Oxidized form) $\xrightarrow{e^- + H^+}$ QH• (Semiquinone – free radical form) $\xrightarrow{e^- + H^+}$ QH_2 (Reduced form)

Fig. 13.15 (a) Ubiquinone or coenzyme Q structure (in the oxidized form), **(b)** Oxidized, semiquinone, and reduced forms of ubiquinone (Q). The semiquinone, $QH^{\cdot}$ can exist as the anion, $Q^{\cdot -}$.R_1, long hydrophobic group; R_2, $-CH_3$; R_3, $-O-CH_3$. The full structure of ubiquinone is given above.

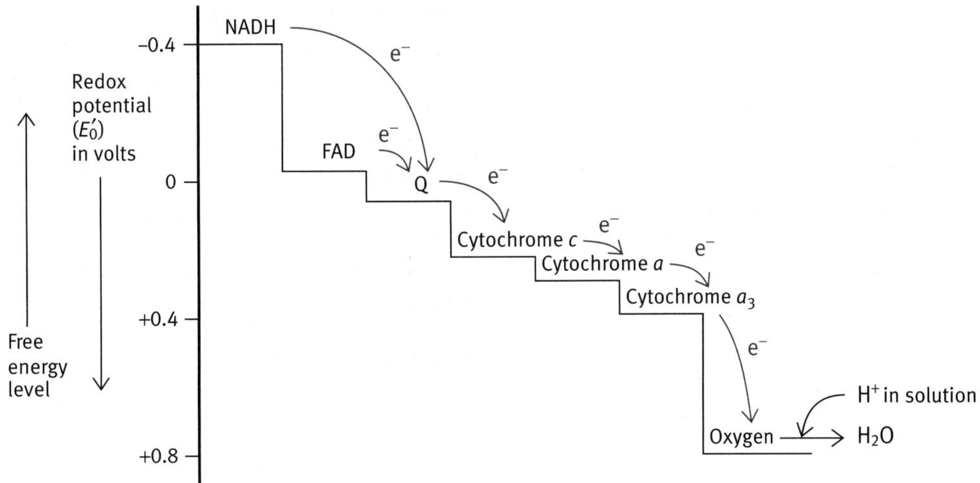

Fig. 13.16 The approximate relative redox potentials of some of the main components of the electron transport system in mitochondria. The arrows indicate electron movements. The role of cytochrome *b* components is shown in Fig. 13.21.

cytochrome *c*. Q takes electrons from complexes I and II and delivers them to complex III. Cytochrome c is the intermediary between complexes III and IV. Complex I carries electrons from NADH to Q. Complex II carries electrons from succinate and other substrates (fatty acids, glycerol phosphate) via $FADH_2$ to Q; complex III uses QH_2 to reduce cytochrome *c*. Complex IV transfers electrons from cytochrome *c* to oxygen. Complexes I, III, and IV are, for convenience, referred to as NADH:Q oxidoreductase, QH_2:cytochrome c oxidoreductase, and cytochrome oxidase, respectively. Complex IV, cytochrome oxidase, is also a multi-subunit structure; electrons are donated to it by cytochrome *c* on the outer face of the inner mitochondrial membrane. Cytochrome oxidase contains the haem proteins, cytochromes *a* and a_3, and copper centres, which participate in the final transference of electrons to oxygen. The final reduction catalysed by cytochrome oxidase is:

$$O_2 + 4e^- + 4H^+ \rightarrow 2H_2O.$$

Let us now see what this electron transport achieves.

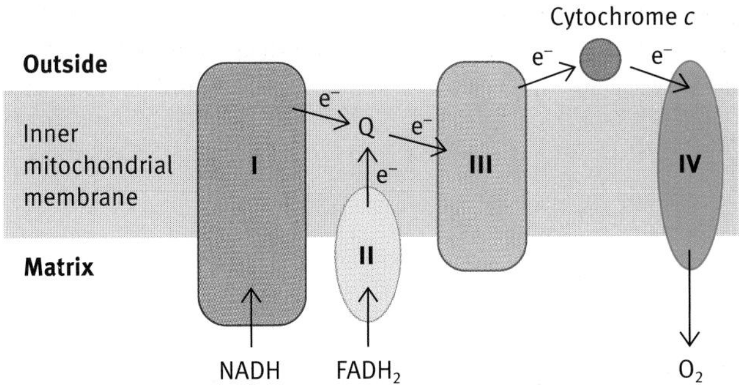

Fig. 13.17 The electron transport chain with electron carriers grouped into four main complexes. Complex I, NADH:Q oxidoreductase; complex II, succinate:Q oxidoreductase; complex III, QH_2:cytochrome *c* reductase; complex IV, cytochrome oxidase. Q, ubiquinone or coenzyme Q. $FADH_2$ is generated in the cycle from succinate by succinate dehydrogenase. Note that the complexes are located in the inner mitochondrial membrane. Q and cytochrome *c* are mobile carriers capable of physically transporting electrons from one site in the membrane to another. Cytochrome *c* is surface-located. Note that $FADH_2$ exists attached to flavoprotein enzymes. The major ones are succinate dehydrogenase and fatty acyl-CoA dehydrogenases involved in fat oxidation; the latter is described in Chapter 14.

Oxidative phosphorylation – the generation of ATP coupled to electron transport

In glycolysis ATP is formed by substrate-level phosphorylation, that is, the phosphorylation (generation of ATP from ADP and P_i) is inseparably linked to the reactions in glycolysis. ATP generation is an intrinsic component of the reactions. The same is true of ATP generation in the TCA cycle via GTP.

In cells, or in intact 'healthy' mitochondria, electron transport results in ATP generation, the two processes are said to be coupled, but, in damaged mitochondria, electron transport occurs readily without ATP generation. This situation perplexed and frustrated biochemists for decades. How is it possible?

The chemiosmotic theory of oxidative phosphorylation

The puzzle was solved by the English biochemist, Peter Mitchell, who in 1961 produced a theory of how electron transport causes ATP synthesis that was so novel that it was at first hardly taken seriously by most and strongly opposed by many. It took a long time for it to be accepted (and for a Nobel Prize to be awarded to Mitchell in 1978). The concept is based on the simple notion that gradients have the ability to do work. A gradient of water pressure can be used to generate electricity, a gradient of air pressure to drive a windmill,

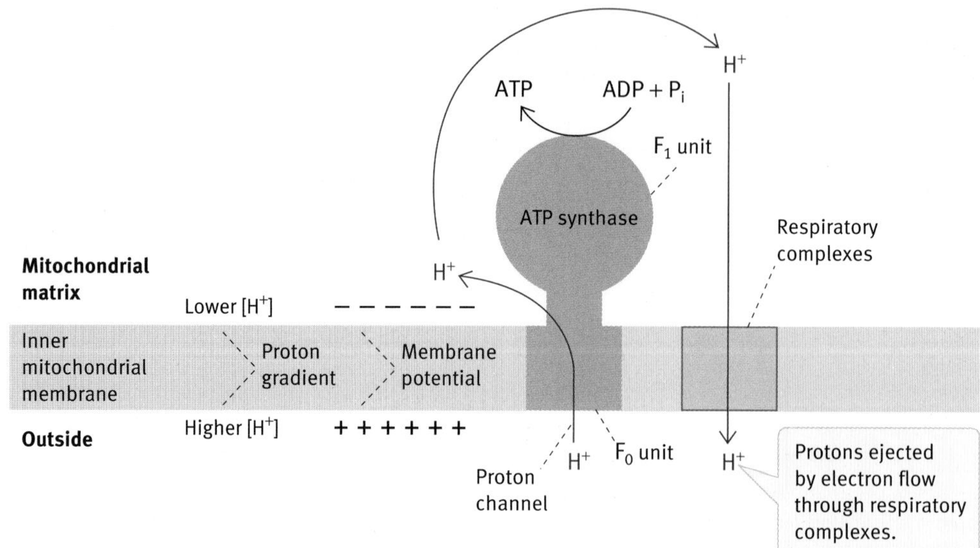

Fig. 13.18 The ATP-generating system of the inner mitochondrial membrane. The entry of H^+ into the matrix via the F_0 unit is discussed later in this chapter.

etc. A chemical gradient is no different. Molecules or ions will migrate from a high concentration to a low concentration and, if a suitable energy-harnessing device is interposed, useful work can be done.

Applying this concept, two things are needed for ATP generation coupled to electron transport: firstly, electron transport must create a gradient of some sort; and, secondly, the gradient must be allowed to flow back through a device that uses the energy of the gradient to synthesize ATP from ADP and P_i. Thirdly, Mitchell's concept required the existence of intact vesicles across whose membrane a gradient could be established. It has to have an inside and an outside, as separate compartments, explaining why damaged mitochondria do not make ATP. The membrane must therefore be impermeable to the solute of which the gradient consists.

Mitchell discovered that electron flow caused protons to be ejected from inside the mitochondrion (or *E. coli* cell) to the outside, thus creating a proton gradient across the membrane (Fig. 13.18); in other words, the pH of the external solution decreased, an exciting moment in the history of biochemistry. A membrane (or charge) potential, negative inside and positive outside, is also generated by the proton expulsion and also contributes to the total energy gradient or **proton-motive force** available for ATP synthesis. The inner mitochondrial membrane itself is virtually impermeable to protons. This is a prerequisite for the system, but inserted into the membrane are special proton-conducting channels. Protons flow from the outside through these channels back into the mitochondrial matrix and the energy of this flow is harnessed to the formation of ATP from ADP and P_i. Mitchell's theory has a majestic simplicity. All aerobic life on this planet is driven by the creation of a pH gradient across a membrane. Dinosaurs roamed on it; from whales to aerobic bacteria, from plants to humans – all are driven by it. It is one of the great concepts in biochemistry.

The proton-conducting channels are knob-like structures that completely cover the inner surface of the cristae. These are, in fact, **ATP synthase** complexes that convert ADP and P_i to ATP, the process being energetically driven by the proton flow. The process will be described shortly.

The overall **chemiosmotic mechanism** in mitochondria is shown in Fig. 13.19. Two main questions can be asked:

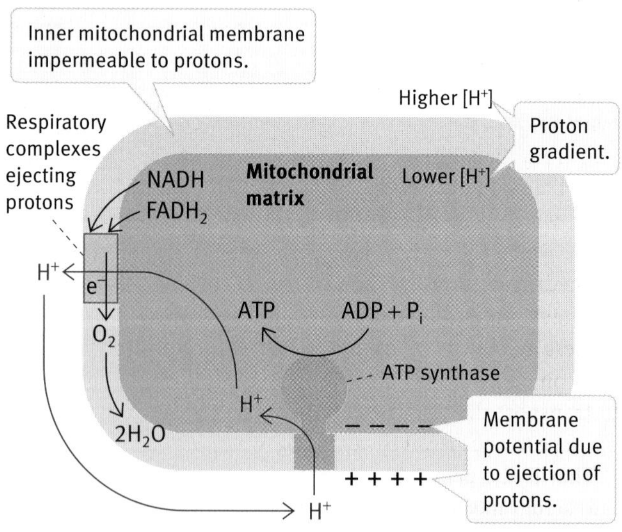

Fig. 13.19 Generation of ATP in mitochondria by the chemiosmotic mechanism. Note that $FADH_2$, produced by the dehydrogenation of fatty acids (described in Chapter 14), enters the same pathway as that utilized by the oxidation of succinate.

- How does the flow of electrons from NADH and $FADH_2$ to O_2 cause protons to be pumped from the matrix side of the inner mitochondrial membrane to outside the membrane?
- How does the flow of protons into the mitochondrion drive the synthesis of ATP from ADP and P_i?

How are protons ejected?

Protons are transferred from the matrix to the cytosolic side of the membrane by three separate complexes. These are I (NADH: Q oxidoreductase), III (QH_2: cytochrome coxidoreductase), and IV (cytochrome oxidase). Complex I is a huge complex with a large domain extending into the matrix where the NADH binds. The mechanism by which complex III ejects protons was the first to be established. The proton-pumping mechanism of complex IV is not fully understood, though the three-dimensional structure of this complex has been obtained.

Complex II, which reduces Q to QH_2 using $FADH_2$ as the reductant (see Fig. 13.17) does not pump protons as the free-energy drop in the process is insufficient. FAD reduction to $FADH_2$ occurs mainly by succinate dehydrogenase and in fatty acid metabolism. The $FADH_2$ has a higher redox potential (lower free energy) than NADH. Proton pumping by complex III is described below, followed by that of complex IV (cytochrome oxidase).

Figure 13.20 illustrates the fact that the energy for proton-gradient formation is derived from the free energy released as electrons are transported down the electron carrier chain.

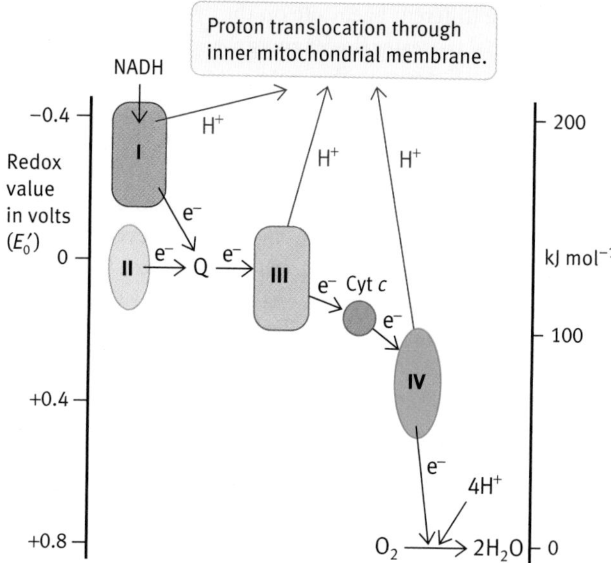

Fig. 13.20 Approximate positions on the redox potential scale of the electron transport complexes. The scale on the right gives the approximate free energy released when a pair of electrons is transferred from components to oxygen. Q, ubiquinone.

The Q cycle in complex III ejects protons from mitochondria

The mechanism by which protons are translocated as a result of electron flow is established in the case of complex III. The *principle* of Mitchell's original idea is as simple as it is ingenious. In essence, hydrogen *atoms* are assembled on the matrix side of the inner mitochondrial membrane, using protons from the matrix and electrons from the transport chain. The H atoms are assembled on Q, forming the reduced form, QH_2, which now diffuses to the opposite face of the membrane where the reverse happens – electrons are stripped off the hydrogen atoms by the electron transport system and the resultant protons escape to the outside. The essentials of the process are the positioning of the responsible catalytic proteins on opposite sides of the membrane and a mobile carrier to transport the hydrogen atoms across the membrane from one face to another. It is called the **Q cycle**.

That is the principle for proton transport in complex III. The actual mechanism is shown in Fig. 13.21; it looks complicated but is in fact very simple with few reactions involved, and achieves what Mitchell's original idea proposed: assemblage of H atoms on one side attached to Q and release on the other side. Note first of all that the red arrows simply represent physical movement of ubiquinone and its derivatives; the second point is that, in effect, two distinct processes are going on, both of which eject a pair of protons. In the membrane there are pools of Q and QH_2.

Now in more detail, let us start with QH_2 formation by complexes I and II, and whose electrons originate from NADH and $FADH_2$, respectively. In this, two protons are taken up from the matrix as shown to the left of the diagram and two electrons from NADH (in complex I) or $FADH_2$ (in complex II) form hydrogen atoms on Q. The QH_2 now migrates to a site, in complex III, *on the external face of the inner membrane*, where one electron is removed and passed on to reduce cytochrome *c*, which transports it to complex IV. Remember that cytochrome *c* is also mobile. The site on complex IV that accepts electrons from cytochrome *c* is exposed on the external face of the inner membrane and cytochrome *c* is also located on this face of the membrane. A proton is ejected, leaving the half-oxidized quinone, QH- (see earlier in this chapter). A further electron is now removed from the latter to form Q, but in this case the electron is handed on, not to cytochrome *c* but to cytochrome *b*, to which we will return shortly. A proton is ejected. That is the end of that half of the story – a molecule of QH_2 from complexes I/II has been oxidized, two electrons passed (one to cytochrome *c* and one to cytochrome *b*), and two protons ejected from the matrix to the outside. The Q so formed now returns to the general pool.

There is another part to the story. A *second* molecule of QH_2 is oxidized in the same way as the first, resulting in the ejection of two more protons. From each of the two molecules of QH_2 we thus have two electrons passing on to complex IV via cytochrome *c* and two to cytochrome *b*. The latter transfers the electrons to a different haem site on the cytochrome *b* whose redox potential is greater (lower energy) and in this way transports the two electrons to the matrix side of the membrane. They are donated here to a molecule of Q and, with a pair of protons from the matrix, hydrogen atoms on Q are assembled as QH_2. The QH_2 now migrates back to the site of the external face and the cycle starts again. (The diagram in Fig. 13.21 shows the same molecule of Q going round the cycle, for clarity, but molecules of Q and QH_2 will enter and exit the pool in a dynamic equilibrium – it amounts to the same thing.) This somewhat convoluted double process actually oxidizes only one molecule of QH_2 (since a molecule of Q is reduced in the whole process), but achieves the ejection of four protons. The net effect of the reactions within complex III can be summarized in the equation:

$$QH_2 + 2H^+(\text{matrix}) + 2\text{cyt}c(Fe^{3+}) \rightarrow Q + 4H^+(\text{outside}) + 2\text{cyt}c(Fe^{2+}).$$

Complex IV also contributes to the proton gradient

Complex IV oxidizes reduced cytochrome *c*, forming water, and it also contributes to the formation of the proton gradient across the mitochondrial membrane. Water is formed by the reaction:

$$4\text{cyt}c(Fe^{2+}) + 4H^+ + O_2 \rightarrow 4\text{cyt}c(Fe^{3+}) + 2H_2O.$$

During the oxidation process, protons are actively pumped out of the mitochondrion by a mechanism not yet fully understood but possibly involving protein conformational changes.

For the oxidation of four reduced cytochrome *c* molecules, four protons are transported out of the mitochondrial matrix (two per electron pair). The protons used to form water are taken from the matrix and this increases the proton gradient. The result of the oxidation of four reduced cytochrome *c* molecules is to remove a total of four protons as water, and four ejected into the cytosolic compartment.

To briefly digress on a separate note of importance, in the process of water formation, electrons are added to oxygen. Although oxygen is a safe molecule, addition of a single electron to an oxygen atom yields a dangerous, highly reactive free radical, superoxide.

Addition of the two electrons forms peroxide, which is also potentially dangerous. Cytochrome oxidase adds four electrons to oxygen, forming water from H^+ but without releasing intermediate species. Superoxide and peroxide are two members of a group of free radicals, in this case, reactive oxygen species (ROS) which can potentially damage DNA, proteins and lipids. Their generation and defence mechanisms of the cells are discussed in Chapter 32.

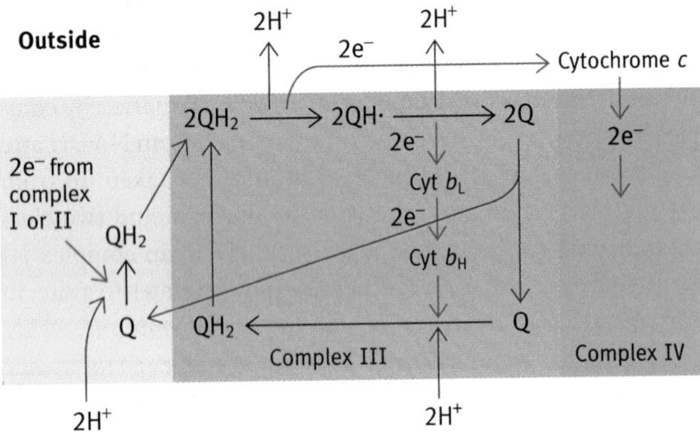

Fig. 13.21 The mechanism of proton translocation by complex III as a result of electron transport in mitochondria. The red arrows represent physical diffusion of components rather than chemical transformations; the latter are indicated by black arrows and electron transport by blue arrows. (However, note that the blue arrow representing electron transport by cytochrome *c* is effected by the physical diffusion of the latter from complex III to complex IV and its return after oxidation.) Cytochrome *b* protein spans the membrane so that electrons are transferred from the outer face to the inner face. The molecules of Q and QH_2 are in equilibrium with a membrane pool of these, but this is not shown in order to simplify the diagram arrangement. The electrons from complex I (as QH_2) arise mainly from NADH generated in glycolysis and the TCA cycle; those from complex II come from the succinate→fumarate step of the cycle via an FAD-protein. As will be described in the next chapter, electrons from fatty acid oxidation also enter complex III via complexes I and II. Q, ubiquinone; QH˙, semiquinone; Cyt, cytochrome. Cyt b_L and b_H refer to haem sites on cytochrome *b* of low and high redox potentials respectively.

ATP synthesis by ATP synthase is driven by the proton gradient

Paul Boyer of UCLA, who won a Nobel Prize for his work on ATP synthase, referred to it as a 'splendid molecular machine' and added 'All enzymes are beautiful, but ATP synthase is one of the most beautiful as well as one of the most unusual and important'; all of which is true.

ATP synthase is the name of the structure with one part visible as a knob (the F_1 unit) projecting into the matrix on the inside surface of the inner mitochondrial membrane and the other anchored in the membrane itself (the F_0 unit; o=oligomycin; see Box 13.2). The knobs are diagrammatically represented in Figs 13.18 and 13.21. A simplified diagram, showing the major components of the ATP synthase is shown in Fig. 13.22. The reaction catalysed by the synthase is:

$$ADP^{3-} + P_i^{2-} + H^+ \rightarrow ATP^{4-} + H_2O.$$

The standard free energy change of this is about +29.3 kJ and so cannot proceed without a large input of energy, which is supplied by the proton gradient established across the membrane by electron transport. Protons flow back into the mitochondrial matrix through the ATP synthase. It is a major metabolic activity.

ATP synthase is found in aerobic organisms wherever the energy derived from electron transport has to be trapped

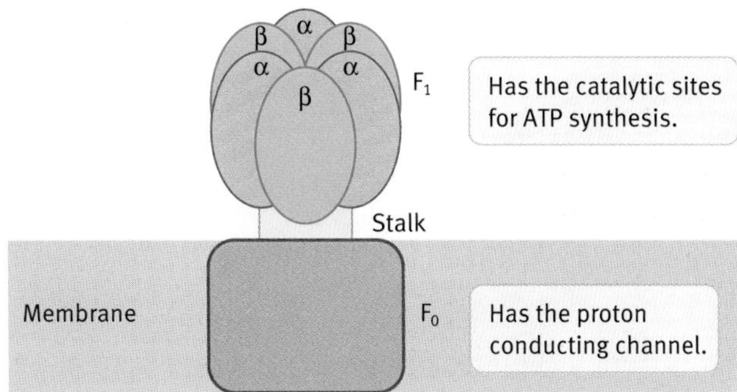

Fig. 13.22 Simplified diagram showing the major components of ATP synthase. The F_0 has multiple subunits and is integral with the lipid bilayer membrane. It is the proton-conducting channel. The F_1 is composed of a hexamer ring of alternating α and β subunits enclosing two central subunits γ and ε, which project downwards and contact the F_0 unit.

as ATP. It is not confined to mitochondria – chloroplasts in plants have the same system. So do *E. coli* cells which are roughly the size of a mitochondrion, the cell membrane in this context being equivalent to the inner mitochondrial membrane. Its ATP synthase units project from the cell membrane into the cytosol and the electron transport system in the membrane creates a proton gradient from outside to inside by pumping protons from the inside to the outside of the bacterial cell. The structures from all sources are essentially the same.

Structure of ATP synthase

The complete structure of ATP synthase is shown in Fig. 13.23 (in this case from *E. coli*, but a very similar arrangement is found in mitochondria). It may look formidable, but we can look at the two parts separately, F_0 in the membrane and F_1 the knob projecting into the mitochondrial matrix, and how they function.

The F_1 unit and its role in the conversion of ADP+P_i to ATP

The F_1 unit is a ring formed by six protein subunits arranged in a barrel-like structure in external appearance more or less like the segments of an orange (Fig. 13.23). All F_1 subunit proteins are designated by Greek letters. The 'knob' consists of a hexamer of three α protein subunits and three β subunits, the two alternating. Each β subunit has a catalytic (enzymic) site which synthesizes ATP from ADP+P_i, so there are three such sites per F_1 unit located at interfaces with the α subunits. The narrow cavity of the barrel is occupied by an elongated asymmetric shaft, the γ subunit, projecting beyond the barrel to form a short 'stalk' which connects the F_1 to the F_0 unit in the membrane. (In Fig. 13.23, the visible part of the γ subunit is shown in yellow and its extension inside the hexamer in a darker shade.) Another subunit called ε forms part of the stalk structure.

Figure 13.25 shows two sections of F_1 as ribbon diagrams, determined by X-ray diffraction, with the asymmetric γ subunit shaft in the centre. We will come to the other subunits in due course, so do not be concerned with them now.

Activities of the enzyme catalytic centres on the F_1 subunit

If we consider for the moment an enzymic site on a *single* **β subunit**, the sequence of events in the synthesis of a molecule of ATP as proposed in Boyer's model is as follows (Fig. 13.24(a)).

- The site is open and nothing is bound to it (the **O state**).
- A conformational change in the protein converts the site to a low-affinity state; ADP and P_i now bind to it loosely but there is no catalytic activity (the **L state**).
- A further conformational change produces a tight-binding state – the ADP and P_i become tightly bound. This is now catalytically active and ATP is formed (the **T state**).
- A conformational change opens up the site, ATP escapes and the site is back to the original open state.

The model postulates that each site progresses sequentially through the three conformations. (It is easy to fall into the error of imagining that the sites rotate – they do not.) ATP synthase is a most unusual enzyme in that catalytic activity is dependent on cooperation between the subunits. At any point in time one of the three β subunit sites is in the O state, one in the L state, and one in the T state (see Fig. 13.24(b)). A site in the T state with a molecule of ATP bound to it will convert to the open O state and release the ATP when the preceding site on the β subunit in the L state becomes occupied by ADP and P_i, which at the same time converts to the T state itself.

The ADP and P_i are now tightly bound and ATP synthesis proceeds.

You may be puzzled that in the third step above, we simply state that ATP is formed from ADP and P_i, which seems energetically wrong. However, the researchers working on this problem found that *when ADP and P_i are firmly bound to the active site of the enzyme* there is little free energy change in the formation of ATP. The equilibrium constant is about one as compared with 10^{-5} for ADP and P_i in free solution (see Chapter 3 if you need to be reminded about equilibrium constants and free energy). This does not conflict with what you have learned about the energetics of ATP, for it applies only to reactants tightly bound to the enzyme. *Energy is needed to release the ATP* so that the conversion of ADP+P_i *in solution* to ATP *in solution* requires the expected energy input. This is supplied by the conformational changes which the enzyme undergoes during the catalytic cycle. These are caused by the rotating asymmetric γ subunit sequentially contacting the F_1 subunits. It is not known how the energy transference occurs, but each β subunit in turn undergoes a conformational change, which puts it into a 'high-energy state' allowing ATP release.

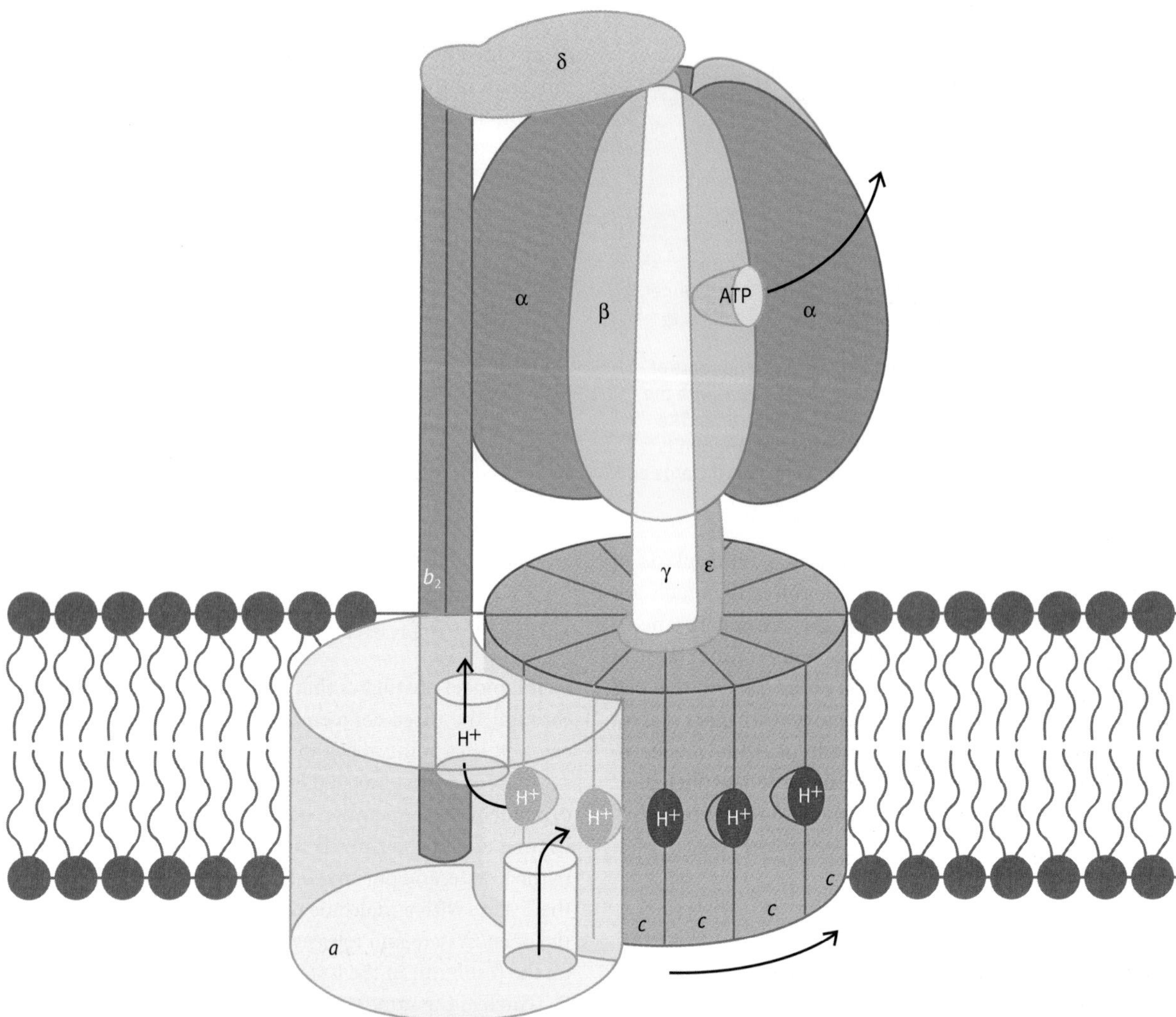

Fig. 13.23 Model of the *E. coli* ATP synthase. All ATP synthases are essentially the same except for a small variation in the number of *c* subunits. We will describe the model in terms of the mitochondrial location since this is most relevant to the text. The F_0 consists of a ring of *c* subunits integrated into the membrane lipid bilayer. Adjacent to it is the *a* subunit also integrated into the bilayer. This has two nonconnecting proton-conducting half channels, one open to the outside of the membrane and the other to the inside of the mitochondrion. Flow of protons from the outside through the F_0 into the mitochondrion matrix causes rotation of the ring of *c* subunits. This drives the rotation of the 'stalk' formed by the γ and ε subunits which project as a central asymmetric 'shaft' into the barrel-like hexamer of subunits constituting the F_1 in the mitochondrial matrix. The F_1 hexamer has three α and three β subunits surrounding the central shaft. Each β subunit has an active site near the interface with the adjacent α subunit in which synthesis of ATP from ADP and phosphate occurs. As the central shaft rotates, it contacts in succession the surrounding subunits and causes conformational changes in the active sites involved in ATP synthesis (see Fig. 13.24). The actual energy-requiring step is the release of ATP from the sites. This is supplied via the conformational changes. The δ and $b_2$subunits have the role of preventing the rotation of the hexamer of the F_1 as the central shaft rotates; it is equivalent to bolting down the casing of an electric motor to stop it rotating as the shaft rotates. A more detailed description of the ATP synthase and the mechanism of its action are given in the text. Fillingame R H; Molecular Rotary Motors; Science; (1999) 286:1687–1688; Reproduced by permission of American Association for the Advancement of Science.

Structure of the F_0 unit and its role

F_0 built into the inner mitochondrial membrane is the motor which is driven to rotate by a flow of protons from outside the inner membrane into the inside of the mitochondrion. It rotates the γ subunit inside the F_1 to which it is connected. The F_0 consists of a ring of *c* subunits (Fig. 13.23; F_0 proteins are denoted by italic letters rather than the Greek ones used for F_1 protein), varying in number from 10 to 14 in the various ATP synthases. Do not worry about the H^+ depicted on the ring of subunits of the F_0 – we will come to that shortly.

Each *c* subunit is a single α helical polypeptide in the shape of a hairpin so that each has two 'arms' spanning the lipid bilayer. The crucially important feature to note is that in the middle of the α helix of one of these arms of each *c* subunit is

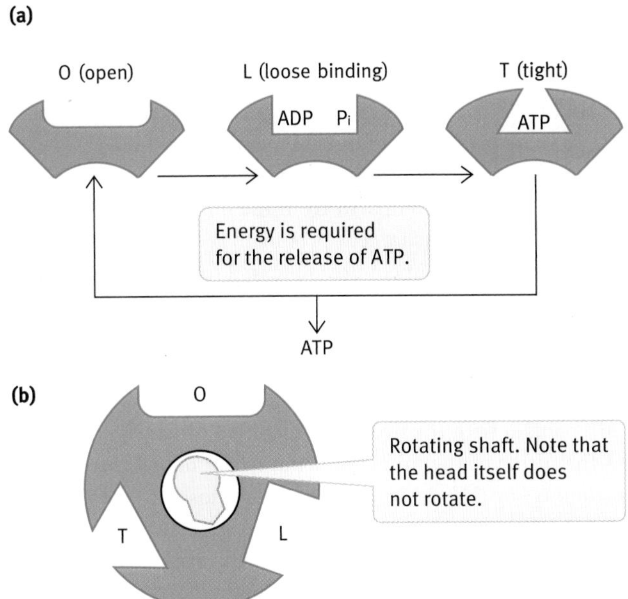

Fig. 13.24 The catalytic sites of ATP synthase as proposed in the Boyer model: **(a)** the changes that occur in a single site of one β subunit of F_1 during the synthesis of ATP. **(b)** The three β subunits work in a cooperative manner and the conversions in one site are coordinated with the other two sites. This means that at any one time an F_1 unit has one subunit in the O state, one in the L state, and one in the T state. The rotating shaft is shown as a notional asymmetric shape to convey that it is believed that it successively interacts with the subunits as it rotates. The actual structure of the shaft within the F_1 barrel is given in Fig. 13.25.

an aspartate residue which is thereby placed in the centre of the hydrophobic lipid bilayer. Adjacent to the ring of *c* subunits is the large *a* protein.

Mechanism by which proton flow causes rotation of F_0

To explain this we will use Fig. 13.26 which gives a different view of the ring of 12 *c* subunits seen in plan from the F_1 side.

It is essential to note that the c ring is surrounded by the hydrophobic lipid bilayer except for the two *c* subunits which interface with the *a* protein. Ten of the subunits will be in the hydrophobic environment of the surrounding lipid bilayer. This energetically requires the central aspartyl residue of each of these to be in the protonated uncharged $-COOH$ state rather than the unprotonated charged $-COO^-$ state. In the case of the two *c* subunits adjacent to the a protein the situation is different because, it is postulated, there are two half channels in the a protein, as shown in Fig. 13.23 as a transparent shape. These expose their aspartyl residues to a hydrophilic environment. They are therefore in the unprotonated $-COO^-$ state. Figure 13.23 shows that two half channels do not make a direct connection between the two faces of the membrane, so that protons cannot flow via them directly across the a protein. One half-channel is open to the

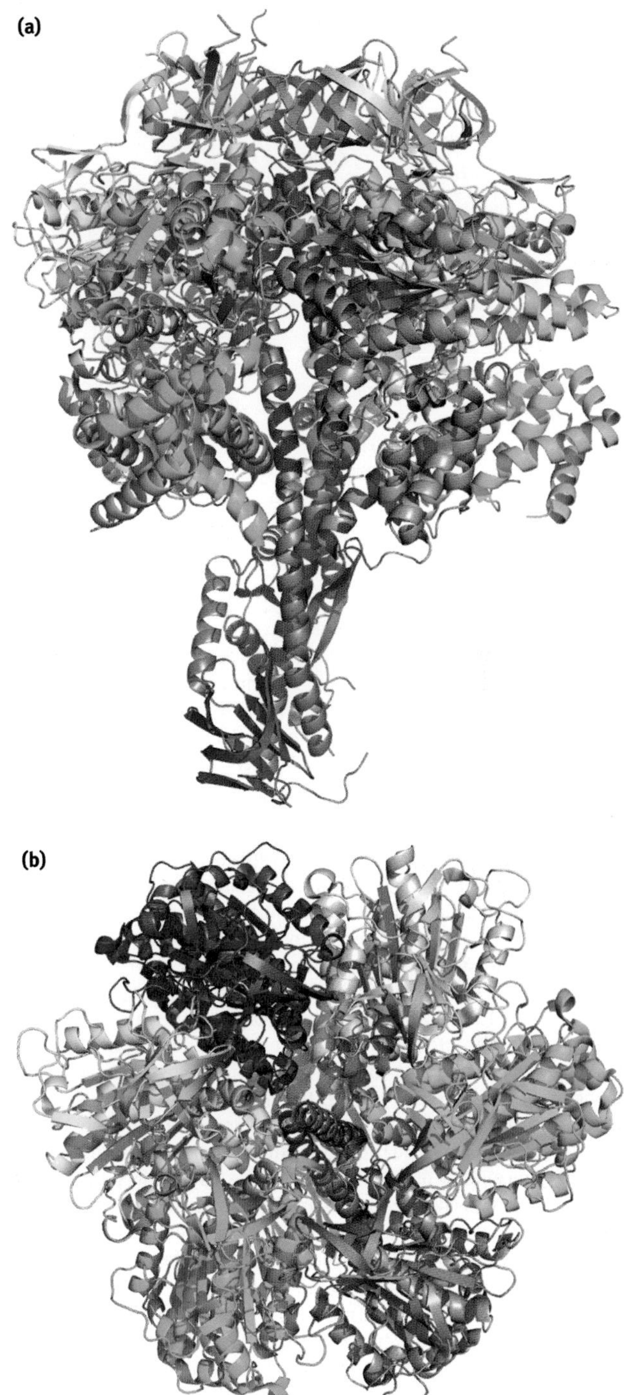

Fig. 13.25 Ribbon diagrams of the three-dimensional structure of the F_1 of ATP synthase (Protein Data Bank Code 1JNV), with the γ subunit in the central cavity (coloured yellow-brown). The ε subunit of the central shaft is coloured purple. In **(a)** the diagram is a longitudinal section of the entire head, showing the conformation of the ε and γ subunits within the *Escherichia coli* F_1 ATPase. In **(b)** a cross-section of the head is shown, giving the relative arrangement of the α and β subunits. Abrahams, J. P., Leslie, A. G. W., Lutter, R, and Walker, J. E; Structure at 2.8 A resolution of F1-ATPase from bovine heart mitochondria; Nature; (1994) 370, 621–8; Nature Publication Group.

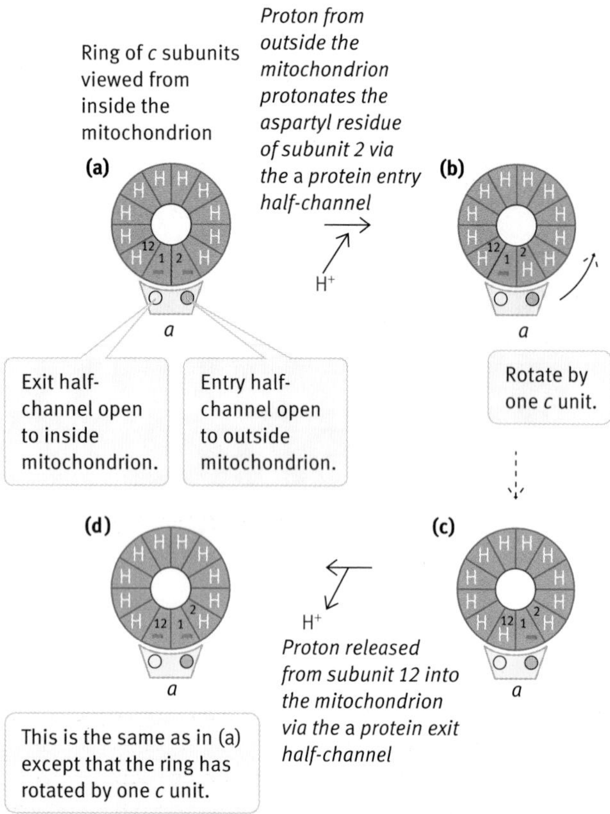
Ring of c subunits viewed from inside the mitochondrion
Proton from outside the mitochondrion protonates the aspartyl residue of subunit 2 via the a protein entry half-channel
(a)
(b)
H⁺
a
a
Exit half-channel open to inside mitochondrion.
Entry half-channel open to outside mitochondrion.
Rotate by one c unit.
(d)
(c)
H⁺
Proton released from subunit 12 into the mitochondrion via the a protein exit half-channel
a
a
This is the same as in (a) except that the ring has rotated by one c unit.

Fig. 13.26 Diagram to illustrate the principle of F_0 rotation. In **(a)** the central aspartyl group in *c* units 1 and 2 are unprotonated and in contact with the hydrophilic environment provided by the two half channels of subunit *a*. If the residue in subunit 2 is now protonated from the ***outside*** of the mitochondrion ***via*** the entry half-channel of ***a*** as in **(b)**, the ring of *c* units will rotate to bring the uncharged residue into hydrophobic contact with the lipid bilayer. At the same time, subunit 12 moves into the hydrophilic region provided by the exit half-channel of *a* as shown in **(c)** and the proton is lost to the in side of the mitochondrion. This restores the situation to that in (a) except that the ring has moved by one subunit as shown in **(d)**. Repetition of this cycle causes stepwise F_0 rotation. The net result is that protons flow from the outside of the membrane to the inside driven by the concentration and charge gradient and in doing so rotate F_0. A molecular model of this diagrammatic figure has been used as the cover picture for this edition (Protein Data Bank code 1C17).

matrix *inside* the mitochondrial membrane (the left-hand one in Fig. 13.26) while the other (on the right) is open to the *outside*.

What causes the ring to rotate? The answer is extremely simple but you will need to follow Fig. 13.26 closely. The state of the aspartyl residue in the centre of each sub unit is shown as a white H for the uncharged state (−COOH) and a green minus sign for the ionized state ($-COO^-$). In Fig. 13.26(a) the aspartyl residues of subunits 1 and 2 are charged since each is exposed to one of the hydrophilic half channels of protein *a*. Those of the other ten of the *c* ring subunits, in contact with the hydrophobic lipid bilayer, are, as stated, uncharged.

The ring cannot move in this state since it would bring the charged group of subunit 2 into the hydrophobic environment (which is thermodynamically 'forbidden'). The aspartyl group of subunit 2 is open to the half-channel, which connects to the *outside* of the mitochondrial membrane where there is a high concentration of protons. This causes its central aspartyl group to become protonated from the *outside pool* thus converting it to the uncharged, protonated −COOH state (Fig. 13.26(b)), which is 'uncomfortable' in a hydrophilic environment. This causes the ring to move by one subunit so that the now uncharged aspartyl group of subunit 2 is thus comfortably placed in a hydrophobic environment. This movement, however, brings the uncharged aspartyl group of subunit 12 into contact with the hydrophilic half-channel which is open to the *inside* of the mitochondrion *where the proton concentration is low* (Fig. 13.26(c)) causing it to lose its proton to the matrix. This produces the state in Fig. 13.26(d), which is the same as in Fig. 13.26 (a) except that the ring has moved by one *c* subunit. Repetition of the same cycle of events causes stepwise rotation of the ring. Each proton joining the aspartyl group from outside the mitochondrion is thus carried round the ring as a passenger on its *c* subunit until after 11 moves it arrives at the exit half-channel of the aprotein (Fig. 13.26(c)) and moves into the mitochondrion (Fig. 13.26(d)). Rotation of the *c* ring relative to the *a* protein has been demonstrated by experiments in which the *a* and *c* proteins were chemically cross-linked.

Thus the proton flow, reinforced by the membrane potential, by this complex pathway from the outside to the inside of the mitochondrial inner membrane generates the force, which rotates the γ subunit inside F_1. The rotation of F_0 is unidirectional. This is the result of much higher proton concentration outside the membrane than inside. This means that protonation of aspartyl residues occurs preferentially from the outside and deprotonation to the inside.

If the F_1 unit is detached from the F_0 its reactions are reversible in the presence of ATP, which is hydrolysed. It is an ATPase. Yoshida's group in Japan has, in an ingenious experiment, directly visualized the reverse-direction rotation of the γ subunit in such detached F_1 units as ATP is hydrolysed.

To summarize, the shaft is asymmetric and sequentially contacts the F_1 hexamer subunits. In some way, not understood, this transmits rotational energy into conformational changes in the F_1 subunits. This supplies the energy to allow ATP release.

It is estimated that for each molecule of ATP synthesized, three protons flow through the F_0, though this is not certain to be the actual figure. The value need not necessarily be a whole number. It is the world's smallest rotary motor and one of the most remarkable enzymes known.

What is the role of the elongated subunit *b* dimer and the *d* subunit on the left of the structure in Fig. 13.23? This is rather interesting. To digress briefly, an electric motor needs to have its outer casing bolted down to a bench or whatever to stop it

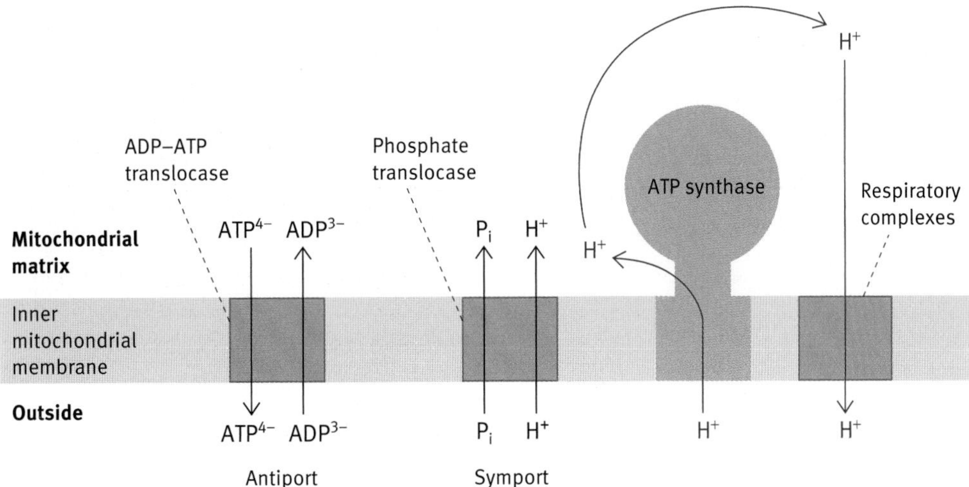

Fig. 13.27 Diagram of transmembrane traffic in mitochondria involved in ATP generation. All of the traffic is *via* specific transport proteins. Other transport systems for metabolites exist.

rotating as the shaft inside it turns, The two protein subunits have the same job in ATP synthesis. They bolt down the F_1 to the membrane to restrain it as the γ subunit rotates.

Transport of ADP into mitochondria and ATP out

The inner mitochondrial membrane is impermeable to most compounds, and to electrons. Special transport systems (translocases) have been developed. Most of the ATP synthesis in most eukaryotic cells occurs in mitochondria while most of the ATP is used outside of the mitochondria. Hence ADP and P_i must enter the mitochondrion and ATP move out. The highly charged molecules cannot diffuse passively across the inner mitochondrial membrane and a special transport mechanism exists. ATP–ADP translocase exchanges ATP inside the mitochondrion for an ADP outside the mitochondrion (Fig. 13.27).

Where does the energy for this ATP–ADP exchange come from? As already explained, electron transport generates not only a pH gradient across the inner mitochondrial membrane but also a membrane potential across the inner mitochondrial membrane, positive outside and negative inside, due to the ejection of H^+ ions (see Chapter 7 if you are not clear what is meant by a membrane potential). ATP carries four negative charges out, while ADP carries only three in. Thus the ATP–ADP exchange tends to neutralize this membrane potential. Therefore the exchange of ATP for ADP costs the equivalent of one proton. The transport of the P_i needed along with ADP for ATP generation is catalysed by a phosphate translocase in the mitochondrial membrane (Fig. 13.28). It carries $H^2PO_4^-$ into the matrix driven by the proton gradient.

Another transport problem occurs in the oxidation of cytosolic NADH generated in glycolysis. Two different 'shuttle' mechanisms exist to cope with this. These will now be described.

Reoxidation of cytosolic NADH from glycolysis by electron shuttle systems

In the aerobic situation, the NADH generated in glycolysis is reoxidized by transferring its electrons into mitochondria. This is the 'normal' route of reoxidation of NADH. NADH itself cannot enter the mitochondrion; there are two systems for transferring its electrons into mitochondria. In these, protons from NADH are transported into the mitochondrion, leaving NAD^+ in the cytosol.

The glycerol phosphate shuttle

The first, the glycerol phosphate shuttle, involves dihydroxyacetone phosphate (generated by the aldolase reaction). An enzyme in the cytosol transfers electrons from NADH to dihydroxyacetone phosphate giving glycerol-3-phosphate (Fig. 13.29). The enzyme is called **glycerol-3-phosphate dehydrogenase**, working in reverse in the above reaction.

Glycerol-3-phosphate can reach the inner mitochondrial membrane (the outer one being highly permeable) where a different type of glycerol-3-phosphate dehydrogenase, *built into the membrane* – with an FAD prosthetic group transfers electrons from glycerol-3-phosphate to the mitochondrial electron transport chain. The dihydroxyacetone phosphate so produced cycles (shuttles) back into the cytosol to pick up more electrons (Fig. 13.29). Note that glycerol-3-phosphate does not have to enter the mitochondrial matrix but its pair of electrons gain access to the electron chain carrying electrons to oxygen, located in the inner mitochondrial membrane. The net effect is to transfer electrons from cytosolic NADH to the mitochondrial electron transport chain.

The malate–aspartate shuttle

Another shuttle, the **malate–aspartate shuttle**, transfers electrons from cytosolic NADH to mitochondrial NAD^+. The

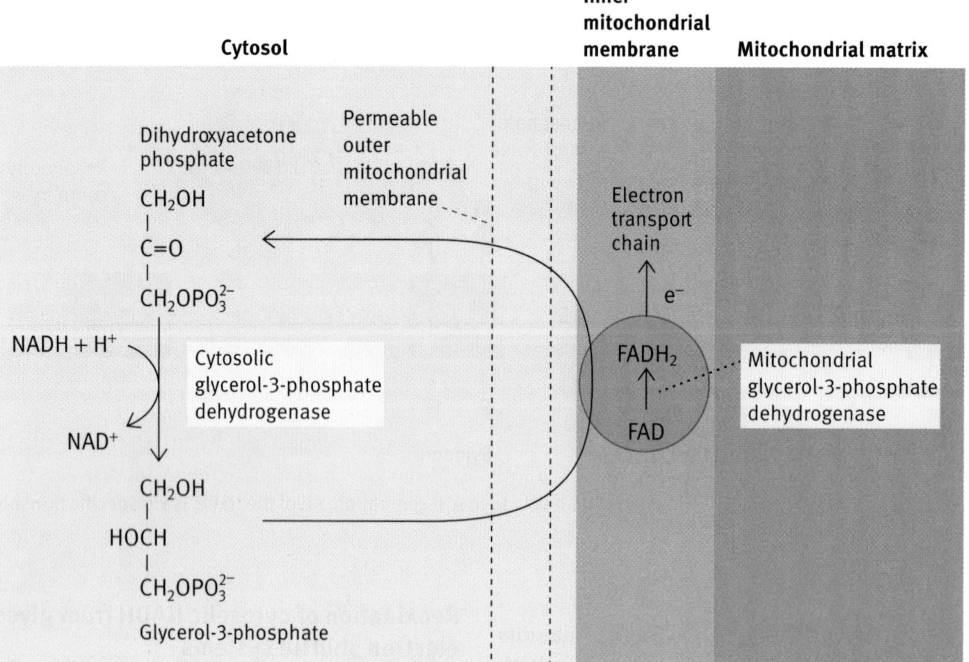

Fig. 13.28 The glycerol phosphate shuttle that transfers electrons from cytosolic NADH to the electron transport chain of mitochondria.

mitochondrial NADH thus generated is then oxidized by the electron transport chain. This shuttle system involves transfer of electrons from NADH to oxaloacetate to form malate in the cytosol which is transported into the mitochondrion, by a specific carrier, where it is reoxidized to oxaloacetate, mitochondrial NAD^+ being reduced (Fig. 13.29). The net effect is that NADH outside reduces NAD^+ inside. This shuttle is a little more complicated in that the oxaloacetate so formed, cannot traverse the mitochondrial membrane to get back to the cytosol. It is converted into aspartate, which is transported, again by a specific carrier, to the cytosol and reconverted into oxaloacetate there; hence the name, the malate–aspartate shuttle. At this stage we will not give the mechanism of aspartate $\rightleftharpoons$ oxaloacetate inter-conversions, since it will be more convenient to do this later when we deal with amino acid metabolism.

Different tissues probably use the two shuttles to different extents. The two differ in a significant way – the glycerol phosphate shuttle results in cytosolic NADH reducing the FAD of the prosthetic group of glycerol-3-phosphate dehydrogenase. The $FADH_2$ has a higher redox potential than NADH (lower energy); it hands on its electrons to the electron transport chain at a point that is further along the chain from that at which NADH hands on its electrons. The net ATP generation from the oxidation of a *cytosolic molecule* of NADH by this glycerol phosphate route is 1.5 molecules. The malate–aspartate shuttle starts with one molecule of cytosolic NADH and ends up with one molecule of mitochondrial NADH whose oxidation generates 2.5 molecules of ATP.

The balance sheet of ATP production by electron transport

It requires an estimated three protons to flow through the ATP synthase to generate one molecule of ATP from ADP and P_i, assuming that the latter two are already inside the mitochondrion. As described, the transport of a molecule of ADP into the mitochondrion and that of one of ATP to the cytosol requires the energy equivalent of one proton entering the mitochondrion. Hence four protons have to be pumped out of the matrix to drive the production of one molecule of ATP made available in the cytosol of the cell. For each pair of electrons transported from NADH to oxygen, the consensus is that 10 protons are pumped out of the mitochondrial matrix (four from complex I, four from complex III, and two from complex IV). Thus the oxidation of one molecule of NADH (located inside the mitochondrion) will produce 2.5 molecules of ATP. For the oxidation of one molecule of $FADH_2$ (that is, a pair of electrons from succinate or fatty acid), six protons are pumped yielding 1.5 molecules of ATP. (Earlier estimates were 3 and 2, respectively.) The values are known as P/O ratios since a pair of electrons reduce one atom of oxygen.

The molecules of NADH produced in glycolysis, located in the cytosol, require separate consideration; these donate their pairs of electrons either to NAD^+ located inside the mitochondrion or to mitochondrial FAD, depending on which shuttle mechanism the cell uses. Thus, a *cytosolic molecule* of NADH may give rise to either 2.5 or 1.5 molecules of ATP.

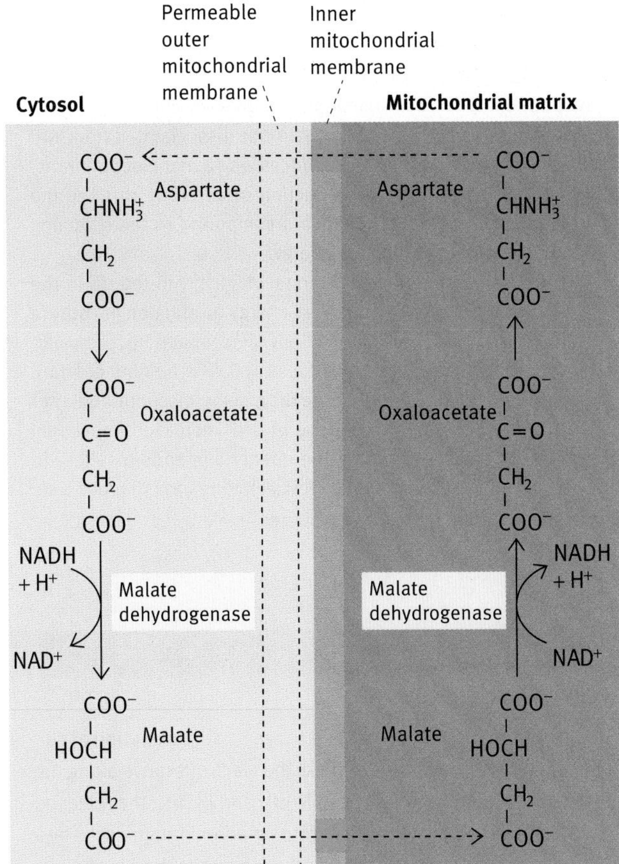

Fig. 13.29 The malate–aspartate shuttle for transferring electrons from cytosolic NADH to mitochondrial NAD^+. The mechanism of the interconversion of oxaloacetate and aspartate is dealt with in Chapter 18. This shuttle, unlike the glycerophosphate shuttle, is reversible, and can operate as shown, bringing NAD^+ inside the mitochondrion, only if the NADH/NAD^+ ratio is higher in the cytosol than in the mitochondrial matrix.

Yield of ATP from the oxidation of a molecule of glucose to CO_2 and H_2O

Starting from free glucose rather than glycogen, the net yield of ATP from the complete oxidation of the molecule is either 30 or 32 depending on which shuttle is used for the cytosolic NADH. To summarize: two from glycolysis at the substrate level (remember that, although four ATP molecules are generated in glycolysis, two were used at the start so the net gain is two). In the TCA cycle, two molecules of ATP (*via* GTP in liver and kidney) are produced per molecule of glucose at the succinyl-CoA stage (one per turn of the cycle but two acetyl-CoA molecules are produced per glucose). Thus we have four molecules of ATP produced at the substrate level; all the rest come from electron transport.

Glycolysis produces, per molecule of glucose, two molecules of NADH, which are located in the cytosol. These will give rise either to a total of five or three molecules of ATP depending on the shuttle used. Per molecule of glucose, pyruvate dehydrogenase produces two molecules of NADH and the TCA cycle, six. Oxidation of these will produce 20 molecules of ATP. Oxidation of the $FADH_2$ generated from the succinate→fumarate step produces a further three ATP molecules.

The total is therefore 2+5 (or 3)+2+20+3=32 (with the malate–aspartate shuttle used) or 30 (with the glycerol-3-phosphate shuttle used). These values are estimates – with substrate-level phosphorylation, ATP generation is always a whole number, but there are not necessarily whole-number relationships between electron transport, proton ejection, and ATP generation.

An *E. coli* cell is equivalent in this context to a mitochondrion, the cell membrane equating to the inner mitochondrial membrane and the bacterial interior to the mitochondrial matrix. In such cells there is no need for shuttle mechanisms to transport NADH electrons to the respiratory pathway. In *E. coli* there is no transport of ATP and ADP needed into the cytosol. The ATP yield from the oxidation of a molecule of glucose in *E. coli* is therefore greater.

Is ATP production the only use that is made of the potential energy in the proton-motive force?

The answer is almost, but not quite. In newborn babies, heat production to maintain body temperature is helped by brown fat cells – brown because they are rich in mitochondria that contain the coloured cytochromes. The generation of ATP in mitochondria is dependent on the inner mitochondrial membranes being impermeable to protons, thus forcing the latter to enter the mitochondrial matrix only via the ATP-generating channels. If you made a hole in the membrane the protons would simply flood through it, effectively acting as a short circuit, no ATP would be generated, and the energy would be liberated as heat. In brown fat cell mitochondria, this is essentially what happens, channels permitting nonproductive (that is, no ATP synthesis) proton flow being made by a special protein, **thermogenin**. Chemicals that transport protons unproductively through membranes (**dinitrophenol** is the classical one) also 'uncouple' oxidation from ATP generation (see Box 13.2).

Bacteria also harness energy by pumping protons across the membrane to the outside of the cell and generating ATP as described by reversed proton flow. However, the proton gradient is also used in uptake of solutes into the cell, the H^+ gradient being used for cotransport (lactose uptake is an example) just as the Na^+ gradient is used in animal cells (see Chapter 7). Remarkably also, the cilia of bacteria are rotated by a flow of protons through the protein machinery that rotates the cilium; it runs on 'proticity' rather than the electricity used by an electric motor. The proton-driven motor of cilia is reminiscent of the F_0 'motor' of ATP synthase.

BOX 13.2 **Inhibitors and uncouplers of oxidative phosphorylation**

The classical inhibitors of ATP generation by the oxidative route are cyanide ions (CN^-), azide ions (N^{3-}), and carbon monoxide. These simply inhibit cytochrome oxidase and thus block the entire respiratory chain. The first two react with the ferric form of the enzyme and CO with the ferrous form. (CO also combines avidly with haemoglobin and deprives cells of oxygen.)

During the period when the respiratory pathway and oxidative phosphorylation mechanisms were being elucidated a variety of natural and synthetic inhibitors were prominent in the literature because they were such valuable tools. Thus amytal and rotenone block electron transfer between NADH and Q; antimycin A blocks the reduction of cytochrome *c* by QH_2; oligomycin blocks transport of protons through the F_0 of ATP synthase thus preventing ATP generation.

We have mentioned that dinitrophenol (DNP) physically transports protons through the mitochondrial membrane, unproductively generating only heat. It dissipates the proton gradient established by respiration. DNP is an uncoupler. Unlike an inhibitor, it does not block the electron transport chain but dissociates it from ATP generation,it 'uncouples' the two processes so that electron transport can take place without production of ATP. One can see the potential marketing appeal of such a compound in the slimming industry. Eat without getting the energy.

DNP had indeed been used pharmacologically in the US in the 1930s in 'slimming pills' as it makes energy production or capture inefficient (i.e. oxidation of fuel without ATP production), the inefficiency increasing with increasing doses of DNP. A number of fatalities from overdosing resulted in the drug being discontinued after about one year of use. As the dose of DNP increased, more fuel needed to be oxidized to meet the energy demands resulting in fatal hyperthermia. The search by the slimming industry for a safe uncoupler, continues but without success so far.

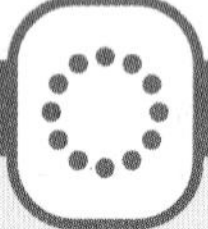

Summary

Glycolysis is Stage 1 in the complete oxidation of glucose or glucosyl units of glycogen. It causes the lysis of the C_6 glucose molecule into the two C_3 molecules of pyruvate (hence the name glycolysis). It occurs in the cytosol. It produces a net gain of only two ATP molecules (three if we start with glycogen) but prepares the glucose for the next stage, the TCA cycle.

In glycolysis there is one oxidation step which reduces NAD^+ to $NADH+H^+$. Since NAD^+ is limited in amount it must be reoxidized *via* the mitochondrion or glycolysis would halt. Under normal conditions the NADH is reoxidized *via* mitochondria. NADH cannot itself enter the mitochondrion but shuttle mechanisms transfer the electrons from NADH either to NAD^+ or to FAD inside the mitochondrion. During vigorous exercise, NADH is formed too rapidly for the normal oxidation route to cope with it. It is rapidly reoxidized by lactate dehydrogenase reducing pyruvate to lactate.

The **TCA cycle** is Stage 2 of the oxidation of glucose. Pyruvate is transported into the mitochondrial matrix where it is converted by pyruvate dehydrogenase into acetyl-CoA, which enters the TCA cycle. The first reaction is the conversion of the acetyl group into citrate by condensation with oxaloacetate catalysed by citrate synthase.

In a single turn of the TCA cycle, the electrons from the acetyl group (plus extra ones originating indirectly from water) are transferred to NAD^+ and FAD, regenerating oxaloacetate. The carbon atoms are removed as CO_2. During the cycle only two ATP molecules are produced per molecule of glucose (in the equivalent form of GTP in liver) but it has generated fuel in the form of NADH and $FADH_2$ for Stage 3, which is the transfer of electrons from these carriers to oxygen, forming water. This is associated with a large release of free energy.

The **electron transport system** is Stage 3 of the oxidation of glucose. It generates most of the ATP. As the electrons move along the hierarchy of electron carriers from NADH and $FADH_2$, the released free energy is used to generate a proton gradient (augmented by a membrane charge potential) across the inner mitochondrial membrane by proton pumping. Protons are pumped out of the mitochondia. This is the celebrated Mitchell chemiosmotic theory. The electron carriers are grouped into four complexes. Proton pumping occurs in complexes I, III, and IV but not II. The mechanism of pumping by complex I is not known. In complex III it is achieved by the Q cycle, Q being ubiquinone, which is a mobile electron carrier. Cytochrome *c* is also mobile and connects complexes III and IV, the latter being cytochrome oxidase, which transfers electrons to oxygen, forming water. Proton pumping here probably involves conformational changes in the protein subunits.

The proton gradient is used to drive ATP synthesis by the molecular machines known as ATP synthase of the inner mitochondrial membrane. They are minute rotating motors driven by proton flow. The rotation causes conformational changes in the ATP synthase subunits, the energy of which drives the condensation of $ADP+P_i$ to ATP. The mechanism of these remarkable rotary machines is now almost fully established. The ATP is transported out into the cytosol by exchange with ADP, the process being driven by the membrane potential. The yield of ATP per molecule of glucose cannot be calculated with absolute precision but approximately 30 molecules are produced from ADP+phosphate, which is lower than previous estimates.

In prokaryotes there are no mitochondria but the cell membrane is equivalent to the inner mitochondrial membrane in the present context.

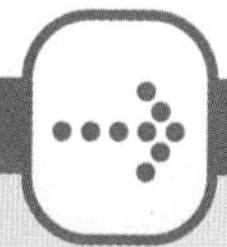

Further reading

To access the further reading, please scan the QR code image or go to http://global.oup.com/uk/orc/biosciences/molbiol/snape_biochemistry/student/reading/ch13/

Problems

1 What is the metabolic significance of the phosphohexose isomerase reaction?

2 What is meant by substrate-level phosphorylation? Give an example of such a system. How does this differ in basic terms from oxidative phosphorylation?

3 The reaction for which the enzyme pyruvate kinase is named never occurs in the cell. Discuss this.

4 How many ATP molecules are generated in glycolysis, from:
(a) glucose
(b) glycosidic unit of glycogen?

5 In calculating how many molecules of ATP are produced as a result of the oxidation of cytosolic NADH, we cannot be sure of the exact answer in eukaryotes. Why is this so?

6 What is the advantage in isocitrate being oxidized before loss of CO_2 occurs?

7 Explain how the TCA cycle acids can be 'topped up' by an anaplerotic reaction. Can acetyl-CoA participate in this?

8 What is the cofactor involved in carboxylation reactions? Describe how it works.

9 Outline the arrangement of respiratory complexes in the electron transport chain.

10 What characteristic do ubiquinone and cytochrome *c* have in common? What are their physical locations in the cell?

11 What is the immediate role of electron transfer in the respiratory chain?

12 It is stated in the text that the yield of ATP from the complete oxidation of a molecule of glucose in eukaryote cells is either 30 or 32 molecules; in the case of *E. coli* it is stated that the yield is greater than this. Why does this difference in statements exist?

13 By means of diagrams and brief notes, explain in outline how the proton gradient generated by electron transport is harnessed into ATP production by ATP synthase.

14 The active sites in the F_1 subunits of ATP synthase are said to be cooperatively interdependent, a most unusual situation. Explain what this means.

15 Give a brief account of the complexes that constitute the electron transport system of the inner mitochondrial membrane with particular reference to the creation of the proton gradient across the membrane.

16 Briefly describe the basic physicochemical principle by which the F_0 unit of ATP synthase is caused to rotate.

17 Which of the following is out of place? Oxaloacetate, malate, GDP, acetyl-CoA, H_2O, NAD^+? Explain your answer.

Chapter 14

Energy release from fat

In the previous chapters, you have seen how energy in the form of adenosine triphosphate (ATP) is obtained in the cell starting with the oxidation of glucose or the breakdown of glycogen.

Fat is the other major source of energy for ATP production. It provides perhaps half of the total energy needs of heart and resting skeletal muscles. By far the largest amount of stored energy occurs as fat for, as mentioned earlier, there appears to be no limit to the amount of neutral fat that can be stored in the body in the adipose cells.

The subject of fat oxidation and concomitant ATP production is greatly simplified because fat oxidation involves the TCA cycle and electron transport system that we have discussed in glucose oxidation. As already described (see Chapter 12), the two systems (glucose oxidation and fat oxidation) converge at acetyl-CoA so all we are mainly concerned with in fat oxidation is the relatively simple task of chopping up fatty acids by removing two carbon atoms at a time as acetyl-CoA.

The processing of fatty acid molecules involves NAD^+ and FAD reduction the reduced forms of which are oxidized by the same pathways we have already discussed. It is efficient to use the same machinery for obtaining energy from all classes of dietary components. Regulation of the pathways in this chapter are dealt with in Chapter 20 on metabolic control.

A few simple points first.

- Fatty acid oxidation occurs inside mitochondria (Fig. 12.4).
- Before oxidation can occur, free fatty acids are released from triacylglycerol (TAG) stores by hydrolysis by the hormone-sensitive lipase (Chapter 20). This also produces a molecule of glycerol per TAG hydrolysed and this is metabolized separately. It is manipulated to enter metabolism in the glycolytic pathway – it is phosphorylated and oxidized to give dihydroxyacetone phosphate of glycolysis fame. Thus, in fasting and starvation, the glycerol moiety *can* give rise in the liver to glucose by the gluconeogenesis pathway (Chapter 16).
- Free fatty acids for oxidation are obtained by peripheral tissues from that released into the blood by adipose cells when the glucagon level is high. They also enter the cells as a result of lipoprotein lipase attack on chylomicrons or very-low-density lipoprotein produced by the liver (VLDL, see Chapter 11). Chylomicrons exist after feeding, while fatty acids are released from adipose cells in fasting/starvation (see Chapter 10).
- Free fatty acids from adipose cells are carried as ionized molecules attached in a freely reversible manner to serum albumin. They readily diffuse into cells so that the amount entering cells increases as their blood level rises as a result of release from adipose cells. Apart from the simple diffusion, a system of fatty acid transporters also exists so there is a saturable and an unsaturable component to fatty acid transport into the cells. As the fatty acids are taken up by cells, more will dissociate from the serum albumin carrier protein to maintain the equilibrium between free and bound fatty acids.
- During conversion into acetyl-CoA, the fatty acid is always in the form of an acyl-CoA. The first stage in oxidation of fatty acids is always to convert the carboxylic acids into the fatty acyl-CoA compounds, a reaction known as fatty acid activation.
- Fatty acids are broken down by removing two carbon atoms at a time as acetyl-CoA, a process known as **β-oxidation**.

Mechanism of acetyl-CoA formation from fatty acids

'Activation' of fatty acids by formation of fatty acyl-CoA derivatives

The term 'activation' of a carboxylic acid refers to the fact that the thiol ester is a high-energy (or reactive) compound. The activation reaction is:

$$RCOO^- + ATP + CoA{-}SH \rightleftharpoons RCO{-}S{-}CoA + AMP + PP_i; \quad \Delta G^{0'} = -0.9\,kJ\,mol^{-1}$$

The free-energy change of this reaction is small (because of the high energy of the thiol ester), but hydrolysis of the inorganic pyrophosphate (PP_i) by the ubiquitous enzyme, inorganic pyrophosphatase, makes the overall process strongly exergonic and irreversible ($\Delta G^{0'} = 32.5\,kJ\,mol^{-1}$). (See Chapter 3 if you have forgotten this point.)

There are three different fatty acid-activating enzymes for shortchain, medium-chain, and long-chain acids, respectively – called **fatty acyl-CoA synthetases** (sometimes called **thiokinases**).

Transport of fatty acyl-CoA derivatives into mitochondria

Activation of fatty acids occurs in the cytosol. The outer mitochondrial membrane is permeable to most metabolites and the fatty acyl-CoA crosses it to enter the intermembrane space but cannot cross the inner mitochondrial membrane to reach the site of conversion into acetyl-CoA, which is the mitochondrial matrix. The acyl group of fatty acyl-CoA is carried through the inner mitochondrial mitochondrial membrane, without the CoA, by a special transport mechanism and then handed over to CoASH *inside* the mitochondrion to become fatty acyl-CoA again. The high-energy nature of the acyl bond is preserved during the transport – otherwise it could not reform fatty acyl-CoA inside the mitochondrion without further energy expenditure. To achieve this, on the external face of the inner membrane of the mitochondrion, the acyl group is transferred to a hydroxylated , nitrogen containing carboxylic acid known as, **carnitine**.

$$H_3C{-}\overset{+}{N}(CH_3)_2{-}CH_2{-}CH(OH){-}CH_2{-}COO^-$$

Carnitine

Although a carboxylic ester is usually of the low-energy type, the structure of carnitine is such that the fatty acyl–carnitine bond is of the high-energy type – the acyl group has a high group-transfer potential. Presumably this is how carnitine has evolved as the carrier molecule in this transport system. The fatty acylcarnitine so formed is transported into the mitochondrial matrix where the reverse reaction occurs – carnitine is exchanged for CoASH and the free carnitine is taken back to the cytosol where it collects another fatty acyl group.

$$H_3C{-}\overset{+}{N}(CH_3)_2{-}CH_2{-}CH(O{-}C(=O){-}R){-}CH_2{-}COO^-$$

Fatty acyl-carnitine

The scheme is shown in Fig. 14.1

The acyl transfer from fatty acyl-CoA to carnitine is catalysed by an enzyme called **carnitine acyltransferase I** located in the outer mitochondrial membrane. The fatty acyl-carnitine

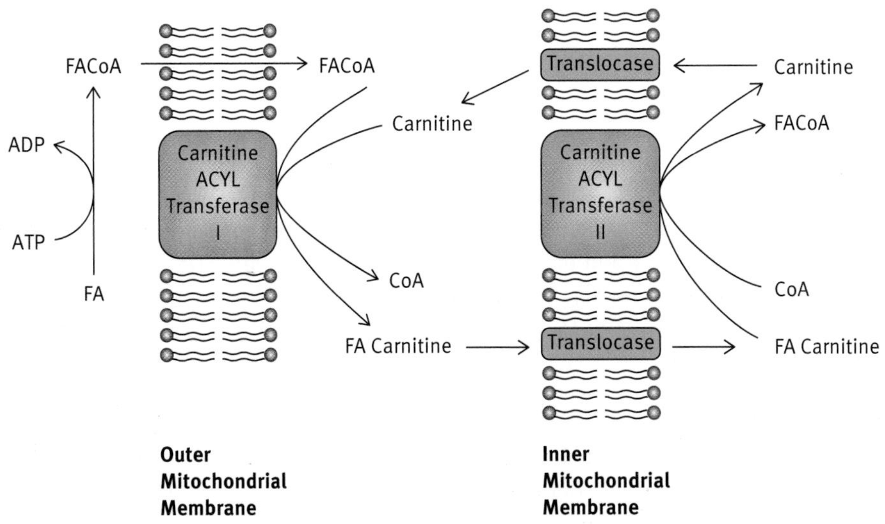

Fig. 14.1 Mechanism of transport of long-chain fatty acyl groups into mitochondria where they are oxidized in the mitochondrial matrix. The acyl–carnitine bond is an unusual ester bond in that it has a high group-transfer potential – the compound is of the high-energy type so that exchange of carnitine for CoASH inside the mitochondrion occurs without need for energy input. See text for structures.

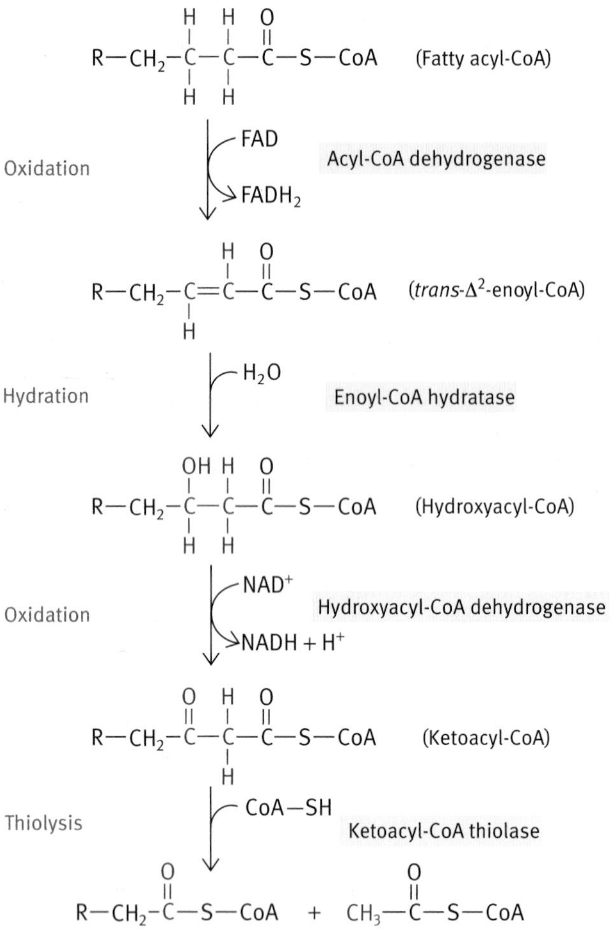

$R{-}CH_2{-}CH_2{-}CH_2{-}CO{-}S{-}CoA$ (Fatty acyl-CoA)

Oxidation — $FAD \rightarrow FADH_2$ — Acyl-CoA dehydrogenase

$R{-}CH_2{-}CH{=}CH{-}CO{-}S{-}CoA$ (*trans*-Δ^2-enoyl-CoA)

Hydration — H_2O — Enoyl-CoA hydratase

$R{-}CH_2{-}C(OH)H{-}CH_2{-}CO{-}S{-}CoA$ (Hydroxyacyl-CoA)

Oxidation — $NAD^+ \rightarrow NADH + H^+$ — Hydroxyacyl-CoA dehydrogenase

$R{-}CH_2{-}CO{-}CH_2{-}CO{-}S{-}CoA$ (Ketoacyl-CoA)

Thiolysis — CoA—SH — Ketoacyl-CoA thiolase

$R{-}CH_2{-}CO{-}S{-}CoA + CH_3{-}CO{-}S{-}CoA$

Fig. 14.2 One round of four reactions by which a fatty acyl-CoA is shortened by two carbon atoms with the production of a molecule of acetyl-CoA. Note the similarity of reaction types in the desaturation, hydration, and ketoacyl formation with the succinate → fumarate → malate → oxaloacetate steps of the TCA cycle.

complex is transferred across the inner membrane by a translocase. Carnitine is released in the matrix and a CoA is transferred to the fatty acid by the enzyme **carnitine acyltransferase II** located in the inner mitochondrial membrane. The free carnitine returns to the intermembrane space by the translocase. Note, Fig 14.1 shows two translocases, this is just for clarity and to show the shuttling of carnitine from the matrix to the intermembrane space and back. Genetic defects are known in which there is a carnitine deficiency or deficiency in the carnitine acyltransferase. In some cases this can manifest itself as muscle pain and abnormal fat accumulation in muscles. Long-chain fatty acids also accumulate in the blood .

Carnitine comes in the diet mainly in meat (hence the name, carne is latin for meat), but it is not a dietary essential, it can be synthesized in the body from the amino acid lysine. Individuals who cannot synthesize carnitine should be given carnitine supplements for life.

Conversion of fatty acyl-CoA into acetyl-CoA molecules inside the mitochondrion by β-oxidation

Let us look at a general phenomenon first. Biological oxidations, in the sense that an oxygen atom is introduced into a molecule, are rarely direct additions of oxygen. Usually they are a series of removal of hydrogen, addition of water and removal of hydrogen. This process is common to fatty acid oxidation and the TCA cycle that you have seen in the previous chapter, that is, the conversion of succinate to oxaloacetate. In both cases there is a **dehydrogenation** by an FAD enzyme, **hydration**, and an **NAD^+-dependent dehydrogenation**. We suggest that you refresh your memory on these by having a quick look at the Conversion of succinate to oxaloacetate Section in Chapter 13: succinate → fumarate → malate → oxaloacetate. The corresponding reactions on fatty acyl-CoA derivatives are shown in Fig. 14.2. The name β-oxidation tells us that the β-carbon atom (C3) is oxidized and is converted into a C=O group, forming a β-ketoacyl-CoA. The ketoacyl-CoA is cleaved by CoASH, splitting off two carbon atoms as acetyl-CoA and forming a shorter fatty acetyl-CoA derivative. Because the molecule is split by the –SH group of CoASH, the enzyme is called a **thiolase**. The thiolase reaction preserves the free energy as a thiol ester of fatty acyl-CoA. When the fatty acyl group has been shortened, by successive rounds of acetyl-CoA production, to the C_4 stage (butyryl-CoA) the next round of reactions produces acetoacetyl-CoA, which is finally split by CoASH into two molecules of acetyl-CoA. A specific thiolase in mitochondria performs this reaction:

$$\underset{\text{Acetoacetyl-CoA}}{CH_3COCH_2CO{-}S{-}CoA} + CoA{-}SH \rightarrow \underset{\text{Acetyl-CoA}}{2CH_3CO{-}S{-}CoA}$$

The reaction sequence in each round involves conversion of saturated (only single bonds in the chain) fatty acyl-CoAs to β-ketoacyl-CoAs. The process is therefore referred to as **β-oxidation of fatty acids**. The NADH and the $FADH_2$ (the latter on the acyl-CoA dehydrogenase) feed electrons into the electron transport chain exactly as already described (see Fig. 13.17 for a summary of this).

Energy yield from fatty acid oxidation

A molecule of palmitic acid (C_{16}), called palmitate at pH 7 in its ionized form, is converted to eight molecules of acetyl-CoA and in this process we generate seven $FADH_2$ molecules and seven NADH molecules. The acetyl-CoA is metabolized by the TCA cycle, and the NADH and $FADH_2$ oxidized by the electron transport chain (see Chapter 13). NADH oxidation generates, per molecule, 2.5 ATP molecules and from $FADH_2$, 1.5 ATP molecules. (Since the NADH is generated in the mitochondrial matrix, it does not have to be carried in by a shuttle mechanism.)

If you count it all up (not forgetting the one GTP per acetyl-CoA from the TCA cycle), the oxidation of one mole of palmitic acid generates 106 moles of ATP from ADP and P_i, allowing for the two ATP consumed in the formation of palmitoyl-CoA. Effectively two ATP molecules are consumed at the beginning though only one directly participates. Due to PP_i release, two high-energy phosphate groups are released as P_i, forming AMP. A kinase phosphorylates this using ATP and thus formation of the acyl-CoA costs the two molecules of ATP converted to ADP. Overall, this represents about a 33% efficiency in trapping the free energy in usable form, based on $\Delta G^{0\prime}$ values.

Oxidation of unsaturated fat

Olive oil and other TAGs have a high content of **monounsaturated fatty acids** with a *cis*-configured double bond (see Chapter 7 for an explanation of this). For example, palmitoleic acid has a *cis*-configured double bond between carbon atoms 9 and 10. So far as oxidation goes, this fatty acid is treated by the cell in exactly the same way as palmitic acid for three rounds of β-oxidation. At this point the product is *cis*-Δ^3-enoyl-CoA:

$$\underset{(4)}{\mathrm{R{-}C}}\mathrm{H}{=}\underset{(3)}{\mathrm{C}}\mathrm{H}{-}\underset{(2)}{\mathrm{CH_2}}{-}\underset{(1)}{\overset{\mathrm{O}}{\overset{\|}{\mathrm{C}}}}{-}\mathrm{S\text{-}CoA}.$$

The double bond in this position prevents the acyl-CoA dehydrogenase from forming a double bond between carbon atoms 2 and 3, as is required in β-oxidation of a saturated acyl-CoA.

An extra isomerase enzyme takes care of this by shifting the existing double bond into the required 2–3 position; it generates the *trans*-isomer in doing so:

$$\mathrm{R{-}CH{=}CH{-}CH_2{-}\overset{O}{\overset{\|}{C}}{-}S{-}CoA} \quad cis\text{-}\Delta^3\text{-Enoyl-CoA}$$

$$\downarrow \text{Isomerase}$$

$$\mathrm{R{-}CH_2{-}CH{=}CH{-}\overset{O}{\overset{\|}{C}}{-}S{-}CoA} \quad trans\text{-}\Delta^2\text{-Enoyl-CoA}$$

This is now on the main pathway of fat breakdown and so the problem of monounsaturated fat oxidation is solved. (If you look at Fig. 14.2, you will see that, in oxidation of saturated acyl-CoAs, it is the *trans*-isomer that is generated.)

Polyunsaturated fatty acids pose additional problems – for example, linoleic acid has two double bonds (Δ^9 and Δ^{12}). In effect, one (Δ^9) is dealt with as described above while the second (Δ^{12}) is handled by using an additional enzyme at the appropriate stage, again putting the molecule on the normal metabolic path of β-oxidation (the steps are not given here).

Oxidation of odd-numbered carbon-chain fatty acids

A small proportion of fatty acids in the diet (for example those from plants) have odd-numbered carbon chains, β-oxidation of which produces, as the penultimate product, a five-carbon β-ketoacyl-CoA instead of acetoacetyl-CoA. Cleavage of this by thiolase produces acetyl-CoA and the three-carbon propionyl-CoA.

$$\mathrm{CH_3{-}CH_2{-}CO{-}CH_2{-}CO{-}S{-}CoA + CoA{-}SH} \rightarrow$$
$$\underset{\text{Propionyl-CoA}}{\mathrm{CH_3{-}CH_2{-}CO{-}S{-}CoA}} + \underset{\text{Acetyl-CoA}}{\mathrm{CH_3{-}CO{-}S{-}CoA}}$$

Propionyl-CoA is carboxylated to succinyl-CoA and hence on to the TCA mainstream by the following reactions. The epimerase catalyses the conversion of D- to L-methylmalonyl-CoA.

$$\mathrm{CH_3{-}CH_2{-}CO{-}S{-}CoA + HCO_3^- + ATP} \xrightarrow[\mathrm{ADP}\,+\,\mathrm{P_i}]{\text{Propionyl-CoA carboxylase}}$$

$$\mathrm{CH_3{-}\underset{H}{\overset{COO^-}{C}}{-}CO{-}S{-}CoA} \xrightarrow{\text{Methylmalonyl-CoA epimerase}} \mathrm{CH_3{-}\underset{COO^-}{\overset{H}{C}}{-}CO{-}S{-}CoA}$$

$$\xrightarrow{\text{Methylmalonyl-CoA mutase}} \underset{\text{Succinyl-CoA}}{\mathrm{{}^-OOC{-}CH_2{-}CH_2{-}CO{-}S{-}CoA}}$$

It is the last reaction that is of great interest, for it involves the most complex coenzyme of all, **deoxyadenosylcobalamin**, a derivative of **vitamin B_{12}**.

Propionate is also formed in the degradation of three amino acids (valine, isoleucine, and methionine) and from the cholesterol side chain. Propionate is a major product of ruminant digestion so the pathway is of particular importance in these animals. It comes from bacterial digestion of plant material in ruminants.

Deficiency of the methylmalonyl-CoA mutase or the inability to synthesize the required cofactor from vitamin B_{12} leads to **methylmalonic acidosis**, usually fatal early in life. Patients typically present at the age of 1 month to 1 year with neurologic manifestations, such as seizures, encephalopathy, and strokes.

Ketogenesis in starvation and type 1 diabetes mellitus

So far we have explained that fatty acids are converted into acetyl-CoA, which then joins the TCA cycle and is oxidized. This is true for tissues in 'normal' circumstances but it has to be qualified for the liver in a particular circumstance.

The body can be in a physiological situation where fat metabolism is the main source of energy. This occurs in starvation after exhaustion of glycogen stores; the same can occur in untreated type 1 diabetes where the inability to metabolize carbohydrate effectively results in an almost analogous glucose 'starvation' irrespective of glucose availability. In this situation, the fat cells are releasing excessive amounts of free fatty acids due to high levels of glucagon and the activation of hormone-sensitive lipase and the liver may produce amounts of acetyl-CoA that cannot be handled by the TCA cycle, as they exceed its capacity.

The liver cell, *in effect*, joins two acetyl groups together, by a mechanism to be described shortly, to form **acetoacetate** ($CH_3COCH_2\ COO^-$), which is partly reduced to β-hydroxybutyrate ($CH_3CHOH\ CH_2\ COO^-$). The two, known as **ketone bodies** by tradition, are released into the blood. They are water soluble and transported in the blood to extrahepatic tissues. (As explained, the term 'ketone bodies' is an historical misnomer - they are molecules, not bodies, and β-hydroxybutyrate is not a ketone anyway.) The liver produces ketone bodies but does not use them itself as metabolic fuel for reasons that we will see shortly.

How is acetoacetate made from acetyl-CoA?

Acetoacetyl-CoA is formed from acetyl-CoA by reversal of the ketoacyl-CoA thiolase reaction:

$$2CH_3CO{-}S{-}CoA \rightleftharpoons CH_3COCH_2CO{-}S{-}CoA + CoA{-}S$$

One would have imagined that free acetoacetate would be formed by simple hydrolysis of the acetoacetyl-CoA (thus pulling the equilibrium over). However, it is not so. Instead, a third molecule of acetyl-CoA is used to form **3-hydroxy-3-methylglutaryl-CoA (HMG-CoA)** by an aldol condensation, followed by hydrolysis of the thiol ester bond to give acetoacetate. The scheme is shown in Fig. 14.3.

HMG-CoA is synthesized by many animal cells. It is a precursor of cholesterol, which is an essential constituent of their membranes. Ketone body formation occurs in mitochondria (where fat conversion into acetyl-CoA occurs). Formation of HMG-CoA for cholesterol synthesis takes place in the cytosol where a separate HMG-CoA synthase isomer occurs attached to the endoplasmic reticulum membrane.

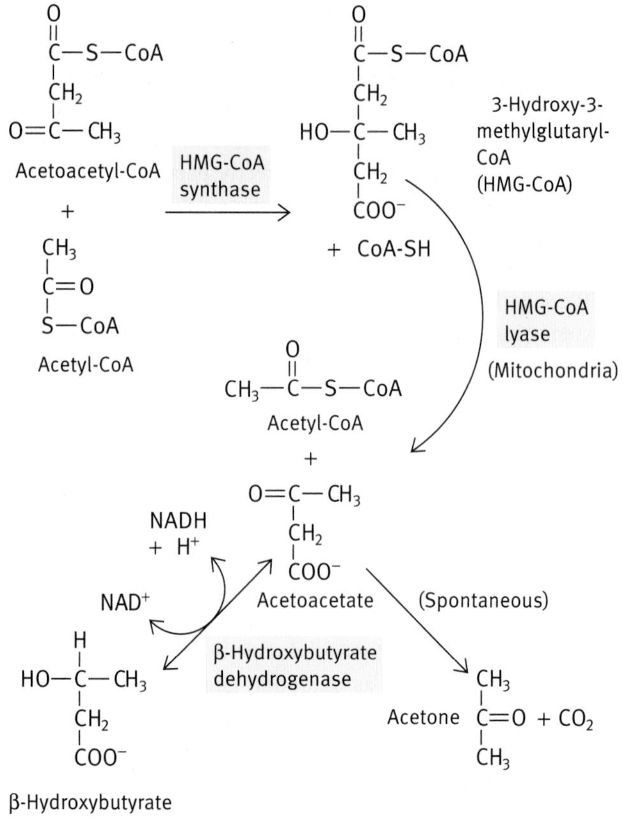

Acetoacetyl-CoA + Acetyl-CoA —(HMG-CoA synthase)→ 3-Hydroxy-3-methylglutaryl-CoA (HMG-CoA) + CoA-SH

HMG-CoA —(HMG-CoA lyase (Mitochondria))→ Acetyl-CoA + Acetoacetate

Acetoacetate + NADH + H^+ —(β-Hydroxybutyrate dehydrogenase)→ β-Hydroxybutyrate + NAD^+

Acetoacetate —(Spontaneous)→ Acetone + CO_2

Fig. 14.3 Ketone body production in the liver during excessive oxidation of fat in fasting/starvation or type 1 diabetes. The process occurs in mitochondria. HMG-CoA is also the precursor of cholesterol but this occurs in the cytosol of liver cells where HMG-CoA synthase also occurs.

Utilization of acetoacetate

Acetoacetate can be used by peripheral tissues to generate energy. (See Chapter 10 for a discussion of the role of ketone bodies in metabolism.) In mitochondria acetoacetate is converted into acetoacetyl-CoA by an acyl-exchange reaction in which succinyl-CoA is converted into succinate:

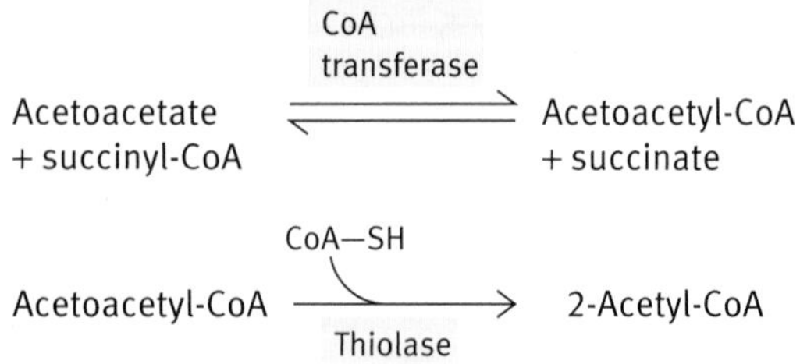

Acetoacetate + succinyl-CoA ⇌ (CoA transferase) Acetoacetyl-CoA + succinate

Acetoacetyl-CoA + CoA—SH —(Thiolase)→ 2-Acetyl-CoA

Acetoacetyl-CoA is cleaved by a thiolase using CoASH to form two molecules of acetyl-CoA. β-Hydroxybutyrate is also utilized by being dehydrogenated first to acetoacetate.

Acetoacetate, being a β-keto acid, tends to decarboxylate spontaneously, and produce acetone (CH_3COCH_3), a volatile solvent, which is exhaled. In untreated type 1 diabetics with high concentrations of ketone bodies, acetone gives rise to a characteristic fruity smell in the breath.

The liver synthesizes but cannot utilize ketone bodies as it has no CoA transferase (also known as thiophorase). In this way the liver does not run a futile cycle, generating and using up ketone bodies at the same time, but they are released into the blood stream and are used by other tissues.

Tissues that possess mitochondria will readily oxidize ketone bodies. The brain can utilize a fair amount of ketone bodies, which relieves the demand for glucose at times when glucose supply is limited (see Chapter 20). Red blood cells clearly cannot use them and they still rely on a glucose supply in fasting and starvation.

Excessive production of ketone bodies leads to **ketoacidosis**, a potentially dangerous condition. Note that ketoacidosis does not happen in fasting/starvation where a small amount of insulin is present and exerts some control over lipolysis but occurs in untreated diabetes type 1 where glucagon acts unopposed and excessive fatty acid degradation occurs despite the presence of high concentrations of glucose in the blood.

Peroxisomal oxidation of fatty acids

The bulk of fatty acid oxidation occurs in mitochondria, but some are oxidized in **peroxisomes**. These are vesicles bounded by a single membrane present in mammalian cells; they cannot synthesize proteins and receive their enzymes from the cytosol by a special transport mechanism. They contain flavoprotein oxidase enzymes, which attack a number of substrates using molecular oxygen generating not water, but hydrogen peroxide.

$$RH_2 + O_2 \rightarrow R + H_2O_2$$

The substrates include some phenols, d-amino acids, and very-long-chain fatty acids ($>C_{18}$), which are not oxidized by mitochondria. The latter are shortened to C_8-acyl-CoA, which is presumably released into the cytosol and transported into mitochondria for conventional oxidation, probably along with acetyl-CoA as well. The fatty acid oxidation is by β-oxidation (see earlier in this chapter), producing acetyl-CoA. The electrons from the first step of the fatty acid oxidation chain (acyl-CoA dehydrogenase) reduce the FAD prosthetic group of the enzyme to $FADH_2$. *In mitochondria*, this is reoxidized by the cytochrome chain to produce ATP. *In peroxisomes*, the $FADH_2$ is reoxidized by molecular oxygen producing H_2O_2. The NADH produced by the later step in the fatty acid β-oxidation (hydroxyacyl-CoA dehydrogenase, Fig. 13.2) is possibly reoxidized by export of the reducing equivalents to the cytosol since there is no cytochrome system in peroxisomes. Peroxisomes do not generate ATP by fatty acid oxidation. There is also evidence that the cholesterol side-chain oxidation required for formation of bile acids occurs in peroxisomes and that synthesis of some complex lipids requires peroxisomes. Biogenesis of these organelles is not fully elucidated. A number of lethal genetic diseases are known in which peroxisome biogenesis is not normal. An example is the **Zellweger syndrome**, often fatal by the age of 6 months; in this there are abnormally high levels of C_{24} and C_{26} long-chain fatty acids and of bile acid precursors.

The metabolic roles of peroxisomes are rather a mixed bag, and they are the least well understood of the organelles, but their essential roles are underlined by the existence of these diseases.

The H_2O_2 generated in peroxisomes is potentially a dangerous oxidant. It is destroyed by the enzyme catalase present in peroxisomes. This is a haem-protein enzyme, which catalyses the reaction:

$$2H_2O_2 + 2H_2O \rightarrow O_2$$

Where to now?

We have so far dealt with energy production from glucose and from fat. The remaining third major food component is in the form of the amino acids. It would be logical to deal with energy production from these next but it is more convenient at this stage to remain with fat and carbohydrate metabolism, then deal with fat and glucose synthesis, and then with the control of metabolism. The reason for this is that energy production from individual amino acids essentially consists of converting them to compounds on the main glycolytic and TCA cycle pathways, so that there is very little information to be given in terms of energy production *per se*. The main biochemical interests of amino acid metabolism lie elsewhere, as will be seen when we deal with this in Chapter 18.

Summary

Energy release from fat involves the oxidation of fatty acids released from triacylglycerol. They are first converted into fatty acyl-CoAs, and the acyl groups transported into the mitochondria. Fatty acyl-CoA cannot enter the mitochondrion but an enzyme of the outer membrane transfers it to carnitine, a curious small molecule, and a transport system transfers the acylcarnitine into the matrix where another enzyme transfers the acyl group back to CoA. In the matrix the fatty acyl-CoA is converted to acetyl-CoA by a process called β-oxidation in which carbon atoms are released two at a time in the form of acetyl-CoA. The fatty acid chain of acyl-CoAs is dehydrogenated by a series of enzymes, which produces β-ketoacyl-CoAs. From these, acetyl units are split off by the enzyme ketoacyl-CoA thiolase, which attaches each to CoASH and releases acetyl-CoA. The fatty acid chain is sequentially shortened until it is completely converted into acetyl-CoA. This is oxidized in the TCA cycle precisely as described for that derived from glucose in Chapter 13.

If fatty acid metabolism is proceeding very rapidly, such as occurs in diabetes or fasting/starvation, acetoacetate and β-hydroxybutyrate are formed from the excess acetyl-CoA; these are water soluble and circulate in the blood to be used by other tissues. They are collectively known as ketone bodies (even though the hydroxybutyrate is not a ketone nor are they 'bodies'). The brain can use them for about half its energy needs, thus reducing its requirement for glucose, which, in starvation, is precious. It has to be synthesized by the liver using amino acids from muscle-protein breakdown (see Chapter 16). Excessive production of the acids can be a serious complication in untreated type 1 diabetes.

Peroxisomal oxidation of fatty acids occurs to some extent. It may be a way of oxidizing fatty acids longer than C_{18}, which are not oxidized by mitochondria. This oxidation does not feed into the electron transport chain but generates hydrogen peroxide.

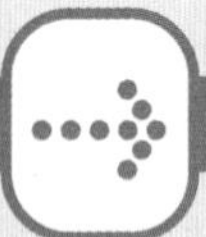

Further reading

To access the further reading, please scan the QR code image or go to http://global.oup.com/uk/orc/biosciences/molbiol/snape_biochemistry/student/reading/ch14/

Problems

1 Peripheral tissues obtain their free fatty acids for oxidation from the blood. Describe three ways in which free fatty acids become available to cells.

2 Which cells of the body do not use free fatty acids for energy supply?

3 In breaking down fatty acids to acetyl-CoA:
(a) What is always the first step?
(b) Where does this occur?
(c) Where does fatty acid breakdown to acetyl-CoA occur in eukaryotes?
(d) How do fatty acid groups reach this site of breakdown?

4 Illustrate similarities in the oxidation of fatty acids to acetyl-CoA with a section of the TCA cycle.

5 What is the yield of ATP from the complete oxidation of a molecule of palmitic acid? Explain your answer.

6 Explain how a monounsaturated fat (Δ^9) is broken down to acetyl-CoA.

7 Is acetyl-CoA, derived from fatty acid breakdown, always fed into the TCA cycle? Explain your answer.

8 HMG-CoA is an intermediate in both acetoacetate synthesis and cholesterol synthesis. Where do these two processes occur?

9 What are peroxisomes and what is their function?

Chapter 15

An alternative pathway of glucose oxidation. The pentose phosphate pathway

A completely different pathway of glucose oxidation exists called the pentose phosphate pathway, but sometimes called the 'direct oxidation pathway' or the 'hexose monophosphate shunt'.

The pentose phosphate pathway is in fact an oxidative pathway that does not provide ATP but meets some other rather specialized metabolic needs:

- It supplies ribose-5-phosphate (a pentose sugar) for nucleotide and nucleic acid synthesis (dealt with in later chapters). Ribose is also a component of the coenzymes NAD^+ and FAD.
- It supplies reducing power in the form of NADPH for synthesis of fat and other compounds, such as cholesterol and steroids.
- It provides a route for excess pentose sugars in the diet to be brought into the mainstream of glucose metabolism.
- It recycles sugars according to the needs of the cell.

The pathway operates mainly in the cytosolic compartment of cells along with glycolysis. The enzymes are most plentiful in tissues with high demands for NADPH where it is used for reductive synthesis, and in rapidly dividing cells, which require ribose-5-phosphate for DNA synthesis. Quantitatively, the main demand is for fatty acid synthesis so that the enzymes of the pathway are plentiful in liver and adipose tissue. Note that in humans, little fatty acid synthesis takes place in adipose tissue, and liver is the main site of fatty acid production. Skeletal muscle by contrast has very little pentose phosphate pathway activity, but since all cells require ribose for nucleic acid synthesis, all tissues probably have some. (This applies to immature erythrocytes; mature ones require NADPH to maintain cell membrane integrity.)

The pentose phosphate pathway has two main parts

In the first part, the oxidative section, glucose-6-phosphate is converted by **glucose-6-phosphate dehydrogenase** into 6-phosphogluconate (*via* 6-phosphogluconolactone), during which $NADP^+$ is reduced to NADPH (Fig. 15.1). Then

Glucose-6-phosphate → (Glucose-6-phosphate dehydrogenase; $NADP^+$ → NADPH + H^+) → 6-Phosphogluconolactone → (6-Phosphogluconolactonase; H_2O) → 6-Phosphogluconate → (6-Phosphogluconate dehydrogenase; $NADP^+$ → NADPH + H^+) → Ribulose-5-phosphate + CO_2 → Ribose-5-phosphate

Fig. 15.1 Oxidative reactions of the pentose phosphate pathway. Control is mainly by availability of $NADP^+$.

6-phosphogluconate dehydrogenase reduces $NADP^+$ and generates a β-keto acid, which is decarboxylated to a ketopentose (**ribulose-5-phosphate**). An isomerase converts the latter into the aldose isomer, ribose-5-phosphate. The oxidative part is irreversible. It produces the two components, ribose-5-phosphate and NADPH.

The rate-limiting step is the first reaction, that of oxidation of glucose-6-phosphate to 6-phosphogluconolactone. The rate of this reaction is tightly coupled to the level of $NADP^+$. This governs the allocation of glucose-6-phospate to the pentose phosphate pathway, rather than the glycolytic pathway.

The oxidative part produces equal amounts of ribose-5-phosphate and NADPH

Tissue demands for the two products, ribose-5-phosphate and NADPH, vary greatly. For example, fat synthesis requires large amounts of NADPH, but production of NADPH also produces ribose-5-phosphate by the reactions given in Fig. 15.1, which may be far more than the cell needs for nucleotide synthesis. Conversely, in a rapidly dividing cell that is not synthesizing fat, large amounts of ribose-5-phosphate are needed to synthesize nucleotides, but there is little requirement for NADPH. The requirements for ribose-5-phosphate and NADPH may, in other cells, be anywhere in between this – equal amounts of the two products or more of one than the other. We will see how the nonoxidative part takes care of this problem. As already implied, control of the oxidative part of the pathway is mainly exercised by the availability of $NADP^+$.

The nonoxidative part interconverts sugars, according to the needs of cells

The nonoxidative part involves the enzymes **transaldolase** and **transketolase**, which together interconvert sugars in accordance with the metabolic needs of the cell. These two enzymes detach, from a ketose sugar phosphate, C_3 and C_2 units, respectively, and transfer them to other aldose sugars (Fig. 15.2). Transketolase is a thiamin pyrophosphate-dependent enzyme as is pyruvate dehydrogenase.

Between them, transketolase and transaldolase can work an almost bewildering range of sugar interconversions. If you put together the reactions of the *oxidative* part given above, the following balance sheet emerges:

$$\text{Glucose-6-phosphate} + 2NADP^+ + H_2O \rightarrow$$
$$\text{ribose-5-phosphate} + 2NADPH + 2H^+ + CO_2$$

In situations where the needs for ribose-5-phosphate and NADPH are *balanced* the nonoxidative section is not required, because the oxidative part produces both products in appropriate amounts.

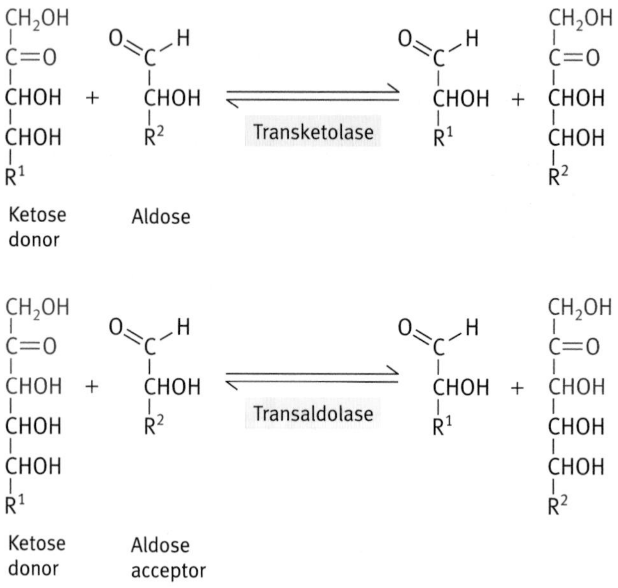

Fig. 15.2 Reactions catalysed by transketolase and transaldolase.

Conversion of surplus ribose-5-phosphate into glucose-6-phosphate

However, if, say, a nondividing fat cell requires *more NADPH than ribose-5-phosphate* the nonoxidative section reconverts (recycles) excess ribose-5-phosphate into glucose-6-phosphate according to the stoichiometry:

$$6\text{Ribose-5-phosphate} \rightarrow 5\text{glucose-6-phosphate} + P_i$$

First, *only part* of the ribose-5-phosphate, an aldose sugar, is converted into **xylulose-5-phosphate**, a ketose sugar, since both transaldolase and transketolase use only ketose sugars as group donors (Fig. 15.3). The remaining part of the ribose-5-phosphate is the aldose sugar acceptor.

The following transformations then occur, reaction 1 being between xylulose-5-phosphate and the remaining ribose-5-phosphate. Fig. 15.4 gives these reactions in full.

(1) $2\,C_5 \rightleftharpoons C_3 + C_7$ Transketolase

Reaction 1, Fig.17.4

(2) $C_7 + C_3 \rightleftharpoons C_4 + C_6$ Transaldolase

Reaction 2, Fig.17.4

(3) $C_5 + C_4 \rightleftharpoons C_3 + C_6$ Transketolase

Reaction 3, Fig.17.4

(4) $2\,C_3 \rightarrow 1\,C_6$

The net effect of the first three reactions is that three molecules of C_5 (indicated in red) are converted into 2.5 molecules of C_6 (blue). The final C_3 compound is glyceraldehyde-3-phosphate, two molecules of which are converted into glucose-6-phosphate by the pathway of gluconeogenesis. Note that this

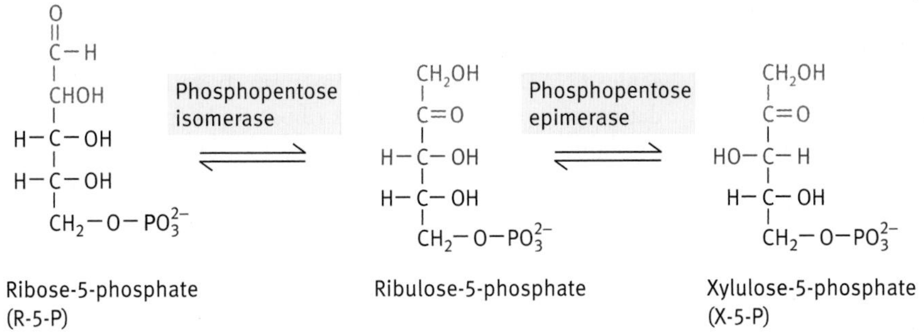

Phosphopentose isomerase

Phosphopentose epimerase

Ribose-5-phosphate (R-5-P)

Ribulose-5-phosphate

Xylulose-5-phosphate (X-5-P)

Fig. 15.3 Conversion of some ribose-5-phosphate into xylulose-5-phosphate. The reason for this is that in reaction 1 of Fig. 15.4, the transketolase requires the donor substrate to be a ketose sugar, which is what xylulose-5-phosphate is. The remaining ribose-5-phosphate (an aldose) acts as the receptor in reaction 1.

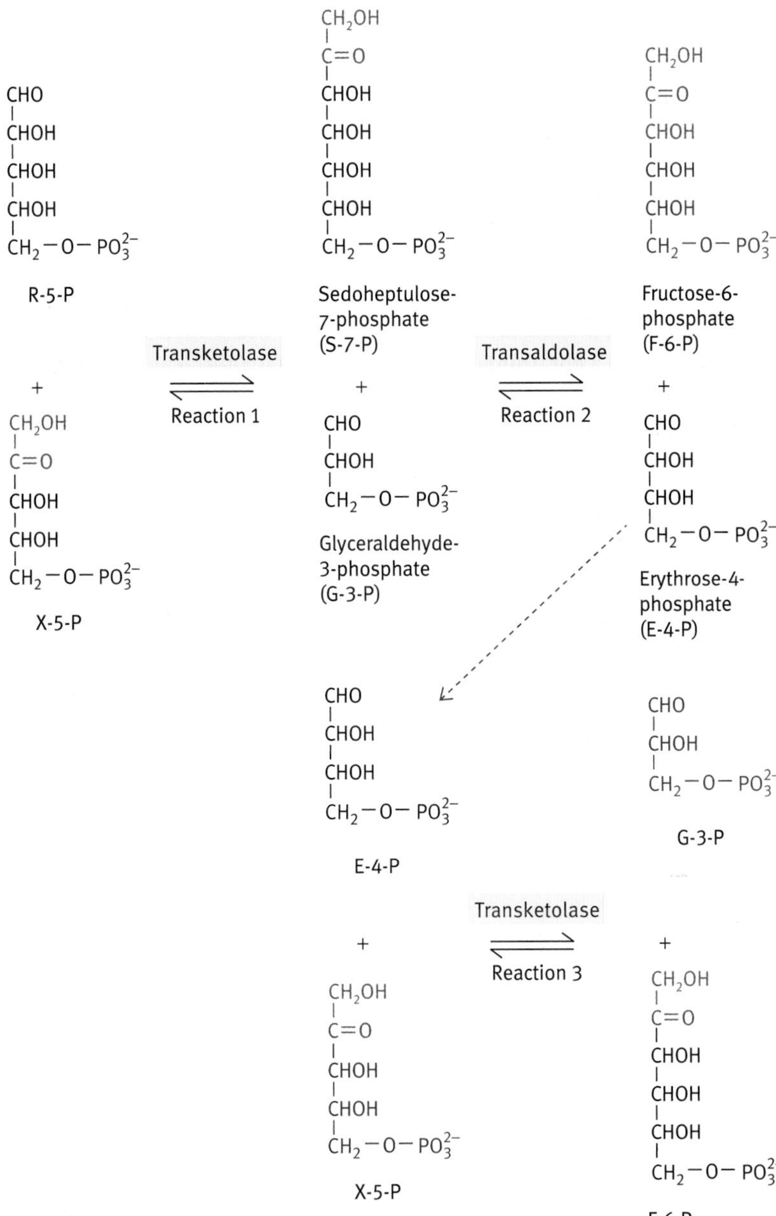

Fig. 15.4 Conversion of ribose-5-phosphate into glyceraldehyde-3-phosphate. Glyceraldehyde-3-phosphate is converted into glucose-6-phosphate by gluconeogenesis reactions. Together, these are the reactions by which six C_5 sugars (three ribose-5-phosphates and three xylulose-5-phosphates) are converted into five glucose-6-phosphates.

set of reactions can also convert dietary ribose into glucose-6-phosphate following its conversion to ribose-5-phosphate by an ATP-requiring kinase.

Two rounds of reactions 1–3 will produce two molecules of glyceraldehyde-3-phosphate, which are converted into fructose-6-phosphate by gluconeogenesis.

$$2\ \text{CHO–CHOH–CH}_2\text{–O–PO}_3^{2-}\ (\text{G-3-P}) \xrightarrow{\text{(Gluconeogenesis reactions)}} 1\ \text{F-6-P} + \text{P}_i.$$

Fructose-6-phosphate is converted into glucose-6-phosphate by phosphohexose isomerase.

The net effect of these reactions is:

$$6\text{Ribose-5-phosphate} \rightarrow 5\text{glucose-6-phosphate} + \text{P}_i$$

Conversion of glucose-6-phosphate into ribose-5-phosphate without NADPH generation

The pentose phosphate pathway is extremely flexible. Consider a different situation where a cell needs ribose-5-phosphate for nucleotide synthesis, but has little demand for NADPH. The following overall process caters for this:

$$5\text{Glucose-6-phosphate} + \text{ATP} \rightarrow 6\text{ribose-5-phosphate} + \text{ADP} + \text{H}^+$$

The mechanism is summarized in Fig. 15.5. In this, glucose-6-phosphate is converted by the glycolytic pathway partly into fructose-6-phosphate and partly into glyceraldehyde-3-phosphate. These are the C_6 and C_3 products of reaction 3. Reversal of the three steps produces xylulose-5-phosphate, which is isomerized into ribose-5-phosphate. This scheme does not involve the oxidative part of the pentose phosphate pathway at all.

Generation of NADPH without net production of ribose-5-phosphate

The oxidative part of the pentose phosphate pathway (see Fig. 15.1) is sometimes referred to as being capable of the direct oxidation of glucose to CO_2, and $NADPH+H^+$. The stoichiometry of the overall process is:

$$6\,\text{Glucose-6-phosphate} + 12\,\text{NADP+} + 7\,\text{H2O} \rightarrow$$
$$5\,\text{Glucose-6-phosphate} + 6\,\text{CO}_2 + 12\,\text{NADPH} + 12\text{H}^+ + \text{P}_i$$

This set of events generates NADPH without net production of ribose-5-phosphate – a situation required in a cell with rapid fat synthesis but no cell division. In fact, a *single* molecule of glucose-6-phosphate is *not* converted into six CO_2 molecules by the pathway. What happens is that six molecules of glucose-6-phosphate can *each* give rise to one molecule of CO_2 and one molecule of ribose-5-phosphate plus one molecule of NADPH by the reactions given above. Now, if nonoxidative reactions convert the six ribose-5-phosphate molecules back into five glucose-6-phosphate molecules, as described, then on a balance sheet it looks as if six glucose-6-phosphate molecules have been converted into five molecules of glucose-6-phosphate plus six molecules of CO_2 and six molecules of NADPH. However, as stated, it has not really oxidized one molecule of glucose completely.

Why is the pentose phosphate pathway so important in the erythrocyte?

Mature erythrocytes do not divide so they have no need for ribose-5-phosphate for nucleic acid synthesis, nor do they synthesize fat. Their energy is derived from anaerobic glycolysis – as they have no mitochondria. They do, however, need a supply of NADPH in order to protect the cell membrane from oxidative damage. The production of NADPH offers a protective mechanism, namely the reduction of glutathione, a thiol compound whose main function appears to be to maintain a reducing situation in the cells by virtue of its –SH group. (For this reason it is abbreviated to GSH.) The presence of glutathione is not restricted to the erythrocyte but this is the function we are going to deal with in this chapter.

Glutathione is a tripeptide of glutamic acid, cysteine, and glycine, the peptide link between glutamate and cysteine being on the γ-carboxyl (Fig. 15.6).

Erythrocytes are particularly susceptible to oxidative damage because of the high oxygen content of the cell giving rise to harmful free radicals or reactive oxygen species (ROS). Erythrocytes

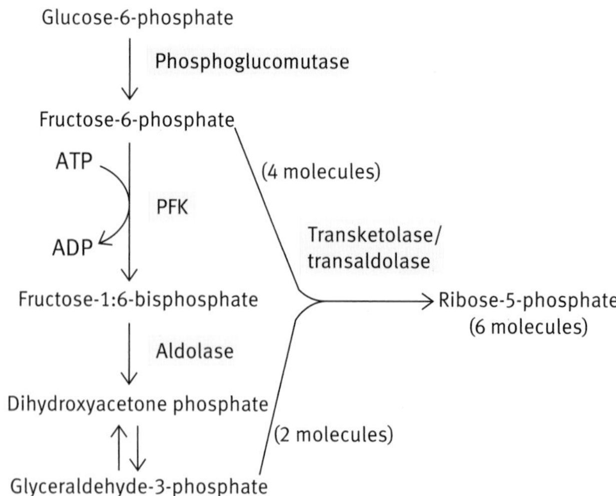

Fig. 15.5 Scheme by which glucose-6-phosphate is converted into ribose-5-phosphate without the production of NADPH. The oxidative part of the pentose phosphate pathway is not involved. PFK, phosphofructokinase.

Glu–Cys–Gly
SH
Reduced glutathione (GSH)

Glu–Cys–Gly
S
S
Glu–Cys–Gly
Oxidized glutathione (GSSG)

Fig. 15.6 Structures of reduced glutathione (GSH) and of the oxidized form (GSSG). Glu, Cys, and Gly are abbreviations for glutamate, cysteine, and glycine, respectively, using the three-letter system.

depend for their integrity on GSH, which reduces any ferrihaemoglobin (methaemoglobin) to the ferrous form, as well as destroying hydrogen peroxide and organic peroxides generated in the erythrocyte (e.g. from an infection, the ingestion of fava beans or through certain drugs) by the action of **glutathione peroxidase**, producing **oxidized glutathione (GSSG)**:

$$H_2O_2 + 2GSH \rightarrow GS{-}SG + 2H_2O$$

The GSH is subsequently regenerated by reduction using NADPH, the reaction being catalysed by **glutathione reductase**:

$$GSSG + NADPH + H^+ \rightarrow 2GSH + NADP^+$$

A continuous supply of NADPH is therefore necessary in the erythrocyte to ensure the integrity of the cell membrane. Defects in the production of NADPH such as glucose-6-phosphate dehydrogenase deficiency, can have dire consequences for the erythrocyte as they result in disintegration of the cell membrane giving rise to **haemolytic anaemia**. As the pentose phosphate pathway is the only means of generation of NADPH/GSH in the cell, the importance of this pathway cannot be overstated.

BOX 15.1 Glucose-6-phosphate dehydrogenase deficiency

G6PD (or G6PDH) deficiency is the most common human enzyme deficiency. It is an X-linked hereditary disorder so, although most individuals with G6PD deficiency are asymptomatic, the symptomatic patients are predominantly male. More than 400 million people in the world are G6PD deficient, with particularly high incidence in African, Middle Eastern, and South Asian people. About 400 different mutations in the G6PD gene are known but not all of them cause clinical symptoms.

Haemolytic anaemia can be triggered in people deficient in G6PD in three main ways, all resulting from presence or production of oxidants in the erythrocyte:

1 A condition known as **Favism** resulting from the ingestion of fava beans (*Vicia faba*) has been known since antiquity. Some forms of G6PDH deficiency, particularly the Mediterranean variant are particularly susceptible to favism. Fava beans, which are a common food in the Mediterranean and the Middle East, contain alkaloids such as **vicine** which are potent oxidants.
2 **Infection** – The inflammatory response to infection can lead to the generation of oxidants, such as free radicals, which enter the erythrocyte causing haemolysis.
3 A number of **oxidant drugs** can cause haemolytic anaemia. Antimalarial drugs, such as pamaquine and choloroquine, can be harmful and individuals should be tested for G6PD deficiency before being prescribed these drugs. Sulphonamides, sulfa antibiotics, and analgesics such as aspirin should also be avoided.

An interesting sidelight on glucose-6-phosphate dehydrogenase deficiency is that mutations leading to a defective enzyme confer a selective advantage in areas where a lethal type of malaria is endemic. Possible explanations for this are that the parasite has a requirement for the products of the pentose phosphate pathway and/or that the extra stress caused by the parasite causes the deficient red blood cell host to lyse before the parasite completes its development. It is interesting to compare the selective advantage conferred in this case with that conferred by the sickle cell trait (see Box 4.3), where a potentially lethal disease can provide a survival advantage because it gives protection against a more lethal disease, malaria.

Summary

The pentose phosphate pathway should not be viewed as a pathway to oxidize glucose; it does not generate ATP nor does it take a molecule of glucose and oxidize it completely. It is a versatile pathway, which supplies three main needs. It produces ribose-5-phosphate for nucleotide synthesis; it supplies NADPH for fat synthesis and other reductive systems; and it provides a route for the metabolism of excess pentose sugars from the diet.

The pathway has an oxidative section converting glucose-6-phosphate to ribose-5-phosphate and produces NADPH. The nonoxidative section manipulates ribose-5-phosphate according to the needs of the cell. If a cell requires equal amounts of ribose-5-phosphate and NADPH, only the oxidative section is required. If it is synthesizing fat and requires a lot of NADPH but little ribose-5-phosphate, the nonoxidative section takes care of excess of the latter by converting it back to glucose-6-phosphate. The pathway is needed in red blood cells to generate NADPH, which is needed to keep the protective molecule glutathione in a reduced state. Without this, certain drugs produce anaemias.

The reaction sequences involved can be complex, involving the interconversion of several sugars. The key reactions in these manipulations are catalysed by transaldolase and transketolase, which between them can effect a range of sugar interconversions.

The pentose phosphate pathway is especially important in the erythrocyte for maintaining the integrity of the cell via production of NADPH. Deficiency of G6PD, a key enzyme in the pathway is the most common human mutation and results in haemolytic anaemia when susceptible individuals are subjected to oxidative stress such as consumption of fava beans, infection, and oxidant drugs.

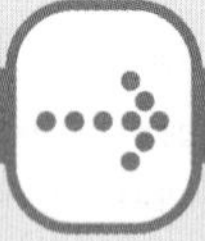

Further reading

To access the further reading, please scan the QR code image or go to http://global.oup.com/uk/orc/biosciences/molbiol/snape_biochemistry/student/reading/ch15/

Problems

1 What are the functions of the pentose phosphate pathway?

2 What is the oxidative part of the pentose phosphate pathway?

3 What enzymes are involved in the nonoxidative part?

4 A nondividing adipose cell requires large amounts of NADPH for fatty acid synthesis but very little ribose-5-phosphate. However, the oxidative section produces equal amounts of the two products. Explain how the nonoxidative reactions deal with this problem.

5 Mature erythrocytes have no need for nucleotide or fatty acid synthesis. Why is the pentose phosphate pathway important in erythrocytes?

Chapter 16

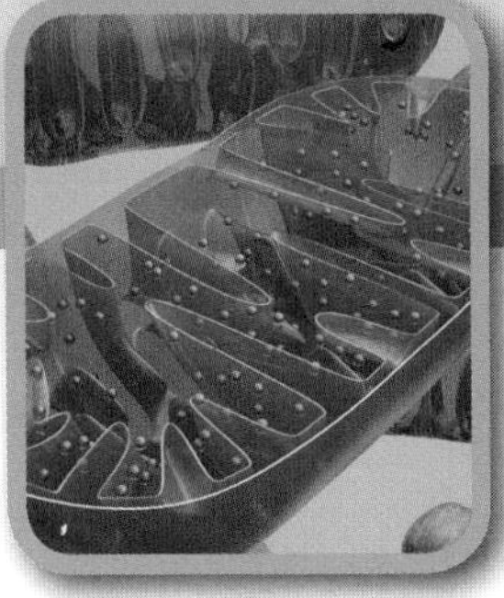

Synthesis of glucose (gluconeogenesis)

The body is able to synthesize glucose from any compound capable of being converted into pyruvate (or any of the intermediates of the TCA cycle), by the process of gluconeogenesis. This excludes fatty acids and acetyl-CoA but includes lactate (which is released into the blood by erythrocytes and by muscle during vigorous exercise), as well as glycerol from triacylglycerols and most of the amino acids, so that a variety of compounds can be converted into glucose and stored as glycogen. However, in addition to this 'routine' metabolic role, gluconeogenesis can literally make the difference between life and death, in starvation.

The brain, as stated several times earlier, requires a constant supply of glucose, which means that the blood glucose concentration must be kept within normal limits or coma and death will ensue. (The brain accounts for only 2% of the body's weight but consumes half the ingested carbohydrate over 24 hours).

The liver's total store of glucose in the form of glycogen is limited and after about 24 hours without food the store is exhausted. Nevertheless, people do not die as a result of one day's fasting. Supply of blood sugar in this situation is dependent on the liver – no other organ, other than the kidney in prolonged starvation, can fill this need. If fasting is prolonged beyond about 24 hours, the liver must synthesize glucose. A minimum of 100 g per day is needed in humans to supply the brain. In early fasting, such as overnight, abut 90% of gluconeogenesis takes place in the liver, but in prolonged starvation the kidney becomes more active and is responsible for up to 40% of the total glucose production. The liver is actually synthesizing glucose at all times except in the fed state, from lactate produced in muscle and erythrocytes, and surplus amino acids, but gluconeogenesis becomes particularly important for survival in fasting and starvation.

Fat mobilization results in ketone body production by the liver, which can supply part of the brain's energy needs and therefore has a glucose-sparing effect as the blood ketone level rises, but it cannot totally replace the requirement for glucose. Although the effects of glucose deprivation on brain are the most dramatic, erythrocytes are also dependent on glucose for energy supply through glycolysis, since they have no mitochondria. The body normally has sufficient fat to supply energy for weeks of starvation, but fatty acids do not pass through the blood–brain barrier in significant amounts and so cannot be used by the brain as fuel.

The main starting point for the pathway of gluconeogenesis in the liver is pyruvate, although the glycerol moiety of triacylglycerol (TAG) and TCA intermediates are also used. To re-emphasize a crucial point, whereas pyruvate can give rise to acetyl-CoA and therefore lead to fat synthesis, the reverse cannot happen. Acetyl-CoA cannot be converted in animals into pyruvate and therefore fatty acids cannot be converted into glucose in a net sense. (It can in *Escherichia coli* and in plants, see later in this chapter.)

We will first describe the pathway by which pyruvate is converted into glucose and after that discuss the broader biochemical implications.

The control of gluconeogenesis is given in Chapter 20.

Mechanism of glucose synthesis from pyruvate

Gluconeogenesis is not a simple reversal of glycolysis. In the glycolytic pathway, there are three reactions, which are irreversible because of thermodynamic considerations:

- phosphorylation of glucose by hexokinase (or glucokinase) using ATP
- phosphorylation of fructose-6-phosphate by phosphofructokinase using ATP
- conversion of phosphoenolpyruvate (PEP) to pyruvate producing ATP.

Glucose is synthesized *via* the intermediate metabolites and reversal of reactions found in glycolysis, but a way is needed to bypass the irreversible reactions.

The first thermodynamic barrier in the process is the conversion of pyruvate to PEP. Because the spontaneous conversion of the enol form of pyruvate to the keto form has a very large negative $\Delta G^{0'}$ value, the PEP $\rightarrow$ pyruvate reaction in glycolysis is irreversible in animal cells (there is no enol pyruvate substrate), but, as mentioned earlier, because of the convention of naming kinases after the reaction involving ATP, it is called pyruvate kinase nonetheless. In animals, a roundabout route involving two reactions is used for the conversion of pyruvate into PEP. Two —Ⓟ groups are used in the process, making the process thermodynamically favourable. The route is as follows:

(1) $\text{Pyruvate} + \text{ATP} + HCO_3^- \rightarrow \text{Oxaloacetate} + \text{ADP} + P_i + H^+$
(catalysed by pyruvate carboxylase)

(2) $\text{Oxaloacetate} + \text{GTP} \rightarrow \text{PEP} + \text{GDP} + CO_2$
(catalysed by PEP carboxykinase, or PEP–CK)

Sum. $\text{Pyruvate} + \text{ATP} + \text{GTP} + H_2O \rightarrow \text{PEP} + \text{ADP} + \text{GDP} + P_i + 2H^+$

The scheme is shown diagrammatically in Fig. 16.1.

GTP is used in the second reaction rather than ATP, which is energetically equivalent. Reaction 1, catalysed by pyruvate carboxylase, which synthesizes oxaloacetate, also has an important role in topping up the TCA cycle (see Chapter 13), quite separate from its role in gluconeogenesis. The second enzyme is usually referred to as **PEP-CK**. Its full name is **phospho-enol-pyruvate carboxykinase** – because in the reverse direction it carboxylates PEP and transfers a phosphoryl group.

Pyruvate carboxylase occurs only in mitochondria, whereas PEP-CK occurs in both mitochondria and the cytosol. Since oxaloacetate cannot be made in the cytosol, there is clearly a problem in converting pyruvate into glucose. There are two solutions, both of which operate. Either PEP is formed in mitochondria and transported out by dedicated transport proteins, or oxaloacetate is reduced to malate and transported out without the need for transport proteins. The malate is dehydrogenated to oxaloacetate in the cytosol and this has the advantage of supplying NADH for gluconeogenesis. Oxaloacetate forms the link between the TCA cycle and gluconeogenesis so that

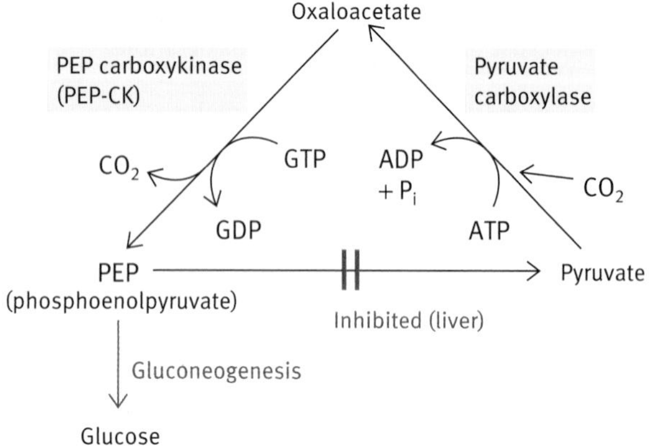

Fig. 16.1 The generation of phosphoenolpyruvate (PEP) for gluconeogenesis from pyruvate in the liver. Note that this scheme makes sense only if the PEP → pyruvate reaction is prevented. How this is done is described in Chapter 20 on metabolic control.

amino acids that can be converted into TCA cycle intermediates, for example, aspartate, can also be used as substrates for gluconeogenesis without the need to be converted into pyruvate first.

Once PEP is formed, the reactions of glycolysis are all reversible until fructose-1,6-bisphosphate is reached. The formation of this in glycolysis from fructose-1-phosphate is irreversible, but the step is bypassed by the simple device of hydrolysing and so removing the phosphoryl group from the 1,6 compound as shown:

Fructose-1,6-bisphosphate

$+ H_2O$

Fructose-1,6-bisphosphatase

Fructose-6-phosphate $+ P_i$

Similarly, when glucose-6-phosphate is reached, the glucokinase (or hexokinase) reaction of glycolysis is irreversible but, in liver, **glucose-6-phosphatase** produces glucose, which is secreted from the cell. Muscle and adipose cells do not have this enzyme and cannot release glucose into the blood. (This is the reason that muscle glycogen stores are used to supply muscle fuel and cannot provide glucose in the blood for use by other tissues.) The complete gluconeogenesis reactions are shown in Fig. 16.2. Note that whether at any given time glycolysis or gluconeogenesis is taking place depends on control mechanisms described in Chapter 20.

There are thus four enzymes involved in gluconeogenesis that do not participate in glycolysis: pyruvate carboxylase, PEP-CK, fructose-1,6-bisphosphatase, and glucose-6-phosphatase. The activities of these enzymes are about 20–50 times greater in rat liver than in rat skeletal muscle, in keeping with the importance of gluconeogenesis in liver.

What are the sources of pyruvate or oxaloacetate used by the liver for gluconeogenesis?

As mentioned above, the main sources of the carbon skeletons for gluconeogenesis are muscle protein amino acids, lactate and the glycerol part of triacylglycerols.

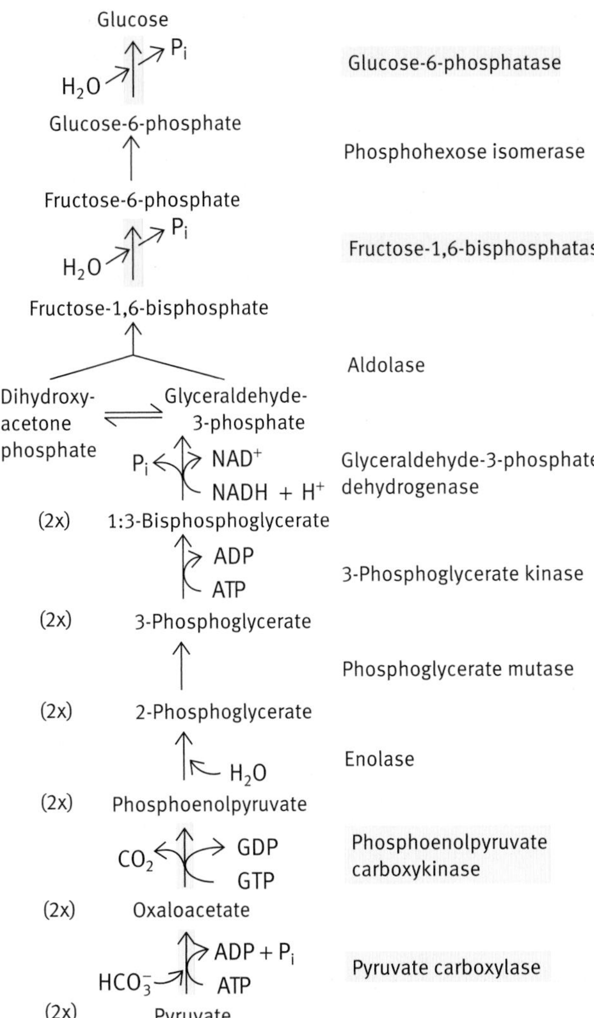

Fig. 16.2 The complete gluconeogenesis pathway from pyruvate to glucose. Reactions in yellow are different from those occurring in glycolysis.

Synthesis of glucose from amino acids

In fasting and starvation, when the glycogen reserves have been exhausted, the main source of pyruvate is the breakdown of muscle proteins. Muscle protein degradation is promoted by the fact that insulin concentrations are low, which means that the inhibition of proteolysis is removed. Glucagon concentrations are high, which stimulates uptake of amino acids by the liver for gluconeogenesis. In addition, the stress hormone **cortisol**, produced during starvation also promotes proteolysis. Hydrolysis of muscle protein produces 20 amino acids. Although the amino acids **alanine** and **glutamine** represent only about 15% of the total muscle protein amino acids, about 50% of the amino acids leaving the muscle to be transported to the liver, are in the form of alanine and glutamine. Alanine and glutamine are both **glucogenic amino acids** – their carbon skeleton is capable of being converted into glucose by the liver and the same is true of most of the other amino acids. (The metabolism of amino acids is separately discussed in Chapter 18, where the mechanisms of the reactions mentioned in this section will be covered.)

How is alanine formed from the protein-breakdown products in muscle? Several of the amino acids give rise to TCA intermediates, which are converted into oxaloacetate via reactions of the cycle. The oxaloacetate can be converted into pyruvate by the reactions shown in Fig. 16.1 and converted into alanine for release into the blood. The amino group of alanine comes from the other amino acids. Note, however, that this formation of alanine in muscle depends on the muscle not oxidizing the pyruvate to acetyl-CoA, which, in starvation, would defeat the whole object of the exercise, which is glucose synthesis. There is a plentiful supply of acetyl-CoA in muscle from the metabolism of fatty acids and ketone bodies, which inhibits the conversion of pyruvate into acetyl-CoA, thus allowing the pyruvate to be channelled into alanine production. Amino groups from other amino acids produced by muscle protein degradation are also transferred to the side chain of glutamate to form glutamine through the action of glutamine synthetase.

The net effect is that many of the amino acids derived from muscle protein breakdown are converted into alanine and glutamine, which are carried in the blood to the liver, where the alanine is converted back into pyruvate, and the glutamine into glutamate and 2-oxoglutarate and thence to glucose. The overall process is summarized in Fig. 16.3.

As the concentration of blood ketone bodies rises during starvation, the brain progressively uses more of these for energy generation and so reduces its utilization of glucose and lessens the demand for gluconeogenesis (which, however,

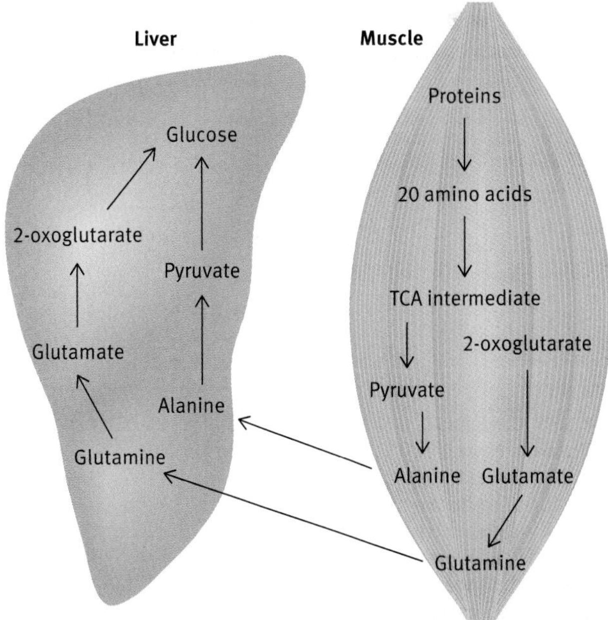

Fig. 16.3 Mechanism by which breakdown of muscle proteins supplies the liver with pyruvate and 2-oxoglutarate for gluconeogenesis.

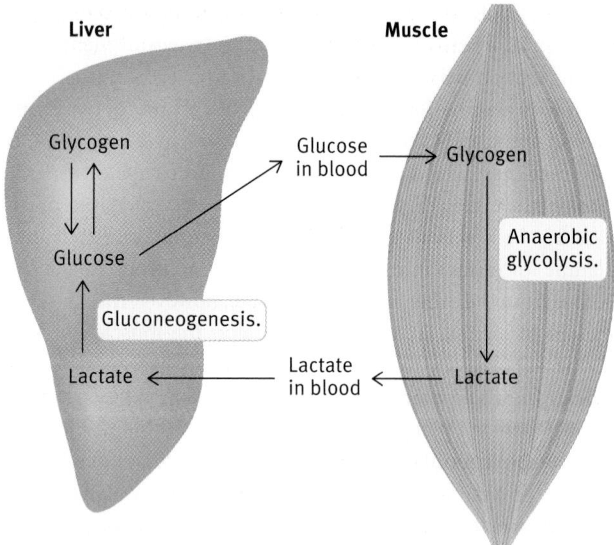

Fig. 16.4 The Cori cycle. Note that this is a physiological cycle involving muscle and liver. The muscle has very low levels of three enzymes essential for gluconeogenesis. Excess lactate is produced in muscle during vigorous action during which glycolysis outstrips the capacity of mitochondria to reoxidize reduced NAD^+ – that is, during anaerobic glycolysis.

always remains essential). This is important for it requires about 2 g of muscle protein to be broken down for each gram of glucose made, a rate of loss that, if continued, would reduce the period that could be survived n starvation.

Synthesis of glucose from lactate

A second source of pyruvate for gluconeogenesis, of less importance in starvation but important in day-to-day normal situations, is **lactate**, produced by the anaerobic glycolysis of glucose or glycogen. Mature erythrocytes have no mitochondria and rely on glycolysis for ATP generation. In normal nutritional situations, a major source of lactate is **muscle glycolysis** – in strenuous muscle activity, the rate of glycolysis within muscle may exceed the capacity of mitochondria to reoxidize NADH and lactate is produced. This travels via the blood to the liver where it is converted into pyruvate and thence to glucose (or glycogen). This constitutes a physiological cycle, called the Cori cycle after its discoverers (Fig. 6.4).

The cycle has two main effects; it 'rescues' lactate for further use and, secondly, it counteracts **lactic acidosis**. Lactic acid in the blood dissociates to lactate and hydrogen ions, and large quantities of the latter may exceed the buffering power of blood and cause a deleterious fall in pH. The synthesis of glucose from lactate involves the uptake of two protons (when $NADH + H^+$ is used for reduction of 1,3-bisphosphoglycerate).

Synthesis of glucose from glycerol

Another metabolite used for gluconeogenesis in the liver is **glycerol** released from **TAG hydrolysis**, mainly in adipose tissue. This is taken up by the liver and converted into glucose by the route shown in Fig. 16.5. The enzyme **glycerol kinase**, which phosphorylates glycerol as the first step in the conversion, is present in the liver, but only in very low amounts in adipose tissue. Glycerol is not converted into glucose (or TAG) in adipocytes but is released into the circulation and enters the gluconeogenic pathway in the liver, allowing the production of glucose in situations where TAG is hydrolysed and the fatty acids used for energy.

During prolonged starvation, after a small initial drop in blood glucose level, the latter remains constant for several weeks and fatty acids supplied by adipose cells likewise remain at a constant level so that the mechanisms are extremely

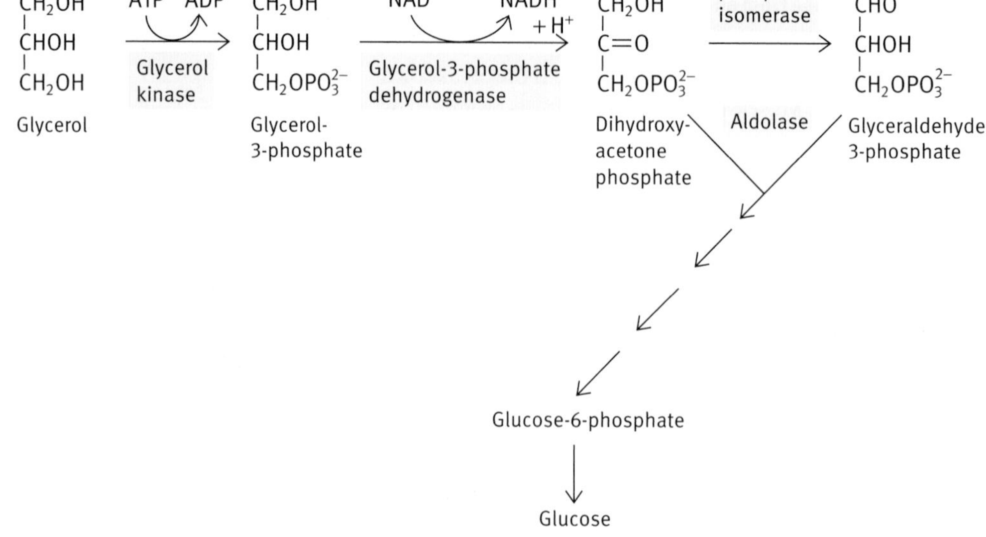

Fig. 16.5 Conversion of glycerol, released by hydrolysis of neutral fat, into glucose in the liver. Much of the glycerol is produced in adipose cells but, since the glycerol kinase is not present there, gluconeogenesis from glycerol occurs in the liver.

effective even though animals cannot use the simple metabolic trick used by plants and bacteria, which allows them to convert fatty acid into glucose, known as the glyoxylate cycle.

Synthesis of glucose from propionate

The synthesis of glucose from propionate is of minor importance except in ruminants, where propionate is a major digestion product. It is converted into succinyl-CoA by the metabolic route for oxidizing odd-numbered fatty acids, present in all animals. Succinyl-CoA, being a component of the TCA cycle, is on the normal metabolic pathways, and is glucogenic.

We will digress briefly to discuss the effect of ethanol metabolism, which has special relevance to gluconeogenesis in the liver.

Effects of ethanol metabolism on gluconeogenesis

Ethanol is oxidized to acetaldehyde, which is further oxidized to acetate, part of which enters the blood to be used by other tissues. Acetate is converted into acetyl-CoA by the acetate-activating enzyme found in most tissues and the acetyl group is then oxidized to carbon dioxide and water by the TCA cycle or diverted into fat synthesis.

The oxidation of ethanol to acetaldehyde occurs in the liver and is mainly due to alcohol dehydrogenase in the cytosol:

$$CH_3CH_2OH + NAD^+ \xrightarrow[\text{Alcohol dehydrogenase}]{} CH_3CHO + NADH + H^+.$$

Oxidation of the acetaldehyde occurs in the mitochondrial matrix:

$$CH_3CHO + NAD^+ + H_2O \xrightarrow{\text{Aldehyde dehydrogenase}} CH_3COO^+ + NADH + 2H^+.$$

A relatively minor route of ethanol catabolism also exists. This is the microsomal ethanol-oxidizing system. Microsomes are small fragments, or vesicles, produced when the endoplasmic reticulum (ER) is disrupted during cell breakage. They do not exist as such naturally but are convenient experimental particles. They contain a cytochrome P450, which uses molecular oxygen, and NADPH to produce acetaldehyde from alcohol:

$$CH_3CH_2OH + NADPH + H^+ + O_2 \rightarrow CH_3CHO + NADP^+ + 2H_2O$$

(It might seem odd to use NADPH in an oxidation reaction, but one atom of oxygen oxidizes the alcohol; the other is reduced to water by the NADPH.) The enzyme is inducible by prolonged heavy intake of alcohol. Many medications are destroyed by P450 s and alcohol may alter the rate of drug metabolism by competing for P450 – something of medical interest. P450 belongs to a class of isoenzymes collectively known as CYP (cytochromes P). They are classified as monooxygenases as they use only one atom of oxygen with the other one being reduced by NADPH to water. Ingestion of alcohol and other foreign chemicals such as pharmaceuticals, pesticides, and herbicides (xenobiotics) increase the synthesis of P450 by upregulation of gene expression (more on P450 in Chapter 32 on special topics).

Effect of ethanol metabolism on the $NADH/NAD^+$ ratio in the liver cell

Alcohol dehydrogenase occurs in the cytosol of liver cells, so the NADH has to be reoxidized *via* the malate–aspartate shuttle, which transports reducing equivalents into the mitochondria. All of this sounds quite harmless, and indeed ethanol produced by microorganisms in the large intestine is absorbed and is a normal metabolite; in humans this can amount to a few grams per day. However, much larger quantities of alcohol may be consumed and the rate of shuttle transfer of reducing equivalents into mitochondria may not keep up with the rate of NAD^+ reduction.

The ratio of NADH to NAD^+ in a liver cell is normally low. With even a moderate amount of alcohol the concentration of NADH is increased. Many of the dehydrogenase reactions in which NAD^+ participates are close to equilibrium in the cell, which means that the normal cytosolic ratios of oxidized and reduced substrates are affected by the changed $NADH/NAD^+$ ratio. In particular, in liver, the lactate dehydrogenase reaction is affected so that the oxidation of lactate arriving at the liver from extrahepatic tissues to form pyruvate is impaired. Remember the lactate dehydrogenase reaction:

$$\underset{\text{Pyruvate}}{CH_3COCOO^-} + NADH + H^+ \rightleftharpoons \underset{\text{La ctate}}{CH_3CHOHCOO^-} + NAD^+$$

The distortion of the normal reduction pattern of NAD^+ affects the metabolism of liver cells.

A serious situation can arise in very heavy drinkers, especially if food intake is restricted during drinking bouts. After 24 hours of food deprivation, the liver glycogen stores are exhausted, and maintenance of blood glucose levels, and therefore brain function, depends on gluconeogenesis. Gluconeogenesis depends on an adequate supply of pyruvate, which as described, is mainly formed from lactate produced by red blood cells and from alanine derived from muscle protein breakdown. However, in the presence of alcohol-induced abnormally high NADH levels in the liver, pyruvate availability is diminished; that formed from alanine may be reduced to lactate, and lactate, arriving from outside, is less efficiently oxidized to pyruvate. Lactate leaks into the blood causing lactic acidosis, and the reduced pyruvate availability impairs gluconeogenesis. In addition, oxaloacetate derived from amino acids that can be deaminated to intermediates of the TCA cycle may be reduced to malate instead of being converted into PEP and channelled into gluconeogenesis. The situation is exacerbated because gluconeogenesis itself is one of the protections against lactic acidosis, two protons being absorbed per molecule of glucose synthesized.

Gluconeogenesis from glycerol could likewise be affected by high NADH/NAD^+ ratios by impairing dehydrogenation of glycerol-3-phosphate.

As a result of the inhibition of gluconeogenesis by alcohol, hypoglycaemia can ensue if alcohol is consumed when the liver is depleted of glycogen. For this reason people are advised not to have alcoholic drinks after heavy exercise or when they have not eaten for a long time.

Alcohol overconsumption is one of the commonest causes of fatty liver (fatty liver disease or FLD). Alcohol dehydrogenase-mediated ethanol metabolism increases the concentrations of NADH in the liver and so decreases the NAD^+/NADH ratio, which in turn inhibits fatty acid oxidation and promotes synthesis of triacylglycerols from the fatty acids which accumulate. The excessive amounts of triacylglycerols are deposited in the liver causing steatosis. Alcohol also promotes the incorporation of fatty acids into cholesterol esters and they also accumulate in the liver. Incorporation into lipoproteins and release into the blood stream can produce a hyperlipidaemia, which is reversible if the consumption of alcohol is reduced.

BOX 16.1 **Alcohol and the Oriental flushing syndrome**

There is considerable genetic variation in the two enzymes that metabolize alcohol to acetate, namely alcohol dehydrogenase and aldehyde dehydrogenase. This variation determines to a large extent individual differences in handling and tolerating alcohol. Acetaldehyde, the product of the reaction catalysed by alcohol dehydrogenase is short lived and does not usually accumulate, as the next enzyme, mitochondrial aldehyde dehydrogenase has a low K_m for acetaldehyde and so 'mops it up' efficiently. A great number of people of East Asian origin have a defective aldehyde dehydrogenase caused by a mutation which results in the substitution of one glutamate by a lysine in the polypeptide chain. The defective enzyme does not catalyse the conversion of acetaldehyde into acetate, which results in increased concentrations in the blood with unpleasant consequences. Even heterozygotes are affected because the defective enzyme forms nonfunctioning dimers with the normal enzyme monomers. The high levels of acetaldehyde cause vasodilation, tachycardia and facial redness, known as the Oriental flush, and can be so unpleasant that a lot of people do not consume alcohol (and do not become alcoholics). About a third to half of people of Chinese and Japanese origin exhibit this flushing after alcohol consumption.

Related to the above is the use of drugs such as sulfiram in the treatment of alcoholics. They inhibit aldehyde dehydrogenase and allow the level of acetaldehyde to rise in the blood producing unpleasant symptoms, which are aimed at discouraging patients from consuming alcohol.

Synthesis of glucose *via* the glyoxylate cycle in bacteria and plants

E. coli can survive quite well with acetate as its sole carbon source. Acetate is converted into acetyl-CoA by an ATP-driven reaction. A net conversion of acetyl-CoA to C_4 acids of the TCA cycle and thence to carbohydrate or any other component of the cell can occur, unlike the situation in animals. In plant seeds, energy stored as TAG is converted to glucose on germination. What allows bacteria and plants to do this?

These organisms possess the normal TCA cycle but, in addition, can bypass some of its reactions by other reactions that do not take place in animals. In the TCA cycle, two carbon atoms are added to oxaloacetate from acetyl-CoA, giving citrate (C_6), but then two carbon atoms are lost as two CO_2 molecules in forming succinate (C_4), so that there is no net conversion of C_2 into TCA cycle intermediates.

The **glyoxylate route** bypasses these losses. Isocitrate is directly split into succinate and glyoxylate – a sort of cycle shortcut:

$$\underset{\text{Isocitrate}}{\begin{matrix}COO^-\\|\\CHOH\\|\\CHCOO^-\\|\\CH_2COO^-\end{matrix}} \xrightarrow[\text{lyase}]{\text{Isocitrate}} \underset{\text{Succinate}}{\begin{matrix}CH_2COO^-\\|\\CH_2COO^-\end{matrix}} + \underset{\text{Glyoxylate}}{\begin{matrix}COO^-\\|\\CHO\end{matrix}}$$

The C_2 glyoxylate now reacts with acetyl-CoA to produce malate – back on the TCA cycle:

$$\underset{\text{Glyoxylate}}{\begin{matrix}CH{-}COO^-\\\|\\O\end{matrix}} + \underset{\text{Acetyl-CoA}}{CH_3{-}\overset{\overset{O}{\|}}{C}{-}S{-}CoA} \xrightarrow[\text{Malate synthase}]{+\,H_2O} \underset{\text{Malate}}{\begin{matrix}OH\\|\\CH{-}COO^-\\|\\CH_2{-}COO^-\end{matrix}} + CoA{-}SH$$

The scheme is shown in Fig. 16.6. The net effect is that acetyl-CoA plus oxaloacetate is converted into malate plus succinate. Succinate and malate can both be converted into oxaloacetate, one molecule then being available for

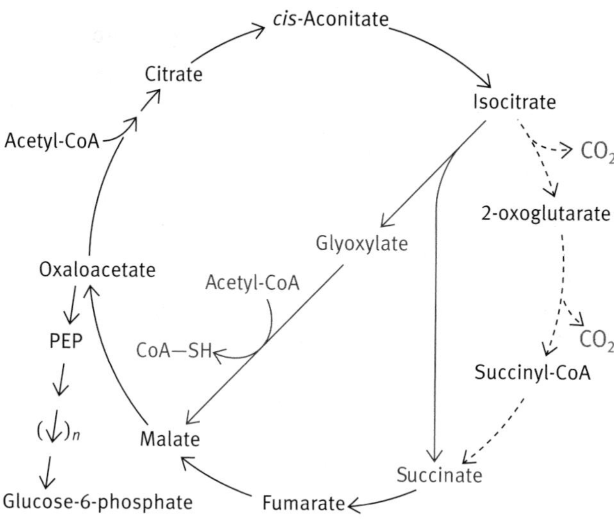

Fig. 16.6 The glyoxylate cycle of plants and bacteria, which permits carbohydrate synthesis from acetyl-CoA. This does not occur in animals. The broken line represents the bypassed section of the TCA cycle; the red lines are reactions peculiar to the glyoxylate cycle. The two CO_2 molecules are highlighted to emphasize that it is these two decarboxylation reactions that must be bypassed. It will be appreciated that, as a result of the glyoxylate reactions, two molecules of oxaloacetate are produced from citrate, one of which is used to form citrate again and one to produce PEP.

conversion to glucose, the other to continue the cycle. In plants, this process occurs in membrane-bounded organelles, called glyoxysomes.

In Chapter 13, we described the enzyme pyruvate carboxylase, which is necessary for topping up the TCA cycle (the anaplerotic reaction). *E. coli* does not possess this enzyme. This is presumably related to the fact that the glyoxylate cycle can generate net increase in cycle intermediates from acetyl-CoA rendering pyruvate carboxylase unnecessary. The same applies to the need to generate oxaloacetate for carbohydrate synthesis.

It should finally be mentioned that by far the greatest amount of carbohydrate synthesis on Earth occurs during photosynthesis in plants. This involves the fixation of CO_2 into another glycolytic intermediate, bisphosphoglycerate, from which glucose synthesis occurs. The mechanism of this is best left until we deal with photosynthesis in Chapter 21.

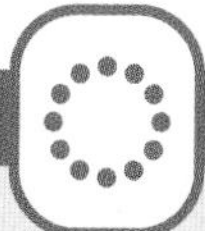

Summary

It is essential, at certain times, for the liver to synthesize glucose by the pathway of gluconeogenesis and release it into the bloodstream. Unless the brain has adequate supplies of glucose it will cease to function normally.

During fasting, after 24 hours of food deprivation, the liver's glycogen stores are exhausted and the essential supplies of glucose to the brain come from glucose synthesized in the liver. The substrate for gluconeogenesis is pyruvate but any compound convertible to pyruvate or an intermediate of the TCA cycle can be gluconeogenic. The main source of this, in starvation, is alanine and glutamine derived from muscle protein catabolism and transported in the blood to the liver. There it is deaminated to pyruvate. The muscle protein breakdown is caused by the high glucagon/insulin ratio and the stress hormone cortisol, which is liberated during starvation.

Gluconeogenesis occurs largely by reversal of glycolytic reactions but there are three irreversible steps in the latter pathway that must be circumvented. Pyruvate is converted into phosphoenolpyruvate *via* oxaloacetate with an input of energy from GTP. The phosphofructokinase and hexokinase reactions are bypassed by phosphatases. The reactions forming phosphoenolpyruvate and the fructose bisphosphatase reaction are important control points in gluconeogenesis (see Chapter 20).

Gluconeogenesis also recycles lactate, produced by muscle (during vigorous exercise) and erythrocytes (continuously), by converting it into glucose or glycogen. The sequence is known as the Cori cycle.

The synthesis of glucose from glycerol arising from triacylglycerol hydrolysis starts with glycerol kinase. It cannot be metabolized by adipose cells, where glycerol kinase activity is very low. It was thought that there was no glycerol kinase in adipocytes, but it has been shown that a glycerol kinase isoenzyme does exist but its activity is 200–600 times less than that of the liver enzyme. Most of the glycerol released in the adipocyte is released into the bloodstream and taken up by the liver. The liver phosphorylates it to glycerol-3-phosphate, then converts it to dihydroxyacetone phosphate, which is on the standard gluconeogenic pathway.

One of the dangers of excessive drinking is when it is allied with fasting; the perturbation of the $NADH/NAD^+$ ratio by alcohol metabolism can impair necessary synthesis of glucose for the brain. Muscle does not release free glucose and has no capacity for gluconeogenesis.

In animals, fats cannot be converted into glucose or C3 metabolites. However, the glyoxylate cycle in plants and bacteria, which is a modified TCA cycle, makes this conversion possible.

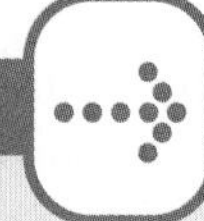

Further reading

To access the further reading, please scan the QR code image or go to http://global.oup.com/uk/orc/biosciences/molbiol/snape_biochemistry/student/reading/ch16/

Problems

1 After 24 hours of fasting, the glycogen reserves of the liver are exhausted but there are still relatively large stores of fat. What is the advantage of synthesizing glucose when large supplies of acetyl-CoA from fatty acids are available for energy?

2 Gluconeogenesis requires production of phosphoenolpyruvate (PEP), but pyruvate kinase cannot form PEP from pyruvate. Why? How is the problem overcome?

3 From PEP, gluconeogenesis in the liver produces blood glucose mainly by reversal of glycolytic steps. However, three enzymes are involved, which do not participate in glycolysis. What are these?

4 Muscle has no glucose-6-phosphatase. What is the metabolic consequence of that?

5 What is the Cori cycle and what is its physiological role?

6 Adipose cells do not have glycerol kinase, the enzyme that converts glycerol to glycerol-3-phosphate, even though adipose cells produce glycerol from triacylglycerol (TAG) hydrolysis. What is the advantage of the presence of glycerol kinase in the liver rather than the adipocyte?

7 Animals cannot convert acetyl-CoA into carbohydrate. Bacteria and plants can. How?

8 Pyruvate kinase is needed to top up the components of the TCA cycle (the anaplerotic reaction). Depletion of the cycle intermediates would impair operation of the cycle. However, *Escherichia coli* does not have this enzyme. Comment on this.

9 In starvation, muscle wasting occurs. What is the relevance of this to gluconeogenesis in the body?

10 Explain how excessive alcohol intake can deprive the brain of glucose if food intake is restricted, as may occur in binge drinking.

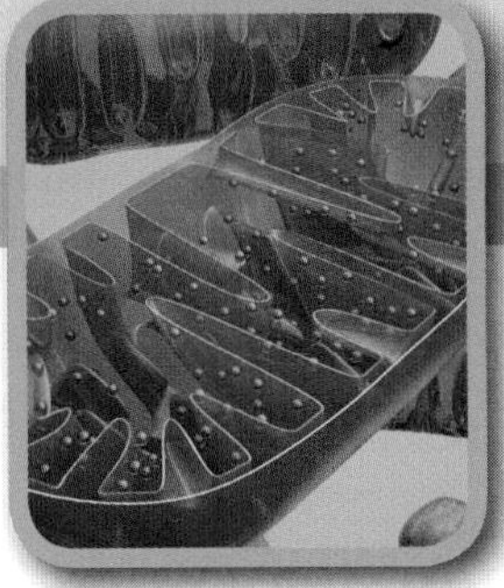

Chapter 17

Synthesis of fat and related compounds

In this chapter we are going to deal with synthesis of fat, or triacylglycerol (TAG), an anabolic process taking place in the fed state, and also with synthesis of related compounds, such as membrane lipids and cholesterol, which take place at all times in cells.

Fat synthesis is an anabolic process taking place in the fed state. Carbohydrate in excess of that needed to replenish glycogen stores is converted into a more suitable form of storage, triacylglycerol. Alcohol and certain amino acids can also be used to synthesize fat. As already explained, if this were not done, and we stored our fuel as glycerol, we would be considerably larger in size as glycerol is stored with water, but fat is compact and anhydrous. Fat synthesis occurs at times of dietary plenty. Conversion of carbohydrate into fat depends on the relative proportions of carbohydrate and fat in the diet. Fat synthesis may be low in mixed diets with a low carbohydrate content, but becomes more important in high carbohydrate diets of adequate or excessive energy value.

Fig. 17.1 Why glucose can be converted into fats but fats cannot be converted into glucose in animals. The reverse pathways in red are not completely the same as the forward pathways and are described in later chapters. In plants and bacteria, fat can be converted into glucose but not by the reversal of the pyruvate dehydrogenase reaction.

Mechanism of fat synthesis

General principles of the process

If you refer to the overall diagram of metabolism (Fig. 17.1), you will see that, as well as fatty acids being converted into acetyl-CoA, acetyl-CoA can be converted into fatty acids. The fact that fatty acids are synthesized from acetyl-CoA, two carbon atoms at a time, explains why most natural fatty acids have even numbers of carbon atoms.

As mentioned, carbohydrate intake in excess can lead to fat deposition. Glucose is converted into pyruvate, which is then converted into acetyl-CoA, which is used to synthesize fatty acids. However, since the step catalysed by pyruvate dehydrogenase is irreversible, acetyl-CoA cannot be converted in a net sense into pyruvate and hence fat cannot be converted into glucose in animals. (We have already seen, in Chapter 16, how bacteria and plants have, a special pathway (the glyoxylate pathway) for converting acetyl-CoA into glucose.)

You will by now be familiar with the concept that in metabolic pathways, at least some reactions are different in the forward and reverse directions. This is also true for fat breakdown and synthesis. For some steps the two pathways use the same reactions (in opposite directions), but there are others that are different in the two directions. In this way both pathway directions are rendered thermodynamically favourable, irreversible, and separately controllable.

Synthesis of malonyl-CoA is the first step

An irreversible reaction is necessary to render synthesis of fatty acids from acetyl-CoA thermodynamically favourable. To refresh your memory, go back to Chapter 3 on 'Energy considerations in biochemistry', since it is essential to appreciate the importance of an irreversible step in order to understand the very first reaction in fatty acid synthesis.

The acetyl-CoA molecule is carboxylated by CO_2, using ATP breakdown as an energy source, to form malonyl-CoA, but, in the next reaction in fat synthesis, the CO_2 is lost. This seems pointless unless you remember that, in this way, the process is thus made thermodynamically irreversible. It is to do with energy, rather than chemical change. Malonic acid has the structure HOOC—CH_2—COOH so malonyl-CoA is HOOC—CH_2—CO—S—CoA. The enzymic reaction below is catalysed by **acetyl-CoA carboxylase** – it adds a carboxyl group to acetyl-CoA, forming malonyl-CoA.

$$CH_3-\overset{O}{\overset{\|}{C}}-S-CoA + ATP + HCO_3^-$$

$$\downarrow$$

$$^-O(O{=})C-CH_2-\overset{O}{\overset{\|}{C}}-S-CoA + ADP + P_i + H^+$$

The enzyme has biotin as a prosthetic group. All carboxylases using ATP to incorporate CO_2 into molecules have this feature. As an intermediate in the reaction, an activated CO_2–biotin complex is formed.

To synthesize fatty acids we have to add two carbon units at a time, starting with acetyl-CoA. *Malonyl-CoA is the active donor of two carbon atoms in fatty acid synthesis*, despite the fact that the malonyl group has three carbons. But before this, a few words of explanation about the acyl carrier protein or ACP.

The acyl carrier protein (ACP) and the β-ketoacyl synthase

When fatty acids are broken down to acetyl-CoA, all of the reactions occur, not as free fatty acids, but as thiol esters with CoA. When fatty acids are synthesized, all of the reactions also occur on fatty acyl groups bound as thiol esters, but instead of CoA being used to esterify the reactants, *half of* the CoA molecule is used. We remind you of the structure of CoA.

Phosphate—pantothenate—NHCH$_2$CH$_2$—SH (in box)
|
Phosphate—ribose—adenine
|
phosphate

Phosphopantotheine moiety in box

The half in the box is 4-phosphopantotheine and this is the 'carrier' thiol used in fatty acid synthesis. It is not free, as is CoA, but is bound to a protein called the ACP. You can think of ACP as a protein with built-in CoA or, alternatively, as a giant CoA molecule with the AMP part of CoA replaced by a protein.

The enzyme, **β-ketoacyl synthase**, also known as **condensing enzyme**, has a reactive thiol. This also is needed for acyl thiol ester formation. In this case the reactive thiol group is that of the amino acid cysteine. The ACP and β-ketoacyl synthase in mammals are part of a large multifunctional complex.

Mechanism of fatty acyl-CoA synthesis

The mechanism described below is that of *E.coli* showing the series of separate enzymes. As mentioned above, there is no separate ACP or other fatty acid synthesizing enzymes in mammals, but they are part of a multi-active site enzyme complex. We will start with the situation shown in Fig. 17.2(a) in which the two thiol groups on the fatty acyl-CoA-ACP synthase complex are vacant.

CH_3CO– is transferred from acetyl-CoA to the ACP thiol by a specific transferase (Fig. 17.2(b)). This is then further transferred to the β-ketoacyl synthase thiol group (Fig. 17.2(c)), leaving the ACP site vacant. Another transferase moves the malonyl group of malonyl-CoA to the latter, forming malonyl-ACP (Fig. 17.2(d)). The synthase now transfers the acetyl group to the malonyl group, displacing CO_2 and forming a β-ketoacyl-ACP; in this initial case it is β-ketobutyryl-ACP (Fig. 17.2(e)). The latter is reduced by activities on the same protein complex (described below) to form butyryl-ACP (Fig. 17.2(f)). The butyryl group is transferred to the β-ketoacyl synthase (Fig. 17.2(g)). The resultant situation is precisely analogous to that shown in Fig. 17.2(c) since both have a saturated acyl synthase complex (acetyl and butyryl, respectively). A series of reactions now ensues, identical to those in Fig. 17.2 (starting with the reaction c → d), except that acetyl- is now butyryl- and the end product is hexanoyl-ACP. Five more rounds produces a palmitoyl-ACP. When C_{16} is reached, a hydrolase releases palmitate (Fig. 17.2(h)),

$$CH_3(CH_2)_{14}CO-S-ACP + H_2O \rightarrow CH_3(CH_2)_{14}COO^- + ACP-SH$$

Note that the β-ketoacyl-ACP synthase reaction leading to stage **(e)** is irreversible because of the energy considerations of the decarboxylation involved.

Palmitoyl-CoA can be further elongated by C_2 units to form long- or very-long-chain fatty acids by Type III fatty acid synthetases (elongases), which are found in the endoplasmic reticulum.

Organization of the process of fatty acid synthesis

The organization of enzymes required to synthesize fatty acids differs among species. Fatty acid synthetase (FAS) can

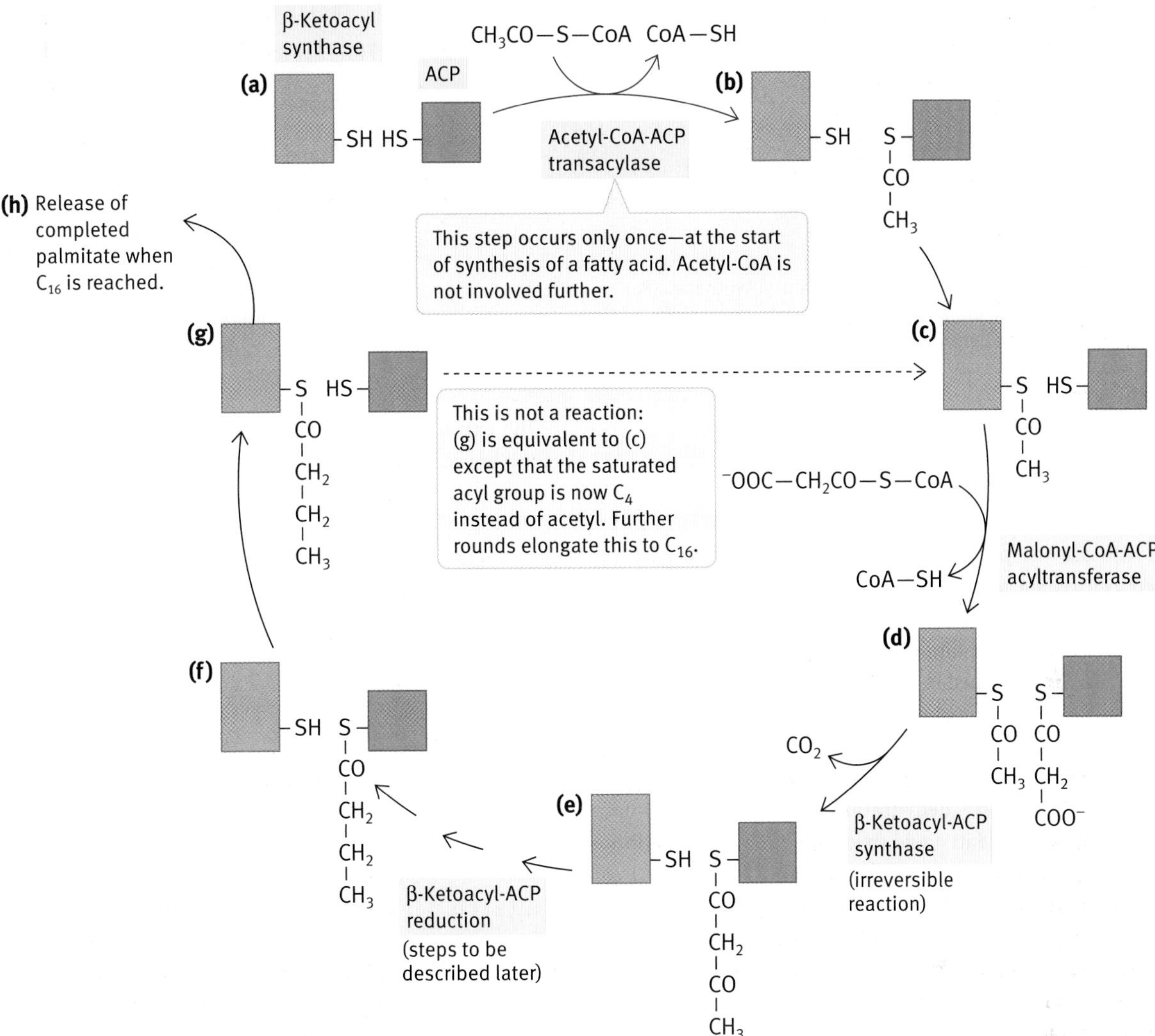

Fig. 17.2 The steps involved in the synthesis of fatty acids. Although the β-ketoacyl synthase of animals is a domain of a single protein molecule, it is shown here as being separate (orange). This is because the functional enzyme unit is a dimer arranged head-to-tail and the transfers between the two —SH groups occur between the domains on the two dimer-constituent protein molecules. Note that the thiol of the ACP (blue) is on the phosphopantetheine moiety and that of the synthase is on a cysteine residue of the protein. After seven successive rounds of reactions, the resultant palmitoyl-ACP is hydrolysed to release free palmitate. ACP, acyl carrier protein.

be divided into two groups according to the organization of their catalytic units:

- Type I FAS systems are multienzyme complexes that contain all the catalytic units as distinct domains covalently linked into one or two polypeptides. Type I systems are mainly eukaryotic. Animal FAS enzymes consist of two homodimers in an X like formation. Mammalian FAS is thought to have evolved through gene fusion.
- In type II FAS systems, the enzymes exist as distinct, individual proteins, each one catalysing a single step in the pathway. This system exists mainly in bacteria and some plants.

In mammals, a pair of such proteins collaborate as the functional dimer complex (Fig. 17.3). The growing fatty acid chain oscillates from the 'ACP'-thiol group to the β-ketoacyl synthase thiol group, but it is never released from the enzyme complex until palmitate synthesis is completed. The elongation step and reductive reactions all occur with the substrate attached to the 'ACP'. The long flexible arm of the 4-phosphopantetheine is presumably needed to permit the different intermediates to interact with the appropriate catalytic centres of the complex. The advantage of a single large complex is that the process of synthesis can be more rapid since each intermediate is positioned to interact with the next catalytic centre rather than having to diffuse away and find the next enzyme.

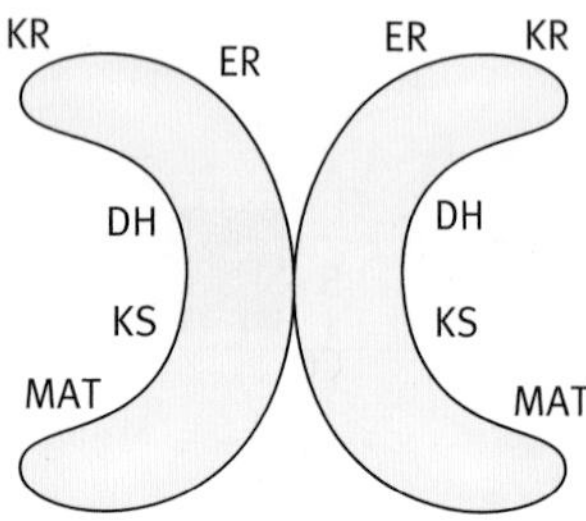

Fig 17.3 Enzyme activities on mammalian fatty acyl synthase. KR, β-ketoacyl reductase; ER, enoyl reductase; DH, dehydratase; KS, β-ketoacyl synthase (condensing enzyme); MAT, malonyl–acetyl-CoA transacetylase.

The reductive steps in fatty acid synthesis

In the above sequence of events involved in saturated fatty acyl-ACP synthesis, the β-ketoacyl group attached to the ACP is reduced by three successive steps shown in Fig. 17.4. The reductant in these steps is **NADPH** (not NADH), the reduced form of nicotinamide adenine dinucleotide phosphate ($NADP^+$), an electron carrier, mentioned in Chapter 15 on the pentose phosphate pathway. (The chemistry of NAD^+ reduction is shown in Chapter 12 and that of $NADP^+$ is the same.)

In summary, the reductive steps in fatty acid synthesis involve hydrogenation, dehydration and hydrogenation, which is the reverse of fatty acid oxidation (dehydrogenation, hydration and dehydrogenation), but using different starting molecules and products (malonyl-CoA as opposed to acetyl-CoA), and cofactors (NADPH as opposed to NAD^+).

$R{-}CO{-}CH_2{-}CO{-}S{-}ACP$ β-Ketoacyl-ACP

Reduction ($NADPH + H^+ \rightarrow NADP^+$) β-Ketoacyl-ACP reductase

$R{-}CH(OH){-}CH_2{-}CO{-}S{-}ACP$ β-Hydroxyacyl-ACP

Dehydration ($\rightarrow H_2O$) β-Hydroxyacyl-ACP dehydratase

$R{-}CH{=}CH{-}CO{-}S{-}ACP$

Reduction ($NADPH + H^+ \rightarrow NADP^+$) Enoyl-ACP reductase

$R{-}CH_2{-}CH_2{-}CO{-}S{-}ACP$ Acyl-ACP

Fig. 17.4 Reductive steps in fatty acid synthesis.

What is $NADP^+$?

NAD^+, we remind you, is:

P—ribose—nicotinamide
|
P—ribose—adenine.

$NADP^+$ is:

P—ribose—nicotinamide
|
P—ribose—adenine
|
P.

The extra phosphoryl group is attached to the 2′-hydroxyl group of the ribose which is attached to adenine.

$CONH_2$ N⁺ O OH OH $^-O{-}P{=}O$ $^-O{-}P{=}O$ O Adenine H OH OPO_3^{2-}

The extra phosphoryl group has no effect on the redox characteristics of the molecule – it is, in fact, purely an identification or recognition signal. An NAD^+ enzyme will not react with $NADP^+$ and *vice versa* (with the odd exception of little significance).

The use of the two coenzymes constitutes an important form of **metabolic compartmentation**. To understand this, remember that in energy release, reducing equivalents are oxidized, while in, for example, fat synthesis, reducing equivalents are used for synthesis. The cell keeps these metabolic activities separate by metabolic compartmentation – one, the oxidative part, uses NADH as electron carrier; the other, the reductive part, uses NADPH.

Separate mechanisms exist to reduce NAD^+ and $NADP^+$. You already know how NAD^+ is reduced in glycolysis, the pyruvate dehydrogenase reaction, the TCA cycle, and fat oxidation. How is $NADP^+$ reduced? This is explained in the next section.

Fatty acid synthesis takes place in the cytosol

The main sites of fatty acid synthesis in humans is the liver and, to a much lesser extent, adipose cells, but some other tissues such as the lactating mammary glands also produce fat. Adipose cells are the main site of storage.

Within a cell, **palmitate synthesis** from acetyl-CoA occurs in the cytosol. This is in contrast to **fatty acid oxidation**, which occurs in the **mitochondria**. The major source of acetyl-CoA for fatty acid synthesis is the pyruvate dehydrogenase reaction (see Chapter 13), which is located in the mitochondrial matrix. Acetyl-CoA cannot cross the mitochondrial membrane to the site of fatty acid synthesis in the cytosol so acetyl residues must be transported from the mitochondria to the cytosol. The acetyl-CoA in the mitochondrion is converted into citrate by the citric acid synthase reaction of the TCA cycle. The citrate is transported by a mitochondrial membrane system into the cytosol where it is cleaved back to acetyl-CoA and oxaloacetate by a separate enzyme, called **ATP-citrate lyase** or the **citrate cleavage enzyme**. This reaction is coupled to hydrolysis of ATP to ADP and inorganic phosphate (P_i), which ensures that it goes to irreversible completion. (Remember that the citrate-synthesizing reaction in mitochondria is irreversible, so a different enzyme is needed for its cleavage.)

$$\text{Citrate} + \text{ATP} + \text{CoA–SH} + H_2O \rightarrow \text{acetyl–CoA} + \text{oxaloacetate} + \text{ADP} + P_i$$

The oxaloacetate cannot get back into the mitochondrion, the membrane of which is impervious to it, and for which no transporter exists. It is reduced in the cytosol to malate by NADH (note, not NADPH); the malate so formed is oxidatively decarboxylated to pyruvate and CO_2 by an enzyme (the **'malic' enzyme**) that uses $NADP^+$ (note, *not* NAD^+), thus producing NADPH, which is needed for fat synthesis (Fig. 17.5). The pyruvate is now transported back into the mitochondrion (Fig. 17.6). The pyruvate transported into the mitochondrion can be converted back into oxaloacetate by the **pyruvate carboxylase reaction**.

$$\text{Pyruvate} + HCO_3^- + \text{ATP} \xrightarrow{\text{(Pyruvate carboxylase)}} \text{Oxaloacetate} + \text{ADP} + P_i + H^+$$

Citrate leaves the mitochondrion only when it is at a high concentration; this occurs when carbohydrate is plentiful. Citrate does not appear in the cytosol at other times.

Thus, the citrate mechanism not only transports acetyl groups out of the mitochondrion, it also generates NADPH for fatty acid synthesis. The reduction in the cytosol of oxaloacetate to malate by NADH and the oxidation of malate to pyruvate by $NADP^+$ constitutes a neat mechanism for transferring electrons from the NADH metabolic 'compartment' into the NADPH 'compartment' used for reductive synthesis reactions. In addition, citrate activates the first reaction committed to fatty acid synthesis – the acetyl-CoA carboxylase, producing malonyl-CoA. Malonyl-CoA only appears in the cell when fatty acid is being synthesized and it constitutes an important regulator of fatty acid synthesis and degradation as we will see in greater detail in Chapter 20.

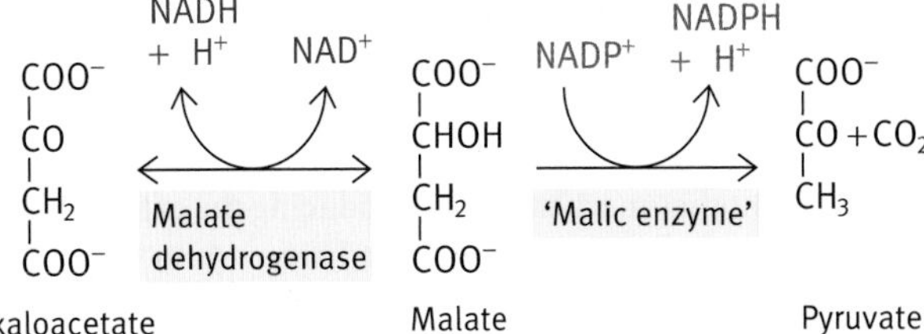

Fig. 17.5 Reduction of $NADP^+$ for fatty acid synthesis. The net effect of the two reactions is to transfer reducing equivalents from NADH to $NADP^+$. The pyruvate so produced in the cytosol enters the mitochondrion. The source of the oxaloacetate is shown in Fig. 17.4.

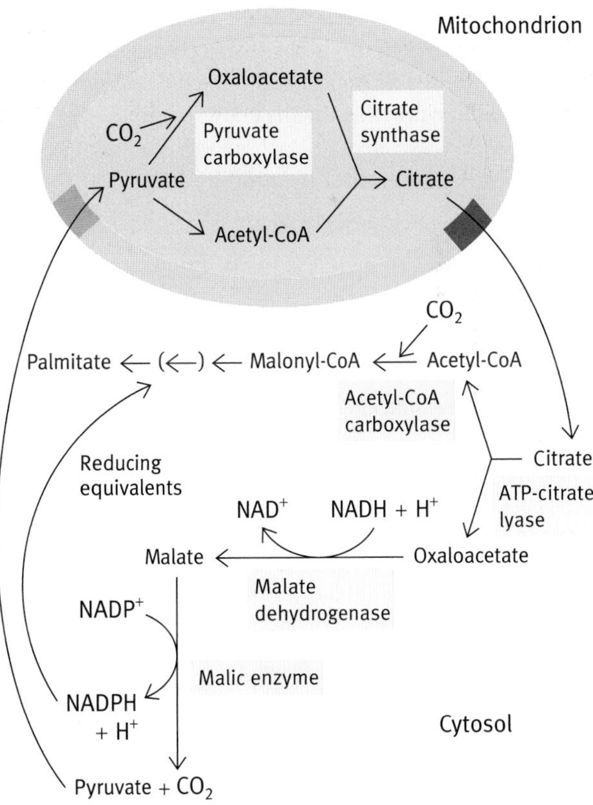

Fig. 17.6 Source of acetyl groups (acetyl-CoA) and reducing equivalents ($NADPH + H^+$) for fatty acid synthesis. The ATP-citrate lyase reaction involves the breakdown of ATP to ADP and P_i. This ensures complete breakdown of citrate.

For every acetyl-CoA molecule produced in the cytosol from citrate, *one* NADPH molecule is generated. However, the formation of each $-CH_2-CH_2-$ group from $-CH_3CO-$ by palmitate synthase requires *two* NADPH molecules (see Fig. 17.4). Yet another mechanism for producing the extra NADPH is required and we have seen this in Chapter 15 on the pentose phosphate pathway.

Synthesis of unsaturated fatty acids

The body requires unsaturated fatty acids for the production of polar lipids for membrane synthesis to achieve lipid bilayer fluidity, and as components or precursors of a number of biomolecules such as prostaglandins and related compounds. An enzyme system exists in the liver which can introduce a single double bond into the middle of stearic acid, generating oleic acid. Δ^9 indicates that the double bond is between carbon atoms 9 and 10 of the fatty acid, with the carboxyl carbon being number one.

$$CH_3(CH_2)_7CH{=}CH(CH_2)_7COOH$$

Oleic acid, 18:1(Δ^9)

However, in animals, the system cannot make double bonds between this central double bond and the methyl end of the molecule although this is possible in plants. Linoleic acid, which has two double bonds, and linolenic acid with three, thus cannot be synthesized by the body.

$$CH_3(CH_2)_4CH{=}CHCH_2CH_2{=}CH(CH_2)_7COOH$$

Linoleic acid, 18:2($\Delta^{9,12}$)

$$CH_3CH_2CH{=}CHCH_2CH{=}CHCH_2CH{=}CH(CH_2)_7COOH$$

α-Linolenic acid, 18:3($\Delta^{9,12,15}$)

Since they are needed for membrane components and for synthesis of eicosanoid regulatory molecules (see later in this chapter), these two acids must be obtained from the diet (see Box 17.1); plants have enzymes that can desaturate in the terminal half of the fatty acids.

The liver can, however, elongate linoleic acid and introduce an extra double bond to give the C_{20} arachidonic acid (20:4, $\Delta^{5,8,11,14}$) with four double bonds. It can also elongate acids to the C_{22} and C_{24} acids involved in the lipids of nerve tissues. All of these transformations occur as the CoA derivatives.

Synthesis of TAG and membrane lipids from fatty acids

TAG is for the main storage form of fat. For the attachment of a fatty acid in ester bond to glycerol, the acid must be activated to the form of acyl-CoA, a reaction catalysed by acyl-CoA synthetase, as already described.

$$RCOOH + CoA{-}SH + ATP \rightarrow RCO{-}S{-}CoA + AMP + PP_i$$

A carboxylic acid ester has a lower free energy of hydrolysis than a thiol ester has and hence activation of the acid makes TAG synthesis from the fatty acyl-CoA exergonic. The acceptor of acyl groups is not glycerol but glycerol-3-phosphate, which arises mainly by reduction of the glycolytic intermediate, dihydroxyacetone phosphate (DHAP; see Fig. 13.29). The main function of glycolysis in adipocytes is the provision of glycerol-3-phosphate for TAG synthesis. The liver can phosphorylate glycerol directly by the action of the enzyme glycerol kinase.

BOX 17.1 The Alpha and the Omega in fatty acids and diet

Another system of nomenclature for unsaturated fatty acids designates the carbon of the end methyl group as ω (omega), and numbers the unsaturated bonds of the carbon chain from there (the carbon atom attached to the carboxyl group is designated as the α carbon atom). What linoleic and linolenic acid have in common, is a double bond in a position less than nine carbon atoms from the omega carbon atom. This nomenclature makes linoleic acid an ω-6 fatty acid and linolenic acid an ω-3 fatty acid.

As mentioned above, linolenic acid is an ω-3 fatty acid, indicating a double bond on the third carbon counting from the ω carbon atom, and linoleic is an ω-6 fatty acid. Sometimes the ω-carbon atom is designated as the n-carbon atom so that linolenic acid is also described as a n-3 fatty acid.

Since mammals cannot make double bonds between Δ^9 and the methyl end of the molecule, fatty acids such as linolenic and linoleic need to be supplied in the diet and they are referred to as **essential fatty acids.** ω-3 and ω-6 acids, such as these two, are found in certain plant seeds, such as linseed, sunflower, corn, and rapeseed (also known as 'canola' from CANadian Oil Low Acidity), while other members of the ω-3 and ω-6 families with longer carbon chains are plentiful in fish oils, such as those of cod and salmon. *ω- or n–3* fatty acids that are important in human physiology are α-linolenic acid (18:3, *n*–3; ALA), eicosapentaenoic acid (20:5, *n*–3; EPA), and docosahexaenoic acid (22:6, *n*–3; DHA).

The potential for the reduction of serum cholesterol levels by increasing the ratio of polyunsaturated fatty acids to saturated fatty acids in the diet is now well known, as is the correlation between high levels of cholesterol and coronary heart disease.

The health benefits of the long-chain omega-3 fatty acids – primarily EPA and DHA are best known. Studies on Greenland Inuit people in the 1970s showed that they consumed large amounts of fat from fish, but had very low incidence of cardiovascular disease. The high amounts of omega-3 fatty acids consumed by the Inuit were associated with low TAG concentrations in the blood, low blood pressure, and low incidence of atherosclerosis.

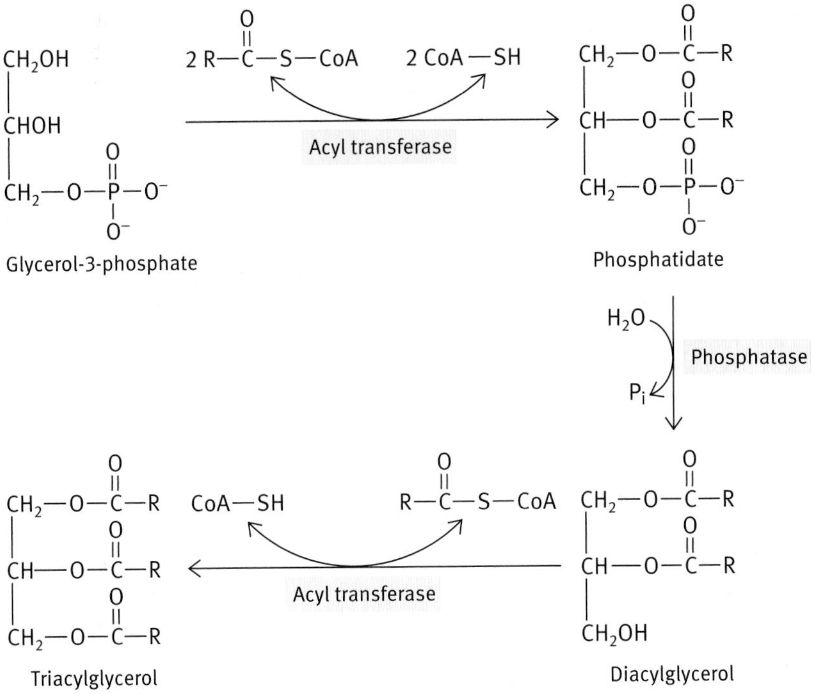

$$\text{Glycerol-3-phosphate } (CH_2OH{-}CHOH{-}CH_2{-}O{-}PO_3^{2-}) + 2\,R{-}CO{-}S{-}CoA \xrightarrow{\text{Acyl transferase}} \text{Phosphatidate} + 2\,CoA{-}SH$$

$$\text{Phosphatidate} + H_2O \xrightarrow{\text{Phosphatase}} \text{Diacylglycerol} + P_i$$

$$\text{Diacylglycerol} + R{-}CO{-}S{-}CoA \xrightarrow{\text{Acyl transferase}} \text{Triacylglycerol} + CoA{-}SH$$

Fig. 17.7 Reactions involved in the synthesis of triacylglycerol (neutral fat) from glycerol-3-phosphate.

The reaction is:

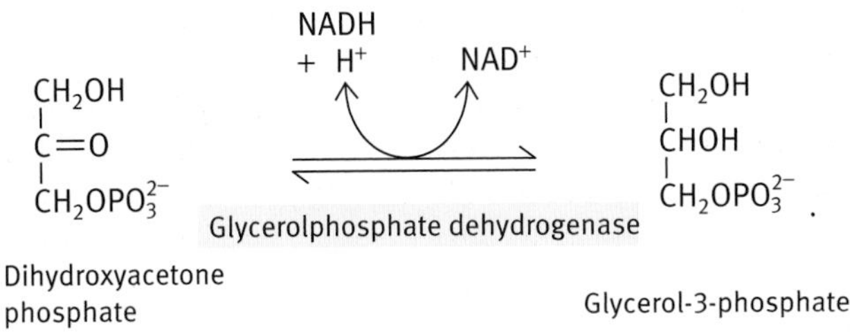

$$\underset{\text{Dihydroxyacetone phosphate}}{CH_2OH{-}C{=}O{-}CH_2OPO_3^{2-}} + NADH + H^+ \underset{\text{Glycerolphosphate dehydrogenase}}{\rightleftharpoons} \underset{\text{Glycerol-3-phosphate}}{CH_2OH{-}CHOH{-}CH_2OPO_3^{2-}} + NAD^+$$

In the fed state, the presence of insulin means that lipoprotein lipase in the adipose tissue capillaries is activated, the hydrolysed fatty acids enter the adipocyte and re-esterification and storage is made possible by the fact that GLUT4 transporters appear on the surface of the adipocyte, which allow the cell to take up glucose and convert it by the glycolytic pathway into DHAP and glycerol phosphate.

The steps in TAG synthesis are shown in Fig. 17.7

Synthesis of new membrane lipid bilayer

There are two aspects in the synthesis of glycerophospholipids, the major components of membranes. First, there are the metabolic reactions by which their synthesis is achieved, which are largely understood. However, there is a different, perhaps more interesting, problem of how these result in the formation of new membrane. Membrane synthesis poses a unique problem. There are many different types of membrane in a eukaryotic cell that have to be extended in order to allow cell growth and division. Membrane synthesis occurs by producing new phospholipids *in situ* in pre-existing membranes. In this section we will first describe the metabolic routes of synthesis that are necessary before we can then discuss the second problem of membrane synthesis.

Synthesis of glycerophospholipids

In Chapter 7 on membrane structure, you have seen that membranes contain glycerol-based phospholipids in which a polar alcohol is esterified to the phosphoryl group of phosphatidic acid.

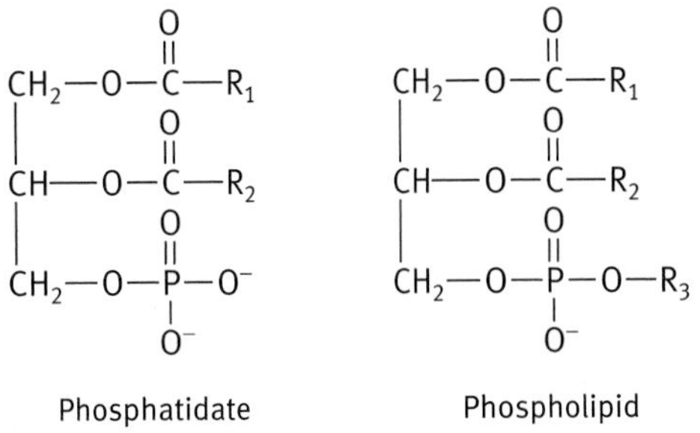

$$\underset{\text{Phosphatidate}}{CH_2{-}O{-}CO{-}R_1,\ CH{-}O{-}CO{-}R_2,\ CH_2{-}O{-}PO_2^-{-}O^-} \qquad \underset{\text{Phospholipid}}{CH_2{-}O{-}CO{-}R_1,\ CH{-}O{-}CO{-}R_2,\ CH_2{-}O{-}PO_2^-{-}O{-}R_3}$$

R_3 may be serine, ethanolamine, choline, inositol, or diacylglycerol. The most prevalent membrane lipids in eukaryotes are phosphatidylethanolamine and phosphatidylcholine so we will consider their synthesis first; ethanolamine is $NH_2CH_2CH_2OH$ and choline is:

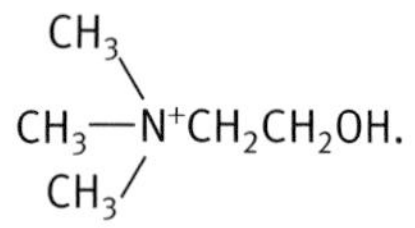

$$(CH_3)_3{-}N^{+}CH_2CH_2OH.$$

Both can be written as the alcohols 'RdOH'. The alcohols are 'activated' to participate in the synthesis; this occurs in two steps. First, a phosphorylation by ATP:

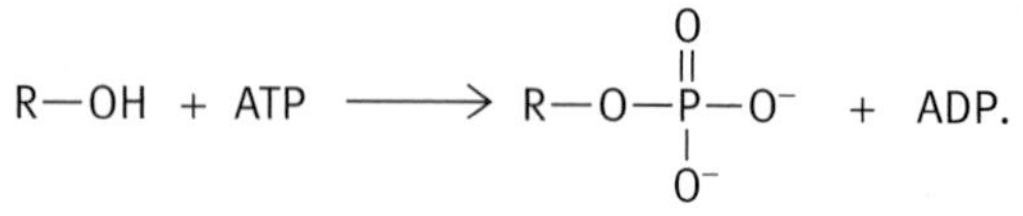

$$R{-}OH + ATP \longrightarrow R{-}O{-}\overset{O}{\overset{\|}{P}}{-}O^- + ADP.$$
$$\qquad\qquad\qquad\quad |\ O^-$$

The second step is a reaction with **cytidine triphosphate (CTP)**. CTP is the same as ATP except that it has the base cytosine (C) in place of adenine (cytosine-ribose is called cytidine). We will come to its precise structure later in the book but it is not needed here. All cells have CTP for it is needed for nucleic acid synthesis (see Chapters 22 and 23). Whether CTP involvement in membrane lipid synthesis rather than ATP is due to an accidental quirk of evolution or whether there is a good reason for it, is unknown, but the fact is that CDP compounds are always the 'high-energy' donors of groups in this area (just as for equally unknown reasons it is always UDP-glucose for processes such as glycogen synthesis in animals). The reaction is, in fact, very reminiscent of UDP-glucose formation from glucose-1-phosphate and UTP (see Chapter 11).

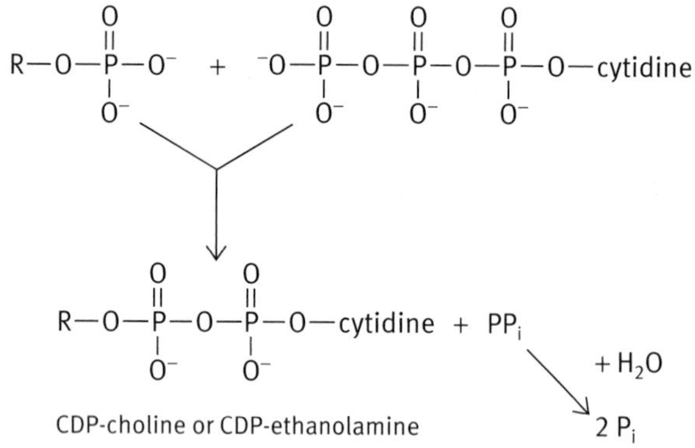

$$R{-}O{-}P(=O)(O^-){-}O^- \;+\; {}^-O{-}P(=O)(O^-){-}O{-}P(=O)(O^-){-}O{-}P(=O)(O^-){-}O{-}\text{cytidine}$$
$$\downarrow$$
$$R{-}O{-}P(=O)(O^-){-}O{-}P(=O)(O^-){-}O{-}\text{cytidine} \;+\; PP_i \xrightarrow{+H_2O} 2\,P_i$$

CDP-choline or CDP-ethanolamine

Hydrolysis of inorganic pyrophosphate (PP_i) to P_i makes the reaction strongly exergonic.

The final reaction in phosphatidylethanolamine and phosphatidylcholine synthesis is as follows:

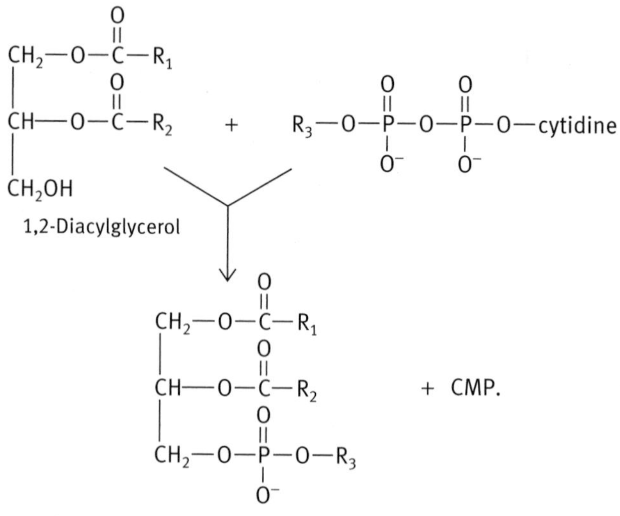

$$\begin{array}{l} CH_2{-}O{-}C(=O){-}R_1 \\ |\\ CH{-}O{-}C(=O){-}R_2 \\ |\\ CH_2OH \end{array} \;+\; R_3{-}O{-}P(=O)(O^-){-}O{-}P(=O)(O^-){-}O{-}\text{cytidine}$$

1,2-Diacylglycerol

$$\downarrow$$

$$\begin{array}{l} CH_2{-}O{-}C(=O){-}R_1 \\ |\\ CH{-}O{-}C(=O){-}R_2 \\ |\\ CH_2{-}O{-}P(=O)(O^-){-}O{-}R_3 \end{array} \;+\; CMP.$$

Phosphatidylcholine or phosphatidylethanolamine

The diacylglycerol comes from hydrolysis of phosphatidic acid by a phosphatase.

In the reactions above we join an alcohol (ethanolamine or choline) to CDP and then transfer phosphoryl alcohol to another alcohol (diacylglycerol); the head group is 'activated'. Energetically, it would be the same to join diacylglycerol to CDP and transfer a diacylglycerol phosphoryl group to ethanolamine or choline – that is, to activate the diacylglycerol instead of the polar alcohol or head group. This is what happens in phosphatidylinositol and cardiolipin (see Chapter 7) synthesis in eukaryotes. In the latter, instead of inositol, the reactant is another molecule of diacylglycerol:

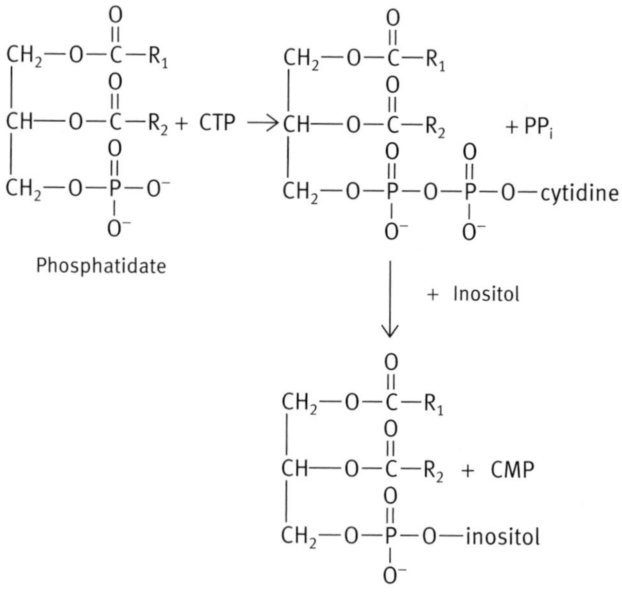

$$\begin{array}{l} CH_2{-}O{-}C(=O){-}R_1 \\ |\\ CH{-}O{-}C(=O){-}R_2 \\ |\\ CH_2{-}O{-}P(=O)(O^-){-}O^- \end{array} + CTP \longrightarrow \begin{array}{l} CH_2{-}O{-}C(=O){-}R_1 \\ |\\ CH{-}O{-}C(=O){-}R_2 \\ |\\ CH_2{-}O{-}P(=O)(O^-){-}O{-}P(=O)(O^-){-}O{-}\text{cytidine} \end{array} + PP_i$$

Phosphatidate

$$\downarrow + \text{Inositol}$$

$$\begin{array}{l} CH_2{-}O{-}C(=O){-}R_1 \\ |\\ CH{-}O{-}C(=O){-}R_2 \\ |\\ CH_2{-}O{-}P(=O)(O^-){-}O{-}\text{inositol} \end{array} + CMP$$

Phosphatidylinositol

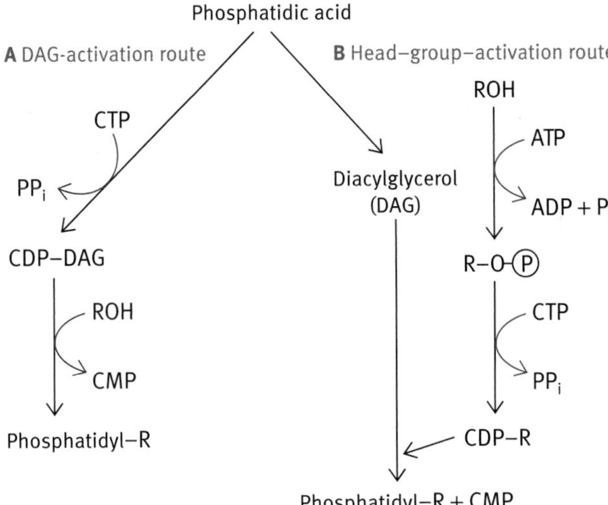

Fig. 17.8 The two pathways (A and B) for synthesis of glycerophospholipid. (ROH, the polar head group to be attached to the phospholipid.) Different cells may use different routes for synthesizing a given phospholipid. In mammals, phosphatidylcholine (PC), phosphatidylserine (PS), and phosphatidylethanolamine (PE) are synthesized by route B; the three are interconvertible by decarboxylation of PS to PE, methylation of PE to PC, and head-group exchange between PE and PS, and free ethanolamine or serine. Synthesis of cardiolipin and phosphatidylinositol in mammals follows route A. In *Escherichia coli* phospholipid synthesis occurs by route A.

The situation is summarized in the scheme in Fig. 17.8. This scheme illustrates the two ways in which glycerophospholipids can be synthesized. It should be emphasized that the field is a complex one in which extensive interconversion of the phospholipids occurs. For example, serine and ethanolamine can be exchanged, phosphatidylethanolamine can be methylated to phosphatidylcholine and phosphatidylserine can be decarboxylated to phosphatidylethanolamine. Phosphatidylcholine can be converted into any of the others by head-group exchange. The particular route of synthesis of a given phospholipid may vary in different organisms.

Sphingolipids (see Chapter 7) are produced from palmitoyl-CoA and serine by a complex set of reactions, which will not be given here.

Synthesis of new membrane lipid bilayer

In eukaryotic cells, the main site of phospholipid synthesis is the membrane of the smooth endoplasmic reticulum (ER). It also takes place in the outer mitochondrial membrane. Fatty acyl-CoAs and glycerophosphate in the cytosol are converted to phosphatidic acid by enzymes located on the cytosolic surface of the ER. The phosphatidate is converted to phosphatidyl derivatives of serine, choline, ethanolamine, and inositol, as described earlier, again by enzymes on the cytosolic surface of the ER. The newly synthesized phospholipids are inserted into the outer leaflet of the lipid bilayer of the ER membrane (possibly at the fatty acyl-CoA stage) and the synthesis completed *in situ*. As stated above, extensive interconversion of the phospholipid head groups results in the formation of the various membrane components. Cells cannot produce new membranes *de novo*. They can only extend existing membranes, which enlarge their structures. At cell division, each daughter cell receives half.

In this way, new lipid membrane is synthesized. The inner and outer layers of a bilayer are different in phospholipid composition (see Chapter 7). The newly synthesized phospholipids are in the outer leaflet of the ER lipid bilayer, and there has to be a transfer of phospholipids to the inner leaflet to maintain the balance (Fig.17.9). Simple flipping of phospholipids from the outer to the inner layer would involve the polar head group traversing the hydrophobic centre of the bilayer, which is energetically unfavourable. A family of phospholipid-transporting enzymes, or so-called flippases, are known. They are driven by ATP hydrolysis. It is not fully clear how the asymmetries in different membranes are achieved and maintained.

There is another problem too. The new membrane synthesized in the ER has to be transported to other sites, such as the plasma membrane, for cell growth and division to occur. There are two possible mechanisms for this; first, membrane vesicles are budded off the ER and migrate to their target membrane and fuse with it thus adding new membrane to the plasma membrane, for example. Phospholipid-transfer proteins are known to exist, which pick up phospholipids from one membrane and transport them to another membrane, and this is the second possible way in which newly synthesized phospholipids may be delivered to target membranes. Membrane synthesis is not completely understood.

Synthesis of prostaglandins and related compounds

The Greek word *eikosi* means 20; this is the basis for the name of a group of compounds called **eicosanoids**, all of which contain 20 carbon atoms and are related to polyunsaturated fatty acids. Although present in the body at very low concentrations, they have a wide range of physiological functions.

There are three main groups named after the cells in which they were first discovered. These are the **prostaglandins**, the **thromboxanes**, and the **leukotrienes**. Prostaglandins were first discovered in semen and so named because it was thought that they originated from the prostate gland (in fact, they originate from the seminal vesicles). However, it is now known that very many tissues produce prostaglandins. Thromboxanes were discovered in blood platelets or thrombocytes and leukotrienes in leucocytes.

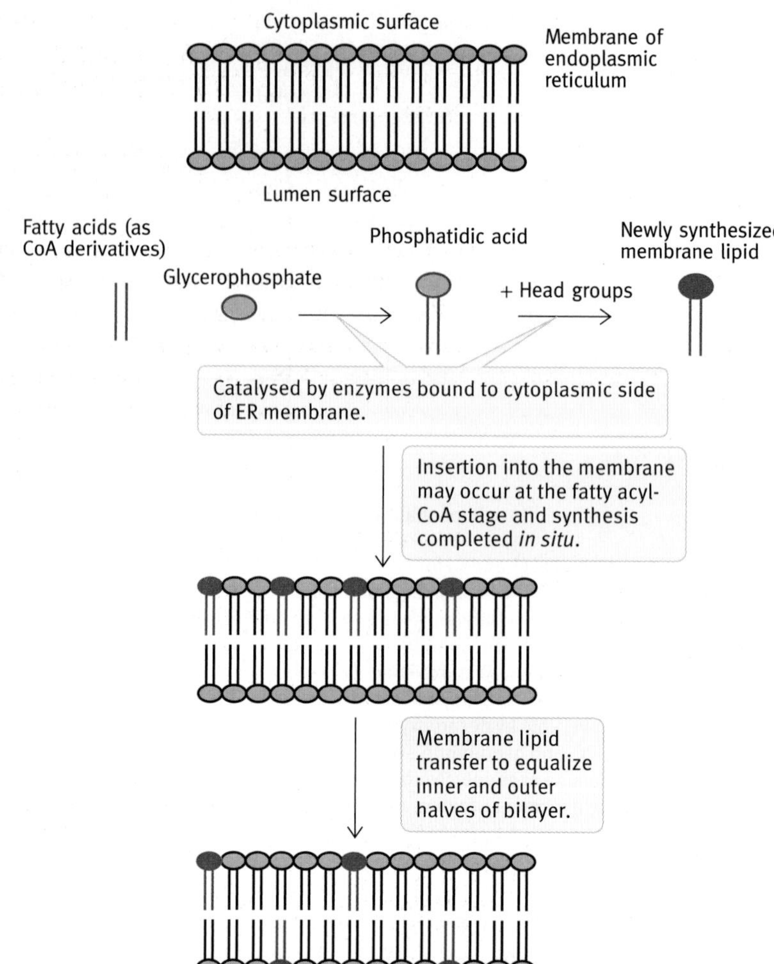

Fig. 17.9 Synthesis of new membrane lipid bilayer. Enzymes which synthesize new membrane lipids are located at the cytosolic surface of the endoplasmic reticulum (ER); the newly synthesized lipids are located in the cytosolic half of the bilayer. Transfer to the inner half to maintain bilayer equality is by a 'flippase'. The new membrane is transported from the ER to other membranes in the cell in vesicle form or *via* phospholipid transport proteins.

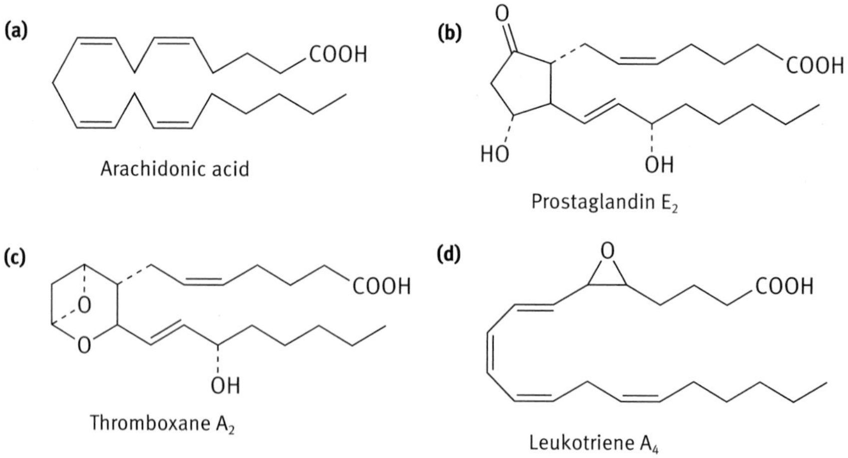

(a) COOH

Arachidonic acid

(b) O COOH HO OH

Prostaglandin E_2

(c) O O COOH OH

Thromboxane A_2

(d) O COOH

Leukotriene A_4

Fig. 17.10 Structure of prostaglandins and related compounds: **(a)** arachidonic acid; **(b)** prostaglandin E_2 (PGE_2); **(c)** thromboxane A_2 (TXA_2); **(d)** leukotriene A_4 (LTA_4). Probably only those specially interested in this area would need to memorize these structures.

BOX 17.2 **Nonsteroidal anti-inflammatory drugs (NSAIDs)**

Pain relief is an area of major medical interest and drugs to achieve pain control are of extreme importance. One of the oldest drugs and still much used is aspirin; as stated in the text, this inhibits the cyclooxygenase (COX) reaction necessary for the conversion of arachidonic acid to the family of prostaglandins.

Prostaglandins have a wide range of protective physiological effects including protection of the mucosal membrane lining of the upper gastrointestinal tract and maintenance of kidney function.

Prostaglandins also cause pain, inflammation, and fever. As well as their normal protective roles, they are formed and liberated at sites of tissue and cell damage. In the arthritic diseases where cell damage at joints result in their production, prostaglandins contribute to the joint pains from which millions of people suffer. There is great interest, therefore, in drugs that inhibit prostaglandin synthesis.

Besides aspirin, many similarly acting drugs, such as ibuprofen, have been produced as anti-inflammatory pain-killing agents. They are about the most commonly used group of drugs and are collectively known as nonsteroidal anti-inflammatory drugs or NSAIDs. These are in contrast to the steroidal drugs such as cortisone and synthetic glucocorticoids (mentioned in Chapter 29 on cell signalling).

NSAIDs can have adverse effects because, as well as suppressing the synthesis of the pain-producing prostaglandins, they also inhibit the normal production of protective prostaglandins. One of the best-known adverse effects is irritation of the lining of the upper gastrointestinal tract leading to bleeding and formation of ulcers. This is especially important in the treatment of rheumatoid arthritis, where high NSAID doses for a prolonged period may be necessary.

There was, therefore, tremendous interest in the development of a new class of NSAIDs, known as **COX-2 inhibitors**, which selectively inhibit the production of the inflammatory pain-producing prostaglandins at the site of cell damage. We will now explain the basis of this development, summarized in Fig. 17.11.

The COX that is involved in prostaglandin synthesis occurs in two isoforms (isoenzymes). **COX-1** is present in almost all cells and

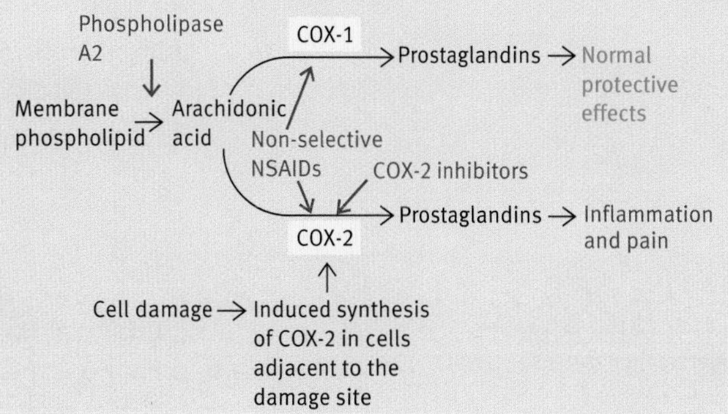

Fig. 17.11 Action of cyclooxygenase isoenzymes and effects of selective inhibition.

is needed for the normal production of prostaglandins. The other form, **COX-2** is normally present in cells in minute amounts or not at all. At the sites of cell damage, adjacent cells are induced to synthesize COX-2 resulting in increased synthesis of prostaglandins causing inflammation and pain.

The 'older' nonselective NSAIDs, such as aspirin, inhibit both COX-1 and COX-2 effectively but, as stated above, may cause the damage to the gastrointestinal mucosal membrane already referred to. COX-2 inhibitors selectively inhibit COX-2 with lesser effects on COX-1 and so while they are effective painkillers, they have reduced undesirable gastrointestinal side effects. However, recent data show that COX-2 selective drugs may cause cardiovascular side effects such as heart attacks or strokes, possibly through inhibition of the synthesis of protective prostacyclins (PGI, a prostaglandin subclass) within the lining of blood vessels. Also COX-2 inhibitors do not stop thromboxane production so the combined effect of reducing prostacyclin production without inhibiting thromboxane production means that they increase the risk of blood clot formation.

As a result, the use of some of the COX-2 inhibitors has now been severely curtailed.

The prostaglandins and thromboxanes

A number of different prostaglandins exist, varying in their detailed structures, with subclasses such as PGE, and PGF. A subscript number indicates the number of double bonds in the side chains attached to the cyclopentane ring structure. Fig. 17.10(b) shows the structure of PGE_2 as an example. In humans the compounds with two double bonds are most important and are synthesized from arachidonic acid, the structure of which is shown in Fig. 17.10(a). Other prostaglandins are derived from related unsaturated fatty acids. Arachidonic acid is present in cells, mainly as the fatty acid component of a phospholipid. The phospholipase, which specifically recognizes the sn-2 acyl bond of phospholipids and catalyses the hydrolysis of the bond releasing arachidonic acid and lysophospholipids is known as phospholipase A2.

The first step in the synthesis of prostaglandins is catalysed by a **cyclo-oxygenase**, which forms the ring structure from arachidonic acid (Fig. 17.10(a)). This enzyme is inhibited by **aspirin** (acetylsalicylic acid), which covalently modifies the enzyme by acetylating an essential serine residue near the active site of the enzyme.

The prostaglandins have a wide variety of physiological effects. They are released immediately after synthesis and act as local hormones on adjacent cells by binding to receptors. They have a number of effects: they cause pain, inflammation, and fever; they cause smooth muscle contraction and are involved in labour; they are involved in blood pressure control; they suppress acid secretion in the stomach. Aspirin, by its inhibition of cyclooxygenase (Box 17.2), suppresses many of these effects.

Thromboxanes (Fig. 17.10(c)) affect platelet aggregation and thus blood clotting. They are formed by further conversion of some prostaglandins, so that aspirin inhibits their formation also; in particular, a low dose (75–100 mg per day) inhibits platelet thromboxane formation resulting in decreased

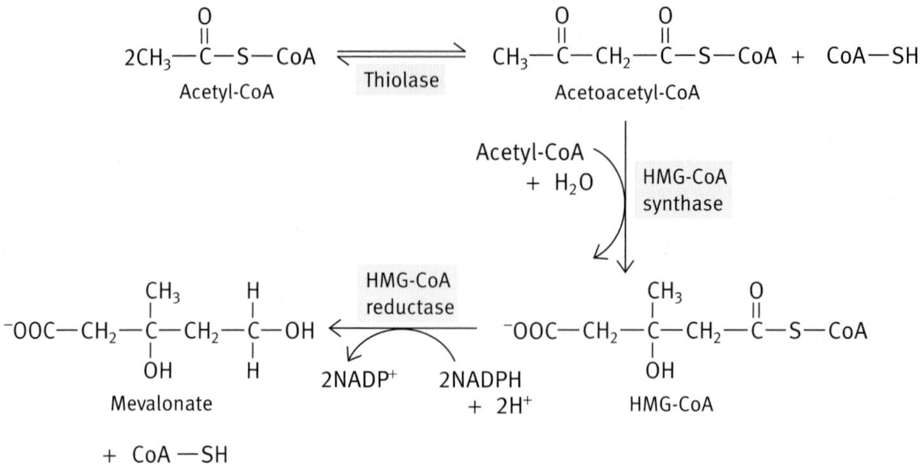

$$2CH_3{-}\overset{O}{\overset{\|}{C}}{-}S{-}CoA \;\underset{\text{Thiolase}}{\rightleftharpoons}\; CH_3{-}\overset{O}{\overset{\|}{C}}{-}CH_2{-}\overset{O}{\overset{\|}{C}}{-}S{-}CoA + CoA{-}SH$$

Acetyl-CoA → Acetoacetyl-CoA

$$\text{Acetoacetyl-CoA} + \text{Acetyl-CoA} + H_2O \xrightarrow{\text{HMG-CoA synthase}} {}^{-}OOC{-}CH_2{-}\overset{CH_3}{\underset{OH}{C}}{-}CH_2{-}\overset{O}{\overset{\|}{C}}{-}S{-}CoA \;(\text{HMG-CoA})$$

$$\text{HMG-CoA} + 2NADPH + 2H^+ \xrightarrow{\text{HMG-CoA reductase}} {}^{-}OOC{-}CH_2{-}\overset{CH_3}{\underset{OH}{C}}{-}CH_2{-}\overset{H}{\underset{H}{C}}{-}OH \;(\text{Mevalonate}) + 2NADP^+ + CoA{-}SH$$

Fig. 17.12 The synthesis of mevalonate from acetyl-CoA. HMG-CoA, 3-hydroxy-3-methylglutaryl-CoA.

platelet aggregation. This therapy has been found to reduce the risk of heart attacks by reducing the danger of blood clot formation blocking the coronary arteries.

Leukotrienes

Leukotrienes are naturally produced eicosanoid lipid mediators, first found in leukocytes, hence the name. They are produced in the body from arachidonic acid by the enzyme 5-lipoxygenase. The structure of one of the leukotrienes is given in Fig. 17.10(d).

One of their effects is to trigger contractions in the smooth muscles lining the trachea. Overproduction of leukotrienes is a major cause of inflammation in asthma and allergic rhinitis. These diseases can be treated either by the use of leukotriene antagonists, which inhibit the synthesis (lipoxygenase inhibitors) or the activity of leukotrienes (blocking their effects on target cells).

Synthesis of cholesterol

Cholesterol is an essential component of cell membranes and the precursor of bile salts and steroid hormones. The liver and, to a lesser extent, the intestine are the most active in its synthesis and from the liver a two-way flux 'equilibrates' the cholesterol of the liver and the rest of the body, as described in Chapter 11.

The starting material for the synthesis of this large molecule is acetyl-CoA. The first few stages of the synthesis are of special interest for this is where control of cholesterol occurs and we will confine this account to the production of mevalonic acid, the first metabolite committed solely to cholesterol synthesis. This involves the reactions shown in Fig. 17.12, which occur in the cytosol.

HMG-CoA (3-hydroxy-3-methylglutaryl-CoA) is also the precursor of acetoacetate, one of the ketone bodies, but ketone body synthesis takes place in the mitochondrion.

The regulation of cholesterol concentrations in cells and blood, and the 'statin' drugs have already been described (see Chapter 11).

Conversion of cholesterol into steroid hormones

The steroid hormones are important signalling molecules. Sex hormones such as testosterone and oestradiol are produced in the gonads and they are involved in determining secondary sex characteristics.

The adrenal cortex produces two main classes of steroid:

- **glucocorticoids**, which have several effects, among which are control of carbohydrate metabolism including gluconeogenesis
- **mineralocorticoids**, which are responsible for maintaining ion balance in the body.

The mode of action of hormones is a major topic dealt with in Chapter 29. All steroid hormones are derived from cholesterol. The outline of production of some of the steroid hormones is shown in Fig. 17.13. The conversion of cholesterol into steroids involves cleavage of the side chain, a reaction in which cytochrome P450 participates, producing pregnenolone, which is the common precursor of all steroids. The structures of testosterone and progesterone are shown in Fig. 29.3.

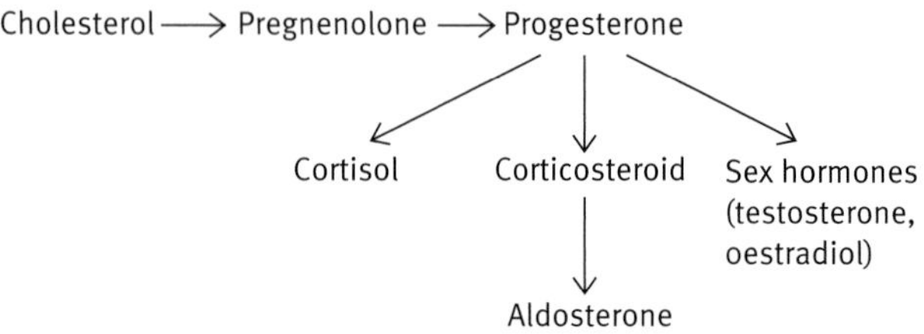

Fig. 17.13 Outline of cholesterol conversion into steroid hormones. Note that individual conversions involve more than single steps.

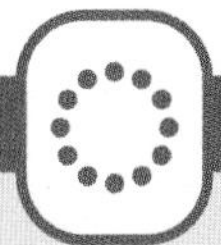

Summary

Fatty acid synthesis occurs in the cytosol using acetyl-CoA as the starting substrate. A multienzyme complex called fatty acid synthase (or palmitate synthase) produces palmitate from acetyl-CoA. The reductant in the process is NADPH (not NADH). Further lengthening of the palmitate is catalysed by a separate enzyme system on the cytosolic face of the ER. Desaturation at selected carbon-carbon bonds also takes place.

The acetyl-CoA from which fatty acids are synthesized is produced from pyruvate inside the mitochondrial matrix. It is not directly transported out to the cytosolic site of synthesis but in the matrix it is converted into citrate by citrate synthase (the first step in the TCA cycle). For fatty acid synthesis, citrate is transported out into the cytosol. There, an ATP-citrate lyase releases acetyl-CoA and oxaloacetate. This reaction depends on the hydrolysis of ATP making it irreversible. NADH is used to reduce the oxaloacetate to malate. Malate is converted into pyruvate by the malic enzyme, which reduces $NADP^+$ to NADPH in the process. NADPH is involved in reductive synthesis in the cell whereas NADH is involved in oxidative processes. Thus the system supplies both acetyl-CoA and the reductant.

Fatty acids are synthesized by the fatty acid synthase cycle. The donor of C2 units in the process is not acetyl-CoA but the C3 molecule malonyl-CoA. This is formed by carboxylation of acetyl-CoA. The CO_2 is released in the donor reaction; its role is to render fatty acid synthesis irreversible. TAGs are synthesized from glycerophosphate and fatty acyl-CoAs.

Membrane lipids are synthesized on the membranes of the ER. Unsaturated fatty acids are produced by separate enzyme systems and used to synthesize prostaglandins and related compounds. Prostaglandins are involved in pain, inflammation and fever. They cause contraction of smooth muscle and have other physiological actions. Prostaglandins and thromboxanes are synthesized from arachidonic acid, with the first step being the formation of a ring structure by cyclooxygenase. Aspirin (acetylsalicylic acid) is a potent inhibitor of cyclooxygenase. It does this by acetylating an essential serine in the enzyme. Thromboxanes are involved in platelet aggregation and blood clotting. Aspirin has a protective action in reducing blood clotting and the blocking of coronary arteries. Leukotrienes involved in smooth muscle contraction in airways and in white cell function are made from arachidonic acid by the action of a lipoxygenase.

Cholesterol is also synthesized from acetyl-CoA; the first committed step is the production of mevalonate by the enzyme HMG-CoA reductase. This is a crucial control point and is the target for the cholesterol-lowering statin drugs. Mevalonate is converted into cholesterol by a long pathway (not described here). Cholesterol is the precursor of steroid hormones.

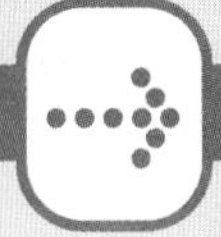

Further reading

To access the further reading, please scan the QR code image or go to http://global.oup.com/uk/orc/biosciences/molbiol/snape_biochemistry/student/reading/ch17/

Problems

1 An early step in fatty acid synthesis is that acetyl-CoA is carboxylated, to be followed immediately by decarboxylation of the product. What is achieved by this carboxylation/decarboxylation sequence?

2 By means of a diagram, outline the steps in a cycle of elongation during fatty acid synthesis, omitting the reductive steps.

3 Discuss briefly the physical organization of the fatty acyl synthase complex in eukaryotes. How does this differ from that in *Escherichia coli*? What is the advantage of the eukaryotic situation?

4 Write down the outline structures of NAD^+ and $NADP^+$. What are the different functions of NAD^+ and $NADP^+$?

5 What are the main sites of fatty acid synthesis in animals?

6 Palmitate synthesis from acetyl-CoA occurs in the cytosol. However, pyruvate dehydrogenase, the main producer of acetyl-CoA used in fat synthesis, occurs inside the mitochondrion. Acetyl-CoA cannot traverse the mitochondrial membrane. How does the fatty acid-synthesizing system in the cytosol obtain its supply of acetyl-CoA?

7 The synthesis of fatty acid involves reductive steps requiring NADPH. What are the sources of NADPH?

8 Describe the synthesis of triacylglycerol (TAG) from fatty acids.

9 What is the role of CTP (cytidine triphosphate) in lipid metabolism?

10 (a) What are eicosanoids?
(b) What are they made from?
(c) Briefly describe their physiological significance.
(d) What is the relevance of aspirin in this area of metabolism?

11 Describe a type of drug that is used to reduce the rate of cholesterol synthesis. What is the rationale for its action?

Chapter 18

Nitrogen metabolism: amino acid metabolism

Digestion of proteins in a normal diet results in relatively large amounts of the 20 different amino acids being absorbed from the intestine into the portal blood, which goes directly to the liver. All cells, except non-nucleated ones, such as mature erythrocytes, use amino acids for protein synthesis and for the synthesis of a variety of essential molecules including membrane components, neurotransmitters, haem, creatine, carnitine, and nucleotides. All cells take up amino acids by selective transport mechanisms since, in their free, ionized form, they do not readily penetrate the membrane lipid bilayer.

As mentioned in Chapter 11, there is no dedicated store of amino acids in the sense that there is no polymeric form of amino acids whose function is simply to be a reserve of these compounds to be called upon when needed. The reserves are in the form of the 'amino acid pool', which includes free amino acids in the cells and in the blood and the rest of the reserves are in the form of functional proteins, the greatest quantity being found in muscles. Muscle proteins are part of the contractile machinery and if broken down in appreciable quantities, for instance to provide amino acids for gluconeogenesis in the liver, muscle wasting ensues.

Humans cannot synthesize 10 of the amino acids. Interestingly enough, the ones that we cannot synthesize are those with many steps in their synthesis and therefore require many enzymes and many genes to code for them – they are, as it were, 'expensive' to manufacture. The amino acids that can be synthesized are those whose carbon skeleton, that is, the relevant oxo-acid can be synthesized. Where that is possible, the amino acid can be produced by the addition of an amino group to the oxo-acid.

When the human diet was from the 'wild', if sufficient food could be obtained to sustain life, energy-wise, it would probably contain all the necessary amino acids. In this situation, loss of the ability to synthesize certain amino acids would not be a disadvantage as one would probably have been more likely to die from insufficient energy sources than from lack of essential amino acids in the food that was available.

With the development of agriculture, however, the vast production of chemical energy, in the form of carbohydrate in cereals, permitted large populations to survive but did not necessarily supply adequate amounts of the essential amino acids, since some plant proteins, such as those in wheat and maize, have low levels of lysine and tryptophan. When humans have a rich source of protein, containing enough essential amino acids to provide for an extended period, there is no method for storing large quantities of them. After immediate needs are satisfied, the surplus amino acids are oxidized or converted into glycogen or fat and the nitrogen excreted as urea.

A condition first described in Ghana in 1932, known as kwashiorkor was thought to be the result of a diet with adequate energy but inadequate protein, often seen in children in rural Africa in times of famine. It is now referred to as oedematous malnutrition and is known to be more complex that a simple protein deficiency, involving low protein intake, increased protein requirement, infection and inadequate antioxidant defences. Kwashiorkor is one end of a spectrum of diseases known as protein energy malnutrition, the other end of the spectrum being marasmus, characterized by emaciation. Marasmus is thought to be the result of inadequate diet both in terms of energy and protein equivalent to adult starvation. Both conditions lead to a compromised immune system, wasting, apathy, inadequate growth and deleterious brain effects. Kwashiorkor is additionally characterized by a low concentration of serum proteins which, by reducing the osmotic pressure of the blood, is a cause of oedema of the tissues, giving the child a plump appearance. Both conditions create a vicious circle since the cells lining the intestine are not functioning properly and as a result when food is available, it is inadequately digested and absorbed. The disease affects developing children more than adults because of their greater demand for protein and energy to support growth.

Nitrogen balance in the body

The nutritive aspects of amino acids can be treated on a generalized level through the concept of **nitrogen balance**. If the total intake of nitrogen (mainly as amino acids) equals the total excretion, the individual is in a state of nitrogen balance. During pregnancy, growth or repair, more is taken in than is excreted – this is positive nitrogen balance. Negative nitrogen balance occurs in fasting and starvation where excretion exceeds intake, when, for example, amino acids of muscle proteins are converted into glucose and the nitrogen excreted. It also occurs in patients with chronic infections or cancer, as glucocorticoids and other stress hormones stimulate protein degradation.

The proteins of animals are continually 'turning over'; they are continuously broken down and resynthesized. Although many of the derived amino acids are recycled back into proteins, something of the order of about 0.3% of total body protein nitrogen per day is converted into urea and excreted. To a much lesser extent, nitrogen excretion also occurs in mammals as ammonia, creatinine, and uric acid.

An essential amino acid is one that, when omitted from an otherwise complete diet, results in negative nitrogen balance or fails to support the growth of experimental animals. Some amino acids are essential without qualification, such as lysine, phenylalanine, and tryptophan (see Table 18.1 for a list of essential and nonessential amino acids). The essential amino acids include a subclass known as conditionally essential. For example, tyrosine is not essential, provided sufficient phenylalanine is available, since phenylalanine is convertible into tyrosine; similarly, cysteine synthesis in mammals requires the availability of methionine, another essential amino acid. The nutritive picture, in this respect, is not completely neat and tidy. 'First-class' proteins are rich in all of the essential amino acids. Plant proteins may be poor in one or two essential amino acids. Since the amino acid compositions of proteins vary, a mixture of plant proteins is needed to ensure adequate amino acid nutrition in vegetarian diets. This is known as the 'complementary value' of proteins and allows populations to survive on a mixture of 'low quality' proteins from plants when they cannot or will not eat animal tissues or products. A time-honoured combination of cereals (short in lysine) and pulses (short in methionine), for example, supplies adequate protein to billions of people worldwide.

Nonessential	Essential
Alanine	Arginine‡
Asparagine	Histidine
Aspartic acid	Isoleucine
Cysteine*	Leucine
Glutamic acid	Lysine
Glutamine	Methionine
Glycine	Phenylalanine
Proline	Threonine
Serine	Tryptophan
Tyrosine†	Valine

Table 18.1 Classification of dietary amino acids in humans.

* Cysteine is produced only from the essential amino acid methionine.

† Tyrosine is produced only from the essential amino acid phenylalanine.

‡ Arginine is required only in the growing stages.

General metabolism of amino acids

The general situation of amino acid metabolism is shown in Fig. 18.1 essentially linking the broad aspects of catabolism in which the integration of amino acid, carbohydrate, and fat metabolism is shown. A physiologically vital aspect is that in starvation, survival depends on the release of amino acids from muscle protein causing muscle wasting. The amino acids travel to the liver and are used for glucose synthesis (Chapter 16).

Aspects of amino acid metabolism

In this chapter we will consider and attempt to answer the following questions:

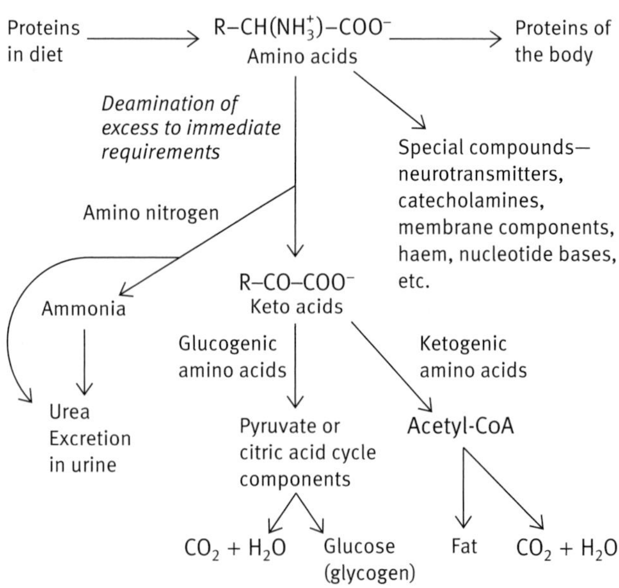

Fig. 18.1 The overall catabolism of amino acids. Note that some amino acids are partly ketogenic and partly glucogenic (see later in this chapter for explanation of these terms).

- How are the amino acids metabolized? How are the amino groups removed? In other words, how does **deamination** occur? Although there are 20 different amino acids most of them are deaminated by a common mechanism.
- How is the amino-group nitrogen, which is removed from amino acids, converted into urea and excreted?
- What happens to the oxo-acids (keto-acids), the 'carbon skeletons', of the amino acids after deamination? In this case, each amino acid has its own special metabolic route, and we will give only a limited treatment of this aspect, dealing with features of special biochemical or medical relevance.
- How are amino acids synthesized? In the animal body this concerns only the nonessential amino acids. If the carbon skeleton (the oxo-acid) is available, synthesis is often the reverse of deamination. We will deal with only a few examples of these amino acids. In bacteria and plants, all amino acids are synthesized by pathways using general metabolic intermediates. Animals depend on this synthetic activity of plants to supply their essential amino acids. The pathways for the synthesis (and breakdown) of all amino acids have been elucidated, and constitute a formidable amount of detailed information. We will deal only with particular cases of amino acid biosynthesis, which are of special biochemical or medical interest.

Finally we will cover aspects of amino acid metabolism involving methyl group transfer, haem synthesis, and transport of ammonia in the circulation.

There are also major topics of biochemistry, such as protein synthesis, and nucleotide synthesis, that have amino acids as their starting points. These are dealt with in the relevant later chapters.

Glutamate dehydrogenase has a central role in the deamination of amino acids

What has dehydrogenation to do with deamination? To explain this we will briefly digress to a piece of simple chemistry. It concerns **Schiff bases**. A Schiff base results from a spontaneous (noncatalysed) equilibrium between a molecule with a carbonyl group (aldehyde or ketone) and one with a free amino group.

$$>C{=}O + H_2N{-}R \rightleftharpoons >C{=}N{-}R + H_2O$$

The reaction is freely reversible so a Schiff base can readily hydrolyse back to a molecule with a carbonyl group.

Let us now look at the involvement of Schiff bases in biological systems and particularly in amino acid metabolism.

In the equation below, going from left to right, we start with an amino acid. If two hydrogen atoms are removed, we have a Schiff base, which can then be hydrolysed to give an oxo-acid and NH_3. It shows how the removal of a pair of hydrogen atoms from an amino acid is a step in **deamination**. The reverse reaction shows how the reactions between an oxo-acid, ammonia and two hydrogen atoms can result in the synthesis of an amino acid.

$$>CHNH_2 \xrightarrow{-2H} >C{=}N{-}H \xrightarrow{H_2O} >C{=}O + NH_3$$

Glutamic acid (glutamate at physiological pH) is of central importance in amino acid metabolism. It is deaminated by **glutamate dehydrogenase**, unusually working with either NAD^+ or $NADP^+$. (The NH_3 is protonated to NH_4^+ at physiological pH.)

$$\underset{\text{Glutamate}}{COO^-{-}CH_2{-}CH_2{-}CHNH_3^+{-}COO^-} + NAD(P)^+ + H_2O \longrightarrow NAD(P)H + H^+ + \underset{\text{2-oxoglutarate}}{COO^-{-}CH_2{-}CH_2{-}C{=}O{-}COO^-} + NH_4^+$$

As described below, glutamate dehydrogenase plays a major role in the deamination, and therefore in the oxidation, of many amino acids. The enzyme is allosterically inhibited by ATP and GTP (indicators of a high-energy charge) and activated by ADP and GDP, which signal that an increased rate of oxidative phosphorylation is needed.

The 2-oxoglutarate produced can feed into the TCA cycle. Since 2-oxoglutarate is converted into oxaloacetate in the cycle (see Fig. 13.10), glutamate can be converted into glucose under appropriate physiological situations. It is a **glucogenic** amino acid. However, there are no corresponding dehydrogenases for the other amino acids. How are they deaminated? A few have special individual mechanisms but many of them have their amino groups transferred enzymically to 2-oxoglutarate, forming glutamate and the corresponding TCA cycle intermediate. The glutamate is then deaminated by the glutamate dehydrogenase reaction given above. The amino acids are thus deaminated by a two-step process.

The first reaction is called **transamination**:

$$R{-}CHNH_3^+{-}COO^- + COO^-{-}CH_2{-}CH_2{-}C{=}O{-}COO^- \longrightarrow R{-}C{=}O{-}COO^- + COO^-{-}CH_2{-}CH_2{-}CHNH_3^+{-}COO^-$$

Say that in the above reaction, R=CH_3, the amino acid, is alanine. Deamination of alanine proceeds as follows:

1. Alanine + 2-oxoglutarate → pyruvate + glutamate
2. Glutamate + NAD^+ + H_2O → 2-oxoglutarate + NADH + NH_4^+

Net reaction: Alanine + NAD^+ + H_2O → pyruvate + NADH + NH_4^+

The two-step process involving transamination and then deamination of glutamate is called **transdeamination** for obvious reasons. Enzymes of the type involved in reaction 1 above are called **transaminases** or **aminotransferases**. A number of these enzymes exist with individual substrate specificities and most amino acids can be deaminated by this route. The one shown is called **alanine transaminase**. Liver damage can cause its concentration in the blood to rise and this may be used as a clinical diagnostic tool.

The reversibility of transamination means that, provided an oxo-acid is available, the corresponding amino acid can be synthesized. When an oxo-acid is not synthesized in the body, as is the case for the oxo-acid equivalent of the essential amino acids, then the amino acid cannot be synthesized and must be taken in the diet. An example of the importance of transamination in the synthesis of nonessential amino acids is in the malate–aspartate shuttle (see Fig. 13.28), in which the following reversible reaction occurs.

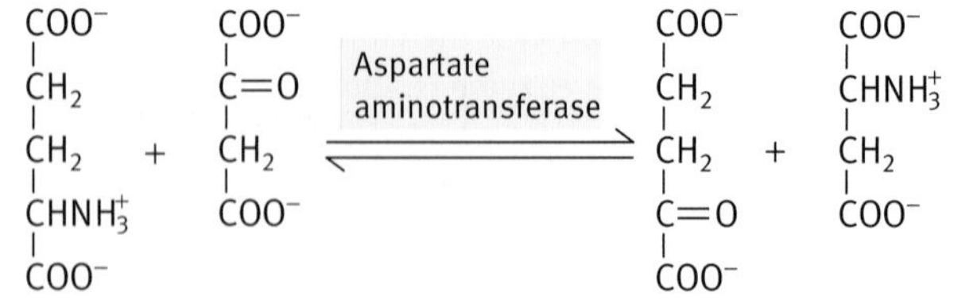

$$\underset{\text{Glutamate}}{COO^--CH_2-CH_2-CHNH_3^+-COO^-} + \underset{\text{Oxaloacetate}}{COO^--C{=}O-CH_2-COO^-} \xrightleftharpoons{\text{Aspartate aminotransferase}} \underset{\alpha\text{-Oxoglutarate}}{COO^--CH_2-CH_2-C{=}O-COO^-} + \underset{\text{Aspartate}}{COO^--CHNH_3^+-CH_2-COO^-}$$

Mechanism of transamination reactions

All transaminases have, tightly bound to the active centre of the enzyme, a cofactor, pyridoxal-5′-phosphate (**PLP**), that participates in the transaminase reaction.

PLP is a remarkably versatile cofactor; it participates as an electrophilic agent in a wide variety of reactions involving amino acids. In simple terms, PLP acts as an intermediary, accepting the amino group from the donor amino acid and then handing it on to the oxo-acid acceptor, both phases occurring on the same enzyme. As so often is the case, the cofactor is a B vitamin derivative. Vitamin B_6 in the diet consists of three interrelated compounds (vitamers), pyridoxine, pyridoxal, and pyridoxamine (Fig. 18.2). They can all be converted into PLP (Fig. 18.2) in the cell. The functional end of the molecule is the —CHO group.

The general reaction catalysed in transamination occurs in two steps:

$$\text{ENZ—PLP} + \text{amino acid 1} \rightleftharpoons \text{ENZ—PLP—}NH_2 + \text{oxo-acid 1}$$

$$\text{ENZ—PLP—}NH_2 + \text{oxo-acid 2} \rightleftharpoons \text{ENZ—PLP} + \text{amino acid 2}$$

where PLP represents pyridoxal phosphate and PLP—NH_2, pyridoxamine phosphate. The mechanism of the reactions is as shown in Fig. 18.3. Both parts of the reaction occur at the active site of the transaminase, the pyridoxamine phosphate remaining attached to the enzyme.

Special deamination mechanisms for serine and cysteine

In addition to the general transdeamination reactions, certain amino acids have their own particular way of losing their amino groups.

Serine is a hydroxy amino acid; cysteine is the corresponding thiol amino acid. Serine can be deaminated by a PLP-requiring dehydratase enzyme (Fig. 18.4). Cysteine can be deaminated by a somewhat analogous reaction in which H_2S

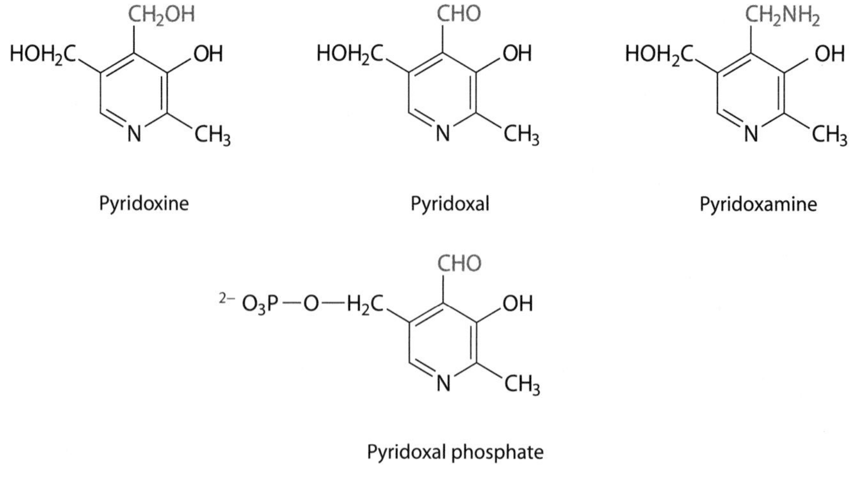

CH_2OH, HOH_2C, OH, N, CH_3 — Pyridoxine

CHO, HOH_2C, OH, N, CH_3 — Pyridoxal

CH_2NH_2, HOH_2C, OH, N, CH_3 — Pyridoxamine

CHO, $^{2-}O_3P—O—H_2C$, OH, N, CH_3 — Pyridoxal phosphate

Fig. 18.2 Structures of vitamin B_6 components and of the transaminase cofactor, pyridoxal phosphate.

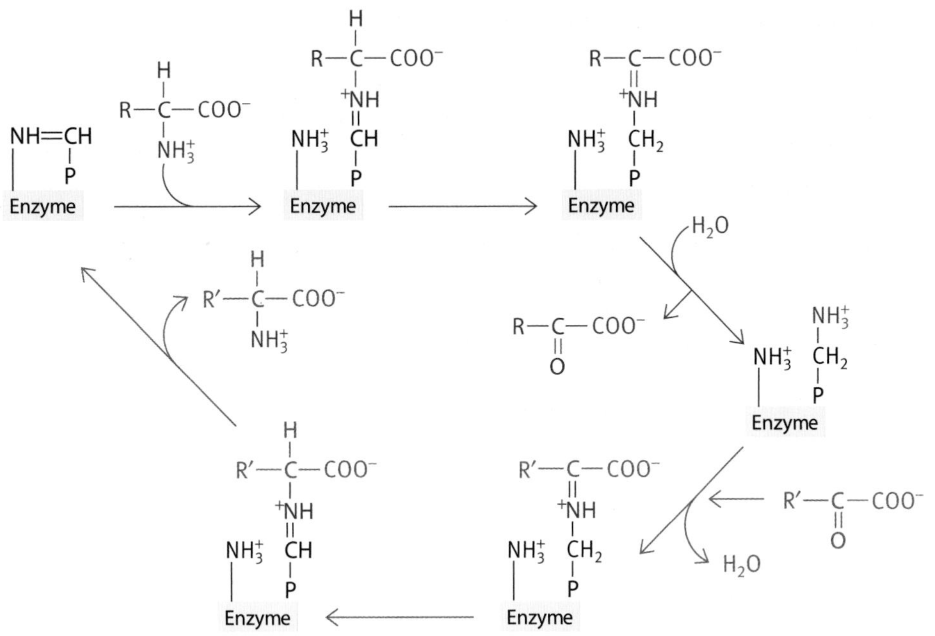

Fig. 18.3 Simplified diagram of the mechanism of transamination. P—CH=NH— in the first structure represents pyridoxal phosphate complexed with a lysine-amino group of the protein. P—CH_2—NH_3^+ represents pyridoxamine phosphate. The forward reaction (red arrows) results in the conversion of an amino acid to an oxo-acid and of pyridoxal phosphate to pyridoxamine phosphate. The reverse reaction (blue arrows) reacting with a different oxo-acid results in transamination between the (red) amino acid and the (blue) oxo-acid.

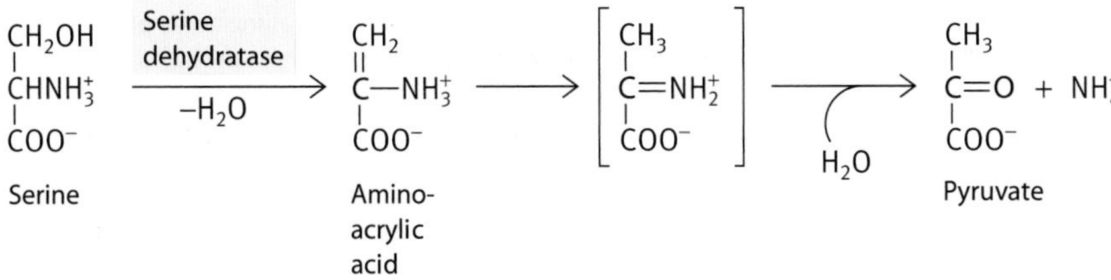

Fig. 18.4 Conversion of serine to pyruvate.

is removed instead of H_2O, but this occurs only in bacteria. In animals, two more complex routes are used involving direct oxidation of the sulphur in one, or transamination and then desulphuration in the other (not shown). The product in both cases is pyruvate.

What happens to the amino group after deamination? The urea cycle

In mammals, the amino groups of catabolized amino acids are excreted mainly as urea, a highly water-soluble, inert, non-toxic molecule. Production of urea prevents accumulation of toxic ammonia produced from amino acids. Urea is produced only in the liver from the guanidino group of arginine by the hydrolytic enzyme **arginase**. The other product is ornithine, an amino acid not found in proteins.

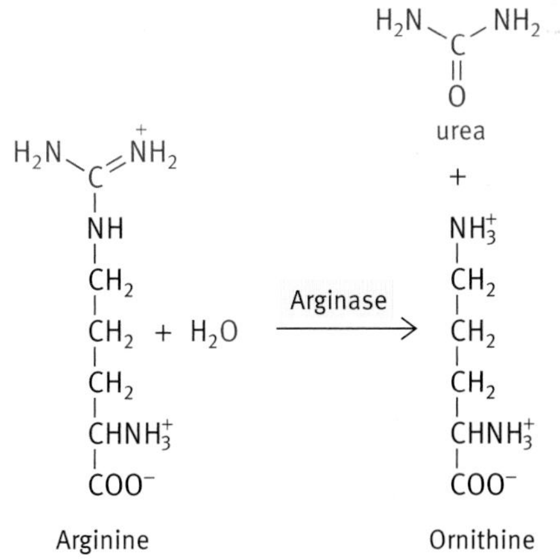

The amino nitrogen of the catabolized 20 amino acids is used to convert ornithine back to arginine. The extra carbon comes from CO_2 (as HCO_3^-). Krebs (who also discovered the TCA cycle) observed together with Henseleit that, when arginine was added to liver cells, the increased urea formation caused by this addition far exceeded the amount arginine added – that is, it was acting catalytically and ornithine did the same, suggesting a cyclical process. It was discovered that citrulline, an amino acid intermediate between ornithine and arginine, is involved, since it also acts catalytically in urea synthesis when added to liver cells. This discovery led to the famous **urea cycle**, the first biochemical cycle to be described. Citrulline, like ornithine, is not an amino acid found in proteins.

$$H_2N-C(=O)-NH-CH_2-CH_2-CH_2-CHNH_3^+-COO^-$$

Citrulline

An outline of the cycle is given in Fig. 18.5. The whole cycle is given in Fig. 18.6.

Mechanism of arginine synthesis

We need first to see how ornithine is converted into arginine. Ammonia, CO_2, and ornithine are the reactants for the first step, that of citrulline synthesis, which occurs in the mitochondrial matrix. Energy is supplied in the form of ATP. Firstly, ammonia and CO_2 are converted into a reactive intermediate, **carbamoyl phosphate**, which then combines with ornithine to give citrulline. Carbamic acid has the structure NH_2COOH and carbamoyl phosphate is therefore $NH_2-\overset{O}{\overset{\|}{C}}-PO_3^{2-}$. It is a high-energy phosphoryl compound, being an acid anhydride. It is synthesized by an enzyme **carbamoyl phosphate synthetase**, catalysing the following reaction:

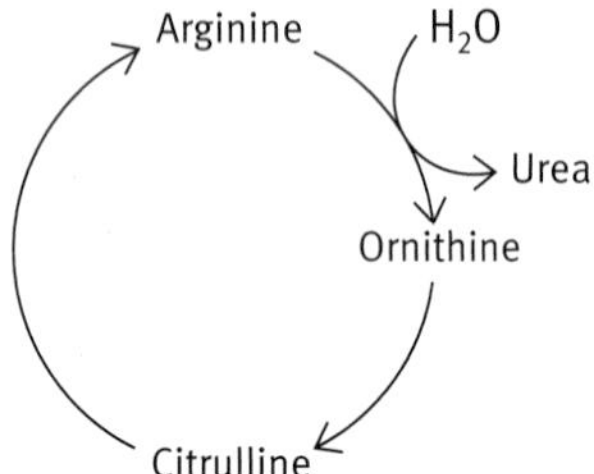

Fig. 18.5 Outline of the arginine–urea cycle. The input of CO_2 and nitrogen into the cycle is dealt with in the section on mechanism of arginine synthesis.

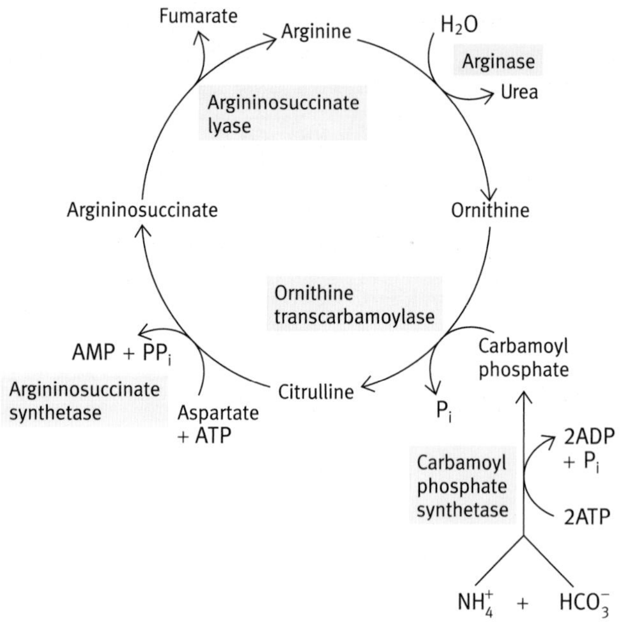

Fig. 18.6 The enzymes of the urea cycle. The levels of urea cycle enzymes are coordinated with the dietary intake of protein. The cycle is allosterically controlled at the carbamoyl phosphate synthetase step. The positive allosteric effector of the enzyme is *N*-acetyl glutamate. The conversion of ornithine to citrulline takes place inside mitochondria while the rest of the cycle occurs in the cytosol.

$$NH_4^+ + HCO_3^- + 2ATP \longrightarrow NH_2-\overset{O}{\overset{\|}{C}}-O-\underset{O^-}{\underset{|}{\overset{O}{\overset{\|}{P}}}}-O^- + 2ADP + P_i + 2H^+$$

Two ATP molecules are used, the first ATP being broken down to ADP and P_i to drive the production of carbamate from ammonia and CO_2 and the second used to phosphorylate the carbamate. It all happens on the surface of the one enzyme. The carbamoyl group of carbamoyl phosphate is now transferred to ornithine by **ornithine transcarbamoylase** giving citrulline. If we represent ornithine as $R-NH_3^+$ the reaction is:

$$\underset{\text{Ornithine}}{R-NH_3^+} + \underset{\text{Carbamoyl phosphate}}{NH_2-\overset{O}{\overset{\|}{C}}-O-\underset{O^-}{\underset{|}{\overset{O}{\overset{\|}{P}}}}-O^-} \longrightarrow \underset{\text{Citrulline}}{\underset{R}{\underset{|}{HN}}-\overset{O}{\overset{\|}{C}}-NH_2} + P_i$$

Conversion of citrulline to arginine

The final step in arginine synthesis is to convert the C=O group of citrulline to the C=NH of arginine. This occurs in the cytosol. Ammonia is *not* used here, but instead the amino group of aspartate is added directly. This is an important point when considering the origin of the two amino groups in a molecule of urea. The first comes from ammonia, that is, deamination of amino acids is necessary. The second is provided directly by the amino acid aspartate without the need for deamination. We will see later how the two arrive at the liver. Ammonia can be used to form glutamate from 2-oxoglutarate and this can generate aspartate by transamination with oxaloacetate. In this way, ammonia, and also the amino groups of most amino acids, can be converted to urea *via* this second stage of the cycle.

First, the enzyme **argininosuccinate synthetase** condenses aspartate with citrulline as follows:

$$\underset{\text{Citrulline}}{R{-}NH{-}C({=}O){-}NH_2} + \underset{\text{Aspartate}}{H_3\overset{+}{N}{-}CH(COO^-){-}CH_2{-}COO^-} \xrightarrow[\text{AMP} + PP_i]{\text{ATP}} \underset{\text{Argininosuccinate}}{R{-}NH{-}C({=}\overset{+}{N}H_2){-}NH{-}CH(COO^-){-}CH_2{-}COO^-}$$

The molecule formed is **argininosuccinate**. The name derives from the fact that the molecule structurally resembles an arginine derivative of succinate. The molecule is now 'pulled apart' by **argininosuccinate lyase** to yield arginine and fumarate (not arginine and succinate).

$$\underset{\text{Argininosuccinate}}{^-OOC{-}CHNH_3^+{-}(CH_2)_3{-}NH{-}C({=}\overset{+}{N}H_2){-}NH{-}CH(COO^-){-}CH_2{-}COO^-} \longrightarrow \underset{\text{Arginine}}{^-OOC{-}CHNH_3^+{-}(CH_2)_3{-}NH{-}C(NH_2){=}NH_2^+} + \underset{\text{Fumarate}}{^-OOC{-}CH{=}CH{-}COO^-}$$

The urea cycle is linked to the TCA cycle as fumarate can be converted into oxaloacetate. Transamination of oxaloacetate forms aspartate which, as we have described above, is available for arginosuccinate synthesis.

Control of the urea cycle

The major control of the urea cycle is by variation in the concentration of enzymes catalysing each step. When there is a large quantity of free amino acids to be metabolized, enzymes are synthesized, and the reverse applies at low quantities. High amino acid concentrations are seen after a rich intake and in starvation when muscle proteins are degraded to supply the gluconeogenic pathway in liver. In addition, carbamoyl phosphate synthetase is inactive without its allosteric activator, *N*-acetyl glutamate, whose cellular level reflects that of amino acid concentrations.

Transport of the amino nitrogen from extrahepatic tissues to the liver

When amino acids are degraded in muscles into their 20 constituent amino acids, these amino acids are not released in the bloodstream in the quantities and proportions that they were found in the original proteins. The predominant, but not the only, amino acids found in the circulation under these circumstances are glutamine and alanine. They are formed by transaminations using the amino groups of the other amino acids present in proteins.

Transport of ammonia in the blood as glutamine

Ammonia produced from amino acids is toxic, and blood ammonia concentrations are kept very low. If ammonia concentrations rise significantly, brain function will be impaired and coma will ensue. Free ammonia is not transported as such from peripheral tissues to the liver, but in the form of the nontoxic amide, glutamine, which is synthesized from glutamate by the enzyme, **glutamine synthetase**.

$$\underset{\text{Glutamate}}{^-OOC{-}CH_2{-}CH_2{-}CHNH_3^+{-}COO^-} + NH_4^+ + ATP \longrightarrow \underset{\text{Glutamine}}{NH_2{-}CO{-}CH_2{-}CH_2{-}CHNH_3^+{-}COO^-} + ADP + P_i + H^+$$

The reaction involves the intermediate formation of an enzyme-bound γ-glutamyl phosphate. This is a high-energy phosphoryl anhydride compound with sufficient energy to react with ammonia.

The glutamate precursor of glutamine can be formed from α-oxoglutarate generated in the TCA cycle, followed by transamination with other amino acids. The glutamine is carried in the blood to the liver where it is hydrolysed to release ammonia, which is used for urea synthesis.

$$\begin{array}{l} CO{-}NH_2 \\ | \\ CH_2 \\ | \\ CH_2 \\ | \\ CHNH_3^+ \\ | \\ COO^- \end{array} + H_2O \xrightarrow{\text{Glutaminase}} \begin{array}{l} COO^- \\ | \\ CH_2 \\ | \\ CH_2 \\ | \\ CHNH_3^+ \\ | \\ COO^- \end{array} + NH_4^+$$

It is thus a safe carrier of two amino groups that can be liberated in the liver and eventually eliminated as urea, which is also nontoxic.

Glutaminase also has a function in the kidney. It liberates ammonia, which acts as a buffer for hydrogen ions so it effectively excretes excessive quantities of acid from the blood. Glutamine is one of the 20 amino acids found in proteins and is involved in the synthesis of several other metabolites.

Transport of amino nitrogen in the blood as alanine

As much as 30% of the amino nitrogen produced by protein breakdown in muscle, is sent to the liver as alanine. The released amino acids from protein breakdown transaminate with pyruvate to yield alanine, which is released into the blood. This is taken up by the liver; the amino group is used to form urea (*via* ammonia and/or aspartic acid). The released pyruvate is converted into blood glucose which can go back to the muscle. The sequence of events is referred to as the glucose–alanine cycle (Fig. 18.7).

This alanine transport from muscle to liver has another important physiological role in starvation. As already explained, after glycogen reserves are exhausted the liver must make glucose to supply the brain and other cells, which have an obligatory requirement for the sugar. The main sources of metabolites for this hepatic gluconeogenesis are the amino acids derived from muscle protein breakdown. Many of the amino acids can give rise to pyruvate in the muscle, which is transported to the liver as alanine. However, the glucose–alanine cycle itself gives no net increase in glucose, and in starvation, the amino acids liberated from muscle protein breakdown must be converted into alanine without utilizing glucose as a source of pyruvate. In that situation, the pyruvate comes from the breakdown of glucogenic amino acids.

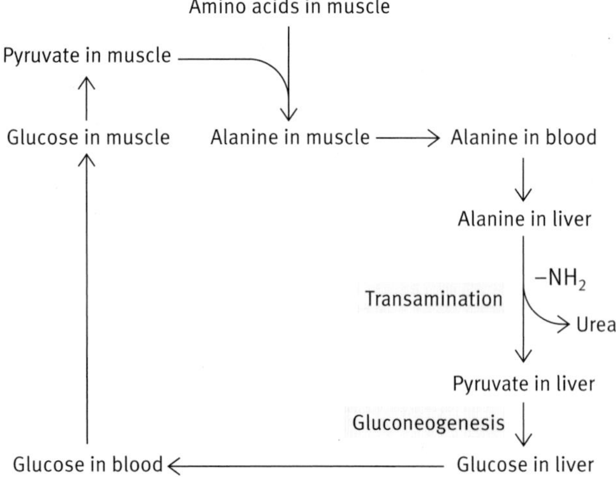

Fig. 18.7 The glucose–alanine cycle for transporting nitrogen to the liver as alanine, and glucose back to the muscles.

Diseases due to urea cycle deficiencies

The urea cycle prevents the accumulation of toxic levels of ammonium ions, therefore, it is not unexpected that deficiencies in enzymes of this pathway can cause diseases. These diseases are very serious but they are fortunately rare.

A variety of such diseases have been identified because of the deficiencies in the synthesis of the activator of carbamoyl phosphate synthetase, *N*-acetyl glutamate, and because of deficiencies of the enzyme itself. Other conditions are associated with separate deficiencies in argininosuccinate synthetase and lyase, and rarely with arginase deficiency.

The diseases vary in severity according to the site of the blockage in the urea synthesis pathway, and according to the fractional loss of the enzyme in question. Many are only partially lost. Severe disease typically is associated with mental retardation and early death. The reason for the toxic effect of ammonia on the brain is not clearly understood.

Hyperammonaemia can also result from liver disease independently of genetic deficiencies of enzymes, for example, because of alcoholism or viral hepatitis.

Various treatments have been devised in different forms of disease to cause alternative routes of nitrogen excretion. Feeding of benzoate and phenylacetate results in excretion of these acids coupled to glycine and glutamine respectively. A more complex metabolic strategy in the case of arginosuccinate lyase deficiency is to provide excess arginine (and a low protein diet of sufficient calorific value). Arginine is converted into urea and ornithine. Ornithine is converted into citrulline (using carbamoyl phosphate), which, with aspartate, forms arginosuccinate. This accumulates and is excreted, carrying with it, and so eliminating, two nitrogen atoms from carbamoyl phosphate and aspartate.

Alternatives to urea formation exist in different animals

In humans, some nitrogen is excreted as uric acid, ammonium ions, and creatinine. However, birds excrete it as a white paste of solid uric acid rather than urea; fish and other animals living in water, excrete it as ammonia. There is an evolutionary logic here. In an aqueous environment, unlimited water means that ammonia can be constantly excreted and dispersed; mammals are intermediate in this respect, and urea, being nontoxic and highly soluble, is the route of choice. Birds have the problem that chicks develop in a closed egg, and accumulation of any soluble form of excretory product could be deleterious. But excreting nitrogen as uric acid, which is almost insoluble, causes no problems as it accumulates.

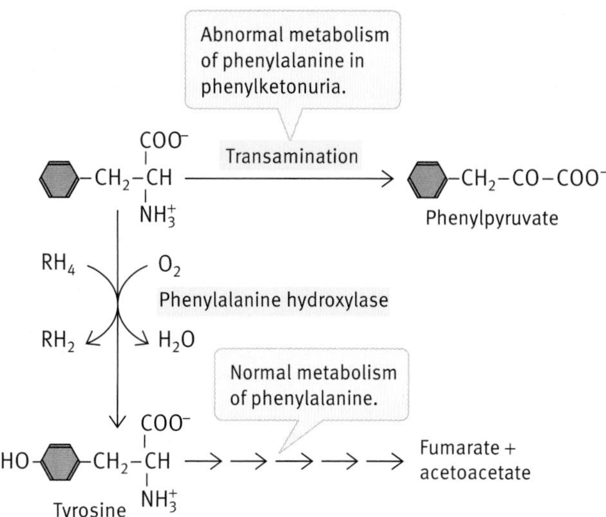

Fig. 18.8 Normal and abnormal metabolism of phenylalanine.

Fate of the oxo-acid or carbon skeletons of deaminated amino acids

Amino acid metabolism links up to the major metabolic pathways of carbohydrate and fat metabolism as already outlined in Fig. 12.10.

As far as general metabolism is concerned, some amino acids are **glucogenic** and some are **oxogenic**. The term 'ketogenic' might seem to imply that an amino acid giving rise to acetyl-CoA results in formation of ketone bodies in the blood, which would conflict with the earlier explanation that ketone bodies arise only in conditions of excess fat metabolism. Acetyl-CoA in normal metabolic situations does *not* give rise to ketone bodies.

The universally used term ketogenic for certain amino acids is an old one, resulting from the use of fasting animals as a test system for the metabolic fate of amino acids since, in such animals, any increase in acetyl-CoA production causes increases in blood ketone bodies that are easily measured. It does not mean that ketogenic amino acids produce these exclusively in normal animals, but simply that they *can* give rise to acetyl-CoA but not to pyruvate. Glucogenic amino acids were detected by increased levels of blood glucose or glucose excretion when administered to diabetic animals, and similar qualifications apply to this term. Whether, in normal animals, the pyruvate goes to glucose formation or is oxidized depends on metabolic controls.

Aspartate, like glutamate, is converted into a metabolite of the TCA cycle (oxaloacetate) by the transamination reaction already described and is therefore glucogenic. Alanine and glutamate, producing pyruvate and 2-oxoglutarate respectively on deamination, are also glucogenic, as are serine, cysteine, and others.

Some amino acids are both ketogenic and glucogenic – for example, phenylalanine produces fumarate (an intermediate in the TCA cycle) and acetyl-CoA. Of the 20 amino acids only two (leucine and lysine) are solely ketogenic. The degradation of four amino acids (isoleucine, methionine, threonine, and valine) is referred to later in this chapter because of its special interest there. The product is propionate, which is metabolized by the route for oxidation of odd-numbered fatty acids, a process involving vitamin B_{12} (see Chapter 14).

Genetic errors in amino acid metabolism cause diseases

Phenylketonuria

Phenylalanine is an aromatic amino acid, an excess of which is normally converted into tyrosine by an enzyme, phenylalanine hydroxylase (Fig. 18.8). This enzyme is interesting in that a pair of hydrogen atoms are supplied by an electron donor – a coenzyme molecule called **tetrahydrobiopterin** (RH_4 in Fig. 18.8; structures of RH_4 and dihydrobiopterin (RH_2) are given below for reference purposes). Oxygen is also needed.

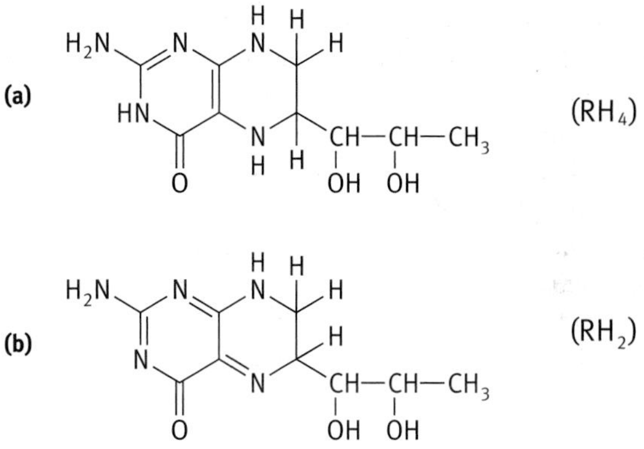

(a) (RH_4)

(b) (RH_2)

It may seem odd that a reaction requires *both* oxygen and a reducing agent but it is a mechanism used in other reactions also, as you will see later. The trick is that one *atom* of an oxygen molecule is used to form an dOH group on the aromatic ring, but this leaves the other oxygen atom to be taken care of.

It is reduced to H_2O by the two hydrogen atoms donated by the tetrahydrobiopterin. A separate enzyme system reduces the **dihydrobiopterin** formed back to the tetrahydro form, using NADPH as the reductant, and so the cofactor acts catalytically.

The phenylalanine hydroxylase belongs to a class of enzymes known as **monooxygenases** (because one atom of O appears in the product) or, alternatively, **mixed-function oxygenases**, because two things are oxygenated – the amino acid and a pair of hydrogen atoms.

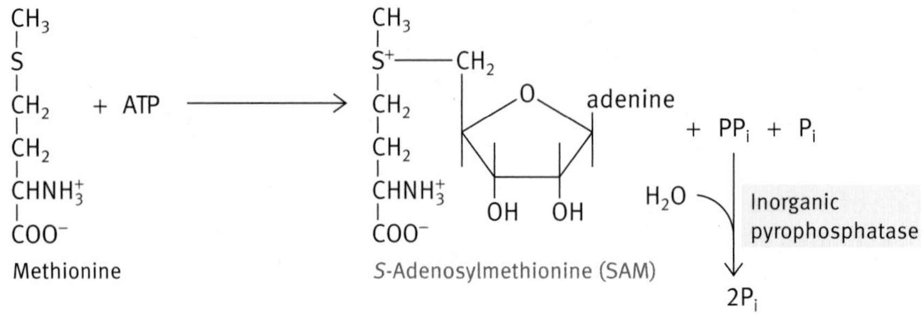

Methionine (CH_3–S–CH_2–CH_2–$CHNH_3^+$–COO^-) + ATP → S-Adenosylmethionine (SAM) (CH_3–S^+(–CH_2–ribose(OH, OH)–adenine)–CH_2–CH_2–$CHNH_3^+$–COO^-) + PP_i + P_i; PP_i + H_2O → (Inorganic pyrophosphatase) $2P_i$

Fig. 18.9 The synthesis of *S*-adenosylmethionine (SAM) from methionine.

Phenylalanine as such is not normally deaminated, being converted into tyrosine, and only then does deamination occur (Fig. 18.8). However, there is a genetic abnormality, occurring in around 1 in 10,000, in which the phenylalanine conversion into tyrosine is impaired or blocked because of enzyme deficiency or, rarely, because of lack of tetrahydrobiopterin. This causes excess phenylalanine to accumulate and, in this situation, it abnormally participates in transamination producing phenylpyruvate (an abnormal metabolite), which spills out into the urine (Fig. 18.8). The disease is called **phenylketonuria** or **PKU**. The consequences of phenylpyruvate in babies are irreparable mental impairment. If diagnosed at birth (by blood analysis, for phenylalanine level), a child with the disease can be given a diet limited in phenylalanine (but adequate in tyrosine) and development is normal. The urine of patients has a mousy colour due to the phenylacetate formed from the phenylpyruvate. Mass screening programmes of neonatals, giving early detection, avoids the tragic consequences of the abnormality. It is not known how phenylpyruvate causes such deleterious brain damage.

Maple syrup disease

Curiously enough, another genetic condition, called **maple syrup disease**, involves accumulation of the oxo-acids of three aliphatic amino acids, valine, isoleucine, and leucine, and this also involves brain impairment. (The name of the disease comes from the oxo-acids in the urine having a characteristic smell.) This disease is much rarer than PKU.

Alcaptonuria

Another much-quoted genetic condition is **alcaptonuria** in which the urine turns black on exposure to air. This is a relatively benign condition but in later years may cause problems in connective tissue. It is due to a block in the tyrosine degradation pathway in which a diphenol intermediate metabolite, homogentisate, is excreted. The diphenol in air oxidizes to form a dark pigment. It was the first metabolic disease for which the genetic pattern of inheritance was worked out.

Methionine and transfer of methyl groups

Methionine is one of the essential amino acids. It has the structure:

$$CH_3—S—CH_2—CH_2—\underset{NH_3^+}{CH}—COO^-.$$

The interesting part is the **methyl group**. Methyl groups are very important in the cell – a variety of compounds are methylated, and methionine is the source of methyl groups that are transferred to other compounds. Methionine is a stable molecule – the methyl group has no tendency to leave; however, if the molecule is converted into **S-adenosylmethionine (SAM)**, a sulphonium ion is created and the methyl group is 'activated' – it has a strong group transfer potential making it thermodynamically favourable for it to be transferred (by transmethylase enzymes) to O and N atoms of other compounds (it cannot form C—C bonds). The reason for the high energy is that when the methyl group is transferred, the sulphonium ion reverts to an uncharged thioether. ATP supplies the energy for SAM synthesis – in this case three –P groups are converted to PP_i+P_i and then the PP_i is cleaved to two P_i (Fig. 18.9).

Transfer of the methyl group of SAM to other compounds generates **S-adenosylhomocysteine**. The latter is hydrolysed to produce **homocysteine** – this is methionine with –SH instead of —S—CH_3 (it is not one of the 'magic' 20 used in protein synthesis). The homocysteine is complexed with serine to form cystathionine, which is hydrolysed to cysteine and 2-oxobutyrate (Fig. 18.10). Deficiency in cystathionine formation leads to accumulation of homocysteine which, for

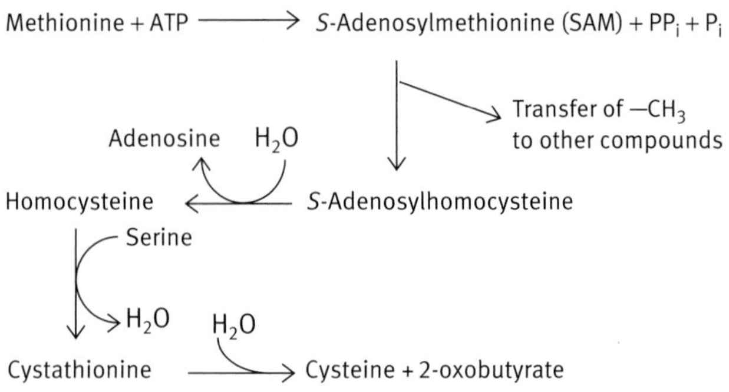

Methionine + ATP → *S*-Adenosylmethionine (SAM) + PP_i + P_i
S-Adenosylmethionine → (Transfer of —CH_3 to other compounds) → *S*-Adenosylhomocysteine
S-Adenosylhomocysteine + H_2O → Homocysteine + Adenosine
Homocysteine + Serine → Cystathionine + H_2O
Cystathionine + H_2O → Cysteine + 2-oxobutyrate

Fig. 18.10 Breakdown of *S*-adenosylhomocysteine to form cysteine and 2-oxobutyrate. Note that the recycling of homocysteine to methionine is dealt with in Chapter 19.

unaccounted reasons, is associated with a variety of childhood pathologies including mental retardation.

Recycling of homocysteine back to methionine requires a folate derivative (see Chapter 19). Hence folate deficiency can cause a build-up of homocysteine, and folate therapy is being tried for patients with raised homocysteine levels. Increased homocysteine concentrations are also believed to be somehow associated with the development of blood vessel blockages and coronary heart disease, and it is currently of considerable medical interest.

There have been numerous studies including meta analysis of various prospective studies with variable results. There is still a controversy as to whether high concentrations of homocysteine are a marker or a causal agent for cardiovascular disease. As hyperhomocysteinaemia can be treated with folate and vitamin B6, clinicians could consider advising supplements to high risk patients (see Chapter 19).

What are the methyl groups transferred to?

In the body the methyl groups of creatine (see Chapter 8), phosphatidylcholine (see Chapter 7), and adrenaline (epinephrine; see later in this chapter) come from *S*-adenosylmethionine, and so do the methyl groups attached to the bases of nucleic acids. Note, however, that the latter do not include that of thymine, which is separately synthesized and is an important topic to be dealt with later (see Chapter 19), as also is the subject of nucleic acid base methylation.

Synthesis of amino acids

In the body, as explained, only the nonessential amino acids can be synthesized, but all are made in plants and bacteria. The pathways of synthesis of all have been long established. Our aim here is to deal only with aspects of special interest and to illustrate how some amino acids are synthesized from glycolytic and TCA cycle intermediates. In fact, five of these intermediates (3-phosphoglycerate, phosphoenolpyruvate, pyruvate, oxaloacetate, and 2-oxoglutarate) together with two sugars of the pentose phosphate pathway are the precursors of all 20 amino acids (in plants and bacteria).

Synthesis of glutamate

As explained earlier, deamination of glutamate occurs *via* glutamate dehydrogenase, an $NADP^+$ enzyme of central importance. This is reversible, and is the route by which glutamate can be formed from 2-oxoglutarate and ammonia. Glutamate formation also occurs in animals *via* transamination of 2-oxoglutarate using other amino acids such as alanine or aspartate as the donor of the amino group. See the reaction for aspartate aminotransferase earlier in this chapter.

Synthesis of aspartic acid and alanine

Aspartic acid and alanine come from transamination of oxaloacetate

$$\left(R = \begin{array}{c} COO^- \\ | \\ CH_2 \\ | \end{array} \right)$$

and pyruvate ($R{=}CH_3$), respectively, with glutamate.

$$\begin{array}{c} R \\ | \\ C{=}O \\ | \\ COO^- \end{array} + \text{glutamate} \longrightarrow \begin{array}{c} R \\ | \\ CHNH_3^+ \\ | \\ COO^- \end{array} + \text{2-oxoglutarate}$$

Synthesis of serine

Serine is formed from the glycolytic intermediate, 3-phosphoglycerate, which is first converted to an oxo-acid, 3-phosphohydroxypyruvate:

$$\begin{array}{c} COO^- \\ | \\ CHOH \\ | \\ CH_2OPO_3^{2-} \end{array} \xrightarrow{NAD^+ \quad NADH} \begin{array}{c} COO^- \\ | \\ C{=}O \\ | \\ CH_2OPO_3^{2-} \end{array}$$

This oxo-acid is transaminated by glutamate to give 3-phosphoserine, which is hydrolysed to serine and P_i.

$$\begin{array}{c} COO^- \\ | \\ C{=}O \\ | \\ CH_2OPO_3^{2-} \end{array} \xrightarrow{\text{Transamination}} \begin{array}{c} COO^- \\ | \\ CHNH_3^+ \\ | \\ CH_2OPO_3^{2-} \end{array} \xrightarrow{\text{Hydrolysis}} \begin{array}{c} COO^- \\ | \\ CHNH_3^+ \\ | \\ CH_2OH \end{array} + P_i$$

Synthesis of glycine

Glycine is the simplest amino acid of all ($H_3N^+CH_2COO^-$). It is formed by a reaction that is completely new, so far as this book is concerned, involving withdrawal of a hydroxymethyl group ($-CH_2OH$) from serine and adding it to a coenzyme, tetrahydrofolate (mentioned in Chapter 9), the function of which is to act as a one-carbon-unit carrier. The one-carbon-unit transfer area is of importance in nucleotide synthesis. It will be more appropriate to go into this thoroughly later (see Chapter 19). (Other sources of glycine exist.)

Haem and its synthesis from glycine

The full structure of haem is given for reference purposes in Fig. 4.19, where it is described in relation to its role in

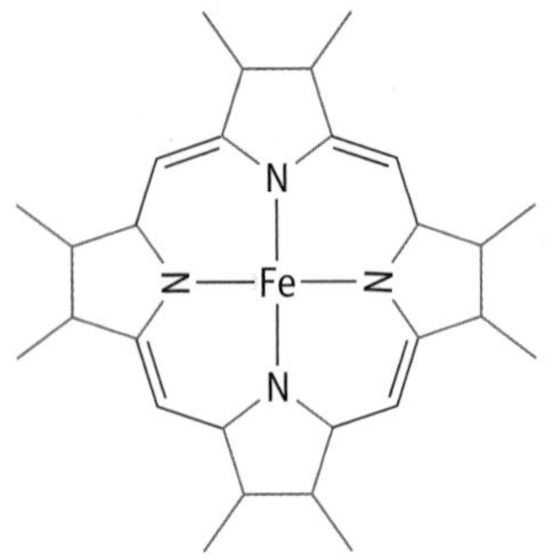

Fig. 18.11 Outline of the haem molecule. (The full structure is given in Fig. 4.19.)

Two molecules of ALA → (ALA dehydratase) → Porphobilinogen (PBG) + 2 H_2O

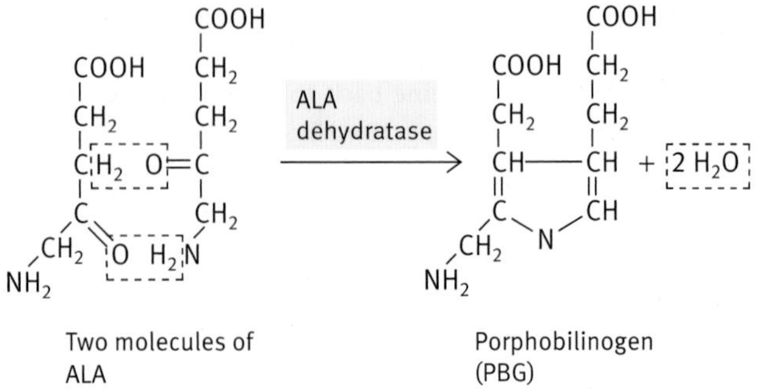

Fig. 18.13 Synthesis of porphobilinogen (PBG) – the aminolevulinate (ALA) dehydratase step. Nonionized structures are given for simplicity. PBG is a monopyrrole; haem is a tetrapyrrole. The conversion of PBG to haem is shown in Fig. 18.14.

haemoglobin, but Fig. 18.11 gives an outline structure. It is a ferrous iron complex with protoporphyrin. Protoporphyrin is a tetrapyrrole, the four substituted pyrroles being linked by methene (=CH−) bridges, such that a **conjugated double-bond system** exists (that is, you can go right round the molecule *via* alternating single and double bonds). This gives protoporphyrin and haem their deep red colour.

In haem, the four pyrrole N atoms are bound to Fe^{2+} leaving two more of the six ligand positions of the Fe^{2+} available for other purposes, as shown in Fig. 4.20.

Red blood cells have the vast majority of the body's haem content, though haem is found in all aerobic cells as the prosthetic group for cytochromes and other proteins.

The synthesis of haem appears to be a formidable task but the essentials are surprisingly simple, requiring in animals only two starting reactants, glycine and succinyl-CoA. You have met succinyl-CoA in the TCA cycle. An enzyme, **aminolevulinate synthase**, or **ALA synthase** (ALA-S), carries out the reaction shown in Fig. 18.12. 5-Aminolevulinic acid (ALA) is the precursor solely committed to porphyrin synthesis. Two molecules of ALA are used to form the pyrrole, **porphobilinogen (PBG)**, the two molecules being dehydrated by ALA dehydratase (Fig. 18.13). The remainder of the haem biosynthetic pathway consists of linking four PBG molecules together, modifying the side groups, and chelating an atom of ferrous iron to form haem. The intermediate tetrapyrroles between PBG and haem are the colourless **uroporphyrinogens** and **coproporphyrinogens** (in which PBG units are linked by methylene bridges) and the red **protoporphyrin** (in which they are linked by methene bridges). The pathway is given in Fig. 18.14; have a quick look at this.

The haem synthesis pathway has a curious feature. The first reaction, synthesis of ALA, occurs inside mitochondria, after which the ALA moves out to the cytosol, but the final three steps also occur in the mitochondria. Why this should be so is not clear.

A group of porphyria diseases is known, each of which is associated with a deficiency of one of the enzymes of the haem biosynthetic pathway (Box 18.1). Each deficiency can cause the accumulation of the metabolite(s) preceding the deficiency and these can have deleterious effects.

Destruction of haem

Red blood cells are destroyed mainly by **reticuloendothelial cells** of spleen, lymph nodes, bone marrow, and liver. Removal of sialic acid groups from the red-cell-membrane glycoproteins is a signal that the cell is aged and ready for destruction. The degraded carbohydrate attaches to cell receptors, which leads to endocytosis of the erythrocyte. The enzyme, haem oxygenase opens up the tetrapyrrole ring, releasing the iron for reuse and forming **biliverdin**, a linear tetrapyrrole (Fig. 18.15). Biliverdin is reduced to **bilirubin**. This is water insoluble but is transported in the blood, attached to serum albumin, to the liver, where it

Succinyl-CoA (CO−S−CoA, CH_2, CH_2, COOH) + $CH_2(NH_2)COOH$ (Glycine) → (ALA synthase) → 5-Aminolevulinic acid (ALA) (CO−CH_2−NH_2, CH_2, CH_2, COOH) + CO_2 + CoA−SH

Fig. 18.12 The first step in haem synthesis, catalysed by aminolevulinate (ALA) synthase. For clarity of illustration, nonionized structures are given.

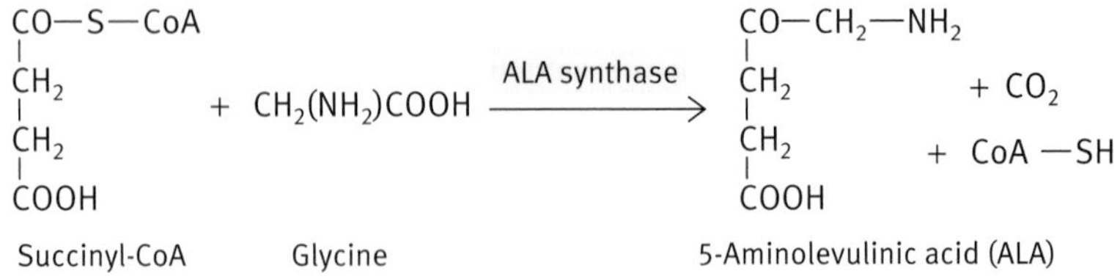

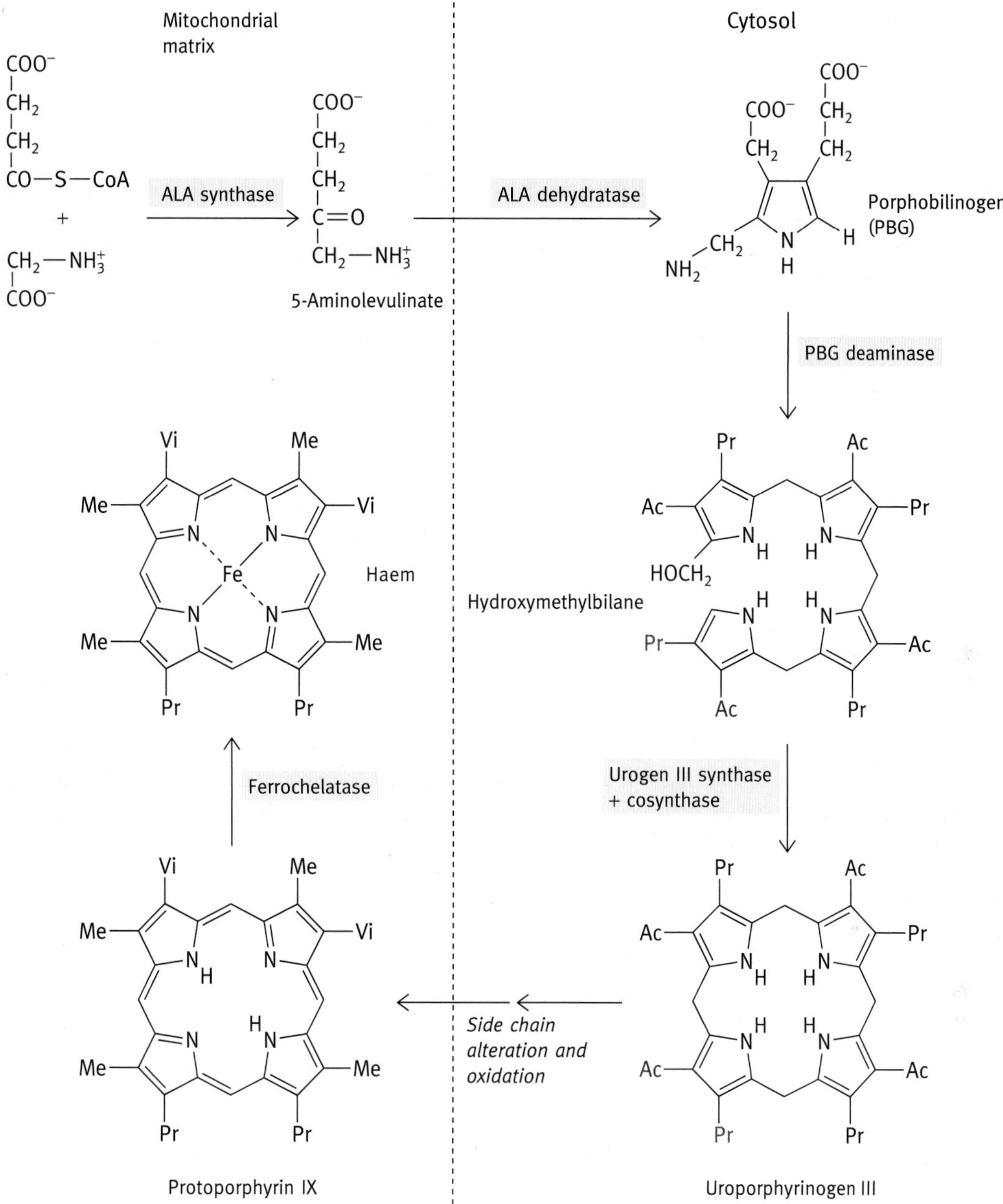

Fig. 18.14 An abbreviated pathway of haem biosynthesis, given for reference purposes. Me, Pr, Ac, and Vi indicate methyl, propyl, acetyl, and vinyl groups, respectively.

is rendered much more polar by the addition of two glucuronate groups (Fig. 18.15) and then excreted, in the bile, into the gut where bacteria convert it to **stercobilin**, giving the brown colour to faeces; modification and partial reabsorption of some bile compounds leads to the yellow colour of urine. **Jaundice** may result from excessive red blood cell breakdown, lack of the glucuronidation enzyme, or blockage of the bile duct.

Synthesis of adrenaline and noradrenaline

Amino acids are used for the synthesis of many nitrogenous molecules including long chain amines, hormones, neurotransmitters, creatine, nucleotides, and others. A single illustrative example (Fig. 18.16) gives the pathway for the synthesis of the hormones **adrenaline** and **noradrenaline**. (Nucleotide synthesis is the subject of Chapter 19.)

BOX 18.1 Acute intermittent porphyria

Acute intermittent porphyria (AIP), the most common type of porphyria disease, though not encountered frequently, is clinically important because it can be life threatening. It is due to a deficiency (about 50%) of the enzyme porphobilinogen deaminase. Most of the time the patient is normal but acute attacks can be precipitated by a variety of drugs such as barbiturates. The triggering agents appear to have in common the ability to induce the synthesis of hepatic cytochrome P450. This is a haem protein that is massively induced by some drugs, barbiturates being the classic ones in this respect. This causes a demand for increased haem synthesis, and in response to this the level of ALA synthase is increased to meet the demand. In acute intermittent porphyrics, the haem biosynthetic pathway cannot handle the increased supply of ALA, resulting in its accumulation and that of PBG, the next metabolite, which spill over into the urine. Such accumulation is associated with the onset of the symptoms of the disease. These are neurological in nature and result in severe abdominal pain and psychiatric abnormalities. It is not known how these effects are caused.

The control mechanisms that underlie the disease are now understood. When cytochrome P450 is induced it causes a reduced haem level. It has been shown that haem controls ALA synthase in three different ways. First it represses transcription of the mRNA for the enzyme in the nucleus; secondly it destabilizes the mRNA resulting in a shorter half-life and reduced synthesis of the enzyme; thirdly, the enzyme is synthesized in the cytosol as a precursor protein, which is transported into the mitochondria where it functions. Haem inhibits this transport. Thus in a normal person, haem biosynthesis maintains a haem level that balances haem production with demand. In a patient with AIP, the impaired haem biosynthesis pathway is inadequate to cope with the extra demand so the ALA synthase is excessively induced, and the ALA and PBG accumulate due to the block.

AIP is known as a hepatic porphyria because its effects originate in the liver. In red blood cells the ALA synthase is coded for by a different gene and haem synthesis control is different in mechanism. The enzyme blockages in erythropoietic porphyrias, as the associated diseases are called, result in accumulations of porphyrins in the skin, giving rise to distressing photosensitivity damage.

For the past few decades the literature has implied that King George III (the 'mad king') had acute intermittent porphyria (and speculated that the attacks may have had some relevance to the American War of Independence), but more recently, opinion has favoured the view that he had variegate porphyria, also of hepatic origin. It has also been implied that Vincent Van Gogh possibly had acute intermittent porphyria (see the May et al. review in the further reading for a fuller account).

In general, metabolic diseases are inherited recessively, because most metabolic pathways can operate on the 50% enzyme level in heterozygotes with one gene deficient. However the porphyria diseases involve rate-limiting enzymes, and a 50% deficiency is sufficient in some circumstances to cause the disease.

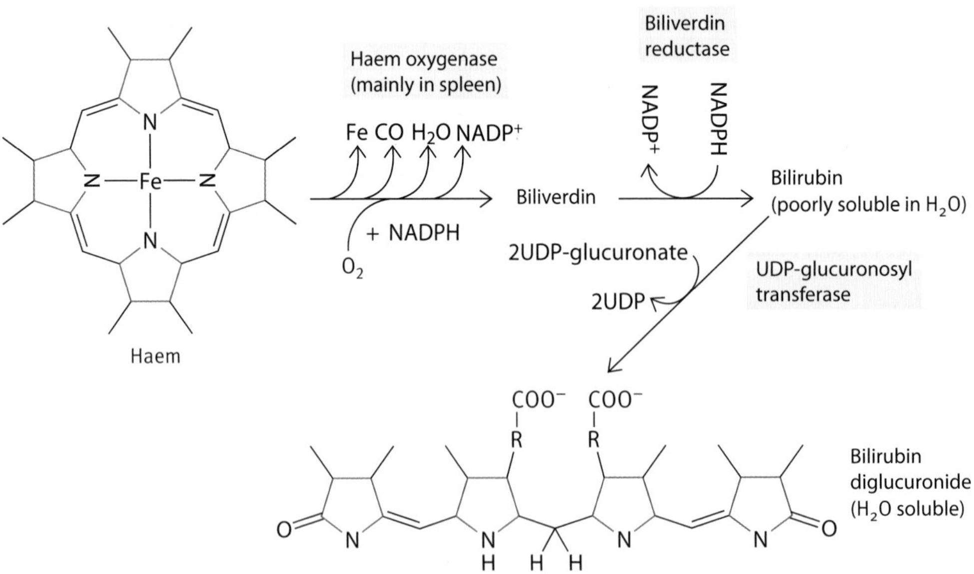

Fig. 18.15 Simplified representation of haem breakdown, the oxygenase reaction, and the reduction of biliverdin (see text). See Fig. 32.5 for UDP-glucuronate formation.

Tyrosine ($CH_2 \cdot CHNH_2 \cdot COOH$) → Tyrosine hydroxylase → Dihydroxyphenylalanine (dopa) ($CH_2 \cdot CHNH_2 \cdot COOH$) → Dopa decarboxylase → Dopamine ($CH_2 \cdot CH_2NH_2$) + CO_2 → Noradrenaline ($CHOH \cdot CH_2 \cdot NH_2$) → Adrenaline ($CHOH \cdot CH_2NH \cdot CH_3$)

Fig. 18.16 Intermediates in the pathway for the synthesis of the catecholamines.

Summary

Amino acids are supplied in the diet from protein hydrolysis in the gut. Proteins in the body are also constantly degraded and resynthesized. The body can synthesize about 10 of the amino acids but the rest must be obtained from the diet. All 20 are needed for protein synthesis. Amino acids are also used to synthesize a wide variety of other molecules.

Amino acids in excess of immediate requirements are deaminated; the amino nitrogen is mainly converted into urea (in mammals) and excreted. The carbon-hydrogen skeletons are oxidized to release energy or converted into fat or glycogen according to the metabolic controls operating at the time and the particular amino acid. Most amino acids are deaminated via transamination with 2-oxoglutarate. The glutamate so formed is deaminated by glutamate dehydrogenase, releasing ammonia. Transaminases are pyridoxal phosphate-containing proteins, the cofactor being essential in the transamination reaction.

A few amino acids such as serine, cysteine, and glycine have different metabolic pathways of their own. Phenylalanine is converted into tyrosine before deamination. If the hydroxylating enzyme is missing, phenylalanine is deaminated to produce phenylpyruvate, which causes mental impairment in children with the disease phenylketonuria. If the disease is detected early, dietary strategies to restrict phenylalanine intake result in normal development. The oxo-acids of branched-chain amino acids also accumulate in a rare genetic disease (maple syrup disease), which also results in mental impairment. Methionine has the special role of providing methyl groups. The latter must be first activated by the formation of *S*-adenosylmethionine.

Haem is synthesized from glycine, the first step being the formation of 5-aminolevulinate by condensation with succinyl-CoA. Haem is destroyed by haem oxygenase, mainly in the spleen, resulting in bilirubin formation, which is excreted as the diglucuronide into the bile. Blockage of the bile duct leads to jaundice.

The ammonia formed by deamination of amino acids is excreted mainly as urea. This is formed in the liver by the Krebs urea cycle. In this, arginine is converted to urea + ornithine by arginase. Arginine is resynthesized from ornithine, using HCO_3^-, ammonium ions, and the amino group of aspartate. Citrulline is an intermediate in the process. The urea cycle is metabolically linked to the TCA cycle.

Urea synthesis is controlled in two ways. The urea cycle is allosterically activated at the carbamoyl synthetase step by *N*-acetyl glutamate whose level reflects that of the amino acids available. In addition, high levels of amino acids cause an increase in the enzymes of the cycle.

Free ammonia from deamination in the tissues is potentially toxic and is transported to the liver as the amide group of glutamine. There the glutamine is hydrolysed by glutaminase to release ammonium ion used in urea synthesis.

The alanine cycle is responsible for transporting amino nitrogen from muscles resulting from muscle protein breakdown to the liver.

Several diseases exist in which urea cycle enzymes are deficient in amount. In some cases these can be treated by dietary strategies.

Alternatives to urea formation exist in different animals. In humans, some nitrogen is excreted as uric acid, ammonium ions, and creatinine. However birds excrete it as a white paste of solid uric acid rather than urea; fish, and other animals living in water, excrete it as ammonia.

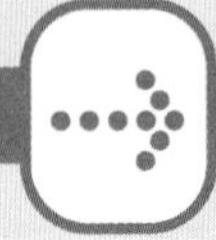

Further reading

To access the further reading, please scan the QR code image or go to http://global.oup.com/uk/orc/biosciences/molbiol/snape_biochemistry/student/reading/ch18/

Problems

1 Explain how an oxidation can result in the deamination of an amino acid.

2 Which amino acid is deaminated by the mechanism referred to in question 1?

3 How are several of the amino acids deaminated, where the reaction in question 2 is involved? Use alanine as an example.

4 What is the cofactor involved in transamination? Give its structure and explain how transamination occurs.

5 Explain how serine is deaminated.

6 What is meant by the terms glucogenic and ketogenic amino acids? Which amino acids are purely ketogenic?

7 Explain the genetic disease phenylketonuria.

8 What is the role of tetrahydrobiopterin in phenylalanine hydroxylation?

9 Methionine is the source of methyl groups in several biochemical processes. Explain how methionine is activated to donate such groups.

10 Describe the first two steps in haem biosynthesis in animals.

11 Outline the reactions of the urea cycle.

12 Why should the level of urea cycle enzymes be increased both in the situation of a high intake of amino acids and in starvation?

13 How are (a) ammonia and (b) amino nitrogen in peripheral tissues transported to the liver for conversion to urea?

14 Several genetic diseases involving the urea cycle are known. What are these and what strategies have been used to ameliorate them?

15 What is the medical relevance of ALA synthase control in liver?

Chapter 19

Nitrogen metabolism: nucleotide metabolism

Several of the preceding chapters have been mainly concerned with energy release from food. In this, ATP occupies the central position, but GTP, CTP, and UTP are also involved in aspects of food component metabolism. However, the involvement of these nucleotides described so far has all been concerned with the phosphoryl groups of the molecules. The nature of the bases, whether A, G, C, or U, has been important only for recognition by the appropriate enzymes but otherwise has not been directly relevant to the metabolic processes. For this reason we have not previously given information about the bases themselves.

We are about to start, in Chapter 22, on the area of **information transfer**, dealing with nucleic acids and protein synthesis. Knowledge of the structures, synthesis, and metabolism of nucleotides is an important prerequisite for understanding the storage and utilizaton of information.

The synthesis and metabolism of nucleotides is also important in understanding several diseases and their treatment including cancer.

Structure and nomenclature of nucleotides

The term 'nucleotide' originates from the name of nucleic acids, originally found in nuclei; you are reminded that a **nucleotide** has the general structure:

phosphate–pentose sugar–base.

A **nucleoside** has the structure:

pentose sugar–base.

Thus, AMP and the corresponding nucleoside, adenosine, have the structures shown:

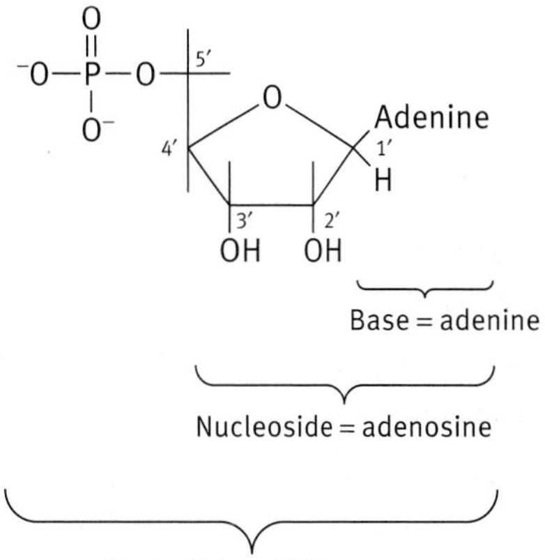

Base = adenine

Nucleoside = adenosine

Nucleotide = AMP

Strictly speaking, the AMP shown here should be written as 5′AMP. The prime (′) indicates that the number refers to the position on the ribose sugar ring, to which the phosphate is attached, rather than to the numbering of atoms in the adenine ring. It is a common practice to assume that the phosphate is 5′ unless specified, since this is the most usual position. Thus 5′AMP is often called AMP, whereas if the phosphate is on the carbon atom 3 of the ribose, this is always specified as 3′AMP. Both are different from cyclic AMP (cAMP, see Chapter 20).

The sugar component of nucleotides

The sugar component of a nucleotide is always a pentose, ribose, or 2′-deoxyribose, which are always in the D-configuration, never the L-form.

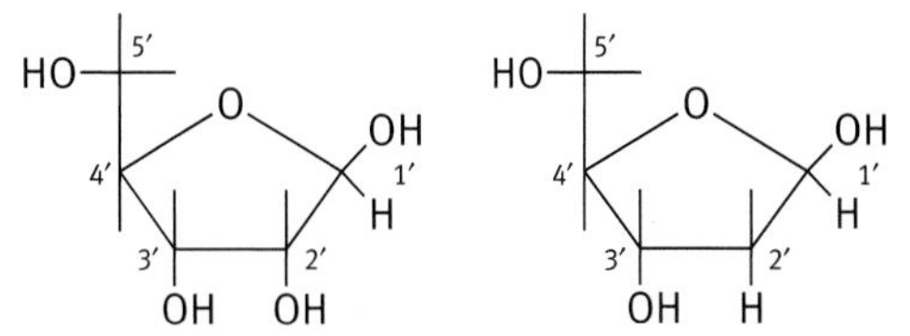

D-Ribose

D-2′-Deoxyribose

In **RNA**, the sugar is always ribose (hence the name, **ribonucleic acid**), and in **DNA**, it is deoxyribose (hence, **deoxyribonucleic acid**). A nucleotide containing ribose is a ribonucleotide but this is not usually specified; unless otherwise stated, a named nucleotide such as AMP is taken to be a ribonucleotide. A deoxyribonucleotide *is always* specified; for example, deoxyadenosine monophosphate or dAMP, etc. (with the one exception mentioned below).

The base component of nucleotides

Nomenclature

We are primarily concerned with five different bases – adenine, guanine, cytosine, uracil, and thymine, all often abbreviated to their initial letter.

A, G, C, and U are found in **RNA**;
A, G, C, and T are found in **DNA**.

The ribonucleotides are AMP, GMP, CMP, and UMP, but older and still used terms are adenylic, guanylic, cytidylic, and uridylic acids, respectively (or adenylate, guanylate, cytidylate, and uridylate for the ionized forms at physiological pH). The deoxyribonucleotides are dAMP, dGMP, dCMP, and dTMP. The latter often is called TMP, or thymidylate, without the d-prefix because T is found only in deoxynucleotides.

When the intention is to indicate a nucleotide without specifying the base, the abbreviations NMP, and 5′NMP, or dNMP, and 5′dNMP for deoxynucleotides are used.

DeoxyUMP exists only as an intermediate in the formation of dTMP; it does not occur in DNA (except as a result of chemical damage to the DNA, when it is promptly removed – see Chapter 23).

Other so-called 'minor' bases exist and are found in transfer RNA (see Chapter 25).

Structure of the bases

The first point is that:

A and G are **purines**
C, U, and T are **pyrimidines**

These names originate from them being formally related to purine and pyrimidine, respectively (neither of which occur in nature).

Purine   Pyrimidine

These rings in future structures will be represented by the simplified forms below.

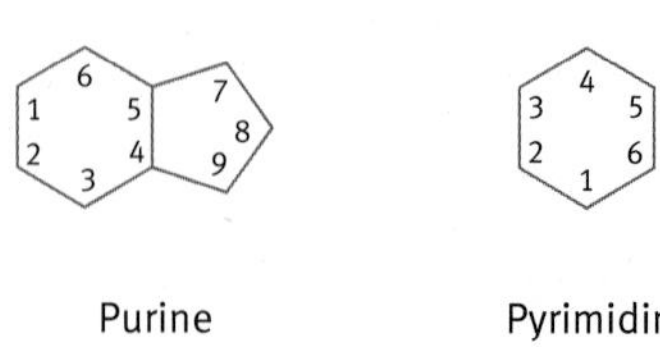

Purine   Pyrimidine

The structures of the nucleotide bases are represented in Fig. 19.1.

Of special importance, note that *T is simply a methylated U.* It will be useful to fix in your mind that T is essentially the same as U except that it is 'tagged' by a methyl group. T is found only in DNA; U only in RNA. The significance of this will be apparent later (see Chapter 23).

Attachment of the bases in nucleotides

The bases are attached to the pentose sugar moieties of nucleotides at the N-9 position of purines and the N-1 position of pyrimidines. The glycosidic bond is in the β-configuration, that is, it is above the plane of the pentose ring. The structures of AMP and CMP are given in Fig. 19.2.

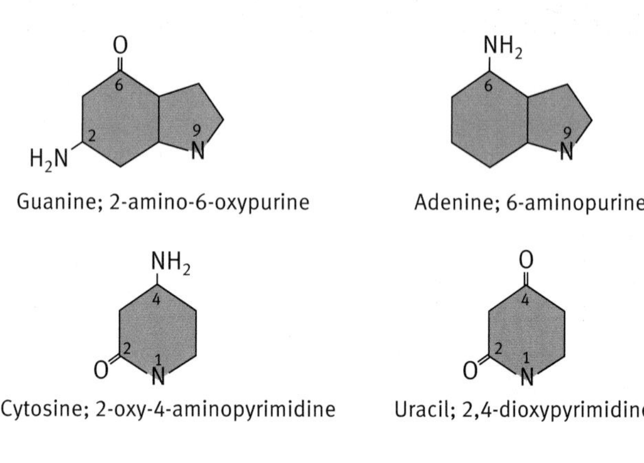

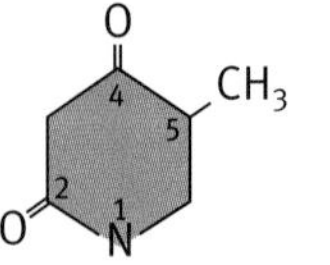

Thymine; 2,4-dioxy-5-methylpyrimidine

Fig. 19.1 Diagrammatic representation of structures of purine and pyrimidine bases found in nucleic acids (full structures in Chapter 23). The oxy and amino groups exist largely, or entirely, in the form shown, rather than the —OH and =NH tautomeric forms. Other minor bases are found in transfer RNA, described in Chapter 25.

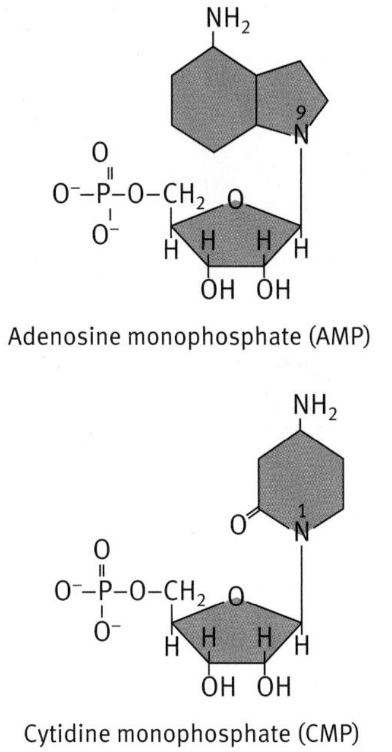

Adenosine monophosphate (AMP)

Cytidine monophosphate (CMP)

Fig. 19.2 Structures of a purine and a pyrimidine nucleotide.

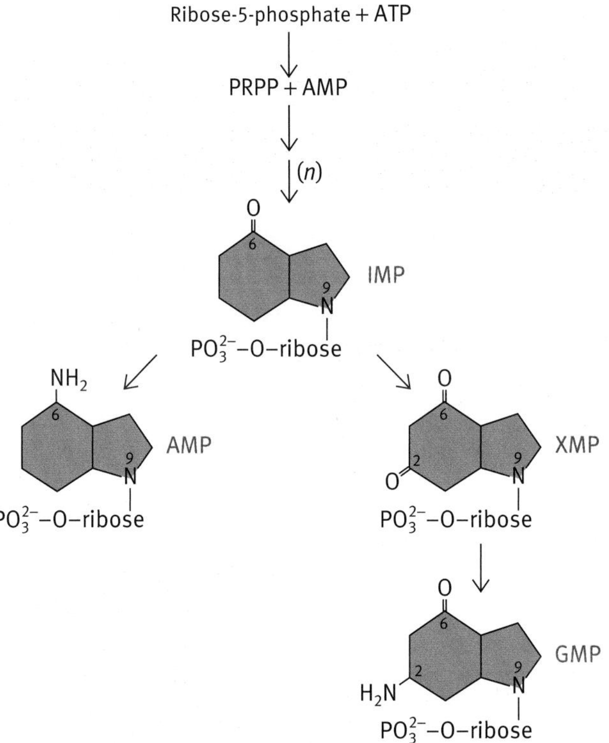

Fig. 19.3 Diagram of the purine *de novo* pathway of GMP, AMP, and XMP synthesis. The base in IMP (inosine monophosphate) is hypoxanthine. The base in XMP (xanthosine monophosphate) is xanthine. PRPP, 5-phosphoribosyl-1-pyrophosphate. The complete pathway of AMP and GMP synthesis can be seen in Figs 19.4 and 19.5.

Synthesis of purine and pyrimidine nucleotides

Purine nucleotides

Most cells can synthesize purine bases *de novo* from smaller precursor molecules. In the *de novo* synthesis of purine nucleotides, bases are not synthesized in the free form but rather the purine ring is assembled piece by piece as a nucleotide starting with an amino group on ribose-5-phosphate. (This refers only to the *de novo* synthesis of purines because free purine bases released by degradation of nucleotides *are* utilized for nucleotide synthesis by the separate **salvage pathway** to be described later.) The mechanism of **ribotidation** (the addition of ribose-5-phosphate to an amino group or to a whole base) is the same in both pathways, as well as in pyrimidine nucleotide synthesis. This brings us to PRPP, the metabolite that is involved in all ribotidation.

PRPP – the ribotidation agent

PRPP is **5-phosphoribosyl-1-pyrophosphate.** It is formed from ribose-5-phosphate (produced by the pentose phosphate pathway; see Chapter 15) by the transfer of a pyrophosphate group from ATP by the enzyme **PRPP synthetase.**

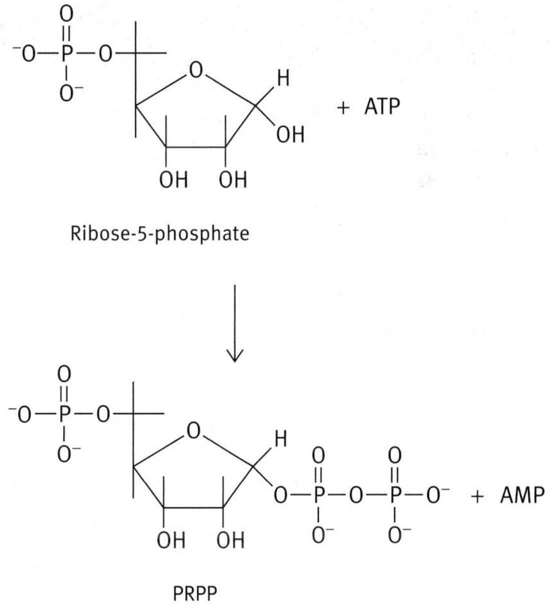

Ribose-5-phosphate + ATP

↓

PRPP + AMP

The PRPP is an 'activated' form of ribose-5-phosphate; appropriate enzymes can donate the latter to an amino group or a whole base forming a nucleotide, and removing —P—P. Hydrolysis of the latter to $2P_i$ drives the reaction thermodymically.

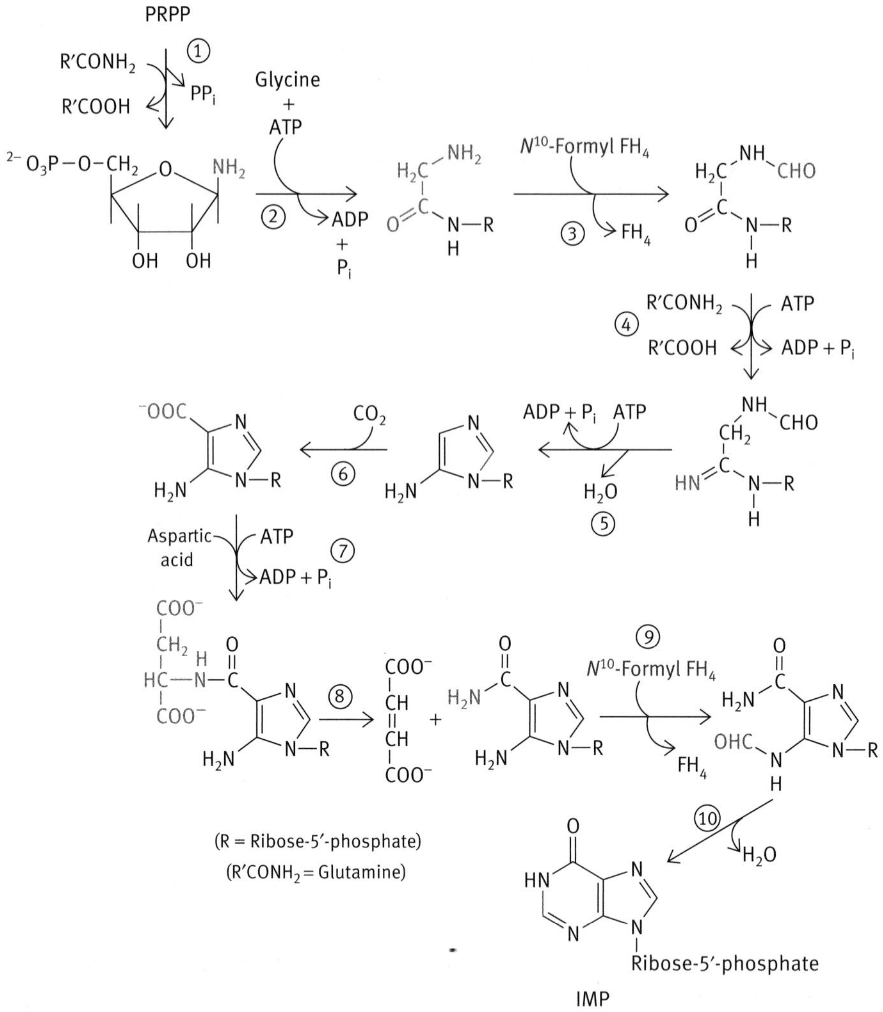

Fig. 19.4 Details of the pathway for the *de novo* synthesis of the purine ring from 5-phosphoribosyl-1-pyrophosphate (PRPP) to inosinic acid (IMP), given for reference purposes. (The circled reaction numbers are referred to in the text.) Blue indicates the structural change resulting from the latest reaction. N^{10}-formyl FH_4, N^{10}-formyltetrahydrofolate.

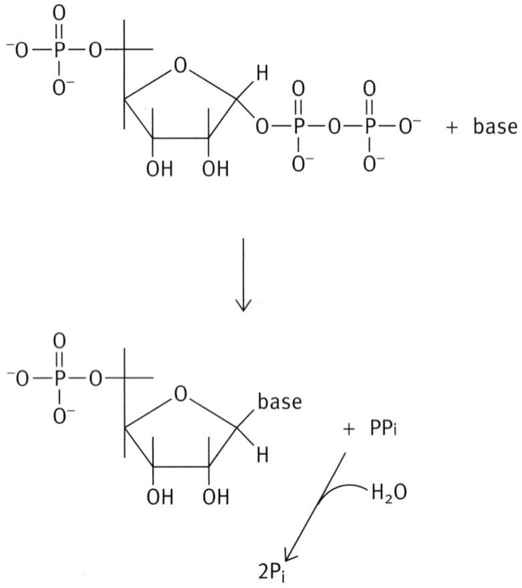

In this reaction the configuration at carbon atom 1 is inverted so that the base is in the required β-position. Note that in the *de novo* pathway (see reaction 1 of this pathway in Fig. 19.4) the 'base' of the starting nucleotide is simply $-NH_2$, derived from glutamine and PRPP giving 5-phosphoribosylamine; in the purine *salvage* pathway it is a purine. We usually associate the term nucleotide with a purine or pyrimidine base but it can be applied to any base attached in the appropriate manner to the sugar phosphate.

The *de novo* purine nucleotide synthesis pathway

To return from this general point to the purine *de novo* pathway in particular, after the formation of 5-phosphoribosylamine, there follows a series of nine reactions resulting in the assembly of the first purine nucleotide in which hypoxanthine is the base (Fig. 19.3). For historical reasons this nucleotide is called IMP or inosinic acid.

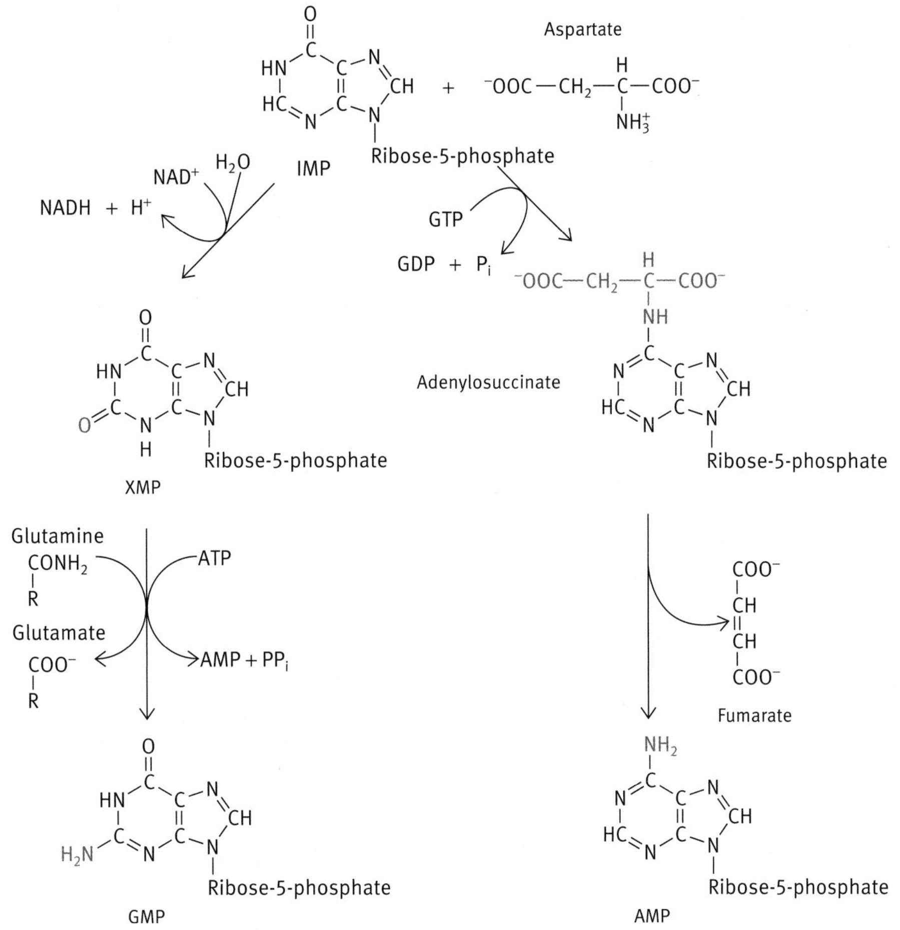

Fig. 19.5 Details of the pathways for the synthesis of GMP and AMP from inosinic acid (IMP), given for reference purposes. The colour shows the change resulting from each reaction. The pathways are summarized in Fig. 19.3. XMP, xanthosine monophosphate.

IMP is a branch-point since its hypoxanthine base may be converted either into adenine or guanine yielding AMP and GMP, respectively. The overall pathway is summarized in Fig. 19.3, but all of the reactions of the pathway are set out in Figs 19.4 and 19.5.

The daunting *de novo* pathway in Fig. 19.4 is given so that we can refer to some reactions of specific interest. Six molecules of ATP are consumed in the synthesis of one purine nucleotide molecule. *The ATP utilization refers only to* —Ⓟ *groups*; there is no loss of the adenine nucleotide of ATP so that the pathway results in a net synthesis of AMP.

Reactions 3 and 9 of this pathway are an important class of reaction and have a general interest and we need to divert to deal with these in some detail. This concerns a type of reaction not dealt with in this book before, and one that is medically important – **one-carbon transfer**.

The one-carbon transfer reaction in purine nucleotide synthesis

Reactions 3 and 9 of the pathway involve the addition of a one-carbon formyl (HCO—) group to intermediates in the pathway (Fig. 19.4). The donor molecule in both cases is ***N^{10}*-formyltetrahydrofolate**, a molecule mentioned in Chapter 9 this book. **Tetrahydrofolate (FH_4, or sometimes THF)** is the carrier in the cell of formyl groups. It is a coenzyme derived from the vitamin **folic acid** (F) or **pteroylglutamic acid**. We suggest that you concentrate only on the relevant parts of this and related structures below, given in blue (not the whole molecules).

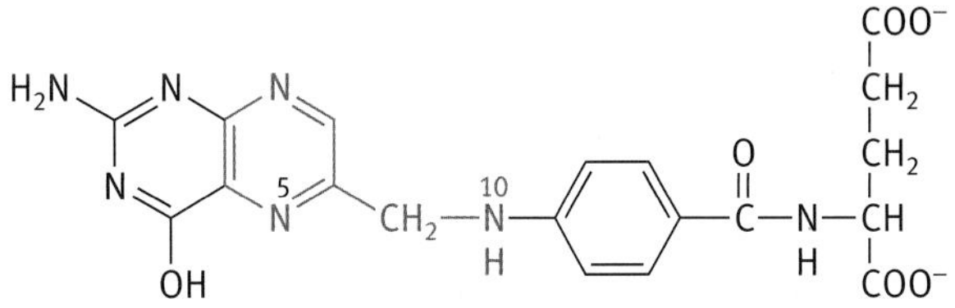

Folic acid (pteroylglutamic acid or F)

FH_4 is, in fact, not only a carrier of formyl groups but of a number of one carbon fragments of different oxidation states. We will return to this point later in this chapter.

The vitamin (F) is reduced to FH_4 by NADPH in two stages.

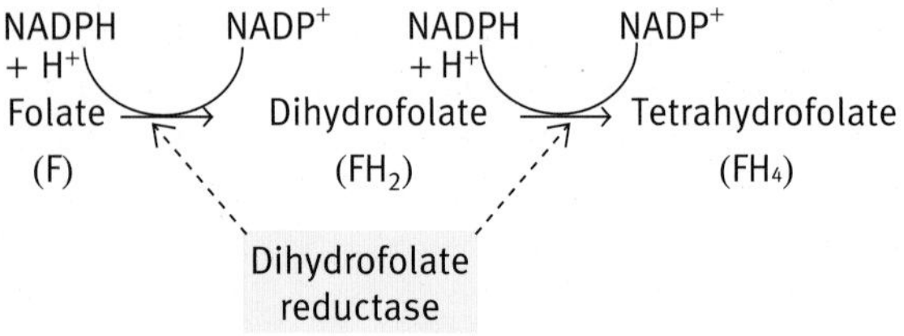

NADPH + H^+ → $NADP^+$; NADPH + H^+ → $NADP^+$

Folate (F) → Dihydrofolate (FH_2) → Tetrahydrofolate (FH_4)

Dihydrofolate reductase

The abbreviated structures of **dihydrofolate (FH_2, or sometimes DHF)** and FH_4 are given below:

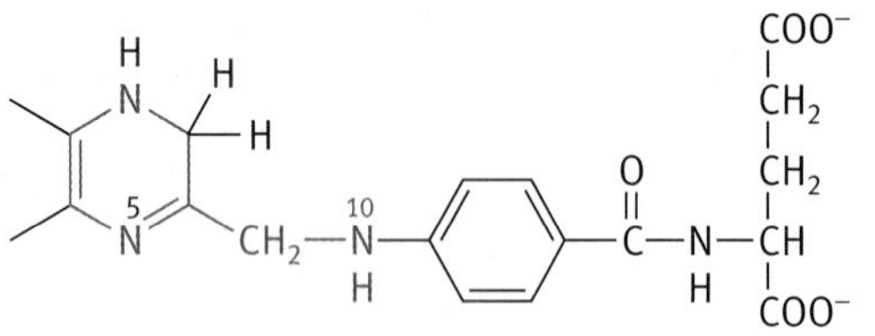

Dihydrofolate (FH_2)

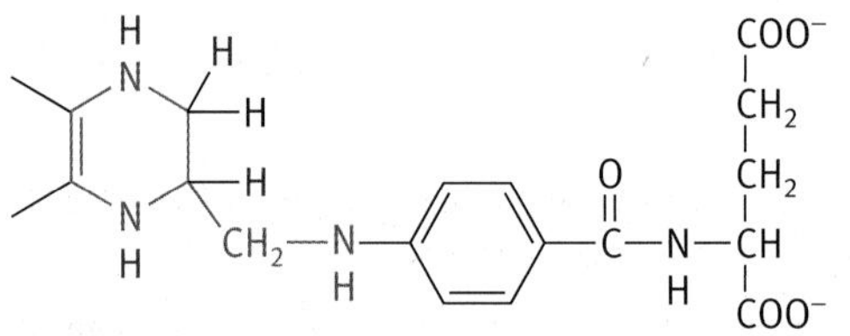

Tetrahydrofolate (FH_4)

If you look at the structure of FH_4 you will notice that the N-5 and N-10 atoms are placed such that a single carbon atom can neatly bridge the gap between them. For our present purposes we can therefore represent FH_4 as:

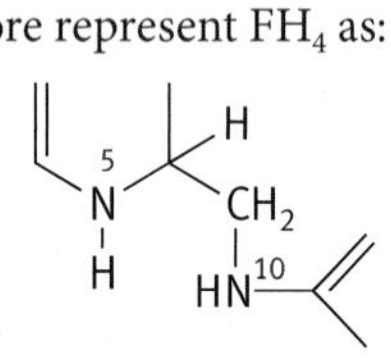

and N^{10}-formyl-FH_4 as:

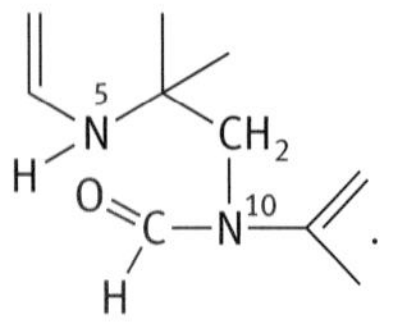

N^{10}-formyltetrahydrofolate (N^{10}-formyl FH_4)

This is the donor of the formyl group in reactions 3 and 9 of the purine biosynthesis pathway; specific formyl transferase enzymes catalyse the reactions.

Where does the formyl group in N^{10}-formyl FH_4 come from?

The answer is the amino acid serine (which is readily synthesized from the glycolytic intermediate 3-phosphoglycerate). An enzyme, **serine hydroxymethylase**, transfers the hydroxymethyl group ($-CH_2OH$) to FH_4 leaving glycine and forming N^5, N^{10}-methylene FH_4.

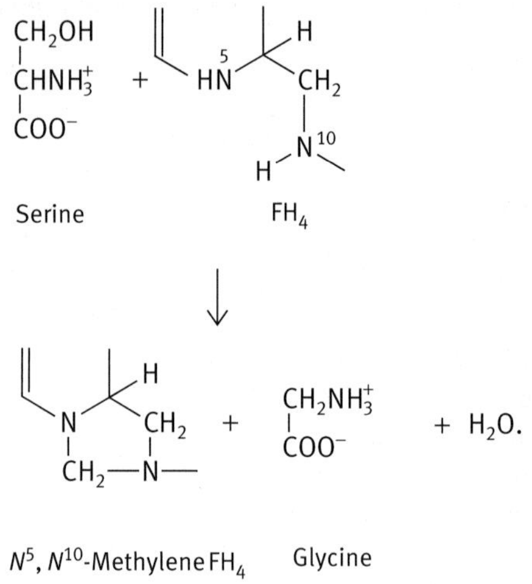

Serine + FH_4 → N^5, N^{10}-Methylene FH_4 + Glycine + H_2O.

The product, N^5, N^{10}-methylene FH_4 is not quite what we want for formylation because the $-CH_2-$ group is more reduced than a formyl group. It can be oxidized by an $NADP^+$-requiring enzyme, forming the methenyl derivative, which is hydrolysed to N^{10}-formyl-FH_4, the formyl group donor.

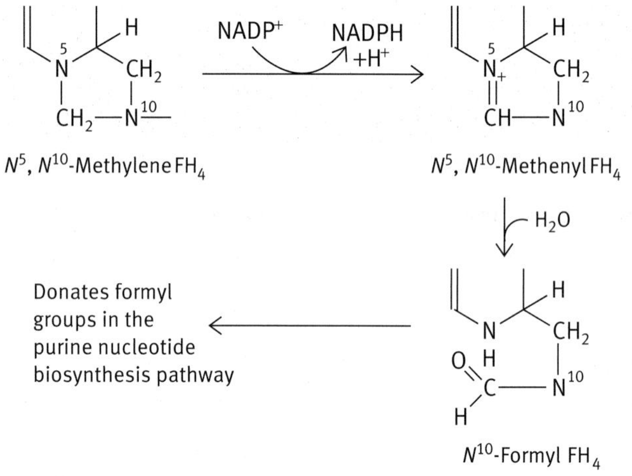

N^5, N^{10}-Methylene FH_4 —($NADP^+$ → NADPH + H^+)→ N^5, N^{10}-Methenyl FH_4 —(H_2O)→ N^{10}-Formyl FH_4 → Donates formyl groups in the purine nucleotide biosynthesis pathway

How are ATP and GTP produced from AMP and GMP?

Most of the synthetic reactions of the cell involve nucleoside triphosphates. As you will see later, these are needed for nucleic acid synthesis. It is a simple but especially important concept that enzymes (kinases) exist in the cell to transfer —Ⓟ groups between nucleotides *at the high-energy level.* There is little free-energy change involved. The main source of —Ⓟ is, of course, ATP, for remember that the energy-generating metabolism constantly regenerates ATP from ADP and P_i. Newly formed AMP and GMP are phosphorylated by kinase enzymes as shown:

$$AMP + ATP \rightleftharpoons 2ADP \quad \text{Adenylate kinase}$$

$$GMP + ATP \rightleftharpoons GDP + ADP \quad \text{Guanylate kinase}$$

GDP+ATP ⇌ GTP+ADP Nucleoside diphosphate kinase

The nucleoside diphosphate kinase has a wide specificity and can use any pair of nucleoside di- and triphosphates.

The purine salvage pathway

We have emphasized that the *de novo* synthesis of purines does not involve free purine bases – purine nucleotides are produced. However, as already indicated, there is a separate route of purine nucleotide synthesis in which **free bases** are converted into nucleotides by reaction with PRPP. The free bases originate from degradation of nucleotides mainly in the liver and supplied to other tissues in the blood – they are salvaged (recycled) and hence the name of the pathway. Two enzymes are involved – these are phosphoribosyltransferases, one of which forms nucleotides from adenine and the other from hypoxanthine or guanine. The latter enzyme, known as **HGPRT** (for **hypoxanthine-guanine phosphoribosyltransferase**), catalyses the reaction:

$$\begin{matrix} \text{Guanine or} & & \text{GMP} & \\ \text{hypoxanthine} + \text{PRPP} & \rightarrow & \text{or} & + \text{PP}_i \\ & & \text{IMP} & \end{matrix}$$

The enzyme salvaging adenine may be of lesser importance than that dealing with guanine and hypoxanthine in humans, for the main routes of nucleotide breakdown produce the free bases hypoxanthine (from AMP) and guanine (from GMP) as shown in Fig. 19.6.

What is the physiological role of the purine salvage pathway?

Since purines are energetically 'expensive' to make, a mechanism for re-utilizing free purine bases is economical since it can reduce the amount of *de novo* synthesis a cell has to carry out. Moreover, certain cells such as erythrocytes have no *de novo* purine synthesis pathway and must rely on the salvage pathway.

The physiological importance of purine salvage is underlined by the rare genetic disease of infants called **Lesch-Nyhan syndrome**, in which the enzyme HGPRT is missing. It is a rare, X-linked inherited disorder. Deficiency of HGPRT results in neurological problems including severe mental retardation and self-mutilation. The disease has a very poor prognosis. Brain possesses the *de novo* pathway only at low levels, so purine nucleotide synthesis is very sensitive to the defects of the salvage pathway. Lack of the salvage pathway reaction leads to a hepatic *overproduction* of purine nucleotides by the *de novo* pathway in these patients because the level of PRPP rises (due to lack of utilization by the salvage reaction) and stimulates the *de novo* pathway. This accounts for the excessive uric acid production which occurs, similar to that in gout (see below), which may result in kidney failure caused by deposition of urate crystals. The connection between the biochemical defect and the neurological symptoms is not clear in Lesch-Nyhan patients. While uric acid overproduction is treatable with allopurinol, the neurological problems are not alleviated. Patients with gout not due to HGPRT deficiency do not develop the neurological symptoms.

The recycling of preformed purine bases has the obvious advantage of energy-saving provided, of course, that the *de novo* pathway synthesis is correspondingly reduced. This is achieved in two ways:

- the salvage pathway reduces the level of PRPP and hence of the *de novo* pathway
- the AMP and GMP produced by the salvage pathway exert feedback inhibition on the pathway.

Formation of uric acid from purines

Nucleotide degradation leads to the production of free hypoxanthine and guanine. Part of this is salvaged back to nucleotides but part is oxidized to **uric acid** (Fig. 19.7). The enzyme, xanthine oxidase, that produces uric acid is present mainly in the liver and intestinal mucosa. **Gout** is a recurrent inflammatory arthritis, as a result of an increased concentration of urate in the blood, leading to the deposition of crystals in tissues, particularly joints and frequently affecting

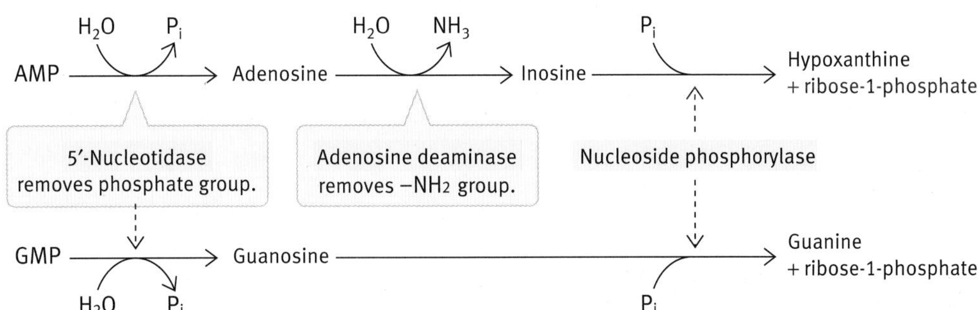

Fig. 19.6 Production of free purine bases hypoxanthine and guanine by nucleotide breakdown. Patients lacking adenosine deaminase in lymphocytes have an immune deficiency that formerly could be treated only by keeping the affected child in a sterile plastic bubble. The disease was the first to be successfully treated by gene therapy in which the normal gene for adenosine deaminase was inserted *in vitro* into bone marrow stem cells and returned to the patient (see Chapter 28).

the metatarsal –phalangeal joint of the big toe. Although gout is traditionally associated with rich living (consumption of alcohol, meat and seafood), the main source of uric acid is in fact excess *de novo* production of purine nucleotides due, in some patients, to a high level of PRPP synthetase activity or an underexcretion of urate. Also, as described above, deficiency of the HGPRT enzyme leads to the overproduction of purine nucleotides. The drug **allopurinol**, used in the treatment of gout, mimics the structure of hypoxanthine. It is a potent xanthine oxidase inhibitor. This inhibition results in xanthine and hypoxanthine formation rather than that of uric acid (Fig. 19.7). These products are more water-soluble than uric acid and more readily excreted, thus preventing the deposition of insoluble uric acid crystals in tissues that results in the clinical symptoms of gout.

Allopurinol

Enol form of hypoxanthine

Control of purine nucleotide synthesis

As with all metabolic pathways, there must be regulation or chemical anarchy would prevail. The *de novo* pathway is a classical example of **allosteric feedback control** (see Chapter 20). The first step of a pathway is a good place for control. In the *de novo* pathway this is the PRPP synthetase. This enzyme is negatively controlled by AMP, ADP, GMP, and GDP. The next enzyme, which catalyses the first *committed* step in the synthesis of purine nucleotides (reaction 2, Fig. 19.4), is inhibited also, as shown in Fig. 19.8. However, this is not quite the end of the story, because the *de novo* pathway produces IMP, and then the IMP is channelled in two directions – to AMP and GMP. They feedback-control their own production. The regulatory loops serve to ensure a balanced production of ATP and GTP since both are required for nucleic acid synthesis (Chapter 23).

Synthesis of pyrimidine nucleotides

Most cells of the body synthesize pyrimidine nucleotides *de novo* but, unlike bacteria, mammals do not appear to have significant pyrimidine salvage pathways for free bases, analogous to those for purines. The nucleoside thymidine, however, is readily phosphorylated to TMP by **thymidine kinase** and, in that sense, salvage of this nucleoside does occur.

The **pyrimidine pathway** is summarized in Fig. 19.9 and given in full in Fig. 19.10. It starts with aspartic acid and produces a ring structure compound, orotic acid. Orotic acid is converted into the corresponding nucleotide by the PRPP reaction and this is converted to UMP. UTP is produced by kinase enzymes much as in the purine pathway. CTP is produced by amination of UTP.

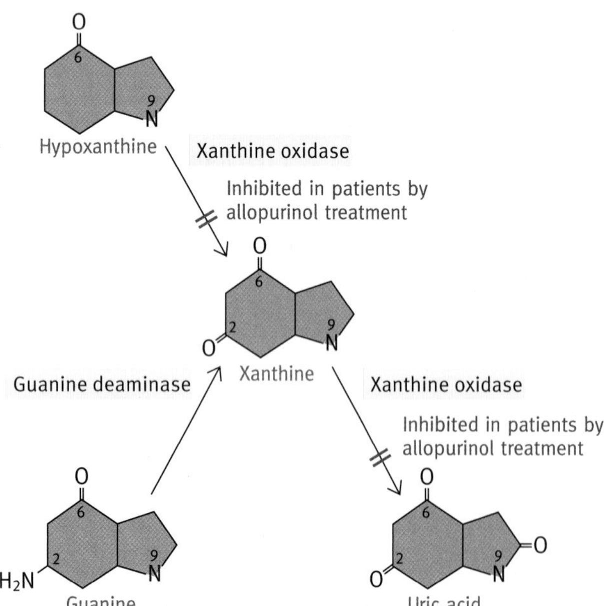

Fig. 19.7 Conversion of hypoxanthine and guanine to uric acid. The drug allopurinol is closely related in structure to hypoxanthine.

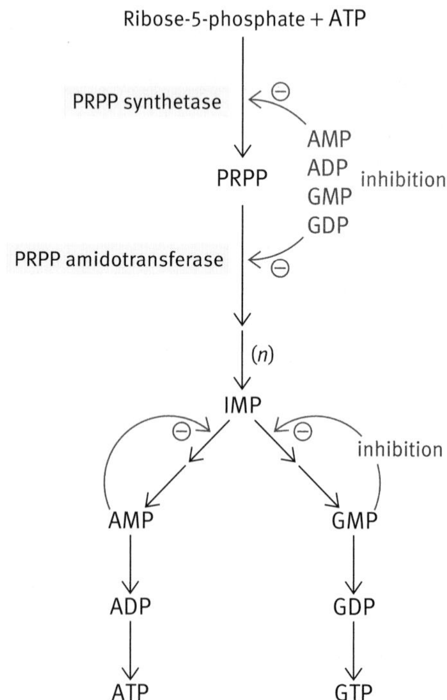

Fig. 19.8 Simplified scheme of the control of the purine nucleotide biosynthesis pathway. PRPP, 5-phosphoribosyl-1-pyrophosphate.

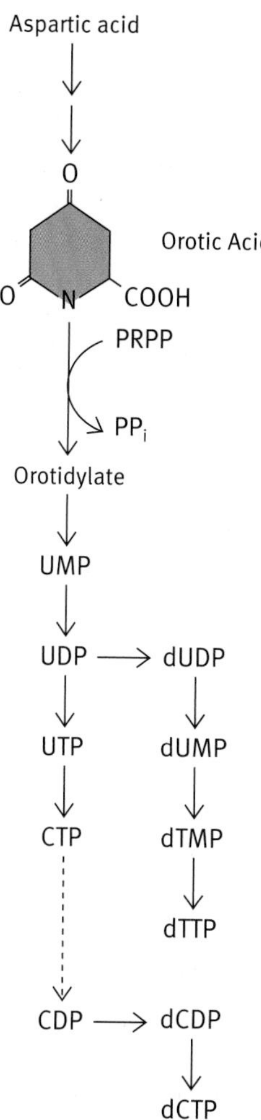

Fig. 19.9 Summary of pyrimidine nucleotide synthesis. (It is assumed here that CTP is converted to CDP.) The formation of deoxynucleotides is described later in this chapter. The complete pathway is given in Fig. 19.10.

In *Escherichia coli*, control of pyrimidine nucleotide synthesis is mainly at the aspartate transcarbamoylase step. In mammals the pathway is controlled at the carbamoyl phosphate synthase step; it is inhibited by pyrimidine nucleotides and activated by purine nucleotides. The latter control serves to keep the supply of all the nucleotides required for nucleic acid synthesis in balance.

How are deoxyribonucleotides formed?

For DNA synthesis dATP, dGTP, dCTP, and dTTP are required (see Chapter 23). The reduction of ribonucleotides to deoxy compounds occurs at the diphosphate level with NADPH as the reductant. The ribonucleotide reductase has a complex mechanism involving formation of a stable radical. There are different, but related, reductases in different organisms.

$$\begin{matrix} \text{ADP} \\ \text{GDP} \\ \text{CDP} \\ \text{UDP} \end{matrix} \xrightarrow[\text{NADPH + H}^+ \;\to\; \text{NADP}^+]{\text{Ribonucleotide reductase}} \begin{matrix} \text{dADP} \\ \text{dGDP} \\ \text{dCDP} \\ \text{dUDP} \end{matrix} + H_2O$$

The resultant dADP, dGDP, dCDP, and dUDP are converted into the triphosphates by phosphoryl transfer from ATP.

However, dUTP is not used for DNA synthesis, since DNA, you will recall, has thymine (the methylated uracil) as one of its four bases, but never uracil. The dUTP is converted into dTTP. This is done in three steps: first, dUTP is hydrolysed to dUMP:

$$\text{dUTP} + H_2O \rightarrow \text{dUMP} + PP_i$$

The dUMP is converted to dTMP and then to dTTP by phosphoryl transfer from ATP. An appropriate system of allosteric feedback controls exists to keep the production of the four deoxynucleotide triphosphates in balance.

Thymidylate synthesis – conversion of dUMP to dTMP

The methylation of dUMP is carried out by the enzyme **thymidylate synthase**; it utilizes N^5, N^{10}-methylene FH_4. In purine synthesis, the methylene group of the latter is oxidized to produce a formyl group. In thymidylate synthesis, the methylene group is transferred and, at the same time, *reduced* to the methyl group of thymine. The reducing equivalents for the reduction come from FH_4 itself, leaving it as FH_2 (note how versatile this coenzyme is). In the scheme below, only the relevant part of N^5, N^{10}-methylene FH_4 is shown.

dUMP (PO_3^{2-}—O—deoxyribose) + N^5, N^{10}-Methylene FH_4

↓

dTMP (PO_3^{2-}—O—deoxyribose) + Dihydrofolate (FH_2)

The FH_2 produced in this reaction is reconverted to FH_4 by **dihydrofolate reductase.** The FH_4 is reconverted to methylene FH_4 by reaction with serine (see earlier in this chapter).

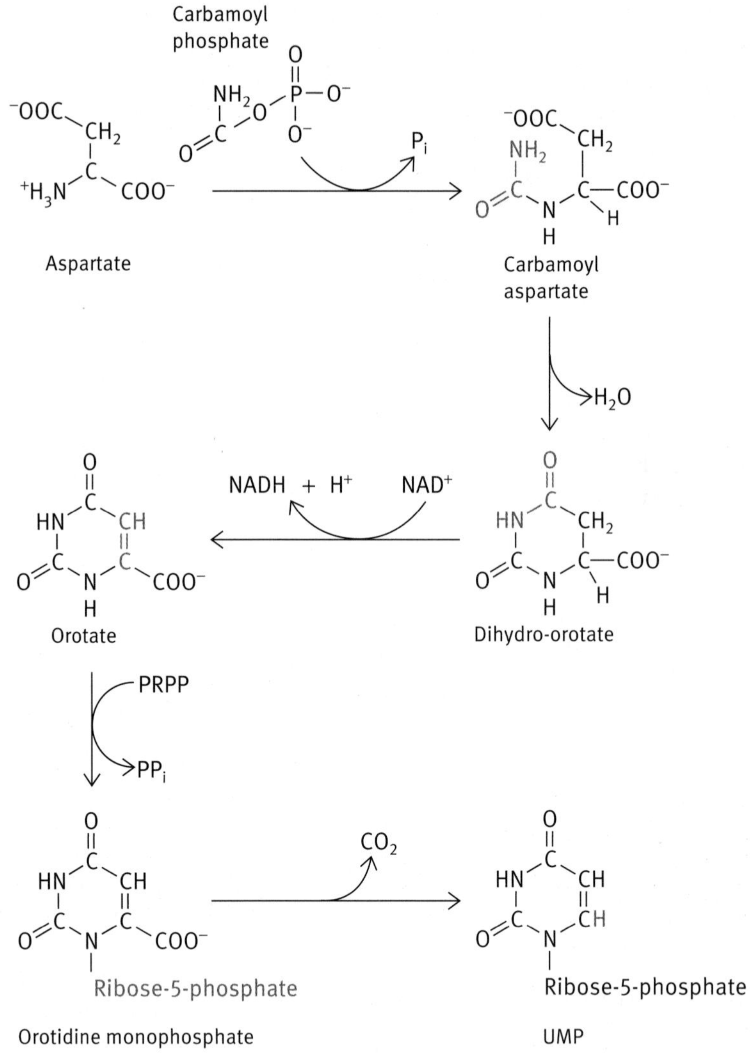

Fig. 19.10 Details of the pathway for the *de novo* synthesis of pyrimidine nucleotides, given for reference. Blue indicates the structural change resulting from the latest reaction.

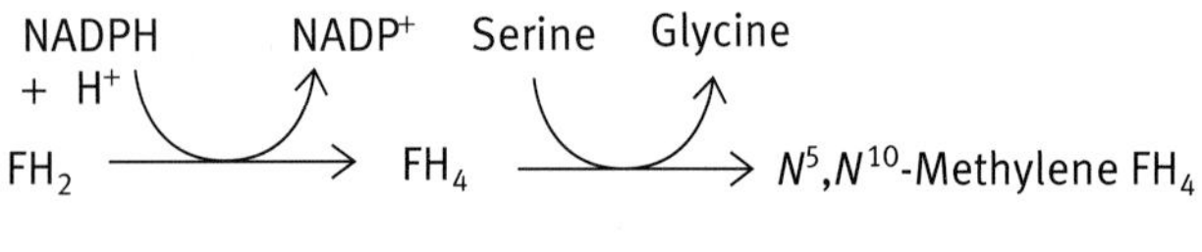

Medical effects of folate deficiencies

Cells must synthesize DNA if they are to divide and they need supplies of all four deoxynucleotides in order to do so; failure to produce any of them in adequate quantities will impair cell division. Some cells, such as erythrocyte precursors and cancer cells, which divide rapidly, are particularly sensitive to restrictions in nucleotide availability.

FH_4 is involved in several metabolic syntheses such as glycine, serine, and methionine formation but these components are usually available from the diet and folate deficiency would not cause a deficiency of these amino acids. Nucleotide synthesis, however, is a different story. Deficiency of folate in the diet leads to a type of **anaemia** known as megaloblastic anaemia, typified by large fragile immature erythrocytes because nucleotides and therefore DNA cannot be synthesized at a rate sufficient for the growth and maturation of these cells. Folate deficiency during pregnancy has also been associated with the birth of infants with neural tube defects such as **spina bifida**, and **anencephaly**. Pregnant women and women planning a pregnancy are advised to take folate supplements.

Thymidylate synthesis is targeted by anticancer agents such as methotrexate

In the synthesis of thymidylate from dUMP and N^5,N^{10} methylene FH_4, described earlier, the products are dTMP and FH_2

since the FH_4 supplies the two H atoms needed for the reduction of the methylene group to a methyl group. The FH_2 must be reduced to FH_4 by dihydrofolate reductase for the molecule to be used catalytically. If this reduction is inhibited, then FH_4 cannot be formed and as the FH_2 is not biologically active an effective folate deficiency is created. The antileukaemic drugs **methotrexate (amethopterin)** and the related **aminopterin** (both known as antifolates) competitively inhibit the dihydrofolate reductase, by mimicking the folate structure. They were the first drugs to be used in this type of cancer treatment known as **chemotherapy**. Cancer cells, like those of leukaemia, require rapid dTMP production to synthesize DNA and are selectively, but not exclusively, inhibited. The scheme is outlined in Fig. 19.11. As methotrexate is not specific for cancer cells only, side effects such as loss of hair also occur in chemotherapy.

The structure of methotrexate (amethopterin) is shown here for interest; aminopterin is similar but lacks the N^{10}-methyl group.

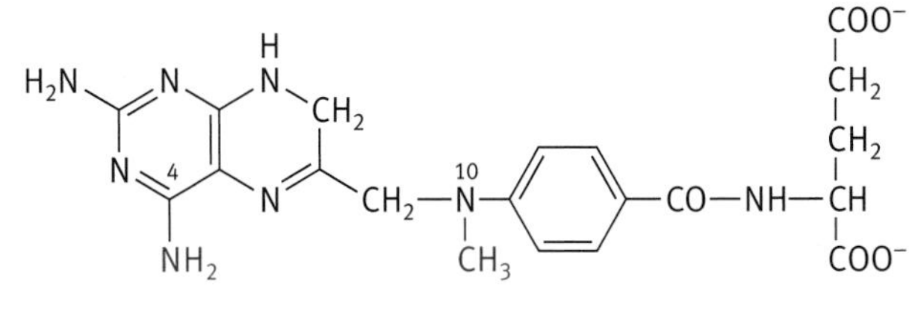

Methotrexate (amethopterin)

Fluorouracil is another agent attacking folate reduction. It is converted in cells into the corresponding deoxynucleotide, a potent FH_2 inhibitor used in some cancer therapies.

Vitamin B_{12} deficiency in cells and the folate methyl trap

Folate, is not just a carrier of formyl and methylene groups. It can carry a number of other one-carbon groups of different oxidation states as shown in Fig. 19.12. All the forms shown are interconvertible but once the N^5,N^{10}-methylene FH_4 is further reduced to N^5-methyl FH_4, the only way to return the folate to the pool of FH_4 derivatives, is to transfer the methyl group to homocysteine and convert it into methionine.

Folate

It is a carrier of 1-C fragments e.g.:

- -CHO N5 formyl FH_4
- -CHO N10 formyl FH_4
- -CH=NH N5 formimino FH_4
- =CH- N5,10 methenyl FH_4
- $-CH_2-$ N5,10 methylene FH_4
- $-CH_3$ N5 methyl FH_4

Fig. 19.12 States of oxidation of one-carbon fragments carried by tetrahydrofolate.

A coenzyme derived from vitamin B_{12} is needed in the transfer of the methyl group of N^5-methyl FH_4 to homocysteine. The reaction is catalysed by **methionine synthase** (Fig. 19.13).

It can be seen that deficiency of B_{12} can cause a functional folate deficiency as in the absence of B_{12}, the folate will be 'trapped' in the methylated form and will be unable to return to the pool of FH_4 for reuse in other reactions in nucleotide synthesis. It is not surprising then, that the haematological picture in vitamin B_{12} deficiency resembles that of folate deficiency, that is, it is a megaloblastic anaemia as in both cases folate cannot be used, The relationship between B_{12} deficiency and inability to use folate is known as the **methyl trap** (Fig. 19.14).

Vitamin B_{12} occurs in most diets and is needed in very small amounts but is not present in plants so that vegan diets are often deficient. A more common cause of B_{12} deficiency is the disease known as **pernicious anaemia**, where a gastric glycoprotein called the **extrinsic factor**, needed for absorption of the vitamin, is missing. In this way the cells of the body lack the coenzyme irrespective of the adequacy of the diet.

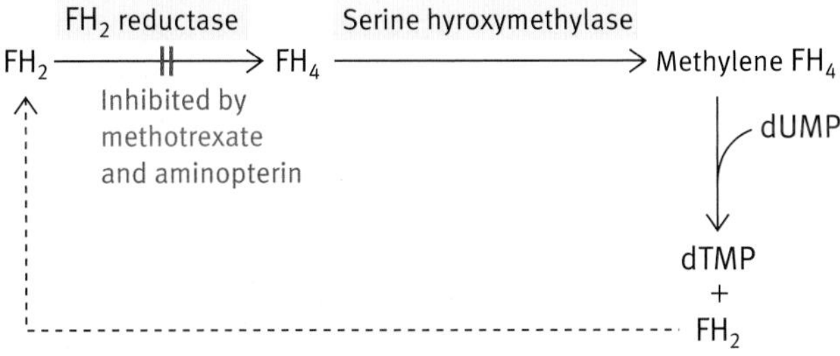

Fig. 19.11 Site of action of the anticancer agent, methotrexate. The relationship of the methotrexate structure to folic acid can be seen by comparing the structure of methotrexate with that of folic acid earlier in this chapter.

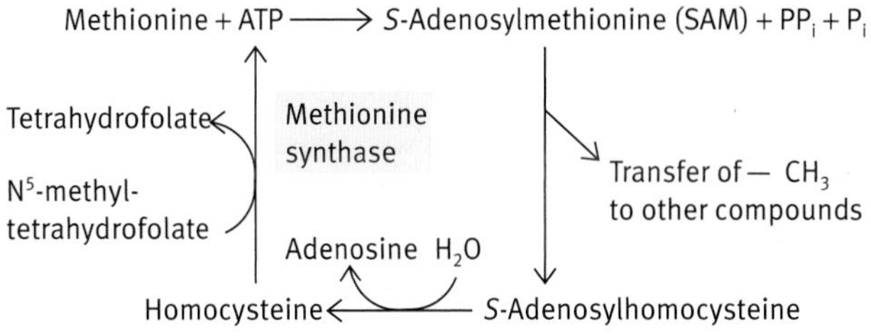

Fig. 19.13 Recycling methionine back from homocysteine. The methyl group of methionine is activated to form *S*-adenosylmethionine (SAM), a major donor of methyl groups. Homocysteine can be recycled to methionine by the transfer of a methyl group from N^5-methyltetrahydrofolate, a reaction that needs a coenzyme derived from vitamin B_{12}.

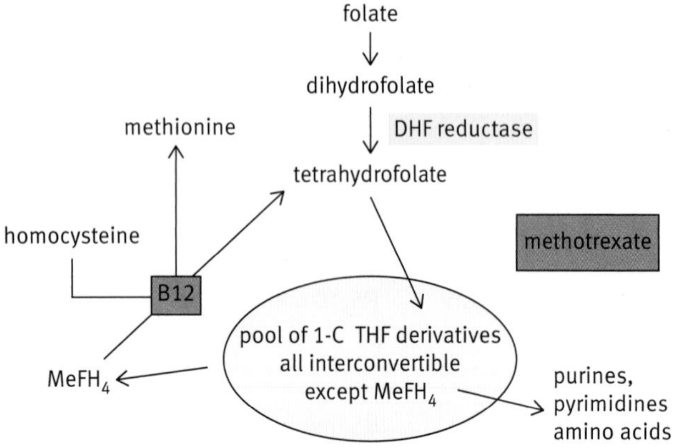

Fig. 19.14 The methyl trap. THF carries 1-C fragments of different oxidation states, all of which are interconvertible except methyl FH_4. To return $MeFH_4$ to the pool the methyl group has to be removed. This requires vitamin B_{12} to transfer it to homocysteine. In vitamin B_{12} deficiency, the FH_4 is 'trapped' in the $MeFH_4$ form resulting in a functional deficiency of folate.

Vitamin B_{12} has no role in nucleotide synthesis as such, but it is involved in other reactions (see Chapter 14), which possibly account for the neurological abnormalities found in pernicious anaemia but not in folate deficiency. Treatment of pernicious anaemia with folate administration does not alleviate the neurological disorders so it is very important to diagnose correctly and treat accordingly. There are pressure groups in various countries lobbying for the supplementation of flour, or other cereal products, with folate, in order to ensure that women of reproductive age are not folate deficient, but in many cases there is strong opposition as folate supplementation might mask an existing B_{12} deficiency in some members of the population. The blood profile would be normal but neurological problems would eventually appear.

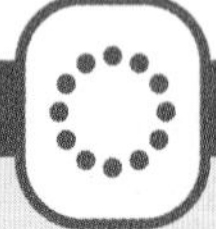

Summary

Nucleotides are synthesized *de novo*. Purine nucleotides are also synthesized by a salvage pathway, by ribotidation of free bases released by the degradation of nucleotides. In mammals the pyrimidine salvage pathway appears to be unimportant.

In the *de novo* synthesis of purine nucleotides, the purine ring is assembled by a series of steps in which the intermediates are already joined to ribose-5-phosphate. 5-Phosphoribosyl-1-pyrophosphate (PRPP) is the universal ribotidation agent. The product of the synthesis pathway is inosine phosphate (IMP), which is converted to form both GMP and AMP. Synthesis of purine nucleotides is allosterically controlled.

The pathway involves two additions of formyl groups, reactions depending on the coenzyme tetrahydrofolate (FH_4). This molecule is formylated using serine as the formyl donor forming glycine and formyltetrahydrofolate.

Dividing cells require deoxynucleotides and so are sensitive to deficiencies in the synthesis pathway. Folate deficiency in the diet can produce anaemias. Deficiency during pregnancy is associated with birth of babies with neural tube defects, such as spina bifida.

The salvage pathway involves two enzymes, one of which catalyses the ribotidation of free adenine, the other of hypoxanthine or guanine, the latter transferase (HGPRT or hypoxanthine-guanine phosphoribosyl transferase) being the more important in humans.

In the Lesch-Nyhan syndrome, infants lack the HGPRT, which leads to mental retardation and self-mutilation; *de novo* synthesis is low in amount in the brain, which therefore is particularly sensitive to deficiency in the salvage pathway. In liver, HGPRT deficiency causes PRPP accumulation possibly because it is not used as much as is normal. This stimulates *de novo* synthesis, leading to overproduction of purines. The excess is converted to uric acid. The uric acid problem does not appear to be related to the neurological symptoms of the Lesch-Nyhan syndrome.

Excessive synthesis of PRPP is a factor causing gout in some patients. Allopurinol is used clinically to inhibit uric acid formation. Pyrimidine nucleotides are formed *de novo* via a different synthesis pathway resulting in UMP synthesis.

The formation of deoxynucleotides needed for DNA synthesis occurs at the diphosphate level, catalysed by a reductase using NADPH. The dNDPs are converted to the triphosphates by phosphoryl transfer from ATP to provide the substrates for DNA synthesis. dUTP is not used in DNA synthesis. dTTP is used. The methylation to form thymidylate occurs at the dUMP level catalysed by thymidylate synthase. The conversion requires FH_4, which, during the formation of the thymine methyl group, is oxidized to dihydrofolate (FH_2). The antileukaemic drugs, methotrexate and aminopterin, inhibit the recycling of the FH_2 back to FH_4 and inhibit cell division and maturation.

Vitamin B_{12} is required for the formation of a cofactor involved in the conversion of homocysteine to methionine. If this vitamin is absent, such as in pernicious anaemia, FH_4 accumulates in the methylated form. This creates a shortage of FH_4 for nucleic acid synthesis. This **methyl trap hypothesis** explains why the red blood cell abnormalities in megaloblastic anaemia resemble those found in folate deficiency.

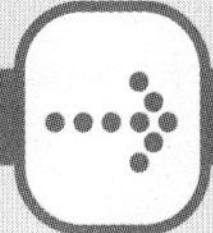

Further reading

To access the further reading, please scan the QR code image or go to http://global.oup.com/uk/orc/biosciences/molbiol/snape_biochemistry/student/reading/ch19/

Problems

1 Describe the method of ribotidation.

2 What is the cofactor involved in the two formylation reactions involved in purine nucleotide synthesis? Give the essential structure of its formyl derivative.

3 What is the origin of the formyl group? Explain how this is generated.

4 What is the function of hypoxanthine-guanine phosphoribosyltransferase (HGPRT)?

5 The Lesch-Nyhan syndrome is a severe genetic disease in children. Discuss its biochemistry.

6 How does the drug allopurinol reduce uric acid formation?

7 Draw a diagram illustrating the main allosteric controls on purine nucleotide synthesis.

8 Several compounds in the body are methylated using methionine as a methyl group source. Is this true of thymidine monophosphate synthesis? Explain your answer.

9 How does the antileukaemic drug methotrexate inhibit cancer cell reproduction?

10 The symptoms of pernicious anaemia resemble, in some respects, those of folate deficiency. What is the reason?

Chapter 20

Mechanisms of metabolic control and their applications to metabolic integration

In this chapter we will try to integrate all the preceding metabolic processes to see how the various pathways work and interact together.

We have collected together metabolic control mechanisms into a separate chapter, rather than dealing with the topic when the pathways were discussed. We can, in this way look at control strategies in general, and see their application to the integration of carbohydrate, fat, and protein metabolism. It is important to look at these pathways because the flow of chemical change (flux) through them is large. In the body, the direction of the fluxes is changed as meals are followed by periods of fasting. The integration of carbohydrate, fat, and protein metabolism illustrates all of the principles of metabolic control. The subject is a fairly complex one, so let us explain how this chapter deals with it. Metabolic control involves control of enzyme activities. We will, therefore, first deal with the strategies by which enzyme activities are regulated. After this we will describe how the pathways are kept in balance by regulatory enzymes that respond to the concentrations of metabolites in other pathways, and adjust their rates of activities accordingly.

There are additional controls as the metabolic activities of cells depend on circulating hormones, which operate in the body as a whole according to current needs. We therefore describe how signals external to the cells – extracellular signals produced by other cells of the body – regulate the metabolic pathways, on top of the more local controls first mentioned.

But before we deal with any of this the first question to consider is the following one.

Why are controls necessary?

When you think of the major pathways that we have dealt with – glycogen synthesis, glycogen degradation, glycolysis, gluconeogenesis, fat synthesis, fat degradation, amino acid synthesis, amino acid degradation, the TCA cycle, electron transport – it is obvious that they cannot all be running at full speed (or even necessarily at all) at the same time. If a metabolic pathway is proceeding in one direction, the reactions in the reverse direction must be switched off, otherwise nothing but heat generation would be achieved. In a single pathway such as glycolysis in muscle for example, its required rate will vary according to the energy needs at the time. The metabolic rate of someone playing squash is about six times greater than that of the person at rest. The regulation of metabolism of the main energy yielding materials achieves at least three goals:

- it avoids the potential problem of futile cycles, or substrate cycles as they are now called (explained in next section)
- it allows response to energy-production needs as energy expenditure varies
- it allows response to physiological needs – metabolic pathways work in different directions after a meal when metabolites are being stored (see Chapter 10) as compared with intervals between meals when stored energy reserves are being utilized. The needs are different again in fasting and prolonged starvation, and in pathological situations such as diabetes mellitus where carbohydrate, fat, and amino acid metabolism are abnormal.

The potential danger of futile cycles in metabolism

The process of gluconeogenesis in the preceding chapter provides a good example for illustrating the potential danger of futile cycles in metabolic systems. Consider the conversion of fructose-6-phosphate into fructose-1,6-bisphosphate in glycolysis and the reversal in gluconeogenesis (Fig. 20.1). In glycolysis, fructose-6-phosphate is phosphorylated using PFK to yield fructose-1,6-bisphosphate, while in the reverse

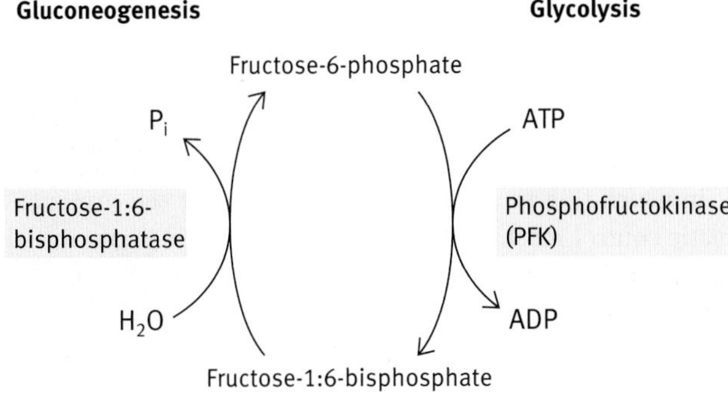

Fig. 20.1 Potential futile cycle at the phosphofructokinase (PFK) step in glycolysis if the reactions were uncontrolled.

direction fructose-1,6-bisphosphatase hydrolyses the product back to fructose-6-phosphate. This cycle of events would, if unchecked, deplete the cell of ATP and achieve no progress in either direction of the two pathways.

Extending this to the whole of glycolysis and gluconeogenesis, if there was no control of forward and reverse reactions these pathways would constitute a giant futile cycle, again achieving nothing but uselessly destroying ATP. The same applies to glycogen synthesis and breakdown, and fat synthesis and breakdown (Fig. 20.2) or, for that matter, to any pathway involving synthesis and breakdown.

It is clear therefore, that the degradative and synthetic reactions in a metabolic pathway must be controlled in a reciprocal manner – activation of one and inhibition of the other. This independent control of the two directions can occur only, as mentioned earlier, at irreversible metabolic steps, for it is here that there are separate reactions in the two directions catalysed by distinct enzymes, which can be separately controlled. A freely reversible reaction is catalysed by the same enzyme in both directions, which cannot be separately controlled. If that enzyme is inhibited, the pathway cannot proceed in either

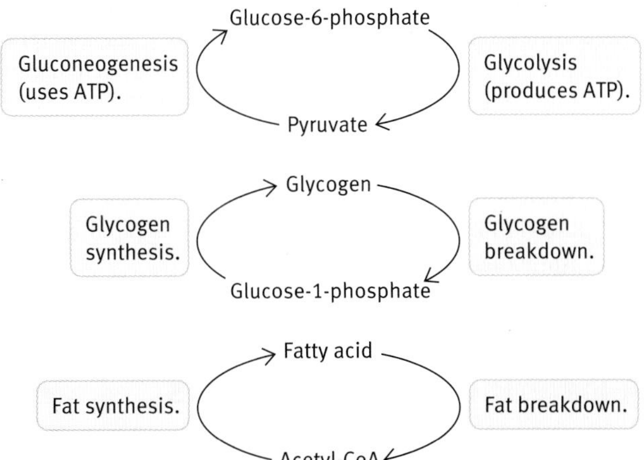

Fig. 20.2 Potential large-scale futile cycles if metabolism were not controlled.

direction. As explained earlier, control of the direction of flow of metabolites in a metabolic pathway is achieved by the fact that key reactions are not exact reversals of one another.

With the realization that controls are operative, the term 'futile cycle' has been replaced by 'substrate cycle' to describe situations such as occurs at the PFK step in glycolysis. Substrate cycling, if allowed to occur at all, may appear to be wasteful. But a limited amount of cycling may be advantageous in making controls on the forward and backward metabolic pathways more sensitive. Suppose you have a pathway in which A is converted to C *via* B and the reaction A→B occurs as a substrate cycle:

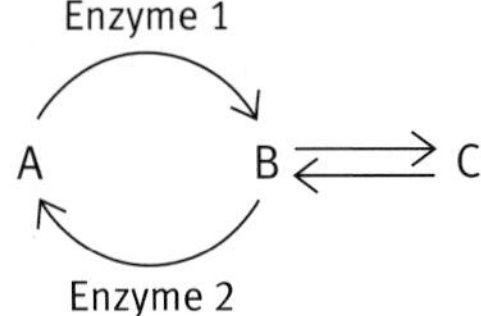

The rate of conversion of A to C can be reduced by inhibiting enzyme 1 or activating enzyme 2. If you do both, the control is more effective; complete shutdown could be achieved without excessively high concentrations of the inhibitor for enzyme 1. Inhibition of the flux from C to A could similarly be achieved by inhibiting enzyme 2 and activating enzyme 1. The fructose-6-phosphate/fructose-1,6-bisphosphate cycle has dual controls of the type described.

How are enzyme activities controlled?

Metabolic regulation of a pathway means regulation of the rate of one or more reactions in that pathway. There are essentially two ways of reversibly modulating the rate of an enzyme activity in a cell. These are:

- a change in the amount of the enzyme
- a change in the rate of catalysis by a given amount of the enzyme, that is, a change in enzyme activity.

(Compartmentation adds a further possibility that availability of substrate may play a role, so that control of transport proteins may be relevant.)

There are ways of *irreversibly* activating enzymes, such as the proteolytic conversion of trypsinogen to active trypsin, but in this chapter we are dealing with control mechanisms that are reversible, since this is essential to meaningful metabolic control.

Metabolic control by varying the amounts of enzymes is relatively slow

The concentration of a protein in a cell can be changed by altering the rate of its synthesis and/or the rate of its degradation.

Proteins in general have variable life spans; the half-life of insulin is 4 minutes and of collagen a few years, but the half-life of enzymes in, say, the liver, might range from about 30 minutes to several days.

This type of control, which is at the gene activation level, is a relatively long-term affair, with effects being seen in hours or days rather than seconds. It is at the level of adaptation to physiological needs. A few examples are given here.

- Lipoprotein lipase concentrations in capillaries (see Chapter 11) are adjusted to the fat demands of the tissues, the levels increasing in lactating mammary glands.
- The liver changes its enzyme content within hours in response to dietary changes – whether it has to cope with high fat or high carbohydrate intake.
- In the fed state, hepatic enzymes involved in fat synthesis increase in amount; a few hours of fasting result in a decrease in the amount of these enzymes.
- Intake of foreign chemicals, such as drugs, results in a rapid increase in hepatic drug-metabolizing enzymes (see Chapter 32).
- A particularly good example is that of enzymes of the urea cycle. Excess nitrogen is converted into urea for excretion into the urine. All of the enzymes of this pathway change in the rat in proportion to the nitrogen content of the diet (see Chapter 19).

Control at the level of enzyme synthesis is spectacular in bacteria. For example, if *Escherichia coli* is exposed to lactose as its sole carbon source, synthesis of the enzyme β-galactosidase, needed to hydrolyse the sugar, starts up immediately; an increase in enzyme activity is detectable in minutes and a 1000-fold increase can be measured within hours. When the stimulus for enzyme synthesis is no longer there, reversal of increases in enzyme concentrations occurs relatively slowly, since they are returned to lower values only by destruction of the proteins (or dilution by cell growth in bacteria since they multiply so rapidly). The half-lives of regulatory enzymes in mammals tend to be short – perhaps 30 minutes to an hour or so. We will return to this subject later when we deal with control of gene expression (see Chapter 26).

Metabolic control by regulation of the activities of enzymes in the cell can be very rapid

For regulation of *activities*, the amounts of given enzymes are not altered, only the rate at which they work. This method of control is very quick, depending on the type of control mechanism.

Which enzymes in metabolic pathways are regulated?

In many or most metabolic pathways, there are certain enzymes with special regulatory properties. Often, but not always, these are found at the first metabolic step, which commits the pathway to the formation of a specific end product.

Consider a metabolic pathway such as:

$$A \rightarrow B \rightarrow C \rightarrow D \rightarrow E \rightarrow \rightarrow \text{utilization for cell synthesis}$$

in which E is an end product needed by the cell. An automatic control mechanism is for the first reaction committed to the synthesis of E to be controlled by the level of E as shown below. This is known as feedback inhibition.

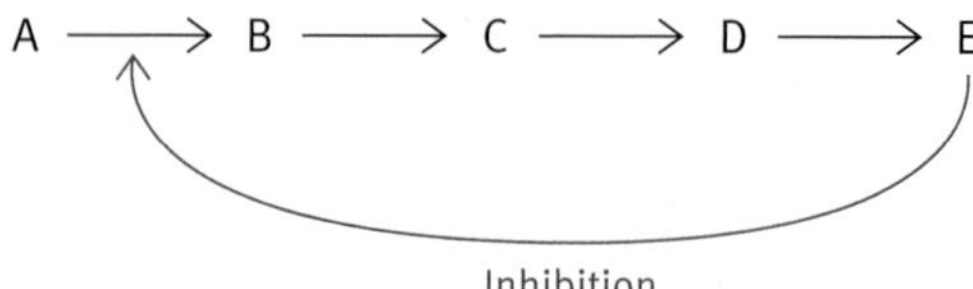

When the level of E is reduced, the inhibition is relieved and E is synthesized. The flux (flow of metabolites through the pathway) is in this sense controlled by the utilization of E, since if more E is produced than can be immediately utilized, its production is automatically diminished.

There are numerous metabolic pathways (such as for the synthesis of amino acids) in which such controls exist. Where the pathway branches, the first enzyme of each branch is often subject to regulation so that formation of each end product is separately controlled. Each end product also partially inhibits the first enzyme of the joint pathway. When it comes to fat and carbohydrate metabolism, the whole system is so complex that the term 'end products' of pathways or 'first' enzymes cannot be defined in quite the same way, so that 'end-product inhibition' is not such a relevant term. Nonetheless the same principle applies, that key intermediates (products) can control metabolically distant enzymic reactions. The control may be feedback, as described, or 'feed-forward' in which metabolites activate downstream enzymes which will be needed to cope with the flow of metabolites coming their way.

The nature of control enzymes

There are essentially two main ways in which the catalytic activity of enzymes is modulated (as distinct from a change in the amount of enzyme).

- **Allosteric control** is effectively instantaneous.
- **Covalent modification** of the protein – mainly by **phosphorylation** and **dephosphorylation** – is rapid but not instantaneous.

Allosteric control of enzymes

Allosteric control of enzymes is of central importance in metabolic regulation. The prefix *allo* means 'other'; it refers to the existence on an enzyme of one or more binding sites other than the active site for substrate binding. The ligands, which bind to the allosteric sites, are called **allosteric activators** or **inhibitors** and collectively allosteric modulators. They may or may not have a structural relationship to the substrates of the enzymes. *At a given substrate concentration*, an allosteric modulator increases or inhibits the activity of the enzyme when it binds at the allosteric site.

Allosteric modulators typically alter the K_m or apparent affinity of the enzyme for its substrates (S). (We have explained earlier (see Chapter 6) that the K_m of an enzyme may not be a true affinity constant.) Most enzymes work in the cell at subsaturating levels of [S], so that an *increase in their affinity will increase their activity and a decrease in their affinity has the reverse effect.* At saturating levels of [S], the activity is unchanged, even if the affinity is changed, but this situation does not (usually) occur in the cell.

The mechanism of allosteric control of enzymes and its reversibility

The mechanism of allosteric control of enzymes is discussed in detail in Chapter 6 using aspartate transcarbamylase as the classic example of an allosteric enzyme.

Allosteric control is virtually instantaneous in both its application and reversibility. The allosteric modulator attaches to its site by noncovalent bonds and when the concentration of ligand is reduced, it dissociates from combination with the enzyme, which reverts to its original state.

Allosteric control is a tremendously powerful metabolic concept

Allosteric modulators, as mentioned, need have no structural relationship to the substrate of the enzyme regulated. They are often from a separate metabolic pathway. This means that any metabolic pathway or area of metabolism can be connected in a regulatory manner to any other metabolic area, the metabolite(s) of one pathway being regulator(s) of another. The systems controlled are complex; glycogen breakdown provides substrates for glycolysis, which feeds pyruvate into the TCA cycle and this, in turn, feeds electrons into the electron transport system, which feeds ATP into the energy-utilizing machinery of the cell. Fatty acid degradation also provides substrates for the TCA cycle, supplying acetyl-CoA whereas acetyl-CoA formed from pyruvate is the precursor for fatty acid synthesis, and so on.

In such a complex system, each part of a pathway detects what is going on by sensing the concentrations of key metabolites. Is there enough or too little ATP? Is the TCA cycle being supplied with too little or too much acetyl-CoA? Is glycolysis too fast or too slow? You can see the chemical chaos that would result unless there was constant second-to-second adjustment of pathways in the light of information reaching each pathway about all other pathways as well as parts of its own pathway. This is why allosteric control is such an important concept – regulatory enzymes can receive signals from anywhere. Thus glycolysis can be 'informed' on how the TCA cycle is doing and how the electron transport system is keeping up ATP supply, the information automatically adjusting activities to make a harmonious chemical machine. As Jacques Monod, the originator of the allosteric concept, pointed out, you simply could not have anything as complex as a cell without it. He described it as the 'second secret of life' (DNA being the first).

Allosteric enzymes often have multiple allosteric modulators

The control network *is* complex. A given enzyme may receive multiple signals from different areas of the metabolic map, each partially inhibiting or stimulating its rate of reaction. Presumably one little evolutionary advantage here and another allosteric signal there has resulted in fine tuning of enzyme activity. Natural selection of the control mechanisms has ensured efficient performance of the whole complicated mass of reactions.

Control of enzyme activity by phosphorylation

The second method of enzyme control is by phosphorylation. To be strictly accurate this section ought to refer to covalent modification of enzymes rather than phosphorylation, because other chemical modifications occur, but phosphorylation is of such overwhelming importance in eukaryotic cells that we will confine ourselves to it. Unlike allosteric control, which is of major importance in both prokaryotes and eukaryotes, phosphorylation is less important in prokaryotes, but in eukaryotes it is of paramount importance in many vital areas quite apart from metabolic control.

Protein kinases and phosphatases are key players in control mechanisms

The principle is simple. Enzymes called **protein kinases** transfer phosphoryl groups from ATP to specific enzymes. When this happens, the target enzyme undergoes a conformational change such that the enzyme (or an enzyme-inhibitor protein) changes its activity (or inhibitory effect). You can imagine that the addition of such a strongly charged group could have an effect on the conformation of a protein molecule. The reverse process is achieved by **phosphoprotein phosphatases** (often abbreviated to **protein phosphatases**), which hydrolyse the phosphate from the protein (Fig. 20.3).

The phosphorylation occurs on the hydroxyl group of a serine or threonine of the polypeptide chain of the enzyme, and is identified by the protein kinase by the neighbouring sequence of amino acids around the target –OH group. (In Chapter 29

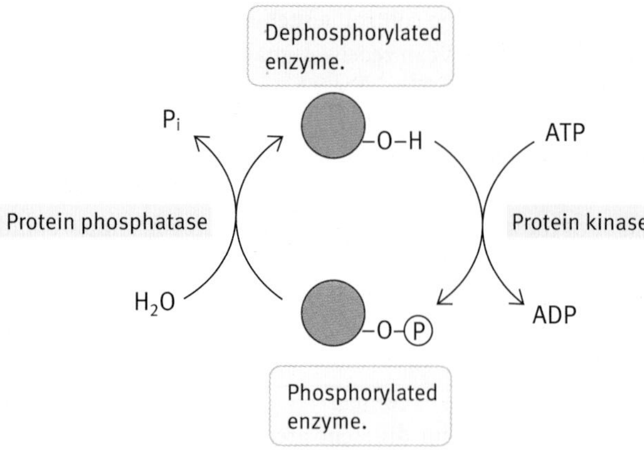

Fig. 20.3 Control of an enzyme by phosphorylation. The phosphorylated enzyme may, in specific cases, be more active or less active than the unphosphorylated enzyme. Note that variations of this may occur in that the activity of an enzyme may be controlled by an inhibitor protein whose inhibitory activity is controlled by phosphorylation.

we describe tyrosine group phosphorylation, which has great importance in gene control as well as the mechanism of action of the hormone insulin.) We show the serine phosphorylation process below.

The control of enzymes by phosphorylation is the second general mechanism by which enzymes are controlled – allosteric control was the first, phosphorylation and dephosphorylation of proteins the second. We will shortly describe how the two methods of enzyme control apply to specific metabolic systems.

Control by phosphorylation usually depends on chemical signals from other cells

Allosteric control gives an immediate regulatory mechanism, but control by phosphorylation requires that the phosphorylation and dephosphorylation processes themselves are controlled. The balance between phosphorylation and dephosphorylation determines the activity of the target enzyme. What, therefore, controls the protein kinases and phosphatases? The answer lies for the most part *not* in intracellular controls but in controlling agents such as hormones, external to the cell. (There are a few individual exceptions to this in which phosphorylation is part of the internal metabolic controls – pyruvate dehydrogenase is one (see Chapter 13) – but the general point applies.)

The internal, allosteric controls coordinate the different pathways and parts of pathways in such a way that metabolic pile-ups and shortages do not occur. As already implied there is a second group of controls in the body, involving hormones that determine the direction of metabolic pathways, such as whether to store fuel, or release it. This is essential if cells are to engage in activities compatible with life. Individual cells cannot regulate the direction of metabolic flow without receiving signals from other cells. Although the overall metabolic control is hormonal, the internal controls determined by allosteric effects and phosphorylation/dephosphorylation reactions are still essential to keep the pathways coordinated avoiding bottlenecks and metabolic complications. A simple parallel is the organization of a navy. Each individual ship (cell) has its own internal (allosteric) system of discipline, which, in all circumstances, maintains it as an organized unit, but what it does as a unit – where it sails and what it does – depends on external signals from higher naval authorities (endocrine glands and the brain, the brain often controlling hormone release from the endocrine glands).

General aspects of the hormonal control of metabolism

In control of carbohydrate , fat, and amino acid metabolism, the hormones of special importance are **glucagon**, **insulin**, and **adrenaline** (also known as **epinephrine**) and these are what we will mainly deal with here. They are the ones which have immediate and rather dramatic effects and which are invoked on a daily basis as we oscillate between eating and fasting periods between meals with occasional adrenaline-releasing events thrown in. Cortisol, released by the adrenal cortex and growth hormone from the anterior pituitary also have effects on metabolism, but their effects are longer term.

Glucagon is produced by the α cells of the islets of Langerhans in the pancreas when blood glucose is low. It causes mobilization and release of stored food components. Insulin is produced by the β cells of the islets of Langerhans in the liver when blood glucose is high and it signals cells to store fuel. Adrenaline, the hormone liberated by the adrenal gland as a result of stressful situations, has a mobilizing effect on food-storage reserves and prepares the body for action. Insulin is the only hormone that lowers blood glucose. All the others mentioned above, glucagon, adrenaline, cortisol and growth hormone increase blood glucose. Insulin is the only true anabolic hormone of the body and its presence signals synthesis and storage.

How do glucagon, adrenaline, and insulin work?

Hormones are chemical signals released into the bloodstream so that all tissues are exposed to them. However, only certain cells, called target cells, respond to a given hormone. What determines whether a cell is a target cell for a particular hormone is the presence of specific receptors for that hormone in that cell. Cells that do not have receptors for a given hormone do not respond to it and are not considered to be target cells. Glucagon, adrenaline, and insulin do not enter their target cells; they combine with membrane receptor proteins, specific for each hormone. These are transmembrane proteins, each

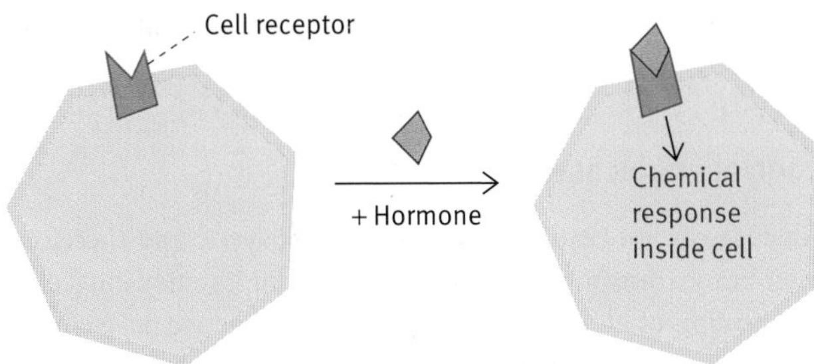

Fig. 20.4 Hormone binding to surface receptor causes chemical events inside the cell.

with an external receptor part displayed like an aerial on the outside of cells ready to combine with its specific hormone or other signalling molecule.

Hormones are quickly eliminated from the blood. Typically their half-life is a few minutes, so unless the source gland of a given hormone is releasing more, the concentration of the hormone in the blood falls, terminating the signal. However, while the hormone is bound to the receptor, changes occur inside the cell (Fig. 20.4). This leads us to the **second messenger** concept.

What is a second messenger?

If we regard the hormone as the first messenger, then binding to its external receptor on a cell leads to the change in the concentration of another signalling molecule within the cell, known as the second messenger – which causes cell responses. Not all hormone signalling involves second messengers, but the ones we are dealing with in this chapter do.

The intracellular second messenger for glucagon and adrenaline is cyclic AMP

Cyclic AMP (cAMP) (not 5′AMP) is adenosine-3′,5′-cyclic monophosphate, a molecule you have not yet met in this book. It is synthesized from ATP by the enzyme **adenylate cyclase** (Fig. 20.5) in response to the binding of glucagon or adrenaline to their cell membrane receptors.

Inside the cell, cAMP is the allosteric activator of a protein kinase (**PKA**) that phosphorylates serine or threonine –OH groups of specific enzymes, thus modulating their activities. The outline of the series of events is shown in Fig. 20.6. The way in which cAMP activates PKA is shown in Fig. 20.7. In the absence of cAMP, PKA is a tetramer with two catalytic subunits (C) and two regulatory subunits (R) and it is inactive. When cAMP binds to the regulatory subunits, the catalytic C subunits are released and become active. PKA then phosphorylates a number of enzymes and activates or inactivates them in this way.

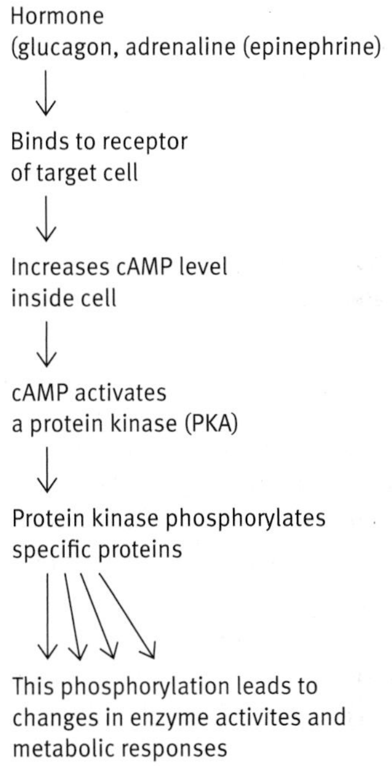

Hormone
(glucagon, adrenaline (epinephrine)

↓

Binds to receptor
of target cell

↓

Increases cAMP level
inside cell

↓

cAMP activates
a protein kinase (PKA)

↓

Protein kinase phosphorylates
specific proteins

↓ ↓ ↓ ↓

This phosphorylation leads to
changes in enzyme activites and
metabolic responses

Fig. 20.6 Steps in the hormonal control of metabolism. Not shown are: (1) that cyclic AMP (cAMP) is continually destroyed; (2) that phosphorylation of proteins is reversed by phosphatases. The metabolic response shown occurs only as long as a hormone is bound to the receptor.

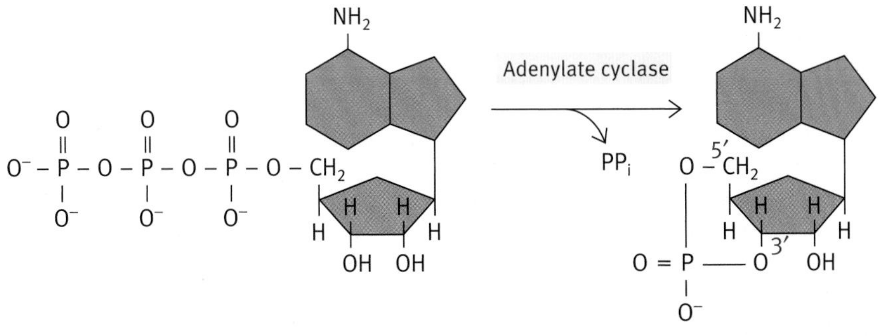

Fig. 20.5 The synthesis of cyclic AMP from ATP by adenylate cyclase.

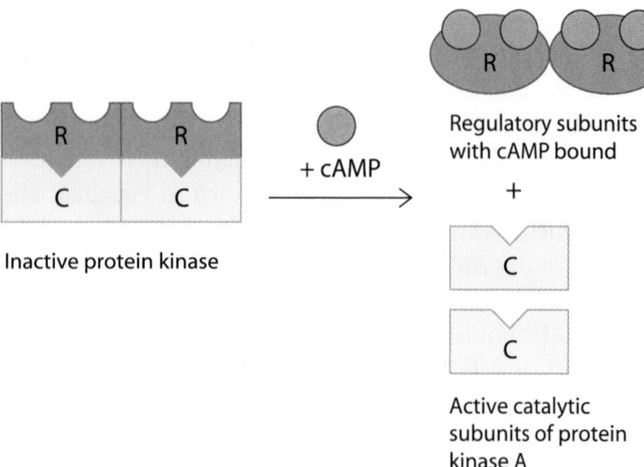

Fig. 20.7 Activation of cyclic AMP (cAMP)-dependent protein kinase (PKA) by cAMP. R, regulatory subunit of PKA; C, catalytic subunit of PKA.

An enzyme, **cAMP phosphodiesterase**, hydrolyses cAMP to AMP (Fig. 20.8) inside the cell so that the activation of cellular PKA depends on continual production of cAMP, which occurs only as long as the hormone is bound to the cell receptor and the adenylate cyclase is activated. Thus everything has the required reversibility essential for a signalling system. If the hormone concentrations fall, cAMP production ceases, existing cAMP is hydrolysed, and everything goes back to the original state and phosphorylated proteins are dephosphorylated by protein phosphatases. Protein phosphatases are usually themselves controlled.

What is missing so far in this account is an explanation of *how* the hormone binding switches on cAMP production and how the latter is switched off. We are deferring description of this until Chapter 29 because it is part of the more general major topic of signal transduction across cell membranes. What we have done so far in this chapter is to give general strategies used by the cell to control metabolism. We can now turn to the application of these to specific pathways.

Control of carbohydrate metabolism

Control of glucose uptake into cells

Glucose cannot readily penetrate lipid bilayers, and therefore membrane transport proteins are needed for its entry into cells. Absorption of glucose from the intestine is active, as we have mentioned (Fig. 7.12). However, transport into other cells of the body is by facilitated diffusion in which a specific transporter protein allows glucose molecules to traverse the membrane, driven only by the glucose gradient across the membrane.

There is a family of such glucose transporters, GLUT1 to GLUT4 and also SGLUT (which specifically transports glucose and sodium across the intestinal cell into the bloodstream), all the isoforms having a common structure characterized by 12 transmembrane sequences. GLUT4 is insulin dependent, and is found in skeletal and heart muscle, and adipose tissue, and others which are noninsulin sensitive, are found in brain (GLUT3 mainly and GLUT1), liver (GLUT2) and red blood cells (GLUT1). They have different affinities for glucose, and occur in different tissues meeting the metabolic needs, responding to hormones under different circumstances. A table showing the types of glucose transporters and their characteristics is shown in Chapter 11 (Table 11.1).

In muscle and adipose cells, where glucose uptake is insulin dependent, there is an intracellular reserve of nonfunctional glucose-transport proteins in the form of membrane vesicles. The binding of insulin to its receptor results in their fusing with the plasma membrane and appearing on the cell surface thus increasing the rate of glucose transport (Fig. 20.9). In the absence of insulin the GLUT4 is mainly present in vesicles in the cytosol. Insulin triggers a complex signalling cascade, causing a translocation of GLUT4 to the plasma membrane. It has been shown in adipose cells that the effect of insulin in recruiting the reserve glucose-transporter protein GLUT4 into the cell membrane is complete in about 7 minutes. If the insulin is removed, the process reverses, the transporters

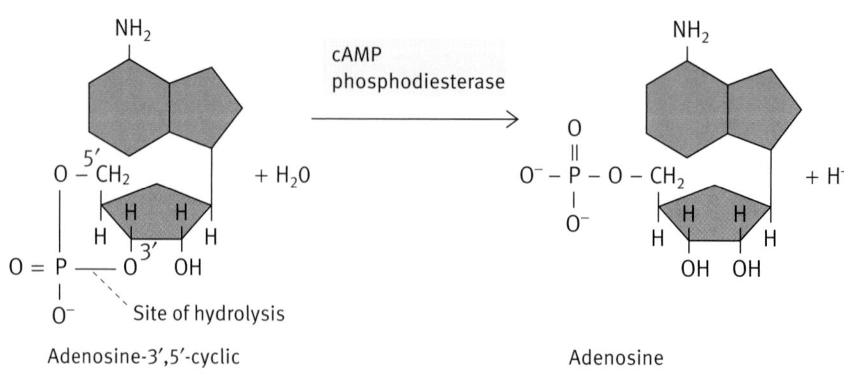

Fig. 20.8 Reaction catalysed by cyclic AMP (cAMP) phosphodiesterase. The name arises from the fact that cAMP contains a doubly esterified phosphoryl group – that is, a phosphodiester.

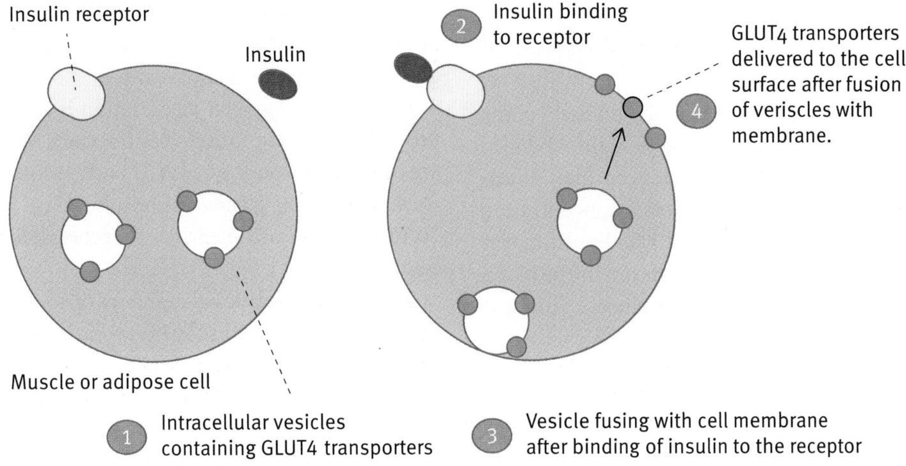

Fig. 20.9 The effect of insulin in mobilizing glucose transporter units (GLUT4) in adipose cells, and heart and skeletal muscle. The glucose transport is of the passive facilitated diffusion type. Note that liver and brain are not insulin responsive in this respect.

being returned to the nonfunctional intracellular reserve in about 20–30 minutes.

Since the transport of glucose is effected by facilitated diffusion not requiring energy, by itself this would simply equilibrate cellular and blood levels, but in the cell the glucose is trapped by phosphorylation and removed by glycogen synthesis or other metabolic pathways and this constitutes a driving force for uptake of the sugar. The first step is the phosphorylation of glucose:

$$\text{Glucose} + \text{ATP} \rightarrow \text{glucose-6-phosphate} + \text{ADP}$$

The enzyme catalysing this reacton is **hexokinase**, but in liver there is an **isoenzyme** known as **glucokinase**, which carries out the same reaction as hexokinase (also discussed in Chapter 11). Isoenzymes are different enzyme forms catalysing identical reactions, but with different characteristics, such as affinity for a substrate or its susceptibility to product inhibition, tailored to the requirements of the tissue in which they are found. Glucokinase has a much lower affinity for glucose than has hexokinase (see Fig. 11.12). This is an important regulatory device. The liver mainly takes up glucose when its concentration in the blood is high and stores it as glycogen. When blood glucose concentration is low, the liver releases glucose into the bloodstream. The low affinity of liver glucokinase minimizes the metabolism of glucose by the liver when glucose is low and allows it only when the blood glucose is high. As the liver glucose transporter is not sensitive to insulin, control of glucose uptake and metabolism in the liver is achieved by GLUT2 having a low affinity for glucose and by the fact that glucokinase has a high K_m, a kind of belt and braces approach. The brain and other cells dependent on glucose have priority for glucose uptake when glucose is low because they have a high-affinity transporter plus a low K_m enzyme, hexokinase. In muscle and adipose tissue uptake is limited to situations of high blood glucose by the transporter being insulin sensitive, but once entry has been achieved, metabolism takes place efficiently as these tissues possess hexokinase.

In addition, hexokinase is inhibited by its product, glucose-6-phosphate, but at physiological concentrations of the latter glucokinase is not. This is physiologically significant. When the blood glucose is high the liver takes it up and converts it into glucose-6-phosphate, reaching concentrations that would inhibit hexokinase but do not inhibit glucokinase and so allowing glucose metabolism and glycogen synthesis to take place. In keeping with its role, the synthesis of this enzyme is increased by insulin.

An important function of insulin is the inhibition of gluconeogenesis in the liver. Insulin is present only when blood glucose is high and inhibition of gluconeogenesis means that no additional glucose is produced and released into the blood stream. Note that the liver is carrying out gluconeogenesis at all times except after a meal. We will see later in this chapter how this control is lost in diabetes mellitus type 1 where no insulin is present. Although the blood glucose is high, the liver keeps producing glucose exacerbating the hyperglycaemia.

Control of glycogen metabolism

Glycogen is degraded by glycogen phosphorylase (see Chapter 11), and synthesized by glycogen synthase (see Chapter 11). Glycogen metabolism control has a vital role in animals, liver and muscle (kidney to a small extent) being the organs involved. The control system is complex and it works on three levels.

- There is the 'routine' ticking-over level, where the controls are of the automatic allosteric type within cells. It keeps the ATP level topped up.
- There is the physiological level depending on the feeding state. After a meal, the blood glucose is high – **insulin** levels will be high, signalling cells to store fuel; **glucagon** levels will be low. As the blood glucose concentration falls, insulin secretion will cease and rapidly disappear from the blood, and glucagon concentration in the blood rises, signalling mobilization of stored glucose from the the liver.
- There is an emergency level, in which a stressful situation results in the brain signalling the adrenal glands to release **adrenaline (epinephrine)**. This also signals for rapid glycogen breakdown as does **glucagon** in the liver. Muscle, lacking receptors, does not respond to glucagon, but does respond to adrenaline. Liver responds to both.

In plants, it is interesting that starch metabolism is much the same as glycogen metabolism in animals, with equivalent enzymes, but they have only the 'routine' level of controls. Plants do not feed at intervals, and do not run away from predators so the 'routine' controls are adequate.

Control of glycogen breakdown in muscle

When a muscle contracts, it hydrolyses ATP to ADP; to regenerate the ATP, glycogen is degraded to provide the energy source. The signal for this in normal 'routine' (nonstressful) situations is AMP (note, not cAMP), which allosterically activates muscle glycogen phosphorylase (Fig. 20.10). Note that metabolism of glycogen in muscle responds to demands for energy by the muscle, whereas glycogen metabolism in the liver responds to concentrations of blood glucose. AMP appears in the muscle cell as the product of a reaction catalysed by the enzyme adenylate kinase when concentrations of ATP in the muscle cell are low.

$$2ADP \rightarrow AMP + ATP$$

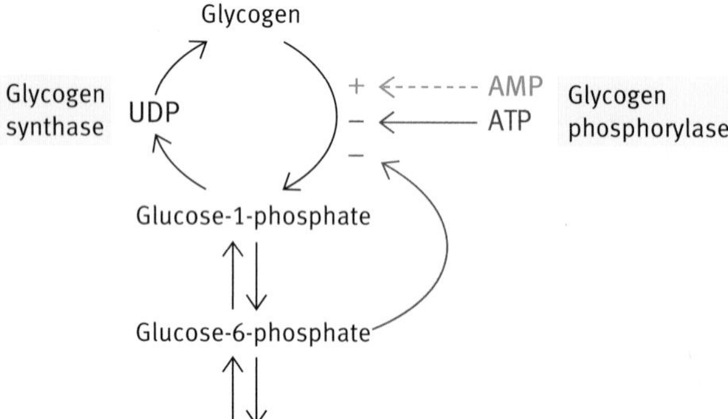

Fig. 20.10 The nonhormonal 'routine' allosteric controls on glycogen metabolism in muscle. See text for explanation. Green lines, allosteric control having positive effects; red lines, negative ones.

AMP is a sensitive indicator of low ATP concentration as it only appears in the cell in this situation, and it activates glycogen phosphorylase allosterically.

In this way glycogen degradation is stimulated and ATP is provided allowing muscle contraction. Most control systems are of the 'push-pull' type. ATP and glucose-6-phosphate allosterically inhibit glycogen phosphorylase in the muscle (Fig. 20.10). If these are plentiful then breakdown of glycogen is switched off.

There is an additional refinement to the control of glycogen metabolism in muscle. In *normal* muscle contraction (not in the stressful situation involving cAMP described below), glycogen breakdown is also *partially* activated by the Ca^{2+} released into the cytosol (see Chapter 8), which triggers the contraction following a motor nerve signal (Fig. 20.11). This mechanism additionally ensures that in muscle contraction, glycogen breakdown keeps pace with energy demand.

There is, however, another situation in which this 'routine' control of glycogen breakdown is overridden. In more stressful situations there is a need is to generate ATP at the maximum possible speed. The first minute or so may make the difference between being eaten and not being eaten, and although glycolysis gives only a low yield of ATP, it can occur very rapidly and does not require oxygen. It is therefore advantageous in an emergency to degrade glycogen at maximum speed to provide glucose-1-phosphate for feeding into glycolysis. This may be important, for the main increased production of ATP by

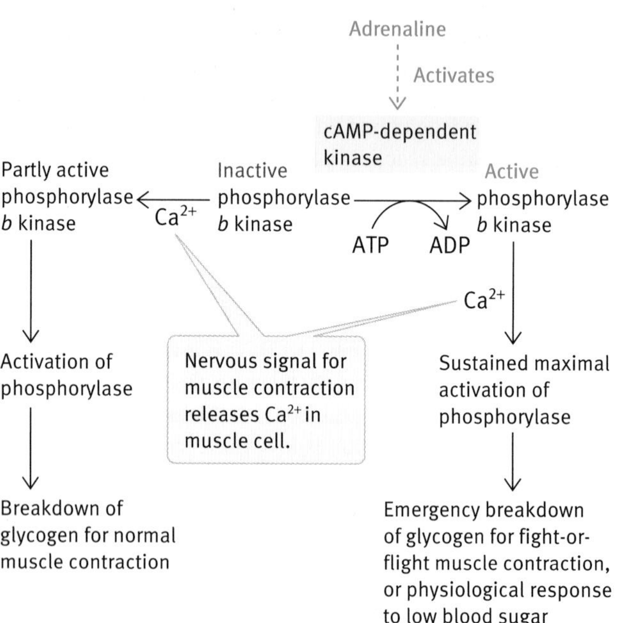

Fig. 20.11 Control of muscle phosphorylase. The enzyme has complex controls. First, it is partially activated allosterically by AMP. Second, it is activated by phosphorylase kinase. Phosphorylase kinase is itself activated in two ways. **(a)** It is partially activated by Ca^{2+} allosterically and **(b)** by a cyclic AMP (cAMP)-dependent protein kinase (PKA) plus Ca^{2+}. The mechanism is described in the text.

oxidative phosphorylation may take a minute or so while the heart speeds up to supply the extra oxygen needed.

This is known as the **flight-or-fight response**. Binding of adrenaline to cells of muscle and liver causes the production of the second messenger, cAMP (note, not AMP). This over-rides other controls, and signals maximum glycogen breakdown. It also stimulates adipose cells to release fatty acids – another energy source.

Mechanism of muscle phosphorylase activation by cAMP

In muscle, in normal situations, phosphorylase exists in a nonphosphorylated form known as **phosphorylase *b***. This is inactive in the absence of the allosteric activator, AMP, and as explained this *partial* activation by the latter is sufficient for routine needs.

In more demanding situations, **phosphorylase *b*** is converted into **phosphorylase *a***, *which is maximally active even without the presence of AMP.* Phosphorylase *b* is converted into the phosphorylated form, phosphorylase *a*, by a cAMP-activated protein kinase (**protein kinase *a*** or **PKA**, Fig. 20.12). Phosphorylase kinase can thus be activated in two ways, one (partially) by allosteric Ca^{2+} activation not involving phosphorylation, as described in Fig. 20.11, and the other by phosphorylation due to hormonal activation of PKA.

The cAMP activation of the phosphorylase kinase, which phosphorylates glycogen phosphorylase *b* is not direct; it activates the protein kinase, **PKA** (A for cAMP), which, in turn, activates phosphorylase kinase (also by phosphorylation), which then activates phosphorylase *b*, converting it into the '*a*' form. The whole scheme from the hormone onwards is shown in Fig. 20.13. What is achieved by having so complicated a mechanism? A cell will have only a relatively small number of hormone molecules attached to its receptors. In an emergency, the body requires a big response to the binding of adrenaline to cells. The attachment of a relatively few molecules of hormone to cell receptors results in a rapid and massive response, glycogen degradation The multiple steps in phosphorylase activation constitute an amplifying or **regulatory cascade**. Suppose that one molecule of hormone activates one molecule of adenylate cyclase (the enzyme producing cAMP – see earlier in this chapter), and that the latter produces 100 molecules of cAMP per minute. In this time period, the amplification is 100-fold. If the cAMP activates a second enzyme which produces an activator at the same rate, the amplification becomes more than 100×100 and so on. In fact, glycogen phosphorylase activation involves four such amplification steps. Regulatory or amplifying cascades are seen whenever a massive chemical response is produced from a minute signal. The highly branched structure of glycogen also contributes to the efficiency of the degradation process. Phosphorylase attacks the end of glycogen chains and the process of producing large numbers of active phosphorylase molecules would be wasted if there were only a few glycogen ends to work on.

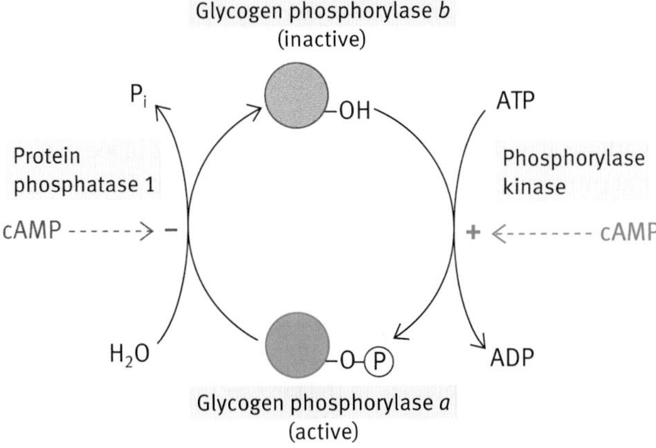

Fig. 20.12 Conversion of glycogen phosphorylase *b* into glycogen phosphorylase *a* by phosphorylase kinase and the reverse by protein phosphatase. It is important to note that the effects of cyclic AMP (cAMP) are not exerted *directly on* the reactions shown – see text. The –OH group is that of a serine residue in the protein.

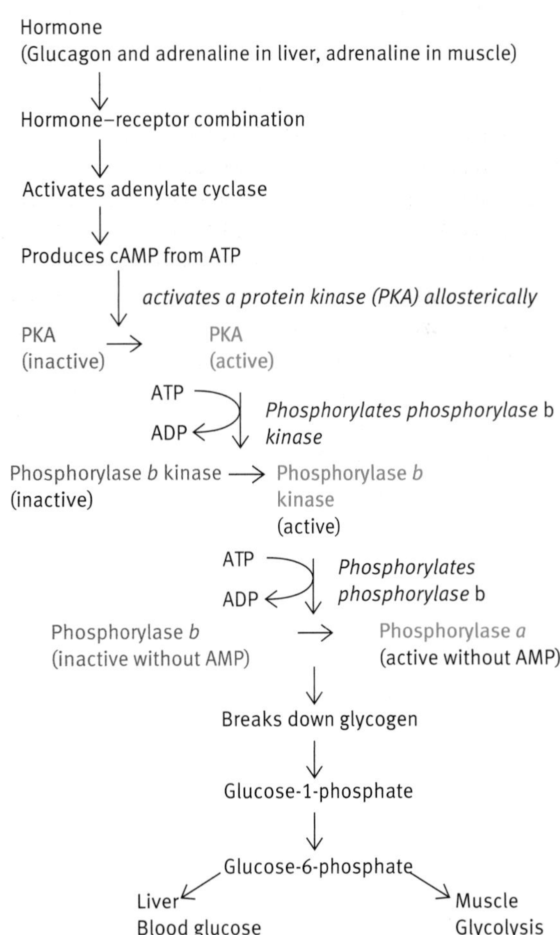

Fig. 20.13 The amplifying cascade mechanism by which hormones activate glycogen phosphorylase.

The numerous branches of the glycogen molecule, allow this to happen. The same is not true of starch but, then, plants do not go into metabolic emergencies!

Control of glycogen breakdown in the liver

As mentioned before, glycogen metabolism in the liver responds to concentrations of blood glucose and in the muscle to demand and supply of energy. Glycogen degradation in the liver leads to the release of glucose into the bloodstream rather than its metabolism by glycolysis as in muscle. Note that muscle lacks the enzyme glucose-6-phosphatase, so it is unable to produce free glucose from glycogen and so channels glucose-6-phosphate into glycolysis in situ. Liver phosphorylase *b* is not allosterically modulated by AMP as is the case in muscle. The glucose-1-phosphate produced by the enzyme is converted into glucose-6-phosphate, which is hydrolysed to free glucose and enters the bloodstream. This occurs during fasting periods and is vital for the metabolism of brain, nerve, and red blood cells, which are obligatory glucose users. The signal for glycogen degradation in the liver is the hormone glucagon, secreted by the pancreas in response to low blood glucose. Since insulin is not produced in this situation, the glucagon/insulin ratio is high and hepatic glycogen breakdown is the predominant event. Glucagon activates the liver phosphorylase by the same mechanism as that triggered by adrenaline in muscle; the second messenger for glucagon is cAMP as for adrenaline.

Liver also participates in the fight-or-flight reaction. It responds to adrenaline by releasing glucose into the blood. In this way it provides muscles with the maximum supply of fuel to generate ATP in the emergency. An added important control is that liver (but not muscle) phosphorylase is allosterically inhibited by free glucose. This makes physiological sense as it means that in the presence of high blood glucose, glycogen is not degraded to provide more glucose.

Reversal of phosphorylase activation in muscle and liver

Metabolic controls must be reversible – there has to be a switch off mechanism. In the case of glycogen phosphorylase in both liver and muscle the *a* form is converted back into the *b* form by dephosphorylation when the cAMP signal is no longer there. The phosphorylase *b* kinase is likewise inactivated.

The dephosphorylation of both is catalysed by **protein phosphatase 1**. However this can happen only when the cAMP signal is no longer there. There is a **phosphatase inhibitor 1**, which inhibits the phosphatase, but the inhibitor is active only when phosphorylated by PKA. What is rather ingenious about the system is that when cAMP is present, it activates PKA, which phosphorylates the inhibitor and activates it. Thus cAMP simultaneously leads to phosphorylase conversion of *b* into the *a* form and prevents its conversion back into the *b* form. When cAMP is no longer present, the inhibitor is dephosphorylated and no longer inhibits the phosphatase which then converts phosphorylase *a* back into the low activity *b* form. In liver, glucose itself plays a part in the inactivation of the phosphorylase. It allosterically induces a conformational change in phosphorylase *a*, which makes it susceptible to dephosphorylation by the protein phosphatase 1. PKA also phosphorylates glycogen synthase, *which inhibits it.* This inhibition will occur probably only so long as cAMP is present. However, glycogen synthase is not activated simply by the absence of cAMP, for there is another inhibitory mechanism; it is this that is removed in the presence of insulin, as described below. The cAMP-dependent inhibition of the synthase is an extra safeguard, which avoids a futile cycle of glycogen breakdown and synthesis.

The switchover from glycogen degradation to glycogen synthesis

Let us look briefly at the physiological context in which these controls are operating. Glycogen metabolism is a balance between glycogen degradation and glycogen synthesis (Fig. 20.14). Apart from the 'routine' controls where everything is ticking over in a balanced way, glycogen degradation will predominate in muscle when adrenaline is present to support vigorous muscular activity. In liver, it will predominate in two circumstances; one when adrenaline is present and two when glucagon is liberated by the pancreas in response to low blood glucose. When adrenaline is no longer present, glycogen breakdown will revert to the routine state in both muscle and liver. In liver, when the blood glucose concentration has been restored and glucagon is at low concentrations, the cAMP signal will disappear. The glucose will make phosphorylase *a* susceptible to conversion into *b*.

The switchover from glycogen breakdown to synthesis occurs after feeding when the blood glucose concentration is high. Glucose enters the pancreatic β-cell and stimulates insulin secretion. Insulin is the signal for the activation of glycogen synthase. It will continue to be secreted until the blood glucose concentration has been lowered to normal levels. The secretion of insulin and glucagon are reciprocally controlled according to the blood glucose concentration and, in the absence of insulin, glycogen synthase is inactivated by PKA. *However, the inactivation of glycogen synthase, which is removed by insulin is different from that produced by PKA in the presence of cAMP.* This will now be explained in the next section.

Mechanism of insulin activation of glycogen synthase

Much of the work on glycogen synthase control has been done on rabbit skeletal muscle but the same may apply to liver.

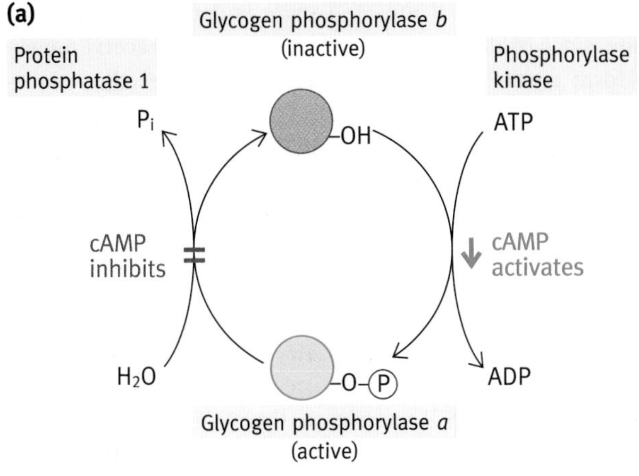

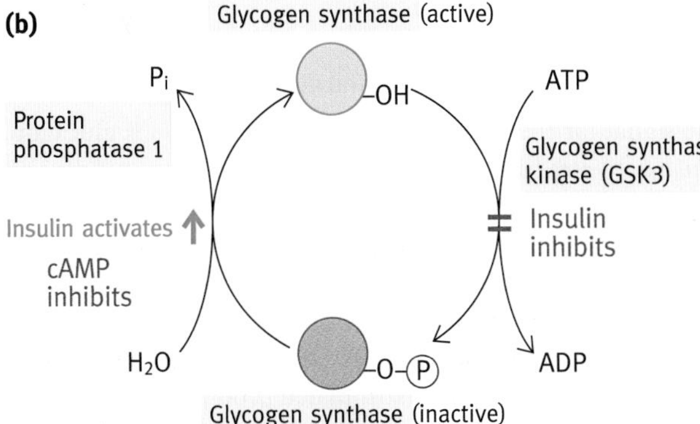

Fig. 20.14 Reciprocal controls on glycogen phosphorylase and glycogen synthase. **(a)** Cyclic AMP (cAMP) causes activation of phosphorylase kinase, which activates phosphorylase by phosphorylating it. The cAMP-stimulated PKA activates the phosphatase-inhibitor protein, which prevents inactivation of the phosphorylase by the phosphatase. It also inhibits synthase activation. **(b)** In the presence of insulin, glycogen synthase kinase 3 (GSK3) is inactivated, thus preventing the latter from inhibiting glycogen synthase. In addition insulin causes activation of the phosphatase, thus activating the synthase. Red, inactive; green, active.

In the situation after feeding, when insulin levels are high and glucagon is low, insulin effects are predominant. There is no cAMP signal when glucagon is low so phosphorylase will be in the relatively inactive *b* form.

Glycogen synthase has multiple phosphorylation sites. One group of sites of special interest to us now is a cluster of three serine residues found in the C-terminal end of the enzyme. A key observation was that when insulin activates glycogen synthase in muscle cells, *these three sites are dephosphorylated.* In the absence of insulin, a specific protein kinase called **glycogen synthase kinase 3** (GSK3) phosphorylates the sites. It inactivates glycogen synthase. *In the presence of insulin, GSK3 is inhibited, and thereby the kinase is inactivated.* At the same time insulin activates protein phosphatase 1, which dephosphorylates glycogen synthase and is thereby activated (Fig. 20.15).

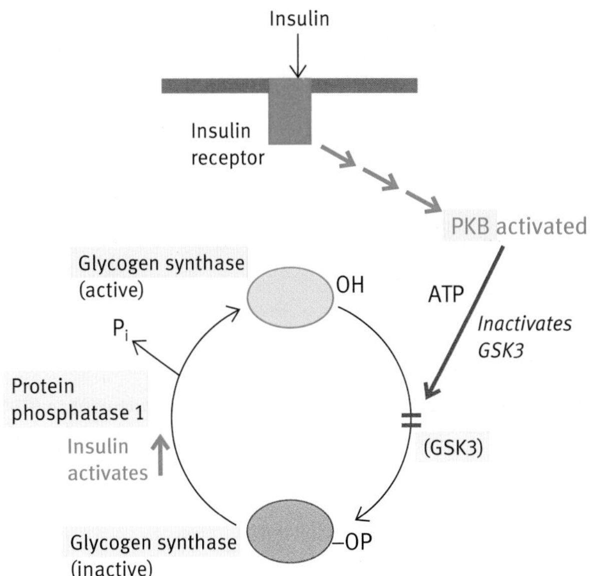

Fig. 20.15 Mechanism of control of glycogen synthase by insulin. The synthase is controlled by phosphorylation, which inactivates it, and by dephosphorylation, which activates it. Both PKA and glycogen synthase kinase 3 (GSK3) phosphorylate the enzyme, at different sites, but it is the phosphorylation performed by GSK3 that is reversed by insulin. It does this by activating PKB, a protein kinase that inactivates GSK by phosphorylating it. A protein phosphatase removes the relevant phosphate groups and thus activates the synthase. The mechanism by which insulin activates PKB is dealt with in the chapter on cell signalling (see Chapter 29).

How does insulin inactivate GSK3?

The mechanism is illustrated in Fig. 20.15. Insulin binds to its receptors on the external surface of target cells. This activates a signalling pathway inside the cell, which results in the activation of yet another protein kinase, **PKB**. (The mechanism of PKB activation by insulin signalling is described in Chapter 29). PKB phosphorylates GSK3, which *inactivates* it. Thus, to summarize, insulin causes inhibition of GSK3 and activation of the protein phosphatase. This latter dephosphorylates and consequently activates glycogen synthase.

The control is reversible. When the insulin level falls, PKB is inactivated by dephosphorylation; this allows GSK3 also to be dephosphorylated, which activates it. The glycogen synthase is now attacked by GSK3 which phosphorylates and inactivates it. In the absence of insulin the protein phosphatase needed to activate the synthase is no longer active, and glycogen synthesis ceases.

Control of glycolysis and gluconeogenesis

Allosteric controls

Figure 20.16 shows the main systems. AMP (not cAMP) indicates an increase in the ratio of ADP to ATP (see earlier in this chapter) so signalling the need to restore ATP levels. As

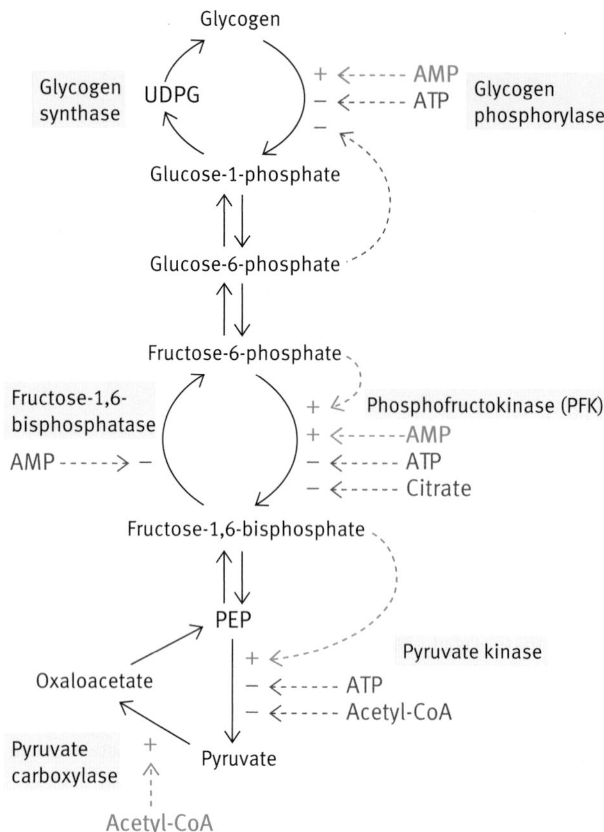

Fig. 20.16 The main intrinsic allosteric controls on glycogen metabolism, glycolysis, and gluconeogenesis. See text for explanation. Green broken lines, allosteric controls having positive effects on activities; red broken lines, allosteric controls having negative effects on activities. UDPG, uridine diphosphoglucose.

well as activating glycogen phosphorylase in muscle, AMP is an allosteric activator of **phosphofructokinase (PFK)**, which is the key controlling enzyme in glycolysis subject to multiple controls. At the same time, AMP inhibits the **fructose-1,6-bisphosphatase**. Activation of glycogen degradation by AMP increases the level of fructose-6-phosphate which also activates PFK. Activation of PFK will, in turn, increase the level of fructose-1,6-bisphosphate, which activates pyruvate kinase, an example of feed-forward control. The activating effects of AMP on PFK are balanced by inhibitory effects of ATP. The cell is thus constantly adjusting the glycolytic speed according to the ATP/ADP ratio (via AMP). In addition, when the ATP level is high, the TCA cycle flux (the passage of metabolites through the pathway) is diminished and citrate accumulates. The latter is transported out of the mitochondria to the cytosol where it allosterically inhibits PFK. This again makes physiological sense since if the TCA cycle is partially shut down, glycolysis, which feeds the cycle, should likewise be inhibited.

The activation of pyruvate carboxylase by acetyl-CoA to produce oxaloacetate needs explanation. Accumulation of acetyl-CoA occurs if the TCA cycle is deficient in oxaloacetate, since oxaloacetate is needed to accept the acetyl moiety to form citrate. Pyruvate carboxylase catalyses the anaplerotic reaction, which tops up the TCA cycle, and counteracts this deficiency (see Chapter 13). The acetyl-CoA thus automatically activates the synthesis of oxaloacetate. Inhibition of pyruvate dehydrogenase by acetyl-CoA helps to ensure adequate pyruvate for oxaloacetate production. This is also an important reaction in gluconeogenesis, which occurs in fasting (see Chapter 16). When glucose concentrations are low, acetyl-CoA produced by fatty acid oxidation activates pyruvate carboxylase and produces oxaloacetate, which will be converted into PEP and eventually into glucose.

However, as with glycogen metabolism, these internal allosteric controls on glycolysis are overridden by extracellular signals from hormones.

Hormonal control of glycolysis and gluconeogenesis

The signal for the liver to produce blood glucose is glucagon, *which activates both glycogen breakdown and gluconeogenesis.* In this way, cAMP (whose synthesis is increased by glucagon) *in liver* switches on glycogen breakdown *and* gluconeogenesis both of which produce glucose. Glycolysis needs to be inhibited otherwise both glucose synthesis and oxidation would take place at the same time (Fig. 20.17(a)). The same applies to the effects of adrenaline-stimulated production of cAMP.

In muscle, adrenaline stimulation of cAMP production increases generation of ATP by glycolysis. As mentioned earlier, adrenaline is released in the fight-or-flight reaction in which vigorous muscular contraction maybe called for. Glycolysis speed-up is required in muscle to prepare for the fight-or-flight reaction. How is it that gluconeogenesis is favoured in liver by an increased cAMP whereas glycolysis is favoured in muscle using the same signalling molecule? There are a number of mechanisms that ensure that metabolism flows in different directions in the two organs.

The muscle cannot produce free glucose from glycogen via glucose-6-phosphate as it lacks glucose-6-phosphatase. It does not carry out gluconeogenesis as it does not respond to glucagon, which activates gluconeogenic enzymes beyond pyruvate carboxylase in the liver. So increased cAMP in muscle, leads to glycogen degradation and the product, glucose-1-phosphate is converted into glucose-6-phosphate and channelled into the glycolytic pathway, which will generate ATP (Fig. 20.17(b)).

In contrast, the liver responds to glucagon, which activates other gluconeogenic enzymes, such as fructose-1,6-bisphosphatase and glucose-6-phosphatase, so that the oxaloacetate produced by activation of pyruvate carboxylase by cAMP is channelled into glucose production and not the TCA cycle. In addition, as we will see shortly, cAMP inhibits some glycolytic enzymes in the liver, but not in muscle, so that glycolysis in the liver is inhibited by cAMP. We will see even more controls on glucose metabolism, which are different in muscle and in liver.

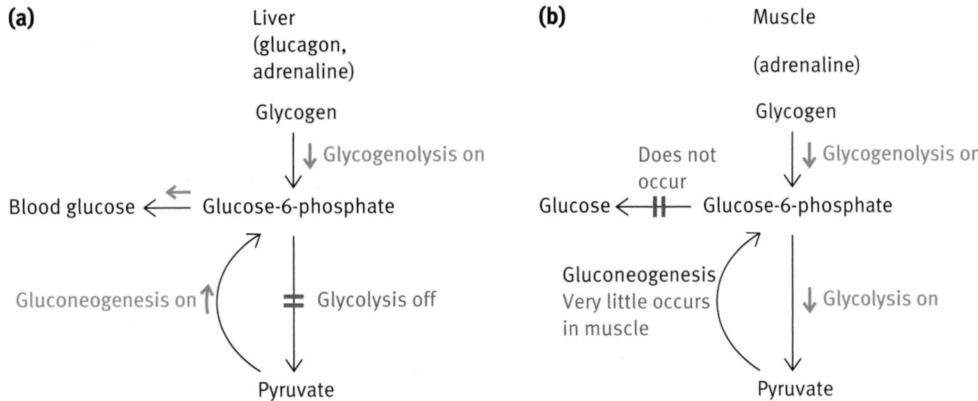

Fig. 20.17 The different control requirements of **(a)** liver and **(b)** muscle in response to glucagon and/or adrenaline. Both hormones have cyclic AMP as second messenger. The term glycogenolysis used here is a synonym for glycogen breakdown.

Control of glycolysis and gluconeogenesis by fructose-2,6-bisphosphate

This is an area of elegant complexity. *Fructose-2,6-biphosphate (F-2,6-BP) is a regulatory molecule for PFK.* It has not previously been mentioned in this book. The reverse of the PFK reaction, catalysed by fructose-1,6-bisphosphatase (FBPase), has to be reciprocally controlled if a futile cycle is to be avoided (Fig. 20.18). PFK is allosterically inhibited by ATP and is inactive unless this is counteracted by F-2,6-BP. *The rate of glycolysis in muscle and liver parallels the concentration of F-2,6-BP* so we must consider what controls the concentration of this molecule. The synthesis of F-2,6-BP is catalysed by an enzyme, which phosphorylates fructose-6-phosphate in the 2 position. It is called **PFK2**, and is bifunctional, with two catalytic sites. One synthesizes F-2,6-BP, the other hydrolyses it back to fructose-6-phosphate. The enzyme does one or the other but not both at the same time. It is phosphorylated by a kinase, PKA activated by cAMP. *In the phosphorylated form PFK2 hydrolyses F-2,6-BP. When the phosphate group is removed it synthesizes it* (Fig. 20.19). *F-2,6-BP stimulates PFK (note, not PFK2) but inhibits the reverse reaction by FBPase.*

Muscle and liver PFK2 enzymes are different

Adrenaline in muscle and liver, and glucagon in liver, as described earlier, cause the production of cAMP as a second messenger. The latter activates PKA in both tissues. The liver and muscle have different isoforms of PFK2. The liver one is phosphorylated by PKA, causing it to switch from synthesizing F-2,6-BP to hydrolysing it. It thus inhibits PFK and, therefore glycolysis, and stimulates FBPase, which is needed for gluconeogenesis.

Muscle PFK2 has no site for phosphorylation (the serine residue present in liver PFK2 that receives the phosphate group is replaced by an alanine residue in the muscle enzyme). Thus, cAMP does not stimulate hydrolysis of F-2,6-BP in muscle. The concentration of F-2,6-BP rises in the presence of an adrenaline signal. It is not known, precisely, how this occurs, but the

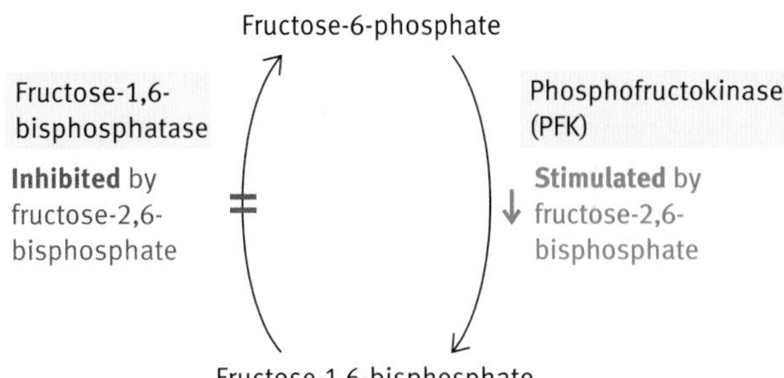

Fig. 20.18 Fructose-2,6-bisphosphate and the control of glycolysis in liver (see text).

most likely explanation is that, by stimulating glycogen breakdown, the supply of fructose-6-phosphate, the substrate of PFK2, increases and stimulates the synthesis of F-2,6-BP.

Control of pyruvate kinase

Pyruvate kinase (Fig. 20.20) is another enzyme responding to glucagon *via* cAMP. In the liver, but not in muscle, cAMP causes phosphorylation of the enzyme, resulting in its inactivation. This is therefore a glycolytic switch off, additional to that at the PFK step. Gluconeogenesis is, as stated, a function of the liver mainly in response to glucagon (though other hormones such as cortisol also have a role; see below). We have described how gluconeogenesis from pyruvate requires the following steps, catalysed by pyruvate carboxylase and phosphoenolpyruvate carboxykinase (PEP-CK) respectively:

$$\text{Pyruvate} + \text{ATP} + \text{HCO}_3^- + \text{H}_2\text{O} \rightarrow \text{oxaloacetate} + \text{ADP} + \text{P}_i + \text{H}^+$$

$$\text{Oxaloacetate} + \text{GTP} \rightarrow \text{PEP} + \text{GDP} + \text{CO}_2.$$

You can see that there would be a futile cycle, as shown in Fig. 20.21, if pyruvate kinase continues to catalyse the reaction:

$$\text{PEP} + \text{ADP} \rightarrow \text{pyruvate} + \text{ATP}$$

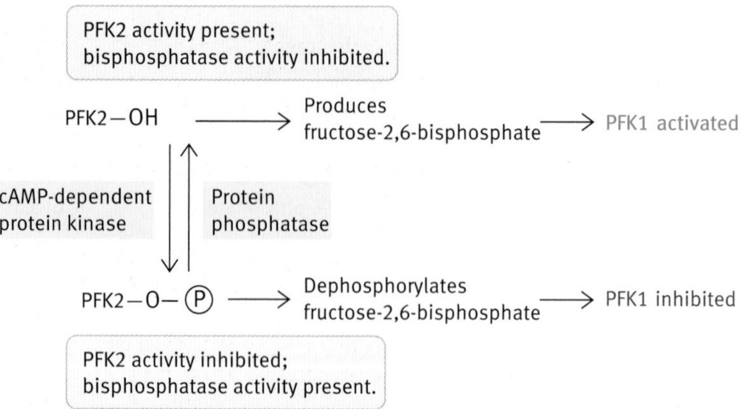

Fig. 20.19 Control of liver PFK2 by phosphorylation by cyclic AMP-dependent protein kinase (PKA). The PFK2 is a double-headed enzyme: (1) it synthesizes fructose-2,6-bisphosphate when unphosphorylated; (2) it dephosphorylates the 2,6 compound when phosphorylated.

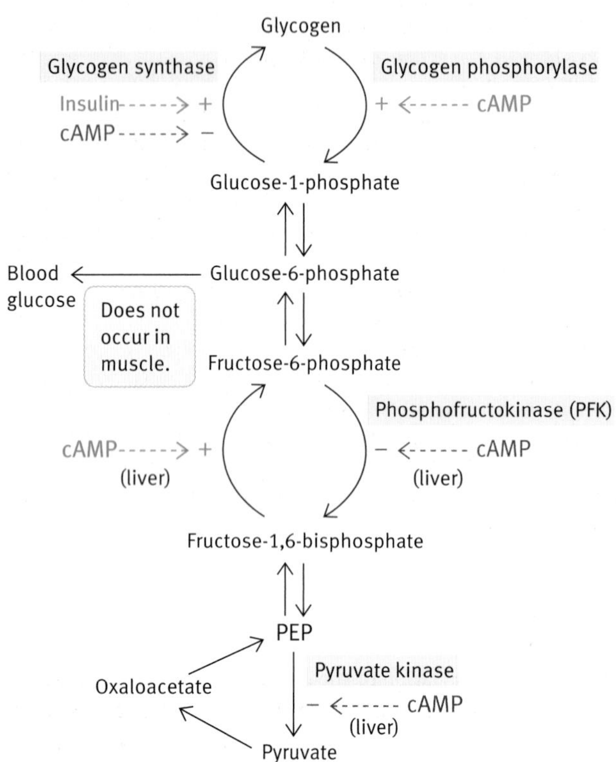

Fig. 20.20 The main external controls in this area of metabolism in liver. Note that 'cAMP' represents the action of glucagon and adrenaline. Its action is not directly on the enzyme being controlled.

For gluconeogenesis, the pyruvate kinase in liver needs to be inactivated if the PEP is to be sent into the gluconeogenic pathway. Inactivation of the liver enzyme by cAMP achieves this.

The net effect of all these controls is that in liver, the PFK in the presence of glucagon is inhibited, the FBPase is activated and gluconeogenesis supplies blood glucose. Figure 20.20 summarizes this. Once again muscle has different needs – it does not synthesize glucose, glycolysis must not be switched off, and its pyruvate kinase is not phosphorylated in response to cAMP elevation by adrenaline.

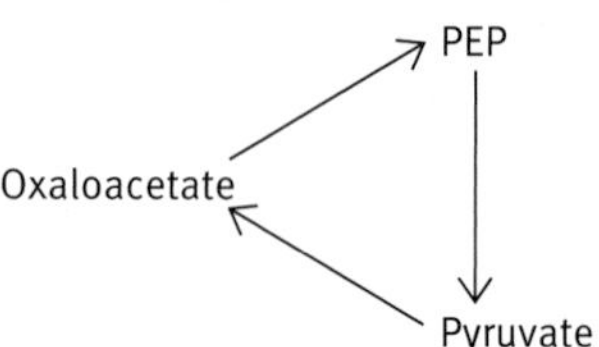

Fig. 20.21 Potential futile cycle. PEP, phosphoenolpyruvate.

Glucocorticoid stimulation of gluconeogenesis

In fasting and starvation, the main substrate for gluconeogenesis in the liver comes from amino acids derived from muscle protein degradation. During periods of stress, the cortex of the adrenal glands liberates steroid hormones called glucocorticoids, the principal one in humans being **cortisol**. It has complex effects in the body including promotion of gluconeogenesis and protein degradation in muscle and other peripheral tissues. In this way, amino acids are supplied to the liver, which uses them as gluconeogenic precursors. Cortisol also affects the activity of the PEP-CK gene needed for PEP synthesis.

Fructose metabolism and its control differs from that of glucose

In Western societies, large amounts of fructose are consumed, largely in the form of sucrose, but also in fructose drinks and other manufactured foods. Fructose is absorbed from the intestine and is metabolized mainly (or entirely) by the liver. Its metabolism is not insulin controlled, and therefore not directly affected by diabetes, and was therefore thought to be suitable for patients with the disease. It is converted in the liver into fructose-1-phosphate by **fructokinase** and this is converted by **aldolase B** into glyceraldehyde and dihydroxyacetone phosphate. (Aldolase B is different from the glycolytic aldolase A.) The glyceraldehyde is phosphorylated to the 3-phosphate, so the products are the same as in glycolysis (glyceraldehyde-3-phosphate and dihydroxyacetone phosphate). Part of this is converted back into glucose.

So, the story so far is not very different from that of glucose metabolism. However, the control situations *are* different and result in fructose being of significance beyond its calorific value in causing increased fat synthesis. The aldolase B reaction by-passes the main glycolytic control of the phosphofructokinase step. The result is that fructose metabolism may swamp the liver cell with NADH reducing equivalents, since the glyceraldehyde-3-phosphate is oxidized by NAD^+. This is similar to what happens with excessive alcohol metabolism (see Chapter 16). This, and the rapid formation of pyruvate from fructose leads to increased fat synthesis, and export from the liver as very-low-density lipoproteins (VLDL) (see Chapter 11). Dyslipidaemia such as hypertriglyceridaemia may be reverted by decreasing the sucrose and/or alchohol content of the diet.

Control of pyruvate dehydrogenase, the TCA cycle, and oxidative phosphorylation

Pyruvate dehydrogenase occupies a strategic position in metabolism as the irreversible reaction producing acetyl-CoA by which pyruvate from glycolysis feeds into the TCA cycle and into fat synthesis. The pyruvate dehydrogenase complex is regulated in several ways. Acetyl-CoA and NADH, which are products of the enzyme reaction, are allosteric inhibitors. CoA–SH and NAD^+, which are substrates, are allosteric activators. In this way the acetyl-CoA/CoA ratio and the $NADH/NAD^+$ ratio control the activity of pyruvate dehydrogenase. This is of great significance in regulating production of acetyl-CoA and its flow into the TCA cycle, but it is also of paramount importance for glucose production by the liver in fasting. Acetyl-CoA and NADH concentrations increase in fasting as they are products of fatty acid oxidation stimulated by glucagon. The reaction catalysed by pyruvate dehydrogenase is irreversible which means that we cannot synthesize glucose from fatty acid as we cannot convert acetyl-CoA into pyruvate. In other words, we cannot use our fat stores to replenish our blood glucose levels (except for a small contribution from glycerol). Body protein has to be degraded in order to provide pyruvate, which will be converted into glucose by gluconeogenesis in the liver, vital for supplying glucose for use by brain, nerve, and erythrocyte. It is important then, that any pyruvate, which has been produced from precious body protein degradation in fasting is not converted into acetyl-CoA, which is plentiful in this situation, but is channelled into gluconeogenesis instead. This is exactly what happens in fasting and this is why the inhibition of pyruvate dehydrogenase by acetyl-CoA is so important.

Of major importance, particularly in muscle, is the negative control by high ATP levels. This, in effect, is monitoring the 'energy charge'. If ATP is low then the TCA cycle activity is increased. If ATP is high, the fuel supply is cut off.The ATP control of pyruvate dehydrogenase is not a direct one. At high ATP/ADP ratios a **pyruvate dehydrogenase kinase** is activated and inactivates the dehydrogenase by phosphorylation of the enzyme; this is reversed by a phosphatase (Fig. 20.22). The kinase is actually part of the pyruvate dehydrogenase complex. The inactivating kinase is additionally allosterically activated by NADH and acetyl-CoA. Usually protein kinases are activated by extracellular signals, but this is one of the exceptions.

In the TCA cycle and electron transport chain, the intrinsic controls are different in that the major internal controls are the availability of NAD^+ and ADP as substrates. If much of the NAD^+ is present in its reduced form (NADH) then the dehydrogenases in the TCA cycle are restricted in their activity by lack of substrate. Since NADH accumulates when electron transport to oxygen cannot cope with available NADH produced by the cycle, the cycle is inhibited. Similarly if the ADP/

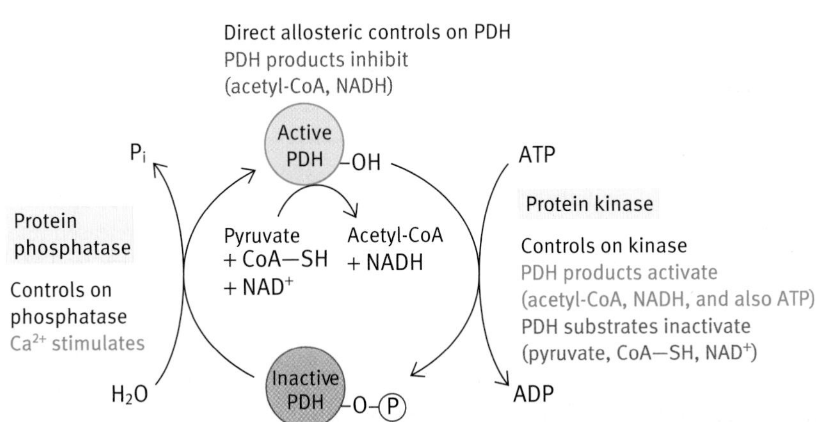

Fig. 20.22 The intrinsic control of pyruvate dehydrogenase (PDH) by direct allosteric control and by phosphorylation and dephosphorylation in the mammalian enzyme complex. The multiplicity of these controls reflects the strategic position of PDH, whose activity is the gateway for pyruvate to enter the TCA cycle and the fat synthesis pathways. Despite their complexity in detail, they largely add up to activation by substrates and inhibition by products. (In this context ATP can be regarded as a product of the result of entry of acetyl-CoA into the TCA cycle.) It is worth noting that control by phosphorylation of proteins is usually associated with extrinsic controls, rather than with the intrinsic ones here.

ATP ratio is low, electron transport is inhibited as oxidation and phosphorylation are tightly coupled (called **respiratory control**). This is of major importance as it allows ATP production to switch off if enough ATP is available. In addition to this control by NAD^+ and ADP availability, the cycle is allosterically controlled at the citrate synthase step (ATP inhibits), the isocitrate dehydrogenase step (ATP inhibits and ADP activates), and the 2-oxoglutarate dehydrogenase step (succinyl-CoA and NADH inhibit). In this way the cycle metabolites are kept in balance with one another and do not lead to accumulation, or shortages, of metabolites.

Controls of fatty acid oxidation and synthesis

Nonhormonal controls

These are illustrated in Fig. 20.23. Acetyl-CoA carboxylase plays a key role here. It converts acetyl-CoA into malonyl-CoA, which enters the fatty acid synthesis pathway. Fatty acid oxidation must be inhibited to avoid a futile cycle uselessly synthesizing fatty acids and degrading them again to acetyl-CoA. The two pathways for fatty acid synthesis and degradation suppress each other. Fatty acyl-CoAs (the product of the first step in fatty acid oxidation) allosterically inhibit acetyl-CoA carboxylase, the first committed step in fat synthesis, while malonyl-CoA inhibits transfer of the fatty acyl groups to carnitine, which prevents them from being transported to the intramitochondrial site of their oxidation (Chapter 14). Acetyl-CoA carboxylase therefore is a key controlling enzyme. It is activated by citrate, which is transported out of the mitochondria only when its concentration is high so allowing fatty acid synthesis to take place rather than increasing the activity of the TCA cycle.

An important control on acetyl-CoA carboxylase is by phosphorylation, which inhibits it. This is effected by the AMP-activated kinase to be described shortly.

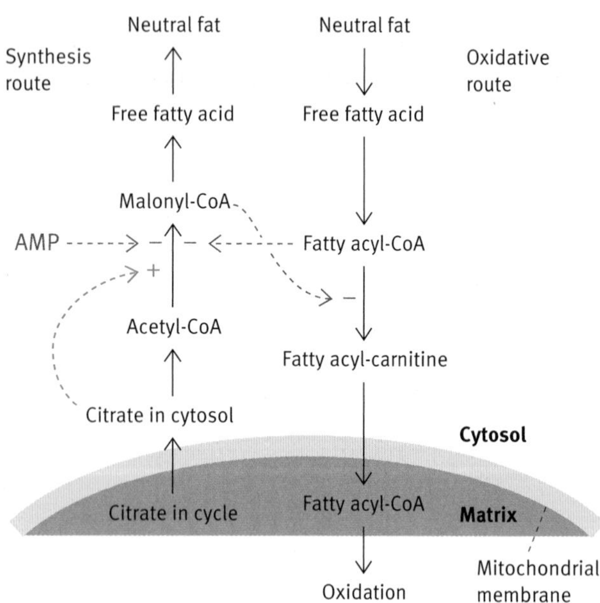

Fig. 20.23 Major intrinsic control points in fat oxidation and synthesis by which the two routes mutually suppress each other. Note that once acetyl-CoA is formed from fat breakdown, its further metabolism is subject to the controls of the TCA cycle, etc., already described. Dashed lines indicate allosteric effects.

Degradation of acetyl-CoA carboxylase is another type of control of fat metabolism

A protein known as TRB3 mediates the degradation of the carboxylase by proteasomes (Chapter 25). TRB3 is induced by cellular stress; it blocks the action of insulin and increases fat oxidation by causing the degradation of acetyl-CoA carboxylase. It shifts the balance of control from synthesis to energy production, much as AMPK (AMP-activated kinase) does, but in a different way.

Hormonal controls on fat metabolism

What determines whether adipocytes store fat as triacylglycerol (TAG) or release free fatty acids from stored TAG into the blood? Insulin is the signal for the former, glucagon and adrenaline for the latter. The concentration of blood glucose is the main controller, for it determines the relative concentrations of the two hormones.

Adipose cells contain a **hormone-sensitive lipase**, which is activated by glucagon and adrenaline (via cAMP). It carries out the reaction:

$$\text{Triacylglycerol} + H_2O \rightarrow \text{diacylglycerol} + \text{free fatty acid}$$

The diacylglycerol is further degraded into glycerol and free fatty acids. The hormone-sensitive lipase is controlled by phosphorylation, the phosphorylated enzyme being active. A cAMP-activated protein kinase carries this out and a protein phosphatase reverses it in the absence of cAMP (Fig. 20.24).

Insulin antagonizes these effects; it stimulates fat synthesis in liver and the high insulin/glucagon ratio restricts release of fatty acids from adipocytes, and increases glucose uptake.

Adipose cells take up fatty acids from VLDL released into the circulation by the liver, and re-esterify them to TAG. Glycerolphosphate is required for this esterification (Chapter 16), Adipocytes cannot phosphorylate glycerol released from TAG hydrolysis by hormone-sensitive lipase (or lipoprotein lipase for that matter) as they lack glycerol kinase, but glycerolphosphate can be supplied by reduction of the glycolytic

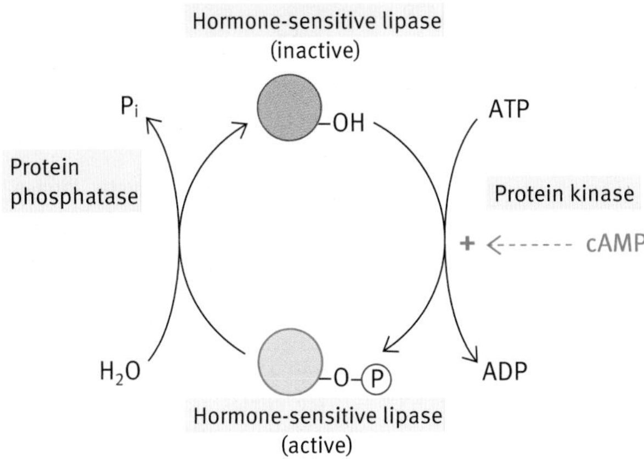

Fig. 20.24 Activation of adipose cell hormone-sensitive lipase by cyclic AMP (cAMP)-dependent phosphorylation. Insulin antagonizes the effects of the catecholamine hormones and glucagon, which increase the cAMP level. Note also that glucagon and adrenaline cause inhibition of fat synthesis in liver and adipose cells, respectively, by preventing dephosphorylation of acetyl-CoA carboxylase.

intermediate dihydroxyacetone phosphate. Entry of glucose into the adipocyte is stimulated by insulin via GLUT4 recruitment to the cell surface. Glucose then enters the cell and the glycolytic pathway producing dihydroxyacetone phosphate, which is reduced to glycerolphosphate. This makes the role of glycolyis in the adipocyte rather different from other tissues as it is not primarily an energy generating process, but one that provides glycerolphosphate for re-esterification of fatty acids and storage of TAG (see Chapter 29 on GLUT4 recruitment)

Effects of leptin and adiponectin on fat metabolism

In dealing with obesity and appetite controls (Chapter 9), we mentioned two hormones produced by adipocytes of the fat depots in the body, leptin and adiponectin. Both are secreted by white adipose tissue and both enhance tissue sensitivity to insulin. Leptin concentrations reflect the size of the adipose tissue. In obesity, leptin concentrations are high, but the obese seem to be leptin resistant, whereas adiponectin concentrations are decreased in obesity. We have seen how leptin has neurogenic effects in the hypothalamus affecting food intake. In addition, both leptin and adiponectin also have a more direct control on fat metabolism. They both activate the AMPK described below and inhibit acetyl-CoA carboxylase by phosphorylation.

Responses to metabolic stress

We have already referred to the release of the hormone cortisol in times of stress. But also there are situations, such as excessive exercise or oxygen deficiency, where cells become deficient in ATP to a dangerous point. In heart cells, ATP shortage could be very serious.

Cells have a general 'emergency' response, essentially shutting down anabolism (synthetic reactions). Two mechanisms are known – one is the AMPK, and another is the response to hypoxia. These will now be described.

Response to low ATP concentrations by AMP-activated protein kinase

We have described how the production of ATP by oxidative phosphorylation is controlled. ATP level in the cell is of critical importance and there is a more general ('global') control in cells, which senses their energy state. **AMPK** is the central effector of this control. Note that this is not AMP kinase (adenylate kinase), which phosphorylates AMP. We have already seen that when AMP is present at increased levels, it indicates a potential deficiency of ATP.

The AMPK is activated by the increased AMP concentration. It closes down non urgent metabolic processes, such as the synthesis of proteins and other cellular components, by phosphorylating key enzymes. It is itself controlled by phosphorylation; another **AMPK kinase** phosphorylates and activates it in the presence of AMP while a protein phosphatase can reverse this. The activated enzyme both shuts down anabolic reactions and increases ATP-generating catabolic processes (Fig. 20.25).

AMPK when activated has a wide range of activities.

- It mobilizes the glucose transporting GLUT4 and increases glucose uptake by cardiac muscle, skeletal muscle and adipocytes.
- It inhibits fatty acid synthesis by phosphorylating acetyl-CoA carboxylase, as described above.
- It activates the glycolytic enzyme PFK2 in cardiac, but not skeletal, muscle in oxygen deficiency.

The mobilization of GLUT4 by AMPK provides an explanation for the observation that glucose uptake by muscle is increased by exercise, without further insulin secretion. It is well known that type 1 diabetics need less insulin to maintain their blood glucose during exercise and need to adjust their insulin dose to avoid hypoglycaemia. By promoting glucose uptake into muscle AMPK may also combat insulin resistance in which glucose transport is deficient so that drug activators of AMPK, which are being developed by the pharmaceutical industry, may have antidiabetic value.

AMPK has some undesirable effects too. Cells in the mass of a tumour tend to be deficient in oxygen and AMPK is believed to protect tumour cells from oxygen depletion and so can assist in cancer development.

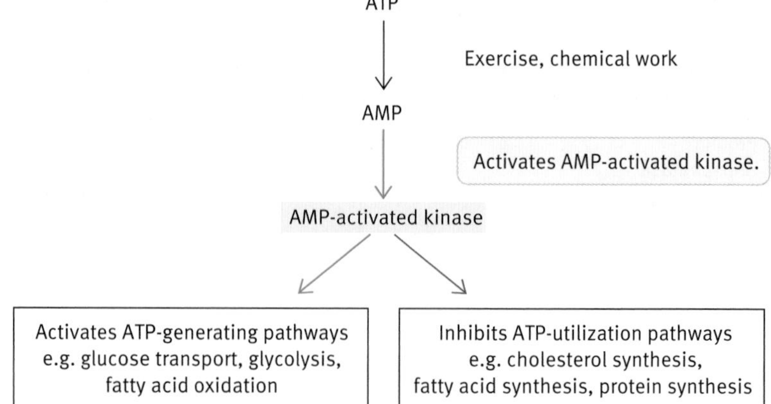

Fig. 20.25 Simplified version of the role of AMP-activated kinase in restricting anabolic reactions when increase in the AMP/ATP ratio indicates that ATP levels are suboptimal.

Response of cells to oxygen deprivation

Protection against hypoxia

Hypoxia refers to a situation in which the oxygen level in a tissue is low. In almost all mammalian cells (except red blood cells since they get their ATP from anaerobic glycolysis), an adequate supply of oxygen is of overriding importance since most of the ATP production depends on it. There are several protective responses to hypoxia, most at the level of gene activation. These include the following:

- increased production of the hormone **erythropoietin** causes increased red blood cell production and hence increases the oxygen-carrying capacity of blood
- some glycolytic enzymes and glucose transporters are induced. This is of physiological significance as glycolysis is the only alternative way of generating ATP in the absence of oxygen
- factors are produced that promote **angiogenesis** (the production of new blood vessels).

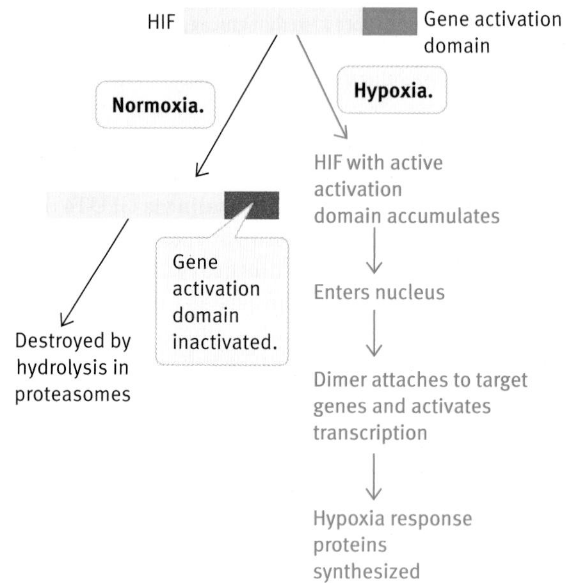

Fig. 20.26 Summary diagram of response to oxygen levels. HIF, hypoxia-inducible factor.

Mechanism of the response to hypoxia

The key to hypoxic gene activation is a family of transcription factors called **hypoxia-inducible factors (HIFs)**. A transcription factor is a protein that enters the nucleus and attaches to the control sections of target genes, which it activates to produce mRNAs and hence enables the synthesis of the proteins coded by those genes. This is the subject of Chapter 26, so here we need only talk in general terms of gene activation. A transcription factor has a domain (region), which promotes specific gene activation.

In normoxia (normal oxygen levels), HIF protein is destroyed rapidly – its half-life in the cell is about 5 minutes, so that its level remains low, therefore HIF has essentially no role to play in normal oxygen conditions. However, in hypoxia its level increases so that it causes the synthesis of protective response proteins. Figure 20.26 gives a general summary of the mechanism.

What is the nature of the switches in hypoxia that prevent proteolytic breakdown of HIF? The control depends on two separate post-translational modifications of the HIF. In normoxia HIF is rapidly destabilized by hydroxylation of two critical proline residues, which leads to proteolytic destruction of the HIF in proteosomes. The proline hydroxylases are rate-limited by oxygen concentration and are believed to be acting as oxygen sensors so that, at low oxygen levels, HIF is not subject to proline hydroxylation and accumulates. An asparagine hydroxylation also occurs.

The hypoxia response mechanism described here has potential medical relevance, because hypoxia in specific tissues is one of the effects of heart attacks, strokes, and other vascular diseases.

It presumably may also have the less desirable effect of helping cells in the relatively oxygen poor centre of tumours to thrive.

Integration of metabolism: the fed and fasting state, and diabetes mellitus

We have seen in this chapter the types of metabolic control mechanisms that deal with fuel homeostasis. We have looked at mechanisms dealing with carbohydrate and fat metabolism separately. In this section we will present four figures, which give a summary of the metabolic pattern in the fed state, the fasting state, prolonged starvation, and we will look into the similarities and differences in the metabolic pattern in starvation and diabetes mellitus type 1. The text will follow the sequence of numbers in the figures to outline each of the metabolic processes and the mechanisms by which they are achieved. We will concentrate mainly on the processes, which are specific to each of the metabolic situations described.

Metabolism in the fed state

Fig. 20.27 shows the metabolic pattern in the fed state. It is characterized by high insulin:glucagon ratio and high blood glucose concentration.

1 In the fed state, 2–4 hours after a meal, there is an increase in the plasma concentrations of glucose, amino acids and TAG in the form of chylomicrons.

2 Brain and nerve rely on glucose as metabolic fuel. Fatty acids are not used to any significant extent as transport across the blood brain barrier is very poor. Glucose enters the brain via the GLUT3 transporter, which has high affinity for glucose and is independent of insulin. Glucose is phosphorylated by hexokinase, which has a low K_m for glucose. The glucose is completely oxidized by glycolysis, the TCA cycle and the electron transport chain and oxidative phosphorylation. Metabolism of glucose in the brain is, in fact, the same in the fed and fasting state as long as adequate glucose is supplied in the bloodstream.

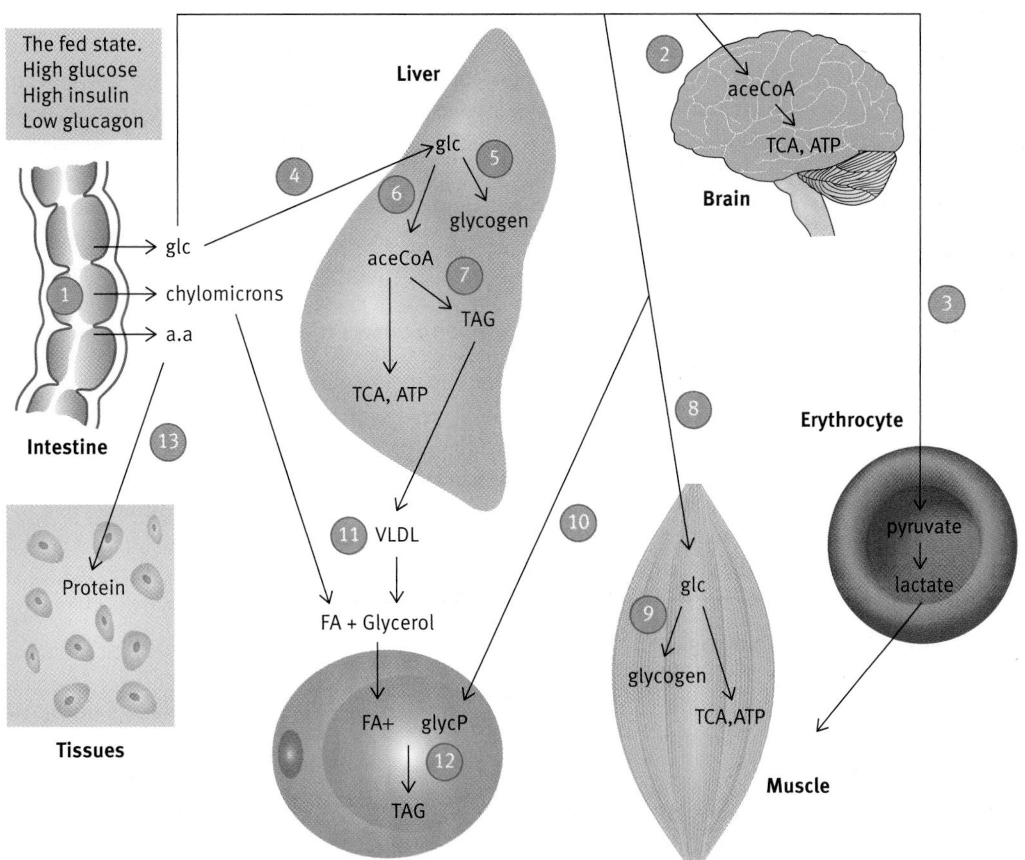

Fig. 20.27 Metabolism in the fed state.

3 Glucose is taken up by the erythrocyte via the GLUT1 transporter, which is similar to the one in brain in having high affinity for glucose and is independent of insulin. The erythrocyte is entirely dependent on glucose for energy as it has no mitochondria and cannot use fatty acids. Glucose is metabolized by glycolysis and the pyruvate is converted into lactate, which enables regeneration of the NAD^+ necessary to continue glycolysis. Metabolism in the erythrocyte, as in the brain, is the same in the fed and fasting state.

4 Glucose is taken up by the liver via the GLUT2 trasporter. This is also independent of insulin but has low affinity for glucose so that the liver only takes it up when glucose concentration in the blood is high. In addition, the liver has glucokinase, which has a much higher K_m for glucose than hexokinase so that it will only metabolize glucose when its concentration inside the hepatocyte is high.

5 Glycogen synthesis is activated in the liver. Glycogen synthase is dephosphorylated and activated whereas glycogen phosphorylase is dephosphorylated and inhibited. Glucokinase is not subject to product inhibition so glucose-6-phosphate is continuously generated and is channelled into glycogen synthesis and glycogen storage.

6 When the stores of glycogen are replete, any excess glucose is channelled into glycolysis in the liver. Note that glycolysis in the liver is not an energy generating process but one that provides acetyl-CoA, the starting material for fatty acid synthesis. Glycolysis is stimulated by the increased activity of glucokinase, PFK and pyruvate kinase. Gluconeogenesis is inhibited.

7 Fatty acid and TAG synthesis are activated. Acetyl-CoA carboxylase, the enzyme that catalyses the rate limiting step in fatty acid synthesis, the conversion of acetyl-CoA into malonyl-CoA, is activated. The product of the reaction, malonyl-CoA inhibits carnitine transferase and in this way any newly synthesized fatty acid cannot be transported into the mitochondrion for oxidation. This ensures that no futile cycle is set up. The fatty acid is available for esterification and production of TAG, which is exported from the liver into the circulation in the form of VLDL.

8 Glucose is taken up by muscle cells. The binding of insulin to its receptors leads to the delivery of GLUT4 transporters to the cell surface. GLUT4 are insulin dependent, which means that the muscle cells can only take up glucose when glucose is high.

9 Once inside the cell, glucose is phosphorylated and glucose-6-phosphate is channelled into glycogen synthesis. As in the liver, glycogen synthase is activated and phosphorylase is inhibited.

10 Glucose is taken up by adipocytes via the insulin sensitive GLUT4 transporter. As is the case with muscle cells, adipocytes only take up glucose when glucose is high. Glucose enters the glycolytic pathway until the stage of dihydroxyacetone phosphate, which is then reduced to glycerolphosphate needed for re-esterification of fatty acids.

11 Dietary TAG in the form of chylomicrons, and endogenously synthesized fat in the form of VLDL are hydrolysed by lipoprotein lipase in the adipose tissue capillaries. Lipoprotein lipase is activated by insulin. The resulting fatty acids enter the adipocyte and the glycerol returns to the liver.

12 The fatty acids delivered from the capillaries to the adipocytes are re-esterified into TAG using glycerol phosphate produced by glycolysis. Hormone-sensitive lipase in the adipocytes is inhibited by insulin and by the fact that glucagon is low, again avoiding a futile cycle that would lead to TAG synthesis and hydrolysis at the same time. TAG is then stored in the adipocytes.

13 Finally, amino acid uptake into muscle, liver, and other tissues is activated by insulin. Synthesis of protein and other N containing molecules takes place.

Metabolism in the fasting state

The fasting state is characterized by low glucose, low insulin, and high glucagon. In the absence of food, stored glycogen, body protein, and TAG are mobilized to supply fuel to the various tissues. There is a specific need for glucose to supply the brain and erythrocytes, and there is a general need for fuel for the rest of the body.The liver maintains blood glucose concentrations at about 4 mM, which are adequate to supply brain and erythrocyte. The main source of energy for the rest of the body is the stored TAG. Fig. 20.28 shows the metabolic pattern in fasting.

1 The first supplier of glucose is liver glycogen. Glucagon activates glycogen phosphorylase and inactivates glycogen synthase by phosphorylation producing glucose 1 phosphate and then glucose 6 phosphate. The liver possesses glucose-6-phosphatase, which hydrolyses glucose 6 phosphate to glucose, which is then released into the bloodstream. The fact that the liver has GLUT2, which is not insulin dependent allows glucose to leave the hepatocyte. Liver glycogen will become totally depleted after 24 hours of fasting.

2 Glucose is taken up by the brain even though the blood glucose is low, as the GLUT3 transporter is not insulin dependent and has high affinity for glucose. Once in the cell, glucose undergoes complete oxidation as happens in the fed state.

3 Glucose is taken up by the erythrocyte as GLUT1 has high affinity for glucose and is not insulin dependent.

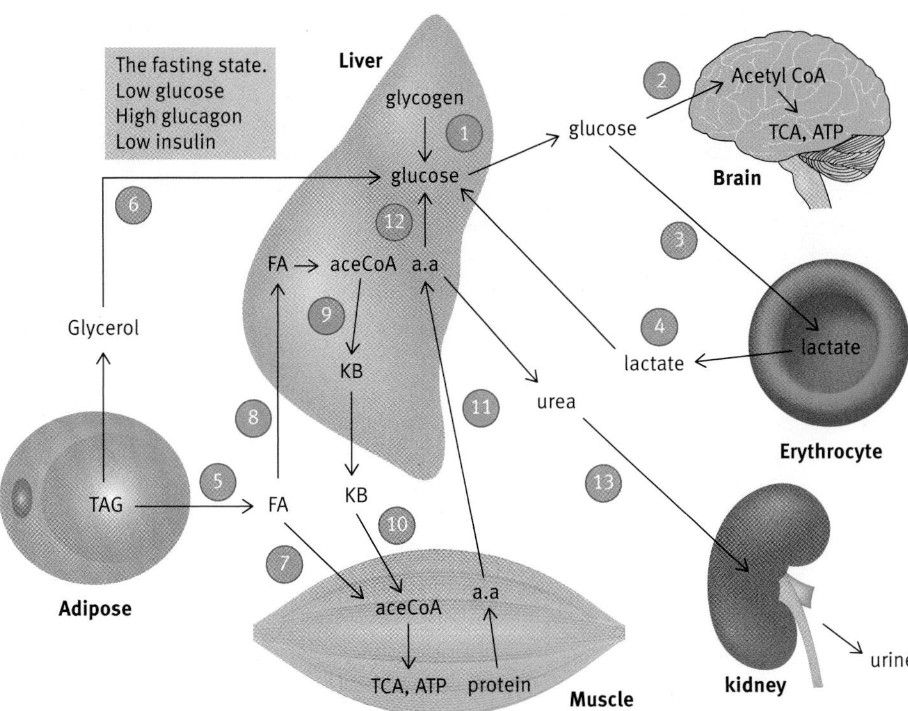

Fig. 20.28 Metabolism in the fasting state.

Glucose is metabolized to pyruvate and lactate as in the fed state.

4 The lactate returns to the liver where it is converted into glucose. Gluconeogenesis is stimulated by glucagon and the lack of insulin. The glucose is released into the bloodstream.

5 In the adipocyte, TAG is hydrolysed into fatty acids and glycerol, which are released into the bloodstream. This is made possible by the fact that hormone-sensitive lipase in the adipocyte is activated by glucagon, and lipoprotein lipase in the capillaries is inactive as insulin is low.

6 Glycerol returns to the liver where it is converted into glucose by gluconeogenesis, which is stimulated in the presence of glucagon.

7 Fatty acids released from adipose tissue travel in the bloodstream bound to albumin and enter the muscle cells where they are oxidized for energy.

8 Fatty acids are also delivered to the liver where they are transported into the mitochondria, as the carnitine shuttle is now active in the absence of malonyl-CoA and are oxidized to acetyl-CoA.

9 The concentrationof acetyl-CoA, the product of fatty acid oxidation exceeds the capacity of the TCA cycle to oxidize it and it is converted into the ketone bodies (KBs), acetoacetate and 2-oxobutyrate.

10 KBs can be used by the muscle as fuel via metabolism to acetyl-CoA. Muscle can oxidize them (and brain to a small extent in early fasting), but the liver cannot oxidize them as it lacks the necessary enzymes.

11 Production of glucose by glycogenolysis is short-lived and by gluconeogenesis from lactate and glycerol insufficient to meet the needs of the brain and erythrocyte. Fatty acids cannot be converted into glucose as the conversion of pyruvate into acetyl-CoA is irreversible, therefore the only other source of glucose is the body's own protein. Glucagon stimulates protein degradation and the amino acids are released into the bloodstream and taken up by the liver. Pyruvate dehydrogenase is inhibited by glucagon and also by the products of fatty acid oxidation, acetyl-CoA, and NADH, so that any pyruvate produced from body protein is not converted into acetyl-CoA, but channelled into glucose production.

12 Most amino acids are glucogenic,and when they are deaminated in the liver their carbon skeletons enter the gluconeogenetic pathway and produce glucose, which is released into the bloodstream.

13 The amino groups are incorporated into urea in the liver and the urea is excreted in urine by the kidney.

Metabolism in prolonged starvation

If the metabolic pattern of fasting would continue in prolonged starvation, body protein would be severely depleted very soon. Only about a third of body protein can be lost without severe or fatal consequences. Some adaptations take place shown in Fig. 20.29.

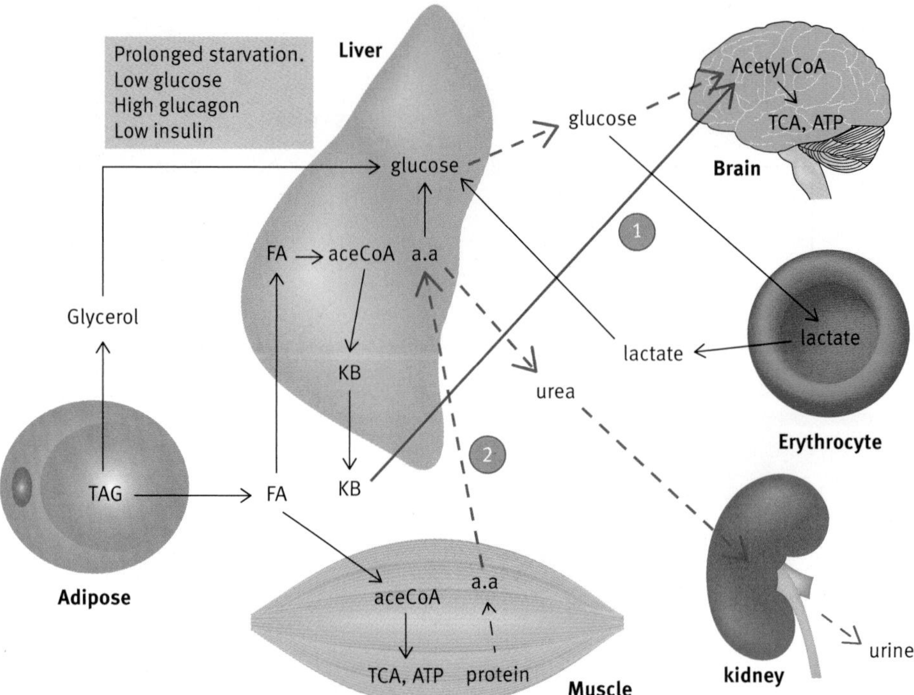

Fig. 20.29 Metabolism in prolonged starvation.

1 Ketone body (KB) production increases in the liver. They can enter the brain and be metabolized because they are able to cross the blood–brain barrier and the brain has the enzymes needed for their oxidation. In prolonged starvation the brain can use KBs to supply up to half its requirement of energy.

2 This means that the brain needs less glucose, therefore the rate of protein breakdown can be substantially reduced preserving precious reserves for longer. This can be observed as a reduced rate of gluconeogenesis, release of glucose into the bloodstream and urea production. Reduced proteolysis is achieved by the fact that KBs act on the β-cells of the pancreas and lead to the release of a small amount of insulin, which is sufficient to dampen proteolysis and lipolysis in prolonged starvation.

Metabolism in type 1 diabetes mellitus

Type 1 diabetes has been described as 'starvation in the midst of plenty'. We are now going to compare the metabolic pattern in starvation and diabetes mellitus and point out the similarities and differences. Fig 20.30 shows the metabolic pattern in diabetes superimposed onto that of starvation. The bold arrows indicate the processes that take place in starvation but are exaggerated in diabetes.

In both starvation and diabetes, glucagon is high. The difference between starvation and diabetes is that in starvation, glucose and insulin are both low, whereas in diabetes, glucose is high and insulin is absent. The fact that insulin is absent means that uptake of glucose by muscle and adipose tissue is insignificant as the GLUT4 transporters are in intracellular vesicles and not on the cell surface. In addition there is uncontrolled glucose production exacerbating the hyperglycaemia.

1 Lipolysis is uncontrolled as it is under the unopposed influence of glucagon.

2 KB production is uncontrolled, again under the unopposed influence of glucagon.

3 KBs in starvation have a dampening effect on proteolysis by stimulating the release of insulin. This control is lost in diabetes as the β cells are incapable of producing insulin. This again means that glucagon is acting unopposed.

4 Gluconeogenesis continues, stimulated by glucagon although blood glucose concentrations are high. This illustrates how important the hormonal control of metabolism is, as high concentrations of the local metabolite, glucose, cannot overcome the effects of glucagon.

The excessive ketogenesis and gluconeogenesis result in hyperglycaemia and ketoacidosis, the hallmarks of type 1 diabetes.

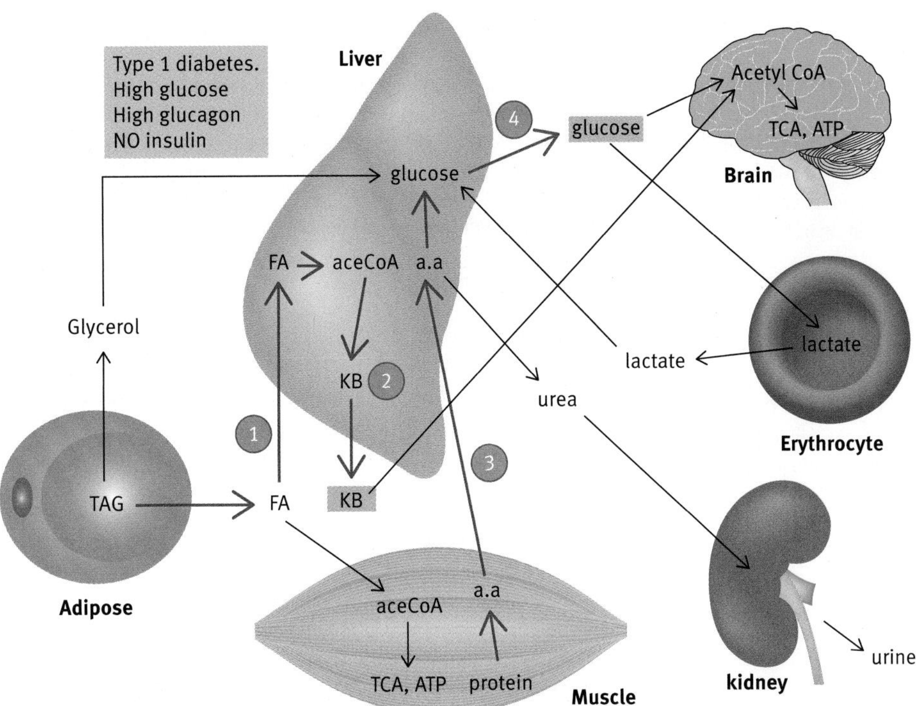

Fig. 20.30 Metabolism in untreated diabetes mellitus type 1.

BOX 20.1 **Diabetes mellitus**

Diabetes mellitus is the commonest endocrinological disorder in the world. It accounts for 90% of all endocrinological disorders and is a major cause of blindness, amputations and early death. There are two forms of the disease:

- **type 1**, which used to be known as **juvenile-onset diabetes** or **insulin-dependent diabetes mellitus,** is caused by an autoimmune destruction of insulin-producing cells of the pancreas, so that insulin production is deficient or absent.
- **type 2**, which used to be called **maturity-onset diabetes** or **noninsulin-dependent diabetes,** where insulin concentrations may be normal or increased, but cells are relatively resistant to the hormone. The onset of this usually occurs after the age of about 35, frequently associated with obesity. The increased incidence of obesity in children seen in the last 20 years or so has led to the earlier appearance of type 2 diabetes, even in childhood.
- As much as 3–4% of the population of economically developed countries are diabetics, 80–90% of whom are type 2 diabetics. The incidence is increasing in less economically developed countries as well, with the biggest increase in type 2 diabetes mainly associated with obesity.

Fig 20.31 shows three typical glucose tolerance curves of one normal and two diabetic subjects with type 1 and 2 diabetes respectively.

The protocol for the tolerance curves requires the subject to fast for 12 hours and then to ingest a glucose load, usually 75 g. Blood glucose concentration is measured at regular intervals (15–30 min) and a graph of glucose concentration against time is plotted.

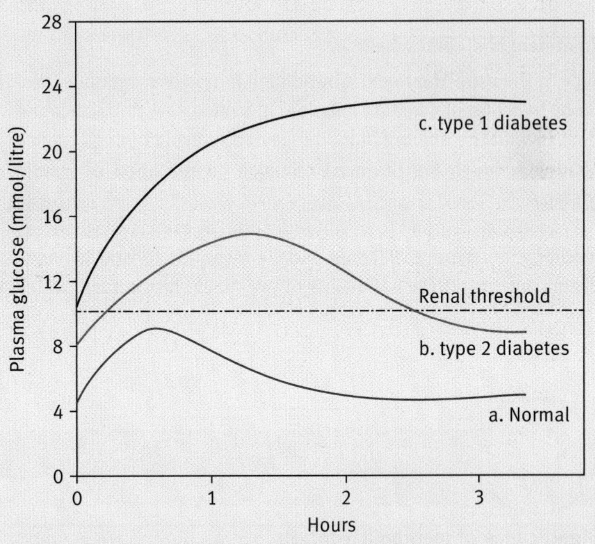

Fig. 20.31 Glucose tolerance curves of normal and diabetic subjects.

The criteria for normality are that the fasting blood glucose is not above 6.2 mM (the concentration below which long-term diabetic complications are not observed), the maximum value does not exceed the renal threshold (the concentration of glucose in the blood

(*Continued*)

Box 20.1 **Continued**

above which the kidney cannot reabsorb it), and that blood glucose returns to normal fasting concentration within 2 hours (curve a).

Curve b shows the results of a type 2 diabetic. The fasting blood glucose is higher than 6.2 mM, but not above 10 mM in this case, the concentration of glucose after the load exceeds the renal threshold, and does not return to normal fasting in 2 hours.

Characteristics of the disorder

Type 1 diabetes is usually of early onset. It is characterized by polyuria (excessive production of urine), polydipsia (excessive thirst and drinking), polyphagia (excessive hunger and eating), fatigue, weight loss, muscle wasting and weakness.The hallmarks are **hyperglycaemia** and **ketoacidosis** (excessive production of KBs and decrease in blood pH). Diabetic ketoacidosis is a medical emergency that can lead to coma and be fatal. Ketoacidosis occurs because of uncontrolled glucagon- induced lipolysis resulting in excessive production of KBs, such as acetoacetate and 2-oxobutyrate, which exceed their rate of utilization. Spontaneous decarboxylation of acetoacetate acid produces acetone, which gives a characteristic fruity smell to the breath of people with untreated diabetes.

Type 2 diabetes is usually milder and is characterized by hyperglycaemia but usually **no ketoacidosis.** The absence of ketoacidosis seems to imply that fat metabolism is more sensitive to insulin than carbohydrate metabolism is and that the amounts of insulin present can inhibit excessive lipolysis even though they cannot control hyperglycaemia. Hyperosmotic, not ketotic coma (HONK) can however be a dangerous complication of severe type 2 diabetes.

Treatment of diabetes mellitus

Type 1: The usual treatment is by administration of exogenous insulin by injection or constant infusion. It is important to balance the dosage with the amount of food ingested to avoid hypoglycaemic incidents, which are the most common complication of insulin therapy.

Type 2: May respond to weight reduction, by exercise and dietary modification. Type 2 diabetes usually responds to oral hypoglycaemic agents. These are mainly of two types: biguanides, which increase the number of GLUT4 transporters and therefore lower blood glucose by increasing glucose uptake by the periphery, and sulphonylureas, which act on the pancreatic β cell to increase insulin secretion.

Chronic complications of diabetes

A number of long-term complications can arise from poor control of diabetes. They include:

- microangiopathy, which is characterized by changes in the walls of small blood vessels seen as thickening of basement membranes affecting circulation in small blood vessels
- retinopathy, which is as a result of microangiopathy in retinal vessels; blindness is 25 times more common in the diabetic than the nondiabetic patient
- nephropathy is also a result of microangiopathy with renal failure, which is 20 times more common in diabetic as opposed to nondiabetic patients
- neuropathy, which is impairment of nerve function
- postural hypotension, impotence and foot ulcers are well known in diabetic patients, resulting from a combination of angiopathy and neuropathy
- the incidence of cardiovascular disease is high in diabetic patients.

For all these reasons, good diabetic control is of the utmost importance. Diabetic control can be monitored by measuring the amounts of a number of glycosylated blood proteins. When glucose concentrations are high over extended periods of time, a number of cell components become glycosylated nonenzymatically and their function is impaired. This is unfortunate but it also provides us with a diagnostic tool to monitor diabetic control. One of the proteins that becomes glycosylated is haemoglobin and measurements of glycosylated haemoglobin known as HbA1c, in a patient's blood provide a measure of diabetic control over the last 3–4 months, which is the life span of a red cell.

Curve c shows the results of a type 1 diabetic. Fasting glucose concentration is above 10 mM , the maximum is very high and it takes a long time to return to the subject's own fasting blood glucose.

Summary

Significance of metabolic controls. Metabolic pathways must be regulated to avoid futile substrate cycling, and to respond to changing physiological needs. Enzyme activities may be controlled by changing the amount of enzyme, and/or by changing the rate of catalysis of enzymes. The first is slow; the second is almost instantaneous. Control points in metabolic pathways are usually at irreversible steps in which the forward and backward reactions can be separately controlled. Regulation of enzymes at these points is by allosteric mechanisms and/or covalent modification.

Allosteric control is a powerful concept essential for cells to exist. Allosteric enzymes are multisubunit proteins with allosteric sites, to which molecules (modulators) attach and affect the activity. The effect of substrate concentration on their reaction rates is sigmoidal rather than hyperbolic. Attachment of modulators usually alters the affinity of the enzyme for its substrate(s) and this affects its rate of activity at a given suboptimal substrate concentration. Two theoretical models that account for their properties are the concerted model and the sequential model. Both involve conformational changes.

Allosteric control coordinates the rates of disparate metabolic pathways and is virtually instantaneous.

Control by phosphorylation involves covalent modification of enzymes by protein kinases. The phosphorylation may activate or inhibit the activity and is reversible by phosphatases. The phosphorylation states of the enzymes are usually regulated by hormones. Glucagon, adrenaline (epinephrine), and insulin are the important ones in our present context. The first two cause the production of a second messenger molecule, cyclic AMP (cAMP), which activates kinases. Insulin control also involves phosphorylation but is more complex.

Control of glucose uptake into cells. Glucose does not diffuse across membranes; it must be transported by proteins. The facilitated diffusion occurs by the movement of transport proteins from within the cell to the membrane. These glucose transporters are GLUT isoforms. This process is not insulin responsive in brain, red blood cells or liver but is controlled by insulin in adipose cells and muscle.

Once glucose enters the cell it is rapidly phosphorylated by hexokinase (most tissues) or glucokinase (liver). The lower affinity of glucokinase means that at times when blood glucose is low and the liver is releasing glucose it does not efficiently take it up again. Hexokinase is inhibited by glucose-6-phosphate at physiological concentrations but glucokinase is not. This allows the liver to take up glucose at high blood sugar levels and to synthesize glycogen using glucokinase to phosphorylate glucose. Not only does glucokinase only operate efficiently at high blood glucose but it is increased by insulin.

Control of glycogen metabolism. Glycogen phosphorylase in the absence of hormonal signals exists in muscle and liver as phosphorylase *b*, which is allosterically activated by AMP and inhibited by ATP. Ca^{2+} released during muscle contraction activates. Adrenaline (and glucagon in the liver) causes cAMP to be formed in muscle and liver. This triggers the conversion of phosphorylase *b* to form *a*, which is fully active without AMP. The conversion is by a catalytic cascade of kinases, which amplifies the response. At the same time cAMP causes inactivation of glycogen synthase so that a futile cycle is avoided. Glycogen synthase is active only when dephosphorylated. The main activating signal is insulin, which causes the dephosphorylation of the synthase at specific sites.

Control of glycolysis and gluconeogenesis. The allosteric controls here fit a logical pattern. If ATP levels are suboptimal, AMP, which is a sensitive indicator of this, activates glycogen breakdown and glycolysis. It also inhibits the reverse pathway involved in gluconeogenesis. ATP inhibits glycolysis as does citrate. They both signal adequate energy levels.

The hormonal controls differ in muscle and liver. A major control in liver is in response to glucagon. The pancreas secretes glucagon when blood glucose levels are low. The liver's response is to inhibit glycolysis, increase gluconeogenesis and channel glucose-6-phosphate into the production of blood glucose by glucose-6-phosphatase. This is achieved by cAMP indirectly regulating the activities of an enzyme PFK2, which controls the level of the fructose-2,6-bisphosphate. The latter compound is a main determinant of the rate of glycolysis. It activates PFK.

PFK2 is an unusual enzyme with two different catalytic sites. In the nonphosphorylated form it synthesizes the 2,6 compound and when phosphorylated it hydrolyses it. cAMP, produced in response to the glucagon signal, activates a kinase, which phosphorylates PFK2. This destroys the 2,6-activating molecule and inhibits glycolysis. The final part of this involved control is that the 2,6 compound inhibits fructose-1,6-bisphosphatase. This enzyme is required for gluconeogenesis so the net effect of destruction of the 2,6 compound in the presence of cAMP is inhibition of glycolysis and activation of gluconeogenesis. Adrenaline has a similar effect to glucagon and is responsible for the flight-or-fight reaction to a threatening situation.

In muscle, adrenaline also produces cAMP but the need here is to maximize glycolysis to produce ATP. In this tissue the PFK2 is a different isoenzyme and is not affected by cAMP. The increased glycogen breakdown caused by cAMP stimulates glycolysis and also, in some indirect way related to this, increases the level of the activating fructose-2,6-bisphosphate. Muscle does not have the gluconeogenesis pathway and does not contribute to blood glucose.

Gluconeogenesis is also promoted in liver by an additional hormonal control. The starting point of the pathway is phosphoenolpyruvate (PEP). When glucagon raises the level of cAMP it inhibits the pyruvate kinase enzyme, thus channelling PEP into the gluconeogenesis pathway. The stress hormone cortisol liberated in starvation also promotes gluconeogenesis (see Chapter 16).

Fructose metabolism and its control differs from that of glucose and leads to increased fat synthesis.

Pyruvate dehydrogenase is controlled by a combination of allosteric controls in which substrates activate and products inhibit. There is however an inbuilt protein kinase. Most protein kinases are subject to hormonal controls but in this case it is inhibited by products of the enzyme reaction and activated by substrates. This makes physiological sense because phosphorylation of the enzyme by the kinase inactivates it.

The TCA cycle is controlled both allosterically and by availability of NAD^+ and ADP. Electron transport is tightly coupled to ATP production and availability of ADP is of prime importance in control.

Fatty acid oxidation and synthesis are reciprocally controlled allosterically so that either oxidation or synthesis is proceeding and a futile cycle is avoided. An AMP-activated kinase also inhibits fatty acid synthesis by phosphorylating acetyl-CoA carboxylase, the first committed step in the synthetic pathway.

This inhibits the enzyme. The malonyl-CoA produced by the carboxylase inhibits the transport of fatty acids into mitochondria, thus inhibiting their oxidation. In fat cells, hormonal control is principally at the level of triacylglycerol (TAG) breakdown and synthesis. Glucagon and adrenaline (*via* cAMP) activate the hormone-sensitive lipase, which liberates fatty acids into the blood while insulin promotes TAG synthesis.

Overall regulation of ATP levels is a safety control mechanism. If the ATP 'charge' is suboptimal, its level is maintained by switching off nonvital synthetic reactions using ATP. This is done by the ubiquitous AMP-activated kinase, which shuts down many processes by specific phosphorylations. This is a response to metabolic stress.

Response of cells to oxygen deprivation is how the body deals with another stress situation. **Hypoxia** is a situation in which the oxygen level in a tissue is abnormally low and protective responses occur. These are to increase erythropoietin production, which increases red blood cell numbers; they increase glycolytic enzymes and glucose transporters. Since glycolysis can produce ATP anaerobically, they increase growth of new blood vessels in the tissue.

These responses require gene activation by a hypoxia-inducible transcription factor (HIF). In normoxia, HIF is inactivated and destroyed due to hydroxylation of proline and asparagine residues. This does not occur in hypoxia in which situation the HIF remains active and induces the protective responses.

The system has medical interest in that heart attacks and other vascular diseases cause hypoxia in specific tissues. It also unfortunately helps hypoxic cells of tumours to survive and to become vascularized thus potentially helping cancers to develop.

Integration of metabolism. The fed state refers to the situation 2–4 hours after a meal. It is characterized by high glucose, amino acid, and fat concentrations in the blood. The hormonal status is that of high insulin and low glucagon. Synthesis and storage of macromolecules is favoured and degradation processes inhibited. Glycogen is synthesized and stored in muscle and liver as glycogen synthase is active and phosphorylase inactive, both through phosphorylation. TAG are hydrolysed by lipoprotein lipase and the resulting fatty acids re-esterified and stored in the adipose tissue. Lipolysis is inhibited in the presence of insulin. Amino acids are taken up by the periphery and protein synthesis is favoured and proteolysis is inhibited.

The fasting state. Glucagon is high and insulin is low. This leads to glycogen breakdown in the liver, supplying glucose in the blood which is used by the brain and erythrocytes. Glycolysis in the liver is inhibited and gluconeogenesis is activated by activation of PEP carboxykinase, fructose-1,6-bisphosphatase and glucose-6-phosphatase. Glycogenolysis and gluconeogenesis from lactate are inadequate to provide glucose for a long time and degradation of body protein occurs, which provides gluconeogenic substrates. Fatty acids are mobilized from the adipose tissue and provide most of the energy of tissues other than brain nerve and erythrocyte. The excess degradation of fatty acids leads to the appearance of KBs, which can be used by muscle as fuel.

Prolonged starvation. A number of adaptations take place which limit the extent of glucagon induced proteolysis and lipolysis. KB production is increased and the brain can satisfy up to 50% of its energy requirements by using KBs. There is less requirement for proteolysis to supply the brain and red cell with glucose. KBs have a regulatory role as well, as they lead to the production of a small amount of insulin, which can reduce the rate of proteolysis and lipolysis so limiting production of glucose and KBs.

Metabolism in diabetes type 1. The metabolic pattern resembles that of prolonged starvation but is exaggerated as glucagon is acting uncontrolled due to the complete absence of insulin. Lipolysis, proteolysis, gluconeogenesis, and KB formation continue unopposed and lead to hyperglycaemia and ketoacidosis, the hallmarks of type 1 uncontrolled diabetes.

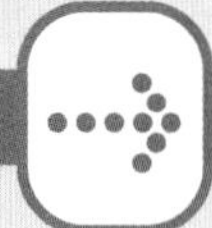

Further reading

To access the further reading, please scan the QR code image or go to http://global.oup.com/uk/orc/biosciences/molbiol/snape_biochemistry/student/reading/ch20/

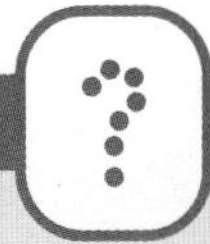

Problems

1 What are the two main ways by which the activities of enzymes may be reversibly modulated?

2 Compare the relationship between substrate concentration and rate of enzyme catalysis in a nonallosteric enzyme and a typical allosteric enzyme.

3 What is the main feature of allosteric control that makes it such a tremendously important concept?

4 What are the salient features of intrinsic regulation and extrinsic regulation by extracellular signals?

5 By means of a diagram, illustrate the main *intrinsic* controls on glycogen metabolism, glycolysis, and gluconeogenesis and explain the rationale.

6 Pyruvate dehydrogenase (PDH) is a key regulatory enzyme. In general, products of the reaction inhibit the reaction. There are three mechanisms of control involved; what are they?

7 What controls the release of insulin and glucagon from the pancreas?

8 What is a second messenger? Name the second messenger for adrenaline and glucagon and explain how it exerts metabolic effects.

9 How does insulin control the rate of glucose entry into fat cells?

10 Explain how cAMP activates glycogen breakdown.

11 Glucagon activates liver phosphorylase via cAMP as its second messenger. Muscle does the same with adrenaline stimulation. However, cAMP has quite different effects on liver and muscle glycolysis. Explain these.

12 Several hormones that elicit different cellular responses nonetheless use cAMP as their second messenger. How can one compound be used for these?

13 Phosphofructokinase is a key control enzyme. What is the major allosteric effector for this enzyme and how is its level controlled in liver?

14 To synthesize glucose in the liver, phosphoenolpyruvate (PEP) is needed. However, production of PEP would achieve little in this regard if it were dephosphorylated to pyruvate by pyruvate kinase. How is this futile cycle avoided? Why would this mechanism not be appropriate in muscle?

15 How does glucagon cause fatty acid release by fat cells?

16 Metabolic pathways often include at least one reaction with a large ΔG value. What advantages accrue from this?

17 The control of glycogen synthase is to a large extent effected by insulin. Explain in outline the nature of this control.

18 What part does glucose play in the regulation of glycogen breakdown in liver?

19 Describe the reciprocal controls that operate so that fatty acids are either synthesized or oxidized but not both at the same time.

20 Why is phosphorylation such a potent way of controlling enzyme activity?

21 Discuss the role of the AMP-activated protein kinase (AMPK).

Chapter 21

Raising electrons of water back up the energy scale – photosynthesis

Chapter 13 describes how ATP generation in aerobic cells depends on transporting electrons of **high-energy potential**, present in food, down the energy scale to end up as the electrons present in the hydrogen atoms of water.

Since the amount of food on Earth is limited, if life in general is to continue indefinitely, a way must exist to raise those electrons back up the energy scale. A minor qualification to this statement is that life forms have been discovered in deep oceans around cracks in the Earth's crust from which compounds such as H_2S emerge. H_2S is a strong reducing agent (that is, of low redox potential) and its electrons could be transported down the energy gradient releasing energy, provided appropriate biochemical systems are there. Such life could presumably exist as long as H_2S and other such agents are generated in the Earth's crust but, for continuation of the vast majority of living organisms, electron recycling is necessary.

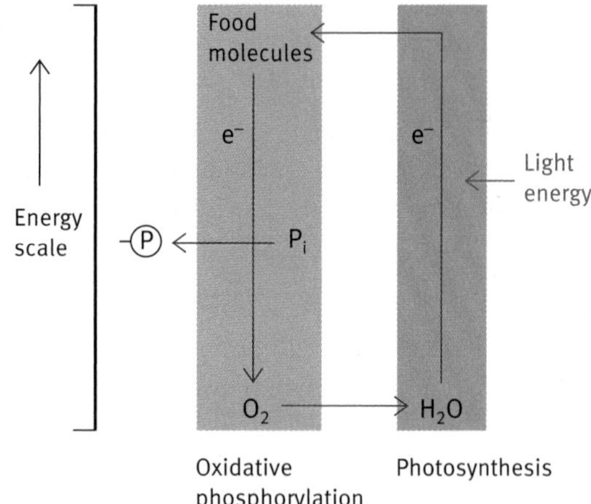

Fig. 21.1 Global 'electron cycling' by oxidative phosphorylation and photosynthesis. Note that the fixation of CO_2 is a process secondary to the raising of electrons from water to a higher energy potential.

Overview

Photosynthesis is the biological process that recycles electrons from water, to produce oxygen and carbohydrate. It has several crucial advantages – water is an inexhaustible source of electrons, sunlight an inexhaustible source of energy, and, in releasing oxygen, an inexhaustible supply of electron sinks (acceptors) is provided to allow the energy to be extracted back from the high-energy electrons of food by life forms in general. The onset of photosynthesis was arguably the most important biological event following the establishment of life. The global energy cycle is shown in Fig. 21.1.

You are probably familiar with the concept of photosynthesis producing carbohydrate (usually starch or sugar) from CO_2 and H_2O; the overall equation (written for glucose production) is

$$6CO_2 + 6H_2O \rightarrow C_6H_{12}O_6 + 6O_2 \; \Delta G^{0'} = 2820 \, kJ\,mol^{-1}$$

To synthesize glucose from CO_2 and H_2O there are two basic essentials, looked at from the point of view of energy. First, there must be a reducing agent of sufficiently low **redox potential** (**high energy**). If you need to refresh your memory on redox potentials, turn to Chapter 7. In photosynthesis, the reducing agent is NADPH. (Gluconeogenesis in animals, as described in Chapter 16 , uses NADH as the reductant, but, in photosynthesis, NADPH is used.) Secondly, there must be ATP to drive the synthesis of carbohydrates.

Light energy is directly involved *only* in transferring electrons from water to $NADP^+$ and in the generation of the proton gradient that drives ATP production.

Site of photosynthesis – the chloroplast

Photosynthesis occurs in the **chloroplasts** of the cells of green plants. They are reminiscent of mitochondria in being

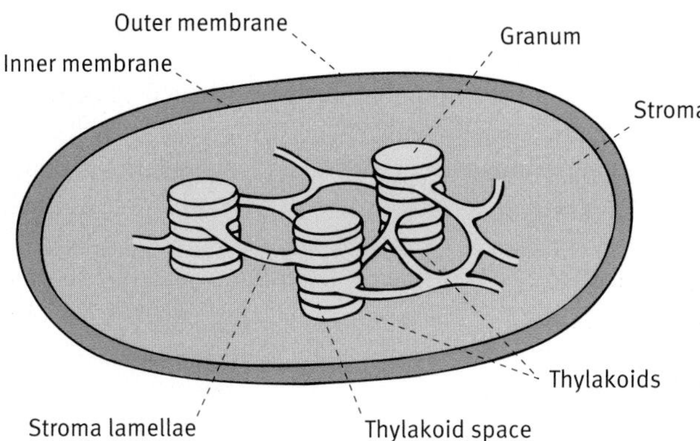

Fig. 21.2 A chloroplast. Grana are stacks of thylakoids.

membrane-bounded organelles in the cytosol with an outer permeable membrane and an inner one impermeable to protons. Like mitochondria they have their own DNA coding for part of their proteins. Their protein-synthesizing machinery is prokaryotic in type and it is believed that they arose from a symbiotic colonization of eukaryotic cells by primitive prokaryotic photosynthetic unicellular organisms.

Unlike mitochondria, however, chloroplasts contain yet another type of membrane-bounded structure – the thylakoids – membrane sacs within the chloroplast. The **thylakoids** are flattened sacs piled up like stacks of coins into grana, which are connected at intervals by single-layer extensions. Inside the thylakoids is the thylakoid lumen; outside is the chloroplast stroma (Fig. 21.2).

All of the light-harvesting chlorophyll and the electron transport pathways are in the thylakoid membranes. The conversion of CO_2 and H_2O into carbohydrate molecules is not itself light dependent and occurs in the chloroplast stroma. The latter processes are referred to as 'dark reactions', not to imply that they only occur in the dark, but rather that light is not involved in them. In fact, the dark reactions occur mainly in the light, when NADPH and ATP generation is occurring in the thylakoid membranes. This is summarized in Fig. 21.3.

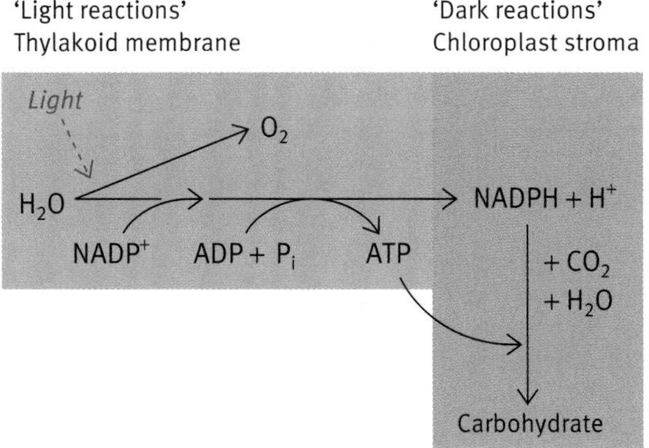

Fig. 21.3 The processes in photosynthesis.

The light-dependent reactions of photosynthesis

The photosynthetic apparatus and its organization in the thylakoid membrane

It would be useful for you to refresh your memory of the electron transport chain in mitochondria for there are considerable similarities between this and the photosynthetic machinery. In the inner mitochondrial membrane there are four complexes (see Fig. 13.17) that transport electrons. Connecting these complexes by ferrying electrons between them are ubiquinone, a small lipid-soluble molecule, and cytochrome *c*, the small water-soluble mobile protein.

In the thylakoid membrane there are three complexes (Fig. 21.4); connecting the first two is the electron carrier plastoquinone the structure of which is very similar to that of ubiquinone (see Chapter 13, Stage 3). Connecting the second two complexes is plastocyanin, a small water-soluble protein that has a bound copper ion as its electron-accepting moiety; this oscillates between the Cu^+ and Cu^{2+} states as it accepts and donates electrons.

The three complexes are named photosystem II (PSII), the cytochrome *bf* complex, and photosystem I (PSI). PSII comes before PSI in the scheme of things; the numbers refer to the order in which they were discovered rather than to their place in the scheme. The function of the whole array shown in Fig. 21.4 is to carry out the overall reaction:

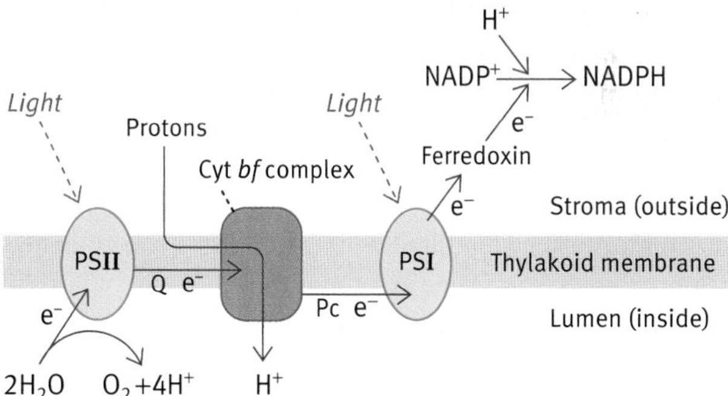

Fig. 21.4 The diagram gives an overall view of the light-dependent part of the photosynthetic apparatus, without reaction details. The feature to note is that the purpose is to take electrons from water and transfer them to $NADP^+$ and in the process to create a proton gradient across the thylakoid membrane that can drive ATP synthesis by the chemiosmotic mechanism. The gradient is created from two sources – the splitting of water (photosystem II, PSII) and the proton pumping of the plastoquinone (Q) – cytochrome *bf* complex. Electrons are transported from PSII to the cytochrome *bf* complex by plastoquinone (Q). The cycle is analogous to that in mitochondria (where 'Q' is used for ubiquinone) that results in four protons translocated per plastoquinol (QH_2) molecule oxidized. Note also that plastocyanin (Pc) is a mobile carrier analogous in this respect to cytochrome *c* in mitochondria and that ferredoxin is located on the opposite side of the membrane.

$$2H_2O+2NADP^+ \rightarrow O_2+2NADPH+2H^+$$

This involves a very large increase in free energy. What is unique in photosynthesis, so far as biochemical reactions are concerned, is that this energy is supplied by light. For each molecule of NADPH produced, four photons are absorbed. During the reduction process a proton gradient is established that is used to generate ATP. The arrangement is rather beautiful for it supplies the two requirements for carbohydrate synthesis from CO_2 and water – a reducing agent and ATP. Added to this, as mentioned, it generates an electron sink – oxygen – which makes it possible for living organisms to recover the energy entrapped in the carbohydrate produced. It is a magnificent system.

We now turn to the light-harvesting machinery present in PSII and PSI.

How is light energy captured?

In green plants, the light receptor is chlorophyll. Other receptor pigments exist in bacteria and algae.

Chlorophyll is a tetrapyrrole similar to haem except that it has a magnesium atom at its centre instead of iron and the substituent side groups are different. One of these side groups is a very long hydrophobic group that anchors it into the lipid layer. As in haem, there is a conjugated double-bond system (alternate double and single bonds right round the molecule) resulting in strong absorption of certain wavelengths of light and thus a strong green colour from wavelengths not absorbed.

In green plants there are two chlorophylls (*a* and *b*) differing in one of the side groups. They both absorb light in the red and blue ranges, leaving the intermediate green light to be reflected. The two chlorophylls have slightly different absorption maxima, which complement each other, so that in the red and blue ranges between them they absorb a higher proportion of the incident light. Chlorophyll *a* is shown below. (Chlorophyll *b* is the same except for the $-CH_3$ side chain (in red), which is $-CHO$ in chlorophyll *b*.)

Structure of chlorophyll *a*

When a chlorophyll molecule absorbs light, it is excited so that one of its electrons is raised to a higher energy state; it moves into a new atomic orbit. In an isolated chlorophyll molecule, after such excitation, the electron drops back to its ground state liberating energy as heat or fluorescence in doing so (and nothing is thereby achieved). But, when chlorophyll molecules are closely arranged together, a process known as **resonance energy transfer** transfers that energy from one molecule to another. In green plants, chlorophyll molecules are packed in functional units called **photosystems**, such that this resonance transfer occurs readily. Thus when a chlorophyll molecule is excited by absorption of a photon, its energy is transferred to another molecule and it drops back to the ground state itself. The excitation wanders at random from one chlorophyll molecule to another (Fig. 21.5).

This has a function, for among large numbers of 'ordinary' chlorophyll molecules there is a special **reaction centre** (an arrangement of a pair of chlorophyll molecules in association with proteins). The properties of this reaction centre are such that the excitation of a constituent chlorophyll molecule by resonance transfer, results in the excited electron being at a somewhat lower energy level, as compared with that of other excited chlorophyll molecules, so that resonance energy transfer *from* this molecule to other chlorophyll molecules does not occur. The excitation energy is, in this sense, trapped in an energetic hole – a shallow hole because the 'trapped' electron is still at a higher energy level than an unexcited electron, sufficient for the excited reaction centre to hand on an electron to an electron acceptor of appropriate redox potential. The latter is the first carrier of an electron transport chain in photosynthesis (to be described shortly). The chlorophyll molecules feeding excitation energy to the centres are called **antenna chlorophylls** (Fig. 21.5).

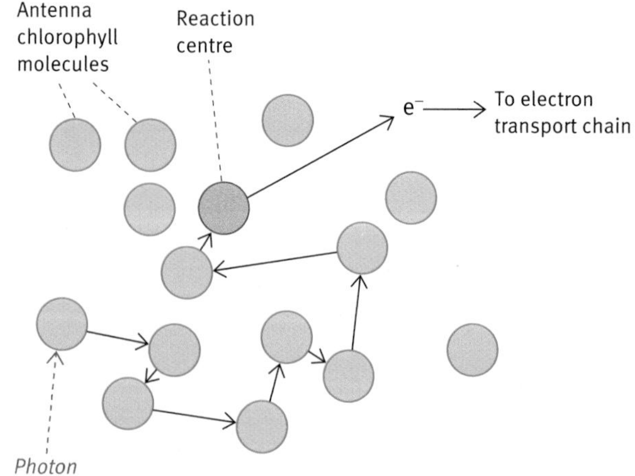

Fig. 21.5 Activation of reaction-centre chlorophyll molecule (red circle) by resonance transfer of energy from activated antenna chlorophyll molecules (green circles). The reaction centre of PSII is called P680 and of PSI, P700. Close packing of the chlorophyll molecules is needed for efficient resonance energy transfer.

A photosystem is therefore a complex of light-absorbing chlorophylls, reaction-centre chlorophyll, and an electron transport chain. In the case of PSII (the first system), the reaction-centre chlorophyll is called P680 because it absorbs light up to that wavelength (in nanometres) and in PSI is called P700 for an equivalent reason.

Mechanism of light-dependent reduction of NADP+

Figure 21.6 shows the 'Z' arrangement of the two photosystems II and I, with the redox potentials of the components indicated. Why two photosystems? A familiar analogy might help at the outset. An electric torch with a bulb (globe) requiring 3 V to light it up often uses two 1.5 V batteries in series. In photosynthesis, lighting of the bulb is represented by NADP+ reduction, and the batteries by the two photosystems operating in series. In fact, as stated, the latter supply more energy than is needed and some of it is sidetracked into ATP generation.

Photosystem II

Let us start with a P680 chlorophyll of a reaction centre of PSII. In the dark, it is in its ground, unexcited state in which it has no tendency to hand on an electron. When energy of a photon reaches it *via* antenna chlorophylls, it is excited such that it has a strong tendency to hand on its excited electron. It is, in fact, a reducing agent and it reduces the first component (a chlorophyll-like pigment lacking the Mg^{2+} atom, called **pheophytin**) of the PSII electron transport chain.

Two molecules of reduced pheophytin then hand on an electron each (one at a time) to reduce **plastoquinone**, the lipid-soluble electron carrier between PSII and the **cytochrome *bf* complex**:

O
H_3C
H_3C
R
O
$2e^-$
$2H^+$
OH
H_3C
H_3C
R
OH
R = a long hydrophobic group

Plastoquinone (Q) Plastoquinol (QH_2)

The latter complex contains two cytochromes and an iron-sulphur centre (see Chapter 13 if you need to be reminded what this is); the complex transports electrons from **plastoquinol** (**QH_2** – reduced plastoquinone) to **plastocyanin** to give the reduced form of the latter (see Fig. 21.4). Plastocyanin

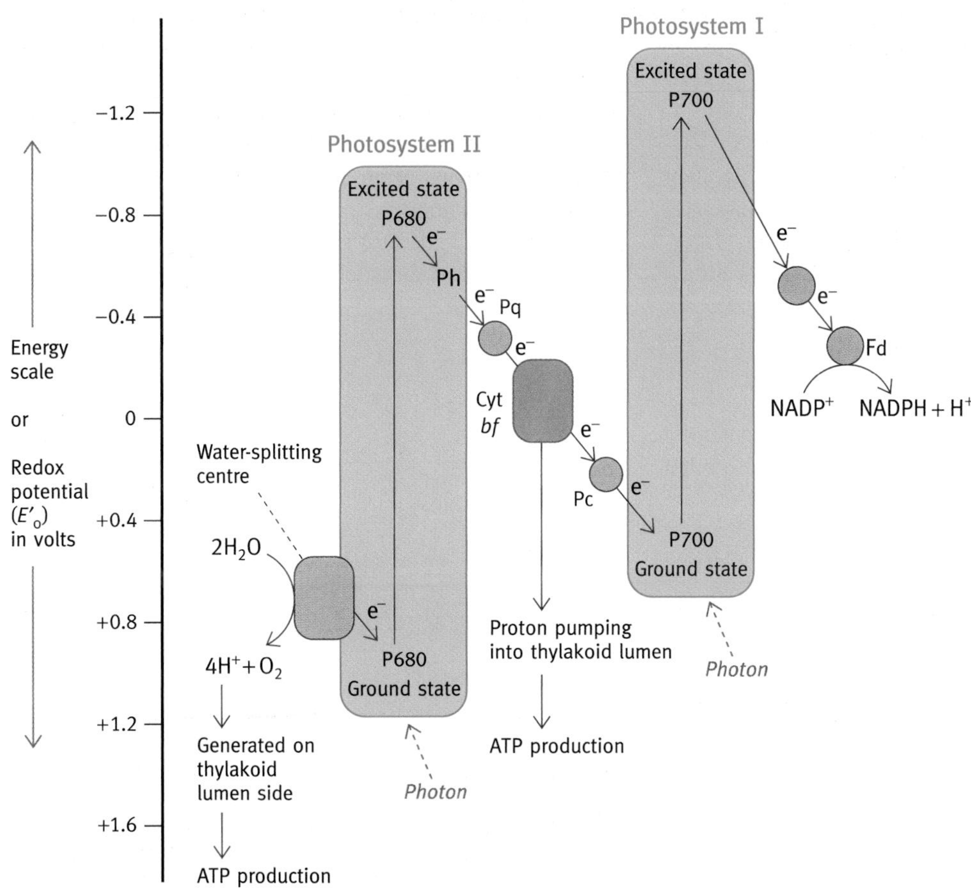

Fig. 21.6 The Z-scheme of photosynthesis. Simplified diagram of the electron flow from H_2O to NADP+. In viewing the diagram, start with a photon of light activating P680 and the transfer of an electron to P700. The electron-depleted P680 then accepts an electron from water and awaits a new round of activation. P700 behaves similarly except that it accepts an electron from the PSII transport chain and its donated electron is transferred to NADP+. Ph, pheophytin; Pq, plastoquinone; Pc, plastocyanin; Cyt *bf*, cytochrome *bf* complex; Fd, ferredoxin.

is a copper-protein complex in which the copper ion alternates between the Cu^{2+} (oxidized) and Cu^{+} (reduced) forms. At this point we will leave PSII and move on to PSI but we will return to the reduced plastocyanin very soon.

Photosystem I

The chlorophyll at the reaction centre of PSI is P700; when activated by a photon arriving from the antenna chlorophylls it becomes a reducing agent. It passes on its electron to a short chain of electron carriers (details of this not given), which reduces **ferredoxin**, a protein with an iron-cluster electron acceptor. Ferredoxin is a water-soluble, mobile protein residing in the chloroplast stroma (that is, outside of the thylakoid membrane). It reduces $NADP^{+}$ by the following reaction, catalysed by the FAD enzyme called **ferredoxin-NADP reductase** (FAD or flavin adenine dinucleotide was discussed in Chapter 13):

$$2\,\text{Ferredoxin}_{RED} + NADP^{+} + 2H^{+} \rightarrow 2\,\text{ferredoxin}_{OX} + NADPH + H^{+}$$

If we summarize what has taken place in PSI, an electron has been excited out of P700 and transported to ferredoxin, which, in turn, has reduced $NADP^{+}$. However, this has left P700 an electron short; it is now $P700^{+}$, an oxidizing agent. It accepts an electron from plastocyanin (Pc), which, you remember, we left after it had been reduced by PSII. The reaction is:

$$P700^{+} + Pc_{Cu^{+}} \rightarrow P700 + Pc_{Cu^{2+}}$$

To go back further, remember that we started with light exciting an electron out of P680 (Fig. 21.6), the reaction-centre pigment of PSII; this leaves $P680^{+}$, which must have its electron restored so that it can revert to the ground state, ready for another photon to start a new round of reactions. The electron comes from water.

The water-splitting centre of PSII

$P680^{+}$ is a very strong oxidizing agent – it has a very strong affinity for an electron (greater than that of oxygen), so that it can even extract electrons from water. Four electrons are extracted from two molecules of H_2O with the release of O_2 and four H^{+} into the thylakoid lumen. It is necessary to extract all four electrons so as not to release any intermediate oxygen free radicals, which are dangerous to biological systems, just as in mitochondria, the addition of electrons to oxygen to form H_2O must be complete. In PSII there is a complex of proteins with Mn^{2+}, known as the **water-splitting centre**, which extracts the electrons from water, with the release of oxygen and protons, and passes them on to $P680^{+}$ molecules, thus restoring them to the ground state (Fig. 21.8(a)). The P680 is now ready to be activated by another photon.

How is ATP generated?

The **cytochrome *bf* complex** of PSII, which uses plastoquinol (QH_2) to reduce plastocyanin, resembles complex III of mitochondria (see Fig. 13.17) in that electron transport through the complex causes translocation of protons from the outside of the thylakoid membrane to the inside. In addition, the water-splitting centre generates protons inside the thylakoid lumen, the two effects producing a proton gradient that reduces the pH of the thylakoid lumen to about 4.5. The uptake of a proton in the reduction of $NADP^{+}$ by ferredoxin in the stroma further contributes to the proton gradient across the membrane.

There is a device that under certain circumstances leads to an increased proton translocation and therefore a greater potential for ATP synthesis. When virtually all of the $NADP^{+}$ has been reduced by ferredoxin, ferredoxin donates electrons to the cytochrome *bf* complex (Fig. 21.7) instead. The passage of these through the complex to plastocyanin leads to increased proton pumping by that complex. The extra ATP production is referred to as **cyclic photophosphorylation** driven by **cyclic electron flow.**

The proton gradient is used to generate ATP from ADP and P_i by the chemiosmotic mechanism described for mitochondria. The whole process of the events described so far is summarized in Fig. 21.8(b).

In mitochondria, the proton gradient is from the outside (high) to the inside (low). The thylakoid membrane is formed from invaginations of the inner chloroplast membrane (cf. the inner mitochondrial membrane), which explains why the proton gradients and the ATP synthases of mitochondria and thylakoid discs look as if they are in the opposite orientation.

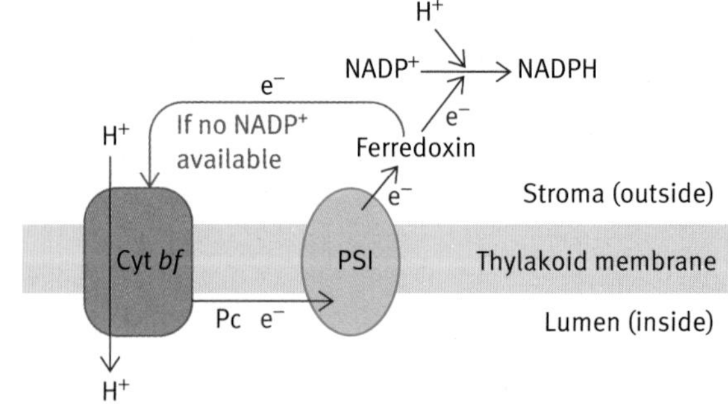

Fig. 21.7 Cyclic electron flow. When all of the $NADP^{+}$ is reduced, ferredoxin transfers electrons back to the cytochrome *bf* complex. Flow of electrons through this leads to increased proton pumping and hence increased ATP synthesis. Pc, plastocyanin.

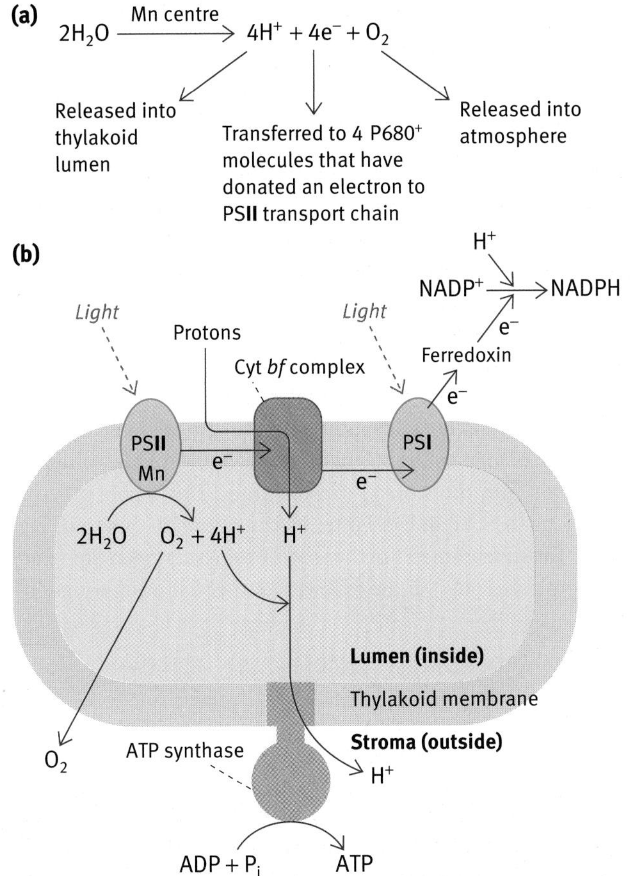

Fig. 21.8 Processes in thylakoid sacs. **(a)** Sum of reactions at the Mn water-splitting centre. **(b)** Routes followed by protons and electrons in thylakoids.

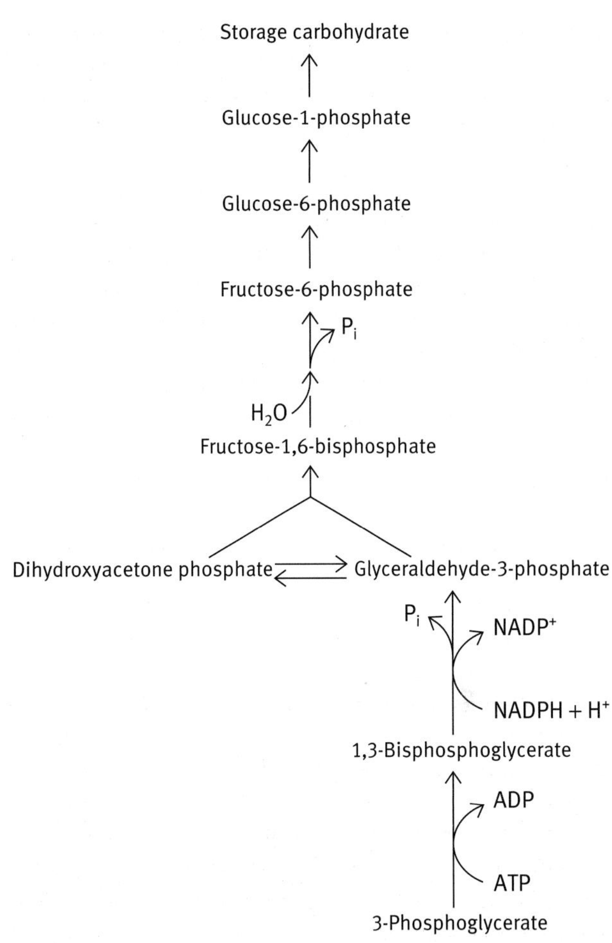

Fig. 21.9 Pathway of starch synthesis in photosynthesis, starting with 3-phosphoglycerate. The process is the same as in gluconeogenesis in liver except that NADPH is the reductant rather than NADH. In starch synthesis, the activated glucose is ADP-glucose rather than the UDP-glucose involved in glycogen synthesis. The special question in photosynthesis is the mechanism by which 3-phosphoglycerate is produced (see text for this).

The 'dark reactions' of photosynthesis – the Calvin cycle

How is CO_2 converted into carbohydrate?

As already emphasized, the aspect of photosynthesis that is fundamentally different from other biochemical processes is the harnessing of light energy to split water and reduce $NADP^+$. From there, although the actual metabolic pathways by which glucose and its derivatives are synthesized are unique to plants, it is nonetheless 'ordinary' enzyme biochemistry and quite secondary to the light-dependent process.

Getting from 3-phosphoglycerate to glucose

Glucose is formed from the metabolite 3-phosphoglycerate by a series of steps that are the same as the process of gluconeogenesis in the liver (see Fig. 16.2) except that NADPH is used instead of NADH as the reductant (Fig. 21.9). This leaves the question of how 3-phosphoglycerate is produced in photosynthesis.

3-Phosphoglycerate is formed from ribulose-1,5-bisphosphate

The most plentiful single protein on Earth is the enzyme called '**Rubisco**' for short, which stands for **ribulose-1,5-bisphosphate carboxylase/oxygenase**. It is the enzyme that utilizes CO_2 to produce 3-phosphoglycerate, one of the first steps in a process commonly referred to as 'carbon fixation'.

To remind you of terminology, ribose is an aldose sugar and ribulose is its ketose isomer. Ribulose-1,5-bisphosphate is cleaved by Rubisco into two molecules of 3-phosphoglycerate; this fixes one molecule of CO_2.

$$\begin{array}{c} CH_2OPO_3^{2-} \\ | \\ C{=}O \\ | \\ CHOH \\ | \\ CHOH \\ | \\ CH_2OPO_3^{2-} \end{array} \xrightarrow[\text{Ribulose-1,5-bisphosphate carboxylase}]{CO_2 \; + H_2O} \begin{array}{c} CH_2OPO_3^{2-} \\ | \\ CHOH \\ | \\ COO^- \\ + \\ COO^- \\ | \\ CHOH \\ | \\ CH_2OPO_3^{2-} \end{array} + 2H^+$$

Ribulose-1,5-bisphosphate Two molecules of 3-phosphoglycerate

The 3-phosphoglycerate is converted into carbohydrate as already described. This leads to the next question.

What happens to ribulose-1,5-bisphosphate ?

The answer to this is very simple, in principle. From six molecules of ribulose bisphosphate (30 carbon atoms in total) and six molecules of CO_2 (six carbon atoms), 12 molecules of 3-phosphoglycerate (36 carbon atoms) are produced. Two of these ($2C_3$) ('the profit') are used to make storage carbohydrate (C_6) by the pathway already outlined in Fig. 21.9.

The remaining 10 phosphoglycerate molecules (30 carbon atoms in total) are manipulated to produce six molecules of ribulose bisphosphate (30 carbon atoms in total). The manipulations involve C_3, C_4, C_5, C_6, and C_7 sugars, and aldolase and transketolase reactions (Chapter 15) reminiscent of the pentose phosphate pathway. The reactions are rather involved and are not presented here; instead they are summarized below.

$C_3+C_3 \rightarrow C_6$	Aldolase
$C_6+C_3 \rightarrow C_4+C_5$	Transketolase
$C_4+C_3 \rightarrow C_7$	Aldolase
$C_7+C_3 \rightarrow C_5+C_5$	Transketolase

In summary,

$$5C_3 \rightarrow 3C_5.$$

The outcome of it all is that six molecules of ribulose bisphosphate, plus six molecules of CO_2 and six of H_2O, are converted into 12 molecules of 3-phosphoglycerate. From there, the six molecules of ribulose bisphosphate are regenerated, plus the dividend of one molecule of fructose-6-phosphate, which is converted into the storage carbohydrate. This whole process, known as the Calvin cycle after its discoverer, is shown in Fig. 21.10. The stoichiometry of the whole business is given (for reference purposes only) in the following rather daunting equation:

$$6CO_2+18ATP+12NADPH+12H^+ +12H_2O \rightarrow$$
$$C_6H_{12}O_6+18ADP+18P_i+12NADP^+.$$

Rubisco has an apparent efficiency problem

At the dawn of photosynthesis there was no oxygen and, it is thought, much higher CO_2 concentrations than now, but, as photosynthesis occurred, oxygen accumulated. It so happens

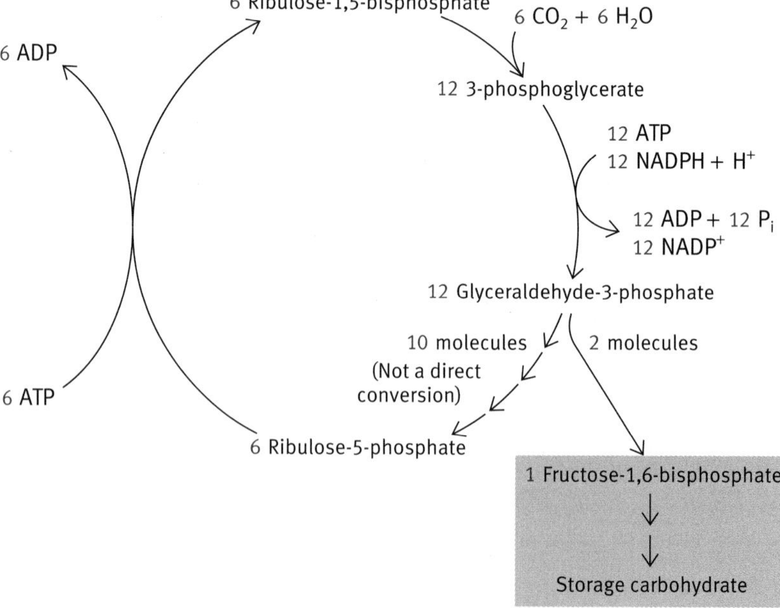

Fig. 21.10 Net effect of the reactions in the Calvin cycle. Note: the diagram is simplified in that the conversion of 3-phosphoglyceraldehyde to ribulose-5-phosphate involves part of it being converted to fructose bisphosphate as an intermediate. The presentation is intended to show the net effect of all the reactions involved. Dihydroxyacetone phosphate, which is in equilibrium with glyceraldehyde-3-phosphate, is also omitted for simplicity.

that Rubisco, in addition to using CO_2, can also react with oxygen, the two competing with one another, which is why the enzyme is called **ribulose bisphosphate carboxylase/oxygenase**. The oxygenation reaction, so far as is known at present, serves no useful purpose and is apparently wasteful in the sense that it degrades ribulose-1,5-bisphosphate and wastes ATP in a reaction pathway known as **photorespiration**.

A molecule of ribulose bisphosphate is degraded to one molecule of 3-phosphoglycerate and one molecule of glycolate + CO_2 + P_i. This release of CO_2 is wasteful to the fixation process and it sacrifices a high-energy phosphoryl group. The glycolate is salvaged by conversion into glycine.

At high temperatures the wasteful oxygen reaction is maximized to the detriment of photosynthetic efficiency. This can reduce CO_2 assimilation by about 30%. Presumably, in earlier times, when there was little oxygen and higher CO_2 levels, this would not have been significant, but today in the presence of high oxygen levels the situation is different. It might have been expected that a new Rubisco that excluded the oxygen use reaction would have evolved but this has not occurred. There may be good reasons for the apparent inefficiency that are not yet appreciated. However, in some plants, a biochemical device has evolved to raise the CO_2 level in cells where Rubisco operates, and thus minimize the oxygenase reaction. This occurs in species of plants that live in high-light and high-temperature environments where the problem of photorespiration would be maximized – in plants such as maize and sugar cane.

The C_4 pathway

C_3 plants are so called because the first stable labelled product that is experimentally detectable, if they are allowed to photosynthesize in the presence of $^{14}CO_2$, is the C_3 compound, 3-phosphoglycerate, produced by the Rubisco reaction. Some plants, however, *initially* fix CO_2 from the atmosphere into oxaloacetate (Fig. 21.11). The latter is a C_4 compound, the process is referred to as **C_4 photosynthesis**, and the plants as **C_4 plants**.

The anatomical or cellular structure of C_4 plant leaves differs from that of C_3 leaves. In the former, the mesophyll cells just below the epidermis cells at the surface, which are exposed to the atmospheric CO_2, do not contain the Rubisco enzyme but do fix CO_2 very efficiently into oxaloacetate by the carboxylation of phosphoenolpyruvate (PEP) by the enzyme **PEP carboxylase**. (*Note that this is not found in animals.*)

$$CO_2 + H_2O + PEP + NADP^+ \rightarrow oxaloacetate + CO_2 + NADPH + H^+$$

PEP carboxylase has a high affinity for CO_2, and there is no competition from oxygen. The oxaloacetate is reduced to malate, which is transported into neighbouring bundle sheath cells where the Calvin cycle occurs. Here the CO_2 is released from malate by the 'malic enzyme' (which we have seen before in fatty acid synthesis).

$$Malate + NADP^+ \rightarrow pyruvate + CO_2 + NADPH + H^+$$

This raises the concentration of CO_2 in the bundle sheath cells 10–60-fold, resulting in a more efficient operation of the Rubisco reaction. The pyruvate returns to the mesophyll cell where it is reconverted into PEP by **pyruvate phosphate dikinase**. This is an unusual reaction (*also absent from animals*) in which two Ⓟ groups of ATP are released:

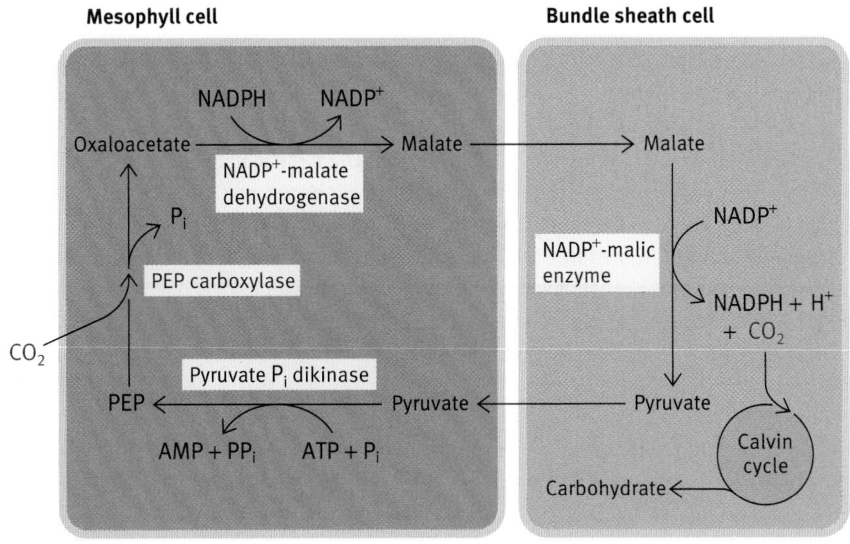

Fig. 21.11 The C_4 pathway for raising the CO_2 concentration for photosynthesis in bundle sheath cells. Important note: the pathway of oxaloacetate formation from pyruvate is quite different from that in animals. Direct conversion of pyruvate to phosphoenolpyruvate (PEP) does not occur in animals and nor does the carboxylation of PEP. It is important to be clear on this for it has a major effect on metabolic regulation in animals. Note also that the malate dehydrogenase that reduces oxaloacetate uses NADPH, unlike that in the TCA cycle, which is NAD^+-specific.

$$CH_3{-}CO{-}COO^- + ATP + P_i \longrightarrow CH_2{=}C(OPO_3^{2-}){-}COO^- + AMP + PP_i + H^+$$

In animals, PEP can be made from pyruvate only *via* oxaloacetate by a quite different route (see Chapter 16). Once phosphoglycerate is made in the bundle sheath cells, the Calvin cycle operates in exactly the same way as in C_3 plants.

The C_4 route incurs an energy price for raising the CO_2 concentration in bundle sheath cells, since ATP is consumed in making PEP and transporting acids. However, at higher temperature and light levels, the C_4 route becomes a considerable advantage. C_4 plants such as maize or sugar cane are prolific producers of carbohydrates.

There is great diversity in the biochemical processes used by different C_4 plants. For example, the C_4 acid labelled in the presence of $^{14}CO_2$ may be aspartate in some species, and not malate. These plants have high levels of aspartate aminotransferase (see Chapter 18) instead of $NADP^+$-malate dehydrogenase. C_4 plants have also evolved three distinct options for decarboxylating C_4 acids in bundle sheath cells, two being located in the mitochondria unlike the $NADP^+$-malic enzyme shown in Fig. 21.11, which is located in the cytosol. They are all mechanisms by which the same end is achieved, that is, the increase of CO_2 levels where Rubisco operates.

Succulent plants living in arid areas save water during the day by closing stomata on leaves and using the C4 pathway. This means that CO_2 cannot be taken in except at night when they fix the CO_2 as malate, which is stored until the sun shines. In light, the CO_2 is then released by decarboxylation for fixation by the Calvin cycle.

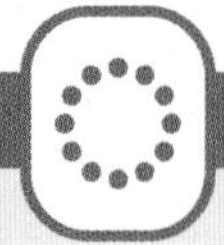

Summary

Photosynthesis occurs in plant cell chloroplasts. The part dependent on light is the splitting of water to generate NADPH. NADPH is used for the reductive synthesis of carbohydrate from CO_2 and water.

Chlorophyll is a green pigment, which receives light energy. It is present in the membrane of organelles called thylakoids. When activated by photons, chlorophyll molecules donate electrons to chains of electron carriers arranged in two photosystems (PSI and PSII). The electrons are finally used to reduce $NADP^+$. The loss of the electrons by chlorophyll makes it a very powerful oxidizing agent capable of accepting electrons from water in the water-splitting centre.

During passage of electrons from one photosystem to the other, ATP is generated by the chemiosmotic mechanism. Thus both NADPH and ATP are produced. Carbohydrate is synthesized using these in the Calvin cycle. The key reaction in this is catalysed by ribulose-1,5-bisphosphate carboxylase/oxygenase (Rubisco) which generates 3-phosphoglycerate from which carbohydrate synthesis proceeds by reversal of glycolytic reactions but using NADPH as reductant.

Rubisco works less efficiently at low CO_2 levels because it can also react apparently wastefully with oxygen in a process known as photorespiration. Oxygen and CO_2 compete for the Rubisco and at higher temperatures the photorespiration reaction is maximized. At the dawn of photosynthesis it may be speculated that this did not matter because the ratio of CO_2/oxygen would have been very much higher. But today, especially in tropical plants such as maize (corn) and sugar cane grown in high temperatures, it becomes a very significant factor.

Such plants, known as C_4 plants, have developed a means of combating this in the form of a device that greatly elevates the concentration of CO_2 available for photosynthesis, thus minimizing the oxygenase reaction of Rubisco. The CO_2 is first incorporated into oxaloacetate (a C_4 acid) in mesophyll cells by a reaction with phosphoenolpyruvate (a reaction that does not occur in animals or *Escherichia coli*). The oxaloacetate is reduced to malate, which migrates into the bundle sheath cells containing the Calvin cycle. There the malate is decarboxylated to pyruvate and CO_2 by the malic enzyme thus greatly raising the CO_2 concentration, which competes more effectively with oxygen in the Rubisco reaction and minimizes photorespiration. Analogous mechanisms using different acids operate in other C_4 plants. Temperate plants that fix CO_2 initially into 3-phosphoglycerate are known as C_3 plants.

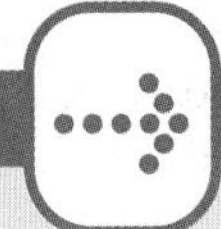

Further reading

To access the further reading, please scan the QR code image or go to http://global.oup.com/uk/orc/biosciences/molbiol/snape_biochemistry/student/reading/ch21/

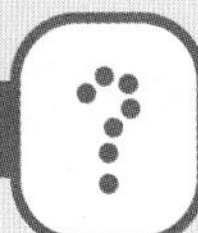

Problems

1 Explain, in general terms, what is meant by the terms 'light' and 'dark' reactions in photosynthesis.

2 What is meant by the term 'antenna chlorophyll'?

3 What is meant by:
(a) photophosphorylation?
(b) cyclic photophosphorylation?

4 The oxidation of water requires a very powerful oxidizing agent. In photosynthesis what is this agent?

5 Proton pumping due to electron transport in photosystem II (PSII) causes movement of protons from the outside of thylakoids to the inside. In mitochondria protons are pumped from the inside to the outside. Comment on this.

6 If a photosynthesizing system is exposed for a very brief period to radioactive CO_2, in C_3 plants the first compound to be labelled is 3-phosphoglycerate. Explain why this is so.

7 Describe the Calvin cycle in simplified terms.

8 The enzyme Rubisco can react with oxygen as well as CO_2, the oxygen reaction being, so far as we know, an entirely wasteful one. At low CO_2 concentrations such as can occur particularly in intense sunlight, the wasteful oxygenation reaction is maximized. What mechanisms have evolved to ameliorate this problem?

9 In C_4 plants, pyruvate is converted into phosphoenolpyruvate by an ATP-requiring reaction. Comment on this, bearing in mind the corresponding process in animals.

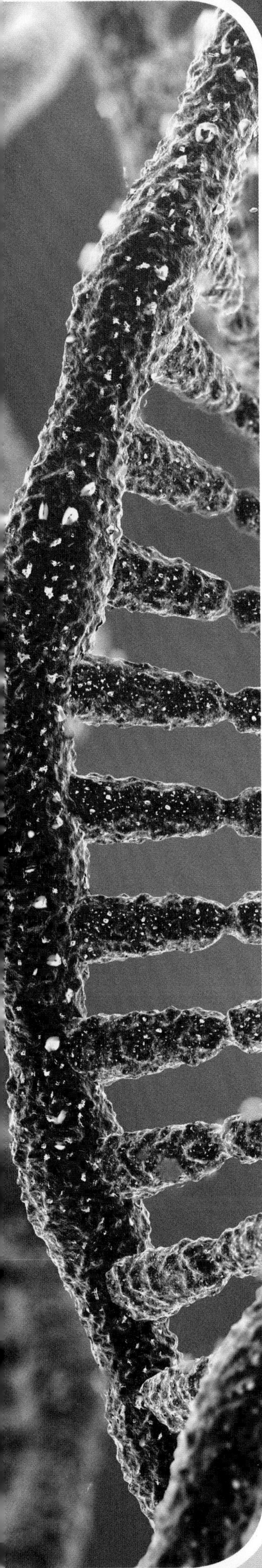

Part 4

Information storage and utilization

Chapter 22

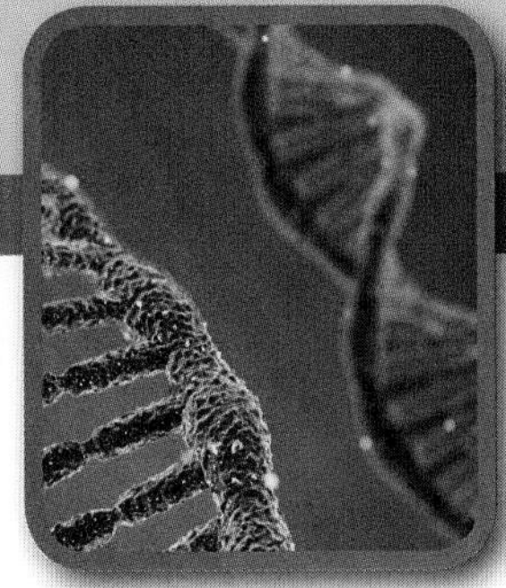

The genome

A brief overview

The genome of an organism contains the information it needs both to carry out its cellular functions and also to reproduce and pass its characteristics on to a new generation. In this chapter we will deal with what genomes are and their typical structure. The genomes of all free living organisms (this does not include viruses) consist of DNA, and the term genome usually refers to all the information coded by the DNA of the cell. In eukaryotes we may distinguish between the nuclear genome and those of mitochondria and chloroplasts. In many viruses, such as influenza and HIV, RNA is the genetic material rather than DNA and it is referred to as the viral genome.

In this chapter we discuss the chemistry of DNA and its associated molecules, its physical state in the cell, how this varies during the cell cycle, the size of the genomes of different organisms and the relationship of genome size to the complexity of an organism. Structures of genes are covered but functional aspects of the genome – how genes work and are controlled – come in the subsequent Chapters 23–25 on DNA synthesis, gene transcription and protein synthesis respectively.

A great advance in our knowledge of the genome has been the complete determination of the base sequence of the human genome (published in draft form in 2001 and updated since) and that of many other species. Knowledge of the components and organization of the genome has been crucial in elucidating how genes work, but has also thrown up many surprises. For instance, the human genome contains far fewer protein-coding genes than scientists expected, but many of the noncoding sequences are now thought to be functionally important, rather than being 'junk' DNA as previously believed. Genome research has also contributed to important technological advances, such as methods for the location and isolation of disease-producing genes and DNA fingerprinting in forensic science. These technologies will be explored in detail in Chapter 28, using the information in this chapter as a foundation.

The structures of DNA and RNA

A basic essential for the existence of life is that organisms must carry genetic information that is replicated and given to their offspring so that they are copies of themselves. At the origin of life this genetic information was almost certainly in the form of RNA, but in all cellular life it is DNA. Some viruses have RNA as their genetic material but viruses are not cells. RNA has been retained in cells as the intermediary between genes and the ribosomes in the form of messenger RNA. Other important roles of RNA will be described in Chapters 24 and 25.

DNA is chemically a very simple molecule

The relatively simple structure of DNA molecules may seem surprising, considering their large size and central role in determining the characteristics of life forms. DNA has only four different 'units', known as nucleotides. Great numbers of these are linked together to form immensely long thin threads. So far as defining amino acid sequences, the sequence of nucleotides is a form of code based on triplets of bases called **codons** that correspond to individual amino acids in polypeptide chains. The information that ultimately specifies the characteristics of complex organisms such as humans is carried in this way.

DNA (like RNA) has the essential characteristic of being able to direct its own replication. At the origin of life there could not have been the elaborate machinery that cells use today to replicate nucleic acids, so life had to start with a basically simple molecule and a basically simple replicative mechanism that would operate before the development of cells. Although much of the detail has been refined by evolution to produce the present day highly controlled replication machinery, the basic principle remains. Life was locked into it at the beginning.

DNA and RNA are both nucleic acids

The term **nucleic acid** arose because DNA was first isolated from cell nuclei. DNA is an acid because of its phosphate groups, which at physiological pH are dissociated to liberate hydrogen ions. Its nucleotide subunits contain a pentose sugar, **2-deoxy-D-ribose** and therefore it is called **deoxyribonucleic acid**, or DNA for short. As already stated, RNA has a similar structure, but differs from DNA in that the pentose sugar in the nucleotide subunits of RNA is **D-ribose, not deoxyribose.** It is therefore called **ribonucleic acid** or RNA for short. The two sugars are shown below. 2-Deoxy-D-ribose lacks the oxygen on the carbon-2 position; it is usually simply referred to as **deoxyribose**.

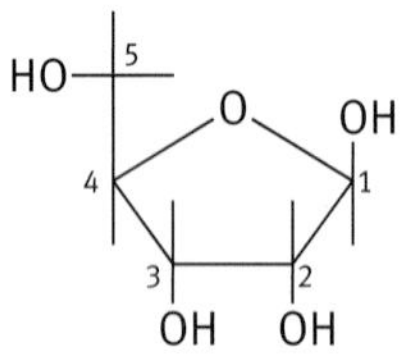

D-Ribose 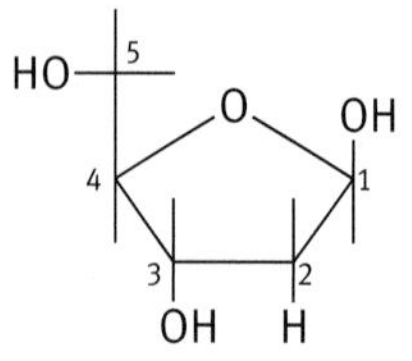 2-Deoxy-D-ribose

Although we have already discussed nucleotides in Chapter 19, which dealt with their synthesis, we will repeat some of the material here both for convenience and because of its importance.

The primary structure of DNA

DNA is a **polynucleotide**. A single nucleotide has the structure:

Phosphate—sugar—base

The structure of a deoxyribonucleotide is shown in the non-ionized form for simplicity:

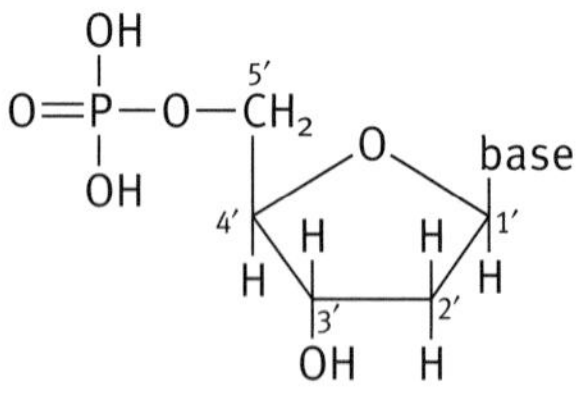

To specify a position in the deoxyribose moiety, a prime (′) is added to distinguish it from the numbering of the base ring atoms. Thus the sugar carbon atoms are 1′, 2′, 3′, 4′, and 5′ (pronounced 'five prime', etc.) and indicated outside the ring. The sugar is in the furanose, five-membered ring form. Since the bond between the phosphate and the sugar is between an acid (phosphoric acid) and an alcohol (the 5′–OH of deoxyribose) it is a phosphate ester or phosphoester. The nomenclature of nucleotides is summarized below, but is described in more detail in Chapter 19.

There are four different nucleotide bases in DNA

The bases are **adenine**, **guanine**, **cytosine**, and **thymine** – abbreviated to **A**, **G**, **C**, and **T**. A and G are **purines**, C and T, **pyrimidines**. The numbering of atoms in the bases is given inside the ring structures.

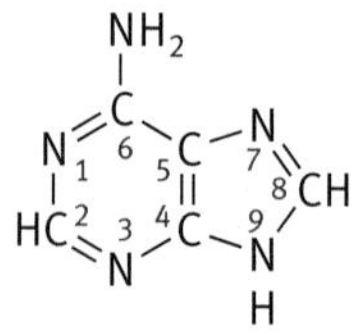

Adenine (A) 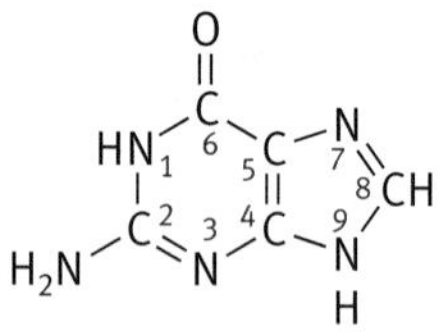 Guanine (G)

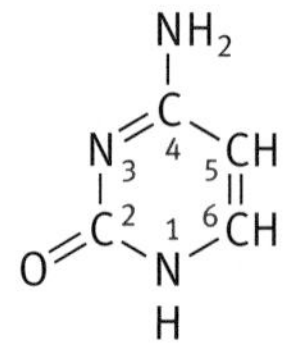

Cytosine (C) 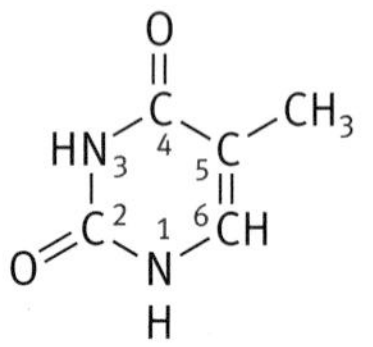 Thymine (T)

Attachment of the bases to deoxyribose

The bases are attached to deoxyribose *via* a glycosidic link between carbon atom 1 of the deoxyribose and nitrogen atoms at positions 9 and 1, respectively, of the purine and pyrimidine rings. The linkage is β (i.e. on the same side of the sugar ring as the 5′ carbon).

The structure, base-sugar, is called a **nucleoside**; if the sugar is deoxyribose it is a **deoxyribonucleoside**. The name of each nucleoside is derived from the name of the base. Although they are usually abbreviated it is useful to learn these names, which are shown in Table 22.1, with both deoxyribonucleosides and ribonucleosides included for completeness.

The structures of the deoxyribonucleosides are shown below in diagrammatic form in which the heterocyclic ring

Base	Deoxyribonucleoside	Ribonucleoside
Adenine	Deoxyadenosine	Adenosine
Guanine	Deoxyguanosine	Guanosine
Cytosine	Deoxycytidine	Cytidine
Thymine	Deoxythymidine *or* Thymidine	–
Uracil	–	Uridine

Table 22.1 Nomenclature of the major bases and nucleosides found in DNA and RNA.

structures are not shown in detail but the side groups that differentiate the bases are given.

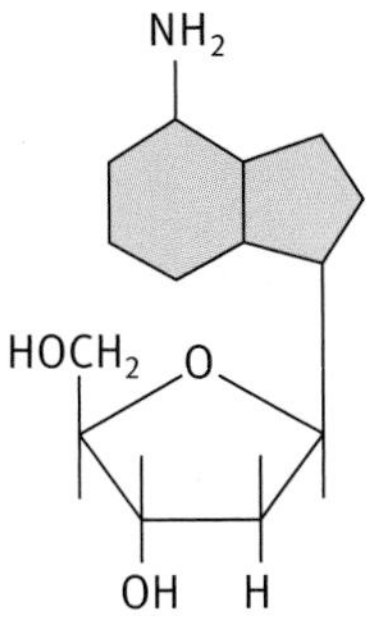

Deoxyadenosine

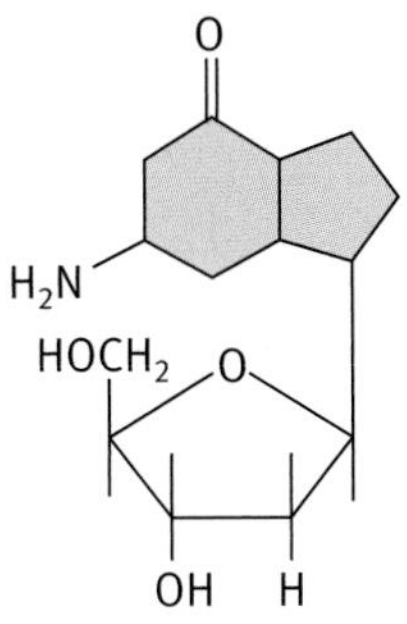

Deoxyguanosine

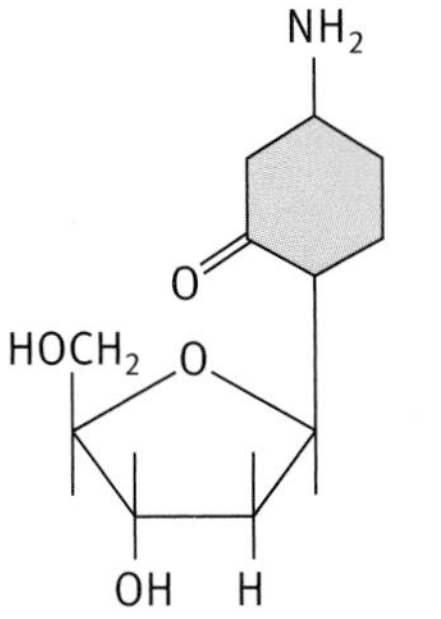

Deoxycytidine

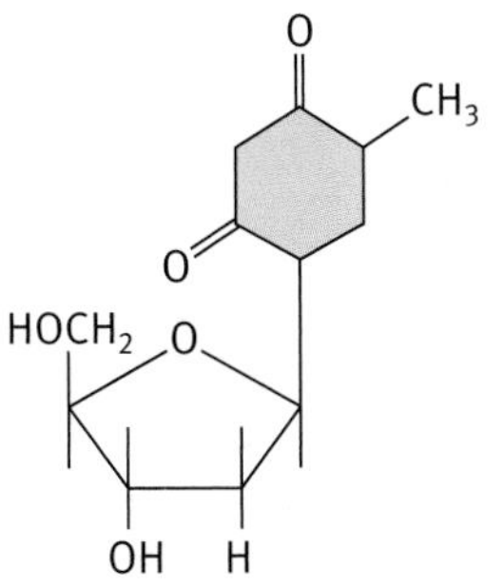

Deoxythmidine
(or, simply, thymidine)

The physical properties of the polynucleotide components

The nucleotides in DNA are the 5′ phosphate compounds, dAMP, dGMP, dCMP, and dTMP. dTMP is often abbreviated simply as TMP, since the ribonucleotide rarely occurs in nucleic acids. (Ribothymidine is found as a modification of uridine in some tRNA molecules).

As noted, the phosphoric acid —OH group of nucleotide components in DNA is ionized at physiological pH and thus has a negative charge. This, together with the hydroxyl groups of the deoxyribose, makes the exterior of the DNA double helix strongly hydrophilic. Importantly, the bases are different in that they are relatively water-insoluble, guanine almost completely so. Their flat faces are essentially hydrophobic so they have a tendency to bind face-to-face because of hydrophobic interactions. At the edge of each, however, there are polar groups with hydrogen bonding potentiality so they can bind edge-to-edge by hydrogen bonds in the core of the double helix to form base pairs.

Structure of the polynucleotide of DNA

A **dinucleotide** consists of two nucleotides linked together by a phosphate group between the 3′-OH of one and the 5′-OH of the second (the nonionized forms are shown here for clarity; see Fig. 23.9 for a description of the reaction).

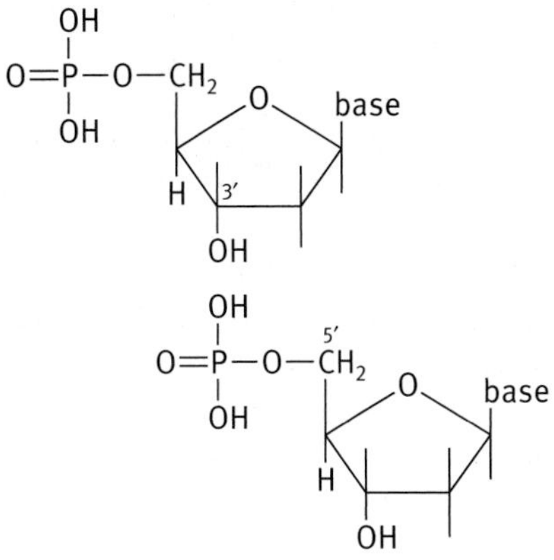

Two nucleotides

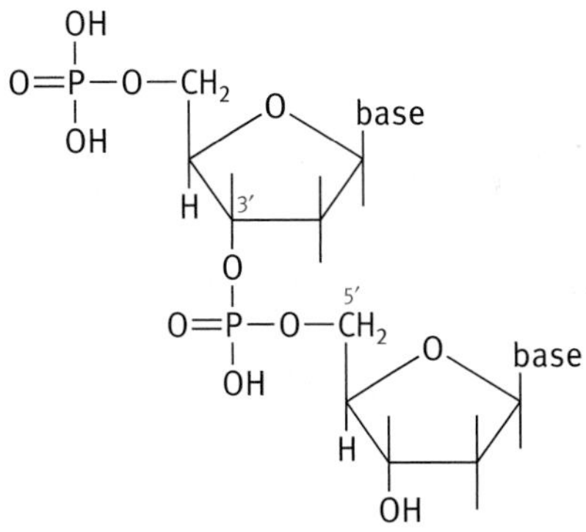

A dinucleotide

In a mononucleotide, the phosphate group is a **primary phosphate ester**, which means that there is only a single ester bond. In the polynucleotide structure, **phosphodiester links** are formed – the phosphate being linked to two deoxyribose moieties by two ester bonds.

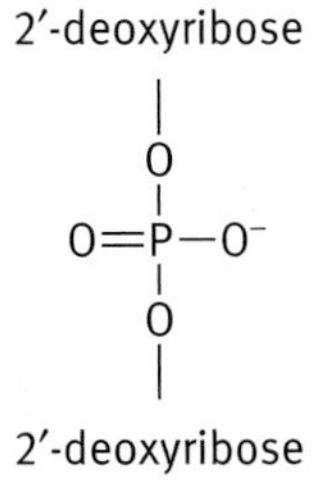

2′-deoxyribose
|
O
|
O=P—O⁻
|
O
|
2′-deoxyribose

In the dinucleotide shown above it is a 3′,5′-phosphodiester link. Nucleotides can be added (by energy-requiring reactions) in the same way indefinitely, giving a polynucleotide. DNA is in its primary structure a polynucleotide of immense length. In dealing with proteins you may recall that there is a polypeptide backbone with amino acid side-chains attached. By analogy a **polynucleotide** has a backbone of alternating sugar-phosphate-sugar groups with a base attached to each sugar residue on the

1′-position; coded information is carried in the sequence of the bases. DNA therefore has the primary structure.

Backbone section — Informational or coding section of the structure

2′-deoxyribose—base
phosphate
2′-deoxyribose—base
phosphate
2′-deoxyribose—base

or, in structural terms,

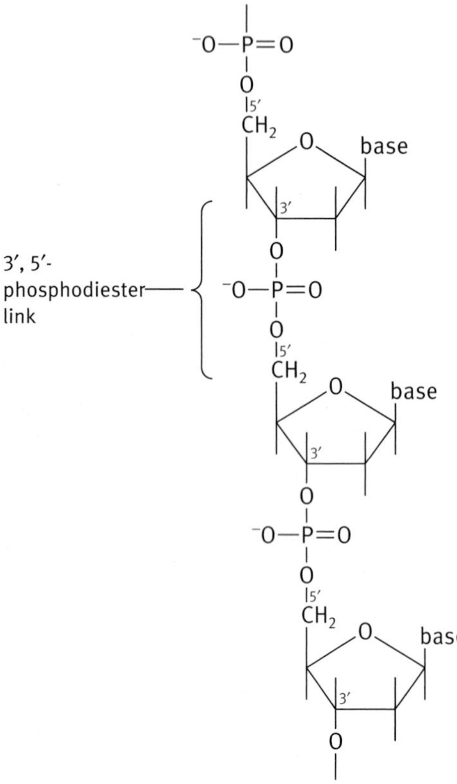

3′, 5′-phosphodiester link

In evolution it is highly probable that ribonucleotides predated deoxyribonucleotides, and RNA predated DNA. The cell nevertheless goes to considerable energetic expense to convert ribonucleotides to the deoxyribonucleotides required for DNA synthesis. The reason for this is that DNA is chemically more stable than RNA. Genetic information gathered over millions of years is stored in chemical form in DNA molecules, but molecules always have some degree of instability – they spontaneously breakdown. The presence of the 2′-OH group of ribose makes a ribopolynucleotide less stable than the corresponding deoxyribose molecule. This is because, as illustrated in the structures below, the 2′-OH group is suitably placed for a nucleophilic attack on the phosphorus atom in the presence of OH^- ions, thus causing breakage of the phosphodiester link. In DNA, lacking the 2′-OH group, this does not happen.

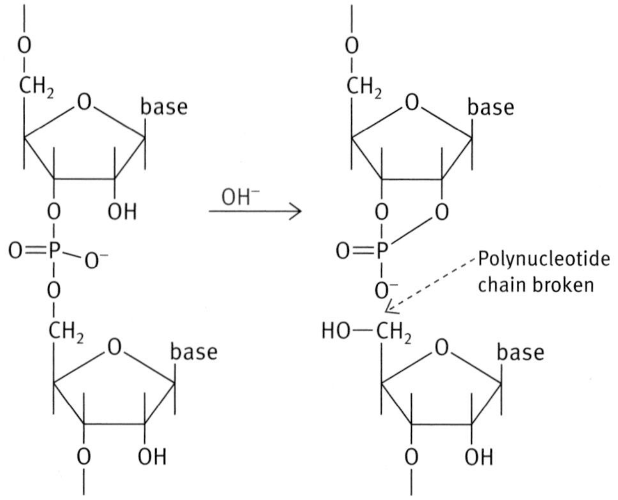

Polynucleotide chain broken

The 2′-OH of ribose facilitates the reaction because it can generate a $2'\text{-}O^-$, which attacks the phosphorus atom and converts the phosphodiester group into a 2′,3′-cyclic nucleotide, thus breaking the polynucleotide chain. Hydrolysis of the cyclic nucleotide produces a mixture of 2′ and 3′ nucleotides at the breakpoint.

The difference in stability is illustrated by the fact that dilute NaOH will completely destroy RNA at room temperature while DNA is unaffected. DNA is therefore a more stable repository of genetic information than is RNA. Nevertheless chemical damage continually occurs in DNA. DNA repair processes are discussed in Chapter 23.

Thymine instead of uracil allows DNA repair

Uracil, found in RNA, and thymine, found in DNA, have very similar structures and can both pair with adenine. The presence of thymine rather than uracil in DNA is explained by the need for genome repair. Cytosine bases in DNA undergo spontaneous deamination to uracil, as illustrated in Figure 19.1, which would lead to genetic mutation if unrepaired. There is a DNA repair enzyme that will rectify the problem (see Chapter 23). If uracil occurred normally the repair process would replace uracils that were part of the normal DNA sequence as well as those generated from cytosine. The occurrence in DNA of thymine, which has the same structure as uracil but with an additional methyl group, disposes of this problem as the repair process recognizes and replaces uracil but not thymine.

The DNA double helix

There will be few readers who have not heard of the double helix. DNA almost always exists as a double strand – only in a few viruses is it not double stranded. In other words, you have two polynucleotide molecules paired together. What holds them together? The answer is **complementary base pairing**.

Complementary base pairing

Complementarity refers to the bases. A/T and G/C in DNA chains are complementary in structure so that, when they are opposite one another in the two chains, hydrogen bonds form between them, two between A and T and three between G and C, attaching the double helices together. Only A—T and G—C pairing takes place in DNA, this being known (after the scientists who elucidated the structure) as **Watson–Crick base pairing**. The geometry of base pairing is shown in Fig. 22.1. The base pairs always include one purine (larger molecule) and one pyrimidine (smaller) so that the pairs are essentially the same size. Because of base complementarity, in different double-stranded DNA molecules the amount of G equals that of C, and the amount of A equals T. Discovery of this 'rule' by Erwin Chargaff was a vital clue in the elucidation of DNA structure. However, DNA genomes from different organisms vary in their percentages of [A+T], and thus, of course, also of [G+C], reflecting their different genetic information. The human genome is about 60% [A+T], and 40% [G+C].

Complementary base pairing is a spontaneous process between closely positioned atoms, requiring no catalysis. This spontaneity is confirmed by the phenomenon of **hybridization**. If a long molecule of DNA in solution is cut up into double-stranded pieces 20 or more nucleotides long, and then the mixture is heated to an appropriate temperature (about 95 °C), the two strands of each piece of DNA will separate – referred to as **DNA melting**, due to heat disruption of hydrogen bonds and associated weak forces. This results in many pieces of single-stranded DNA of different base sequence. However, if the solution is cooled, the pieces

```
                        H
                         \
           H   N          N—H--O        CH3
            \C≠ 7\       /      \\      /
             8    C—C          C—C
             |9  //5 6\\      /4  5\\
  Adenine    N—C4     1N--H—N3     6C—H    Thymine
      Sugar/    \3 2/         \2 1/
   phosphate     N=C           C—N
                  \           //  \
                   H          O    Sugar
                                   phosphate
                              H
                             /
           H   N       O--H—N      H
            \C≠ 7\    //     \    /
             8    C—C          C—C
             |9  //5 6\       //4 5\\
  Guanine    N—C4     1N—H--N3     6C—N    Cytosine
      Sugar/    \3 2/         \2 1/
   phosphate     N=C           C—N
                    \         //   \
                    N—H--O         Sugar
                   /               phosphate
                  H
```

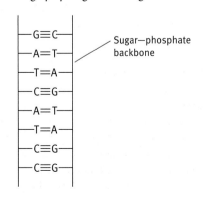

Fig. 22.1 Hydrogen bonding in the Watson–Crick base pairs.

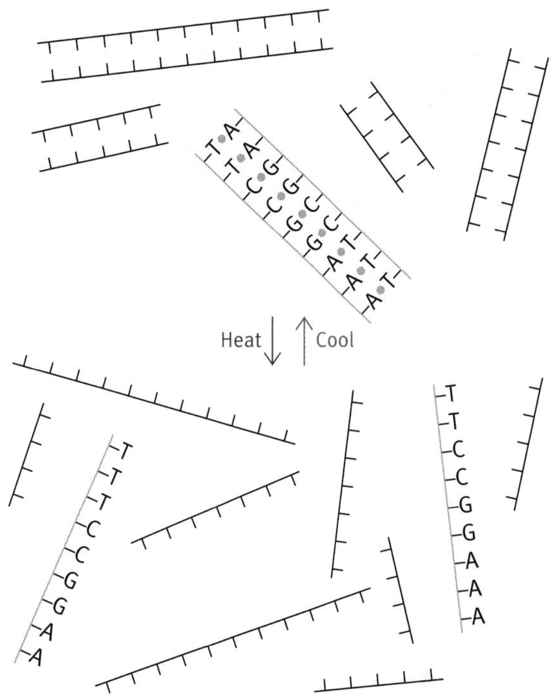

Fig. 22.2 Spontaneous hybridization of pieces of complementary DNA. Base sequences are shown on a single piece of DNA to illustrate the fact that hybridization depends on them.

will 'find' their complementary partners and reassociate, a process known as hybridization (Fig. 22.2). At low temperatures there is a thermodynamic driving force for hybridization, since the formation of hydrogen bonds releases energy. The most stable state in which free energy is minimized is that in which the bases are paired, since this gives the maximum number of hydrogen bonds. In a piece of DNA rich in G+C, the two strands will be more strongly held together than in a piece rich in A+T and the melting temperature is therefore higher. This hybridization, sometimes called **annealing**, is of practical use in many gene manipulation techniques (see Chapter 28).

For the first approximation then, a stretch of DNA is commonly represented as shown below, the long solid line representing the sugar-phosphate backbones and the attached bases interacting by hydrogen bonding.

```
|—G≡C—|
|—A=T—|   Sugar—phosphate
|—T=A—|   backbone
|—C≡G—|
|—A=T—|
|—T=A—|
|—C≡G—|
|—C≡G—|
```

However, this straight ladder structure does not occur under normal solution conditions, for it violates a structural requirement as shown in the diagram below.

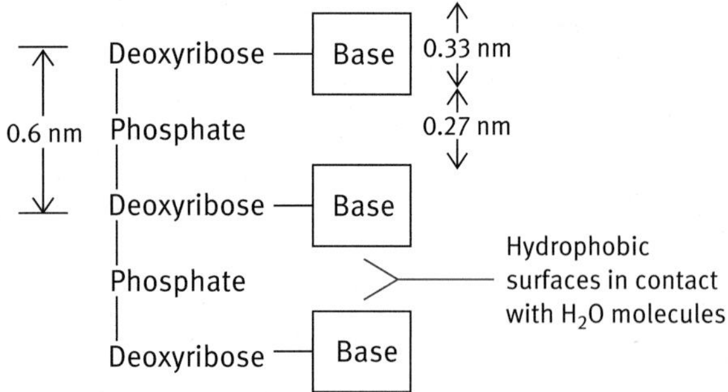

The length of the phosphodiester link is 0.6 nm (1 nanometer (nm) = 10^{-9} m), while the bases are about 0.33 nm thick, so that in the straight ladder-like structure there would be a gap between the bases. The faces of bases are hydrophobic, and in the 'straight' structure above, they would be exposed to H_2O molecules, an unstable situation. Instead, each of the two DNA chains forms a helix, with the bases inside and the hydrophilic sugar and phosphate groups outside (Fig. 22.3(a), (b)). The sloping of the chains as they follow a double helical pathway collapses the bases together and minimizes the exposure of the hydrophobic faces to water, as illustrated in Fig. 22.3(c).

The base pairs still lie almost flat, stacked on top of one another – a phenomenon known as **base stacking**. They interact with each other via van der Waals interactions, which contribute to the stability of the helix. The hydrogen-bonding face at the edge of the bases is still free so that it can bond to its partner strand. In the DNA double helix, the stacking of bases is not exactly vertical as shown in Fig. 22.3(c). Successive base pairs rotate slightly relative to one another, as illustrated in Fig. 22.3(d), such that approximately 10 base pairs are required to rotate through one complete turn, and there is a slight cross-helical slope called 'tilt'.

The helices are right-handed – as you move along a strand or a groove you continually turn clockwise; alternatively, imagine a right-handed person driving in a screw. The turning motion gives the direction of twist. The structure of the double helix is such that there are major and minor grooves (see Fig. 22.3(a)). Any base pair can be viewed from both the major and the minor grooves but only their edges are visible. The major grooves have more atoms accessible for bond formation and hence provide easier access for proteins to 'recognize' (by which we mean attach to) the base pair edges. A given base pair 'looks' quite different when viewed from the two grooves (Fig. 22.4), the significance of which will become apparent in Chapter 26, where gene regulation is discussed.

The DNA conformation described above is known as the **B form** (Fig. 22.5) and is the normal form that exists in cells. Watson and Crick proposed a B form structure with a rotation of 36° between each base pair, that is, with 10 base pairs in each 360° turn and the 'pitch' of the helix (length of one 360° turn) being 34 nm. The updated structure differs slightly from this model, as it has 10.5 base pairs per turn and therefore a pitch of 36 nm (the actual dimensions may vary slightly, depending for instance on the base pair composition of the molecule), but the Watson-Crick model is essentially correct. The helix diameter is approximately 2 nm.

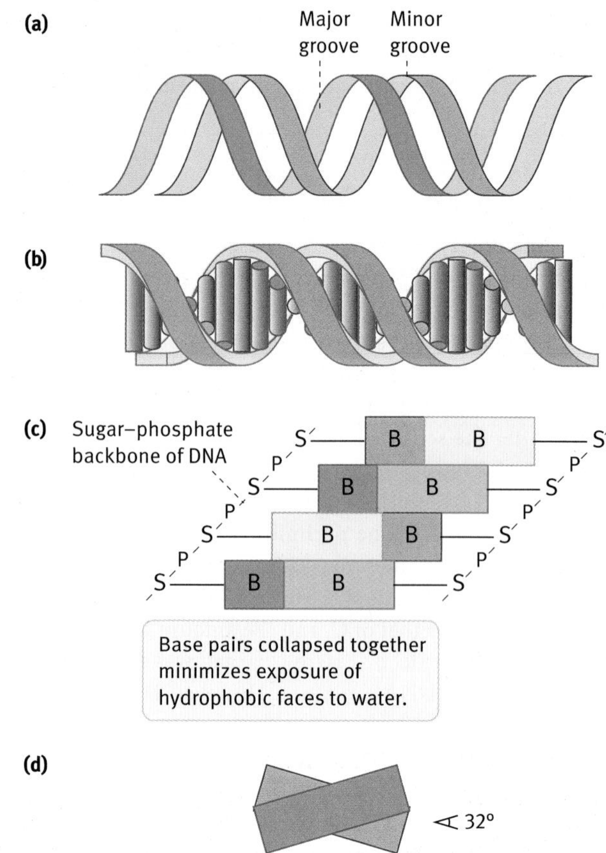

Fig. 22.3 **(a)** Outline of the backbone arrangements in the DNA double helix. **(b)** As (a), but showing the base pairs in the centre of the helix. (Note that each coloured band is a base *pair*.) **(c)** How a skewed arrangement of the ladder collapses the bases together. **(d)** Diagram of two successive base pairs in a double helix showing the twist imposed by the double helix. A more realistic model corresponding to (b) is shown in Fig. 22.5. Figs 2.4a & 3.1 from Calladine, C. R. and Drew, H. W. (1992). Understanding DNA; Academic Press Inc, Elsevier, © 1992.

DNA *can* adopt different configurations in special circumstances. When dehydrated, the double helix is more squat and the bases are more tilted; this is known as the **A form**. Another form is known as **Z** (because the polynucleotide backbone zigzags); in this form the double helix is left-handed. Z DNA has been observed to occur in short synthetic DNA molecules with alternating purine and pyrimidine bases provided the solution is of high ionic strength. The biological significance of A and Z DNA is not known, but is the subject of investigation. In particular, localized sections of the genome may adopt the Z form as a means of regulating gene expression.

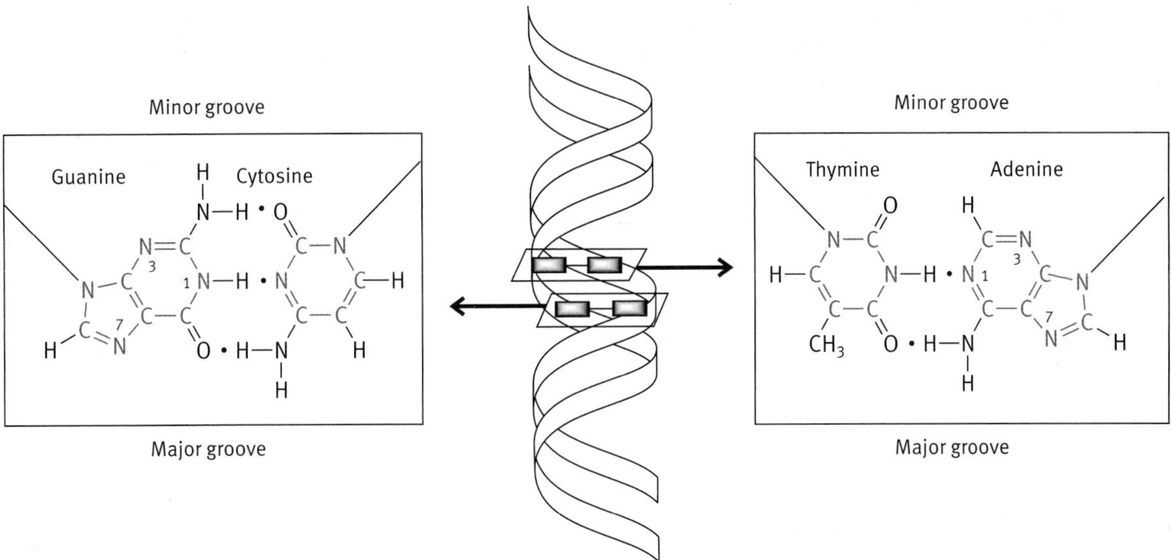

Fig. 22.4 The edges of a given base pair in DNA look different when viewed from the major and minor grooves. DNA-binding proteins designed to recognize specific sequences of base pairs in DNA can identify (bind to) the characteristic chemical groupings of different base pairs without unwinding the DNA. The importance of this is discussed in Chapter 26, which deals with gene control. Ptashne, M (1992); A Genetic Switch: Phage 1 and Higher Organisms, 2nd edition; Reproduced by permission of John Wiley & Sons.

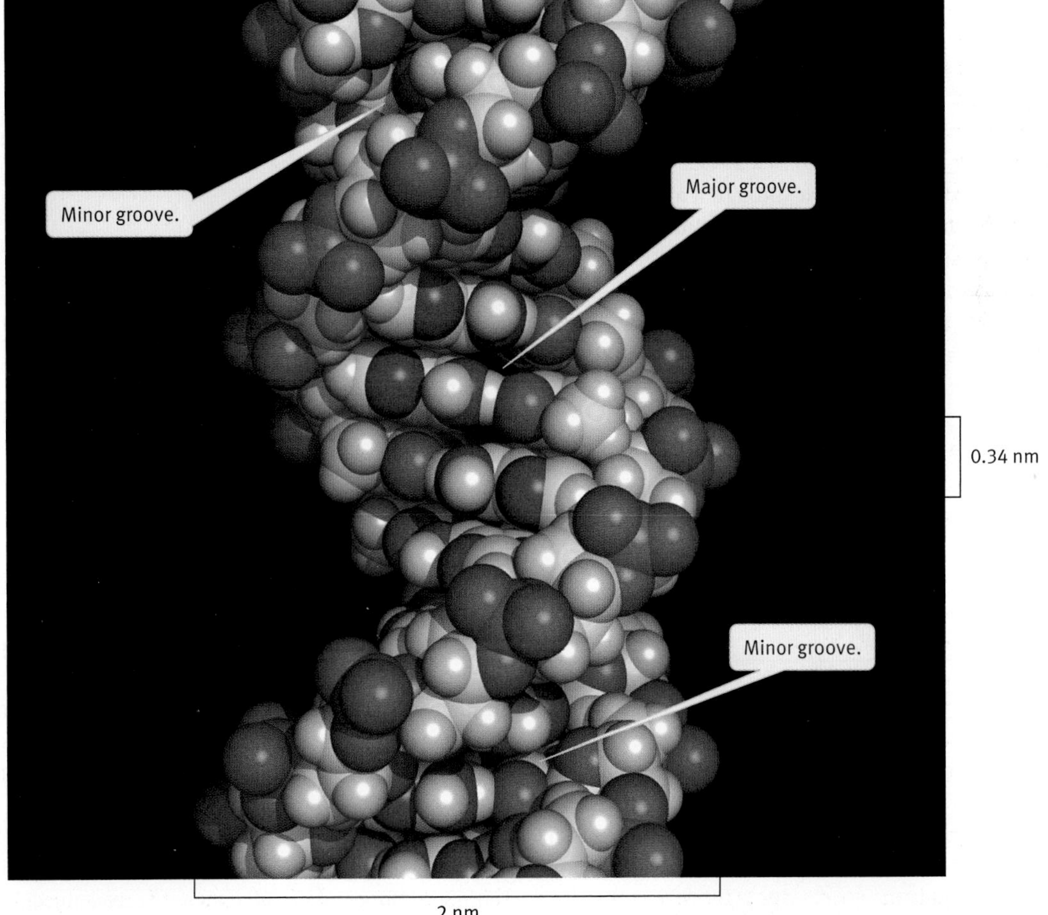

Fig. 22.5 A model of B DNA, Protein Data Bank code 1BNA. Space-filling atomic model of a DNA segment with one major groove and part of two minor grooves.

An important property of the double helix structure is that it can bend. A molecule of eukaryotic DNA is vastly longer than the widest dimension of a nucleus; to pack it in, flexibility is necessary to allow for all the coiling and folding required.

While DNA is almost always in the double helix form, single-stranded DNA does occur, for example, in certain bacterial viruses. In such situations the molecule takes up very complex internally folded structures to satisfy thermodynamic considerations. However, even in these viruses the life cycle involves a double helix form of DNA so that the basic principles of genetic information with complementary base pairing are the same in all cases.

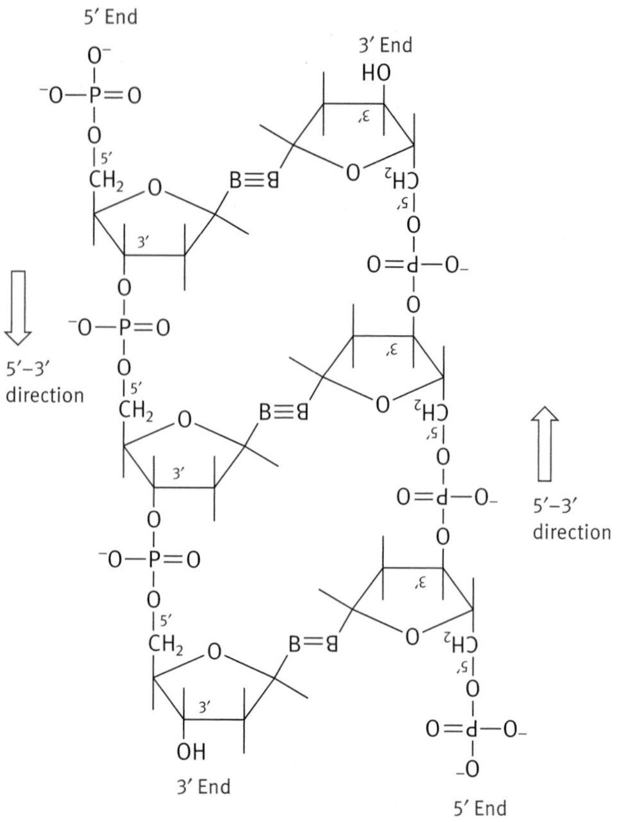

DNA chains are antiparallel; what does this mean?

By antiparallel we mean that the two chains of a double helix have opposite polarity – they run in opposite directions. It may not be immediately clear what is meant by the polarity or direction of a DNA strand. In DNA we are talking about the direction in which a sequence of nucleotides is read, and not polarity in the sense of a bond between oppositely charged ions. It is worth spending a little time on this so that you are comfortable with the concept, because a lot of biochemistry requires an understanding of it. Two antiparallel strands of DNA are illustrated in Fig. 22.6.

Any single linear strand of DNA has (obviously) two ends. One has a 5′-OH group on the sugar nucleotide that is *not* connected to another nucleotide, which may have a phosphate on it; this is the 5′ end. The other end has a 3′-OH group that is *not* connected to another nucleotide (though it also may have a phosphate group on it); this is the 3′ end. At one end of a linear piece of DNA double helix, there is always one 5′ end and one 3′ end. If a piece of DNA is circular, as in bacteria, there are no free ends, but there is still inherent polarity in the individual strands, as you can see from the deoxyribose moieties.

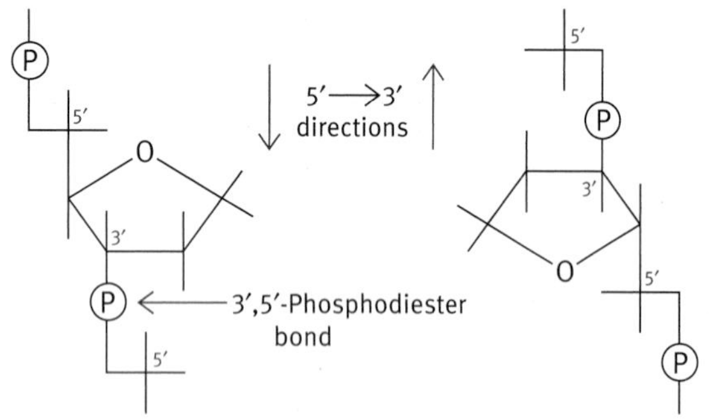

5′⟶3′ directions

3′,5′-Phosphodiester bond

Thus the structure on the left runs 5′⟶3′ down the page and that on the right 5′⟶3′ up the page. In Figure 22.6 showing double-stranded DNA, in each strand the 5′⟶3′ direction runs from the 5′ carbon of the deoxyribose towards the 3′ one of the same sugar.

There are conventions in writing down the base sequence of DNA. It is usual to represent a polynucleotide structure

5′ End

3′ End

5′–3′ direction

5′–3′ direction

3′ End

5′ End

Fig. 22.6 Two antiparallel strands of DNA. GC base pairs are linked by three hydrogen bonds and AT base pairs by two hydrogen bonds. B, base.

simply by a string of letters representing the bases of component nucleotides, but sometimes the phosphodiester link is indicated by the letter p inserted between the bases (for example, CpApTpGp, etc.). Suppose we have a piece of double-stranded DNA whose base sequence is:

5′ CATGTA 3′

3′ GTACAT 5′.

Sometimes it is useful to write both strand sequences, but usually it is not necessary to write both sequences since, given one, the complementary sequence is automatically specified. So you will find that the structure of a gene is often given as a single base sequence, despite there being two strands. There is a convention that a single base sequence is written with the 5′ end to the left, and it is not always necessary therefore to specify the 5′ and 3′ ends of a sequence. Thus, if the structure illustrated above is part of a gene, it would be written as CATGTA. Note also that if it is part of a protein-coding sequence, the sequence given would conventionally be that of the 'coding strand' (see Chapter 24, Coding and noncoding strands).

Base pairing in RNA

Although RNA molecules are generally single rather than double stranded, RNA can of course undergo base pairing. Intra-chain base pairing, where an RNA molecule folds back on itself, is common. Indeed, certain classes of RNA molecule such as transfer RNA (tRNA) and ribosomal RNA (rRNA) depend on their folded three-dimensional structure for function (see Chapter 25). Stem loop structures are often found such as those seen in tRNA (see Fig. 25.2). In RNA, A pairs with U and G with C, but nonWatson-Crick GU base pairs are also found with some frequency, even though they are less stable. You will meet GU base pairing again in the context of 'wobble base pairing' in protein synthesis in Chapter 25, but it is also found in the stems of stem loops and other folded RNA structures.

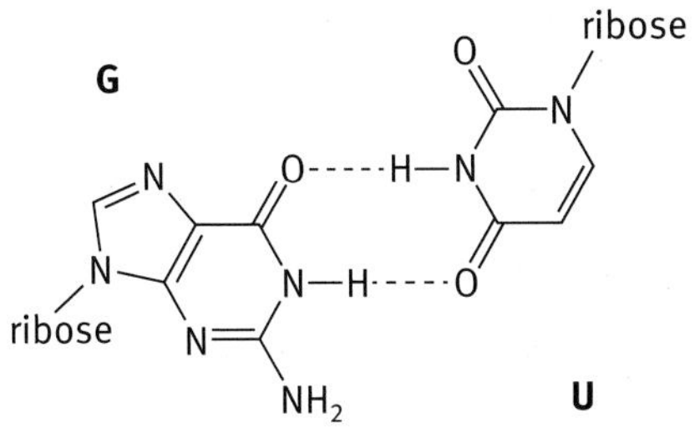

Stretches of base paired RNA form double helices, but the additional hydroxyl group on the ribose sugar prevents formation of the B form of helix and the structure resembles the A form of DNA.

Genome organization

The prokaryotic genome

The genome of a typical prokaryote such as *Escherichia coli* consists of a single closed circle of double-stranded DNA. There is no nuclear membrane surrounding it so that the DNA is in direct contact with the cytosol. The cell grows continuously in suitable nutritive conditions because the chromosome remains in an active state throughout the life cycle of the cell. After the DNA has been duplicated and a critical size reached, cell division occurs with each daughter cell receiving one chromosome. DNA replication and cell division are coordinated, but there are no separate phases in the life cycle – cell growth and replication of DNA go on continuously.

Plasmids

Besides their single chromosome, bacteria such as *E. coli* often contain additional, much smaller, circular double-stranded DNA **plasmids**. The main chromosome in *E coli* is 4.6 million base pairs (4.6 Mbps) in length, while plasmids typically contain a few thousand base pairs. Whereas the main chromosome carries the 'housekeeping' genes needed for the basic processes of life, plasmids may carry genes that confer additional nonessential properties, such as antibiotic resistance. They also include genes required for their own replication, and a single bacterial cell may contain multiple copies of the same plasmid. Plasmids are of medical importance, because they can be transmitted not just **vertically** (to daughter cells when the bacterial cell divides) but also **horizontally**, from one bacterial cell to another, thus contributing to the spread of antibiotic resistance in pathogenic organisms. They are also of great practical use in gene manipulation, since additional DNA sequences can be engineered into them and then replicated many times within the host cell.

The eukaryotic genome: chromosomes

Eukaryotic genomes are differently organized from those of prokaryotes. The genome of each species is divided into a characteristic number of double-stranded **linear** DNA molecules, the chromosomes. In most stages of development of sexually reproducing eukaryotes the cells are **diploid** (gametes excepted), which means that they have two sets of chromosomes, one derived from each parent. Humans, for example, have 46 chromosomes consisting of 22 **homologous pairs** known as **autosomes** and two **sex chromosomes**, X–Y in the male and X–X in the female. The two members of a homologous chromosome pair each have the same genes in the same order, so eukaryotes have two copies of each autosomal gene. The base sequences of the two homologous chromosomes are therefore almost the same, but the two copies of a gene may differ slightly in base sequence. Different forms of the same gene are known as **alleles**, and the inheritance of different alleles contributes to genetic variation. For example, a gene on human chromosome 4 encodes a protein, glycophorin A, that spans the membrane of red blood cells. Two alleles of the gene, the M and N alleles, encode proteins that differ at just two out of 131 amino acids, leading to individuals having different MN blood groups depending on which alleles they inherit.

During formation of the gametes, homologous chromosomes exchange sections of their DNA sequence by the process of crossing over (see Chapter 30 on Cell division). This further increases genetic variation from generation to generation.

Unlike prokaryotes, eukaryotes contain their chromosomes within a nuclear membrane. When a gene is active, a messenger RNA (mRNA) is transcribed (copied) from the DNA, processed, transported through the nuclear membrane and is then translated by ribosomes into proteins. This separation of gene transcription from mRNA translation in time and space is an important difference between eukaryotes and prokaryotes. By contrast, in *E. coli* the mRNA is translated immediately, beginning even before it has been completely synthesized and released from the DNA.

The mitochondrial genome

Besides the nuclear chromosomes, eukaryotic cells contain extra genomes in mitochondria (and chloroplasts in plant cells). It has been explained in Chapter 2 that mitochondria originated in the evolutionary sense from the engulfment of a prokaryote cell by the precursor of modern eukaryotic cells. Mitochondria reproduce by division.

Most mitochondrial proteins today are coded for by genes in the nucleus, synthesized in the cytosol and transported into the organelle, but a small proportion are still made in the organelle. The mitochondrial DNA is double stranded and circular with, in humans, 16,569 base pairs. Each mitochondrion contains multiple copies of the genome. The protein-coding sequences are tightly packed along the length of the DNA, an arrangement that reflects the prokaryotic origin of the genome and is unlike that in eukaryotic nuclear DNA (see next section). The processes of gene transcription and mRNA translation in mitochondria are also prokaryote like.

Mitochondrial inheritance is different to that of nuclear genes. Sperm have very little cytoplasm, so mitochondria are transmitted by the egg and inherited only from the mother. Thus the mitochondrial genome's inheritance is described as 'maternal'. Since the proteins encoded by the mitochondrial genome function in ATP synthesis, maternally inherited mitochondrial genome mutations, though rare, can have profound consequences for human health.

The structure of protein-coding genes

What is a gene?

It is surprisingly difficult to give a single, simple definition of a gene. For a geneticist it is a unit of heredity, which can be tracked through generations as it determines a particular characteristic or **trait** of the organism. In physical terms, it is a particular sequence of DNA, found at a certain place, or **locus**, in the genome, yet a chromosome is a continuous DNA molecule, and defining where one gene begins and another ends can be difficult. Biochemists often think of a gene as a stretch of DNA that carries coded information for the sequence of amino acids in a single polypeptide chain, but this is also an oversimplification. Protein-coding sequences are flanked by regulatory DNA sequences that, though not themselves encoding amino acids, are crucial for determining when and where the protein is made and may also be considered part of the gene. Additionally a number of genes are transcribed to produce nonprotein-coding RNA molecules. In the section below, we discuss protein-coding genes as DNA sequences that are transcribed as mRNA, while the regulatory sequences of genes are considered further in Chapter 25.

Protein-coding regions of genes in eukaryotes are split up into different sections

In prokaryotes the genes are arranged close together on the DNA with short 'spacer segments' between them. The coding region that specifies the amino acid sequence of a protein is continuous. In eukaryotes this is not the case. The coding region is interrupted by segments of DNA that do not code for amino acid sequences. The interrupting sequences are called **introns** while the coding sections are called **exons** (see Fig. 24.12). There can be 1–500 introns in a gene, which can each vary from 50 to 20,000 base pairs in length. Exons are smaller, usually around 150 base pairs. Thus, typically, the total length of the exons of a eukaryotic split gene is very much smaller than the total of its introns. In the human genome exons total only about 1.6% of the DNA, while the transcribed sequences of genes, including introns make up around 25% of the genome. When split genes are transcribed into RNA, the sections of the latter corresponding to the introns are removed and the coding region corresponding to the exons joined together to produce a continuous messenger. The process is called '**splicing**'. It is complex and best left to Chapter 24.

The evolutionary origin and significance of split genes is the subject of debate. Soon after their discovery in the 1970s, the Nobel prizewinner Walter Gilbert hypothesized that primitive genes were short protein-coding sequences that became fused together, with **intergenic** or **intervening sequences** (hence the term introns) between them, to encode more complex proteins. This is the **exon theory** of gene origin, or **introns early** model. An alternative proposal is that introns were inserted into eukaryotic genes later in evolution (the '**introns late**' model), with the suggestion that introns are invasive 'parasitic' sequences since they seem at first glance to have no useful function. There is still no clear resolution to this debate. Introns and the splicing mechanism for their removal have been found in all eukaryotes studied, suggesting that they were present in the early stages of eukaryote evolution, although the density of introns varies considerably between different phyla. In contrast, all prokaryotes studied to date lack introns and the splicing machinery. If introns were also present in the ancestors of present day prokaryotes, then they may have been lost through an evolutionary drive towards rapid protein synthesis since splicing of mRNA is unnecessary for intronless genes.

How is it then that introns have survived in eukaryotic genomes? One possibility is that they may have facilitated evolution of proteins. Individual exons often code for discrete structural and functional protein domains. New proteins are believed to have evolved particularly rapidly by '**domain shuffling**' – the concept is that the same domain is used repeatedly, combined with various other domains, so that new families of proteins are assembled from pre-existing domains. The separation of the parts of a gene into exons coding for discrete protein domains could facilitate this process as new genes could

(a) Exon shuffling

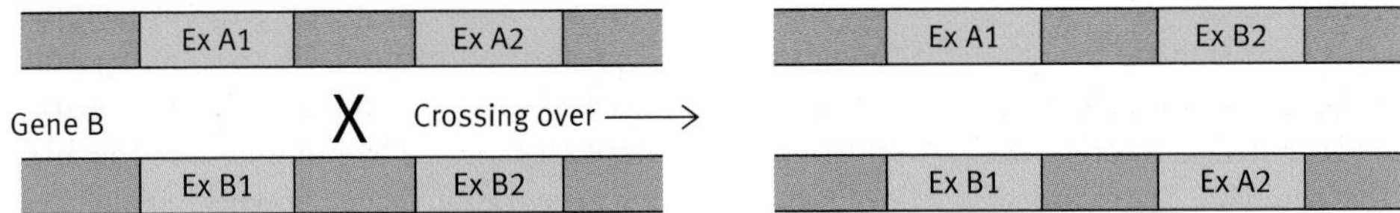

(b) Gene duplication

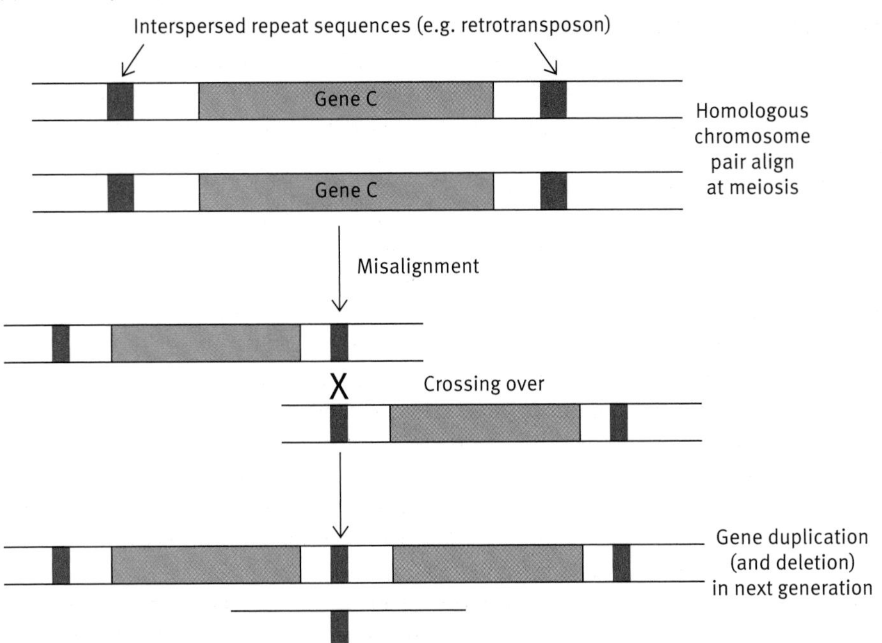

Fig. 22.7 Exon shuffling and gene duplication can occur as a result of misalignment during homologous recombination. **(a)** Exon shuffling can result from recombination between two genes whose introns share the same DNA sequence. **(b)** Gene duplication can occur when repetitive DNA sequences misalign.

be assembled by **exon shuffling** involving genetic homologous recombination (see Chapter 23, Homologous recombination). Introns would help in the process because they are zones where breaks and joins could take place without disrupting exons and hence without destroying protein domains (Fig. 22.7a). The process would provide a more rapid means of producing novel proteins by recombination events than would point mutations in DNA leading to single amino acid changes.

An additional reason for intron survival could be the added flexibility provided by **alternative splicing** (see Chapter 24). Here the removal of varying sections of sequence from RNA during splicing can increase the number of proteins encoded by a single gene, a process that is believed to contribute to the complexity of 'higher' eukaryotes such as humans.

Gene duplication facilitates evolution of new genes

Besides contributing to evolution through exon shuffling, homologous recombination also leads to the generation of new genes through gene duplication (Fig. 22.7b). Duplicate genes allow one copy to be modified by accumulating mutations while the other retains the essential function of the original gene. A good example is that of serine proteases of the chymotrypsin family, which have evolved to carry out diverse functions including digestive enzymes and blood coagulation factors.

Most of the human genome does not encode proteins

One might have expected that to code for an organism such as a human, the genome would have an organized look about it. However, when the human genome project was completed and the base sequence of the DNA available, the reverse was found. It looks more like a disorganized mess thrown together haphazardly. As one distinguished worker in the field put it: 'The general arrangement of the genome provides another startling jolt. In some ways it may resemble your garage/bedroom/refrigerator/life – highly individualistic and unkempt, with little evidence of organization and much accumulated clutter. Virtually nothing is ever discarded and valuable items are scattered indiscriminately, apparently carelessly'.

The valuable items referred to in this quotation are the protein-coding genes and the clutter is DNA that until recently was presumed to have no function and was called 'junk' DNA. It was presumed to be useless DNA that could not be discarded; it constitutes over half of the total in humans. Genes occupy only a fraction of the human genome, and the actual protein-coding sequences (excluding introns) constitute only 1.6% of the total DNA. The estimate for the number of protein-coding genes in the human genome has been revised downward, from a suggestion of 30,000–40,000 in the 2001 Human Genome publication, to around 21,000 in 2011. In the sections below we briefly mention the major nonprotein-coding classes of DNA in the human genome, and indicate why the term 'junk' DNA may no longer be considered appropriate.

Mobile genetic elements: transposons and retroviruses

Around half of the human genome derives from mobile genetic elements (also called transposable elements or transposons). These are sequences that can (or could in the past) move from one part of the chromosome to another. Transposons are found in many species, both prokaryotes and eukaryotes. Two main classes are found in the human genome: DNA transposons and retrotransposons. This classification is based on the mechanism by which the elements move around the genome. **DNA transposons**, which make up around 3% of the genome, are those in which the transposon sequence is cut out of the chromosome and becomes inserted in some other place. The transposon codes for a transposase enzyme, which facilitates the process. **Retrotransposons** are not cut out of the DNA sequence. Instead they replicate by their DNA being transcribed into RNA copies, which are then copied back into double-stranded **complementary DNA (cDNA)** by a retrotransposon encoded enzyme, **reverse transcriptase** (see Fig. 23.26). The cDNA copies insert themselves into new positions in the genome. Retrotransposons are further subdivided into two classes. Firstly, there is a class with DNA sequence and mode of replication similar to retroviruses (see Chapter 23). These have distinctive repeated sequences termed long terminal repeats, or LTRs, at each end. The second class, non-LTR-retrotransposons, also encode reverse transcriptase, but use a simpler though not yet fully understood mechanism for generating and inserting new copies. Retrotransposons with LTRs are often termed endogenous retroviruses and comprise around 8% of the genome sequence, while nonLTR retrotransposons and their derivatives make up a massive one-third of the genome.

We have been careful to talk about transposable elements *and their derivatives*, because the majority of transposon sequences in the human genome have undergone mutations that mean they are no longer able to move around. DNA transposons and LTR retrotransposons (endogenous retroviruses) have little, if any, mobility in the present day genome, although the small proportion of nonLTR retrotransposons that can still move around contribute to human genome variation and occasionally cause disease by jumping into the middle of protein-coding genes or into gene regulatory sequences. NonLTR retrotransposon remnants (LINES and SINES) make up a high percentage of the repetitive DNA sequences discussed below.

Retroviruses such as the human immunodeficiency virus (HIV) differ from LTR-retrotransposons because they have an extra gene (the *env* gene) that allows the RNA intermediate to escape the host cell and infect a new one, while retrotransposons stay within the original cell. Present day retroviruses are probably evolutionarily derived from LTR-retrotransposons, while some of the LTR-transposons in the current genome may derive from infecting retroviruses that lost their *env* gene.

Repetitive DNA sequences

About half the DNA of the human genome is made up of repetitive sequences of different types, a high proportion of which have evolved from transposable elements. 'Interspersed' repeat sequences derived from transposons are scattered through the genome, while some other repeat sequences are clustered in particular regions. Although the origin and function of much of the repetitive DNA in the human genome is uncertain, once there it has contributed to our evolution, as repeated sequences form sites where unequal crossing over, as illustrated in Figure 22.7, can generate duplicated genes.

- **LINES (Long Interspersed Elements)** derive from non-LTR-transposons. They are a few thousand base pairs long and together make up around 21% of the genome. The major class is the LINE-1 or L-1 element, 6,000 bp long, of which there are more than 500,000 copies. Fewer than 100 of these are thought to be actively mobile.
- **SINES (Short Interspersed Elements)** are also nonLTR retrotransposon derivatives, but are only 100–500 base pairs long and make up 13% of the human genome. The main class is the Alu family (so called because the sequence contains the recognition site for the restriction endonuclease *AluI*). Alu elements are also found in mammals other than humans. A typical Alu element is around 300 bp long, and there are more than 1 million of them in the genome. SINES seem to be derived from endogenous RNA molecules that at some stage in evolution were copied into cDNA and inserted in the genome. They have multiplied because they contain sequences that allow them to be further transcribed into RNA copies, which then 'hijack' the enzymes encoded by L1 elements to make more DNA and insert back into the genome. Many of the *Alu* elements in the

human genome no longer move around, but they have been active in our recent evolutionary history, and some are still mobile.

- **Simple sequence repeats** or **tandemly repeated sequences**: the other main category of repetitive DNAs, make up at least 3% of the genome and consist of sequences of short nucleotide units arranged in head to tail or 'tandem' arrays. An example is five bases in tandem, TTCCA/TTCCA/TTCCA repeated dozens or thousands of times. Repeats of this style are found particularly around the centromeres (the region of the chromosome that attaches to the spindle at cell division) and telomeres (ends of the chromosomes), but also at other places in the genome. Their function, if any, is unknown. Most tandem repeated sequences are highly **polymorphic**. That is, despite their being located at characteristic sites (**loci**) on the chromosomes, the number of repeats at any particular locus varies between individuals. This gives them some important practical applications in forensic science and as genetic markers that can assist in locating human disease genes. We will deal with this aspect in Chapter 28.

RNA-coding genes

Besides genes that are transcribed into mRNA in the first step of protein synthesis, the genome also contains sequences that encode RNA that is not translated (nonprotein-coding RNAs or ncRNAs). For example **transfer RNA** (tRNA) and **ribosomal RNA** (rRNA) have specific functions in protein synthesis (Chapter 25). rRNA is the most abundant RNA in the cell and the genome contains multiple copies of the rRNA genes (rDNA), to allow rapid synthesis. In humans rDNA encoding three of the four rRNA molecules is arranged in tandem arrays on five homologous chromosome pairs, and these sections of the chromosomes cluster together to form a recognizable subnuclear structure, the nucleolus, which acts as a 'factory' for rRNA production.

A recent (and surprising) discovery has been the widespread occurrence of **microRNAs**: short RNA molecules that are not translated, but play multiple roles in regulating gene expression. We discuss these further in Chapter 26. The extent to which DNA previously thought of as 'junk' actually encodes miRNAs is the subject of current research. Sensitive detection methods suggest that over 90% of the nucleotides in the human genome are transcribed in some cell type at some time, but what proportion of these rare transcripts are actually functional is not known.

Pseudogenes

Pseudogenes are DNA sequences that are similar to those of coding genes, but no longer encode a functional product. They are evolutionary 'relics' that have often arisen as a result of gene duplication: one duplicate copy of the gene accumulates sequence mutations while the other continues to function. They may also originate from mRNA that is copied into cDNA and inserted into the genome by the retrotransposon machinery. The human genome contains thousands of such pseudogenes.

Genome packaging

The prokaryotic genome is compacted in the cell

One of the most intriguing problems in biology is that of packaging the long DNA genome into the microscopic cell. The *E. coli* bacterium is approximately 2 μm by 0.5–1 μm. In order to fit into this tiny volume, the circular chromosome, 1 mm in length, is bound by positively charged molecules including small basic proteins, which counteract the negatively charged phosphate groups of the DNA and allow compaction of the chromosome into the nucleoid structure. The DNA is also **supercoiled** (see Chapter 23). Less is known of the details of prokaryotic chromosome packaging than that of eukaryotes, discussed in the next section.

How is eukaryotic DNA packed into a nucleus?

The packaging of eukaryotic genomes seems even more amazing than that of prokaryotes: for example, the human cell nucleus contains about 2 metres of chromosomal DNA, and this is packed into a sphere about 10 μm in diameter. DNA in eukaryotic cells exists as **chromatin** – a DNA–protein complex. The main proteins are **histones**: these are small basic proteins rich in arginine and/or lysine, giving them positive charges that form ionic bonds with the negative charges on the phosphate groups on the outside of the DNA. The amino acid sequences of eukaryotic histones are highly conserved throughout evolution. One of the histones differs only in two amino acids in the entire molecule between peas and cows and the changes are conservative (valine for isoleucine, lysine for arginine). The extreme conservation illustrates the fundamental importance of histones for cell function, as it indicates that changes in their structure would be lethal or sufficiently deleterious for natural selection to eliminate.

We will start with the four histones called H2A, H2B, H3, and H4. H2A and H2B are encoded by different genes, but are more closely related in sequence to each other than to H3 and H4. Two molecules of each of these histones form an octamer protein complex around which 146 bp of DNA are wrapped, forming just under two complete turns around the octamer. The octamer and its associated DNA form a unit called a **nucleosome**, which has the shape of a disc about

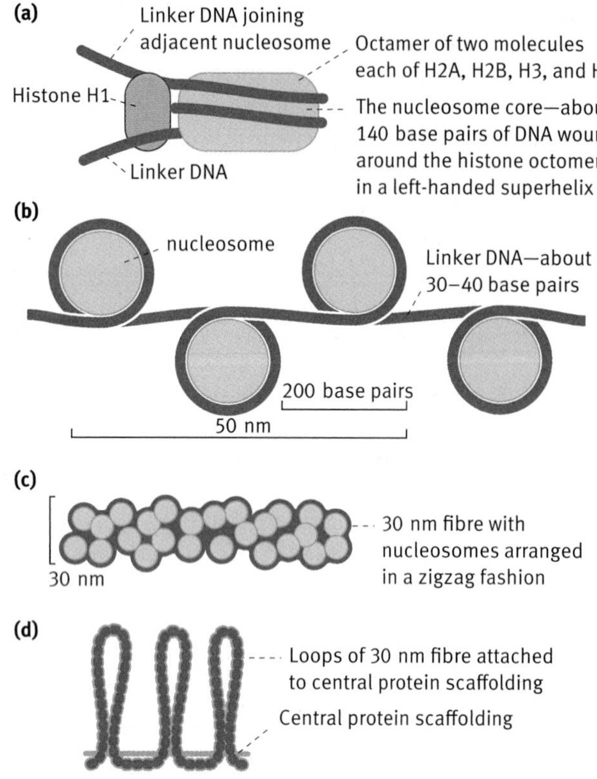

Fig. 22.8 Order of chromatin packing in eukaryotes. **(a)** Diagram of a nucleosome. **(b)** Beads on a string form, the 10 nm fibre. **(c)** A 30 nm fibre of chromatin (see Fig. 22.9 for an electron micrograph of a 30 nm fibre). **(d)** Loops of the 30 nm fibre are attached to a central protein scaffold in a 360° array. It is believed that these loops are yet further condensed, perhaps by supercoiling and ultimately into the extremely compact metaphase chromosome. The latter condensation stage is not illustrated.

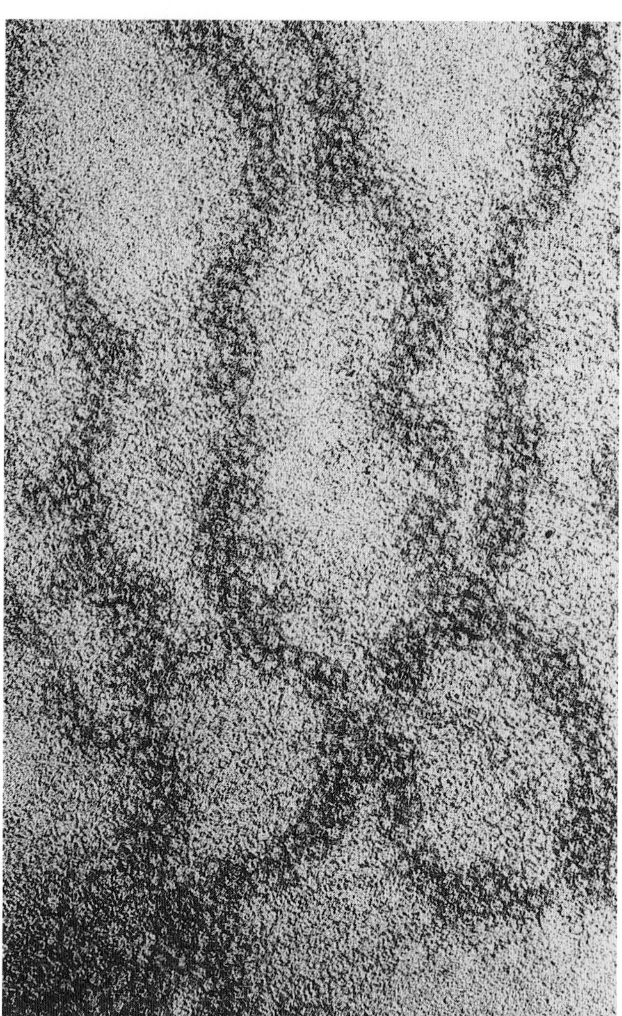

Fig. 22.9 Electron micrograph of a 30 nm fibre of chromatin. Fig. 28.19 in Lewin B. (1994) Genes V, Oxford University Press, Oxford. Photograph was provided by Prof. B. Hamkalo.

10 nm wide and 5 nm thick. Short linker sequences of DNA join successive nucleosomes, arranged like beads on a string (Fig. 22.8(a),(b)). This arrangement condenses (packages) the 2 nm diameter DNA double helix into the 10 nm fibre shown in Fig. 22.8(b). Histone H1 is different from the other four, being larger and less evolutionarily conserved. It binds the DNA as it enters and leaves the nucleosome (Fig 22.8(a)) and plays a role in condensing the nucleosomes together. This allows condensation of the 10 nm fibre to 30 nm in diameter, illustrated in Fig. 22.8(c). An electron micrograph of such a fibre is shown in Fig. 22.9. The exact structure of the 30 nm fibre is not known and may be variable. It could be a zigzag, as shown Fig. 22.8(c), or a coil or solenoid with the H1 histones in the centre.

Further packaging of chromatin takes place through formation of long loops that are attached to a central chromosomal nonhistone protein scaffolding (Fig. 22.7(d)). This looped structure can form yet more densely packed structures, not fully understood, involving folding and/or coiling and achieving a 10,000-fold compaction of the length of the original DNA.

The tightness of DNA packaging changes during the cell cycle

Eukaryotic cells have a strictly controlled cell cycle (see Chapter 30). **Mitosis** (nuclear division involving segregation of the chromosomes) and **cytokinesis** (division of the cytoplasm) take place during **M phase** of the cycle, which alternates with **interphase**. During interphase, cell growth and DNA replication take place (see Fig. 23.4). The degree of compaction of chromatin is variable, depending on the transcriptional activity of the genome. The highest degree of compaction is seen during mitosis, when the genome is not transcriptionally active and the replicated chromosomes must be separated without tangling. In interphase, chromatin is less tightly packed during RNA and DNA synthesis to allow enzymes and regulatory molecules to access the DNA. Thus, using the light microscope, mitotic chromosomes are visible, while individual chromosomes are not distinguishable in interphase nuclei.

The tightness of DNA packing can regulate gene activity

To use information encoded in the base sequence of DNA, the molecule must be accessible to enzymes and other proteins and not concealed in tightly condensed structures. The majority of the chromatin during interphase is in this accessible state even if it is not actively transcribed at all times. This less compact chromatin, which is observed to stain lightly in microscopy studies, is termed **euchromatin** (where 'eu' means true). Chromatin that remains tightly compacted in interphase is identifiable as dark staining regions known as **heterochromatin**. Regions of the chromosomes that have no coding function, such as highly repetitive sequences at the centromeres and telomeres, are always compacted and termed constitutive heterochromatin, while regions that remain compacted only under certain circumstances (for instance in a particular cell type or at a particular developmental stage) are known as facultative heterochromatin.

The exact degree of packing of transcriptionally active chromatin is difficult to determine experimentally, but it is clearly highly dynamic, perhaps alternating between 10 nm and 30 nm fibre structures with short sequences of the DNA becoming transiently free of nucleosomes to allow access by other factors. Additionally, higher order chromatin structures allow spatial association of different parts of the genome so that specific regions can be clustered together in nuclear 'compartments' (similar to clustering of rDNA in the nucleolus), facilitating their co-ordinated regulation. Chromatin structure and organization is thus of great importance for regulating eukaryotic gene transcription, as discussed further in Chapter 26.

BOX 22.1 Size of genomes related to complexity of organisms

When we come to discuss quantitative aspects of the genome, it is customary always to refer only to the haploid genome (a single copy of each chromosome) whether the organism in question is haploid, diploid or polyploid. This makes comparisons possible. You might predict that the size of the genome would correlate with the 'complexity' of the organism, with a larger genome allowing a greater repertoire of cellular functions. In prokaryotes this prediction roughly holds true, but in eukaryotes it is incorrect, as illustrated by the estimates for a selection of organisms given below. Given that such a small proportion of the DNA is devoted to protein-coding genes it is not surprising perhaps that base pair numbers can show little correlation with coding gene numbers. Nor does the eukaryotic gene number obviously correlate with complexity very well.

The smallest known cellular genome is that of *Mycoplasma genitalium* with 580,000 base pairs (per haploid genome) and 485 protein-coding genes. *Escherichia coli* has about 4.5 million base pairs and just over 4,000 coding genes. The yeast *Saccharomyces cerevisiae* has 12 million base pairs and about 6,000 genes, while another single celled eukaryote, *Trichomonas vaginalis*, is, amazingly, the current record holder for eukaryotic gene number, with 60,000 genes in its 160 million base pair genome. The fruit fly, *Drosophila*, has 170 million base pairs and 14,000 genes, but the much simpler nematode roundworm *C. elegans* with only about a thousand cells and a smaller genome has 18,000 genes. Humans, with about 10 trillion cells, have over 3 billion base pairs and an estimated 21,000 protein-coding genes, not so different from the roundworm. *Arabidopsis*, a cress plant that is used as a model organism by plant geneticists (partly because it has, for a plant, a small genome), has about 125 million base pairs and 25,000 genes. Current thinking is that the more complex organisms make more efficient use of genes, for instance by alternative splicing and microRNA regulation, to generate additional complexity.

It should be noted that the cited numbers of protein-coding genes in eukaryotes are estimates, with a considerable degree of uncertainty despite the elucidation of the sequence of whole genomes. This is because of the split structure of eukaryotic genes; it is not necessarily a simple matter to deduce gene numbers from the genome sequence. Bacterial genes are easier to identify and count from the sequenced genomes because all that needs to be looked for is a stretch of DNA long enough to code for a protein before running into a stop codon (see Chapter 25).

Summary

DNA is a polynucleotide consisting of strands of deoxynucleotides linked by 5′→3′ phosphodiester links between the sugar residues. Although RNA probably preceded DNA in early life as the genetic material, DNA is chemically more stable because of the lack of the oxygen atom on the 2′ position of the deoxyribose and this is probably the reason for its present dominance in life, though RNA genomes exist in viruses.

DNA is a double helix of two antiparallel strands (they run in opposite directions) held together by complementary base pairing. It contains four bases, A, T, G, and C, but no U. T is the same as U except that it has a methyl group. Complementary base pairing by hydrogen bonding occurs between A and T, and G and C; this holds the two strands together. High temperatures cause strand separation due to hydrogen bond breakage. Reversal occurs on cooling, the phenomenon being known as strand hybridization or annealing. This is highly specific for pieces longer than about 15–20 nucleotides and is at the centre of DNA technologies (see Chapter 28).

The bases in a double helix point to the inside of the molecule, and the phosphate–sugar backbone to the outside, with the edges of the bases visible in the two grooves of the double helix. This is known as the B form. The bases themselves have flat hydrophobic faces and are stacked on one another. Most DNA is in the B form but A and Z forms are possible in certain circumstances, their biological significance, if any, being unknown.

Each eukaryotic cell has about 2 m of DNA packed into the microscopic nucleus. There it is in the form of chromatin in which the double helix is wrapped around nucleosomes; these are octets of histone proteins. Between nucleosomes, and linking them together, are stretches of DNA 30–40 base pairs in length, forming a 'beads on a string' structure. This nucleosome–DNA fibre is further packed in complex looping arrangements. The maximum condensation occurs in mitotic chromosomes during cell division in which form the DNA is genetically inert. In *Escherichia coli* the chromosome is circular and the packing is not so highly structured as in chromatin and lacks nucleosomes.

In eukaryotes cell division occurs by mitosis (except for germ line cells). The process involves prophase, metaphase, anaphase and telophase with the resulting cells segregating by cytokinesis.

The role of DNA is to store information on the amino acid sequences of proteins, in the form of base sequences of genes. A gene is a small section of DNA, which is part of the large molecule of DNA constituting a chromosome and has no physical independence. The base sequences distinguish one gene from another. In addition to protein-coding genes, there are mobile genes in chromosomes as well as genes coding for ribosomal and transfer RNAs.

The mitochondrial genome is small and circular. It codes for only a small number of mitochondrial proteins, the rest being coded for by nuclear genes.

The sizes and organization of genomes have produced surprises. The amount of DNA is not proportional to the complexity of an organism. Amphibia have more DNA than humans per cell. Humans have about 21,000 genes, not the 40,000 previously estimated. *E. coli* has 4,000, while the cress plant has 25,000.

The human genome has been sequenced. Only about 1.6% of the DNA actually codes for protein sequences and genes in total occupy only 20%, most of this being due to noncoding introns.

Half the DNA is repetitive and was regarded as 'junk' DNA of no known function. A change in thinking has occurred about junk DNA since large numbers of microRNA-coding genes have been discovered in it. There is evidence that the small RNA transcripts play essential parts in the determination of the phenotypes of complex organisms.

Further reading

To access the further reading, please scan the QR code image or go to http://global.oup.com/uk/orc/biosciences/molbiol/snape_biochemistry/student/reading/ch22/

Problems

1 Write down the structure of a dinucleotide.

2 Ribonucleic acid (RNA) almost certainly evolved before deoxyribonucleic acid. How do you think DNA evolved?

3 The flat faces of the bases of DNA are hydrophobic. Explain the structural repercussions of this fact on the structure of double-stranded DNA.

4 What is the main form of double-stranded helical DNA called? Is it a right- or left-handed helix? Approximately how many base pairs are there in a stretch of DNA that completes one rotation of the helix?

5 Explain what is meant by DNA chains being antiparallel in a double helix.

6 Explain in everyday language what is meant by a $5' \rightarrow 3'$ direction in a linear DNA molecule.

7 If you see a DNA structure simply written as CATAGCCG, what exactly does this means in terms of

a double-stranded structure and the polarity of the two chains? Explain your answer.

8 What are Alu sequences?

9 All genes are sections of chromosomes that code for the amino acid sequences of proteins. Comment briefly on how this statement has to be qualified, especially in the light of recent discoveries.

10 Which of the following is out of place – adenine, guanine, thymine, cytosine, uracil?

11 Describe the broad differences between the typical prokaryotic and eukaryotic genomes.

Chapter 23

DNA synthesis, repair, and recombination

DNA synthesis is simple in concept, but complex in practice. The mechanism was initially studied mainly in *Escherichia coli* but sufficient is now known of synthesis in eukaryotic cells to be sure that the processes are basically the same in both, although they differ in absolute detail.

Each time a cell divides, its DNA must be duplicated or, as is more usually stated, the chromosome(s) must be replicated, so that a complete complement of DNA can be given to each daughter cell. A human cell has about 6 billion base pairs in its total DNA (3.2 billion per haploid genome). The magnitude of the task of faithfully replicating these needs no emphasis. Even a single incorrect base in a gene may cause a protein with impaired function to be produced. The unavoidable minute proportion of errors that are not repaired are the feedstock of evolution, and, unfortunately, of genetic diseases.

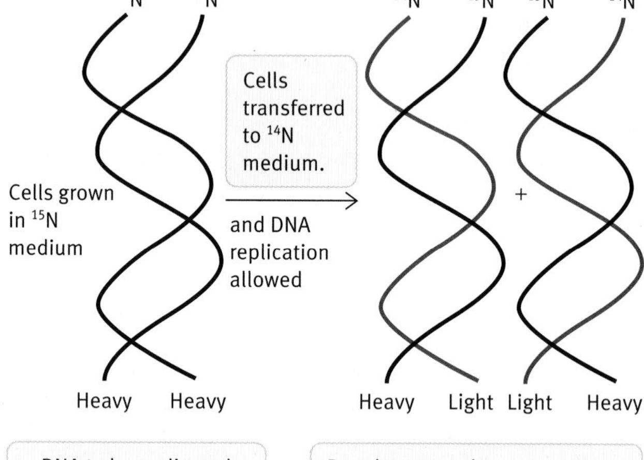

Fig. 23.1 Demonstration of semiconservative DNA replication by Meselson and Stahl. The DNA of cells was labelled by growing them in a medium in which the nitrogen source was ^{15}N, so that both strands of DNA were 'heavy'. They were then transferred to ^{14}N medium so that all subsequent DNA chains synthesized would be 'light'. The density gradient analysis indicated that, one generation after the transfer, each DNA molecule contained one 'heavy' and one 'light' strand. This is known as semiconservative replication. Continuation of the experiment for further generations confirmed the result. The red strands are newly synthesized.

Overall principle of DNA replication

We will go into the question of *how* DNA is synthesized in due course, but for the moment let us look at it at a general level.

A chromosome is double-stranded DNA. Its replication is described as **semiconservative** in that the two original strands, called parental strands, are separated and each acts as a template to direct the synthesis of a new complementary strand; each new double helix has one old and one new strand. This was established in the classic experiment shown in Fig. 23.1.

The basis of the replication is that of complementarity in that a G will base pair with C, and A with T, so that a base on the parental strand specifies which base is to be incorporated into the new strand as its partner. Since this copying process depends on Watson–Crick hydrogen bonding of base pairs, it follows that strand separation is essential to unpair the bases in double-stranded DNA, and make them available for base pairing with incoming nucleotides.

DNA replication does not start just anywhere in the genome. In the circular *E coli* chromosome of approximately 4.6 million base pairs, the strands are initially separated at one particular sequence called the **origin of replication**. *Two* **replication forks**, moving in opposite directions, synthesize DNA at a maximum rate of around 1000 base pair copies per second, with separation of parental DNA and synthesis of new DNA occurring at the same time (Fig. 23.2). The two forks meet at the opposite side of the circle.

Eukaryotic chromosomes are linear, and longer than the *E. coli* chromosome. This means that although the basic mechanism is the same, the organization of eukaryotic DNA replication has to be slightly different. The entire *E coli*

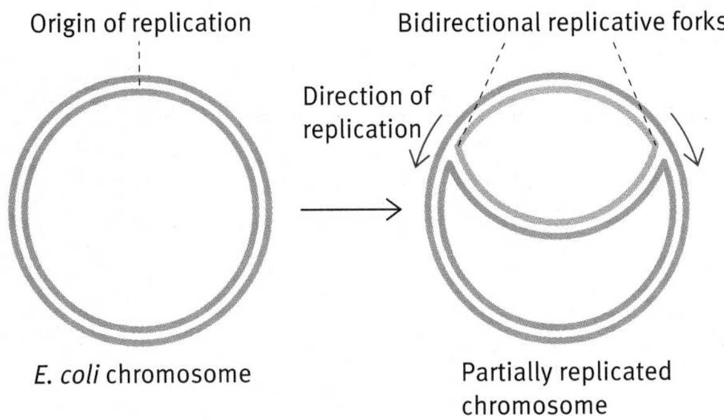

Fig. 23.2 Bidirectional replication of the *Escherichia coli* chromosome. Parental strands are blue; newly synthesized strands are red.

chromosome is replicated from a single origin and is therefore termed a single **replicon**, but in eukaryotes the rate of DNA synthesis (on average 50 base pairs copied per second) is too slow by far for a single replicon to synthesize a whole chromosome in the time available. To cope with this, there are hundreds of origins of replication along the chromosome from which replication forks work in both directions (Fig. 23.3). A vital requirement is that each section of DNA replicates once, and once only, in a given cell division cycle.

Control of initiation of DNA replication in *E. coli*

Before cell division occurs, there must be a complete duplication of the chromosome. Exactly how cell division and DNA replication are coordinated in *E. coli* is not understood. Protein synthesis and a critical enlargement of the cell are required. As already seen (Fig. 23.2), in *E. coli* there is a single point of origin of DNA synthesis called *oriC* at which replication commences bidirectionally.

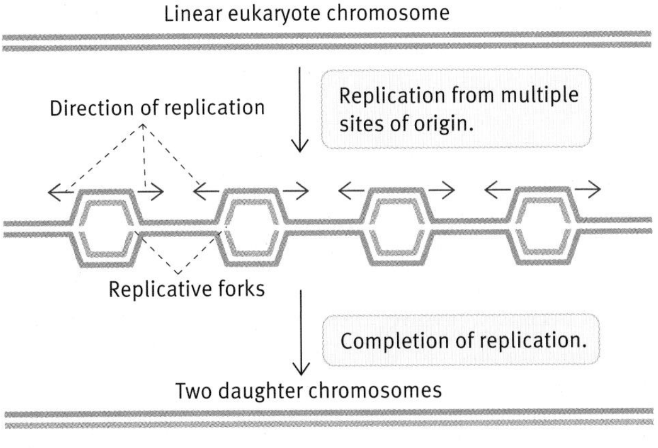

Fig. 23.3 Diagram of multiple bidirectional replication forks in a eukaryotic chromosome.

The origin of replication has a specific base sequence, very rich in A—T pairs, presumably to facilitate strand separation. (Remember that A—T pairs have two hydrogen bonds and G—C pairs three and, therefore, the former are less tightly bound together.) At the time of initiation, a protein referred to as **DnaA** (the protein coded for by the gene *dnaA*) binds in multiple copies to this region and causes strand separation. This permits the main unwinding enzyme, **helicase** (or **DnaB**, the protein coded for by the gene *dnaB*), which works at each replication fork, to attach and begin progressive unwinding of the strands in both directions. The helicase is believed to move along one strand using ATP hydrolysis as the source of energy needed to break hydrogen bonds, thus displacing the other strand and unwinding the DNA. There is a mechanism, which will be discussed in the following section, to stop the two strands coming together again prematurely.

Initiation and regulation of DNA replication in eukaryotes

DNA synthesis is confined to a period in the eukaryotic cell cycle called the S (for synthesis) phase. Different cell types vary but, in cultured animal cells, the **S phase** typically takes about 8 hours out of a total cycle of 24 hours. Before the S phase is the **G_1 phase**, (G for gap), and afterwards the **G_2 phase** (Fig. 23.4). To proceed to cell division, a mammalian cell requires a mitogenic (mitosis- or cell division-producing) signal from other cells. This takes the form of protein signalling molecules that attach to surface receptors and transmit the signal to the interior of the cell. The latter process (that of cell signalling) is the subject of Chapter 29, and eukaryotic cell-cycle control is dealt with more fully in Chapter 30.

Unwinding the DNA double helix and supercoiling

DNA strand separation by helicase presents topological problems. (Topology refers to the arrangement in space of components relative to each other.) To explain these we must deal with the subject of DNA supercoiling.

Since duplex DNA has two strands in the form of right-handed helices, it has an inherent degree of twist, there being one turn of the helix per approximately 10 base pairs. A short piece of linear DNA that is free to rotate on its own long axis adopts this strain-free configuration, known as the **relaxed state**. Suppose instead that you clamped one end of the duplex so that it was not free to rotate and you gave an extra twist to the other end, so that the coil of the double helix is tightened – the number of turns of a given length of

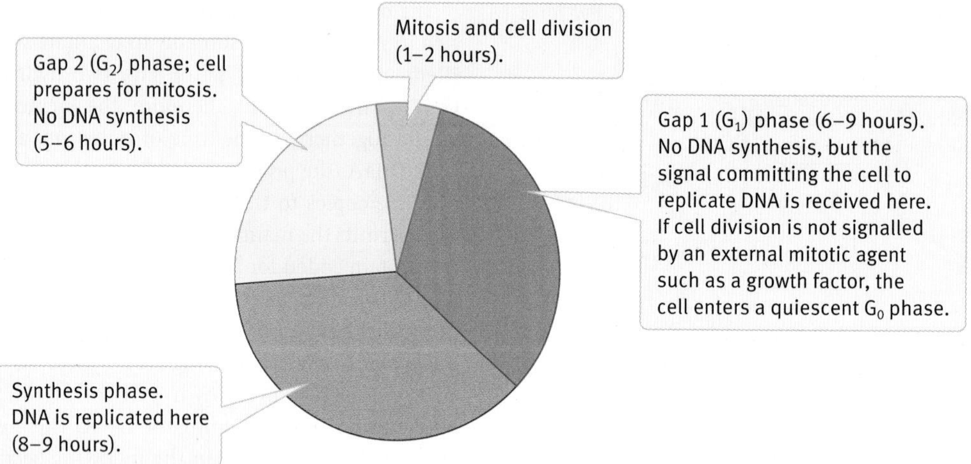

Fig. 23.4 The eukaryotic cell cycle. The duration of the cell cycle varies greatly between different cell types. The times given here are for a rapidly dividing mammalian cell in culture (24 hours to complete the cycle). See Chapter 30 for a more detailed account of the cell cycle.

DNA is increased; that is, the number of base pairs per turn is decreased. It is now **positively supercoiled** or **overwound**. If you twisted in the opposite direction, the coil would be opened up – the number of turns per unit length would be reduced, or the number of bases per turn increased. The DNA would be **negatively supercoiled** or **underwound**. Both the underwound and overwound states are under tension and one way of accommodating the strain is for the DNA double helix to coil upon itself forming a **coiled coil** or **supercoil** (Fig. 23.5). You can illustrate supercoiling with a piece of double-stranded rope. Have someone hold one end or clamp it somehow, so that the rope cannot rotate freely, and twist the rope on its axis. Coils will form to take up the twisting strain. If you release the end of the supercoiled rope it will spin back to the relaxed state. To determine whether a coil is positive or negative, look along the coil from either end. If the uppermost strand is turning to the left it is positive, if to the right it is negative (see Fig. 23.7).

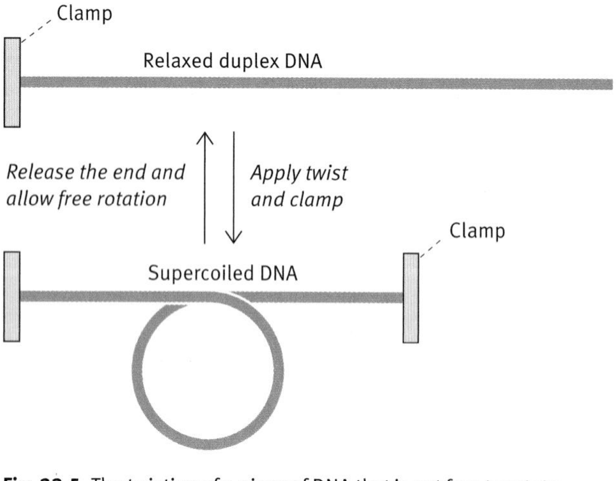

Fig. 23.5 The twisting of a piece of DNA that is not free to rotate induces supercoiling to accommodate the twisting strain. Cellular DNA is effectively clamped and is not free to rotate. If, somehow, free rotation is allowed, the supercoil will relax. If the applied twist is in the direction of unwinding, the supercoil will be negative; it will be positive if the applied twist is in the opposite direction.

What has this to do with DNA replication? DNA in the cell is not free to rotate on its own long axis; in *E. coli* the closed-circle chromosome effectively 'clamps' the DNA. In eukaryotes, the DNA is of such vast length, arranged in fixed loops (see Fig. 23.8) and attached to protein structures, that once again free rotation is impossible. But, separation of DNA strands demands that the duplex rotates. This causes over-winding – it generates positive supercoils ahead of the replication fork and, as the helix tightens, further strand separation is resisted. If unrelieved, the tension would bring strand separation and DNA replication to a halt.

A simple experiment will convince you of this. If you take a short piece of double-stranded rope or string and pull the ends apart, the rope will spin, thus preventing the accumulation of positive supercoils, and the rope strands will separate completely. Now take a long piece of the same rope coiled on the floor, or have someone hold one end of a reasonably long piece, so that it cannot freely rotate, and try to pull the strands apart. Positive supercoils will snarl-up and oppose the separating process and prevent any further separation. This would be the situation in DNA replication in the cell if something were not done about it.

It follows that, for DNA synthesis to proceed, the positive supercoils ahead of the replication fork must be relieved and this, of necessity, involves the transient breakage of the polynucleotide chain.

How are positive supercoils removed ahead of the replication fork?

A group of enzymes, known as **topoisomerases**, catalyse the process. They act on the DNA and isomerize or change its topology. There are two classes of topoisomerase, types I and II. We will deal with the principles of their mechanisms first, and then explain their roles in DNA replication.

In type I, the enzyme breaks *one* strand of a supercoiled double helix, which permits the duplex to rotate on the single phosphodiester bond of the partner strand, effectively introducing a swivel, or point of free rotation, into the DNA. After rotation has occurred, the enzyme reseals the duplex (Fig. 23.6). The enzyme *does not hydrolyse the phosphoester bond it attacks* – it transfers the bond from the deoxyribose-3′-OH to the —OH of one of its own tyrosine side chains. Since little energy change is involved, the process is freely reversible. The enzyme does not use ATP. A **topoisomerase type I** can relax only supercoiled DNA.

A **type II topoisomerase** breaks *two* strands of the DNA double helix, again by transferring phosphoester bonds to itself, making the breakage of the polynucleotide chains freely reversible. The enzyme physically transfers the DNA duplex of a supercoil through the gap. (You can imagine that untangling string or a fishing line would be helped if you could cut a loop, transfer a strand through the cut and magically rejoin the ends without a knot.) Conformational change in the protein is involved, generated by ATP hydrolysis. In *E. coli*, the topoisomerase II is called **gyrase**; it introduces negative supercoils into the DNA. The mechanism of this is illustrated in Fig. 23.7. In this figure we use the removal of a single positive supercoil from a circular DNA molecule for purposes of illustration. In Fig. 23.7(b) the supercoil is positive. The gyrase cuts both strands of the lower duplex to form a gap (Fig. 23.7(c)) through which the front strand is transferred. It then reseals the cut, now at the front (Fig. 23.7(d)), creating a negative supercoil. ATP hydrolysis supplies the energy required in the physical transfer process. Gyrase actively inserts negative supercoils and, since this will relax positive supercoils, permits DNA synthesis to proceed.

Thus, in prokaryotic and eukaryotic DNA replication, the potential snarl-up of strand separation through the accumulation of positive supercoils is averted.

When DNA is carefully isolated from cells it is found to be negatively supercoiled. In relaxed DNA, the double helix has one turn per 10.5 base pairs; in cellular DNA it has about one turn per 12 base pairs – it is underwound. The degree of supercoiling is roughly comparable in the DNA of different cells, which suggests that it is of importance and that its generation is controlled. The reason for this is possibly that, in such a state, DNA strand separation occurs more readily than in the relaxed or positively supercoiled state. In *E. coli* the degree of underwinding will be a balance between topoisomerase I relaxing negative supercoils, and gyrase inserting them.

Eukaryotic DNA, like that of prokaryotes, is underwound or negatively supercoiled in the cell. However, unlike the situation in prokaryotes, no eukaryotic topoisomerase is known that can actively insert negative supercoils into DNA. How then is the underwinding achieved?

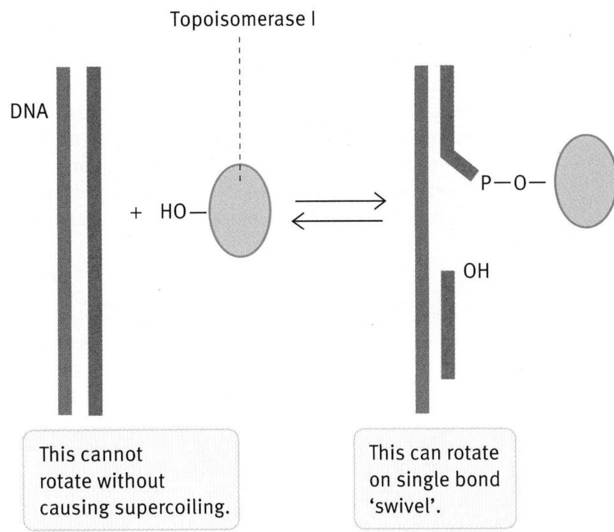

Fig. 23.6 Mechanism of topoisomerase I action. The enzyme breaks a phosphodiester link in the backbone of one strand of DNA by transferring a phosphoester bond to a tyrosine –OH group in its own protein. The DNA can then be allowed to rotate on the single-bond 'swivel' in the partner strand. After rotation, the enzyme restores the original phosphoester bond to remake the phosphodiester link. Note that the phosphoester bond is transferred to the enzyme; it is not hydrolysed so the process is freely reversible.

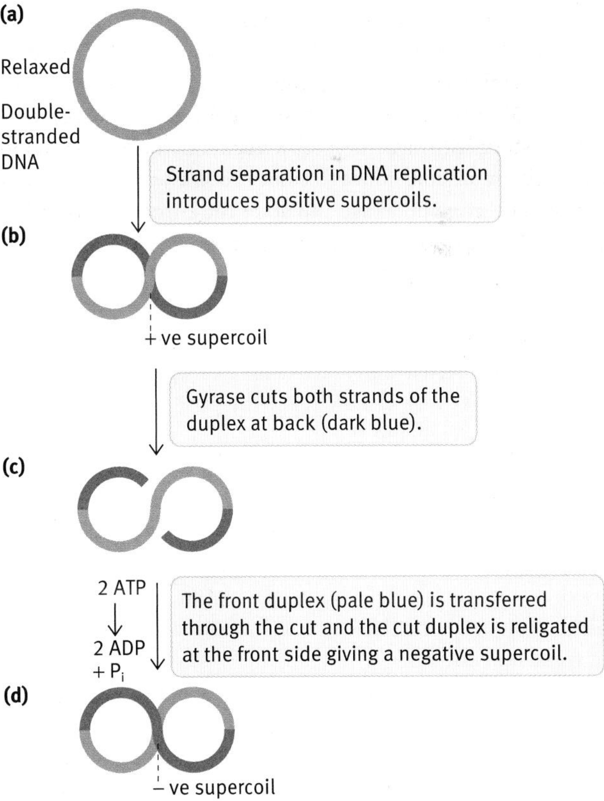

Fig. 23.7 Simplified mechanism by which gyrase in *Escherichia coli* neutralizes positive supercoiling by insertion of negative supercoils. Unlike topoisomerase I, ATP is required to provide energy for transfer of the duplex through the cut.

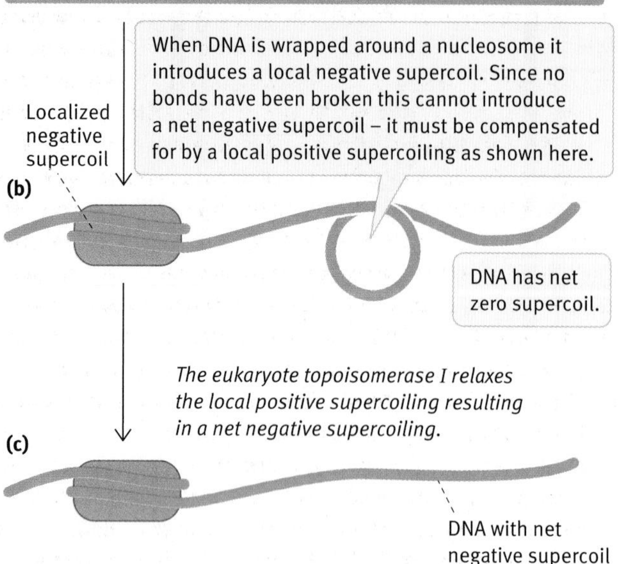
(a) DNA of chromosome (zero supercoil); note that this strand is not free to rotate.
When DNA is wrapped around a nucleosome it introduces a local negative supercoil. Since no bonds have been broken this cannot introduce a net negative supercoil – it must be compensated for by a local positive supercoiling as shown here.
Localized negative supercoil
(b)
DNA has net zero supercoil.
The eukaryote topoisomerase I relaxes the local positive supercoiling resulting in a net negative supercoiling.
(c)
DNA with net negative supercoil

Fig. 23.8 A mechanism by which eukaryotic DNA becomes negatively supercoiled despite the absence of any enzyme capable of actively inserting negative supercoils such as the prokaryotic gyrase. Steps (a)–(c) are referred to in the text.

When chromatin is assembled, it is believed that the DNA winds around nucleosomes in such a manner that, in the local region in contact with the protein, it is in an underwound state. This is achieved by left-handed coiling around the nucleosome core. Since this nucleosome winding does not involve any bond breakage and since the chromosomal DNA cannot freely rotate, it follows that there cannot have been any *net* change in the supercoiling of the DNA. Therefore, the local negative supercoiling at the nucleosome must be compensated for by positive supercoiling elsewhere so that the net change in the structure is zero (Fig. 23.8(b)). The eukaryotic topoisomerases I or II now relax the positively supercoiled section, thus achieving the insertion of a negative supercoil (Fig. 23.8(c)). A prokaryotic type of gyrase is thus not needed; the fact that prokaryotes do not have the nucleosome structures correlates with the need for their own type of gyrase. The antibiotic, **nalidixic acid**, inhibits bacterial gyrase, and is used to treat certain infections resistant to other antibiotics. Since humans do not have gyrase, nalidixic acid can be used to treat patients.

So far we have dealt with the broad aspects of DNA replication – its semiconservative nature based on Watson–Crick base pairing, with the cell cycle, with the initiation of replication, and with the mechanism of unwinding. We now turn to the mechanism of DNA synthesis. The enzyme(s) that catalyse this are called **DNA polymerases**; they polymerize nucleotides into DNA, using deoxynucleotide triphosphates as substrates.

The basic enzymic reaction catalysed by DNA polymerases

A series of facts first:

- There are three DNA polymerases in *E. coli*, called Pol I, II, and III – named in order of their discovery.
- The DNA synthesis occurring in the replication fork is catalysed by Pol III or its eukaryotic equivalents, but Pol I also plays an essential role in DNA replication as well as in repair. Less is known of Pol II, but it is believed to be associated with certain types of DNA repair.
- The substrates for DNA polymerases are the four deoxyribonucleoside triphosphates dATP, dCTP, dGTP, and dTTP. These are synthesized in the cell as described in Chapter 19. The regulatory mechanisms in their synthetic pathways ensure that they are produced in adequate and coordinated amounts.
- The polymerase must have a DNA template strand to copy. 'Copy' is used in the complementary sense – a G on the template strand is 'copied' into a C in the new strand, and likewise A into T, C into G, T into A.
- A most important fact to fix in your mind: *a DNA polymerase can only elongate (add to) a pre-existing strand called a primer.* This **primer** may only be 20 nucleotides long but without it nothing happens. **DNA polymerases cannot start a chain – they cannot join together two free nucleotides**. The priming mechanism is described below.
- As illustrated in Fig. 23.9, the polymerase attaches a nucleotide to the 3′ free OH group at the end of the primer or newly synthesized strand, liberating inorganic pyrophosphate (PP_i). Hydrolysis of PP_i increases the negative $\Delta G^{0\prime}$ value for the synthesis thus helping to drive the reaction. Incorporation of a nucleotide into the new strand of DNA involves the formation of hydrogen bonds with its template partner, with the liberation of energy, thus adding to the thermodynamic drive of the process.
- Which of the four deoxyribonucleoside triphosphates is accepted by the DNA polymerase is determined by the base on the parental strand being copied.

DNA synthesis always proceeds in the 5′→3′ direction with respect to the growing strand. Be sure that you know what this means – that the growing DNA chain is being elongated in the 5′→3′ direction – a nucleotide is added to the free 3′-OH of the preceding terminal nucleotide. Note that when we talk of synthesis being in the 5′→3′ direction we always refer to the direction of elongation – the polarity of the *new* strand. We are *not* referring to the template strand, which has the opposite

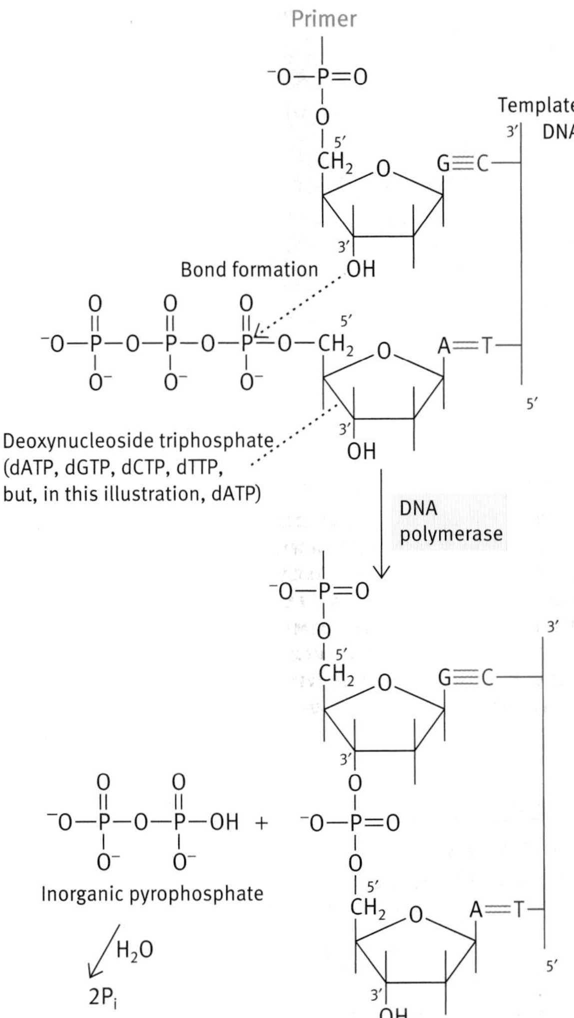

Fig. 23.9 The elongation reaction catalysed by DNA polymerase. The diagram shows the addition of an adenine deoxynucleotide from dATP to the 3′ end of the primer DNA strand, the base selected for addition being determined by the base on the template strand. Note that the synthesis is in the 5′→3′ direction; the chain is being lengthened in the 5′→3′ direction. The dotted line with arrow shows the attack of the 3′-OH on the α-phosphate.

polarity. The polarity of DNA strands has been explained in the previous chapter.

How does a new strand get started?

DNA polymerases cannot initiate new chains and yet, at each origin of replication, new chains must be initiated. The solution to the question posed above is rather surprising in that DNA chains are initiated (primed) by RNA. The structure and synthesis of RNA are described in detail in the next chapter, but for now it is enough to know that RNA has the same structure as single-stranded DNA except that the sugar is ribose and the base thymine (T) is replaced by uracil (U), which like thymine pairs with adenine. RNA is synthesized by RNA polymerases by essentially the same basic chemical mechanism as outlined above for DNA except that ATP, CTP, GTP, and UTP, are used. In the present context, the vital difference is that RNA polymerases *can initiate new chains.* A template strand is needed to direct the sequence, but unlike DNA polymerase, RNA polymerases can take two nucleotides and link them without needing an existing polynucleotide 3′ end to add to.

When a small piece of RNA primer (perhaps 10–20 nucleotides) has been synthesized by a special RNA polymerase called **primase**, DNA polymerase takes over and extends the chain. Primers are later removed.

We now come to yet another problem due to the antiparallel nature of DNA.

The polarity problem in DNA replication

DNA is synthesized at each replication fork, which steadily progresses along the chromosome. In *E. coli*, there are two DNA polymerase III molecules involved in each fork, one for each strand, the two molecules, each with enzymatic activity, being linked together into a single asymmetric holoenzyme dimer. As shown in Fig. 23.10, the polymerase dimer molecules that are replicating the two strands must physically move in the same direction (that is, up the page as it were). To synthesize a new DNA strand in the 5′→3′ direction, DNA polymerase must move along the template DNA strand from its 3′ to its 5′ end. With overall movement of the polymerase dimer being towards the replication fork, this works fine for one parental strand (the left-hand one in Fig. 23.10), but not

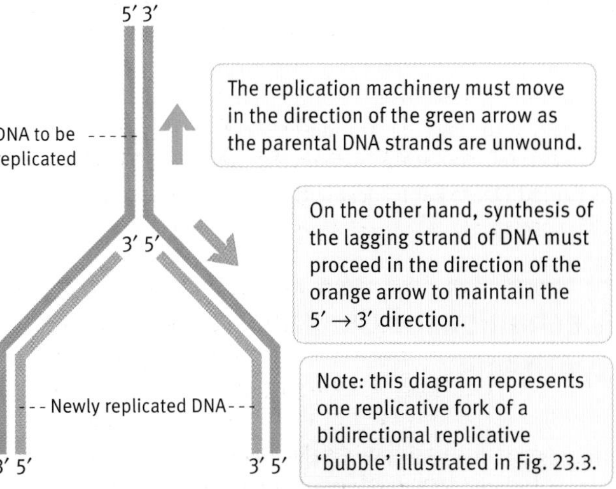

Fig. 23.10 The polarity problem in DNA replication.

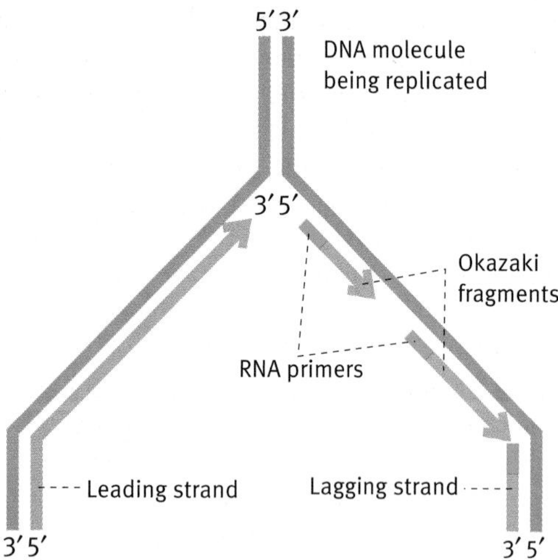

Fig. 23.11 Diagram of a replication fork. The leading strand is synthesized continuously, while the lagging strand is synthesized as a series of short (Okazaki) fragments.

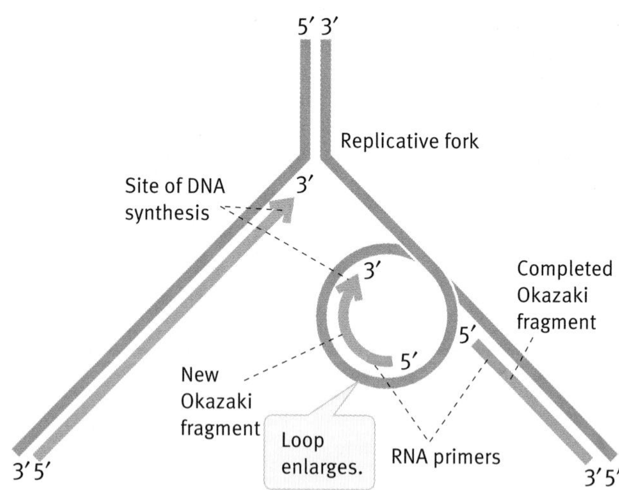

Fig. 23.12 Principle of Kornberg's 'loop' model for Okazaki fragment synthesis. (Pink arrowheads indicate DNA synthesis.) This model requires the loop to fall away when the new Okazaki fragment meets the old one. A new loop has then to be made. Both strands can be synthesized in this model in the required 5′→3′ direction as the replication machinery moves in the direction of the fork. It is not possible to specify the precise mechanical details of the looping mechanism. A more detailed model is shown in Fig. 23.13.

for the other. The left-hand strand, with no problems, is called the **leading strand** and the other the **lagging strand**.

How can the lagging strand initiate? In the case of the leading strand, primase lays down a single primer at the origin of initiation and the DNA polymerase proceeds from there until replication is complete, but the same will not suffice for the lagging strand. The solution is that, as the DNA unwinds, there is repeated initiation by primase, each primer being extended by the polymerase into a short stretch of DNA synthesis, 1,000–2,000 bases long in *E. coli* and 100–200 in eukaryotes. The net result is illustrated in Fig. 23.11. The short stretches of DNA attached to RNA primers on the lagging stand are called **Okazaki fragments** after their discoverer (Reiji Okazaki). (As a brief aside, note that in Figs. 23.10–23.13, for convenience only one half of the replication 'bubble' is shown, but do not be misled by this into forgetting that a linear chromosome is not replicated from the end, but from several points in the middle of the sequence).

This still leaves the original problem of how a DNA polymerase can synthesize DNA away from the replication fork while moving towards it – a physical or topological problem. It also leaves the lagging strand as a series of disconnected short pieces attached to RNA primers that must be made into uninterrupted DNA. Let us deal with the physical problem first.

Mechanism of Okazaki fragment synthesis

The basic principle in solving the physical or topological problem of how the lagging strand is synthesized is simple. The lagging template strand is looped, so that for a short distance it is oriented with the same polarity as the leading-strand template. The replicative machinery can therefore proceed in the direction of the fork and synthesize both new strands.

Although the principle is simple, the mechanical problem of how the loop system can move along the entire length of the parental strand and permit the synthesis of Okazaki fragments is not simple because it requires that the loop is reformed and enlarged at regular intervals, and a new RNA primer laid down at each stage.

This model, put forward by Arthur Kornberg, is shown in Fig. 23.12. As the loop enlarges and the replicative machinery moves forward, the polymerase will meet the 5′-RNA end of the previous Okazaki fragment. At this point the polymerase detaches and the loop falls away and a new one is started. To understand this model we must first discuss the replicative machinery at the replication fork.

Enzyme complex at the replication fork in *E. coli*

The functional complex of proteins and protein subunits at the replication fork is illustrated in Fig. 23.13. The key enzymes are the helicase to unwind the double helix, attached to which is the primase, which synthesizes RNA primers at intervals on the lagging strand as it moves along, and the extremely complex Pol III. The helicase unwinding activity is ATP driven and moves along a DNA strand and, in doing so, separates the two strands of the double helix. The primase and helicase (*E. coli*) form a complex in the replication fork known as a **primasome**. Finally, a single-strand binding protein (**SSB**), which has a high affinity for single-stranded DNA, but with no

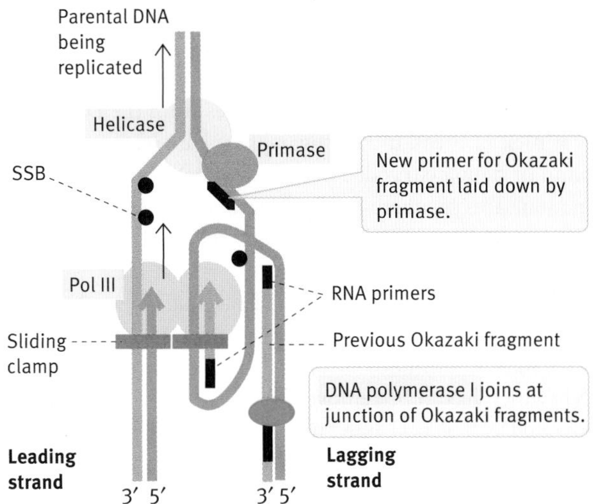

Fig. 23.13 A more detailed loop model. This is to solve the problem of how the polymerase can move forward (upwards on the page) but still synthesize DNA in the required 5′→3′ direction when the template strand polarity demands that synthesis is in the opposite direction (downwards on the page). The action of DNA polymerase I is described in Fig. 23.16 and in the text below. The sliding clamps are described in the text and shown in Fig. 23.14. SSB, single-strand binding protein.

base sequence specificity, binds to the separated DNA strands and stabilizes the single strands. As indicated, in *E. coli* there are two connected molecules of Pol III in the replication fork, one synthesizing the leading strand and the other the lagging strand. They have the same core enzyme, but the holoenzyme dimer is asymmetric with extra subunits present on the lagging strand side. Pol III has high **processivity** – that is, once it locks on to a DNA template it does not fall off but can go on adding more nucleotides without dissociating, allowing it to rapidly replicate long stretches of DNA. A special mechanism prevents premature dissociation. We will describe this now.

The DNA sliding clamp and the clamp-loading mechanism

The sliding clamp is a ring-shaped multisubunit protein structure surrounding the DNA. The ring has a hole big enough for double-stranded DNA to slide through it, but it cannot fall off the DNA (Fig. 23.14). This clamp structure is found both in *E. coli* where it is known as the **β protein** and in eukaryotes, where its name, **proliferating cell nuclear antigen (PCNA)**, describes its function as a protein in the nucleus necessary for cell proliferation. (The term antigen here reflects its mode of discovery). Although the overall structures of the two clamps look the same (Fig. 23.14), the *E. coli* clamp is a dimer whereas PCNA has a three-subunit structure, and they have little protein sequence homology. The convergent evolution of different proteins for the same function emphasizes its vital role. In both *E. coli* and in eukaryotes an additional protein complex is needed to load the clamp onto the DNA. In *E. coli* this is known as the **γ complex**. As explained, at the initiation of replication a special RNA polymerase, the primase, lays down a short stretch

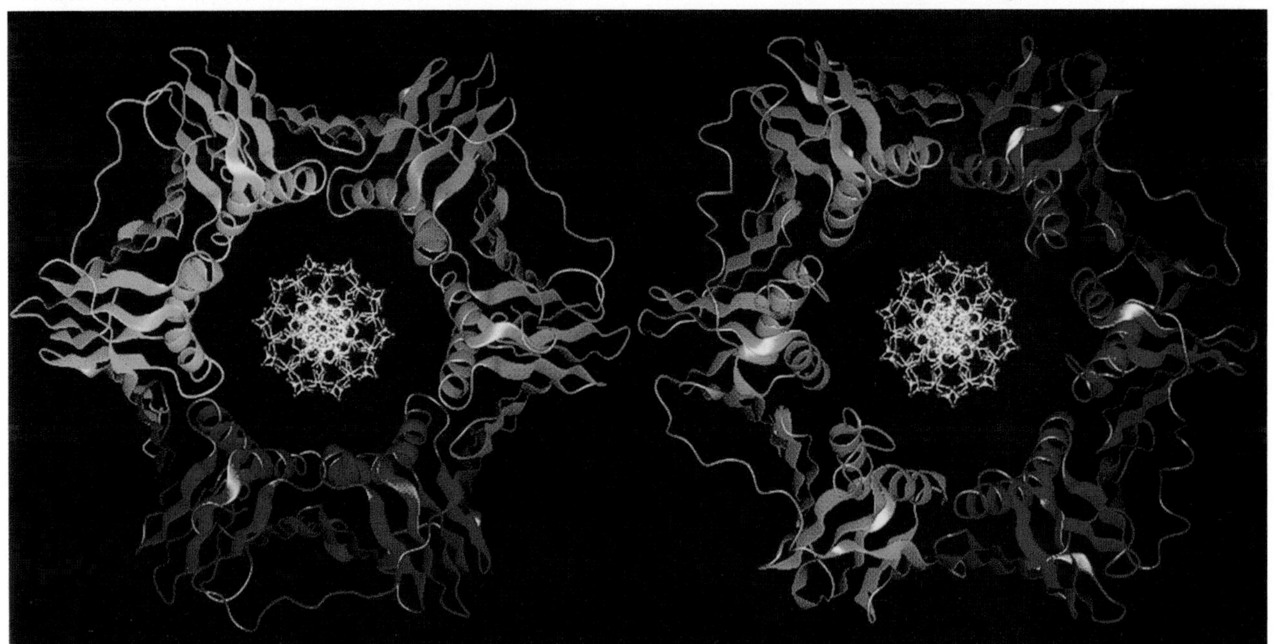

Fig. 23.14 Ribbon representations of the yeast and *Escherichia coli* sliding 'clamps'. **(a)** The yeast clamp (PCNA) that confers processivity on DNA polymerase δ is a trimer. **(b)** The *E. coli* clamp (β protein) that attaches to DNA Pol III is a dimer. The individual subunits within each ring are distinguished by different colours. Strands of β sheet are shown as flat ribbons and α helices as spirals. A model of B DNA is placed in the centre of each structure to show that the rings can encircle duplex DNA. Krishna et al; Crystal structure of the eukaryotic DNA polymerase processivity factor PCNA; Cell; (1994) 79, 1233–43; Elsevier.

of RNA against the template DNA strand. The γ complex, with a bound molecule of ATP, recognizes the short stretch of RNA primer/DNA hybrid and attaches a sliding clamp around it. It does this by seizing a circular clamp from solution, which it opens and places around the RNA primer/DNA hybrid. The ring then snaps shut, this step being associated with ATP hydrolysis and release of the loading γ protein. The face of the clamp has a site for binding the Pol III DNA polymerase, which is recruited from solution. The Pol III is now firmly attached to the DNA by the clamp so that it cannot fall off but is free to move along the DNA and replicate the template strand (Fig. 23.15). The clamp-loading complex places a clamp wherever there is an RNA primer laid down.

When the Pol III synthesizing the lagging strand reaches the primer for the next Okazaki fragment, it must detach and re-initiate at the next RNA primer laid down by the primase. We turn now to the question of how the Okazaki fragments are tidied up into continuous DNA.

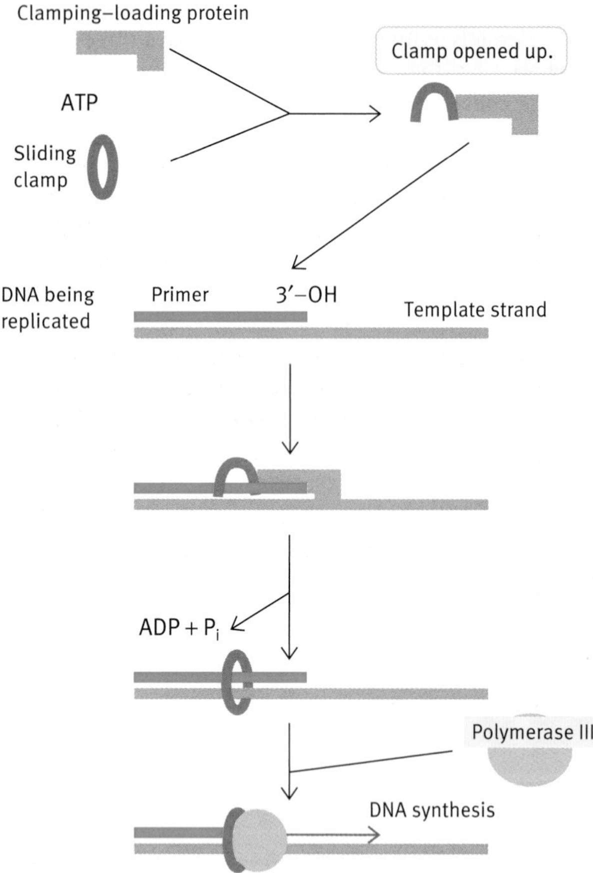

Fig. 23.15 The steps in loading a sliding clamp for DNA synthesis in *Escherichia coli*. The clamps exist in solution as complete rings. The clamp-loading protein, in the presence of ATP opens up the ring, binds to the DNA wherever a primer laid down by primase awaits elongation, and snaps the clamp around the DNA to which a Pol III molecule attaches. Adapted from Fig 1 in Kelman, Z., and O'Donnell, M.; Annu Rev Biochem; (1995) 64, 171; Reproduced by permission of Annual Reviews Inc.

Processing the Okazaki fragments

In *E. coli*, when the Pol III reaches the RNA primer of the preceding Okazaki fragment, it disengages from the DNA, leaving a **nick** (a break in the sugar-phosphate backbone of one strand of double-stranded polynucleotide) at the DNA/RNA junction. This is where DNA polymerase I (Pol I) comes in. Pol I is an astonishing enzyme with three separate catalytic activities on the same molecule.

If we look at the problem, as illustrated in Fig. 23.13, the separate pieces of DNA, the Okazaki fragments, must be converted into a continuous DNA molecule. Each piece starts with RNA which has to be removed, replaced with DNA, and the separate DNA pieces joined up. The Pol I attaches to the nicks between successive Okazaki fragments, and adds nucleotides to the 3′-OH of the preceding fragment, moving, as with Pol III (or any DNA synthesis), in the $5' \rightarrow 3'$ direction. Since, as Pol I moves, it encounters the RNA of the next Okazaki fragment, the nucleotides of this are hydrolysed off. Thus, as it were, the front end of Pol I removes RNA nucleotides, and a site further back adds DNA nucleotides to fill the gap with DNA. The 'front' activity is a **5′→3′ exonuclease** activity – 'exo' because it works on the end of the molecule, 'nuclease' because it hydrolyses nucleic acids, and $5' \rightarrow 3'$ because it nibbles away at the 5′ end of the RNA and moves in the direction of the 3′ end of the molecule. Note that Pol III does not have a $5' \rightarrow 3'$ exonuclease activity, so cannot chop out the RNA primer when it meets the preceding Okazaki fragment. As stated, it disengages from the DNA at this point and hands over the job to Pol I.

The DNA Pol I has (unlike Pol III) *low processivity* – it does not hold on to the DNA template strand firmly and detaches relatively soon after the RNA has been replaced. It does not have the ring-shaped clamp to hold it on to the DNA. This is essential for otherwise it would go on replacing long stretches of the newly synthesized Okazaki fragments. When it detaches, a nick is left in the chain, which is healed by a separate enzyme, called **DNA ligase**.

DNA ligase catalyses formation of a phosphodiester bond between the 3′-OH of one DNA fragment and the 5′-phosphate of the next, a process requiring energy. In some prokaryotes and all eukaryotes, ATP supplies this. The three-step mechanism is that the enzyme (E in the scheme below) accepts the AMP group of ATP, liberating pyrophosphate, and then transfers AMP to the 5′-phosphate of the DNA. Finally the DNA-AMP reacts with the DNA-3′-OH, releasing AMP and sealing the break.

$$\text{E+ATP} \rightarrow \text{E—AMP+PP}_i$$
$$\text{E—AMP+(P-) 5'DNA} \rightarrow \text{E—AMP—(P-)—5'DNA}$$
$$\text{DNA-3'-OH+AMP—(P-)—5'DNA} \rightarrow \text{DNA-3'-O—(P-)—5'DNA+AMP.}$$

Linkage of AMP to the enzyme and then to the DNA is *via* its 5′-phosphate group. The linkage to the enzyme is to a lysine side chain, forming an unusual phosphoamide bond.

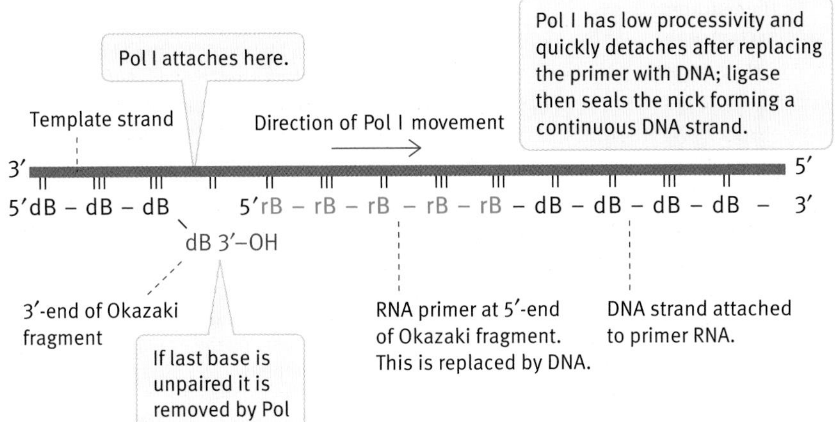

Fig. 23.16 Pol I actions in processing Okazaki fragments. dB, deoxynucleotide; rB, ribonucleotide. Removal of the last base, if unpaired, is described in Fig. 23.19.

In *E. coli* NAD^+, rather than ATP, is the AMP donor. This is most unusual role for NAD^+, which you have met only as an electron carrier. However, NAD^+ can donate an AMP group just like ATP (look at the structure of NAD^+ in Chapter 12, NAD^+ – an important electron/hydrogen carrier?) and, for some reason, *E. coli* uses this route.

What happens then, in summary, is the following. The DNA Pol I binds at the attachment site shown in Fig. 23.16 – at the nick. The polymerase adds DNA nucleotides to the 3′ end of the fragment on the left side of the diagram, and moves in the 5′→3′ direction. The RNA, and some DNA, is nibbled away and the gap replaced by DNA. The Pol I detaches and a ligase joins the two fragments of DNA together. Thus, a series of Okazaki fragments becomes a continuous new DNA strand. Pol I also proofreads the nucleotide additions (see Exonucleolytic proofreading, later in this chapter).

The machinery in the eukaryotic replication fork

The general, principles of DNA replication in *E. coli* also apply to eukaryotes, but there are differences in detail. A large number of eukaryotic DNA polymerases have been identified (15 in humans) and not all have had their roles clearly elucidated. Three in particular, Pol α, Pol δ, and Pol ε, are involved in replication of nuclear chromosomes, while Pol γ replicates the mitochondrial genome. DNA polymerase α is a multisubunit enzyme, part of which has *primase* activity. Thus, Pol α is responsible for initiating replication of the leading strand and the Okazaki fragments on the lagging strand by synthesis of RNA primers, to which it then adds about 30 nucleotides using its DNA polymerase activity, before handing over to Pol δ and Pol ε. The roles of these two enzymes have been the subject of debate. Pol δ was believed to be the other eukaryotic polymerase, besides Pol α, that is essential for genome replication, but recent research suggests that Pol δ primarily replicates the lagging strand, while Pol ε replicates the leading strand. Both Pol δ and Pol ε are clamped to the template DNA by PCNA and are therefore more processive than Pol α.

Other eukaryotic DNA polymerases function in DNA repair and recombination. It seems that several of them play multiple and overlapping roles, suggesting that as DNA synthesis is so crucial to survival some redundancy has evolved in their functions.

When eukaryotic chromosomes are replicated the nucleosomes (see Chapter 22) must be displaced or otherwise coped with at the replication fork as the polymerase moves along. This process is not fully understood, but it seems that the parental histones are somehow 'shared out' to both daughter DNA molecules. Immediately behind the fork nucleosomes are fully reassembled, incorporating new histones as necessary, so that the replicated DNA immediately regains the normal chromatin structure.

Telomeres solve the problem of replicating the ends of eukaryotic chromosomes

The linearity of eukaryotic chromosomes poses a problem not encountered in the replication of circular chromosomes of *E. coli*.

Consider the replication of the chromosome represented in Fig. 23.17 as a very short one for diagrammatic convenience. It is shown as being replicated (in a bidirectional manner) from a single initiation site in its centre (but remember that a real

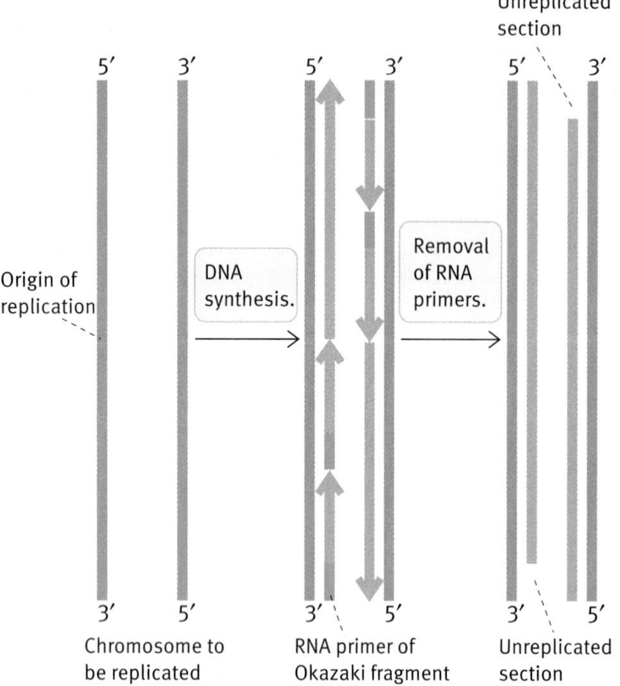

Fig. 23.17 The shortening of linear chromosomes by replication. For diagrammatic convenience the bidirectional replication of a very short piece of DNA is represented. It should be noted that primer removal from Okazaki fragments is a continuous process – it is represented here, for clarity, as occurring as a separate event. A typical chromosome will have multiple origins of replication. The pink lines represent new DNA synthesis; the green lines the RNA primers of Okazaki fragments.

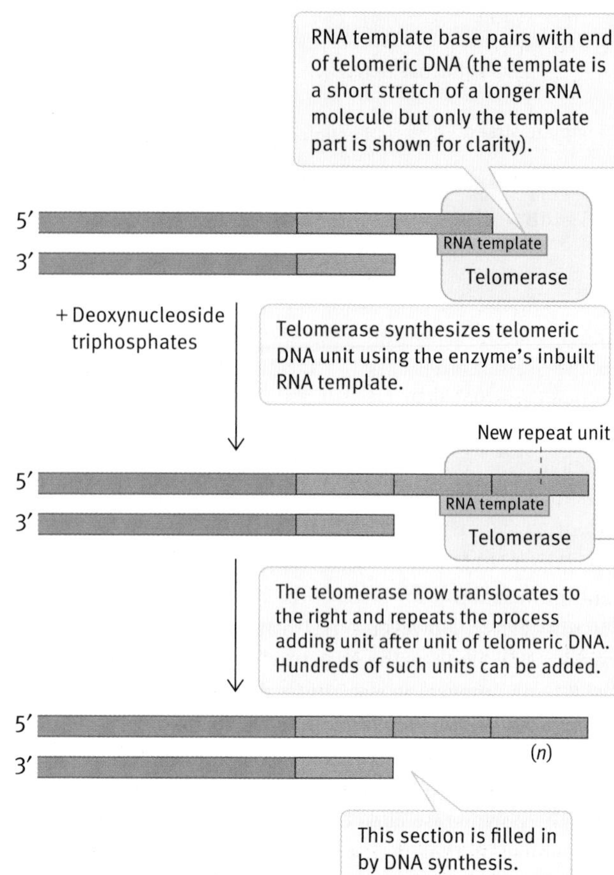

Fig. 23.18 Mechanism by which telomerase synthesizes telomeric DNA. The blue lines indicate the chromosomal or informational DNA (the 'real' chromosome) and the red lines the pre-existing telomeric DNA at one end of a chromosome. The telomerase has an inbuilt short RNA molecule that contains the sequence complementary to the repeating unit characteristic of the species. The enzyme becomes positioned with the RNA pairing with the terminal bases of the pre-existing telomere and adds one repeating unit of TTAGGG (in the case of humans) one base at a time to the G-rich strand. Synthesis is, as always, in the 5′→3′ direction. The enzyme moves so that the RNA template is now paired with the end bases of the new repeating unit and a further unit is added, and so on. The newly synthesized telomeric DNA acts as the template for filling in the opposite strand, using conventional RNA priming, so that the telomere is double stranded.

eukaryotic chromosome has many such sites). The 5′ ends of each strand are fully replicated by leading-strand synthesis. This is not true of the lagging 5′ ends, because the synthesis of the end Okazaki fragments requires RNA primers to be laid down, as shown, on the 3′ ends of the template strands. When the primers are removed it leaves these ends unreplicated and no mechanism exists by which they could be replicated by the DNA synthesizing machinery that we have described so far, since all DNA synthesis requires a starting primer. To fill in the missing parts would require RNA primers but there are no templates against which they could be laid down. This means that, at each cell division, chromosomes would become progressively shorter, on average in a vertebrate, by about 100 nucleotides per cell division. A more potentially disastrous situation could hardly be imagined.

The mechanism of DNA synthesis means that incomplete replication of linear double-stranded DNA *cannot* be avoided. The solution adopted is that, at the ends of eukaryotic chromosomes, stretches of special DNA called **telomeric DNA** are attached, which have no informational role; the ends of the chromosomes containing it are called **telomeres** (Fig. 23.18). The lagging-strand ends will still not be replicated by the DNA synthesis machinery so far described, but it no longer matters for only a short piece of telomeric DNA is lost. For added protection, in many rapidly dividing cells, the telomere is elongated at each round of replication so that repeated loss of the ends does not put the important chromosomal sequences at risk. However, this telomere lengthening does not continue through the full lifetime of the organism (see following sections).

How is telomeric DNA synthesized?

A telomere consists of repeating short stretches of bases – the repeated sequences vary between species. In humans there are hundreds of repeats of the TTAGGG sequence. The enzyme

telomerase adds these sequences one after the other to the 3′ end of pre-existing telomeric DNA.

Telomerase has two remarkable features:

- it uses RNA as the template for DNA synthesis – it is a **reverse transcriptase** (see later in this chapter)
- it carries its own RNA template in its structure.

This RNA carries the sequence complementary to 1.5 repeats of the telomeric sequence. It hybridizes to the end of the overhang (Fig. 23.18), creating a template that can be used to elongate the overhang using the 3′ end of the overhang as a primer. The telomerase protein has the polymerase activity that catalyses this process. When the RNA template has been copied the telomerase moves along and hybridizes to the end of the new repeating unit and thus the telomere is constructed in a discontinuous manner. Telomerase extends only one strand of the DNA, which is made double stranded by conventional lagging-strand DNA synthesis. Removal of the final RNA primer still results in the lagging strand being shorter than its partner, so there is always a residual overhang, but the added telomere repeats protect the end of the chromosome.

The necessity for telomeres has been demonstrated by the use of **yeast artificial chromosomes (YACs)**. These contain the three types of DNA essential for chromosome replication – centromeres, sites of origin, and telomeres. It was shown that YACs are correctly maintained for generations when inserted into yeast cells but that, when they lack telomeric ends, they disappear in time from the cells.

Telomeres stabilize the ends of linear chromosomes

Quite apart from the problem of replicative shortening of chromosomes, linear chromosomes face another problem. Free DNA ends are likely to be mistaken by the cell for damaged DNA and may be attacked by repair systems or nucleases. The ends have to be protected. Studies of mammalian telomeres suggest that a stretch of the double-stranded telomere loops back on itself and the single-stranded overhang is tucked into the double-stranded DNA, and base pairs with an earlier copy of the repeat. The loop structure thus formed is stabilized by binding specific proteins, which give further protection.

Telomere shortening correlates with ageing

In vertebrates telomerase is active in rapidly dividing cells such as germ cells (the cells that give rise to gametes) and early embryonic cells. Lengthening of the telomeres is necessary here to ensure that the daughter cells produced by repeated cell divisions maintain viable chromosomes. However, in somatic cells where cell division occurs only to replace dead cells or heal wounds addition to the telomeres does not occur. When such cells are isolated in culture they undergo a limited number of divisions before undergoing senescence and dying. It is proposed that gradual shortening of the telomeres ensures a limited lifespan for somatic cells, perhaps to protect the organism from faults such as errors in genome sequence that could build up as their cells age. Thus, shortening of telomeres is thought to contribute to human ageing.

If telomerase is activated in cultured cells, they often escape senescence and become immortal. It is significant that telomerase is reactivated in most cancer cells, which undergo uncontrolled division and do not senesce despite an accumulation of genome mutations. Telomerase is thus a potential target for cancer therapy.

How is fidelity achieved in DNA replication?

The 4.6 million base pairs of the *E. coli* genome must be copied very rapidly, as the genome can be duplicated in 40 minutes. Yet, despite this speed, faithful replication of the DNA sequence is critical for survival of the organism. Human DNA polymerases can work at a slower rate, because of the large number of replication origins and the longer duration of a cell cycle, but a human diploid cell has to replicate over 6 billion base pairs for each cell division. With a large genome, even a very accurate replication process may be insufficient to avoid occasional errors, and a single mistake, if an incorrect base is incorporated into a critical region of a critical gene, can cause a genetic disease. The cells achieve an error rate of <1 in a billion. How is this done?

When a deoxynucleotide triphosphate (dNTP) enters the active site of a DNA polymerase, it has to pair with the template nucleotide in a Watson–Crick fashion. If it does not it is incorrect. Given the emphasis that is placed on the specificity of base pairing, it may be a surprise to learn that other hydrogen bonding between nonWatson–Crick pairs can, in principle, occur. Indeed, unusual base pairs are important in protein synthesis (see the wobble mechanism in Chapter 25). So how is Watson–Crick base pairing achieved in DNA and how are 'illegitimate' base pairs excluded? The free-energy difference between the pairing of a correct base and an incorrect one is not large enough to give sufficient discrimination. There has to be something else selecting the correct dNTP. The structure of the polymerase protein plays a large part, as in enzyme-substrate binding in general. The two main factors which operate in the polymerases are **geometric selection** and **conformational changes** in the polymerases.

Geometric selection. The geometry of nucleotide Watson–Crick base pairs (A/T and G/C) are almost identical in their shape, the distance apart, and angle of their glycosidic links. The 'illegitimate' base pairs have a different geometry from Watson–Crick pairs. An incorrect dNTP pairing with the base in the template strand will therefore not have the correct geometric shape to fit the polymerase active site.

Conformational changes. When a correct dNTP pairs with the template nucleotide a large conformational change occurs in the polymerase active site. The DNA polymerase has an open structure in the absence of substrate. The entry of a correct dNTP causes the enzyme to close it up around the base pair and place it in the appropriate position for catalysis of phosphodiester bond formation. The conformational changes are 10,000 times slower with the entry of an incorrect dNTP.

The selectivity achieved by these factors is high, with an error rate of one in a million, but this would still give an unacceptable mutation rate. Further improvement in selectivity is needed. The next stage is for the polymerase itself to check the correctness of each addition.

Exonucleolytic proofreading

Many DNA polymerases including *E. coli* polymerases I and III and their eukaryotic counterparts have a catalytic activity that we have not mentioned so far. They have 3′→5′ exonuclease (backward) activity, which can chop off the last added nucleotide from the growing DNA chain. Note that this is quite different from the forward-acting 5′→3′ exonuclease by which Pol I removes the RNA portions of Okazaki fragments. However, this backward chopping activity only occurs if the last added nucleotide was incorrect. Each time the enzyme adds a nucleotide it checks if it was correct; if not, it is removed and it has another try to replace it with a correct one. It is a **proofreading** mechanism found in many DNA polymerases.

The mechanism of proofreading has been examined in DNA Pol I. The synthesis site and exonuclease sites are sufficiently close on the enzyme surface for the newly formed end of the DNA chain to slide from one site to another (Fig. 23.19). An incorrect base is more likely to become detached from its template partner and slide from the synthesis into the exonuclease site than is an end with the correct base. The incorrect base is removed by hydrolysis of the phosphodiester bond and the correctly paired template-primer complex is returned to the synthesis site. Proofreading occurs in the synthesis of the leading and lagging strands and in the processing of the Okazaki fragments. However, in eukaryotes Pol α lacks exonuclease and hence proofreading activity. Since Pol α mainly synthesizes the RNA primers, and only short initial stretches of DNA, this does not matter too much.

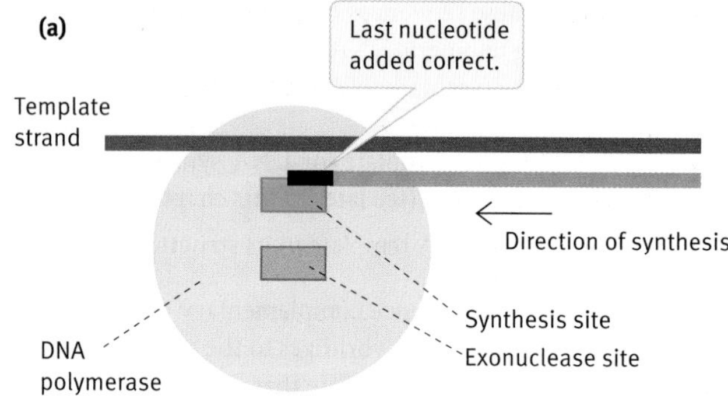

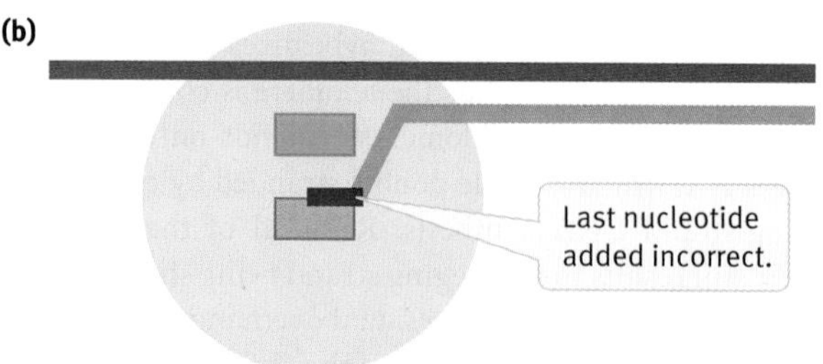

Fig. 23.19 Simplified diagram of the mechanism of exonucleolytic proofreading by DNA polymerase I. **(a)** Situation if last addition was correct. **(b)** Situation if last addition was not correct. An incorrect base on the growing DNA chain (mauve) in the lower diagram increases the chance of it detaching from the template base and swinging to the exonuclease site so that the error is removed. The polymerase can then replace it with the correct nucleotide when the chain swings back. If the last added base is correct, it is less likely to detach from the synthesis site.

Methyl-directed mismatch repair

The number of mismatches that slip through into the newly synthesized DNA would still give an unacceptable rate of mutation. The cell therefore has a backstop mechanism to replace faulty bases even after the newly synthesized DNA has been released from the polymerase. The final fidelity mechanism in *E. coli* is called **methyl-directed mismatch repair**. It increases fidelity. If a mismatch has escaped the polymerase proofreading correction, the error will cause a distortion in the duplex chain illustrated diagrammatically below.

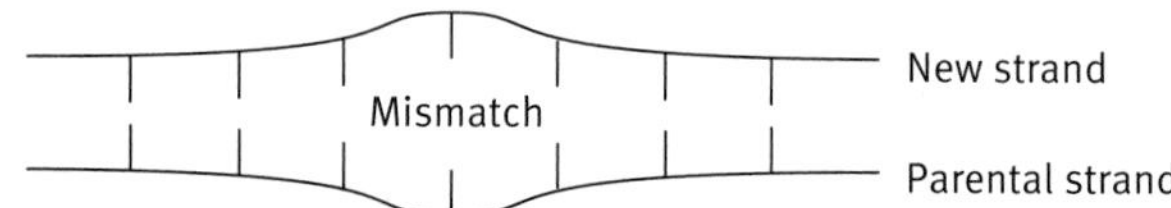

The repair system can recognize the distortion in the DNA, but it must also discriminate between the parental (template strand), which by definition has the correct sequence, and the new strand, which is incorrect. It must remove the base from the new strand and replace it, rather than replacing the correct base in the template, as that would perpetuate the mutation. How is this strand discrimination made? In *E. coli*, wherever there is a GATC sequence in the DNA, the adenine of this sequence is **methylated** by an enzyme in the cytosol. This does not affect base pairing or DNA structure. It takes some time after its synthesis for the new strand to be methylated and so, for this brief period, it is unmethylated. Thus the parental strand is methylated but the just-synthesized new strand is not. The repair system involves three proteins designated Mut S, Mut H and Mut L. First **Mut S** recognizes the mismatch distortion in the double helix. The role of **Mut H** is to bind to the DNA at *unmethylated* GATC sites, i.e. those that have recently been replicated. When **Mut S** binds to a mismatch, it is itself bound by **Mut L**, which then also binds to the nearest **Mut H. Mut H** is thus stimulated to nick the unmethylated GATC (to 'nick' DNA is to break a phosphodiester bond without removing any nucleotides) (Fig. 23.20). The GATC nicked by Mut H may be some distance from the mismatch, so helicase, SSB, and an exonuclease cooperate to remove the newly replicated strand of the DNA from the nick to beyond the mismatch and Pol III synthesizes a replacement strand, in doing so correcting the error. DNA ligase completes the repair by sealing the nick. To correct one base, thousands of nucleotides may be replaced (Fig. 23.20). This error-correction system increases the fidelity of replication so that there is a final error rate of less than or equal to 10^{-10}.

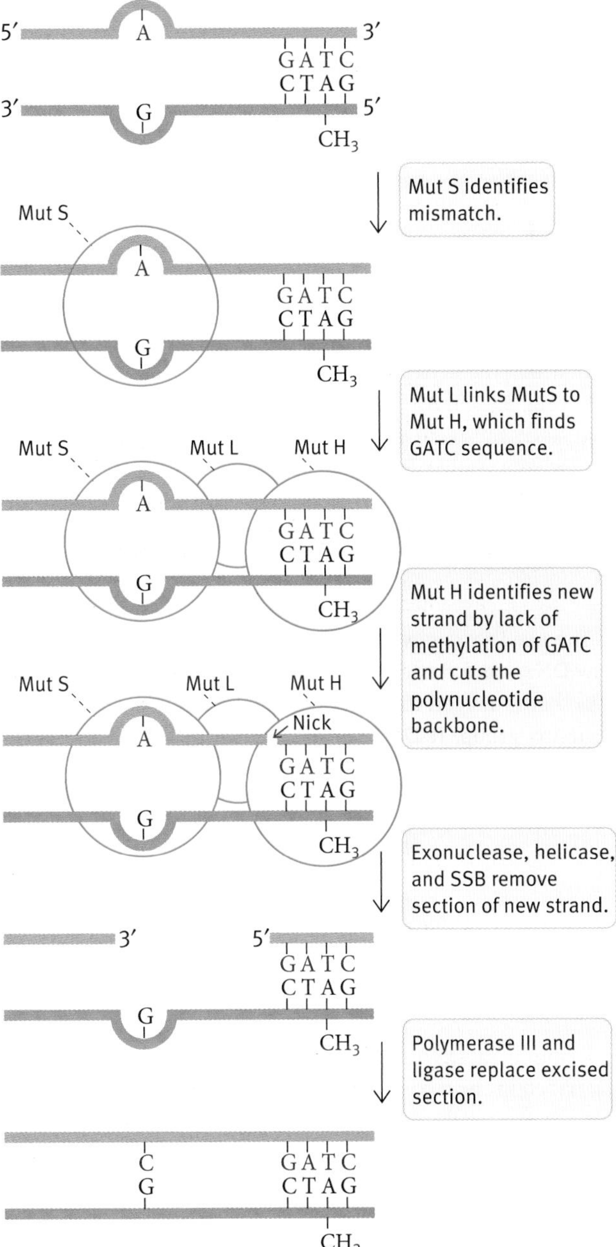

Fig. 23.20 Methyl-directed pathway for mismatch repair. If the GATC sequence is distant from the error, bending of the DNA could bring the two into proximity. Mut proteins are coded for by mutator genes whose inactivation increases DNA synthesis error rates.

Mismatch repair also occurs in eukaryotes. In humans, proteins corresponding to Mut S and Mut L proteins of *E. coli* are known, but nothing corresponding to Mut H has been found, as mechanisms other than methylation are used for strand identification. Mutations in the genes encoding the Mut S and Mut L homologues are associated with increased risk of colon cancer, evidence of the importance of mismatch repair in humans.

Repair of DNA damage in *E. coli*

The mechanisms described above ensure that DNA is replicated with the degree of accuracy needed to ensure continuity of cellular life. However, chemical changes that damage DNA occur at a rate that would result in large numbers of mutations per day, in each cell, if there was not constant repair. Some of the damage changes are spontaneous. The glycosidic link that binds the bases to deoxyribose is somewhat unstable (for purines more than pyrimidines) so that depurination and depyrimidation occur spontaneously. Numbers of purines and a lesser number of pyrimidines are hydrolysed from the DNA every day in a human cell, creating apurinic or apyrimidinic (AP) sites. In addition, cytosine and adenine occasionally chemically deaminate to become uracil and hypoxanthine, respectively (see following structures in text and Fig. 19.7).

The DNA of all cells is also subject to 'insults' by a wide variety of agents and many of these cause carcinogenic mutations. An important cause of damage is oxygen free radicals generated in cells. Free radicals, and the mechanism by which they damage biological molecules, are described in Chapter 32, together with the protective mechanisms developed against them. Reactive free radicals are also generated by ionizing radiation. UV light is well known to cause cancers by cross-linking adjacent pyrimidine bases. The best known are thymine dimers, but all four types of pyrimidine dimer can be formed. A variety of other abnormal molecules can be produced by UV light. Besides this, certain chemicals such as aflatoxins (fungally produced carcinogens that often contaminate food crops), form reactive molecules that attack and modify DNA.

When other molecules such as proteins are damaged they are simply destroyed, but DNA must be repaired at all costs (or, if not, in a complex animal the whole cell must be destroyed by apoptosis in case it should develop into a cancer; see Chapters 30 and 31). The importance of DNA repair is shown by the variety of systems that exist to achieve it, making it a complex topic that could be the subject of an entire book. Here we can only summarize the mechanisms discovered in *E. coli*, and draw a brief comparison with eukaryotes.

Many DNA repair mechanisms rely on the rarity of damage occurring at the same place to *both* strands of a duplex DNA molecule (though it does happen). If only one strand is damaged the other strand can act as a template for repair.

- **Direct repair.** Exposure of DNA to UV light can result in the covalent linking of two adjacent pyrimidine bases (on the same strand), forming a dimer, often of thymines (**T dimer**). (The structure is shown in two dimensions, side by side for clarity, rather than one on top of the other.)

T dimer

In *E. coli*, thymine dimers can be repaired directly, without the need for synthesis of replacement DNA. The abnormal bonds between the two bases are cleaved by a light-activated enzyme called photolyase. Another direct repair system involves a 'suicide enzyme' that removes alkyl groups from bases. The alkyl (e.g. methyl and ethyl groups) are added by mutagenic chemicals and would cause the altered bases to pair abnormally at the next round of replication, introducing a mutation. The repair enzyme removes an alkyl group onto its own structure, but in doing so destroys its own activity. It is more of a specific protein reagent than an enzyme since it is changed in the process, unlike a true catalyst.

- **Nucleotide excision repair.** Lesions that distort the double helix, including T dimers, are repaired by the removal (excision) of a short stretch of the DNA strand that includes the lesion, followed by its correct replacement, using the opposite strand as the template. In *E. coli*, four proteins called UvrA, UvrB, UvrC and UvrD (where Uvr stands for UV repair), cooperate to cut the DNA on either side of the lesion and remove 12–13 nucleotides (Fig. 23.21). The nuclease activity of the Uvr protein complex is termed excision endonuclease or **excinuclease**. DNA polymerase I adds nucleotides to the 3′ end of the cut chain and once the gap has been filled ligase heals the remaining nick. The system depends on it being possible to recognize which strand of the DNA is faulty.
- **Base excision repair and AP site repair.** Deamination converts cytosine to uracil and adenine to hypoxanthine. As neither of these bases occur in normal DNA, DNA glycosylase enzymes recognize them and hydrolyse

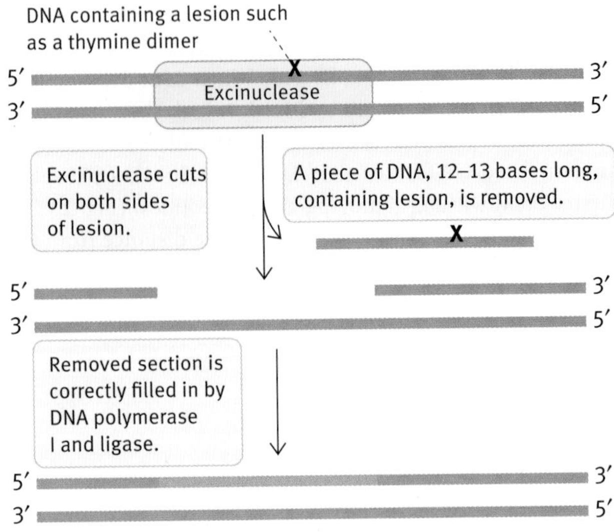

Fig. 23.21 The pathway of nucleotide excision repair in *Escherichia coli*.

them off, leaving AP (apurinic or apyrimidinic) sites in which the deoxyribose moiety has no base attached to it (Fig. 23.22). AP sites can also be formed spontaneously, since the purine-deoxyribose link, especially, is somewhat unstable. Repair of AP sites involves nicking of the polynucleotide chain adjacent to the lesion followed by replacement of the damaged section by DNA polymerase I and sealing by ligase.

The need to remove uracil formed from cytosine explains why DNA has T, instead of U. Remember that T is, in essence, a U that is tagged in DNA for identification purposes with a methyl group. If DNA *normally* contained U, it would be impossible to distinguish between a U that should be there and an 'improper' U, formed by deamination of C.

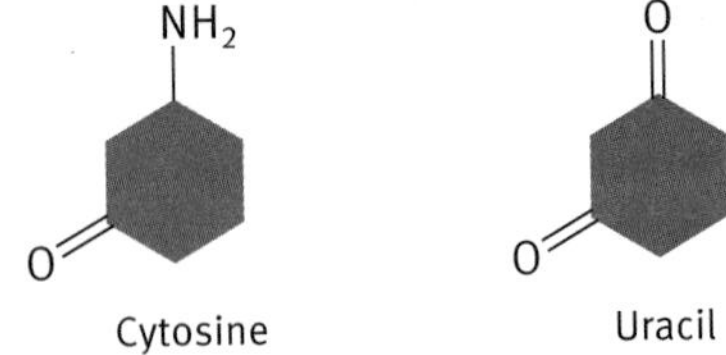

Using T in DNA, instead of U, solves the problem. (As described in the next chapter, U can be used in RNA, because RNA has a relatively short lifetime, is much smaller, and errors do not have the same long-term consequences. Hence RNA is not repaired.)

Repair of double-strand breaks

Double-strand breaks in DNA are caused by ionizing radiation and other agents. They pose an extremely dangerous repair problem because there is no undamaged partner strand to act as a template for the repair. Two methods

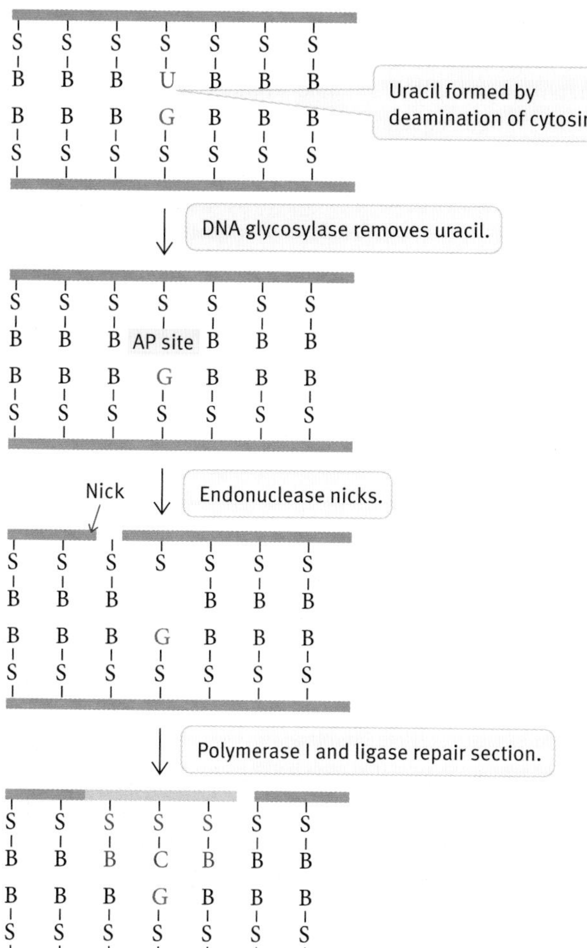

Fig. 23.22 AP site formation and repair. In the example given, the site is created by removal of a uracil by a glycosylase, but sites are also formed by spontaneous hydrolysis of purine bases (and, to a lesser extent, of pyrimidine bases) from the nucleotide. S, sugar; B, base.

have evolved. In the **end-joining** mechanism, the cut ends are simply ligated together. This is a risky process as it may change the sequence due to nucleotides being trimmed off at the cut ends.

The other method is more accurate. It uses **recombination** with a homologous undamaged sequence to direct the repair. For instance, if a double-strand break occurs on one branch of a replication fork, the homologous sequence on the other branch can be used as the template for repair by recombination. **Homologous recombination** has important functions outside of DNA repair: it generates genetic diversity by allowing exchange of DNA sequences between chromosomes, and the mechanism is outlined later in the chapter.

Translesion synthesis

If DNA repair fails, there is a 'last resort' mechanism, that allows the cell to complete DNA replication by synthesizing a new strand across the lesion. This is a highly error prone process as the lesion cannot act as a template for addition of the correct nucleotides. Instead, a specialized translesion polymerase moves along a template strand that carries an unrepaired lesion such as an AP site or thymine dimer, but instead of using base pairing, the polymerase itself selects the nucleotides for incorporation at that point. In *E. coli*, this is known as the SOS response. In eukaryotes, several DNA polymerases that are not used in normal replication are able to carry out translesion synthesis.

DNA damage repair in eukaryotes

For the most part, analogous DNA repair mechanisms to those in *E. coli* are present in eukaryotic cells, including humans, although we and other placental mammals lack the photolyase system. Inherited mutations affecting proteins needed for DNA repair cause a number of genetic diseases that predispose to cancer. For example, **xeroderma pigmentosum** can be caused by a mutation affecting any one of seven proteins involved in nucleotide excision repair. Sufferers develop skin cancer because they cannot repair DNA damage caused by sunlight. Genes for proteins corresponding to the mismatch-repair proteins, Mut S and Mut L of *E. coli*, have been found, and their mutations are associated with **hereditary nonpolyposis colorectal cancer** (HNPCC).

Homologous recombination

Genetic recombination involves the rearrangement of DNA sequences. Recombinant DNA technology, as used in genetic engineering, is discussed in Chapter 28, but recombination also occurs naturally in living cells. The main type is **general** or **homologous recombination** between separate chromosomes at a region where their base sequences are homologous (largely, but not necessarily entirely, the same). There is another quite different chromosomal rearrangement process known as **site-specific recombination** that is involved in antibody production (see Chapter 33).

Homologous recombination creates diversity through **reassortment** of genes: it results in individual organisms with new combinations of genes that are then subject to natural selection, hence driving evolution. In eukaryotes homologous recombination takes place during formation of gametes by meiotic cell division (see Chapter 30). Bacteria contain only a single chromosome, but they can temporarily acquire a homologous section of DNA from another bacterium through a process called **conjugation**, giving an important opportunity for recombination. As mentioned above, homologous recombination is also a mechanism for the repair of double-strand breaks in DNA.

The result of homologous recombination is illustrated in Fig. 23.23; in this, you see two chromosomes (both are DNA duplexes) with a stretch of homologous DNA in the base sequence in the two. They come together at this point and exchange sections of duplex DNA as shown in the diagram. The recombinant process produces a very small patch of **heteroduplex** DNA (double-stranded DNA containing some mispaired sequence) if the two homologous pieces are

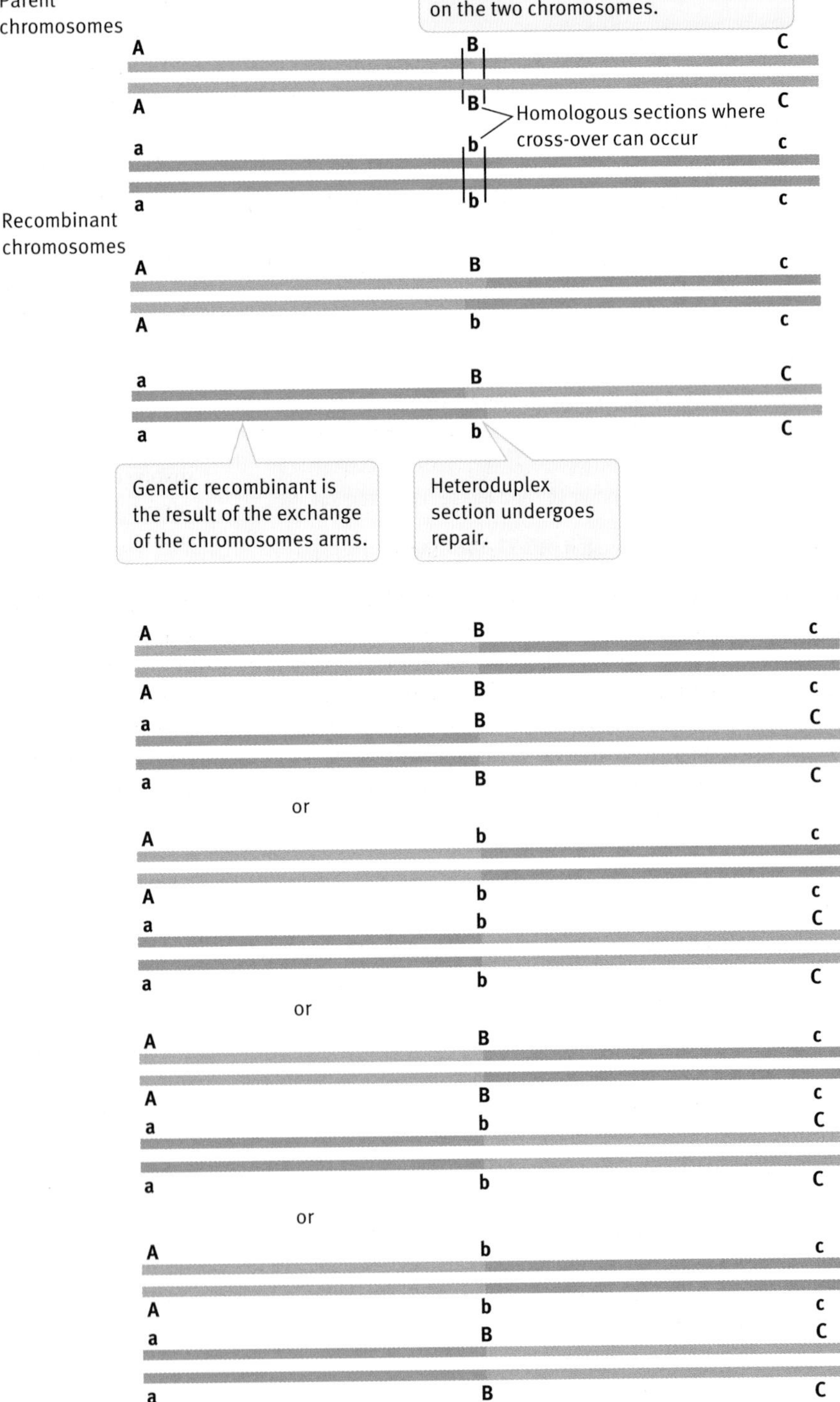
Parent chromosomes
A, B, C, a, b, c represent alleles of genes on the two chromosomes.
Homologous sections where cross-over can occur
Recombinant chromosomes
Genetic recombinant is the result of the exchange of the chromosomes arms.
Heteroduplex section undergoes repair.
or
or
or

Fig. 23.23 Homologous recombination, which occurs *via* cross-over junctions as described in the text. Genetic recombination is the resultant reciprocal exchange of chromosome arms or sections. Gene conversion is the nonreciprocal change in DNA sequence that may result from repair of the short heteroduplex section. A, a, B, b, C and c represent allelic genes on the two chromosomes. (Gene alleles represent the same gene but differ in base sequence such that the proteins expressed are slightly different, or are expressed differently.)

slightly different at the cross-over point. This hybrid DNA patch could indeed cause a nonreciprocal change of sequence when the mismatches are repaired (a phenomenon known as gene conversion), but the main event is that the two arms on either side of the cross-over point are reciprocally exchanged, producing extensive swapping of genes between the two parent chromosomes.

Mechanism of homologous recombination

The process of homologous recombination can be summarized as follows:

- Homologous DNA duplexes pair up due to the similarities in their sequence. Single or double-stranded breaks in the DNA allow one strand of each duplex to invade and base pair with the other, forming a hybrid structure of both DNA duplexes called a Holliday junction. The junction can migrate along the DNA allowing more extensive sequence exchange.
- Further breakage and rejoining of the DNA occurs to resolve the Holliday junction and separate the DNA duplexes, which have now exchanged sequence and hence recombined.

Formation of cross-over junctions by single-strand invasion

A model to account for homologous recombination was put forward by Robin Holliday in 1964. This has now been modified slightly as his model envisaged single-stranded nicks to the DNA to allow strand invasion. Although it is now known that most recombination events, whether in DNA repair or meiotic crossing over, involve double-stranded breaks to the DNA with end processing and replication of a short stretch of DNA, the Holliday model is often shown, as in Fig. 23.24, because it allows a relatively simple view of events at the cross-over site. In this model, single-stranded nicks in both of the homologous DNA molecules allows the nicked strands to cross-over, each invading the other molecule and forming a short stretch of heteroduplex DNA. The nicks are sealed by ligation, resulting in the formation of the cross-over **Holliday junction** shown in Fig. 23.24. As explained in the figure, the cross-over junction can move along the chromosomes as long as there is homology, increasing the length of the heteroduplex sections and moving the cross-over along (known as **branch migration**).

Resolution of the cross-over junction

Resolution of the Holliday junction starts with three-dimensional rearrangement of the junction, a process known as **isomerization**. The result is a structure in the form of a cross. The noninvading strands are then cut and religated as shown in Fig. 23.24, to separate the DNA molecules. There are two options for cutting and ligation, and it seems to be random which one occurs. The first option creates a patch of heteroduplex DNA, which can undergo mismatch repair. The second option results in reciprocal exchange of the sequences on either side of the junction.

Molecular mechanism of homologous recombination in *E. coli*

The main function of homologous recombination in bacteria is to repair double-stranded breaks (DSBs). The molecular mechanism of homologous recombination is more fully established in *E. coli* than in eukaryotic cells, but sufficient is known to make it likely that the two classes of cells have much in common.

The initial phase involves **single-strand invasion** (Fig. 23.25). A DNA duplex containing a DSB is processed by exonuclease activity to produce single-stranded 3′ ends. An intact homologous duplex is required, and can be provided by the other branch of the replication fork if the repair takes place during replication. The single-stranded regions of the broken duplex 'invade' the intact homologous partner, each base pairing with complementary sequence to form an open structure called a D-loop. Specialized proteins are involved in processing the DSB ends and invasion: a crucial component is the recombinase enzyme; in *E. coli* this is RecA. Multiple RecA molecules bind to the single invading DNA strand, causing it to have an extended conformation that make its bases available for hybridization. The single strand invades the duplex and RecA facilitates its search for a complementary sequence with which it base pairs, forming the D-loop structure. Synthesis of new DNA occurs to replace the sequence digested during end processing. Two Holliday junctions are formed, which undergo branch migration and resolution as in Fig. 23.24. Note that crossing over will occur only if the two junctions are resolved differently, as shown in Fig 23.25. If the two junctions are both resolved in the same way a 'patch' of heteroduplex DNA is formed, which may result in gene conversion, but there is no crossing over.

Recombination in eukaryotes

In **meiosis**, in which haploid gametes are produced (see Chapter 30), chromosomes linked by **chiasmata** (strand cross-overs) are seen. These are the sites of homologous recombination that generate genetic diversity. Two eukaryotic recombinase proteins, Dmcl and Rad 51, have been identified that show similarities in structure and function to *E. coli* RecA. Dmcl is believed to be specific for meiotic recombination while Rad51 is required for double-strand break repair. In both yeast and mice, mutations affecting Rad51 function cause hypersensitivity to ionizing radiation (hence the designation 'Rad'), highlighting the conservation of the recombination and repair mechanisms throughout evolution.

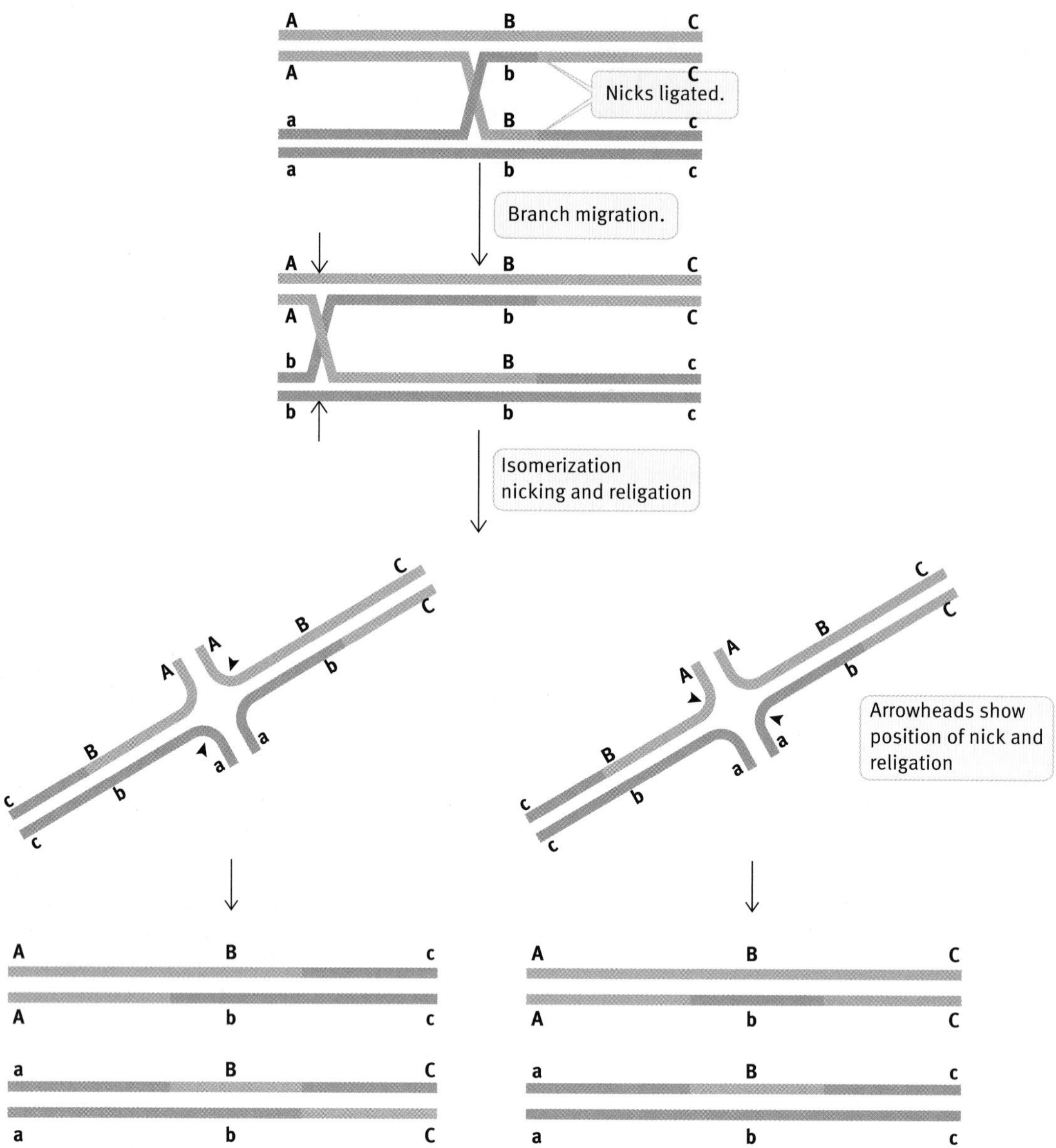

Fig. 23.24 A Holliday junction after mutual strand invasion and ligation of nicks. Note that once a limited amount of exchange has occurred, as shown in the top figure, the extent of the cross-hybridization can extend, so long as the sections exchanged are homologous so that hybridization can occur. In doing so, the cross-over junctions are moved; this is known as branch migration. The migration can occur to the limits of the homologous sections. The Holliday junction is resolved by nicking and religating the DNA. If this occurs in the way shown here, it results in recombination between the two original chromosomes as shown in Fig. 23.23. The short stretch of heteroduplex DNA formed at the junction site may undergo mismatch repair, resulting in gene conversion.

Replication of mitochondrial DNA

In humans each double-stranded circular mitochondrial genome has 16,600 base pairs that code for 24 RNAs (tRNAs and rRNAs) and 13 proteins. As cells contain many mitochondria and each mitochondrion contains multiple copies of the genome an individual cell may contain hundreds or even thousands of genome copies. Replication of mitochondrial genomes is not synchronized with nuclear genome replication and seems to occur randomly during the cell cycle.

Mitochondrial DNA is replicated by a special DNA polymerase, Pol γ. Like most mitochondrial proteins Pol γ is encoded by the nuclear genome and imported into the mitochondrion. Human mitochondrial DNA is observed to accumulate sequence changes and hence evolve significantly more rapidly than the nuclear genome. It was thought this was caused by a lack of DNA repair in mitochondria.

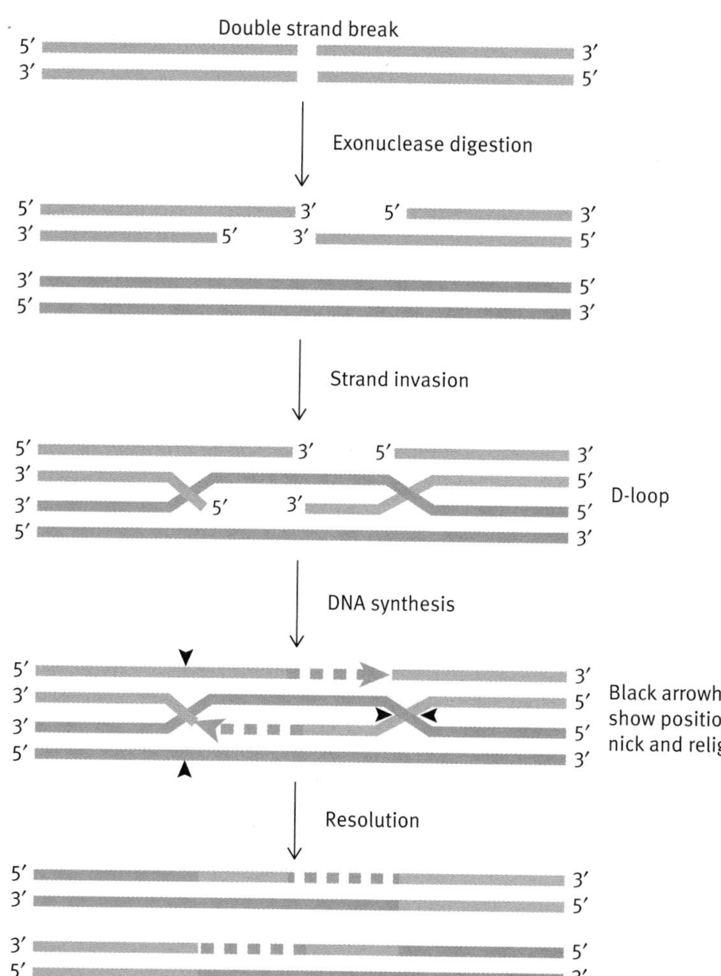

Fig. 23.25 Steps in homologous recombination. A DNA double-strand break is processed by exonuclease, leaving single-stranded tails each with a 3′-OH group. The single-stranded tails invade homologous duplex DNA forming the D-loop structure. This process requires multiple proteins including RecA in *E. coli* and Rad51 in eukaryotes. New DNA is synthesized using the invaded strands as the template, forming two Holliday junctions, which are resolved as shown in Fig. 23.24. From Björklund S, Gustafsson CM. Trends in Biochemical Sciences (2005) 30(5):240–4.

However, it is now clear that Pol γ does carry out proofreading, and that mitochondria have at least one DNA repair system, base excision repair, though they lack others. Nevertheless, the rapid rate and multiple rounds of DNA replication that occur in each cell may contribute to accumulation of mitochondrial mutations during the lifespan of the organism, and it is suggested that this may be a factor in our ageing process.

Inheritance of mitochondrial mutations is a significant cause of genetic disease. Since mitochondria are transmitted only from the egg, mitochondrial genome disorders show an unusual inheritance pattern, termed maternal inheritance.

DNA synthesis by reverse transcription in retroviruses

We will finish the chapter with a brief look at a fundamentally different mechanism of DNA synthesis, that is used by retroviruses to copy their single-stranded RNA genome into DNA. This involves an enzyme whose discovery caused initial disbelief, to be followed by the award of a Nobel Prize in 1975 to its discoverers, David Baltimore, Renato Dulbecco, and Howard Temin; it is called **reverse transcriptase**. Before this discovery, it was, of course, known that DNA directs RNA synthesis, but the accepted dogma was that the reverse never happened. Besides the medical significance of retroviruses, in particular the human immunodeficiency virus (HIV) that causes AIDS, reverse transcriptase has important uses in recombinant DNA technology (see Chapter 28). It is now known also that retrotransposons found in eukaryotic genomes use reverse transcriptase to replicate; thus the human genome contains genes that encode reverse transcriptase.

An outline of retrovirus replication is shown in Fig. 23.26. The RNA genome and reverse transcriptase are carried in the virus particle, but replication takes place once they enter the host cell. Viral reverse transcriptase is a versatile enzyme; it has polymerase and RNAse activity and the polymerase can copy both RNA and DNA. Initially, it copies a short

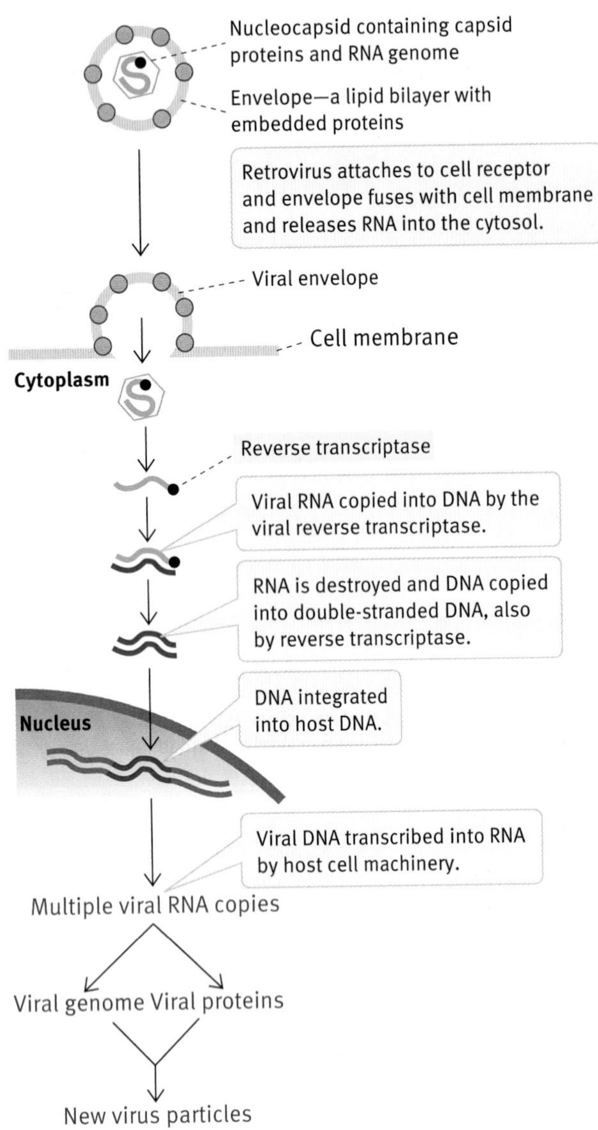
Nucleocapsid containing capsid proteins and RNA genome
Envelope—a lipid bilayer with embedded proteins
Retrovirus attaches to cell receptor and envelope fuses with cell membrane and releases RNA into the cytosol.
Viral envelope
Cell membrane
Cytoplasm
Reverse transcriptase
Viral RNA copied into DNA by the viral reverse transcriptase.
RNA is destroyed and DNA copied into double-stranded DNA, also by reverse transcriptase.
Nucleus
DNA integrated into host DNA.
Viral DNA transcribed into RNA by host cell machinery.
Multiple viral RNA copies
Viral genome
Viral proteins
New virus particles

Fig. 23.26 Replication of a hypothetical retrovirus.

stretch of the RNA template into DNA using, as a primer, a transfer RNA (tRNA) molecule that is 'borrowed' from the host cell. The 3′ end of the tRNA, base pairs to the template strand and is extended as DNA by the reverse transcriptase. The viral genome contains a sequence that is repeated at each end, and the first short DNA copy made includes this repeat sequence. The short copy transfers to the other end of the genome next, pairing with the repeat sequence, and is extended to make a full DNA copy of the genome. The RNA/DNA hybrid thus formed is converted to single-strand DNA by RNA hydrolysis using the RNAse activity also present in reverse transcriptase, leaving a short stretch of RNA to act as a primer for second strand synthesis. The single-stranded DNA is then copied by reverse transcriptase to form double-stranded DNA (called **proviral DNA**) (Fig. 23.26), which is incorporated into host DNA by a separate enzyme, integrase, also carried in the virus. Once in, it is replicated along with the host DNA chromosome. For the production of new retrovirus particles, the proviral genes (that is, viral genes in the host chromosomes) are transcribed into RNA transcripts that direct synthesis of the proteins needed for new virus particle assembly.

Summary

DNA synthesis is catalysed by DNA polymerases, which require four deoxynucleoside triphosphates, a template or parental strand to copy, and a primer. The primer is a short RNA copy of part of the parental strand to which deoxynucleotides are added to the 3′ end. Synthesis starts at a site of origin on the chromosome where strand separation occurs.

E. coli has a single site of origin while eukaryotic chromosomes have hundreds. As the polymerase proceeds, a helicase separates parental strands, producing supercoiling ahead of it. Supercoils are removed by topoisomerases. A polymerase dimer replicates both DNA strands while moving towards the replication fork. The problem of maintaining the 5′→3′ direction of synthesis of both strands is solved by discontinuous synthesis of the lagging strand followed by processing the separate Okazaki fragments into a single chain.

The mechanism of synthesis of DNA means that linear eukaryotic chromosomes are shortened on each replicative round. To prevent loss of coding DNA, the ends are protected by telomeres, consisting of repetitive DNA added by the enzyme telomerase. This enzyme is not present in somatic cells so their telomeres shorten with age, limiting the number of cell divisions possible. Stem cells (and cancer cells) that replicate indefinitely have telomerase to maintain telomere length.

Fidelity of replication is achieved by several means. DNA polymerase selectively accepts triphosphates that form Watson–Crick base pairs with the template nucleotides, but the free-energy difference between the formation of a correct and an incorrect pair is not enough to give sufficient discrimination. The shape of the correct base pairs, being different from incorrect ones, plays an important part in the selection process. The DNA polymerases also proofread by removing and replacing the last added nucleotide if it is incorrect.

DNA of cells is subject to continual insults from radiation and chemical instability, for which there is a variety of repair mechanisms, including direct repair, base excision and nucleotide excision repair.

Genetic recombination involves the *in vivo* rearrangement of the DNA of chromosomes within cells. The main type is **general** or **homologous recombination** between separate chromosomes (DNA duplexes) at a point where there are homologous sections of DNA. The continual reassortment of genes by homologous recombination (coupled with selection) means that new combinations of genes may be assembled to be tested by evolution.

Retroviruses such as HIV have an RNA genome but on infection replicate their RNA as DNA using reverse transcriptase.

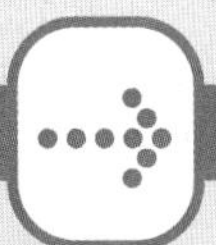

Further reading

To access the further reading please scan the QR code image or go to http://global.oup.com/uk/orc/biosciences/molbiol/snape_biochemistry/student/reading/ch23/

Problems

1 What is a replicon?

2 In separating the strands of parental DNA during replication, what topological problem occurs

3 How is the problem referred to in the preceding question solved, both in *E. coli* and eukaryotes?

4 By means of diagrams, explain the actions of topoisomerases I and II.

5 Eukaryotes have no topoisomerase capable of inserting negative supercoiling into DNA and yet eukaryotic DNA is negatively supercoiled. Explain how this is brought about.

6 What are the substrates for DNA synthesis?

7 Why is TTP used in DNA synthesis – why not UTP as in RNA?

8 (a) Can a DNA chain be synthesized entirely from the four deoxytriphosphate substrates? Explain your answer.

(b) In which direction does DNA synthesis proceed? Explain your answer so as to be totally unambiguous.

9 What are the thermodynamic forces driving DNA synthesis?

10 *E. coli* polymerase I is a complex enzyme. Describe its different activities and explain their roles in DNA synthesis.

11 Discuss the mechanism by which DNA polymerase III of *E. coli* achieves a high standard of fidelity in DNA synthesis.

12 The proofreading activity of *E. coli* polymerase III is important, but insufficient to give a sufficiently high fidelity rate. If an improperly paired nucleotide is incorporated giving a mismatch, it has to be replaced. This demands that the repair system recognizes which of the two bases in the mismatch is wrong. How is

this done and how is the problem fixed? Does this mechanism exist in humans?

13 Explain what a thymine dimer is, how it is formed, and how it is repaired.

14 Explain how eukaryotic chromosomes become shortened at each round of replication.

15 Explain how the DNA-shortening problem in replication is coped with.

16 What is meant by processivity? Which DNA polymerase does not have it and why?

17 Which of the following components does not belong to the series? CTP, UTP, DNA, ATP, GTP, RNA.

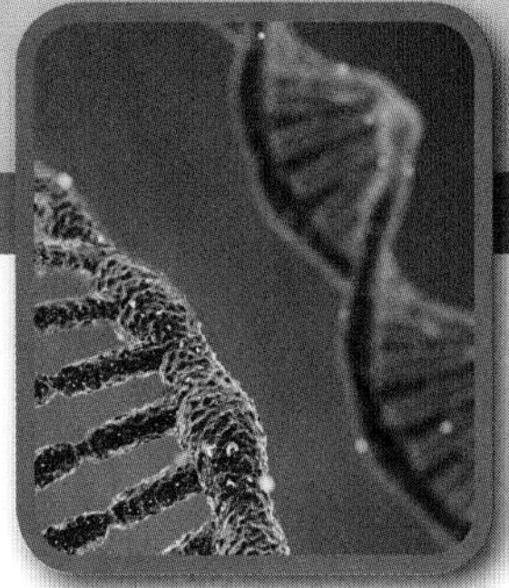

Chapter 24

Gene transcription

Information in DNA, encoded in the sequence of the four bases, is used to direct the assembly of the 20 standard amino acids in the correct sequence to produce a protein, or, in the case of nonprotein-coding genes, to produce the correct RNA sequence. A gene does not participate directly in protein synthesis; indeed in eukaryotes the DNA is enclosed inside the nuclear membrane while the protein-synthesizing machinery is outside in the cytosol, so the two never meet. How then does a gene direct protein synthesis? It does so by sending out RNA copies of its coded information to the cytosol. Three main classes of RNA are involved in protein synthesis. Ribosomal RNA (rRNA) and transfer RNA (tRNA) have specialized functions that are described in detail in Chapter 25. It is the **messenger RNA (mRNA)** that carries the sequence information of protein-coding genes to the cytosol.

Messenger RNA

The structure of RNA

RNA stands for ribonucleic acid. It is a polynucleotide essentially similar to DNA but with these differences:

- The sugar is D-ribose, not the deoxyribose of DNA. Ribose has an —OH in the 2′ position.

D-Ribose 2′-Deoxy-D-ribose

- mRNA is single stranded, not a duplex of two molecules as is DNA. mRNA is a copy of only one of the two strands of the DNA of a gene.
- Its four bases are A, C, G, and U. There is no T. U and T have identical base-pairing properties and thus they both pair with A.

Were it not for these differences, the structure of a single strand of DNA, shown in Chapter 22, could be that of single-stranded RNA. There are the same $3' \rightarrow 5'$ phosphodiester bonds between successive nucleotides.

How is mRNA synthesized?

The building-block reactants for RNA synthesis are ATP, CTP, GTP, and UTP, which are produced in all cells. In *E. coli*, all RNA is synthesized from these by a single enzyme called DNA-dependent RNA polymerase, or **RNA polymerase**. In eukaryotes three RNA polymerases exist, known as polymerases I, II, and III, or, often, as Pol I, II, or III. Eukaryotic mRNA is synthesized by polymerase II.

The synthesis first requires that the duplex DNA strands are separated to provide a single-stranded template for directing the sequence of nucleotides to be assembled into mRNA. The two strands are transitorily separated over a short sequence at the site of mRNA synthesis, and then come together again after the polymerase has passed. In effect, a separation 'bubble'

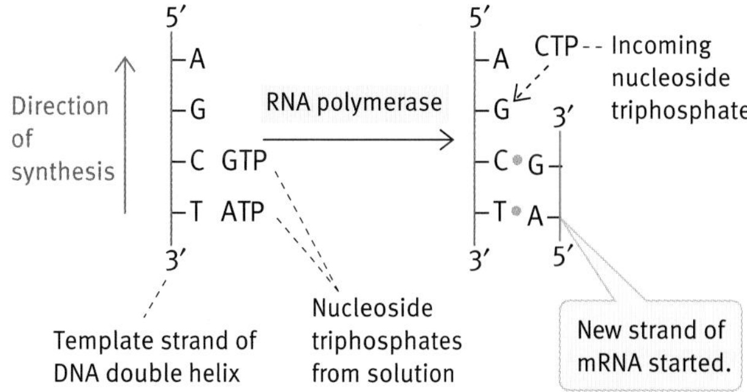

Fig. 24.1 Copying mRNA from a DNA template strand. The nontemplate strand is not shown. Note that the separation of the two strands is transitory. A bubble of DNA strand separation moves along the DNA as the polymerase progresses along it.

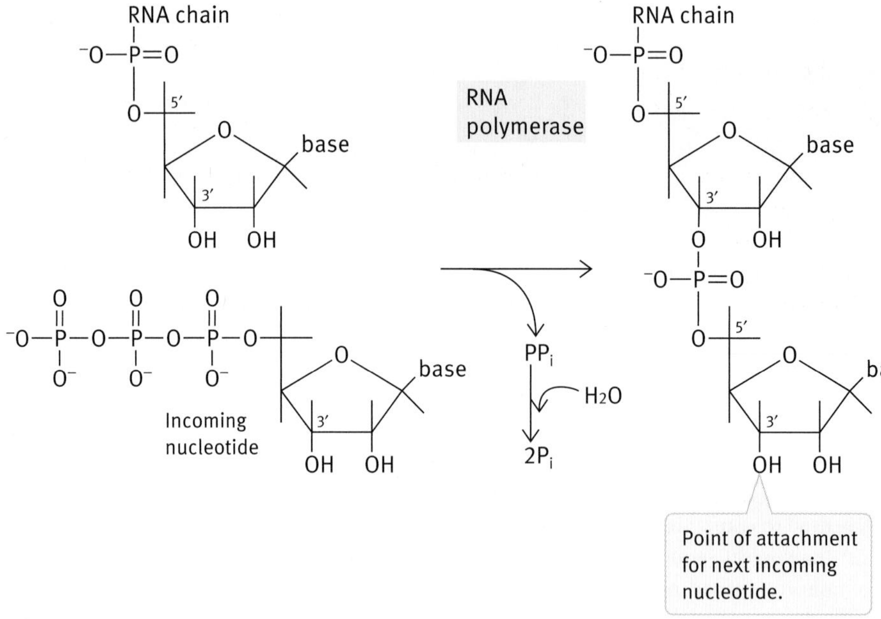

Fig. 24.2 The reaction catalysed by RNA polymerase.

moves along the DNA. The basic process of synthesis, called **gene transcription**, is much the same as in DNA synthesis in that the base of the incoming ribonucleotide is complementary to the base on the DNA template (Fig. 24.1) but, unlike DNA synthesis, only one strand is formed.

The RNA polymerase works its way along the template, joining together the nucleotides in the correct order as determined by the DNA template. Like DNA synthesis *RNA synthesis is always in the 5′→3′ direction*. That is, new nucleotides are added to the 3′-OH and so the chain elongates in the 5′→3′ direction. The template is antiparallel, running in the opposite (3′→5′) direction. The chemical reaction catalysed by the polymerase is very much like that of DNA synthesis in that it involves the attachment of the α-phosphoryl group (the first one attached to the ribose) of the nucleotide triphosphates to the 3′-OH of the preceding nucleotide, splitting off inorganic pyrophosphate (PP_i). PP_i is hydrolysed to two P_i molecules, making the reaction, shown in Fig. 24.2, strongly exergonic, again like DNA synthesis.

An important point of difference, however, is that RNA polymerase *can initiate new chains* – it does not need a primer; it can synthesize the entire mRNA molecule from the four nucleoside triphosphates, provided a DNA template is there. This is quite different from DNA polymerase, which can only elongate existing chains.

Some general properties of mRNA

In a typical chromosome there are thousands of different genes. An mRNA molecule is a copy of a single gene (or, in prokaryotes, often a small group of genes). mRNA molecules are therefore tiny in comparison to the length of chromosomal DNA, and while the cell contains a relatively small and fixed number of chromosomes, it contains large numbers of different mRNA molecules depending on which proteins are required. Multiple copies of each required mRNA are made, increasing the complexity of the cell's mRNA content.

DNA is immortal in cellular terms, but mRNA is ephemeral, with a half-life of perhaps 20 minutes to several hours in mammals and about 2 minutes in bacteria. Thus, for expression of a gene (i.e. when the protein coded for is actually being synthesized), a continuous stream of mRNA molecules must be produced. The gene, as it were, 'stamps out' copy after copy, RNA polymerase being the stamping machinery. This might seem wasteful but it gives the important benefit of permitting control of the expression of individual genes. Once mRNA synthesis ceases and the mRNA already made breaks down, synthesis of that protein stops.

In prokaryotes a single mRNA molecule may carry the coded instructions for the synthesis of several proteins: prokaryotes are said to make **polycistronic mRNA** (from the now rather historic term **cistron**, which is more or less synonymous with the term gene). Genes that are clustered on the prokaryotic chromosome often encode a set of proteins that are all involved in the same metabolic pathway. Thus, making a single polycistronic mRNA is a way of coordinating protein expression. In eukaryotic cells, an mRNA almost always specifies a single polypeptide. i.e. eukaryotic mRNA is monocistronic. However, a single eukaryotic gene may give rise to different mRNAs by alternative splicing of the primary transcript (see later), and the gene may therefore encode more than one polypeptide.

Some essential terminology

Transcription and translation

The flow of **information** in gene expression is:

(transcription) (translation)
DNA - - -→ mRNA - - -→ protein

(The broken arrows represent *information* flow, not chemical conversions. DNA cannot be converted into RNA nor RNA into protein.)

The 'language' in DNA and RNA is the same – it consists of the base sequences. In copying DNA into RNA there is transcription of the information. Hence mRNA production is called **transcription**, and the DNA is said to be transcribed. The RNA molecules produced are called transcripts, and, in the case of those which are yet to be modified, primary transcripts. The 'language' of the protein is different – it consists of the amino acid sequence the structures and chemistry of which are quite different from those of nucleic acid. The synthesis of protein, directed by mRNA, is therefore called **translation.** If you copy this page in English you are transcribing it. If you copy it into Mandarin characters, you are translating it.

Coding and noncoding strands

We have so far talked of mRNA synthesis as 'copying' the DNA. The mRNA copy that is produced has the same base sequence (apart from U replacing T) as the DNA strand that is not used as the template for transcription. Therefore, when looking at the double-stranded DNA sequence of a gene, the strand that is not used as the template is termed the **coding strand**, or the **sense** strand, while the template strand is the noncoding strand (Fig. 24.3). As in DNA replication the sequence of mRNA is determined by Watson–Crick base pairing of the incoming nucleotides with the template, so the mRNA has the reverse complementary sequence to that of the template. 'Reverse' refers to the antiparallel 5′→3′ directions of the complementary polynucleotide chains.

5′ and 3′ ends of a gene: upstream and downstream sequences

A gene has two strands of DNA of opposite polarity, and duplex DNA has therefore no intrinsic polarity. Yet we often refer to '**the 5′ end**' of the gene, and when doing so we are referring to the coding strand, which has the same 5′→3′ polarity as the mRNA.

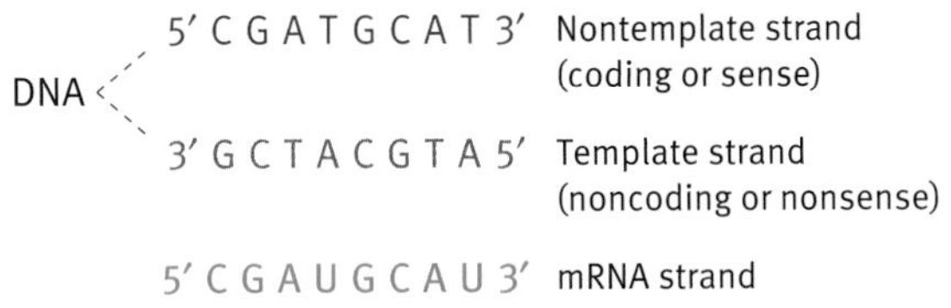

DNA
5′ C G A T G C A T 3′ Nontemplate strand (coding or sense)
3′ G C T A C G T A 5′ Template strand (noncoding or nonsense)

5′ C G A U G C A U 3′ mRNA strand

Fig. 24.3 Relationship of transcribed mRNA to template and nontemplate strands. The mRNA is the sense of the information (see text). In viruses a frequently used terminology is: template, minus (–) strand; nontemplate, plus (+) strand.

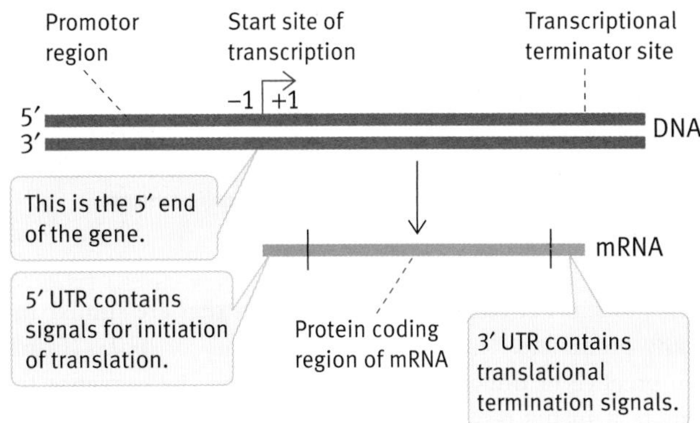

Fig. 24.4 Geography of a prokaryotic gene and its mRNA. The '5′ end' of a gene refers to the nontemplate strand or sense strand of the DNA. Note that the Figure is not to scale as the transcribed sequence of a gene is usually much longer than the promoter.

As discussed in Chapter 22, one definition of a gene is a specific sequence of DNA that is transcribed into RNA. However, DNA adjacent to the transcribed sequence plays important roles in the transcription process, and must also be considered, There is a region of DNA adjacent to the 5′ end of the gene, called the **promoter** that is essential for transcription of the gene, but is not itself transcribed into RNA. At the opposite (3′) end is a **terminator region** necessary for termination of transcription. A typical gene is illustrated in Fig. 24.4. The first template nucleotide is given the number +1 and the nucleotide 5′ to this –1 and so on. The start site is illustrated by an arrow (→) that indicates the direction of transcription. By analogy to the flow of a river, nucleotides 5′ to this are referred to as 'upstream' and 3′ to this as 'downstream'.

A final point to note from Figure 24.4 is that an mRNA molecule has sections at each end that are not translated into protein. The 5′ **untranslated regions (UTRs)** contain encoded signals necessary for initiation of translation and the 3′ UTR signals for its termination (see Chapter 25).

Apart from the basic chemistry already described, gene transcription in prokaryotes and eukaryotes are rather different processes. We will deal first with *E. coli* and then with eukaryotic cells.

Gene transcription in *E. coli*

Phases of gene transcription

There are three phases – **initiation, elongation,** and **termination**.

Initiation of transcription in *E. coli*

Initiation takes place when RNA polymerase locks on to the gene at the promoter. How does this occur? The polymerase

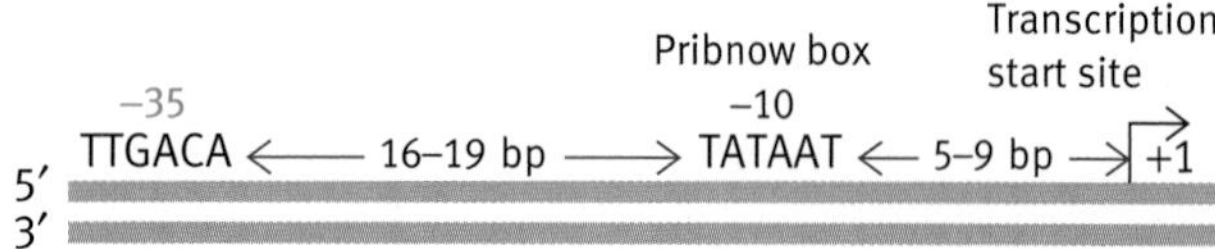

Fig. 24.5 Consensus sequences of *Escherichia coli* promoter elements.

binds to short stretches of base sequences that are similar in many promoters. These are called 'elements' or 'boxes'(from the practice of drawing a rectangle around the sequence to denote it in figures; there is no physical entity corresponding to the 'box' in the cell). In a typical *E. coli* promoter there are two boxes – the **Pribnow box** (named after its discoverer, David Pribnow), centred at nucleotide −10, and the other centred at −35. The consensus sequences of the boxes are shown in Fig. 24.5. A **consensus sequence** is obtained by comparing the sequence of, for example, the Pribnow box in a number of different genes, for they are often not exactly the same. You then look at the first nucleotide position in the box and count up which base is most often used by the different genes and so on for all the other positions. The consensus sequence itself may be found in few if any genes, but the sequences actually found will be recognizably similar with only small variations from it.

When we quote a DNA sequence we often give the sequence of only one strand, conventionally that of the coding strand in the 5′→3′ direction but, of course, in a gene there is also the second DNA strand. Thus, although we say the Pribnow box has the sequence TATAAT, it really is:

5′−TATAAT−3′
3′−ATATTA−5′.

It is the *double strand* that is recognized by the proteins required for transcription.

Correct initiation of transcription is obviously important. Synthesis of an mRNA needs to commence at the correct nucleotide on the template and on the correct strand. The −35 and Pribnow boxes are the signals for positioning the RNA polymerase. RNA polymerase of *E. coli* is a large complex of several protein subunits. The '**core**' enzyme (two copies of an α subunit, two related subunits termed β and β', and an ω subunit) has an affinity for DNA, and can even catalyse RNA synthesis from the template strand, but it cannot recognize the correct initiation site until it is joined by another protein from the cytosol, the sigma subunit (σ) or **sigma factor**. With this attached, the polymerase binds to the −35 and Pribnow boxes and initiation of transcription can start. (We remind you that, although the DNA bases are Watson–Crick paired in the centre of the duplex and largely concealed, their edges are 'visible' in the DNA grooves and can be recognized, that is to say, contacted, by proteins; see Fig. 22.4.) This aligns the enzyme in the correct starting position, and the correct orientation.

Elongation

The polymerase can now separate the DNA strands as it moves, thus making the template strand bases available for pairing with incoming bases of NTPs. The enzyme synthesizes the first few phosphodiester bonds from nucleoside triphosphates and initiation is thus achieved. At this point sigma factor protein detaches (to be used again) and the polymerase, now released, moves down the gene synthesizing mRNA. The polymerase moves at the rate of 50–100 nucleotides per second (compare with DNA polymerase at up to 1,000 nucleotides per second) and unwinds the DNA ahead. The DNA rewinds behind it forming a temporary unwound 'bubble'(about one and a half turns of the helix in length), which passes along the gene with the polymerase (Fig. 24.6).

An incorrect base in an mRNA molecule could lead to synthesis of an abnormal protein, but as mRNA molecules are short lived, and many copies are transcribed from a single gene, fidelity is not quite so crucial in transcription as it is in DNA replication. Nevertheless, RNA polymerase does carry out some proofreading, as it pauses at and can remove the latest nucleotide added if it is incorrect.

Termination of transcription

At the end of many transcribed prokaryote genes are sequences that result in the transcribed RNA having a **stem loop structure**. This needs to be explained. Although mRNA is single stranded, it maximizes base pairing within itself to achieve the lowest free energy. The newly synthesized mRNA is attached to the DNA template strand by base pairing but only for about 12 nucleotides because the polymerase deflects the mRNA from the template. Once free from the template, the newly formed mRNA can undergo internal base pairing. Near the end of the gene, the sequence of bases in the mRNA produced is such that the stem loop structure shown in Fig. 24.7 forms because the base sequence here permits formation of G—C pairs, a stable structure due to triple

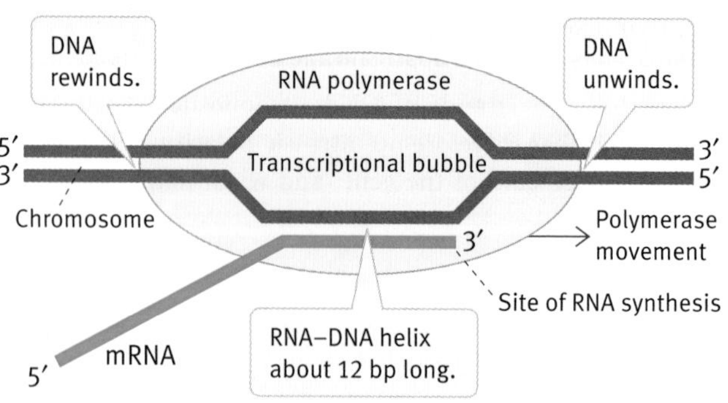

Fig. 24.6 DNA transcription by *Escherichia coli* RNA polymerase. The polymerase unwinds a stretch of DNA about 17 base pairs in length forming a transcriptional bubble that progresses along the DNA. The DNA has to unwind ahead of the polymerase and rewind behind it. The newly formed RNA forms an RNA–DNA double helix about 12 base pairs long.

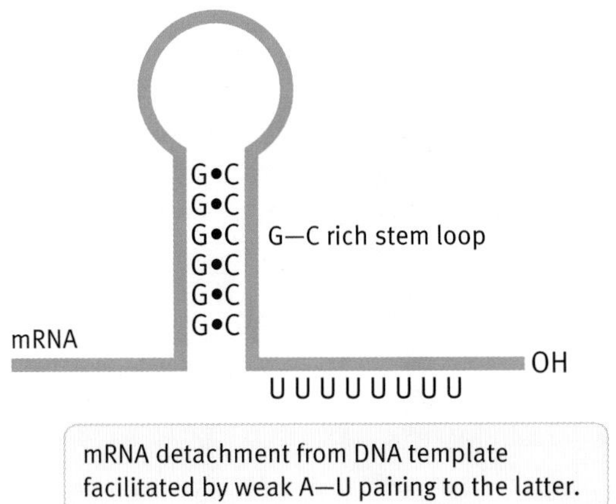

Fig. 24.7 The stem loop structure of an RNA transcript involved in the Rho-independent termination of gene transcription (see text).

hydrogen bonding between G and C. The stem loop structure somehow disrupts the elongation process. Perhaps it prevents binding of the mRNA to the template at this point, since, if its bases are preferentially internally paired, they cannot pair with the template DNA. Immediately following the stem loop structure in the mRNA transcript is a string of around 8 U residues, giving weak bonding of the RNA to DNA because of weaker double A—U hydrogen bonding. This facilitates complete detachment of the mRNA and hence terminates transcription.

There is an alternative method of termination of transcription in many prokaryote genes. This requires an additional protein called the **Rho** factor, which attaches to the newly transcribed mRNA and moves along it behind the RNA polymerase. At the termination site, the polymerase pauses, possibly because of a difficult to separate G—C-rich sections of the DNA, and this pause allows the Rho factor to catch up with the polymerase. The Rho factor has an unwinding (helicase) activity for unwinding the RNA–DNA duplex formed by transcription. ATP breakdown is involved. Unwinding releases mRNA and terminates transcription. Rho dependent termination ensures that the mRNA ends, appropriately, after the end of a protein-coding sequence. In *E. coli* the mRNA starts to direct protein synthesis before the full mRNA molecule is completed because transcription occurs in contact with the cytosol. In fact, the ribosomes follow closely behind the polymerase and can thus block Rho factor from binding mRNA. After the end of the protein-coding sequence is reached the mRNA is naked of ribosomes, allowing Rho to bind. Furthermore, if for some reason translation cannot keep up with transcription, for instance if the cell is starved of amino acids so that ribosomes are moving slowly along a message, Rho can stop the cell wasting energy as it binds untranslated mRNA and by terminating transcription prevents further synthesis of unusable message.

The rate of gene transcription initiation in prokaryotes

It is clearly important for cells to express genes only where and when their products are needed. We are mostly saving for Chapter 26 the question of how gene expression is selectively controlled. Many genes, however, are said to be **constitutively expressed** as they are 'switched on' all the time. Such genes code for enzymes and other proteins that are needed at all times and in amounts that do not vary from time to time. However, among these constitutive proteins some will be required in larger amounts than others. The major control on the level of gene expression in bacteria is the rate of mRNA production and this is largely determined by the frequency of initiation of transcription of a given gene. This varies because genes have promoters of different 'strengths'. A 'strong' promoter will initiate transcription frequently and cause many mRNA transcripts of the gene to be made and hence a lot of the specific protein. A 'weak' promoter has the reverse effect. The strength of a promoter is a function of the precise base sequence of the Pribnow and –35 boxes, the distance between them, and the nature of the bases in the –1 to –10 region. The greater the affinity of these regions for the polymerase, the stronger the promotion, though this may not be the sole determinant.

Control of transcription by different sigma factors

One mechanism for controlling selective gene transcription in prokaryotes will be described in this chapter, because it depends on a factor that is intrinsic to the basic process of transcription, the sigma factor. Under certain conditions, the usual sigma protein (termed σ^{70} for its molecular mass, or σ^{D}) is replaced by one that causes the RNA polymerase to initiate at a different set of genes that have different promoter sequences. This is often a reaction to environmental stress, for instance after a heat shock (a sudden rise in temperature) *E. coli* transitorily increases the synthesis, stability, and activity of a sigma protein, σ^{32} or σ^{H}, that normally is present at a nonfunctional level. This factor directs the transcription of genes for a set of 'heat-shock proteins' that protect the cell against the consequences of the heat shock. Other specialized sigma factors similarly direct transcription of groups of genes that are required for specific functions, such as cell motility, or to respond to particular environmental stimuli.

Gene transcription in eukaryotic cells

Eukaryotic RNA polymerases

The basic enzymic reaction by which RNA is synthesized in eukaryotes is the same as in prokaryotes (see Fig. 24.1).

However, in eukaryotes there are three different RNA polymerases, designated I, II, and III, which are responsible for transcribing different classes of gene.

- RNA polymerase I (Pol I) transcribes the major rRNA transcript.
- RNA polymerase II (Pol II) transcribes mRNA.
- RNA polymerase III (Pol III) transcribes 'small' RNAs: tRNAs, 5S rRNA and small nuclear RNAs (snRNAs), a class of nonprotein-coding RNA molecules with varied functions such as RNA splicing (see Split genes and RNA splicing later in this chapter).

We will refer to transcription by Pol I and Pol III later in this chapter, but our main focus will be on Pol II, as it transcribes essentially all protein-coding genes.

Gene expression in eukaryotes is a highly regulated process. A key difference from prokaryotes is that the mRNA is not translated as soon as it is produced. Instead *the primary transcript is greatly modified before it is a functional mRNA.* The major modifications are:

- addition of a modified 'cap' nucleotide to the 5′ end
- splicing to remove introns
- addition of a polyA tail to the 3′ end.

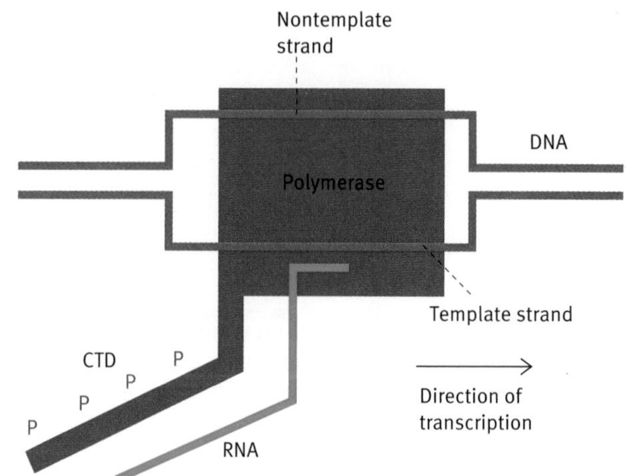

Fig. 24.8 Diagrammatic representation of RNA polymerase II transcribing a gene. It is designed to represent the enzyme melting the DNA as it progresses; the template strand is in a groove of the enzyme containing the active centre while the other strand takes a separate path. The two reassociate behind the enzyme. The RNA forms a hybrid with the template strand for eight or nine bases but is then diverted to a separate exit from the enzyme in the direction of the phosphorylated C-terminal domain (CTD). It has been postulated that the latter has enzymes attached for capping and splicing the RNA transcript as it is synthesized, as well as packing the mRNA for transport into cytosol.

Each of these modification steps is described in more detail later in this chapter. However, capping and splicing occur simultaneously to transcription so Pol II has to accommodate this. Eukaryotic **RNA polymerase II**, is a large multisubunit enzyme. The major subunits are homologous to those of the core prokaryotic RNA polymerase, but the largest subunit has a feature not found in prokaroyotes, an elongated **carboxy terminal domain (CTD)** containing multiple repeats (52 in mammals) of the seven amino acid sequence Tyr—Ser—Pro—Thr—Ser—Pro—Ser. The CTD of Pol II is involved in regulating transcription and recruiting factors that cap and splice the primary transcript as transcription continues. It also recruits factors involved in adding the polyA tail. The CTD is itself subject to regulation through phosphorylation of the repeated serine residues.

During transcription, the template strand of the DNA is positioned in a groove in the polymerase with the nontemplate strand outside. When transcription begins, the first eight nucleotides of the RNA transcript forms a duplex with the template strand after which a lobe of the enzyme diverts the transcript along a groove so that it exits in the direction of the CTD. The template DNA strand exits behind the main body of the polymerase and then reassociates with the nontemplate strand (Fig. 24.8). The groove of the enzyme in which the template strand resides is a claw-like structure, which closes and firmly attaches the DNA to the enzyme. This is essential because the gene transcribed may be enormous in length, and if it prematurely detached, there is no mechanism for it to reattach. One function of the phosphorylated CTD is to position processing enzymes so that they can act on the emerging RNA transcript.

Eukaryotic RNA polymerase must somehow negotiate histones, as these are not totally stripped away from the DNA template before transcription. However, the mechansism by which this occurs is not yet understood.

How is transcription initiated at eukaryotic promoters?

As in prokaryotes, eukaryotic genes have promoter sequences at their 5′ ends, However, there is an added layer of complexity due to the organization of eukaryotic genes in tightly packed chromatin. For transcription of a particular gene to begin, the DNA sequence must be made accessible to RNA polymerase and other proteins through localized 'opening' of the chromatin. As this is an aspect of gene regulation, further discussion of the process of chromatin opening will be left until Chapter 26. For now we will assume that the promoter sequences are available for binding.

Type II eukaryotic gene promoters

Figure 24.9 shows the typical DNA components of a type II promoter. There are two sections.

The term **basal elements** refer to the **initiator (Inr)** , a short pyrimidine-rich sequence at the start site, and the **TATA box**,

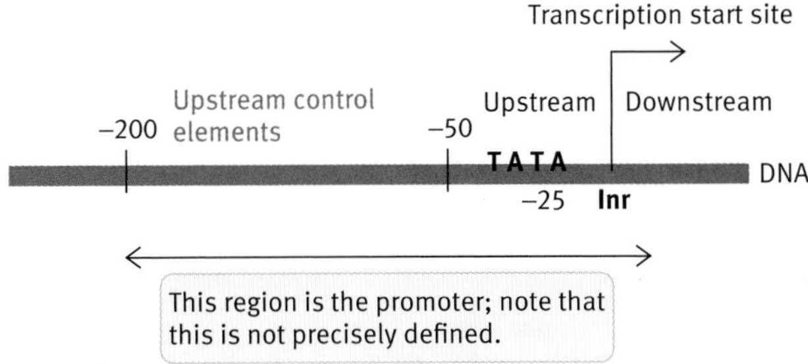

Fig. 24.9 Eukaryotic type II gene control elements. Examples of upstream common control elements are the CAAT box and the GC box. Inr, initiator (a pyrimidine-rich stretch on one of the strands).

centred usually around −25 base pairs from the start site, with a consensus sequence TATAAAA reminiscent of the Pribnow box of prokaryotes. Many, but not all, type II genes have the TATA box. Where there is no TATA box, the Inr and a short section called the **downstream promoter element (DPE)** position the polymerase. The DPE is about −25 base pairs downstream from the start site.

The **upstream control elements** are found at variable positions within the range of about −50 to −200 or so base pairs. Those most commonly found are the CAAT (pronounced CAT) box with a consensus sequence GGCCAATCT, and the GC box (GGGCGG). Eukaryotic genes usually have at least one of such common control elements but different gene promoters are quite different with respect to which are present. In addition to the CAAT and GC boxes there may be any number of additional control elements, which vary greatly from gene to gene as they are concerned with the complexities of gene regulation in a multicellular organism.

Although the TATA box and Inr are analogous to the Pribnow and −35 boxes in prokaryotes there is a difference in that they are not recognized by the RNA polymerase itself. In eukaryotes a set of **general transcription factors** are required to position Pol II correctly on the promoter and open up the DNA duplex. These factors are designated TFIIA, TFIIB etc, where II denotes a transcription factor required for Pol II-transcribed genes, The exact functions of all the TFII proteins and the sequence of events when the **basal initiation complex** of general factors and RNA polymerase is assembled on the DNA is not known for certain: the order of assembly may vary. However, a large complex called **TFIID** is a key component in committing a gene to transcription. The heart of the TFIID complex is the **TATA-box-binding protein (TBP)**, which attaches TFIID to the TATA box. The other components are known as **TAFs** (**TBP**-Associated Factors; Fig. 24.10). TBP binding causes localized bending and distortion of the DNA, which makes it recognizable by other factors. RNA polymerase joins up to the TFIID complex on the TATA box and several other general initiation factors bind, such as TFIIB which plays a role in linking up the polymerase to the TFIID. The position of the TATA box relative to the start site

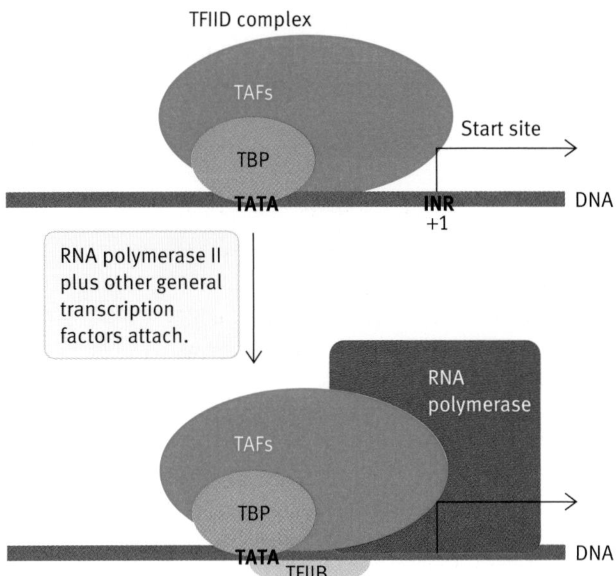

Fig. 24.10 Diagrammatic representation of components of the basal initiation complex. TFIID is a complex of the TATA-box-binding protein (TBP) and a number of TAFs (TBP-associated factors). RNA polymerase attaches to the preinitiation complex and forms the basal transcription complex.

means that TFIID ensures that Pol II is positioned at the start site and pointing in the right direction. In genes that do not have a TATA box it is the **Inr** and the **DPE** that position the basal transcription machinery.

Elongation of the transcript requires Pol II modification

After the initiation complex is formed, Pol II must be able to travel along the DNA template. For this to occur one of the general transcription factors TFIIH, which has protein kinase activity, phosphorylates serines in the CTD of Pol II. This enables Pol II to leave the initiation complex and bind a new set of factors associated with elongation and processing of the RNA transcript. Many of the initiation factors are removed from RNA polymerase and the promoter, but TFIID remains bound to the TATA box, thus facilitating a further round of initiation. TFIID will leave the promoter only after the need for transcription is over.

Capping the RNA transcribed by RNA polymerase II

The RNA of the eukaryotic gene primary transcript immediately undergoes a modification at its 5′ end, called **capping**. At the 5′ end of the primary RNA transcript there is a triphosphate group because the first nucleotide triphosphate simply accepts a nucleotide on its 3′-OH. The terminal phosphate of this is removed and a GMP residue is added from GTP (Fig. 24.11). The 5′–5′ triphosphate linkage is unusual. The G

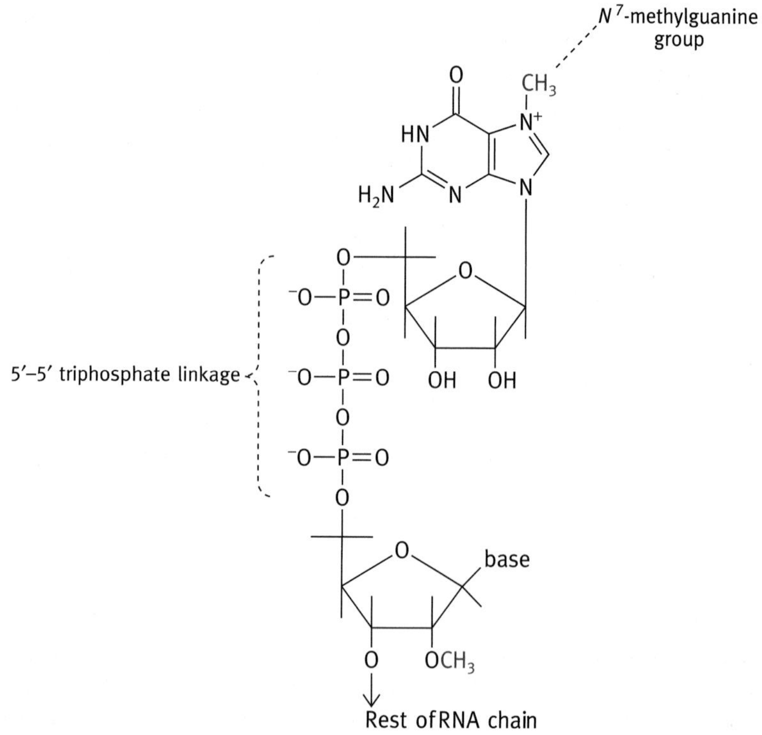

Fig. 24.11 Structure of the 5′ cap in eukaryotic mRNA. The terminal nucleoside triphosphate of the primary RNA transcript is converted to a diphosphate followed by a reaction with GTP in which pyrophosphate is eliminated. This is followed by methylation reactions. (A third methyl group may be added to the 2′-OH of the next nucleotide of the primary transcript.) The capped primary transcript is then processed to mRNA – see text for details.

is then methylated in the N-7 position as is also the 2′-OH of the second nucleotide. The cap protects the end of the mRNA from exonuclease attack and it is involved in initiation of translation as described in the next chapter.

This is not the only modification of the primary transcript, because many eukaryotic genes are split genes.

Split genes and RNA splicing

The DNA coding for a given protein is split up into several parts, linked together by intervening stretches that do not code for amino acid sequences. The latter sections with no protein-coding content are called **introns**, and the coding stretches, **exons**. There can be 1–500 or so introns in a gene (Fig. 24.12(a)), which can typically vary in length from about 50 to 20,000 base pairs (or sometimes longer). Exons vary in size but usually are smaller than introns, around 150 base pairs in length on average. The primary transcript is processed to eliminate the introns and link together the exons into one mRNA molecule. This is known as **RNA splicing** (Fig. 24.12(b)).

Mechanism of splicing

Removing the unwanted RNA introns of a primary transcript and joining up the exons into mRNA looks a formidable task but the key to it is the **transesterification** reaction. In this, a phosphodiester bond is transferred to a different —OH group. There is no hydrolysis and no significant energy change during the bond rearrangements.

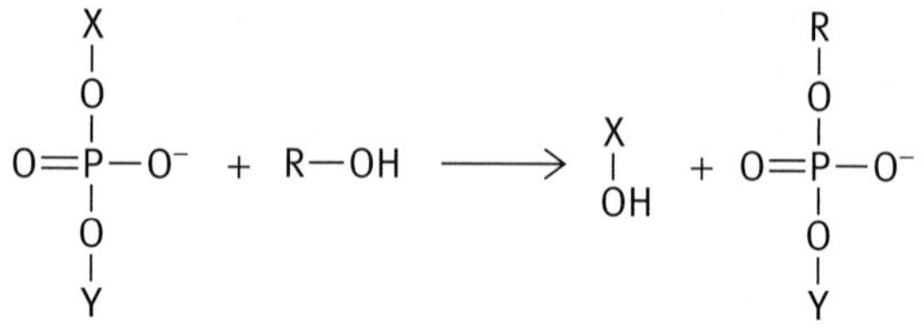

$$X{-}O{-}P(=O)(O^-){-}O{-}Y + R{-}OH \longrightarrow X{-}OH + R{-}O{-}P(=O)(O^-){-}O{-}Y$$

If X–Y is an RNA chain, it would be broken.

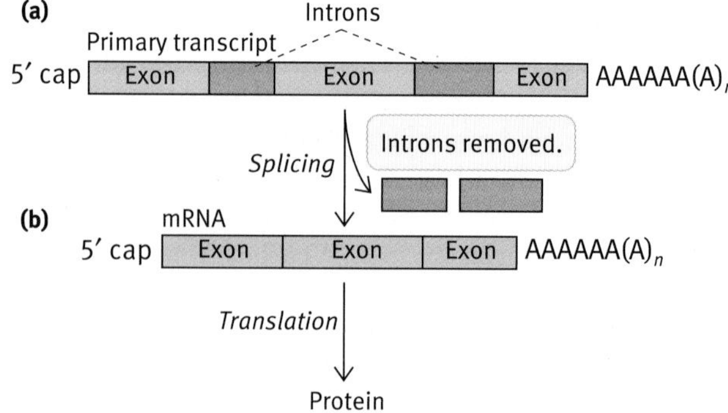

Fig. 24.12 Primary polymerase II transcript of a eukaryotic gene: **(a)** with introns after capping and addition of polyA tail; **(b)** excision of introns to form the mature mRNA is called splicing. Note that in reality intron sequences are often much longer than exons.

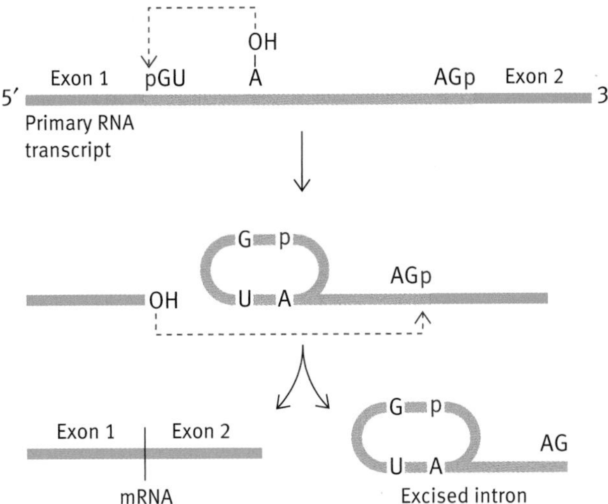

Fig. 24.13 Mechanism of RNA splicing.

In RNA splicing, the exon–intron junctions are 'marked' by consensus sequences called the **5′ and 3′ splice sites**; all introns begin with GU and end with AG (though the full consensus sequences are longer than this). The ROH of the diagram in Fig 24.13 is the 2′-OH of an adenine nucleotide in a short sequence (seven bases long in yeast) of the intron chain, known as the **branch site**. The reason for this name will be seen from the structure in Fig. 24.13. The 2′-OH group attacks the 5′ phosphate of the G nucleotide at the splice site, forming a lariat structure. This breaks the chain at the 3′ end of exon 1, thus producing a free 3′-OH and the 3′-OH attacks the 5′ end of exon 2, joining the two exons and releasing the lariat.

In most cases, the splicing reaction in the nucleus is catalysed by very complex protein-RNA structures called **spliceosomes**. They are complexes of about 300 different proteins and also five RNA molecules, 100–300 bases long in higher eukaryotes, called **small nuclear RNAs (snRNAs)**. These are associated with proteins in structures known as **small ribonucleoprotein particles (snRNPs**, pronounced 'snurps'), each containing multiple protein subunits. There are five snRNPs in a spliceosome, known as **U1**, **U2**, **U4**, **U5**, and **U6**, (see **Rino** review in the Further reading online). Additional proteins known as **splicing factors** are also needed. The U1 and U2 snRNPs bind by base pairing to the 5′ splice site and the branch site respectively and then associate with each other. A trimer of U4, U5, and U6 then associates with the complex to form the spliceosome.

The splicesome undergoes various rearrangements during different phases of the splicing reaction. These involve changes in the base pairing of snRNAs with the transcript and with each other. Several of the interactions and rearrangements require ATP hydrolysis, so although the transesterification itself does not involve energy changes, splicing is an energy requiring process. U6 is the snRNP that actually catalyses the transphosphoesterification reactions but initially its catalytic centre is masked by U4. Release of U4 allows U6 to displace U1 from the 5′ splice site and base pair with U2, this linking the 5′ and 3′ ends of the intron and forming the active site of the splicesome. It is notable that the active site is formed by RNA.

Faulty splicing can lead to genetic diseases. In **β-thalassaemia** (Box 4.3), the β subunit of haemoglobin is not produced in normal amounts, because the G at the 5′ splicing sequence of an intron is mutated to an A, and primary transcripts are therefore not properly processed to mRNA.

Alternative splicing or two (or more) proteins for the price of one gene

There is one well-established advantage that split genes confer – alternative splicing. In typical splicing, all the exons of the primary RNA transcript are linked together to form the mature mRNA leading to the formation of a single protein. However, there are many known cases where the splicing can occur in different patterns so that a particular group of exons form one mRNA, whereas a different group from the same gene transcript forms another leading to different proteins. The mechanism is often employed to produce variant forms (isoforms) of a protein required in different tissues or at different times, such as occurs in antibody production (see Chapter 33). A simple example is that of the human calcitonin gene, which encodes a transcript that is spliced differently in thyroid and neuronal cells, producing either calcitonin or calcitonin gene-related peptide (CGRP) (different peptide hormones with distinct functions). Alternative splicing may partly explain why the human genome has a surprisingly small number of protein-coding genes, since it seems that most human primary transcripts can be alternatively spliced.

Ribozymes and self-splicing of RNA

The spliceosome is, as described, a complex with scores of protein components and several RNA components and a very elaborate mechanism. The biochemical world was shocked when it was discovered in the 1980s that a few RNA gene transcripts accurately self-spliced without any help from proteins. It was the first case known of a specific biochemical reaction occurring as the result of catalytic activity brought about by a macromolecule other than a protein. In the protozoan *Tetrahymena*, one of the rRNAs is made as a precursor transcript containing an intron which has to be spliced out to produce the mature rRNA. The two exons on either side of the intron become joined together by the splicing to form the mature molecule. The mechanism of the self-splicing in *Tetrahymena* is shown in Fig. 24.14. No protein is required. Guanosine or a guanosine nucleotide is needed as a co-factor, The guanosine 3′-OH (G–OH) attacks the phosphodiester bond thus releasing the 5′ exon with a 3′-OH group. The latter now makes an attack on the second phosphodiester

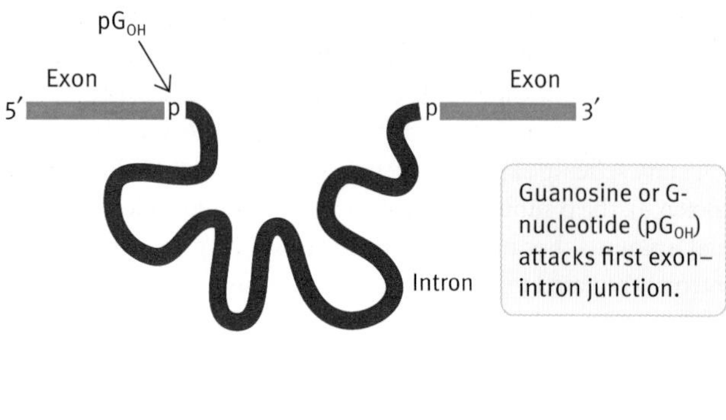

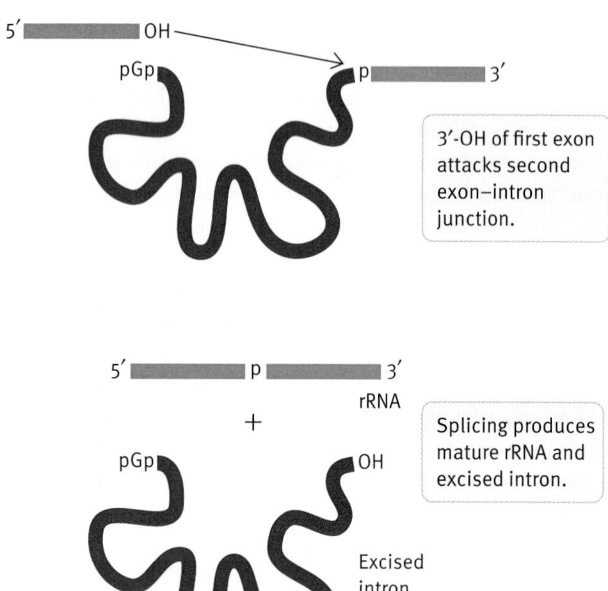

Fig. 24.14 Mechanism of the self-splicing reaction of the *Tetrahymena* ribosomal RNA (rRNA) precursor. G_{OH} represents guanosine, GMP, GDP, or GTP.

bond releasing the intron and splicing the two exons. The chemistry of the two reactions involved is as illustrated in the diagram of transesterification (see Mechanism of splicing section). The internal sequence of the intron is important as it folds to form a three dimensional structure that provides a binding site for the guanosine, and also positions the exons by base pairing.

This self-splicing is not a true catalytic reaction because the molecule itself is changed. However, RNA can also act as a true catalyst. An enzyme called **ribonuclease P**, discovered in *E. coli* but subsequently also found in eukaryotes, processes tRNA precursors by a specific hydrolytic reaction. This enzyme has an RNA component attached to a protein, but the RNA by itself is capable of catalysing the hydrolysis. Because of its similarity to an enzyme, the term **ribozyme** was coined. A number of ribozymes are known but they are not common and seem to occur for the most part unpredictably. Most spectacularly, a reaction of central importance in protein synthesis is catalysed by an RNA component of the ribosomes (see Chapter 25) of all species so far as is known. It seems likely that ribozyme-catalysed reactions are relics that have been retained from the 'RNA world' before proteins existed.

As they can be targeted to their substrates by base pairing, ribozymes can be engineered to cleave specific RNA sequences and thus they have potential as therapeutic agents, for example against viral infection.

Termination of transcription in eukaryotic cells: 3′ polyadenylation

Termination of transcription in eukaryotic cells is less understood than the prokaryotic mechanism. Most, though not all, eukaryotic mRNAs end in a string of up to 250 adenine residues known as a 3′ polyA tail. The polyA tail is not directly encoded by the gene, but its position is directed by a polyadenylation signal (AAUAAA) that is encoded by the gene and transcribed in the primary transcript. The polymerase transcribes a short distance beyond the polyadenylation signal and then, somehow, terminates. The RNA is cleaved near the polyadenylation signal by a specific endonuclease. Then polyA polymerase, which is not dependent on a DNA template, uses ATP as the source of adenine nucleotides to form a polyA tail. The polyA tail increases the stability of the mRNA and increases the efficiency of translation.

The phosphorylated CTD of RNA polymerase II is involved in each stage of mRNA processing, because it sequentially brings the factors involved in 5′ capping, splicing, and polyadenylation to the transcript. After transcription is completed, the CTD is dephosphorylated and the polymerase returns to the promoter for another round of transcription. Mature, processed mRNAs are exported from the nucleus through the nuclear pores (see Chapter 27).

Editing of mRNAs

In a few cases, mRNAs are 'edited' after transcription so the final coding sequence of the mRNA is not exactly that specified by the DNA template. Although not nearly as common as alternative splicing, RNA editing can add yet again to the diversity of proteins encoded by the human genome. For example the gene for apolipoprotein B has two mRNA transcripts. One of these has a DNA-coded C enzymatically deaminated to a U to form a translational stop codon (see next chapter). Thus, two proteins with different roles in lipid transport and metabolism are encoded by the same gene. Other forms of mRNA editing include deletion of insertion of nucleotides. For instance in the mitochondria of trypanosomes (parasitic protozoans that can cause disease), mRNA transcripts are produced that have several extra U residues inserted at specific places to produce the correct coding sequence for proteins.

Transcription of nonprotein-coding genes

Nonprotein-coding genes that are transcribed by eukaryotic RNA polymerase I and III encode RNA molecules that are stable end products with specific cellular functions. Most of the RNA content of a cell is ribosomal RNA (rRNA), which forms the core of ribosomes. As it is needed in large quantities, the eukaryotic genome contains multiple copies of the rRNA-encoding genes. There are four different rRNA components in eukaryotic ribosomes (three in *E. coli* ones); these are described in the next chapter. Three of the four rRNAs are transcribed from a single gene giving a large primary transcript, which is subsequently processed by cleavage into the mature rRNA molecules. This major rRNA precursor transcript is synthesized by Pol I. The smallest rRNA molecule (5S rRNA) is transcribed from a separate cluster of genes by Pol III, which is also responsible for synthesis of other 'small' RNAs i.e. tRNAs and snRNAs including those involved in the splicesome.

Pol I and Pol III transcribed genes have characteristic promoter sequences that are different from those of protein coding genes. The polymerases each require their own set of transcription factors, but both are guided to the correct position on the template DNA by TBP, the TATA- binding protein. In this case TBP is not part of the TFIID complex: it is recruited to the promoter by different mechanisms that do not rely on TATA box recognition.

RNA transcripts synthesized by Pol I and Pol III undergo extensive processing, including cleavage of the molecule (in tRNA by the ribozyme ribonuclease P) and chemical modification of certain bases. However, they are not capped and polyadenylated. Although Pol I and III have homology with prokaryotic RNA polymerase and Pol II they lack the CTD and hence lack the site where mRNA capping and polyadenylation occur.

Gene transcription in mitochondria

Mitochondria are replicating organelles of eukaryotic cells, with their own DNA and protein-synthesizing machinery (the same is true of plant chloroplasts). The mitochondrial genome was discussed in Chapter 22. Mitochondrial genes are transcribed by a special monomeric (i.e. nonmulti-subunit) RNA polymerase, which is encoded by the nuclear genome. In mitochondria *both* strands of the DNA are completely transcribed into single long transcripts (the polymerase molecules moving in opposite directions). The single primary transcripts are then processed into mRNAs, tRNAs, and rRNAs. Mitochondrial transcription has a mixture of prokaryotic and eukaryotic characteristics: in mammalian mitochondria the mRNAs are polyadenylated (eukaryotic) but not capped (prokaryotic), and there are no introns in the genes (prokaryotic).

Summary

For genes to be expressed (to cause production of a protein), one strand of the DNA of a gene, the template strand, is copied (transcribed) into single-stranded RNA. The coding region of a prokaryotic gene is a continuous stretch of DNA so that the primary transcript is a messenger RNA (mRNA), while in eukaryotes the primary transcript is processed to form mature mRNA.

Each gene has adjacent to its 5′ end a promoter region, which controls its transcription at the level of initiation. While the basic mechanism of RNA synthesis is the same in both prokaryotes and eukaryotes, the mechanism of initiation and its control are different.

In prokaryotes, RNA is transcribed by a single RNA polymerase using ATP, CTP, GTP, and UTP as building blocks. RNA polymerases do not require a primer. The promoter contains specific DNA sequences known as the Pribnow and −35 boxes. The RNA polymerase directly attaches at these sites, which position the enzyme correctly. The rate of mRNA production is determined by the affinity of the polymerase for the promoter. As the polymerase moves along the DNA it separates the two DNA strands forming a transcription bubble and copies the template strand into RNA. Termination of transcription at the end of the coding region of the gene can occur in either of two ways. A stem loop in the mRNA formed by G–C base pairing followed by a string of U residues causes detachment from the DNA template. Alternatively a Rho factor, which is a helicase, unwinds the mRNA from the DNA, causing detachment.

In *E. coli*, groups of genes called operons often occur and are transcribed together forming polycistronic messengers, while in eukaryotes mRNA is monocistronic.

Eukaryotic transcription is carried out by three RNA polymerases: Pol II transcribes protein coding genes, while Pol I transcribes a large rRNA precursor and Pol III transcribes 'small RNAs' (5S rRNA, tRNAs and snRNAs). Transcription by Pol II depends on transcription factors such as TFIID. TFIID binds the TATA box in the promoter and positions the polymerase on the DNA.

In eukaryotes the protein-coding region of most genes is split into exons separated by noncoding introns. The primary transcript is processed as transcription proceeds. It is 'capped' at the 5′ end by addition of a methylated GMP moiety. The transcript is spliced by spliceosomes to remove introns and link the exons into a continuous mRNA molecule. Alternative splicing of the transcript may produce different mRNAs, and hence a single eukaryotic gene can encode more than one protein. Most eukaryotic transcripts have a polyA tail added at the 3′ end. Some are edited after transcription. The mature mRNAs are transported into the cytosol *via* pores in the nuclear membrane.

In rare cases, transcribed RNAs splice themselves without the aid of proteins, instead using catalytic RNAs (ribozymes), which were first discovered through their role in this self-splicing process.

Ribosomal and transfer RNAs essential for protein synthesis are transcribed by Pol I and Pol III. The transcripts undego extensive processing but are not capped and polyadenylated. The small circular genome of mitochondria is transcribed by a dedicated RNA polymerase that is encoded by the nuclear genome. Mitochondrial transcription has a mix of prokaryotic and eukaryotic characteristics.

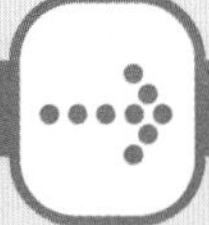

Further reading

To access the further reading, please scan the QR code image or go to http://global.oup.com/uk/orc/biosciences/molbiol/snape_biochemistry/student/reading/ch24/

Problems

1 In what ways does RNA synthesis differ from DNA synthesis?

2 By means of notes and diagrams, describe the components of an *E. coli* single gene and its associated flanking regions.

3 Describe the process of initiation of transcription of a gene in *E. coli*.

4 Describe two methods by which, in *E. coli*, gene transcription is terminated.

5 What factors determine, in a constitutive gene of *E. coli*, the strength of a promoter?

6 Describe, in broad terms, the main ways in which the formation of mRNA in eukaryotes differs from that in prokaryotes.

7 By means of a diagram, explain the mechanism of splicing. What possible biological significance does the existence of introns have?

8 In what ways does eukaryotic initiation of transcription differ from that in prokaryotes?

9 Discuss the two viewpoints on the nature of introns.

10 What are proteomes and genomes? Are they fixed in size?

11 There are estimated to be about 21,000 genes in humans. Does this mean that the maximum number of human proteins is also about 21,000?

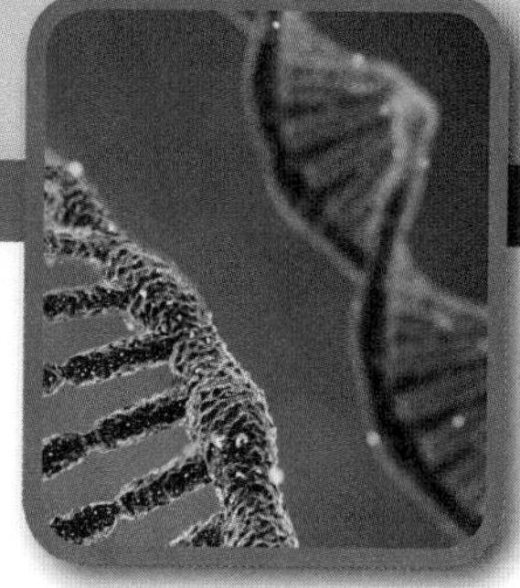

Chapter 25

Protein synthesis and controlled protein breakdown

In Chapter 24 we dealt with the production of messenger RNA (mRNA). In this chapter we deal with the way in which mRNA directs the assembly of amino acids into the finished protein coded for by the gene from which the mRNA was transcribed. This process is known as **translation** since the 'language' of nucleic acid bases is translated to the 'language' of protein amino acids. Amino acids are brought to the mRNA by adaptor tRNA molecules, which recognize the codon sequences in the mRNA by base pairing.

Translation is a highly ordered process. It is organized on **ribosomes**. These are complex RNA and protein structures present in large numbers in the cytosol and also in mitochondria and chloroplasts (since these organelles synthesize proteins encoded by their own genomes). An *Escherichia coli* cell has about 20,000 ribosomes, constituting about 25% of its dry weight. Prokaryotic ribosomes and those in mitochondria and chloroplasts are smaller than eukaryotic ones, with about 55 proteins and >80 proteins respectively. However, their structures are basically similar, each having a small and a large subunit, each made of many molecules.

Ribosomes are the molecular machines that move along mRNA and assemble the amino acids into the sequences that constitute proteins. Ribosomes work quickly and accurately; an *E. coli* ribosome adds up to 20 amino acid residues per second at 37 °C with about one mistake in the addition of 10^4 amino acids. Although an incorrect amino acid residue in a protein could mean a nonfunctional molecule an error does not have long-term consequences, as in DNA synthesis, because the faulty protein is simply destroyed. Evolution seems to have resulted in an error rate that allows most proteins of a few hundred amino acids to be mainly error-free, with an acceptable error rate in the largest proteins. Greater accuracy might have made protein synthesis too slow.

We have earlier referred to the 'RNA world', now believed to have predated the DNA world. Ribosomes were probably originally only RNA, and proteins were a refining addition later in evolution. As pointed out earlier, one of the essential properties of nucleic acid molecules is that they can, due to specific base pairing, direct their own replication. A 'useful' RNA molecule could have been directly replicated in primitive circumstances but nothing analogous to this is known for proteins. While protein-containing ribosomes today synthesize proteins, if this were the case at the origin of life a logical impasse arises – how can a structure dependent on proteins be required for the synthesis of proteins? An ancient RNA-only ribosome explains this.

Protein synthesis is basically the same in prokaryotes and eukaryotes but there are sufficient differences to require some separate treatment. We will first outline the concepts that apply to both, then deal with protein synthesis in *E. coli*, followed by a discussion of differences in eukaryotic systems.

At the end of the chapter we will deal with the targeted breakdown of proteins. It may seem that the breakdown of proteins is a mundane subject analogous to the simple digestion of proteins. Nothing could be further from the truth, for targeted breakdown inside cells is fundamental to many aspects of cellular mechanisms, including the control of cell division in eukaryotes. The mechanisms involved are elegant and have been conserved throughout much of evolution. The key player in this endgame is a cellular structure known as the **proteasome**, yet another astonishing molecular machine.

Essential basis of the process of protein synthesis

mRNA is a long molecule with four different bases in its component nucleotides; A, U, G, and C. Proteins are synthesized from 20 different amino acids (ignoring a relatively rare additional one – selenocysteine; see later in this chapter). They are listed in Table 4.1. The sequence of bases in the messenger specifies the sequence of amino acids in the protein. A one-base code (in which a single base represents an amino acid) could code for only four amino acids; a two-base code could code for 16; still not enough. A three-base code will do it with some excess capacity left over, and this is the situation.

A triplet of bases called a **codon** represents each amino acid on the RNA. With four possible bases at each site, 64 different triplets or codons (4×4×4) are possible.

The genetic code

The identification of which codon triplets correspond to which amino acids constitutes the **genetic code**. The code is virtually universal, with exceptions that in mitochondria, derived from symbiotic early prokaryotes, there are one or two minor variations with corresponding variations in their translation apparatus and the same is true of some protozoans.

Three codons have been reserved as 'stop' signals, which indicate to the protein-synthesizing machinery that the protein is complete. These three codons (UAA, for example, is one) have no amino acids assigned to them in the genetic code. If, by chance, a mutation produces a stop signal in an mRNA coding region (known as a **nonsense mutation**), the completed protein will not be produced from that gene. However, with only three stop triplets, the chance of this is much less than if there were 44 of them. Of the remaining 61 codons, all code for amino acids, which means that an amino acid is likely to have several different codons, giving what is known as a **degenerate code**. The assignment of codons to amino acids, the genetic code, is shown in Table 25.1. Note that the codon sequence is the mRNA sequence, and hence (with T instead of U) the same sequence is found on the coding strand of the DNA.

Only two amino acids, methionine and tryptophan, have single codons (AUG for methionine). The rest have more than one – leucine, arginine and serine have six. Codon assignment is not random. Where several codons exist for one amino acid, they tend to be closely related and vary mainly in the third base. For example, those for isoleucine are AUU, AUC, and AUA. Not only that, but codons for similar amino acids tend to be similar. For example isoleucine and leucine are very similar aliphatic hydrophobic amino acids; their codons include CUU (leucine) and AUU (isoleucine). This has important genetic consequences, since it means that many mutations involving single-base-changes have relatively little effect on the protein synthesized (changing AUU to AUC still represents isoleucine) or else substitute a very similar amino acid (changing CUU to AUU substitutes isoleucine for leucine). Isoleucine and leucine are so similar in size and hydrophobic properties that the substitution may not impair the function of the protein. Thus the arrangement of the genetic code provides a 'genetic buffering' action whereby the effects of many single-base-change mutations on the proteins synthesized are minimized.

If you know the base sequence of an mRNA you can work out the amino acid sequence of the protein it codes for. The reverse is not completely true because we cannot be sure in the case where an amino acid has several codons which of these was in the mRNA.

5′ base	Middle base				3′ base
	U	C	A	G	
U	UUU Phe	UCU Ser	UAU Tyr	UGU Cys	U
	UUC Phe	UCC Ser	UAC Tyr	UGC Cys	C
	UUA Leu	UCA Ser	UAA Stop*	UGA Stop*	A
	UUG Leu	UCG Ser	UAG Stop*	UGG Trp	G
C	CUU Leu	CCU Pro	CAU His	CGU Arg	U
	CUC Leu	CCC Pro	CAC His	CGC Arg	C
	CUA Leu	CCA Pro	CAA Gln	CGA Arg	A
	CUG Leu	CCG Pro	CAG Gln	CGG Arg	G
A	AUU Ile	ACU Thr	AAU Asn	AGU Ser	U
	AUC Ile	ACC Thr	AAC Asn	AGC Ser	C
	AUA Ile	ACA Thr	AAA Lys	AGA Arg	A
	AUG Met†	ACG Thr	AAG Lys	AGG Arg	G
G	GUU Val	GCU Ala	GAU Asp	GGU Gly	U
	GUC Val	GCC Ala	GAC Asp	GGC Gly	C
	GUA Val	GCA Ala	GAA Glu	GGA Gly	A
	GUG Val	GCG Ala	GAG Glu	GGG Gly	G

Table 25.1 The genetic code

* Stop codons have no amino acids assigned to them.

† The AUG codon is the initiation codon as well as that for other methionine residues.

A preliminary simplified look at the chemistry of peptide synthesis

Some (or many) students find protein synthesis a difficult subject to learn. It might help if we give a very simplistic description of the essence of the chemistry and energy needs of the synthesis of a polypeptide devoid of other complicating aspects. If you are completely familiar with this central chemistry (which is slightly counter-intuitive), it will be easier to understand all the other activities going on in the ribosome. These other activities organize the process and ensure that the genetic code of the messenger RNA being translated is accurately followed; that is where the real complexity lies.

To form a peptide bond two amino acids must become joined together by a CO—NH link. This requires energy so the first step is that each amino acid that will participate in protein synthesis is activated. An activated amino acid has sufficient energy to form a peptide bond. The activation reaction also links each amino acid to a transfer RNA (tRNA) molecule. tRNAs are adaptor molecules that enable the codon sequence of the mRNA to direct the order in which amino acids are linked to form a protein. The activation process consists of attaching an amino acid to a cognate (correct) tRNA molecule by an ester bond between a hydroxyl group on the ribose at the 3′ terminus of the tRNA and the -COOH of the amino acid. It is catalysed by a family of enzymes called **aminoacyl-tRNA synthetases**, each of which activates a specific amino acid and recognizes specific tRNA molecules. The overall activation reaction is shown below using nonionized structures for simplicity (tRNA-OH signifies the 2′- or 3′-OH of the 3′ terminal ribose of a tRNA molecule).

$$\text{tRNA—OH} + \text{HOOC—}\underset{\substack{|\\ \text{R}}}{\text{CH}}\text{—NH}_2 + \text{ATP} \longrightarrow$$

Amino acid

$$\text{tRNA—O—OC—}\underset{\substack{|\\ \text{R}}}{\text{CH}}\text{—NH}_2 + \text{AMP} + \text{PP}_i.$$

Aminoacyl—tRNA

Activation takes place in the cytosol at the expense of ATP hydrolysis. The inorganic pyrophosphate (PP_i) is hydrolysed to two inorganic phosphate (P_i) molecules so that the process has a large negative ΔG value.

Protein synthesis always involves these activated amino acids (called **aminoacyl-tRNAs**), never free amino acids. For simplicity we will in the scheme below represent them as tRNA-AA. ***Note that the growing polypeptide chain also has its carboxyl group esterified to the 2′ or 3′ hydroxyl group at the 3′ end of a tRNA, and it is this carboxyl that reacts with the free NH_2. group of the activated amino acid. Thus, protein synthesis consists of extending the −COOH end of the peptide not the amino terminal end.*** Fig. 25.1 shows that *the peptide (in red) is transferred from its tRNA to the NH_2 of the incoming aminoacyl-tRNA (in blue).* This gives us a peptide, one amino acid longer, attached to the last arrived tRNA.

$$\text{tRNA}_1\text{−peptide} + \text{tRNA}_2\text{−AA} \rightarrow \text{tRNA}_1 + \text{tRNA}_2\text{−AA−peptide}$$

The transfer is effected by the $-NH_2$ of the incoming aminoacyl-tRNA attacking the carbonyl carbon of the peptide-ester linkage to its tRNA; the *peptide* is transferred to the $-NH_2$ of the incoming activated amino acid.

This mechanism means that in the synthesis of a protein 200 amino acids in length, the last peptide bond synthesis involves a polypeptide of 199 amino acids being transferred to the amino end of the 200th aminoacyl–tRNA. One might intuitively expect that each incoming aminoacyl group would simply be added to the peptide amino end but, as stated, it is not like that. To use a fanciful image, to complete the dog, you do not add the tail to the dog, you add the dog to the tail, or, if you prefer it, in completing a wall you do not add the last brick to the wall, you add the wall to the last brick. The amino terminal end of the protein emerges from the ribosome first.

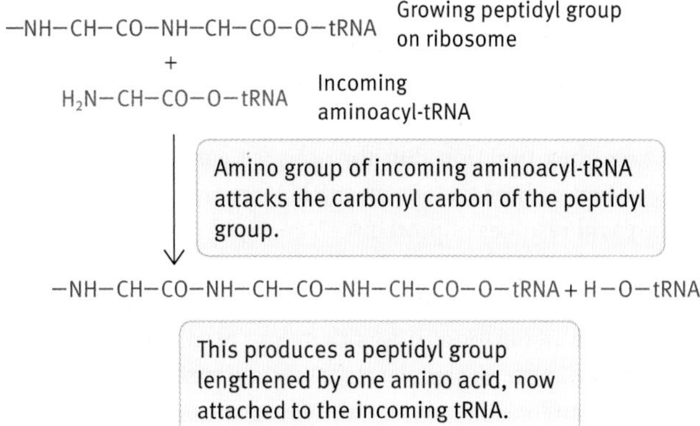

Fig. 25.1 Illustration of the reaction by which the growing peptide chain on the ribosome is elongated by one amino acid unit. Note that the peptidyl group is transferred from its transfer RNA (tRNA) attachment to the newly arrived aminoacyl-tRNA; the amino group remains free. This extends the carboxyl terminal end of the peptide chain by one aminoacyl residue. The same process is repeated over and over again until the protein chain is completed. Note particularly that after each round the peptide is attached to the most recent incoming aminoacyl-tRNA.

The above is the essence of the *chemistry* of polypeptide synthesis. The actual process on the ribosome is more complicated but, as already implied, the complications are to ensure that the process is organized to ensure that translation of messenger starts at the right place, that the codons on the mRNA are faithfully translated, that the ribosome moves along the mRNA one triplet at a time for each amino acid inserted, and that the finished polypeptide is successfully released.

ATP and GTP hydrolysis in translation

We have explained that ATP is used to activate each amino acid, and no further energy input is needed for the reaction forming each peptide bond. You are familiar with this type of energy utilization for bond formation. However, on the ribosome, each time an amino acid is added to the growing peptide chain, two molecules of GTP are hydrolysed to GDP and P_i, but the energy from the hydrolysis is not used for bond formation. It looks like a waste of energy, but what happens is that the hydrolysis causes a conformational change in the protein to which the GTP is attached. In each case the protein attaches to the ribosome in the GTP form, and cannot be released except in the GDP form. Hydrolysis occurs only after a slight delay, which allows an essential step to happen. This GTP/GDP switch mechanism is a crucial part of the organization of events on the ribosome, as will become evident.

We will now turn back to dealing with the complexities of protein synthesis but from this preliminary account you should understand what it is all aimed at – simply to synthesize one peptide bond after another.

How are the codons translated?

It is important at this stage to remember that protein synthesis occurs only on **ribosomes** – we will see shortly what these are and how they function. But for now, we can concentrate on the concept of how codons on mRNA are translated into the amino acid residues of proteins.

There is no physical or chemical resemblance or relationship between an amino acid and its codon that could lead to their direct association. It was therefore predicted that there must be **adaptor molecules** to associate amino acids with particular codons and, since the hydrogen-bonding potential of codons was most likely to be of importance, the adaptors were postulated to be small RNA molecules. Almost at the same time these were discovered as the **transfer RNA (tRNA)** molecules.

Transfer RNA

These are small RNAs, less than 100 nucleotides in length. When depicted diagrammatically they have a cloverleaf structure (Fig. 25.2(a)). Internal base pairing forms the stem loops. The important parts (from our present viewpoint) are the three unpaired bases, which form the **anticodon** (explained below), and the 3′-CCA terminal trinucleotide flexible arm to which an amino acid can be attached. In order to link two amino acids, two of these tRNA molecules with attached amino acids have to be positioned side by side on the ribosome with their anticodons hydrogen-bonded to adjacent codons on the mRNA. In three dimensions the tRNA molecules are folded up into quite a narrow shape as shown in Figure 25.2(b).

The anticodon is a triplet of bases complementary to a codon (Fig. 25.3(a)). It is located at a hairpin bend of the tRNA so that the three bases are unpaired and available for hydrogen bonding. Thus if a codon on the mRNA is 5′UUC3′ (coding for phenylalanine), the anticodon corresponding to this on a tRNA molecule will be 5′GAA3′. This particular tRNA molecule must have *only* phenylalanine put on to it and not any other amino acid; we will come to how this is done shortly. Since there are 61 codons each representing an amino acid (the other three are the stop codons), to translate these, it might be expected that there are 61 different tRNA molecules each with its own anticodon complementary to one codon and each accepting the one amino acid represented by its codon. In fact, there are fewer than 61 tRNA species – obviously there must be at least one for each of the 20 amino acids, but some tRNA molecules can recognize more than one codon differing from each other in the 3′ base. In these cases each of the codons recognized by a single tRNA must, of course, represent the same amino acid. The arrangement means that the cell needs to make fewer tRNA molecules. How is this achieved? The answer is **wobble pairing**.

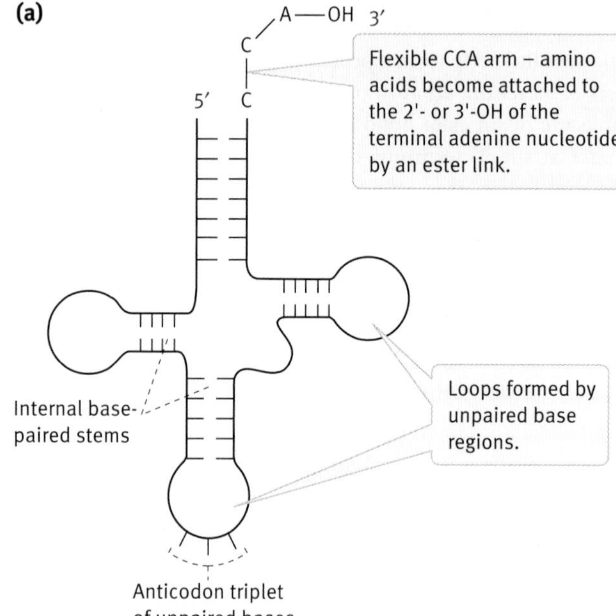

Fig. 25.2 (a) The cloverleaf structure of transfer RNA. **(b)** The folded structure of tRNA molecules.

The wobble mechanism

In view of the importance of complementarity in DNA replication and transcription where Watson–Crick base pairing is a vital essential, it may be disconcerting that, in codon-anticodon base pairing, the rules are bent a little. This applies *only* to the first (5′) base of the anticodon; that is to say, the *third (3′)* base of the codon. It is known as wobble pairing; a U in this position will pair with A or G on the codon, and a G with C or U.

One needs to be careful when thinking about the 'first base of the anticodon' to allow for the antiparallel nature of base pairing. The tRNA molecule *on its own* as in Fig. 25.2

is shown following the usual convention for nucleic acid sequences with the 5′ end to the left, but when it is base paired on a codon, it is flipped over as in Fig. 25.3a. Thus, the anticodon CGG will base pair in the normal Watson–Crick fashion as shown, the first (5′) base of the latter being printed in red:

3′-CGG-5′ anticodon
5′-GCC-3′ codon

The wobble mechanism permits the same anticodon to pair 'improperly', as follows:

3′-CGG-5′ anticodon
5′-GCU-3′ codon

Since GCC and GCU both code for the amino acid alanine, wobble pairing does not alter the amino acid sequence of the protein synthesized, but it enables a single tRNA to translate both. In short, as stated, it permits the cell to manage with fewer species of tRNA molecules.

Another mechanism that allows wobble pairing is to have the nonstandard base **hypoxanthine** (see Fig. 19.3) as the 5′ base of the anticodon. This will pair with C, U, or A in codons (Fig. 25.3). The nucleoside formed when hypoxanthine is bonded to a ribose sugar is called **inosine**, so the nonstandard base found in several tRNA anticodons is often referred to as inosine rather than hypoxanthine.

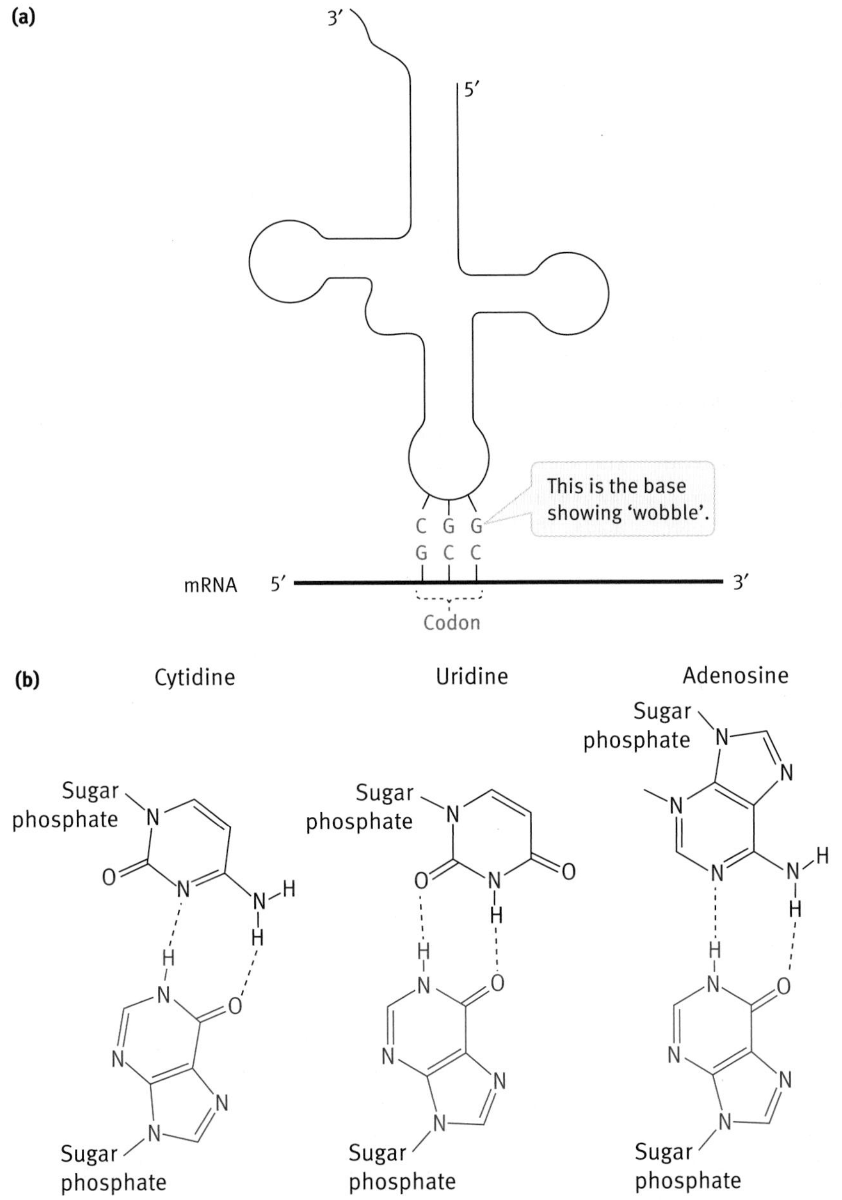

Fig. 25.3 **(a)** Base pairing of an anticodon of a transfer RNA (tRNA) molecule to a messenger RNA (mRNA) codon. To achieve antiparallel pairing the tRNA molecule is flipped over. This is why a tRNA structure is presented in one way in Fig. 25.2 (with 5′ by convention to the left) but in this paired form in the reverse way; the mRNA is written with the 5′ end to the left. In the example shown, the tRNA can base pair with codons GCC and GCU, both of which code for alanine. **(b)** Base pairing by the unusual base hypoxanthine, found in the 5′ position of certain anticodons, allows wobble pairing with three different 3′ codon bases.

How are amino acids attached to tRNA molecules?

The system depends on a tRNA molecule having attached to it the particular amino acid specified by the codon which is complementary to the anticodon on that tRNA molecule. Thus a tRNA molecule with the anticodon GAA must be 'charged' only with phenylalanine, since UUC and UUU are the codons for that amino acid. If any other amino acid were to be attached to that tRNA, a mistake would be made in the synthesis of protein molecules since phenylalanine would be replaced by the other amino acid. This has been demonstrated by experiments which showed that a noncognate ('incorrect') amino acid chemically attached to a tRNA for a different amino acid was inserted into a protein as if it was the correct one specified by that tRNA. Accuracy depends on the synthetase recognizing its own (cognate) tRNA.

tRNA specific for phenylalanine is depicted as $tRNA^{Phe}$, and so on for each of the 20 amino acids, using the accepted three letter abbreviations of amino acid names (see Table 4.1). (Note that $tRNA^{Phe}$ specifies only the tRNA; it does not mean that it has Phe attached. For the latter situation the term Phe-$tRNA^{Phe}$ is used.) Enzymes that attach amino acids to tRNAs are called **aminoacyl-tRNA synthetases** or, sometimes, **aminoacyl-tRNA ligases**. Most organisms have a different aminoacyl-tRNA synthetase for each of the 20 amino acids found in proteins. Each enzyme recognizes both a specific tRNA and its cognate amino acid, and joins them together. There is no general pattern for the way in which a synthetase recognizes its tRNA – the anticodon is recognized in some cases, but in others a specific base or several bases elsewhere in the molecule are recognized.

The overall activation reaction is repeated here for convenience:

$$\text{Amino acid} + \text{tRNA} + \text{ATP} \rightarrow \text{aminoacyl-tRNA} + \text{PP}_i + \text{AMP}.$$

However, the reaction on the enzyme actually occurs is two stages, as shown below:

$$\text{Amino acid} + \text{ATP} \rightarrow \text{aminoacyl-AMP} + \text{PP}_i;$$
$$\text{Aminoacyl-AMP} + \text{tRNA} \rightarrow \text{aminoacyl-tRNA} + \text{AMP}$$

In the aminoacyl-AMP the carboxyl group of the amino acid is bonded to the phosphoryl group of AMP as shown below.

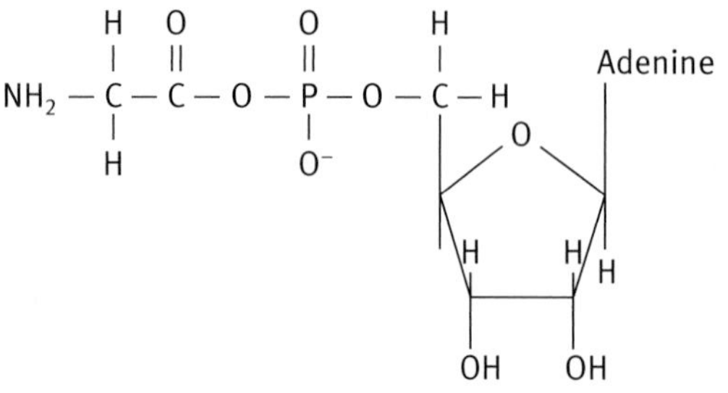

```
      H   O      O       H
      |   ||     ||      |          Adenine
NH2 — C — C — O — P — O — C — H        |
      |          |        |    O       |
      H          O-       |  /   \     |
                          |/       \   |
                          \H      H/ H
                           |        |
                          OH       OH
```

The PP_i is hydrolysed to $2P_i$ and this drives the reaction to the right.

Proofreading by aminoacyl-tRNA synthetases

There is more to the above reaction. The synthetase, as well as selecting a correct tRNA, must also select the correct amino acid. As stated, the accuracy of the translation of mRNA into protein depends on the accurate selection by these enzymes of the correct amino acid. Protein synthesis on the ribosome depends on codon-anticodon base pairing to achieve the correct amino acid sequence. The ribosome has no mechanism for checking that a particular tRNA is carrying the correct amino acid, so it is very important that the enzyme attaching the amino acid to the tRNA does not have a high error rate. The initial selection is on the active site of the synthetase selecting its correct amino acid. It is relatively easy for an enzyme to distinguish between amino acids with markedly different structures, but much harder to do this between amino acids such as valine and isoleucine that are very similar.

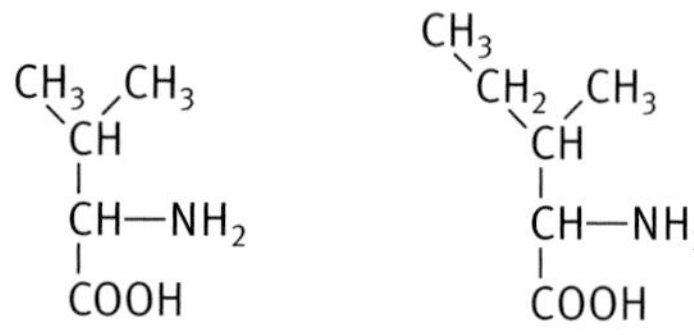

```
                          CH3
CH3  CH3                    \
   \ /                      CH2  CH3
   CH                          \ /
   |                           CH
   CH—NH2                      |
   |                           CH—NH2
   COOH                        |
                               COOH

  Valine                    Isoleucine
```

As valine is smaller than isoleucine it fits into isoleucine binding site on the aminoacyl-tRNA synthetase specific for isoleucine and the difference in binding energies due to the single extra -CH_2 group in isoleucine is not enough to give a sufficiently high degree of selectivity between the two amino acids. The isoleucyl-tRNA synthetase would attach valine to $tRNA^{Ile}$ at a rate that would result in an unacceptable rate of errors in mRNA translation unless there was a corrective mechanism. A 'proofreading' mechanism has evolved on many aminoacyl-tRNA synthetases, involving the aminoacyl-AMP intermediate. As well as the catalytic site, the enzyme has a nearby editing site. Valyl-AMP fits in this site on the isoleucyl-tRNA synthetase and is then hydrolysed and released as valine and AMP, while isoleucyl-AMP is too big to fit in the editing site, and proceeds to the second stage of the synthetase reaction. Similarly, threonyl-tRNA synthetase hydrolyses seryl-AMP. In contrast, the tyrosine specific enzyme has no such mechanism because tyrosine is sufficiently different from all other amino acids that the initial selection is adequate.

The tRNA molecule has a terminal 3′ trinucleotide sequence of CCA, the terminal adenosine having free 2′ and 3′-OH groups on the ribose moiety (Fig. 25.4(a)). It is to one of these that the amino acid is attached by an ester bond (Fig. 25.4(b)). (There are two classes of synthetases, each class having its own group of tRNAs. One class attaches its amino acid to the 3′ of the ribose, and the other class to the 2′.) This CCA trinucleotide forms a flexible arm, which can position the aminoacyl group on the appropriate reactive site on the ribosome. The ester bond formed by the aminoacyl group is at a slightly higher energy level than that of a peptide bond so there is no thermodynamic problem in transferring the aminoacyl-ester group to the $-NH_2$ of another aminoacyl group to form a peptide bond. In other words, the energy required for the formation of a peptide bond is inserted into the process by the aminoacyl-tRNA synthetase using ATP as the energy source.

Ribosomes

Ribosomes derive their name from the content of RNA or ribonucleic acid, which accounts for about 60% of the dry weight. A ribosome consists of two subunits – in *E. coli*, a large one containing two molecules of RNA (23S and 5S; see below for an explanation) and a small subunit with one RNA molecule (16S). Eukaryotic ribosomes are somewhat larger. You have already met mRNA and tRNA; **ribosomal RNAs (rRNAs)** are different again. Because of internal base pairing, they assume highly folded compact structures. Fig. 25.5 illustrates the complex secondary structure of an RNA molecule of an *E. coli* ribosome. The rRNAs are associated with many proteins, forming solid particles. The rRNA and protein composition of prokaryotic and eukaryotic ribosomes is summarized in Table 25.2.

With very large structures such as a ribosome, their sizes are measured in terms of the rate at which they sediment in an ultracentrifuge – expressed as Svedberg units or S values. (An ultracentrifuge spins so fast and therefore generates such a high G force that large molecules in solution move towards the bottom of the tube.) An *E. coli* ribosome is 70S, the subunits being 50S and 30S (the S values are not simply additive, since S depends both on mass and shape).

The overall *principle* of protein synthesis can be given very simply. The ribosome becomes attached to the mRNA near the 5′ end and then moves down the mRNA towards the 3′ end, assembling the aminoacyl groups of charged tRNA molecules into a polypeptide chain according to the sequence of codons in the mRNA, which is read in the 5′→3′ direction. The *N*-terminal amino acid of the polypeptide is the first to emerge

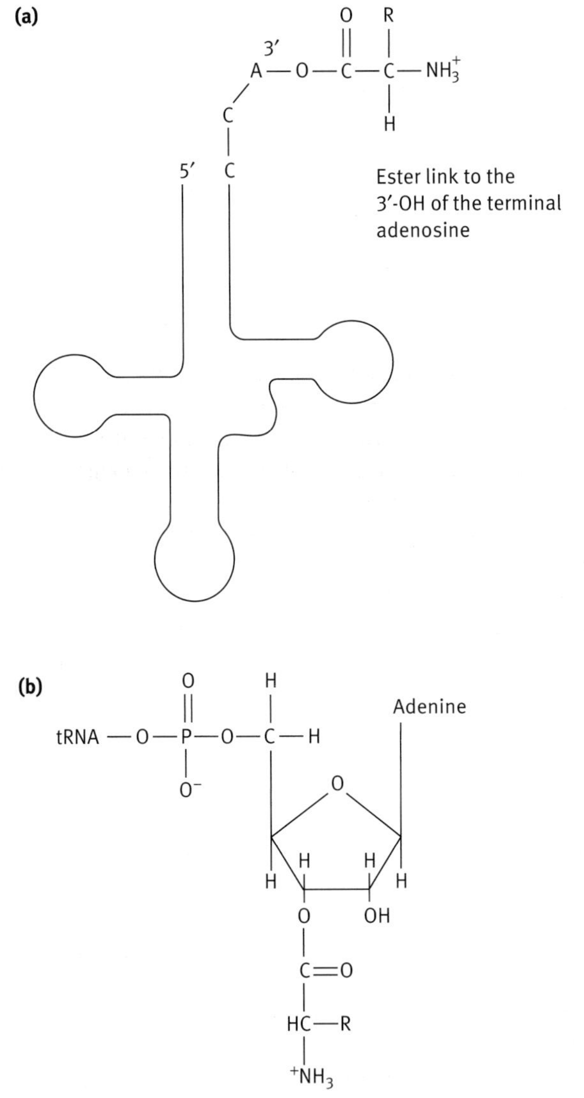

(a)

3′ A—O—C(=O)—C(R)(H)—NH_3^+

5′ C C

Ester link to the 3′-OH of the terminal adenosine

(b)

tRNA—O—P(=O)(O⁻)—O—CH₂ ... Adenine

O, H, H, H, H, O, OH

C=O

HC—R

$^+NH_3$

Fig. 25.4 (b) Diagram of transfer RNA molecule showing the CCA base sequence at the 3′ end, where the amino acid is attached by ester link. **(b)** Structure of the terminal nucleotide with the attached amino acid. In some cases the ester link is on the 2′-OH of the ribose.

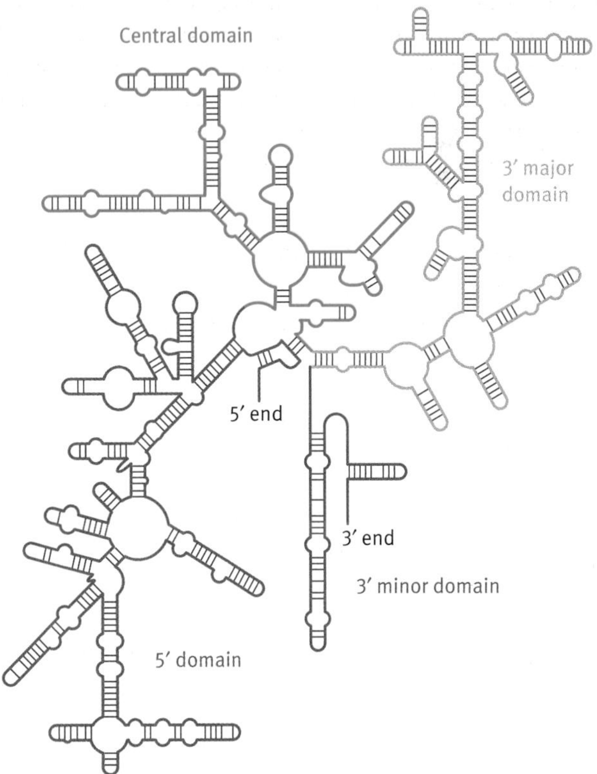

Fig. 25.5 16S RNA. Fig 9.5 from Lewin, B. (1994) Genes V, Oxford University Press, Oxford.

	Subunits	Subunit composition
Prokaryotic ribosome 70S MW=2,500,000	**50S** MW=1,600,000	5S rRNA (120 nucleotides) 23S rRNA (2,900 nucleotides) ~34 proteins
	30S MW=900,000	16S rRNA (1,540 nucleotides) 21 proteins
Eukaryotic ribosome80S MW=4,200,000	**60S** MW=2,800,000	5.8S rRNA (160 nucleotides) 5S rRNA (120 nucleotides) 28S rRNA (4,700 nucleotides) 49 proteins
	40S MW=1,400,000	18S rRNA (1,900 nucleotides) ~33 proteins

Table 25.2 The rRNA and protein composition of prokaryotic and eukaryotic ribosomes

from the ribosome's exit tunnel through the large subunit and the C-terminal one is the last. At the end of the mRNA, the ribosome meets a stop codon, the protein is released, the ribosome detaches from the mRNA, and dissociates into its two subunits. Note that there are no 'special' ribosomes. In a given cell, any ribosome can use any mRNA; a slight qualification is that mitochondrial and chloroplast ribosomes are different from those in the cytosol of eukaryotic cells, the former being prokaryotic in type.

That is the principle – to understand the process we need to give more detail. Synthesis of a protein molecule can be divided into three phases – **initiation, elongation,** and **termination**.

Initiation of translation

It is important that a ribosome begins its translation of an mRNA at exactly the correct point – in other words, that it initiates translation correctly. mRNAs have 5′ and 3′ untranslated regions and the coding region lies in between. The ribosome therefore must recognize the *first* codon of the coding sequence, which is not at the beginning of the mRNA, and start translating at that site. Absolutely precise initiation is essential, for this puts the ribosome in the correct **reading frame**.

This may need explanation. Suppose the *coding region* of an mRNA starts with the sequence:

5′-AUGUUUAAACCCCUG------3′.

The first five amino acids are specified by the codons AUG, UUU, AAA, etc. There is nothing to indicate what constitutes a codon, other than that the first three bases of the coding part of the mRNA encountered constitute codon 1, the next three are codon 2, and so on. There are no commas or full stops between them. The message therefore depends on starting to read *exactly* at AUG. Suppose an error of one base is made and a U is deleted from the codon UUU at the second base. The codons would then read:

5′-AUG,UUA,AAC,CCC,UG------3′

The codons translated would be totally different and the amino acids inserted would correspondingly be totally different. This error is known as a **reading frameshift**; mutations deleting or adding one or two bases from an mRNA also cause reading frameshifts and result in the amino acid sequence of the polypeptide synthesized after the frameshift being completely different from that required, the process being halted when a stop codon arising randomly as a result of the frameshift is reached. The resultant polypeptide is useless garbage instead of the one specified by the gene. This contrasts with a deletion of a three nucleotide codon in which case only one amino acid would be missing from an otherwise accurate sequence.

Initiation of translation in *E. coli*

First the ribosome must be positioned at the correct starting place on the message. As shown in Fig. 25.6, there is a purine-rich sequence of 3–8 nucleotides long 5′ to the translational start site on the mRNA known as the **Shine-Dalgarno sequence**, which is complementary to a section of the 16S rRNA. Binding of the two by base pairing correctly positions the mRNA on the small ribosomal subunit. A ribosome has three sites on it, each of which can accommodate a tRNA molecule. These are the **A**, **P**, and **E** sites (for **acceptor** or **amino-acyl**, **peptidyl**, and **exit** sites), the roles of which will become apparent shortly. The sites extend into both ribosome subunits. The mRNA is positioned such that the first and second codons are aligned with the P and A sites respectively (Fig. 25.6). Many bacterial mRNA molecules are polycistronic; the *lac* mRNA discussed in the next chapter is one such example, with three regions in the one mRNA molecule coding for

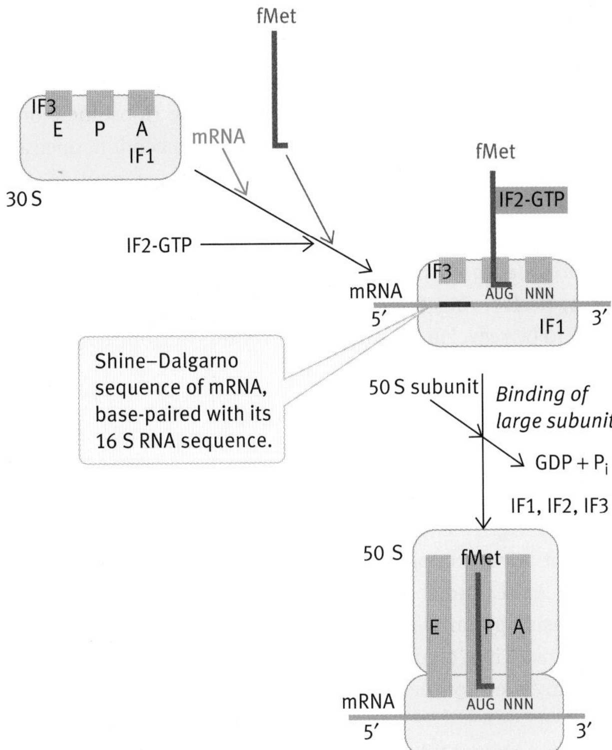

Fig. 25.6 Initiation of translation in *Escherichia coli*. The initiating transfer RNA, $tRNA_f$, is represented by the blue line, the anticodon being the horizontal short line. The binding of fMet-$tRNA_f$ to the 30S subunit requires IF2. NNN represents any codon (N for any nucleotide).

three different proteins. Each cistron has a Shine-Dalgarno sequence adjacent to it so that each can be initiated independently (Fig. 25.7).

Since initiation of translation is *not* at the 5′ end of the mRNA but at a distance along the molecule after the Shine-Dalgarno sequence, the first codon must be identified in some way. The start site is the codon **AUG**, but there is a problem. An AUG triplet can occur anywhere in the mRNA, since it is needed to code for each internal methionine. Also, AUG codons are likely to occur at random in reading frames other than the correct one, as illustrated below.

5′-AUGUUUAAAUGGCCUGAAG------3′.

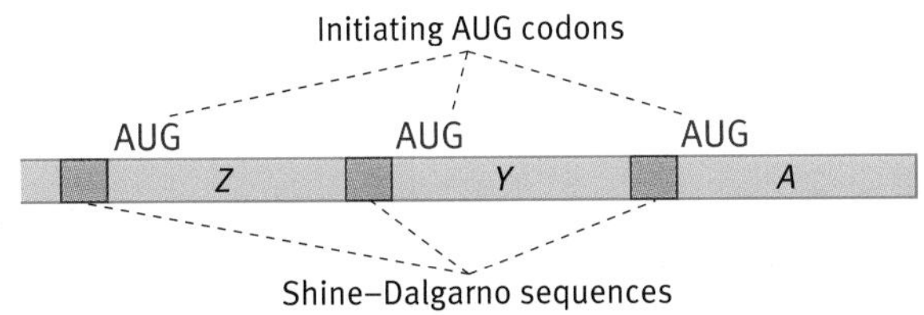

Fig. 25.7 The structure of a polycistronic prokaryote messenger RNA. *Z*, *Y*, and *A* refer to coding regions of the *lac* mRNA (see Fig. 26.2).

Translation of this sequence in the reading frame starting at the first AUG gives Met-Phe-Lys-Trp-Ile-Lys- etc., but if the ribosome initiates translation at the second AUG in this sequence, translation will occur in a different reading frame to give Met-Ala-Glu- etc.

So how is the absolutely crucial step of identifying the correct initiating AUG carried out? The answer is that the Shine-Dalgarno sequence centred approximately 10 nucleotides upstream of the AUG base paired with the 3′ end of the 16S RNA of the small ribosomal subunit (30S ribosome), so positioning the AUG in the P site that is present in the 30S ribosome. Thus the recognition signal is effectively longer than just the AUG.

Now that the AUG is in the correct position, the initiating tRNA can be placed so that its anticodon 3′-UAC-5′ can base pair with the AUG initiation codon on the mRNA. There are two different types of tRNAs specific for methionine, both with the same anticodon, but one is used exclusively for initiation and the other exclusively for inserting methionine internally into the growing polypeptide during the elongation process. The two differ in that the initiating methionyl-tRNA has structural features that are recognized by an **initiation protein factor (IF2)**. Binding of this to the methionyl-tRNA makes it the only aminoacyl-tRNA of all the aminoacyl-tRNAs that can directly enter the P site that is present on the small ribosomal subunit. All the other aminoacyl-tRNAs that are involved in *elongation* of the polypeptide following initiation are recognized by a different cytosolic factor (EF-Tu; described below), which delivers them to the complete 70S ribosome (large 50S subunit joined with the small 30S subunit), but exclusively into the A site formed by both subunits. EF-Tu does not bind to the initiating tRNA. Possibly the requirement to get the first aminoacyl-tRNA directly into the P site is why the first step of initiation always takes place on the small ribosomal subunit alone.

There is another difference between the two methionyl-tRNAs in bacteria that does not occur in eukaryotes – the methionine that becomes attached to the initiating tRNA in prokaryotes is formylated on its $-NH_2$ group by a transformylase using N^{10}-formyltetrahydrofolate (see Chapter 19) as a formyl donor. Prokaryotic proteins are synthesized with *N*-formylmethionine (fMet) as the first amino acid residue, though the formyl group, and frequently the methionine also, are removed before completion of the synthesis.

That then is how the initiating methionyl-tRNA is distinguished from the methionyl-tRNA used in elongation, the most important factor being that they are recognized by different protein factors, IF2 and EF-Tu respectively. The initiating tRNA can be called $tRNA^{fMet}$ (f for formyl), and the charged version fMet-$tRNA^{fMet}$, while the tRNA for methionine involved in elongation is called $tRNA^{Met}$.

There is a slight complication in that the initiation codon for some prokaryotic proteins is GUG rather than AUG. Even though GUG normally encodes valine, when it is in the initiation position downstream of the Shine-Dalgarno sequence, it is recognized by the initiator tRNA^{fMet}, which can bind to it through wobble base pairing, and thus fMet is incorporated as the first amino acid as normal.

Initiation factors in *E. coli*

To summarize the initiation procedure, in the cytosol there is a pool of 30S and 50S ribosomal subunits. There are three cytosolic initiation factors that participate in the process, and then are released for further use.

- **IF1** binds to the 30S subunit and prevents an aminoacyl-tRNA entering the A site before initiation is complete.
- **IF2**, as explained above, binds to fMet-tRNA^{fMet} and delivers it to the 30S subunit. IF2 also binds GTP. It is in fact a GTPase enzyme that hydrolyses GTP to GDP and P_i, but the reaction does not occur until the initiation complex is fully formed and the 30S and 50S subunits are bound together with the mRNA and fMet-tRNA^{fMet} (Fig. 25.6).
- **IF3** binds to the 30S subunit and prevents it from associating prematurely with the 50S subunit.

In the presence of mRNA, fMet-tRNA^{fMet}, and GTP, a complex of these with a 30S subunit (with its attached IFs 1 and 3) is formed. The fMet-tRNA^{fMet} is delivered to the P (for peptidyl) site on the subunit with its anticodon paired with the AUG codon, by IF2 (Fig. 25.6). This complex now associates with a 50S subunit. The event is accompanied by the hydrolysis of GTP and the release of GDP, P_i, IF1, IF2, and IF3 (Fig. 25.6).

We now have a complete 70S ribosome positioned on the mRNA with the fMet-tRNA^{fMet} in the P site with its anticodon base paired with the initiating AUG codon. The A site is vacant, awaiting delivery of the second amino acid on its tRNA. Initiation is complete.

Once initiation is achieved, elongation is the next step

Elongation factors in *E. coli*

Two soluble protein factors in the cytosol are involved in elongation. They cannot both be attached to the ribosome at once. They bind alternately, perform their task for each round of peptide bond formation, and detach. In both cases they can attach only when bound to a molecule of GTP. Like IF2, both are latent GTPases, active only when ribosome-bound. Hydrolysis of GTP to GDP causes them to detach. In the cytosol their GDP is exchanged for GTP so that the factors are then ready to participate in a new round of elongation.

The two factors are **EF-Tu** (elongation factor, temperature unstable) and **EF-G**, also known as **ribosomal translocase.** EF-Tu has the task of delivering the incoming aminoacyl-tRNA to the ribosome, EF-G of moving the ribosome along the mRNA in the $5' \rightarrow 3'$ direction to the next codon, once an aminoacyl group has been added to the growing peptide (see below).

It is useful to keep in mind this central concept of a pair of factors *alternately* hopping on to the ribosome in their GTP form, performing their tasks, and detaching in their GDP form to be recycled for subsequent rounds of elongation.

Mechanism of elongation in *E. coli*

We have already given the chemistry of elongation in principle (earlier in this chapter). We suggest that you follow the steps in Fig. 25.8 as you read the next part. Starting with the initiation complex (state a), we have an fMet-tRNA^{fMet} in the P site, and the A site is vacant. It might be worth re-emphasising that *only* in the initiation process does the P site accept a tRNA charged with an amino acid – in this case *N*-formylmethionine; *all* subsequent aminoacyl-tRNAs enter the A site.

Aminoacyl-tRNAs (other than the initiating species) are complexed in the cytosol with the elongation factor EF-Tu, carrying a molecule of GTP bound to it. This factor attaches to the ribosome only if both GTP and an aminoacyl-tRNA molecule are bound to it. The EF-Tu-GTP-aminoacyl-tRNA complex binds to the ribosome such that the aminoacyl-tRNA occupies the A site with its anticodon positioned at the mRNA codon (Fig. 25.8, state b). EF-Tu exists in high concentration in the cytosol, sufficient to bind all of the aminoacyl-tRNA. Hydrolysis of its bound GTP molecule by EF-Tu releases the latter in its GDP form due to a conformational change, and frees the ribosome to proceed with the next step in elongation (Fig. 25.8, state c).

The aminoacyl groups on the two tRNA molecules on the P and A sites are in the vicinity of the catalytic site of **peptidyl transferase**, which transfers the fMet group from the tRNA in the P site to the free amino group of the incoming aminoacyl-tRNA in the A site, producing a dipeptide attached to the tRNA (state d). Despite its name, *'peptidyl transferase' is an RNA molecule with catalytic properties*, part of the 23S RNA of the large ribosomal subunit, and is not a protein; it is a ribozyme. The peptidyl transferase reaction is shown, using the first peptide bond synthesis as an example:

$$\underset{\text{(P site)}}{\text{tRNA—O—CO—fMet}} + \text{NH}_2\text{—}\underset{\text{R}'}{\underset{|}{\text{CH}}}\text{—CO—O—}\underset{\text{(A site)}}{\text{tRNA}} \longrightarrow$$

$$\underset{\text{(P site)}}{\text{tRNA—OH}} + \text{fMet—CO—NH—}\underset{\text{R}'}{\underset{|}{\text{CH}}}\text{—CO—O—}\underset{\text{(A site)}}{\text{tRNA}}.$$

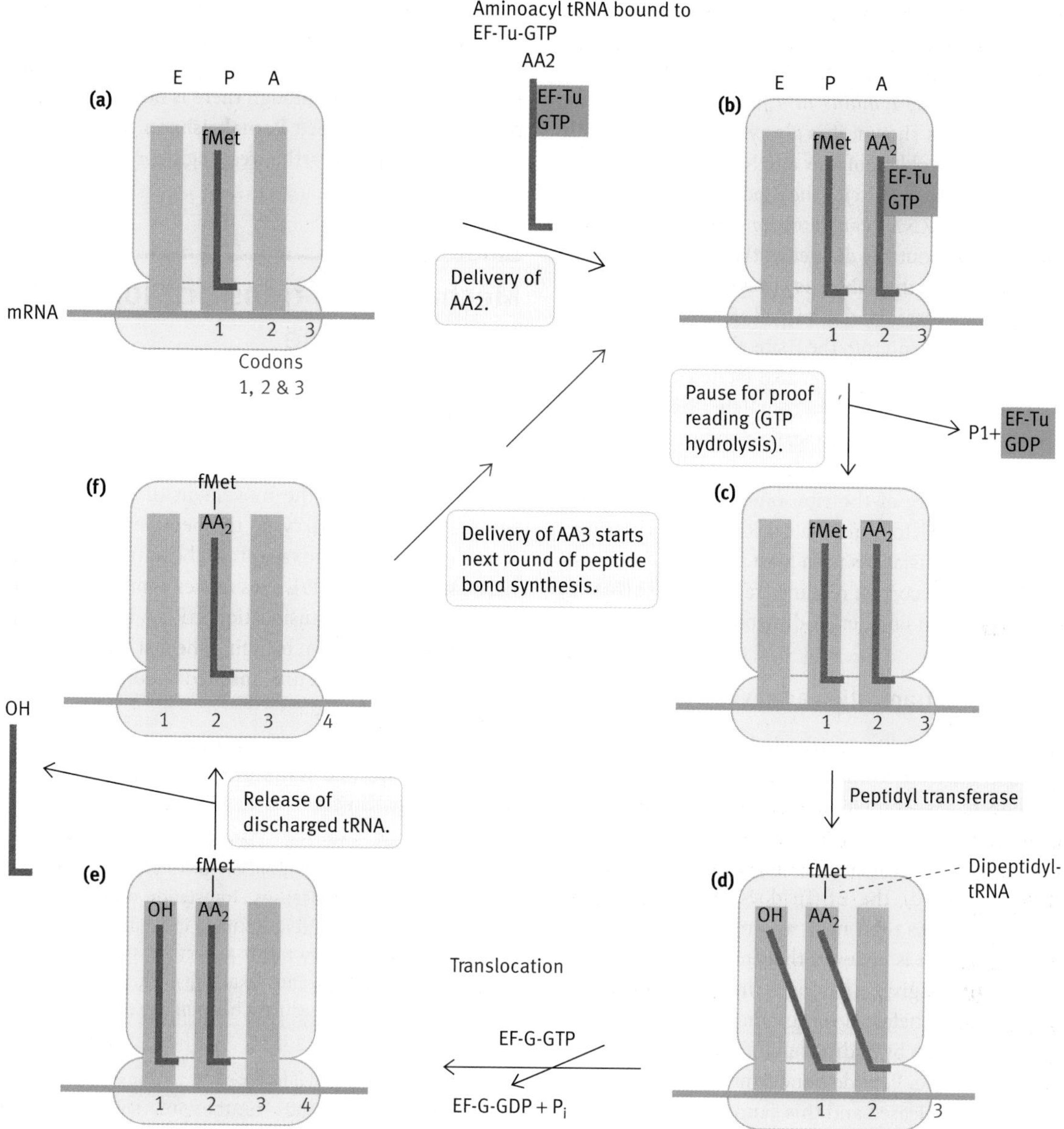

Fig. 25.8 The elongation process in protein synthesis in *Escherichia coli* following translational initiation. Transfer RNAs are shown as blue lines, the anticodon being represented by the short horizontal section; AA2 and AA3 represent amino acids and fMet represents formylmethionine. The positioning of EF-Tu-GTP on the tRNA is purely diagramatic but see Fig. 25.9(c). The reason for the naming of the enzyme peptidyl transferase is not evident from the diagram, but if you do the next round of synthesis you will see that in all subsequent rounds of synthesis, it is a peptide that is transferred to the incoming aminoacyl-tRNA as is also explained in the text. E, exit site; P, peptidyl site; A, acceptor site. Note movement of mRNA.

The aminoacyl group on the tRNA in the P site is transferred to the free amino group of the aminoacyl-tRNA in the A site. As the synthesis of the polypeptide proceeds, the aminoacyl group in the P site is actually the partially completed polypeptide chain and this is transferred to the incoming amino acid. This is why the process is called the **peptidyl transferase reaction**. Adjacent to the peptidyl transferase site there is the opening to a tunnel through the large ribosomal subunit. As the polypeptide is synthesized it is fed into this tunnel to emerge from the subunit, the amino terminal end first (see Fig. 25.11).

During the peptidyl transfer process, each tRNA becomes bound to two sites on the ribosome (Fig. 25.8(d)). The discharged tRNA straddles the P and E (exit) sites – the anticodon end of the molecule is still in the P site but the other end is in the E site. Similarly, the tRNA in the A site (now carrying the peptide) straddles the A and the P sites. This was shown by chemical protection studies.

The model proposed to account for this attachment of tRNAs to two sites at once (Fig. 25.8) envisages that one end of the aminoacyl-tRNA swings over as shown (c→d), peptide transfer occurs at the same time, and the discharged tRNA swings one end to the E site.

In this model an important feature is that, during the process, the *nascent peptide remains in a fixed position relative to the large subunit*, as shown; it is always in the P site. This would eliminate the problem of how a tRNA physically moves with a polypeptide attached to it. The model also removes the problem of how the tRNAs can move on the ribosome during translocation without the danger of them diffusing away, since at least one end of the tRNAs is always attached to a site. Translocation now causes the discharged tRNA to move with respect to the small subunit into the E site, the tRNA exits, and we have the situation shown in Fig. 25.8 (state f). This moves the ribosome to the next codon and completes the first round.

An interesting comment on elongation comes from one of the leading laboratories in this field, to quote 'Perhaps most impressive is the ability of the ribosome, together with the elongation factor, EF-G, to translocate mRNA and tRNAs over very large molecular distances with high speed and accuracy, while maintaining the correct reading frame' (see **Korostelev and Noller** reference in Further reading online).

How is accuracy of translation achieved?

As discussed above, fidelity of translation in the ribosome is due to codon-anticodon interaction, but exactly how this selection is effected is a problem. The binding-energy difference between a correct and an incorrect codon-anticodon base pairing alone is insufficient to account for the translational fidelity achieved. The EF-Tu does not 'know' which aminoacyl-tRNA has to react next, and must deliver them to the A site randomly. It is believed that the 'pause' for hydrolysis of GTP by EF-Tu gives time for an incorrect aminoacyl-tRNA to diffuse away before it reacts since it will not be held in the A site as strongly by hydrogen bonding to the codon as a correct one would be. Peptide bond synthesis cannot occur until EF-Tu-GDP is released and this cannot happen until the GTP is hydrolysed (Fig. 25.8(c)).

It seems, however, that this is unlikely, by itself, to achieve the degree of fidelity observed of an error rate between 1 in 10^3 and 1 in 10^4. It is believed that the ribosome may play a part in the accuracy of codon-anticodon recognition and this is a function of the 16S RNA molecule in the small subunit of the ribosome. There are three bases, two adenines and one guanine, in the 16S sequence, which, if altered, render *E. coli* nonviable due to inaccurate protein synthesis. When a cognate tRNA binds to the anticodon, it is believed that these three bases flip out of their orientation in the rRNA helix and interact with the codon-anticodon base pairs. They check that the first two base pairs of the codon-anticodon action are *bona fide* Watson–Crick base pairs rather than aberrant pairs. The third (wobble) pair is not so checked; it is the first two that are the main specificity determinants. If the codon-anticodon pairing is correct it triggers the hydrolysis of GTP attached to the EF-Tu that delivered the aminoacyl-tRNA to the A site. This causes the release of EF-Tu-GDP and in turn this allows peptide bond formation. It thus appears that the ribosome itself plays an active role in achieving fidelity of translation, though there is not a full understanding of the mechanism. (See **Ramakrishnan**, and also **Zaher and Green** references in Further reading.)

Mechanism of translocation on the *E. coli* ribosome

Translocation of peptidyl-tRNA from the A site to the P site is catalysed in *E. coli* by **EF-G (ribosomal translocase)**. It attaches to the ribosome after each peptide synthesis is completed and catalyses the movement of the tRNA straddling the A/P sites (now carrying the peptide) completely into the P site, leaving the A site vacant. At the same time the discharged tRNA straddling the P/E sites moves completely to the exit site and leaves. During translocation mRNA moves with the newly formed peptidyl-tRNA, bringing the just translated codon into the P site and the next untranslated codon into the A site.

BOX 25.1 Effects of antibiotics and toxins on protein synthesis

Antibiotics are the chemical missiles that microorganisms throw at each other in the competition for survival. They attack critical points in cellular processes. Translation offers many such targets. Many of the antibiotics used in medicine are specific for bacteria because they target aspects of translation that differ between prokaryotes and eukaryotes. Thus, they can target the pathogenic organism without harming the patient.

In prokaryotes:

- **chloramphenicol** inhibits peptidyl transferase
- **erythromycin** binds to the 50S subunit and inhibits translocation
- **kirromycin** and fusidic acid prevent EF-Tu release
- **streptomycin** (an aminoglycoside) affects initiation and causes misreading of codons; **neomycin** and **kanamycin**, which are structurally related, act similarly
- **tetracyclines** inhibit the binding of aminoacyl-tRNAs to the A site of the ribosome
- **puromycin** affects both prokaryotes and eukaryotes: by mimicking an aminoacyl tRNA in the A site, it gets added to the nascent polypeptide and causes premature chain termination.

In addition to these examples of antibiotic action, **diphtheria toxin** inhibits EF_2, the eukaryotic translocase (equivalent to EF-G in bacteria). **Ricin**, the toxin of the castor bean, is an *N*-glycosidase, which removes a single adenine base from one of the eukaryotic rRNAs and inactivates the large subunit. One molecule of ricin can destroy a cell containing tens of thousands of ribosomes.

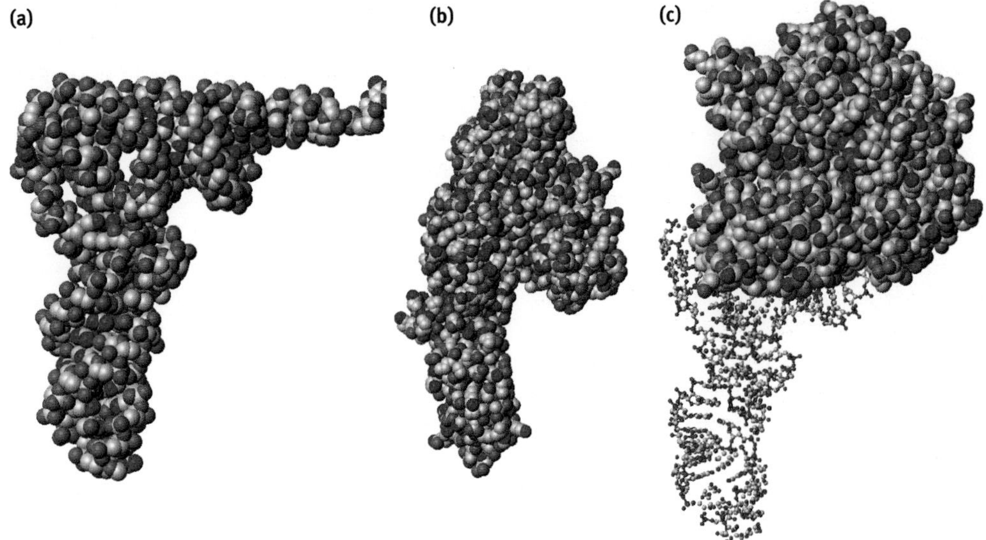

Fig. 25.9 Space-filling models of **(a)** yeast transfer RNA^{Phe} (Protein Data Bank code 4TNA), **(b)** EF-G in complex with GDP (Protein Data Bank code 1DAR), and **(c)** the ternary complex of Phe-tRNA^{Phe}, EF-Tu, and a GTP analogue (Protein Data Bank code 1TTT), showing the protein tRNA mimicry described in the text.

EF-G has attached to it a molecule of GTP and the translocation is associated with the hydrolysis of this and release of the EF-G-GDP complex. The mechanism by which EF-G acts involves **protein mimicry**. It mimics the shape of tRNA. Figure 25.9 shows space-filling models of (a) tRNA^{Phe}, (b) the protein EF-G in complex with GDP, and (c) the ternary complex of Phe-tRNA^{Phe}, EF-Tu, and a GTP analogue. You can see that a domain of EF-G resembles in shape that of the tRNA stem helix carrying the anticodon, and that the shape of the whole EF-G molecule is a close match to the complex between EF-Tu and tRNA. Possibly the EF-G-GTP inserts itself into the A site and physically pushes the peptidyl-tRNA into the P site, and the discharged tRNA from the P site into the exit site. During translocation GTP is hydrolysed into GDP, and, in this form, the EF-G-GDP is able to detach. It is not fully understood, however, how translocation is achieved.

Termination of protein synthesis in *E. coli*

At the end of each mRNA coding section there is a stop codon (Table 25.1). In *E. coli* there are two release factors, **RF1** and **RF2**, which recognize the stop codons UAG and UGA respectively, while both recognize UAA. There is a case of protein mimicry here as the primary amino acid sequences of RF1 and RF2 suggest that these proteins also mimic the shape of the tRNA as occurs in EF-G. The release factors carry a molecule of water to the ribosome that hydrolyses the ester bond between the now completed polypeptide chain and the final tRNA. This releases the polypeptide from the ribosome. A third release factor, **RF3**, yet another GTPase, triggers dissociation of RF1 or RF2 from the A site, and once GTP is hydrolysed to GDP, RF3 itself dissociates.

After this the remaining ribosome complex is disassembled into the two subunits and the remaining bound tRNAs and mRNA released. The dissociation procedure requires a **ribosomal recycling factor (RRF)**, **EF-G** and IF3. The ribosome recycling factor is a protein, which also accurately mimics the shape of tRNA. By binding the A site it triggers the recruitment of EF-G, which in turn triggers release of the remaining tRNAs.The two ribosomal subunits and mRNA dissociate. IF3 attaches to the small subunit, and prevents reassociation with the large one.

Physical structure of the ribosome

In the preceding diagrams we have used simple shapes for the ribosome for the purpose of clarity, but a good deal is known of the structure, visible in the electron microscope and recently elucidated through the crystal structure of the prokaryotic 50S and 30S subunits, and the intact 70S ribosome.

The ribosome has a distinctive shape that reflects its function, as shown in Fig. 25.10. The tRNAs bind to the face of the large subunit, which faces the small subunit, but at the contact faces there is a hollow big enough to accommodate the tRNAs and form a channel. The aminoacyl-tRNAs enter the channel on the right of the diagram and the discharged tRNAs emerge on the left. The peptidyl transferase catalytic region is located on the face of the large subunit within the cavity,

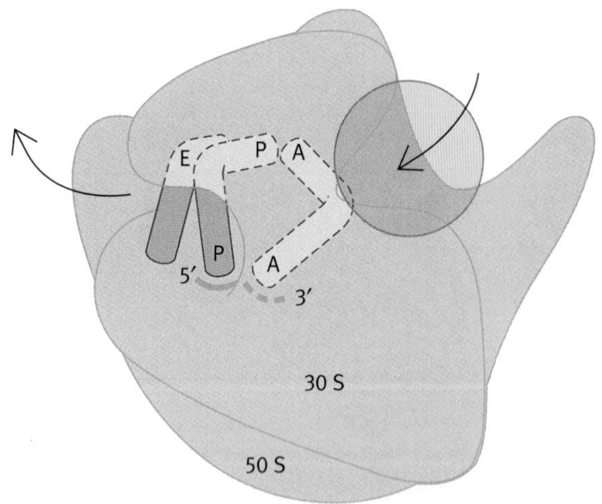

Fig. 25.10 Diagram of a ribosome with plausible locations for ribosome-bound transfer RNAs in the A/A, P/P, and E states (see Fig. 25.8 for the terminology). The shaded area at the right shows the approximate site of interaction of EF-Tu. The polarity of a fragment of messenger RNA containing the A- and P-site codons is shown. The arrows indicate the likely path of a tRNA as it transits the ribosome. Adapted from Fig. 1 in Noller, H. F.; Annu rev Biochem; (1991) 60, 193; Reproduced by permission of Annual Reviews Inc.

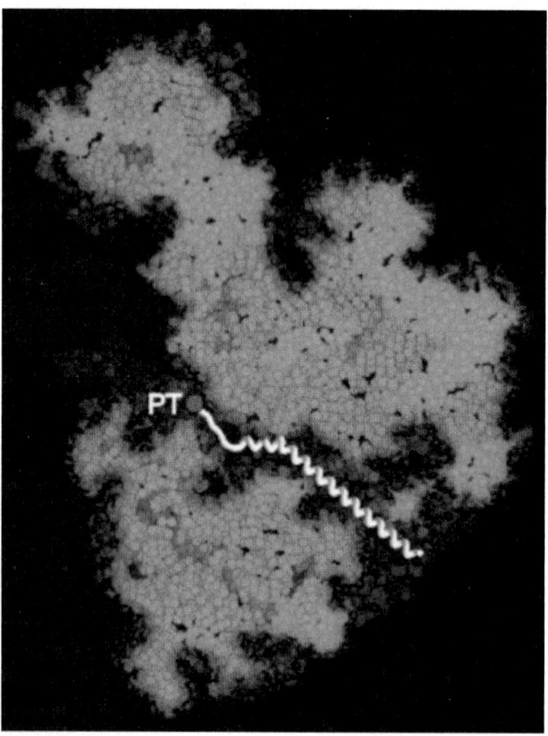

Fig. 25.11 The modelled nascent polypeptide shown schematically in the exit tunnel of the large ribosomal subunit. PT, peptidyl transferase centre. Buried RNA atoms are shown in grey, and protein atoms in green. N N Ban et al; The complete atomic structure of the large ribosomal subunit at 2.4 A resolution; Science; Aug 11, 2000. Reproduced by permission of American Association for the Advancement of Science.

and a tunnel for the exit of the synthesized polypeptide chain opens next to this. The chain is fed into it and emerges from the large subunit, $-NH_2$ end first. The mRNA binds to the face of the small subunit facing the cavity. Fig. 25.11 shows a cross-section of the large ribosomal subunit with the nascent peptide in the exit tunnel.

What is a polysome?

It takes about 20 seconds for a ribosome to synthesize a polypeptide of average length in *E. coli*, which adds about 20 amino acid residues per second. If only a single ribosome at a time moved along the mRNA molecule, the latter could direct the synthesis of the protein at the rate of 1 molecule per 20 seconds. However, as soon as an initiated ribosome has got under way and has moved along about 30 codons, another initiation can occur. One ribosome after another hops on to the mRNA. They follow one another down the mRNA, each independently synthesizing a protein molecule – a typical case would be about five ribosomes per mRNA molecule, but this varies with the length of the mRNA. This greatly increases the rate of protein synthesis. The term **polysome** is thus a shortened version of **polyribosome** and refers to the complex of multiple ribosomes associated with an in mRNA molecule in the process of translation.

Protein synthesis in eukaryotes

The major point at which protein synthesis in eukaryotes differs from that in prokaryotes is in initiation of translation. It is similar in that it requires the initiating methionyl-tRNA to bind into the P site of a small ribosomal subunit that positions itself correctly so that the initiating AUG is in that site. Following the formation of this complex the large subunit joins it and elongation proceeds as in bacteria. Initiation in eukaryotes involves over 12 eIFs (**eukaryotic initiation factors**) and one of these (**eIF3**), the largest and earliest discovered, has 13 subunits. However, despite this complexity, there are counterparts of IF1, IF2 and IF3. As in prokaryotes the IF1 and IF3 counterparts bind to the small (40S) ribosomal subunit and prevent it from associating prematurely with the large (60S) subunit.

One difference between prokaryotes and eukaryotes is that the initiating amino acid is methionine, not *N*-formylmethionine, although it still has a special initiatior tRNA, designated $tRNA_i^{Met}$, rather than the regular $tRNA^{Met}$ to bind to.

Another difference is that there is no equivalent to the Shine-Dalgarno sequence to direct the ribosome to the first AUG. Instead the small (40S) ribosomal subunit, in this case already complexed with Met-$tRNA_i^{Met}$ in the P site, recognizes and binds near the cap structure at the 5′ end of the mRNA. There are two GTP-binding proteins, eIF2 and eIF5b, that do the work of prokaryotic IF2, bringing the Met- $tRNA_i^{Met}$ to the ribosome (Fig. 25.12a). Strange though it may seem, the 3′

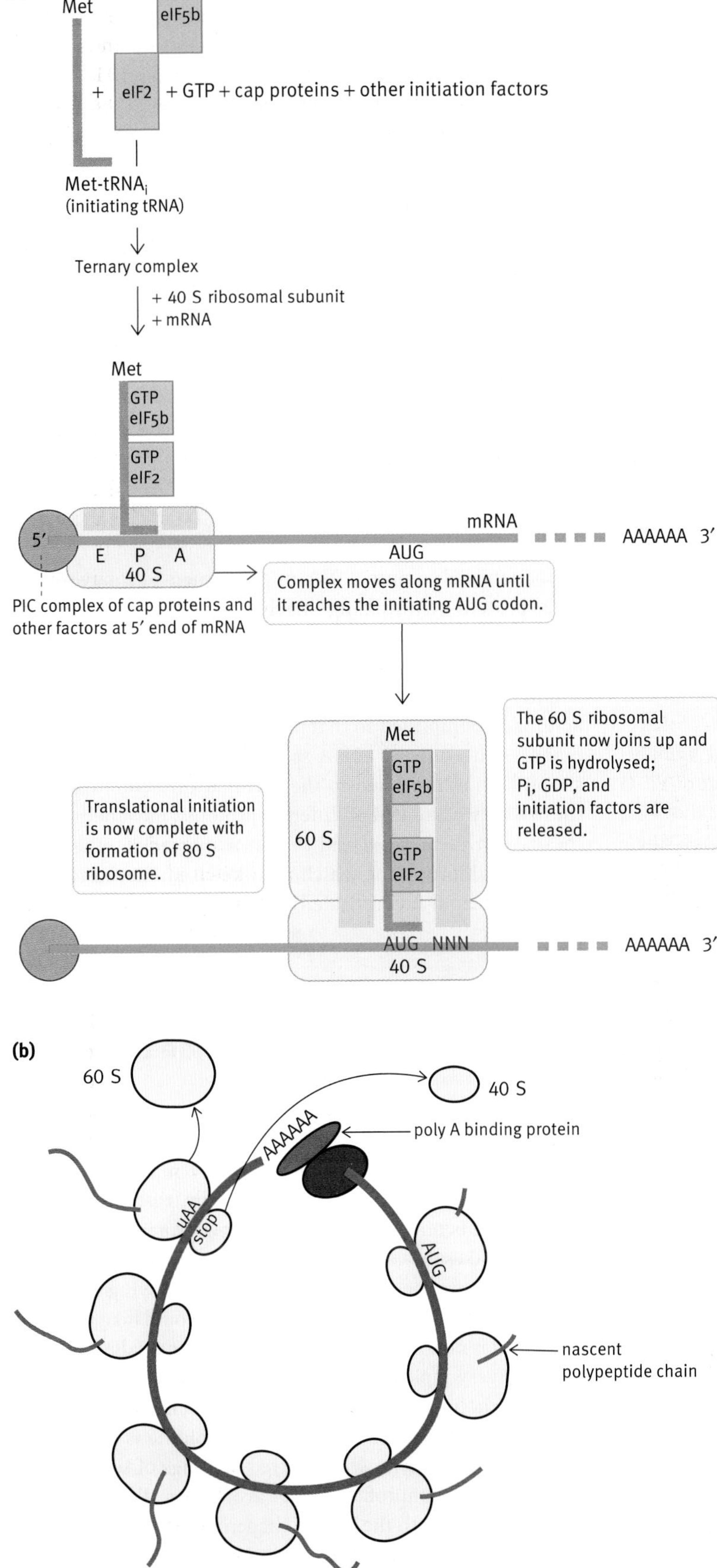

Fig. 25.12 **(a)** Simplified diagram of intiation in eukaryotes. Note that several eukaryotic initiation factors, besides eIF2 and eIF5b, are involved. eIF2 and eIF5b are the eukaryotic initiation factors corresponding to IF2 in prokaryotes. They select the initiating Met-$tRNA_i^{Met}$ and deliver it to the P site on the small ribosomal subunit with mRNA bound. The anticodon cannot base pair with the mRNA until the ribosome has found the initiation codon. PIC, preinitiation complex. NNN represents the second codon, N representing any nucleotide. **(b)** Interaction between protein factors bound at the 5′ and 3′ ends of the mRNA circularizes the mRNA during translation, so that ribosomes completing translation are positioned appropriately to restart.

polyA tail of the mRNA is also involved in initiating translation. A polyA binding protein attached to the tail is needed for binding of the 40S subunit into the **preinitiation complex (PIC)**. This arrangement means that the eukaryotic mRNA is held in a loop structure as it is translated, with its 5′ and 3′ ends close together. The ribosomal subunits that dissociate from the 3′ end as they complete translation are thus well positioned to begin again at the 5′ end (Fig.25.12(b)).

The next stage is to position the PIC at the AUG start codon of the mRNA with the Met- $tRNA_i^{Met}$ anticodon base paired with it. The PIC scans the message by moving along it until it finds the correct AUG. This is usually simply the first one encountered although there is often a characteristic set of bases adjacent, called the **Kozak sequence**, that aids recognition by PIC and hence increases translational efficiency. This process of 5′ cap binding and scanning for the first AUG by the PIC explains why eukaryotic mRNAs are usually monocistronic, only encoding a single protein. Once the PIC reaches the first AUG the large 60S ribosomal subunit joins to the small subunit forming the 80S complex, GTP on eIF2 and eIF5b is hydrolysed, and the eIF-GDPs and all the other initiation factors are released as initiation is complete. The eIF-GDPs are recycled back to the GTP state in the cytosol ready for further use.

The next stage is **elongation** – the synthesis of the complete polypeptide chain of the protein. This proceeds as in prokaryotes involving two cytosolic elongation factors. The counterpart of EF-Tu in eukaryotes is called EF1a, and that of EF-G (translocase) is EF2.

Termination is similar to that in *E. coli* except that a single release factor, eRF1, rather than the separate RF1 and RF2 factors, recognizes all three stop codons.

Incorporation of selenocysteine into proteins

Francis Crick invented the phrase, the 'magic 20', to indicate the amino acids for which there are codons in the mRNAs and hence are incorporated into proteins. However there is a 21st amino acid called **selenocysteine** that is incorporated by a 'freak' mechanism into a few specific proteins (encoded by specific mRNAs) during their synthesis on the ribosome.

$$H{-}Se{-}CH_2{-}CH(NH_2){-}COOH$$

There is no codon for this amino acid but a specific UGA *stop codon* is used as a substitute. A messenger RNA for a selenocysteine containing protein has a long stem loop section near to a UGA codon, which identifies it as the one to be used. The stem loop is called the **selenocysteine insertion sequence (SECIS)**. A special tRNA with an anticodon complementary to the UGA codon is used in the incorporation. The aminoacyl synthetase that recognizes this particular tRNA attaches *serine* to it. The seryl tRNA is then converted by a reaction with **selenophosphate** (produced enzymically from the metal and ATP) to **selenocysteinyl-tRNA**. This is delivered to the ribosome by a protein factor, which recognizes only this aminoacyl-tRNA. The incorporation involves a SECIS binding protein and an associated SECIS specific elongation factor. While the mechanisms of selenocysteine incorporation in eukaryotes and in *E. coli* are similar, the details of the process are less well elucidated in eukaryotes.

Selenium is toxic in large amounts, but is still an essential trace metal. There are not many selenocysteine proteins – the family of glutathione peroxidase enzymes (Chapter 32) is an example. They occur in both prokaryotes and eukaryotes; about 25 are known in humans.

Protein synthesis in mitochondria

Mitochondria contain DNA of their own and have their own protein-synthesizing machinery. The ribosomes of mitochondria are prokaryote-like and use fMet-$tRNA^{fMet}$ for initiation. They have other features of interest, such as a slightly different genetic code, and codon-anticodon interactions are simplified so that mammalian mitochondria can manage with only 22 tRNA species. The possibility of such simplification is related to the fact that mitochondria synthesize only a handful of different proteins, most being synthesized in the cytosol and then transported in. The same is true of chloroplasts, which are likewise believed to have originated from incorporated (in this case, photosynthetic) prokaryotes.

Folding up of the polypeptide chain

The newly synthesized polypeptide chain emerges from the channel in the ribosome in a denatured (unfolded) state. Proteins are not active in this state; they must acquire the three-dimensional structure specified by the amino acid sequence. It is not fully understood how proteins fold up so rapidly. It takes the cell anything from one second to two minutes. To achieve the correct conformation by trying all variations until the lowest free energy is found would take millions of years. It is now believed that certain sections, called 'molten globules' very rapidly fold to give the main features of a secondary structure, which is followed by side chain adjustments to give the final tertiary structure. The folding is probably a stepwise procedure in which correct foldings of sections are preserved, and nonprofitable ones avoided or allowed to correct themselves with the help of chaperones.

Chaperones (heat shock proteins)

As a newly synthesized polypeptide emerges from the ribosome, hydrophobic groups, which will ultimately be buried in the interior of the folded native protein, are exposed to the aqueous medium of the cytosol. Unless something is done to prevent it, these groups will form nonspecific hydrophobic associations to whatever other hydrophobic groups are available, either on the same or adjacent polypeptides. Such random, 'improper' associations could prevent the polypeptide from folding (unless corrected).

A family of proteins collectively referred to as **chaperones** guard against this. They are highly conserved in evolution and are present normally in all cells. They were discovered when it was observed that in cells subjected to temperatures higher than normal, certain proteins increased in amounts and were called **heat shock proteins (Hsps)**. It was later discovered that these Hsps are chaperones. What does heat shock have to do with protein synthesis? Heat denatures (unfolds) native proteins; newly synthesized polypeptides are also unfolded as they emerge from the ribosome. The heat denatured proteins have improperly exposed hydrophobic groups, and if they are to be salvaged must be refolded. So the problem is similar to that of newly synthesized proteins and the cell increases production of chaperones as an emergency response to heat.

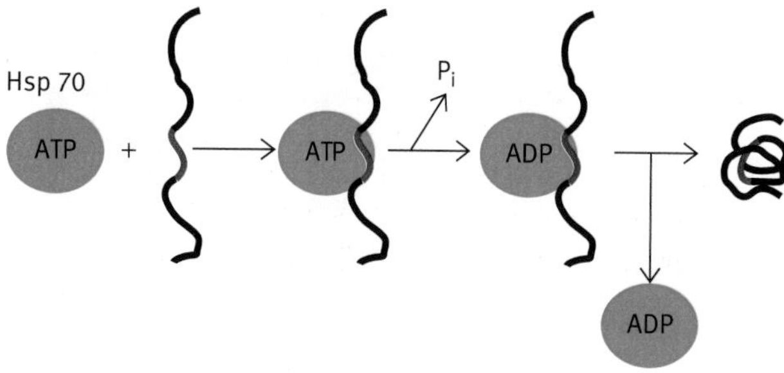

Fig. 25.13 Simplified diagram of *Escherichia coli* chaperone Hsp70 action in assisting folding of a polypeptide. The participation of other cochaperone proteins in the process has been omitted for simplicity. The essence of the process is that the chaperone is a slow ATPase and alternates between an ATP-bound form and an ADP-bound form, the former having a high affinity for the polypeptide and the latter a low affinity. The red line represents a hydrophobic patch on the polypeptide. Hsp70 can attach to nascent proteins emerging from the ribosome.

Mechanism of action of molecular chaperones

The Hsps are classified into three groups. The two best known are the **Hsp70** and the **Hsp60** groups.

Hsp70 attaches to the hydrophobic groups of nascent polypeptides (Fig. 25.13). It has a molecule of attached ATP, in which form it has a high affinity for the unfolded chain. Hsp70 is a slow ATPase. After the chaperone attaches to a polypeptide, the ATP is hydrolysed to ADP and a conformation change of the chaperone encloses the substrate, holding it in its unfolded state. ADP is then exchanged for ATP, triggering the reopening of the chaperone so it releases the polypeptide, allowing it to fold. The ATPase/ADP exchange is thus a timing mechanism to determine how long the chaperone remains attached to the unfolded polypeptide. By attaching, the Hsp prevents incorrect hydrophobic associations occurring until polypeptide synthesis is likely to be completed. When it detaches, the correct folding is given the opportunity to occur.

Probably the Hsp70 type of chaperone assistance is all that is required for many proteins; even large ones probably have several essentially structurally independent domains which fold sequentially as the polypeptide emerges. However, a number of proteins use a different form of assistance for completion, a molecular chaperone of the Hsp60 class (sometimes known as **chaperonins**). The best known of these is a protein complex in *E. coli* known as **GroEL**, together with a 'lid' structure known as **GroES**. GroEL is a multisubunit structure with two back-to-back rings of seven subunits. In Fig. 25.14 all such details of subunit structure are omitted for simplicity. A space-filling model of the chaperonin, together with a vertical cross-section, is given in Fig. 25.15.

We mentioned earlier that in Anfinsen's experiment, in which ribonuclease was refolded *in vitro*, the denatured protein was at low concentration which favoured the refolding because there is less chance of aggregation with other molecules. GroEL provides what is sometimes known as an 'Anfinsen cage' because it literally encloses a polypeptide so that it can fold in a hydrophilic box secluded from all the other proteins in the cell. It gives the protein a private folding chamber and then ejects it. If it has failed to fold it can try again because its hydrophobic groups are still exposed and will reattach to the GroEL. Hsp60 is needed typically for proteins imported, in an extended form, into the mitochondrial matrix where the concentration of proteins is extremely high. As illustrated in Fig. 25.14 the unfolded, or partially unfolded, protein attaches to the Hsp60 entry point by its hydrophobic groups. At this stage the chamber has a lining of exposed hydrophobic groups to which the unfolded protein can attach. ATP attaches to the central domain of the barrel and this results in a dramatic allosteric conformational change in the lining of the cavity containing the unfolded protein. This causes hydrophilic groups to be exposed instead of the hydrophobic ones, the cavity changes its shape and enlarges, and the GroES cap attaches and seals in the protein.

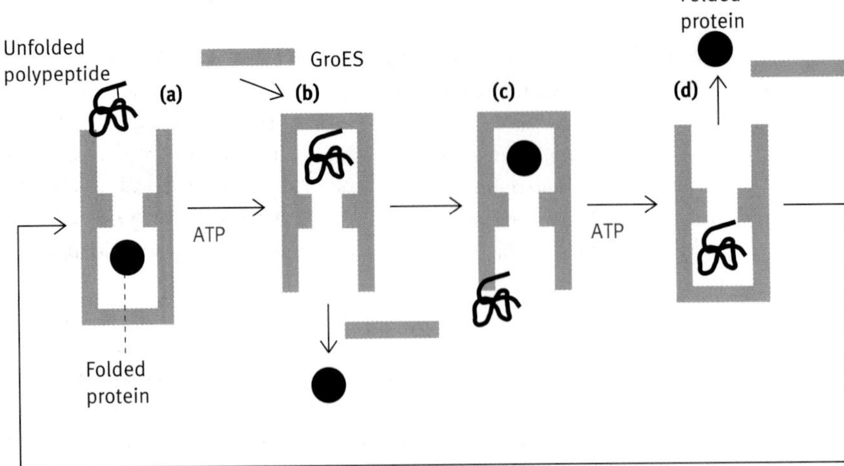

Fig. 25.14 Simplified diagram to illustrate the principle of GroEL action in *E. coli*. The 'lid' structure is known as GroES. The chaperonin has two folding chambers, which in this model are postulated to work alternately. In **(a)** an unfolded polypeptide has attached *via* its hydrophobic groups to GroEL. The lower cavity has a folded polypeptide represented as a solid circle waiting for release (this corresponds to the situation in the upper chamber in (c)). In **(b)** the unfolded polypeptide has entered the hydrophilic cavity and a 'lid' seals it in. Meanwhile the lower cavity has released its folded protein. In **(c)** and **(d)** the situation is just the same as (a) and (b) but upside down. Note that the diagram does not show the ring structure of the chaperonin nor the conformational changes that occur. These involve the changing of the lining of the cavity accepting the protein from hydrophobic to hydrophilic. The mechanism involves the hydrolysis of seven molecules of ATP at the steps indicated. The next figure shows a more realistic structural representation. The mammalian Hsp counterpart may function as a single ring.

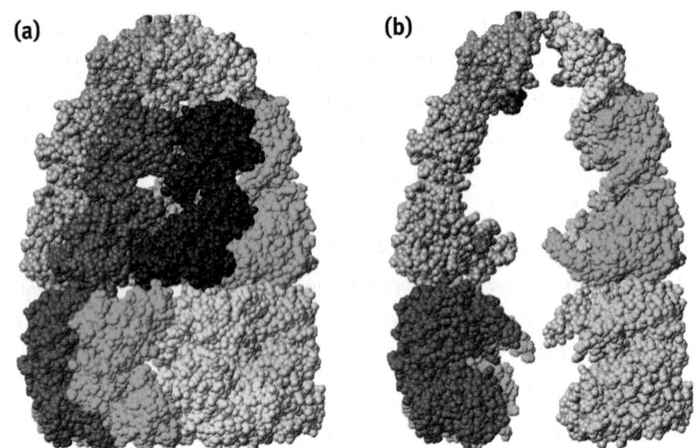

Fig. 25.15 **(a)** Space-filling model of GroEL with a GroES cap (Protein Data Bank code 1AON). This state corresponds with state (b) in the previous figure. **(b)** A cross-section of the same.

The hydrophilic box is an ideal environment for refolding to occur. Hydrophobic groups of the folding protein become hidden and its hydrophilic ones exposed, as is required in a folded protein. After a short interval the ATP is hydrolysed, the cap detaches, the folded protein is ejected and the conformational changes of the chaperonin reverse. If the folding was not successful hydrophobic groups will still be exposed and it can have another try. Note that the GroEL provides no folding guidance to the protein but just optimal conditions for it to fold itself. It is the amino acid sequence which determines the folded configuration. A main driving force for the folding is the hiding of hydrophobic groups from a hydrophilic environment. GroEL has two cavities and it has been suggested that they act alternately as shown in Figure 25.14. The yeast Hsp chaperone is believed to also act using two chambers alternately, but the mammalian one is thought to function as a single chamber.

Enzymes involved in protein folding

The chaperone assistance in folding is not the end of the folding story; there are two additional problems, this time requiring enzymic intervention.

The first is by **protein disulphide isomerase (PDI)**. This enzyme 'shuffles' S–S bonds in polypeptide chains; it rearranges them. If an incorrect S–S bond were to be formed, being covalent it would not break spontaneously and would lock the polypeptide in an incorrect configuration. PDI, by breaking and reforming S–S bonds between different cysteine residues, permits the folding to correct itself. It transfers an S bond from one disulphide bridge to another, using itself as the intermediary attachment site. There is little or no free energy change in the process and so it is freely reversible. High concentrations of PDI are found inside the endoplasmic reticulum involved in folding proteins destined for secretion, many of which have disulphide bridges.

Another enzyme is **peptidyl proline isomerase (PPI)**. Peptide bonds between amino acids commonly form in the *trans* rather than the *cis* configuration (see The peptide bond is planar section in Chapter 4), as steric factors favour the *trans* form. The unusual structure of proline, however, means that the *cis* and *trans* conformation of the peptide bond between proline and another amino acid can both occur and they are not easily interconverted. The 'incorrect' conformation can hold up folding of the protein. The PPI plays the role of 'shuffling' proline residues between the conformations in order to permit the whole protein to fold correctly.

Protein folding and prion diseases

Prion diseases are an unusual group of fatal neurological degenerative diseases that affect humans and animals. In sheep, the disease is known as **scrapie** because the animals scrape off their wool by rubbing against fence posts; while cattle are affected by **bovine spongiform encephalopathy (BSE)**, commonly known as **mad cow disease**. Human prion diseases include **Creutzfeldt–Jakob disease (CJD)** and **kuru**. Kuru used to be transmitted in certain New Guinea tribes by cannibalism. The diseases can be transmitted by consumption of infected tissue or, rarely, can be an inherited trait. The only known case of a prion disease being transmitted from animals to humans is mad cow disease, which in humans is known as **new variant CJD**. An outbreak of nvCJD in the 1990s, mainly in the United Kingdom, was caused by consumption of meat products from animals that had been fed infected material. The spread of disease has now been controlled by stricter legislation.

Prion diseases were initially believed to be due to a 'slow virus'. However, no nucleic acid-containing infectious agent was found. It is now known that the diseases are associated with an abnormal form of a normal protein found in brain. The *disease-producing unit* is called a **prion** (for **proteinaceous infectious particle**) and the harmful protein itself is called **PrP^{Sc}** (for the prion protein, scrapie, although it applies to human prion protein not just sheep). The *normal* counterpart of **PrP^{Sc}** is called **PrP^{c}** (for the normal constitutive prion protein) and is of unknown function. The PrP^{c} and PrP^{Sc} proteins have identical polypeptide amino acid sequences *and are coded for by the same gene*, but their folded conformations are different. PrP^{c} has four α helices and no β sheet content, is soluble and protease sensitive. In PrP^{Sc} two of the four α helices are folded instead into four β strands that make up a single β sheet, which make it resistant to proteases.

The question arises as to how an improperly folded protein can be infectious. No mechanism is known by which a protein molecule can direct its own replication. Instead, what happens is that PrP^{Sc} somehow causes PrP^{c} (the normal protein) to convert to the abnormal form. This has been demonstrated *in vitro* by incubating the two together. It is believed that it occurs by a 'seeding' mechanism where PrP^{Sc} proteins form aggregates to which PrP^{c} proteins attach and refold giving more PrP^{Sc}. The conversion *in vivo* of PrP^{c} to PrP^{Sc} is a rare event in the absence of infection by PrP^{Sc}, so spontaneous occurrence of the disease is rare. It is believed that mutations in the gene for the normal PrP^{c} may increase the probability of incorrect folding, which might explain the hereditary origin of some cases of the disease. Once some PrP^{Sc} is formed, it would then trigger the autocatalytic formation of more.

It is not known how prions cause disease. It is thought to be associated with accumulation in the brain of the long aggregate fibres which are known as **amyloids**. It is now known that other proteins are capable of forming similar β strand rich aggregates also often called amyloids and associated with diseases such as Alzheimer disease. In the case of Huntington disease (see Chapter 28) the protein aggregates are associated with an abnormal protein encoded by an inherited mutated gene.

Programmed destruction of protein by proteasomes

Introduction

There are three main ways in which proteins are broken down in the human body. The most obvious is in digestion, which is not an intracellular process. The second is intracellular, in the lysosome system (see Chapter 27) in which material to be destroyed is enclosed in a vesicle to which are delivered vesicles of destructive enzymes for all classes of materials. The third is by the **ubiquitin–proteasome system**, which is our present topic. This destroys individual selected protein molecules in cells. It is tightly controlled in an elaborate and sophisticated way.

Controlled protein breakdown in eukaryotic cells is of almost astonishing importance. For example, the cell cycle control (see Chapter 30) depends on the specific breakdown of cyclin proteins at precisely the correct time to allow the cycle to progress from one phase to the next. Failure to destroy them as appropriate would cause an improperly controlled cell cycle with disastrous consequences. Enzymes and other proteins are destroyed in a selective manner; some proteins have, in humans, a half-life of hours and others of days. Despite chaperones and the other methods of guarding against faulty folding of polypeptides errors still occur and unless these faulty proteins were removed they would accumulate and cells would be loaded down with them. As will be described in Chapter 27, many proteins are transported into the endoplasmic reticulum (ER) as a linear polypeptide where they fold up into the mature form. Mistakes happen and unfolded proteins are sensed and transported back out of the ER lumen into the cytosol where they are degraded in proteasomes. As a final example of its importance, as will be explained later (see Chapter 33), a major part of the immune protection against viruses is dependent on the proteasomal destruction of proteins.

The structure of proteasomes

Proteasomes are organelles which provide a cavity in which proteins destined for destruction are segregated from the rest of the cell in an unfolded form and degraded to small peptides. These organelles are large protein structures about 2 million Daltons in size. They are present in all eukaryotic cells in both the cytosolic and nuclear compartments in large numbers and are visible in the electron microscope. A model of the structure is shown in Fig. 25.16. There is a central **20S core** (see Ribosomes Section earlier in this chapter for an explanation of S values) consisting of a barrel-shaped cylinder made of four annular rings of protein subunits; the end ones, known as α rings, sandwich the two β rings. The latter contain proteolytic enzymes on the inside of the cylinder; the α rings have no known enzyme activity. At both ends of the 20S core are **19S caps**, also known as the regulatory units since they control the selection of proteins to be admitted and unfold the proteins so that they can enter the cavity of the 20S core where the actual hydrolysis takes place. The unfolding by the caps is ATP-driven. The dimensions of the proteasome cavity are such that extended polypeptide chains can be accommodated, but not folded proteins. Here the polypeptide is cut into small peptides, which emerge into the cytosol. It is more or less like a tree trunk being sliced up into short logs. This may seem a trivial detail, but the correctly 'sized' peptides are vital in the immune system.

Proteasomes occur in a limited number of modern bacteria (eubacteria) but are found in the archaea, which often live in hostile environments such as sulphur hot springs at 80 °C and pH 2. These proteasomes are different in that they lack the end caps, and have only the 20S core, virtually identical in appearance to the core of those in yeast, but with fewer types of subunits in the rings. It is remarkable that the basic 20S core structure has been conserved in archaebacteria, yeast, and humans.

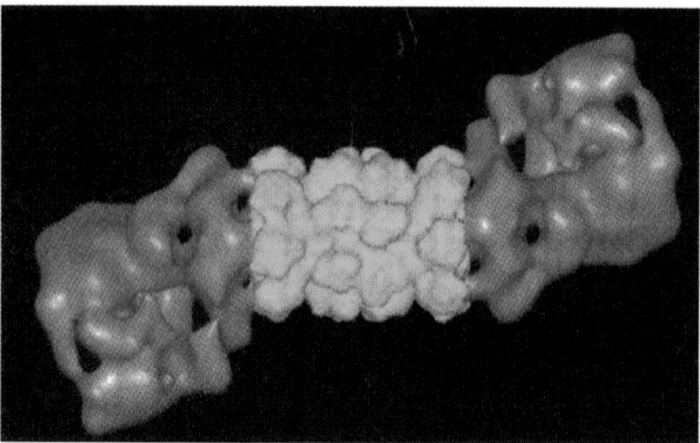

Fig. 25.16 Model of the proteasome. The yellow represents the structure of the 20S proteasome core and the blue, the 19S caps. Baumeister, W et al; The proteasome: paradigm of a self-compartmentalizing protease; Cell; (1998) 92, 367-80; Elsevier.

Proteins destined for destruction in proteasomes are marked by ubiquitination

Ubiquitin is a small protein of 76 amino acids found universally in eukaryotes but not in prokaryotes. It has been highly conserved throughout evolution. Its amino acid sequence is identical in all animal species and differs from that in yeast and plants by only three, very conservative, amino acid changes.

Mechanism of ubiquitination of proteins

Three enzymes are involved (Fig. 25.17).

- Step 1: the ubiquitin is activated, its terminal –COOH becomes attached by a thioester link to the –SH group of enzyme 1 (**E1**) the **ubiquitin activating enzyme**. This requires ATP breakdown to AMP and PP_i because the thioester is a high-energy bond.
- Step 2: E1 transfers its attached ubiquitin to an –SH group on **E2**. (The latter exists in the cell complexed with **E3**.)
- Step 3: E3 is the final enzyme. It exists as a complex with E2. The E2/E3 complex is the enzyme **ubiquitin ligase** (the active form of E3). It transfers the ubiquitin from E2 to the ε amino group of a lysine residue on the target protein.
- Step 4: The ubiquitin attachment is the 'death ticket' for a protein but rather oddly the attached ubiquitin itself becomes ubiquitinated. About four molecules are optimal for proteasomal destruction. The multiubiquitinated protein molecule binds to the proteasome caps, is unfolded, and fed into the proteasome core containing the proteases. Prior to this, the attached ubiquitin molecules are released by enzymes and recycled.

Selection of target proteins for ubiquitination

This is a critical question for unless there are rigorous selection criteria there would be mayhem from random destruction. The selection system is not fully understood but a good deal is known. One criterion is the *N*-terminal amino acid of a protein. Some proteins with very short half-lives in yeast (less than 1 hour) are characterized by arginine, lysine, or aromatic *N*-terminal amino acids. However, there are other criteria. Some destabilizing signals are masked and revealed by conformational changes induced by a ligand. Phosphorylation by a protein kinase is also believed in some cases to trigger recognition of a target protein by the E3 ligase. Denaturation may reveal destabilizing signals that are normally hidden. With all the different signals to recognize there needs to be a variety of different ligases and there are. Several hundred different E3 enzymes and a variety of E2s exist so that there are a great many different ubiquitin ligases. There must also be a special mechanism whereby for example cyclins in the cell

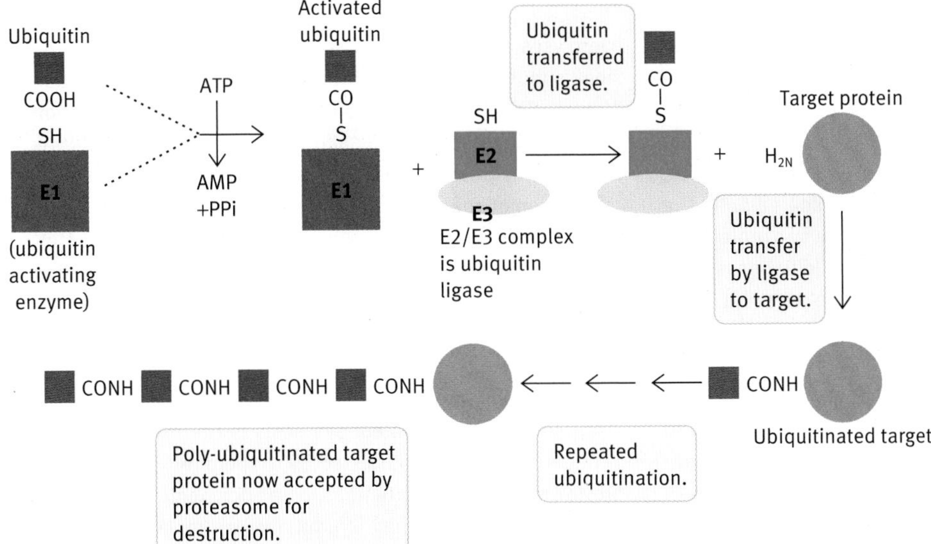

Fig. 25.17 Sequence of events in ubiquitinating target proteins destined for proteasomal destruction. Ubiquitin is first activated by attachment to E1 by a thioester link at the expense of ATP breakdown. It is then transferred to ubiquitin ligase, which is a complex of E2 and E3. The ligase transfers it to a lysine amino group on the target protein. This is repeated to give a polyubiquitinated target, which is then accepted by a proteasome for destruction.

cycle are triggered to be degraded at the end of each cell cycle phase (see Chapter 30). The complexity of the organization that must go into the safe running of this system is somewhat mind-boggling.

The role of proteasomes in the immune system

Somatic cells are at risk of virus infection. The body defends itself by destroying infected cells thus aborting virus multiplication, but it must detect those cells carrying foreign proteins and ignore normal ones. Their destruction is brought about by cytotoxic killer T cells (see Chapter 33), which recognize infected cells by a remarkable mechanism in which proteasomes play an essential part. All somatic cells continually hydrolyse 'samples' of cytosolic proteins into short peptides and display these on their surface for inspection by the killer cells, which ignore displayed peptides originating from normal (self) proteins, but recognize those from a foreign source, usually a viral protein, synthesized in the cytosol of an infected cell. The cell is then attacked and destroyed. A more mechanistic account of this system can be found in Chapter 33.

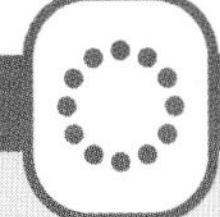

Summary

Protein synthesis is the joining together of amino acids in the correct sequence to form a polypeptide chain. It involves three phases, initiation, chain elongation, and termination.

Messenger RNA (mRNA) contains information of the amino acid sequence of the protein in the form of triplets of bases known as codons. Each codon corresponds to a specific amino acid, the correlation between them comprising the genetic code.

The sequence of codons in an mRNA molecule is translated by cytosolic ribosomes into the sequence of amino acid residues in a polypeptide chain. Ribosomes have a large and a small subunit each containing RNA and many proteins.

As well as ribosomes, many cytosolic proteins are needed, and also a group of transfer RNAs (tRNAs). The tRNAs act as adaptors between the amino acids and the codons on the messenger being translated by the ribosome. Each tRNA has an unpaired triplet of bases known as the anticodon. There are tRNAs specific for each amino acid; protein synthesis depends on amino acids being placed on their own cognate tRNA – the one whose anticodon is complementary to the codon specifying it on the messenger. The amino acids are attached to the tRNAs by aminoacyl-tRNA synthetases that recognize their appropriate tRNA and the cognate amino acid, and join them together by ester linkage. ATP is broken down to supply the energy to link the amino acid to the tRNA. This supplies the energy for the subsequent formation of a peptide bond.

The initiation codon is AUG, which specifies methionine, but it is also used for other methionine residues within the protein. Initiation has to distinguish the first methionine. There is a special $tRNA_i^{Met}$ specific for methionine that is recognized by a protein that delivers methionyl-tRNA only to the initiation codon. In *E. coli* the first methionine is formylated. The essence of what follows is that the next amino acid on its tRNA is delivered to the second codon on the ribosome and the first methionine is transferred to the incoming aminoacyl group to form a dipeptide.

The ribosome has three tRNA sites, A, P, and E (acceptor or aminoacyl, peptidyl, and exit). Initiation in *E. coli* involves formation of a preinitiation complex (PIC) of formyl-methionyl-tRNA, mRNA and initiation factors on the small ribosomal subunit. The formyl-methionyl-tRNA is delivered to the P site by a protein factor, which recognizes only the initiation tRNA. Initiation is completed by joining the large ribosomal subunit to the small one.

Chain elongation involves delivery of aminoacyl-tRNAs to the A site by an elongation factor bound to GTP, where they bind by codon-anticodon hydrogen bonding. The P site is already occupied by the peptidyl-tRNA. The peptidyl group is transferred to the amino group of the latter thus elongating the nascent peptidyl chain in the amino to carboxyl direction.

Translocation transfers the elongated peptidyl-tRNA to the P site while the discharged tRNA moves to the exit site. The vacant A site accepts the next incoming aminoacyl-tRNA and a new round of elongation ensues. Translocation requires the cytosolic translocase (EF-G in *E. coli*) to attach to the ribosome and GTP hydrolysis is involved. The synthesis of new peptide bonds is effected by peptidyl transferase, which is a ribozyme formed by folded rRNA. Probably the primitive ribosome was entirely RNA and proteins a refining addition. This avoids the origin of life 'chicken-and-egg' problem of how proteins could be essential for the synthesis of the first proteins.

At the end of the mRNA is the termination site, which comprises a stop codon where a protein factor releases the polypeptide. GTP hydrolysis plays a role in inducing conformational changes in proteins attached to ribosomes during various phases of translation.

To achieve fidelity of translation, the aminoacyl-tRNA synthetase attaching the amino acid to tRNAs in some cases hydrolyses incorrect amino acid loadings. Additionally the small subunit rRNA checks the codon – anticodon interaction Watson–Crick base pairing before the peptidyl transfer reaction occurs. A pause mechanism allows incorrect aminoacyl-tRNAs to leave the ribosome.

Initiation of eukaryotic translation differs in that protein factors, the small ribosomal subunit and (non-formylated) Met-$tRNA_i^{Met}$ assemble at the mRNA 5′ cap site. The preinitiation complex moves down the mRNA until it encounters the AUG initiation codon; initiation is then completed by attachment of the large subunit. Elongation and termination are similar in eukaryotes and prokaryotes.

The final stage of protein synthesis is the folding of polypeptide chains. This is assisted by chaperones, proteins that provide maximum opportunities for the polypeptide to fold correctly either by preventing unprofitable hydrophobic associations or providing an optimum environment for correct folding. Incorrect folding of the prion protein causes diseases such as mad cow disease (bovine spongiform encephalopathy). Other protein misfolding diseases exist.

Selenocysteine is found in a few proteins. It is incorporated by a special mechanism where a codon used as a stop codon in other proteins is recognized by the tRNA carrying this unusual '21st amino acid'.

Targeted protein degradation is of central importance in many aspects of the life of the cell. Prominent among these is the degradation of cyclins (see Chapter 30), which control the eukaryotic cell cycle. The selection of proteins for destruction is effected by attachment of ubiquitin proteins, which direct the protein to proteolytic destruction chambers known as proteasomes. The selection of proteins for ubiquitination is a complex process involving multiple ubiquitin ligase enzymes.

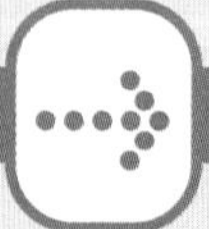

Further reading

To access the further reading, please scan the QR code image or go to http://global.oup.com/uk/orc/biosciences/molbiol/snape_biochemistry/student/reading/ch25/

Problems

1 There are 64 codons available for 20 amino acids. Why do you think 61 codons are actually used to specify the 20 amino acids?

2 Despite the facts stated in question 1, there are fewer than 61 tRNA molecules. Explain how this is so.

3 In diagrams, when a tRNA molecule is shown base paired to a codon, the molecule is shown flipped over as compared with the same tRNA shown on its own. Why is this so?

4 At which points in protein synthesis do fidelity mechanisms operate?

5 Describe the participation and, where known, the role of GTP in protein synthesis.

6 Studies have indicated that, in *E. coli*, tRNA molecules (with their aminoacyl or peptidyl attachments) straddle A, P, and E sites on the ribosome. Explain why this occurs.

7 The mechanism of initiation of translation in eukaryotes is not compatible with polycistronic mRNA. Explain why.

8 Explain the role of chaperones in protein synthesis.

9 What diseases are associated with improper protein folding?

10 (a) If you are given the base sequence of the coding region of a mRNA can you deduce the amino acid sequence of the protein it codes for?

(b) If you are given the amino acid sequence of a protein can you deduce the sequence of the coding region of the mRNA that directed its synthesis? Explain your answers.

11 Consider an mRNA that codes for a protein 200 amino acid residues in length. What would the resultant polypeptide be from translation of this messenger if codon number 100 was mutated so that its first base was deleted or if the first and second bases were deleted? What if all three bases were deleted? Explain your conclusions.

12 In a ribosome, is the RNA there simply as an inert scaffold on which to hang proteins? Discuss two pieces of evidence that bear on this problem.

13 Explain the difference between transcription and translation.

14 Describe a proteasome. What are its roles in cells? Give some examples of the latter. What is the evidence for their great importance in cells?

15 How are proteins targeted to the proteasomes?

16 Codons and anticodons specifically base pair on the ribosome. Is the formation of hydrogen bonds itself sufficient to account for the observed fidelity of protein synthesis?

17 Explain briefly how an Hsp70 chaperone works. Why should chaperones be given the prefix Hsp in their names?

Chapter 26

Control of gene expression

In Chapters 24 and 25 we have described the processes by which genes are expressed: transcription and translation. Some mention has been made of gene control, or regulation of gene expression, but as this is an important and complex topic extensive coverage has been saved for this chapter. First, we need to consider why gene control is necessary. We have seen that protein synthesis uses cellular resources and energy: this means that evolution will favour systems that conserve resources by synthesizing particular proteins only when they are needed. For instance, bacteria have evolved to adapt to changes in the availability of particular nutrients in their environment by synthesizing only the enzymes needed to metabolize those nutrients, rather than constantly synthesizing all the enzymes encoded by their genome. Eukaryotic cells also have to respond to environmental changes, such as hormonal signals or the availability of particular metabolites. A second driver for gene regulation is that multicellular eukaryotes consist of multiple cell types, all deriving from a single cell, the zygote. This cellular specialization also requires controlled gene expression so that the appropriate proteins are made in each cell type.

Levels of regulation

As protein synthesis takes place in multiple stages there are many levels at which it can be regulated: mRNA transcription, processing and nuclear export of mRNA (in eukaryotes) and translation. Examples of each of these are described below, but if we consider gene regulation as an energy saving device it should come as no surprise that the bulk of regulation takes place at the level of transcription, and mostly at the initiation step. There is generally no point in expending energy to begin a process you are not going to complete. However, regulation at later stages does take place, and is often a fine tuning or rapid response mechanism.

Gene control in *E. coli*: the *lac* operon

We have already seen in Chapter 24 how *E. coli* can regulate transcription in response to environmental stress via different sigma factors. In this chapter we consider a different mechanism of transcriptional regulation, that occurs in the *E. coli lac* operon. This is a good example of gene control in response to the availability of a particular carbon and energy source, the disaccharide lactose present in milk. In *E. coli* glucose-metabolizing enzymes are made **constitutively** (at all times). Since glucose is the most common sugar and other sugars are shunted on to the glucose metablic pathways, glucose-metabolizing enzymes are always needed so the promoters of the genes coding for them have no 'on' and 'off' switches – they are always 'on'. In contrast, the enzymes required for utilization of lactose are **inducible**: that is in the absence of lactose they are made only at very low **basal** levels, but when lactose is abundant and glucose scarce the *E. coli* cell responds rapidly by greatly increasing their expression. The regulation is at the level of transcription initiation.

In order to explain how transcription is regulated in response to lactose, it is first necessary to remind you that many prokaryotic mRNAs are **polycistronic**: it is common to find a group of genes encoding enzymes required for a single metabolic pathway clustered together in the genome, so that they can be transcribed as a single polycistronic mRNA. This allows coordinated regulation of their expression by a single promoter, as is the case for the three enzymes required for utilization of lactose in *E. coli*. The term used for such a cluster of genes and their promoter is an **operon**.

The key enzyme needed to utilize lactose is **β-galactosidase**, so called because lactose is a β-galactoside and must be hydrolysed to free galactose and glucose before it can be metabolized (Fig. 26.1). A transport protein, **lactose permease** or **β-galactoside permease** is needed to transport the lactose into the cell. A third protein, **galactoside transacetylase**, is believed to be involved in protection of the cell against nonmetabolizable, potentially toxic β-galactosides that may be imported, though less is known of this enzyme and its role than of the other two proteins encoded by the operon. The three proteins are normally made in minute amounts (basal levels) because they are not required unless lactose is encountered. When lactose is encountered as the sole energy source, there is an almost instant burst of synthesis of the

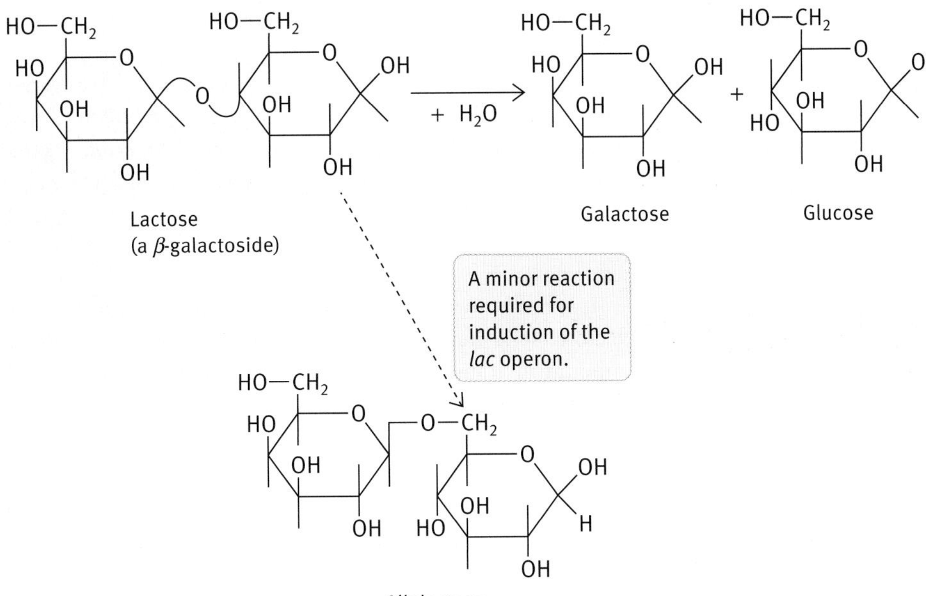

Fig. 26.1 The reactions catalysed by β-galactosidase. See the Structure of the *E. coli* lac operon Section for an explanation of allolactose formation. Hydrolysis of lactose to galactose and glucose also occurs in the human digestive system, where the enzyme is called lactase.

three proteins. The cell can then use the lactose as a carbon and energy source. However, if, in addition to lactose being present, there is also glucose, then production of the three enzymes would be wasteful since this merely leads to production of more glucose inside the cell when there is plenty of it available anyway. The cell therefore 'ignores' the lactose signal and does not produce the lactose metabolizing enzymes.

Structure of the *E. coli lac* operon

The approximate arrangement of the operon is shown Fig. 26.2. The three genes encoding β-galactosidase, lactose permease, and transacetylase are called the *lacZ*, *lacY* and *lacA* genes, respectively. There is also a separate ***I* gene** (*I* for inducibility)

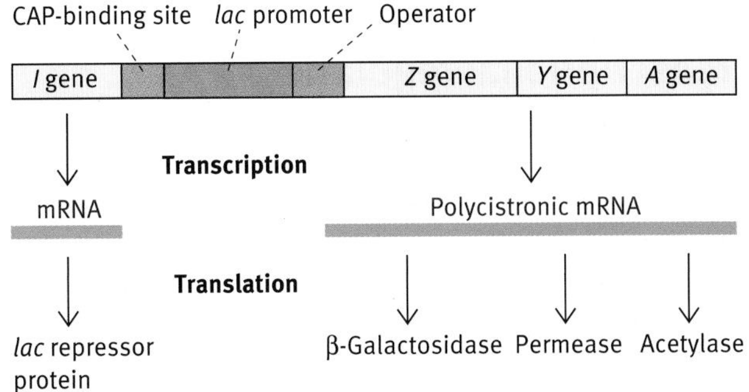

Fig. 26.2 Diagram of the *lac* operon. Note that the *I* gene is an independent gene that codes for the *lac* repressor protein. Similarly, there is a completely independent gene producing the catabolite gene-activator protein (CAP) to which cyclic AMP can bind.

that codes for a protein called the ***lac* repressor**, and there is a stretch of DNA called the **operator region** to which the *lac* repressor protein can bind. The **promoter**, where RNA polymerase binds to begin transcription, is flanked by two regulatory DNA sequences, each of which is recognized and bound by a different regulatory protein. The lac repressor binds to the **operator** sequence, which lies just downstream of (and partially overlapping) the promoter, while just upstream of the promoter there is a stretch of DNA to which a cyclic AMP (cAMP) receptor protein (**catabolite gene-activator protein – CAP**) can bind. CAP is given this general name because, unlike the lac repressor it is involved in regulating a number of genes that encode enzymes involved in metabolizing substrates other than glucose. CAP is part of the cell's response to a shortage of glucose.

We can now systematically go through the control mechanism noting the following points. The *lac* promoter *on its own* is a weak one so that the RNA polymerase does not readily bind to it and initiate transcription. Without extra help the *lac* operon is not transcribed except at a low basal level. The extra help in polymerase binding is given by the attachment of the protein, **CAP**, to the adjacent site. When CAP is attached to the DNA, the promoter becomes a strong one. However, CAP does not attach unless cAMP is bound to it – it is an allosteric protein and *cAMP is abundant in the cell only when glucose levels are low.* The binding of cAMP to CAP is freely reversible. When glucose is at a high level, cAMP is scarce, CAP does not bind DNA, the RNA polymerase therefore does not bind effectively, and *lac* operon transcription is minimal and so is the production of β-galactosidase.

Does this mean that when glucose is scarce and cAMP-CAP assists RNA polymerase binding to the promoter the *lac* operon is always transcribed? The answer is no. As explained, there is no point in doing so unless lactose is present. In the absence of lactose the ***lac* repressor** protein is attached to the operator and as it sits between the promoter and the protein-coding sequences it blocks the RNA polymerase from transcribing the genes. Like CAP, the *lac* repressor is an allosteric protein. In the absence of lactose in the environment, the repressor has strong affinity for the operator stretch of DNA, but if lactose is present the repressor is allosterically altered and no longer binds the operator. Lactose is often referred to as the **inducer** of the *lac* operon. However, this is not strictly speaking true: when the bacterial cell encounters a high level of lactose in the environment a small amount is able to enter the cell via the basal level of permease and as this lactose is hydrolysed by the basal level of β-galactosidase, a proportion is converted to the lactose isomer, **allolactose**, which is the true inducer (see Fig. 26.1). Allolactose binds reversibly to the repressor protein, causing an allosteric change, which causes the repressor to dissociate from the operator and unblocks the operon. With the unblocking of the operator, the RNA polymerase is now free to move down the operon, producing the polycistronic mRNA this is translated to produce the three enzymes.

It is not known why allolactose is the inducer rather than lactose. It is formed by β-galactosidase catalysing a small amount of transglycosylation in which a galactosyl residue is reversibly transferred to the 6-OH of glucose instead of to water (Fig. 26.1). Conceivably it is a mechanism for preventing the firing off of wasteful induction when only minimal 'uneconomic' amounts of lactose are present.

Figure 26.3 illustrates regulation of the *lac* operon in three different situations, showing how the CAP and lac repressor proteins enable an appropriate respose to different environmentaly level of glucose and lactose.

The *lac* operon was the first understood example of prokaryotic gene control, but similar mechanisms apply to operons associated with other metabolic pathways. The *E. coli* ***gal* operon**, encoding enzymes involved in metabolism of galactose, is regulated by CAP (thus responding, like the lac operon, to low glucose levels) and a specific gal repressor, which prevents expression unless the inducer, galactose, is present. The ***trp* operon**, which contains the five structural genes needed to produce enzymes involved in synthesis of the amino acid tryptophan, is controlled by a ***trp* repressor protein**, but in this case the repressor does not bind to the operator unless the repressor is allosterically modified by tryptophan. This reflects the function of the trp operon as synthesis of tryptophan is not necessary if the amino acid is already abundant. The *trp* operon is additionally regulated by a process known as **attenuation**, which is described later in this chapter.

Situation (a). High glucose; no cAMP; no lactose; no transcription of *lac* operon.

Start site of transcription

CAP site | Promoter | Operator | *Z* gene | *Y* gene | *A* gene

lac repressor protein binds to operator in absence of inducer.

CAP

RNA polymerase

CAP does not bind in absence of cAMP

Does not bind efficiently if CAP is not bound to CAP site.

Situation (b). Low glucose; high cAMP; CAP–cAMP complex binds CAP site; RNA polymerase can now bind to promoter; no lactose; repressor protein blocks operator; no transcription.

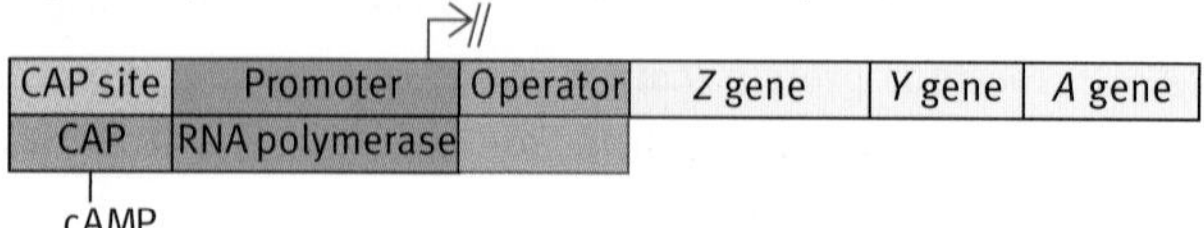

Situation (c). Low glucose; high cAMP; lactose present; repressor protein–allolactose complex detaches from operator; transcription of operon proceeds.

Transcription

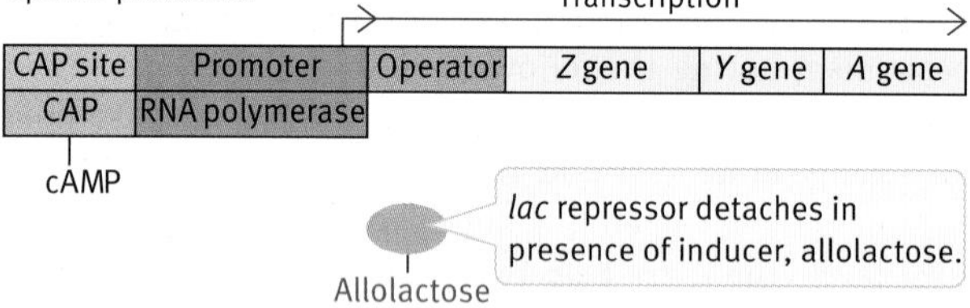

Fig. 26.3 Expression of the ***lac*** operon. **(*a*)** In the presence of high glucose there is no cyclic AMP (cAMP) to cause catabolite gene-activator protein (CAP) to bind, and this binding is necessary for the attachment of RNA polymerase to the promoter. With no lactose the lac repressor is bound to the operator. **(*b*)** With low glucose but no lactose, although CAP binds and assists the RNA polymerase to bind, transcription still does not occur because the *lac* repressor is bound to the operator, blocking polymerase movement. **(*c*)** As glucose is low CAP has assisted RNA polymerase to bind. In the presence of lactose, a small amount is converted to allolactose, which acts as the inducer. Allolactose binds to the repressor, causing its release from the operator and transcription can proceed.

Transcriptional regulation in eukaryotes

A general overview of the differences in the initiation and control of gene transcription in prokaryotes and eukaryotes

Although gene control in eukaryotes is also largely effected at the level of transcriptional initiation, and involves activator and repressor proteins similar in function to CAP and the *lac* repressor, the situation is considerably more complex than in prokaryotes. One reason for this is the packaging of eukaryotic genomes into chromatin. A second is the large number of different cell types and functions associated with multicellularity. We will consider the requirements of multicellularity first.

An *E. coli* cell has about 4,000 genes. During the lifetime of a single cell it may be necessary to transcribe all of those genes. For the most part what is required is an 'on-off' switch for regulated genes such as those of the *lac* operon or, for constitutive genes, a permanent 'on' condition. The rate of initiation depends in both cases on the affinity of RNA polymerase for the promoter, while in regulated genes RNA polymerase is also 'helped' or 'hindered' by activator and repressor proteins that bind specific DNA sequences such as the CAP-binding site and the operator. In a eukaryote, the situation is analogous, in that regulatory proteins known collectively as **transcripton factors (TFs)** bind to specific DNA sequences and control the rate of transcription initiation by RNA polymerase. However, it is much more complex. A human cell has about 21,000 protein-coding genes. There are many different types of cell: liver, muscle, brain, epithelial, blood, bone, and so on. Many proteins are common to all – those for glycolysis are an example, encoded by what are known as **housekeeping genes** – but each type of cell also has its own cohort of proteins needed for specific cell functions. Liver cells have liver-specific proteins not present in brain and muscle cells, and *vice versa*. However, the DNA of all cells is the same (ignoring gametes and special cases such as the gene rearrangements in B and T cells of the immune system described in Chapter 33, Generation of antibody diversity section). In a liver cell, genes for liver-specific proteins must be activated while those coding for brain-specific proteins must be 'ignored' by the transcription apparatus in that cell. All cells of the body arise from a single fertilized egg cell and differentiation into specific cell types involves gene control.

Even in mature cells when the differentiation into cell types has been achieved, another set of control problems exist. The activities of each cell must be such that they correspond to the needs of the organism as a whole. An obvious instance of this is that cell division should not proceed independently (as happens in cancer): the cell must receive one or more signals from other cells before it proceeds to division. Additionally, the rate of synthesis of individual proteins varies over short time periods according to needs. For example, after feeding, the level of enzymes devoted to breakdown or storage of foodstuffs increases resulting from hormonal activation of specific genes. The activities of many cells are controlled by a whole battery of hormones, growth factors, and cytokines (see Chapter 29), which regulate appropriate genes. A given gene in a cell may be simultaneously regulated by a multiplicity of signals from hormones or other factors.

A consequence of this complexity is that there can be a large number of short regulatory DNA sequences associated with each eukaryotic gene. A plethora of different TFs in eukaryotic cells interact with these sequences. It has been estimated that more than 5% of human genes code for TFs. Tissue-specific expression of genes depends on the presence or activity of particular TFs in the cells. Some TFs are activated only when the cell receives the appropriate signal such as from a hormone; a wide variety of external signals work in this way, by activating specific TFs. This system gives some of the very great flexibility needed for gene control in eukaryotes.

After this general preview, it is evident that crucial elements of transcriptional regulation in eukaryotes that we need to consider are the regulatory DNA sequences and their cognate transcription factors. 'Cognate' means 'related to': here meaning that a particular TF relates to a regulatory DNA sequence by having an affinity for and binding specifically to it. DNA binding and transcriptional activation by TFs is considered later in this chapter. Initially, we will focus on the DNA. We will be discussing protein-coding genes, transcribed by RNA polymerase II, since these are most subject to complex control.

DNA elements involved in eukaryotic gene control

Eukaryotic gene regulation involves several classes of DNA element, which are illustrated in Fig. 26.4 and described in more detail below.

Promoters

We have already introduced Type II eukaryotic promoters in Chapter 24. As a reminder, the region around the start site of transcription contains the basal elements: typically some combination of the initiatior (Inr) sequence, the TATA box and the downstream promoter element (DPE). Unlike in prokaryotes, where RNA polymerase itself recognizes the promoter, general transcription factors such as the TFIID

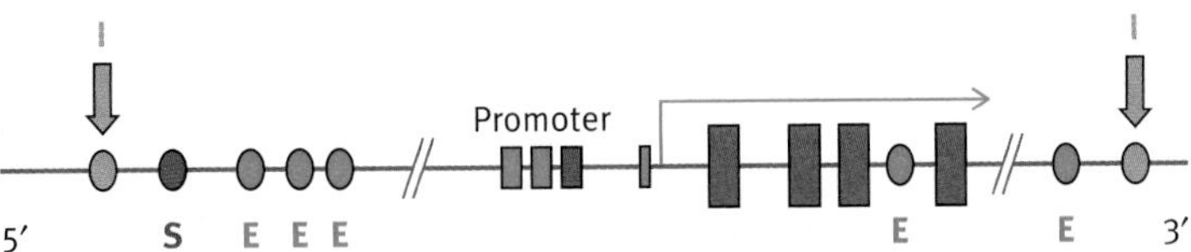

Fig. 26.4 DNA elements involved in eukaryotic gene control. Transcription is indicated by the green arrow. The promoter contains both positive (green) and negative (red) regulatory elements. Enhancer elements (E) are typically thousands of base pairs away from the transcription start site, and may be upstream (5′) or downstream (3′) of, or even with an intron of the transcribed sequence. Distant regulatory elements that repress transcription are termed silencers (S). Insulators (I) are sequences that demarcate the regulatory unit and prevent the regulatory sequences within it from influencing adjacent genes.

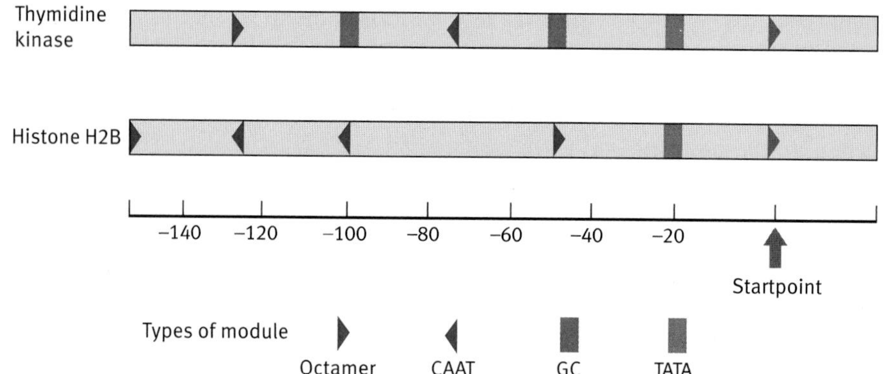

Fig. 26.5 The elements of two eukaryotic gene promoters. Promoters contain different combinations of TATA boxes, CAAT boxes, GC boxes, and other general control elements. Adapted from Fig. 29.10 in Lewin B. (1994) Genes V, Oxford University Press, Oxford.

complex are required to recognize and bind these sequences and position RNA polymerase II correctly. These basal elements make up the **core promoter**; the minimal sequence of DNA that can correctly initiate transcription. However, with only these basal elements, transcription is slow and inefficient. Additional **upstream control elements** such as the CAAT and GC boxes (see Type II eukaryotic gene promoters in Chapter 24) are also present. These and other common control elements are present in the promoters of many genes, in variable numbers and at various positions. They are recognized and bound by TFs that are common to many cell types, and generally increase the rate of transcription.

Figure 26.5 shows examples of the variety of upstream control elements found in promoters. In addition to these, there may be a number of control elements concerned with tissue-specific expression, hormonal control, and control by many other factors, as appropriate to the particular gene. To give a single example, the promoters of genes encoding globins have multiple regulatory elements containing the sequence GATA. A specific TF found only in developing red blood cells binds the GATA sequence and is required for activation of the globin genes. This factor is known as GATA1, named for the DNA sequence it binds. A tissue such as muscle does not contain GATA1, and therefore globin genes are not expressed there.

The eukaryotic gene promoter is usually taken to be roughly in the region of 200 base pairs upstream (5′) of the start site of transcription. However, fully regulated gene expression often involves much more distant sequences, known as enhancers.

Enhancers

Enhancers can greatly increase the expression of the gene, sometimes by as much as 200 times. Like promoters they contain clusters of short control sequences, many of which bind the same TFs that bind upstream control elements in promoters. The remarkable thing is that a given enhancer may be thousands of base pairs distant from the gene that is affected and may be upstream or downstream of the gene. Whereas a gene promoter can only function when correctly oriented with respect to the direction of transcription, it does not matter which orientation the enhancer section is in. Enhancer operation is often tissue specific. For example, enhancers both upstream and downstream of the globin genes contain GATA1 binding sites, which result in efficient globin gene expression only in developing red blood cells.

The obvious problem is how the enhancer can act at such a distance from the gene it affects. This is achieved by looping of the DNA between the promoter and its enhancer so that the two can be brought into proximity with one another (see Figure 26.10). Some of the proteins that bind to the enhancer cause bending of the DNA and thus bring the cluster of protein factors on the enhancer to a position of interacting with and activating the transcriptional complex on the gene promoter.

Some distant regulatory sequences do not activate transcription, but instead repress it in circumstances where the gene product is not required. It is obviously not appropriate to call these sequences enhancers, so an alternative term, **silencers**, is used.

Insulators

Since enhancers can act at a distance and in either direction, they can potentially influence the expression of any gene that comes within their range. To guard against this the genome contains a rather different class of regulatory sequences called **insulators** . The proteins that bind insulators are not TFs. Instead they divide the chromosome into sections; an enhancer can only act on promoters that are not separated from it by a protein bound insulator sequence.

DNA binding by transcription factors

From what has been said in this chapter so far, it is clear that the capacity of transcription factors to bind to specific DNA sequences is of central importance to gene control. We will now look in more detail at DNA binding by these proteins.

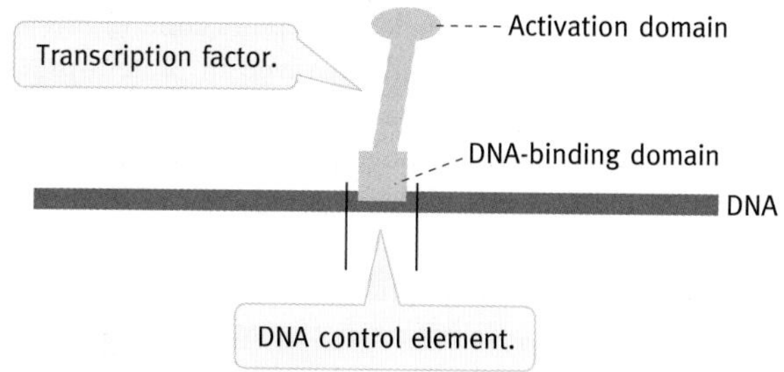

Fig. 26.6 Transcription factor domains.

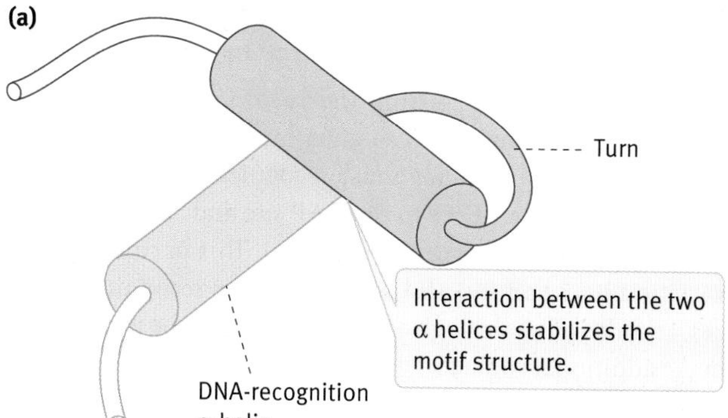

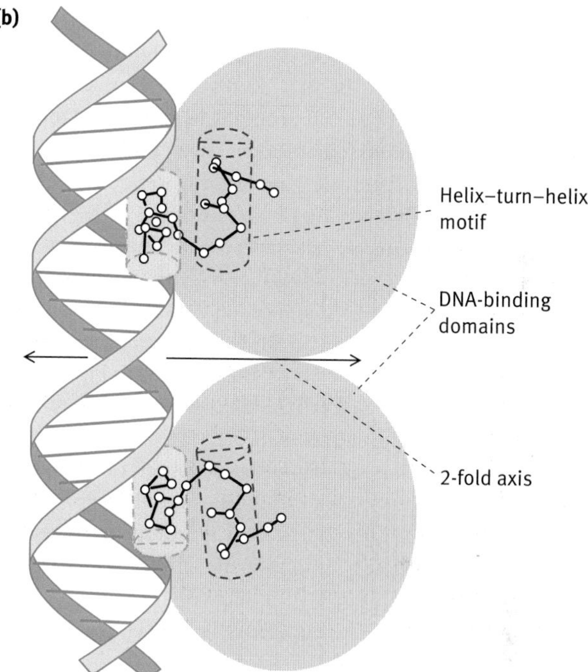

Fig. 26.7(a) The helix–turn–helix motif of a DNA-binding protein monomer. **(b)** Dimer helix–turn–helix protein binding to DNA in major grooves. Douglas H. Ohlendorf et al.; Many gene-regulatory proteins appear to have a similar α-helical fold that binds DNA and evolved from a common precursor; Journal of Molecular Evolution; 1983, 19, 2; Reproduced by permission of Springer.

Transcription factors have a domain structure, in which one domain binds DNA and another has the actual activation role (Fig. 26.6). They bind to double-stranded DNA in which the bases are already Watson–Crick hydrogen bonded to each other, so the binding proteins have to recognize the exposed edges of the bases visible in the grooves of the double helix (see Fig. 22.4). Sequence-specific binding to a double helix involves noncovalent bond formation between the amino acid side chains of the protein and the exposed groups of the DNA bases, though additional stabilizing bonding to the sugar-phosphate-sugar backbone may occur. In many cases, the contact is made by a 'recognition helix' of the protein, an α helix that fits into the major groove of the DNA. The edge view of attachment groups of the base pairs is more variable here than in the minor groove and thus has more potential recognition specificity. The DNA sites that the TFs recognize are usually sections of about 20 base pairs in length.

There are large numbers of different TFs but they can be grouped into a small number of families on the basis of characteristic structures of the protein 'motifs' involved in the recognition. Some of these DNA-binding motifs are found in both prokaryotic and eukaryotic factors. The DNA-binding motifs are small parts of the whole TF, since other domains are involved in different aspects of its function such as the activation of transcription and interaction with additional regulatory molecules. We will now describe four major classes of TF classified by DNA-binding motif: helix-turn-helix, zinc finger, leucine zipper and helix-loop-helix proteins. These four classes do not cover all TFs, but provide a brief overview of some of the most common types. A more general term than TF, **DNA-binding proteins**, is often used as a blanket term for proteins in these classes, since some other proteins that bind DNA using these motifs are not actually concerned with transcriptional regulation. Note, however, that the term DNA-binding protein implies sequence specificity. It is not used for proteins such as histones that bind any DNA sequence,

Helix–turn–helix proteins

This was the first type of sequence-specific DNA-binding protein to be identified and the most studied; it occurs commonly in prokaryotes and eukaryotes. We will use a prokaryotic transcription factor from bacteriophage lambda, known as **Cro**, whose binding mechanism was first elucidated, as an illustrative example. Cro is involved in controlling the lysogenic/lysis switch in the phage life cycle. The cAMP CAP and lac repressor of *E. coli* are also members of this class, as are the **homeodomain proteins** of eukaryotes. The latter are a group of TFs of great importance in embryonic development.

The helix–turn–helix (HTH) motif is a small section of the protein that makes the binding contact to the recognition sequence on the DNA. The motif has two α helices linked by a β turn; one of the two (the recognition helix) sits in the major groove of DNA (Fig. 26.7(a)). The Cro protein is a dimer, the recognition helices of the two HTH motifs fitting into adjacent binding sites (Fig 26.7(b)). The use of a dimer rather than a monomer presumably gives tighter binding.

Different HTH proteins must recognize and bind different DNA sequences (CAP to the CAP site and lac repressor to the lac operator sequence, for example). This occurs through variation in the amino acid sequence of the recognition helices, and also often through adjacent regions of the protein that make additional contacts with the DNA.

Zinc finger proteins

The zinc finger motif is found very commonly in eukaryotic DNA-binding proteins. The name derives from a finger-like projection of the polypeptide chain, which binds inside the major groove of DNA. The first discovered, 'classic' type, is a finger structure of about 30 amino acid residues stabilized by a zinc atom bound to four residues, two cysteines and two histidines, but in others the zinc-binding residues are four cysteines. Several of these fingers often occur as a group in DNA-binding proteins so that successive fingers can bind adjacent major groove sections of the DNA giving firm attachment of the protein. For instance, the transcription factor TFIIIA, which regulates the 5S RNA gene, has nine zinc fingers. The classic type of zinc finger is illustrated in Figure 26.8. In all zinc fingers one side of the structure is an α helix that recognizes the binding site on the DNA.

An important class of regulatory protein that contains zinc fingers with four cysteines are the steroid receptor family (dealt with in Chapter 29). Steroid receptors are members of the **nuclear receptor** family of zinc finger transcription factors, which also includes receptors for thyroid hormone, vitamin D, and retinoic acid. (Retinoic acid is important in

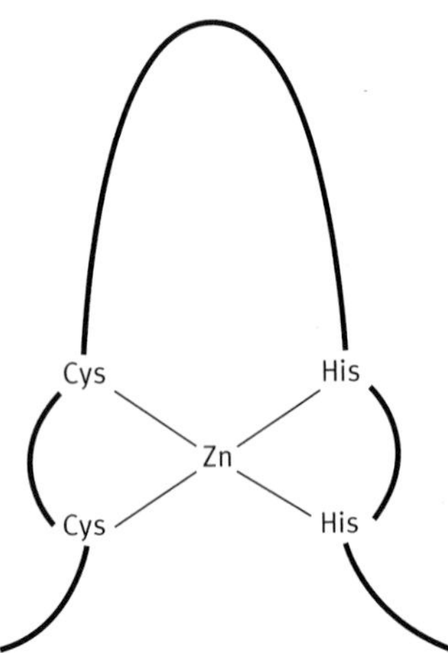

Fig. 26.8 Diagram of one type of zinc finger structure. This motif inserts into the major groove of DNA and binds to five base pairs by its recognition *α* helix, which makes up one side of the finger structure. Different proteins have variable numbers of zinc fingers arranged sequentially which bind into successive regions of the major groove in the DNA giving firm attachments.

embryonic development). Nuclear receptors contain two zinc fingers, one of which is actually involved in protein-protein rather than protein-DNA interactions. The receptors homodimerize or heterodimerize through protein interactions of one of their two zinc fingers, and then bind to their DNA recognition sequences by the other.

Another family of transcription factors, the GATA factors, are also zinc finger proteins, but not members of the nuclear receptor family. One member of the GATA family, GATA1, has already been mentioned as a cell type specific TF in developing red blood cells.

In recent years, genetically engineered zinc fingers have been designed to target specific DNA sequences. When artificially coupled to nucleases rather than as part of transcription factors they are potentially of use to introduce targeted mutations or new sequence into DNA, for instance in gene therapy (see **Klug** reference in Further Reading).

Leucine zipper proteins

The leucine zipper motif is found in many eukaryotic transcription factors. However, unlike HTH and zinc finger motifs, leucine zippers are not actually DNA-binding motifs. Leucine zipper transcription factors cannot bind DNA unless they first dimerize, and the leucine zipper is the protein-protein interaction motif (dimerization domain) that allows this.

Leucine zipper proteins contain a long α helix in which one end is the dimerization domain while the other binds DNA. In the dimerization domain, every seventh amino acid is leucine. Since there are 3.6 amino acid residues per turn, the leucines all appear on the same side of the α helix forming a row of hydrophobic faces. Two leucine zipper monomers therefore bind each other by hydrophobic forces between the leucine side chains (Fig. 26.9(a)). The term 'zipper' is really a misnomer resulting from the initial belief that the leucine residues interdigitated. The DNA-binding region of the α helix is a region rich in positively charged arginine and lysine residues. The dimer attaches to the DNA in a 'scissor' grip, the two arms being in adjacent major groove sections of the duplex (Fig. 26.9(b)). The dimers may be of the same or different monomers; heterodimers recognize different adjacent base pair sequences thus giving flexibility of control.

Helix–loop–helix proteins

Another family of eukaryotic transcription factors characterized by dimerization rather than DNA-binding domain is the **helix–loop–helix** (HLH) family. The HLH motif should not be confused with the helix–turn–helix. As in leucine zipper proteins, a single α helix allows both dimerization and DNA binding, and the latter cannot occur without the former. However, the α helix of each monomer is interrupted by a polypeptide loop that gives flexibility for the short helices of the dimer to associate with each other (Fig. 26.9(c)).The adjacent DNA-binding region is again rich in basic amino acids.

The muscle-specific TF MyoD is a famous member of this family. It plays a similar role to that of GATA1 in red blood

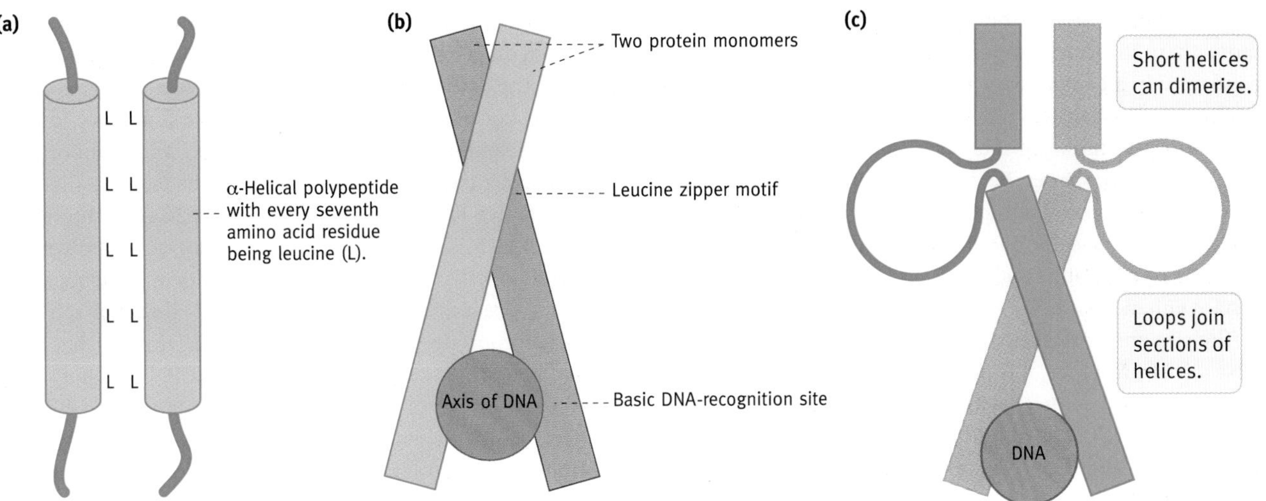

Fig. 26.9 **(a)** The leucine zipper motif. The hydrophobic leucine residues are opposed, not interdigitated as in a zipper. **(b)** The leucine zipper protein attached to DNA, looking down the axis of DNA. The two arms of the protein lie in adjacent sections of the major groove. The attachment sites are rich in basic amino acids. **(c)** The helix–loop–helix is a different type of dimerization in that the flexibility of the loop allows the short helical sections to bind to each other.

cells in driving tissue specificity, in this case by activating expression of multiple genes required for the development of contractile muscle cells. Like leucine zipper proteins, HLH proteins such as MyoD can homo- or heterodimerize, allowing flexibility in gene control.

How do eukaryotic transcription factors influence transcription?

Unlike the DNA-binding domains of TFs, activation domains are not easily identifiable by typical structural motifs. Activation domains have wide range of structures and therefore they are not used to classify transcription factors. They also work in a variety of ways, which we will now explore.

Interaction with the **basal initiation complex** is a common mechanism. You will recall that prokaryotic TFs such as CAP interact directly with RNA polymerase. Eukaryotic TFs rarely, if ever, do so. You will also recall that eukaryotic RNA polymerase II is recruited to the promoter by general or basal transcription factors, of which the TFIID complex, including TBP (the TATA binding protein) is key in initiating the process. The activation domain of a specific TF often interacts with TFIID or another basal TF. Through doing so, it may increase the binding of the basal TF to the promoter, and/or increase the activity of the already bound factor.

The mediator

Besides the general transcription factors such as TFIID, another complex of proteins termed the **mediator** was discovered, initially in yeast, to act as a link between TFs and RNA polymerase II. Roger Kornberg, who was awarded the Nobel prize in 2006, studied yeast transcription because it is somewhat simpler than transcription in multicellular eukaryotes. On the basis of what was known of the mechanism of transcription by polymerase II, it was possible to assemble *in vitro* the known essential components for transcription in a partially purified form. Kornberg and co-workers found that the assembled components transcribed genes at a basal level, but did not respond to addition of activators (TFs). Something was missing. Addition of crude (unpurified) yeast extracts raised the basal level of transcription and the system now responded to activators, suggesting that these extracts contained a factor that permitted TFs to control Pol II. This factor was given the name of **mediator** because it apparently mediated the transmission of regulatory information between TFs and the polymerase within the initiation complex (Fig. 26.10).

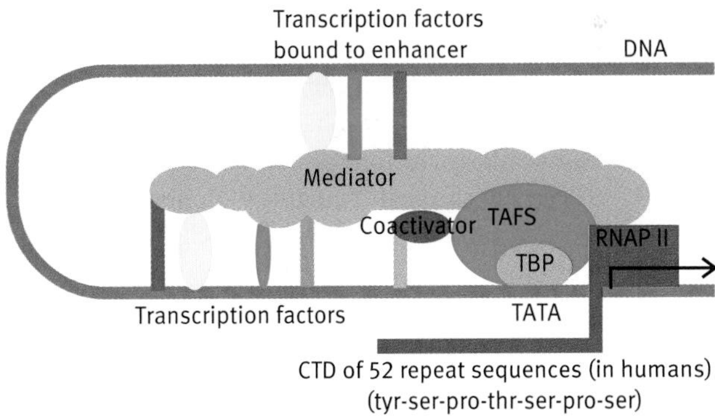

Fig. 26.10 Cartoon of the transcription initiation complex illustrating the role of the mediator. The sizes, positions and interactions of the various components are conjectural. They represent the binding of the general transcription factors (TBP and associated TAFS of the TFIID complex) to the TATA box, which positions the RNA polymerase II (RNAPII) at the correct site. The upstream sequence-specific transcription factors bound to the promoter interact with the mediator as do those bound to the enhancer. The mediator is a complex of 30 proteins in humans; its representation is based on the fact that it does not bind to DNA but forms a physical connection between the polymerase and other components of the initiation complex. It is believed to convey 'instructions' from the various regulatory elements to the polymerase.

The mediator turned out to be an immense complex of more than 20 protein subunits. Its presence has now been confirmed in all eukaryotes. Homologues of most of the yeast mediator subunits are found in humans. The yeast and human mediators are similar in their arrangement in the complexes and have similar shapes. Mediator proteins do not themselves bind DNA, but make multiple contacts with basal TFs, Pol II itself, and with specific TFs bound to both upstream control elements of promoters and, through bending of the DNA, to distant enhancers.

The size and complexity of the mediator makes it difficult to study. For instance, it is not known to what extent it forms a pre-assembled complex with Pol II and basal TFs (a 'holo-enzyme') before the basal initiation complex assembles on the promoter. Indeed, this may vary depending on the cell type and situation. The extent to which some components of the mediator vary and therefore contribute to cell type specific transcription, and transcriptional responses to cell signalling, is also the subject of current research.

Coactivators

Rather than act directly, the activation domains of many TFs bind to coactivator proteins, which then take part in further interactions. Coactivators are not TFs and do not themselves bind to DNA but are essential for the activity of certain TFs. A well-studied example of a coactivator that works with a number of different TFs is **CBP**. CBP stands for CREB-binding protein, CREB being the first TF found to have CBP as a coactivator. **CREB** in turn stands for **cAMP-response-element-binding protein**. CREB is TF that activates specific genes in response to cAMP signalling, as discussed further in Chapter 29.

When CBP is recruited to a promoter by binding to CREB or another TF, it then activates transcription by more than one mechanism. It can bind to components of the basal initiation complex. However, it also has an enzyme activity that catalyses modification of histone proteins. Thus, the coactivator CBP takes part in chromatin modification, another important aspect of eukaryotic gene control, which we explore in more detail below.

Most transcription factors themselves are regulated

The TFs that bind to common upstream elements such as the CAAT and GC boxes are present and active in most cells. TFs that are involved in more complex gene control often exist in an inactive form in the cell and cannot stimulate transcription until they are activated. Activation may be by phosphorylation or another modification that causes a conformational change in the protein, allowing it to bind its target DNA sequence. Activation may also be associated with the movement of the TF from the cytosol to the nucleus where it can then bind to its cognate DNA elements.

The activation is usually the result of signals arriving at the cell from other cells. Figure 26.11 gives a few examples, in outline, of the activation mechanisms involved: in Fig. 26.11(a) a steroid hormone is shown to enter the cell directly (due to its lipid solubility) and on binding to a soluble receptor protein causes a conformational change in the latter so that it is now an active TF for cognate genes; Figure 26.11(b) shows that cAMP, which is elevated as a result of the action of certain hormones, activates protein kinase A, which phosphorylates an otherwise inactive transcription factor – the latter activates genes appropriate to the hormone signal. This is what happens when cAMP causes activation of CREB. Phosphorylated CREB is then able to bind and recruit is coactivator, CBP. Fig. 26.11(c) shows the general concept of the way in which many hormones bind to membrane receptors and induce a signal cascade inside the cells. This results in activation of specific TFs often by their phosphorylation. The regulation is effected largely by signals from other cells in the form of hormones, cytokines, and growth factors. This type of cell signalling control lies at the heart of cell regulation and is much more fully dealt with in Chapter 29. Inappropriate activation of TFs is important in the generation of cancer because genes are activated when they should not be. The same is true of overproduction of certain TFs, which leads to improper stimulation of cell growth as covered in Chapter 31.

Transcription repressors

Before moving on to chromatin modification, we will briefly consider transcription repressors. We have so far described eukaryotic TFs mainly as activators of transcription. However, as emphasized in the introduction, eukaryotic gene transcription is typically regulated by multiple signals to a given promoter, and these signals may be either negative or positive. The combination of multiple signals gives finely balanced regulation that can respond to complex situations. Negative control is achieved by TFs acting as repressors.

Repression by transcription factors can occur through a wide variety of mechanisms, some essentially equivalent to those operating in activation, that is by interaction of a repression domain or a corepressor with the basal initiation complex and/or the mediator, or by repressive chromatin modification. The refinement of eukaryotic gene control is often achieved by transcriptional activators and repressors competing for the same or overlapping binding sites on DNA, and in this case the repressor may act simply by blocking access of an activator to the promoter. Whether the gene is activated or repressed will therefore depend on the relative levels or activities of the repressor and the activator in the cell. The thyroid hormone nuclear receptor can even switch between working as an activator and a repressor, depending on whether it binds a coactivator or a corepressor, which in turn depends on the presence or absence of its ligand, thyroid hormone (Fig. 26.12).

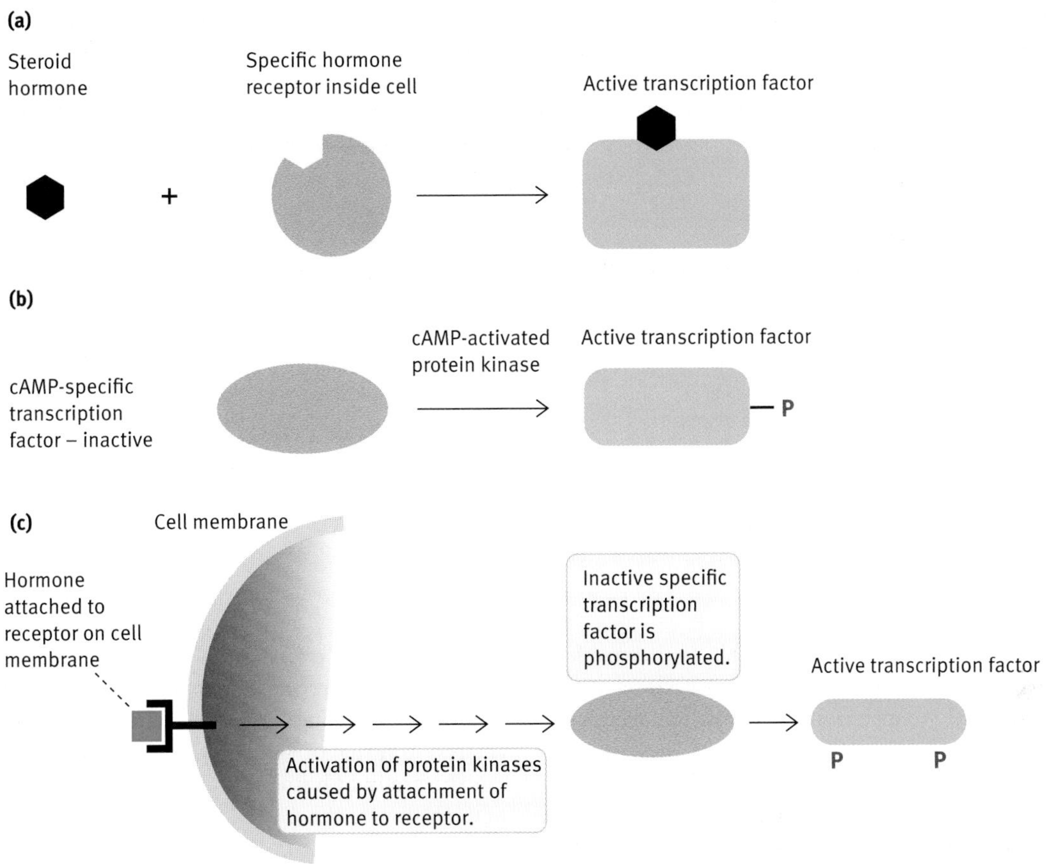

Fig. 26.11 Examples of transcription factor activation. **(a)** A steroid hormone enters the cell; it attaches to a receptor specific for that hormone and causes a conformational change in the receptor protein, which is now an active transcription factor. This activates the gene(s), which are controlled by the particular hormone. A family of steroid hormone receptors, which are zinc finger proteins, exists. **(b)** Cyclic AMP (cAMP) is produced as a result of adrenaline binding to cell receptors. The cAMP activates a protein kinase, which phosphorylates the inactive transcription factor, which is activated. **(c)** Protein hormones such as insulin do not enter the cell but bind to receptors on the cell surface. This results in a sequence of events that ends in the phosphorylation of the appropriate inactive transcription factors and activates them. Note that there are many different hormones that bind to different specific receptors and activate different transcription factors. In many cases activation causes transport of the transcription factors from the cytosol into the nucleus. The transcription factors bind to specific response elements of different genes. Thus each hormone can exert control over appropriate genes (activation of transcription factors inside cells by steroid binding and receptor-mediated signal transduction is dealt with in Chapter 29 on cell signalling).

The role of chromatin in eukaryotic gene control

As discussed in Chapter 22, eukaryotic genes *in vivo* are in the form of the protein–DNA complex known as **chromatin**, not as naked DNA. DNA is wrapped around nucleosomes made of octamers of histone proteins, making just under two turns around each nucleosome. Individual nucleosomes are separated by linker DNA, which varies somewhat in length in different species but averages about 50 base pairs, so the whole length of DNA per nucleosome is about 200 base pairs (see Fig. 22.8). Chromatin used to be regarded as an inert structure with the sole function of condensing the DNA to fit into the nucleus. However, it is now known to be dynamic, with changes in its structure and organization reflecting transcriptional activity.

It is not difficult to envisage that histones and other chromatin proteins affect the accessibility of DNA to RNA polymerase and transcription factors. The 'default' state of chromatin (the state in the absence of any action to counteract it) is a 'shut-down' condition – the genes are inactive. The reason is that gene promoters are blocked by nucleosomes that prevent assembly of the basal initiation machinery on the DNA.

Transcriptional activation in eukaryotes involves 'opening up' or unblocking of the promoters. It requires modification of the nucleosome structure, a process known as **chromatin remodelling** (Fig. 26.13). It is not known whether a nucleosome physically leaves the DNA or just changes its attachment so as to permit the transcription complex to assemble on the promoter. The term 'remodelling' avoids the implication of what exactly is happening in molecular terms. It is, however, known that nucleosome remodelling is carried out by protein factors and is an energy-requiring ATP-dependent process.

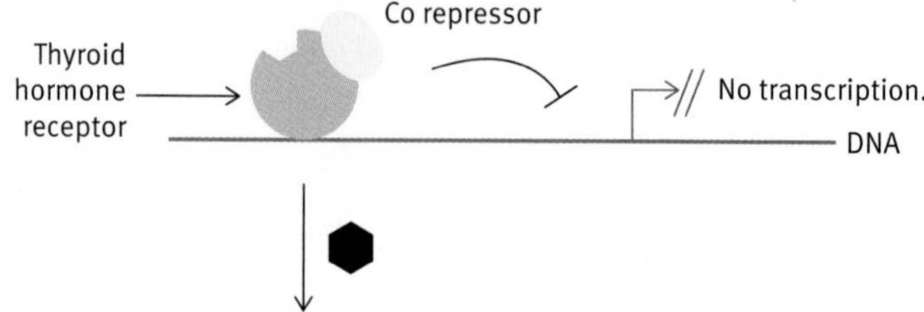

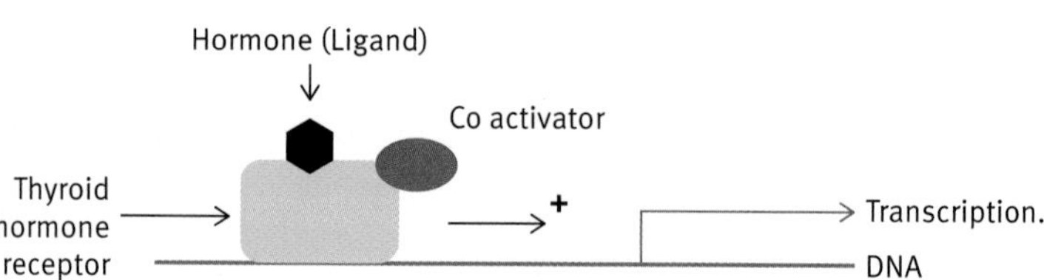

Fig. 26.12 The thyroid hormone receptor (a nuclear receptor zinc finger protein) is converted from a repressor to an activator of transcription when bound by its ligand, thyroid hormone. Hormone binding changes the conformation of the receptor causing release of its corepressor and binding of its coactivator.

Another form of chromatin modification is the covalent modification of histones. Chemical groups such as acetyl (ethanoyl) or methyl groups may be added or removed from the *N*-terminal 'tails' of the histone proteins. These regions of histones are rich in basic amino acids such as lysine and arginine, which being positively charged favour interaction with the negative sugar-phosphate backbone of DNA. As the addition of an acetyl group neutralizes the positive charge (Fig. 26.14), it was suggested that acetylated histones are less tightly bound to DNA, making it accessible for transcription. The situation is actually more complex than this relatively simple model suggests, as histone acetylation also affects interactions between nucleosomes, and between histones and other proteins. Nevertheless, histone acetylation is generally associated with 'loosely wound' chromatin and therefore with transcriptionally active genes.

Nucleosomes
DNA
TATA
Chromatin remodelling exposes promoter.
Nucleosomes
TATA

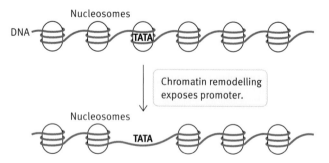

Fig. 26.13 Chromatin remodelling. The principle is that the promoter in chromatin is blocked by nucleosomes. Gene activation requires exposure of the promoter; this may require the physical removal of one or more nucleosomes, or it could be some change in the relationship of the nucleosome(s) to the DNA which effectively gives accessibility to the promoter. Use of the term 'chromatin remodelling' reflects the current uncertainty about exactly what happens at the molecular level.

Histone acetyl transferases (HATs) and **histone deacetylases (HDACs)** are the enzymes that add or remove acetyl groups from histones (Figs. 26.14 and 26.15). HAT enzymes catalyse transfer of the acetyl group of acetyl-CoA to the ε-NH_3^+ group of lysine residues in the *N*-terminal domains of the histones. These domains are exposed on the surface of the nucleosomes like short tails so that they are accessible to HAT activity. It was mentioned above that the transcriptional coactivator CBP works in part by modifying chromatin, and in fact CBP has HAT activity, as do a number of other coactivators. Conversely, some corepressors have HDAC activity.

How is DNA first made accessible to transcription factors?

It may not have escaped your notice that in discussing the roles of transcription factors and chromatin modification in gene control we have presented a bit of a 'chicken-and-egg'

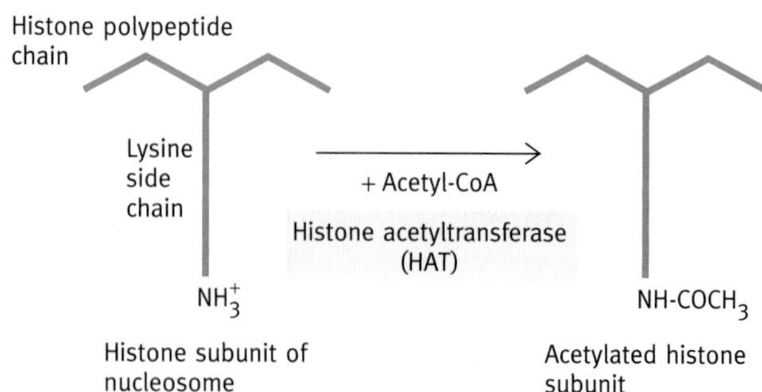

Fig. 26.14 Acetylation of the lysine residues of the *N*-terminal tails of subunits of the histone octamers of nucleosomes. The acetylation reduces the positive charge on the protein and is believed to result in lessening the attachment of DNA to the nucleosome leading the chromatin remodelling described in the text.

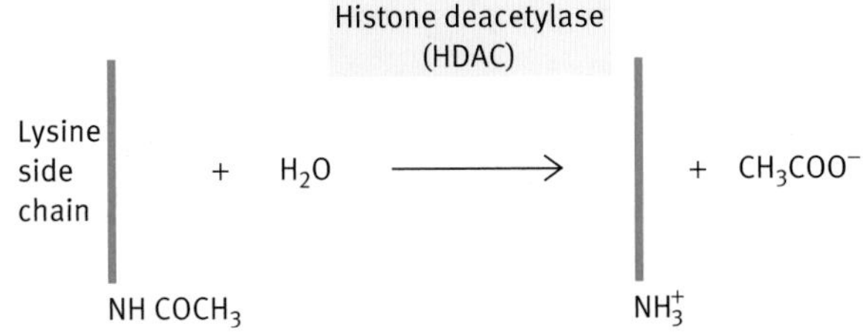

Fig. 26.15 Reaction catalysed by histone deacetylase.

dilemma. Chromatin has to be opened up to allow transcription factors access to DNA, yet DNA-bound TFs are responsible for recruitment of chromatin modifying enzymes. How can one process begin without the other. The probable answer is that certain TFs, recently termed **pioneer factors**, are able to bind to their DNA elements even before chromatin remodelling has occurred. This leads to remodelling of the promoter, which allows all the other factors to access and assemble on the promoter.

DNA methylation and epigenetic control

Methylation of bases of DNA occurs in both bacteria and eukaryotes. In bacteria it is concerned with preventing restriction enzymes from hydrolysing the cell's own DNA; the restriction enzymes are a protection against bacteriophage infection as discussed in Chapter 28.

In mammals methylation of cytosine has a different role, transcriptional regulation. The presence of 5-methylcytosine, shown below, is associated with transcriptionally inactive genes.

Cytosine   5-Methylcytosine

It is not easy to determine whether DNA methylation causes repressive modification of chromatin or whether cause and effect are the other way round and chromatin proteins regulate DNA methylation. However, in at least some cases DNA-binding proteins that specifically recognize 5-methylcytosine recruit histone deacetylases, thus causing the condensation of chromatin and making the DNA inaccessible to transcription factors.

Cytosine methylation occurs specifically where there is a cytosine 5′ of a guanine in the DNA sequence. This is often shown as CpG, where 'p' denotes the linking phosphate (to distinguish it from a hydrogen bonded CG base pair). Gene promoters are often relatively rich in CpG, creating so called CpG islands in the genome. Given that DNA methylation is associated with transcriptional repression, it is unsurprising that CpG islands tend to carry low levels of methylation in active promoters.

A key feature of cytosine methylation is that the methylation pattern of a DNA sequence can be preserved through DNA replication and cell division. To understand this, we have to consider the responsible enzymes, DNA methyltransferases. There are two classes of methyltransferases both using *S*-adenosylmethionine as methyl donor (see Methionine and transfer of methyl groups in Chapter 18). ***De novo* methyltransferases** methylate previously unmethylated DNA, while **maintenance methyltransferases** recognize methylated cytosine in CpG on the template DNA strand, and add a methyl group to the cytosine paired with the guanine of the CpG on the newly replicated strand (Fig. 26.16). The importance of this is that methylation patterns associated with particular patterns of gene expression and hence with particular differentiated cell types are preserved when the cell divides. This kind of change in DNA, which is heritable in the sense of being preserved through cell division but does not involve a mutation or change of base

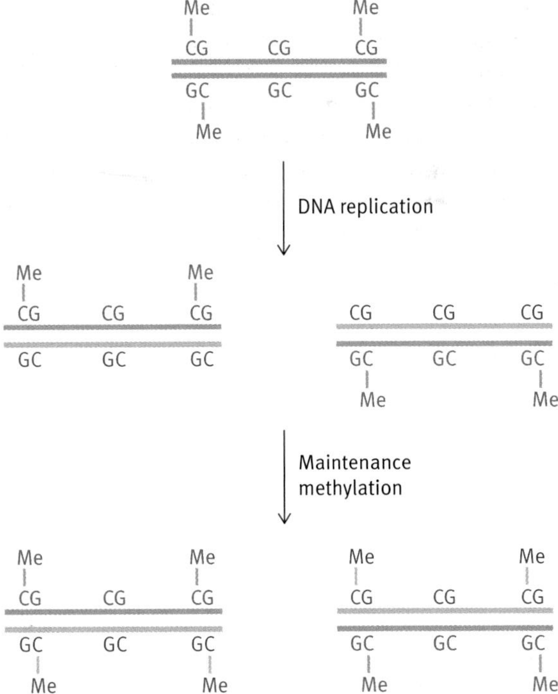

Fig. 26.16 Differential DNA methylation is preserved after DNA replication by the action of maintenance methyltransferases. The methyltransferase enzyme recognized 5-methylcytosine in CpG islands on the template DNA strands, and methylates corresponding cytosines on the newly synthesized strands.

sequence, is termed an **epigenetic** modification. The study of epigenetics is becoming increasingly important, in part because abnormal changes in the epigenetic modification of DNA, and hence with gene expression, are found in many cancers.

Epigenetic modifications of DNA are not usually passed from generation to generation, largely due to a global wave of DNA demethylation in the fertilized egg. Methylation patterns are gradually re-established during embryonic development as cells and tissues differentiate. However in a restricted set of genes the methylation pattern is locked in and is inherited by offspring, a phenomenon known as **genomic imprinting**.

BOX 26.1 **Genomic imprinting disorders**

Mammals inherit one copy of each autosomal (non-sex chromosome) gene from their mother and one from their father. In the vast majority of cases genes are transcribed, when expression is required, from both the maternally and the paternally inherited copy of the gene. However, a restricted subset of mammalian genes (fewer than 100 are known so far in humans) are **imprinted**. That is, transcription is restricted to either the maternally or the paternally inherited copy and the other copy is stably repressed. It is a characteristic of the gene itself whether it is maternally or paternally imprinted, and the repression of one parental copy can be transmitted through cell division and from generation to generation. We do not understand why some mammalian genes are imprinted, and nor do we fully understand how differential gene expression depending on parental origin is achieved and maintained. However, we know that differential DNA methylation at sequences termed **imprinting control regions** is important.

Imprinted genes are often concerned with growth and development, and disruption of imprinted genes or their regulation causes developmental disorders that puzzled clinicians before the phenomenon of imprinting was discovered. A cluster of genes on human chromosome 15 that contains both maternally and paternally imprinted genes is prone to deletion, and the same deletion was observed to cause two entirely different syndromes in different individuals. It was eventually realized that **Prader-Willi Syndrome** is caused by a deletion in the paternally inherited chromosome, resulting in functional loss of paternally expressed genes. The loss cannot be compensated by the remaining maternal chromosome because those genes are silenced on it. **Angelman Syndrome** is the 'reciprocal' disorder caused by deletion of the same region of the chromosome, but from the maternally inherited copy with functional loss of a different gene that is only maternally expressed. Another imprinting disorder associated with an imprinted cluster of genes on chromosome 11, **Beckwith-Wiedemann syndrome**, causes overgrowth of the fetus. In some cases it is caused by the child inheriting both copies of the chromosome from its father, resulting in genes that promote fetal growth being expressed at twice the normal level, as they are normally active only on the paternal and silenced on the maternally inherited copy of the chromosome.

Gene control after transcription is initiated: an overview

Although gene control mainly occurs at the first stage of gene expression, there are many examples of regulation at later stages, just some of which are discussed below. Once transcription is initiated it is sometimes terminated before synthesis of the RNA is complete, as in the prokaryotic examples of attenuation and riboswitches. In eukaryotes, alternative splicing of mRNA (see Chapter 24) is also a potential control point: the relative quantities of differently spliced mRNAs (and thus different proteins) produced from a single gene may vary in different cell types or in response to changed conditions. The stability of mRNAs is also regulated; the longer the lifetime of a message the more rounds of translation it can undergo and hence the more protein is made. Translation may also be regulated.

Gene control post-transcription initiation in prokaryotes

Attenuation in the *E. coli trp* operon

We have already seen that the *trp* repressor regulates initiation of transcription at the *E. coli trp* operon, so that enzymes required to synthesize tryptophan are made only when the cellular level of the amino acid is low. The operon is under a second level of regulation known as attenuation, which allows a more finely graded response to tryptophan levels that occurs once transcription has begun. The mechanism is logical but a little complex (Fig. 26.17). As this is a prokaryotic system, translation of the *trp* operon mRNA initiates before the mRNA is complete. Translation begins with the synthesis, not of one of the enzymes, but of a short 'leader' peptide encoded by the 5′ end of the message. The leader peptide contains codons encoding tryptophan. If there is tryptophan in the cell, then tRNA charged with tryptophan will be available and ribsomes will move rapidly along the sequence encoding the leader peptide. The effect of this is to leave part of the mRNA sequence free to base pair with itself, forming a stem-loop transcription termination signal so the mRNA is never completed and the enzymes needed for tryptophan synthesis are not made. The leader peptide has no other function but to play this regulatory role by its translation, and is rapidly degraded.

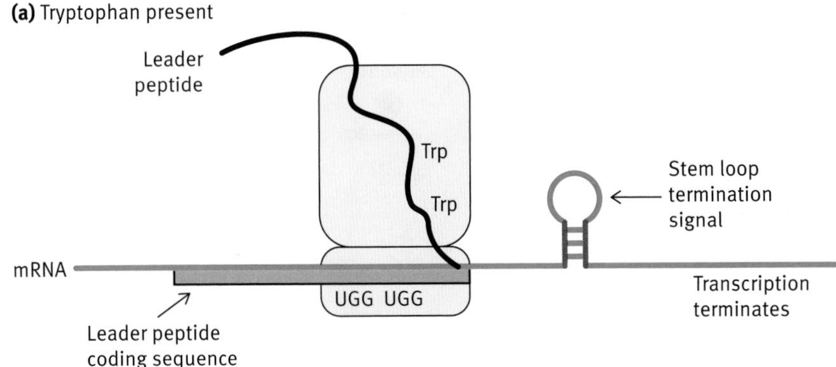

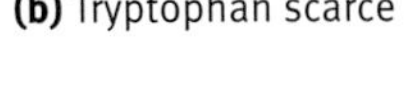

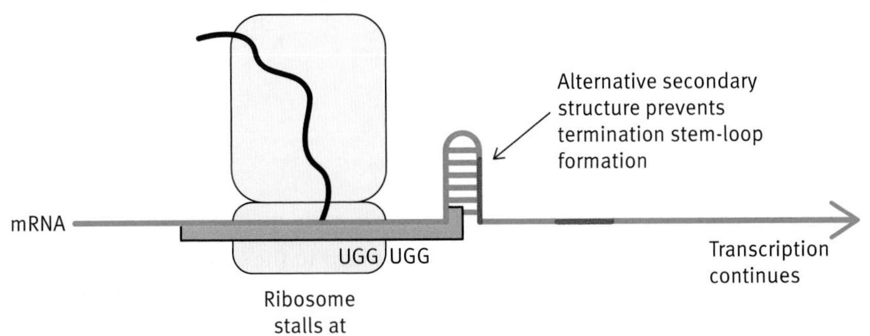

Fig. 26.17 Attenuation of transcription in the trp operon. **(a)** Tryptophan present: the ribosome translates the full leader peptide, reading through the *trp* codons. A stem-loop termination signal forms in the mRNA and transcription terminates, leaving a truncated mRNA, which does not encode the enzymes required to synthesize more tryptophan. **(b)** Tryptophan scarce: the ribosome stalls at *trp* codons of the leader peptide, allowing formation of an alternative mRNA secondary structure which prevents formation of the termination signal. Transcription continues, giving full-length mRNA that encodes enzymes required for tryptophan synthesis.

If tryptophan and therefore trp-charged tRNAs are scarce, the ribsome translating the leader peptide stalls at the *trp* codons, and in doing so prevents formation of the premature termination stem-loop structure. The complete operon mRNA is transcribed and the enzymes needed for tryptophan synthesis are made. Thus, the combination of *trp* repressor and attenuation gives a fine tuned system, with high levels of trp blocking transcription via the repressor, intermediate levels allowing just a few full-length transcripts to form while most are prematurely terminated by attenuation, and low levels of trp allowing the synthesis of many full-length transcripts.

Attenuation is not confined to the trp operon. Operons encoding enzymes for the synthesis of histidine and leucine are also controlled in this way, with leader peptides containing multiple *his* and *leu* codons respectively.

Bacterial riboswitches

Riboswitches regulate bacterial protein synthesis by a mechanism somewhat analogous to the attenuation system described above, in that they depend on mRNA forming secondary internally base-paired structures in response to the levels of key metabolites. The difference is that it is the metabolite itself rather than ribosomes binding to mRNA that determines whether or not the secondary structure forms. Riboswitches regulate the production of enzymes required for the synthesis of flavin mononucleotide, thiamine pyrophosphate, certain amino acids, purines, and *S*-adenosylmethionine. The principle is that mRNAs coding for the synthesis of enzymes involved in the production of the metabolite contain a sequence in the untranslated leader sequence known as an **aptamer** (from the Latin *aptus* to fit), which specifically binds its cognate metabolite. When the metabolite binds to its aptamer the RNA undergoes a conformational change, which inhibits the expression of the gene(s) coding for enzyme(s) involved in the synthesis of the metabolite. The control may be achieved in one of two ways (Fig. 26.18).

- The conformation of the aptamer when the metabolite is bound allows formation of a premature termination stem-loop structure which aborts transcription of the mRNA (Fig. 26.18(a)).
- Alternatively it may cause the masking of the Shine–Dalgarno sequence on the mRNA and so prevent translation (Fig. 26.18(b)).

In either case, the effect is to prevent the synthesis of enzymes that are not needed because the relevant metabolite is already abundant in the cell.

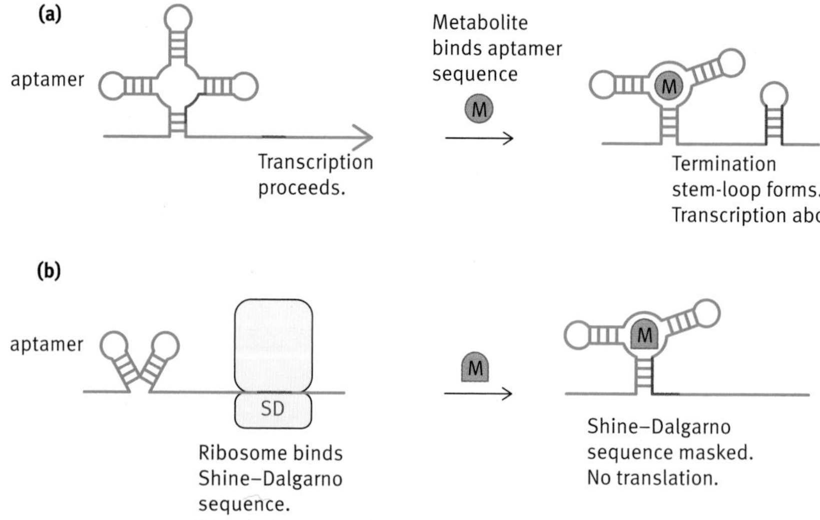

Fig. 26.18 Mode of riboswitch action. Metabolites when present in abundance bind aptamer sequences in mRNA, and hence repress expression of genes required for their own synthesis (i.e. negative feedback control).

mRNA stability and the control of gene expression

Although regulation of gene transcription is of overriding importance in the control of gene expression, the stability of individual mRNAs is also significant. In general the rate of synthesis of a protein is a reflection of the mRNA level for that protein although there are some cases of specific control of translation (discussed below). The level of an mRNA in a cell is a function of its synthesis and breakdown rates. At a given rate of mRNA production, a message with a longer half-life will be present at a higher steady-state level within the cell than a less stable one, resulting in a higher rate of synthesis of the cognate protein. Mechanisms that determine the half-life of an mRNA can thus provide a way of regulating gene expression.

Prokaryotic mRNAs in general have an ephemeral existence, with half-lives of 2–3 minutes. In mammals, the half-lives of individual mRNAs range from about 10 minutes to 2 days, with 3–4 hours being the average. The mRNA for globin, a stable one, has a half-life of about 10 hours. Regulatory proteins, such as those involved in the cell cycle (see Chapter 30) tend to be coded by short-lived messages so that changes in the rate of transcription of their genes have rapid effects on synthesis of the proteins; the relevant messages usually have half-lives of less than 30 minutes. This reflects the need for the cellular level of regulatory proteins to change rapidly.

Determinants of eukaryotic mRNA stability and their role in gene expression control

Almost all eukaryotic mRNAs have a polyA tail added to their 3′ ends (Fig. 26.19(a)) before they emerge from the nucleus, histone mRNAs being exceptions. They all have the methylguanosine cap structure added to the 5′ end. You will recall from Chapter 24 that the polyA tail actually loops around and physically interacts with the 5′ end of the message via polyA binding proteins. This looping increases translation of the message, but it also stabilizes it. Both the polyA tail and the 5′ cap protect the mRNA from degradation by cellular exonucleases, while looping protects the cap from removal by a 'decapping' enzyme.

Degradation of mRNA begins with the removal of the polyA tail (deadenylation) by $3' \rightarrow 5'$ exonuclease activity. This occurs gradually as the mRNA ages. Once the polyA tail is below a critical length it cannot interact with the 5′ end of the message, and the 5′ cap is exposed to attack by the decapping enzyme. Removal of the cap exposes the 5′ end of the mRNA to attack by a $5' \rightarrow 3'$ exonuclease, while $3' \rightarrow 5'$ digestion can also continue.

mRNAs with different half-lives undergo deadenylation at different rates. What determines the rate at which the polyA tail of an mRNA is broken down? In many cases it is specific sequences in the 3′ untranslated region (3′ UTR), and proteins that interact with them. mRNAs with short half-lives contain characteristic AU-rich elements (AREs) while very stable mRNAs often contain a polypyrimidine (C rich) element (PRE). ARE-binding proteins recruit deadenylation enzymes to the message, while the PRE-binding protein is protective.

Mammalian histone mRNAs are unique in that they lack a polyA tail. Their stability is determined by a stem loop at the extreme 3′ end of the molecules (Fig. 26.19b)) and is regulated to match the requirement for new histone proteins only in the S phase of the eukaryotic cell cycle (see Chapter 30) when DNA synthesis and nucleosome assembly are occurring. Histone genes are transcribed during S phase but in the next,

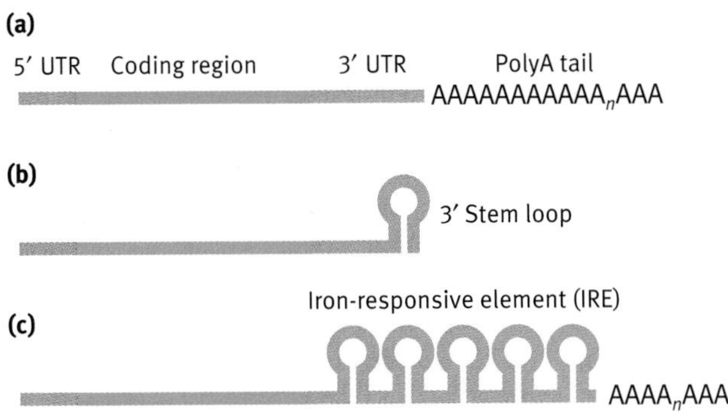

Fig. 26.19 Some structures present in the 3′ untranslated regions (UTRs) of mammalian mRNAs that influence the half-lives of the molecules in the cell: **(a)** the polyA tail found in most eukaryotic mRNAs; **(b)** the G–C rich 3′ stem loop found in histone mRNAs; **(c)** the iron-responsive element (IRE) of transferrin-receptor mRNA. Adapted from Fig. 4 in Ross, J.; Microbiol Rev.; (1995) 59, 423; American Society for Microbiology.

G2, phase of the cycle the level of histone mRNAs rapidly falls. The fall is partly due to reduced transcription but also the half-life of the mRNAs is reduced from 40 to 10 minutes. This change is dependent on the 3′ stem loop, which regulates both histone mRNA translation and mRNA stability through a complex mechanism, the details of which are not completely understood.

The synthesis of the **transferrin-receptor protein** is another case where mRNA stability regulates synthesis of the protein and is itself regulated. The receptor is responsible for the transport of iron into cells by endocytosis (see Figure 27.12) and more receptor protein is therefore required when iron is scarce, to increase the efficiency of iron import. In the 3′ untranslated region of the mRNA is a group of five stem loops called an **iron-responsive element** (**IRE**; Fig. 24.19(c)). In the absence of iron, an **IRE-binding protein (IRP)** attaches to the IRE and stabilizes the mRNA, thus increasing receptor synthesis with a consequent increase in the import of iron into the cell. In iron abundance, iron complexes with the IRP, which then no longer binds to the IRE. The IRE is thus exposed to attack by a specific endonuclease which cleaves the message, leading to its degradation (Fig. 26.20(a)). The result is a reduced import of iron into the cell. mRNAs encoding other proteins involved in iron metabolism also contain the IRE, but in these messages the IRE is in the 5′ UTR, and has the opposite effect: IRP binding increases translation of these proteins, which are required when iron is abundant (Fig. 26.20(b)).

While discussing mRNA stability, we should also mention that the class of regulatory small RNAs known as **microRNAs (miRNAs)** act by base pairing with the 3′ UTR sequences of their target mRNAs. This can block translation or lead to degradation of the target message. MicroRNAs are discussed further in the last section of this chapter.

Translational control mechanisms in eukaryotes

Regulation of gene expression by changing mRNA stability is often found where a rapid change in proten level is needed, either during the cell cycle, as for histones, or due to a change in cell conditions, as in the response to iron. The same is true of gene control at the level of translation. We will continue looking at iron homeostasis, a well-studied example of translational regulation in mammals.

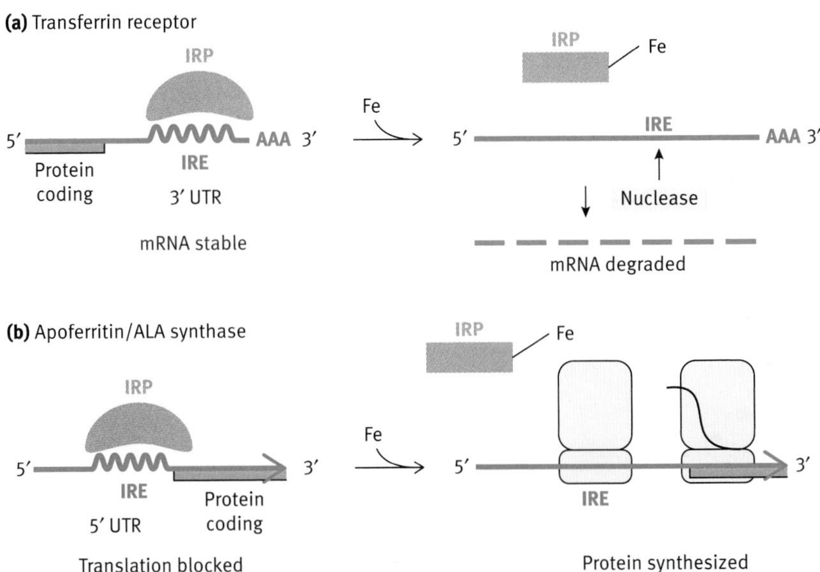

Fig. 26.20 Regulated synthesis of proteins involved in iron homeostasis and haem synthesis through mRNA stability and translation. **(a)** mRNA encoding transferrin receptor is stabilized by binding of the IRE-binding protein (IRP) to the 3′ UTR when iron is scarce. Transferrin receptor is made, increasing iron uptake into cells. When iron is abundant iron binding to IRP releases IRP from the mRNA and hence mRNA is degraded. **(b)** Translation of mRNA encoding apoferritin or ALA synthase is blocked by IRP binding the 5′ UTR when iron is scarce. When iron is abundant IRP is released, allowing ribosomes to translate the mRNA. Ferritin prevents iron toxicity by sequestering excess iron. ALA synthase catalyses synthesis of haem, which requires iron.

Translational control in iron homeostasis and haem synthesis

Iron is transported in the blood plasma as a complex with a protein, transferrin, produced in the liver. The complex is taken up into cells by receptor-mediated endocytosis (see Chapter 27), where iron is needed by iron-requiring enzymes. The transferrin receptor involved has previously been discussed as its level is regulated by stability of its mRNA. Excess iron is toxic, and therefore is stored in liver cells (hepatocytes) as ferritin – a complex of a protein, apoferritin, and inorganic iron. The role of hepatocytes in iron homeostasis is illustrated in Figure 26.21. A balance between the levels of the transferrin receptor and ferritin must be maintained. When iron is scarce, the transferrin receptor is synthesized but ferritin synthesis is repressed to prevent ferritin sequestering scarce iron. When iron is abundant transferrin-receptor synthesis is repressed to stop cells taking up toxic levels of iron, and ferritin is made to soak up the excess.We have already described how the iron-responsive element (IRE) and the iron-responsive element binding protein (IRP) regulate the stability of the transferrin-receptor mRNA. The apoferritin mRNA also contains an IRE, but in the 5′ UTR rather than the 3′ UTR. While IRP binding to the 3′ UTR of the transferrin-receptor mRNA stabilizes the message and therefore increases translation, IRP binding to the 5′ UTR of the apoferritin mRNA blocks translation by the ribosome (Fig 26.18 (b)). Thus, when iron is scarce the transferrin receptor is made and ferritin is not. Conversely, when iron is abundant it binds the IRP and stops it binding to the mRNA. Apoferritin mRNA can now be translated while the transferrin-receptor message is degraded. Ferritin is made and the transferrin receptor is not.

The control of haem biosynthesis is of unusual interest due to its medical importance and is intimately tied up with regulation of iron uptake and storage in the cell. The rate-limiting enzyme in haem production is aminolevulinate synthase (ALA synthase) (see Fig. 18.12). *Synthesis* of ALA synthase is controlled by iron levels in the cell, by the same translational regulation that increases ferritin synthesis in response to iron (Fig. 26.20(b)). Like the apoferritin message, the mRNA that encodes ALA synthase has an **iron-responsive element** in its 5′ UTR that blocks translation unless iron is present at a high level. Thus, haem is only made if the necessary iron is available.

Haem biosynthesis is dealt with further in Chapter 18. Intermediates of haem biosynthesis, if allowed to accumulate in excess of needs, cause diseases, the best known of which is **acute intermittent porphyria** (see Box 18.1). The porphyria diseases are summarized, and a summary of the control area given, in the May et al. reference in the further reading resource on line.

Regulation of globin synthesis

The two components of haemoglobin, the protein globin and the porphyrin haem, need to be produced in the correct relative amounts if excess of one or the other is to be avoided. A coordinating regulatory link exists. In reticulocytes (developing red blood cells), in the absence of haem, a protein kinase phosphorylates the translation initiation factor eIF2. This halts the initiation of translation and thus prevents globin synthesis. In the presence of haem, the kinase is inactivated

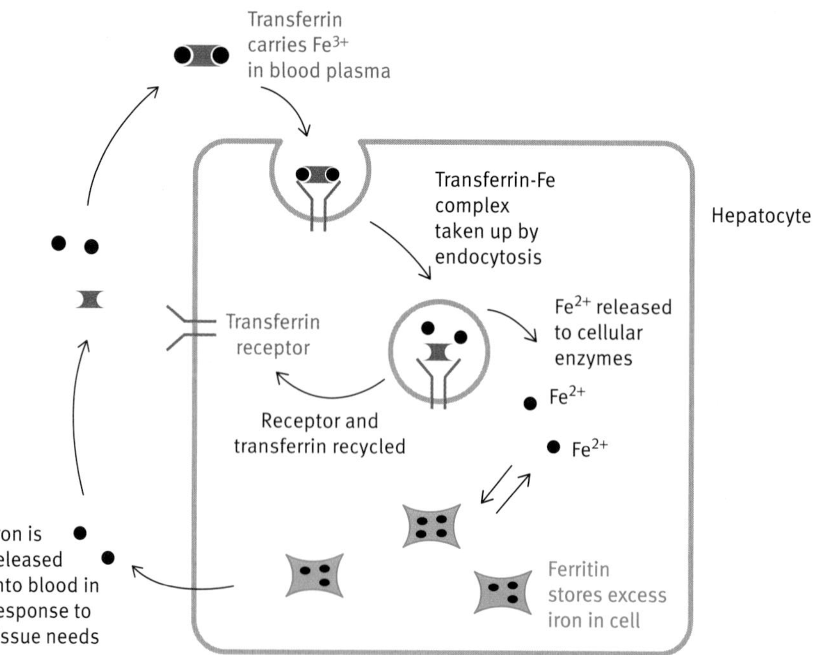

Fig. 26.21 Simplified scheme illustrating the role of liver cells (hepatocytes) in iron homeostasis. The levels of ferritin and the transferrin-receptor are controlled by post-transcriptional regulation of their gene expression, as described in the text.

and a phosphatase hydrolyses the phosphate off eIF2 and restores the normal activity of the initiation factor. Thus globin synthesis can proceed only when haem is present. This system of repressing a factor *generally* needed for translation of all mRNAs in order to regulate globin synthesis works in red blood cells only because globin mRNA is essentially the one type of mRNA they contain.

Small RNAs and RNA interference

We have already discussed the existence in biochemical processes of species of RNA other than messenger RNA, such as ribosomal, transfer and small nucleolar RNAs (snoRNAs) (see Chapter 25). Their functions in protein synthesis have been known for a long time. The more recent discovery of widespread eukaryotic gene regulation by several classes of small RNA was a surprise. Small regulatory RNAs have already been mentioned briefly in the context of controlling RNA stability. They also exert their effect through regulating RNA translation and in some cases by regulating transcription. This phenomenon of gene silencing by **RNA interference (RNAi)** has been the focus of enormous recent attention. Although its extent and overall importance are still unclear, more examples are being discovered every day. It is also of great interest because it offers a powerful experimental and potentially therapeutic tool for controlling gene expression.

Classes and production of small RNAs in eukaryotes

Three main classes of small regulatory RNAs are recognized: small interfering RNAs (siRNAs), microRNAs (miRNAs) and piwi-interacting RNAs (piRNAs). The latter class is confined to germ cells where they seem to prevent the activation of transposons. This is a relatively specialized role and we will not consider piRNAs further here. **siRNAs** can be exogenous or endogenous in origin. In plants and in some invertebrates such as the fruit fly *Drosophila*, which lack sophisticated immune systems, RNA interference can act as protection against infectious viruses. In these cases the triggering siRNAs are derived from the viral RNA, so are exogenous in origin. However, in siRNA-mediated gene regulation, which is now found to be widespread in plants, invertebrates and vertebrate species including mammals, siRNAs are endogenous, encoded by the organism's own genome. **MicroRNAs**, which are produced and processed slightly differently from siRNAs, are genome-encoded and are involved in developmental gene regulation in plants and animals.

All these classes of small RNA are 20–30 nucleotides in length, and although they ultimately act as single-stranded RNA, they are produced via a double-stranded RNA (dsRNA) precursor (Fig. 26.22). In the case of siRNA, the dsRNA may be viral RNA, or provided experimentally (exogenous siRNA), or it may be derived from transcription of repetitive elements such as transposon derived sequences in the genome (endogenous siRNA). MicroRNA precursors known as primary miRNAs (pri-miRNAs) are generally transcribed by RNA polymerase II. Some miRNA genes are clustered in the genome, while others are found within the introns of protein-coding genes. The sequence of pri-miRNAs allows them to fold back on themselves forming 'hairpin' loops through internal complementary base pairing. A pri-miRNA is generally several thousand bases long and is initially 'trimmed' in the nucleus by a ribonuclease enzyme called **Drosha** to leave a small hairpin loop termed a pre-miRNA, which is exported to the cytosol.

Once the pre-miRNA is in the cytosol, it joins a pathway that seems to be common for both miRNA and siRNA processing. A second ribonuclease, related to Drosha and called **Dicer**, cleaves the hairpin loop off the pre-miRNA leaving the double-stranded stem. Dicer also cleaves double-stranded siRNA precursurs into shorter duplexes around 22 bp in length. The miRNA or siRNA duplex then starts to assemble with a complex of proteins termed **RISC (RNA-induced silencing complex)**. During formation of RISC, one strand of the duplex (the **passenger strand**) is removed and degraded, while the other, the **guide strand**, is retained. The guide strand is the mature regulatory RNA, which will direct RISC to its complementary target sequence. Although it is not fully understood how the passenger and guide strand are selected for degradation and retention, respectively, an enzyme called **argonaute**, which forms part of RISC, is involved in the separation of the two. Argonaute proteins are fascinating multifunctional proteins, and as you will see they play several different roles in RNAi.

Molecular mechanism of gene silencing by RNAi

Most small regulatory RNAs inhibit gene expression post transcription

Most small regulatory RNAs silence gene expression by base pairing of the guide strand with a target mRNA in the cytosol. What happens next appears to be determined by the degree of complementarity between the guide strand and its target. If it is perfect or nearly so (as is the case for most siRNAs and some miRNAs) the guide strand base pairs with the translatable section of the mRNA. **Argonaute** comes into play again, this time cleaving the mRNA, which is thereby cut into two pieces. As one piece of the mRNA is no longer protected by a 3′ polyA tail, and the other lacks a 5′ protective cap, they are subject to further digestion by exonucleases (Fig. 26.23(a)). The regulatory RNA, on the other hand, remains intact and can target another mRNA molecule, making RNAi a highly efficient process.

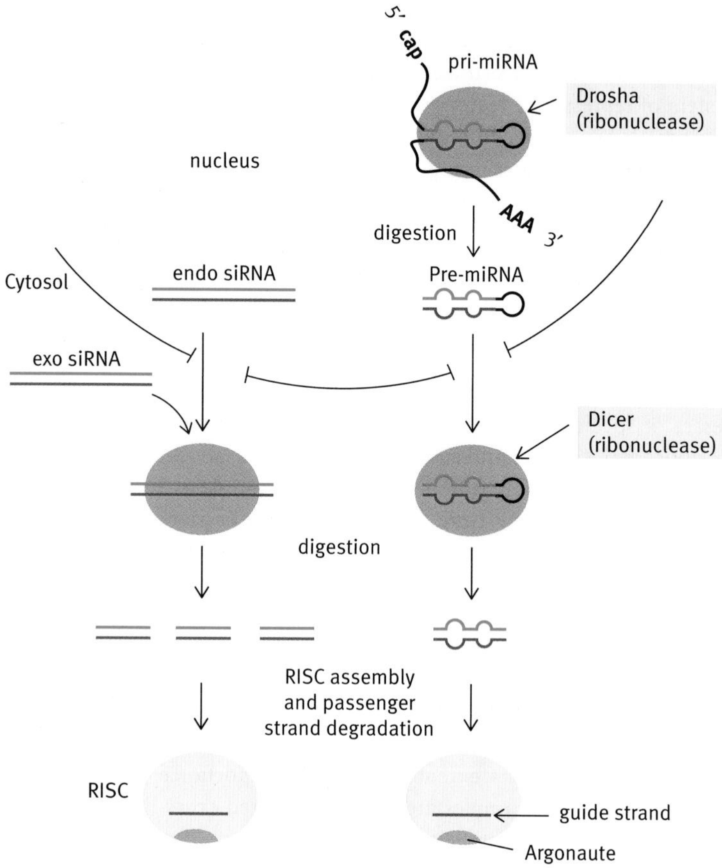

Fig. 26.22 Processing of short regulatory RNAs. Double-stranded siRNA precusors can be endogenous (transcribed from genomic sequences) or exogenous (e.g. viral) in origin. Pri-miRNAs are transcribed as single-stranded RNA, which forms double-stranded regions via internal base pairing. Pri-miRNAs are cleaved by Drosha ribonuclease to produce small hairpin pre-miRNAs, which are exported to the cytosol. Both siRNA precursors and pre-miRNAs are cleaved to smaller sizes (20–30 bps) by Dicer, which removes the hairpin loop of miRNAs leaving the partially base-paired stem. Passenger strands are degraded and guide strands are loaded onto RISC (RNA-induced silencing complex), which includes the Argonaute protein. The order of events during passenger strand digestion and RISC assembly is not known for certain; it appears that Argonaute is involved in separation of the guide and passenger strands.

For most miRNAs the complementarity match between the guide strand and the mRNA is only partial, and in these cases RISC attaches to the 3′ untranslated region (3′ UTR) of the mRNA. There appear to be multiple mechanisms by which inhibition can then occur (Fig. 26.23(b)). When there is a partial mismatch the mRNA is not cleaved by Argonaute, but in some cases RISC binding accelerates digestion of the 3′ polyA tail and therefore causes destabilization and nuclease digestion of the mRNA. In other cases RISC binding prevents translation by ribosomes. The RISC attachment site on mRNA is downstream of the stop codon where the ribosomes detach and translation terminates, so it is not clear exactly how translation is inhibited: the Argonaute component of RISC may be involved here in yet another role. A domain of the Argonaute protein has homology with one of the eukaryotic translation initiation factors, and it may compete with this factor for binding to the 5′ cap of the mRNA, thus reducing the efficiency of translation.

There is much yet to be understood about the precise mechanism of gene silencing by miRNAs *in vivo*. Often inhibition is not complete: protein synthesis from the inhibited gene is reduced but not absent. Partial silencing is called **gene knockdown**. To complicate matters further, there are also a few examples where miRNA regulation has been shown to *increase* gene expression, and more research is needed to understand how this comes about and what determines which miRNAs have which effect.

Some RNAi acts at the level of chromatin

Although most RNAi prevents translation, some classes of siRNA, particularly in plants, act at the level of chromatin and therefore inhibit transcription. Here the guide strands bind either to complementary DNA sequence in the genome, or to nascent mRNA transcripts. Binding directly to the DNA or to mRNA while it is still being transcribed triggers recruitment of chromatin modifying enzymes to the target gene. This causes formation of compacted heterochromatin, hence shutting down transcription. This RNAi mechanism seems particularly important in shutting down transcription from retrotransposons, hence preventing the movement of mobile genetic elements and protecting the integrity of the genome.

In vivo functions and importance of noncoding RNA

Gene silencing triggered by RNA was first described in plants when it was observed that RNA virus infection of plants led to them developing an immunity to the virus. After this, it

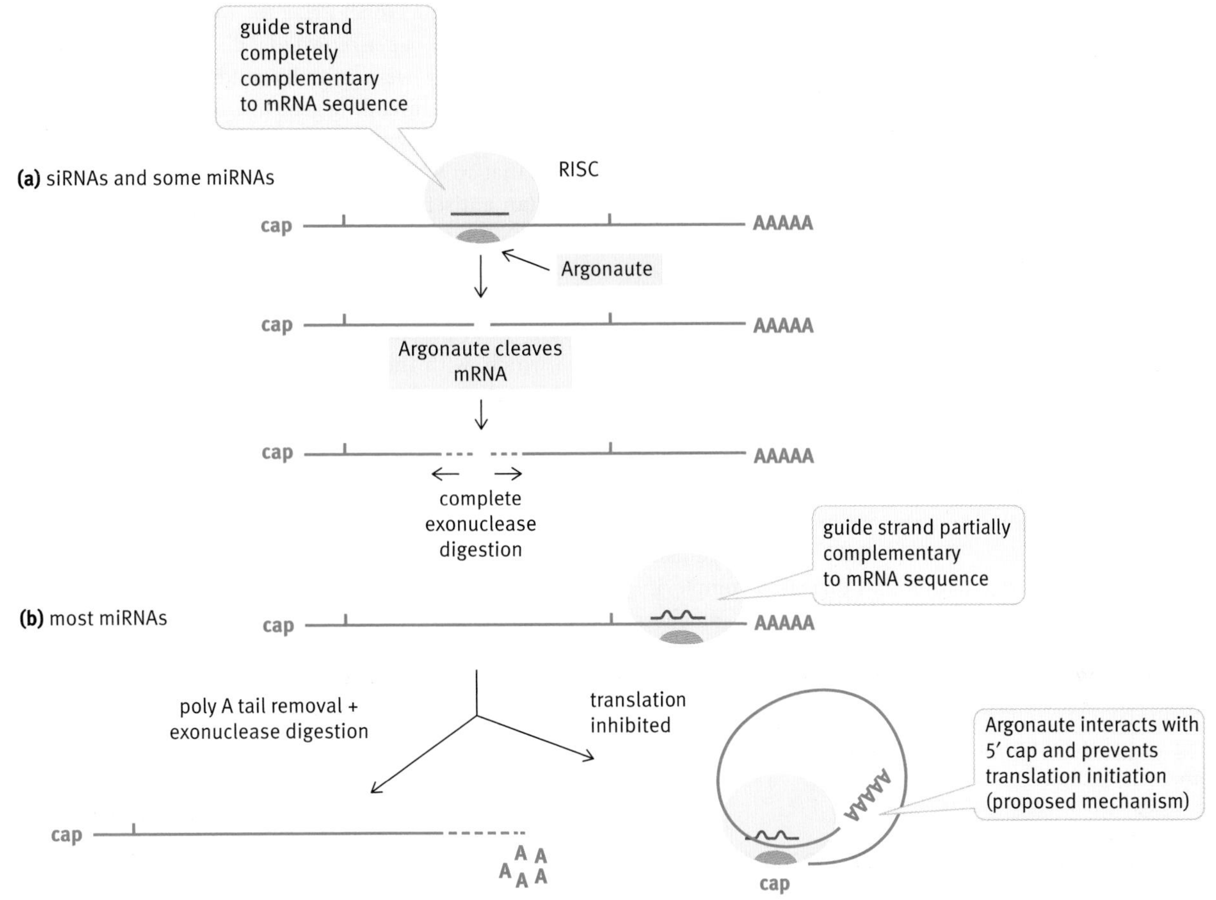

Fig. 26.23 Gene silencing by RNAi. **(a)** siRNAS and the small number of miRNAs that are perfectly complementary to their mRNA target sequence base pair to the protein-coding region. The mRNA is cleaved by the endonuclease activity of Argonaute, leaving the cut ends exposed to exonuclease digestion. The mRNA is completely broken down. **(b)** Most miRNAs are partially complementary to their target mRNAs and bind the 3′ UTR. RISC either accelerates polyA tail breakdown, exposing the 3′ end of the mRNA to exonuclease digestion, or inteferes with translation. The mechanism shown here with Argonaute binding to the 5′ cap, and hence blocking the translation initiation factor access, is speculative as the details of the process are not fully understood.

was discovered in the small roundworm *Caenorhabditis elegans* that a gene essential for its development (called *lin-4*) did not code for a protein but for a small noncoding RNA that silenced the expression of the gene *lin-14*. *lin-14* encodes a protein essential for the development of the worm that regulates the timing of certain developmental events, and *lin-14* is itself regulated by the *lin-4* noncoding RNA. *lin-4* turned out to be the first characterized miRNA. The discovery of other miRNAs in many species including humans followed. More than 1,000 human miRNAs have now been reported and it seems likely that more will be discovered.

Before the discovery of miRNA genes, most thought that the basic rules of genome function had been established despite gaps in our understanding. We still accept that protein-coding genes determine more or less all heritable characteristics, since the proteins that are produced determine the chemistry and assembly of organisms. However, there is an increasing body of evidence pointing to microRNAs and other classes of noncoding RNA as regulators of a wide range of the most fundamental processes of life, including development, cell growth, and apoptosis. Gene silencing by RNAi can be viewed as an example of epigenetic regulation, as it can be stably transmitted through cell division. siRNA expression and RNAi induced chromatin modification have even been shown, at least in *C. elegans*, to pass through the germ line from one generation to the next.

In order to understand the scale and importance of regulation by noncoding RNAs and RNAi, it seems that what is now needed is a genome-wide insight, both to identify regulatory RNAs and to explore their functions. To this end an international collaborative research effort has been set up with the aim of identifying the functional elements in the human genome and their transcriptional activities. It is called ENCODE (Encyclopaedia of DNA Elements). In its initial phase, ENCODE set out to examine a total of 1% (about 30,000 kilobases) of the human genome in great detail. To make up the total it sampled 44 different sections of the genome taken from different representative areas, such as areas rich in protein-coding genes and areas hitherto believed to be transcriptionally inert 'junk DNA'. It was found, remarkably, that despite mostly not being protein-coding genes, about 90% of the bases in the genome are found in RNA transcripts. It could have been that this low level RNA transcription was background 'transcriptional noise' that had no function. However, it was found that a number of the noncoding transcripts were conserved across mice and humans, showing that these at least almost certainly have an important role. As well as microRNAs and endogenous siRNAs, long noncoding RNAs have also been found that regulate gene expression via a variety of mechanisms.

Perhaps the most profound outcome of this new field of research is the possibility that noncoding RNAs appear to constitute another layer of control that may orchestrate the function of the genomes of humans and complex organisms generally. It may be that evolution has increased complexity by having large numbers of regulatory RNAs that somehow network a limited number of protein-coding genes into producing greater phenotypic complexity than could be otherwise achieved. This could potentially solve the puzzle posed by finding surprisingly few protein-coding genes in the human genome.

The potential medical and practical importance of RNAi

The discovery of gene silencing by RNAi has also generated huge excitement because of its potential as an experimental and therapeutic tool. In 2006 Andrew Fire and Craig Mello were awarded the Nobel Prize for work that opened up the field. Their 1998 paper reported a discovery that was almost accidental. They were initially studying the possibility of silencing a protein-coding gene in *C. elegans* using **antisense RNA** chemically synthesized in the laboratory. The concept here is that an RNA molecule *complementary* to a messenger RNA (an antisense RNA), present in excess, would bind to the mRNA by Watson–Crick pairing and block the translation of the latter, and thus inhibit expression of the gene. An antisense preparation when injected into the worm was effective, but paradoxically so was a sense strand preparation which, with the same base sequence as the messenger, could not have hybridized with the latter. It was obvious that a different explanation for the silencing must be involved. The explanation was that both preparations were contaminated with *double-stranded RNA*, which was what did the silencing.

Because RNA with defined sequence can be synthesized in the laboratory, RNAi is a relatively easy method by which expression of a particular gene can be knocked down experimentally, usually in cultured cells. Double-stranded siRNAs are introduced directly into the cells, or for a longer-term effect DNA sequences encoding short hairpin RNAs that mimic microRNAs are introduced. Observing the phenotypic effects of reduced expression of the target gene gives insight into its normal function.

The possibility that RNAi could be used therapeutically, for instance to treat cancer by knocking down expression of cancer-causing oncogenes, is also exciting. The main difficulty is that of delivering the therapeutic RNAs to the patient's cells and tissues. Gene therapy approaches may be used to introduce DNA encoding therapeutic microRNAs.

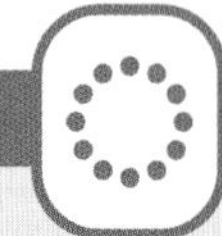

Summary

Gene expression can be regulated at each stage of protein synthesis, but the majority of regulation is of transcription. In *E. coli*, groups of genes called operons often occur and are transcribed together forming polycistronic messengers. In the *lac* operon, comprising three genes, a repressor protein effects control by blocking an operator region at the initiation site of transcription. The repressor is an allosteric protein and in the presence of lactose it detaches and allows transcription of genes required to form enzymes needed to utilize the sugar. The actual inducer is allolactose formed from lactose. This type of operon control is prevalent in other metabolic systems.

Eukaryotic protein-coding gene control is also mainly at the level of transcription initiation but the process is more complex. Eukaryotic cells of an animal have to respond to batteries of signals from hormones, growth factors, and cytokines, so that a given gene is likely to be instructed by a multiplicity of signals that the initiation process interprets. Eukaryotic genes are in a closed-down default state – they are inactive until something is done to change this. The keys to eukaryotic gene control are transcription factors. These are proteins that bind to gene regulatory sequences upstream of the start site; they open up specific promoters. Some genes have enhancers, large distances away from the gene, to which transcription factors also attach and enhance gene activity. Sections of DNA, called insulators, confine enhancer effects to their intended target gene(s).

The recognition of specific DNA sequences by the various transcription factors is a crucial aspect of transcriptional regulation. There are a number of characteristic motifs found in DNA-binding proteins. These include helix–turn–helix, zinc finger, leucine zipper and helix–loop–helix proteins. Many act as dimers that recognize two DNA sequences in the major grooves of the double helix. Eukaryotic transcription factors activate transcription by a variety of mechanisms that include interacting via the mediator with the basal initiation complex, recruitment of coactivators, and interaction with chromatin. Transcription factors are themselves activated and inactivated by specific signals and may act as repressors rather than activators.

Chromatin is important in eukaryotic gene regulation as nucleosomes control access of transcription factors to DNA. Chromatin is modified by histone acetylases and deacetylases and other remodelling enzymes, which may be recruited by transcription factors. Methylation of mammalian DNA on cytosine also regulates gene transcription and is an example of epigenetic control as it is passed on stably through cell division.

Gene regulation after transcription initiation involves regulation of mRNA stability and of translation. The coordinated synthesis of mammalian iron binding proteins in response to iron levels provides a good example of post-transcriptional gene regulation involving both these mechanisms.

RNA interference (RNAi) is an ancient natural mechanism of silencing protein-coding genes. It is widely found in eukaryotes, but not in prokaryotes. Its function was probably initially to protect against invasion by RNA viruses, but it is also widely involved in regulating gene expression, making it an important topic of current research. Gene silencing by RNAi also has great potential as an experimental and therapeutic tool.

Two classes of small regulatory RNAs, short interfering RNAs (siRNAs) and microRNAs (miRNAs) are made in the cell from dsRNA intermediates, or siRNA may come from exogenous sources such as viruses. Most RNAi silencing occurs though base pairing of the regulatory RNA to mRNA, which causes degradation of the message or blocks translation. Some siRNA acts on chromatin to shut down transcription from retrotransposons, which prevents them from moving around and protects genome integrity.

Noncoding RNAs are now thought to be involved in orchestrating gene expression in many fundamental cellular processes. This potentially changes our view of 'junk' DNA as a high percentage of nonprotein-coding sequences in the human genome appear to be transcribed. It seems that microRNAs enable increasing complexity of life forms without a corresponding increase in protein-coding genes, though much remains to be learnt on this subject.

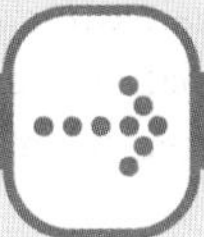

Further reading

To access the further reading, please scan the QR code image or go to http://global.oup.com/uk/orc/biosciences/molbiol/snape_biochemistry/student/reading/ch26/

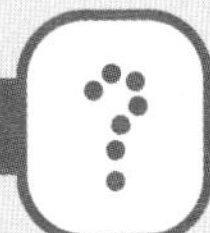

Problems

1 Describe how the *lac* operon is controlled.

2 In terms of structural motifs, there are several families of transcription factors. What are these?

3 What part does acetyl-CoA play in the initiation of eukaryotic gene transcription?

4 What is one probable way by which an activated eukaryotic gene is inactivated?

5 The control of gene initiation by the acetylase and deacetylase implies that these enzymes are somehow targeted to specific gene promoters. How is this done?

6 Explain how the red blood cell ALA synthase level is coordinated with the availability of iron.

7 What is a noncoding RNA? Give examples.

8 Where are miRNA genes located in the genome?

9 What is the evidence that miRNAs represent meaningful transcription and not just background transcriptional 'noise'?

10 What is believed to be the significance of miRNAs?

11 Describe briefly the mechanism of RNA interference (RNAi).

12 What is one medical potential of RNAi?

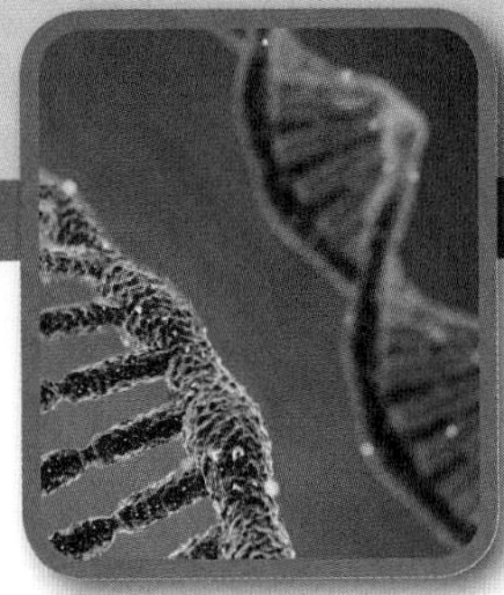

Chapter 27

Protein sorting and delivery

As explained in Chapter 25, the majority of proteins made in eukaryotic cells are synthesized by cytosolic ribosomes. The only sites of protein synthesis other than the cytosol are the mitochondria and chloroplasts, and these produce only a handful of proteins that are encoded by the organelles' own genomes. Which protein a given cytosolic ribosome synthesizes at any one time is solely a function of the mRNA that it happens to be translating, but, once synthesized, proteins have a number of different destinations.

So far as delivery goes, cytosolic proteins present no problems – they are synthesized in the cytosol, released from the ribosome, and stay there. However, proteins destined for other cellular compartments present intriguing problems. How do the integral proteins of the plasma membrane and other membranes get to be there? How are blood serum proteins selectively released by liver cells to their exterior? The same question applies to release of any of the many other extracellular proteins such as digestive enzymes, insulin from the pancreas, and connective tissue proteins from fibroblasts. Most mitochondrial proteins are encoded by nuclear genes and are hence synthesized in the cytosol. How are they selectively transported into the mitochondria? Lysosomes and peroxisomes are membrane-bounded vesicles full of specific enzymes, but they cannot synthesize proteins. Again, how are the different enzymes transported into the correct compartment? The nucleus has its own cohort of proteins such as the enzymes responsible for synthesizing and transcribing DNA, but these are synthesized in the cytosol. There is also traffic of proteins (and RNA) out of the nucleus. How is the two-way traffic across the nuclear membrane organized? As you will learn from Chapter 29 on cell signalling, many gene-control proteins exist in the cytosol but, on receipt of extracellular signals, they enter the nucleus to regulate transcription of genes.

It is not just a question of how proteins are able to cross membranes but also how specific proteins are selected from the whole mixture of proteins in the cell to be delivered to, and transported across, or into, the correct membrane. Many of the mechanisms of protein targeting have been substantially elucidated, though much detail of the processes remains to be understood.

A preliminary overview of the field

An overview summary without any details may be useful. An important distinction when considering proteins that are destined for cellular compartments other than the cytosol is whether they are synthesized on free cytosolic ribosomes, or on ribosomes that attach to the endoplasmic reticulum (ER), which acts as a 'staging post' from which proteins are despatched to further destinations.

- **Cytosolic proteins** are released from the ribosome on completion of their synthesis and stay in the cytosol (Fig. 27.1).
- Proteins destined for **mitochondria, peroxisomes, or the nucleus** are released from the free cytosolic ribosomes and are then transported into the appropriate organelle, but by a different mechanism in each case; this is known as **posttranslational transport** (Fig. 27.1).
- The synthesis of **extracellular (secreted) proteins, lysosomal proteins, proteins that function in the lumen of the endoplasmic reticulum** (ER), and all **integral membrane proteins** *commences* on free ribosomes, but these become transitorily attached to the ER membrane and the proteins are transported into the ER lumen or ER membrane *as they are synthesized*. This is known as **cotranslational transport**.

Once inside the ER lumen, proteins move along to the smooth ER and then are transported to the Golgi apparatus. In the ER and Golgi, proteins have carbohydrates added. The Golgi sorts out and packages proteins into transport vesicles, which deliver their cargo to appropriate target membranes or

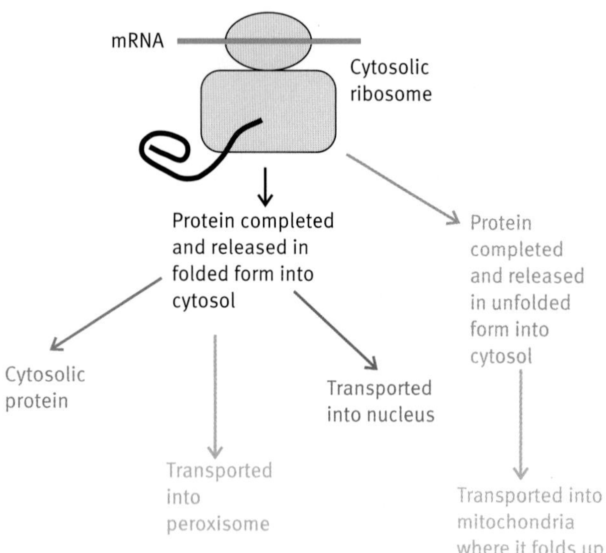
mRNA
Cytosolic ribosome
Protein completed and released in folded form into cytosol
Protein completed and released in unfolded form into cytosol
Cytosolic protein
Transported into nucleus
Transported into peroxisome
Transported into mitochondria where it folds up
(The method of transport into the three organelles is different in each case.)

Fig. 27.1 Preliminary overview summary of events in posttranslational targeting of proteins to the cytosol, peroxisomes, nucleus, and mitochondria. The essence of the process is that free ribosomes synthesize complete polypeptide chains, release them in the cytosol, and these proteins are then transported to their target destinations but by different mechanisms in each case. This contrasts with cotranslational transport, depicted in Fig. 27.6, in which proteins are transported across the endoplasmic reticulum lipid bilayer as they are synthesized.

compartments. Secretory vesicles eject their contents from the cell by exocytosis (see Chapter 7), and transport vesicles deliver their contents to endosomes (see later in this chapter) to form lysosomes. Vesicles carrying integral proteins in their membrane fuse with their target to produce new membrane complete with proteins (Fig. 27.2).

Some additional information on the ER and the Golgi apparatus will be useful since these remarkable organelles play such a central role.

Structure and function of the ER and Golgi apparatus

The ER is a membranous structure that pervades eukaryotic cells to varying degrees. It is a complex of linked sacs so that its lumen is one continuous cavity separated from the cytosol by the single ER membrane. Its size varies enormously in different cells, depending on the metabolic functions and state of the cell. Part of the ER when seen in the electron microscope is studded with attached ribosomes and is the **rough ER**; the rest, **smooth ER**, has no attached ribosomes. The rough and smooth ER are not physically discrete membrane structures but are different regions of a continuous structure (Fig. 27.3) that have different functions. The ER lumen is continuous with the space enclosed by the double membrane surrounding the nucleus.

The ribosomes are not a permanent fixture on the rough ER but are those which just happen, at the time, to be translating an mRNA that codes for a protein destined for secretion, inclusion in a lysosome, or incorporation as an integral membrane protein. When the ribosome on the ER has completed the synthesis of a protein molecule it detaches and its subunits re-enter the general cytosolic pool. There are no 'special' ribosomes for this role. Polypeptides are transported into the ER lumen as they are synthesized and are folded in the lumen. The proteins progress through the smooth ER, in which they often have carbohydrates added. To leave the smooth ER, proteins are enclosed in small transport vesicles that bud off and transport them to the **Golgi apparatus** (Fig. 27.4).

The ER and the Golgi are completely closed structures, with no physically evident entry or exit sites. The smooth ER is also the major site for synthesis of new lipid bilayer, a function which correlates with the formation of transport vesicles.

The Golgi apparatus consist of 4–6 (more in plant cells) membranous flattened structures enclosing spaces known as **cisternae**. They resemble a stack of large plate-like vesicles placed near the nucleus. The side facing the ER is the transport vesicle reception area known as the *cis* cisternae (cis=near to; cisternae=chambers). Transport vesicles carrying newly synthesized proteins are budded off from the smooth ER, move to fuse with the *cis* membranes, and deliver their contents to the Golgi cisternae. Proteins move through the Golgi stacks and are modified by different enzymes in successive cisternae, progressing towards the *trans* (trans=distant) side where they are sorted, packaged into vesicles and despatched to their destinations. The Golgi thus takes newly synthesized proteins arriving from the ER, modifies them, packages them into membrane vesicles addressed to their proper destinations, and finally despatches them. It is what a mail-sorting office is to posted mail. There is even a 'return-to-sender' service; proteins that are needed in the ER, such as chaperones and enzymes involved in polypeptide folding, are sent back from the Golgi in appropriately addressed transport vesicles.

There is still some controversy over how proteins move through the Golgi from the *cis* to the *trans* side. In one model they are carried between the cisternae in transport vesicles, while in another the cisternae themselves go through a maturation process, starting as *cis* and ending as *trans* cisternae. In the latter model, Golgi enzymes must be returned to their starting points via transport vesicles while proteins for onward transport remain in the cisternae as they mature.

That is the end of the general overview, and we can now turn to the molecular processes. We start with ER-mediated processes. These will be followed by the transport of proteins into mitochondria and peroxisomes (in which the ER and Golgi are not involved), and finally with transport of proteins across the nuclear membrane, which is a quite different story from all the rest. However, the following section describes a feature of some of the molecular processes that is worth spending a little time on first.

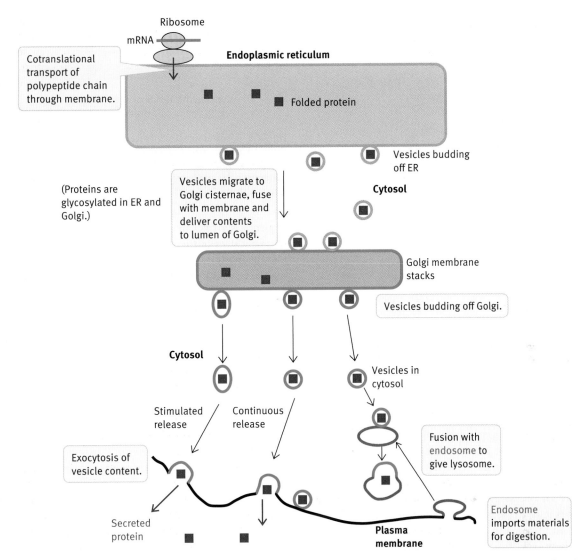

Fig. 27.2 Preliminary overview of how proteins are secreted from cells and how enzymes are delivered to lysosomes. The initial transport of proteins through the lipid bilayer into the lumen of the endoplasmic reticulum (ER) is cotranslational – the polypeptide traverses the membrane as it is synthesized. This is quite different from the transport into peroxisomes, mitochondria, and nuclei which occurs posttranslationally (see Fig. 27.1). Proteins are glycosylated as they pass through the ER and Golgi. Targeting of new membrane proteins to their appropriate sites is basically by a similar mechanism in each case; the proteins are inserted into the ER membrane as they are synthesized and then sections of membrane, complete with the new proteins, are packaged as vesicles. These migrate to and fuse with the target membrane, thus delivering both new lipid bilayer and membrane proteins.

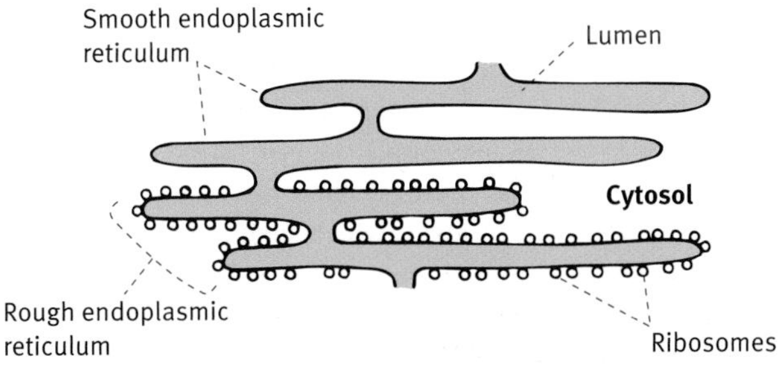

Fig. 27.3 Diagrammatic representation of the endoplasmic reticulum.

The importance of the GTP/GDP switch mechanism in protein targeting

As we go through the protein-targeting mechanisms, GTP hydrolysis to GDP + P_i will be encountered several times. You are used to ATP breaking down to perform chemical or other work, but GTP is often broken down by **GTPase proteins**, apparently without any useful work being performed. This is not a waste of energy. The hydrolysis of the protein-bound GTP to GDP causes an allosteric conformational change in the protein. This acts as a switch to allow the next step in a process

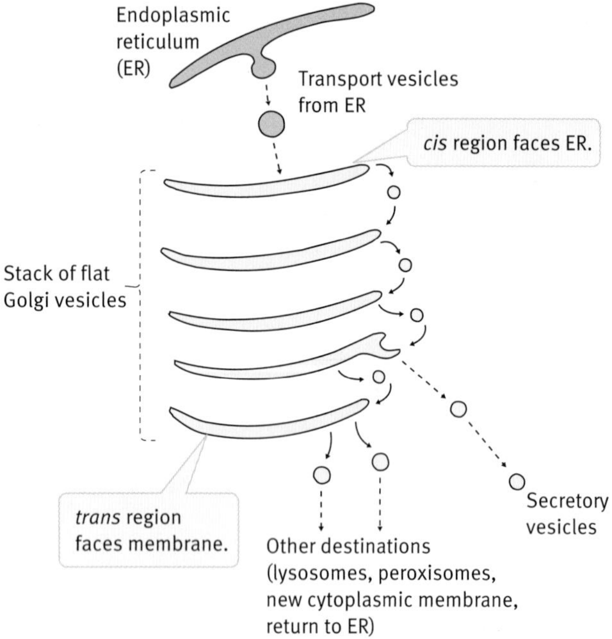

Fig. 27.4 The central role of the Golgi apparatus in posttranslational sorting and targeting of proteins. In addition, newly synthesized membrane lipids are transported to the appropriate destination by transport vesicles.

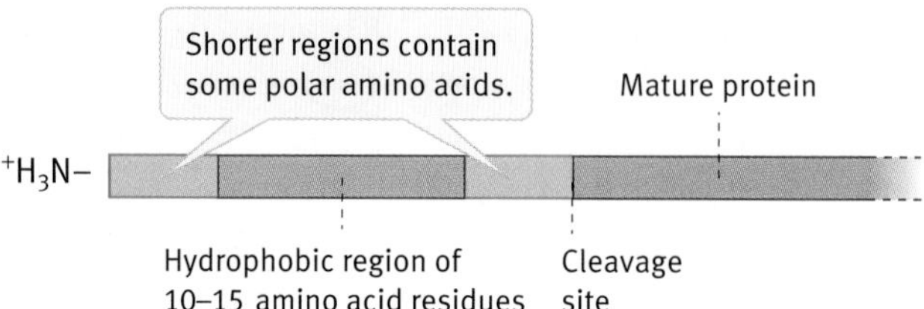

Fig. 27.5 A typical signal sequence at the *N*-terminal of a protein destined to be transported through the endoplasmic reticulum membrane. Such signal peptide sequences show the same general pattern of polar and hydrophobic amino acids but no specific amino acid sequence. Acidic amino acids are not present.

to occur, and, since the hydrolysis is irreversible, it confers a unidirectionality on the process. Note that there is believed to be a slight delay between the initial trigger and GTP hydrolysis by these proteins (they are often called 'slow' GTPases), which gives time for one stage in a process to be completed, and then moves it on to the next stage. After the GTP hydrolysis, the GDP on the protein is exchanged for GTP, restoring the original form. GTPase-activating proteins (GAPs) and guanine nucleotide exchange factors (GEFs) are often involved in regulating this type of switching, which is important in many processes. It occurs in microtubule dynamics (see Chapter 8), protein synthesis (see Chapter 25), cell signalling (see Chapter 29), transfer of proteins into the ER, nuclear–cytosolic transport and vesicle transport (this chapter).

Translocation of proteins through the ER membrane

Gunther Blobel of the Rockefeller Institute, New York, was awarded a Nobel Prize in 1999 for his work on protein transport and localization. He and coworkers established that proteins that are to be secreted or end up in the external plasma membrane, inside lysosomes, or in the lumen of the ER itself, have at their *N*-terminal end a **signal peptide sequence** of about 29 amino acids. If the coding information for such a signal sequence is artificially added to the mRNA for a protein that normally stays in the cytosol, the protein will be transported into the ER as it is synthesized. The signal peptide amino acid sequences of different proteins have a pattern, rather than a fixed sequence, as shown diagrammatically in Fig. 27.5. They have a short, positively charged *N*-terminal section and a central hydrophobic region 10–15 amino acid residues in length. Artificial substitution of a single charged residue into the hydrophobic central region is enough to render the signal sequence inactive. After translocation the signal sequence is cleaved off by a **signal peptidase** on the inner face of the ER membrane, so the mature protein does not have the sequence.

Mechanism of cotranslational transport through the ER membrane

It will help if you follow each numbered step in Fig. 27.6 as you read the following section. You will notice that GTP/GDP switching is involved, and ensures that the steps occur in the correct order. A free ribosome in the cytosol translating an mRNA for an ER-targeted protein synthesizes the signal sequence first (since it is at the *N*-terminal end of the polypeptide). The signal sequence is recognized by a **signal-recognition particle (SRP)**, an RNA-protein complex in the cytosol that binds to the nascent signal peptide when it emerges from the ribosomal tunnel and arrests further elongation of the polypeptide chain (step 1 in Fig. 27.6). The SRP is a GTP-binding protein.

The loaded ribosome migrates to the ER membrane (step 2 in Fig. 27.6), which has on it **SRP receptors** or **SRP docking proteins**. The receptors are found only on the ER membrane and, like the SRP, they are GTP-binding proteins. The SRP of the cytosolic ribosome-SRP complex attaches to the SRP receptor (step 3 of Fig. 27.6), and positions the ribosome on the membrane. In the ER membrane there are protein assemblies known as **translocons**. These are channels, formed by several subunits of a protein complex, which span the membrane. In the absence of a ribosome the translocon is in an effectively closed condition. A ribosome attaching to the SRP receptor on the membrane becomes associated with a translocon channel.

Opening of the translocon channel by the ribosome and transfer of the signal sequence from the SRP to the channel

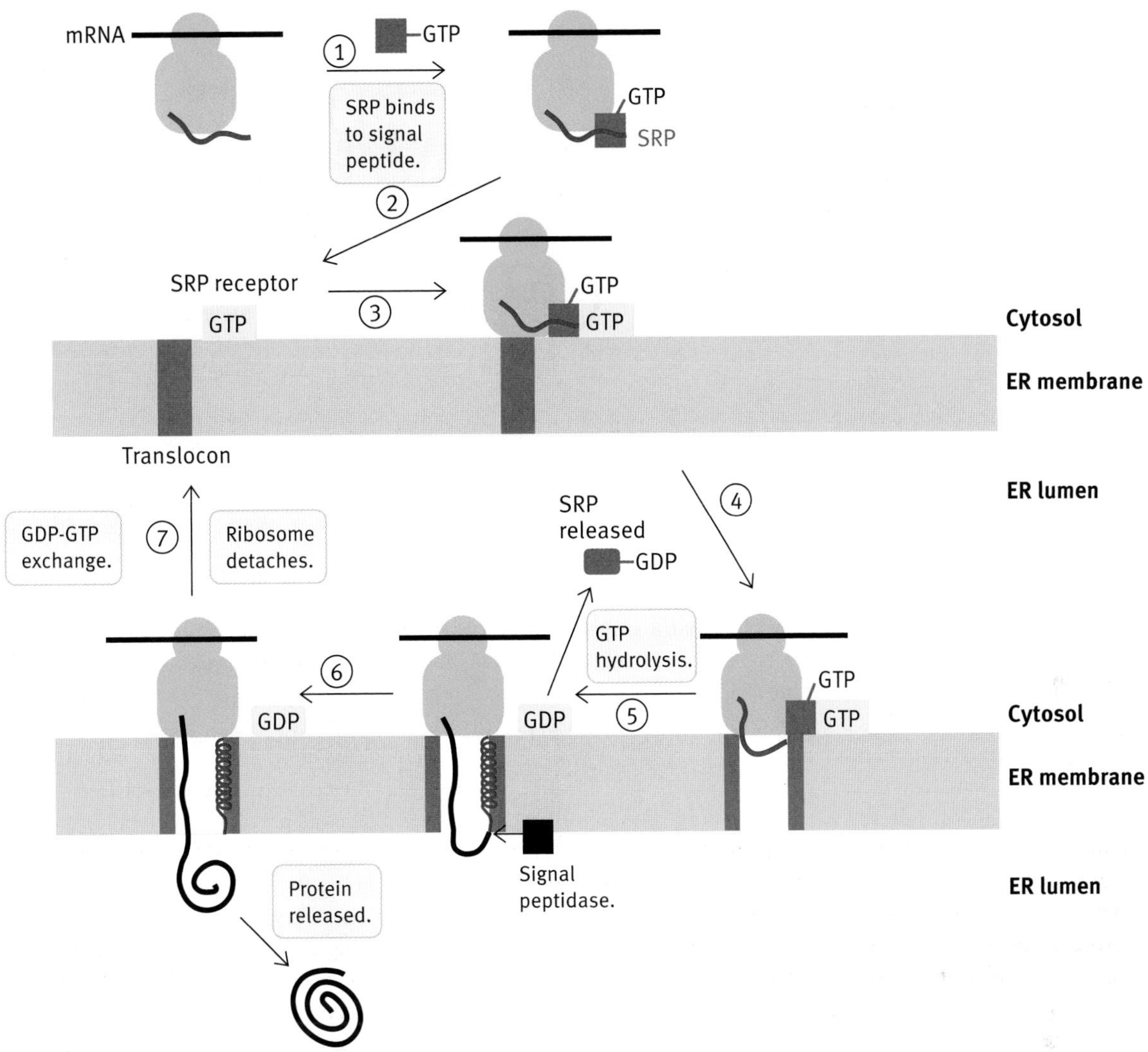

Fig. 27.6 Sequence of events by which proteins are cotranslationally transported into the lumen of the endoplasmic reticulum (ER). The numbered steps are referred to in the text. The signal peptide is shown in red and the polypeptide that constitutes the mature protein is shown in black. SRP, signal-recognition particle.

now takes place (step 4 of Fig. 27.6). Next, GTP hydrolysis occurs; SRP and SRP receptor GTPases are activated by conformation changes that occur when the two form a heterodimer, *but GTP hydrolysis to GDP occurs only once the ribosome-signal sequence complex is transferred from the SRP to the translocon channel* (step 5 in Fig. 27.6), The GDP-bound forms of the SRP and SRP receptor do not bind each other efficiently, so the SRP is now released into the cytosol for further use.

The ribosome recommences synthesis of the polypeptide, which traverses the membrane via the translocon channel as it is synthesized. The signal peptide is postulated, in this model, to remain in the channel but the cleavage site is exposed at the internal face of the membrane as illustrated in step 5. Other models show the signal peptide going straight through. The signal peptidase, which is responsible for the cleavage (step 6 in Fig. 27.6), has a hydrophobic patch that attaches it on the membrane so that as the signal peptide cleavage site emerges from the membrane it encounters the peptidase.

On completion of the polypeptide synthesis, the polypeptide is released into the lumen of the ER. The signal peptide is presumed to be destroyed and the translocon, it is postulated, becomes closed. The ribosome dissociates into its subunits to rejoin the cytosolic pool for further use. The SRP and SRP receptor undergo exchange of GDP for GTP so that they are ready for reuse (step 7 of Fig. 27.6).

Synthesis of integral membrane proteins

Integral membrane proteins are synthesized on the rough ER, integrated into the membrane and transported *in situ* by vesicles, as new membrane, to the plasma or other target membranes. How are the proteins fixed in the ER membrane

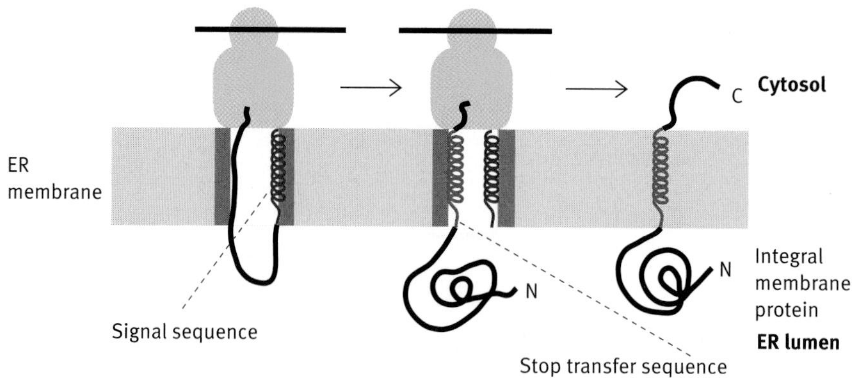

Fig. 27.7 Insertion of integral membrane proteins into the membrane of the endoplasmic reticulum (ER). The initial steps are as depicted up to step 5 in Fig. 27.6 but the ribosome synthesizes another sequence, shown in blue, which is a stop transfer or anchor sequence. This arrests the further movement of the polypeptide so that when the protein is fully synthesized and exits the channel, the protein is left as an integral membrane protein. As described in the text, models have been proposed for a mechanism by which the integral protein may be oriented across the membrane in the opposite orientation, with the *N*-terminus in the cytosol. There must presumably be a mechanism for releasing the protein laterally from the translocon.

instead of going right through it? One model is that a **stop transfer sequence** or **anchor sequence** is translated, which anchors the polypeptide into the membrane (Fig. 27.7). Once it is fully synthesized the protein is presumed to exit laterally from the translocon into the lipid bilayer.

This sequence of events produces a transmembrane protein with its *N*-terminus inside the ER (Fig. 27.8(a)). Other orientations occur (Fig. 27.8(b),(c)). A possible mechanism for achieving the reverse orientation (b) is shown in Fig. 27.9. Here a **noncleavable signal peptide** positioned within the polypeptide also acts as a stop transfer signal and is therefore called a **signal anchor sequence**. It is envisaged that the signal sequence inserts into the channel in a hairpin-looped fashion, resulting in the situation shown in Fig. 27.9. Release of the protein from the channel would produce an integral protein with the C-terminal end inside the ER and the *N*-terminal end in the cytosol. The synthesis of serpentine proteins such as the G-protein coupled receptors discussed in Chapter 29, which criss-cross the membrane several times, is believed to involve a succession of stop transfer sequences.

Folding of the polypeptides inside the ER

The lumen of the ER contains **chaperones** (see Chapter 25) that attach to unfolded and partially folded polypeptides. Their function is to hold the chain in a conformation that prevents the polypeptides from going down an unproductive folding route leading to aggregation. A key component in this system is the ER isoform of the chaperone **Hsp 70**, which interacts with the polypeptide chain as it emerges from the translocon channel so that the incoming polypeptide folds correctly. The lumen also contains **protein disulphide isomerase** and **peptidylproline *cis-trans*-isomerase** whose role in assisting correct folding has already been dealt with in Chapter 25. Proteins that misfold and are not corrected are not allowed to accumulate. They are transported back into the cytosol, and degraded in proteasomes.

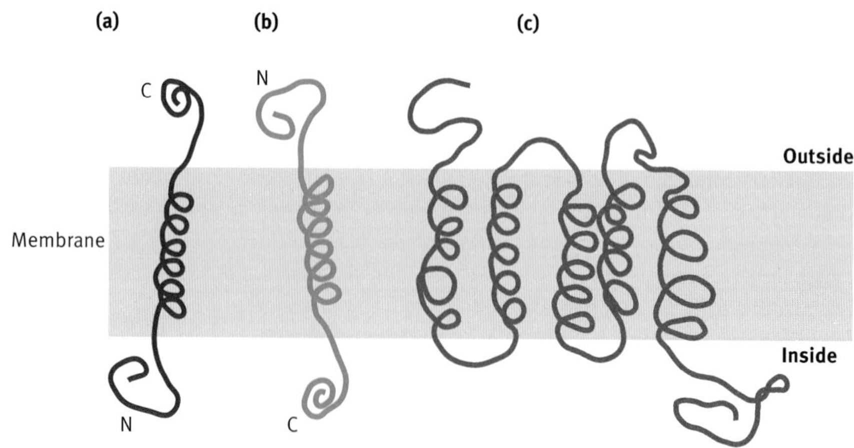

Fig. 27.8 Different orientations of integral membrane proteins. See text.

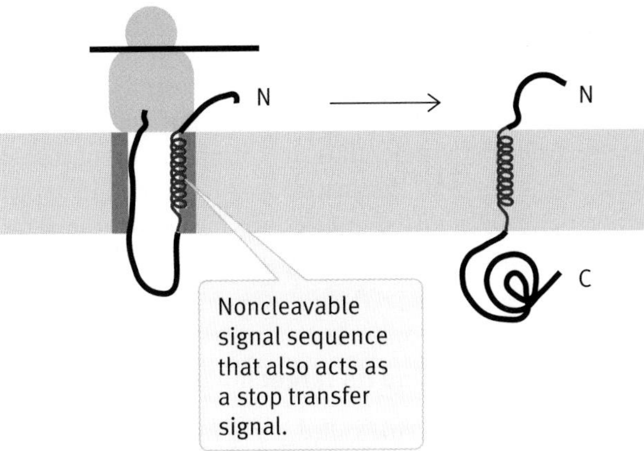
N
N
C
Noncleavable
signal sequence
that also acts as
a stop transfer
signal.

Fig. 27.9 How a transmembrane protein may be synthesized with the orientation shown. See text for explanation. The signal sequence is also a stop transfer signal. In this model it inserts into the channel in a looped fashion.

Glycosylation of proteins in the ER lumen and Golgi apparatus

In Chapter 4 we described proteins, particularly membrane and secreted proteins, which have complex oligosaccharides added to them. The attachment points are either the amide $-NH_2$ of asparagine side groups (*N*-glycosylation) or the $-OH$ of serine and threonine residues (*O*-glycosylation). *N*-glycosylation takes place inside the ER. The first step involves a 'core' oligosaccharide of 14 sugar units, which is assembled in the cytosol and transported through the membrane attached to a long hydrophobic chain called **dolichol phosphate**. A transferase enzyme on the inside of the ER membrane transfers the oligosaccharide group to the nascent polypeptide chains as they enter the ER lumen. *O*-glycosylation of proteins occurs in the Golgi cisternae.

Proteins for lysosomes

In the case of enzymes destined for inclusion in lysosomes, the signal is on the carbohydrate part of the glycoprotein. In the Golgi the carbohydrate attachment is enzymically modified so that it terminates in a mannose-6-phosphate residue, which attaches to membrane receptors and leads to their inclusion in lysosomal delivery vesicles.

All types of cell components, such as proteins, nucleic acids, carbohydrates, and lipid components, are destroyed by lysosomal digestion. The importance of this is underlined by the existence of a group of genetic diseases known as **lysosomal storage disorders** that arise because of the lack of one or more specific lysosomal enzymes (see Box 27.1). Their cognate target material in the lysosomes scheduled for destruction is not destroyed and the organelles become overloaded with it, sometimes with fatal results.

BOX 27.1 **Lysosomal storage disorders**

In one family of these genetic diseases, called sphingolipidoses, specific lysosomal enzymes are missing, which impair degradation of gangliosides (see Chapter 7) of the cell membranes. A classic example is Tay–Sachs disease. Children with this devastating condition become progressively paralysed, deaf and blind, and usually die by the age of four.

In I-cell disease all of the lysosomal enzymes are missing so that all manner of molecules accumulate within the vesicles. It is caused by a deficiency in the tagging of lysosomal enzymes with mannose-6-phosphate so they are not packaged into transport vesicles, and are secreted from cells into the plasma, rather than ending up in lysosomes. The disease is often fatal before 10 years of age.

In Pompe's disease, one of a series of glycogen storage diseases, the deficiency is of a lysosomal enzyme which hydrolyses the $(1\rightarrow4)$-α links of glycogen. The physiological significance of lysosomal glycogen breakdown is not known, as glycogen breakdown for energy occurs in the cytosol. Nevertheless, lack of this process is fatal. In the absence of the enzyme, there is a massive accumulation of glycogen in the cells usually causing death in infancy.

Proteins to be returned to the ER

Somewhat surprisingly, proteins such as chaperones and protein disulphide isomerase that function in the ER may not be initially retained there, after synthesis. They are moved to the Golgi and then returned to the ER in transport vesicles. The 'return address label' that distinguishes these proteins is a C-terminal peptide sequence of four amino acids, Lys–Asp–Glu–Leu (KDEL in the single-letter abbreviation system for amino acids; see Table 4.1), which is recognized by specific receptors in the Golgi membrane.

Proteins to be secreted from the cell

The Golgi packages proteins for secretion into COP-coated (see below) vesicles that migrate towards the plasma membrane (see Fig. 27.4). There are two types of secretion. Some proteins are released continuously as produced, while others are released periodically as required. Serum proteins, for example, are released continuously from the liver. The transport vesicles fuse with the cell plasma membrane as they arrive there and release their contents by exocytosis. In the case of digestive enzymes, however, these are released from the pancreas only when food enters the gut. In this case the vesicles from the Golgi containing these enzymes are larger secretory vesicles also known as **secretory granules**. These store the enzymes until they are needed, at which point a neuronal or hormonal stimulus causes release by exocytosis. There are many other examples where release of a secreted protein depends on a specific signal – release of insulin being a familiar one.

Proteins are sorted, packaged, and despatched from the ER and Golgi by vesicular transport

As already stated, movement of proteins from the ER and the Golgi and between Golgi cisternae occurs via membrane transport vesicles that bud off from the organelles. The vesicles move to their target membranes to which they fuse and hence deliver their cargo. Vesicles also transport proteins from the Golgi to the cell membrane as new membrane proteins or for secretion by exocytosis, to lysosomes, or back to the ER. There are two main classes of transport vesicle; COP-coated vesicles (COP stands for **co**at **p**rotein complex), which are involved in transport between the ER, the Golgi, and the cell membrane; and clathrin-coated vesicles, which transport enzymes from the Golgi to lysosomes and also function in endocytosis. We will deal with each class in turn.

Mechanism of COP-coated vesicle formation

Most transport vesicles are of this type. Two varieties exist, COPI and II, which are involved in transport from the Golgi and from the ER respectively. Figure 27.10 illustrates formation of a COPI-coated vesicle, in which a cytosolic GTP-binding protein, Arf, recruits **coatomer proteins** to the membrane. (The name Arf comes from another role of the protein, not relevant in the present context.) Initially, GDP-bound Arf binds the Golgi membrane at a point where there is a GEF embedded in it, causing Arf to become activated by exchanging GDP for GTP. The coatamer proteins recruited by GTP-bound Arf cause the membrane to deform so that the vesicle buds off, containing its 'cargo'. During transport of the vesicle, Arf's GTPase activity causes hydrolysis of GTP to GDP, and this triggers release of both Arf and coatamer proteins from the vesicle, thus uncoating it. The uncoated vesicle continues to its destination and fuses with its target membrane. COPII-coated vesicles are formed in the same way, but the recuiting GTPase and coatamer subunits are different proteins.

How does a vesicle find its target membrane?

The vesicle has a molecule called a v-SNARE (v for vesicle) on it that binds to a complementary t-SNARE (t for target) on the target membrane (Fig. 27.11). There are specific v- and t-SNAREs for particular vesicles and particular targets, ensuring that the cargo proteins are delivered to the appropriate destination. SNAREs are long helical proteins, which can associate strongly and bring the two membranes together, ready for fusion. A number of other proteins assist in the process.

If the target membrane is the plasma membrane, then vesicle contents are secreted by exocytosis. This method of targeting vesicles to its destination is a general one, not just confined to protein secretion. In nerve-impulse transmission, when a signal arrives at a synapse, vesicles containing neurotransmitter substances fuse with the presynaptic membrane and eject their contents into the synapse. This occurs also by courtesy of v-SNAREs attaching to membrane t-SNAREs. The **tetanus** and **botulinus** neurotoxins, two of the most deadly substances known, have protease activity which snip off these SNARES and interfere with nerve-impulse propagation, since neurotransmitters are not then released.

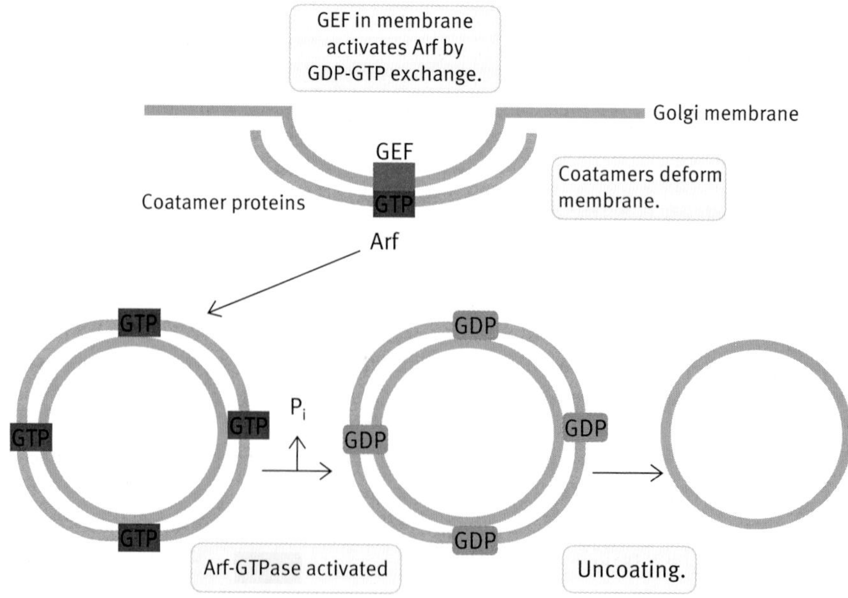

Fig. 27.10 Formation of COPI-coated transport vesicles from the Golgi membranes. Arf–GTP recruits coatamer proteins to the Golgi membrane. During transport, the Arf–GTPase is activated, giving Arf–GDP. This triggers the uncoating of the vesicle followed by fusion with its target membrane. In the case of a secretory vesicle the contents are ejected by exocytosis.

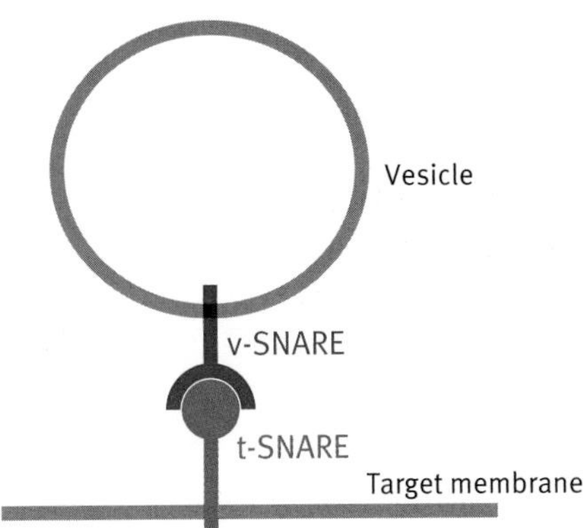

Fig. 27.11 Binding of v-SNAREs and t-SNAREs leads to uncoating of the vesicles followed by fusion of the two membranes. This is a general mechanism for vesicle targeting.

Clathrin-coated vesicles transport enzymes from the Golgi to form lysosomes

Lysosomes are membrane-bounded organelles found in the cytosol of all eukaryotic cells. They are bags of destructive enzymes formed by the fusion of **endosome vesicles** and **lysosomal enzyme transport vesicles**. Both of these are clathrin-coated vesicles. Clathrin is the equivalent of the coatamer proteins for this class of vesicle, and like the COPI coatamers it is recruited to the Golgi membrane by Arf, forming a basketwork-like coating that deforms the membrane. As stated earlier, a mannose-6-phosphate tag on the lysosomal enzymes selectively directs them to the transport vesicles.

An endosome is formed by receptor-mediated endocytosis described in outline for the uptake of LDL in Fig. 11.22. A particle to be delivered to a cell binds to specific protein receptors on the cell membrane. These are transmembrane proteins with the ligand-binding domain exposed to the outside and a cytosolic domain exposed on the inside. A protein known as **adaptin** combines with the cytosolic domains of a number of the ligand-bound receptors and clusters them together forming a depression in the membrane. **Clathrin** attaches to the adaptin molecules, the depressions being known as **clathrin-coated pits** (Fig. 27.12). The pits invaginate and another protein called **dynamin** attaches to the neck of the invagination causing a vesicle to be nipped off as a **clathrin-coated vesicle**. Inside the cell, the coated vesicle is uncoated and the coat molecules recycled back to the membrane. The interior of the vesicle (now called an **endosome**) is acidified by proton pumps in the membrane, forming a late endosome.

A lysosomal enzyme transport vesicle, similarly uncoated, now fuses with the endosome and delivers the hydrolytic enzymes. The result is a **lysosome** in which the material initially endocytosed is hydrolysed into its component parts. The cell is protected from destruction by the lysosomal enzymes because they are segregated by the lysosomal membrane, and also because the enzymes require a pH between 4.5 and 5.0 for activity maintained inside the vesicle by the ATP-dependent proton pumps in the membrane. If a lysosome were to rupture, the buffering of the cytosol would maintain the pH at 7.3 or so, at which lysosomal enzymes are inactive.

Intracellular components such as aged mitochondria are also destroyed by enveloping them in vesicles known as **autophagosomes** followed by lysosomal enzyme delivery.

Posttranslational transport of proteins into organelles

To remind you, the transport of proteins across membranes described so far has been cotranslational. However, transport into mitochondria, (and chloroplasts in plants), peroxisomes, and the nucleus is posttranslational. The polypeptides are completely synthesized, released by cytosolic ribosomes and then transported. Different mechanisms are used for transport into the different organelles.

Transport of proteins into mitochondria

A mitochondrion contains hundreds of proteins, only 13 of which (in humans) are coded by mitochondrial genes and synthesized within the mitochondrion. All of the mitochondrially encoded proteins are subunits of the large oxidative phosphorylation complexes of the inner mitochondrial membrane, such as cytochrome oxidase and ATP synthase. However, the majority of mitochondrial proteins are coded by genes in the nucleus, the mRNAs for which are translated on free ribosomes and the polypeptides released into the cytosol. There they have to be selected from other cytoplasmic proteins, delivered to receptors on the mitochondrial membrane and transported to their destinations.

The transport of proteins into mitochondria from the outside is quite complex because of the several compartments to which the proteins have to be targeted – the mitochondrial matrix, the inner and outer mitochondrial membranes and the intermembrane space. In some cases alternative routes to the same compartment have evolved.

Mitochondrial matrix proteins are synthesized as preproteins

Proteins entering the matrix have to cross both membranes; this happens at points where the inner and outer membranes become close together. Most matrix proteins are synthesized as **preproteins** with targeting presequences that are removed once they reach their destination. One of the best characterized

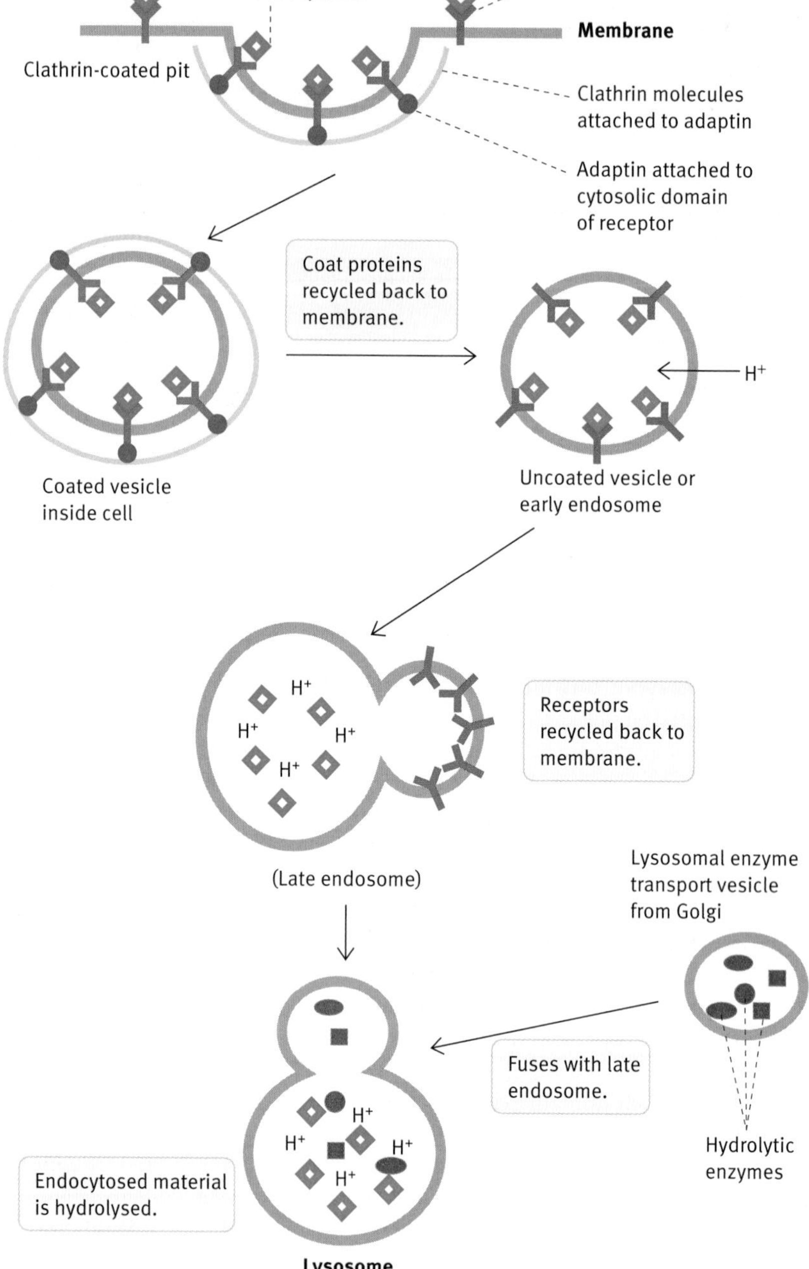

Fig. 27.12 Receptor-mediated endocytosis involves clathrin-coated vesicles. Ligand molecules bind to membrane receptors. Adaptin molecules attach to receptor domains inside the cell and cluster them into a coated pit. Clathrin molecules bind to adaptin and the pit invaginates into a coated vesicle in the cytosol. The latter is uncoated and the coat molecules are recycled. The uncoated vesicle, now an endosome, is acidified. Acidification releases the receptors (which in many cases are recycled back to the cell membrane) producing a late endosome. A Golgi transport vesicle containing hydrolytic enzymes fuses with the latter, forming a lysosome.

targeting presequences is an *N*-terminal sequence of 15–35 amino acids, which is variable, but usually consists of hydrophobic, hydroxylated, and basic amino acids that form an amphipathic α helix with hydrophobic side chains on one side and basic, positively charged groups on the other. However, other matrix proteins have internal recognition signals that are not removed.

As the polypeptide is synthesized on the ribosome, it is held in the unfolded form by Hsp 70 chaperones attached to the extended chain (Fig. 27.13). The polypeptide–chaperone complex docks with a mitochondrial receptor, which is part of the **translocase** of the **outer *m*itochondrial membrane (TOM)**. This is a multiprotein complex forming a channel through which the preprotein traverses the outer membrane. The TOM complex is involved in the transport of virtually all proteins that enter the mitochondrion.

In the case of proteins destined for the mitochondrial matrix, the preprotein is transported in an extended form (but

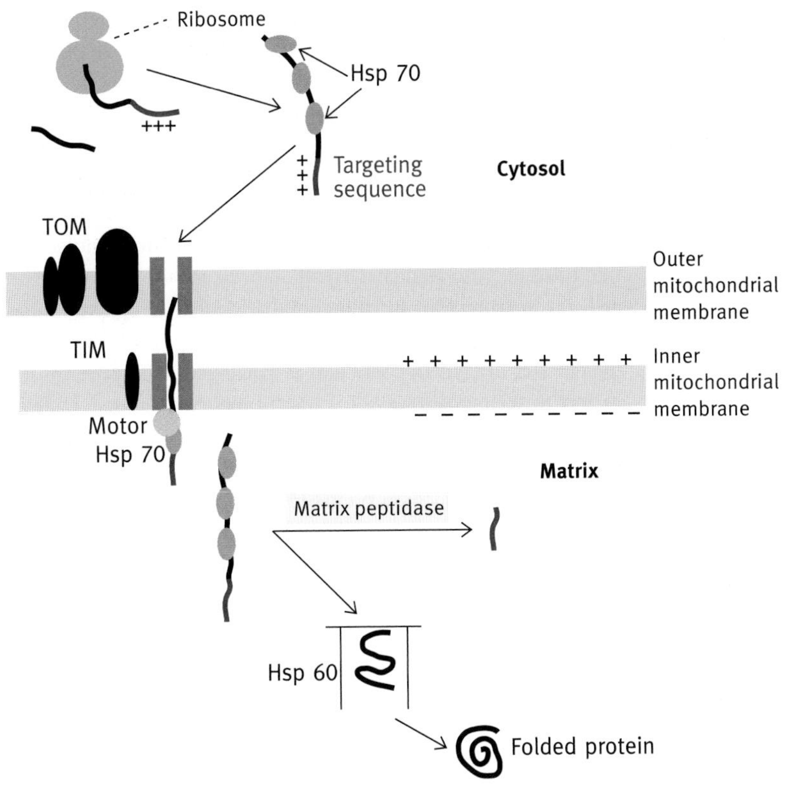

Fig. 27.13 Targeting of proteins to the mitochondrial matrix. Proteins are synthesized as preproteins with an amphipathic targeting sequence at the *N*-terminal (red). They are released into the cytosol but held in the unfolded state by Hsp 70 chaperones (blue). The preprotein attaches to the TOM complex receptors. It is transported through to the TIM complex and into the matrix where a peptidase cleaves off the targeting sequence and the protein folds, assisted by the chaperone Hsp 70 and the chaperonin Hsp 60. The Hsp 60 provides an isolated cage amongst the densely packed matrix proteins inside which the transported polypeptide can fold. The TOM and TIM complexes contain a variety of protein subunits. Import of the polypeptide into the matrix is assisted by Hsp 70 and other subunits that form the translocation motor. The inner membrane must have a charge potential for protein transport through it.

without the chaperones) through the outer membrane, and then across the **intermembrane space**. It is delivered to the ***t*ranslocase** complex of the ***i*nner *m*itochondrial** membrane **(TIM)**. For transport by TIM the inner membrane must have a charge potential (negative inside), generated by electron transport, which attracts the positively charged targeting sequence through the membrane. As the preprotein enters the matrix it meets an ATP-driven translocation motor, composed of the mitochondrial isoform of Hsp 70 plus other TIM subunits, which drives the import process to completion. A **matrix peptidase** hydrolyses off the targeting peptide sequence and the protein becomes fully folded, a process involving the Hsp 70 chaperone and/or the Hsp 60 chaperonin. The latter forms a secluded chamber amongst the extremely densely packed proteins of the matrix inside which the protein can fold properly. (A description of chaperones, chaperonins and their mechanisms can be found in Chapter 25.)

Delivery of proteins to mitochondrial membranes and intermembrane space

Integral proteins of the inner membrane are delivered by a number of different routes, but the first, common stage is that they are transported across the outer membrane by the TOM complex. One route involves the proteins transferring from TOM to the same TIM complex that transports matrix proteins. Some of them then move laterally from the TIM complex into the inner membrane, but others are actually transported initially into the matrix. Once there, removal of the preprotein signal sequence exposes another leader sequence, which targets the protein to the inner membrane via an export protein complex. (The same export complex also transports proteins synthesized within the mitochondria to the inner membrane.) An alternative route by which proteins made in the cytosol reach the inner membrane involves proteins that contain internal targeting sequences rather than *N*-terminal presequences. These are transported from TOM across the intermembrane space to a different TIM complex, from which they move laterally into the inner membrane.

Proteins for the outer membrane have internal signals that the TOM complex recognizes. As with inner membrane proteins, a number of mechanisms exist for their delivery. Some are transported by TOM into the intermembrane space and from there they return to the outer membrane. Others are transferred from TOM to an alternative outer membrane complex, which inserts them into the membrane, apparently without prior transport into the intermembrane space.

Proteins are delivered to the intermembrane space also by a variety of routes. One of these is that the protein is first inserted into the inner membrane and then cleaved to release the external part into the space. Another is that as it emerges from TOM the protein is recognized by a receptor protein complex that directs it to remain and complete its folding in the intermembrane space. These mechanisms have only recently been discovered and are still not fully understood.

Plant **chloroplasts** also import about 90% of their total proteins from the cytosol. The mechanism is similar to that for mitochondria with the targeted proteins synthesized on cytosolic ribosomes as preproteins but the targeting peptide, known as the **transit peptide**, is 30–100 amino acids long and is not positively charged as is the mitochondrial targeting sequence. Possibly correlating with this, there is no requirement for a negative membrane potential on the inner membrane as there is for mitochondria.

Targeting peroxisomal proteins

Peroxisomes and their metabolic functions have been described earlier (see Chapter 2). They are small organelles with a single membrane and have no DNA or protein synthesizing machinery. The peroxisomal matrix contains about 50 different enzymes, which are synthesized on free cytosolic ribosomes and transported into the organelle. The proteins become fully folded in the cytosol and, unusually, they are transported into the peroxisome in this form. This is in marked contrast to mitochondrial import, described above, where the polypeptides are kept unfolded.

There are two known **peroxisome-targeting signals (PTS1 and PTS2)** on proteins to be transported. PTS1, the most common, is a C-terminal tripeptide Ser–Lys–Leu (SKL in the single-letter abbreviations) which is not removed. PTS2 is nonapeptide located near the *N*-terminus. A variety of cytosolic proteins called **peroxins** are needed for transport of the targeted proteins. Some of these act as receptors that recognize the PTS-containing proteins and bring them to docking complexes on the peroxisomal membrane. The exact nature of the translocon for peroxisomal proteins has still not been elucidated. Since folded proteins are transported into the peroxisome, a very large membrane pore is implied, but how the transport occurs is not known, nor whether there are separate translocons for PTS1 and PTS2 mediated import.

Interest in the field is heightened by the existence of genetic diseases that are due to defects in various aspects of peroxisome biogenesis, including matrix protein import. In one of these, **Zellweger syndrome**, death often occurs before 6 months of age. Yeast genetic studies have revealed 33 PEX genes involved in peroxisome biogenesis, many of which are conserved in humans. Of these, 13 have been shown to be associated with peroxisomal biogenesis disorders.

Nuclear–cytosolic traffic

The eukaryotic nucleus is surrounded by inner and outer membranes, the space between them being continuous with the ER lumen. The existence of a separate nuclear compartment means that there is intensive traffic across the nuclear membrane, different in nature from anything dealt with in preceding sections. Moreover, we are dealing with two-way traffic of proteins and protein-RNA complexes. For instance, during DNA replication histone mRNA is exported to the cytosol for translation, while histone proteins are imported from the cytosol to allow nucleosome assembly. The nuclear membrane is studded with pores of huge size and elaborate construction. It has been calculated that each pore must transport 100 histone molecules per minute. The nucleus manufactures all the RNA of the cell (apart from that in mitochondria and chloroplasts) most of which is required in the cytosol so that it has to be transported out. mRNA in particular is 'packaged' with proteins for export to the cytosol. rRNA is not transported as such; instead the ribosomal proteins are transported in, the ribosomal subunits assembled and transported out. By contrast, in the case of **snRNPs** (ribonucleoproteins involved in mRNA splicing in the nucleus), the snRNA component is transported out of the nucleus to be equipped with its protein in the cytosol, and then the complete snRNP is transported back into the nucleus.

Why is there a nuclear membrane?

This is a slight, but relevant, diversion from the mechanism of nuclear-cytosolic traffic. An interesting question is why evolution has led to eukaryotes sequestering their DNA inside a nuclear membrane. An *E. coli* cell manages perfectly well without a nucleus and has none of the associated transport problems. The eukaryotic nucleus separates the act of transcription of DNA, in both time and space, from translation of the mRNA in the cytosol. This provides a time-gap for the mRNA to be modified before translation; the most important of these modifications in evolutionary terms may have been splicing, which, in allowing the existence of split genes, facilitated exon shuffling (see Chapter 22). In addition, differential splicing allows the production of different proteins from a single gene. In *E. coli*, ribosomes begin translating mRNA even before the synthesis of the latter has been completed. It is not easy to see how splicing could occur in this circumstance.

Much of eukaryotic gene control occurs through hormones and cytokines arriving at the cell. This often involves specific proteins being transported from the cytosol into the nucleus on arrival of a signal (Fig. 27.14). Transcriptional control and cell signalling are major topics dealt with in Chapters 26 and 29 respectively.

After that diversion we will turn to the mechanism of nuclear–cytosolic traffic.

The nuclear pore complex

Each nucleus has several thousand pores forming aqueous channels between the cytosol and nucleoplasm (Fig. 27.15). Nuclear pores are huge structures with a total size of 10^8

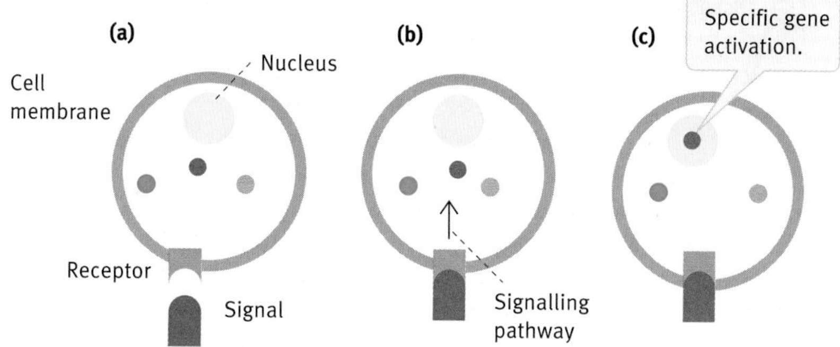

Fig. 27.14 Outline of the role of nuclear import in the control of genes by cell signalling. **(a)** A hormone or other chemical signal arrives at a cell surface and binds to its specific receptor. **(b)** The receptor triggers a series of events that selects a specific cytosolic protein (blue) to migrate into the nucleus. **(c)** The protein inside the nucleus causes specific gene activation. It should be noted that this is an exceedingly simplified scheme to give the bare outlines of gene control by hormones etc. as related to nuclear transport. Although it applies to many signalling pathways, lipid-soluble hormones such as steroids, thyroxine, and vitamin D diffuse through the membrane and bind to cytosolic receptors or to receptors already in the nucleus. Nonetheless most signalling molecules do not enter the cell. More detailed information is given in Chapter 29.

Daltons that are built up of over 30 species of protein subunits present in multiple copies. The proteins are called **nucleoporins**. Pores have been isolated free of the membrane by detergent treatment and their structure studied. They consist of concentric inner and outer rings of protein subunits at the nuclear and cytosolic ends of the pore, and an additional ring of integral membrane proteins situated where the inner and outer nuclear membranes are continuous at the site of the pore (Fig. 27.16). The pore complex has eightfold rotational symmetry, with the nucleoporin proteins present in eight or multiples of eight copies. In addition to the main pore structure spanning the membrane, filaments extend from the nuclear and cytosolic rings into the nucleoplasm and cytosol respectively. The filaments guide proteins to be transported toward the pore. A distal protein ring links the ends of the nucleoplasmic filaments to form a basket-like structure.

Molecules up to about 40,000 Daltons can enter the nucleus simply by diffusion, though proteins in the upper ranges of this do so more slowly than smaller ones. Anything greater than this size must be specifically accepted by the transport machinery of the pore.

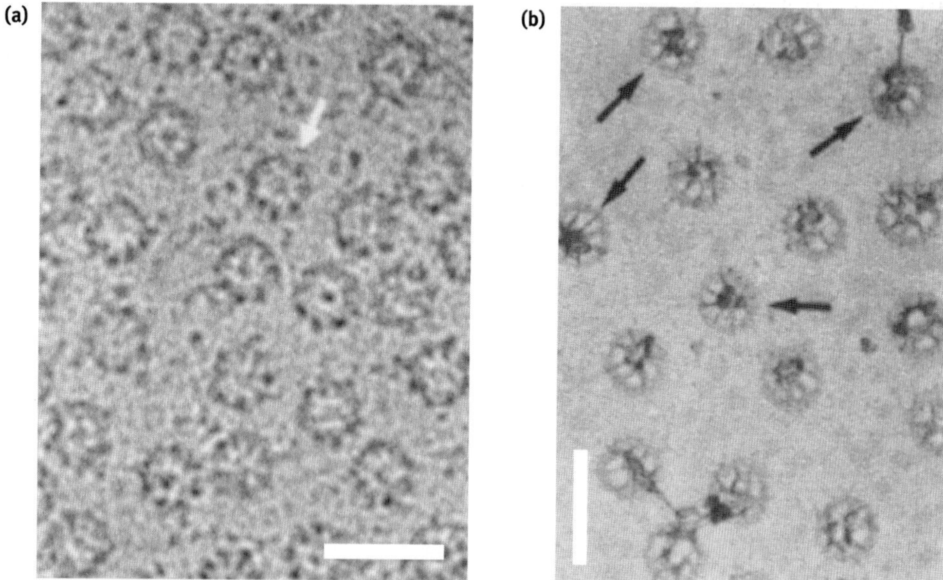

Fig. 27.15 Spread *Xenopus* oöcyte nuclear envelopes (NEs) prepared for transmission electron microscopy (TEM). **(a)** Electron micrograph of chemically unfixed and unstained *Xenopus* oöcyte nuclear pore complexes (NPCs) embedded in a thick (i.e. ~250 nm) amorphous ice layer. **(b)** Nuclear face of quick-frozen, frozen-hydrated, and metal-shadowed NPCs revealing well preserved nuclear baskets (see arrows). Scale bars, 200 nm. Photographs courtesy Professor Ueli Aebi, University of Basel, Switzerland.

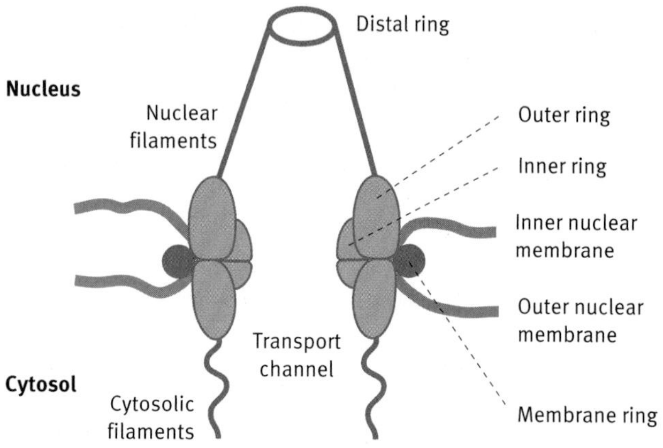

Fig. 27.16 Simplified diagram of the main body of a nuclear pore. See text for the mechanism of transfer through the pore.

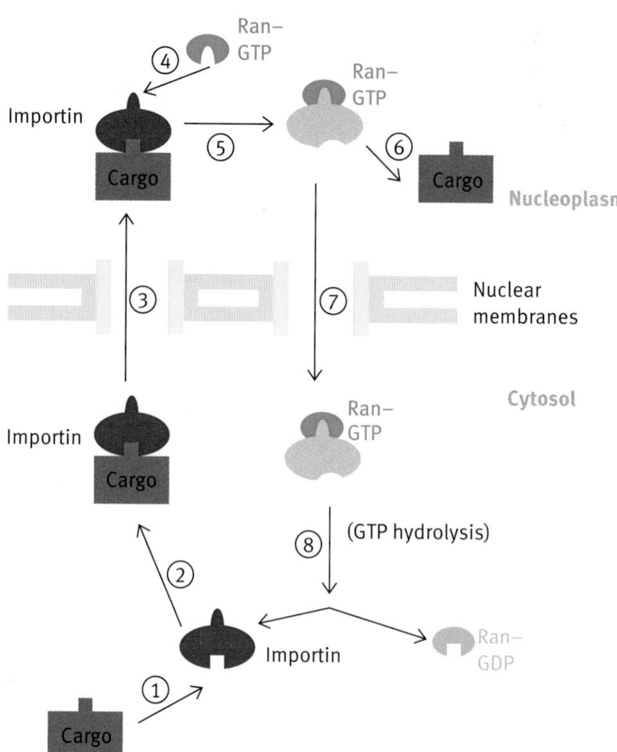

Fig. 27.17 Mechanism of protein import into the nucleus. See text for explanation. The deep red colour represents importin in its form capable of attaching to a protein to be transported; the pale red form cannot do so. Note that Ran–GDP is transported back into the nucleus (not shown) to complete the cycle. Adapted from Fig. 2 in Mattaj, I. W., and Engelmeier, L.; Annu Rev Biochem; (1998) 67, 265; Reproduced by permission of Annual Reviews Inc.

Nuclear localization signals

Large (> 40,000 Daltons), nuclear-localized proteins need a **nuclear localization signal (NLS)**. Many nuclear proteins contain NLSs that are rich in basic arginine and lysine residues, which may be located anywhere in the polypeptide chain. The 'prototype' of this class of NLS is that found in a protein known as the 'T antigen' of the SV40 virus, which is transported into the nucleus as part of the infective process. The T antigen NLS is PKKKRKV (using single-letter abbreviations of amino acids). Other members of this class of basic NLS are **bipartite**; an example is found on nucleoplasmin, a chromatin assembly protein. The sequence of this is KR, followed by a 10 amino acid spacer, then by KKKK. Mutation of an NLS results in a normally nuclear-located protein remaining in the cytosol, whereas artificial addition of such a signal to a normally cytosolic protein results in it being transported into the nucleus. Most transcription factors are examples of proteins with NLS sequences. As well as NLS signals, **nuclear export signals (NESs)** have been identified and are discussed later in this section.

Variations on these 'classical' signals occur, for instance in the protein that binds mRNA in the nucleus and transports it *via* the nuclear pore into the cytosol as a **hetero-ribonucleo-protein complex (hnRNP)**. mRNAs are relatively short-lived in the cytosol (20 minutes to a few days in mammals), so the mRNA binding protein shuttles back into the nucleus to pick up more. It has a signal of 38 amino acid residues, which can act as both an import and an export signal.

Importins combine with nuclear localization signals on proteins to be transported into the nucleus

Importins, also known as **nuclear import receptors** or **karyopherins**, constitute a family of soluble cytosolic proteins that recognize the NLS on 'cargo' molecules (step 1 in Fig. 27.17). Each importin has two binding sites; one binds to the NLS on the 'cargo' proteins to be transported and the other to receptor nucleoporins on the filaments of the pore complex (step 2). Many of the nucleoporins of the pore (about 30% of the total) contain short stretches of clustered hydrophobic amino acids known as FG repeats (F and G are one letter abbreviations for phenylalanine and glycine respectively) interspersed by hydrophilic regions. The FG repeats are binding sites for the importins (carrying the cargo). It is believed that movement through the pore (step 3) involves progressive binding and detachment of the importin/cargo complex from the FG repeats so that it moves inward from one to the next.

GTP/GDP exchange imparts directionality to nuclear–cytosolic transport

The nuclear pore has no ATPases or other known direct energy input mechanisms. The driving force for the transport comes indirectly from a gradient created by GTP hydrolysis, which will now be described.

The importin undergoes a conformational change when a small protein, **Ran**, a GTP-binding protein, combines with it. **Ran is bound to GTP in the nucleus**, and when it attaches to the arriving importin–cargo complex (step 4 in Fig. 27.17) the conformational change in the importin causes it to release its cargo (steps 5 and 6). **Ran–GTP**, coupled to the importin, then recycles back to the cytosol *via* the nuclear pore (step 7). Ran has GTPase function, but this requires a **GTPase-activating protein (GAP)** that is found only in the cytosol, not in the nucleus. Thus, in the cytosol Ran–GTP is hydrolysed to **Ran–GDP** (step 8), resulting in release of the importin in a conformation able to pick up a new cargo molecule to transport into the nucleus. Ran–GDP is recycled back into the nucleus by a nuclear transport factor, where a **guanine nucleotide exchange factor (GEF)** stimulates its conversion back to Ran–GTP. The exchange factor does not occur in the cytosol. *The system depends on the Ran protein being in the GTP form in the nucleus and in the GDP form in the cytosol.* To reemphasize the crucial point, this is achieved because of the asymmetric compartmentalization of the exchange factor and the GTPase-activating protein.

Exportins transfer proteins out of the nucleus

The reverse transport of proteins carrying a **NES** out of the nucleus into the cytosol occurs by a cycle (Fig. 27.18) that is the mirror image of the import cycle. In the nucleus there is a family of **exportin** proteins (also known as **nuclear export receptors** or **karyopherins**), each specific for proteins carrying the appropriate NES. Attachment of the exportin to the NES of its target cargo (step 1) occurs only when Ran–GTP is attached to the exportin (steps 2 and 3). Note that this is the reverse of the situation with importin, which releases its cargo when Ran–GTP attaches. The Ran–GTP–exportin–cargo is transported through the pore (step 4) into the cytosol where, as before, the GAP activates the Ran–GTPase (step 5) causing hydrolysis of the GTP and dissociation of the complex into Ran–GDP and exportin. This releases the cargo (steps 5 and 6). The Ran–GDP is recycled back to the nucleus (step 7). The exportin is also recycled back to the nucleus.

The establishment of the gradients of Ran–GTP and Ran–GDP across the nuclear membrane drives nuclear transport, and the contrasting properties of importin and exportin give the opposing directionality.

Regulation of nuclear transport by cell signals and its role in gene control

As already explained, in much of eukaryotic gene control, regulatory proteins are transported into the nucleus on receipt of cell signals. The proteins must each have an NLS so the question arises of why they are not carried into the nucleus in the absence of any signal. Such proteins have their NLS rendered ineffective until an extracellular signal arriving at the cell causes changes in the protein that make it available for transport.

There are several mechanisms that bring this about. One is to mask the NLS on a given protein by the binding of another protein molecule (Fig. 27.19(a)). An example of this is to be found in Chapter 29, describing how certain steroid hormones

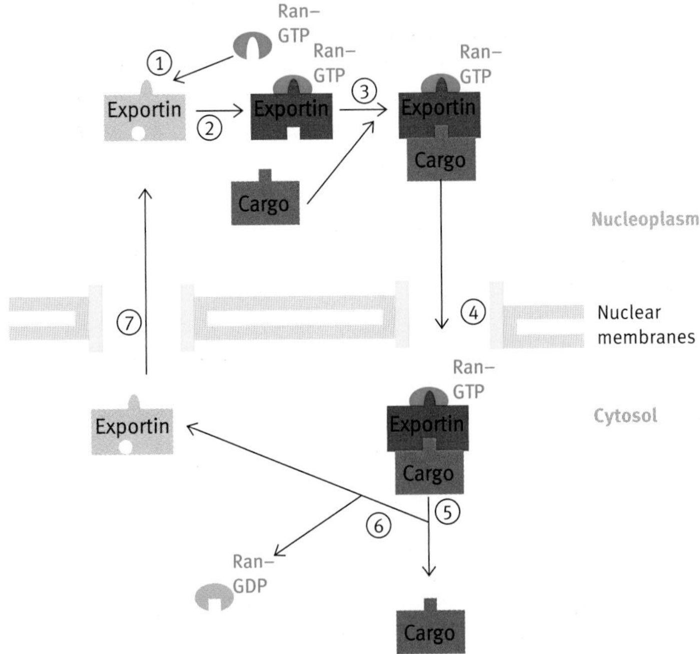

Fig. 27.18 The mechanism of export of proteins from the nucleus. See text for explanation. The deep red colour represents exportin in a form capable of carrying a cargo; the pale red form cannot do so. Adapted from Fig. 2 in Mattaj, I. W., and Engelmeier, L.; Annu Rev Biochem; (1998) 67, 265; Reproduced by permission of Annual Reviews Inc.

(a) Masking of NLS by another protein

Masking protein
NLS
Detached masking protein
Cell signal.
NLS
Cytosol
Transport into nucleus.
Nuclear membrane
Nucleus
NLS
Transcription factor, activates genes.

(b) Masking of NLS by phosphorylation

P P
NLS
Importin does not bind; protein is not imported into nucleus.
P_i P_i
Signal
NLS
NLS
Gene activation.

(c) Enhancement of NLS by phosphorylation

NLS
Importin binds only weakly; protein is only slowly imported into nucleus.
Signal
P P
NLS
Phosphorylation increases importin binding affinity.
Protein is now transported in at an enhanced rate.
P P
NLS
Gene activation.

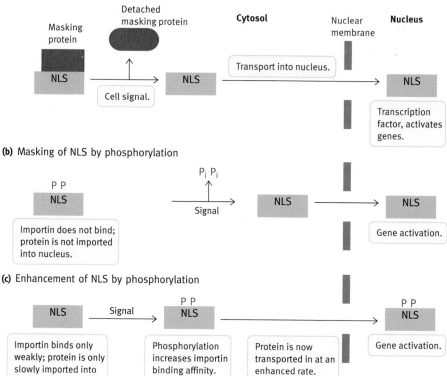

Fig. 27.19 Mechanisms of signal-mediated regulation of import of proteins into the nucleus. Extracellular signals to cells are of central importance in gene control, and nuclear import regulation plays a vital role in this. Proteins required for specific gene control reside in the cytosol until a signal causes them to be imported into the nucleus. This requires that the proteins in question are not transported into the nucleus (or at a slow rate) until a signal is received. There are several mechanisms for this, three of which are illustrated. NLS, nuclear localization signal. Adapted from Australian Biochemist, (1998) 29, 5–10; reproduced with permission of Professor D.A. Jans.

combine with their cognate receptor protein, causing a conformational change in the receptor that results in detachment of a masking protein. The NLS of the receptor is thereby made available for binding to the importin, resulting in import of the receptor into the nucleus where it acts as a transcription factor for those genes that are activated by the steroid hormone.

An alternative mechanism is to mask the NLS by phosphorylation (Fig. 27.19(b)). A third possibility (Fig. 27.19(c)) is that some NLSs have a low binding affinity for importins, which is increased by phosphorylation of the protein. In this case dephosphorylation by protein phosphatases provides a way of reversing the process when the cell signal is terminated.

The major role played by nuclear-cytosolic transport in cell signalling and gene control obviously adds greatly to its importance and interest as a cellular process.

Summary

Proteins are synthesized on cytosolic ribosomes but function in different cellular locations. They are targeted to their destinations by a number of mechanisms. GTP-binding proteins with 'slow' GTPase activity play an important part in many protein transport processes, acting as molecular switches. The proteins undergo allosteric changes when GTP replaces GDP or bound GTP is hydrolysed.

Proteins destined for the lumen of the rough ER traverse the ER membrane as they are synthesized (cotranslational transport). Free cytosolic ribosomes begin translating mRNAs for these proteins. The initial sequence of polypeptide produced is called a signal peptide. A SRP binds to this, and halts polypeptide synthesis. The SRP docks the ribosome to a translocon site in the ER membrane. The signal peptide is postulated to open the translocon channel and as translation resumes it leads the nascent polypeptide through the channel. On reaching the ER lumen the signal peptide is cleaved off, the synthesis of the protein is completed and it is released into the lumen where it folds.

Synthesis of integral membrane proteins also occurs in the ER. In this case, in addition to the signal peptide, the polypeptide chain contains a stop transfer or anchor signal that fixes the protein in the membrane. Vesicles transport new membrane sections complete with proteins to specific existing membranes.

Proteins imported into the ER lumen move along through the smooth ER, processed en route by attachment of carbohydrates, and are transported in vesicles to the Golgi. Here the proteins are modified further. Amino acid or carbohydrate sequences act as 'address labels' for particular cellular destinations and allow proteins in the Golgi to be sorted and transported by vesicles to lysosomes, the plasma membrane or back to the rough ER.

Membrane transport vesicles are of two major types; COP-coated and clathrin-coated, depending on their origin and destination. Clathrin-coated vesicles are involved in endocytosis and lysosome formation. Vesicles are targeted to their destination by v-SNARES, protein complexes that bind to complementary t-SNARES on their target membrane.

Transport of proteins into mitochondria is posttranslational. On completion of synthesis, the polypeptide is held in an unfolded form by a chaperone. It is delivered to the receptor of the transport complex, known as TOM (translocase, outer membrane) on the outer membrane of the mitochondrion. Mitochondrial matrix proteins are translocated from TOM to TIM (translocase, inner membrane) and then into the mitochondrial matrix where they fold up aided by chaperones. Proteins for the outer and inner mitochondrial membranes and the intermembrane space are also delivered unfolded, by a variety of mechanisms.

Peroxisomal proteins are transported in a fully folded form into the organelle. Two peroxisomal targeting amino acid signals are known, but the mechanism of translocation is not yet understood.

Nuclear transport occurs through elaborate pores in the nuclear membrane. Proteins containing NLSs in their amino acid sequences are imported complexed to importin proteins. Conversely, export of proteins from the nucleus depends on exportins that recognize NESs on target proteins and transport them through the pores. Nuclear traffic also depends on the GTPase protein, Ran. Compartmentation of a GEF to the nucleus and a GTPase-activating protein (GAP) to the cytosol results in Ran being GTP-bound inside the nucleus and GDP-bound in the cytosol. Ran–GTP causes cargo release from importin in the nucleus while hydrolysis of GTP causes cargo release from exportin in the cytosol. The Ran–GTP/Ran–GDP gradient so established across the nuclear membrane gives directionality to nuclear pore transport.

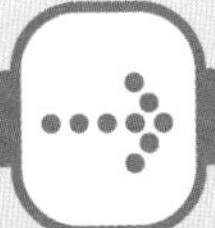

Further reading

To access the further reading, please scan the QR code image or go to http://global.oup.com/uk/orc/biosciences/molbiol/snape_biochemistry/student/reading/ch27/

Problems

1 What is meant by cotranslational and posttranslational transport of a protein across a membrane? Give examples.

2 In protein targeting, GTP hydrolysis is often involved although this is not associated with any chemical synthesis or performance of other obvious work. What is the function of this?

3 What are the functions of lysosomes?

4 What are lysosomal transport vehicles and where are they produced?

5 What evidence is there to indicate that lysosomal digestion is an essential process?

6 Why is Pompe's disease somewhat of a biochemical puzzle?

7 The transport of proteins through nuclear pores is critically dependent on the asymmetric distribution of two specific protein activities between the nucleoplasm and cytosol. Explain this.

8 What are FG repeats of the nuclear pore?

9 Briefly summarize the basic differences between the targeting of proteins for secretion, for import into mitochondria, nucleus, and peroxisomes respectively.

Chapter 28

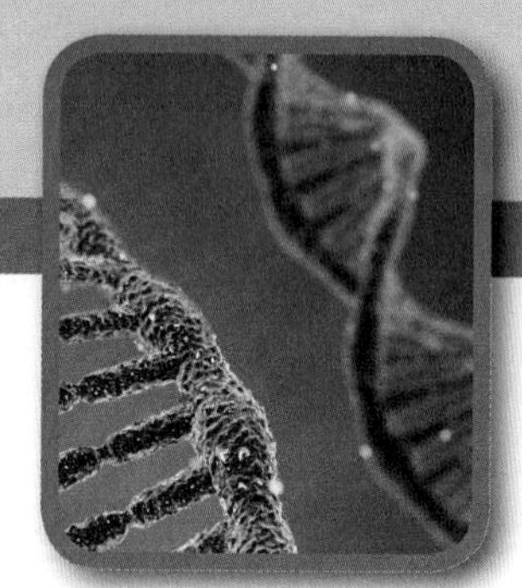

Manipulating DNA and genes

DNA manipulation techniques are a major part of a revolution in medical and biological sciences in general. Areas referred to as **recombinant DNA technology**, **gene technology**, **genetic engineering**, and **biotechnology** depend on DNA manipulation. At one time, working with DNA seemed extremely difficult. Other than their size, different DNA molecules have similar physical and chemical properties, being made up of sequences of just four nucleotides. Cellular DNA molecules are also huge and therefore difficult to handle. However, it has turned out that DNA can be manipulated more easily than other macromolecules, such as proteins or complex polysaccharides. There are two main factors that make this the case. The first is the capacity of DNA to base pair specifically with DNA and RNA of complementary sequence. This property of hybridization is utilized repeatedly in working with DNA. The second factor is the availability of a multiplicity of enzymes that nature uses in defense, DNA repair, and replication. This means that DNA can be cut, the pieces joined together, duplicated, sequenced, and detected by labelled complementary DNA probes. The realization that these enzymes, many of which we have met in earlier chapters, could be harnessed for use in the laboratory was the beginning of recombinant DNA technology.

When two molecules of DNA of different origin are covalently linked together, end to end, the resultant molecule, which does not occur naturally, is known as a **recombinant DNA** molecule. This is different from natural genetic recombination in cells.

DNA manipulations are very often the most powerful approach available in biological and medical sciences. To give a few examples, the technologies make it possible to isolate individual genes, to determine their base sequences, to manipulate the base sequences in any desired way, to transfer genes back to their original source or from one species to another, and to detect genetic abnormalities. It is possible to produce proteins such as human hormones or other factors using bacteria, or other cells loaded with extra genes, as protein factories. The proteins can be produced in virtually unlimited amounts even though they may occur in the body only in very small amounts. Amino acid sequences of proteins may be modified by changing the DNA coding for them. Minute amounts of DNA can be rapidly amplified to produce multiple copies, a technique used in forensic science in which human DNA is 'fingerprinted' for identification purposes. The base sequence of the coding region of genes permits deduction of the amino acid sequence of proteins (and is often the easiest way to determine them). Evolutionary relationships can be traced from DNA sequences and since DNA has some degree of stability under favourable circumstances, studies on ancient Egyptian mummies and even extinct fossilized organisms are possible.

Recent advances such as the technology of DNA microarrays, which enable the expression of large numbers of genes to be studied simultaneously, and modern 'next generation' sequencing can generate vast quantities of data very quickly. This would be of little use without mechanisms for storing, retrieving, and analysing it. This brings us to the international DNA databases in which information of all kinds related to DNA is recorded as it becomes available from research. The databases are freely available on the Internet, together with software programs for retrieving and interrogating the data in a wide variety of ways. In fact, 'mining the databases' has become an important research activity in its own right. Computational skills as well as a knowledge of molecular biology are required for this. The DNA databases complement the protein databases already described (see Chapter 5). Use of the information in databases is referred to as **bioinformatics** and increasingly there are specialized courses in this subject. More information on this aspect is given at the end of this chapter.

Basic methodologies

Some preliminary considerations

One aspect of DNA manipulation that is potentially confusing is that it is usual to talk about a 'fragment', a 'molecule', or a 'piece' of DNA being sequenced, cloned, or digested. In fact these manipulations require multiple copies of the DNA being worked on. Even the minute quantities of DNA that are handled in laboratories usually contain billions of individual molecules. In fact, the production of a DNA **clone**, described later in this chapter, is done with the purpose of producing potentially unlimited copies of a particular DNA sequence with a view to later manipulation. A newer and often easier technique, the polymerase chain reaction (PCR), is also used for the same purpose.

Another aspect to bear in mind is that genes and other specific DNA sequences form parts of long continuous molecules within the cell. This means that it is impracticable to isolate a gene or other piece of DNA directly from a cell lysate as one does for proteins. The first step in working with genomic DNA, therefore, is to obtain fragments of a size that can be handled (in practice up to a few thousand base pairs in length). Genomic DNA can be broken into smaller pieces by physical methods, but is more common to use enzymes as they give greater control over the process.

Cutting DNA with restriction endonucleases

The only thing that distinguishes a gene or other DNA section from all the other DNA is its base sequence. The discovery of a group of enzymes in different bacteria that hydrolyse DNA at specific sequences was the vital first development that opened the way to manipulating DNA.

The class of enzymes called **restriction enzymes** recognize short sequences of bases so as to make cuts at precise points in both strands of a DNA molecule. Different bacteria have different restriction enzymes that recognize different base sequences in the DNA and therefore have different cutting sites. The bacteria have these enzymes as protection against bacterial DNA viruses (bacteriophages). When these viruses infect a cell, they insert their own DNA, which takes over the targeted cell, and directs its biochemical activities to synthesizing new virus particles, which are then released. The cell's restriction enzymes cut the invading DNA, which destroys the phage. There will be large numbers of short base sequences in the infected cell's own genome identical to the sequence targeted by the restriction enzyme. So why does the latter not destroy the cell's own DNA? The bacteria protect themselves by methylating A or C bases in the target sites in their own DNA; the enzymes no longer recognize the methylated sites. As different species and strains of bacteria produce different enzymes with different target sequences, DNA replicated in one strain of bacteria will be recognized and digested should it be transferred to another strain, thus infections by foreign DNA are 'restricted'. Over 100 restriction enzymes cutting at specific base sequences are now known so that we now have great control over where we cut DNA. To illustrate this point, an enzyme from *Escherichia coli*, *Eco*RI, cuts double-stranded DNA at the sequence:

5′G↓AATTC3′

3′CTTAA↑G5′

and that from *Bacillus amyloliquefaciens*, *Bam*HI, cuts at the sequence:

5′G↓GATCC3′

3′CCTAG↑G5′

These two enzymes make staggered cuts in the two DNA strands as indicated by the arrows, producing overhanging ends, but others make straight-through cuts. *Hae*III from *Haemophilus aegyptius* is an example that produces 'blunt-ended' fragments:

5′GG↓CC3′

3′CC↑GG5′

Different restriction enzymes recognize sequences four to eight base pairs in length, which are typically palindromic (i.e. they read the same 'forwards' on one strand as 'backwards' on the complementary strand). As a specific four base sequence will be present by chance more frequently in a DNA molecule than a specific six or eight base sequence, enzymes with longer recognition sequences have fewer target sites in a given DNA molecule and therefore cut the DNA into longer fragments. For cutting genomic DNA enzymes with different length recognition sequences can therefore be exploited for different purposes depending on the length of DNA fragments required.

Restriction enzymes are named after the bacterium (and bacterial strain) of their origin and a Roman numeral where more than one enzyme occurs in that species. Thus the three enzymes mentioned above are called *Eco*RI, *Bam*HI, and *Hae*III, respectively. *Eco*RI was the first to be isolated from *E. coli* strain R. The target sites in a DNA molecule are often termed restriction sites. Note that restriction enzymes cut double-stranded, not single stranded, DNA so the existence of the complementary strand is implied when stating that an *EcoRI* restriction site has the sequence GAATTC.

Restriction enzymes make it possible to cut DNA at base sequences with precision, producing defined fragments that can be characterized in a preliminary way simply by their size. Many different restriction enzymes are now available commercially, cutting at different DNA sequences. The choice of recognition sequence and also of whether the enzyme produces overhanging or blunt ends is often an important consideration for further manipulation of the DNA.

Separating DNA pieces

DNA pieces can be separated by **gel electrophoresis**, a technique that has already been described in Chapter 5 in the context of protein investigation. Each phosphate group in DNA carries a negative charge so that molecules migrate to the anode. The regular repeat of the phosphate groups means that all DNA molecules have the same charge to mass ratio, so the basis of size separation is through molecular sieving, with smaller molecules moving faster than large ones due to experiencing less resistance from the gel. For shorter pieces of DNA, up to about 1,000 base pairs in length, the pores in **polyacrylamide** gels are sufficiently large to allow the DNA molecules to migrate through them and the apparatus is basically the same as used for proteins. DNA molecules differing in length by a single nucleotide are separated.

For longer pieces of DNA **agarose** gels are used because it is not possible to create stable polyacrylamide gels with sufficiently large pores. Agarose is a passive uncharged seaweed polysaccharide. A typical agarose gel electrophoresis apparatus is shown in Fig. 28.1.

Visualizing the separated pieces

DNA that has been separated by electrophoresis may be detected by staining with fluorescent DNA-binding dyes such as ethidium bromide, which intercalates between the bases, and viewing the fluorescence in UV light. The technique is not sensitive enough to detect single DNA molecules, but where multiple copies of the same piece of DNA are present they all migrate to the same point in the gel and produce a stained 'band' that can be visualized. For samples containing just a few different sized pieces of DNA this analysis may be sufficient. However, you can appreciate that a sample derived from human chromosomal DNA cut with a restriction enzyme, for example, will contain so many DNA fragments spread all through the gel that staining in this way will give a meaningless blur. For these situations detection of specific sequences depends on the base-pairing properties of DNA, as described below.

Detection of specific DNA fragments by nucleic acid hybridization probes

In Chapter 22 we described the self-annealing properties of DNA strands. If a mixture of different pieces of double-stranded DNA is heated, the hydrogen bonds holding the Watson–Crick base pairs together are disrupted and the strands separate. On cooling, the pieces find their original partners again by base pairing, so that only correctly base-paired double strands are reformed. This theme of specific base pairing is central to recombinant DNA work.

The specificity of hybridization makes it a sensitive tool for identifying specific sequences irrespective of how many other molecules are present, so that a single-gene sequence for example can be detected in the DNA from an entire genome. To do this a **hybridization probe** is needed; a probe is a piece of DNA complementary to the one you wish to find, and with some sort of label incorporated in it, such as radioactivity or fluorescence. A probe optimally needs to be at least 20 nucleotides long to give sufficient hydrogen-bonding attachment and is often hundreds of bases long. The DNA being probed is first rendered single stranded by heat or NaOH. The hybridization is carried out in defined conditions of temperature and ionic strength. The sensitivity of hybridization to detect DNA sequences can be varied. If you want only perfect base-pair matching to occur, a temperature just below the melting point of a double helix is used where a single mismatch will prevent hybridization (these conditions are known as high stringency). At lower temperatures hybridization will occur despite imperfect matching. The degree of stringency becomes important in using single nucleotide polymorphisms

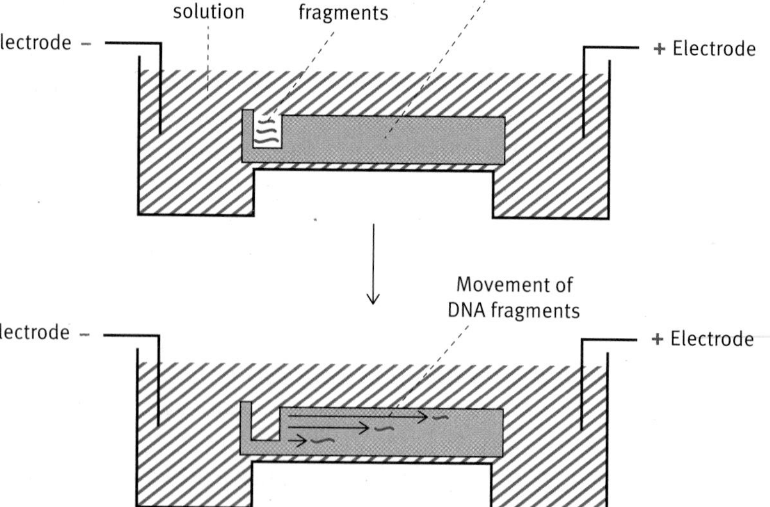

Fig. 28.1 Agarose gel electrophoresis of DNA. The apparatus is shown in cross-section and side on. It differs from the polyacrylamide gel electrophoresis apparatus in that the gel is horizontal rather than vertical. DNA fragments move through the gel towards the anode, but are impeded by the polymeric agarose, with larger DNA fragments moving more slowly than the more manoeuvrable smaller ones.

(SNPs) as genetic markers for locating disease-causing genes, as described later in this chapter.

The probe may be labelled by enzymically adding a radioactive phosphoryl group (^{32}P) or synthesizing the probe from radioactive nucleotides. Fluorescent tags are often preferred now to avoid using radioactivity. Methods for obtaining a suitable probe will vary with the individual experiment.

Southern blotting

Hybridization probing is often used to visualize DNA molecules that have been separated by electrophoresis. Probing cannot be done on either agarose or polyacrylamide gels, which are too fragile to undergo the process, so it is necessary to first transfer the DNA fragments to a more robust medium, typically a nylon membrane, while retaining their pattern of distribution. This technique, illustrated in Figure 28.2, is known as Southern blotting, or Southern hybridization, after its inventor, Edwin Southern. The DNA in the gel is first made single stranded by exposure to dilute alkali, and then transferred by capillary action to the nylon membrane, which is laid over it. The relative positions of the different DNA fragments are thus preserved on the membrane. (The term 'blotting' is used for this process by analogy with the use of blotting paper to soak up excess ink when laid over a document written with an old fashioned fountain pen.) The membrane is then soaked in buffer containing the labelled probe, which base pairs with the DNA fragment of interest. The position of that fragment is then determined by autoradiographic (exposure of the membrane to X-ray film), or fluorescent detection of the hybridized probe.

Adaptations of the blotting method have been given names that play on the name 'Southern'. In **Northern blotting**, mRNA, which can also be size separated by electrophoresis, is detected using a labelled DNA hybridization probe. A further application is called **Western blotting**, but here proteins

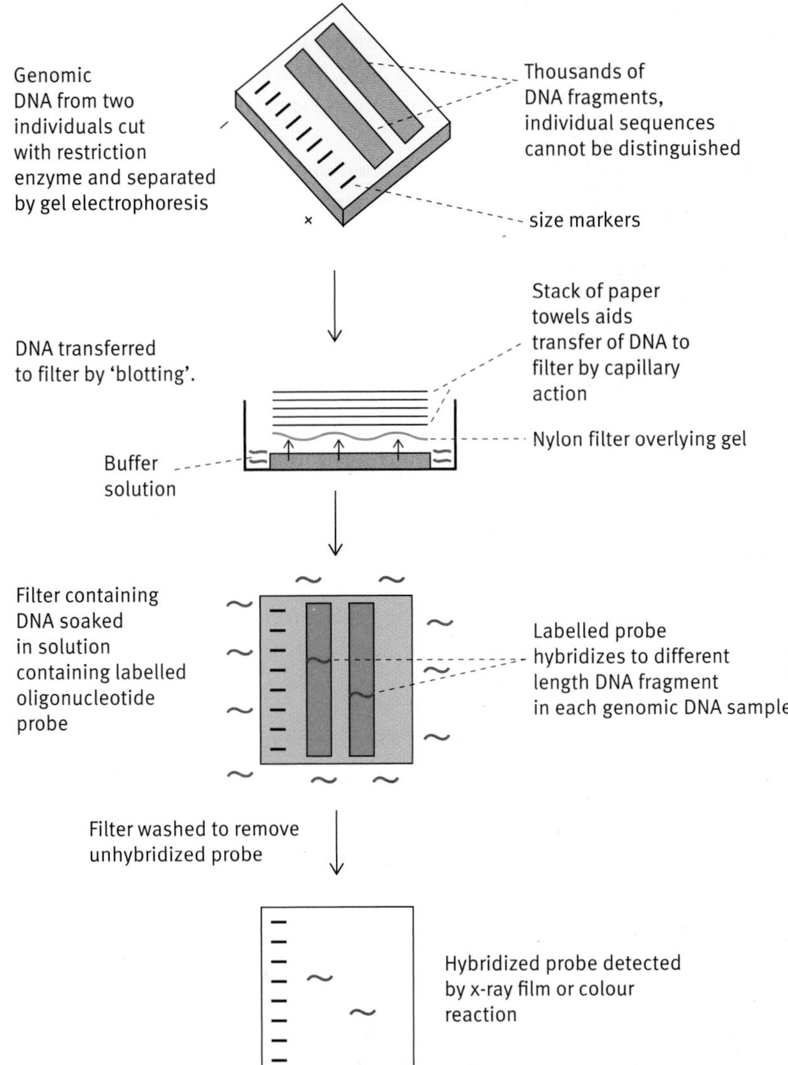
Genomic DNA from two individuals cut with restriction enzyme and separated by gel electrophoresis
Thousands of DNA fragments, individual sequences cannot be distinguished
size markers
DNA transferred to filter by 'blotting'.
Stack of paper towels aids transfer of DNA to filter by capillary action
Nylon filter overlying gel
Buffer solution
Filter containing DNA soaked in solution containing labelled oligonucleotide probe
Labelled probe hybridizes to different length DNA fragment in each genomic DNA sample
Filter washed to remove unhybridized probe
Hybridized probe detected by x-ray film or colour reaction

Fig. 28.2 Southern blotting procedure. See text for details.

rather than nucleic acids are detected using specific antibodies, not hybridization probes.

Chemical synthesis of DNA

We have made several mentions of the use of hybridization probes from various sources. While probes may be prepared from previously cloned sections of DNA, it is also common to synthesize them in the laboratory. Automated synthesizers can be programmed to make DNA molecules of defined sequence by a chemical method that does not require replication of a template strand. Molecules up to 30–40 nucleotides long can easily be made with a good degree of sequence accuracy. Synthetic **oligonucleotides** of 20–30 bases are long enough to hybridize specifically to unique sequences in genomic DNA, and can be used as probes and also as primers for DNA replication in the sequencing and PCR methods described below. Probes may be designed based on known peptide rather than nucleic acid sequence, and in this case allowance must be made for the degeneracy of the genetic code. A mixture of oligonucleotide sequences is synthesized that contains all possible codons that might be used to encode the amino acids in the peptide sequence.

Sequencing DNA

Determination of the base sequence of DNA is of central importance. The standard method that has been in use since the 1970s is based on enzymic replication of DNA. It is called the **dideoxy method** or **chain-termination method** or often 'Sanger sequencing' after its developer, Fred Sanger, winner of two Nobel prizes.

The principle of DNA sequencing by the chain-termination method

The sequencing procedure requires that the piece of DNA is copied, *in vitro*, by DNA polymerase. This requires, apart from the four deoxynucleoside triphosphates (dNTPs), that the DNA to be copied is single stranded and that a primer is hybridized to the start site because DNA polymerase cannot initiate new chains. Duplex DNA is rendered single stranded by treatment with NaOH or heat. The single-stranded DNA to be sequenced is incubated with a primer, a suitable DNA polymerase, and the four nucleotides (dATP, dGTP, dCTP, and dTTP). The products are separated by polyacrylamide gel electrophoresis to separate DNA molecules. Each addition of a nucleotide alters the migration of a DNA piece so that chains form separate bands, each being one nucleotide different in length from the next.

The next point is the crucial one; it concerns **dideoxy** derivatives of nucleoside triphosphates (**ddNTPs**), because if

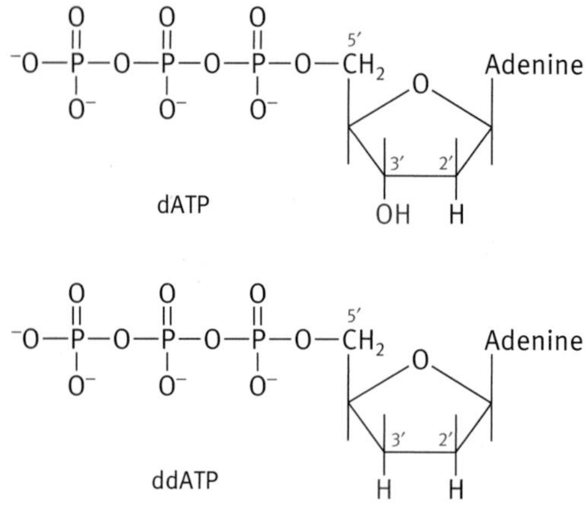

Fig. 28.3 The structures of deoxyATP (dATP) and dideoxyATP (ddATP). The absence of the 3′-OH group means that when a ddNTP is added to a growing DNA chain the chain is terminated.

one of these molecules is added to a growing DNA chain, synthesis stops. This is because DNA polymerase adds a nucleotide to the 3′-OH of a growing DNA chain and ddNTPs lack the 3′-OH group (Fig 28.3). Hence although they can still be added to a chain *via* their 5′-phosphate, the chain is then terminated.

We remind you that when we talk of a 'piece' of DNA being sequenced, multiple copies of that piece are involved in the experiments. Even a minute amount of DNA contains a large number of individual molecules. Suppose that we have in the copying process all four dNTPs plus a small amount of *one* ddNTP. If we take ddATP as an example, every time addition of an A is specified by the template strand, most of the new chains will have a normal adenine nucleotide added, and will continue to grow, but a fraction will, by chance, have a dideoxy form of the adenine nucleotide added, thus terminating those particular chains. (The fraction terminated will depend on the relative proportions of dATP and ddATP.) This ratio is adjusted so that the terminated chains are sufficient in number to be detected as a separate band on an electrophoretic gel by autoradiography. The rest of the molecules will go on being added to until another A is due to be added and the same will happen again.

How is this interpreted as a base sequence?

Suppose the piece of DNA being sequenced has a sequence with Ts placed as shown:

3′X–X–T–X–X–X–X–T–X–X–T 5′

If, during the copying, ddATP is present (along with dATP) such that, at the addition of each A, a small proportion of the growing chains are terminated, the copying will produce the following population of chains (attached to the primer):

(a) 5′X–X–ddA 3′

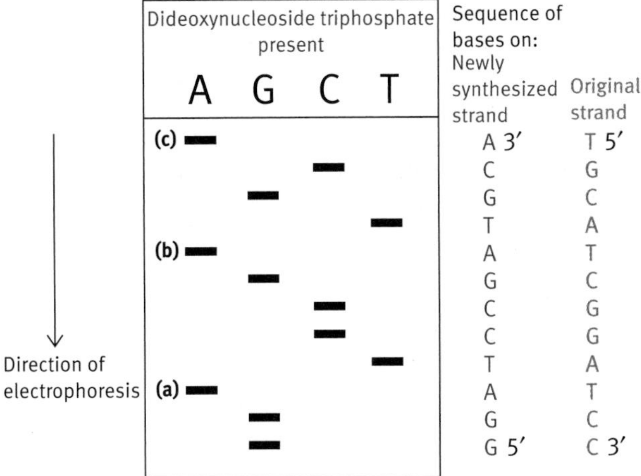

Fig. 28.4 An autoradiograph sequencing gel. The sequence is read from the bottom to the top. **(a)**, **(b)**, and **(c)** are the bands produced in the presence of dideoxyATP. See text for explanation. Manual sequencing has been largely replaced by the automated method that is also described in the text, but the basic principle remains the same.

(b) 5′X–X–A–X–X–X–X–ddA3′

(c) 5′X–X–A–X–X–X–X–A–X–X–ddA3′.

On a sequencing gel, these bands will be seen as the bands in Fig. 28.4 (left-hand column).

If a second incubation contains ddTTP instead of the ddATP, an analogous set of chains terminating in T will be produced and, similarly, terminating in C and G if ddCTP or ddGTP, respectively, are present in duplicate incubations. All four incubations are run side by side on a gel, giving the bands shown in Fig. 28.4. From this the base sequence of the piece of DNA can be read off from the bottom of the gel upwards as the sequence of the newly synthesized chain (and therefore of the partner to the template strand being sequenced) rather than of the template strand itself. This gives the sequence in the 5′→3′ direction, since synthesis always proceeds in that direction.

Automated DNA sequencing

At first sequencing was done manually with radioactive labels. Subsequently an automated procedure was introduced but the basic principle is the same. The four ddNTPs are labelled each with a differently coloured fluorescent label. Electrophoretic separation of the chains is done in fine capillaries containing a fluid matrix rather than a gel. Resultant separations are automatically scanned for the different colours, and a computer prints out the base sequence. Figure 28.5 shows a section of a sequencing run from an automated sequencer. It is now usual for research workers to send a sample of DNA with a primer in a small plastic tube to a sequencing service and wait for the answer.

The development of automated sequencing made it possible for the international Human Genome Project consortium to complete the sequence of the entire 3.2 billion bases of human DNA in 2003. Many other genomes including *E. coli*, yeast, *Drosophila*, mouse, the rice plant, *Arabidopsis* (a favourite plant for genetic studies), and others are also available. Today, because of the Human Genome Project and other programmes, DNA work is therefore often performed using sequence information already available in databases.

Next generation DNA sequencing

Since the completion of the Human Genome Project the focus of human genomic studies has switched to sequencing individual genomes in order to study human genetic variation. To make this a practical proposition, still faster and cheaper sequencing methods were needed. A number of these 'next generation sequencing' (NGS) technologies are now in use.

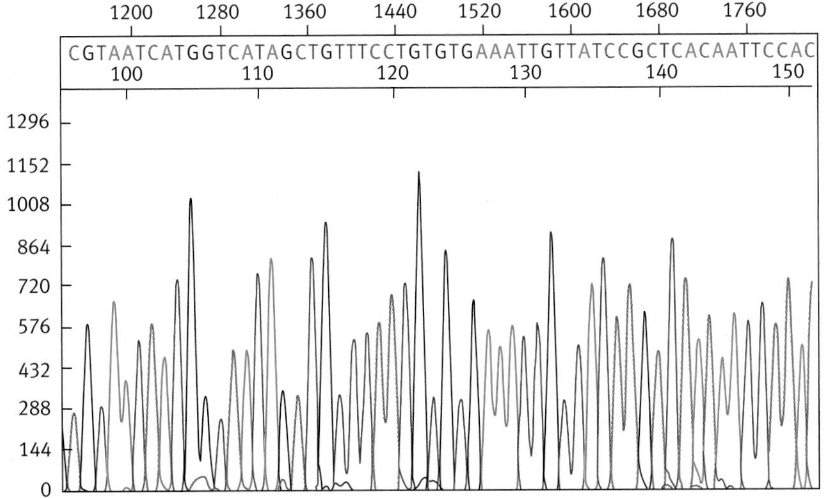

Fig. 28.5 Graph from automatic sequence analyser. Fluorescence detection of fragments produced by the dideoxy method, the fluorescence being of a different colour for each of the four ddNTPs. Kindly provided by Arthur Mangos and Dr Z. Rudzki, Molecular Pathology, Institute of Medical and Veterinary Sciences, Adelaide, Australia.

Most of these continue to utilize DNA replication *in vitro*, but the process has been speeded up in two key ways. The first of these is that the number of sequencing reactions that can be carried out simultaneously is vastly increased, and the second is that capillary electrophoresis has been dispensed with in the analysis. Whereas a Sanger sequencing machine could analyse 96 reactions; NGS technologies allow simultaneous sequencing of hundreds of millions of DNA fragments, a process known as **massively parallel sequencing**. In many of these technologies the process starts in an oil–water emulsion in which each minute droplet acts as a separate reaction compartment. Alternatively, the millions of fragments to be sequenced are attached at one end to a glass slide, and reactants are allowed to flow over them.

A commonly used detection method based on glass slide technology is that marketed by Illumina. Here the single-stranded DNA fragments to be sequenced are replicated one nucleotide at a time, by washing a mixture of all four dideoxyNTPs over the slide. Each of the four ddNTPs (A, C, G and T) has a different-coloured fluorescent dye attached to it. Thus, after the first reaction each DNA fragment has incorporated just one nucleotide into its growing second strand, and the glass slide is photographed at high resolution to detect which colour and hence which base has been added at each position on the slide. The fluorescent dye and the dideoxy block further replication and are then chemically removed from the first incorporated nucleotides, the replication step is repeated so that a second nucleotide is added to each sequence, the slide rephotographed and so on.

While NGS technologies can carry out millions of sequencing reactions simultaneously, the length of each DNA piece that can be successfully sequenced in a single reaction is typically shorter than that achieved by Sanger sequencing. While Sanger sequencing produces individual sequence 'reads' of several hundred bases, NGS technologies often produce reads of tens of bases only. In order to compile complete genome sequences NGS relies on powerful computer programmes to line up the data from multiple sequencing reactions, and also on the existence of the 'reference' Human Genome Sequence for comparison. Nevertheless, these are hugely powerful technologies that are transforming many areas of biological and medical research.

Amplification of DNA by the polymerase chain reaction

This technique has assumed enormous importance in DNA studies of all kinds because it enables a stretch of DNA to be picked out and quickly amplified exponentially. It is so sensitive that a few molecules of DNA among millions of others can be detected selectively and amplified. The essential requirement is that the sequences flanking each end of the selected section are known so that complementary DNA primers can be obtained. The specificity of selecting a particular section of DNA from all the others depends on this. The principle of the method is that the selected section, primed on both strands of the DNA, is enzymically replicated from the four dNTPs and then this is repeated with the old and the new strands all being replicated from the added primers giving an exponential increase of the section. The incubation mixture contains a heat-stable DNA polymerase and excess primers, as well as the dNTPs, so that no more components need to be added throughout the cycles. Progression through the cycles is determined solely by the temperature changes. The heat-stable DNA polymerases required to remain active through the process are derived from bacterial species that survive in high temperature environments, such a volcanic hot springs. An example is *Thermus aquaticus* from which the often used *Taq* polymerase is derived.

To describe the method let us consider a stretch of DNA that you want to amplify. This is shown in green in Fig. 28.6(a). The parent strands are separated by heating (Fig. 28.6(b)). For simplicity we will illustrate in Fig. 28.6 only what happens to one of the two parent strands since the same applies to both.

Two chemically synthesized DNA primers, about 20 nucleotides in length, complementary to the 3′ flanking sequences on the two template strands are added in large excess (because each round incorporates the primers into the new strand). These attach on cooling to priming sites (one shown in Fig. 28.6(c)). Replication to the end of the piece occurs, producing a 'long' product (Fig. 28.6(d)). *That is the end of the first cycle.*

The next cycle is started by heating to separate the strands (Fig. 28.6(e)) and after cooling, primers attach to both strands and replication from these occurs. The DNA polymerase used is heat stable and does not need to be added again. The original strand once again gives rise to a long product but the *first* long product will be copied in the opposite direction to produce a 'short product' (Fig. 28.6(f)) *which is the desired copy.* At the end of the second cycle we therefore have the original strand and two long products and one short product. *That is the end of the second cycle.*

The third cycle is started by again heating to give strand separation; again primers attach on cooling (Fig. 28.6(g)). Replication produces another short strand giving a duplex of the latter (Fig. 28.6(h)). From now on the short strands increase exponentially as the number of cycles increases but the long products increase only arithmetically. The latter therefore becomes a minute proportion of the total and can be ignored.

The entire process is automated with the heating and cooling performed in as many cycles as you wish, each cycle taking only a few minutes. Amplification of many millions of fold is easily achieved (2^n fold where n=number of cycles).

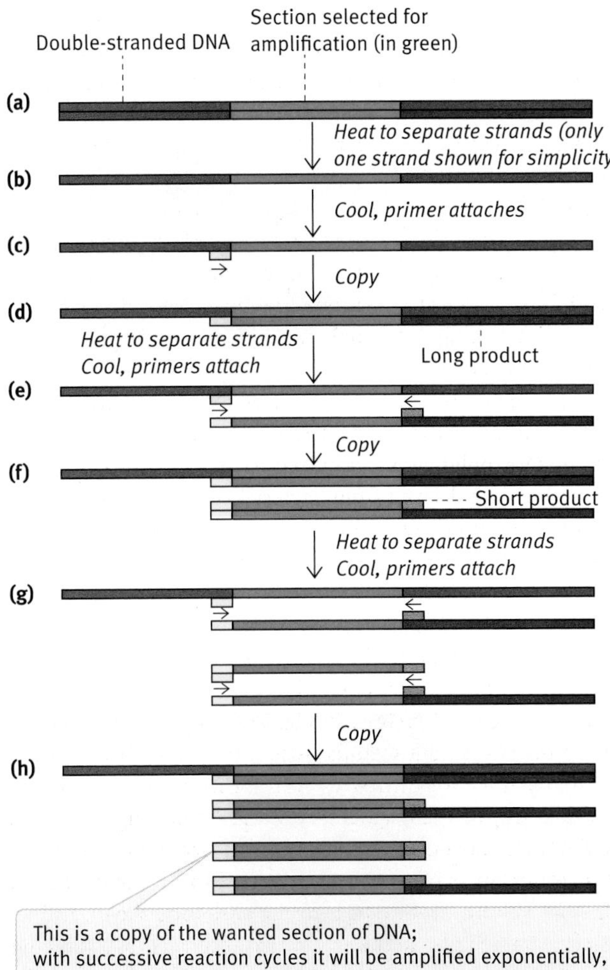

Fig. 28.6 The basic principle underlying amplification of a DNA section by the polymerase chain reaction (PCR). Incubations contain four dNTPs, a heat-stable DNA polymerase, and primers. **(a)** The green bars represent the section chosen for amplification by the use of appropriate primers (arrows), each complementary to the 3′ flank of the section in one strand to be amplified. **(b)** Heating separates the strands and, from this point, to keep the diagram to a manageable size, amplification of only one strand is shown; the process amplifies both strands. Cool so that the primer appropriate for this strand binds. DNA is replicated (in this case beyond the end of the desired piece). Heating separates the strands. **(f)** The new strand is primed. **(g)** The new strand is replicated. **(h)** The separated wanted section. You can see that we now have reproduced our selected piece. With about 25 rounds of replication, heating, cooling, priming, and synthesis, the piece can be amplified a millionfold. Note that, every time priming and copying occurs, all of the original and new molecules are so primed and copied that a detailed diagram of what happens becomes rather involved. The essential point is that an enormous amplification of the selected piece occurs.

Analysis of multiple gene expression in cells using DNA microarrays

One of the newer additions to the armoury of molecular biology tools are the **DNA microarrays**, colloquially referred to as **DNA chips**.

Northern blotting, as described earlier, can be used to detect the transcripts of individual genes by hybridizing mRNA to single-stranded DNA probes of the gene from which they were transcribed. A Northern blot can give information on the size of a particular transcript and by analysing samples from different tissue and cell types, on when and where it is expressed. Genomics however is concerned with the *level* of transcription of many genes, even thousands, at any one time and in any one tissue so that overall patterns of gene expression can be studied. The interest stems from the realization that the interactions of many genes determine the life of cells and organisms. If the degree of expression of many genes could be determined it could for example throw new light on gene control in normal and disease states. As an example, comparing total patterns of gene transcription in normal and cancer cells of the same tissue can lead to better classification of the cancers and more specifically tailored therapeutic regimes. A human has about 21,000 genes whose expression might be looked at. How could this be done for large numbers of them at any one instant of time? DNA microarray technology performs this seemingly impossible task.

The principle is that an array of pieces of DNA, which are specific targets, are spotted on to a DNA-binding surface called a chip, most often of glass. (*Note that these are referred to as the probes.*) The spots are minute in size and are applied robotically or synthesized *in situ*, the location of each spot being identifiable and each sequence corresponding to one in a known specific gene. Many thousands of such sequences can be placed on a postage-stamp-sized chip. The specific target sequences in the spots may be copies of the coding sequences of genes or parts of genes amplified by PCR, or synthetic oligonucleotides synthesized on the glass *in situ* by robotic means. The availability of whole genome sequences means that the sequence of any identifiable gene is known. Ready-made microarrays specific for groups of genes, such as those relevant to cancer, are available commercially.

Suppose your aim is to compare gene expression in cancer cells with that in normal cells of the same type, using an appropriate DNA microarray (Fig. 28.7). Total mRNA is isolated from the two samples. Although in some cases mRNA is analysed directly, it is usually first copied to make complementary cDNA (see later in this chapter), creating for each cell type a mix of cDNA copies of each of the transcripts that were present in the mRNA preparation. The two separate sets of cDNAs are labelled with fluorescent dyes, red in the case of the cDNA made from cancer cells and green for the normal control cell samples. The strands are separated by heat and allowed to hybridize with the target probe sequences on the microchip. The chip is washed and scanned automatically so that the intensity of the different-coloured fluorescence on each spot is quantified. Genes whose expression is increased in the cancer cells as compared with normal ones appear as red fluorescent spots because more cDNA has hybridized to the gene probe than has cDNA from the normal cells. In the reverse situation, where expression of a gene in the cancer

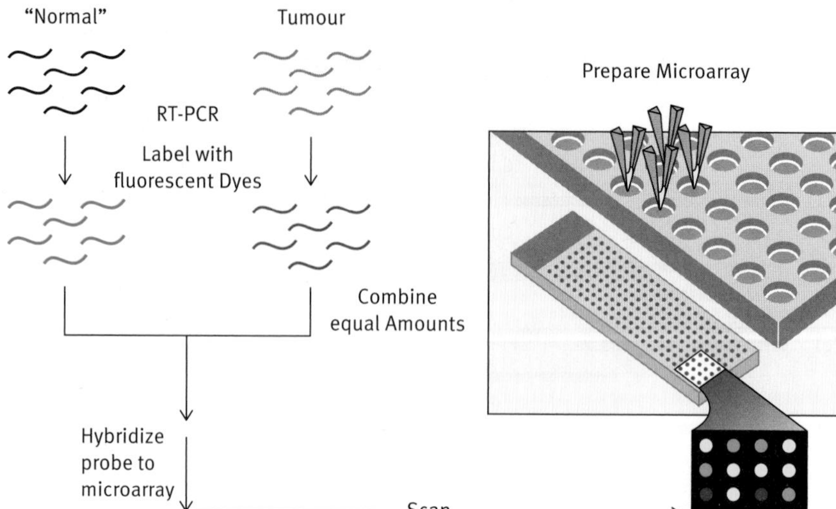

Fig. 28.7 Use of a DNA microarray to compare gene expression in normal and tumour cells. The use of RT-PCR to create cDNA copies of messenger RNA is explained later in this chapter.

cells is decreased relative to the normal control, the spot is green, and if there is no difference between the two types of cells the spot is yellow (red and green combined). Dark spots indicate little or no expression in either cell type. A computer records the results in terms of quantity of each specific mRNA detected by a spot on the microarray. Since each of these corresponds to a known gene the method gives a global picture of the expression of all the genes at any one time in the cancerous and normal tissues.

Figure 28.8 shows a comparison of expression when a microarray was probed with labelled cDNA generated from RNA extracted from different cell lines. Different patterns of expression are found.

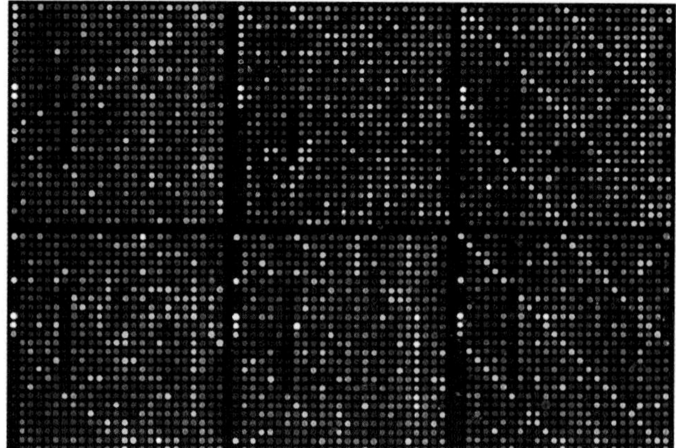

Fig. 28.8 Gene expression analysis using a DNA microarray (or DNA chip). A library of about 19,000 oligonucleotide probes for human gene coding regions was hybridized with complementary DNA from two cell lines, labelled with a green and red fluorescent dye respectively. A small section of it was enlarged for this figure. See text for explanation provided. We are grateful to Mark Van der Hoek, of the Adelaide Microarray Facility, Institute of Medical and Veterinary Sciences, University of Adelaide, Australia, who kindly supplied this photograph.

Joining DNA to form recombinant molecules

One of the central techniques of DNA manipulation is to create new DNA molecules by joining together pieces that are not found as such in nature. The new molecules are recombinants. There are various ways of doing this and which is used will depend on the actual experiment. One of the simplest methods is to make use of the staggered cut and cohesive or sticky ends made by restriction enzymes such as *Eco*RI. These short single-stranded overhangs will base pair with complementary sequences under suitable conditions. Although it is the same physical process as probe hybridization, base pairing of overhanging ends created by restriction digests is often termed annealing (or re-annealing) of DNA.

—X—X—G↓A—A—T—T—C—X—X—
—X—X—C—T—T—A—A↑G—X—X—

*Eco*RI ↓

—X—X—G　　　　A—A—T—T—C—X—X—
—X—X—C—T—T—A—A　　　　G—X—X—

Sticky ends

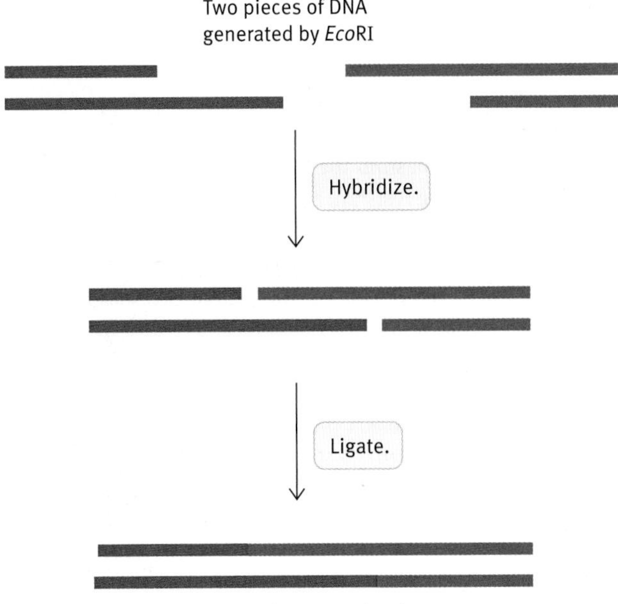
Two pieces of DNA generated by *Eco*RI
Hybridize.
Ligate.
Recombinant molecule

Fig. 28.9 Principle of construction of recombinant DNA molecules by annealing of overhanging ends.

If DNA from different sources is cut with the same restriction enzyme (of the type which makes a staggered cut), the fragments will have complementary overhanging ends. When two pieces of DNA with complementary ends are mixed, they join together by base pairing as indicated in Fig. 28.9. The single-stranded 'nicks' in the sugar-phosphate backbones are covalently sealed by a DNA ligase (the enzyme which has the *in vivo* role of joining Okazaki fragments during DNA replication); this synthesizes a bond between the 3′-OH of one piece and the 5′-phosphate of the other.

Sometimes it is necessary to join pieces of DNA with blunt ends. This can also be done using the ligase enzyme. Joining blunt ends is less efficient and less controlled than joining overhanging ends as it is not aided by the annealing step, but it has the advantage that blunt-ended DNA from any source can be joined to any other blunt-ended DNA molecule.

Cloning DNA

As mentioned earlier, for DNA manipulation, it is necessary to have large numbers of individual molecules of the DNA sequence of interest. One very efficient method of achieving this is to amplify it by PCR. However, an earlier and still frequently used approach is to introduce the sequence into a host (usually bacterial) cell where it can be replicated. To do this the DNA is covalently attached to the DNA of a **cloning vector**, which can be transferred into the host.

Cloning in plasmids

There are several different vectors to choose from. They differ mainly in the size of the piece of DNA that they can accommodate but some have other useful characteristics. The most commonly used vectors are bacterial plasmids for they are the easiest to handle and have long been the backbone of recombinant DNA technology. They can handle DNA inserts to be cloned up to approximately 10 kb pairs in length.

The *E. coli* cell has a single major circular chromosome that constitutes most of the cell's genetic make up. In addition, however, there are tiny separate minichromosomes or plasmids, often in multiple copies in the cytosol – they are circular DNA molecules carrying a handful of genes that usually have a protective role in the cell. Typically they carry genes conferring antibiotic resistance on the cell. Each plasmid has an origin of replication, which is required for duplication of the plasmid in the cell. Fig. 28.10 shows the general principle of constructing a recombinant plasmid. The plasmid is cut at a specific cloning site using a restriction enzyme giving

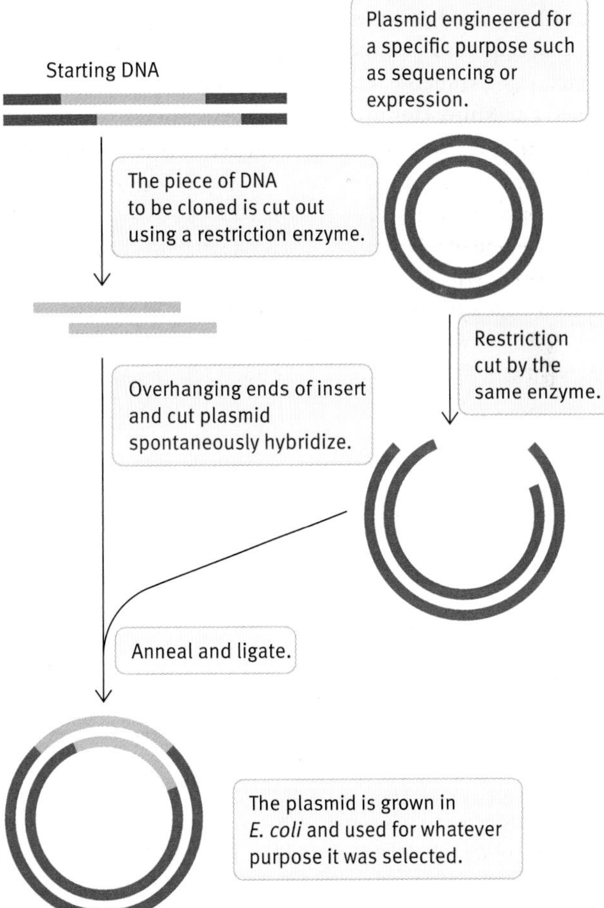
Starting DNA
Plasmid engineered for a specific purpose such as sequencing or expression.
The piece of DNA to be cloned is cut out using a restriction enzyme.
Restriction cut by the same enzyme.
Overhanging ends of insert and cut plasmid spontaneously hybridize.
Anneal and ligate.
The plasmid is grown in *E. coli* and used for whatever purpose it was selected.

Fig. 28.10 Method of constructing a recombinant plasmid for cloning a DNA insert.

overhanging ends. The DNA piece to be cloned is generated by cutting the source DNA with the same enzyme so that the pieces and the cut plasmid have complementary ends, which anneal together. The cut plasmids and the pieces are annealed and enzymically ligated (covalently joined) together. The next step is to introduce the plasmids into *E. coli* cells, which we will take as our example. Plasmids, naturally, are weakly infectious but *E. coli* cells can be made 'competent' by chemical and heat treatment to take up plasmids more readily, a process known as **transformation**.

In order to understand the further steps of a cloning exercise it is necessary to bear in mind that we are handling, not individual source DNA molecules and plasmids, but many millions of them and millions of *E. coli* cells. Each stage of the process is much less than 100% efficient, so that selection steps are needed to make sure we end up with what we want. First of all, a proportion of the plasmids in our ligation reaction will simply religate back together to reform the original plasmid without acquiring the source DNA (i.e. in the language of cloning they 'lack an insert'). Secondly, transformation is inefficient. This can act in our favour as it is rare for even one plasmid to enter a bacterial cell. For two to enter one cell is so rare that it can be ignored as a possibility. This ensures that each host cell that acquires a plasmid with an insert will contain just one cloned DNA sequence, even if our original source DNA contained a mixture. However, the inefficiency of transformation also means that the majority of *E. coli* cells that go through the transformation step do not pick up a plasmid at all.

Naturally occurring plasmids have been engineered for use as cloning vectors so that they allow selection, first of *E. coli* cells that have acquired the plasmid, and secondly of *E. coli* cells containing a plasmid that has a cloned insert. The inclusion of an antibiotic resistance gene in the plasmid allows the first. The transformation step is carried out by mixing the ligated DNA with *E. coli* cells in liquid culture, as illustrated in Figure 28.11. The *E. coli* are then spread on an agar plate containing the antibiotic, for example, ampicillin. Only *E. coli* that picked up a plasmid will be able to grow and replicate, with each individual cell growing up into a colony containing large numbers of identical cells, each carrying multiple copies of the same plasmid with its specific DNA insert.

Often, further selection is used in order to select *E. coli* containing a plasmid that has picked up an insert, rather than plasmid that religated back to itself. Plasmid vectors such as pUC18, illustrated in Figure 28.12, have been designed to allow visual inspection to tell whether a bacterial colony has received a plasmid with an insert (known as blue/white selection). Fig. 28.12 shows the construction of the plasmid. It has a gene for ampicillin resistance and a gene for the *N*-terminal 146 amino acids of the enzyme β-galactosidase, a gene you may remember from our discussion of the *lac* operon in Chapter 26. The gene also has several

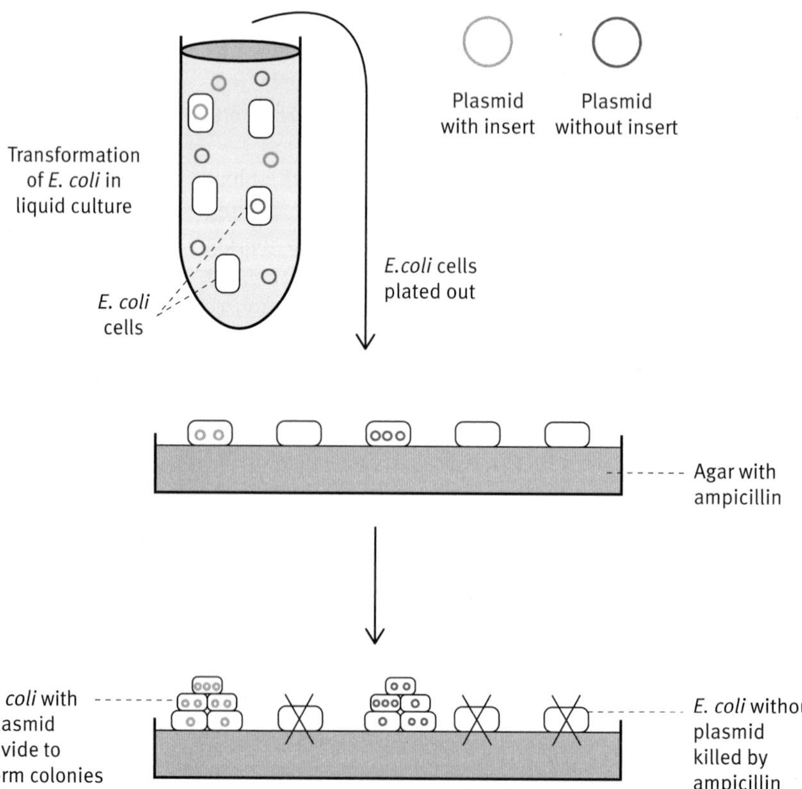

Fig. 28.11 Transformation of *E. coli* with recombinant plasmids. Following cloning of a DNA insert into a plasmid vector, competent *E. coli* cells are mixed with the products of the ligation reaction. A proportion of the cells take up plasmid DNA, one copy per cell. The *E. coli* cells are then plated on nutrient agar containing ampicillin. Only cells containing plasmid are ampicillin resistant and can form colonies. *E. coli* cells may acquire plasmid with an insert (orange) or without one (blue).

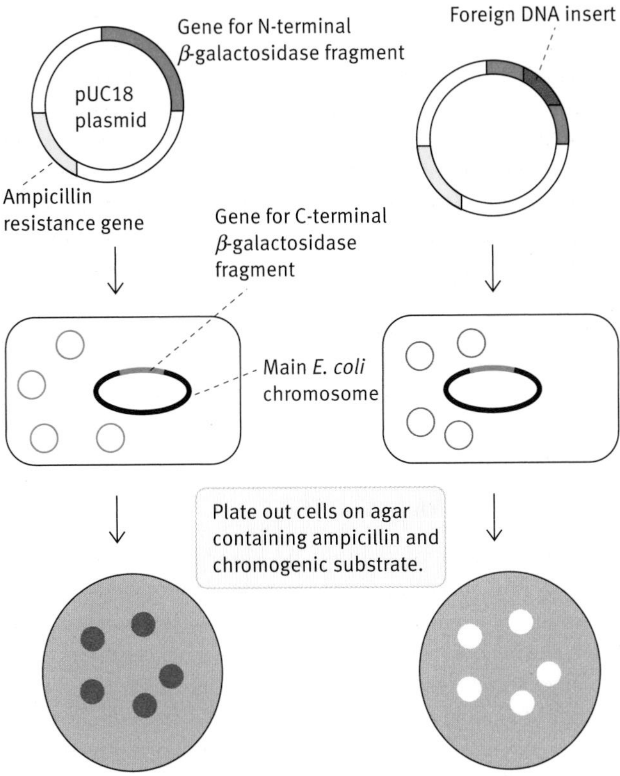

Fig. 28.12 Diagram of pUC18 plasmid used to clone a piece of foreign DNA (red). The insertion of the foreign DNA inactivates the gene coding for one part of the β-galactosidase protein. The *Escherichia coli* cells used have been engineered to contain the gene for the rest of the β-galactosidase protein. The plasmids without the insert allow the synthesis of the fragment that complements the latter so that active enzyme is produced. This hydrolyses a chromogenic substrate taken up from the medium and turns the bacterial colonies blue. The plasmids with the foreign insert do not produce active enzyme and the colonies appear white.

restriction enzyme sites, which can be used as cloning sites, situated within the gene for the *N*-terminal fragment of β-galactosidase. Following transformation, only cells with a plasmid will form colonies on the ampicillin containing agar plate, but most of the incorporated plasmids will not have received the DNA insert. This is where the gene for the *N*-terminal fragment of β-galactosidase come in. If the part enzyme it encodes joins to the missing C-terminal portion of the enzyme the two reconstitute the active enzyme. An engineered *E. coli* strain is used for the cloning into whose main chromosome the gene for the missing C-terminal section of β-galactosidase has been inserted. Any plasmid that had the DNA fragment inserted into it cannot synthesize the *N*-terminal β-galactosidase fragment because the insertion disrupts the coding region. A plasmid with a successful insert will therefore not give rise to active β-galactosidase. On the other hand, cells with plasmids lacking the DNA insert will produce both the *N*-terminal and the C-terminal enzyme fragments and generate active enzyme.

The cells are plated on a medium containing ampicillin and a chromogenic (colour-generating) substrate which enters the cells and gives rise to a blue colour if hydrolysed (by β-galactosidase if it is present). Colonies of cells carrying the plasmid without a DNA insert produce the enzyme and turn blue. Cells carrying a recombinant plasmid do not produce active enzyme and the colonies are white (Fig. 28.12).

Cloning libraries

In some cases, the source DNA in the type of cloning exercise described above is DNA that has already been cloned or amplified by PCR, and is being prepared for further work. The source DNA will therefore consist of a single DNA sequence (though remember always that it will contain multiple copies of that sequence). However, in other exercises the source DNA consists of a mixture of sequences, for instance the genomic DNA of an organism of interest. In this case each plasmid that receives an insert at the ligation stage will receive a different fragment of DNA, and each *E. coli* colony that receives a plasmid with insert will receive a different cloned portion of the genome. The collection of clones created is known as a **library**, and in this case because the source was genomic DNA, a **genomic library**. The goal is that each DNA sequence in the genome should be represented in the library.

To go back to our blue/white colour selection, if we are creating a library we know that each white *E. coli* colony contains a plasmid with an insert, but not which of the many cloned sequences each colony contains. To find a particular clone containing a sequence of interest, the library must be **screened**. The method used for screening depends greatly on what we are looking for and what we know about it, but in many cases we have some knowledge of at least part of the desired sequence. For instance, if we are interested in DNA that encodes a protein for which we have even part of the amino acid sequence, we can work backwards to predict part of the DNA sequence we are looking for and synthetically create an appropriate oligonucleotide hybridization probe. Screening by hybridization proceeds rather in the manner of Southern blotting described earlier. It involves overlaying the colonies with a DNA-binding membrane, which when removed has taken a 'print' of the colonies. The print will contain some of the cells from each colony; the *E. coli* cells are lysed (opened up) by treatment with NaOH and the DNA is fixed to the membrane in single-stranded form. The membrane is subjected to hybridization with the probe, allowing detection of which colony contained complementary DNA, and the corresponding colony can then be identified on the original plate. Once the colony of interest is known, it is then a routine matter to grow cells from it in liquid culture and produce as many copies of the plasmids with its DNA insert as you want. The plasmids are easily purified from the cells because of their small size compared with the main *E. coli* chromosome. Your desired piece of DNA can be released from the plasmid by cutting with the same restriction enzyme that was used to construct the recombinant plasmid in the first place.

It may not seem obvious in the example given why we would want to clone DNA encoding a protein for which we already have some amino acid sequence. However, we may have only managed to obtain the sequence of a few amino acids from the purified protein, so our DNA clone, up to several thousand base pairs in length, will provide us with extra sequence on either side of that used for the oligonucleotide probe. Remember also that only a few tens of base pairs can be chemically synthesized, so if we want to do any further work with DNA encoding our protein we need to obtain it by the cloning method, and that once the DNA is contained in a cloning vector we have a convenient method of producing more copies of it at will.

Cloning vectors for larger pieces of DNA

Plasmids, as mentioned, can be used for cloning DNA fragments up to 10 kb pairs in length, which is convenient for many routine DNA manipulations, but it may be necessary to clone longer pieces. In practice genomic libraries are rarely constructed in plasmid vectors because the number of different clones needed to accommodate say the entire human genome sequence in 10 kb pair fragments would be entirely unmanageable. Alternative vectors are available that can take larger inserts.

Lambda phage (bacteriophage λ) can be used for inserts up to 20 kb pairs. Bacteriophage λ is a bacterial virus, which infects *E. coli*. It is comprised of a shell of protein molecules (the head) enclosing its genome (a single molecule of DNA), and a tail which in effect is a device for injecting the DNA into the cell (see Fig. 2.8) where it replicates. One of the remarkable properties of lambda phage is that the DNA is automatically packaged into the heads by self-assembly of the head and this can be done *in vitro* using 'packaging kits' with the necessary protein components. The lambda phage genome can be engineered in a similar way to plasmids for use as a cloning vector, so that it contains suitable restriction sites into which inserts can be ligated. The inserted foreign DNA replaces most of the phage's own genome sequence, but leaves sufficient for it to be packaged into infective phage. *E. coli* cells are thus infected by recombinant phage rather than transformed by naked plasmid DNA. The cells are then grown on an agar plate. In this case there is no antibiotic selection, so all the *E. coli* grow and cover the plate with a continuous lawn of bacteria. Phage is replicated in each infected cell, causing it to lyse and infect neighbouring cells so that clear plaques are created in the opaque lawn of bacteria. The plaques can be screened for the wanted insert in a similar manner to that used for screening colonies containing plasmids. Infectious phage containing the cloned sequence of interest can then be recovered from the plaque and grown in culture.

Cosmids are hybrids of lambda phage and plasmids used for cloning pieces of DNA 40–50 kb pairs in length. For cloning pieces of DNA 500 kb pairs or longer, **yeast artificial chromosomes** (YACs) are used. These are artificial DNA chromosomes with telomeres, centromeres, and origins of replication and are reproduced in the host yeast cells. **Bacterial artificial chromosomes (BACs)** are also used. Libraries produced in BACs were used to clone the human genome in preparation for sequencing.

Applications of recombinant DNA technology

Working with RNA and cDNA

In our discussion of cloning and library production so far we have talked about using genomic DNA as the source of clones. It would often be more convenient to work with mRNA, since we would only be working with the portions of the genome that actually encode proteins. mRNA can be purified from cells, but RNA is much less stable than DNA and cannot be ligated into cloning vectors and propagated in host cells. It is therefore very common to copy mRNA sequence *in vitro* to make **complementary DNA (cDNA)**. The method of doing so makes use of the viral enzyme reverse transcriptase and is illustrated in Fig. 28.13. Since eukaryotic mRNA has polyA tails, synthetic oligonucleotides containing

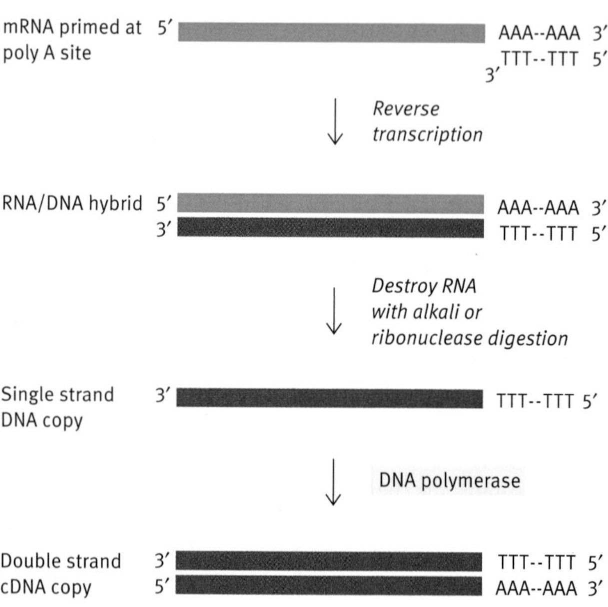

Fig. 28.13 Preparation of a complementary DNA (cDNA) library from eukaryotic messenger RNA (mRNA) (green). The reverse transcriptase is primed with oligo-dT and a DNA copy (red) is synthesized. The RNA is destroyed by alkali and the single DNA strand duplicated by DNA polymerase.

only T bases (oligo-dT primers) are hybridized to the mRNA and used as primers by reverse transcriptase. This gives an RNA-DNA duplex. The RNA strand is destroyed and the single-stranded DNA is copied to give double-stranded DNA by an exonuclease-free version of DNA polymerase I. The collection of double-stranded cDNAs can be ligated into plasmid vectors to created a cDNA library, which can be screened as described earlier. Note that the population of mRNA from which cDNA is copied varies from tissue to tissue and from time to time depending on which genes are being expressed at the time, so that cDNA libraries prepared from various tissues of the same organism will vary in their content, unlike genomic libraries.

cDNA has other uses besides making libraries. It is often used as a more robust surrogate for mRNA in analysing gene expression. Use of cDNA for this purpose has already been mentioned in the context of microarrays. RT-PCR is another common technique for detecting the presence of a particular mRNA transcript in a cell or tissue type. Here the RT stands for 'reverse transcription', as the mRNA is first copied to make just a single cDNA strand, which is then subjected to PCR using primers specific for the transcript of interest. If the mRNA was present in the cell an amplified double-stranded DNA PCR 'product' of the expected size will be made, and if it was absent no PCR product will be seen. RT-PCR methodology has been successfully adapted to make it quantitative (qPCR or Real-Time PCR) so that the level of gene expression, not just the presence or absence of a transcript, is measured.

Production of human proteins and proteins from other sources

This has advantages over isolation of proteins from cells in which they occur naturally. A protein of therapeutic value such as human growth hormone, which is a therapy for some types of dwarfism, used to be isolated from human pituitary glands obtained from cadavers. Quite apart from the difficulty of obtaining the source material, there was the risk of transferring to the patient an infectious agent such as a prion. In treating diabetes, insulin previously obtained from beef pancreas sometimes produced an immunological reaction, which is less of a risk if human insulin produced in bacteria (or more commonly today in yeast) is used. A major advantage of the technology is that it can produce large amounts of proteins, which may occur in the body only in small amounts. Tissue plasminogen activator (see Chapter 32) is an example. The reason is, of course, that the amount of a protein produced naturally is a function of the promoter of the gene. In recombinant DNA technology a powerful promoter is supplied.

To produce eukaryotic proteins in *E. coli*, cDNA is used because bacterial genes are not interrupted by introns and *E. coli* therefore cannot splice the RNA transcripts as in eukaryotes. In a eukaryotic cell, an mRNA has already had the introns spliced out so that cDNA copied from it can be transcribed and translated into protein in *E. coli*, provided a promoter and the necessary bacterial translational signals are added. A starting point is a cDNA library from which the clone containing the cDNA of interest may be isolated.

Expressing the cDNA in *E. coli*

To produce eukaryotic proteins in *E. coli*, plasmids engineered so that production of the wanted protein is maximized, so called **expression vectors**, are available commercially. Built into such a plasmid are a strong bacterial promoter and also a sequence that transcribes into a ribosome-binding (Shine–Dalgarno) site on the mRNA. A plasmid containing the cDNA of interest is introduced by transformation into bacterial cells where mRNA is produced and translated into the wanted protein (Fig. 28.14).

Bacteria do not glycosylate proteins or carry out some of the other post-translational modifications that occur in eukaryotic cells. Some proteins may function sufficiently without these modifications to be used therapeutically or in research. If post-translational modification is required, eukaryotic cells

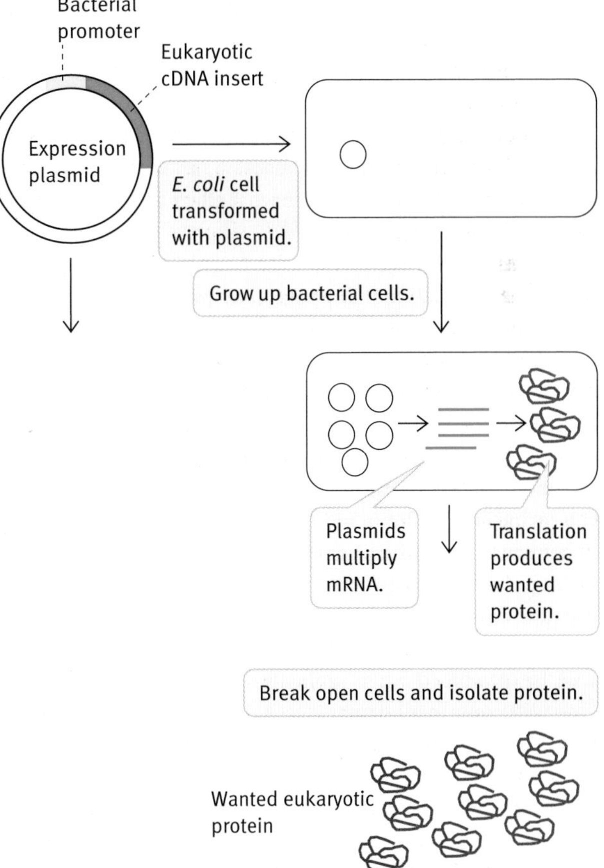

Fig. 28.14 Expression of a eukaryotic protein in *Escherichia coli* using an expression plasmid. See text for explanation.

and appropriate expression vectors may be used. A wide variety of human proteins are now produced in *E. coli*, yeast, and other cells. Examples include insulin, tissue plasminogen activator (used for blood-clot removal), and human growth hormone. The viral hepatitis B protein used as a vaccine against the disease is also produced in this way.

When you transform *E. coli* with the expression plasmid, you want to produce as much protein as possible so a strong promoter is used. However, the energy requirements of protein synthesis mean the cell cannot grow rapidly and produce lots of foreign protein at the same time. To cope with this, an inducible promoter is used (somewhat similar to the *lac* operon promoter, which only functions in the presence of its inducer allolactose). The bacteria grow rapidly without inducer and then when sufficient cells are formed inducer is added to the culture to switch on synthesis of the desired protein.

Another practical problem is that eukaryotic proteins produced in bacteria may not fold correctly, and unfolded protein produced in large quantities precipitates out as 'denatured' inclusion bodies. Procedures have been developed by which, in some cases, these inclusion bodies can be dissolved and refolded properly *in vitro*, though larger proteins may present difficulties.

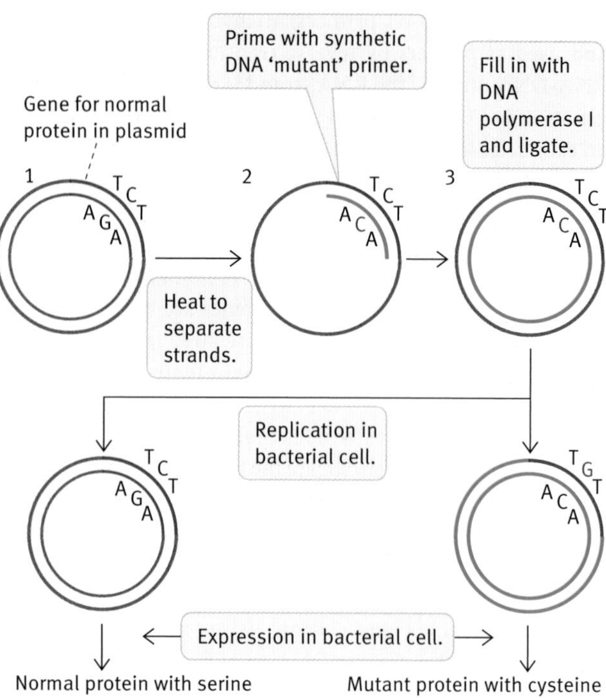

Fig. 28.15 Steps in site-directed mutagenesis to replace a single amino acid of a protein. See text for explanation.

Site-directed mutagenesis

This enables one or more selected amino acid residues in a protein to be altered and makes it possible to investigate protein structure and mechanism of action. For example in studying the mechanism of an enzyme reaction, an amino acid in the active site can be changed to see what effect the change has on activity.

We start with a bacterial plasmid carrying a cloned cDNA encoding the enzyme. Suppose that you wish to change a selected serine residue to a cysteine whose coding triplets in the DNA are, for purposes of illustration, AGA and ACA respectively (the code for these amino acids is degenerate). The two strands of the duplex are separated by heating and a synthetic primer about 20 nucleotides in length containing the mismatched nucleotide (shown in red in Fig. 28.15) at the serine codon site is added. The primer is long enough to hybridize despite the mismatch. The primer is extended by DNA polymerase I to form a double-stranded plasmid and ligation completes the plasmid, which is introduced into a bacterial cell where it replicates. Two versions of plasmids will be produced after replication from the two strands. One will contain a normal cDNA and the other the mutated cDNA. The mutated cDNA can be isolated and expressed to produce the mutated protein. An alternative is to replace a section of DNA with a 'cassette' carrying the desired sequence. A section of the coding sequence in an expression plasmid is excised by two restriction enzyme cuts and replaced by a synthetic DNA piece coding for the desired amino acid sequence.

PCR in forensic science

DNA data are now widely accepted as evidence in criminal trials and paternity determinations. Each human being has a large number of polymorphic loci in their chromosomes consisting of short repeated sequences known as **microsatellites**, or **short tandem repeats (STRs)** (see Box 28.1). The term locus (plural loci) refers to a specific region within a genome; use of the term avoids stating whether the region is actually a gene or not (remembering that most of the genome sequence is not actually genes). A polymorphism is a part of the genome sequence that frequently differs between individuals due to normal human variation. In STRs the polymorphism is in the number of repeats present at each locus, and this variation in length can be readily detected by PCR.

In forensic work the microsatellites with tetra- or pentanucleotide-repeat units (such as GATA, GATA, GATA, GATA, etc.) are usually used. We will use the example of a GATA repeat in the subsequent account. The number of repeats in different chromosomal loci typically varies from 4 to 40 in number giving rise to different alleles at the same locus amongst individuals in the population. In a pair of homologous chromosomes, one derived from the father and one from the mother, the repeat number in a given locus is by chance likely to be different in the two chromosomes (Fig. 28.16). If you select, say, nine microsatellite loci in a sample of DNA and examine the repeat number in both chromosomes of homologous pairs, you will obtain a pattern of (2×9) 18 numbers. *(Note that in Fig. 28.16 we show only three loci being used for*

BOX 28.1 Repetitive DNA sequences

About half the DNA of the genome is made up of repetitive sequences of different types. See Chapter 22, for long interspersed repeated sequences (LINES) and short interspersed repeated sequences (SINES).

Tandemly repeated sequences are the other main category of repetitive DNAs. These consist of sequences of short nucleotide units arranged one after the other. An example is five bases in tandem, TTCCA/TTCCA/TTCCA (with their complementary strands) repeated dozens or thousands of times. Repeats of this style are found particularly in the tightly packed heterochromatin of centromeres, in patches near the centromeres and near the ends of the chromosomes. They constitute perhaps 10% of the total DNA. Nothing is known of their function, if any, though it has been suggested that they may originate from transposons.

Most tandem repeated sequences are highly polymorphic. This gives them some important practical applications in human genetics. The term 'polymorphism' refers to the situation in which, within a population, there are several or many variants of a sequence at a given locus. In other words the sequence at this locus is highly likely to be different between individuals. Within an individual there will be two versions (alleles), one each on the maternal and the paternal chromosomes. A common class of polymorphism is where the number of repeats of a tandem repeated sequence unit at a given chromosomal locus varies between two homologous chromosomes. The number is usually different between the maternal and paternal chromosomes. As an example, at a particular locus there could be five repeated units on one chromosome and 30 on its homologous partner. Geneticists have identified large numbers of such polymorphisms throughout the human genome. These are important as genetic markers; for example to locate a disease-producing gene and as the basis for DNA fingerprinting in forensic science. The longer ones are known as variable number of tandem repeats (VNTRs), but for experimental convenience the shorter tandem repeats known as STR polymorphisms (STR=short tandem repeats) are most used. They are also known as microsatellites.

Genetic markers are very important in locating disease-producing genes and in forensic science. VNTRs have been commonly used, but now SNPs have replaced them, except in the case of DNA fingerprinting where, for special reasons, STRs are preferred.

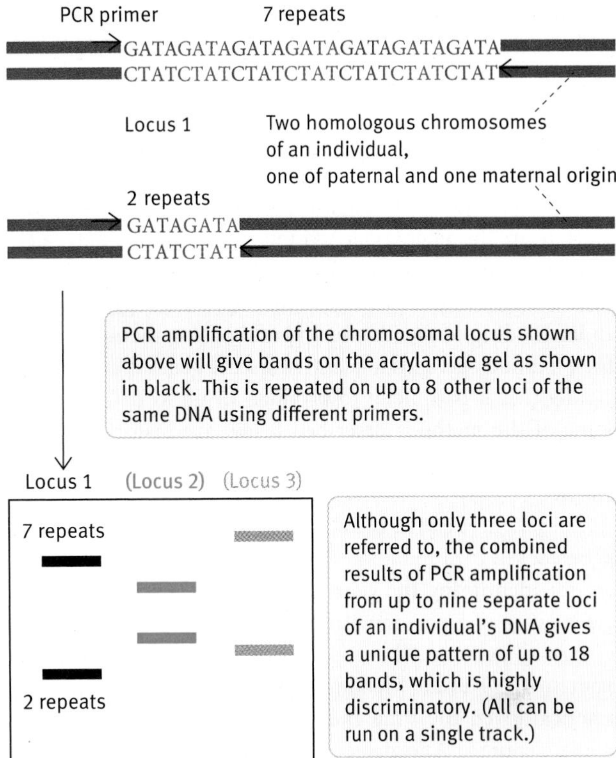

Fig. 28.16 Diagram illustrating the principle of DNA fingerprinting in forensic science. A microsatellite locus on homologous chromosomes of an individual is shown. Polymerase chain reaction (PCR) products give two bands. This process is repeated on up to eight more selected loci on the DNA of the same individual so that altogether a pattern of up to 18 bands will result that is unique to the individual. The process is repeated on the forensic sample of DNA to see if the patterns match. (Note that loci 2 and 3 are not illustrated but the bands that might be produced from them are shown. Also, although we show only three locus results, in practice up to nine may be used to increase the certainty of identification. Note also that the smaller pieces run faster than the larger ones, so that different repeat numbers are separated into discrete bands.)

the purpose of clarity in illustrating the principle.) This pattern is not quite unique to each individual but the chance of another *unrelated* individual (apart from an identical twin) having the same pattern is one in billions. Near certainty exists if a pattern is found in an individual to be different from the forensic sample for then they are beyond reasonable doubt not the same. Therefore DNA profiling is an extremely decisive method for excluding suspects in a case.

How is this put into practice? From genetic chromosome mapping studies and the human genome sequence, the location and flanking sequences of large numbers of microsatellite loci are known. Using a pair of synthetic primers that flank a given locus the repeat section is amplified by PCR. Note that primers that would hybridize to the core GATA sequence itself would be useless – the primers must be complementary only to the *flanking* sequences, as these are locus specific. The length of the product from each locus will depend on the number of GATA repeats and so the products from the two chromosomes will usually give rise to two bands on an electrophoresis gel. From the three different loci, the combined PCR products will give a pattern of up to six bands as shown in Fig. 28.16 and up to 18 from nine loci. (It will not be 6 or 18 if some loci are homozygous.) In current profiling, all of the loci of interest are PCR'd together in one mix, known as a '**multiplex**'. The pattern obtained from the DNA of a suspect is compared with that obtained from a forensic sample. Because of the sensitivity of PCR, minute traces of DNA in bloodstains, semen samples, swabs wiped on the steering wheel of a car, or a single hair follicle, for example, is sufficient. The procedure can also establish paternity since the mother and father will have different alleles at many STR loci, and these are inherited by the rules of classical genetics.

Locating disease-producing genes

In situations where nothing is known about a genetic disease, apart from the phenotypic symptoms, it can be extremely valuable if the gene responsible for the disease can be located and isolated. Studies of the gene may open the way to a better understanding of the disease and of devising therapies for it. The methodology used to identify disease genes is evolving so rapidly that it is difficult to produce an up to date account. Some of the methods described below are historic, but are included for interest or to enhance understanding of current practice.

If a genetic disease is known, by its inheritance pattern, to be caused by a mutation in a single gene, a sensible strategy for identifying the gene involved might be to sequence the genome of a person with the disorder and compare it to a 'normal' genome. This approach was of course impossible prior to the Human Genome Project, and is still less than practical given the time and cost involved in sequencing a whole genome. Affordable individual genome sequencing for medical purposes may, however, be not far off. One of the first individual genomes to be sequenced (the 'reference' human genome being a composite) was that of James Watson, one of the discoverers of DNA structure, whose genome was sequenced in 2007, taking 4 months and costing around US$ 1 million. Such is the progress of sequencing technology that a prize of $10 million has been offered to the first organization to sequence 100 human genomes in 10 days at a cost of less than $10,000 per genome, and the foundation offering it anticipates awarding the prize in 2013. However, even with fast and affordable genome sequencing another challenge to disease gene identification remains – how to distinguish a disease-causing mutation from the natural harmless polymorphisms that are scattered throughout the genome. What is needed to help with both the cost and time difficulties of sequencing and the latter problem is a way to narrow down the location of the disease-causing mutation so that the sequencing effort and subsequent analysis can be concentrated on the correct region. An older strategy known as **positional cloning** makes this possible.

Positional cloning was used, prior to the availability of the genome sequence, to identify the genes for **cystic fibrosis** and **Huntington disease**, for example. The principle of the positional cloning strategy is based on genetic linkage. If two loci are close together on a chromosome they tend to be inherited together – they are tightly genetically linked – while if far apart they tend to be inherited independently as illustrated in Fig. 28.17. This is due to genetic recombination in meiosis in which pairs of homologous chromosomes align, and two of the four chromatids swap sections generating (natural) recombinants; the probability of a recombination event occurring in the DNA section between loci is greater the further they are apart.

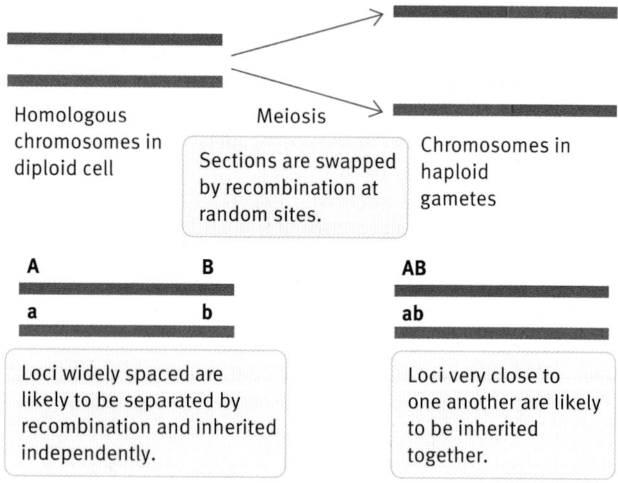

Fig. 28.17 Diagram to illustrate the principle of genetic linkage.

Genetic studies have identified polymorphic loci spaced at intervals across the human genome. These can be used as genetic 'markers' since they can be detected experimentally by DNA methodologies. In order to find an approximate location for a disease gene the inheritance of these polymorphic markers is tracked through individual families in which the disease occurs. If it is found that family members with the disease tend to share a particular allele of the polymorphic marker, that tells us *not* that that allele causes the disease, but that the marker locus is close to the disease gene on the chromosome and tends to be inherited with it. Thus the location of a co-inherited marker points to the location of the disease gene.

Several different types of polymorphic markers have been used, one replacing the other as techniques for their detection became easier. The earliest were **restriction fragment length polymorphisms (RFLPs)**. These are based on the fact that a single nucleotide change can change the restriction sites on a piece of DNA so that a different pattern of restriction fragments are produced on enzyme digestion, and detectable by Southern blot analysis. The method was successful in isolating the genes for cystic fibrosis and Huntington disease. However, restriction analysis takes several days and the amount of work in gene isolation was enormous. Newer markers, particularly the microsatellites already described above, have largely replaced it. These markers can be detected in hours using PCR amplification followed by electrophoresis. The latest type of genetic markers of increasing importance are **single nucleotide polymorphisms (SNPs)**. These are, as the name implies, changes in one base pair so that A=T becomes G≡C for example. SNPs have several advantages. They occur very frequently in the genome so that several millions of SNPs and their positions are known in the human genome. The next important thing is that they can be detected by DNA hybridization even though only a single base change is involved. For this a probe is made to the sequence containing the SNP and the hybridization done at high stringency. DNA microarrays are used so that huge numbers of SNPs can be studied simultaneously and automatically.

Currently in the first phase of analysis about 300 widely spaced SNPs are looked at spread over the entire genome. Finding SNPs that are co-inherited with a disease identifies the chromosome and broad region within it where the disease gene is located. Concentrating on this region by using the same technique of linkage analysis but with more closely spaced polymorphic markers gives a narrower localization. When the region of interest has been narrowed down sufficiently the reference human genome sequence may be helpful in identifying genes in that region that are already known (though perhaps not yet associated with any disease), open reading frames (sequences long enough to code for proteins without any stop codons), and other sequence features such as promoters that indicate the presence of protein coding genes. Other useful database information may be available such as expression data for genes in that region; for instance a gene known to be expressed in muscle (either in humans or in a related species such a mouse) would be a good candidate gene for a disease where muscle is affected. Sequencing the candidate gene in patients and their unaffected relatives can now take place, and finding the same mutation, or different mutations in the same gene, in several families with the disease is good evidence that the correct gene has been identified.

The microarray method of detecting SNPs is assuming great medical importance, not only in identifying genes for so called 'single-gene disorders' as described above, but also for **Genome Wide Association Studies** (Fig 28.18) that aim to find genetic causes for common complex diseases such as type 2 diabetes and coronary heart disease. Susceptibility to these diseases is thought to be influenced by multiple genes as well as environmental factors, each of which contributes to a small extent to the probability of an individual developing the disease. The aim of GWA studies is to analyse large populations and correlate particular polymorphisms with development or not of the disease(s) in question. It is important to realize that, just as with single-gene disorders, inheritance of a particular polymorphic marker by individuals who develop a disease does not mean that that polymorphic locus causes the disease; it simply points to a region of the chromosome where genes associated with susceptibility are found and acts as a trigger for further research. Nevertheless these studies are already enhancing our understanding of the biological causes of common complex disease.

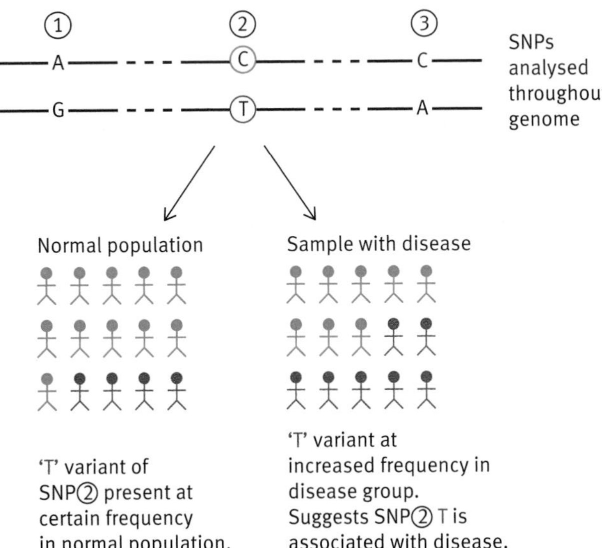

Fig. 28.18 Diagram to illustrate the principle of a Genome Wide Association study used to identify correlations between genome sequence polymorphisms and a common disease.

Knockout mice

Mutants have long played an important role in biochemistry and molecular biology. For decades prokaryotic mutants have provided information on the function of genes and of the proteins they encode. Mutants are comparatively easy to obtain in bacteria because huge numbers of cells can be chemically mutated and the desired mutant selected by screening procedures. Obtaining mammalian mutants is a much more difficult task for obvious reasons but procedures have been developed for obtaining strains of mutant animals that specifically lack a known gene. Such animals are useful medically in that they can act as models of human diseases caused by loss of function of a given gene. Mice are the favourite experimental mammal for such work and the term **knockout mice** is an accepted one. (Mice with an inserted gene are called knockin mice.) Libraries of knockout mice as models of different human diseases are being developed. As an example a mouse model of the human disease familial hypercholesterolaemia has been made by knockout of the gene for the low-density lipoprotein receptor. The production of knockout mouse strains requires that every cell in the body should carry the mutation and this requires that the mutation is introduced into germ line cells. **Embryonic stem (ES) cell** technology achieves this.

The embryonic stem (ES) cell system

There are various groups of **stem cells** in the body known as **adult stem cells**, which provide replacement cells for a limited range of cell types. Thus there are adult stem cells for blood cells (see Fig. 2.7), sperm, skin, bone, and epithelial cells. **Embryonic stem (ES) cells**, on the other hand, are **pluripotent** – they can give rise to all types of cells. After fertilization, a mammalian egg divides to form a solid ball of cells, which develops into a blastocyst. This is a sphere of cells containing a cavity and an inner cell mass (Fig. 28.19). The latter will give rise to all cells of the mature animal, but at this stage individual cells are not committed to becoming any

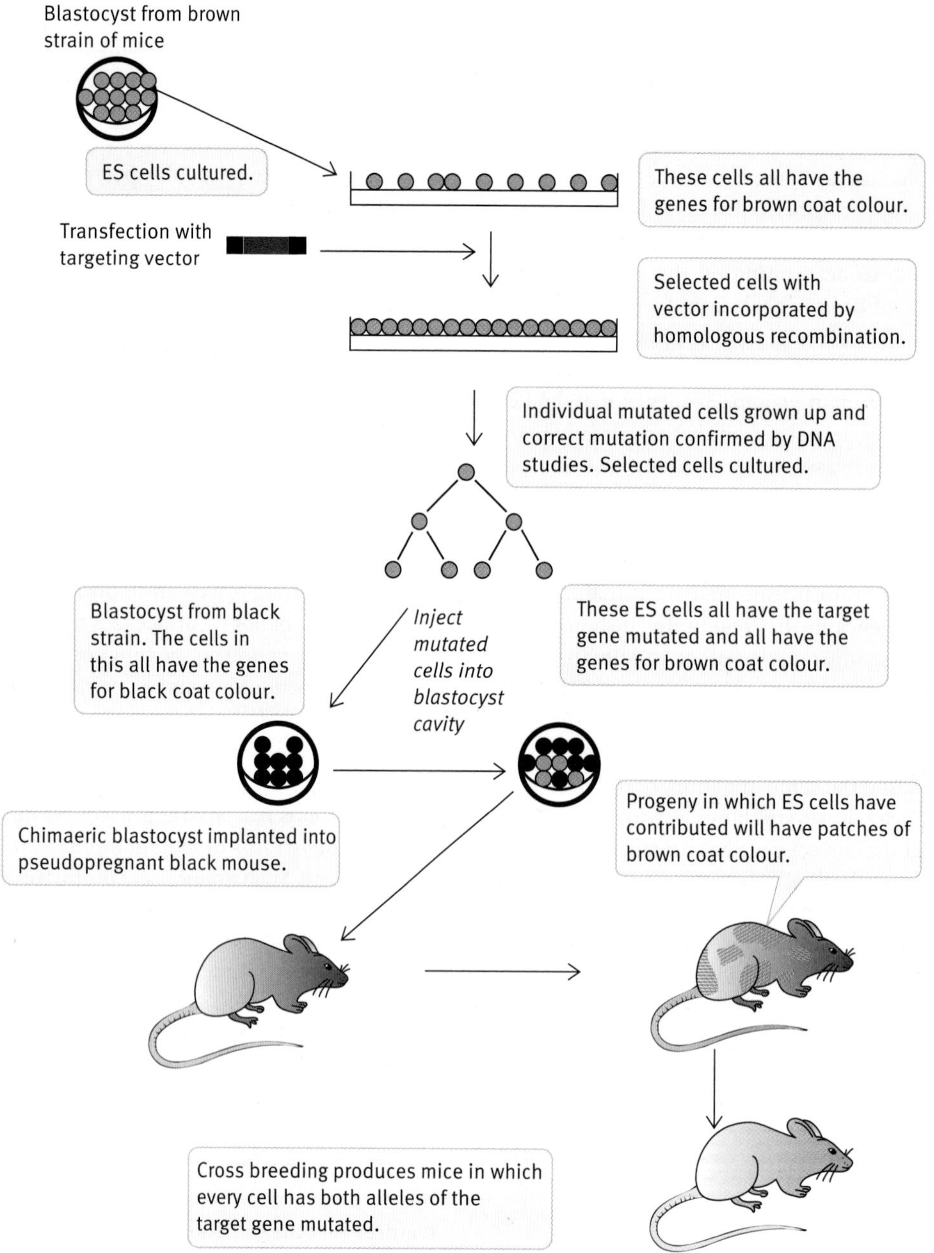
Blastocyst from brown strain of mice
ES cells cultured.
These cells all have the genes for brown coat colour.
Transfection with targeting vector
Selected cells with vector incorporated by homologous recombination.
Individual mutated cells grown up and correct mutation confirmed by DNA studies. Selected cells cultured.
Blastocyst from black strain. The cells in this all have the genes for black coat colour.
Inject mutated cells into blastocyst cavity
These ES cells all have the target gene mutated and all have the genes for brown coat colour.
Chimaeric blastocyst implanted into pseudopregnant black mouse.
Progeny in which ES cells have contributed will have patches of brown coat colour.
Cross breeding produces mice in which every cell has both alleles of the target gene mutated.

Fig. 28.19 Outline of procedure for obtaining knockout mice by gene-targeted mutation. The diagram illustrates how coat colour provides a convenient guide in selecting progeny. ES: embryonic stem cells.

particular type in the adult. The inner cell mass cells can be propagated in long-term culture, and under appropriate conditions they do not differentiate but retain their pluripotency. These cultured pluripotent cells are the ES cells that can be injected back into the cavity of a blastocyst. When the latter is re-introduced into a mouse foster mother it develops into progeny in which the introduced ES cells contribute to all tissues, including germline cells. In any one tissue of progeny mice some of the cells will be derived from the cells of the 'natural' blastocyst and some from the (foreign) injected ES cells, which were obtained from a different blastocyst. This provides a route for gene targeting, which is described in the next section.

Gene targeting

One method of achieving a functional gene knockout is to partially or fully replace the selected gene with a piece of foreign DNA. The basis of the targeting is that at both ends of the section to be incorporated there are regions of homology with the section it is to displace. Homologous recombination at each end of the DNA to be inserted specifically targets the selected gene.

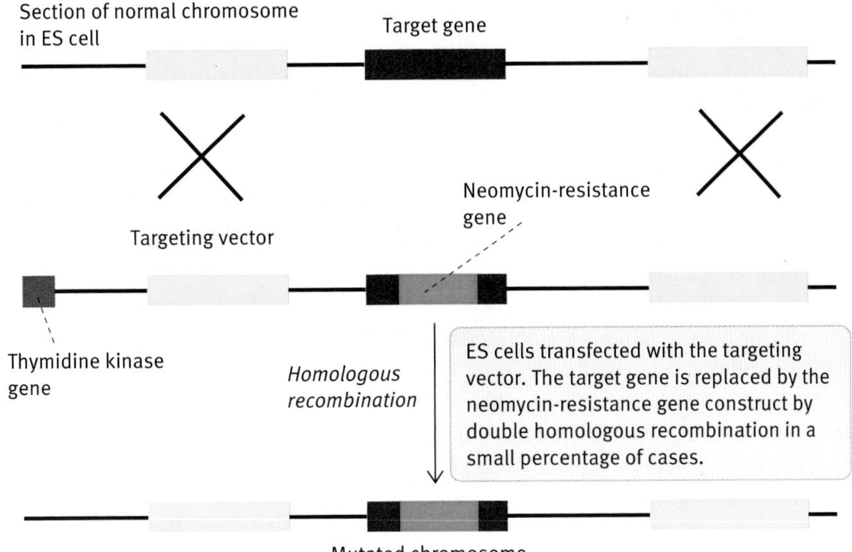

Fig. 28.20 Gene targeting by homologous recombination in mouse embryonic stem (ES) cells. Cultured ES cells are transfected with a targeting vector constructed by recombinant DNA methods. In this, the target gene is replaced by a NRG flanked by sequences homologous to the normal gene. Homologous recombination replaces the target gene with the NRG construct in a small number of cases. Stably transfected cells are cultured and cells that have incorporated the foreign gene by homologous recombination are identified by recombinant DNA methods and cultured. The mutant cells are used to obtain knockout mice as described in the text and illustrated in Fig. 28.19. The yellow rectangles represent normal genes flanking the target gene. The blue rectangle represents the addition of a viral thymidine kinase gene which is placed outside of the region of homology between the vector and targeted chromosome. This is to permit negative selection of cells in which the entire vector is incorporated randomly rather than a section of it by homologous recombination.

The first step is to construct a **targeting vector**. In the example that we will use here, the target gene is inactivated by replacing it with a gene conferring resistance to the antibiotic neomycin, which is toxic to eukaryotic cells. In the vector, the **neomycin-resistance gene (NRG)** has flanking sequences homologous to the targeted gene (Fig. 28.20) so that recombination at each end of the replacement section causes the new piece of DNA with NRG to replace the targeted gene. The procedure for obtaining knockout mice is given in Fig. 28.19. We suggest that you follow the steps in the figure.

ES cells are obtained from a mouse blastocyst and cultured *in vitro.*

- The cells are **transfected** by the targeting vector, usually by **electroporation.** This involves giving an electric shock to the cells, which transitorily forms holes in the cell membrane and allows the vector to enter. Most frequently, in mouse cells, the introduced DNA integrates into the host chromosomes in a random fashion, which is not what is wanted, but a low but useful percentage (0.5–10%) integrate by homologous recombination at the target site.
- Stably transfected cells are now resistant to the neomycin due to expression of the NRG so that these cells can be selected by growing them in the presence of the antibiotic. Cells without the NRG fail to grow. After that, it is necessary to select those cells in which integration has occurred by homologous recombination rather than by random insertion. A rather clever trick facilitates this. In constructing the targeting vector, a gene for **thymidine kinase**, derived from the herpes simplex virus, is inserted at one end of the construct, well outside the region of homology to the target gene. If the DNA is incorporated randomly, the thymidine kinase gene goes in with it and is expressed; if the incorporation is by homologous recombination the kinase gene is not incorporated (because it is outside the homology region). If the kinase gene is expressed, it makes the cell sensitive to **gancyclovir,** since this drug has to be phosphorylated by the viral thymidine kinase to become toxic to the cell. Thus, in the **positive-negative selection,** as it is called, only those cells with the NRG but without the thymidine kinase gene will grow, and these are the desired homologous recombinants.
- The cells that survive this test are separately grown up as clones, and individual clones of cells are analysed to confirm that the replacement gene has been correctly inserted by homologous recombination. This is done using DNA manipulation techniques, previously Southern blotting, or more recently by PCR and sequencing.
- Cells in which the targeted gene has been replaced are grown up and injected into the cavity of a mouse blastocyst derived from an animal whose coat colour is different from that of the mouse from which the injected ES cells originated (see below).

- The hybrid blastocyst is placed in a pseudopregnant foster mother (pseudopregnancy results from mating with a sterile male). The injected, mutated, cells participate in formation of tissues as the embryo develops so that in each tissue of the progeny mouse, some cells will have originated from the original inner cell mass and some from the injected transfected cells. In other words, the mouse will be a chimera.
- As a ready guide to what is occurring genetically, coat colour is used. Let us suppose that the original blastocyst from which the ES cells for manipulation were obtained had a brown coat colour and the recipient blastocyst into which modified cells were injected came from a black mouse strain. The mouse resulting from implantation of the latter into the foster mother will be a patchy black/brown mixture as shown in Fig. 28.19. These are not yet the end of the story for only some of their cells are mutated.
- What is needed are mice in which every cell in the body is a knockout (null) mutant with respect to the target gene, including the germline cells. To obtain these, selective breeding is used to produce progeny some of which are homozygous for the knockout with both copies of the gene mutated in every cell.

Stem cells and potential therapy for human diseases

Work on ES cells and the identification of human adult stem cell populations inspired the hope of curing human degenerative disease and restoring damage caused by certain traumatic injuries by the introduction of stem cells to replace lost or damaged ones. As an example a heart attack causes sections of cardiac muscle to die and, in principle, stem cells could replace these. **Human ES cells (hESCs)** were first isolated and grown in 1998, seventeen years after the first derivation of mouse ES cells. However, there are strong impediments to the development of hESC-based therapy. Firstly, ES cells would be immunologically different from the patient's cells and therefore subject to rejection. Secondly, to obtain embryological stem cells involves the destruction of a human embryo and this raises ethical considerations that have led to legal restrictions on research in many countries. Despite these issues, a limited number of human ES cell lines have been derived. In 2012 a paper in *the Lancet* reported preliminary results of a clinical trial in which retinal cells derived by *in vitro* differentiation of hESCs were transplanted into two patients suffering from macular degeneration, a leading cause of blindness. The cells persisted with no sign of forming tumours or being rejected, although there was limited improvement in the patients' vision.

It is likely that the transplanted cells in this first human clinical trial were not rejected by the eye because this organ is relatively immunoprivileged (protected from the immune response). For stem cell therapy of other organs the ideal situation would be if a patient's own differentiated adult somatic cells could be reprogrammed from the differentiated state back to the pluripotent state of ES cells. This could also overcome ethical objections to stem cell therapy. It was once thought that differentiation of cells in which pluripotency was lost was irreversible but the famous cloning of Dolly the sheep showed that the differentiated nucleus could be reprogrammed to the stem cell state by injecting it into an egg, which had had its own nucleus removed. This is called **somatic cell-nuclear transfer (SCNT)**. However, the proposal that donor human embryos could be used as recipients for SCNT to produce autologous hESCs cells (that is hESCs immunologically the same as the patient's cells), still raises both practical and ethical objections.

A recent exciting advance has been the production of **induced pluripotent stem cells (iPSCs)** from adult differentiated cells, first from mouse and then from humans. iPSCs are generated by introducing DNA expressing four key transcription factors into the cells using retroviral vectors and then selecting for pluripotency. Mouse iPSC lines have been obtained, which resemble ES cells in having the ability to generate all cell types when injected into blastocysts of early embryos, which are then allowed to develop in pseudopregnant mice. Human iPSC lines have been derived from cells from patients with a range of diseases including Parkinson Disease and β-thalassaemia. In some cases these cells have proved promising as *in vitro* models of the disease that can be used to screen new drugs. iPSCs have also been tested as therapies in mouse and rat models of human disease. However, much remains to be done before the technique is applicable to human therapy.

Gene therapy

The development of sophisticated DNA manipulation technology, as well as the generation of mice in which genes can be knocked out (or knocked in) in a controlled fashion, raises hopes of treating human disease by **gene therapy**, which in its simplest form involves correction of a gene deficiency by inserting a normal copy of the gene into the patient's cells.

Initially, gene therapy was considered primarily for the treatment of diseases in which there is a known single-gene deficiency, such as β-thalassaemia, cystic fibrosis, or Duchenne muscular dystrophy. The first gene-therapy clinical trial was performed in 1990. This was on patients with **severe combined immunodeficiency disease (SCID)** in which the gene for **adenosine deaminase** is deficient in lymphocytes. This leads to a build up of adenosine, the metabolic repercussions of which are to cause a toxic build up of dATP and inhibition of DNA synthesis in the lymphocytes. This impairs the immune response. Bone marrow cells of patients were treated *in vitro* with a retroviral vector carrying a functional adenosine deaminase gene and the cells returned to the patient. This gave encouraging results with a number of children.

In 2001, a French group used gene therapy to cure children with a very severe and fatal immune defect called **X-linked, severe combined immunodeficiency** (then treatable only by isolation in a 'bubble' to prevent infection). This disease, found almost exclusively in boys, is different from that caused by adenosine deaminase deficiency. It is caused by a mutation in a lymphocyte protein that is common to a number of interleukin receptors. These are necessary for the immune cells to signal each other to raise appropriate immune responses. The researchers used a retroviral vector carrying a corrective functional receptor protein gene. This was inserted *in vitro* into the patients' bone marrow haematopoietic stem cells. They were then transplanted back into the patients. Remarkably, 10 out of 11 patients benefited significantly from the therapy with most appearing to be cured. Unfortunately, within 3 years of the treatment, two of the patients had developed leukaemia, which appears to have been due to insertional mutagenesis. That is, the insertion of the retroviral gene vector into the patients genome had activated a protooncogene (a cancer causing gene, see Chapter 31). It appeared that rather than inserting at a random position, the vector had a high chance of inserting next to this protooncogene because it is in a genomic region that is transcriptionally active, and hence in an 'open' accessible chromatin conformation, in haematopoietic stem cells. Because X-linked SCID is such a severe condition, it seemed worth persisting with the development of gene therapy despite this setback, and work is now focused on modifying the retroviral vector to avoid it causing leukaemia.

Most gene therapy trials to date have been disappointing not because of harmful side effects, but through lack of efficacy. Despite success in a few cases, gene therapy has unfortunately proved more difficult than was anticipated. Nevertheless, hundreds of clinical trials are in progress, many of them aimed at treating cancer, where the unusual genetic make up of tumour cells provides hope that they can be targeted without harming the patient. Another possible use of gene therapy is in correcting genetic defects in patient-derived iPSCs, so that they can be replaced therapeutically into the patient, providing an immunological match. This type of therapy is, however, many steps away from realization.

Transgenic organisms

Transgenic or genetically modified organisms (GMOs) are produced by the introduction of foreign genes into the organism's cells or genome. Although it is not very efficient, it has been found that DNA injected into an animal cell's nucleus may be incorporated into its genome and expressed. This technique has been successfully used to alter the genome of animal somatic cells and then combined with somatic cell-nuclear transfer to produce transgenic embryos, which are transplanted into a surrogate mother. This methodology has given us transgenic farm animals including sheep and goats that produce human proteins, for therapeutic use, in their milk.

Genetic modification has been more widely used in agricultural plants. Foreign genes can be inserted successfully into plant chromosomes by using, as a cloning vector, a naturally occurring (but suitably altered) plasmid, called the Ti (or tumour-inducing) plasmid, which is normally found in the pathogenic soil bacterium *Agrobacterium tumefaciens.* Alternatively, DNA molecules may be literally shot into plant cells, with a gun-type instrument that fires a cloud of fine shot loaded with DNA. In this way, crop plants are being engineered with specified phenotypic characteristics – for example, to be resistant to herbicides. The purpose is to control weeds by blanket spraying with the herbicide – only the resistant crop plant survives. Other plants have been genetically modified to alter their nutritional properties, for example 'golden rice', which is designed to produce β-carotene and hence combat vitamin A deficiency. The introduction of GM crops has met fierce opposition from some quarters because of environmental concerns and questions about their ownership by for-profit companies, but they are gradually gaining acceptance.

DNA databases and genomics

DNA databases are the essential partners to the protein databases described in Chapter 5. Genomics refers to the study of large numbers of genes. It is the partner to **proteomics** in which large numbers of proteins are studied together. The computational use of the protein and DNA databases is collectively known as **bioinformatics**. The databases are now of great importance in biochemistry and molecular biology, medicine, and virtually all biological sciences.

When a gene or other section of DNA has been isolated and sequenced, the information about it is recorded in one of the international databases in the public domain. Free usage of software is available to interrogate the bases and analyse the data in ways appropriate to the type of questions being asked. When an unidentified gene or other DNA sequence is isolated and fully or partially sequenced, a search of the databases for matching sequences will reveal whether information on that piece of DNA, or a closely related one, already exists. The complete sequence may then be available and its location in the chromosomes relative to other genes identifiable. The search might also, for example, reveal that it is part of a family cluster of related genes; its function may be known or perhaps an association of it with a disease may have been determined from other work. It is also possible to analyse a DNA sequence for the presence of open reading frames, potential splicing patterns or transcription factor binding sites, to give but a few examples.

The databases resulting from the Human Genome Project and many other major research projects such as ENCODE are now revealing information of importance to medicine, biotechnology, or basic science. This can speed up a research project to an almost unimaginable degree. The uses are almost limitless and 'mining the databases' is an important branch of research in many areas of biological science.

Use of the DNA databases is mainly a research activity for it requires computational skills and considerable knowledge of biochemistry and molecular biology. Specialist courses on bioinformatics are given in many departments. Further information is given in the bioinformatics overview section in Chapter 5. Some relevant website addresses are given in Box 5.1.

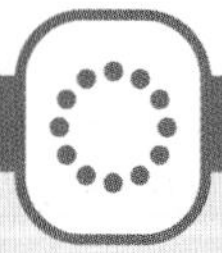

Summary

The technology of DNA manipulation has become the most powerful approach to many biological and medical problems. It permits the isolation of genes, determination of their nucleotide sequences, detection of abnormal genes, and production of human and other proteins in unlimited amounts in hosts such as yeast and bacteria. It can also produce proteins with specific amino acid substitutions.

The techniques are many and varied but a number of basic principles apply.

- DNA can be cut with precision at known sequences using a battery of restriction enzymes.
- DNA sequences can be identified by hybridization with probes obtained by isolation or synthesis.
- Recombinant DNA molecules in which different pieces of DNA are joined together can be produced in a number of ways.
- Recombinant molecules or individual sections of DNA can be amplified and purified by cloning techniques.
- The polymerase chain reaction (PCR) can amplify a selected stretch of DNA millions of fold in an hour or so. Pieces to be amplified can be selected using sequence information from genome projects. RNA can be copied to cDNA and similarly amplified by RT-PCR.
- DNA has been sequenced using the dideoxy method, which is now fully automated. Next generation sequencing has greatly increased the speed and reduced the cost of generating massive amounts of sequence data.

Polymorphic 'markers' are known that are spaced throughout the human genome. Polymorphic microsatellite loci can be selectively amplified by PCR. This is the basis of forensic DNA fingerprinting. Polymorphic markers, including microsatellites and SNPs can also be used to locate disease-producing genes. Genome Wide Association studies of the linkage between SNPs and common complex diseases form an expanding field. Microarray technology permits rapid analysis of SNP variation and also of gene expression on a large scale.

Genetic engineering combined with stem cell technology has been harnessed to generate genetic knockout mice, which can provide animal models of human disease. Human gene and stem cell therapies are still in the developmental stage. However, transgenic animals that produce therapeutic human proteins in their milk and genetically modified crop plants are in production.

DNA databases are essential parts of DNA technology. These, with the help of appropriate software for retrieving and analysing the vast amount of information, are a vital part of bioinformatics, a discipline that is assuming great importance. Mining the databases is now an important part of molecular biology.

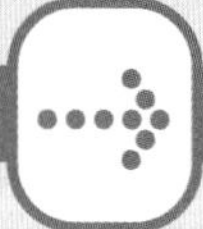

Further reading

To access the further reading, please scan the QR code image or go to http://global.oup.com/uk/orc/biosciences/molbiol/snape_biochemistry/student/reading/ch28/

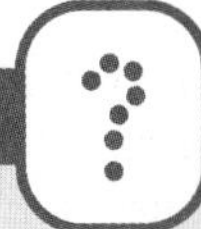

Problems

1 How does a restriction enzyme differ from pancreatic DNase?

2 *Eco*RI cuts at a hexamer base sequence; such a sequence must occur many times in *E. coli* DNA. Why does the enzyme not destroy the cell's own DNA?

3 What is meant by the term 'overhanging ends', as applied to DNA molecules?

4 What is a genomic clone? What is a cDNA clone? How do they differ? Refer to eukaryotes in your answer.

5 What is a dideoxynucleoside triphosphate? What is the precise function of these in the Sanger technique of DNA sequencing?

6 Explain in a few sentences the importance and principle of the polymerase chain reaction, specifying the requirements for it to be used.

7 What is a bacterial plasmid expression vector?

8 What is restriction analysis?

9 What is a stem cell and why are they of great current interest?

10 Describe in conceptual outline how mammalian genes can be targeted to produce specific mutants, generally known as knockout mice.

11 In the polymerase chain reaction (PCR):
(a) Why are two primers added?
(b) Why are the primers added in large excess?
(c) Which enzyme is used and what are its special characteristics for use in the PCR?

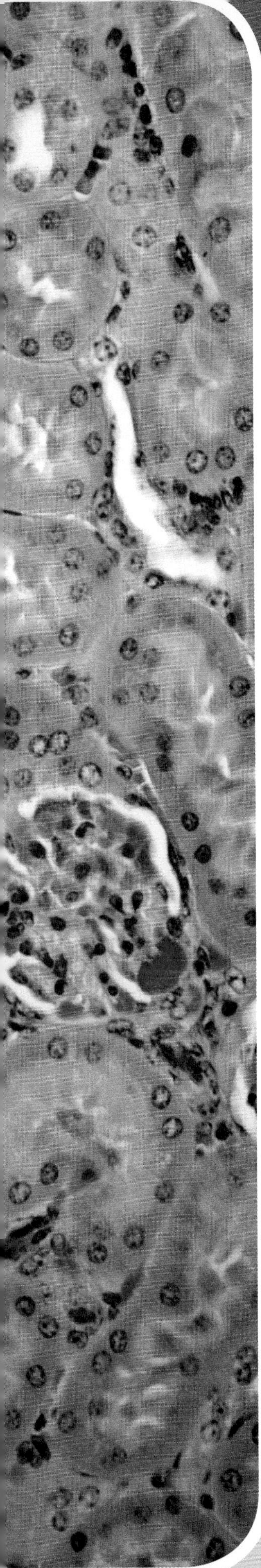

Part 5

Cells and tissues

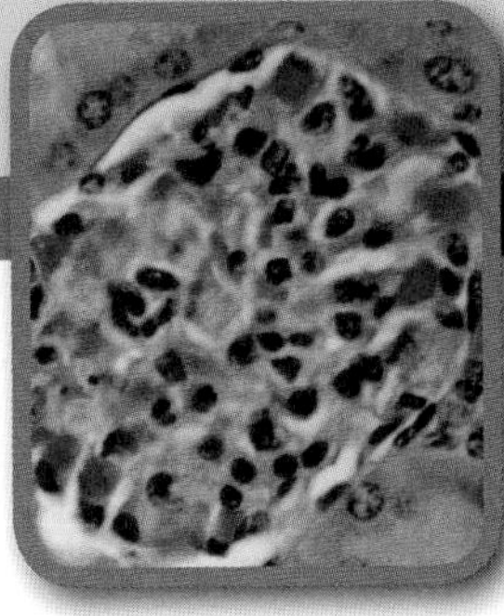

Chapter 29

Cell signalling

Cell signalling brings together several major biochemical concepts that have been dealt with separately elsewhere in the book. Important in this category are conformational changes in proteins, protein kinases, second messengers (see Chapter 20), gene control in eukaryotes, transcription factors (see Chapter 24), and the cell cycle (see Chapter 30). In this chapter there are brief recapitulations of essentials where these seem useful and there are references to fuller treatments of individual topics.

Overview

There is hardly a topic in the molecular aspects of life more important than cell signalling in an organism as complex as a mammal. A human being is a society of about 10^{13} individual cells, which have to be coordinated so that their activities correspond to those that are needed to maintain function of the body as a whole. This is achieved by chemical signalling between cells. There is no room for independent cellular enterprise. Cancer is the end result of a cell replicating, irrespective of whether or not it should do so. It is a disease involving aberrant signalling systems (see Chapter 31). The number of cells in a human body far exceeds the total human population so the organizational task is large and in recent years it has emerged that the controls operating on mammalian cells are far more complex than was ever imagined. There is no parallel to this in prokaryotes.

The most fundamental controls are those on cell replication, differentiation, and apoptosis, or programmed cell death (see Chapter 30), which is a normal part of life. Metabolic events must also be regulated in the interests of survival (see Chapter 20). Cells in the body are bombarded with instructions on all manner of things by cell signalling. As well as coordination of cell growth, replication, differentiation, and death, there are ever-changing physiological requirements such as responding to the state of nutrition, inflammatory signals, and adapting to a changing environment. Cell signalling by hormones and other factors is the mechanism by which cells communicate with each other about these metabolic needs.

There are, in the broadest sense, three avenues of signal delivery.

Neurotransmitter molecules are released by nerve endings signalling the next neuron, or activating muscle contraction. They are usually ligands controlling gated ion channels present on all cells, but of special importance in nerve-impulse conduction. We have described the mechanism of these in Chapters 7 and 8.

The second delivery system is *via* **hormones** (see Chapter 20). These are produced and secreted into the bloodstream by cells specialized for their production and aggregated into endocrine glands. The insulin-secreting cells of the pancreas are a typical example. Hormones are released into the circulation and reach distant **target cells** – those able to receive the signal by virtue of having receptors for the hormone.

Neurological control by transmitters and hormonal control issued by 'central bodies' (the brain and the endocrine glands) are easy concepts to grasp. The control is hierarchical – similar to how human societies are organized.

The third broad class of signalling control involves mutual control by cells. The signalling molecules are known as **cytokines** or **growth factors**; there is no absolute distinction between the two terms and what they are called often depends on how they were discovered. Factors involved with blood cells are usually called cytokines and growth factors were often detected by their effects on cell growth and replication but they may have other effects on different cells or the same cells at different times. Here are a few examples of the function of cytokines and growth factors.

- Normal (noncancerous) cells will not grow on nutritive plates unless supplied with growth factors (like those in fetal calf plasma).
- When stem cells differentiate into specific somatic cells, cytokines/growth factors direct the process. Production of various classes of blood cells from adult stem cells is a

typical example of a battery of different cytokines controlling the process.

- When a wound heals, growth factors direct cell replication.
- In the immune system, cytokine signals are needed to trigger B cells to differentiate into antibody-producing cells.
- Interferons produced by virus-infected cells signal neighbouring cells to take protective actions.

From these few examples you can see that they are of fundamental importance and are intensely researched, especially as they are of major medical interest.

Identification of the many growth factors and cytokines and elucidation of their signalling pathways has revealed much about the nature of cellular controls and of diseases such as cancer. It is not understood, however, how local collections of cells can between themselves coordinate the activities of all the cells in, for example, a tissue. Tissues and organs, such as the liver, essentially remain the same size in adults because cell division and cell death balance each other. Cell signalling somehow achieves this. For example, if as much as two-thirds of the liver of a rat is removed, cell division is increased until – in about 2 weeks – normal liver size is again achieved. It then reverts to the normal slow rate needed to compensate for cell death. How this is achieved is not known, nor is it understood how each organ reaches its appropriate size, and then remains constant.

All eukaryotic signalling molecules combine with protein receptors. Most of these are on the outside of cells, though a handful of lipid-soluble signals enter the cell before they combine with **intracellular** receptors (Fig. 29.1 (a)). A cell without a receptor for a given signal cannot respond to it. The chemistry of the signalling molecules has no known intrinsic significance because they do not enter into any reactions. All they do is to bind to their specific receptors. It is the relationship between the receptor and the signal, which is important. Nothing happens to the signalling molecule apart from its ultimate degradation. When a signalling molecule binds to a **transmembrane** receptor, conformational changes occur in the receptor that result in its cytosolic domain changing its shape. The signal is thus conveyed (transduced) across the membrane by the transmembrane receptor protein.

In the case of gene-activating signals, the change in the receptor domain results in a chain of events, which triggers the relay of the message to the nucleus. In the case of hormones affecting metabolism, the action may be primarily in the cytosol (Fig. 29.1(b)). In both cases, the whole chain of events from the membrane inwards are called **signal transduction pathways** . There are various types of these pathways. Some, like the Ras pathway (see below), involve a chain of proteins; others produce a **second messenger**, which passes on the message to other proteins. The second-messenger term (see Chapter 20) is used for a transduction pathway component, which is a

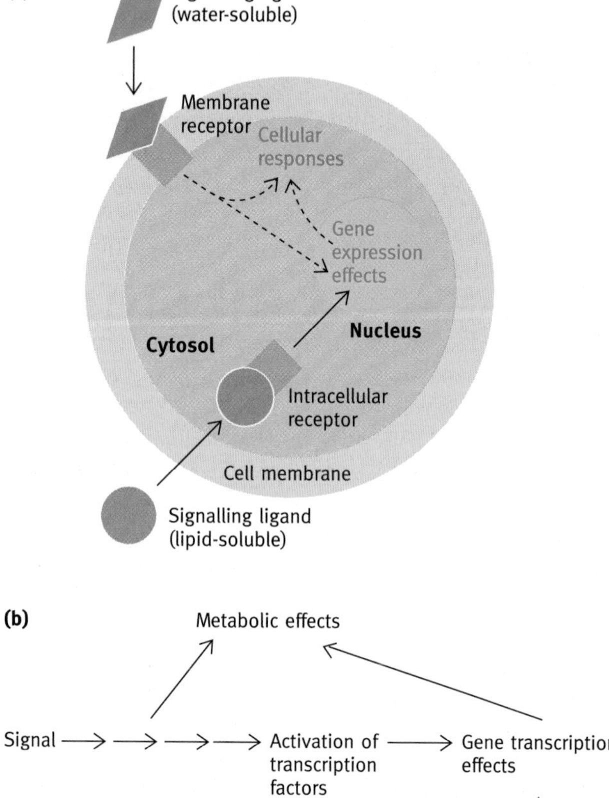

Fig. 29.1 (a) Outline of receptor-mediated signalling. Water-soluble signalling molecules cannot pass through the lipid bilayer. They bind to external domains of receptors. The lipid-soluble signalling molecules such as steroids and thyroxine enter the cell directly and bind to intracellular receptors. Note that some intracellular receptors reside in the nucleus as described later; cytosolic steroid receptors move into the nucleus upon ligand binding. **(b)** A more detailed account of cellular responses to signals.

small molecule rather than a peptide, but what is important is its shape and binding properties. It fits to its target component and relays the signal.

With so many signals there have to be different receptors and signal transduction pathways, just as a landline telephone system depends on separate lines to take signals to the correct house. Variation in the structures of the protein components involved gives a multiplicity of receptors and pathways.

It adds up to a mind-boggling complexity but fortunately for our understanding there is a relatively small number of reasonably direct pathways used for individual receptors and transduction pathways. For example, the different Ras pathways have variations in individual components of the pathway and in this way they can route signals from different receptors to their correct molecular destinations.

Organization of this chapter

In the sections that follow, we will initially present the types of receptor and their general characteristics. First,there is the class of **intracellular** receptors, the prototype of which is the glucocorticoid-specific one. Then, of the **extracellular** receptors, most fall into two main categories – these are **tyrosine kinase receptors** and **G-protein receptors**. A number of examples of the rest of the transduction pathways are then given.

The prototype of the tyrosine kinase pathways is the **Ras** pathway, an ancient highly conserved signalling pathway present in all animals from fruit flies to humans. Its relative, the **JAK/STAT** pathway, that follows, has a different type of tyrosine kinase receptor and pathway, and is used by many cytokines. How insulin acts (involving another tyrosine kinase receptor) is also a topic of obvious medical interest.

Turning to the second major class, the G-protein-associated signal transduction pathway, of which adrenaline (epinephrine) signalling is the prototype. This illustrates a different type of pathway activation, which does not involve tyrosine phosphorylation of receptors. G protein associated receptors form a very large group but they all share a common basic mechanism. 'G-protein receptors' refers to the fact that in all cases the signal is transduced by a GTP-binding protein. The phosphatidylinositide cascade pathway is a more complicated example of a G-protein pathway. G-protein receptors are of great medical significance. Half of all pharmaceutical drugs are designed to target them.

We are also going to look at some different receptor mechanisms, for example, how light signals are handled in the visual process and how the simplest signal of all, nitric oxide (NO), acts, a powerful vasodilator involved in many physiological processes.

What are the signalling molecules?

As stated in the overview, the signalling molecules do not enter into chemical reactions – they are simply molecules of the right shape and properties for binding with great specificity by noncovalent bonds to their receptors. Briefly, to illustrate their variety, in chemical terms they include **proteins**, **large peptides**, **small peptides**, **steroids**, **eicosanoids** (see Chapter 17), **thyroid hormone**, **adrenaline (epinephrine)**, **nitric oxide**, and **derivatives of vitamins A and D**. Examples of the first five groups respectively are **insulin**, **glucagon**, **vasopressin**, **sex hormones**, and **prostaglandins**.

A biological classification is as follows (nitric oxide being in a class of its own):

- neurotransmitters
- hormones
- growth factors and cytokines
- vitamin A and D derivatives.

This classification is important in terms of nomenclature and physiology, but they are all signalling molecules that bind to cellular receptors and elicit responses. They have common basic functions and we will now describe the basis of their classification.

Neurotransmitters

A variety of neurotransmitters are involved in nerve function but we will mention only a few relevant to our immediate topic. The sympathetic (involuntary) nervous system, which innervates fat cell depots for example, secretes adrenaline (epinephrine) and noradrenaline (norepinephrine) on receipt of a nerve impulse. Thus, a fat cell may be regulated by adrenaline from the adrenal glands *via* the blood (see Chapter 20) or from nerve endings. The difference is that the latter delivery route is faster and the signal is released precisely at the target cell site. The motor neurons innervating voluntary striated muscle trigger contraction by the release of acetylcholine from the nerve endings. All of them work by binding to external cell receptors, which usually control the opening of ion channels.

The transmitters are rapidly destroyed and the neurons become ready for the next impulse.

Hormones

These are the 'classic' signalling molecules, most of which have been known for a long time. They are important both in metabolic control and in control of the expression of specific genes. Hormones are produced by endocrine glands, which secrete them directly into the bloodstream. They reach their target cells by circulating round the whole body and so reach cells distant from the secreting gland. Only the **target cells** have specific **receptors** capable of picking up the signal. A large number of hormones are known and their biological effects well documented. Table 29.1 lists some of the principal hormones, for reference purposes.

Much of the endocrine system is under a hierarchical control system with the **hypothalamus** being at the top of the chain of command. The hypothalamus is a small part of the brain that produces hormones that stimulate release of anterior pituitary hormones, known as tropic hormones because they cause target endocrine glands (thyroid, adrenal cortex, and the gonads) to release *their* hormones, the 'final' ones in the chain, which elicit responses from cells in general. Feedback loops in which end products (the final cell-targeted hormones) inhibit the first step (release of hypothalamic hormones) maintain appropriate levels of circulating hormones. There are exceptions to this control system, a notable one being release of the pancreatic hormones, insulin and glucagon, which is more directly controlled by the level of blood glucose.

Secreting organ	Hormone	Target tissue	Function
Hypothalamus	Hormone-releasing factors	Anterior pituitary	Stimulation of circulating hormone secretion
	Somatostatin (also from pancreas)	Anterior pituitary	Inhibits release of somatotrophin
Anterior pituitary	Thyroid-stimulating hormone (TSH)	Thyroid	Stimulates T_4 and T_3 release
	Adrenocorticotrophic hormone (ACTH)	Adrenal cortex	Stimulates release of adrenocorticosteroids
	Gonadotrophins (luteinizing hormone [LH]) and follicle-stimulating hormone (FSH)	Testis and ovary	Stimulates release of sex hormones and cell development
	Somatotrophin (growth hormone)	Liver	Stimulates synthesis of insulin-like growth factors, IGFI and IGFII
	Prolactin	Mammary gland	Required for lactation
Posterior pituitary	Antidiuretic hormone (ADH) or vasopressin	Kidney tubule	Promotes water resorption
	Oxytocin	Smooth muscle	Stimulates uterine contractions
Thyroid	Thyroxine (T_4), triiodothyronine (T_3)	Liver, muscles	Metabolic stimulation
Parathyroid	Parathyroid hormone	Bone, kidney	Maintains blood Ca^{2+} level, stimulates bone resorption
	Calcitonin (also from thyroid)	Bone, kidney	Inhibits resorption of Ca^{2+} from bone
Adrenal cortex	Glucocorticosteroids (cortisol)	Many tissues	Promotes gluconeogenesis
	Mineralocorticosteroids (aldosterone)	Kidney, blood	Maintains salt and water balance
Adrenal medulla	Catecholamines (adrenaline, noradrenaline)	Liver, muscles, heart	Mobilize fatty acids and glucose into bloodstream
Gonads	Sex hormones (testosterone from testes, oestradiol, progesterone from ovaries)	Reproductive organs, secondary sex organs	Promote maturation and function in sex organs
Liver	Somatomedins (insulin-like growth factors: IGFI, IGFII)	Liver, bone	Stimulate growth
Pancreas	Insulin	Liver, muscles, adipose	Stimulates glycogenesis, lipogenesis, protein synthesis
	Glucagon	Liver, muscles, adipose	Stimulates glycogen breakdown, lipolysis

Table 29.1 The principal hormones

Cytokines and growth factors

It has been suggested that cytokines and growth factors should be regarded as developmental regulatory factors. **Growth factors**, for example, are more understandable if regarded as signals that, depending on the cell type and the circumstances, may induce cell division or inhibit it. They may control differentiation or instruct the cell to undergo programmed cell death (**apoptosis**). The cytokines and growth factors control such fundamental processes because they control specific gene transcription. Many such factors exist and this is an intensively researched area and of great medical interest. There are no 'authorized' definitions that distinguish between cytokines and growth factors, and the terms are sometimes used interchangeably. The many factors controlling blood cell development, including those involved in immune responses (Chapter 33) together with the interferons (see later in this chapter), are referred to as **cytokines**. Cytokines and growth factors are regulatory proteins or peptides secreted by many cell types that, unlike the cells producing hormones, are not specialized for producing signals, but are typical of whichever tissue they belong to, such as hepatocytes and lymphocytes. Most cytokines/growth factors are **paracrine** in their action – they diffuse short distances to act only on local cells and are rapidly destroyed – while some are **autocrine** in action (Fig. 29.2). These, such as **interleukin 2**, which stimulates T cell proliferation (Chapter 33), act on the same type of cells that secrete them. The localization of their action to closely neighbouring cells may be a biologically important function of their action. The converse can also be the case; the cytokine **erythropoietin** produced by the kidney medulla controls the proliferation of erythrocytes. It is released when the oxygen tension in the blood is low and is different in that it is released

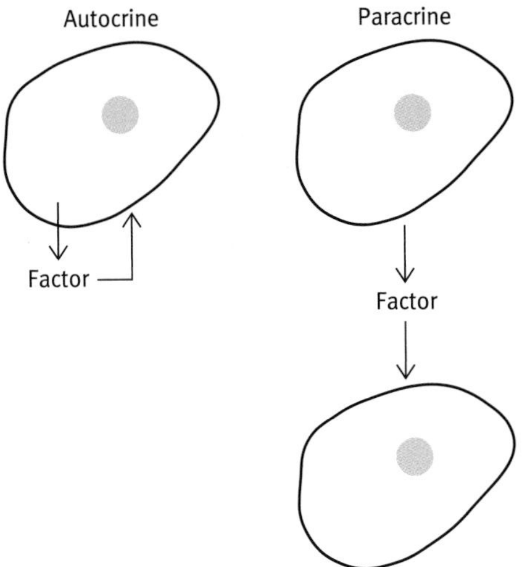

Fig. 29.2 Autocrine signals affect the cell producing them; paracrine signals diffuse only a short distance to affect nearby cells.

into the general circulation of the blood so it is also classified as a hormone.

The names given to cytokines and growth factors usually depend, as stated, on the way they were discovered. One of the earliest known growth factors was **platelet-derived growth factor (PDGF)**. Blood platelets lyse at the sites of damage in blood vessels initiating clotting and the released PDGF stimulates cell division and repair. However, other cells also produce the factor so its name does not fully define its role. **Epidermal growth factor (EGF)** stimulates the growth of skin cells. **Interleukins** are produced by leucocytes (white blood cells) and affect other leucocytes. **Colony-stimulating factors (CSFs)** are so named because they were discovered in experiments in which they stimulated the growth of colonies of white cells on culture plates. Some are used clinically to control white cell production. For example, people with leukaemia given bone marrow transplants first have their bone marrow cells ablated (number reduced) by radiation and chemotherapy. After the transplant, there is a period during which there are insufficient neutrophils, with the risk of infections, since these white cells combat bacteria. Treatment with **granulocyte-colony-stimulating factor (G-CSF)** stimulates neutrophil (granulocyte) production thus reducing the risk period.

Growth factors/cytokines and the cell cycle

The eukaryotic cell cycle is described in Chapter 30 but, briefly, for present purposes, eukaryotic cell division involves a progression of phases from one division to the next. DNA synthesis is confined to a definite period of time called the S (for synthesis) phase (see Fig. 23.4), typically lasting about 8 hours out of the total cycle of 24 hours. Before and after the S phase are the G_1 and G_2 phases respectively (G denoting gap in DNA synthesis). The crucial checkpoint known as the **restriction point** (in mammals) controls the transition from the G_1 to the S phase. Once past this checkpoint the cell is committed to duplicate its chromosome and proceed to mitosis. If a mitogenic (replication) signal is not received in the G_1 phase, the cell is shunted into a quiescent (G_0) phase until a signal arrives. This is the state of most cells in a mature tissue. The mitogenic signal is delivered by a growth factor or cytokine, which thus assumes critical importance in the control of cell division.

Vitamin D and retinoic acid

We usually think of vitamins as being enzyme cofactors, or components of these systems, but vitamins A and D are different (see Chapter 9). Retinoic acid, derived from vitamin A, is important as a signalling molecule in embryonic development and normal cell growth, and also in vision. Calcitriol (1,25-dihydroxyvitamin D, the active form of vitamin D) has an important role in control of genes involved in calcium absorption from the intestine.

Having considered many of the 'signals' and their target processes, we will now move on to the question of how cells detect signals and transduce them to the interior of the cell.

Responses mediated by intracellular receptors

As already mentioned, a handful of hormones are lipid soluble and easily traverse the cell membrane. These include steroid hormones, thyroid hormones, vitamin D, and retinoic acid. They meet their receptors *inside the cell* in contrast to most hormones and other signals, which are hydrophilic and do not enter the cell but meet their receptors exposed on the outside surface. The lipid-soluble hormones regulate expression of specific genes in target cells at the level of initiation of gene transcription. Examples of the steroids include **glucocorticoids**, **oestrogen**, and **progesterone** (see list of hormones in Table 29.1) The structures of a number of the lipid-soluble signalling molecules are shown in Fig. 29.3 for reference. A superfamily of related steroid/thyroxine receptors exists, suggesting an ancient, common ancestor protein.

We will use the glucocorticoid receptor as an example. Glucocorticoid hormones have diverse effects on metabolism (see Box 29.1), including increasing gluconeogenesis (see Chapter 16). The receptor exists in the cytosol, attached to a complex of heat shock proteins (Hsps, see Chapter 25), which mask its nuclear localization signal (NLS), a peptide sequence on the protein (see Chapter 28). When a glucocorticoid hormone binds to a specific site on the receptor, the receptor undergoes a conformational change and the Hsps dissociate from it, revealing the NLS so that the receptor is transported into the nucleus. There, it combines as a dimer at the specific

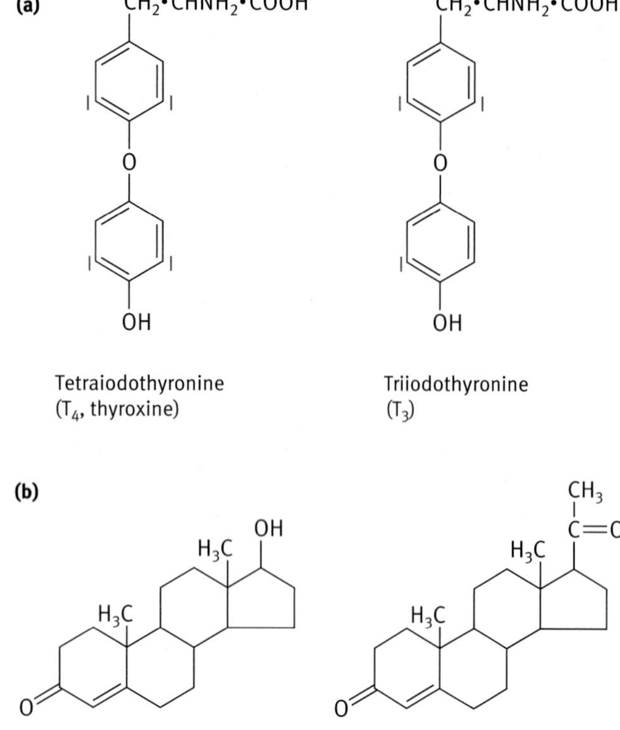

Fig. 29.3 **(a)** The structures of the thyroid hormones; **(b)** the structures of two steroid hormones.

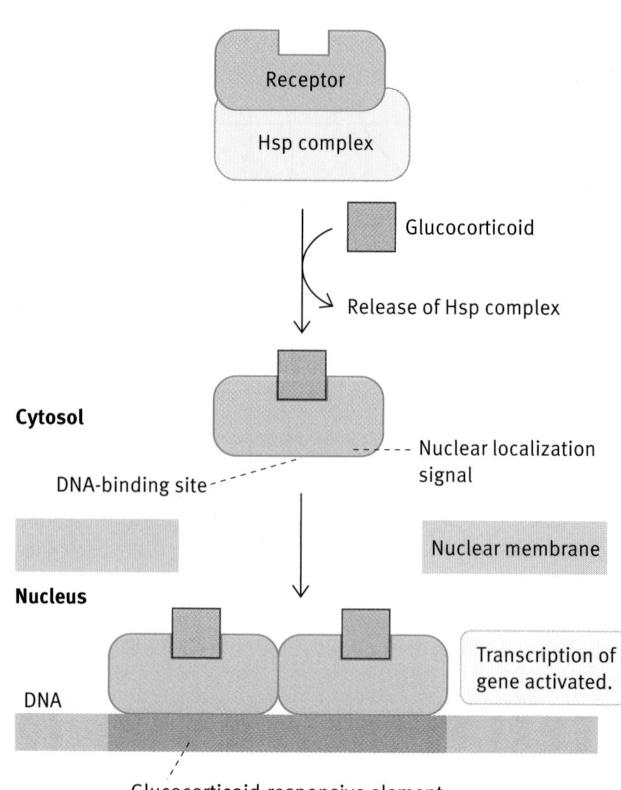

Fig. 29.4 Mechanism of action of a glucocorticoid hormone. The receptor is bound to the HSp complex. Binding of the glucocorticoid to its receptor leads to the release of the Hsp and the exposure of the NLS and DNA binding site on the receptor. The hormone-receptor complex binds the glucocorticoid responsive element on DNA causing activation of selected genes. Hsp, heat shock proteins; NLS, nuclear localization signal.

glucocorticoid response element (Fig. 29.4) on the DNA causing activation of appropriate genes.

In the case of the steroid family of receptors, which reside in the nucleus (those for thyroid hormones, and retinoic acid), the NLS is not obscured by the Hsp complex so that the receptor – Hsp complex is transported as a unit into the nucleus soon after it is synthesized. However, until the Hsp is released by hormone attachment, it does not become an active transcription factor and does not modulate genes. The principle of the control is the same as with glucocorticoids despite these differences.

Another signalling molecule, nitric oxide, is also lipid soluble with an intracellular receptor but is in quite a different category and for convenience will be dealt with later in this chapter.

Responses mediated by receptors in the cell membrane

For the water-soluble hormones, most of the signalling in the body occurs *via* cell membrane receptors. There are large numbers of them, but all have an external domain to which the signal binds, a transmembrane domain, and a cytosolic domain. When the hormone binds its specific receptor, the cytosolic domain undergoes a conformational change that activates a signalling pathway. This in turn activates specific genes and/or modulates metabolic systems.

There are three main types of membrane-bound receptors

Metabotropic receptors are G-protein coupled receptors and binding of the ligand to them mainly affects metabolic events in the cell. Examples are glucagon and adrenaline receptors

Catalytic receptors are tyrosine kinase linked receptors and they mainly (but not exclusively) affect gene expression, and cell growth and differentiation. Examples include the insulin receptor and cytokine receptors. Tyrosine kinase receptors are known not only to be important regulators of normal cellular processes but also to be involved in the development and progression of many cancers.

Ionotropic receptors are ligand-gated ion channels, an example of which is the nicotinic acetylcholine receptor. They

BOX 29.1 The glucocorticoid receptor and anti-inflammatory drugs

Several synthetic glucocorticoids are among the most effective suppressors of inflammation, their activity being mediated by the glucocorticoid receptor described below. The inflammatory response provides immediate protection against infections and assists in tissue repair after injury, but inappropriate or unregulated inflammation is a key element in the development of **rheumatoid arthritis** and other autoimmune diseases such as **inflammatory bowel disease**, the skin disorder **psoriasis**, and **multiple sclerosis**.

One important element in inflammation is the release by phagocytic white blood cells (neutrophils and macrophages) of protein mediators called chemokines (they attract white cells) and cytokines (see earlier in this chapter), of which tumour necrosis factor-α (TNF-α) and interleukin-1 (IL-1) are two important examples. These attach to cell-surface receptors and stimulate the inflammatory response. They do so by initiating a cascade of intracellular events, leading ultimately to the activation of the transcription factor family **NF-κB**. This is present in the cytosol of all cells but in an inactive form due to being complexed with **IκB** inhibitory proteins. The TNF-κ signal causes the activation of kinases that phosphorylate the inhibitor protein marking it for degradation by proteasomes. This exposes the nuclear localization signal (NLS) on the NF-κB, which is then transported into the nucleus where it activates many genes, leading to the inflammatory response.

The anti-inflammatory glucocorticoid drugs attach to the glucocorticoid receptor in the cytosol occupying the position that the glucocorticoid hormone would normally occupy as shown in Fig. 29.4. The receptor/drug complex also moves into the nucleus where it attaches to and inhibits the activated NF-κB protein, thus helping to inhibit the inflammatory response.

Glucocorticoids remain important drugs in the treatment of inappropriate inflammation. Recently, however, new classes of anti-inflammatory drugs have been developed. These include inhibitors of the inflammation-associated enzyme cyclooxygenase 2 (**COX-2 inhibitors**), and the use of monoclonal antibodies or soluble cytokine receptors, which mop up cytokines such as TNF before they have a chance to bind their cell-surface receptors (TNF inhibitors). Finally, small-molecule drug candidates targeting components of the intracellular signalling cascades are also under development.

This is a large and complex system, not fully elucidated, and has been given in brief outline here. It is an area of intense medical research. See Box 17.2 for more on COX-2 inhibitors, which are now of restricted use because of deleterious side effects.

are mainly involved in ion transport processes and neurotransmission. The mechanism has been described in Chapter 7 so in this chapter we will concentrate on the G-protein coupled and the tyrosine kinase linked receptors.

The tyrosine kinase-coupled receptors

Most tyrosine kinase receptors are monomeric but some exist as multimeric complexes, for example, the insulin receptor, which is a tetramer where the four subunits are held together by three disulphide bridges in the absence of the ligand. Fig. 29.5 shows typical examples of tyrosine kinase receptors and their ligands and Fig. 29.6 shows the special features of the insulin receptor.

When specific ligands bind to monomeric tyrosine kinase receptors, they dimerize by lateral movement in the lipid bilayer (Fig. 29.7a). This brings the cytosolic domains of a pair together. These domains are themselves tyrosine protein kinases and they phosphorylate each other on tyrosine groups by the reaction shown in Fig. 29.8.

There are a number of proteins in the cytosol, which are involved in signalling processes, by binding to phosphorylated receptors and acting as adaptor molecules. In this way they link the phosphorylated receptors to their specific signalling pathways. Tyrosine kinase signals are usually (but not always) conveyed to the nucleus. (We will see how in the case of the insulin receptor a number of adaptor molecules direct

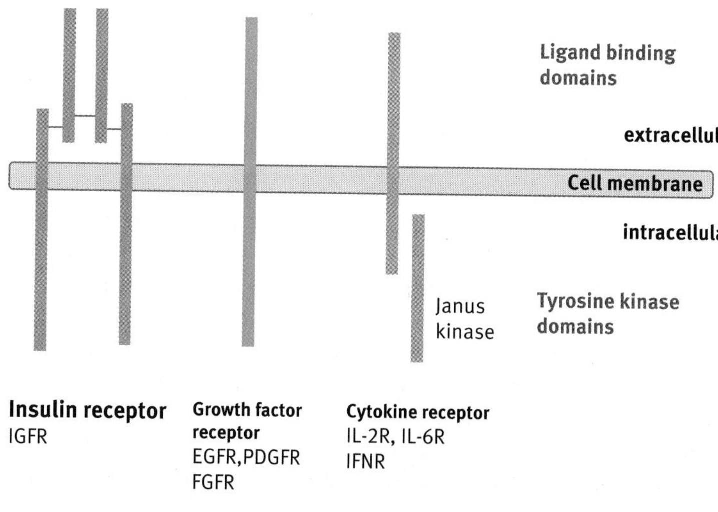

Fig. 29.5 Types of tyrosine kinase receptor structures. Simplified diagram of the main types of tyrosine kinase receptor structures. There are three main types: (a) the insulin receptor type, a heterotetramer of two α (extracellular) and two β (intracellular) subunits. Ligands include insulin and insulin-like growth factor (IGF); (b) the growth factor receptor type, a monomer whose ligands include epidermal growth factor (EGF), platelet-derived growth factor (PDGF), and fibroblast growth factor (FGF); (c) cytokine receptor type. The tyrosine kinase is not part of the receptor as for the two above but a separate protein known as Janus kinase (JAK). Ligands include interleukins 2 and 6 (IL-2R, IL-6R) and interferon production regulator (IFNR)

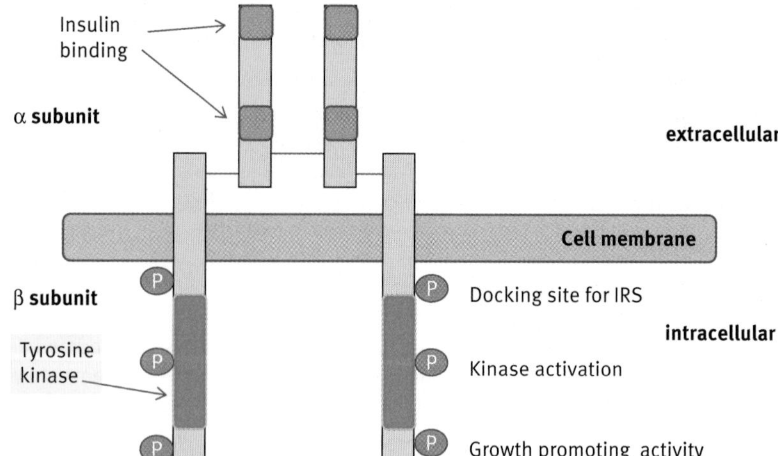

Fig. 29.6 The structure of the insulin receptor. It is a heterotetramer composed of 2 α- and 2 β-subunits, the α being extracellular and the β spanning the membrane. Insulin binds to the α subunits and the tyrosine kinase domain is in the β subunits, on the cytosolic aspect of the membrane.

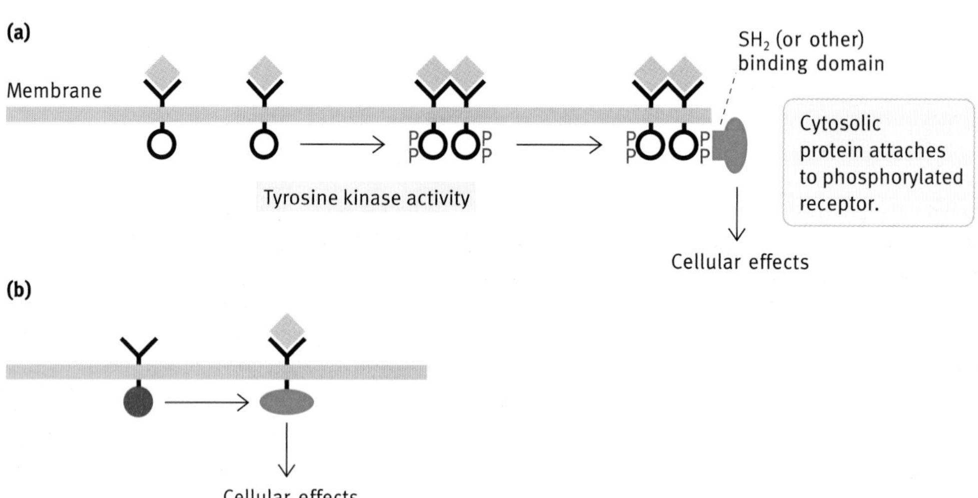

Fig. 29.7 **(a)** Diagram illustrating the tyrosine kinase type of receptors; the phosphorylation of the receptors causes attachment of the appropriate cytosolic protein by its SH2 domain; the attached protein then activates a signalling pathway. (SH2 domains are common, but other binding domains exist.) **(b)** In this type of signalling, the receptor is not phosphorylated but attachment of the signal molecule causes a conformational change in the cytosolic domain of the receptor. This leads to the activation of one or more signalling pathways. The G-protein-coupled receptors are of this type.

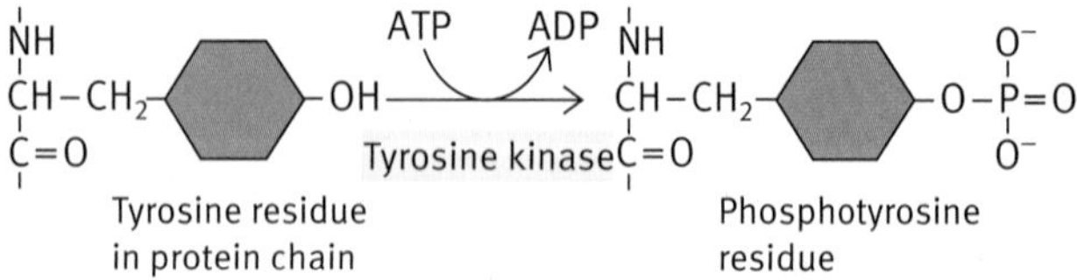

Fig. 29.8 Tyrosine phosphorylation by tyrosine kinase.

the signal to different pathways, some cytosolic and metabolic, and others affecting gene expression).

Around the phosphorylation sites on a receptor are amino acid sequences that are recognized by these adaptor proteins, so that the correct adaptor proteins bind to the correct receptor sites and thus the correct signalling pathway(s) are activated (see section on Binding domains). A given receptor may be connected to a number of signalling pathways and thus have multiple control effects in the cell.

The G-protein-coupled receptors

We have earlier described the control system of the GTP/GDP switch (see Chapter 27), which is important in protein synthesis, protein delivery mechanisms and elsewhere. The principle, to recapitulate very briefly is that the G-control proteins have GTP bound to them. They are monomeric **GTPases** and hydrolyse the attached GTP to GDP and phosphate causing a conformational change in the protein. In the case of G-coupled receptors this occurs as a result of the binding of the specific hormone or other signal to the receptor, which causes an allosteric change in the cytosolic domain. In turn this leads to a response to the signal (Fig. 29.7b). A very wide range of hormones have receptors of this type. We will describe later in detail how the 'prototype' of G-protein receptors, the adrenaline-specific one operates.

General concepts in cell signalling mechanisms

Protein phosphorylation

The interconversion of phosphorylated and dephosphorylated forms of proteins plays a crucial role both in metabolic control (see Chapter 20) and in signalling, and there are large numbers of different protein kinases in eukaryotes for phosphorylation of proteins. The principle is simple. Addition of a highly charged phosphate group to a protein can have a major effect on the conformation of a protein, which results in the protein participating in the relevant signalling pathways. Reversal of this by protein phosphatases terminates the action. Fig. 29.9 illustrates how phosphorylation is involved in signalling both in the control of metabolic processes and control of gene expression. The figure shows that the protein kinase and the phosphatase are themselves regulated by a certain signal so that highly integrated control is possible. In this type of signalling system, phosphorylation is usually on serine and/or threonine —OH groups.

Binding domains of signal transduction proteins

Cell signalling in mammals is complex and includes proteins recognizing other proteins, or small molecules, and joining together to form chains or clusters, which relay the signals, one to another, to their destinations. The same recognition domains are found in different proteins building up signalling pathways reminiscent of the sections in a child's building blocks that click together. In Chapter 4, we described how domains of proteins are found over and over in different proteins, an evolutionary concept known as domain shuffling.

Such a domain, called SH2, is found in a wide variety of regulatory proteins. The name SH2 means Src (pronounced sarc) homology domain, region 2. This refers to a kinase found in the Rous sarcoma virus, which causes tumours. One of the Src kinase domains recognizes a small amino acid sequence. SH2 proteins typically bind to phosphorylated tyrosines on the cytosolic face of membrane receptors. As mentioned before, the phosphorylated tyrosines have adjacent to them different amino acid sequences in different receptors, and different SH2 domains of signalling proteins recognize these, as well as the tyrosine phosphate group. In the human genome many hundreds of genes coding for hundreds of proteins with variant SH2 domains have been identified.

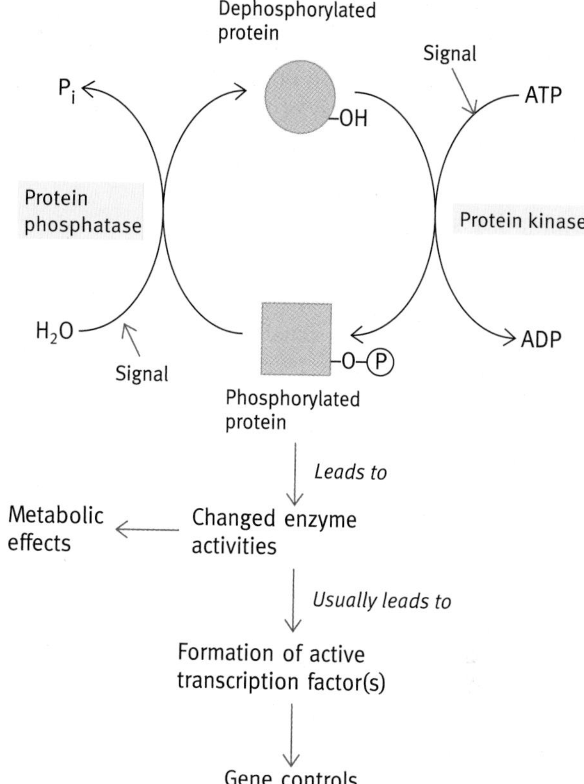

Fig. 29.9 Central principle of control by many extracellular signals. The diagram illustrates protein kinase and protein phosphatase action. Phosphorylation is on serine or threonine residues of proteins. The phosphorylation brings about a conformational change in the protein that changes its activity in some way, resulting in a cellular response. The process is reversed by removal of the phosphoryl group by a protein phosphatase action. In the case of some membrane receptors, described later in the text, phosphorylation of their cytosolic domains on tyrosine —OH groups of protein side chains is involved in their activation. Note that an arrow may represent several steps.

In addition to SH2, the prototype Src kinase has an **SH3 domain**, which binds to proline-rich sequences around phosphorylated tyrosines and to proline-rich regions of other proteins. These are found in many signalling proteins often in addition to the SH2. Thus an adaptor protein may bind to its phosphorylated receptor by the SH2 domain and also connect to other proteins in signalling pathways *via* its SH3 domain.

The existence of variants of these domains provides for the specific recognition of large numbers of different activated receptors by adaptor proteins.

Another class of proteins involved in signalling pathways binds to membrane areas enriched in inositol-containing second messengers (see Chapter 20) by the **pleckstrin homology** or **PH domain**, first found in a blood platelet protein (pleckstrin). It has since been found in over 60 signalling proteins downstream from the receptor in signalling pathways.

In summary, the use of a few main types of recognition domains with large numbers of variants giving many signalling proteins allows evolution of great flexibility in receptor recognition and the assembly of transduction pathways in a modular fashion.

Terminating signals

Regulatory systems in the body must be reversible, otherwise a signal once given could not be switched off. Cancers are often the result of runaway signalling pathways (see Chapter 33). Much signalling involves phosphorylations and there are families of protein phosphatases, which reverse these phosphorylations. It is estimated that 2–3% of human genes code for phosphatases. Some are specific for serine/threonine phosphates and others for tyrosine phosphates and may themselves be regulated by phosphorylation. In addition, some signalling pathways have individual restraining mechanisms, which will be dealt with when we describe specific examples.

The rest of this chapter illustrates examples of the main classes of signalling pathways. To help you keep track of the text as we deal with them, they are listed below as a reference guide. There is a bewildering collection of abbreviations referring to these molecules, the story made even more confusing because their name denotes the organism, tissue, or cell where they were discovered. So we have to resign ourselves to the fact that a protein first discovered in a virus that causes cancer in a mouse (i.e. Ras), is a perfectly normal protein involved in, for example, perfectly normal insulin signalling. Table 29.2 may be of help here. It gives a summary of proteins and peptides involved in signalling pathways with a short description of the origin of their name and their characteristics.

Examples of signal transduction pathways

- Pathways from tyrosine kinase-associated receptors:
  - the Ras pathway
  - the phosphatidylinositide 3-kinase (PI 3-kinase) pathway (encountered in insulin signalling)
  - JAK/STAT pathways.
- G-protein-associated pathways:
  - cyclic AMP (cAMP) pathway
  - phosphatidylinositol cascade pathway
  - vision – the light-transduction pathway.
- Signalling pathways mediated by cyclic GMP as second messengers:
  - membrane receptor-mediated pathways
  - nitric oxide signalling.

Signal transduction pathways from tyrosine kinase receptors

The Ras pathway

This widespread signalling pathway from membrane receptor to genes in eukaryotes is the mechanism of action of a wide variety of growth factors. **Ras**, the protein that gave its name to this pathway, is found in all eukaryotic cells. It is part of a major signalling pathway from growth factor-specific tyrosine kinase receptors, to modulation of gene transcription factors. Note that there are no low-molecular-weight second messengers in this pathway – all of the components are proteins.

Ras, and other components of the pathway, were known to be important in cell regulation, because mutations resulting in abnormal forms of some of the proteins are **oncogenic** – they are associated with cancer. Ras was discovered as the oncogenic protein coded for by the **rat sarcoma virus**, which causes muscle tumours (sarcomas) in rats. Its normal counterpart is found in all eukaryotic cells and a mutated form of Ras (not of viral origin) occurs in many human cancers.

The signalling pathway involves the receptor stimulating a cascade of three cytosolic protein kinases, the final one of which causes gene activation. Fig. 29.10 gives a simplified outline of the pathway to provide a preliminary orientation.

Mechanism of the Ras signalling pathway

We will use the receptor for the **epidermal growth factor (EGF)** to illustrate this (Fig. 29.11). The receptor for EGF is a monomer with external binding, single transmembrane and cytosolic domains. The latter is a tyrosine kinase. When EGF binds, the receptors dimerize in the membrane, bringing their kinase domains together and they phosphorylate each other as shown in Fig. 29.7a.

In the cytosol is a **growth factor receptor-binding protein (GRB)**, an adaptor protein. It binds to the phosphorylated receptor *via* its SH2 domain, but not to the nonphosphorylated form. The next component in the pathway is **SOS**, initially found in a *Drosophila* (fruit fly) genetic mutant where it was named 'son of sevenless'. SOS is activated by the receptor-bound GRB (but not by free GRB). The activated GRB-SOS complex in turn activates Ras, the mechanism of which we will now describe.

Concept of the GTP/GDP switch mechanism, as seen in the Ras pathway

Ras has a switch mechanism widely used in control pathway proteins; it is a GTPase that belongs to a family known as **small monomeric GTPase proteins**. (The term **G-protein**, which is reserved for the trimeric GTPase components of G-protein receptors, is described later (see Fig 29.25). They have a molecule of GDP *or* GTP bound to them. Ras is active when GTP is bound to it and inactive when GDP is bound (Fig. 29.12), the nucleotides causing conformational changes in the protein. To

Term	Description and characteristic
Ras	Oncogene product(first described in rat sarcoma virus, which causes tumours). A small anchored GTPase involved in tyrosine kinase receptor signalling and mainly associated with pathways involving gene expression.
SH2	Src homology domain, a domain on a kinase first described in Rous sarcoma virus. Binds phosphorylated tyrosines.
SH3	Similar to SH2 but binds proline rich areas surrounding phosphorylated tyrosines.
PH domain	Pleckstrin homology domain in an adaptor protein. Binds to inositol rich areas near phosphorylated tyrosines. First found in the blood protein pleckstrin.
GRB	Growth factor receptor binding protein. An adaptor protein that binds to phosphorylated tyrosines through its SH domains
SOS	Protein operating in signal cascade of Ras/MAPK pathways. Stands for 'son of sevenless' as it is a protein downstream from the gene 'sevenless', mutations of which lead to failure to develop the seventh central photoreceptor in drosophila eye.
GAPs	Proteins that associate with a G protein, i.e. a GTP binding protein, and increase the rate of degradation of GTP thus switching off the signal. Involved both in Ras and cAMP pathways.
Raf	a MAP kinase kinase kinase. Raf phosphorylates MEK, which is a MAP kinase kinase. A serine/threonine kinase (name comes from rapidly accelerated fibrosarcoma, a retroviral oncogene).
MEK	MAP kinase kinase. Phosphorylated and activated by Raf. Phosphorylates ERK or MAPK. A tyrosine/serine/threonine kinase.
MAPK	Mitogen activated protein kinase. The last kinase in the Ras pathway, which activates a transcription factor. Also known as ERK (extracellular signal regulated kinase). Phosphorylated by MEK or MAPKK
ERK	Another name for MAPK. A serine/threonine kinase involved in gene expression by phosphorylating and activating specific transcription factors.
PI 3 kinase	Phosphatididyl inositol kinase. Involved in metabolic effects of insulin via activation of Akt/PKB.
Akt/PKB	Serine/threonine kinase activated by PI 3 K and PDK1 and involved in phosphorylating a number of proteins and mediating the effects of insulin.
PDK1	A kinase, which together with PI 3 kinase activates Akt/PKB.
SHC	A protein with SH domain involved in the activation of Ras and initiating the effects of insulin etc. on gene expression.
JAK	Janus kinase possesses two tyrosine kinase activities and phosphorylates and activates cytokine receptors and transcription factor STAT (JAK named after double faced Roman god Janus).
STAT	Trancription factor (Signal Transducer and Activator of Transcription). Involved in cytokine signalling.
SOCS	Suppressors of cytokine signalling. Proteins that bind to phosphorylated JAKs at the active site thus inhibiting the kinase action.
PIAS	Protein inhibitors of activated STATS. Proteins that bind phosphorylated STAT dimers and prevent them from switching on genes.
G-protein	Heterotrimeric protein on the cytosolic side of cell membrane, associated with a receptor. Subunits are alpha, beta, and gamma. The alpha subunit is a GTPase, which can activate adenyl cyclase, which produces cAMP.
CREB	CAMP responsive element binding protein. Dimerize and become active transcription factors when phosphorylated by PKA. Bind and activate CREs, which are cAMP responsive elements in promoter regions of genes.
PKA	cAMP activated protein kinase A.
GRK	G-protein receptor kinases. Phosphorylate G-protein coupled receptors and inactivate them.
arrestin	Protein that binds activated G-protein receptors and inactivates them by uncoupling them from G-proteins.
Phospholipase C	Hydrolyses PIP2 into diacylglycerol and IP3 both of which are second messengers.
PKC	Protein kinase C. Activated by calcium ions and DAG (diacylglycerol).
calmodulin	Calcium binding protein, mediating the effects of calcium as a third messenger by modulating the activity of various enzymes by activating a number of kinases.
transducin	A heterotrimeric G-protein involved in vision.

Table 29.2 Proteins/peptides encountered in signalling pathways (listed in order of appearance in the text)

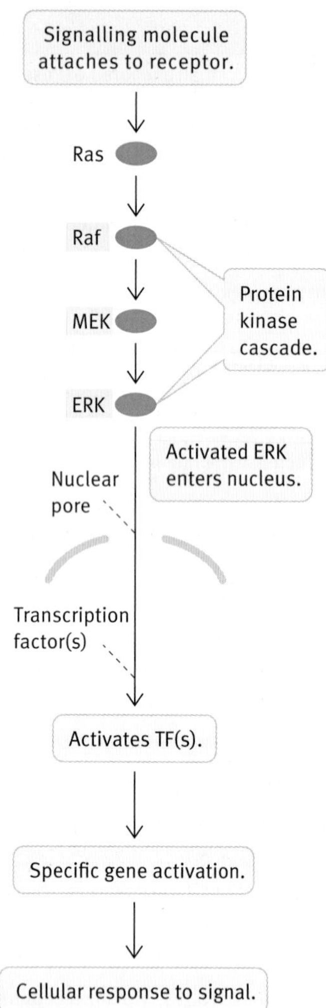

Fig. 29.10 Simplified overview of the Ras signal tranduction pathway. Raf, MEK, and ERK are protein kinases, each of which phosphorylates the next component of the pathway. The nomenclature and further details of the pathway are given in the text and subsequent figures. TF, transcription factor. Arrows represent activation.

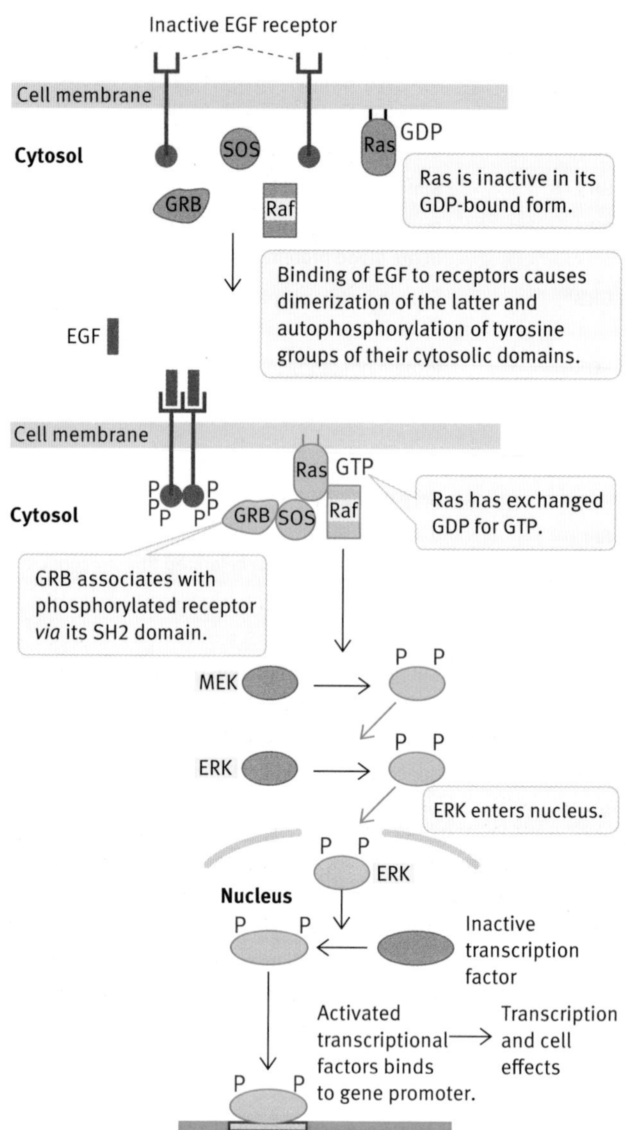

Fig. 29.11 The Ras pathway of signal transduction. See text for explanation. In this diagram signalling by epidermal growth factor (EGF) is used as an example. Activation of the various proteins involves changes in conformation but these are not indicated to keep the presentation reasonably simple. The nomenclature is explained in the text. GRB, growth receptor-binding protein.

activate Ras (for Ras to activate the next component), the GDP must be exchanged for GTP, but this exchange occurs only when Ras is in contact with GRB-SOS bound to the receptor (which occurs only when a signalling molecule is attached to the receptor).

When the EGF is no longer present (because it has been degraded), the signal has to be terminated. The Ras protein is a slow GTPase that switches itself off by hydrolysing its attached activating GTP molecule to GDP + P_i. If the receptor is still activated, Ras will be reactivated by a repeat of the GDP/GTP exchange process; if not, Ras will be inactive until more EGF arrives. One gets a glimpse here of the complexity of signalling controls, because other proteins called **GTPase-activating proteins (GAPs)** stimulate the rate of GTP hydrolysis by the Ras-type proteins and speed up the switch-off. GAPs provide a modulation of the control. This accounts for the fact that GAP mutations can be oncogenic. If the GTPase activity of Ras is not stimulated by a functional GAP, Ras does not switch itself off.

The MAP kinase cascade in the Ras pathway

Ras, in its GTP-bound form, activates a cascade of three protein kinases, collectively called **mitogen-activated protein kinases (MAP kinases)**. It activates **Raf** by combining with it and inducing a conformational change. Raf is the first protein kinase in the Ras pathway downstream from Ras. The name Raf derives from a viral oncogene; after activation by Ras-GTP, it phosphorylates **MEK**, the second protein kinase,

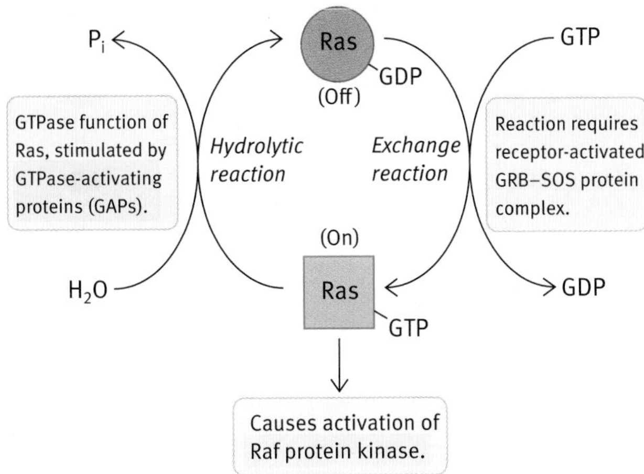

Fig. 29.12 The control of the Ras protein.

which in turn phosphorylates **ERK**, the last of the three protein kinases (nomenclature explained above). The cascade is as shown in Fig. 29.11, the first kinase (Raf) activating the second by phosphorylation, and the second phosphorylating the third, which migrates into the nucleus. This succession of phosphorylations amplifies the signal. Amplifying cascades were discussed in glycogen breakdown (see Chapter 20) and blood clotting (see Chapter 32).

When the phosphorylated ERK is transported into the nucleus it phosphorylates target transcription factors, which results in the transcription of specific genes, the synthesis of their related proteins, and the desired cellular response to the EGF signal, such as cell proliferation. Raf and ERK are serine/threonine-type-specific kinases. MEK, uniquely, has dual specificity for these and tyrosine.

To summarize, the activation sequence, also illustrated in Fig. 29.11, is as follows:

- EGF attaches to its receptor; the receptor dimerizes and is autophosphorylated on its tyrosine -OH groups
- GRB-SOS attaches to the phosphorylated receptor by the SH2 domain of GRB
- receptor-bound GRB-SOS activates Ras; Ras activates Raf, a protein kinase
- Raf phosphorylates and thus activates MEK, also a protein kinase
- MEK phosphorylates ERK, the final protein kinase
- ERK migrates into the nucleus and phosphorylates target transcription factor(s); the latter activates gene transcription; specific proteins are synthesized that promote cell proliferation.

Nomenclature of the protein kinases of the Ras pathway

Raf, MEK, and ERK are collectively known as **m**itogen-**a**ctivated **p**rotein kinases (**MAP** kinases), a mitogen being something that stimulates cell division. Since there are several parallel pathways involving MAP kinase cascades (see below), *individual* MAP kinases in a cascade are given identifying names. In the case of the Ras pathway, these are Raf, MEK, and ERK. MEK stands for MAP kinase/ERK; that is a MAP kinase whose substrate is ERK. **ERK** stands for **extracellular signal-regulated protein kinase**. Since related, but different, MAP kinase variant cascades have kinases analogous to Raf, MEK, and ERK, but with different specificities, these also have individual names.

It is useful to be able to collectively refer to all kinases of the Raf type, and similarly to all protein kinases of the MEK type, and also to all of the ERK type found in the different signalling pathways. The terminology seems daunting at first sight, but is really quite simple. In the literature the terms **MAPKKK**, **MAPKK**, and **MAPK** are used; to understand them it is best to start with the last and work backwards. Thus ERK is a MAP kinase or MAPK. MEK is a kinase that phosphorylates ERK (a MAP kinase), and therefore is a MAP kinase-kinase or MAPKK. Raf is a kinase that phosphorylates MEK (a MAP kinase-kinase), and therefore is a MAP kinase-kinase-kinase or MAPKKK. (See terms in brackets in Fig. 29.14; these are the generic names for MAP kinases in Ras-type parallel pathways.)

Inactivation of the Ras pathway

The removal of EGF or other factors from receptors terminates the activation of the receptor but the whole pathway needs to be switched off also. Ras is inactivated by its intrinsic GTPase activity. How are the activated MEK and ERK kinases inactivated? A family of **protein phosphatases** is known which inactivate MAP kinases and other signalling proteins by dephosphorylating them. Somewhat surprisingly, the protein kinases and their related phosphatases are physically bound together (Fig. 29.13), implying that no sooner has a phosphorylation taken place than it is reversed. A competing push-pull situation is less likely to get out of control – perhaps analogous to continually touching the brakes of a car going down a steep hill, and, as stated, a signalling pathway out of control is characteristic of cancer cells.

Some of the protein phosphatases are induced by the same signal that activates the pathway, so that negative feedback loops are established. The importance of protein phosphatases is demonstrated by the toxins that increase or decrease protein dephosphorylation (Box 29.2).

Scaffold proteins prevent cross-talk between pathways

There are multiple Ras-type pathways operating in parallel using protein kinases corresponding to Raf, MEK, and ERK in other pathways and targeted at different transcription factors (Fig. 29.14) or cytosolic effectors.

These multiple pathways convey signals from different receptors to the nucleus so that activation of specific genes appropriate to the particular signal occurs. It is necessary that the different pathways do not 'cross-talk' *inappropriately* just

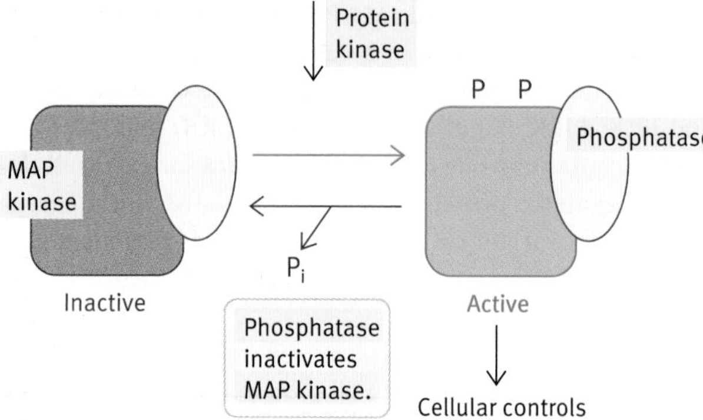

Fig. 29.13 The association of a MAP kinase with a phosphatase provides a rapid molecular switch mechanism. The MAP kinase is activated by phosphorylation and the associated phosphatase reverses the activation by removing the phosphoryl groups. The implication is that MAP kinase pathway signalling involves the rapid oscillation of the kinase components between the active and inactive states. This oscillation provides stability of control.

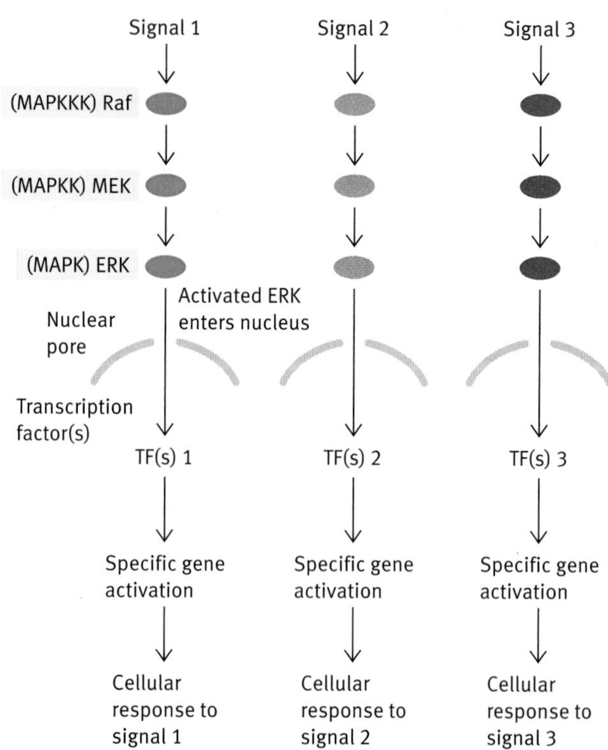

Fig. 29.14 Multiple signal pathways of the Ras type. Raf, MEK, and ERK are shown as green shapes. The orange and blue shapes represent protein kinases corresponding in function to Raf, MEK, and ERK but are protein kinases of other signalling pathways. The terms in brackets are the generic names for MAP kinases in the Ras-type parallel pathways. TF, transcription factor.

BOX 29.2 **Some deadly toxins work by increasing or inhibiting dephosphorylation of proteins**

The crucial importance of protein phosphatases is emphasized by the fact that they are targets of some biological toxins. One of current environmental interest is the phosphatase-inhibitory toxin produced by blue-green algae. This is a cyclic peptide, called microcystin, containing seven residues. It is a powerful liver toxin and a dangerous cancer-producing agent. Inhibition of dephosphorylation would prevent switching-off of signalling pathways. The appearance of toxic algal blooms in rivers is a world-wide problem, which accounts for the increased production of microcystin. The cause is the nitrification of rivers, which occurs because of fertilizers leaking from agricultural areas, coupled with the reduced water flows caused by irrigation and also the high concentrations of phosphates leaking into rivers from household cleaning products. Harmful algal blooms (HAB) have been associated with various kinds of shellfish poisoning. Shellfish are filter feeders and so accumulate a lot of toxins produced by green blue algae.

One of the poisons produced by shellfish, called okadaic acid, is an inhibitor of serine/threonine-specific phosphatases of the type found in the Ras pathway. This toxin is of medical concern, as well as a problem for the shellfish industry.

as you do not want telephone lines to different houses to allow cross-talk when they should be conveying separate messages to their different destinations.

However, different MAP kinase cascades may contain a component common to two or more pathways. This would seem to allow a mix up of signals from different receptors and negate the whole point of having specific receptors transmitting their signal to specific endpoints. A solution to the problem is found in scaffold proteins, which bind together in a series, the components of specific cascades. This is illustrated in Fig. 29.15. It is thus possible to use the same MAP kinase in two pathways since the scaffold protein does not allow cross-talk between pathways.

Signal sorting

A more difficult problem exists in that it is known that two independent signals may activate the same component of different signalling pathways and yet the signals have to remain separate and reach different specific final targets. For example it is known that Ras can be activated both by growth promoting mitogens and by hormones, and yet their effects on the cell are quite different. As an example, EGF and insulin can both activate Ras, but the responses to the two signals are quite different. Somehow the cell sorts out signals from receptors to the correct end points.

Let us look at an example studied in yeast. Two receptors were studied, one activated by a mating pheromone and the other by high salt concentrations. The responses to the two signals are totally different. They both activate different signal cascades, each attached to a scaffold protein, but are activated by a single protein kinase, which interacts with a MAP kinase common to both pathways. The single activating kinase is attached to the cell membrane. In the absence of a signal to

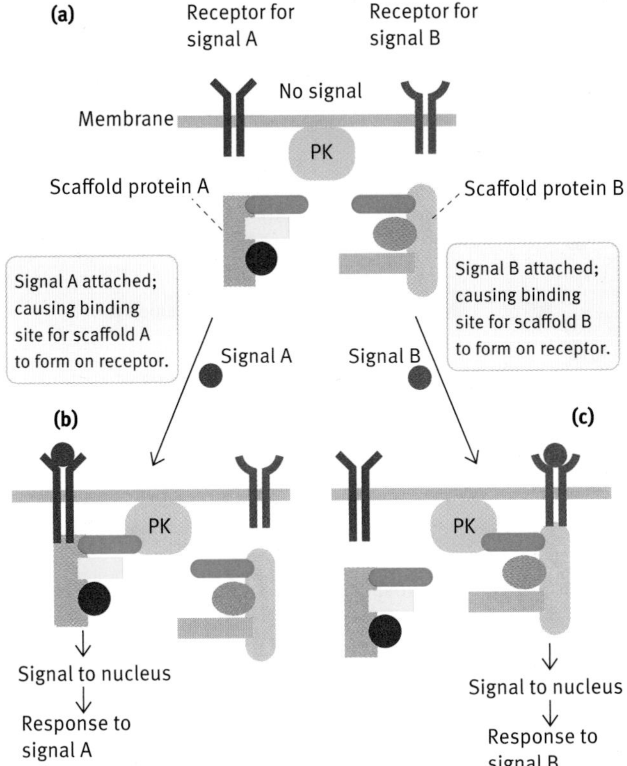

Fig. 29.15 Simplified diagram to give the overall concept of how two receptors can route signals to two signalling cascades *via* a common protein kinase (PK) in a generalized way. In some cases a domain of the scaffold protein itself constitutes the middle MAP kinase. The figure also illustrates how scaffold proteins allow independent pathways to have components in common (here seen in dark green) without cross-talk occurring.

either receptor, the MAP kinase systems on their scaffold proteins are detached from the membrane protein kinase so that neither signal pathway is activated. When specific ligand binds to a receptor there is conformational change in its cytosolic domain, which creates a binding site for its specific scaffold protein. The latter binding brings the first MAP kinase on the scaffold into contact with the kinase that activates it, thus firing off the signal pathway. When the signal leaves the receptor the conformational change reverses, the scaffold protein detaches and the signal is terminated. In this way signals from the different receptors are routed down separate pathways to the correct endpoints. The mechanism is presented in Fig. 29.15 in a generalized way to give the overall concept without details specific to yeast

The problem cited above, of different signals activating Ras but the correct messages getting to the intended destinations seems much more difficult in mammalian systems and the mechanism by which the cell sorts the signals so efficiently is not understood. However, there are aspects to Ras that may ultimately be found to be relevant. In mammals, four largely homologous Ras isoforms are known, which activate different signalling pathways within the cell, and there is considerable speculation on their compartmentation.

It is speculated that **lipid rafts** may play a role in this area. These are small islands of membrane lipids rich in sphingolipids and cholesterol, which do not mix with the rest of the lipid bilayer and remain discrete as slightly thickened areas due to the length of the hydrocarbon tails of the sphingolipids. It is thought that possibly these lipid rafts are sites for the specific attachment of regulatory proteins of signalling pathways, though here again no definite examples of how this may function in signal sorting are known. We are a long way away from the first fluid mosaic model of membrane structure that postulated that lipids in membranes were randomly distributed in the plane of the membrane. We know that cholesterol and sphingolipids will coalesce in areas of cell membranes, but after more than 10 years of research into this topic, there still seems to be no agreement amongst scientists that the mechanism of segregation of signals by a cell is through lipid rafts.

The phosphatidylinositide 3-kinase (PI 3-kinase) pathway and insulin signalling

This is another example of the class of signalling pathways involving tyrosine kinase-associated receptors. The PI 3-kinase pathway has a bewildering range of control roles involved in cell proliferation, differentiation, inhibition of apoptosis, and other cellular activities including metabolic control; it is activated by many different receptors responding to a range of hormones, growth factors, and neurotransmitters. We will use **insulin** to illustrate the pathway noting that only some of the effects of insulin are mediated through this pathway.

The **insulin receptor** is unusual in structure in that it resembles two receptors of the EGF type, covalently dimerized. Fig 29.16 shows a summary of the effects of insulin. They can be divided into three categories: metabolic effects, gene expression effects, and mitogenic effects. All these effects, which vary in time course, must be explained by the binding of insulin to its receptor. We now know that two signalling pathways are activated by the binding of insulin to its receptor.

The metabolic effects result from a signalling cascade involving the enzymes **PI 3-K** (phosphoinosite 3-kinase) and **Akt/PKB**, both kinases. **Akt** is the mammalian homologue of the retrovirus Akt 8, the 't' standing for thymoma. The alternative name is PKB ,protein kinase B. The gene expression and mitogenic effects result from activation of the **Ras** signalling cascade, in a similar way to that of EGF described earlier.

For the sake of clarity we are going to look at the metabolic effects first and then at the gene expression and mitogenic effects remembering that they are taking place simultaneously. Insulin signalling is a good example of how one ligand (insulin) binding one receptor, can have such diverse effects as glycogen synthesis and inhibition of apoptosis.

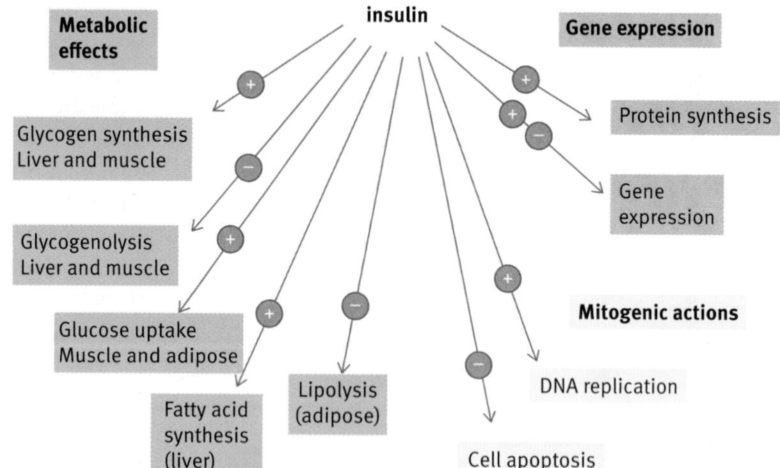

Fig. 29.16 Cellular effects of insulin. The cellular effects of insulin can be divided into three categories: metabolic effects, gene expression effects, and mitogenic effects. + denotes activation of the process by insulin; – denotes inhibition.

The insulin receptor is a tyrosine kinase. If you look at Fig. 29.6 you will notice that the tyrosine kinase domain is on the cytosolic part of the receptor. Furthermore, there are three phosphorylation sites. When insulin binds the receptor, the latter becomes autophosphorylated on specific tyrosine OHs. They become docking sites for a number of proteins involved in the signalling cascade. The phosphorylated tyrosine nearest to the membrane is involved in events through phosphorylation of insulin receptor substrates (IRS 1/2) and activation of PI 3 K and Akt/PKB proteins. The phosphorylated tyrosine in the centre is involved in activation of various kinases and the one furthest away from the cell membrane is involved in pathways of gene expression and mitogenic activity.

Let us follow the events shown in Fig. 29.17, which shows the first steps in the metabolic effects (i.e. activation of Akt/PKB):

1 On binding of insulin, the receptor becomes autophosphorylated.
2 Insulin receptor substrates, IRS 1/2, attach to the phosphorylated receptor by virtue of their SH2 domains and become phosphorylated by the receptor tyrosine kinase.
3 The enzyme, **PI 3-kinase**, binds to the phosphorylated IRS and is brought in an activated form close to the plasma membrane. PI 3-kinase consists of two subunits, P85 and P110, one regulatory and one catalytic.
4 The substrate of PI 3-kinase is a membrane component, **phosphatidylinositol-4,5-bisphosphate** (**PI(4,5)P$_2$** or **PIP$_2$**). As a result of the PI 3-kinase action, a **second messenger**, **PI(3,4,5)P$_3$** (or **PIP$_3$**), is formed (Fig. 29.18).
5 The membrane section enriched with PI(3,4,5)P$_3$ attracts *inactive* cytosolic Akt/PKB to attach to the membrane using its PH (pleckstrin homology) domain, which binds to PI(3,4,5)P$_3$.
6 There it encounters a membrane-located kinase known as **PDK1**, which phosphorylates the Akt/PKB, and activates it.

Akt/PKB has many protein targets, as we will soon see, which result in the metabolic effects of insulin. We will look

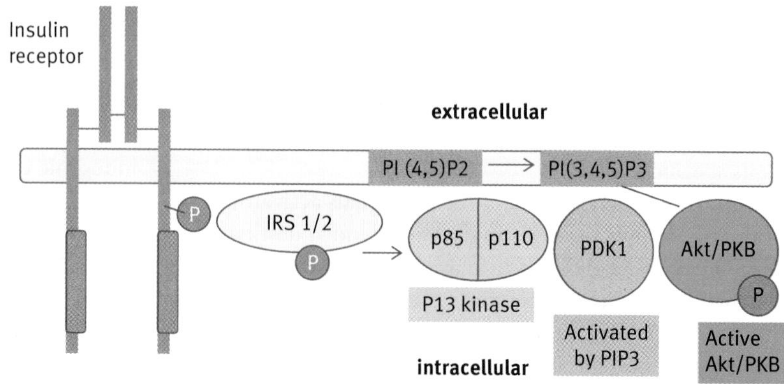

Fig. 29.17 The activation of Akt/PKB following the binding of insulin to its receptor. IRS bind the phosphorylated receptor with their SH2 domain and are themselves phosphorylated and acivated. The phosphorylated IRS phosphorylates and activates PI3-kinase, which is attracted to the membrane by virtue of its PH domain. PI3-kinase phosphorylates PI (4,5)P2 to produce PI(3,4,5)P3 in the membrane which activates PDK1, which in turn phosphorylates and activates Akt/PKB. SH2, Src homology; PH, pleckstrin homology; PDK1, phosphoinositide dependent protein kinase.

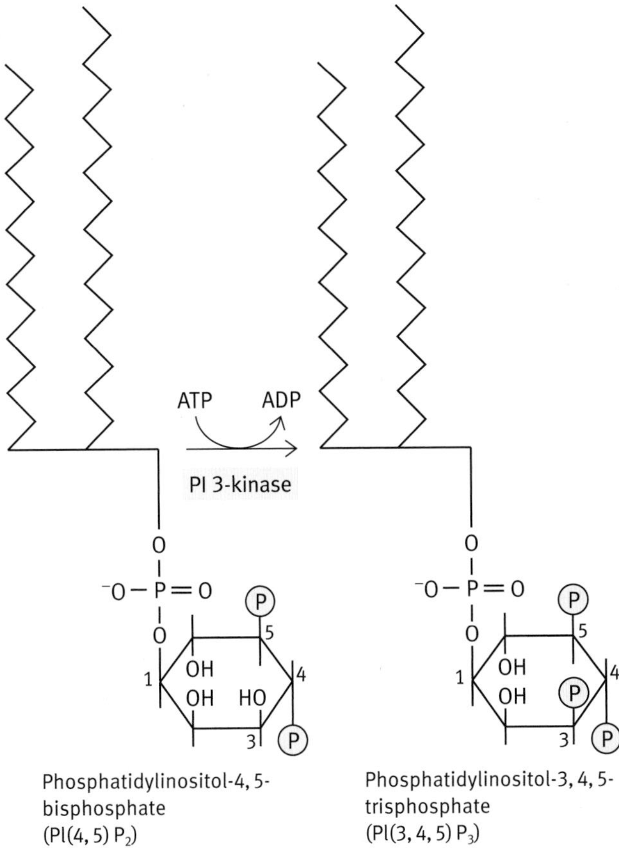

Fig. 29.18 Production of the second messenger $PI(3,4,5)P_3$ from the membrane component $PI(4,5)P_2$ by PI 3-kinase, the latter being insulin-activated.

at three in particular, glucose uptake by muscle and adipose cells, promotion of glycogen synthesis in liver and muscle, and inhibition of lipolysis in the adipose tissue.

Fig 29.19 shows how insulin stimulates glucose uptake by cells of muscle and adipose tissue and glycogen synthesis in liver and muscle cells.

Liver and adipose tissue cells take up glucose through GLUT4 transporters, which are insulin sensitive. When glucose is low and insulin is low, the transporters are found on intracellular vesicle membranes. Active Akt/PKB resulting from binding of insulin to its receptor activates the translocation of GLUT4 vesicles to the cell membrane where they fuse with the membrane and allow the appearance of the transporters on the membrane thus allowing glucose entry into the cell. The glucose is metabolized in both cells. In the adipocyte it provides glycerol phosphate for esterification of fatty acids.

In muscle and liver insulin activates glycogen synthesis. When glucose and insulin are low, glycogen synthase kinase (GSK) phosphorylates and inhibits glucogen synthase (GS3) (see Chapter 20) so no glycogen is formed when glucose is low. In the presence of insulin, the active Akt/PKB phosphorylates glycogen synthase kinase and inhibits it. It therefore cannot phosphorylate the GS, which remains active and allows glycogen synthesis.

Fig. 29.20 shows the inhibition of lipolysis in adipocytes in the presence of insulin. When glucose is low and insulin is low, cAMP increases in the cell through activation of adenylate cyclase by glucagon (see later in the chapter). cAMP activates protein kinase A (PKA), which phosphorylates hormone sensitive lipase in the adipocytes and activates it so TAG is

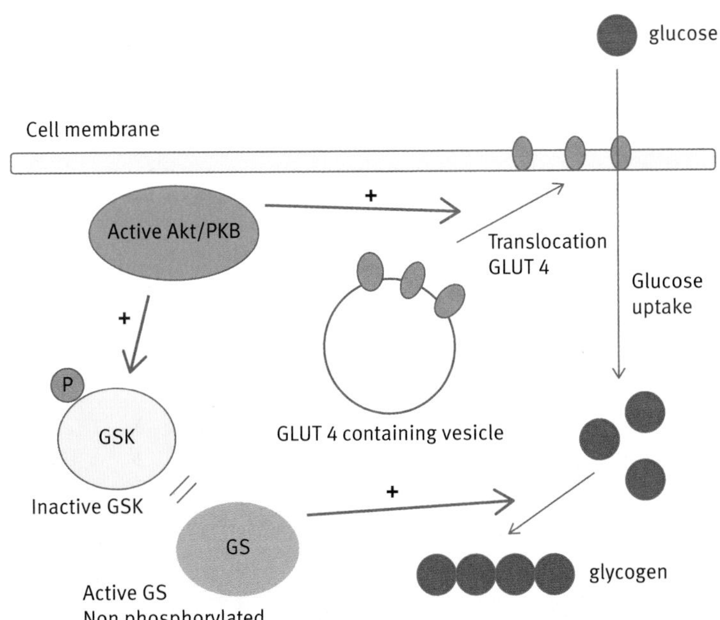

Fig. 29.19 Translocation of GLUT4 transporters to the cell membrane (muscle and adipose) and activation of glycogen synthesis (muscle and liver) by insulin. The active Akt/PKB phosphorylates GSK and inhibits it. It therefore cannot phosphorylate the GS, which remains active and allows glycogen synthesis. GSK, glycogen synthase kinase; GS, glycogen synthase.

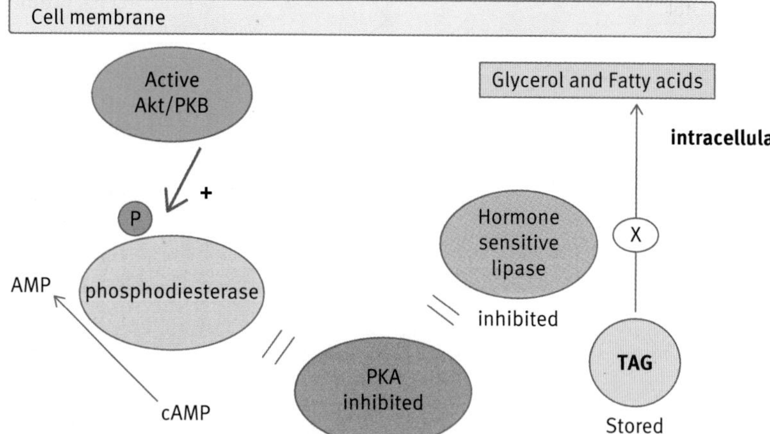

Fig. 29.20 Inhibition of lipolysis in adipose tissue by insulin. In the presence of insulin, active Akt/PKB phosphorylates and activates a phosphodiesterase, which removes cAMP by converting it into AMP. This means that PKA is inactivated and cannot phosphorylate and activate hormone sensitive lipase, which leads to inhibition of TAG hydrolysis. PKA, protein kinase A; HLS, hormone sensitive lipase.

hydrolysed, and glycerol and fatty acids are released into the bloodstream. In the presence of insulin, active Akt/PKB phosphorylates a phosphodiesterase, which removes cAMP by converting it into AMP. This means that PKA is inactivated and cannot phosphorylate and activate hormone sensitive lipase, and TAG hydrolysis is inhibited.

As mentioned above, insulin has effects on gene expression and cell differentiation, which are also mediated through its binding to its receptor but the signalling pathway is different and more similar to the EGF signalling pathway described earlier (i.e through Ras and MAPK).

Fig 29.21 outlines the effect of insulin on gene expression. SHC is an adaptor protein (containing Src homology at the carboxyl terminus); it is another substrate for the insulin receptor kinase. On phosphorylation it associates with GRB and SOS and activates the MAPK pathway. In a similar manner to EGF signalling cascade, Raf is activated, phosphorylates MEK kinase which phosphorylates and activates MAPK which in turn phosphorylates specific transcription factors and affects gene expression.

Having a membrane site, which entices other signal transducers to bind to particular locations where they can be phosphorylated by kinases lying in wait, is a very interesting arrangement, which may have wide applications. It introduces the possibility of cellular controls involving microlocalization of components.The clinical importance of these proteins involved in MAPK signalling pathways is made obvious by the fact that mutations in a number of them are associated with a great number of human cancers.

The JAK/STAT pathways: another type of tyrosine kinase-associated signalling system

A wide variety of genes are controlled by JAK/STAT receptors. The first discovered was involved in the control of genes by the antiviral agent **interferon** (produced in response to an

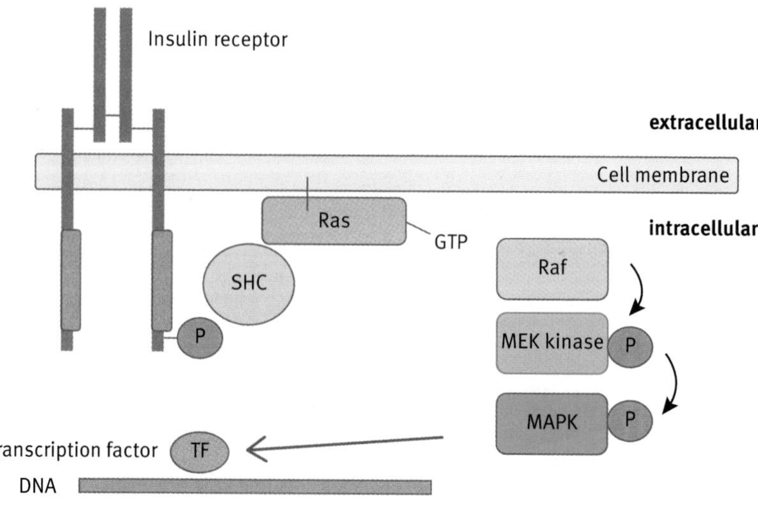

Fig. 29.21 The effect of insulin on gene expression through Ras and MAPK. In the presence of insulin the adaptor protein SHC docks at the phosphorylated tyrosine furthest away from the membrane and activates the anchored protein Ras by phosphorylation. There follows a cascade of Ras phosphorylating and activating Raf, which in turn phosphorylates and activates MEK kinase, which phosphorylates and activates MAPK, which phosphorylates specific transcription factors leading to effects on gene expression, and cell differentiation and proliferation.

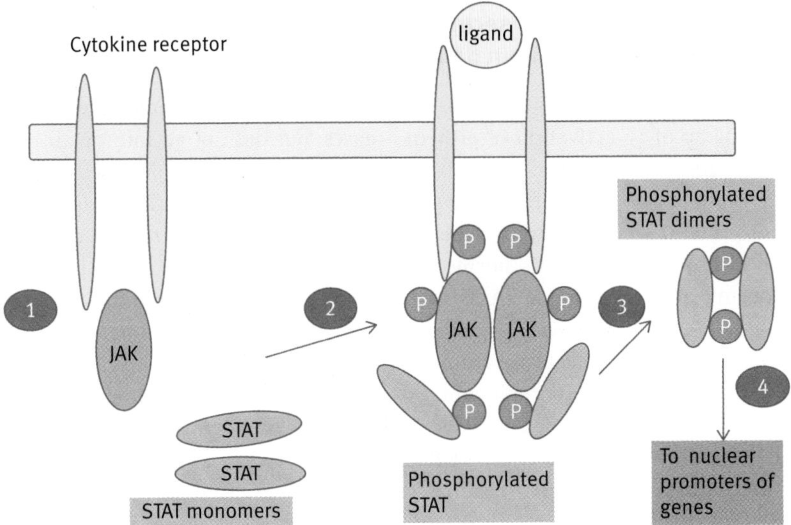

Fig. 29.22 Cytokine signalling through JAK/STAT pathway. The cytokine receptor is not a tyrosine kinase as the insulin receptor or EGF receptor. The tyrosine kinase is a separate protein, JAK with two tyrosine kinase domains. On ligand binding the receptor recruits JAK, which phosphorylates the tyrosine receptor, itself, and monomeric proteins known as STAT. The phosphorylated STAT dimer is a transcription factor, which travels to the nucleus and binds promoter regions of selected genes. JAK, Janus kinase; STAT, signal transducer and activator of transcription.

infection), but they are now known to be involved in the signalling pathways of a wide variety of cytokines and growth factors. The remarkable feature of JAK/STAT signalling pathways is that instead of the multiplicity of intermediates in the pathway the signal is carried from receptor to nucleus by a single protein dimer (Fig. 29.22).

The receptors in the cell membrane exist as monomers. The cytosolic domains have no tyrosine kinase activity themselves but depend on the JAK family of tyrosine kinases to phosphorylate and activate downstream proteins involved in their signal transduction pathways. The receptors exist as paired polypeptides, exhibiting two intracellular signal-transducing domains. JAKs associate with a proline-rich region in each intracellular domain, which is adjacent to the cell membrane. (JAK name derives from Janus the two-faced Roman god; JAKs have two kinase sites). In the absence of an activating signal the kinases are inactive. On binding of a cytokine, the receptors in the membrane dimerize and their associated JAKs are activated by autophosphorylation. They phosphorylate tyrosine groups on the receptor and also on cytosolic proteins known as **STATs** (STAT, **signal transducer and activator of transcription**). STATs bind to the phosphorylated receptor via their SH2 domains and are themselves phosphorylated by the JAKs on tyrosine sites. The phosphorylated STATs leave the receptor and form phosphorylated dimers, each dimer being an active transcription factor, which is transported into the nucleus. There it binds to its target gene-control elements and switches on specific genes. In the case of interferon activation, they protect the cell against viruses.

The pathways have great versatility. There are three classes of receptors and four different kinases (JAKs 1, 2, and 3 and TYK2) that bind to different phosphorylated receptors with high specificity. There are, in mammals, seven different STAT proteins, which can form homodimers or heterodimers. Many different signalling pathways can thus be assembled.

Termination of JAK/STAT signalling pathways

As emphasized previously, signalling pathways have to have a mechanism for switching off as well as a switching on otherwise a signal would last forever. The JAK/STAT pathways have a multiplicity of checks. There is a phosphatase that reverses the activation by dephosphorylating proteins. Cytokines attached to receptors may be internalized and degraded. There are also **suppressors of cytokine signalling (SOCS)** proteins, which bind to and inhibit the JAKs attached to the receptors. The SOCS proteins inhibit the protein kinase domain by blocking its active site.

The STAT dimers activate the genes needed to respond to the cytokine signal but also activate SOCS genes giving a feedback-control system. The cytokine thus switches on the positive and negative responses at the same time. Gene-knockout mice (Chapter 28) lacking SOCS, when given interferon, show adverse physiological effects as they are unable to terminate the signal. Additionally there are protein inhibitors called **protein inhibitors of activated STATs (PIAS)**, which bind to activated STAT proteins and prevent them from switching genes on.

G-protein-coupled receptors and associated signal transduction pathways

Overview

We now turn to the second major class of signal transduction pathways one that does not involve tyrosine kinases but does involve G-protein coupled receptors. They constitute a very large superfamily of receptors, examples of which are found in a great variety of organisms from yeast and insects to humans. The signals include hormones, growth factors, odours, and light.

Each receptor has associated with it on the cytosolic side of the cell membrane, a **heterotrimeric G-protein**, made up of three subunits, α, β, and γ. All the G-proteins have a common pattern. The α subunit of the trimeric *inactive* G-protein has a molecule of GDP bound to it (hence the name, G-protein). On binding of the signal molecule to the receptor, a conformational change in the associated G-proteins occurs and GDP is replaced by GTP. This in turn causes activation of a signal pathway involving the formation of second messengers, which bring about the cellular responses as outlined earlier (see Fig. 29.7b).

A note on terminology: Ras also is a GTP/GDP-binding protein and has a switching on and terminating mechanism much the same as that to be described for G-proteins. The difference is that Ras belongs to the class known as **small monomeric GTPase proteins** rather than G-proteins, *a term used only for the heterotrimeric class*. Let us now look at this class of signalling system in greater detail:

Structure of G-protein-coupled receptors

All are proteins whose polypeptide chain crosses the membrane seven times (Fig. 29.23); this type of protein is called serpentine or polytopic. In the case of adrenaline (epinephrine), the hormone binds to a cleft formed by the transmembrane helices while the heterotrimeric G-proteins associate with a polypeptide loop on the cytosolic side of the membrane. In real life, the transmembrane domains are clustered together, not extended as in the figure.

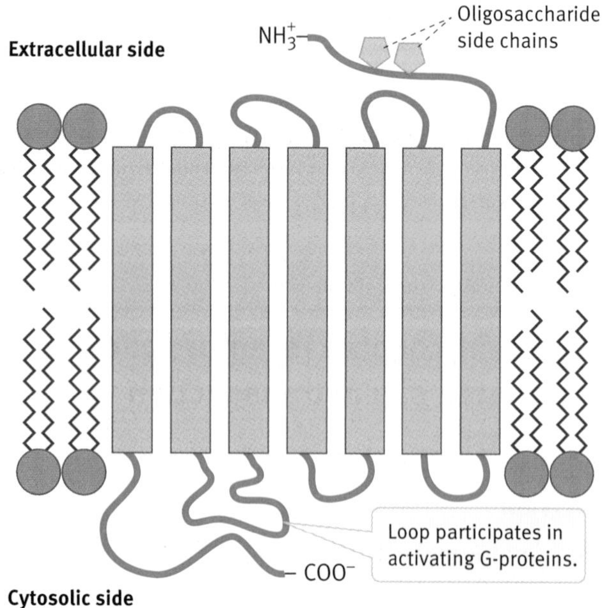

Fig. 29.23 The structure of the β_2 adrenergic receptor.

cAMP as second messenger: adrenaline signalling – a G-protein pathway

In Chapter 20 we dealt with control of metabolism by cAMP activation of protein kinases, but did not go into detail. We will deal with this now and also with the mechanism by which cAMP regulates gene expression. cAMP is the *second messenger*, for a wide array of hormones, including adrenocorticotrophic hormone (ACTH; or corticotrophin), antidiuretic hormone (ADH; or vasopressin), gonadotrophins, thyroid-stimulating hormone (TSH), parathyroid hormone, glucagon, the catecholamines adrenaline (epinephrine)and noradrenaline (norepinephrine), and somatostatin (the latter negatively controls cAMP levels). This list, which is not exhaustive, illustrates that cAMP has different effects in different cells and the response to cAMP in a given cell is appropriate to the signal that that cell recognizes via its receptors. Thus, in cell A, signal X elevates cAMP levels. This produces cellular responses appropriate to signal X. Cell B does not have receptors for X, but does for signal Y, which also elevates cAMP. In cell B this causes responses appropriate to signal Y. For example, cAMP mobilizes glycogen breakdown in liver cells and triacylglyceride breakdown in fat cells.

Control of cAMP levels in cells

cAMP is produced from ATP by adenylate cyclase (Fig. 29.24), an enzyme which is an integral cell membrane protein (Fig. 29.25 a). A typical example of a pathway that activates adenylate cyclase is that using the **β_2-adrenergic receptor**. Associated with the cytosolic face of the receptor is a G-protein. The α subunit has a site on it that can have bound to it *either* GTP or GDP. When the site on the α subunit is occupied by GDP, nothing happens. This is the situation in the absence of hormone, illustrated in Fig. 29.25 (a).

When a molecule of adrenaline binds to the receptor, the receptor undergoes a conformational change, causing conformational change in the α subunit of the G-protein bound to it, which exchanges its GDP for GTP. *This exchange cannot take place unless the G-protein is attached to a receptor to which the hormone is bound.* The α-GTP complex detaches, migrates, and binds to an adenylate cyclase enzyme molecule, which now becomes activated and produces cAMP (Fig. 29.25(b,c)) by the reaction shown in Fig. 29.24(a).

The hormone switches on cAMP production using the G-protein as an intermediary. Activation of cAMP production by the α-GTP complex must be limited in time – it must be switched off, for otherwise a single hormone stimulus would last indefinitely, long after the response would be appropriate.

The α subunit of the G protein in its GTP-bound form is a GTPase. It hydrolyses its attached GTP molecule to GDP (Fig. 29.25 (d)). The activity is low so that the hydrolysis occurs only after a delay. As soon as GDP is formed, the α subunit reverts to its original state; it detaches from the adenylate cyclase, which is now inactivated. The α subunit rejoins the β and γ subunits,

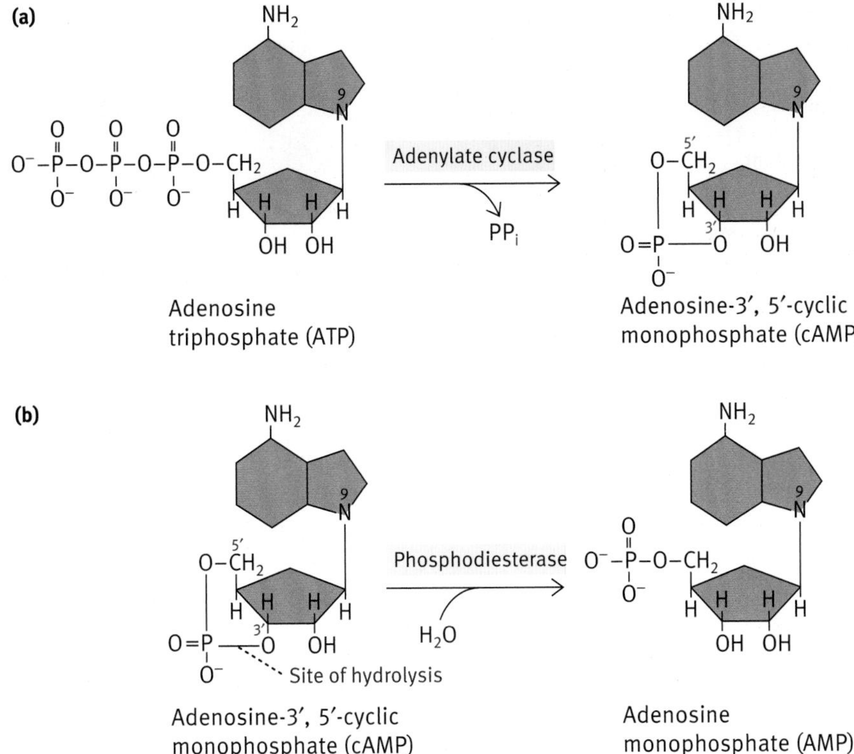

Fig. 29.24 (a) Formation of cyclic AMP (cAMP) by adenylate cyclase; **(b)** hydrolysis of cAMP by phosphodiesterase. Adapted with permission from Dohlman, H. G., Caron, M. G., and Lefkowitz, R. J.; Biochemistry; 26, 2660. Copyright 1987 American Chemical Society.

and the G-protein–GDP trimeric complex is reassembled, in contact with the receptor (Fig. 29.25 (e)). If the hormone is still bound to the receptor, the whole cycle can start again (back to Fig. 29.25 (b)). If the hormone has already dissociated from the receptor, (back to Fig. 29.25(a)), the process comes to a halt. Hydrolysis of cAMP into AMP by phosphodiesterase in the cell completes the termination of the signal (see Fig. 29.24(b)). Thus, to continue cAMP production, the α subunit moves back and forth from receptor to adenylate cyclase, the duration of its stay on the adenylate cyclase being that required for the hydrolysis of its attached GTP molecule. The situation is rather like that of a timed light switch on a staircase where you press a button, the light goes on, and the button slowly comes back out and switches off the light after a minute or two. You have to keep on pressing the button at intervals for the light to stay on. The G-protein can be thought of as a timing device, which limits the period of activation of adenylate cyclase. The system has an amplifying effect, since one molecule of hormone, on binding to a receptor, causes the synthesis of many molecules of cAMP, each of which furthers the activation.

The importance of terminating the signal through GTP hydrolysis becomes obvious when we look at a condition in which this cannot take place, cAMP activates different processes in different cells. In the liver it leads to glycogen degradation but in the gut it activates Na^+ secretion from the gut mucosal cell into the intestinal lumen.

In **cholera** the secretion of Na^+ and water into the intestinal lumen becomes uncontrolled. This happens because the cholera toxin, an enzyme, inactivates the GTPase activity of the G-protein α subunit by ADP ribosylating the subunit, that is by transferring an ADP-ribose molecule to the GTPase, blocking the active site. So, once the adenylate cyclase is activated, it cannot be switched off. The α subunit cannot hydrolyse GTP and the prolonged cAMP production results in massive loss of Na^+, accompanied by water molecules, causing severe diarrhoea and possible death from fluid and electrolyte loss.

GTPase-activating proteins (GAPs) regulate G-protein signalling

We have already described GAPs acting on Ras-type signal transduction proteins. GAPs controlling the heteromeric G-proteins also exist, acting on the α subunit, whose GTPase activity can be enhanced more than 2000-fold.

Different types of G-protein receptor

In the case given in Fig. 29.25, the GTP-α subunit stimulates adenylate cyclase activity and is called G_s (s for **stimulatory**). Another type of receptor for adrenaline (epinephrine) (known as the α_2 receptor) operates similarly, except that the GTP-α subunit (called G_i **for inhibitory**) inhibits adenylate cyclase. Examples of G_i receptors are those for **angiotensin** and **somatostatin**. Thus, a hormone can exert different effects on different cells, according to the type of receptor present. It illustrates the way in which different G-proteins associated with particular receptors can control different signal transduction pathways. Since there are multiple

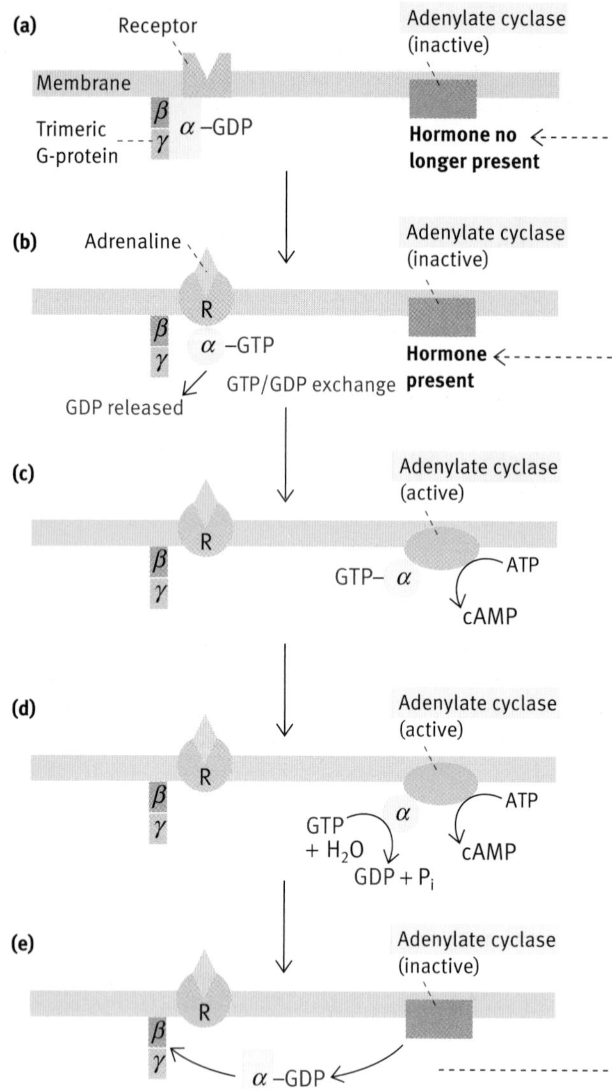

Fig. 29.25 The control of adenylate cyclase activity by a hormone such as adrenaline. The geographical locations of the subunits shown are hypothetical, the essentials being that the α-GTP complex activates adenylate cyclase. The steps (a)–(e) are referred to in the text.

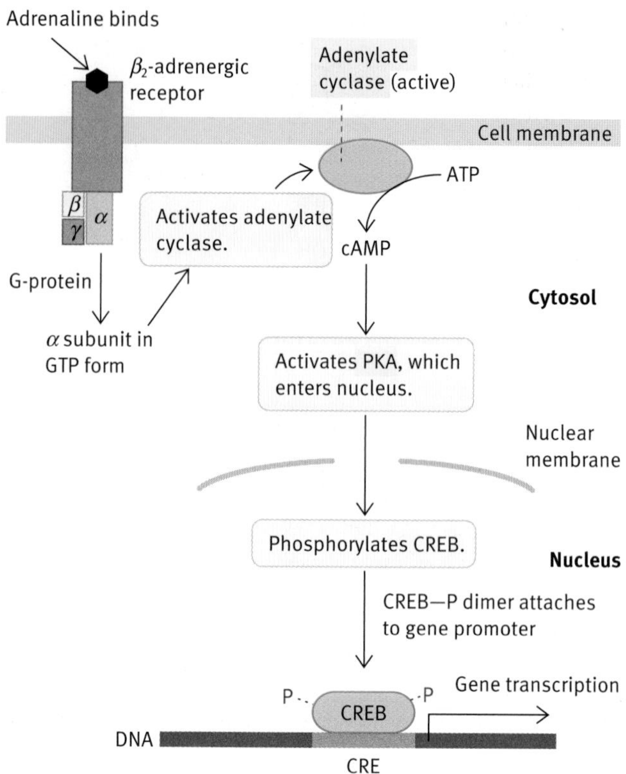

Fig. 29.26 The β_2-adrenergic receptor function. Binding of adrenaline to the receptor causes the heterotrimeric G-protein to convert to the GTP form, detachment of the α subunit – GTP, which activates adenylate cyclase. The cyclic AMP (cAMP) produced activates protein kinase A (PKA), which is transported into the nucleus. There, the PKA phosphorylates cAMP-response-element-binding protein (CREB), which dimerizes and in this form is a transcription factor for specific genes. This activates the genes, resulting in the appropriate responses to the adrenaline. CRE, cAMP-response element of gene promoter.

forms of each of the G-protein heterotrimer subunits, different combinations are possible giving flexibility of control options.

How does cAMP control gene activities?

In Chapter 20, we described how cAMP activates **protein kinase A (PKA)**. PKA, as well as being involved in metabolic control is part of a pathway of gene control, as shown in Fig. 29.26. Activated PKA is transported into the nucleus. The promoters of several cAMP-inducible genes contain **cAMP-response elements (CREs)**. A response element is a section of DNA promoter to which a transcription factor binds and activates the gene. A **CRE-binding protein (CREB)** , when phosphorylated by PKA, dimerizes and becomes an active transcription factor. Mutations that permanently activate the α subunit of the G-protein lead to excessive PKA activation and may be oncogenic.

Desensitization of the G-protein receptors

Typically, cellular responses to extracellular signals diminish with prolonged exposure to the signal molecule. In some cases synthesis of the receptor is reduced, and/or the receptors are endocytosed and degraded in lysosomes, as happens for example with the insulin receptor. The reduction in receptor numbers is known as **downregulation**. An alternative mechanism of diminishing the cellular response to a signal is desensitization in which the receptor is inactivated when the signal is prolonged. Many of the G-protein-coupled receptors are desensitized by a family of enzymes called **G-protein receptor kinases (GRKs)**, which phosphorylate the receptors inactivating them. Then an inhibitory protein, **β-arrestin**, binds to the phosphorylated site.

This is not to be confused with the tyrosine kinase receptors in which activation is coupled with phosphorylation. The GRKs phosphorylate only *activated* G protein receptors to which their specific signal molecules are bound – they only inactivate receptors, which are in the activated state. There is evidence that the GRKs are themselves subject to controls.

The phosphatidylinositol cascade: another example of a G-protein-coupled receptor that works *via* a different second messenger

Other G-protein-coupled receptors exist that use quite different signalling pathways from that described for the adrenergic receptors. In the signalling system we will now describe, a different enzyme is activated by the α-GTP (α subunit in the GTP form), leading to the formation of a second messenger other than cAMP. Examples of signals that use the mechanism include acetylcholine and vasopressin (ADH).

We have seen how the cell membrane component, **phosphatidylinositol-4,5-bisphosphate (PI(4,5)P_2)** is involved in the insulin signalling pathway through PI 3-kinase. The signalling pathway we are about to describe is different from the PI 3-kinase although they both start with PI(4,5)P_2 (see Fig. 29.18). Once again a G-protein links the receptor to the intracellular signalling pathway. Binding of the hormone to a receptor causes the GTP-α subunit to exchange its GDP for GTP; it then migrates to, and activates, a membrane-bound enzyme, **phospholipase C (PLC)**, which hydrolyses PI(4,5)P_2 into **inositol trisphosphate (IP_3)** and **diacylglycerol (DAG)** (Fig. 29.27). The IP_3 leaves the membrane and causes release of Ca^{2+} from the lumen of the endoplasmic reticulum (ER) where the ion is stored in high concentration relative to that in the cytosol. It opens IP_3-gated Ca^{2+} channels in the membrane, allowing the ion to be released into the cytosol. The signal is reversed when the hormone is no longer attached to the receptor, when GTP is hydrolysed, IP_3 is degraded, and Ca^{2+} returns to the lumen by a Ca^{2+}/ATPase pump. Thus, the combination of a hormone, or other signal, with a receptor associated with the phosphatidylinositol cascade results in increases of intracellular DAG and Ca^{2+} (Fig. 29.28).

DAG is the physiological activator of **protein kinase C (PKC)**, which is also activated by Ca^{2+}. Cytosolic PKC is attracted to the membrane by DAG and activated. It is a protein kinase with multiple target proteins and also multiple versions of PKC exist. PKC is involved in phosphorylating

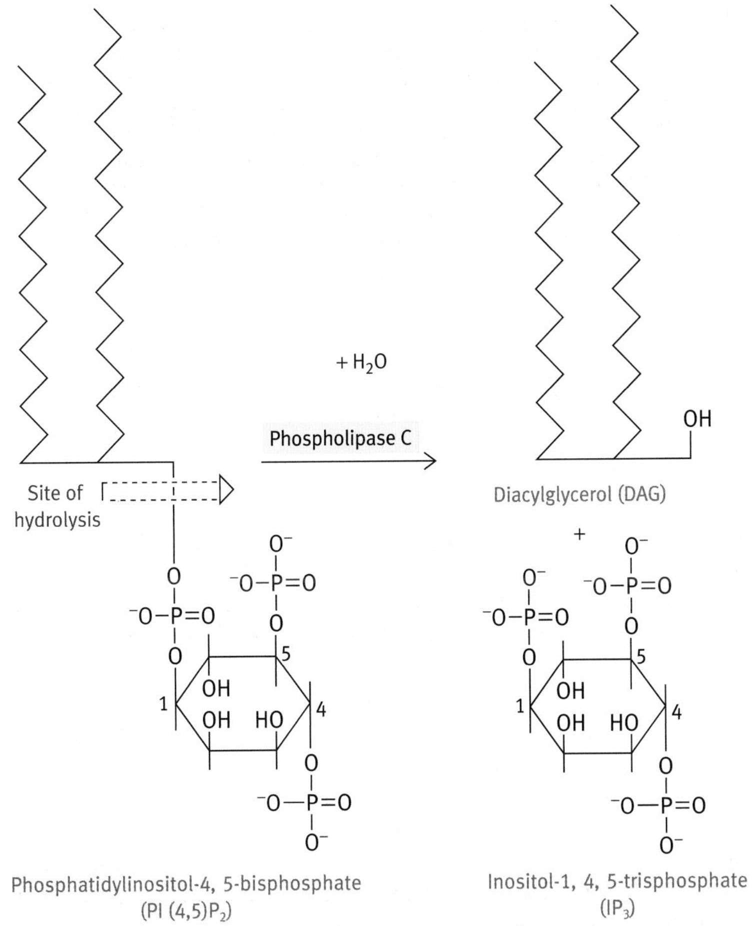

Fig. 29.27 Hydrolysis of phosphatidylinositol-4,5-bisphosphate (PI(4,5)P_2) to diacylglycerol (DAG) and inositoltrisphosphate (IP_3).

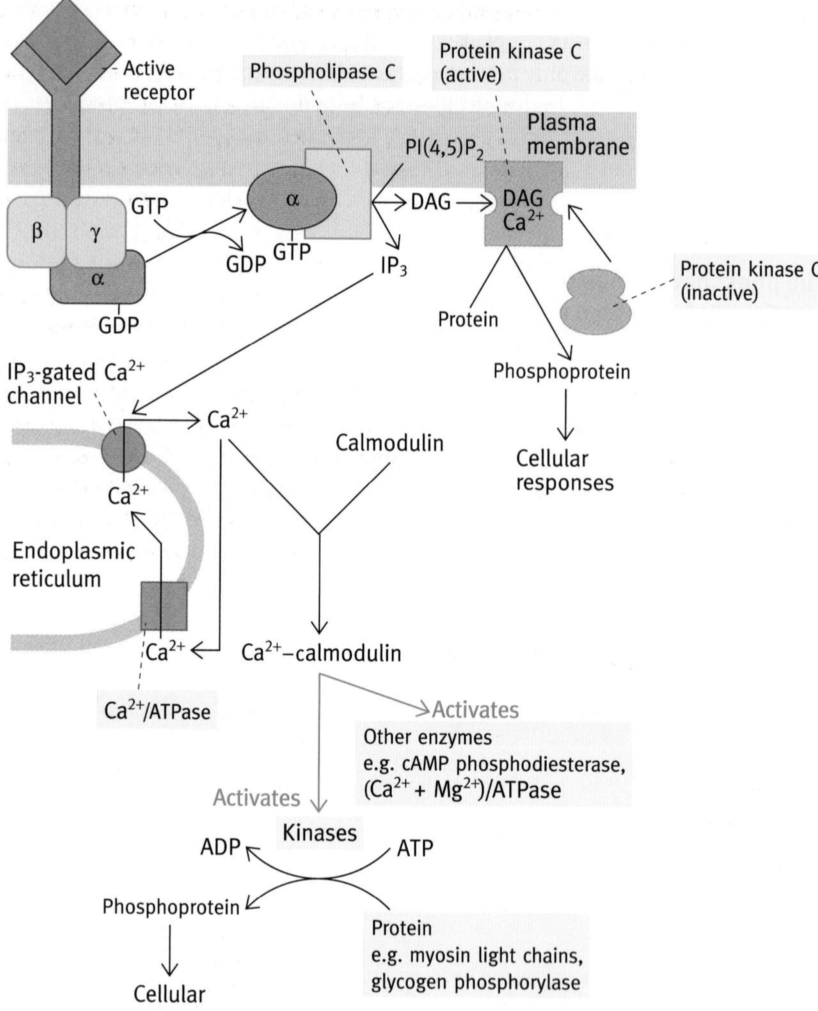

Fig. 29.28 The phosphatidylinositol cascade: interactions of diacylglycerol (DAG), inositoltrisphosphate (IP_3), and Ca^{2+} as second messengers. Note that the G_α protein binds to phospholipase C (PLC) only in the GTP-complexed form. On hydrolysis to GDP the process reverses. This illustrates the versatility of G-protein-associated receptors. Different receptors are associated with different G-proteins whose α subunits, when in the GTP form, control the activities of different enzymes – in this case G_α activates phospholipase C. See earlier in this chapter for a description of what happens. (Compare with Fig. 27.19, where the G-protein α subunit activates adenylate cyclase.)

transcription factors in the nucleus and it also has a controlling role in the regulation of cell division as is illustrated by the tumour promoting effect of **phorbol esters.** These are analogues of DAG (Fig. 29.29) capable of activating PKC, leading to cell division. It may seem incongruous that DAG, a normal cellular signalling molecule, has the same effect in activating PKC as does a promoter of tumour formation, but DAG is rapidly destroyed and activates PKC only when required, whereas phorbol esters are longer-lived and deliver an inappropriately prolonged signal. DAG and Ca^{2+} are both needed for maximal activation of PKC but quite apart from this, Ca^{2+} is an important second messenger on its own.

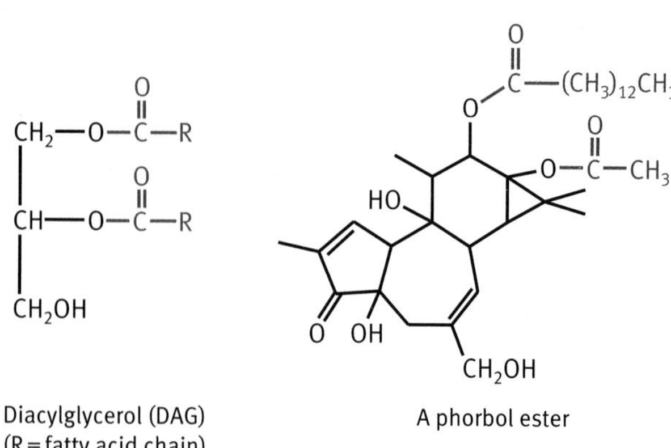

Fig. 29.29 Phorbol esters are analogues of diacylglycerol, the natural activator of protein kinase C. (The complete structure of the phorbol ester is given only to illustrate this point.)

Other roles of calcium in regulation of cellular processes

Ca^{2+} ions control a wide variety of cellular processes. The human body contains about 1 kg of calcium, with about

99% of it as a structural component of bones and teeth, and about 1% in the blood and extracellular fluid; only a tiny proportion is intracellular. The cytosolic concentration of Ca^{2+} is very low and this is achieved by Ca^{2+}/ATPases, which pump Ca^{2+} either to the outside of the cell, the mitochondria, the ER lumen, or in the case of skeletal muscle, into a special sac called the **sarcoplasmic reticulum** (see Chapter 8). Appropriate signals open gated calcium channels (see Chapter 7), which release the ion back into the cytosol where it acts as a regulator. The steep concentration gradient across the membranes means that there is an instant delivery of Ca^{2+} to the cytosol.

A general second-messenger role of the Ca^{2+} involves combination with a widely distributed protein called **calmodulin.** It has four sites for binding Ca^{2+} with high affinity, causing a conformational change in the protein. Calmodulin is sometimes found in association with the enzymes it controls, or it may be free and attach to enzymes in its Ca^{2+}-bound form, depending on the enzyme. The Ca^{2+} causes a conformational change in the calmodulin, which alters the activity of the enzyme it is associated with. A number of calmodulin-Ca^{2+}-activated protein kinases exist and Ca^{2+} can, via this route, exert multiple cellular effects. The target proteins of the calmodulin-activated kinases include glycogen phosphorylase, and myosin light chains (see Chapter 8) but these are only a few of many examples.

Vision: a process dependent on a G-protein-coupled receptor

The versatility of G-protein signalling pathways using different specific G-proteins is illustrated by this example, in which light is the signal. The challenge here is to convert the stimulus of light photons into chemical changes, which result in impulses in the optic nerve carrying signals to the brain, manifested as vision.

The vertebrate retina has two types of cells for light detection: **rods** for black-and-white and dim-light vision, and **cones** for colour vision. The rod cell (which is the one that we will discuss) has two segments. The inner segment has the mitochondria, nucleus, and the synthetic machinery of the cell. The inner end of the cell makes a synapse with a bipolar cell that connects with the optic nerve. At the other end is a cylindrical rod-shaped section in which there is a stack of membranous discs (as many as 2,000) embedded in the cytosol (Fig. 29.30). These discs contain the light-detection machinery.

Transduction of the light signal

The process of light detection is complex. Here we are going to deal only with the essential principles. In the dark, the rod cells have a relatively high level of **cyclic GMP (cGMP)**, analogous to cAMP, synthesized by a **guanylate cyclase** (see Fig. 29.34). In this instance, *cGMP is not strictly a second messenger,*

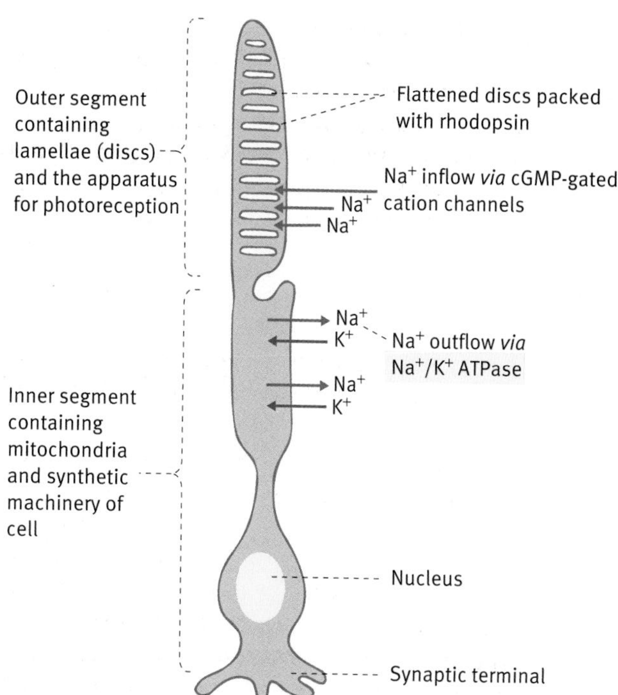

Fig. 29.30 Structure of a rod cell.

since it is not produced as a result of receptor activation. In the cell membrane there are ligand-gated cation channels (see Chapter 7), which are kept open by cGMP binding to them as their controlling ligand.

In the dark, the constant inflow of Na^+ through these channels gives the cell membrane a potential across it at equilibrium of –30 mV (see Chapter 7 if you want to refresh your memory on membrane potentials).

The light receptor in the discs is **rhodopsin**, a complex of the protein opsin and the visual pigment, **11-*cis*-retinal**, synthesized from dietary retinol (vitamin A) or beta-carotene (pro-vitamin A). The 11-*cis*-retinal is linked to the ε–NH_2 amino group of a lysine residue in the rhodopsin. Absorption of a photon converts rhodopsin to meta-rhodopsin II. As with other G-protein-associated receptors it has seven transmembrane helices. Light causes activation of rhodopsin by a conformational change in the visual pigment to become **all-trans-retinal** (Fig. 29.31). This causes a conformational change in the cytosolic domain of rhodopsin. The associated heterotrimeric G-protein, called **transducin**, exchanges its bound GDP for GTP. The α-GTP complex detaches and activates a membrane-bound enzyme, in this case cGMP phosphodiesterase, which degrades cGMP (Fig. 29.32). This enzyme has a similar action to the cAMP phosphodiesterase shown earlier in Fig. 29.24(b). *Lowering* the cGMP concentration results in closure of the cation channels, the inflow of Na^+ ceases and the membrane potential increases to –70 mV – it becomes hyperpolarized. This triggers a nerve impulse in the optic nerve to flow to the visual centre in the brain.

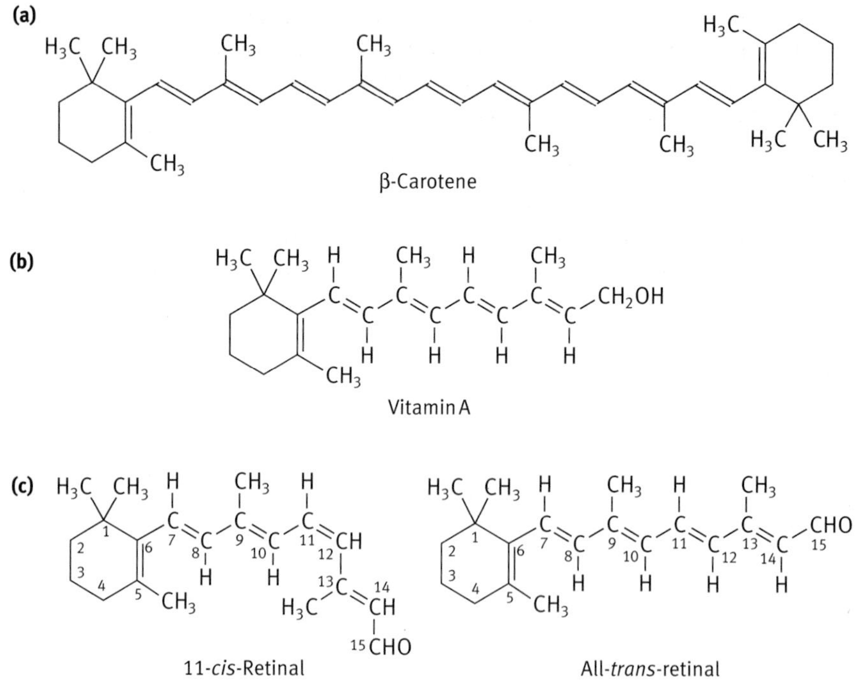

(a) β-Carotene

(b) Vitamin A

(c) 11-*cis*-Retinal All-*trans*-retinal

Fig. 29.31 **(a)** Structure of β-carotene, which is the principal accessory photosynthetic pigment in plants and a precursor of vitamin A in animals. **(b)** Structure of vitamin A. **(c)** Structures of 11-*cis*-retinal and all-*trans*-retinal. These structures are given for reference purposes only.

The recovery of the cell after illumination is complex. A primary event is the inactivation of the α-GTP subunit by its GTPase activity, which results in the reassembly of the trimeric transducin. Recovery of the cell involves inactivation of the activated rhodopsin, lowering of the Ca^{2+} concentration in the cell, which stimulates cGMP synthesis, and recycling of the rhodopsin by a complex pathway.

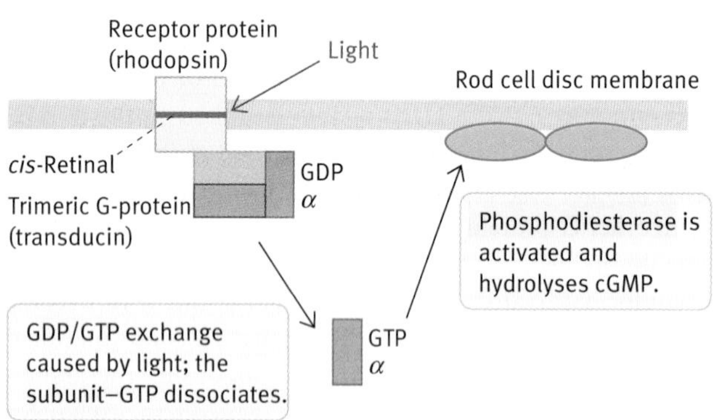
Receptor protein (rhodopsin)
Light
Rod cell disc membrane
cis-Retinal
GDP
α
Trimeric G-protein (transducin)
Phosphodiesterase is activated and hydrolyses cGMP.
GDP/GTP exchange caused by light; the subunit–GTP dissociates.
GTP
α

Fig. 29.32 The G-protein-coupled receptor involved in vision. The receptor consists of seven transmembrane helices with 11-*cis*-retinal as the chromophore.The cytosolic domain on light stimulation causes the trimeric G-protein transducin to exchange its GDP for GTP. The α-GTP subunit migrates to the membrane and activates a phosphodiesterase which hydrolyses cyclic GMP (cGMP). This results in closure of cation channels and cessation of Na^+ inrush. The consequent increase in membrane polarization is converted into a signal in the optic nerve.

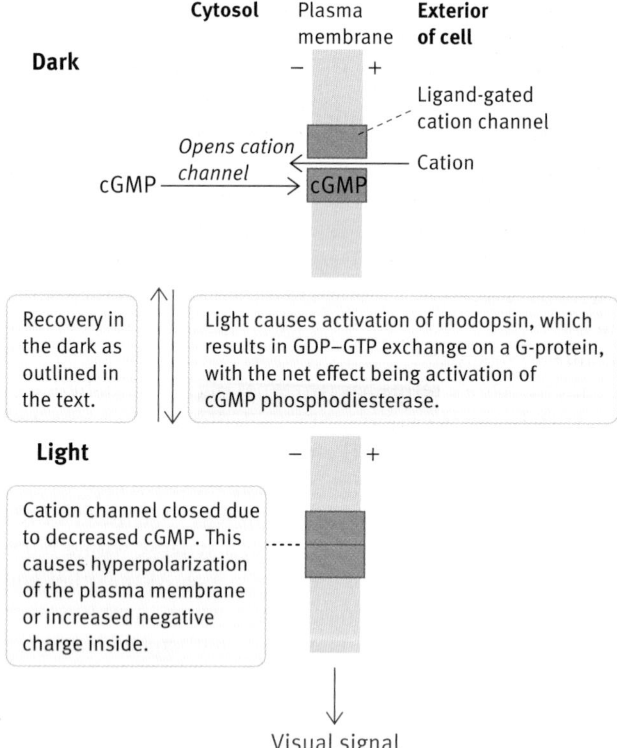
Cytosol
Plasma membrane
Exterior of cell
Dark
–
+
Ligand-gated cation channel
Opens cation channel
Cation
cGMP
cGMP
Recovery in the dark as outlined in the text.
Light causes activation of rhodopsin, which results in GDP–GTP exchange on a G-protein, with the net effect being activation of cGMP phosphodiesterase.
Light
–
+
Cation channel closed due to decreased cGMP. This causes hyperpolarization of the plasma membrane or increased negative charge inside.
Visual signal

Fig. 29.33 Simplified diagram of the visual process. The action of light on rhodopsin results in the activation of cyclic GMP (cGMP) phosphodiesterase. Decreased cGMP results in the closure of a cation channel, hyperpolarization of the cell membrane, and an optic nerve impulse.

To summarize the whole process (Fig. 29.33), in light, cGMP concentrations decrease, cation channels close and hyperpolarization of cell membrane leads to a visual signal in the optic nerve. After illumination, cGMP levels are restored, Na^+ channels open, and the system is restored to readiness for the next photon.

Colour vision in the cone cells is due to the presence of three different visual pigments all proteins with 11-*cis*-retinal attachments, and each more than 95% homologous in amino acid sequences to rhodopsin. The three pigments are responsible for red, green and blue vision respectively.

We now come to two different pathways where cGMP is acting as a true second messenger.

Signal transduction pathway using cGMP as a second messenger

Membrane receptor-mediated pathways

The heart produces a neuropeptide hormone which regulates salt balance and affects blood pressure. It acts via cGMP as a second messenger. The reaction for the formation of cGMP is shown in Fig. 29.34. The hormone, **atrial natriuretic peptide**, produced by endothelial cells, combines with its membrane receptor on kidney cells whose *inner domain is a guanylate cyclase* and this is activated by a conformational change resulting from the attachment of the neuropeptide to the receptor (Fig. 29.35). The raised cGMP concentration mediates cell responses by activating protein kinases resulting in the appropriate cellular effects, including increased Na^+ excretion by the kidneys. Hydrolysis of cGMP by a phosphodiesterase reaction analogous to that for cAMP (see Fig. 29.24(b)) confers reversibility.

There is a second control system producing cGMP as a second messenger. (cGMP is, as stated, not definable as a second messenger in the visual process.)

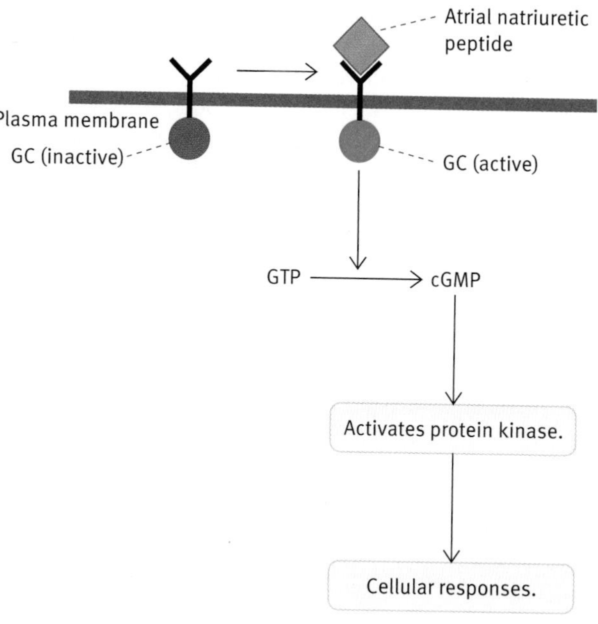

Fig. 29.35 Production of the second messenger cyclic GMP (cGMP) by a membrane receptor (R), activated by the atrial natriuretic peptide (blue). It has a single polypeptide chain with a membrane-spanning sequence and a guanylate cyclase catalytic site (GC) on the cytosolic face. On attachment of the signal peptide to the receptor, it undergoes a conformational change responses, which activates the cyclase and leads to the production of cGMP from GTP. This, *via* the activation of a protein kinase, leads to the cellular responses.

Nitric oxide signalling – activation of a soluble cytosolic guanylate cyclase

The second guanylate cyclase is present in the cytosol. This has a haem molecule as its prosthetic group to which binds a signalling molecule of surprising simplicity – **nitric oxide (NO)**. The haem molecule functions as a detector of NO at concentrations as low as 10^{-8} M and transduces the signal into cGMP synthesis from GTP. NO is produced from the arginine guanidino group by **nitric oxide synthase**, in endothelial cells lining parts of the vascular system and

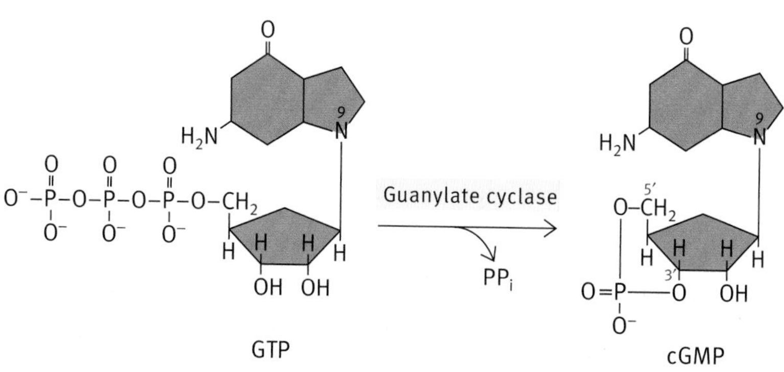

Fig. 29.34 Formation of 3′,5′-cyclic GMP by guanylate cyclase.

elsewhere. The NO diffuses into the smooth muscle of blood vessels, causing cGMP production, which, in turn, causes muscle relaxation and vessel dilation. The NO has a lifetime of a few seconds. NO is also produced in response to shearing forces exerted by blood flow on the endothelial cells lining the vessels. This results in vasodilation. Since NO is oxidized to higher oxidized states of nitrogen in seconds, it is a very locally acting (paracrine) hormone; being lipid soluble it escapes from cells producing it and enters adjacent cells. Trinitroglycerine, a drug long used in the treatment of angina, slowly produces NO thereby relaxing blood vessels (including coronary vessels) and reducing the workload of the heart. NO is part of a complex regulatory system with multiple physiological effects. The phosphodiesterase, which destroys cGMP is inhibited by the drug sildenafil (**Viagra**). This potentiates the effect of NO, production of the latter being increased by sexual stimulation. Dilation of blood vessels in the penis aids erection.

It has been suggested recently that control of the soluble guanylate cyclase is more sophisticated than hitherto believed. It seems that, in resting states, smooth muscle tone is maintained by a low concentration of NO that combines with the haem prosthetic group and partially activates the cyclase. When there are bursts of NO production, such as caused by liberation of acetylcholine, the higher concentration of NO combines at a nonhaem site and gives a transitory full activation of the cyclase. This results in immediate smooth muscle relaxation.

Overview and summary

Figure 29.36 gives a simplified overview of the signalling pathways dealt with in this chapter with emphasis on the protein kinases and second messengers involved.

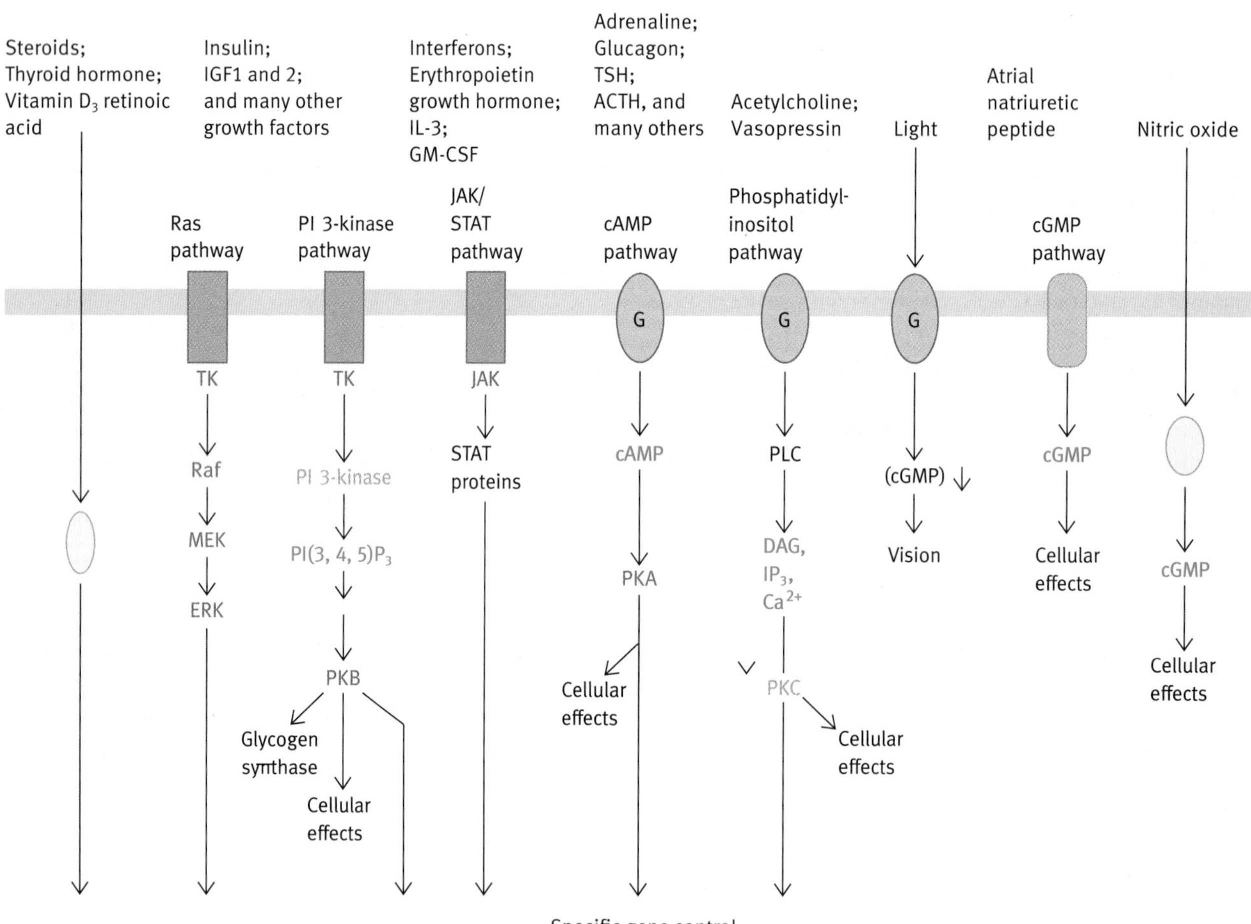

Fig. 29.36 Simplified summary diagram of the signal tranduction pathways. The figure shows only the kinases (red), which are activated by the receptors, and second messengers formed (green). Yellow receptors are intracellular. Receptors shown in red are phosphorylated on binding of the signal (TK, tyrosine kinase; JAK, Janus kinase). G indicates that the receptor is of the heterotrimeric G-protein-coupled type. The type of receptor shown in orange is unusual in that the guanylate cyclise, which forms the second messenger is part of the receptor itself. Note that the pathways are of extraordinary complexity with multiple parallel pathways. Note also that the light receptor is an exception in that its effect is to reduce the level of cGMP, which is not a second messenger in this context. EGF, epidermal growth factor; IGF, insulin-like growth factor; IL-3, interleukin 3; GM-CSF, granulocyte–macrophage colony-stimulating factor; TSH, thyroid-stimulating hormone; ACTH, adrenocorticotrophic hormone. The lists of growth factors, cytokines and hormones using the different pathways are illustrative examples only, not exhaustive lists.

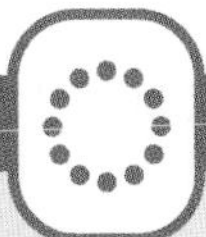

Summary

Cells have receptors that receive signals from other cells. In unicellular organisms such as yeast and bacteria this is limited in scope, but in animal cells such as mammalian ones, the number of signals needed to coordinate their activities is large. They control the most fundamental aspects of gene control, cell survival, programmed cell death, and cell division as well as metabolism. Cancer usually involves malfunction of signalling pathways.

The signals are variously proteins, peptides, steroids and other lipid-related molecules, and nitric oxide. They are hormones, neurotransmitters, growth factors, and cytokines.

The signalling molecules bind to specific receptors of target cells and activate signalling pathways, which result in gene-control events in the nucleus and/or more direct effects on metabolism.

Steroids enter the cell directly and attach to intracellular receptors but water-soluble signals, the predominant class, bind to membrane-bound receptors exposed on the cell exterior. They span the membrane and are exposed also on the inside where they cause an allosteric change in their cytosolic domains. This leads to activation of signalling pathways within the cell.

There are two main classes of membrane receptor, tyrosine kinase-associated and G-protein-linked. The tyrosine kinase-associated receptors (apart from the insulin receptor) dimerize in the membrane on signal binding and tyrosine –OH groups of the cytosolic domains are phosphorylated. Protein phosphorylation is the predominant process in this class of signalling. Adaptor proteins bind to tyrosine phosphate groups by specific domains such as SH2 and link the receptors to other proteins of signal transduction pathways, again by specific domains such as SH3. Variable forms of these domains are found on large numbers of different proteins. Chains of signal-transducing proteins are assembled into pathways that transmit the signal to the nucleus where they control gene activities.

The Ras pathway, universally found, is the prototype tyrosine kinase-associated pathway. It transmits signals *via* a serine/threonine kinase-amplifying cascade to gene-control proteins (transcription factors), which enter the nucleus. The Ras protein, which is one of the components of the signalling pathway, has a GTP/GDP switch. Ras is activated only when the receptor has a signal molecule attached to it. It involves a bound molecule of GDP being exchanged for GTP. However, the bound GTP molecule is slowly hydrolysed to GDP thus inactivating Ras and blocking the signal transduction pathway. This is a timing device, which limits the effect of a signal activation event. It can be reactivated again only if the signal is still attached to the receptor. The signal is destroyed so that the pathway remains activated only so long as the signal is being produced, usually by other cells.

When the signalling molecule is no longer present, phosphatases inactivate the phosphorylated proteins. Overactivity of the Ras protein, because of mutation impairing the GTP/GDP switching off, is associated with a number of human cancers, a fact that is true of many of the signalling pathways. In another type of tyrosine kinase pathway, the JAK/STAT pathway, activated receptors cause phosphorylation of cytosolic STAT proteins that enter the nucleus and act as gene controllers. This is a direct pathway in contrast to the multicomponent Ras pathway. Controls exist to reverse the signal so that runaway activation of the pathway does not occur. Insulin operates via another pathway involving PI 3-kinase present in the membrane.

G-protein-linked receptors are many and versatile; over half of the pharmaceutical drugs are targeted to them. The classical G-protein-associated pathway is that activated by adrenaline. The membrane receptors are linked on the cytosolic side to a heterotrimeric protein with α, β, and γ subunits. On receipt of a signal, the α subunit exchanges GDP for GTP and migrates to, and activates, a membrane-bound adenylate cyclase that produces the second messenger, cAMP. (see Chapter 20 for a fuller treatment of second messengers.) The signal is automatically cancelled after a brief period because the α subunit slowly hydrolyses the GTP to GDP. The subunit then detaches and rejoins its partners in the G-protein. The GTP/GDP switch is a molecular timing mechanism. In cholera, GTP hydrolysis is inhibited, thus inappropriately indefinitely extending the formation of cAMP, which causes the intestinal symptoms of the disease, diarrhoea and dehydration.

Other G-protein-associated receptors activated by their specific signalling molecules cause the formation of different second messengers, the phosphatidylinositol cascade being an example. Here the second messengers are inositol triphosphate (IP_3), which causes Ca^{2+} release into the cytosol and diacylglycerol formation, which activates protein kinase C. The Ca^{2+} is important in many cellular control systems.

Visual receptors in which photons are the signal also are G-protein-linked systems but in this case the signal causes a reduction in cGMP. This results in a signal in the optic nerve to the visual centre in the brain.

cGMP production is the second messenger of a different type of membrane receptor, which when activated is itself a guanylate cyclase.

Nitric oxide signalling is different again in that it causes cGMP production but the receptor is located in the cytosol. Signals are terminated by the hydrolysis of cGMP by a phosphodiesterase.

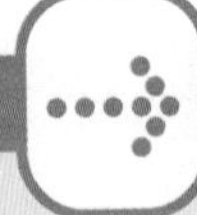

Further reading

To access the further reading, please scan the QR code image or go to http://global.oup.com/uk/orc/biosciences/molbiol/snape_biochemistry/student/reading/ch29/

Problems

1 How is cAMP production controlled? In particular, describe the role of GTP in the process. What is the relevance of the latter to cholera?

2 How does cAMP exert its effect as second messenger?

3 How does nitric oxide exert its controlling effect?

4 cAMP and cGMP are not the only second messengers. Describe another system.

5 Make a general comparison of the activation mechanism of three types of receptor: (a) the adrenaline receptor activating adenylate cyclase; (b) the EGF receptor of the Ras pathway; and (c) the interferon receptor.

6 What are the classes of signalling molecules between cells in the mammalian body?

7 Lipid-soluble signalling molecules directly enter cells but lipid-insoluble ones do not. Does this mean that the two types act in totally different ways? Explain your answer.

8 If a protein is found to have an SH2 domain, what is its likely function in the cell?

9 Outline and compare the salient features of a Ras signalling pathway with a JAK/STAT pathway.

10 G-protein-associated receptors are tremendously versatile in the signals the different receptors respond to. Outline the events in G-protein signalling and explain why it can be so versatile.

11 Can cGMP be properly regarded as a second messenger in the visual system? What about the nitric oxide signalling system?

12 Most tyrosine kinase-associated receptors dimerize on receipt of a signal. What is the exception to this?

13 What is the role of GTP/GDP switches? Illustrate your answer with examples from cell signalling, protein targeting and protein synthesis respectively.

14 Ras is a GTPase but it is not called a G protein. Explain why.

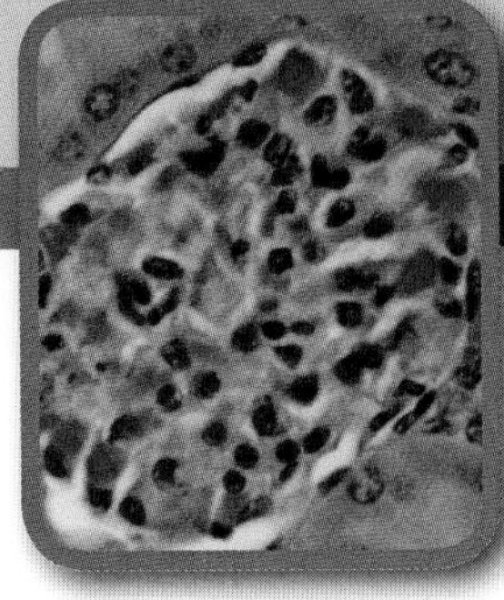

Chapter 30

The cell cycle, cell division and cell death

The eukaryotic cell cycle

All life depends on the self-reproduction of cells, which has gone on continuously for billions of years. Cells multiply by dividing into two and, before the next cell division, the daughter cells generally grow and double in size so that normal cell size is maintained. In complex eukaryotes such as mammals cell signalling provides the stimulus for rapid somatic cell division when increased cell numbers are needed, for instance during embryonic development, but once the need for growth is over cell division is restricted to replacing dead cells. Each cell division requires the total DNA in the nucleus to be replicated exactly so that there are two sets of chromosomes, no less and no more. The two sets are then segregated into the two daughter cells at cell division. This must be done with absolute precision or both daughter cells would be genetically abnormal. It is not surprising that the eukaryotic cell cycle has an elaborate system of controls and checkpoints. The process has been tightly conserved, so that the cycle and its regulation are essentially the same in all eukaryotes.

The cell cycle is divided into separate phases

In a typical replicating animal cell in laboratory culture, from the completion of one cell division to the next takes about 24 hours depending on the cell type (Fig. 30.1). The cell division phase or **M phase**, in which the duplicated chromosomes are separated (**mitosis**) and the cell divides in two (**cytokinesis**), takes about an hour. During mitosis the chromosomes are in a highly condensed phase and are inactive – they are not transcribed to make messenger RNA. When the daughter cells are formed by cytokinesis, the chromosomes are unravelled from the highly compacted state to the more extended form pervading the nucleus. In the period between cell divisions, known as **interphase**, the genes can be active and the cell more or less continuously synthesizes most of the proteins and other components needed for cell growth. DNA synthesis is, however, confined to the **S phase**, which has a duration of about 8 hours out of a 24-hour cycle in mammalian cells. In some cells, such as the early embryonic cells of amphibians, M phase and S phase simply alternate with no pauses in between, so that the embryo increases its cell number with no overall growth. However, in most cell types M and S phases are separated by G or 'gap' phases. Prior to S phase there is a **G_1 phase**, in which the cell grows and prepares to enter S phase. After completion of S phase, in which the genome is completely duplicated, the cycle progresses through the second gap phase or G_2 in which the cell prepares for mitosis. Interphase therefore consists of G_1+S+G_2. The latter leads into the mitosis (M) phase. While the G_2 phase is usually quite brief, the length of G_1 is typically around 10 hours in a 24 hour cycle, but is variable depending on cell type. For instance, human intestinal epithelial cells, which undergo a lot of wear and tear, complete a full cycle in around 10 hours, while the cycle in adult pancreatic β cells lasts for months. Cells in G_1 may enter a quiescent phase, termed **G_0**, in which they will not proliferate unless they receive a specific signal that causes them to re-enter the cell cycle.

The cell cycle phases are tightly controlled

The cell cycle must be controlled to avoid genetic abnormalities. For this reason progression through the phases of the cycle is subject to controls and checkpoints to ensure that everything is in order. There are several vital requirements to be met.

- If the DNA of a cell is damaged and not repaired, the cell must not be allowed to proceed to mitosis to avoid the production of genetically abnormal, potentially cancerous, daughter cells.
- The DNA must be completely replicated before mitosis for the same reason. Equally important, and again for much the same reason, the DNA must be replicated once and once only. This means that the replicons, the sections of DNA under the control of single centres of origin, must fire only once per cell cycle.

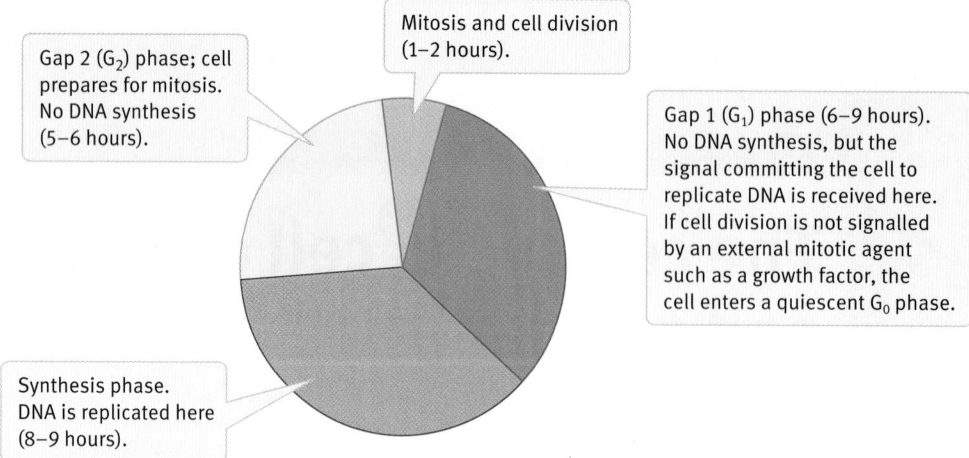

Fig. 30.1 Overview of the eukaryotic cell cycle. The duration of the cell cycle varies greatly between different cell types. The times given here are for a rapidly dividing mammalian cell in culture (24 hours to complete the cycle).

- At the metaphase stage of mitosis (see later) the duplicated chromosomes must be correctly positioned at the equator and attached to spindle fibres. If this were not so, genetically abnormal daughter cells would result from the imperfect segregation of the chromosomes.
- Each phase of the cycle must be completed before the next phase is initiated.
- Replication of the DNA must not proceed in a mammalian cell unless a mitogenic signal to proceed to cell division is received from other, neighbouring cells. In an animal as complex as a mammal, individual cells are part of a vast community that exists for the welfare and reproduction of the organism as a whole. They must fit in with the needs of other cells, and not proceed to cell multiplication independently. Uncontrolled, independent cell division occurs in cancer. To prevent it in the normal way, cell signalling pathways coordinate cell multiplication with the requirements of the body as a whole. The effectors of this control are intercellular signals, proteins known as **growth factors** and **cytokines** that were introduced in Chapter 29.

Cell cycle controls

Cytokines and growth factor control in the cell cycle

The cytokines and growth factors are signalling molecules that bind to cell-surface receptors and activate pathways that control genes. Deficiencies in this control can lead to chaotic cell multiplication and cancer.

In the present context, it is the mitogenic (mitosis-stimulating) effects of cytokines and growth factors that are important. If a mitogenic signal is not received by a cell in early G_1 phase the cycle does not proceed, and the cell enters G_0, in which it metabolizes normally but does not undergo mitosis. Most somatic cells are in G_0 most of the time. On receipt of a mitogenic signal, however, the G_0 cell can re-enter the cycle again at G_1.

The intercellular cytokine/growth factor signals mainly come from neighbouring cells. These signals maintain correct organ cell numbers, for once adult size is reached, cell multiplication largely ceases except to replace dead cells and for wound healing. Although much is known of the mechanisms of individual signal transduction pathways, how these collectively add up to coordination of the mass of cells in tissues is unclear. The mitogenic control of cell division operates mainly in the G_1 phase of the cell cycle as described later.

Cell cycle checkpoints

In addition to control by mitogenic factors, there are other checks related to safety requirements. Towards the end of G_1, G_2, and M phases, there are checkpoints at which the cycle is halted if one of the safety requirements is not met. The arrest gives an opportunity for the defect to be rectified, in which case the cycle can proceed to the next phase. If the fault is not corrected, the cell may self-destruct by activating a chain of events leading to **apoptosis** (programmed cell death – see below). Figure 30.2 shows the checkpoints in the mammalian cell cycle. We will come to the nature of these checkpoints shortly.

Cell cycle controls depend on the synthesis and destruction of cyclins

In the cell cycle, as in so many cellular control systems, protein kinases are of overwhelming importance. The cycle kinases are however of a unique kind; they are without activity on their own and require the binding of other proteins

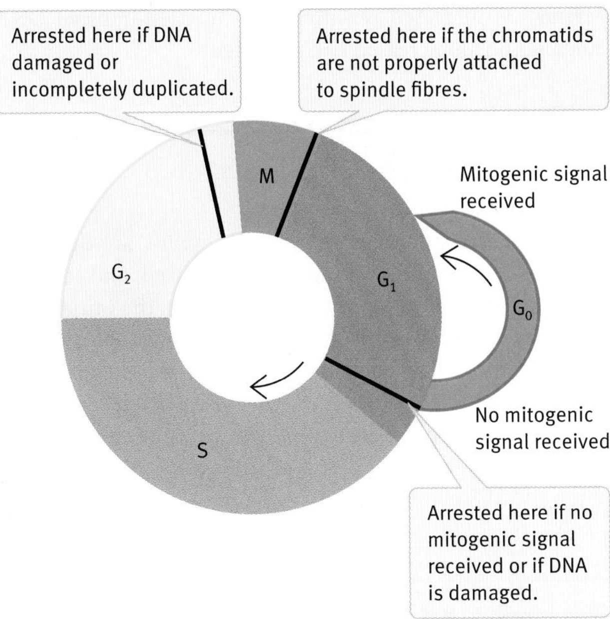

Fig. 30.2 The eukaryotic cell cycle, showing checkpoints in G_1, G_2, and M phases.

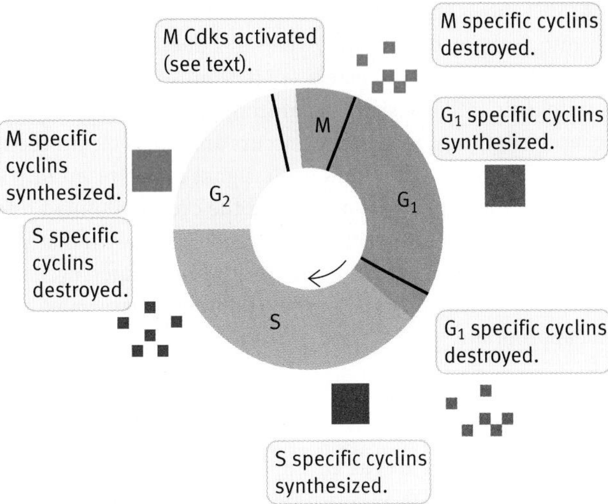

Fig. 30.4 Cyclin–Cdk cycle. Simplified diagram of cell cycle control by synthesis and destruction of phase-specific cyclins that activate phase-specific Cdks. The M phase Cdks are inactive until just prior to M phase when they are rapidly activated by dephosphorylation (see text). Note that more than one cyclin and Cdk may act in a given phase but this is not shown for simplicity.

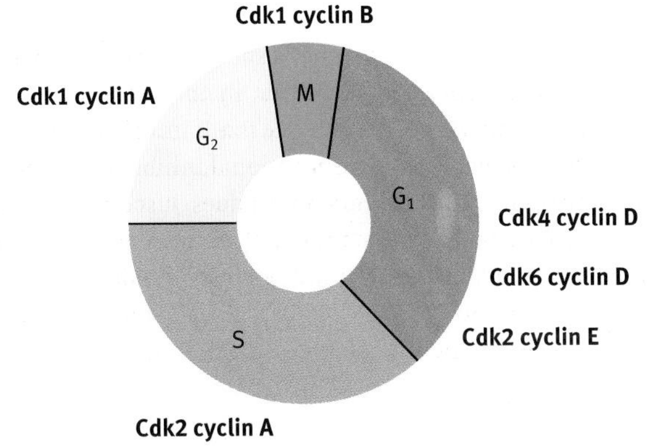

Fig. 30.3 Diagram showing proposed cyclin-Cdk complexes and their appropriate phases of operation in the mammalian cell cycle.

known as **cyclins** before they have activity. They are therefore known as **cyclin-dependent kinases** or **Cdks**. A complex variety of Cdks operate in a mammalian cell cycle, each working in its own designated phase. The amounts of the Cdks remain essentially constant in the cell during the cycle, but cyclins are destroyed at the end of each phase and new ones are synthesized to enable progression to the next phase; this is the basis of the main control of the cell cycle. The sequence of cyclin involvement is quite complex but Fig. 30.3 shows the generally accepted cyclin-Cdk combinations. You might expect that a different cyclin-Cdk pair would operate in each phase of the cycle, but the system is not that simple; in some cases the same Cdk is activated in different phases by different cyclins. For example, as shown in Fig. 30.3, Cdk2 acts in G_1 when bound and activated by cyclin E, but in S phase it is partnered with cyclin A. In each phase, Cdk2 acts on different target proteins, depending on which cyclin it is partnered with. Because of the complexity of the system, subsequent figures will refer to the cyclins simply as G_1, S, G_2 and M cyclins.

The mechanism of Cdk activation by cyclins is known. In the absence of a cyclins, the active site of a Cdk is blocked due to the conformation of the protein. Combination with a cyclin causes a conformational change, which partly unblocks the site. There are, however, further levels of control. Complete activation requires phosphorylation of the Cdk on a threonine residue. This is done by a Cdk activating kinase (CAK). In the case of Cdk1, still further control exists in relation to progression to M phase (see later). There are also a number of Cdk inhibitor proteins (Ckis) that interact with the cyclin-Cdk complexes to regulate the cycle in response to checkpoints.

At the start of each cycle phase, genes have to be activated so that the appropriate cyclins are synthesized. If this does not happen, the cycle cannot proceed through that phase. At the end of each phase, the cyclins are destroyed by proteasomes and new cyclin synthesis specific for the next phase is needed (Fig. 30.4). This may seem an expensive way to achieve control but it is a decisive procedure leaving no room for partial inactivation or reversibility.

Controls in G_1 are complex

The progression through G_1 to S phase involves multiple gene controls. At the end of the preceding M phase, all cyclins are destroyed so that at the start of G_1 there are no active Cdks.

Fig. 30.3 shows that cyclin D is needed for the progression of G_1 into S phase, but cyclin D synthesis does not begin automatically when the cell enters G_1. It requires the presence of a mitogenic (cell division) signal from growth factors. If this is received, G_1 cyclin synthesis occurs and the cycle proceeds towards S phase. If the mitogenic signal is not received, G_1 cyclin synthesis does not occur, and the cycle is shunted into G_0 (as occurs commonly in somatic cells). It is known that several human cancers are associated with unregulated cyclin D synthesis, which allows G_1 to progress into S phase when it is not appropriate.

The G_1 checkpoint

The G_1 checkpoint acts as an important 'gateway' in the cell cycle. In mammals it is known as the **restriction point**, and once past it the cell is committed to proceed right through to M phase. Chief players in the checkpoint mechanism are two proteins that are described more fully in Chapter 31 because of their roles as 'tumour suppressors'. These are **p53** and the **retinoblastoma protein** or **Rb**. Rb halts the cell cycle in the absence of mitogenic signals, while p53 is a key regulator in the best-known check; that is, if the DNA of a cell is damaged it is not allowed through the checkpoint. Briefly, in the presence of damaged DNA, p53 increases in amount in the cell and is activated. It is a transcription factor that sets up a train of events causing the cell cycle to halt so that DNA repair can be attempted. If repair fails the cell is signalled to self-destruct by apoptosis to avoid the risk of genetically abnormal cells being produced.

How is DNA damage detected?

The mammalian genome is inevitably subject to damage by, for example, ionizing radiation, reactive oxygen species and other agents. Fortunately, the repair mechanisms, described in Chapter 23, detect and repair many types of lesions. If these are not repaired the cycle must not be allowed to proceed. DNA damage results in exposure of single-stranded sections. For example, a replication fork stalled for some reason will have stretches of single-stranded DNA, and double-stranded DNA breaks may also have some terminal single-stranded DNA. A protein known as **replication protein A (RPA)** attaches to the single-stranded DNA and this attracts a complex of protein kinases to assemble. This is the signal that is detected by p53, which halts the cycle to enable the DNA lesion to be repaired. If repair is not achieved p53 initiates apoptopic destruction of the cell.

One of the protein kinases in the complex that assembles in response to RPA is mutated in the disease **ataxia telangiectasia**. The protein is called **Ataxia telangiectasia mutated (ATM)**. Normally, activation of ATM by double-stranded DNA breaks leads to blocking of the $G_1 \rightarrow S$ phase transition, but in ataxia telangiectasia this does not occur, and hence the person has increased sensitivity to radiation-induced DNA damage and an increased risk of cancer.

Progression to S phase

Once past the G_1 checkpoint, the G_1 cyclins are destroyed by proteolysis, S phase cyclins are synthesized, and the cycle enters S phase. DNA replication is initiated. The initiation of duplication in each replicon is effected by a complex of proteins and the mechanism, controlled by S phase cyclin-Cdk, ensures that each complex fires only once per cell cycle. Once committed to S phase, the cycle advances to the checkpoint in M phase at the end of G_2, the S phase cyclins being destroyed at the end of S phase.

Progression to M phase

The mitotic-related cyclins are synthesized and accumulate in the cell during S and G_2 phases, and combine with the relevant Cdks. Cyclin-Cdk complexes, as stated above, require addition of a phosphoryl group to a specific threonine residue on the Cdk for activation. However, in the case of M phase Cdk1, when this is done, the enzyme is not immediately active because other kinases add two additional inhibitory phosphoryl groups to two other amino acid residues. Just before mitosis the two additional phosphoryl groups are removed by a phosphatase enzyme, causing activation of the Cdk. The reason for this convoluted process is probably that it permits build up of the triply phosphorylated, inactivated, cyclin-Cdk1 and then a very rapid activation by simple hydrolytic dephosphorylation just prior to mitosis. However, before entering M phase, the G_2 checkpoint must be passed. This arrests the cycle if the DNA has not been completely replicated or is damaged. Unless the fault is corrected, the cell destroys itself by apoptosis.

M phase

In M phase the cell undergoes dramatic changes in which the nuclear membrane disappears, the chromatin condenses into compact mitotic chromosomes, the spindle fibres develop and the duplicated chromosomes become positioned at the equator of the cell ready for segregation. A single M phase Cdk initiates all of these cellular events, which are described in more detail later. A checkpoint during mitosis ensures that all chromosomes are correctly positioned so that they will be distributed equally between the two new daughter cells. Any chromosome not attached correctly to the spindle fibres is a signal to halt the cycle. The protein complex responsible for

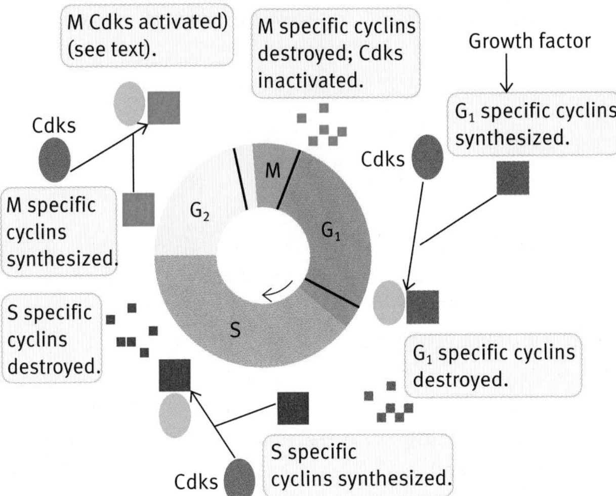

Fig. 30.5 Simplified diagram of control mechanism. In mammalian cells the cell cycle cannot pass the restriction point in G_1- unless a mitogenic signal is received, which activates the synthesis of G_1-specific cyclins. The latter are required to activate Cdks necessary for activation of genes involved in the progression into S phase at which stage the G_1- cyclins are degraded. Activation of Cdks by mitosis-specific cyclins is required for progression to the M phase. The different cyclins target the Cdks to different substrates as appropriate for the different phases of the cell cycle. Red represents inactive Cdks and green the activated forms. Note that in the interests of clarity the figure does not show the multiplicity of Cdks or of the activating cyclins involved. The details of these do not affect the essentials of the control scheme.

progression to the next phase is the **Anaphase promoting complex** (**APR**), and once past these checks the cell enters anaphase of mitosis where the duplicated chromosomes are drawn apart. When chromosome segregation is complete the spindle is disassembled, the nuclear membrane reforms and the cell completes cytokinesis (division of the cytoplasm). The APR also triggers proteolytic destruction of the M phase cyclins so that the cell is now ready to commence a new cell cycle.

A summary of the Cdk-cyclin control of the cell cycle is given in Fig. 30.5.

Cell division

Mitosis

Eukaryotic cell division, in all but germ line cells which give rise to sperm and eggs, involves **mitosis** (Fig. 30.6), in which replicated chromosomes are equally partitioned into the two daughter cells. As discussed in Chapter 22, during interphase the DNA of the chromosomes is packaged with histones and other chromatin proteins but in a relatively diffuse manner. In this state the genes can be transcriptionally active.

As the cycle enters the first stage of mitosis (**prophase**), duplicated chromosomes become condensed. At this stage each duplicated chromosome is referred to as consisting of two 'sister' **chromatids** (Fig. 30.7). The chromatids are identical double-stranded DNA molecules produced by DNA replication; in each chromatid one of the strands is newly synthesized, and the other is the parental strand. The pair of chromatids is held together along their length by **cohesin** proteins. In this state the genes are in a shut down condition. The cell now undergoes a major structural change; the nuclear membrane disintegrates, allowing the **mitotic spindle**to invade the area. The spindle is an arrangement of microtubules (Chapter 8) originating at the **centrosomes**, a complex of proteins located at each pole of the cell. At **metaphase** the chromosomes are arranged at the spindle equator with each chromatid attached to spindle fibres at the **kinetochore**, a protein complex assembled at the **centromere** of each chromatid. The centromere is a specialized DNA sequence that directs assembly of the kinetochore proteins. In **anaphase** the chromatids separate due to the activation of an enzyme (a protease) that breaks down the cohesin proteins holding the chromatids together. The separated chromatids, now called **daughter chromosomes**, move towards the spindle poles and the poles move further apart. The molecular mechanisms involved in these movements are described in Chapter 8. This segregation process is completed at **telophase** when the chromosomes reach the spindle poles. The mitotic apparatus is disassembled and nuclear membranes reform. **Cytokinesis** or cytosolic division is now completed, with each daughter cell receiving one full set of chromosomes. The chromosomes decondense into the active interphase state.

Meiosis

Germ line cells divide to produce gametes (eggs and sperm in mammals). The crucial genetic difference between gametes and somatic cells is that gametes are **haploid** while somatic cells are **diploid**. As a reminder, diploid cells contain homologous pairs of chromosomes, one member of each homologous pair inherited from the mother and one from the father. Each member of a homologous chromosome pair has the same genes arranged in the same order, so that (apart from genes on the X and Y chromosome in males), somatic cells have two copies of each gene. The two copies may be completely identical, or minor sequence differences may cause an individual to have two different **alleles** of a single gene. These allelic differences contribute to genetic variation between individuals. At meiosis the gamete receives just one member of each homologous chromosome pair. Diploidy is restored when the sperm and egg fuse at fertilization, and this results in the offspring inheriting a set of alleles that is a mix of those originally present in each parent.

Mitosis and meiosis have much in common so far as the cellular mechanism goes. The difference is that, in mitosis,

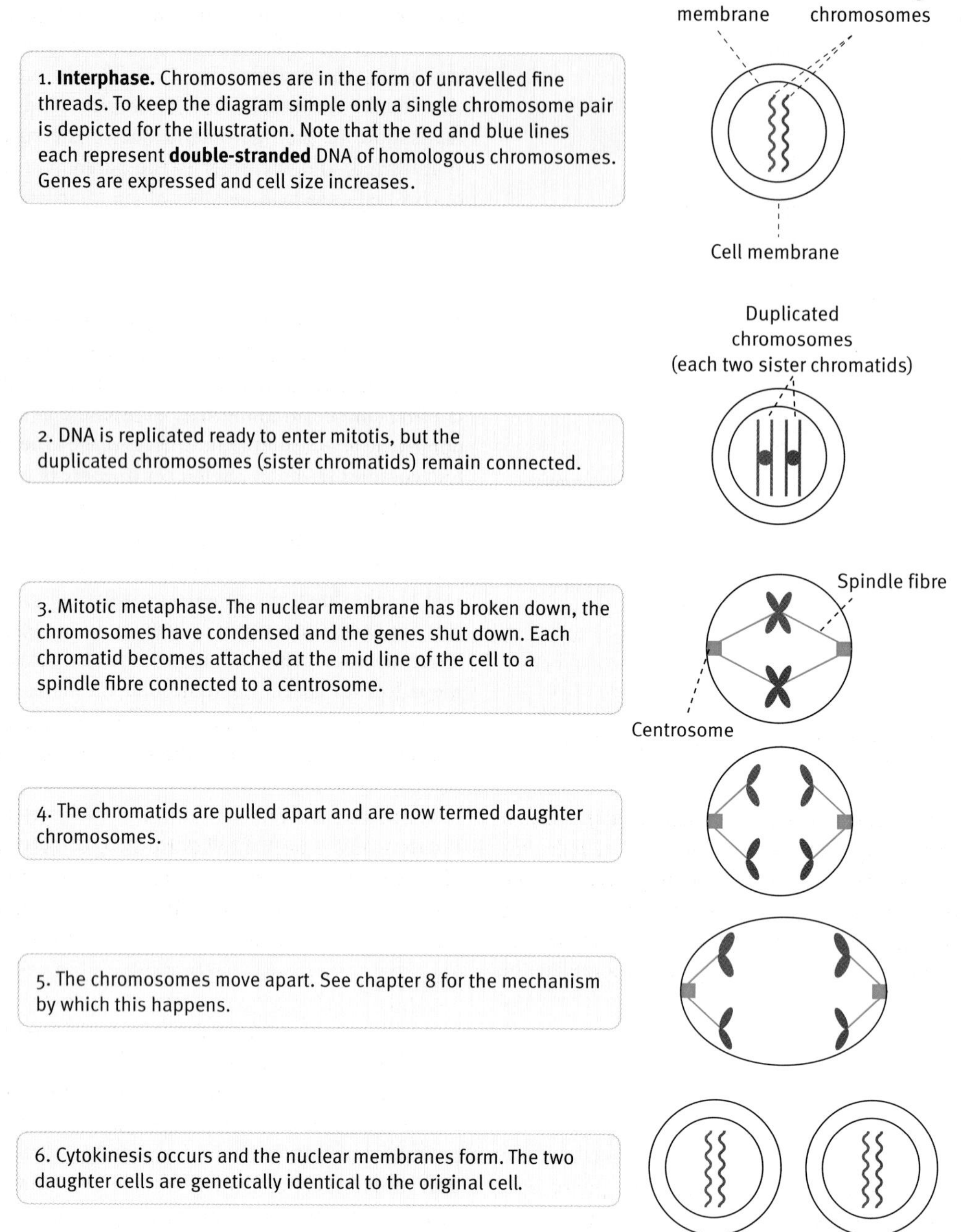

Nuclear membrane
Two homologous chromosomes
1. **Interphase.** Chromosomes are in the form of unravelled fine threads. To keep the diagram simple only a single chromosome pair is depicted for the illustration. Note that the red and blue lines each represent **double-stranded** DNA of homologous chromosomes. Genes are expressed and cell size increases.
Cell membrane
Duplicated chromosomes (each two sister chromatids)
2. DNA is replicated ready to enter mitotis, but the duplicated chromosomes (sister chromatids) remain connected.
Spindle fibre
3. Mitotic metaphase. The nuclear membrane has broken down, the chromosomes have condensed and the genes shut down. Each chromatid becomes attached at the mid line of the cell to a spindle fibre connected to a centrosome.
Centrosome
4. The chromatids are pulled apart and are now termed daughter chromosomes.
5. The chromosomes move apart. See chapter 8 for the mechanism by which this happens.
6. Cytokinesis occurs and the nuclear membranes form. The two daughter cells are genetically identical to the original cell.

Fig. 30.6 A simplified diagram of mitosis.

DNA is replicated once and a **single** cell division occurs. In meiosis the DNA is also replicated once but **two** cell divisions occur. Exactly as in mitosis, in meiosis DNA replication and chromosome condensation produce compacted chromosomes consisting of pairs of sister chromatids (Fig. 30.8(a)); at this stage, as in mitosis, the nuclear membrane disappears. Up till now the process has been the same as in mitosis but now meiosis is different in that the two members of a homologous pair of chromosomes align themselves together in what is called **synapsis** and swap small sections of DNA by **crossing over** (Fig. 30.8(b)). This process, termed recombination, is described in detail in Chapter 23. Recombination by crossing over at meiosis creates further genetic diversity by creating single chromosomes that contain a mix of alleles from the original paternal and maternal members of a homologous chromosome pair.

Now cell division occurs but of a type that does not occur in mitosis. Instead of the chromatids separating, the two members

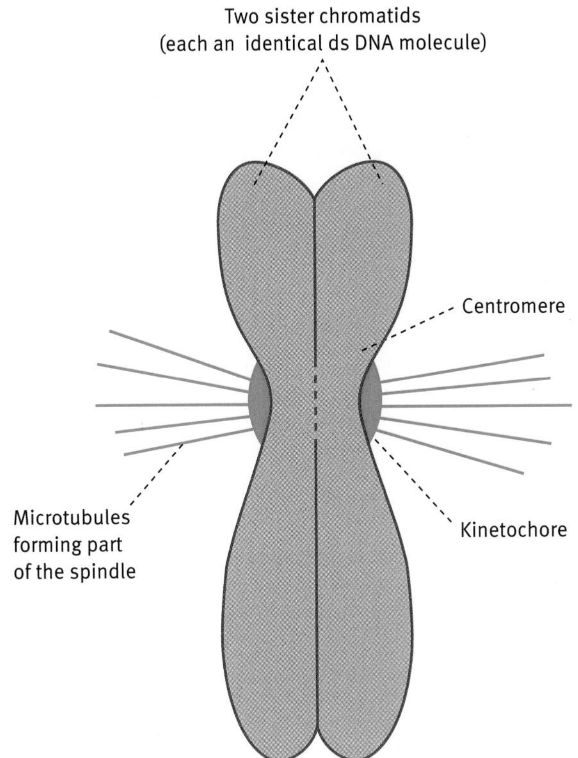

Fig. 30.7 Chromosome at mitotic metaphase consisting of two sister chromatids.

of the homologous chromosome pair are separated in the first meiotic cell division (Fig. 30.8(c)), one going into each daughter cell (Fig. 30.8(d)). These immediately undergo a second cell division (*without prior DNA replication*) with the chromatids now separating (Fig. 30.8(e)) as in mitosis. This gives four haploid cells (gametes) from the original diploid parent cell (Fig. 30.8(f)).

Apoptosis

The name **apoptosis** is derived from the Greek prefix 'apo' meaning detached, and 'ptosis' meaning falling, with allusion to falling leaves. Leaf fall is a form of programmed death essential to the tree. Apoptosis is programmed cell death, essential to the life of multicellular animals.

Apoptopic death is not the only form of cell death within living animals; there is also **necrosis**. Necrosis occurs as a result of crude mechanical damage or oxygen deprivation to cells, which stop ATP production. The cell contents leak out of ruptured membranes and can set up inflammation in the surrounding tissue. When severe damage occurs, such as long-term oxygen deprivation to cells, pores open up in cell membranes, there is a large influx of calcium ions, the cells swell and burst. Necrotic destruction of cells is not a regulated process; it is essentially an accidental one. This is a fundamental difference from apoptosis, which is brought about by a highly regulated elaborate multipathway process. In apoptosis the cell is killed but does not rupture the membrane. The cell is systematically degraded to a shrunken remnant by fragmentation of the DNA, and blebbing off of small vesicles, which are phagocytosed by white cells. The phagocyte recycles the degraded material. The mechanism avoids the inflammation caused by necrosis.

What is the function of apoptosis?

The importance of deliberate programmed cell death in the life of an animal, and its scale and complexity, initially came as a big surprise to many. It has two broadly different roles, the first connected to normal processes. In embryonic development it is essential that, at the appropriate stage, certain cells are removed. An obvious example is the disappearance of the tail of the tadpole on development into a frog. Also, development of the nervous system involves death of neurons that fail to make profitable connections. In adult animals a constant amount of cell death occurs. For example, in the bone marrow and thymus, development of cells of the immune system involves the destruction every day of large numbers, probably billions, of B and T cells that would cause autoimmune attacks if allowed to survive. These are essential normal processes in which the cells self-destruct on receipt of signals.

The second reason for apoptosis is connected to the potential for malfunction of cellular processes. An obvious example is that a single cell that escapes normal controls on cell division and/or divides with irreparably damaged DNA can result in a cancer that could be fatal. Another major reason for programmed self-destruction is that one of the main strategies of the immune system to combat viral or other infection of cells (Chapter 33) is for activated killer T cells of the cellular immunity system to deliver a signal to the cell to self-destruct thus aborting virus replication. The complexity of cell growth regulation is such that virtually all cells have an elaborate mechanism in them ready to cause self-destruction should such irreparable trouble in the life cycle become apparent. The tumour suppressing protein p53, (see Chapter 31), plays a central role by triggering self-destruction if damage to the genome is detected that cannot be repaired. It is striking that p53 is deficient in a high percentage of all human cancers, and this deficiency is probably the reason that apoptosis has not been initiated, so that the 'corrupt' cell is allowed to survive and divide when it should have been destroyed.

Cells thus contain a self-destruction 'kit' poised to be activated, a very delicate situation that has to be counter-balanced by a system to prevent accidental activation. There is a constant fine balance between proteins that promote apoptosis and others that prevent it, as described in the following section.

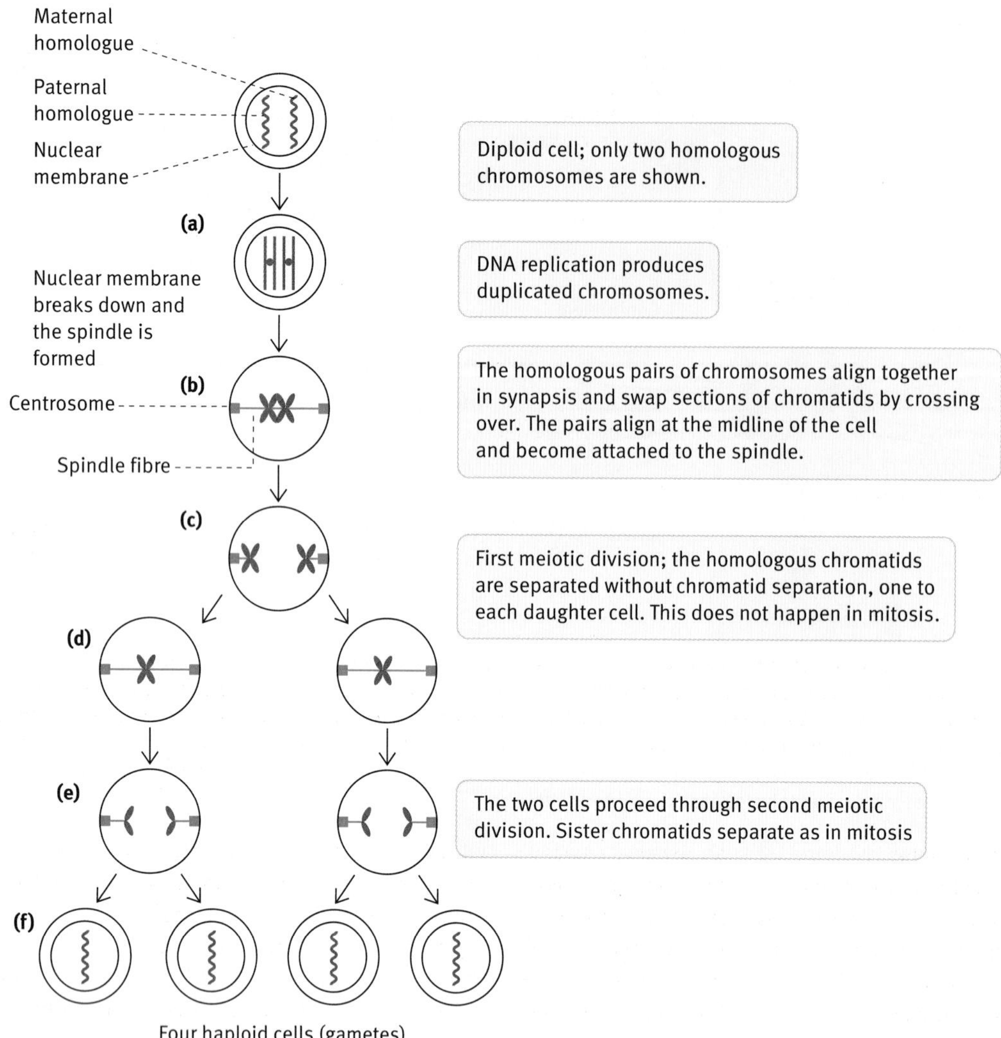

Fig. 30.8 A simplified diagram of meiosis. The process of meiosis to generate haploid germ cells occurs mechanistically by stages similar to those of mitosis. The diagram therefore just shows the changes that occur to the chromosomes ignoring spindle formation and the chromatid separation process. (See Chapter 8 for this mechanism.) The main difference between mitosis and meiosis is that, in meiosis, the chromosomes are replicated once but two cell divisions occur. Thus the daughter cells (the gametes) receive only one member of each homologous chromosome pair. In the figure a single pair of homologous chromosomes is used for illustrative purposes.

There are two main pathways that initiate apoptosis

The two pathways that initiate apoptosis are called **intrinsic** and **extrinsic**. The intrinsic pathway follows from a wide variety of events that happen inside the cell. These events may be the result of external agents such as radiation causing damage to DNA or short-term oxygen deprivation, but there is no external signal delivered to the cell actually instructing it to destruct. Instead internal mechanisms such a p53 detecting DNA damage trigger the apoptotic pathway. Another form of cellular 'stress' that can induce apoptosis is called endoplasmic reticulum (ER) stress, in which proteins cotranslationally transported into the ER lumen fail to fold properly so that the unfolded polypeptides accumulate in large quantities. These are transported into the cytosol and lead to apoptosis.

The extrinsic pathway of apoptopic induction, on the other hand, depends on the existence of protein **death receptors** produced by cells and displayed on their outside surface. When specific ligands (generally expressed on the surface of other cells such as killer T cells) combine with these receptors the cell self-destructs.

Caspase enzymes are the effectors of apoptosis

Both the intrinsic and the extrinsic apoptotic pathways ultimately lead to the activation of **caspase** enzymes. These are so named because they have a cysteine residue at their active

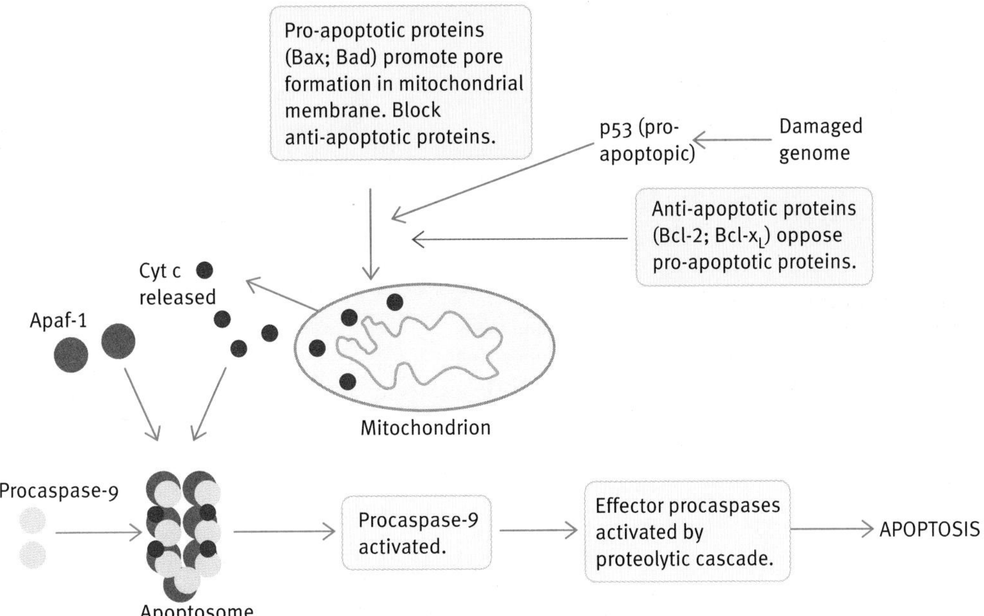

Fig. 30.9 Highly simplified diagram of the main features of the intrinsic pathway of apoptosis induction. The intrinsic pathway is a response to a variety of cell stress events. The stress stimulus leads to the formation of pores in the mitochondrial membrane and the release of cytochrome c. Cytochrome c forms a complex with the Apoptopic protease-mediating factor (Apaf-1), the apoptosome, to which procaspase-9 attaches and is activated. The process is controlled by opposing proteins of the Bcl-2 family. Bcl-2 and Bcl-x$_L$ inhibit apoptosis, while Bax and Bad promote it. The role of p53 is greatly simplified in the diagram. Green indicates proapoptopic pathway; red indicates antiapoptopic pathway.

site and attack proteins, cleaving after an **asp**artate residue (**c-aspases**). There is a sequence of caspase enzyme activation that leads to the final effector caspases targeting large numbers of proteins for cleavage. For instance, cleavage of a cellular DNase inhibitor protein activates the DNase enzyme, which then takes the cell's DNA apart. Caspases are present in the cell as inactive **procaspases**, which are activated by the apoptotic pathways as described below.

Mechanism of an intrinsic pathway of apoptosis

Given the central benign role of mitochondria in producing most of the cell's ATP, the idea that they become deadly killers in intrinsic apoptosis is almost difficult to accept, yet such is the case. We remind you that one of the electron transfer proteins in oxidative phosphorylation is **cytochrome c**. Unique among the cytochromes, it is not an integral membrane protein, but only loosely attached by noncovalent bonds to the inner mitochondrial membrane. This protein, vital to ATP production, if released into the cytosol, triggers the intrinsic apoptotic pathway.

The initial 'stress' to the cell which, as mentioned, can be caused by a wide variety of events such as a damaged genome or accumulation of unfolded polypeptides first results in the formation of pores in the mitochondrial membranes. Cytochrome c leaks out into the cytosol. There it forms a complex with the protein **Apoptotic protease-mediating factor (Apaf-1)** and an inactive procaspase enzyme, **procaspase-9**. The resulting protein complex is called an **apoptosome** (Fig. 30.9). In humans the apoptosome probably has seven Apaf-1 and seven cytochrome c molecules. The Apaf-1 subunits have a domain to which the procaspase-9 binds. As a result of the binding, procaspase-9 is activated as a proteolytic enzyme. Most procaspases are activated by proteolytic cleavage, but the initiator procaspase-9 is not cleaved. Instead, incorporation into the apoptosome apparently causes a conformational change in the protein that results in activation. The active **caspase-9** then activates other procaspases present in the cytosol in a proteolytic cascade, culminating in the cleavage and activation of the effector procaspases.

Regulation of the intrinsic pathway of apoptosis by Bcl-2 proteins

The intrinsic apoptotic pathway is tightly regulated in mammalian cells by two groups of proteins, one favouring apoptosis and the other inhibiting it. Confusingly, they all share at least one homologous domain and thus belong to a single protein family, the **Bcl-2** family. **Bax** and **Bad** are pro-apoptopic members of the family and are believed to be involved in the initial formation of pores in the mitochondrial membrane. **Bcl-2** (the first discovered member) and **Bcl-x$_L$** are important antiapoptotic proteins that prevent pore formation and cytochrome c release. Bcl-2 and Bcl-x$_L$ block the pro-apoptotic activity of Bax and Bad in nonapoptotic cells. Whether a stimulus that causes cell stress produces cell destruction will

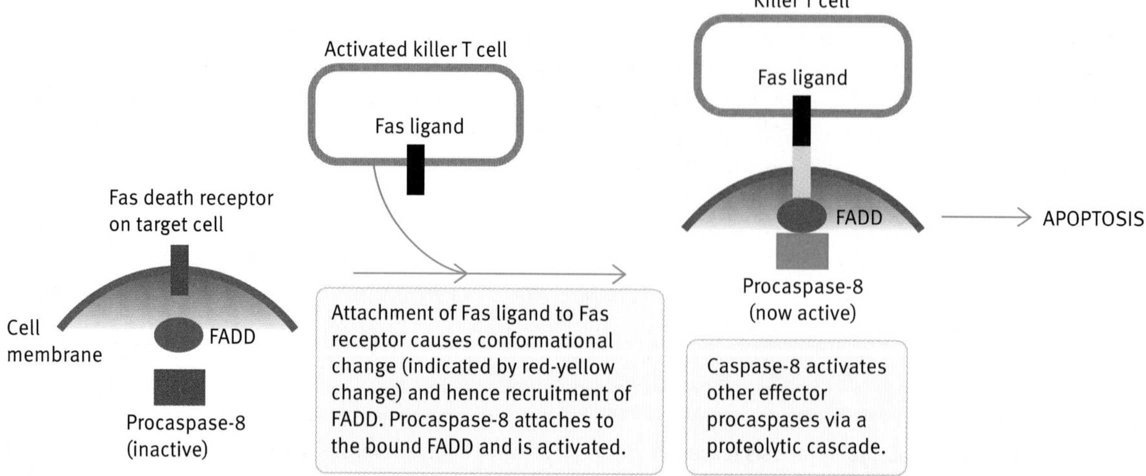

Fig. 30.10 The extrinsic pathway of apoptosis (described in the text).

therefore depend on the balance between the pro-apoptopic and antiapoptotic proteins.

Mechanism of the extrinsic pathway of apoptosis

The extrinsic pathway is different in its induction from that of the intrinsic pathway, but ultimately the pathways merge at the stage of activation of effector caspases. The extrinsic pathway is based on signals being delivered to protein death receptors, which somatic cells produce on their surface. Several of these are known but we will use the one called **Fas** for illustration. Fas protein is a member of the **TNF (tumour necrosis factor)** receptor superfamily. The Fas receptor is the one that killer T cells of the immune system use to deliver a self-destruct signal to cells that it has recognized to be infected with a virus. The other death receptors have a similar mechanism of action. Such killer T cells produce on their surface a protein called the **Fas ligand**, which specifically recognizes, and binds to, the Fas receptor on the infected cell. The binding triggers the target cell to self-destruct by apoptosis and in this way aborts virus production in that cell.

The Fas receptor is a transmembrane protein. Attachment of Fas ligand to the external domain causes the cytosolic domain to recruit an adaptor protein called **Fas-associated protein with a death domain (FADD)** to the cytosolic domain of the receptor (Fig. 30.10). Procaspase-8 then binds as a cluster to the FADD–receptor complex and is activated by cleavage. Once active caspase-8 is formed a proteolytic caspase cascade ensues, leading to activation of the same effector caspases as in the intrinsic pathway.

An alternative extrinsic delivery of an apoptopic signal to cells has been discovered. Killer T cells attach to target cells and release proteases from granules in their cytosol which the target cell takes up. They activate cellular procaspases directly by proteolysis. The enzymes have been christened **granzymes**. The killer cells also release **perforin**, which causes holes to form in the target cell membrane by which the granzymes enter.

Understanding of the complex mechanism(s) of apoptosis and its controls has opened up the hope that targeted drugs to induce apoptosis may attack cancer cells. Inhibitors of apoptosis may possibly also be used to treat or manage degenerative diseases.

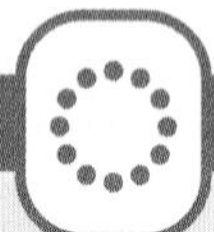

Summary

The eukaryotic cell cycle is divided into phases. These are the first gap phase (G_1), the DNA synthesis phase (S), the second gap phase (G_2), and the mitotic or cell division phase (M), occurring in that sequence. Progression through the phases depends on the synthesis of cyclin proteins specific for different phases, and at the end of each phase the cyclins are destroyed by proteolysis. The cyclins are required to activate different cyclin-dependent protein kinases (Cdks) and also to determine which substrates a given kinase works on in each phase of the cycle.

Cyclin synthesis in G_1 requires the receipt by the cell of a mitogenic signal from a growth factor or cytokine; in its absence the cell enters a quiescent G_0 phase, which is the state of most somatic cells. A mitogenic stimulus reverts it to G_1. To advance from G_1 to S phase, the cycle passes a checkpoint. This halts the cycle if DNA is damaged and, if it is not repaired, the cell is given an apoptopic signal to self-destruct. Once past this checkpoint the cycle is committed to proceed through DNA replication and division, However, entry to M phase requires passing a further checkpoint at the end of G_2; if the DNA has not been replicated completely, or is damaged, the cycle is halted. After entering M phase, a further check is made to establish that all of the chromosomes are correctly placed on the mitotic spindle. If not, the cycle is halted to prevent production of daughter cells with an abnormal chromosome complement. Once past this checkpoint cell division is completed, all cyclins are broken down and the cell enters G_1 where it again awaits a mitogenic signal.

In eukaryotes cell division occurs by mitosis (except in germ line cells). The chromosomes have been duplicated by DNA replication and mitosis involves segregation of the resulting daughter chromosomes to each end of the cell, followed by cytoplasmic division or cytokinesis. Each daughter cell receives a full diploid chromosome complement. Germ line cells divide by meiosis so that the resulting gametes are haploid and after fertilization a diploid cell is reformed. In meiosis there are two cell divisions following a single round of DNA replication. At the first meiotic division duplicated homologous chromosomes pair up and exchange DNA between them by crossing over. Each daughter cell then receives a single member of the homologous chromosome pair. At the second meiotic division the duplicated chromosomes are separated into two daughter chromosomes as in mitosis.

Apoptosis is a process by which eukaryotes dispose of unwanted cells. It occurs on a large scale in normal development and in response to stresses on the cell. The cell effectively dismantles itself without lysis, the remnant body being phagocytosed. This avoids necrotic lysis that could cause inflammation and toxic effects. Cells contain inactive proteolytic enzymes called caspases that are activated to destroy the cells. The activation may arise basically in two ways. In the intrinisic pathway internal events such as DNA damage cause the release of cytochrome *c* from mitochondria. This causes aggregation and activation of an initiator caspase, which triggers a proteolytic cascade leading to activation of effector caspases. These target various cellular proteins leading to autodestruction of the cell. In the alternative extrinsic pathway, an external death signal is delivered, for instance by a killer T cell recognizing a virally infected cell. This activates a death receptor on the target cell surface, which in turn activates an initiator caspase, leading ultimately to autodestruction.

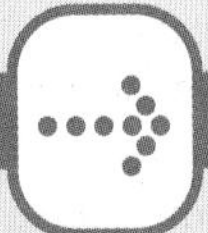

Further reading

To access the further reading, please scan the QR code image or go to http://global.oup.com/uk/orc/biosciences/molbiol/snape_biochemistry/student/reading/ch30/

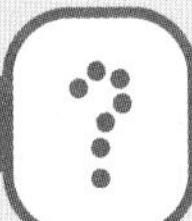

Problems

1 The restriction point in the G_1 phase of the cell cycle is of major importance. Discuss this with reference to the need for a growth factor signal for cell division to proceed.

2 Discuss the role of cyclins in the eukaryotic cell cycle.

3 If a mitogenic (growth factor) signal is not received in G_1 the cell enters the G_0 phase. Discuss this.

4 What checks are made at the G_1 restriction checkpoint and the mitotic checkpoint?

5 Summarize the differences between mitosis and meiosis.

6 What are caspases? Describe their role and the mechanisms by which they become involved in it.

7 What is the purpose of apoptosis?

8 What initiates apoptosis in cells (apart from the killer T cell death signal)?

9 What is the broad mechanism by which cytochrome c causes apoptosis?

10 What regulates apoptosis in mammalian cells?

Chapter 31

Cancer

General concepts

We will first deal with the general cell biology of cancer and then with some of the basic molecular aspects.

Development of cancer involves loss of control of a number of processes; above all, it involves the uncontrolled multiplication of cells. Although rapid cell division occurs normally throughout life in tissues such as those producing blood cells in the bone marrow, somatic cells for the most part stop dividing except for the small numbers of divisions needed to replace dead cells. Many tissues, however, retain the ability to divide rapidly if they are given the requisite signals to do so. Wounding of the skin, for example, fires off cell replication to heal the wound, but when this is achieved normal cell division ceases. If two-thirds of a rat liver is surgically removed, cell division restores the original size in a week or two but then the division ceases. Signalling pathways from mitogenic growth factors and cytokines control the replication process so that it is appropriate to the needs of the body as a whole.

A cancer cell is aberrant in that it replicates without mitogenic signals from other cells or with a diminished requirement for such signals. Moreover, cancer cells may fail to self-destruct by apoptosis after DNA damage, or other cell cycle mishap, has occurred, and by continuing to divide, they may perpetuate and exacerbate the damage in their daughter cells. This difference between normal and cancerous cell division can be illustrated by tissue-culture experiments. Adherent mammalian cells of epithelial and mesenchymal origin will divide on nutritive medium in plastic dishes if they are supplied with growth factors, usually provided by fetal calf serum. The cells multiply and spread out to cover the surface of the plate in a single layer until they are in contact with each other and they then stop dividing. This has been called **contact inhibition**. Unlike normal cells, cancer cells keep on dividing after the plate is completely covered, the cells piling up on one another to form a solid mass, which is essentially the equivalent of a tumour (Fig. 31.1).

Most normal cells can divide only a limited number of times

Normal somatic cells cannot be repeatedly subcultured in nutritive medium indefinitely, but cancer cells can. A favoured hypothesis for what limits the lifespan of normal cells relates to the **telomeres**, the repeated sequences found on the end of eukaryotic chromosomes. In Chapter 23 we described how a linear chromosome is shortened after each round of DNA replication. To cope with this DNA loss, each eukaryotic chromosome has **telomeres** at its ends, extensions of each DNA strand consisting of repeating short units of nucleotides put on by the enzyme **telomerase**. The telomeric DNA protects the chromosome ends from the attention of repair enzymes that might otherwise detect them as sites of DNA damage and push the cell into an inappropropriate DNA damage response. It is also there to be progressively lost on chromosome replication and, in being sacrificed, protect the 'real' chromosomal DNA carrying the genes. In rapidly dividing embryonic cells and stem

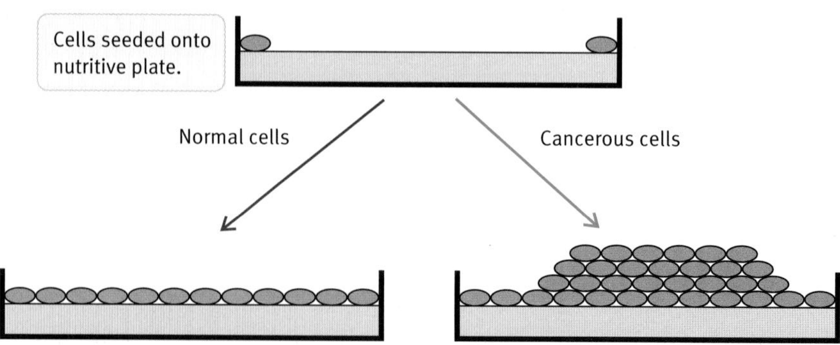

Fig. 31.1 Growth characteristics of normal and some cancerous animal cells in culture. Normal cells show contact inhibition – they stop dividing when they are in contact with each other. Cancer cells do not.

cells telomerase replenishes the lost telomeric DNA so that no matter how many times the chromosome is replicated it is protected from shortening. However, somatic cells do not contain telomerase so that once the telomere is reduced to a critical length cell division ceases. It would seem that somatic cells have a 'ration' of telomeric DNA for so many cell divisions and that is it. Thus, their lifespan is finite. Nonetheless, cancer cells derived from those same normal somatic cells are immortal.

Cancer cells have no limitation on the number of cell divisions they can make

There is no intrinsic limitation on the ability of cancer cells to divide. The immortalization of cancer cells has been shown to depend on telomere maintenance. Most, though not all, cancerous cells have been found to have telomerase activity, unlike the normal cells from which they developed. This does not imply that reactivation of telomerase causes cancer, rather that it is a corequisite for the immortalization of cancer cells, and this has suggested that telomerase could be a target for cancer therapy.

It is found that 10–20% of human tumours do not switch on telomerase synthesis. They are nevertheless immortal because of the adoption of alternative mechanisms of lengthening telomeres (ALT) in which the telomere of one chromosome acts as the template for another.

Types of abnormal cell multiplication

There are two broad classes of abnormal cell growth in the body. Some cells grow and form a solid tumour, which simply gets bigger but remains as a single 'lump'; this is a **benign tumour**, so called because it does not spread to adjacent tissues. It may not pose a threat to life other than from the effects of physical compression, or its own biochemical activity, if it, for example, is a hormone-producing tumour. Such tumours may often be relatively easy to treat surgically. The other class is **malignant cancer cells**; they invade neighbouring tissues and ultimately may break off and migrate to lymph nodes and elsewhere and set up further cancer centres. This spread is known as **metastasis**. Such metastasizing cancers are the dangerous ones; the cells have properties over and above their immortality, which enable them to spread into vitally important organs.

Cancers arising from epithelial cells (the most common) are called **carcinomas**, and from muscle cells, **sarcomas**. **Leukaemias** are 'cancers of the blood' in which large numbers of abnormal white blood cells continue to multiply and circulate without giving rise to, and often at the expense of, normal functional white blood cells.

Cancer development involves a progression of mutations

A cancer is a clone of cells originating from a single somatic cell that outgrows its neighbours. It is initiated by **genetic mutation**, which may be a factor in why the incidence of cancer increases with age; there has been more time for initiating mutations to accumulate. However, single genetic changes are not sufficient to cause cancer. Additional multiple genetic events have to occur, often over years. Cancerous cells undergo cytological changes that are diagnostically important; they tend to de-differentiate; that is to say, they lose the characteristic phenotype of the original cell from which they were descended.

Uncontrolled growth alone does not constitute a malignant cancer, which usually starts as a benign cell growth due to one or a few genetic changes. During this abnormal cell division further genetic changes accumulate in the multiplying cells. The changes do not occur in a fixed progressive series, so that different cancers and even individual cancers of the same type may have different histories of mutations. Some of the genetic changes give variant cells growth advantages over neighbouring tumour cells. For example, tumours can promote angiogenesis (development of blood vessels which sustain development of the tumour mass so that the cells are not restricted in growth by oxygen deprivation). A tumour cell may develop the ability to break loose from its starting clone of cells and escape to colonize other parts of the body. This metastasis requires several changes in a cell, incompletely understood, but which include cell-surface changes enabling it to detach from other cells in the tumour, and secretion of enzymes to break down connective tissue basal lamina layers so that the cell enters the lymphatics or bloodstream. The comprehensive review by **Hanahan and Weinberg** (see Further reading online) discusses the cellular abnormalities that are found in most human cancers.

Development of colorectal cancer

A classic example of the progression of cancer development is colorectal cancer. This originates from a single cell in the gut epithelial layer causing the formation of a precancerous mass of cells known as a polyp (Fig. 31.2). If not removed (as is possible relatively simply during colonoscopy), the polyp cells go on multiplying and cell division provides the opportunity for accumulation of further mutations, which might lead to the evolution of a corrupted 'cancer genome' over a period, usually of several years. The polyps often leak blood into the gut, and this can be detected by a simple widely available faecal test. Removed polyps can be checked cytologically to see whether any malignancy had developed. Figure 31.2 is a cartoon depiction of the genetic changes conferring growth advantages, ultimately resulting in cells that invade adjacent tissues and metastasize to other locations in the body.

Mutations cause cancer

The causes of cancer are basically at the genome level and they include the following:

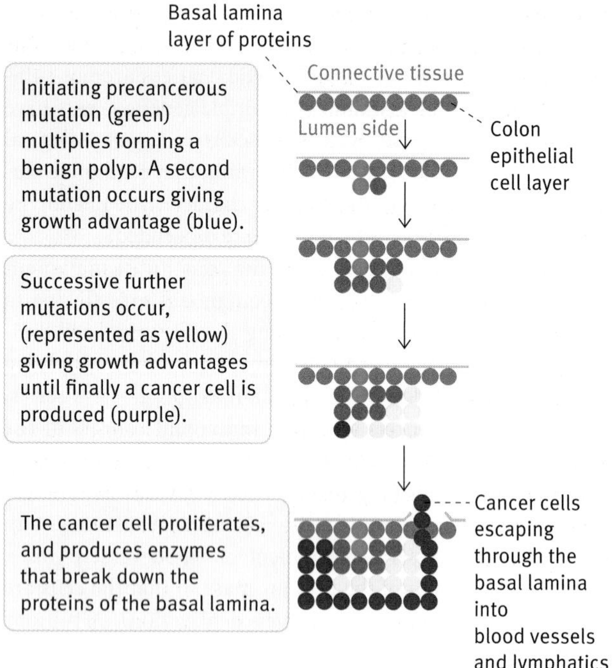
Basal lamina layer of proteins
Connective tissue
Lumen side
Colon epithelial cell layer
Initiating precancerous mutation (green) multiplies forming a benign polyp. A second mutation occurs giving growth advantage (blue).
Successive further mutations occur, (represented as yellow) giving growth advantages until finally a cancer cell is produced (purple).
The cancer cell proliferates, and produces enzymes that break down the proteins of the basal lamina.
Cancer cells escaping through the basal lamina into blood vessels and lymphatics

Fig. 31.2 Cartoon of development of a colorectal cancer. An initial mutation in an epithelial cell causes it to proliferate into a benign polyp. A series of further mutations occur giving the mutant cells growth advantages until the accumulation of mutations results in the development of a cancer cell (purple). This multiplies and metastasizes by invading the epithelial layer and breaking down the basal lamina proteins. The cancer cells escape into the lymphatics and blood vessels and colonize other tissues of the body. Only four mutations are depicted here to keep the figure to a reasonable size. The progression from initial mutation to malignancy usually takes several years. Large numbers of cells are obviously formed at each stage but could not be shown here.

- Carcinogenic chemicals that react with and covalently modify the DNA. The cytochrome P450s (see Chapter 32) activate some of the procarcinogens to actual carcinogens. An example of a natural carcinogen is the **aflatoxins**, which are liver carcinogens. These may be present in foods such as peanuts that have been fungus-infected due to improper storage.
- Ionizing radiation causes chromosome breakages and rearrangements.
- UV light is well known to cause melanomas. Other skin lesions are triggered by UV exposure in patients with defective nucleotide excision repair (xeroderma pigmentosum, see Chapter 23).
- In addition to the above external agents, reactive oxygen radicals, superoxide, and particularly hydroxyl radicals generated in cells (see Chapter 32) covalently modify DNA, and set up destructive chain reactions.
- Point mutations may be produced due to faulty DNA replication. Mutations predisposing to cancer may occur at mitosis, or they may be inherited due to mutations in germline cells. The mutation rate is greatly increased if DNA-repair mechanisms are deficient.
- Viruses can also produce cancers in animals. **Papilloma virus**, which causes cervical cancers, is one of a small group that cause human cancers. **RNA retroviruses** are of great interest from a molecular biology viewpoint and are known to cause cancer in animals (see later), but they are rarely associated with human cancers. To date only one form of human leukaemia is definitively established to be caused by a retrovirus. Another human retrovirus, **HIV** is associated with the development of cancers but this arises from the suppression of the immune response rather than from the virus itself causing cancer. It seems that continuous immunological surveillance is utilized to destroy cancer cells, presumably because of their surface changes. Without this surveillance, cancers would occur more frequently than they do, and its loss leads to cancer in patients with HIV.

The types of genetic change involved in cancer

As stated, cancers result from multiple mutations acquired possibly over years and accumulating gradually in successive generations derived from an initial clone of cells. Some of the mutated genes in a tumour may have little to do with the generation of cancer and be incidental to it, while others may be indirectly involved in conferring a selective growth advantage. Much is yet to be discovered about the mechanisms by which cancer cells acquire their particular characteristics, such as the capacity to metastasize. However there are two broad classes of cancer-associated genes, the mutation of which is known to be of critical importance and about which a good deal is known.

One broad class is the **oncogenes**. These give abnormal signals leading to uncontrolled cell division (*onco*=Greek for mass or tumour). Their effect is positive; they do something that actively causes or facilitates the development of cancer. Put colloquially, they are the 'bad' genes. They are most often derived from normal, essential, genes in control pathways, described below. The second broad category of genetic changes that contribute to cancer development is due to the loss of functional genes that protect cells against uncontrolled cell division, the **tumour-suppressor genes**. They are in colloquial terms the 'good' genes. Mutation of these genes does not itself lead to uncontrolled cell growth, but removal of their protective effect means that cells are more likely to make the progression to the cancerous state. These are also described below.

A third class of genes that are often mutated in cancer are **DNA-repair genes**. Xeroderma pigmentosum, arising from loss of nucleotide excision repair, was mentioned earlier. In Chapter 23 we also described the mechanism of methyl-directed mismatch repair involving Mut H, S, and L proteins in *Escherichia coli*. As mentioned, it has been found that proteins corresponding to Mut S and L (but not H) are present in eukaryotes, mutations of which are associated with a form of heritable bowel cancer in humans, **hereditary nonpolyposis colorectal cancer (HNPCC)**. Since mismatched bases (if not corrected) can lead to mutations and possibly to the formation of oncogenic forms of normal genes, it follows that DNA-repair systems are normally protective and defects in them increase the chance of cancer-causing mutations.

We will now describe the **oncogenes** and the **tumour-suppressor genes** in turn.

Oncogenes

An oncogene is an abnormal gene that can initiate uncontrolled cell divisions or, as in the case of Bcl-2 (see later), cause inappropriate survival of a damaged cell. Oncogenes may be detected by isolating and fragmenting cancer cell DNA and introducing the pieces into normal cells (transfection), where they cause abnormal cell division. There are, however, a great many genes that are not known to have any potential to cause cancer when mutated. What makes particular genes liable to initiate cancer if mutated?

This is the point at which several major topics come together. In Chapter 29 we discussed cell signalling in which the predominant emphasis was on gene control and the way in which cytokines, growth factors, and hormones from outside the cell control specific genes in the nucleus to produce the required responses. Taking the Ras pathway (see Chapter 29) as an illustrative example, signals from other cells attach to receptors on the outside of the target cell and the signalling pathway conducts the instruction to the nucleus via a series of intermediate proteins. Mutation of genes encoding the proteins that contribute to the signalling pathway are associated with cancer development. Cancer in fact typically involves at least one signalling pathway becoming faulty.

How can a signalling pathway become faulty? Consider a cascade of events such as shown below in which A–D are components of the signalling pathway (arrows represent activations, not conversions, please note).

$$\text{Mitogenic factor} \rightarrow \text{cell receptor} \rightarrow \text{A} \rightarrow \text{B} \rightarrow \text{C} \rightarrow \text{D} \rightarrow \text{cell division}$$

If any of the protein components of the chain were to be altered as a result of gene mutation so that it is in a permanently active state, it will be the same as if a mitogenic growth factor or cytokine is present, even in the absence of the latter (Fig. 31.3). The nucleus is erroneously given the signal to activate genes leading to cell division. The pathway is active without a mitogenic signal having been received by the cell receptor so that, in terms of cell cycle controls, the cell is 'permitted' to proceed through G_1 in the absence of an appropriate signal. This is in contrast to a *normal* cell, which goes into G_0 phase in the absence of a signal. A gene in a signalling pathway of this type has the potential to become an oncogene, and is therefore called a **protooncogene**. Oncogenes arise from normal essential genes; it is the normal regulatory role of the protooncogenes that makes them potentially oncogenic.

A somewhat different type of oncogene is exemplified by antiapoptotic Bcl-2 proteins (overexpressed in many cancers), which favour cancer progression. Excess Bcl-2 protein blocks protective apoptosis signals and thus allows cells (e.g. with damaged DNA) to survive and become cancerous. If, as in normal cells, the apoptotic signal were not blocked, the cell would self-destruct and prevent progression to cancer.

How are oncogenes acquired?

Protooncogenes may be converted to oncogenes in several ways. The simplest is that a point mutation in a protooncogene may result in the production of an abnormal hyperactive protein that causes abnormal cell cycle control. The **Ras** protein encoding gene, one of the early components of the Ras signalling pathway, is a protooncogene. It has the automatic GTPase activity, which limits the period of its activation by hydrolysing bound GTP to GDP. The Ras protein has, in effect, to continually ask the receptor whether the signalling molecule is still there. If it is not, in normal cells, the whole pathway is switched off. The oncogenic form of the Ras protein lacks the ability to hydrolyse GTP and so cannot operate the GTP/GDP-off switch. The Ras signalling pathway therefore fires off uncontrollably once activated. Many human cancers have an abnormal Ras protein, in some cases due to a point mutation in the gene. It is also known that the gene encoding the next component in the pathway, **Raf**, acts as an oncogene in many human cancers. The oncogene encodes a form of the Raf protein kinase that lacks the normal regulatory domain and continually activates the signalling pathway. This concept of intermediate signalling pathway components being abnormally active is illustrated diagrammatically in Fig. 31.3.

Chromosomal translocations can result in oncogene formation by fusing a section of a protooncogene with a different gene. The resultant fusion protein may be overproduced and/or be hyperactive. For example, in **Burkitt's lymphoma**, a cancer of human immune system B cells, the coding region of the *c-myc* protooncogene is placed under the control of a strong immunoglobulin gene promoter by a chromosomal

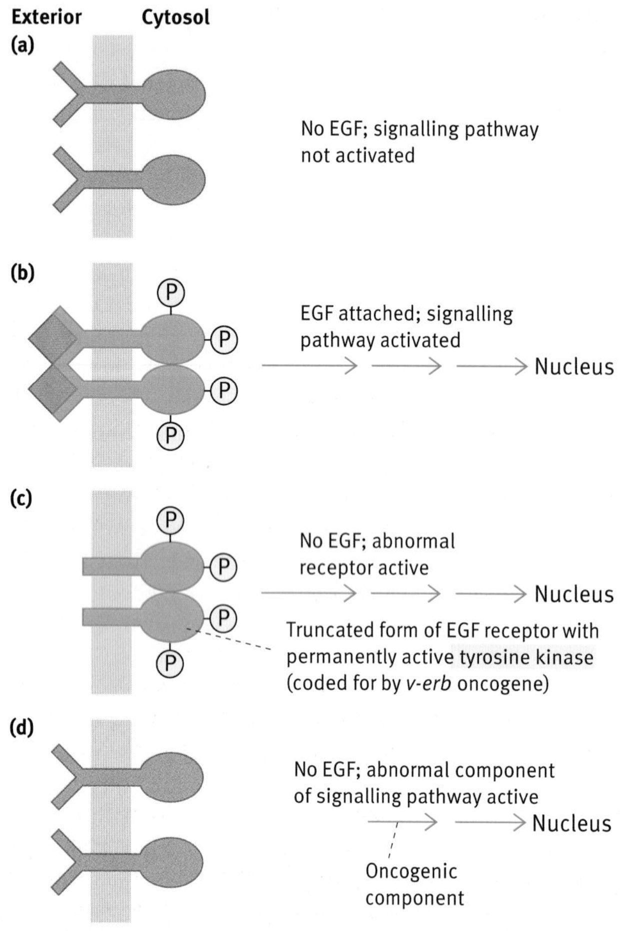

Fig. 31.3 Oncogene products as components of signalling pathways from the membrane to the nucleus – the epidermal growth factor (EGF) receptor is used as an example: **(a)** normal receptor – not activated; **(b)** normal receptor – activated; **(c)** permanently active truncated EGF receptor; **(d)** normal receptor – pathway component permanently active. Examples of the latter in human cancers are oncogenic forms of Ras and Raf in the Ras signalling pathway.

breakage and rearrangement, resulting in the formation of excessive amounts of the transcription factor **Myc**. Since Myc is a transcription factor that activates genes involved in cell growth and cell replication, this leads to excessive gene activation and to uncontrolled multiplication of the lymphocytes.

Like all human genes that are located on autosomes (non-sex chromosomes), protooncogenes are present in the cell in two copies. The nature of oncogenic mutations, as described above, results in a protooncogene *acquiring* an active function, such as abnormal activity in the absence of a signal or inappropriate overexpression. This means that oncogenes are **dominant**. In other words, conversion of only one of the cell's two copies of the protooncogene to its oncogenic form is needed for the cancer-causing effect.

Retroviruses can activate or acquire cellular protooncogenes

Retroviruses are RNA viruses known to cause cancers in animals and in one human leukaemia. When a retrovirus infects a cell, its reverse transcriptase copies its RNA genome into DNA, which then randomly inserts itself into the host DNA as a permanent genetic passenger. The retrovirus carries strong promoter elements, and if these insert upstream of a protooncogene they may cause it to be overexpressed, giving it oncogenic activity. Alternatively, the retrovirus may insert into a tumour-suppressor gene (see later) and disrupt normal production of protein coded for by that gene. In some cases, a retrovirus acquires an oncogene from the host genome as it is reproduced in a cell. In this case the retrovirus becomes the infectious transmitter of a **viral oncogene**, which becomes inserted into a new host cell's DNA on infection.

Many cellular protooncogenes were in fact originally discovered in their retroviral oncogene form, and this has influenced their nomenclature. Retroviral oncogenes are named (in italics) after the retroviruses carrying them. Thus, the *ras* gene was originally found in the rat sarcoma virus. The viral oncogene is identified by the letter 'v' (for virus) and the protooncogene by 'c' (for cellular) so that we have v-*ras* and c-*ras* etc. The expressed proteins are written as Ras, Raf, etc. (not in italics). As an example, the v-*erb* oncogene codes an abnormal truncated form of the receptor for epidermal growth factor (EGF; Fig. 31.3). It is permanently in the active state. If this gene is inserted into a host cell by a virus, the abnormal receptor is produced and is present in the cell membrane. It remains permanently in the phosphorylated active state even though there is no EGF. It thus continually, and improperly, activates the Ras pathway, which causes cell division in the absence of EGF signalling, which normally would be required for division to occur.

Tumour-suppressor genes

We have emphasized that oncogenes commonly affect signal transduction pathways and gene control. When we come to tumour-suppressor genes the emphasis shifts to cell cycle control and particularly to the restriction point in G_1 phase of the cycle, which determines whether the cycle may proceed into S phase. Several tumour-suppressor genes are known, mutation of which has been found in particular groups of cancers. The two best known tumour suppressors are **p53** and the **retinoblastoma** gene encoding the **Rb** protein. The retinoblastoma gene is so named because a mutant form was first discovered in a rare retinal cancer. It has since been found to be mutated in a number of other cancers.

Tumour-suppressor genes encode proteins that protect against the development of cancer. It is therefore mutations that cause loss or loss of function of the tumour-suppressor protein that are associated with cancer formation. This means that tumour-suppressor gene mutations tend to be recessive in that both copies of the gene must be mutated, unlike oncogene mutations, which act dominantly.

Mechanism of protection by the p53 gene

Defective p53 genes are involved in the development of over half of all human cancers and p53 is described as the guardian of the genome. Consider a normal cell in early G_1 phase; appropriate mitogenic factors, if present, will activate genes leading to the production of the G_1-specific cyclins, which activate the cyclin dependent kinases (Cdks). In the *normal* cell, the amount of p53 protein is low and inactive and the cycle can proceed. If the DNA is damaged, however, the p53 gene is activated and more p53 protein is produced. p53, in the presence of damaged DNA, becomes phosphorylated on a serine residue, which both activates it and decreases its rate of destruction. p53 is a transcription factor that activates over sixty other genes, including for example that encoding p21, one of the Cdk inhibitors (Ckis) mentioned briefly in Chapter 30. p21 inactivates the G_1-specific Cdk activity, thus arresting the cycle at the restriction point and giving time for the DNA to be repaired. If repair is successful the p53 gene is no longer activated and the level of p53 protein returns to its low level and the cycle continues normally. If the DNA damage is not repaired, the p53 causes the cell to self-destruct by the apoptopic mechanism. In the absence of functional p53 genes, the cycle is not arrested and the apoptosis signal is not delivered, so a cell with abnormal DNA is allowed to replicate, thus increasing the chance of cancer developing.

Mechanism of protection by the retinoblastoma gene

This is another well-known tumour-suppressor gene. The protein encoded by the **retinoblastoma** gene (**Rb**), like p53 also blocks the cycle at the G_1 checkpoint. Rb combines with a transcription factor called **E2F** and prevents E2F from activating genes required for progression into S phase. In *normal* cells, when a mitogenic stimulus is received, the genes for G_1 cyclin synthesis are activated so that the Cdk activity increases. The Cdk phosphorylates and inactivates the Rb protein, preventing it from combining with E2F so that the cycle can advance to S phase. However if the cycle is proceeding in an aberrant way due to an oncogene bypassing the need for a mitogenic signal, since the normally required Cdks are not active, the Rb is not phosphorylated and remains bound to E2F, arresting the cycle. Rb if present and active can therefore restrain the cell cycle from proceeding abnormally. Retinoblastoma gene defects are known to be associated with certain types of cancer such as retinoblastoma, osteosarcoma, and small cell lung cancer.

Molecular biology advances have potential for development of new cancer therapies

Advances in understanding the molecular biology of the cell may provide new and more specific targets for attacking the causes of cancer. The membrane receptor tyrosine kinases, which are of central importance in signalling pathways and are often mutated or overexpressed in cancer, have attracted much attention as potential therapeutic targets. Therapeutic monoclonal antibodies have been developed that interact with and block the function of these receptors. An alternative approach is to target toxins to specific cancer cells using immunological methods based on the fact that the cancer cell surface is changed from the normal. Strategies to boost the immune system to increase immunological surveillance are also being attempted.

There are other approaches such as developing agents aimed at preventing vascularization of tumours that could help lead to their demise since the mass of cells become short of oxygen without an increased blood supply. The discovery of RNA interference (RNAi, see Chapter 26) and its applicability to mammalian cells is very recent, but it raises the possibility of the process being used against cancer. The small interfering RNAs (siRNAs) are easily synthesized chemically. Their sequence specificity means that, in principle, any messenger RNA can be targeted to silence (or knockdown) expression of specific genes. Oncogenes are an obvious target. An interesting advance has been to use a viral vector to deliver to cultured cells a sequence of DNA, which is transcribed into a specific siRNA.

The completion of the human genome project and the resulting boom in genomic technology and information facilitates the identification and detection of disease-producing genes. A problem in understanding and treating cancer is that individual cancers have different characteristics, which makes their classification difficult and this complicates devising specifically tailored therapeutic regimes. The development of DNA microarrays, which make it possible to study the simultaneous expression of large numbers of genes, has the potential to classify cancers in a way that may help the establishment of improved therapy regimes. By comparing the expression of blocks of genes in normal and cancerous tissues, new understanding of the disease may emerge. Drugs can be tested on cancer cell lines to investigate their effects on cancer gene expression. A similar promise is inherent in the related technology of proteomics.

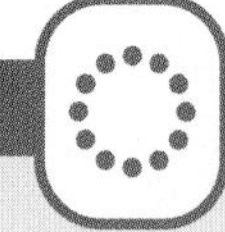

Summary

Cancer cells grow uncontrollably when they should not, often due to abnormalities of cellular signalling systems. They become independent of mitogenic signals and proceed to S phase in the absence of such a signal. They evade signals to self-destruct and, because they acquire the ability to lengthen their telomeres, can divide indefinitely. They progressively mutate, which gives them growth advantages, and finally the acquisition of mutations may enable them to metastasize to invade other tissues and spread throughout the body.

Development of cancer is a process where successive mutations confer growth advantages on individual cells in a tumour. While a single initiating mutation may cause cell division, development of an invasive cancer requires a series of mutations, which may take several years. The mutations do not occur in a fixed order, and cancers of the same type may have different histories and mutations. A typical example is colorectal cancer in which first a benign polyp develops. Usually only after several years does it become a cancer. Removal of the polyp in colonoscopy will prevent it developing into a malignant form.

Major causes of mutations are carcinogenic agents, such as chemicals, certain viruses, and ionizing radiations. Mutations may result also from faulty DNA replication or repair, or from large scale chromosomal rearrangements such as translocations.

A major class of cancer-causing mutations result in the formation of oncogenes from normal protooncogenes. Protooncogenes are usually genes coding for proteins in signalling pathways that control replication. Oncogenic mutations cause inappropriate activity of the signalling pathways and hence inappropriate cell division.

Alternatively mutations may affect the activities of tumour-suppressor genes, which normally guard against cancer; p53 and the retinoblastoma gene encoding the Rb protein are prime examples. About half of all human cancers are associated with deficient p53 protein. p53 works by halting the cell cycle if the genome is damaged. If the damage is not repaired p53 delivers an apoptopic signal for the cell to self-destruct, thus avoiding development of a potentially cancerous cell. If both copies of the p53 gene are deficient this safeguard is not present and an oncogenic mutation is more likely to progress to a cancer. The Rb protein can prevent the cell cycle from progressing from G_1 into the S phase if appropriate mitogenic signals are not present. Thus, loss of active Rb allows inappropriate cell division to proceed in cancer cells.

Many new targets for possible cancer drug therapies are arising from understanding the molecular biology of cancer.

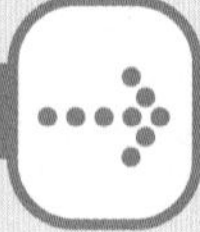

Further reading

To access the further reading, please scan the QR code image or go to http://global.oup.com/uk/orc/biosciences/molbiol/snape_biochemistry/student/reading/ch31/

Problems

1 Explain what a protooncogene is. What are the mechanisms by which a protooncogene can become an oncogene?

2 How does the p53 gene protect against the development of cancer?

3 What are oncogenes and tumour-suppressor genes?

4 When somatic cells become cancerous they often develop telomerase activity. Explain the significance of this.

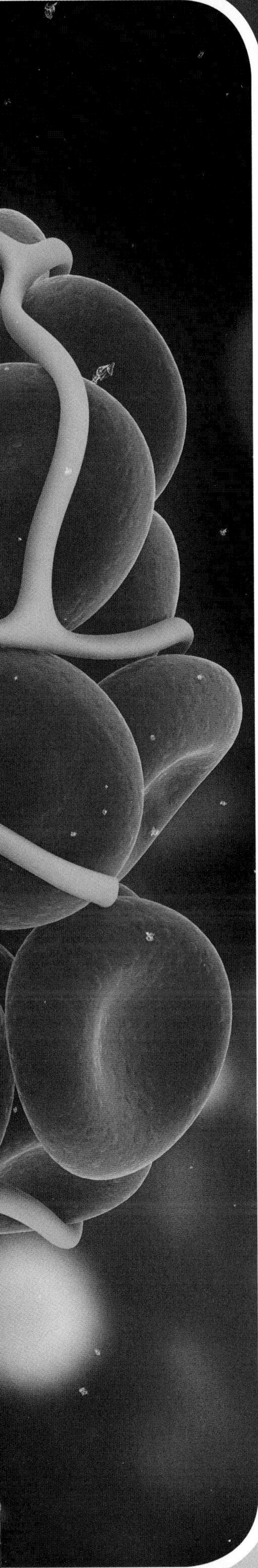

Part 6

Protective mechanisms against disease

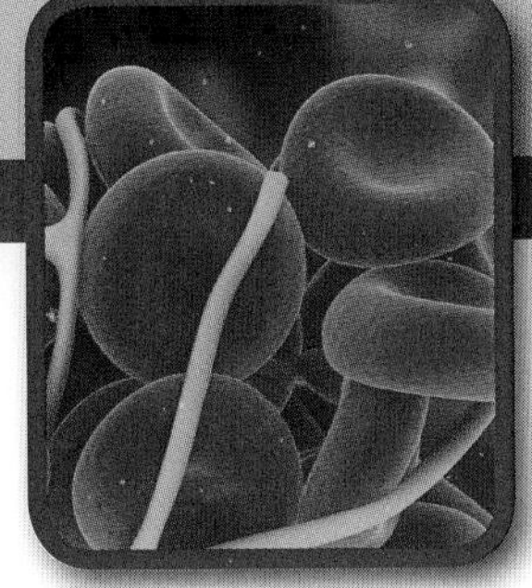

Chapter 32

Special topics: blood clotting, xenobiotic metabolism, reactive oxygen species

At the molecular level, life is a dangerous business, particularly in organisms as complex as mammals. There is a delicate balance between having chemical mechanisms, without which life would be impossible, and the dangers inherent in these mechanisms. It is, for example, efficient to circulate oxygen to all the cells of the body by a high-volume flow of blood, but this means that the body has to be ready to form a clot at the site of a wound instantly to prevent death by bleeding. Unfortunately, inappropriate blood clotting (thrombosis) also constitutes a life-threatening hazard, and this must also be guarded against without impairing the ability of rapid wound healing.

It is also efficient to generate usable energy from foodstuffs by transferring electrons to oxygen, and provided the oxygen molecule is reduced to water the process is benign. No process is perfect, however, and incompletely reduced **reactive oxygen species (ROS)**, often **free radicals**, are generated, which are destructive to biological molecules. There are also other causes of ROS generation such as ionizing radiation and certain chemicals.

We need to take in a wide variety of foods to obtain essential nutrients but this means the intake of a variety of potentially toxic molecules, which if not disposed of would be dangerous.

In this chapter we have collected together the enzymic mechanisms by which the body protects itself from such hazards. It is more usual to find these attached to other chapters where they have mechanistic relevance, but here they are presented as biological topics in their own right.

Other protective mechanisms fit into specific chapters. The immune system is the major protective mechanism against attack by disease-causing pathogens but this is a subject in its own right and is dealt with in Chapter 33. DNA repair to cope with damage caused by ionizing radiation, ultraviolet (UV) light, and mutations is also a protective mechanism but this has already been covered in Chapter 23 in which DNA synthesis is dealt with. Similarly, tumour suppression genes, which protect us from cancer, are best left to Chapter 31 where cancer is discussed.

Blood clotting (thrombus formation)

Coagulation of blood is necessary for haemostasis, the process of cessasion of blood loss from a damaged blood vessel. The response has to be rapid and substantial while the initial signal in chemical terms is small. A large amplification of the signal is needed – the response must, in quantitative terms, be much greater than the signal.

We have seen earlier that biochemical amplification is achieved by means of a **reaction cascade**. In this, an enzyme is activated that then activates another enzyme and so on. The fact that the enzymes activated are themselves catalysts means there is an amplification at each step. In case this is not clear, if a single enzyme molecule activates 1,000 molecules of the next enzyme in 1 minute and each molecule of the second enzyme does the same for the next enzyme and so on, in a cascade of four steps, in a very short time vast numbers of active molecules of enzyme number four are created. This enzyme at the end of the cascade can rapidly catalyse a massive response.

Blood clotting can be divided into two parts. There is the cascade resulting in the activation of the enzyme that forms the clot, and there is the mechanism of clot formation itself by that enzyme. The cascade process is based on **proteolysis**, that is, hydrolysis of peptide bonds of inactive precursor proteases, which activates them (*cf.* trypsinogen activation to trypsin in digestion). All of the necessary inactive precursor proteins are present in the blood and they are activated in response to the signal of damaged blood vessels.

What signals the necessity for clot formation?

When a wound occurs, the endothelial cell layer lining the blood vessels is damaged, exposing the structures underneath, such as collagen fibres. These have a negatively charged or 'abnormal' surface. The blood-clotting (or thrombus formation) response is a localized reaction around the site of damage. Initially, a temporary plug is formed by the aggregation of blood platelets around the hole. Liberation of ADP and thromboxane A_2 activates other blood platelets to aggregate on the wound. For the formation of a clot, a small group of proteins in the blood are attracted to the abnormal surface, the net result of which is that two proteases mutually activate each other. One of them is called factor XII. (The nomenclature is slightly confusing in that some 'factors' are enzymes and some are cofactors for enzymes and they are not numbered according to the sequence of their appearance in the process.) Factor XII activates a cascade of three steps resulting in active factor X (another protease). Factor X activates prothrombin to the active protease, thrombin. (The prefix 'pro' or the suffix 'ogen' refer to the inactive molecules, which are converted into the active form.) Thrombin causes clotting (or thrombus formation). (We usually expect enzyme names to end in 'ase' but several of the classic proteases end with 'in' – for example, thrombin, pepsin, trypsin, chymotrypsin, and others.)

The process of blood clotting described above is known as the **intrinsic pathway**. If blood is put into a glass vessel, the glass surface triggers clotting. No external factors need to be added, blood clotting is an inherent property of blood and for this reason, the process is called **intrinsic**.

There is also an **extrinsic pathway**. This is triggered by the release of a protein complex called tissue factor from damaged cells and tissues. Since something has to be added to blood, it is called the *extrinsic* pathway and it is shorter than the intrinsic one. A protease is activated and this activates the same factor X as occurs in the intrinsic pathway, resulting again in active thrombin formation. The two pathways are set out in Fig. 32.1.

The intrinsic pathway, being longer, is slower to cause clot formation when measured *in vitro*, than the extrinsic pathway, also measured *in vitro*. However, in the genetic disease haemophilia A, in which blood clotting fails to occur, it is the intrinsic pathway that is deficient due to the absence of factor VIII required for factor X proteolytic activation. However, it seems that, for normal physiological clotting, both pathways function as one and both are essential. Interactions between the two pathways have been identified to be involved in physiological clotting.

How does thrombin cause thrombus formation?

In the circulating blood there is a protein called **fibrinogen**. The basic molecular unit consists of short rods made up of three polypeptide chains; two of these rods are joined together by S–S bonds near their *N*-terminal ends, forming the fibrinogen monomer. As shown in Fig. 32.2, at their joining points

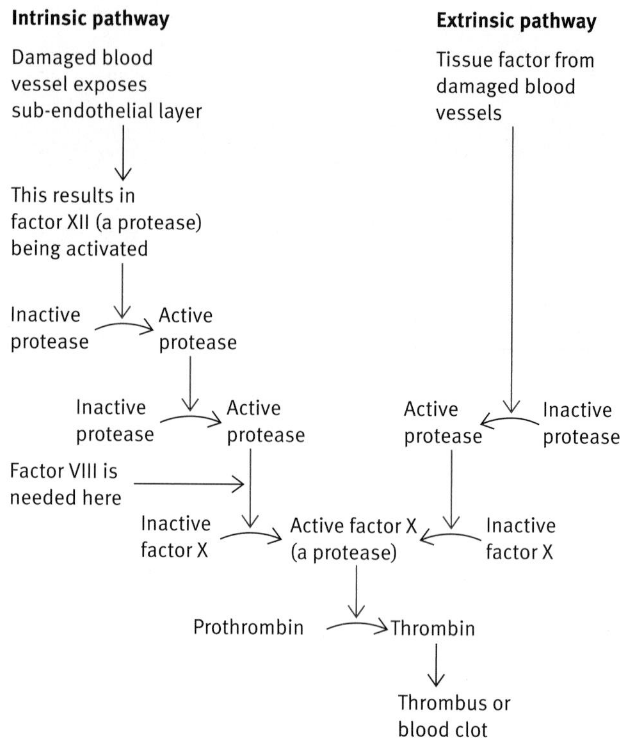

Fig. 32.1 Simplified diagram of intrinsic and extrinsic pathways of blood clotting. The names of the various proteases and other factors involved are omitted for simplicity. (Thirteen factors, numbered I-XIII, are in fact known.) The proteases listed are specific for their particular substrate. Factor VIII is the protein missing in patients with **haemophilia A**.

two of the three chains in each short rod project as negatively charged peptides, called **fibrinopeptides**. The negative charges on the monomers repel each other and prevent association.

Thrombin cleaves the fibrinopeptides off, giving a **fibrin monomer**. The fibrin monomer is now able to polymerize spontaneously by noncovalent bond formation. The sites at the end of the fibrin monomer are complementary to sites at the centre of adjacent molecules so that a staggered arrangement forms from the polymerization (Fig. 32.3). This gives a so-called 'soft clot'. A more stable 'hard clot' is formed by subsequent covalent cross-linking between the side chains of adjacent fibrin molecules.

The covalent crosslinks are curious in that a glutamine side chain on one monomer is joined to a lysine side chain on the next in an enzymic transamidation reaction:

$$\underset{\text{(Glutamine side chain)}}{-\text{CONH}_2} + \underset{\text{(Lysine side chain)}}{\text{H}_3\text{N}^+-} \rightarrow \underset{\text{(Cross-link)}}{-\text{CO}-\text{NH}-} + \text{NH}_4^+.$$

The fibrin strands entangle blood cells, forming a blood clot.

Keeping clotting in check

Blood clotting is potentially dangerous unless limited to local sites of bleeding. Once started, there is the danger of

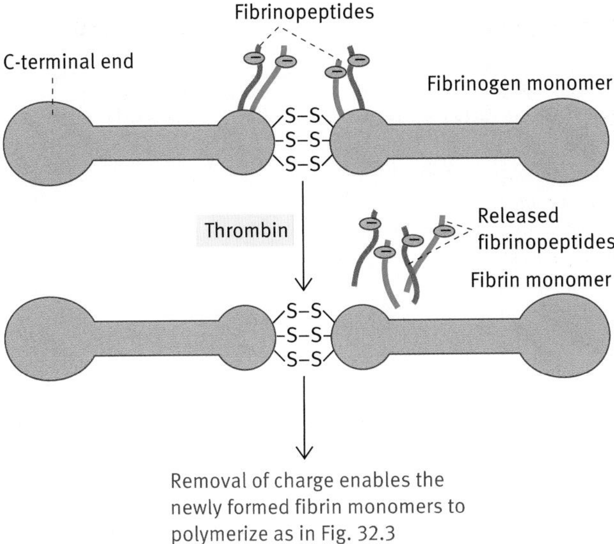

Fig. 32.2 Fibrinogen monomer and its conversion to fibrin monomer. Each half of the fibrinogen monomer is composed of three polypeptide chains, two of which terminate in the negatively charged fibrinopeptides.

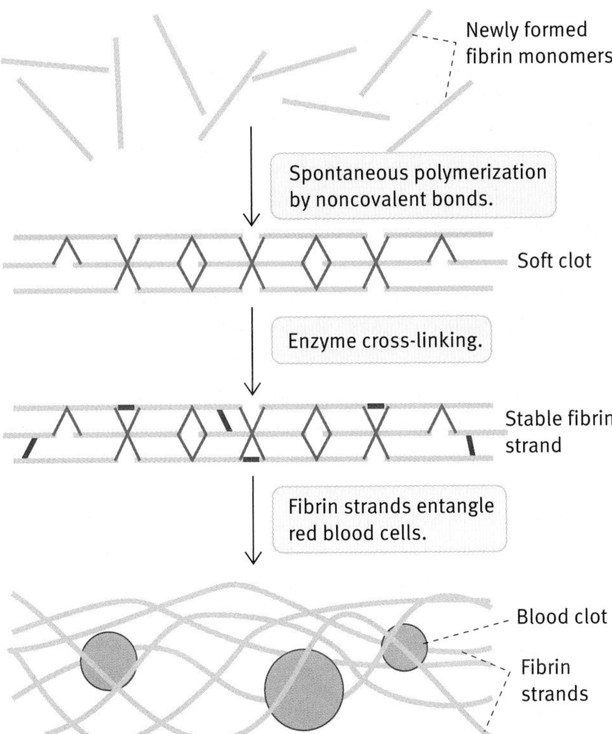

Fig. 32.3 Spontaneous polymerization of fibrin monomers and their enzymic cross-linking to form a stable fibrin strand. The fibrinogen cannot polymerize until thrombin hydrolyses off the fibrinopeptides because (1) the negative charges prevent association and (2) the central sites with which the ends of the fibrin monomers associate are masked by the fibrinopeptides. Noncovalent bonds are shown in blue; the covalent crosslinks, shown in red, are arbitrarily represented in position and number.

an autocatalytic process like this getting out of hand, causing inappropriate clotting. An elaborate series of safeguards exists. Proteinase inhibitors (for example, antithrombin) in blood 'dampen down' and prevent the clotting reactions from spreading; heparin, a sulphated polysaccharide (a glucosaminoglycan; see Chapter 4) present on blood vessel walls, increases this inhibitory effect; another protease, **plasmin**, dissolves blood clots. Plasmin itself is formed from inactive **plasminogen**; activation of plasminogen is effected by a protein, **tissue plasminogen activator** (**TPA**, or **t-PA**), released from damaged tissues. Tissue plasminogen activators are used therapeutically to prevent thrombosis. Although TPA is present in tissues in minute amounts, the gene coding for it and the corresponding cDNA have been isolated and used for commercial production of the protein for injection (the technology is described in Chapter 28). This, and other inhibitory mechanisms, limit the reaction to the damaged surface area, the latter being required for initiation of the process. Blood-clotting control is complex; inappropriate clotting (thrombosis) is responsible for large numbers of deaths.

Blood clotting requires aggregation of blood platelets. This is facilitated by thromboxane A_2 (see Chapter 17), which is synthesized by cells lining blood vessels. The synthesis is selectively inhibited by low doses of aspirin (75 mg/day – see Box 17.2). This is commonly used medically to help prevent vascular problems.

Rat poison, blood clotting, and vitamin K

The widely used rat poison, **warfarin**, kills by preventing blood clotting so that the rodents die from unchecked internal bleeding from minor lesions that normally occur. Warfarin is used clinically, for example, in patients with atrial fibrillation, in which areas of stagnant blood due to defective atrial contractions raise the risk of clotting. It is structurally similar to **vitamin K** (K from the Danish *Koagulation*) and acts as a competitive inhibitor. It competes with the vitamin for an enzyme site and inactivates it (have a quick look at the two structures in Fig. 32.4). Vitamin K is needed for **prothrombin** conversion to **thrombin**; it acts as a cofactor in an unusual enzyme reaction that adds an extra —COOH group, using CO_2, to several glutamic acid side chains of prothrombin (γ carboxylation). The carboxyglutamate is highly negatively charged and binds Ca^{2+}, which is essential in the prothrombin→thrombin activation process. Half the human body's vitamin K is from the diet, and half formed from gut bacteria (see Chapter 9).

$$\begin{array}{c} | \\ CH_2 \\ | \\ CH_2 \\ | \\ COO^- \end{array} + CO_2 \longrightarrow \begin{array}{c} | \\ CH_2 \\ | \\ HC—COO^- \\ | \\ COO^- \end{array}$$

A glutamic acid side chain of prothrombin — Carboxyglutamate — This group efficiently binds Ca^{2+}

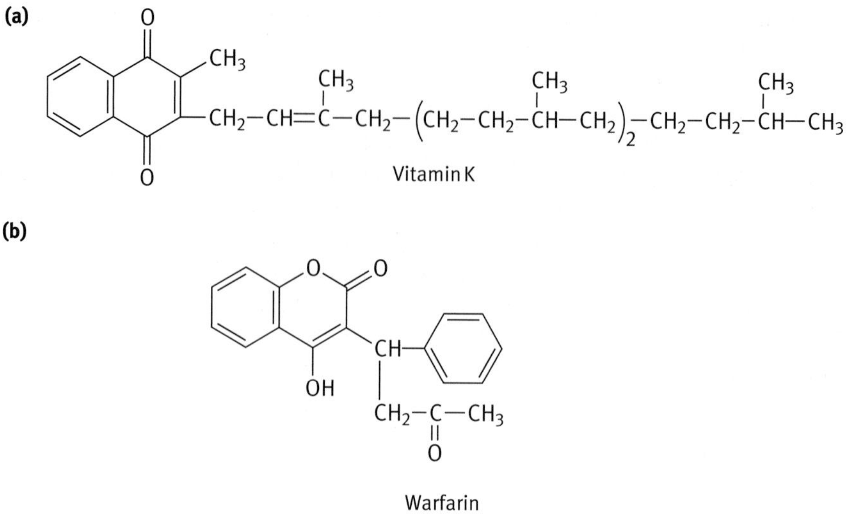

Fig. 32.4 Comparison of the structures of **(a)** vitamin K and **(b)** warfarin. Vitamin K is needed for blood clotting. Warfarin antagonizes vitamin K and prevents blood clotting.

Protection against ingested foreign chemicals (xenobiotics)

Large foreign molecules such as proteins are dealt with by the immune system (Chapter 33). Small foreign molecules (which are not indicative of invasion by a living pathogen) are dealt with by different systems, enzymic in nature.

Human beings ingest large numbers of different foreign chemicals, collectively referred to as **xenobiotics** (*xeno* meaning foreign). These include pharmaceuticals, pesticides, and herbicides, as well as complex structures of plants. Many of these are relatively insoluble in water, but soluble in fats, and they therefore tend to partition into the lipid part of membranes and the fat globules of adipose cells rather than being excreted in the urine or bile. Unless they are rendered more polar and therefore more water soluble, they will accumulate in the body with deleterious consequences. The excretion of foreign chemicals is facilitated by their metabolism and the **cytochrome P450 (P450)** system is of central importance in this process. P450 is a haem–protein complex. The name comes from P for pigment and 450 from the absorption maximum (in nanometres) of the complex formed with carbon monoxide. (CO is not involved in the reaction – it just happens to give a complex with a spectrum that makes measurement of the amount of P450 easy.) A typical reaction of the P450 system is to add a hydroxyl group to an aliphatic or aromatic grouping. This is known as **phase I** of xenobiotic metabolism. In **phase II** different enzymes add various highly polar groups, such as glucuronate, to the hydroxyl groups, thus increasing their water solubility and facilitating excretion in the urine.

Let us look at the P450 system to start with.

Cytochrome P450

There are multiple isoforms or isozymes of P450 s in the body. A nomenclature system has been adopted based on amino acid sequence homologies. All are given the name **CYP** followed by a number indicating families (>40% homology), then a capital letter indicating subgroups (>55% homology), and then a number defining an individual isoform. An example is CYP2B4. The nomenclature is relevant because of the medical importance of the enzymes but for the purposes of this discussion only the term P450 will be used.

P450 enzymes have two roles. The first is the involvement in normal metabolic processes, such as the conversion of cholesterol to other steroids, which explains why the adrenal glands are rich in P450 enzymes. The second role is in xenobiotic metabolism.

The remarkable thing about the P450 system is the large number of different compounds that it tackles, including some that living organisms could not have encountered before. It may be that the collection of compounds, such as terpenes, alkaloids, etc. found in plants were developed as protection for the plants against attacks (for example, grazing) by animals. Having a detoxifying system would confer an evolutionary advantage to animals as it would allow them to survive by coping with almost anything present in plants and other food. The basis of this versatility is that a P450 enzyme has a wide specificity – it attacks a variety of related structures – and different P450 enzymes exist with different but overlapping specificities.

P450 enzymes are found in most tissues, the liver having the largest amounts. They are anchored into the smooth endoplasmic reticulum (ER) facing the cytosol. P450s collectively can bring about a surprising variety of reactions, including dehalogenations and desulphurations, but the most important are hydroxylations. In hydroxylations, a foreign compound, AH, is attacked according to the reaction:

$$AH + O_2 + NADPH + H^+ \rightarrow A{-}OH + H_2O + NADP^+$$

It is called a **monooxygenase reaction** because it uses only one atom of oxygen from each O_2 molecule. NADPH is used to reduce the other oxygen atom to water. It is also called a **mixed-function oxygenase** because it both hydroxylates AH and reduces O to H_2O. The electrons from NADPH are transferred, one at a time, to the Fe^{3+} in the haem of P450 by a P450 reductase enzyme also present in the smooth ER membrane.

Secondary modification – addition of a polar group to products of the P450 attack

This is known as **phase II** of xenobiotic metabolism. The products of the P450 attack are converted into a more water-soluble form for excretion in the urine. There are several reactions for doing this but we will describe the most important ones only. These are **glucuronidation** and conjugation with **glutathione**.

The glucuronidation system

This is present in the smooth ER and it transfers the highly hydrophilic **glucuronate** group from UDP-glucuronate to the hydroxyl group generated on a foreign chemical by P450 (Fig. 32.5). UDP-glucuronate is produced by oxidation of UDP-glucose. Glucuronidation facilitates excretion in the urine and bile by increasing water solubility of the chemical. The same system is used for the excretion of endogenous products such as bilirubin arising from haem degradation (see Chapter 18).

The glutathione *S*-transferase system

Glutathione is a tripeptide of glutamic acid, cysteine, and glycine, the peptide link between glutamate and cysteine being on the γ-carboxyl (Fig. 32.6). It is present in large amounts in liver, muscle, and other tissues. Its main function is to maintain a reducing environment in the cells by virtue of its –SH group. For this reason it is abbreviated to GSH.

Glutathione S-transferases add the sulphur to xenobiotics, which include halogenated molecules, epoxide metabolites of carcinogens, and others. The transferases are present in the smooth ER. The reaction catalysed is as follows:

$$RX + GSH \rightarrow RSG + HX$$

where R is an electrophilic xenobiotic.

The reaction is an important defence against reactive carcinogenic molecules for it prevents them reacting with DNA and causing genetic damage. The product, RSG, of the transferase reaction is modified before excretion. The glutamyl and cysteinyl groups are removed by hydrolysis and the cysteinyl amino group acetylated to form a **mercapturic acid**, which is excreted.

(a) Glucose-1-phosphate

↓ UTP → PP_i

UDP-glucose

↓ H_2O, 2 NAD^+ → 2 NADH + H^+

UDP-glucuronate

↓ R–OH (UDP-glucuronosyl transferase)

R–glucuronate + UDP

(b)

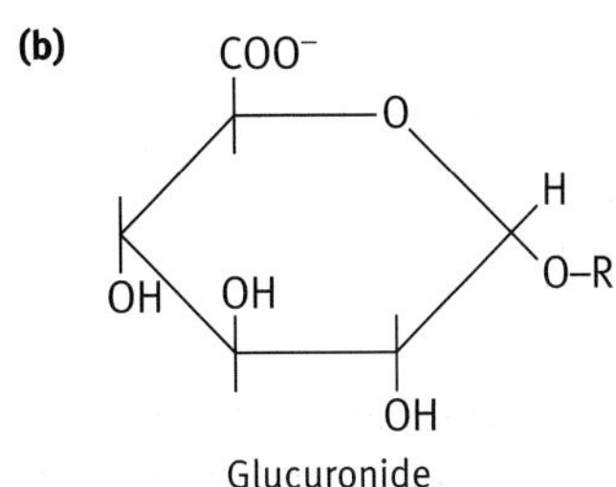

Glucuronide

Fig. 32.5 (a) The glucuronidation system. **(b)** Structure of a glucuronide.

Glu–Cys–Gly
SH
Reduced glutathione (GSH)

Glu–Cys–Gly
S
S
Glu–Cys–Gly
Oxidized glutathione (GSSG)

Fig. 32.6 Structures of reduced glutathione (GSH) and of the oxidized form (GSSG). Glu, Cys, and Gly are abbreviations for glutamate, cysteine, and glycine, respectively, using the three-letter system.

Glutathione is involved in quite separate important protective reactions against peroxides.

Medical significance of P450s

Many pharmaceutical drugs are metabolized by P450 so that the half-life of a drug in the body is related to the rate it is metabolized and excreted. The amount and activity of a given enzyme may vary from one individual to another because of genetic variations so that the correct dose of a drug may vary from person to person. Another aspect is that many P450s are induced by drugs. Thus if barbiturates are fed to a rat, there is a massive proliferation of the smooth ER and of the P450s in it. Once the inducing drug is disposed of, the smooth ER returns to normal. This can complicate drug therapy because if a patient on a correct dose of one drug is given another that induces the P450 that attacks the first drug, then the dose of the first one may now become inadequate. An illustrative case is warfarin the anticlotting agent, the dose of which is carefully calibrated. If the patient is also given a P450-inducing drug the warfarin dose could now be inadequate.

The P450s are not always beneficial in their attacks on xenobiotics. An ironic twist to the story is that the oxidation of some substances by the P450 increases their carcinogenic effect.

Multidrug resistance

Another form of protection of cells against toxic chemicals is by lowering their concentration inside cells. Many cells, including those of human tissues, express a P glycoprotein (P for permeability) in their cell membranes, which is an ATP-driven multidrug transporter of drugs out of the cell. It is one of a very large family of **ABC (ATP-binding cassette) transporters** with common structural features (see Chapter 7). They are found both in bacteria and eukaryotes. A remarkable range of chemicals is transported, among them several of the anticancer drugs used in chemotherapy, and many other pharmacological agents and cytotoxic chemicals. Multidrug resistance can occur after prolonged administration of a drug due to P glycoprotein(s) being induced. The acquired resistance might include drugs different from the original one that caused the resistance in the first place.

P glycoprotein does not only transport xenobiotics out of cells but also normal metabolites. An important biological role is the secretion of steroids from the cells that synthesize them. This accounts for the fact that adrenal cortical cells are rich in P glycoprotein. The system also transports cholesterol out of cells (see Chapter 11) and participates in the formation of HDL particles, as well as transport materials across the blood-brain barrier. The molecules transported by P glycoprotein have no remarkable chemical similarities but all are amphipathic compounds preferentially soluble in lipids.

Protection against reactive oxygen species

As stated in the introduction, oxidation of foodstuffs may sometimes lead to incompletely reduced ROS. ROS may be compounds such as hydrogen peroxide (H_2O_2) and they may also be free radicals, which have an unpaired electron, such as superoxide ($O_2^{\cdot-}$) and the highly reactive hydroxyl radical ($OH^{\cdot}$)

Formation of the superoxide anion and other reactive oxygen species

As described in Chapter 13, oxygen is an ideal electron sink for the energy-generating electron transport system. Its position on the redox scale means that, in energy terms, electrons from NADH have a long way to 'drop', meaning that the negative free-energy change of the overall oxidation of NADH to produce water is large. O_2 accepts four electrons and four protons to give H_2O, the final product of the electron transport chain:

$$O_2 + 4e^- + 4H^+ \rightarrow 2H_2O$$

O_2 is relatively unreactive and therefore itself does no chemical damage, and the product is H_2O. During evolution, the switch from energy generation by anaerobic metabolism to the use of oxygen as the electron sink was one of the most important events. There is, however, a darker side to the story, for O_2 has the potential to be dangerous in the body.

One way in which the danger occurs is when a single electron is acquired by the O_2 molecule to give the **superoxide anion**, a free radical that is a reactive corrosive chemical agent:

$$O_2 + e^- \rightarrow O_2^{\cdot-}$$

Although free radicals such as the superoxide anion are present in minute amounts, they set up chain reactions of chemical destruction in the body. Unpaired electrons are usually very reactive and seek a partner. The unpaired electron of $O_2^{\cdot-}$ acquires a partner by attacking and destroying a covalent bond of some cell constituent. In doing so it acquires a partner electron but also generates a new free radical species with an unpaired electron from the attacked molecule that, in turn, attacks yet another molecule of cell constituent producing yet another free radical, and so on. The destruction initiated by the free radicals is thus a self-perpetuating chain of harmful reactions.

Superoxide is formed in the body in several ways. In the electron transport chain, the final enzyme that donates electrons to oxygen, cytochrome oxidase, does not release partially reduced oxygen intermediates in any significant amounts ensuring that an O_2 molecule receives all four electrons resulting in H_2O formation. However, in the respiratory chain, it is inevitable that small amounts of superoxide are formed. Moreover, mutations in mitochondrial DNA may block electron transport pathways and deflect electrons into formation of reactive oxygen species. Mitochondria have few or no DNA-repair systems and so mutations accumulate in them.

In addition, there are other oxidation reactions in the body that produce small amounts of dangerous oxygen species. Spontaneous oxidation of haemoglobin (Hb) to methaemoglobin (Fe^{3+} form) is another source; as a rare event, the oxygen in oxyhaemoglobin (HbO_2) instead of leaving as O_2

and leaving behind haemoglobin in the Fe^{2+} form, leaves as O_2^- with the formation of methaemoglobin, the Fe^{3+} form.

There are other sources of ROS. Ionizing radiation, by its interaction with water generates ROS, which can lead to free radical attack on biological components. Excessive amounts of neutrophils attracted to irritated joints may lead to superoxide release and contribute to arthritic damage.

ROS have defensive roles as well. When phagocytes ingest a bacterial cell there is a rapid increase in oxygen consumption, which is used to oxidize NADPH *via* a mechanism that generates superoxide anions. These are shed into the vacuole and converted into H_2O_2, which helps to destroy the contained bacterial cell. Some oxidases that directly oxidize metabolites using molecular oxygen, O_2, and quite distinct from the respiratory pathways we have described, generate H_2O_2.

Flavoprotein oxidases, with FAD as their prosthetic group, generally catalyse reactions of the type:

$$AH_2 + O_2 \rightarrow A + H_2O_2$$

Xanthine oxidase, involved in purine metabolism (see Chapter 19), is of this type. H_2O_2 is potentially dangerous because, in the presence of metal ions such as Fe^{2+}, it can generate the highly reactive hydroxyl radical ($OH^{\cdot}$), not to be confused with the hydroxyl anion (OH^-).

$$H_2O_2 + Fe^{2+} \rightarrow Fe^{3+} \rightarrow OH^{\bullet} + OH^-$$

(H_2O_2 generation is the result of two electrons being added to O_2; when three electrons are added, $OH^{\cdot}$ and OH^- are formed.) The hydroxyl radical can attack DNA and other biological molecules.

The biological injuries caused by free radical damage are not precisely established but it has been suggested they contribute to ageing, cataract formation, the pathology of heart attacks, and other problems.

There are essentially two protective strategies – one chemical and the other enzymic. The two protective enzymes, catalase and superoxide dismutase, are important here. Another enzyme that destroys H_2O_2 is glutathione peroxidase, described below. Since brain has little catalase (the enzyme that decomposes H_2O_2 to H_2O), glutathione peroxidase is possibly the main enzyme responsible for protection against H_2O_2 in that organ.

Mopping up oxygen free radicals with vitamins C and E

The chemical method of dampening or **quenching** the chain reactions, initiated by superoxides, is the use of **antioxidants.** The requirement for a quenching reagent is that it should itself be attacked by a free radical such as the superoxide anion but generate a free radical insufficiently reactive to perpetuate the chain reaction. The main biological quenching agents are **ascorbic acid** (vitamin C) and **α-tocopherol** (vitamin E). The former is water soluble, the latter is lipid soluble, and so, between them, they can provide protection for both phases of the cell. These two vitamins are not the only antioxidants – some normal metabolites such as uric acid are effective; **β-carotene** (see Fig. 29.31) is also an antioxidant. Certain polyphenolic compounds (**procyanidins**) found in red wine and other food sources have antioxidant properties (see Box 32.1)

Bilirubin, produced from haem breakdown by haem oxygenase, is also an effective antioxidant.

BOX 32.1 Red wine and cardiovascular health

Epidemiological studies in the 20th century revealed what became known as the French Paradox – despite high consumption of saturated fat, the French population was found to have relatively low rates of coronary heart disease.

Comparing figures for heart disease in men aged 55–64 years in Europe, North America, and Australasia, it was also found that the highest number of deaths were in traditional beer and spirit-drinking countries, while France had the lowest number, and the highest wine consumption. One of the reasons proposed for the lower CV incidents in France was red wine.

Recent epidemiological studies have shown that the moderate consumption of red wine might reduce the risk of death from cardiovascular disease.

The substances in red wine that may have this effect are polyphenols found in the grape skin and seeds. The polyphenol content of wine varies with grape variety, cultural conditions, and the wine making process. The most abundant polyphenols in red wine are the procyanidins, made up of repeating units of two to six molecules of the polyphenol, catechin (oligomeric procyanidins – OPCs). Some other sources of OPCs are cocoa solids, such as those found in dark chocolate, and in several fruits, particularly cranberries.

Procyanidins are antioxidants. They quench oxygen free radicals and prevent damage by these agents. In blood vessels they appear to protect low-density lipoproteins (see Chapter 11) from free radical-mediated oxidation. Plaque formation in arteries involves LDL oxidation, and heart attacks ensue when plaques in arteries burst; blood clots form and cause blockages.

Procyanidins may have an additional protective effect by increasing the synthesis and release of nitric oxide (see Chapter 29) by the vascular endothelium, thus promoting vasodilatation, and lowering the blood pressure.

Experiments with cultured endothelial cells identify the most potent vasoactive polyphenols in red wine as OPCs. A wide range of red wines across the world was tested and the OPC content of each wine correlated with the suppression of synthesis of endothelin-1, a powerful vasoconstricting peptide.

Enzymic destruction of superoxide by superoxide dismutase

Most or all animal tissues contain the enzyme **superoxide dismutase** and, most appropriately, it is found in mitochondria. It is also present in lysosomes (see Chapter 27) and peroxisomes (see Chapter 17) as well as in extracellular fluids, such as lymph, plasma, and synovial fluids. Superoxide dismutase catalyses the reaction:

$$2O_2^{-} + 2H^+ \rightarrow H_2O_2 + O_2$$

It is an oxidoreduction reaction between two superoxide anions. The hydrogen peroxide is destroyed by **catalase**:

$$H_2O_2 \rightarrow 2H_2O + O_2$$

The glutathione peroxidase–glutathione reductase system

Glutathione we have already met. It is a thiol tripeptide (γ-glutamyl-cysteinyl-glycine), which is abbreviated to GSH (Fig. 32.6). As stated, it is found in most cells where, because of its free thiol group, it functions as a reducing agent. It keeps proteins with essential cysteine groups in the reduced state. This reaction with proteins is nonenzymic, but another protective action of GSH is to inactivate peroxides via the action of **glutathione peroxidase**, producing **oxidized glutathione (GSSG)**:

$$H_2O_2 + 2GSH \rightarrow GS \quad SG + 2H_2O$$

Organic peroxides (R–O–OH) generated by free radical attack on membrane lipids are also destroyed in this way. The GSSG is subsequently reduced by NADPH, the reaction being catalysed by **glutathione reductase**:

$$GSSG + NADPH + H^+ \rightarrow 2GSH \quad NADP^+$$

Erythrocytes depend for their cellular integrity on GSH, which reduces any ferrihaemoglobin (methaemoglobin) to the ferrous form, as well as destroying peroxides. This explains why the pentose phosphate pathway is so important in erythrocytes (see Fig. 15.1) as it supplies the NADPH needed for the reduction of GSSG. Usually, patients with defective glucose-6-phosphate dehydrogenase, the first enzyme in the pentose phosphate pathway, have enough enzymic activity for normal function. However, when extra stress is placed on the cell, for example, by the accumulation of peroxides due to the action of the antimalarial drug **pamaquine** (a glycoside found in fava beans) and other drugs, the supply of NADPH can no longer be maintained (see Box 15.1). The integrity of the cell membrane is impaired and haemolysis results in such patients from taking the drug. People who are homozygous for a mutant gene that leads to the absence of glucose – phosphate dehydrogenase (G6PDH) may die of haemolytic anaemia after the ingestion of fava beans, a condition known as favism. As with sickle cell anaemia, people who are heterozygous for G6PDH deficiency may have some protection from lethal forms of malaria.

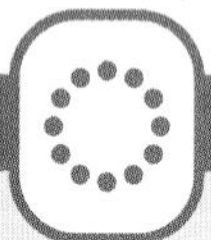

Summary

There are a number of processes in the body that are essential protective devices against different hazards.

Blood clotting involves two separate pathways of proteolytic enzymes, which converge at active factor X. This activates prothrombin to thrombin, a proteolytic enzyme that converts fibrinogen to fibrin. Fibrin is a fibrous complex that entraps blood cells into a soft clot, which is then stabilized by cross-link formation between the strands.

One pathway is triggered by a wound exposing an 'abnormal' surface, such as collagen. This triggers the activation of a long proteolytic cascade in which each activated component activates the next. The purpose is to amplify a minute signal into a massive response. It is called the intrinsic pathway.

The shorter extrinsic pathway results from the release of a factor from damaged cells. Both pathways are needed for efficient clotting.

Vitamin K is needed for clotting. It is involved in the carboxylation of a number of glutamate side chains of prothrombin. The carboxyglutamate binds Ca^{2+} efficiently, which is needed for the activation of prothrombin to thrombin. The same applies to several other clotting factors. The agent warfarin antagonizes vitamin K and prevents clotting. It is used therapeutically, in carefully graded amounts, as a guard against inappropriate clotting and also as a rat poison.

A complex system is important in protecting against inappropriate clotting since all the components are present in blood; inappropriate clotting causes many deaths.

Xenobiotics are foreign chemicals, including pesticides, pharmaceuticals, and many others. They would accumulate in the body unless excreted. Since most are lipid soluble, they must be rendered more hydrophilic for excretion in the urine.

The cytochrome P450 family of enymes typically hydroxylate a wide variety of the compounds, followed by a secondary addition of polar attachments. Multiple P450s exist, each with wide overlapping specificities so that a vast range of chemicals can be attacked.

The P450 systems are of great medical importance since they can interfere with drug therapy regimes.

Reactive oxygen species are formed in the body by several mechanisms and are destructive agents. Superoxide radicals are converted to hydrogen peroxide by the enzyme superoxide dismutase. The enzyme catalase destroys the peroxide. another system involving glutathione protects red blood cells against peroxides. Vitamins C and E are antioxidants. They protect against free radicals as quenching agents – they terminate the destructive reaction chains initiated by free radicals (but can have side effects if ingested excessively).

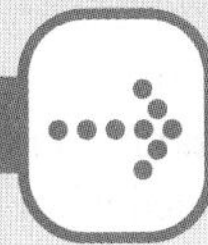

Further reading

To access the further reading, please scan the QR code image or go to http://global.oup.com/uk/orc/biosciences/molbiol/snape_biochemistry/student/reading/ch32/

Problems

1 Blood clotting involves cascades of enzyme activations. What are the advantages for having such a long process?

2 Explain how thrombin triggers the formation of a blood clot.

3 The spontaneous polymerization of fibrin monomers forms a 'soft clot'. How is this converted into a more stable structure?

4 What is the role of vitamin K in blood clotting?

5 What is the function of cytochrome P450?

6 How is NADPH involved in an oxygenation reaction?

7 What is the function of UDP-glucuronate in disposing of water-insoluble compounds?

8 What is multidrug resistance?

9 What is a superoxide?

10 What mechanisms exist for guarding against the deleterious effects of superoxide?

Chapter 33

The immune system

Overview

The immune system is a large subject area. Apart from its medical importance, it is one of immense relevance to molecular biology. In this chapter we will give an outline of the basic molecular mechanisms involved with some cell biology background to put the subject into a meaningful context.

The body is constantly under the threat of invasion by pathogenic organisms, such as bacteria and viruses. The immune system is the main protection against these invading organisms. Its importance is obvious and illustrated by the consequences of a compromised immune system either due to genetic defects, such as adenosine deaminase deficiency (see Chapter 19) or to infection by HIV (the AIDS virus). It also helps to protect against abnormal cells growing within the body. Immunological surveillance against cancer cells is believed to be an important role of the immune system.

There are two distinct immune systems. One is the **innate system**, which is mainly a protection against infection by microorganisms. It is less protective than the **adaptive immune system** but has the advantage that it is instantaneous, while the adaptive response usually takes a week or more to become fully effective. Bacterial infections can multiply very rapidly so that the timing of the defence mechanism is important. The innate response is also necessary for the triggering of the adaptive response.

This chapter is mainly about the adaptive response and we shall give only a brief outline of the innate system.

The innate immune system

Phagocytic white cells known as **macrophages**, present in tissues, and **neutrophils** in blood, engulf and destroy pathogens, and an inflammatory process ensues. They secrete **chemokines**, which attract more white cells to the site of infection, and also **cytokines**, which cause cells to produce an inflammatory response (see Chapter 29). The phagocytes recognize components of the infective prokaryotes that never occur in the body. Typical examples are the bacterial cell wall components such as lipopolysaccharides – carbohydrates anchored into the membrane lipid bilayer by fatty acid hydrocarbon chains (see Chapter 7). There are no counterparts of these molecules in an animal and so the phagocytes safely attack them as foreign. Prokaryotic protein synthesis (see Chapter 25) is initiated by *N*-formylmethionine while eukaryotes do not have formyl groups attached. Such formylated proteins are powerful chemokines which attract macrophages to the site of infection. The innate immune system has natural killer cells (NK cells), which destroy pathogen-infected cells, and possibly also some types of cancer cells, but the recognition procedure is different from that adopted by the killer cells in the adaptive immune system (see later).

The innate immune response does not confer any protection against subsequent infection by the same pathogen – there is no lasting immunity as happens with the adaptive response.

The adaptive immune response

This will be referred to simply as the immune response. In evolutionary terms it is a relatively recent development confined to vertebrates. It depends on the recognition of one or more specific macromolecules, mainly proteins, as foreign to the body. A foreign macromolecule in the body is a warning signal that leads to a defensive response against an invader. A molecule that elicits an immune response is known as an **antigen**. A patient recovering from a disease such as measles, is protected, often for a long time, from a further measles infection but not from other pathogens.

The adaptive response requires the help of cells of the innate system known as **antigen-presenting cells** (**APCs**). These are **dendritic cells** and macrophages, but not neutrophils. Vaccination involves injection of killed pathogens or harmless proteins derived from pathogens in order to trigger the immune response. The innate immune response is stimulated by injecting an immunostimulant to start the process. This is usually given in the form of an adjuvant, a preparation of microbial origin, which mimics infection. It stimulates the activity of APCs to initiate the immune response.

The immune system responds to macromolecules. Proteins are the most important class that elicits an immune response. Carbohydrate attachments to proteins are also often antigenic, as are some polysaccharides themselves. The immune system does not protect against small foreign molecules, such as drugs, although antibodies can develop to small molecules if they become attached to proteins. There are enzymic protection systems that deal with unattached small foreign molecules (see Chapter 32).

The problem of autoimmune reactions

The body has thousands of proteins differing from foreign proteins only in the detail of amino acid sequences and yet the distinction between 'self' and foreign proteins is made, otherwise the immune system would attack the components of the body in what is known as an **autoimmune reaction**. The avoidance of autoimmune attack is of overriding importance, and it is a complicated system that enables this distinction between self and foreign. The protection is not absolutely perfect, as evidenced by the existence of **autoimmune diseases**. An example of an autoimmune disease is **myasthenia gravis** in which an autoimmune reaction destroys the acetylcholine receptors of muscle, thus preventing nervous stimulation of contraction. **Diabetes mellitus type 1** is the result of an autoimmune attack on the insulin-producing β-cells of the pancreas. **Rheumatic fever** is an autoimmune attack on heart cells caused by *Streptococcal* infections. (Antibodies against bacterial antigens happen to cross-react with a heart muscle protein.)

The cells involved in the immune system

All blood cells have their origin in bone marrow stem cells, which continually divide (see Fig. 2.7). Two of the principal players in the immune system are **B cells** and **T cells**, collectively known as **lymphocytes**. They express specialized protein receptors on their surface that can specifically bind to antigens from the pathogens, resulting in activation of the cell to produce an immune response. On B cells these receptors are antibodies, on T cells they are called T cell receptors (TCR) (see later). Although B and T cells both originate from the same stem cells in the bone marrow, there is a major difference; B cell precursors multiply and undergo their primary maturation in the bone marrow, but the cells committed to becoming T cells migrate to the thymus gland, a small structure behind the breastbone, where they multiply and mature – hence the name T (for thymus) cell.

B cells are responsible for producing antibodies that bind specifically to antigens, but only after an elaborate preparation. To do this effectively they require the cooperation of a class of T cells known as **helper T cells** – they help the B cells to carry out their function. Two separate classes of T cells are generated in the thymus – helper T cells and **cytotoxic T cells**.

The phagocytic APCs are an essential requirement for long-lasting immunity. **Dendritic cells** are the most important of these. They engulf and digest pathogens, hydrolysing the antigen proteins into short peptides. These peptides are placed in grooves of special proteins, which are then displayed on the outside of the phagocytic cell and are inspected by passing T cells. This process is known as antigen presentation. The T cells are activated only by peptides presented by an APC – they do not react directly with a foreign antigen. The binding results in the activation of the T cells whose function will be described shortly.

What does the adaptive immune response achieve?

There are two main outcomes. First, there is the production of **antibodies**. These are soluble proteins in the blood and tissue fluids that combine with the foreign **antigens**. Antigens are macromolecules, normally foreign to the body, which can lead to **anti**body **gen**eration. The attachment of the antibody to antigen results in protective events described later. This type of immunity is called **humoral immunity**, for, in times past, body fluids were referred to as humors.

The second type of immunity is known as **cell-mediated immunity**. In this, the **cytotoxic T cells** recognize cells infected with intracellular pathogens, such as viruses, and also some cancer cells. The cytotoxic T cell attaches to the infected cell and kills it, thereby stopping the infection from spreading. Direct contact is made between the cytotoxic cell and its target cell – hence the term, cell-mediated immunity. Thus the two mechanisms complement each other. Using a viral infection, as the example, the humoral system is extracellular and prevents a virus in the blood or in a mucous membrane from infecting a cell. The antibodies cannot reach the virus once it is inside a cell so a different system is employed. The cell-mediated immune system destroys the host cell after it has become infected. The cytotoxic T cell delivers a 'death signal' to the infected cell, which triggers events inside the targeted cell that lead to its destruction.

Where is the immune system located?

There is no single organ housing the immune system. Instead, there are large numbers of separate cells, which, if put together, would be equivalent to a large organ. They are distributed in spleen, lymph nodes, intestinal, and mucosal lymphoid tissue, and about 10% in blood and lymph circulation. Lymph is a solution of soluble blood proteins and small molecules – it is essentially a blood 'filtrate' lacking red blood cells that leaks out of the blood and bathes cells. It is drained into thin-walled lymphatic vessels, which return the lymph to the blood *via* two main ducts. The lymphocytes are able to squeeze through certain capillary walls and so can migrate from blood to lymph and also from blood directly into the lymph nodes. In being carried through the blood and lymph, they are able to rapidly reach sites of infection in the body. The lymph is propelled by the movements of the body. On the route of the lymph returning to the blood, the lymphatic vessels expand at places into **lymph nodes**, known as **secondary lymphoid tissues** (bone marrow and thymus being primary), in which high concentrations of B and T cells are found. The dendritic cells that present antigen are situated in the secondary lymph nodes. They are continuously visited by the T and B cells as they circulate around the body. If any of the lymphocytes recognize the antigen presented by a dendritic cell, they will stay in the lymph node and undergo the differentiation processes, which initiate an immune response to the foreign antigen. Lymphoid tissues are found throughout the body, for example in the armpit, groin, tonsils, adenoids, and intestine. Dendritic cells, which normally reside in the tissues ready to phagocytose invading pathogens, can migrate easily through the lymph to the lymphoid tissues and present the pathogen antigens to the circulating lymphocytes. Activation of helper and cytotoxic T cells by specific antigens presented by the dendritic cells in the lymphoid tissue, is a crucial step in the generation of both types of immune response by mechanisms to be described shortly.

After that general introduction, we will deal with immunity under two main headings – humoral or antibody-based immune protection, and then cell-mediated immunity.

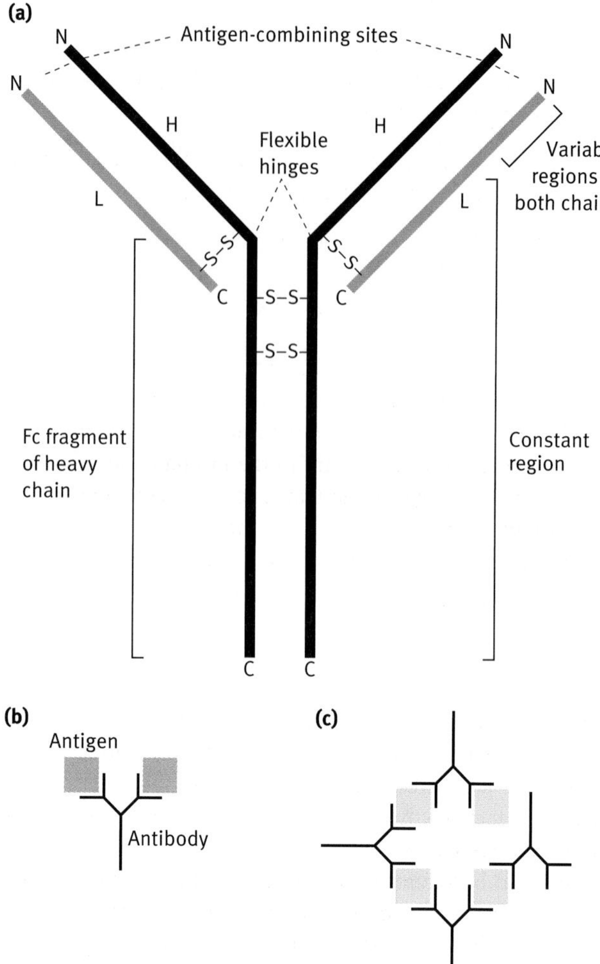

Fig. 33.1 **(a)** An immunoglobulin G (IgG) molecule. IgM and IgA molecules differ in the Fc fraction, but are similar in the variable part. H, heavy chain; L, light chain. N and C refer to the N-terminal and C-terminal ends of the polypeptides, respectively. The Fc fragment is the C-terminal half of the two H chains, bound covalently into a single fragment by —S—S— bonds. The name arises from crystallizable (c) fragment (F), one of the products of papain hydrolysis of the antibody molecule at the two peptide bonds at the flexible hinges. **(b)** Cross-linking of antigen with single specific epitopes. **(c)** Antigens with multiple antigenic determinants form a cross-linked insoluble cluster with antibody.

Antibody-based or humoral immunity

Structure of antibodies (immunoglobulins)

Antibodies are a class of proteins known as **immunoglobulins**. There are several different classes of these proteins. Immunoglobulin G (**IgG**) is the most abundant as the immune response develops. The basic structure of all immunoglobulins, typified by IgG, can be thought of as a Y-shape, made up of two identical **light (L) polypeptide** chains and two identical **heavy (H) chains** held together by disulphide bonds (Fig. 33.1).

The ends of the Y arms have hypervariable regions on both the heavy and light chains so that in a collection of IgG molecules they will all have the same constant regions (Fig. 33.1), but each will have a different version of the variable regions. It is these hypervariable regions of the two chains that form the antigen-binding site on each of the Y arms. On each arm the light and heavy variable regions form a single binding site to which an antigen might fit. Because of the variability, vast numbers of binding-site specificities exist, one per immunoglobulin molecule, the sites on the two arms being identical on each individual molecule. The antigen binds by noncovalent bonds. An antibody thus has two identical antigen-binding sites, which means that it can bind more avidly to an antigen and can also cross-link two identical antigen molecules. At the fork of the Y there is a flexible 'hinge' region, which increases the antibody's ability to cross-link two antigen molecules. This helps to make aggregates of pathogens, such as bacteria, which aids their removal from the body.

Although an individual antigen might be quite large, a specific antibody binds to only a small part of the antigen – in the case of a protein antigen to only a few amino acids. The specific part of the antigen that is recognized by an antibody is called an **epitope**. Thus a given protein antigen provokes the production of a large number of different antibody molecules, each combining with a specific epitope. This binding of multiple antibodies to different regions on the same antigen also helps to remove the antigen from the body (see below).

What are the functions of antibodies?

By binding to antigens on the surface of pathogens, antibodies have three main functions.

- They act to prevent the binding of the pathogens to their target cells (e.g. viruses, bacteria, or their toxins).
- They greatly increase the ability of neutrophils to phagocytose the pathogens.
- By forming aggregates of antigens and antibodies, they activate the **complement system**. This is a group of proteins in blood that bind to the antigen-antibody complexes through an enzyme cascade pathway, ending with the deposition of a lytic complex on the surface of the pathogen. This can lead directly to the killing of some bacteria. In addition, some activated complement components strongly attract neutrophils to the site of infection, and complement-coated bacteria, or other antigens, are more easily phagocytosed and digested by the neutrophils.

There are different classes of antibodies

Different classes of immunoglobulins differ in the **constant regions** of their H chains – these regions are constant only within each class. These differences have nothing to do with the antigen-binding sites but are related to the precise role of the antibody (see later).

When the body responds to an antigen, the antibody produced first is **IgM**, a multisubunit, or pentameric, form of antibody, with five of the basic structures joined together to give 10 antigen-binding sites. IgM has fairly low affinity for binding to its specific antigen but is particularly efficient at binding and cross-linking viruses and bacteria into aggregates, because of its multiple binding sites, and in activating complement. However, it is not effective at promoting phagocytosis because IgM is not recognized by neutrophils.

A few days after antigen exposure, **IgG** antibodies start to appear; these are monomeric. They bind to the same antigen with a much higher affinity than IgM and are very effective at complement activation and aiding phagocytosis by neutrophils The neutrophils carry surface receptors for the Fc portion of IgG (Fig 33.1). Antigens coated in IgG antibodies are said to be **opsonized** for phagocytosis. The B cells producing IgM, switch to producing IgG during their activation by specific antigen in the lymph node, aided by helper T cells. This switching of classes takes several days. Class switching in a B cell changes the constant region of the heavy chain but not the variable regions of the H and L chains. The antigen-binding specificity therefore does not change but the cell switches the class of antibody it produces. (The mechanism of gene segment rearrangements is discussed below.)

IgA, another class, is important as a first line of defence for mucosal tissues. It is transported through the epithelial cells that line the intestine and lungs, by combining with a special secretory polypeptide, which transports it into the mucus of the intestine and respiratory tract. It protects, for example, against the cholera bacillus infecting intestinal cells in humans previously exposed to the disease. It is also a component of breast milk, which helps protect the neonate from intestinal infections.

IgE, another class of antibody, binds to receptors on mast cells (a type of innate cell found in tissues) and triggers the release of cytokines and histamine. IgE antibodies are important in immunity to parasitic helminth worm infections, but also are responsible for allergies such as **hay fever**.

Generation of antibody diversity

The body is potentially exposed to vast numbers of different pathogen antigens and it is potentially capable of producing an adequate number of different antibodies that can specifically bind to these antigens. Each newly developed B cell can produce, in terms of antigen specificity, only a single antibody species, but each cell, as released from the bone marrow, produces a different one. Given the enormous numbers of cells involved, there will be an antibody that by sheer chance binds to whatever antigen comes along.

What is the source of this variability? It has been estimated that there are between 10^{11}–10^{25} different epitopes and more than 10^{11} different lymphocyte specificites. This poses a problem; humans have only about 30,000 genes so it is not possible to have a different gene for each heavy and light chain specificity. Clearly, some special mechanism must exist, which gives rise to the enormous diversity of immunoglobulin genes. The genomic arrangement is truly impressive. There is a cluster of linked gene segments for the heavy chain and a different cluster for the light chain. These gene segments are rearranged during B cell development to form a single gene for each chain. This rearrangement defines the specificity of the variable region of the chain, and for the class of the heavy chain, and allows for almost unlimited diversity from a relatively small starting number of gene segments. We are used to cells having constant and stable genes in their protein-coding capacity but here we have exactly the opposite – a mechanism that ensures a tremendous variability in genes coding for the antigen-binding sites of antibody proteins.

To understand how this generation of diversity is achieved, let us look first at the L chain of immunoglobulin molecules. As outlined, it has two sections – a terminal variable region that participates in the antigen-binding site and a constant region that is identical in all immunoglobulin molecules of a particular class. Consider a stem cell in the bone marrow *before* it has become committed to becoming a B cell. At this stage it has not assembled its gene for the immunoglobulin L chain, but it has sections of DNA, known as gene segments, from which a gene coding for an antibody L chain will be assembled. This is a very exceptional process. There is one gene segment that codes for the constant (C) domain, and two adjacent clusters of gene segments from which the variable domain will be selected. The first of these clusters contains at least 300 different variable (V) gene segments, and the second contains four different joining (J) gene segments.

When the bone marrow stem cell is committed to becoming a B lymphocyte, a rearrangement of these DNA sections occurs so that one V gene segment is selected and aligned next to one of the four J gene segments, which aligns next to the C region gene segment. A functional L chain gene is produced by site-specific or somatic recombination (explained later) in which the intervening V and J gene segments are completely removed from the DNA. This recombination results in a new composite gene different in each of the B lymphocytes, as shown in Fig. 33.2. *The DNA recombination is a completely random process.* Thus, in the final mRNA coding for an IgG L chain, any one of 300 V sections is joined to any one of the four J sections giving 1200 different combinations. The recombination, known as **somatic**, or **site-specific**, recombination, involves excision of a piece of DNA and rejoining of the remaining sections. The recombination is brought about by two enzymes, which cleave the DNA at specific sites, thus excising a section of DNA. The double-stranded breaks so created are end-joined by other enzymes in a sloppy, imprecise way creating more diversity. Yet further diversity is created by an enzyme, which adds random nucleotides to the cut ends before joining. The number of possible L chain variants is greatly increased by this joining variability.

The H chain gene is assembled in a similar random manner from variable sections giving large numbers of possible H chain gene variants. The variability is greater in the H chain assembly due to the greater diversity in joining events than occurs with L chain assembly. This is because the H chain gene complex also contains a number of diversity (D) gene segments between the V and the J gene regions, so that the variable region of the H chain gene is composed of a single V-D-J, which binds the C_H gene segment, giving rise to an IgM antibody. As the antigen-binding site is made from the combination of the variable parts of the H and L chains, and there are large numbers of different H and L genes, the number of different binding sites coded by the assembled L and H genes is sufficient to account for the immune system's ability to produce vast numbers of different antibodies. In summary, each developing B cell assembles, at random, a gene coding for one antibody with its specific antigen-binding site. T cells undergo a similar process during their development to generate TCRs. These are also composed of two polypeptide chains that contain a variable and a constant region.

Since the B cells are diploid, there will be both maternal and paternal DNA sections so that production of more than a single gene coding for antibodies might be expected. However, a process of allelic exclusion ensures that only one version of H and L chains is assembled from maternal or paternal sections so that a B cell is monospecific in its antibody-binding. Similarly, there is allelic exclusion of the TCRs so each T cell is also monospecific for its antigen.

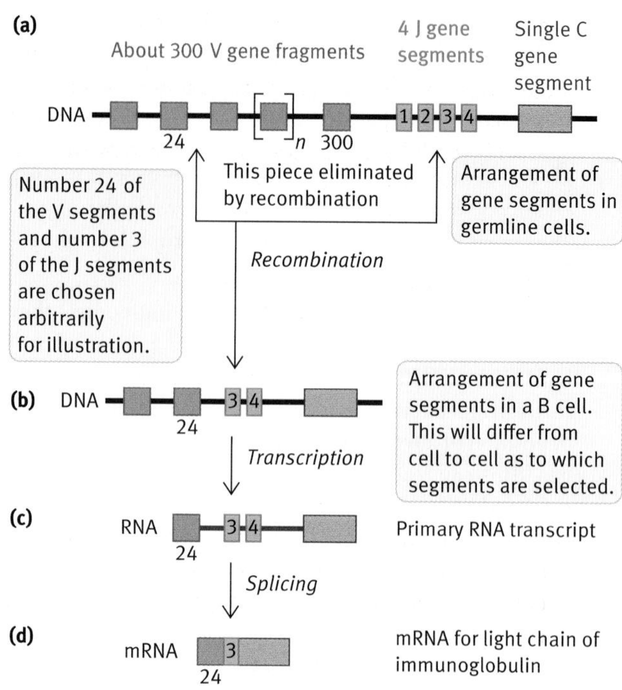

Fig. 33.2 The process of rearrangement leading to a functional L chain immunoglobulin gene. **(a)** Arrangement of gene segments in the stem cell where the immunoglobulin genes are not being expressed. **(b)** The randomly chosen V gene (V_{24} in this example) is joined to one of the J genes (J_3 in this example), the intervening DNA being excised. **(c)** Transcription begins at the V_{24} gene segment, the J genes (J_3 and J_4) also being transcribed. **(d)** After messenger RNA splicing to remove transcripts of J_4 and introns, those corresponding to V_{24}, J_3, and *C* now make up the mRNA. Allelic exclusion ensures that only one of the pair of alleles becomes a functional immunoglobulin gene so that a given cell produces only one immunoglobulin, not two.

Activation of B cells to produce antibodies

This involves an elaborate sequence of events. A preview of each step, without going into detail, may help you keep track of what is happening in the subsequent description.

- In the bone marrow, as the immature B cells develop, each B cell produces a limited number of copies of its own particular antibody molecules, the binding specificity of which is determined by the random gene assembly process described above. These antibody molecules are not released at this stage but are fixed in the membrane with the antigen-binding site displayed on the outside.
- In the bone marrow, any immature B cell that binds an 'antigen', which is a self-component, is eliminated (clonal deletion).
- The survivors, which are likely to be specific only for foreign antigens, develop into mature B cells and are released into the circulation.

- If a released B cell now meets and binds its antigen, it multiplies into a clone of identical cells that secrete IgM antibodies as a first line of defence against the antigen. These cells do not produce antibodies effectively without a helper T cell, which itself has been activated by encountering a peptide derived from the same antigen (which has been presented to it by an **APC**, usually a dendritic cell).
- After activation by the helper T cell, the B cell multiplies further and switches class to IgG and differentiates into an IgG antibody-secreting plasma cell.

We will now describe how these events happen.

Deletion of potentially self-reacting B cells in the bone marrow

The immature B cells formed in the bone marrow have between them the potential to produce antibodies to attack just about every macromolecular component in the body. Any B cell whose antibody, if produced, would act against a self-component should be eliminated otherwise autoimmune diseases will develop (see later). This elimination takes place during the development of the B cells in the bone marrow. There the immature B cells are exposed to the self-antigens they are ever likely to meet.

Each immature B cell in the bone marrow produces about 10^5 copies of its antibody and displays them on its surface (fixed into the membrane), where they will serve as the antigen receptor for the mature B cell. The antibody is not released from a B cell until it has matured and has been activated by its specific antigen. If, during development in the bone marrow, an immature B cell encounters an antigen, which binds to its displayed antibody, it is assumed (as it were) that it is a self-component, which must be tolerated and not attacked, by the immune system. Any immature B cells whose antibodies bind to an antigen will die without leaving the bone marrow. After this selective elimination , the surviving B cells should respond only to foreign antigens. They finish their development to become mature, but naïve, B cells and enter the circulation. Peter Medawar in the UK and Macfarlane Burnet in Australia shared the 1960 Nobel Prize for the theory explaining the basis of immunological tolerance to self-components.

The theory of clonal selection

When the B cell survivors of the elimination process in the bone marrow emerge into the circulation, they include a vast number of cells each with its own randomly designed antibody receptor exposed, so that, when a foreign (unprocessed) antigen is encountered, a few of these B cells will by sheer chance bind to the antigen. The number of cells happening to do so will be small, but the body does mount a large-scale response requiring a large number of B cells specific for that antigen. How does this happen? The answer is, by **clonal selection**, a mechanism proposed by Macfarlane Burnet. The principle is simple. An antigen, by binding, selects the B cells carrying a displayed antibody specific for itself and initiates a train of events, which leads to a multiplication of those particular cells. The displayed antibody is an antigen receptor, and binding activates signalling pathways that ultimately lead to clonal expansion (increase in the numbers of that particular cell). A foreign invader (or other antigen), in short, arranges for its own destruction. The principle is illustrated in Fig. 33.3.

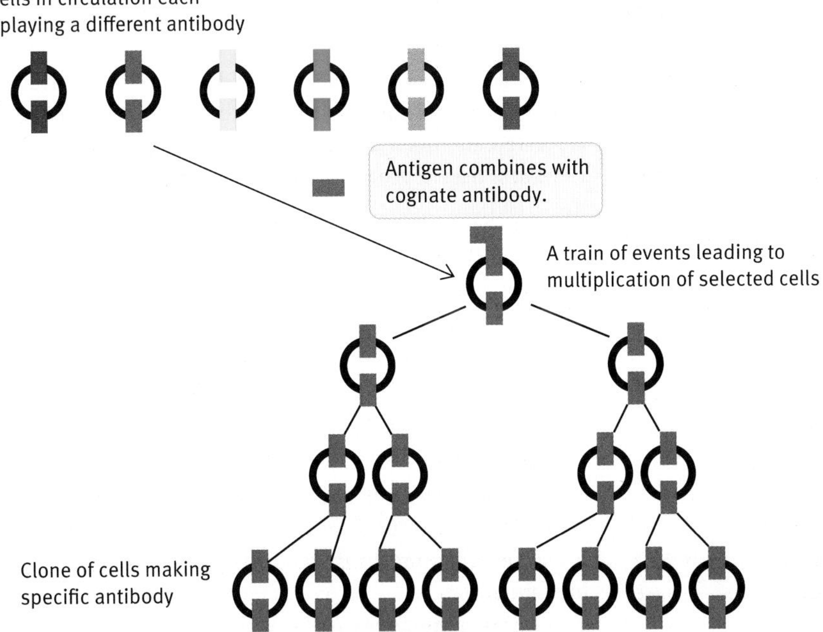
B cells in circulation each displaying a different antibody
Antigen combines with cognate antibody.
A train of events leading to multiplication of selected cells
Clone of cells making specific antibody

Fig. 33.3 The principle of clonal selection. The released population of B cells contains a vast array of cells, each of which makes a different antibody, which at this stage is displayed as a receptor in the membrane (coloured blocks). An antigen binds to its cognate receptor and triggers a train of events leading to multiplication of the selected cell forming a clone. The antigen thus engineers its own destruction by selecting for amplification the clone of cells most suited to attack it. This binding of the antigen is only the initial signal leading to multiplication, which requires a complex set of events to be described. Note that at this stage the antibody is not released but is an integral membrane protein with its binding sites displayed as receptors. See text.

B cells must be activated before they can develop into antibody-secreting cells

The selection process described above results in a clone of B cells, which recognize an antigen. During this process the binding of the antigen to the B cell's surface antibody activates the B cell. It internalizes the bound antigen, hydrolyses it into peptides and displays these on the outside of the cell (how this is done is explained shortly). The B cell is now functioning as an APC (Fig. 33.4).

You can regard the B cell at this stage as being on a 'standby' basis waiting for a further stimulus. This is given by a **helper T cell** produced in the thymus. Each helper T cell has on its surface multiple copies of its receptor, TCR. They bind to specific antigenic peptides displayed on the surface of antigen-presenting cells. During T cell development in the thymus, any T cell that would bind to such a displayed peptide from a 'self' antigen is destroyed (again avoiding the development of autoimmune disease). T cells that survive and enter the circulation should recognize only peptides from foreign antigens.

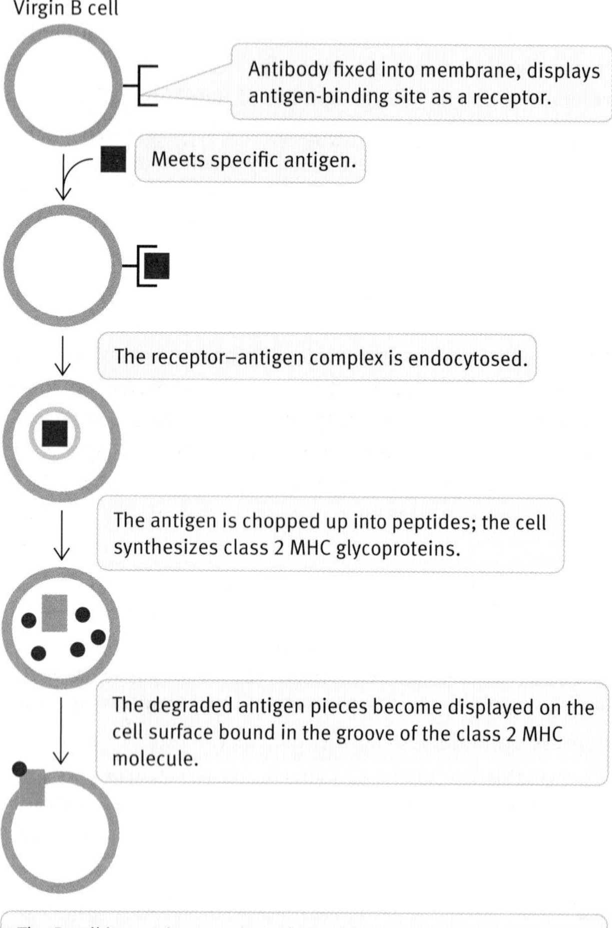

Virgin B cell
Antibody fixed into membrane, displays antigen-binding site as a receptor.
Meets specific antigen.
The receptor–antigen complex is endocytosed.
The antigen is chopped up into peptides; the cell synthesizes class 2 MHC glycoproteins.
The degraded antigen pieces become displayed on the cell surface bound in the groove of the class 2 MHC molecule.
The B cell is now in an activated state; it does not produce antibodies unless it binds to a helper cell that has been activated by an antigen-presenting cell displaying the same antigen. If this happens, the B cell matures into an antibody-secreting plasma cell and also multiplies due to cytokine release by the helper cell. (This is illustrated in Fig. 30.6.)

Fig. 33.4 Conversion of a virgin B cell into an activated state but not yet to an antibody-producing cell.

The new helper T cells from the thymus require activation themselves before they can 'help' a B cell. This is done by **dendritic cells**; these **APCs** are widely distributed in the body. They phagocytose foreign pathogens, hydrolyse them into peptides and display these on their external surface. If the helper T cell recognizes and binds to its specific target peptide on the dendritic cell, this means that that there is a foreign invader to be tackled. This interaction with the dendritic cell delivers signals, which activate the T cell (Fig 33.5), which then produces a self-stimulating cytokine (autokine) that induces multiplication of the helper T cell into a clone of activated cells.

Now that the helper T cell is activated, if it recognizes its specific peptide on the surface of a B cell, cytokines are released, which cause the B cell to multiply and differentiate into a clone of antibody-producing cells. These are known as plasma cells and reside for long periods (several years in humans) in the bone marrow and other lymphoid tissues. The antibody is secreted into the blood and tissue fluids where it can bind to its target antigen. Before activation by antigen and help from T cells, the B cells do not secrete antibodies but only display them in the cell membrane. The fully activated **plasma cell**, no longer expresses the antibody in the cell membrane because differential splicing (see Chapter 24) at this stage eliminates two exons from the mRNA, which code for a transmembrane polypeptide segment that fixes the antibody into the membrane at the earlier stage (Fig. 33.6). The loss of this transmembrane region allows the antibody to be secreted from the cell rather than becoming embedded in the membrane.

The requirement for an activated helper cell to activate the B cells is, in effect, a double check that the target of the B cell is foreign and not a self-protein (since *both* B and T cells are survivors from a process that eliminated those cells responding to self-antigens).

There are cases where helper T cells are not required for the production of antibodies. Polysaccharides of the type found typically in the lipopolysaccharides attached to the outer membrane of some bacteria are antigenic. The antigens in this case have multiple recognition sites occurring one after another in the carbohydrates. Possibly because of this, they give signals to the B cells that are sufficient to cause their differentiation into antibody-secreting cells without helper T cells being involved. It is believed that a special class of B cells is involved. This type of response does not, however, produce IgG antibodies or memory cells, so there is no effective or lasting immunity.

Fig. 33.7 summarizes the process of antibody production.

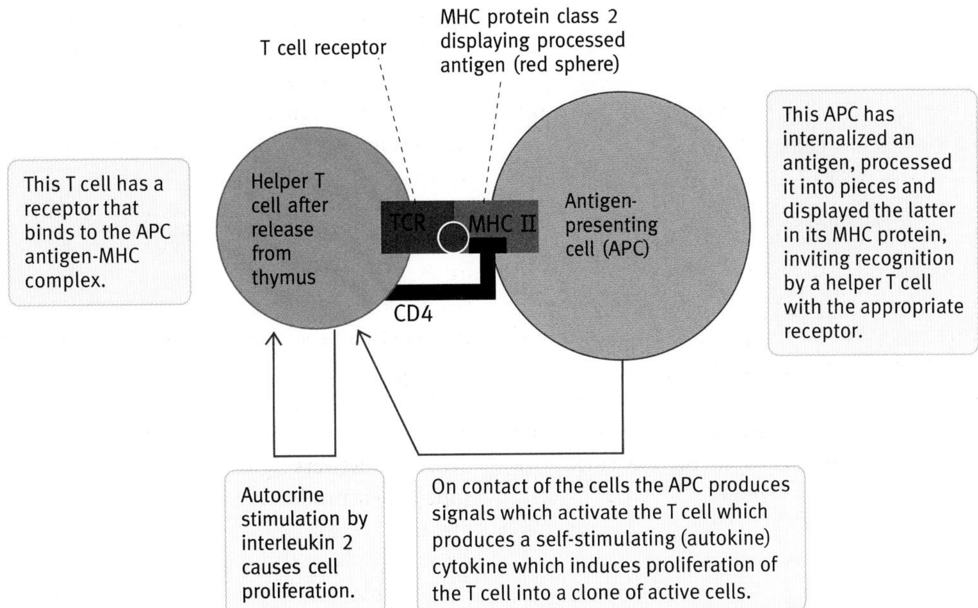

Fig. 33.5 Activation of a helper T cell by an antigen-presenting cell (APC), a dendritic cell. The autocrine stimulation of the helper T cell allows cell multiplication to occur after separation of the cells. CD_4 is a glycoprotein on the T cell that interacts with the MHC 2 protein of the APC. This interaction is necessary in addition to the binding of the T cell receptor (TCR) with the APC MHC-antigen complex. The CD_4 protein is the receptor by which the AIDS virus (HIV) infects the helper T cell.

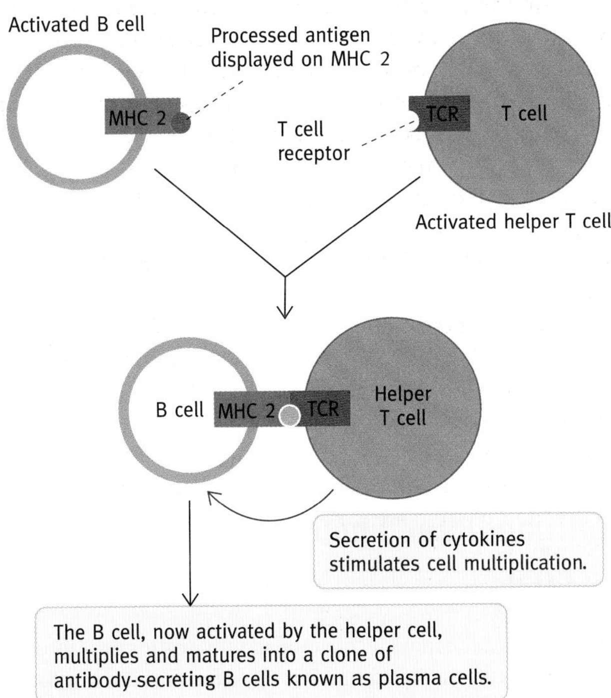

Fig. 33.6 A helper T cell activating a B cell to become a clone of plasma cells secreting antibody. The cytokines are growth factors that stimulate local cells to divide (paracrine action). On activation of a B cell to become a plasma cell, differential processing of RNA transcripts eliminates the membrane-anchoring section of the antibody. TCR, T cell receptor.

Affinity maturation of antibodies

The B cell(s) were initially selected for multiplication by an antigen that combined with the cell's displayed antibody. However, the latter's binding site was generated by random variation and the antigen happened to fit to it by chance. The fit is not likely to be particularly good, so the antibody produced is of low binding efficiency and therefore of low protective effect. During B cell multiplication, after antigen binding, rapid site-specific mutations occur in the variable region DNA that produce yet more variations in the antibody-binding sites of the multiplying B cells. This is coupled with a selection of those cells with improved binding affinity for the relevant antigen. Cells with poor binding ability that do not capture an antigen molecule, die. As the number of antigen molecules falls, owing to their removal by phagocytosis, the competition for them increases. The result is that during an immune response there is a progressive Darwinian evolutionary improvement in the antibody quality, as only the best B cells are able to capture an antigen molecule and survive. The process of **affinity maturation** takes place in special centres (germinal centres) in secondary lymphoid organs. The antibody class-switching that occurs as the immune response progresses (see earlier in this chapter) is also an important aspect of antibody quality improvement. Helper T cells are essential for this affinity maturation and class switching; babies born with rare genetic defects in their ability to make helper T cells can produce lots of low affinity IgM antibodies but are unable

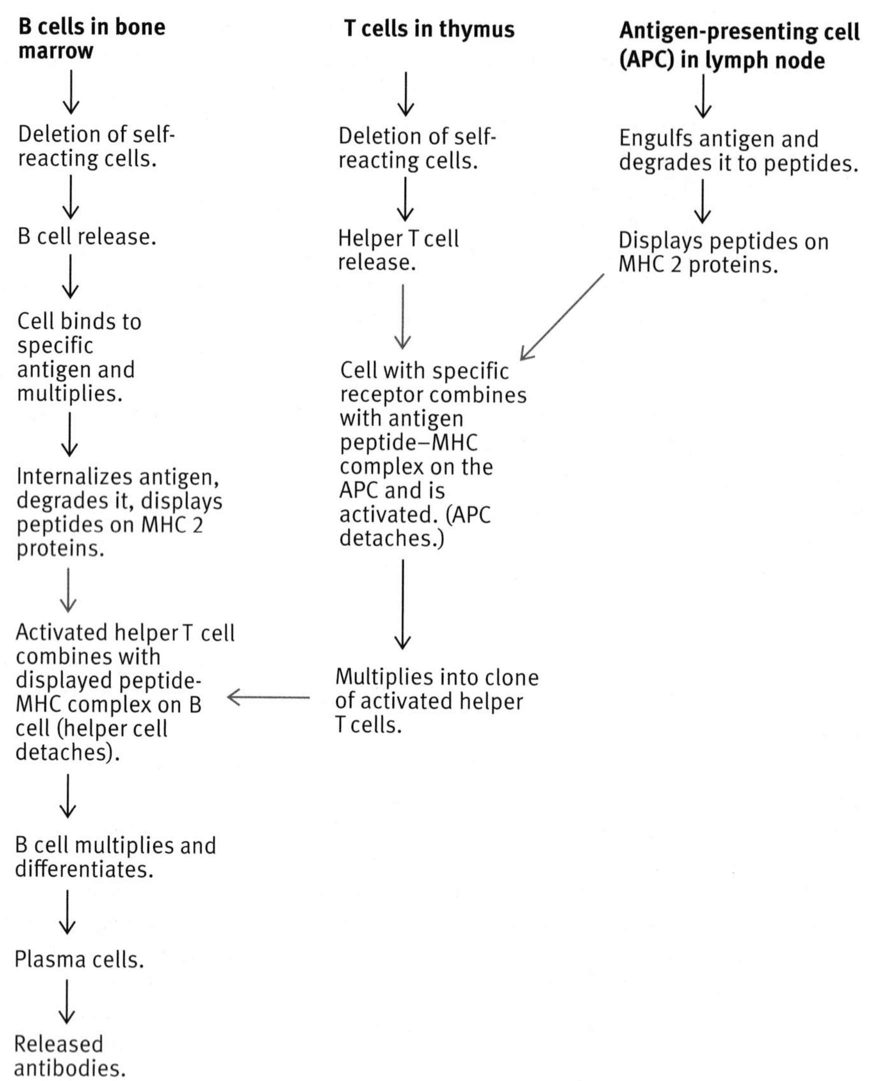

Fig. 33.7 Summary scheme of antibody production by B cells. The crucial point to note is that a B cell cannot proceed to release antibodies until it is given the signal to do so by a helper T cell. The helper cell is activated by combining with an antigen-presenting cell (APC) that has engulfed the *same* antigen that was endocytosed by the B cell. The APC displays the antigen peptides on its MHC proteins and this is recognized by the specific helper TCR. The helper T cell thus activated has to see the identical MHC-peptide complex on the B cell. You will notice that in this sense the B cell is acting as an antigen-presenting cell. Cytokines are liberated at activation steps and play an essential part in the process (see Fig. 33.5). The red arrows are to draw attention to cell–cell interactions.

to fight bacterial infections, which require high affinity antibodies of other classes.

Memory cells

When B cells are activated and proliferate, not all of the resultant clone of cells mature into antibody-secreting plasma cells. A proportion become long-lived memory cells that may circulate for years or an animal's lifetime. They are the basis of long-term immunity from repeat infections. If an appropriate antigen is encountered at a subsequent time, the immunological response is very rapid. Hence, vaccination can often give long-term protection against an infective disease.

Cell-mediated immunity (cytotoxic T cells)

This is the second major arm of the adaptive immune response. It is called cell-mediated because the effector cells, known as **cytotoxic T cells**, make physical contact with their target cells. This is in contrast to the humoral response, where the effector agent is an antibody that can attack its antigen target, and cell contact is not involved in the actual attack process.

Having a second arm of the adaptive immune response is very useful. The humoral response protects against pathogens in the blood and tissue fluids before they have infected host cells. Once inside the cells, the antibodies cannot reach them so a different mechanism is needed. The cytotoxic T cells attack and trigger the destruction of the infected cell, and this aborts multiplication of the pathogen. But how does the cytotoxic T cell 'know' which cells are infected? A remarkable mechanism achieves this. Somatic cells display a group of proteins on their surface membranes that are collectively known as the major histocompatibility complex (MHC). These proteins are synthesized in the cell, but in order to be folded into the correct shape for transport to the plasma membrane they must first incorporate a small peptide that is produced from other cytosolic proteins in the cell. Usually this is derived from misfolded proteins, which are hydrolysed into peptides in proteasomes in the process of their removal (see Chapter 23). But if the cell becomes infected with a virus or other pathogen it will also hydrolyse some of the pathogen's proteins and these will then also become incorporated into MHC molecules and displayed on the cell surface. If a cytotoxic T cell recognizes the foreign peptide displayed in the MHC on the cell surface, it will then kill the cell. It will not kill cells displaying MHC containing peptides derived from the body's own proteins.

The cytotoxic T cells, like helper T cells, develop in the thymus and undergo the same selective process, which occurs with B cells in the bone marrow (see earlier). During their development they express TCR on their surface that can recognize MHC molecules containing peptides on the somatic cells. T cells will only recognize MHC containing foreign peptides. Any developing T cells in the thymus whose TCR binds to a self peptide in the MHC are programmed to die. It has been estimated that about 95% of developing T cells in the thymus die without reaching maturity because their TCR either cannot recognize MHC at all, or because it recognizes MHC containing self peptides. This highlights the random nature of gene recombination in TCR and antibody generation, and the need to remove any self-reactive cells before they mature and can cause autoimmune damage.

How are cytotoxic T cells activated after release from the thymus? This again is largely the job of the dendritic cells. The dendritic cells are able to engulf the pathogens without being infected by them, hydrolyse their proteins into peptides and display these within the MHC molecules on their surface (see below). If a cytotoxic T cell recognizes its specific target peptide among them it binds to it on the surface of the dendritic cell, which then delivers essential signals that activate it. Only the dendritic cell can deliver this activation signal to the T cell. It cannot be directly activated by binding to an infected somatic cell. This mechanism prevents the inappropriate activation of self-reactive T cells that might have escaped the selection process during their development in the thymus. As a result of activation, and with the stimulus of cytokines and help provided by the helper T cells, the cytotoxic T cells divide to become a clone of active 'killer' cells. These are now able to leave the dendritic cell and attack and kill any somatic cell displaying the same foreign peptide-MHC complex. Once they have targeted one infected cell for destruction, they can move on and target the next infected cell, sparing the ones that have not been infected and so do not display pathogen peptides in their MHC molecules.

Mechanism of action of cytotoxic T cells

There are two mechanisms by which target cells are killed. The cytotoxic T cell makes contact with the cell and either delivers an apoptotic death signal (see Chapter 30), instructing the cell to self destruct, or it liberates **perforin**, which renders the cell permeable and allows entry of proteases known as granzymes, which the T cell also secretes. These also trigger apoptotic cell death.

The role of the major histocompatibility complexes (MHCs) in the displaying of peptides on the cell surface

The peptides to be displayed on the outside of antigen-presenting cells (B cells, helper cells, and dendritic cells) are placed in a groove of newly synthesized MHCs, which are then transported to the cell surface. The receptors on helper cells and cytotoxic T cells recognize the complex of their target peptides in MHC grooves, not the peptide alone or the MHC alone. There are two main classes of MHCs called **class 1** and **class 2**.

Most somatic cells of the body express class 1 on their surface; cytotoxic T cells recognize only peptides held in class 1. Therefore most somatic cells of the body are subject to cytotoxic T cell surveillance for infection. The somatic cells display peptides produced by proteasomes in the cytosol on newly synthesized class 1 MHCs.

The antigen-presenting cells are different. They phagocytose antigens and fragment them into peptides in endosomes (see Chapter 27), not proteasomes as in class 1. Peptides produced in this way are routed to be displayed on newly synthesized class 2 MHCs. However, some of the protein antigens phagocytosed by the dendritic cells are degraded into peptides and uploaded into MHC class I molecules by processes that are still not fully understood, but might involve degradation by proteasomes if the antigens escape into the cytosol and might also involve specialized subsets of dendritic cells (see Joffre 2012 in Further reading online). Whatever the mechanism, dendritic cells are able to present peptide antigens in the grooves of both MHC class I and class II without needing to be infected by the pathogen and thus they are able to activate both the helper and cytotoxic T cells with TCR specific for the particular peptide-MHC combination.

Some viruses might elude cytotoxic T cell attack by inhibiting the display of MHC-peptides. Remarkably, the NK cells of the innate immune system recognize cells with deficient display and have mechanisms to kill them, making it more difficult for the viruses to escape detection.

CD proteins reinforce the selectivity of T cell receptors for the two classes of MHCs

The binding of T cell receptors to their targets is primarily determined by recognition of peptide-MHC complexes, but this is reinforced by additional proteins associated with the TCR on the T cell surface. Cytotoxic T cells express a protein called CD8, which binds to a *constant* region of MHC class 1 proteins; helper T cells express another protein, CD4, which binds to class 2 MHCs. (CD means cluster of differentiation; they are cell surface markers.) This extra binding is necessary for effective binding of the different cells to their targets and so confines the two types of T cells to their proper targets (Figs. 33.5 and 33.8). The binding also delivers one of the signals necessary for T cell activation. The CD4 is the route by which the AIDS virus enters and destroys the helper T cells thus removing both humoral and cell-mediated protection of the body with disastrous effects.

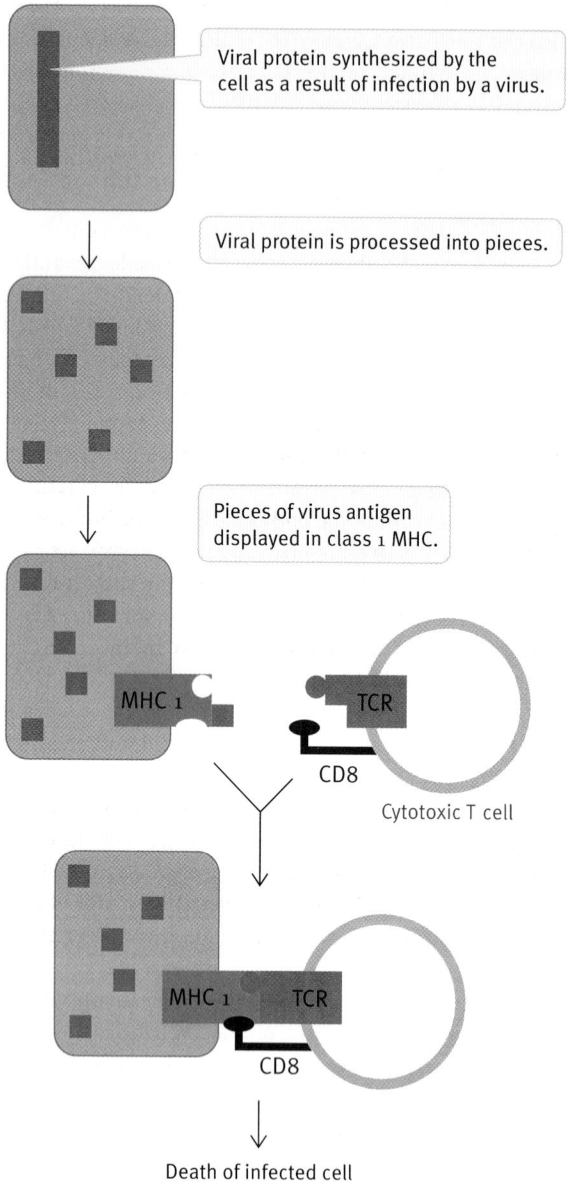

Fig. 33.8 Sequence of events in cell-mediated immune reaction to a foreign antigen synthesized inside cells (for example, as a result of virus infection). A proportion of the clone of cells develops into memory cells. The infected cell may be killed by perforation of its cell membrane due to the release of the protein perforin, or death may be due to apoptosis. TCR, T cell receptor.

Why does the human immune system reject transplanted human cells?

Tissue grafts and organ transplants are recent developments, so there was no known evolutionary mechanism to guard against them. The main cause of rejection is the presence of MHC proteins on cells. Each individual has several genes coding for a number of variants of the MHC proteins. In addition, the genes for these are highly polymorphic (see Chapter 22) within a population – they vary from one individual to the next. It is unlikely, therefore, that the MHC proteins of one individual are exactly the same as those of the next. Those which are different will therefore be antigenic if transferred to another person. The evolution of such variation in the MHC molecules of the population serves as a protection for the species as a whole. Viruses and other pathogens employ devices to outwit the immune system. Suppose that, by chance, a virus evolves so that its processed antigen is not displayed on a host MHC protein or not recognized by the cytotoxic T cells. The cell is not labelled as virally infected and is not attacked. The virus might then multiply and kill its host. However, the next individual it infects is unlikely to have exactly the same MHC genes and the chance of the same evasive device being successful again is reduced. The disease is therefore less likely to sweep through the whole population.

Monoclonal antibodies

If an animal is injected (immunized) with an antigen, antibodies appear in the blood against the antigen and from this, an immune serum can be obtained and used for a variety of purposes such as experimentally to detect a particular protein. Such sera are not specific, as they contain many different antibodies specific for the different epitopes on the antigen, each produced by a different clone of B cells. The relatively small amount of immune serum that can be produced in a single animal and the variability of the sera between animals meant that this approach had only limited use experimentally and almost none clinically. It was realized that a 'pure' antibody, in which every immunoglobulin molecule is identical and specific for the antigen of interest, would be a powerful tool in biochemistry and medicine.

The achievement of this earned Niels Jerne, Georges Kohler, and Cesar Milstein, a Nobel Prize in 1984. A given B cell, as explained, produces one antibody in terms of binding specificity and one only (though many copies of it). If a B cell that produces the wanted antibody could be isolated and grown in culture, a pure antibody would be produced. It is possible to isolate such a specific B cell, but B cells will not grow in culture for more than perhaps a few days. Only transformed cancerous cells will grow indefinitely in culture.

Pure clones of cancerous B cells are available in tumours called **multiple myelomas**; these are derived from single B cells, which multiply uncontrollably and will grow indefinitely in culture. Most multiple myelomas produce large amounts of pure immunoglobulin that is not of a wanted antibody specificity, but some have been found that produce no immunoglobulin of their own, and these cells can be used to generate monoclonal antibodies by fusing them to the antibody-producing B cells. This is an effective way of producing an immortalized B cell that secretes the wanted antibody.

In brief outline, in the original technique, mice were immunized with an antigen and all the B cells from the spleen (a secondary lymphoid organ were fused with the myeloma cells (this could be done simply by using a common chemical, polyethylene glycol). The fused cells are called **hybridoma cells**, but there would also be large numbers of unwanted cells, which had not fused and also hybridomas that were not producing the appropriate antibody. The unfused cells were eliminated by culturing the mixed population in a selective medium (not described here), which did not allow them to survive. To select for hybridomas producing the wanted antibody, they were cultured in small wells in a tissue culture plate, in which only a single hybridoma cell was initially placed per well. After suitable growth, the wells were tested for the presence of antibody reacting to the antigen of interest. Once identified, cells could be grown up in unlimited quantities producing the monoclonal antibody also in unlimited amounts. The cells could be stored indefinitely in liquid nitrogen and grown up whenever more of the antibody was wanted.

Monoclonal antibodies have become the basis of a biotechnology industry, so powerful are they as a tool. Almost any protein can be detected by developing the appropriate monoclonal antibody. They are used in clinical laboratories for measuring specific human proteins or for HIV or other virus detection. This is the basis of **ELISA (enzyme-linked immunosorbent assay)**, widely used in clinical medicine; polyclonal antibodies are also used in this technique in some cases but monoclonal antibodies give more reproducible results. Suppose you want to test for the presence of a hormone in urine or of a virus in some biological fluid. The sample is adsorbed to the bottom of a plastic well. The excess is washed away and any remaining sites on the plastic well that could adsorb proteins are swamped by nonspecific protein. An antibody specific for the antigen of interest that has had conjugated with it an enzyme, such as horseradish peroxidase, is added to the test well. If the virus or hormone is present, the antibody will attach. The excess is washed away. A colourless substrate for the peroxidase is added which the attached enzyme converts to a coloured compound. This is quantitated colorimetrically and indicates the amount of the antigen of interest present in the test sample. Monoclonal antibodies give absolute specificity. There are several variations which can increase the sensitivity. In one of these, the initial (rabbit) antibody used has no enzyme attached. A second antibody, such as goat antirabbit immunoglobulin, specific for the first antibody (with attached enzyme in this case) binds to it; this increases the amount of enzyme fixed since multiple copies bind, and so increases the sensitivity of the assay. For example, to detect antibodies to HIV the patient's blood is added to a well containing adsorbed HIV antigen. If an antibody is present in the blood it will bind. After appropriate washing the well is treated with an antibody to human immunoglobulin (raised in another species, often in goats), to which peroxidase has been bound. Production of a colour from added substrate indicates that there is an antibody to the HIV in the patient's blood.

Experimentally, monoclonal antibodies can be labelled with fluorescent compounds and used to identify specific protein production, for example in cells of embryos (Fig. 8.13 was obtained in this way). They also have the potential to target therapeutic agents to specific sites in the body, such as tumours, an area of great interest because cancerous cells may have antigenic 'markers' on them to which specific antibodies could attach. The antibody could for example be loaded with a therapeutic agent against the cancer (see Firer & Gellerman 2012 in Further reading online).

Humanized monoclonal antibodies

The use of antibodies to combat diseases has been used for about a century. Tetanus infection was treated by injection of horse serum from animals that had been injected with the tetanus toxoid. Subsequent injections carried the risk of potentially fatal immune responses to the horse serum proteins.

Mononoclonal antibodies specific for a given target protein do not have the contaminating proteins but since monoclonal antibodies are generated in mice, they still have the drawback of a possible immune response to the mouse proteins, which at best could diminish the effectiveness of the treatment.

To overcome this drawback, 'humanized' monoclonal antibodies have been developed, in which the variable region of the mouse antibody is retained (the specific 'business ends' of the molecule) while other constant protein components are either removed or replaced by human versions of the proteins. The latter is achieved by genetic engineering in such a way that the DNA of human antibody heavy and light chain genes replaces the relevant variable regions from the mouse and this recombinant DNA is then cloned into cells that can be grown in culture indefinitely to secrete large amounts of the humanized antibody. This approach has had considerable success, and a number of diseases have been therapeutically treated with such monoclonal antibodies, including rheumatoid arthritis and some forms of cancer. The use of humanized antibodies reduces the risk of unwanted immune reactions.

Recently this has culminated in the development of a strain of mice (known as the xenomouse) by genetic engineering that produce totally human monoclonal antibodies. Such an antibody produced in this way has received FDR approval in 2006 for use in humans (see article by C. T. Scott in Further reading).

Summary

The immune system of animals protects against pathogenic invaders. There is an innate immune system in which phagocytes engulf invading pathogens. This gives immediate protection, and helps to trigger the adaptive response.

The adaptive immune system is the most important and the one mainly discussed in this chapter. It responds to macromolecules, mainly proteins and complex polysaccharides, but not to small foreign molecules unless attached to large ones. Foreign macromolecules are indicative of an invader. They are called antigens because antibodies are generated in response to them. There are two arms of the adaptive immune systems, the humoral in which antibodies prevent pathogens in the blood and tissue fluids from entering cells and aid their removal from the body, and the cell-mediated immunity in which cytotoxic T cells destroy infected cells.

Cells of the immune system, known as lymphocytes, recirculate through the blood and lymphoid tissues to the sites of potential infection. They originate in the bone marrow. B cells undergo their primary maturation there and when released are the generators of antibodies. T cell precursors migrate from the bone marrow to the thymus where they undergo their primary differentiation. When released and activated they become either cytotoxic T cells or helper T cells. Helper T cells are essential to 'help' B cells to form antibodies.

Antibodies are proteins; several classes exist for different roles but immunoglobulin G illustrates essential features of all. They are composed of two heavy and two light chain polypeptides, the *N*-terminus of each containing the hypervariable region, which forms the specific antibody-binding site. Each B cell makes a single antibody in terms of binding specificity and different B cells make different ones.

Each B cell in the bone marrow assembles a different variable region by random recombination of pre-existing gene segments for the the heavy and light chain genes, and displays the resultant antibody on the outside of the cell. If it combines with a protein in the bone marrow it is likely to be self-reacting and undergoes self-programmed death (apoptosis; Chapter 30), so avoiding the development of an autoimmune disease. T cells undergo a similar process in the thymus.

Survivors after release, on binding to a (presumptively foreign) antigen, multiply. The antigen thus selects cells capable of protecting against itself, a process known as clonal selection.

The bound antigen is engulfed by the B cell and processed into peptides, which are displayed, on class 2 MHC proteins, on the cell surface. The selected B cells must interact with an activated helper T cell specific for the same antigenic peptide before it differentiates into an antibody-secreting cell. The helper T cell is activated by an APC, usually a dendritic cell, displaying the same foreign peptide in its class 2 MHC proteins as displayed on the B cell it will activate. The activation processes involve the production of cytokines, which promote T cell division and are of pivotal importance in the entire system of immunity. The AIDS virus destroys helper T cells, thus impairing antibody production. During the interaction between the activated helper T cell and its cognate B cell the helper T cell produces cytokines that complete the signal for the B cell to multiply and secrete antibodies.

The humoral response protects against infections in the blood and tissue fluids but once cells are infected the antibodies cannot reach them. The cell-mediated immune response deals with this situation by destroying the infected cell. Cytotoxic T cells have receptors (TCR) for foreign antigen peptides displayed on class 1 MHC proteins on the infected cells. Before it can kill the infected cell the cytotoxic T cell must be activated by binding to the same foreign peptide displayed on class 1 MHC proteins of the dendritic cells. The use of two classes of MHC proteins is a form of compartmentation. It ensures that cytotoxic cells do not attack B cells displaying the same foreign peptide, since in this case it is on a class 2 MHC protein.

In organ transplantation, immune rejection results from the foreign MHC proteins. MHC molecules are highly polymorphic and vary between individuals. This provides some degree of protection for the species as a whole against an infection.

Monoclonal antibodies are preparations obtained from a clone of a single activated B cell that has been fused with an immortal cell line so that the antibody molecules are all identical and can be produced in large amounts, as opposed to antibodies isolated from blood, in which there is a wide variety of antibodies and a finite supply.

Monoclonal antibodies are used as tools to detect specific proteins in cells and may have application in targeting chemotherapeutic agents specifically to cancer cells. They, and polyclonal antibodies, have an important use in the ELISA method for clinical detection of antigens and antibodies.

If mononoclonal antibodies generated in mice are used for therapy they may induce an immune response to the mouse immunoglobulin. 'Humanized' monoclonal antibodies in which the variable region of the mouse antibody is retained while other constant protein components are replaced by human versions of the proteins is achieved by genetic engineering.

Further reading

To access the further reading, please scan the QR code image or go to http://global.oup.com/uk/orc/biosciences/molbiol/snape_biochemistry/student/reading/ch33/

Problems

1 Describe the structure of an IgG molecule.

2 Explain how the body can have genes that code for so many different antibodies.

3 What are the different classes of lymphocyte involved in immune protection and what is their function?

4 What is an antigen-presenting cell?

5 When a host cell displays, for example, a viral antigen on its surface, which class of MHC molecule is it displayed on? Which class of MHC molecule does a cytotoxic T cell recognize?

6 What is meant by the term clonal selection theory?

7 How is autoimmunity avoided, or how does the immune system become self-tolerant?

8 Production of an antibody by a B cell after stimulation (to become a plasma cell) by a helper T cell does not change the antigen-binding site of the antibody, but the antibody has to be released instead of residing in the membrane. How is this achieved?

9 When a B cell (plasma cell) starts to secrete antibody, the latter may have a relatively poor affinity for its antigen but this is rapidly improved. How is this achieved?

10 Which immune systems have the glycoprotein CD4 and which have CD8? What are their roles in the immune reaction? What is the relevance of CD4 to the AIDS virus?

11 Why are there two classes of MHCs (major histocompatibility antigens)?

12 What are humanized monoclonal antibodies?

Answers to problems

Chapter 1

1 A high entropy level means a relatively low-energy level and a low entropy level means a relatively high-energy level.

2 It is difficult to define what it is though everyone feels that they know what it is. What one knows is that there are different types of energy, which can be converted into one another.

3 There is potential energy, which may be gravitational potential energy like a rock about to fall, or the chemical potential energy of a molecule about to react and liberate energy. There is the kinetic energy of a moving object; gravitational potential energy becomes kinetic energy when the rock falls. There is also heat energy, but this can do work only if there is a temperature gradient.

4 Molecules such as starch are long strings of glucose units linked together like a string of the same letter. Proteins and DNA are made of different units linked together so the sequence of these can vary and convey a message or instruction. They are more like meaningful sentences. The information in the sequences of these molecules is the basis of life.

5 It is an RNA molecule that delivers the message of genes from the DNA to the ribosomes, which translate the message into proteins. The message is in the form of a copy of the sequence of bases that make up the DNA of the gene.

6 The structures of both DNA and RNA are such that because of their hydrogen-bonding capability they can direct their own replication. This would have been essential to establish life in the primitive environment before any of the elaborate replication mechanisms of the modern cell had been developed. Without accurate self-replication, there can be no life.

Chapter 2

1 A living cell has to take in molecules that it needs from outside and get rid of waste molecules. This means that there is a high volume of traffic across the membrane usually requiring transport mechanisms in the membrane. The traffic has to be adequate to sustain the needs of the cell; the bigger the cell the more traffic is required. The ratio of membrane area to volume of the cell diminishes with increasing size so that cells must remain small enough for this ratio to remain adequate for the needs of the cell. Also molecules must reach all parts of the cell. Diffusion is slow so a small size is also required for this.

2 There has to be sufficient room to accommodate the molecules needed for life. If you were building a doll's house but had to use full size bricks this would set a lower limit to how small you could make it.

3 Prokaryotes are the bacteria. They are very small and are surrounded by a rigid cell wall. They have no internal membranes and thus lack a defined nucleus. Their DNA is in contact with the cytosol. They have a single circular main chromosome but in addition small circular plasmids of DNA may also be present, which replicate independently. Prokaryotes in general have specialized in rapid cell division as their survival strategy. Prokaryotes are typically haploid; their DNA is segregated at cell division in a relatively simple way without obvious physical changes. Eukaryotic cells are those of plants and animals. They are typically about 1000 times larger in volume than bacteria; they have no cell wall outside of their plasma membrane (plants excepted). The most important characteristic is the possession of a 'true' nucleus surrounded by a membrane and containing the chromosomes. They also have a variety of internal membranes in the form of other organelles. Eukaryotic chromosomal DNA molecules are linear and, except for stem cells and cancer cells, mammalian cells are limited in the number of cell divisions they can undergo. Eukaryotic cells are typically diploid (gametes excluded); segregation of DNA involves the elaborate physical events of mitosis.

4 Somatic cells are the 'ordinary' cells of most tissues. They are differentiated into liver cells, muscle cells, etc., and at cell division give rise only to their specific type. Their ability to divide is limited due to telomere shortening and they only divide to repair damage and replace dead cells. Stem cells are of two broad types, embryonic and adult. The former are pluripotent – they give rise to all types of cells in the body. At cell division the two daughter cells can either differentiate into somatic cells or remain as stem cells and thus are constantly renewed. Telomeres are maintained and cell division is unlimited. Adult stem cells are multipotent but not pluripotent. They renew themselves but can give rise also to some somatic cells. An example is bone marrow stem cells, which can give rise to any type of blood cell.

5 It is best to use an organism that gives the best chance of getting an answer, usually the simpler the better. The system works because the fundamental processes of life are basically the same in all life forms.

6 Antibody protection against influenza is directed at the haemagglutinin protein. This is continually mutated but, because there are many epitopes on the molecule, the immune protection loss is partial and gradual. This antigenic drift leaves people susceptible to infection but residual protection means that it is mild. If by a recombination between different strains a totally new haemagglutinin is present (antigenic shift), a lethal pandemic can result.

Chapter 3

1 The amount of ATP that can be synthesized using 5000 kJ of free energy is 5000/55, which equals 90.91 mol. The weight of disodium ATP produced is thus 551 × 91 g daily, or 50,141 g, which is equal to 72% of the man's body weight. The reason why this is possible is that ATP is being continuously recycled to ADP + P_i and back again to ATP.

2 The $\Delta G^{0\prime}$ value refers to standard conditions where ATP, ADP, and P_i are present at 1.0 M concentrations. In the cell, the concentrations will be very much lower, the actual ΔG value for ATP synthesis will be different from the $\Delta G^{0\prime}$ value according to the following relationship:

$$\Delta G = \Delta G^{0\prime} + RT\,2.303 \log_{10} \frac{[\mathrm{ADP}][\mathrm{P_i}]}{[\mathrm{ATP}]}.$$

3 ATP and ADP are high-energy phosphoric anhydride compounds, whereas AMP is a low-energy phosphate ester. The factors that make hydrolysis of the former more strongly exergonic include the following:

Release of phosphate relieves the strain caused by the electrostatic repulsion between the negatively charged phosphate groups. The released phosphate ions fly apart. A factor also contributing to the exergonic nature of the hydrolyses is the resonance stabilization of the phosphate ion, which exceeds that of the phosphoryl group in ATP. Hydrolysis of AMP causes little increase in resonance stabilization.

4 (a) The nonpolar molecule cannot form hydrogen bonds with water molecules so that those of the latter surrounding the benzene molecule are forced into a higher-order arrangement in which they can still hydrogen bond with each other (so the total bonding does not change). This increased order lowers the entropy and increases the energy of the system; insertion of the benzene molecules into the water is therefore opposed and they are forced into a situation with minimal benzene/water interface as spherical globules and then as a separate layer. The effect is known as hydrophobic force.

(b) The polar groups on glucose can form hydrogen bonds with water.

(c) The Na^+ and Cl^- ions become hydrated. The energy release as a result of this exceeds that of the ionic attraction between them. The separation also has a large negative entropy value.

5 The enzyme AMP kinase transfers a phosphoryl group from ATP to AMP by the following reaction:

$$\mathrm{ATP} + \mathrm{AMP} \rightarrow 2\mathrm{ADP}$$

Hydrolysis is not involved so there is no significant $\Delta G^{0\prime}$ change in the reaction.

6 In the cell, PP_i will be hydrolysed to $2P_i$, whereas with a completely pure enzyme there will be no inorganic pyrophosphatase present. In the former case, the $\Delta G^{0\prime}$ of the total reaction will be −32.2 − 33.4 + 10 kJ mol^{-1} = −55.6 kJ mol^{-1}. For the completely pure enzyme, the $\Delta G^{0\prime}$ will be −22.2 kJ mol^{-1}.

7 Ionic bonds, hydrogen bonds, and van der Waals forces have average energies of 20, 12–29, and 4–8 kJ mol^{-1}, respectively. The energies of activation for formation of weak bonds are very low and so can occur without catalysis. Weak bonds in large numbers can confer definite structures on molecules, but nonetheless can be broken easily, resulting in flexibility of structures.

8 (a)

pH0	pH4	pH9	pH14
$HO{-}P({=}O)(OH){-}OH$	$HO{-}P({=}O)(OH){-}O^-$	$HO{-}P({=}O)(O^-){-}O^-$	$^-O{-}P({=}O)(O^-){-}O^-$

(b) $\mathrm{pH} = \mathrm{p}K_a + \log \frac{[\mathrm{salt}]}{[\mathrm{acid}]}$

The relevant pK_a is 7.2.

$$\mathrm{pH} = 7.2 + \log \frac{[\mathrm{Na_2HPO_4}]}{[\mathrm{NaH_2PO_4}]}$$
$$= 7.2 + \log 1$$
$$= 7.2.$$

(c) Histidine with a pK_a value of 6.5.

9 The binding of the molecules is *via* noncovalent bonds. Since these are weak, several are needed to make the binding effective. Noncovalent bonds are short range and, in the case of hydrogen bonds, highly directional. Put these considerations together and only molecules that exactly fit together can make the required number of bonds. This is the basis of biological specificity. The fact that the bonds are weak (low ΔG) means that they can form spontaneously, without catalytic assistance, and importantly can dissociate to make the associations reversible, which is also essential in many situations. In the case of antibody-antigen binding in which irreversibility is needed for immune protection, very large numbers of bonds are involved. Many structures in the cell, in some cases very elaborate ones, depend on the same principle of specific protein-protein interactions to bring about their self-assemblage.

10 In chemistry, a high-energy bond is one requiring a large amount of energy to break it, which is the reverse of the intended biological concept. Most importantly, however, it is

incorrect to envisage the energy of a molecule to be located in a particular bond but rather, it resides in the molecule as a whole. It is the free energy fall in converting ATP to its products that provides the energy for work. A fall in free energy is a property of a complete reaction, not only of the breakage of a particular bond. This does not alter the fact that Lipmann's concept was important in that ATP behaves *as if the* energy is located in the bond. For this reason it can still be found in textbooks.

11 It is the relentless drive of the universe to achieve maximum stability or lowest energy state. The most stable forms of atoms are when the outer valence electron shell is fully occupied by eight electrons (the octet rule). Atoms with this structure are the inert gases such as helium. Atoms on earth cannot change their electron numbers but by sharing electrons such as occurs in covalent bond formation they can mimic or approach the noble gas structure and thus achieve greater stability.

12 Nothing can be exempt. To assemble a living cell from smaller environmental molecules, energy is needed. This is obtained by the breakdown of food molecules such as glucose to CO_2 and water, which leave the cell. In doing so this increases entropy to a greater degree than the assembly of the cell lowers entropy. Therefore the total entropy of the universe is increased and the second law is obeyed. Life is thus an ordinary process mechanistically and is not magical.

13 Free energy is a term that applies to chemical reactions not a property of individual molecules. One can talk of the free energy of formation of a molecule because that refers to reactions. In a chemical reaction there is a liberation of energy but only part of this is available to do work such as being coupled to the synthesis of molecules with a higher energy content. This is the fraction known as free energy. It is expressed as ΔG – that is the change in free energy occurring in a reaction. This is the all-important thermodynamic term in considering the reactions of life. The part of the total energy change that is not available for utilization in work goes on satisfying the second law of thermodynamics, which specifies that any happening must increase the total entropy of the universe.

14 Entropy is the degree of randomness in any system and in the universe. A system of high entropy is at a lower energy than one at a lower entropy so that increasing entropy level increases stability. The 'aim' of the universe is to achieve maximum entropy and this appears to be the driving force of everything. The second law of thermodynamics demands that all happenings raise the entropy of the universe. It is a somewhat elusive concept to grasp. One common way to raise total entropy is to release heat from a chemical reaction, which increases the random movement of gas molecules in the air. All reactions must contribute to entropy increase, otherwise they would be defying the second law; this part is not available for work. What is left is free energy.

Chapter 4

1 It refers to the amino acid sequence of the polypeptide chain. The amino acids are linked together by peptide (—CO—NH—) bonds. The structure of such a chain is given in Chapter 4.

2 The polypeptide chain of a protein is folded into a specific, usually compact shape, the precise folding being dependent on noncovalent bonds and covalent disulphide bonds between the amino acid residues. Heat disrupts the former bonds, causing the polypeptide chains to unravel and become entangled into an insoluble mass, devoid of biological activity. This is known as protein denaturation.

3 Examples of all are given at the beginning of Chapter 4.

4 **(a)** Around 4; **(b)** around 10.5–12.5; **(c)** around 6.5.

5 Aspartic and glutamic acids, lysine, arginine, and histidine. The carboxyl and amino groups of all other amino acids are bound in peptide linkage (excluding the two end ones).

6 Primary, secondary, tertiary, and quaternary. Illustrated in Fig. 4.2.

7 It lies in the four-way desmosine group shown in Fig. 4.12(b).

8 Collagen fibres are designed for great strength. To achieve this, three polypeptides are associated in a closely packed triple superhelix. The pitch of the supercoil is such that there are three amino acid residues per turn and where the polypeptides touch, there is always a glycine residue. The side group of a single hydrogen atom permits close association whereas bigger side groups would prevent this.

9 (a) Proteins are functional only in their native state, which means their correct three-dimensional folded state and, in most proteins, is due to noncovalent bonds. These bonds are easily disrupted by mild heat, so the polypeptides unfold and become irreversibly tangled up into aggregates.
(b) A number of proteins such as insulin have a few disulphide bonds (three in the case of insulin) which, being covalent, are not destroyed by heat. Although noncovalent bonds are still important in insulin structure, the disulphide bonds confer a degree of increased stability. The extreme case is keratin where large numbers of disulphide bonds produce a very stable structure. You can boil hair as long as you like without producing any permanent structural changes.

10 Particularly in the larger proteins, when the three-dimensional structure is determined, separate regions of the protein are often seen (see Fig. 4.7(d)) separated by unstructured polypeptide. The impression is that these are essentially independent globular sections, which could exist on their own. Indeed, in a number of cases, domains have been isolated from multi-functional enzymes and found to have partial catalytic activity on their own. This is sometimes of great

practical use in that it permits the preparation of an enzyme fragment with a wanted activity, but is free from other activities of the complete enzyme. It is essential for a protein domain to be formed from a given section of a chain; it cannot leave the domain and then double back into it, because then the structure would not be stable on its own if separated from the rest of the protein. Domains are of very great interest because it is likely that they have played an important role in evolution by a process of domain shuffling. This is described in detail in later chapters but for the moment it is often found that domains are coded by sections of genes, which are believed to be shuffled about to form new genes. When expressed, these give rise to new functional proteins. Evolution of new enzymes by this means would be much faster than gradually altering proteins by point mutations.

11 Phenylalanine and isoleucine are hydrophobic and are to the maximum possible extent shielded from water in the hydrophobic interior of globular proteins. Aspartic acid and arginine have highly charged side chains at physiological pH and therefore need to be on the outside exposed to water. Inside a protein molecule these groups would tend to have a destabilizing effect because they cannot have their bonding capabilities satisfied in the hydrophobic environment. Formation of bonds releases energy so that only when bonding is maximized do you get the most stable structure.

12 The charges on the GAGs cause the molecule to adopt an extended conformation and the GAG units permit the formation of very large molecules so that the total volume of water trapped is large.

13 The polypeptide backbone has hydrogen-bonding potential, which must be satisfied if structures of maximum stability are to be achieved. The backbone cannot avoid crossing the hydrophobic interior of proteins, and the hydrophobic side chains of the amino acid residues in these regions offer no possibility of hydrogen-bond formation. The solution is for backbone groups to hydrogen bond to other backbone groups and this is what happens in both the α helix and the β-pleated sheet structures.

14 As shown in Fig. 4.21, myoglobin has the higher affinity and the curve is hyperbolic as compared with the sigmoid curve of haemoglobin. Myoglobin functions purely as an oxygen reserve in muscle. The high affinity means that it maximizes the storage. It is released only when the O_2 tension in the muscle falls. Haemoglobin has to take up the maximum amount of O_2 in the lungs and release the maximum amount in the capillaries. The sigmoid curve helps this. It releases O_2 most rapidly in the capillaries because the O_2 tension there corresponds to the steep part of the curve.

15 On binding of oxygen to the haem iron, Fe moves into the plane of the porphyrin ring. This requires the slightly domed tetrapyrrole to flatten. The Fe is attached to the F8 histidine of the protein and the molecule re-arranges itself. This allows the tetrapyrrole to flatten. The movement affects subunit interactions causing the tetramer to change its conformation from the T to the R state (see Fig. 4.27).

16 Adult haemoglobin in the deoxygenated state has a cavity that binds 2,3-bisphosphoglycerate (BPG) by positive charges on the protein side chains. Binding of BPG is possible only in the deoxygenated state and therefore binding increases unloading of oxygen – it reduces affinity of the haemoglobin for oxygen. Foetal haemoglobin has a γ subunit instead of the adult β one. It lacks one of the BPG charge-binding groups. BPG therefore binds to foetal haemoglobin less tightly, thus raising its affinity for oxygen above that of the maternal haemoglobin.

17 It is necessary for the major transport of CO_2 as HCO_3^-, as illustrated in Fig. 4.29.

18 The statistical chance of a polypeptide with a random primary sequence folding up into a functional domain is vanishingly small. It is observed that many different proteins have folded structures, which are strikingly similar but performing different roles. This can be explained by domain shuffling in which a polypeptide primary sequence that folds up to form a functional entity is used over and over again in different domain combinations and with variations that adapt it to new roles. This would lead to more rapid evolution than depending on single amino acid residue changes due to point mutations.

19 The disease was the first understood case of a 'molecular disease'. In this a single amino acid residue change in haemoglobin from glutamate to valine results in deoxyhaemoglobin crystallizing out as long rods, which distort the red blood cell leading to serious vascular problems. The disease has apparently been positively selected in those areas with a vicious form of malaria, against which the abnormal haemoglobin gives a protection, which outweighs the deleterious effects of the disease.

20 Neutrophils secrete elastase in the mucous lining of lung tissue that destroys elastin, the elastic structural material of lungs. This, unchecked, converts the minute alveoli into much larger structures with a greatly reduced surface area for gas exchange. α_1-Antitrypsin in the blood diffuses into the lung and keeps elastase in check and thus prevents damage. Smoking has two effects: (a) it inactivates α_1-antitrypsin by converting a crucial methionine side chain to a sulphoxide ($S \rightarrow S{=}O$); (b) by irritating the lungs it attracts neutrophils resulting in more elastase being liberated.

21 The peptide bond is a hybrid between two structures.

```
(1)  —C      H          (2)  —C      H
       \    /                  \    /
        C—N                     C=N+
       //   \                  /    \
      O      C—               O-     C—
```

The link is approximately 40% double bonded. This restricts rotation of the peptide bond.

22 Proline. It is an imino acid rather than an amino acid. It is also an α-helix breaker while the others are good α-helix formers.

Chapter 5

1 The original Sanger technique was to label the *N*-terminal amino group of a peptide derived from partial hydrolysis of the protein. The label is coloured or fluorescent for ease of detection and the labelling is stable to acid hydrolysis. The peptide is hydrolysed and the products separated by chromatography. This identifies the first (*N*-terminal) amino acid residue. Sequencing a protein by this procedure is very laborious, because the exercise has to be performed on many peptides and the complete sequence determined from overlapping peptides. This method would now not be used for anything but very small peptides, if that. The Edman procedure is to label the peptide with a reagent, which under appropriate conditions causes the *N*-terminal amino acid to detach, still labelled. This exposes the next amino acid residue to be labelled. The procedure is automated and can be used for sequences of up to about 30 residues. The labelled amino acids detached in turn are identified by chromatography. The method is laborious if a complete protein has to be sequenced and requires the disruption of the protein into oligopeptides, which are sequenced separately. Mass spectrometry can determine the sequence of oligopeptides very rapidly and promises to be of increasing importance. It should not be forgotten that often the quickest method of sequencing a protein is to isolate the DNA responsible for coding the amino acid sequence of the protein. The understanding of this statement may have to wait until later sections of the book have been studied.

2 It refers to the total collection of proteins present in cells. This varies from time to time and from cell to cell as proteins are synthesized and destroyed.

3 Proteomics refers to the simultaneous study of large numbers of proteins at once. For example it may be a study of the changing protein profile during development or between normal and cancerous tissues. To do this work, it is necessary to be able to identify proteins very rapidly and from small amounts such as in single spots on two-dimensional gels. Mass spectrometry has been the main development in doing this. Proteins may be sufficiently characterized by the method to search for them in protein databases in minutes.

4 The SDS molecules insert into the proteins by their hydrophobic tails and denature them. The large numbers of SDS molecules that become inserted with their strong negative charge outside, swamp all charges of the native protein. This results in separations based purely on molecular sieving and therefore of molecular size. Nondenaturing gels by contrast separate proteins both by molecular sieving and the charges on the proteins. SDS can also solubilize proteins.

5 Molecular sieving depends on packing matrices with different pore sizes. Entry of a protein into the beads retards it as compared with a larger protein that flows around the beads. Ion-exchange chromatography involves the use of packings with different ionic groups. Proteins are absorbed selectively and may be eluted selectively by buffers. Reverse-phase chromatography uses a hydrophobic packing to which proteins absorb selectively and can be eluted selectively by solutions of different hydrophobicities or ionic strengths. Affinity chromatography involves attachment to the column packing of groups, which are known to specifically bind to the wanted protein. It can be eluted by a solution of the affinity molecule.

6 The protein can be digested with, say, trypsin and a peptide mass analysis by MALDI-TOF spectrometry performed on it. The pattern is usually sufficient to identify the protein in a database particularly if about five peptides are obtained. An alternative is to obtain limited sequence data by tandem mass spectrometry, which unambiguously identifies a protein.

7 Mass spectrometry involves ionized molecules in the gas phase, which suited many organic molecules but not proteins as they are large and nonvolatile. In the 1980s two methods for obtaining suitable ions from proteins were developed. MALDI (matrix-assisted laser-desorption ionization) involves a UV-light-absorbing matrix mixed with the protein. Laser light causes formation of charged-protein-derived ions. Electrospray, involving spraying a solution at a high electrical potential, also achieves this.

8 Protein databases have been established in various international centres to store information on proteins as it is reported. They contain vast amounts of information on all aspects of protein structure and function. This includes amino acid sequences, tertiary structures and much else. The databases are freely available as well as software for analysing the information in many ways. Mining the databases is itself a research activity. They can enormously speed up research because if a protein can be identified with one in a database, information is then immediately available. They are of constant use in proteomics in which large numbers of proteins have to be identified.

9 Tandem mass analysers separated by a collision chamber are used. The first analyser selects the peptide to be sequenced. It enters the middle chamber and collides with argon gas molecules, which fragments the peptides in a predictable way. The fragmentation products are separated in the second mass analyser and the results displayed as a spectrum from which the amino acid sequence can be deduced.

10 The first method is X-ray diffraction of proteins labelled by heavy metal isomorphous replacement. A newer version of this is to use synchrotron radiation. For this proteins labelled with incorporated selenomethionine can be used. These are conveniently made in *Escherichia coli* or other expression systems of cDNA molecules. A different method, currently applicable for small proteins is nuclear magnetic resonance

(NMR). This requires high protein concentrations but crystallization is not needed. However, synchrotron radiation can use extremely small crystals, which are easier to obtain than those needed for laboratory X-ray diffraction.

Chapter 6

1 Very little of significance because the plot will be arbitrary depending on the time period selected for the measurement of activity. This is because, at higher temperatures, destruction of the enzyme is likely to occur and, at any given temperature, the amount of destruction will be proportional to the time the enzyme is exposed to that temperature. If information is sought on the heat stability of the enzyme it is much better to expose it for a standard time at the different temperatures and, after cooling, to measure the activity in each sample at the normal incubation temperature.

2 The $\Delta G^{0\prime}$ value of a reaction determines whether a reaction may proceed, but says nothing about the rate at which it does proceed (if at all). The latter is determined by the energy of activation of the reaction and the rate at which the transition state is formed.

3 This can involve several factors:
(a) the active site binds the transition state much more firmly than it does the substrate and, in doing so, lowers the activation energy
(b) it positions molecules in favourable orientations
(c) it can exert general acid-base catalysis on a reaction
(d) it may position a metal group, which facilitates the reaction.

4 An affinity constant for a substrate binding to an enzyme represents the equilibrium $E+S \rightleftharpoons ES$. A K_m, which is a measure of the concentration of S at which the enzyme is working at half maximum velocity, is derived from a measure of the rate of product formation. This involves the catalytic rate at which the enzyme converts ES to product. For a K_m to represent a true affinity constant, it is necessary that the rates of the reaction $E+S \rightarrow ES$ and the reverse $ES \rightarrow E+S$ are very large compared with the rate of removal of ES by conversion to product, so that the latter can be ignored in the substrate-binding equilibrium.

5 (a)

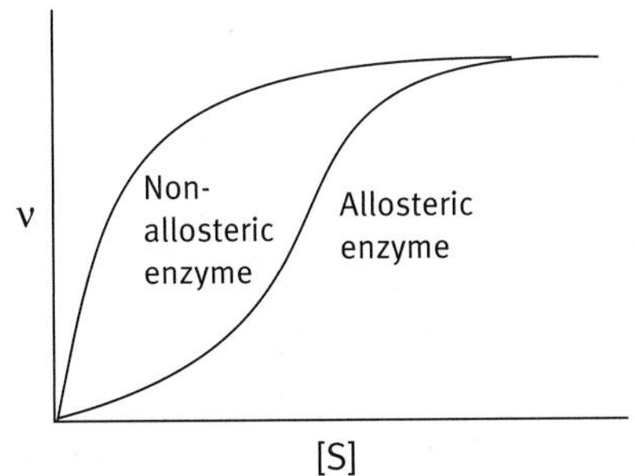

(b) Allosteric modulators (with a few exceptions in which V_{max} is changed) work by changing the affinity of an enzyme for its substrate. This requires that the substrate concentrations for such enzymes are subsaturating, which is usually the case. A positive allosteric modulator moves the sigmoid substrate/velocity curve to the left and a negative effector moves it to the right (see Fig. 6.8). A sigmoidal relationship amplifies the effect of such changes on the reaction velocity and so increases the sensitivity of control. This can be seen in Fig. 6.7.

6 They would have no effect since the modulators change the affinity of the enzyme for its substrate. At saturating concentrations this cannot be used to increase the rate of the reaction.

7 A double reciprocal plot (Lineweaver–Burke plot) is needed. A noncompetitive inhibitor will reduce the V_{max} at infinite substrate concentration (the intercept at the vertical axis) but will not alter the K_m. A competitive inhibitor on the other hand has no effect on the V_{max} at infinite substrate concentration but does change the K_m. This is illustrated in Fig. 6.11.

8 A transition state binds more tightly to the active centre of an enzyme than does the substrate molecule. This is the fundamental basis of the mechanism of enzyme catalysis. The affinity in one recorded case was a thousand or more times greater than that of the substrate. Therefore, there is considerable interest in the potential of stable transition state molecules as specific enzyme inhibitors.

9 It is essential to make sure that you are measuring initial velocities in which the amount of reaction catalysed is linear with time. Otherwise other limiting factors may occur such as inhibition by product, exhaustion of substrate or denaturation of enzyme. In these situations the activity measured is not a true reflection of the amount of enzyme.

10 The serine –OH is perfectly positioned next to a histidine group, which readily accepts the proton thus freeing the oxygen atom to make a bond with the target carbon atom.

11 The aspartate $-COO^-$ group forms a strong hydrogen bond with the histidine side chain and holds the imidazole ring in the tautomeric form and most favourable orientation to accept the proton from serine (see Fig. 6.14). Amidation of the carboxyl group greatly reduces its hydrogen-bonding potential.

12 The enzymes have different active sites for binding their respective substrates. Chymotrypsin has a large hydrophobic pocket to accommodate the large aromatic amino acid residues. Trypsin has an aspartate residue at the binding site to which the basic amino acid residues characteristic of trypsin substrates attaches. Elastase accepts only peptides with small amino acid residues because a pair of valine and threonine residues restricts entrance to the site.

13 A thiol protease is similar to a serine protease except that cysteine replaces the active serine and the intermediate acyl enzymes are thiol esters. Papain is such an enzyme. Aspartyl proteases have a pair of aspartyl residues, which alternate in acting as proton donors and acceptors in the catalytic mechanism. One such enzyme is essential for replication of the AIDS virus.

Chapter 7

1 The basic structure of all biological membranes is the lipid bilayer. It is formed from components that have amphoteric structures, comprised of a hydrophilic head and two long hydrocarbon tails, which are hydrophobic. The molecules arrange themselves so a double layer is formed producing a two-dimensional sheet, with the hydrophilic heads on the outside and the hydrophobic tails pointing inwards, towards each other, sandwiched between them. The centre of the lipid bilayer is therefore hydrophobic. The structure is held together by noncovalent forces. The adjacent heads are exposed to water, and the arrangement maximizes van der Waals forces between the hydrophobic tails. The arrangement is thermodynamically optimal and therefore stable.

2 It acts as a fluidity buffer by preventing the polar lipids from associating too closely towards the centre of the bilayer, but also acts as a wedge at the surface layer.

3

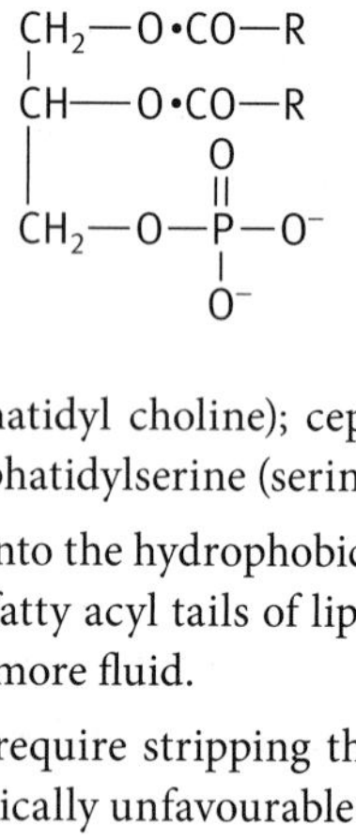

```
CH2—O•CO—R
|
CH——O•CO—R
|      O
|      ||
CH2—O—P—O−
       |
       O−
```

4 Lecithin (phosphatidyl choline); cephalin (phosphatidyl ethanolamine); phosphatidylserine (serine).

5 They put a kink into the hydrophobic tail, which prevents close packing of the fatty acyl tails of lipid bilayers, and thus make the membrane more fluid.

6 To do so would require stripping the molecules of H_2O molecules, an energetically unfavourable process.

7 It involves a membrane protein forming a hydrophilic channel, which permits solutes to traverse the membrane in either direction, according to the direction of the concentration gradient. No energy input is involved. The anion transport of red blood cells is an example; it allows Cl^- and HCO_3^- ions to pass through in either direction.

8 A triacylglycerol is a neutral molecule with no hydrophilic groups. The molecule is therefore water insoluble and to minimize contacts with water the molecules are forced to become spherical globules or form a separate layer. A polar lipid is amphipathic with a charged group at one end and hydrophobic tails at the other. It can therefore assemble into a bilayer structure with hydrophobic tails in contact with each other and shielded from water. The charged group can interact at the surface with water giving the bilayer a stable minimum free energy arrangement of the molecules.

9 This is done by symport and antiport systems. In the case of glucose active transport (a symport), the glucose molecule is cotransported into the cell with sodium ions whose concentration is much higher outside than inside. The latter are ejected from the cell by the Na^+/K^+ ATPase that maintains the ion gradient. Thus ATP indirectly supplies the energy for glucose transport. An example of an antiport system is the pumping of Ca^{2+} from a cell, driven by the co-transport of Na^+ into the cell. The energy once again comes from the ATP-dependent Na^+/K^+ ATPase.

10 It is a membrane channel formed by proteins, which opens and closes on receipt or cancellation of a signal respectively. A ligand-gated channel opens on binding of a specific molecule. The acetylcholine-gated channel in nerve conduction is one. A voltage-gated channel opens in response to a change in membrane potential. The Na^+ and K^+ voltage-gated channels in neuron axons are examples.

11 The drug 'freezes' the Na^+/K^+ pump in the phosphorylated form when applied to the outside face of the protein in the presence of K^+. This inactivates the pump and results in the intracellular concentration of Na^+ rising, which lowers the steepness of the gradient of this ion across the cell membrane. This causes an increase in the intracellular Ca^{2+} concentration, which increases the strength of the heart muscle contraction. The increase in Ca^{2+} is due to the fact that an antiport system driven by the Na^+ gradient exports calcium from the cell. Lowering the Na^+ gradient reduces the outflow of Ca^{2+}.

12 A shell of water molecules surrounds ions in solution. The hydrated ions are too large to pass through the channel. Eight molecules of water surround a potassium ion; normally there is an energy barrier to the removal of these, but the selectivity filter is lined by peptide carbonyl groups, which bind to the ion exactly mimicking, in a thermodynamic sense, the water molecules. The potassium ion is able to slip from solution into the channel without any thermodynamic barrier. It can move through the pore from one binding site to the next and as it leaves the pore is once again hydrated. A sodium ion is smaller but in its hydrated form too large to traverse the pore. The dehydrated ion, however, is too small to bind to the carbonyl groups lining the pore and so there is a thermodynamic barrier to it doing so.

Chapter 8

1 Acetylcholine stimulation of the muscle receptor leads to liberation of Ca^{2+} from the sarcoplasmic reticulum into the myofibril. On the actin filaments are tropomyosin molecules and, in turn, these have a Ca^{2+}-sensitive troponin complex attached to them. When Ca^{2+} binds, a conformational change occurs; it was believed that the tropomyosin blocked the attachment of myosin heads to actin. This is now questioned. A Ca^{2+} ATPase returns the Ca^{2+} to the sarcoplasmic reticulum and terminates the contraction.

2 In smooth muscle myosin heads there is a p-light chain that inhibits the binding of the myosin head to the actin fibre, preventing contraction. Neurological stimulation opens channels allowing Ca^{2+} to enter the cell. The Ca^{2+} combines with calmodulin regulatory protein, which activates a myosin light chain kinase. Phosphorylation of the light chain abolishes its inhibitory effect and contraction occurs.

3 In one case ATP is synthesized from ADP + P_i. The synthesis on the enzyme surface involves little or no free-energy change, but energy is required for release of ATP. This is mediated by a conformation change in the proteins of the ATP synthase head. In the case of myosin, ATP is split into ADP+P_i but on the enzyme surface there is little free-energy change involved. It is on the release of the P_i and ADP that the power stroke of contraction occurs. This is mediated by a conformational change in the myosin head. In neither case is there a covalent intermediate in the synthesis or breakdown of the ATP.

4 It used to be thought that the myosin head swings at its point of attachment to the coiled-coil rod of myosin so that the angle of the head to the actin filament changed. This is now known to be incorrect. The angle of attachment of the head to the actin filament does not change but an α helix of the head known as the lever arm swings to cause the power stroke. The diagram in Fig. 8.7 provides an illustration of this.

5 Actin is involved in movement of nonmuscle cells. Polymerization and depolymerization of actin microfilaments allow cells to extend and 'crawl' over underlying surfaces. Contraction also occurs in such cells. Actin filaments anchored in the cell membrane provide the means for small bundles of myosin molecules to exert a contractile force. A second role is for actin filaments to form a transport track along which special minimyosin molecules move. The latter have a myosin head but the rod-like structure is replaced by a short tail to which vesicles may be attached.

6 Hollow tubes formed by the polymerization of tubulin protein subunits. They have a definite polarity with (+) and (–) ends. The tubulin monomers have a bound molecule of GTP. The end GTP-monomer added protects the tubule from collapse. However, the tubulin has low GTPase activity and hydrolyses the GTP to GDP + P_i after an interval. Unless GTP-monomers are added before this, the tubule collapses.

7 The microtubule-organizing centre (MTOC) protects the (–) end. The (+) ends *of growing* tubules are protected by a tubulin-GTP cap. The GTP is hydrolysed slowly, removing the protection. Unless a new tubulin-GTP molecule is added before this, the microtubule collapses. When the microtubule reaches a 'target' structure, it is then protected.

8 They are molecular motors that move along microtubule tracks and can pull a load with them. Kinesin and dynein travel in opposite directions (in terms of microtubule polarity) on the microtubule track. Myosins move by the swinging-lever mechanism as in muscle. Kinesin and dynein move by a stepwise movement involving swivelling of the two heads.

9 No, they cannot contract. It is believed that the shortening is due to depolymerization but precisely what causes chromosome movement is not certain.

10 Filaments 10 nm in diameter, which are intermediate in this respect between microtubule filaments (20 nm) and actin filaments (6 nm). In specialized cases, they form the structural basis of hair. They may be associated with conferring toughness on structures such as neurofilaments and the Z discs of sarcomeres. Lamins of the nuclear membrane are disassembled in cell division – phosphorylation is involved in this.

11 They are both monomers with a definite polarity and both have an associated molecule of nucleotide triphosphate, ATP in the case of actin and GTP in the case of tubulin. They both polymerize into fibres, actin filaments and microtubules respectively. In both cases, the ATP or GTP is hydrolysed to the diphosphate after polymerization. The triphosphate state in both cases favours polymerization and the diphosphate state, depolymerization. In a growing actin filament or microtubule, it is necessary to add a new subunit before the previously added one is converted to the diphosphate state in order to prevent polymer collapse.

Chapter 9

1 It is a source of the essential amino acids, which are needed for protein synthesis and as precursors of other molecules, and which the body cannot synthesize.

2 It is likely that the bulk of the diet would be made of fat, so ketosis would result. Also a carbohydrate free diet would not provide enough protein from which to make glucose, and body protein would be compromised.

3 Excessive salt intake is related to hypertension and cardiovascular disease and excessive intake of sucrose is related to dental cavities.

4 They aid the correct functioning of the gut and decrease the transit time of the stool in the intestine. Societies whose intake of nonstarch polysaccharides is low, show increased incidence of diverticular disease and colon cancer

5 Leptin is produced by adipose cells. Its concentration in the blood reflects the size of the body's fat reserves. It inhibits appetite and has been found to be effective in controlling obesity in mutant mice and in a small number of children who lack the hormone. Most obese humans are not deficient in the hormone and its levels may in fact be high. The obese are often leptin resistant.

6 Ghrelin, produced by the empty stomach, stimulates appetite via a centre in the brain. Leptin and adiponectin, produced by fat cells (greatest when fat reserves are high) and a peptide produced by the intestine in the presence of food have the reverse effect. Insulin has a leptin-like effect. Cholecystokinin produced by an extended stomach also gives a feeling of satiety.

Chapter 10

1 Pepsin as pepsinogen; chymotrypsin as chymotrypsinogen; trypsin as trypsinogen; elastase as proelastase; carboxypeptidase as procarboxypepidase.The proteolytic enzymes are potentially dangerous in that they could attack proteins lining the ducts. There are no components that amylase might attack. Mucins coating the gut cell lining of the intestine protect cells against proteolytic attack.

2 In the stomach, the acid pH causes a conformational change in pepsinogen that activates molecules to self-cleave the extra peptide in pepsinogen that inactivates the enzyme. Once some pepsin is formed, it activates more pepsinogen so that an autocatalytic activation cascade occurs. In the small intestine, enteropeptidase, an enzyme produced by gut cells, activates trypsinogen to produce trypsin. This, in turn, activates the other zymogens.

3 Pancreatitis, or inflammation of the pancreas, due to proteolytic damage of pancreatic cells.

4 Because the gut cells can absorb only monomers (plus monoacylglycerols). Fat, protein, polysaccharides, and disaccharides cannot be absorbed.

5 Milk contains lactose, which must be hydrolysed to glucose and galactose by the enzyme lactose. After infancy, many individuals lose the capacity to produce lactase. The lactose is not absorbed and is fermented in the large intestine. It attracts water into the gut by its osmotic effect, leading to diarrhoea.

6

Primary ester

$CH_2O \cdot CO—R$

$CHO \cdot CO—R$

$CH_2O \cdot CO—R$

Primary ester

TAG or triacylglycerol (sometimes called triglyceride, but this is chemically incorrect). The central bond is a secondary ester, the other two are primary.

7 The fat is emulsified by intestinal movement and by the monoacylglycerol and free fatty acids produced by initial digestion together with bile salts, so that the lipase can attack the emulsified substrate. The products of digestion (free fatty acids and monoacylglycerol) are carried to the intestinal cells as disc-like mixed micelles with bile salts, which have a high carrying capacity for such products. Probably at the cell surface, the micelle breaks down.

8 The free fatty acids and monoacylglycerol are resynthesized into TAG. The TAG itself cannot traverse cell membranes but it is incorporated into lipoprotein particles, called chylomicrons. These have a shell of stabilizing phospholipid, some proteins, and a centre of TAG and cholesterol and its ester. The chylomicrons are released by exocytosis into the lymphatics and eventually discharge into the blood as a milky emulsion via the thoracic duct.

9 The osmotic pressure of glucose precludes its storage as a monomer. The osmotic pressure of a solution is related to the number of particles of solute in it. By polymerizing thousands of glucose molecules into a single glycogen molecule, the osmotic pressure is correspondingly reduced.

10 TAG is a more concentrated energy source – it is more highly reduced and is not hydrated. If the energy in fat were present as glycogen, the body would have to be much larger. Glycogen stores in the liver are sufficient to last through 24 hours of starvation only, but there may be sufficient TAG for weeks.

11 Yes – glucose can be converted into acetyl-CoA and thence to fat.No – acetyl-CoA cannot be converted into pyruvate as the reaction catalysed by pyruvate dehydrogenase is irreversible, so the breakdown product of fat, acetyl-CoA, cannot be converted into the metabolite that can be converted into glucose.

No – there are no special amino acid storage proteins (excluding milk) in the body but there is an 'amino acid pool', which is the total amount of free amino acids distributed in cells and extracellular fluid.

12 (a) It is the glucostat of the body. It stores glucose as glycogen in times of plenty; it releases glucose in fasting to keep a constant blood sugar level, which the brain needs.
(b) During prolonged starvation it synthesizes glucose, which can be used by the brain. It converts fats to ketone bodies, which the brain (and other tissues) can use for part of its energy supplies and so conserve glucose.
(c) It synthesizes fat and exports it to other tissues.
(d) It is the site of urea production from amino acid metabolism.

13 No, they do not cross the blood–brain barrier effectively.

14 They store fat in the fed state; they release fatty acids in fasting.

15 They are completely dependent on glucose, which they convert to lactate. Since mature red blood cells lack mitochondria,

they are unable to produce ATP by any metabolic route except glycolysis.

16 When blood glucose is high, insulin is released. This is the signal for tissues to store food. As the blood glucose levels fall, insulin levels fall and glucagon levels rise. The latter is the signal for the liver to release glucose and for adipose cells to release free fatty acids for the tissues to utilize. The brain is not insulin-dependent for its uptake of glucose, so this can proceed even in starvation, when insulin levels are extremely low. In addition, adrenaline can override all controls, causing massive turnout of glucose and fatty acids to cope with an emergency requiring the 'fight-or-flight' reaction.

17 Ketone bodies are produced in excess in uncontrolled type I diabetes. However, in prolonged fasting, when glycogen sources are exhausted, they supply muscles with a ready supply of metabolizable substrate, which is preferentially utilized as an energy source. This minimizes the utilization of glucose. In starvation, keeping up the blood glucose level assumes top priority because the brain must be supplied with it. Nonetheless, during starvation the brain can adapt to use ketone bodies for perhaps half of its energy needs, thus economizing on glucose. Preserving glucose in this situation is important. The pyruvate, which the liver uses to synthesize glucose, comes from breakdown of muscle proteins stimulated by glucagon and cortisol and it is clearly advantageous to keep muscle wasting to the minimal level consistent with keeping the brain supplied with glucose.

Chapter 11

1 $\text{G-1-P} + \text{UTP} \rightarrow \text{UDPG} + \text{PP}_i$.

$\text{PP}_i + H_2O \rightarrow 2\text{P}_i$.

$\text{UDPG} + \text{glycogen}(n) \rightarrow \text{UDP} + \text{glycogen}(n+1)$.

The hydrolysis of PP_i makes the overall reaction strongly exergonic and irreversible.

2 Only the liver (and the less quantitatively important kidney) does so. These alone possess the enzyme, glucose-6-phosphatase.

$$\text{Glucose-6-phosphate} + H_2O \rightarrow \text{glucose} + \text{P}_i$$

3 Glucose phosphorylation. Glucokinase has a much higher K_m than hexokinase. During starvation the liver turns out glucose into the blood, primarily to supply the brain (and red blood cells). The first reaction in the uptake of glucose into brain and liver is phosphorylation of glucose. The lower affinity of glucokinase for glucose, as compared with hexokinase, means that the liver does not compete with brain for blood sugar. It efficiently takes up glucose only when blood glucose levels are high. Since it is not as sensitive to inhibition by glucose-6-phosphate as is hexokinase, it can take up glucose and synthesize glycogen even when cellular levels of glucose-6-phosphate are high.

4 Many glycolipids and glycoproteins contain galactose. Removal of galactose from the diet does not prevent synthesis of these molecules because UDP-glucose can be epimerized to UDP-galactose.

5 In the capillaries, lipoprotein lipase hydrolyses TAG into glycerol and fatty acids and the latter immediately enters adjacent cells.

6 VLDL are lipoproteins, resembling chylomicrons, that carry cholesterol and endogenously synthesized TAG from the liver to peripheral tissues.

7 Conversion to bile salts in the liver. High cholesterol concentration in the blood is a risk factor in cardiovascular disease and a cause of heart attacks. Removal of cholesterol is therefore of importance.

8 The enzyme lecithin: cholesterol acyltransferase (LCAT) transfers fatty acyl groups from lecithin to cholesterol.

9 TAG is *not* released; a hormone-sensitive lipase, stimulated by glucagon (or adrenaline in emergency situations), hydrolyses the TAG in the fat cells and the resulting fatty acids are carried in the blood to the tissues attached loosely to serum albumin. By contrast, fat transport from the liver to other tissues occurs *via* VLDL.

10 In familial hypercholesterolaemia the patient lacks LDL receptors and so cannot remove cholesterol-rich LDL from the blood.

11 Adipose cells liberate unesterified fatty acids, which are carried attached to serum albumin. This dissociates from the latter and enters cells. Liver exports fat as VLDL. In extrahepatic tissues, lipoprotein lipases release free fatty acids from the VLDL and these enter the adjacent cells to be metabolized.

12 Cholesterol is transported to peripheral cells from the liver in the form of VLDL, initially, and then taken up by extrahepatic cells in the form of LDL formed by TAG depletion of VLDL. Cholesterol in excess of requirements is transported out of cells by the ABC1 transport system. It is taken up by HDL and converted into cholesterol ester, which is transferred to LDL. Part of the latter is taken up by the liver resulting in the delivery of cholesterol back to the liver. This is the reverse flow. It is important because the liver converts part of the cholesterol to bile acids which are secreted into the intestine and is one route of avoiding cholesterol excess in the blood. HDLs are associated with protection against arterial blockages and heart attacks.

Chapter 12

1 Glycolysis, the TCA cycle, and the electron transport chain; cytosol mitochondrial matrix, and the inner mitochondrial membrane, respectively.

2 Structures in 'Overview of glucose metabolism', Chapter 12.

$$AH_2 + NAD^+ \rightleftharpoons A + NADH + H^+$$

$$B + NADH + H^+ \rightleftharpoons BH_2 + NAD^+$$

3 FAD is another hydrogen carrier; it is reduced to $FADH_2$. It is not a coenzyme but a prosthetic group attached to enzymes.

4 In aerobic glycolysis, glucose is broken down to pyruvate. The NADH produced is reoxidized by mitochondria. In anaerobic glycolysis, the rate of glycolysis exceeds the capacity of mitochondria to reoxidize NADH. This can occur, for example, in the 'fight-or-flight' reaction. Since the supply of NAD^+ is limited, if NADH were not reoxidized sufficiently rapidly to NAD^+, glycolysis and its ATP production would cease. An emergency mechanism reoxidizes NADH by reducing pyruvate to lactate.

$$\underset{\text{Pyruvate}}{CH_3COCOO^-} + NADH + H^+ \underset{\text{Lactate dehydrogenase}}{\rightleftharpoons} \underset{\text{Lactate}}{CH_3CHOHCOO^-} + NAD^+.$$

5 Structures in 'What is coenzyme A', Chapter 12.

-31 kJ mol^{-1} as compared with -20 kJ mol^{-1} for the carboxylic ester – that is, the thiol ester is a high-energy compound.

The vitamin, pantothenic acid, does not participate in the reactions involving CoA, unlike the situation with other coenzymes, where the vitamin is the 'active part' of the molecule.

6 Pyruvate + CoA–SH + $NAD^+ \rightarrow$ Acetyl–S–CoA + NADH + H^+ + CO_2

$$\Delta G^{0-} = -33.5\ \text{kJmol}^{-1}$$

This reaction is irreversible. It makes it possible to convert carbohydrate into fat but makes it impossible to convert fat into carbohydrate. That means that in fasting, fat cannot supply blood glucose although it can be used for energy. The glucose needs of the body need to be met by muscle protein breakdown.

7 The acetyl group is fed into the TCA cycle.

8 They are reoxidized by the electron transport chain producing H_2O and generating ATP in the process (though this is achieved indirectly via a proton gradient).

9 The Nernst equation relates $\Delta G^{0\prime}$ and ΔE_0 values as follows. $\Delta G^{0\prime} = -nF\Delta E_0$ (F is the Faraday constant = 96.5 kJ V^{-1} mol^{-1}) and ΔE_0 is the difference between the redox potentials of the electron donor and acceptor. In the example given ΔE_0 equals -1.035 V ($-0.219 - 0.816$ V).

Therefore, $\Delta G^{0\prime} = -2(96.5\ \text{kJ V}^{-1}\ \text{mol}^{-1})(-1.035\ \text{V}) = -193\ (-1.035) = 194.06\ \text{kJ mol}^{-1}$.

10 Breakdown of fatty acids to acetyl-CoA by β oxidation.

11 Yes. Glucose is converted to pyruvate and pyruvate dehydrogenase converts this to acetyl-CoA. The latter can be used to synthesize fat.

12 No. Fatty acids are degraded to acetyl-CoA. To synthesize glucose, pyruvate is needed. The pyruvate dehydrogenase reaction is irreversible. There is no net conversion of acetyl-CoA (and therefore of fatty acids) to glucose in animals.

Chapter 13

1 The division of the C_6 molecules into two C_3 molecules occurs by the aldol reaction catalysed by aldolase. This requires the structure:

```
        R
        |
        C=O
        |
  R'—C—R''
        |
  H—C—OH
        |
        R
```

The conversion of glucose-6-phosphate to fructose-6phosphate produces the aldol structure capable of dividing in this way. In straight-chain formulae they are as follows.

Glucose-6-phosphate	Fructose-6-phosphate
CHO	CHOH
CHOH	CO
CHOH	CHOH
CHOH	CHOH
CHOH	CHOH
$CH_2OPO_3^{2-}$	$CH_2OPO_3^{2-}$

2 It occurs when a 'high-energy phosphoryl group' transferable to ADP is generated in covalent attachment to a substrate. Oxidation of glyceraldehyde-3-phosphate by the mechanism given in Figure 13.6 is an example.By contrast, electron transport in mitochondria generates a proton gradient which drives ATP synthesis. No phosphorylated intermediates interpose between ADP + P_i and ATP.

3 Pyruvate kinase works in the reverse reaction but conventionally the nomenclature of kinases always derives from the reaction using ATP. The reaction is irreversible because the product of the reaction, enolpyruvate, spontaneously isomerizes into the keto form, a reaction with a large negative $\Delta G^{0\prime}$.

4 Two and three, respectively. Phosphorolysis of glycogen using P_i generates glucose-1-phosphate, convertible into

glucose-6-phosphate. To produce the latter from glucose, a molecule of ATP is consumed.

5 Cytosolic NADH itself cannot enter the mitochondrion; instead the electrons must be transported in by one of two alternative shuttles, and which is used affects the outcome. The malate-aspartate shuttle (Fig. 13.29) reduces mitochondrial NAD^+, but the glycerol phosphate shuttle reduces FAD attached to glycerol phosphate dehydrogenase of the inner mitochondrial membrane. The redox potential of the former is more negative than that of the latter and the two enter the electron transport chain at different respiratory complexes. The yield of ATP from the glycerol phosphate shuttle is lower.

6 Oxidation forms a β-keto acid that readily decarboxylates.

7 The reaction is catalysed by pyruvate carboxylase.

$$CH_3{-}CO{-}COO^- + HCO_3^- + ATP \longrightarrow {}^-OOC{-}CH_2{-}CO{-}COO^- + ADP + P_i + H^+$$

Acetyl-CoA cannot participate. Although acetyl-CoA is involved in citric acid formation, both carbon atoms are lost as CO_2 in one turn of the cycle. Therefore there can be no increase in cycle acids. Also acetyl-CoA cannot be converted into pyruvate in animals. (In bacteria and plants, acetyl-CoA can produce C_4 acids via the glyosylate cycle; see Chapter 16.)

8 Biotin, a B-group vitamin. Using ATP, it forms a reactive carboxybiotin that can donate a carboxy group to substrates. The reaction is:

$$\text{Biotin–}(CH_2)_4\text{–CO–Enzyme} + HO{-}CO{-}O^- \xrightarrow{ATP \rightarrow ADP + P_i} {}^-O{-}CO{-}N\text{(biotin)–}(CH_2)_4\text{–CO–Enzyme}.$$

9 This is shown in Fig. 13.17.

10 They are both mobile electron carriers. Ubiquinone connects respiratory complexes I and II with III, and cytochrome c connects complexes III and IV. The former exists in the hydrophobic lipid bilayer; the latter in the aqueous phase loosely attached to the outside of the inner mitochondrial membrane.

11 To pump protons from the mitochondrial matrix to the outside of the inner mitochondrial membrane and so generate a proton and charge gradient that can be used to drive ATP synthesis.

12 In the eukaryote, NADH generated in the cytosol has to have its electrons transferred to the electron transport pathway. If the glycerol-3-phosphate shuttle is employed for this, we lose one ATP generated per NADH that has to be so handled. Since there are two NADH molecules generated per glucose glycolysed, this leads to a potential loss of two ATP molecules. In *E. coli* there is no such problem. In addition, in *E. coli* there is no expenditure of the energy used in eukaryotes to exchange ATP and ADP across the mitochondrial membrane.

13 The proton flow back into the matrix causes a rotary motion in the F_0 unit in the inner mitochondrial membrane. This drives a shaft inside the F_1 unit, which somehow causes conformation changes in the F_1 subunits. The energy stored in the conformation changes brings about the synthesis of ATP from ADP+P_i but it is important to note that it is the release of ATP which requires the energy. The reaction forming ATP from ADP+P_i at the enzyme surface involves little free-energy change (a concept which takes a little getting used to). The diagram in Fig. 13.21 would be suitable in this question.

14 There are three sites capable of synthesizing ATP in each F_1 unit but they progress through different stages referred to as open (O state), loose binding (of ADP and P_i – the L state), and tight binding (T state). ATP synthesis occurs at the last stage. Cooperative interdependence means that at any one time only one of the three is in the O state, one in the L state, and one in the T state. The individual sites change simultaneously. Thus the L state can change to the T state only if the preceding O state changes to the L state and so on. Figure 13.27 would be useful here.

15 There are four complexes I–IV. I accepts electrons from NADH and transports them to ubiquinone (Q). Q carries electrons to III. II accepts electrons from $FADH_2$ and also transports them to Q. The FAD is the prosthetic group of succinate dehydrogenase and the fatty oxidation pathway and of the fatty acyl-CoA dehydrogenase. III transports the electrons to IV (via cytochrome c), which finally delivers them to oxygen to form water.

I pumps protons to the outside in a manner not yet understood. II does not pump protons; the free energy fall in the transference of electrons from $FADH_2$ to Q is insufficient for this. III pumps protons by the Q cycle in which Q is reduced using protons from the matrix and oxidized on the opposite face of the membrane, liberating protons to the outside. IV pumps protons but the mechanism is not fully elucidated. Figure 13.20 would be useful here.

16 It depends on the aspartate residues in the ring of c subunits. In the unprotonated charged state, it is thermodynamically obliged to reside in a hydrophilic environment. When protonated, if free to do so, it will move to a hydrophobic environment in the lipid bilayer. In short the principle is that the minimum free-energy state is when charged groups are in a hydrophilic environment and protonated uncharged ones in a hydrophobic environment, and if free to do so will move to this situation.

17 Acetyl-CoA. The rest are involved in the TCA cycle. Acetyl-CoA enters the cycle but is not part of the cycle itself.

Chapter 14

1 (a) From free fatty acids carried by the serum albumin in the blood; these originate from the adipocytes.
(b) From chylomicrons – lipoprotein lipase releases free fatty acids and glycerol from TAG.
(c) From VLDL produced by the liver – released in the same way as (**b**).

2 The brain and red blood cells (the latter have no mitochondria).

3 (a) Activation to convert them into acyl-CoA derivatives.
(b) On the outer mitochondrial membrane.
(c) In the mitochondrial matrix.
(d) As carnitine derivatives, as shown in Fig. 14.1.

4 The reactions in Fig. 14.2 are analogous to the succinate→fumarate→malate→oxaloacetate reactions in the TCA cycle, both in the reactions and in the electron acceptors involved.

5 One molecule of palmitic acid produces eight acetyl-CoAs and generates seven $FADH_2$ and seven NADH molecules. NADH oxidation is estimated to produce 2.5, and 1.5, molecules of ATP per molecule of NADH and $FADH_2$ respectively. If you add to this the yield of ATP from the oxidation of eight molecules of acetyl-CoA (ten per molecule), the total is 108 molecules of ATP (counting the GTP from the TCA cycle as ATP). From this must be subtracted two used in the activation reaction, giving a net yield of 106 molecules.

6 After two rounds of β-oxidation, the *cis*-Δ^3-enoyl-CoA is isomerized to become the *trans*-Δ^2-enoyl-CoA.

7 In situations of rapid fat release from adipose cells such as occurs in fasting or diabetes type 1, the liver converts acetyl-CoA into ketone bodies, which are released into the blood. These are preferentially utilized by muscle, thus conserving glucose; most importantly, the brain can obtain about half its energy needs from ketone bodies.

8 Acetoacetate synthesis occurs in the mitochondrial matrix and cholesterol synthesis in the cytosolic compartment, the process occurring on the endoplasmic reticulum membrane.

9 These are membrane-bound vesicles in the cytosol that contain oxidases. These enzymes oxidize a variety of substrates, using oxygen, and generate H_2O_2.

Catalase degrades the hydrogen peroxide. Substrates include those not metabolized elsewhere; for example, very-long-chain fatty acids are shortened. Oxidation of the cholesterol side chain to form bile salts is believed to occur here. The essential role of peroxisomes is underlined by a fatal genetic disease in which some tissues lack peroxisomes.

Chapter 15

1 It supplies ribose-5-phosphate for nucleotide synthesis; it can supply NADPH for fat synthesis; and it provides for the metabolism of pentose sugars.

2 Glucose-6-phosphate is converted into ribose-5-phosphate + CO_2 and $NADP^+$ is reduced. The reactions are given in Fig. 15.1.

3 Transaldolase and transketolase are the main enzymes whose reactions are given in Fig. 15.2. (Other enzymes of the glycolytic pathway may also participate.)

4 Part of the ribose-5-phosphate is converted to xylulose-5-phosphate, a ketose sugar (since transaldolase and transketolase must have a ketose sugar as donor). The following manipulations now occur.

$$(1)\ 2C_5 \rightarrow C_3 + C_7\ \text{(transketolase)}$$
$$(2)\ C_7 + C_3 \rightarrow C_4 + C_6\ \text{(transaldolase)}$$
$$(3)\ C_5 + C_4 \rightarrow C_3 + C_6\ \text{(tranketolase)}$$

The $2C_5$ in reaction 1 are ribose-5-phosphate and xylulose-5-phosphate; the final C_3 compound is glyceraldehyde-3-phosphate. This is converted into glucose-6-phosphate with loss of P_i. The net effect is that six molecules of ribose-5-phosphate are converted into five molecules of glucose-6-phosphate + P_i. Thus the cell can produce NADPH with no net increase in ribose-5-phosphate.

5 NADPH is required to reduce glutathione, a molecule necessary for the protection of the red blood cell. Patients lacking the enzyme are sensitive to the consumption of fava beans and also to some drugs such as the antimalarial drug, pamaquine, resulting in a haemolytic anaemia.

Chapter 16

1 The brain cannot use fatty acids; it needs to use glucose. So must red blood cells, which have no mitochondria and can generate energy only from glycolysis.

2 The substrate of pyruvate kinase is the enol form of pyruvate, but the keto/enol equilibrium is overwhelmingly to the keto form and hence the enzyme has no substrate. The solution lies in a metabolic route in which two high-energy phosphate groups are expended.

$$\text{(1) Pyruvate}+\text{ATP}+\text{HCO}_3^-\rightarrow\text{Oxaloacetate}+\text{ADP}+\text{P}_i+\text{H}^+$$
$$\text{(catalysed by pyruvate carboxylase)}$$
$$\text{(2) Oxaloacetate}+\text{GTP}\rightarrow\text{PEP}+\text{GDP}+\text{CO}_2$$
$$\text{atalysed by PEP carboxykinase, or PEP}-\text{CK)}$$

3 Fructose-1,6-bisphosphatase, producing fructose-6-phosphate, and glucose-6-phosphatase.

4 Only the liver and kidney produce free glucose.

5 In normal nutritional situations (not in starvation) strenuous muscular activity can generate lactate by anaerobic glycolysis. This travels in the blood to the liver where it is converted back to glucose. Release of glucose into the blood and its uptake by muscle completes the cycle (see Fig. 16.4).

6 Glycerol kinase of liver is required to convert glycerol to glucose by the route shown in Fig. 16.5. Glycerol release occurs in starvation where a prime concern is to produce blood glucose. Since only the liver can do this it makes sense for the glycerol to have to travel to the liver rather than being metabolized by adipose cells that cannot release blood glucose.

7 *Via* the glyoxalate cycle shown in Fig. 16.6. The principle is that the two decarboxylating reactions of the TCA cycle are bypassed.

8 In animals, pyruvate carboxylase synthesizes oxaloacetate (C4) from pyruvate (C3). However, in organisms with the glyoxylate cycle, an extra molecule of acetyl-CoA is converted in the net sense to malate and therefore no topping-up reaction is needed.

9 The liver must be supplied with a suitable substrate for gluconeogenesis. Muscle wasting produces free amino acids, which are converted largely to alanine. The alanine migrates to the liver where it is converted into pyruvate.

10 In starvation, the liver must synthesize glucose to supply the brain since glycogen stores are rapidly exhausted. The pyruvate for gluconeogenesis arises from lactate produced by red blood cells and from alanine coming from muscle. Alcohol raises the ratio of reduced to oxidized NAD^+ in the liver; this can impair conversion of lactate to pyruvate because the equilibrium between these is easily shifted by increased NADH levels and also may cause reduction of pyruvate formed from alanine to lactate. Thus the liver may be deprived of pyruvate needed for gluconeogenesis.

Chapter 17

1 Fatty acids are synthesized two carbon atoms at a time, but the donor of these is a three-carbon unit, malonyl-CoA. Acetyl-CoA is converted to the latter by an ATP-dependent carboxylation. The subsequent decarboxylation results in a large negative $\Delta G^{0\prime}$ value. In other words, the point of the carboxylation and decarboxylation is to make the process of adding two-carbon-atom units to the growing fatty acid chain irreversible.

2 This is given in Fig. 17.2.

3 In eukaryotes, all of the enzyme reactions are organized into a single protein molecule with the enzymic functions catalysed by separate domains. The functional unit is a dimer with the two molecules cooperating as a single entity. In *E. coli* the different activities are catalysed by separate enzymes. The advantage of the eukaryotic situation is that the intermediates are transferred from one active centre to the next. In *E. coli* the components must diffuse to the next enzyme so that the process is slower.

4 See 'Overview of glucose metabolism' (Chapter 12) and 'The reductive steps in fatty acid synthesis' (Chapter 17) for structures. NAD^+ is used in catabolic reactions – it accepts electrons for oxidation and energy generation. $NADP^+$ is involved in the reverse – in reductive syntheses. The existence of the two is a form of metabolic compartmentation that facilitates independent regulation of the processes.

5 The main site in the human body is the liver.

6 The acetyl-CoA in the mitochondrion is converted to citrate; the latter is transported into the cytosol where citrate lyase cleaves citrate to acetyl-CoA and oxaloacetate. This is an ATP-requiring reaction that ensures complete cleavage.

$$\text{Citrate}+\text{ATP}+\text{CoA}-\text{SH}+\text{H}_2\text{O}\rightarrow$$
$$\text{acetyl-CoA}+\text{oxaloacetate}+\text{ADP}+\text{P}_i$$

7 The oxaloacetate from the citrate lyase reaction is reduced to malate by malate dehydrogenase, an NADH-requiring reaction. The malate is oxidized and decarboxylated to pyruvate by the malic enzyme, an $NADP^+$- requiring reaction. This scheme effectively switches reducing equivalents from NADH to NADPH. The pyruvate generated returns to the mitochondrion, as illustrated in Fig. 17.6.

This generates only one NADPH per malonyl-CoA produced, while the reduction steps in fatty acid synthesis require two. The rest is generated by the glucose-6-phosphate dehydrogenase system, described in Chapter 20.

8 The scheme is shown in Fig. 17.7. Glycerol phosphate and fatty acyl-CoAs are used.

9 In the synthesis of glycerol-based phospholipids there are two routes. In one, phosphatidic acid is joined to an alcohol such as ethanolamine (see Fig 17.8). For this the alcohol is activated; in phospholipid synthesis, the activated molecule is

always a CDP-alcohol. For some glycerophospholipid syntheses, the diacylglycerol component is activated (see Fig. 17.8). Again, this is by formation of the CDP-diacylglycerol complex. The situation is reminiscent of the use of UDP-glucose whenever an activated glucose moiety is wanted.

10 (a) Eicosanoids have 20 carbon atoms. They include prostaglandins, thromboxanes, and leukotrienes.
(b) All are related to, and synthesized from, polyunsaturated fatty acids.
(c) Prostaglandins cause pain, inflammation, and fever. Thromboxanes affect platelet aggregation. Leukotrienes cause smooth muscle contraction and are a factor in asthma, by constricting airways.
(d) Aspirin inhibits cyclooxygenase, an enzyme involved in their synthesis, and thus it can suppress pain and fever, and also inhibit blood clotting (see Box 17.2).

11 Mevalonic acid is the first metabolite committed solely to cholesterol synthesis. Structural analogues of mevalonic acid have been found to inhibit HMG-CoA reductase, which is the enzyme responsible for mevalonate production. The drugs act in the body as competitive inhibitors of HMG-CoA reductase.

Chapter 18

1 Oxidation results in the formation of a Schiff base, hydrolysable by water.

$$\text{CHNH}_2 \xrightarrow{-2\text{H}} \text{C=NH} \xrightarrow{\text{H}_2\text{O}} \text{C=O} + \text{NH}_3$$

2 Glutamic acid.

3 Transdeamination is the most prevalent mechanism. The amino group of many amino acids are transferred to 2-oxoglutarate, forming glutamate. The latter is deaminated by glutamate dehydrogenase. As an example:

1 Alanine + 2-oxoglutarate → pyruvate + glutamate.

2 Glutamate + NAD^+ + H_2O → 2-oxoglutarate + NADH + NH_4^+

Net reaction: Alanine + NAD^+ + H_2O → pyruvate + NADH + NH_4^+.

4 Pyridoxal phosphate (Fig. 18.2). The mechanism of transamination is given in Fig. 18.3.

5 By removal of H_2O, to result in a Schiff base.

6 A glucogenic amino acid is one that, after deamination, can give rise to pyruvate (or phosphoenolpyruvate). This may be indirect – any acid of the TCA cycle is glucogenic. A ketogenic amino acid is one that cannot give rise to the above but rather gives rise to acetyl-CoA. Only leucine and lysine are purely ketogenic but some, such as phenylalanine, are mixed. Ketogenic amino acids produce ketone bodies only in circumstances appropriate to this such as starvation. Otherwise the acetyl-CoA is oxidized normally.

7 Phenylalanine is not normally transaminated; it is converted to tyrosine and then metabolized (see Fig. 18.8). If the phenylalanine conversion to tyrosine is defective, phenylalanine does transaminate, producing phenylpyruvate, which causes irreparable brain damage to babies and early death.

8 To supply the reducing equivalent for the formation of H_2O from one atom of the oxygen molecule used in the hydroxylation reaction.

9 By formation of 5-adenosylmethionine (SAM); this has a sulphonium ion structure conferring a strong leaving tendency on the methyl group. The formation of SAM is illustrated in Fig. 18.9.

10 5-Aminolevulinate synthase (ALA) carries out the reaction shown in Fig. 18.12 and ALA dehydratase produces the pyrrole, porphobilinogen, by the reaction in Fig. 18.13.

11 This is shown in Fig. 18.5.

12 Both situations call for high rates of deamination of amino acids (in starvation muscle proteins break down to allow glucose synthesis). The amino nitrogen must be converted into urea.

13 (a) Ammonia is converted into glutamine, which is transported to the liver and hydrolysed.

$$\text{Glutamate} + \text{ammonia} \xrightarrow{\text{ATP} \rightarrow \text{ADP} + \text{P}_i} \text{glutamine}$$

$$\text{Glutamine} + \text{H}_2\text{O} \longrightarrow \text{glutamate} + \text{ammonia}$$

(b) Amino nitrogen is transported from muscle as alanine. The alanine cycle is shown in Fig. 18.6.

14 Any blockage in the cycle leads to ammonia toxicity. Deficiencies are known to occur in the level of the enzyme that synthesizes *N*-acetyl glutamate, in argininosuccinate synthetase and lyase. Usually the deficiency is only partial but in severe cases mental retardation and death can occur. Attempts to cause supplementary excretion of ammonia are used to treat the condition. Feeding of benzoate or phenylacetate in large amounts causes excretion of the glycine and glutamine conjugates respectively. In the case of argininosuccinate lyase deficiency, feeding of excess arginine and a low-protein diet results in excretion of argininosuccinate. The reason is that the arginine is converted to urea (which is excreted) forming ornithine, which is converted to argininosuccinate, using up a molecule of ammonia.

15 Increased activity in liver is associated with acute intermittent porphyria (maybe variegate porphyria), though the connection between 5-aminolevulinic acid (ALA) production and the neurological effect is not understood.

Chapter 19

1 PRPP or 5-phosphoribosyl-1-pyrophosphate is the universal agent. Its formation from ribose-5-phosphate and ATP and its mode of action are given in 'Synthesis of purine and pyrimidine nucleotides' (Chapter 19).

2 Tetrahydrofolate (FH_4).

Formyl-FH_4 =

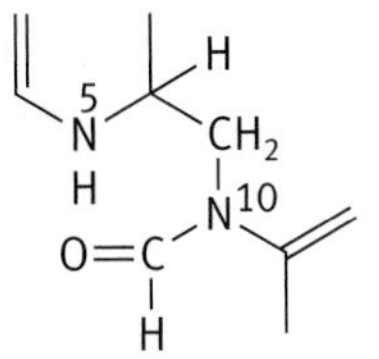

3 Serine. Serine hydroxymethylase transfers $-CH_2OH$ to FH_4 leaving glycine and forming N^5,N^{10}-methylene FH_4. This is oxidized to N^5,N^{10}-methylene FH_4 by an $NADP^+$-requiring reaction. Hydrolysis of this produces formyl-FH_4.

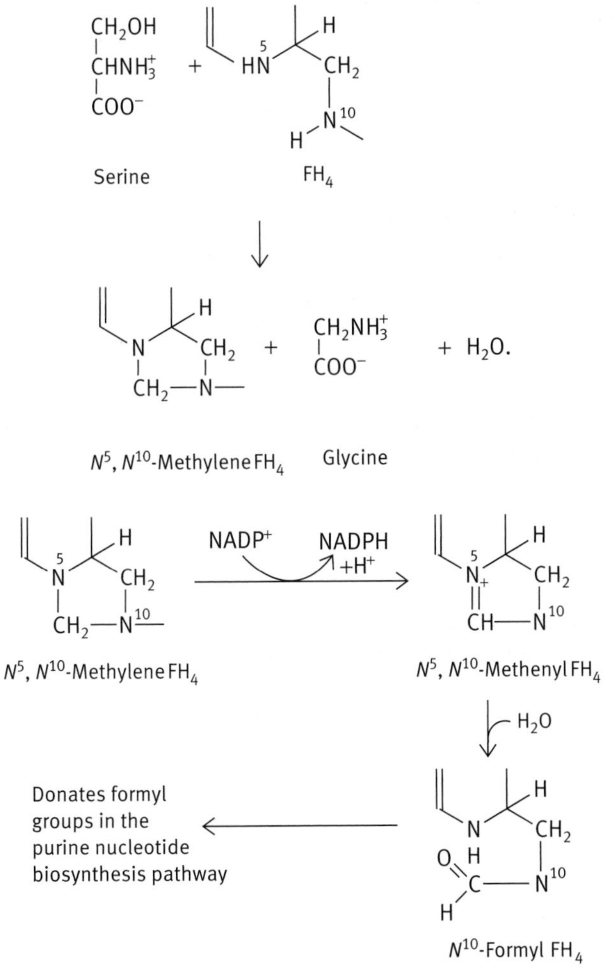

Serine + FH_4 → N^5, N^{10}-Methylene FH_4 + Glycine + H_2O.

N^5, N^{10}-Methylene FH_4 —($NADP^+$ → NADPH + H^+)→ N^5, N^{10}-Methenyl FH_4 —(H_2O)→ N^{10}-Formyl FH_4

Donates formyl groups in the purine nucleotide biosynthesis pathway ← N^{10}-Formyl FH_4

4 It ribotidizes guanine and hypoxanthine in the purine salvage pathway.

5 HGPRT is missing so that purine salvage cannot occur. However, the brain has the direct pathway and it may be that overproduction of purine nucleotides *de novo* occurs due to increased levels of PRPP (because it is not used by salvage). In such patients, excess uric acid is formed but prevention of this by allopurinol does not relieve the neurological symptoms. And gout patients do not suffer the latter, so these symptoms are unexplained.

6 It mimics the structure of hypoxanthine. It is a potent xanthine oxidase inhibitor. This inhibition results in xanthine and hypoxanthine formation rather than that of uric acid.

7 As in Fig.19.8. It is based on feedback inhibition.

8 No. Thymidylate synthetase transfers a C_1 group from methylene FH_4 to dUMP and at the same time reduces it to a methyl group. The H atoms for this are taken from FH_4 so that FH_2 is a product as shown in 'Thymidylate synthesis' (Chapter 19).

9 It inhibits reduction of FH_2, generated by thymidylate synthase, and so prevents dTMP production essential for cell multiplication. FH_4 is essential for the thymidylate synthase reaction.

10 Vitamin B_{12} is required for the methylation of homocysteine to methionine, the methyl donor being methyl FH_4. Lack of vitamin B_{12} causes FH_4 to be 'trapped' as the methyl compound and unavailable for the other folate-dependent reactions.

Chapter 20

1 By allosteric control; or by covalent modification of the enzyme, the chief mechanism for this being phosphorylation.

2

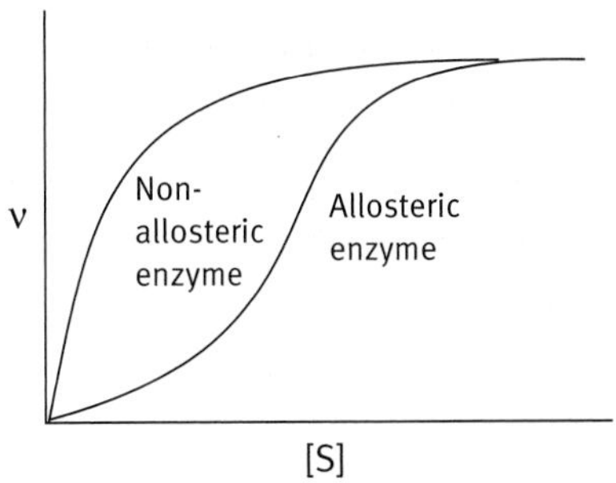

3 It is that an allosteric modulator need have no structural relationship to the substrate of the enzyme. This means that completely distinct metabolic systems can interact in a regulatory fashion.

4 Intrinsic regulation is usually allosteric and can apply to a single cell. It keeps the metabolic pathways in balance.

However, it cannot determine the overall direction of metabolism of a cell – whether, for example, it will store glycogen and fat or release these. These are determined by extrinsic controls (hormones, etc.), which direct the activities of cells to be in harmony with the physiological needs of the body.

5 See Fig. 20.16. AMP activates glycogen phosphorylase and phosphofructokinase while ATP inhibits the latter. The salient feature is that high ATP/ADP ratios stop glycolysis while a lower ATP level (resulting in increased AMP) speeds it up. High citrate levels also logically slow the feeding of metabolite via glycolysis into the TCA cycle. High acetyl-CoA levels may be an indication that oxaloacetate is low – hence activation of pyruvate carboxylase to perform its anaplerotic reaction. At the same time high acetyl-CoA levels indicate that the supply of pyruvate by glycolysis is adequate and inhibition of glycolysis at the PEP step is therefore logical.

6 Direct allosteric controls, controls on a kinase that phosphorylates and inactivates PDH, and controls on protein phosphatase that reverses the phosphorylation.

7 High blood glucose concentrations stimulate insulin release; low glucose concentrations stimulate glucagon release.

8 A hormone, such as adrenaline, is a first messenger; in combining with a cell receptor it causes an increase in a second molecule that exerts metabolic effects. This is a second messenger. For the two hormones mentioned, this is cAMP. It allosterically activates a protein kinase, PKA, whose activity has diverse metabolic effects.

9 By mobilizing glucose transporters into a functional position in the cell membrane.

10 It causes an amplifying cascade of activation, starting with PKA activation as shown in the scheme in Fig. 20.13.

11 In the liver, free glucose is produced, which is released into the bloodstream. In liver, glycolysis is inhibited; in muscle, it increases in rate as the glucose-6-phosphate produced cannot be dephosphorylated but enters the glycolytic pathway.

12 Different cells have receptors for different hormones. Thus, cell A has a receptor for hormone X; cAMP in cell A has effects appropriate to hormone X. Cell B does not have a receptor for hormone X but does for hormone Y. cAMP in cell B elicits responses appropriate to hormone Y but not those appropriate to hormone X.

13 Fructose-2,6-bisphosphate. cAMP decreases its level in liver.. The mechanism is complex, involving a second phosphofructokinase, PFK2.

14 Gluconeogenesis is switched on by glucagon whose second messenger is cAMP. The latter activates a kinase that phosphorylates and inactivates pyruvate kinase. In muscle, adrenaline produces cAMP as the second messenger. The aim of adrenaline here is to maximize glycolysis so inactivation of pyruvate kinase here would be inappropriate.

15 cAMP activates a hormone-sensitive lipase that hydrolyses TAG.

16 Such reactions are irreversible in the cell and act as one-way valves in the pathway; they also ensure that the pathway goes to completion. The carboxylation of acetyl-CoA to malonyl-CoA in fatty acid synthesis is an example in which the reaction serves no apparent purpose other than to confer irreversibility due to the subsequent decarboxylation reaction. However, many pathways have to be reversed; an example is glycolysis, which has to operate in the reverse direction for gluconeogenesis. In such situations, it is necessary to reciprocally control the different directions or the pathway would be simultaneously breaking down glucose units and synthesizing them. To separately control a pathway in the two directions, different enzymes are required. At the irreversible steps, it is necessary to have a different reaction for the reverse direction and this is often a control site. PFK is a typical such step in glycolysis. The reverse step is catalysed by fructose-1,6-bisphosphatase, the two steps being reciprocally controlled.

17 The default state is that glycogen synthase is inactivated by GSK3. There are several kinases phosphorylating the synthase in different positions but insulin activation involves removal of the –P groups added by GSK3 and these are the sites mainly involved in synthase control. To activate the synthase, the GSK3 has therefore to be inhibited and, in addition, the protein phosphatase which dephosphorylates the synthase and thereby activates it is activated by insulin. PKB is the enzyme which inactivates GSK3 by phosphorylating the latter. PKB is activated by a signal pathway activated by binding of insulin to its cell membrane receptor.

18 It has an important role in the control of liver phosphorylase. Glucose binds to phosphorylase *a* (the active phosphorylated form) and induces a conformational change, which makes the enzyme more susceptible to attack by the protein phosphatase 1. This converts the phosphorylase *a* into the relatively inactive '*b*' form and preserves glycogen stores.

19 Malonyl-CoA, the first metabolite committed solely to fat synthesis, inhibits the conversion of fatty acyl-CoAs to the carnitine derivative. The latter is essential for transporting the acyl-CoAs into mitochondria where fatty acid oxidation takes place.

20 The strongly charged phosphate group is very effective at producing a conformational change in a protein, which in most cases is responsible for altering the activity of the enzyme.

21 This is a broadly acting control, which shuts down ATP-consuming synthesis reactions and activates catabolic ATP-generating pathways. The rationale is that AMP is a sensitive signal of reduced phosphorylation charge – that ATP supplies are low. There is only a relatively small amount of ATP in a cell and life depends on its very rapid regeneration. ATP shortage would be much more dangerous than shutting down anabolic reactions temporarily. The AMPK achieves this.

Chapter 21

1 Light reactions involve the splitting of water using light energy, and the reduction of $NADP^+$ to NADPH. Dark reactions mean the utilization of that NADPH to reduce CO_2 and water to carbohydrate. The term 'dark' implies that light is not essential, not that it occurs only in the dark. In fact, dark reactions will occur maximally in bright sunlight.

2 Photosystems contain many chlorophyll molecules. When excited by a photon, one of the electrons in the latter is excited to a higher energy level. Resonance energy transfer permits this excitation to jump from one molecule to another until it becomes trapped in special reaction-centre molecules called antenna chlorophyll molecules. The excitation is insufficient for resonance energy transfer, but sufficient for an electron to enter the electron transport chain shown in Fig. 21.6.

3 (a) Electrons passing along the carriers of photosystem II result in ATP synthesis by the chemiosmotic mechanism as they traverse the cytochrome *bf* complex (see Fig. 21.6).
(b) When all of the $NADP^+$ is reduced, electrons passing along the carriers of photosystem I are diverted to the cytochrome *bf* complex and so generate more ATP (see Fig. 21.7).

4 It is the chlorophyll $P680^+$; that is, the reaction-centre pigment of photosystem II that has been excited by resonance energy transfer from antenna chlorophyll molecules to donate an electron to pheophytin, the first component of the photosystem II electron transport chain. $P680^+$ is an electron short and has a strong tendency to accept an electron; that is, it is a strong oxidizing agent.

5 Thylakoids are formed by invagination of the inner chloroplast membrane (*cf.* inner mitochondrial membrane), which explains the apparent opposite orientation of proton pumping.

6 The enzyme Rubisco (ribulose-1,5-bisphosphate carboxylase) splits ribulose-1,5-bisphosphate into two molecules of 3-phosphoglycerate, a molecule of CO_2 being fixed in the process.

$$\begin{array}{c} CH_2OPO_3^{2-} \\ | \\ C{=}O \\ | \\ CHOH \\ | \\ CHOH \\ | \\ CH_2OPO_3^{2-} \end{array} \xrightarrow[\text{Ribulose-1,5-bisphosphate carboxylase}]{CO_2 \;+\; H_2O} \begin{array}{c} CH_2OPO_3^{2-} \\ | \\ CHOH \\ | \\ COO^- \\ + \\ COO^- \\ | \\ CHOH \\ | \\ CH_2OPO_3^{2-} \end{array} + 2H^+$$

Ribulose-1,5-bisphosphate   Two molecules of 3-phosphoglycerate

7 As in Fig. 21.10.

8 If CO_2 is fixed first into 3-phosphoglycerate by Rubisco, the plant is known as a C_3 plant. These are subject to full competition between oxygen and CO_2 in the Rubisco reaction. However, in high-temperature, high-sunlight areas, in C_4 plants, CO_2 is first fixed into oxaloacetate by pyruvate phosphate dikinase and PEP carboxylase (see Fig. 21.11). This is reduced to malate, which is transported into the underlying bundle sheath cell where the Calvin cycle occurs. Decarboxylation of the malate occurs so that the ratio of CO_2/O_2 in the cell is greatly increased – the concentration of CO_2 in bundle sheath cells may be raised 10–60-fold by this means. The whole scheme is given in Fig. 21.11. Variations in detail of the scheme given occur in different C_4 plants but all are devoted to the same basic strategy.

9 In animals this cannot happen directly – pyruvate kinase cannot form PEP from pyruvate. However, the enzyme in plants is pyruvate phosphate dikinase in which two phosphoryl groups of ATP are used, making the reaction thermodynamically feasible.

Chapter 22

1 The structure can be found in Chapter 12 under The primary structure of DNA.

2 Genetic material needs to be as chemically stable as possible. DNA is more stable than RNA. This is because, the 2′-OH of RNA can make a nucleophilic attack on the phosphodiester bond making RNA less chemically stable than DNA.

3 The phosphodiester bond is longer than the thickness of a base; a straight-chain structure of DNA would leave hydrophobic faces of the bases exposed to water. Because of hydrophobic forces the bases are collapsed together by sloping the phosphodiester bond as shown in Fig. 22.3(c).

4 B DNA; right-handed; 10 base pairs.

5 One chain runs 5′→3′ in one direction and the other 5′→3′ in the other direction. Thus in a linear DNA molecule each end has a 5′ end of one chain and a 3′ end of another chain.

6 A 5′→3′ direction means that you are moving from a terminal 5′-OH group to a 3′-OH group of the polynucleotide chain.

5′CATAGCCG3′
3′GTATCGGC5′

7 Watson–Crick base pairing explains the complementary sequence. By convention a single sequence is written with the 5′ end to the left.

8 Repetitive DNA; an Alu sequence is a few hundred bases long but repeated almost exactly hundreds of thousands of times; scattered throughout the human chromosome.

9 The statement is true to a large extent but not completely so. It has been known for a long time that a few genes such as

those for ribosomal and transfer RNAs do not code for protein but very recently it has been found that in DNA previously sometimes referred to as 'junk' DNA, there are large numbers of 'micro genes', coding only for small RNA sequences, which in some way seem to play a part in the expression of conventional genes and may determine characteristics of eukaryotic organisms.

10 Uracil. The others are components of DNA. Uracil occurs in RNA but not DNA.

11 The prokaryotic genome is haploid and circular and not enclosed by a membrane. It does not have the defined nucleosome structure found in eukaryotes. The eukaryotic genome consists of linear chromosomes, terminating in telomeres and is surrounded by a nuclear membrane. Eukaryotic cells are typically diploid. At cell division the chromosomes condense into a tightly packed form unlike prokaryotic cell division where this does not occur.

Chapter 23

1 A section of DNA whose replication is initiated at a single origin of replication.

2 Ahead of the replicative fork positive supercoils occur.

3 In *E. coli* gyrase, a topoisomerase II, introduces negative supercoils. In eukaryotes a topoisomerase I relaxes positive supercoils.

4 As shown in Figs 23.6 and 23.7.

5 In winding DNA around a nucleosome, a local negative supercoil is introduced, but, since no bonds are broken, there cannot be a net negative supercoiling. The local negative supercoil is compensated for by a local positive supercoiling elsewhere. This is relaxed by topoisomerase I, leaving a net negative supercoiling in the DNA.

6 dATP, dGTP, dCTP, and dTTP.

7 Cytosine readily deaminates to uracil; if uracil were a normal constituent of DNA it would be impossible for repair enzymes to recognize and correct the mutation. Use of T instead of U removes the problem.

8 (a) No. An RNA primer is required. DNA polymerase cannot initiate new chains.

(b) Synthesis proceeds in the $5' \rightarrow 3'$ direction; by this it is meant that the new chain is elongated in the $5' \rightarrow 3'$ direction, new nucleotides being added to the free 3′OH of the preceding nucleotide. It does not refer to the template strand, which runs antiparallel.

9 Hydrolysis of inorganic pyrophosphate; base pairing of the nucleotides and breaking of the high-energy phosphate bond of dNTPs, liberating pyrophosphate.

10 Polymerase I is the enzyme that processes Okazaki fragments into a continuous lagging strand. It has a $5' \rightarrow 3'$ exonuclease and a $3' \rightarrow 5'$ exonuclease, and is a DNA polymerase with low processivity. Its activities are summarized in Fig. 23.16.

11 Correct base pairing of the incoming nucleotide triphosphate with the template nucleotide is the primary essential. The enzyme also has a proofreading activity. It has a $3' \rightarrow 5'$ exonuclease activity that removes the last incorporated nucleotide if this is improperly base paired.

12 The methyl-directed mismatch repair system is described in Fig. 23.20. There is evidence that proteins similar to *E. coli* proteins exist in humans and that, where these are deficient, there is an increased risk of cancer.

13 Thymine dimers are formed when DNA is subjected to UV irradiation. Two adjacent thymine bases become covalently linked. Repair can either be direct by a light-dependent repair system that disrupts the bonds and restores the separate bases; or it can be subject to excision repair (Fig. 23.19).

14 As shown in Fig. 23.17, removal of the 3′ Okazaki fragment primer leaves an unreplicated section.

15 The answer lies in telomeric DNA synthesized as described in Fig. 23.18. The telomerase is a reverse transcriptase.

16 Processivity is the ability of a DNA polymerase to duplicate the coding strand of DNA for the very long stretches involved in DNA replication without falling off. The sliding clamps achieve this. These are annular proteins that are placed around the DNA to be copied, at the site of an RNA primer; the polymerase attaches to the clamp and so is prevented from falling off. DNA polymerase I has low processivity because its job is to remove the RNA primer from Okazaki fragments. If it had high processivity it would remove the newly synthesized DNA of the fragments and replace them unnecessarily. It is best that soon after it has reached the DNA part, it detaches to allow the ligase to heal the nick.

17 DNA. All the rest are the precursors of RNA.

Chapter 24

1 RNA polymerase uses ATP, CTP, GTP, and UTP as substrates (cf. dATP, dCTP, dGTP, and dTTP). Most importantly, RNA polymerase can initiate new chains; DNA polymerase requires an RNA primer. RNA synthesis never includes proofreading.

2 This is illustrated in Fig. 24.4. In more detail the promoter has a –10 (Pribnow) box and a –35 box.

3 RNA polymerase attaches nonspecifically to the DNA, but, when joined by a sigma factor molecule, it binds firmly to a promoter. The Pribnow box and the –35 box position orientate the polymerase correctly. The enzyme synthesizes a few phosphodiester bonds and then the sigma protein flies off and the polymerase progresses down the gene transcribing mRNA.

4 One is by the stem loop structure shown in Fig. 24.7, formed by G—C base pairing. In this, internal G—C base pairing in the mRNA prevents bonding of the mRNA to the template DNA strand. Following this, a string of uracil nucleotides further weakens the attachment (only two hydrogen bonds in a U—A pair), facilitating detachment. There is still some uncertainty as to why this is a reliable terminator. The second method depends on the Rho factor, a helicase that unwinds the mRNA–DNA hybrid. At the termination site, the polymerase pauses (possibly due to a region in the DNA rich in G—C base pairs with triple bonds to be broken) and the Rho factor catches up and detaches the mRNA.

5 The precise base sequence of the –10 and –35 boxes, their distance apart, and the bases in the +1 to +10 region.

6 The primary transcript in eukaryotes has introns that must be spliced out; the 5′ end is capped; the mechanism of termination is unknown; the mRNA is, with few exceptions, polyadenylated at the 3′ end.

7 The mechanism is shown in Fig. 24.13. It is based on a consensus sequence that determines where splicing occurs and on a transesterification reaction involving little change in free energy. Split genes may have facilitated evolution by promoting exon shuffling. Differential splicing can also result in a single gene giving rise to different proteins.

8 The eukaryotic polymerase II does not attach to the DNA directly but rather attaches to a protein complex that assembles on the DNA. In addition, any number of transcription factors may be associated with the complex.

9 The 'introns early' view is that they have always been present in modern genes. The latter are postulated to have been formed from primitive mini-genes, which fused together, the introns being the fused nontranslated regions. On this view, uninterrupted prokaryotic genes are due to all excess DNA having been discarded in the interests of rapid cell division.

The alternative view is that introns are late additions to prokaryotic genes, which are regarded as primitive, and the introns may almost be regarded as parasitic DNA. It is generally accepted that introns may have been important in evolution by facilitating exon shuffling. The argument on intron origins still rages.

10 The proteome is the complete collection of proteins present in a cell at any one time. The genome is the complete collection of genes. The proteome varies according to which genes have been activated so that the number of proteins varies in cells from different tissues and from time to time according to physiological controls. The genome is essentially fixed if we exclude occasional examples of gene amplification.

11 No; alternative splicing may produce multiple versions of proteins from a single gene transcript.

Chapter 25

1 If only 20 codons were used, there would be 44 not coding for any amino acid. Any mutation in a gene-coding region would then be highly likely to inactivate the gene by prematurely introducing a stop codon. Instead, by using 61 codons for amino acids, a base change would either cause no change or would substitute a different amino acid. Because of the arrangement of the genetic code, many of these substitutions would be conservative and cause minimal change to the change to the protein structure.

2 The answer lies in the wobble mechanism described in Chapter 25.

3 The convention is that RNA is shown with the 5′ end to the left. When mRNA is shown in this manner, a tRNA anticodon is base paired in an antiparallel fashion. The tRNA molecule is therefore shown with the 5′ end to the right.

4 In the case of some amino acids, at the stage of attaching the amino acid to tRNA. In all cases at the stage of elongation; a pause in GTP hydrolysis on the EF-Tu gives time for incorrectly paired aminoacyl-tRNAs to leave the ribosome.

5 In general GTP hydrolysis to GDP and P_i is believed to result in conformational changes in the relevant proteins. GTP is involved in assemblage of the *E. coli* initiation complex. It is involved in the delivery of aminoacyl-tRNAs to the ribosome by EF-Tu. It is involved in the translocation step.

6 Figure 25.8 is necessary for this explanation. The straddling has the advantage that in moving from one site to the other the tRNA is never completely detached. It also means that the peptidyl group does not have to physically move relative to the ribosome.

7 Figure 25.12(a) explains this. The scanning mechanism of selecting the start AUG means that there can be only one start site.

8 They bind to nascent polypeptides and prevent premature, improper folding associations. Appropriate release of chaperones facilitates correct folding though the process is not fully understood. Other chaperonins provide a suitable folding box.

9 The prion diseases described in Chapter 25, Protein folding and prion diseases. In these, the faulty protein is due to different folding of the identical polypeptide. They are typified by mad-cow disease.

10 (a) Yes, you can deduce the amino acid sequence of the protein that an mRNA will code for because each codon unambiguously specifies an amino acid.

(b) No, you cannot deduce the base sequence of an mRNA from the amino acid sequence of the protein it codes for. This is because of the degenerate nature of the genetic code, or redundancy. Since several codons may code for a given amino acid it is impossible to deduce which of the alternatives was actually used by the cell in translating the messenger.

11 If one or two bases were deleted the reading frame would be shifted so that improper codons would be read thereafter. The result of this would be meaningless polypeptide after the first 99 amino acids. Or, if the new reading frame encountered a termination codon, the polypeptide would be prematurely terminated. If all three bases were deleted, the only effect would be to delete one amino acid residue and it is possible that the protein would be functional, depending on the effect of the deletion on the structure of the protein.

12 The peptidyl transferase has been shown to be active even after extraction of proteins from the ribosome. It is in fact not an enzyme but a ribozyme. Also, deletion of a single adenine base from one of the ribosomal RNAs by ricin inactivates the ribosome.

13 Transcription is the synthesis of mRNA coded for by the gene; translation is the synthesis of polypeptide coded for by the mRNA. The terms derive from the fact that transcription implies copying in the same language (DNA and RNA are both polynucleotides). Translation implies copying in a different language (mRNA is a polynucleotide, proteins are polypeptides).

14 Proteasomes are protein organelles consisting of a central core made of rings of protein subunits with caps at both ends. Inside the cavity of the central core are proteolytic enzymes, which hydrolyse proteins into peptides and amino acids. The caps act as the gateway for entry of proteins destined for destruction. The proteasomes selectively destroy individual protein molecules, which are targeted into the cavity. The vital role is selective destruction of proteins such as regulatory proteins in cell cycle control. They also play a vital part in the immune system by producing peptides, for example from viruses to be displayed on the outside of cells inviting attack by killer T cells of the immune system. The importance of proteasomes is underlined by the fact that they have been highly conserved throughout evolution. Mutations causing nonfunctional proteasomes in yeast are lethal. It has been realized that protein breakdown in cells, especially by proteasomes, has very great fundamental importance, so the field has become one of the most intensively studied – a revolution in the image of the field.

15 The entry ticket is the attachment of the small protein ubiquitin to target proteins. Polyubiquinated proteins are allowed to enter the proteasome cavity for destruction. The ubiquitin is removed during the entry process and recycled. There is still much to learn about what determines the selection of proteins for ubiquitination but certain N-terminal sequences are known to be a factor.

16 No. The free-energy release that occurs with a correct base pairing and an incorrect one are not sufficiently different to give the required discrimination. However the ribosome participates in the process in that a triplet of bases in the *E. coli* small subunit RNA checks that the first two anticodon-codon base pairing involves genuine Watson–Crick hydrogen bonds before peptide synthesis is allowed to occur.

17 Hsp 70 with a bound molecule of ATP binds to exposed hydrophobic groups on nascent polypeptide chains emerging from ribosomes and prevents unprofitable (improper) hydrophobic associations, which could hinder correct folding. The chaperone detaches on ATP hydrolysis and allows the polypeptide to fold.

Hsp means heat-shock protein. These are formed when cells such as *E. coli* are subjected to sudden temperature increases. They are in fact chaperones whose function is to assist refolding of proteins denatured by the heat. In this, their role is the same as helping to fold the (unfolded) nascent polypeptide chains emerging from the ribosomes.

Chapter 26

1 The operon is blocked by a repressor protein. In the presence of lactose, the repressor detaches (see Fig. 26.3).

2 Helix–turn–helix proteins, leucine zipper proteins, helix–loop–helix proteins, zinc fingers proteins, and homeodomain proteins.

3 In the default condition, eukaryotic genes are shut down. This is due to nucleosomes blocking the promoter sites. Chromatin remodelling unblocks the site by removal of the nucleosomes. This is done by the enzyme histone acetyltransferase (HAT), which acetylates the tails of the histones that constitute the nucleosome octamer. This has the effect of removing the positive charge on the histone lysine side chains. Somehow this results in chromosome remodelling.

4 The enzyme histone deacetylase is believed to reverse the activation process started by the histone acetyltransferase.

5 In the case of HAT, the enzyme is part of the coactivator. The latter binds to the activating transcription factor(s) and brought into proximity to the nucleosomes on the promoter selected by the transcription factors. Negative control may be effected in essentially the same way, a repressing transcription factor binding to the promoter and attracting the deacetylase.

6 It is based on an iron-dependent inhibition of mRNA translation as shown in Fig. 26.20.

7 It is an RNA transcript, which is not translated into a protein, as is messenger RNA. Several types have been known for decades. Examples are ribosomal RNAs, transfer RNAs, snoRNAs. These act as 'infrastructure' supporting the expression

of protein-coding genes. The new type is microRNAs (miRNAs) produced from microgenes, about 75 nucleotides long.

8 They are very widely distributed including most of the areas formerly known as junk DNA. This includes the introns of protein-coding genes. The ENCODE project has shown that at least 80% of the bases in the genome of eukaryotes are transcribed into RNAs. The protein-coding section of conventional genes occupies only a small percentage of the genome. Microgenes exist in very large numbers.

9 The ENCODE project and other studies have shown that microgenes are conserved across different species such as humans and mice. Such conservation indicates that they must have a function naturally selected.

10 There are several. Overall it is believed that they were used in the evolution of complexity in eukaryotes by exerting epigenetic control over protein-coding genes rather than by increasing the number of the latter. There is no clear correlation between complexity and conventional gene number. Prokaryotes have remained relatively simple; there is little non-coding DNA and microgenes appear to be comparatively unimportant so far as is known.

It is also believed that miRNAs by making it possible to silence protein-coding genes protected eukaryotes from excessive transposon proliferation.

11 Microgenes are transcribed into miRNAs about 75 bases long, which adopt a hairpin form. These are processed into small interfering double stranded RNAs (siRNAs), which attach to RISC complexes. One strand is selected and guides the RISC complex to complementary sequences in mRNAs which are then either destroyed or silenced.

12 It lies in the possibility of specifically silencing protein-coding genes. Theoretically a wide variety of diseases could be treated. For example cancer might be treated by targeting an oncogene. Such silencing has been demonstrated in tissue culture. Or a virus could be targeted. The beauty of the system would be that siRNAs can be cheaply made and potentially specific. However the field is in its infancy and problems of delivering the siRNAs to target tissues or cells is an obvious problem.

Chapter 27

1 In cotranslational transport, the polypeptide is transferred through the target membrane as it is synthesized. This occurs in the transport of proteins into the ER. Posttranslational transport is where the protein is fully synthesized in the cytosol, released, and then transported through its target membrane. Examples are mitochondrial proteins, nuclear proteins, and peroxisomal proteins.

2 In almost every case, GTP hydrolysis occurs to produce a conformational change in a protein. Commonly the hydrolysis occurs only after a slight delay, this essentially being a timing device to permit other happenings to occur. An example of this is in the transport of proteins through the ER (see Fig. 27.6). There are other examples (see Q7).

3 To destroy unwanted molecules and structures imported into the cell by endocytosis and to destroy components of the cell destined for destruction.

4 They are vesicles produced by the Golgi sacs containing an array of hydrolytic enzymes. The enzymes are acid hydrolases with an optimum pH of 4.5–5.0. Proton pumps in the membrane maintain this pH inside the vesicle. They fuse with endosomes to form the lysosomes.

5 A variety of genetically determined lysosomal storage disorders exist in which the absence of a specific hydrolytic enzyme causes the lysosomes to become overloaded with material normally disposed of.

6 In Pompe's disease, a fatal genetically determined one, there is a lack of a lysosomal α-1,4-glycosidase that normally degrades glycogen. The lysosomes become overloaded with glycogen. However, it is not clear why disposal of glycogen by this means should be needed. It is not part of the usual accounts of glycogen metabolism.

7 (It will probably be necessary to refer to Figs. 27.17 and 27.18 to follow this explanation.) The Ran-GDP/Ran-GTP exchange enzyme exists only inside the nucleus. The function of Ran-GTP is to effect the release of the cargo from the importin-cargo complex arriving from the cytosol. The Ran–GTP–importin complex migrates back to the cytosol. It is now necessary for the GTP to be hydrolysed to allow the Ran-GDP to dissociate and release the importin. The Ran–GTPase, however, must be activated by GAP, the GTPase-activating protein. This is found only in the cytosol, not in the nucleus (the Ran-GDP itself must return to the nucleus). The released importin picks up another cargo and carries it into the nucleus to start the whole cycle again.

8 They are cluster of hydrophobic amino acids in the proteins lining the nuclear pore. (F and G are the one-letter abbreviations for phenylalanine and glycine respectively). The repeats are binding sites for importins, which move through the pore from one repeat to the next.

9 Proteins destined for secretion are transported cotranslationally into the endoplasmic reticulum and from there to the cell membrane by transport vesicles. Proteins destined for the mitochondria are delivered in an extended form to receptors on the mitochondrial membrane. Nuclear proteins are delivered to the nuclear pore by importins. Peroxisomal proteins are delivered to receptors on the organelle in a fully folded form.

Chapter 28

1 Pancreatic DNase randomly cuts DNA. A restriction enzyme cuts only at specific short sequences of bases.

2 The adenine bases in all of the relevant hexamer sequences of *E. coli* strain R DNA are methylated so that the enzyme does not recognize it. Invading foreign DNA will not be so protected.

3 It refers to ends resulting from a staggered cut made by many restriction enzymes.

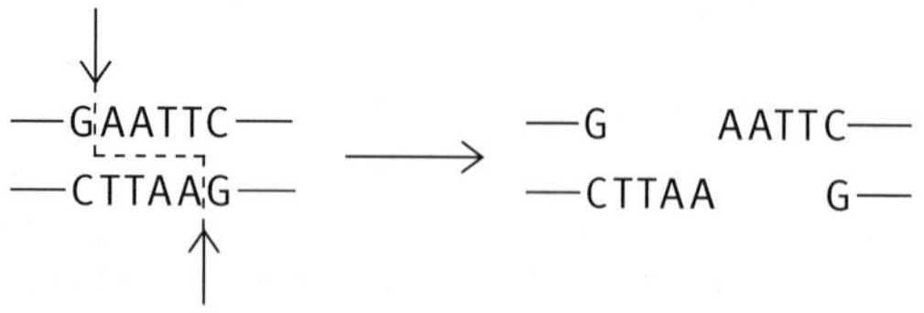

```
   ↓
—G|AATTC—              —G      AATTC—
  └----┐      ——→
—CTTAA|G—              —CTTAA      G—
        ↑
```

The overhanging ends will automatically base pair and 'stick' together.

4 A genomic clone refers to a cloned section of DNA identical to the sequence of DNA in a chromosome. A cDNA clone means complementary DNA obtained from a eukaryotic mRNA. It refers to cloned DNA identical to the mRNA for a gene. The cDNA lacks introns; eukaryotic genomic clones have them.

5 The nucleotide lacks a 3′OH and therefore, when added by DNA polymerase to a growing DNA chain, terminates that chain. The application of this to sequencing is described in Chapter 28, Sequencing DNA.

6 A specific section of DNA on a chromosome can be amplified logarithmically. It involves copying the section of DNA and copying the copies *ad infinitum*. Its importance is that a minute amount of DNA, too small for any studies, can be amplified at will. The essential, apart from enzymes and substrates, is that you must have primers for copying the section in both directions and this means knowing the base sequences at either end of the piece to be amplified.

7 It is an engineered plasmid with a convenient insertion site for, say, a cDNA clone for a specific protein, but which also contains appropriate bacterial DNA transcriptional signals (a promoter) and translational signals. The DNA is transcribed and the mRNA translated in a bacterial cell. It can be used to produce large quantities of specific proteins.

8 It is a technique for detecting gene abnormalities. If DNA is cut by a pattern of restriction enzymes and the resultant pieces separated on an electrophoretic gel, a pattern of bands can be visualized by hybridization methods. Mutations may produce or remove restriction sites so that the pattern may be altered. The method was important, but is largely replaced by PCR.

9 Stem cells divide and the progeny cells can either proceed to develop into mature terminally differentiated cells or divide into more stem cells. This can occur indefinitely, unlike somatic cells, which have a limited potential to divide. Thus a stem cell population keeps up a continual supply of replacement cells. Adult stem cells are committed to become a certain class of cells; bone marrow stem cells differentiate into many types of blood cells. Embryonic stem cells found in blastocysts are pluripotent; they can give rise to any type of the cell in the body. They are of especial interest because they can be cultured *in vitro* and retain this pluripotency. Apart from their use in producing knockout mice, they have the potential to be of enormous medical importance in that they might be developed in controlled *in vitro* conditions into replacement cells for the human body. Production of nerve cells to repair neurological damage is but one possible example.

10 (The procedure is complex and you will probably have to follow Figs 28.19 and 28.20 to understand this answer.) The principle is to inactivate a specific gene by homologous recombination. For this a targeting vector is constructed so that the normal gene is replaced by an inactivated one. The method uses embryonic stem cells obtained from a mouse blastocyst. These are mutated by the vector and cells with the targeted gene correctly replaced are grown up and injected into the blastocyst of another mouse. The blastocyst is implanted into a mouse to develop. The offspring contains cells derived from both the mutated cells and the recipient blastocyst cells. By appropriate cross-breeding, mice which are homozygous to the mutated gene can be obtained. In the procedure, the targeted stem cells are taken from a blastocyst from a mouse of a coat colour that is different from that from which the recipient blastocyst is taken. This enables the process to be followed by coat colour of the progeny.

11 (a) To replicate the DNA, synthesis in both directions is required and a different primer is needed for each direction.
(b) Each time a section of DNA is replicated the primer becomes incorporated into the new strand. Since an enormous amplification is required sufficient primers must be available to support the indefinite numbers of replications.
(c) DNA polymerase; since at each cycle the mixture has to be heated to separate strands, a heat-stable enzyme preparation from a thermophilic bacterium (*Taq* polymerase) is used.

Chapter 29

1 The production is described in Fig. 29.25. GTP hydrolysis has a timing function. In cholera the GTPase is inactivated so that cAMP production remains activated once stimulated.

2 It allosterically activates protein kinase A (PKA). This can have metabolic effects depending on phosphorylation of key enzymes but, in addition, there are genes with cAMP-response elements (CREs). Inactive transcriptional factors can be activated by phosphorylation by PKA. Therefore cAMP can have extensive gene-control effects.

3 It activates a cytosolic guanylate cyclase that produces cGMP. The latter is a second messenger in some systems.

4 The phosphatidylinositol cascade involves the receptor-mediated activation of membrane-bound phospholipase C. This releases inositol trisphosphate (IP_3) and diacylglycerol (DAG). The former increases cytosolic Ca^{2+}; the latter activates a protein kinase, PKC. Thus, IP_3 and DAG are second messengers. The scheme is shown in Figs 29.27 and 29.28.

5 (a) Involves an allosteric change of the cytosolic domain on binding of the hormone.

(b) Involves receptor dimerization and activation of self-phosphorylation of tyrosine residues by an intrinsic kinase.

(c) Involves receptor dimerization and association with a separate tyrosine kinase that phosphorylates the receptor (see Figs 29.9 and 29.22).

6 Neurotransmitters; homones, cytokines, and growth factors; vitamin D_3 and retinoic acid; nitric oxide is in a class of its own.

7 In both cases the signal combines with a receptor protein and this triggers a response which in terms of gene control usually causes a protein to enter the nucleus and effect gene control. In the case of a lipid-soluble signal such as glucocorticoid, it combines with a cytosolic receptor protein, which enters the nucleus and functions as a transcription factor. In the case of a water-soluble signal such as EGF, it combines with a membrane receptor, and this results ultimately in a kinase entering the nucleus and activating a transcription factor. The two systems are fundamentally similar, the differences being in the detail. (In some cases the lipid signal receptor resides in the nucleus, but the same general principle applies.)

8 It is likely to be a component of a signalling pathway which specifically binds to an activated membrane receptor of the tyrosine kinase type.

9 Both have tyrosine kinase-associated receptors. The Ras-associated receptors are themselves tyrosine kinases, whereas the JAK/STAT pathways depend on cytosolic kinases associating with the receptors. The Ras pathway is activated by many hormones whereas the other pathway is best known for its association with cytokine signals such as interferon. The Ras receptor activates a long cascade of proteins terminating in a kinase that migrates into the nucleus and activates transcription factors. The JAK/STAT receptors bind the JAKs, which phosphorylate tyrosine groups of the receptor. The STAT proteins are attracted from the cytosol to bind to these and are themselves phosphorylated by the double-headed JAKs. The phosphorylated STAT proteins migrate into the nucleus where they activate genes. Thus the latter pathway is a much more direct system from receptor to gene whereas the Ras pathway is very long. This is probably to allow amplification of the signal. Less certainly, it may also give more opportunity for cross-talk between Ras and other pathways and provide opportunities for extra controls.

10 All G-proteins are heterotrimeric proteins associated with membrane receptors. The latter do not become phosphorylated but on the binding of the cognate ligand undergo a conformational change. This in turn causes a change in the conformation of the attached G-protein such that the GDP attached to the α subunit is exchanged for GTP. This α-GTP subunit detaches from its partner β and γ subunits and migrates to a target enzyme located on the membrane. In the case of a typical G-protein, the one associated with the β-adrenergic receptor, it activates adenylate cyclase, which produces a second messenger cAMP. The activation is terminated by the slow intrinsic GTPase activity of the α subunit, which acts as a timing device. The α subunit in its GDP form migrates back to the receptor to reform the heterotrimeric G-protein. If the receptor is still activated, the process is repeated. The versatility is due to the fact that different receptors have different G-proteins whose activated α subunit activates or inhibits different enzymes involved in production of second messengers. Thus stimulation of receptors by adrenaline may activate or inhibit cAMP production depending on the nature of the particular G-protein subunit associated with the receptor; in the phosphatidylinositol system, the G-protein subunit activates phospholipase C as one example of the different target enzyme specificities that exist. In this case the second messengers are DAG and PI(3,4,5)P_3. (Note that other GTPase proteins exist, but the term G-protein is restricted to the type described above.)

11 Not really; cGMP is continually produced in the rod cells to keep the cation channels open. The effect of light is to cause the activation of the enzyme, which destroys cGMP. In the case of the nitric oxide signalling system cGMP is a second messenger as the signal activates either a membrane receptor or an intracellular receptor to produce the cGMP.

12 The insulin receptor does not dimerize but it structurally resembles a covalently permanently dimerized receptor of the EGF type.

13 The breakdown of GTP to GDP+P_i may seem to have no purpose in that no chemical bonds are formed or obvious work done. However the breakdown is often a timing device. The G protein activation of cAMP synthesis is an example in which the α subunit–GTP activates adenylate cyclase but GTP hydrolysis switches this off. In protein targeting, it is postulated that GTP hydrolysis by the signal receptor protein reduces its affinity for the peptide and allows the latter to insert into the translocon channel. In protein synthesis GTP hydrolysis allows the EF-Tu to leave the ribosome but only after a slight delay to allow incorrect amino acyl–tRNAs to detach. There are other examples such as the uncoating of COP-coated vesicles and the auto inactivation of Ras.

14 Ras is called a small monomeric GTPase. The term G-protein refers only to the heterotrimeric signalling proteins such as the adrenaline receptor.

Chapter 30

1 In G_1, a growth factor signal is necessary to activate genes that code for the G_1-specific cyclins needed to activate the kinases, which push the cell cycle through the restriction point. In the absence of growth factor signal the cycle is arrested in the G_0 phase awaiting the arrival of a signal.

2 Cyclins are proteins which are necessary for the cyclin dependant kinases (Cdks) to act. They not only activate Cdks but also direct their activity to be appropriate to the needs of the different phases of the life cycle. Cyclins are specific to particular phases of the life cycle and are completely destroyed at the end of their appropriate phase of the cycle. New cyclins are synthesized as the cycle progresses from one phase to the next.

3 At the end of cell division the cyclins are destroyed and new G_1 specific cyclins are synthesized. Without this the cell cannot progress to the G_1 checkpoint. If no mitotic signal is received the cell enters G_0 in which it functions normally in interphase but does not divide. Most somatic cells are in this state most of the time. If a signal is received it can enter the G_1 phase again.

4 If the DNA is damaged in any way the cell is not allowed to proceed to the S phase. Damage is detected by the protein p53, which halts the cycle progression. In addition the retinoblastoma protein can also halt the cycle at this point (described in Chapter 31). At the mitotic checkpoint each duplicated chromosome must be properly aligned on the mitotic spindle. If each chromatid is not attached to a kinetochore fibre the cycle is halted.

5 In mitosis, chromosomes are replicated and a copy of each chromosome is segregated into each of the two daughter cells, which like the parent cell are diploid (2*n*). In meiosis, haploid sperm and egg cells are produced (*n*). In meiosis two cell divisions are involved but only before the first is DNA replicated. At the first cell division each daughter cell receives one chromosome (as a pair of sister chromatids) of each of the parental homologous pairs, randomly assigned. DNA synthesis does not occur before the second cell division and the chromatids of the chromosome duplexes are pulled apart and segregated into the two daughter cells, which each thus receives a single copy of each parental chromosome. The copy has been derived either from the mother or the father of the parent organism. The gametes are haploid (*n*). At the first meiotic division the two members of each homologous chromosome pair swap sections of non-sister chromatids through crossing-over, leading to greater genetic diversity

6 They are proteolytic enzymes, with a cysteine residue at their active centre, which attack a wide variety of proteins at sites adjacent to an aspartate residue. They exist in multiple copies in all cells in an inactive state and become activated in response to an apoptopic signal for the cell to destroy itself. They exist in cells as inactive procaspases. One form of apoptopic stimulus releases cytochrome c from mitochondria; another is a death signal from a killer T cells. In both instances the procaspases are caused to aggregate probably resulting in mutual activation by partial proteolysis. The cascade of proteolysis, which ensues, destroys the cell.

7 It has a vital role in destroying unwanted cells. An obvious example is the destruction of huge numbers of B and T cells of the immune system to eliminate those that would cause an autoimmune reaction. The importance of apoptotic destruction is that it is a neat non-messy way of cell disposal without necrosis that could cause problems. It also has a vital protective role in eliminating potentially cancerous cells. Another role is that killer T cells of the immune system destroy infected cells by delivering a signal for the cell to self-destruct.

8 A wide variety of 'insults' to cells can induce it, including radiation or a damaged genome. p53 plays a vital part in this. If a cell is damaged to the point where an abnormal potentially cancerous state is present, p53 triggers its destruction. In fact about half of human cancers are associated with p53 deficiency. It is not totally understood how p53 induces apoptosis but an early event is the release of cytochrome c from mitochondria into the cytosol, which is a death sentence.

9 Cytochrome c causes procaspases, which are inactive proteolytic enzymes, to aggregate. Activation of these occurs and the caspases then destroy the cell. The activation process is not fully understood, but it may be that the procaspases have a low activity, and aggregation results in mutual activation.

10 Cells are delicately balanced between life and death. There are opposing controls of pro- and anti-apoptosis. Broadly there are two groups of proteins within the Bcl-2 family. Bax and Bad are pro-apoptopic and are believed to initiate cytochrome c release. Bcl-2 and Bcl-x_L are important anti-apoptopic proteins, which prevent cytochrome c release from mitochondria. It is the balance between the two groups, which will determine the fate of a cell. There are additional direct controls inhibiting caspases.

Chapter 31

1 A protooncogene is a normal gene, which can become converted into an oncogene. It is a gene coding for a protein, which is involved in cellular control processes most commonly those that affect cell replication. They were discovered when it was found that oncogenes of retroviruses, known to be cancer-producing, have almost identical counterparts in

normal cells. Conversion of a protooncogene to an oncogene usually causes excessive signalling – either the gene is over-expressed or the gene product (the signalling molecule) has inappropriately prolonged life. The change may be caused by a single base mutation or by the gene being placed under the control of an excessively active promoter. In the case of Burkitt's lymphoma chromosomal rearrangements places a protooncogene under the control of an immunoglobulin gene promoter. A very active retroviral promoter may be inserted at a point in a chromosome so as to control the expression of the protooncogene. Alternatively a mutation may produce an excessively long-lived transcription factor. In the case of the *Ras* oncogenes, mutations eliminate the GTPase activity so that the signalling pathway cannot be switched off.

2 The p53 gene is normally expressed at a low level but this increases if the DNA is damaged. The resultant high levels of p53 protein activate genes that cause the inhibition of the cyclin-dependent kinases, which are needed for the cell to proceed through the restriction point in G_1. The delay allows time for the DNA to be repaired. When this is done, the p53 expression is no longer stimulated, the p53 protein diminishes and the cell cycle proceeds. If the DNA is not repaired the p53 gene can also induce apotosis of the cell.

3 An oncogene is a gene involved in the control of cell division that has become mutated so that it gives inappropriate mitogenic signals leading to uncontrolled cell division. They are the 'bad' genes. Tumour-suppressor genes protect against cancer. The best known is p53. It is activated by damaged DNA and arrests the cell cycle at the G_1 checkpoint. If the damage is not rectified, it signals the cell to die by apoptosis. Mutation of both alleles of a tumour-suppressor gene makes a cell with an oncogene more likely to progress to cancer.

4 A somatic cell shortens its telomeres at every round of DNA replication and this limits the number of cell divisions it can undergo. A cancer cell has to be able to undergo unlimited numbers of divisions and the continual lengthening of the telomeres by telomerase or ALT permits this.

Chapter 32

1 The initial stimulus for blood clotting is, in quantitative terms, very small. To obtain a sufficiently rapid response amplification is needed. Cascades are the biological method of amplification. The final enzyme activation is that of prothrombin to thrombin, an active proteolytic enzyme (see Fig. 32.1).

2 Fibrinogen monomer proteins are prevented from spontaneous polymerization by negatively charged fibrinopeptides (see Fig. 32.2) that cause mutual repulsion. Removal of these by thrombin permits the association of fibrin monomers, illustrated in Fig. 32.3.

3 Covalent cross-links are formed between monomers in the polymer by an enzymic transamidation process between glutamine and lysine side chains:

$$-CONH_2 + H_3N^+- \rightarrow -CO-NH- + NH_4^+$$

4 In the conversion of prothrombin to thrombin, a glutamic acid side chain is carboxylated; the carboxyglutamate binds Ca^{2+}. Vitamin K is a cofactor in the carboxylation reaction.Other conversions in the proteolytic cascade involve this conversion.

$$\begin{array}{c} | \\ CH_2 \\ | \\ CH_2 \\ | \\ COO^- \end{array} + CO_2 \longrightarrow \begin{array}{c} | \\ CH_2 \\ | \\ HC-COO^- \\ | \\ C-COO^- \end{array}$$

5 It is involved in the conversion of hydrophobic xenobiotics to more soluble ones. A typical reaction is:

$$AH + O_2 + NADPH + H^+ \rightarrow AOH + H_2O + NAD^+$$

6 To reduce one oxygen atom to H_2O. It is known as a mixed-function oxygenase.

7 The glucuronyl group is donated to an –OH on the foreign molecule, rendering it much more polar (see Fig. 32.5).

8 It involves ATP binding cassette transporters (ABC transporters) that remove a variety of compounds from cells. It transports steroids in steroid-secreting adrenal cortical cells, but all transported compounds (including anticancer drugs used in chemotherapy) are lipid-soluble amphipathic compounds, indicating a more general role. It has recently been shown to transport the cholesterol to the outside of the cells to be removed to form high-density lipoprotein (HDL).

9 It is an oxygen molecule that has acquired an extra electron.

$$O_2 + e^- \rightarrow O_2^{\overline{\cdot}}$$

Superoxide is extremely reactive, causing damage.

10 Most eukaryotic cells have superoxide dismutase and catalase:

$$2O_2^{\overline{\cdot}} + 2H^+ \rightarrow H_2O_2 + O_2 (\text{dismutase})$$
$$2H_2O_2 \rightarrow 2H_2O + O_2 (\text{catalase})$$

Secondly, there are antioxidants such as ascorbic acid (vitamin C) and α-tocopherol (vitamin E). Superoxide is dangerous because it attacks a molecule and generates another free radical, causing a self-perpetuating chain of reactions. Antioxidants are quenching agents. When attacked by superoxide, the free radical produced from these is insufficiently reactive to perpetuate the chain of reactions.

Chapter 33

1 As in Fig. 33.1.

2 The principle is that each B cell assembles its own immunoglobulin genes randomly from a selection of DNA sections as described in Fig. 33.2.

3 B cells produce antibodies (after activation and maturation into plasma cells). Helper T cells are (in most cases) required for B cells to do this. Cytotoxic or killer T cells bind to host cells displaying a foreign antigen and kill them by perforating their membrane or inducing apoptosis.

4 It is a type of phagocyte (usually a dendritic cell) that engulfs a foreign antigen, processes it into pieces, and displays it in combination with its MHC molecules. If an inactive helper T cell combines with the antigen-MHC complex, it is activated (see Fig. 33.6). B cells after their initial encounter with an antigen also are antigen-presenting cells.

5 In both cases MHC class 1.

6 Each B cell produces a different antibody. For any given antigen there will be very few B cells specific for it, but, when the antigen binds to the displayed antibody, the B cell proliferates into a clone. Thus the antigen automatically selects which B cells are to proliferate.

7 During *primary* maturation of B cells in the bone marrow or of T cells in the thymus, if an antigen binds, the cell is eliminated, the principle being that such antigens will be 'self'. After release from the bone marrow or thymus, the binding to an antigen activates the cells to multiply, but further activation by dendritic cells is required.

8 Differential splicing of introns; at the 3′ end of the immunoglobulin gene is an exon that codes for an anchoring polypeptide sequence. At the onset of secretion a switch in the splicing eliminates this during mRNA formation.

9 The onset of secretion is accompanied by rapid somatic mutation in the cells, which modifies the variable site. Since binding of the antigen causes cell proliferation this constitutes a progressive selection mechanism for those cells producing a 'better' antibody. This is known as affinity maturation.

10 CD4 is present on helper T cells. The protein binds to a constant protein component of MHC class 2 molecules and thus confines interactions to B cells, helper T cells, and antigen-presenting cells. Cytotoxic T cells have CD8, which binds to MHC class 1 molecules; this restricts the interaction of the killer T cells to host cells. The CD4 protein is the receptor by which the AIDS virus infects helper T cells.

11 It is a form of compartmentation that directs immune cells to their correct targets. Most somatic cells have Class 1, for which killer T cells are specific. Helper T cells and B cells are Class 2. This means that killer T cells will not attack them and helper T cells will not combine with somatic cells.

12 They are monoclonal antibodies produced in mouse systems in which the constant mouse protein groups are replaced by their human counterparts. They are used for therapeutic injection into humans to eliminate or reduce immunological reactions generated by mouse proteins. A new strain of mice has been engineered to produce completely human versions of the antibodies.

Index of Diseases and Medically Relevant Topics

Index

B

C

D

E

F

G

H

I

N

O

P

Q

R

S

T